Kasuga M, Liu Q, Miura S, Yamaguchi-Shinozaki K, Shinozaki K (1999) Improving plant drought, salt, and freezing tolerance by gene transfer of a single stress-inducible transcription factor. Nat Biotechnol 17:287–291

Kerby TA, Buxton DR (1978) Effect of leaf shape and plant population on rate of fruiting position appearance in cotton. Agronomy J 70:535–538

Kerby TA, Buxton DR, Matsuda K (1980) Carbon source-sink relationships within narrow-row cotton canopies. Crop Sci 20:208–213

Kitajima S, Sato F (1999) Plant pathogenesis-related proteins: Molecular mechanisms of gene expression and protein function. J Biochem 125:1–8

Kohel RJ (1974) Influence of certain morphological characters on yield. Cotton Grow Rev 51:281–292

Kosmas SA, Argyrokastritis A, Loukas MG, Eliopoulos E, Tsakas S, Kaltsikes PJ (2006) Isolation and characterization of drought-related trehalose 6-phosphate-synthase gene from cultivated cotton (*Gossypium hirsutum* L.). Planta 223:329–339

Kramer PJ (1980) Drought, stress, and the origin of adaptation. John Wiley and Sons, New York

Kumar A, Singh DP (1998) Use of physiological indices as a screening technique for drought tolerance in oilseed *Brassica* species. Ann Bot 81:413–420

Kuznetsov VV, Rakitin VY, Zholkevich VN (1999) Effects of preliminary heat-shock treatment on accumulation of osmolytes and drought resistance in cotton plants during water deficiency. Physiologia Plantarum 107:399–406

Lacape M, Nguyen TB, Hau B, Giband M (2003) Targeted introgression of cotton fiber quality QTLs using molecular markers. International Workshop on Marker Assisted Selection: A Fast Track to Increase Genetic Gain in Plant and Animal Breeding, Turin, Italy, October 17–18, 2003

Lande R, Thompson R (1990) Efficiency of marker-assisted selection in the improvement of quantitative traits. Genetics 124:743–756

Le Houerou HN (1996) Climate change, drought and desertification. J Arid Environ 34:133–185

Leidi EO, Lopez M, Gorham J, Gutierrez JC (1999) Variation in carbon isotope discrimination and other traits related to drought tolerance in upland cotton cultivars under dryland conditions. Field Crops Res 61:109–123

Lu YT, Dharmasiri MAN, Harrington HM (1995) Characterization of a cDNA encoding a novel heat-shock protein that binds to calmodulin. Plant Physiol 108:1197–1202

Lu ZM, Rundel PW, Zeiger E, Rasoul Sharifi M, Chen JW, Percy RG (1996) Genetic variation in carbon isotope discrimination and its relation to stomatal conductance in Pima cotton (*Gossypium barbadense*). Aust J Plant Physiol 23:127

Lu ZM, Zeiger E (1994) Selection for higher yields and heat-resistance in Pima cotton has caused genetically-determined changes in stomatal conductances. Physiologia Plantarum 92:273–278

Lubbers EL, Chee PW, Paterson AH, Smith CW (2006) Fiber quality of a near-isogenic introgression line series from an Upland by Pima interspecific cross. Beltwide Cotton Conferences, San Antonio, Texas, January 3–6, 2006

Ludlow MM, Santamaria JM, Fukai S (1990) Contribution of osmotic adjustment to grain-yield in *Sorghum bicolor* (L) Moench under water-limited conditions. II Water-stress after anthesis. Aust J Ag Res 41:67–78

McKersie BD, Leshem YaY (1994) Stress and stress coping in cultivated plants. Kluwer Academic Publishers, Dordrecht, The Netherlands

Meloni DA, Oliva MA, Martinez CA, Cambraia J (2003) Photosynthesis and activity of superoxide dismutase, peroxidase and glutathione reductase in cotton under salt stress. Environ Exp Bot 49:69–76

Meloni DA, Oliva MA, Ruiz HA, Martinez CA (2001) Contribution of proline and inorganic solutes to osmotic adjustment in cotton under salt stress. J Plant Nutrition 24:599–612

Meredith WR (1984) Influence of leaf morphology on lint yield of cotton – enhancement by the sub okra trait. Crop Sci 24:855–857

Morgan JM (1995) Growth and yield of wheat lines with differing osmoregulative capacity at high soil-water deficit in seasons of varying evaporative demand. Field Crops Res 40:143–152

NCC (2006) World Cotton Database National Cotton Council of America www.cotton.org/econ/cropinfo/cropdata/index.cfm Cited 8 Oct 2006

Niles GA, Feaster CV (1984) Breeding. In: Kohel RJ, Lewis CF (eds) Cotton. ASA/CSSA/SSSA, Madison, WI, pp 202–229

Paterson AH (1995) Molecular dissection of quantitative traits – Progress and prospects. Genome Res 5:321–333

Paterson AH, Damon S, Hewitt JD, Zamir D, Rabinowitch HD, Lincoln SE, Lander ES, Tanksley SD (1991) Mendelian factors underlying quantitative traits in tomato – Comparison across species, generations, and environments. Genetics 127:181–197

Paterson AH, Deverna JW, Lanini B, Tanksley SD (1990) Fine mapping of quantitative trait loci using selected overlapping recombinant chromosomes, in an interspecies cross of tomato. Genetics 124:735–742

Paterson AH, Jiang CX, Wright RJ, Saranga Y, Menz M (2003) QTL analysis of genotype x environment interactions affecting cotton fiber quality. Theor Appl Genet 106:384

Paterson AH, Lan TH, Reischmann KP, Chang C, Lin YR, Liu SC, Burow MD, Kowalski SP, Katsar CS, DelMonte TA, Feldmann KA, Schertz KF, Wendel JF (1996) Toward a unified genetic map of higher plants, transcending the monocot-dicot divergence. Nat Genet 14:380–382

Paterson AH, Lander ES, Hewitt JD, Peterson S, Lincoln SE, Tanksley SD (1988) Resolution of quantitative traits into mendelian factors by using a complete linkage map of restriction fragment length polymorphisms. Nature 335:721–726

Percy RG, Wendel JF (1990) Allozyme evidence for the origin and diversification of *Gossypium barbadense* L. Theor Applied Genet 79:529–542

Peterschmidt NA, Quisenberry JE (1982) Plant water status among cotton genotypes. Beltwide Cotton Prod Res Conf. National Cotton Council of America, Memphis, TN, pp 108–111

Pettigrew WT, Heitholt JJ, Vaughn KC (1993) Gas-exchange differences and comparative anatomy among cotton leaf-type isolines. Crop Sci 33:1295–1299

Pettigrew WT, Meredith WR (1994) Leaf gas-exchange parameters vary among cotton genotypes. Crop Sci 34:700–705

Quisenberry JE, Jordan WR, Roark BA, Fryrear DW (1982) Exotic cottons as genetic sources for drought resistance. Crop Sci 21:889–895

Reinisch AJ, Dong J, Brubaker CL, Stelly DM, Wendel JF, Paterson AH (1994) A detailed RFLP map of cotton, *Gossypium hirsutum* X *Gossypium barbadense* – Chromosome organization and evolution in a disomic polyploid genome. Genetics 138:829–847

Richards RA (2006) Physiological traits used in the breeding of new cultivars for water-scarce environments. Agric Water Management 80:197-211

Richards RA, Rebetzke GJ, Condon AG, van Herwaarden AF (2002) Breeding opportunities for increasing the efficiency of water use and crop yield in temperate cereals. Crop Sci 42:111–121

Rong JK, Abbey C, Bowers JE, Brubaker CL, Chang C, Chee PW, Delmonte TA, Ding XL, Garza JJ, Marler BS, Park CH, Pierce GJ, Rainey KM, Rastogi VK, Schulze SR, Trolinder NL, Wendel JF, Wilkins TA, Williams-Coplin TD, Wing RA, Wright RJ, Zhao XP, Zhu LH, Paterson AH (2004) A 3347-locus genetic recombination map of sequence-tagged sites reveals features of genome organization, transmission and evolution of cotton (*Gossypium*). Genetics 166:389–417

Rosenow DT, Quisenberry JE, Wendt CW, Clark LE (1983) Drought tolerant sorghum and cotton germplasm. Agric Water Management 7:207–222

Rosielle AA, Hamblin J (1981) Theoretical aspects of selection for yield in stress and non-stress environments. Crop Sci 21:943–946

Ruan Y, Gilmore J, Conner T (1998) Towards *Arabidopsis* genome analysis: Monitoring expression profiles of 1400 genes using cDNA microarrays. Plant J 15:821–833

Sanità di Toppi L, Pawlik-Skowrońska B (eds) (2003) Abiotic stresses in plants. Kluwer Academic Publishers, Dordrecht, The Netherlands

Saranga Y, Flash I, Yakir D (1998) Variation in water-use efficiency and its relation to carbon isotope ratio in cotton. Crop Sci 38:782–787

Saranga Y, Jiang CX, Wright RJ, Yakir D, Paterson AH (2004) Genetic dissection of cotton physiological responses to arid conditions and their inter-relationships with productivity. Plant Cell Environ 27:263–277

Saranga Y, Menz M, Jiang CX, Wright RJ, Yakir D, Paterson AH (2001) Genomic dissection of genotype x environment interactions conferring adaptation of cotton to arid conditions. Genome Res 11:1988–1995

Schena M, Shalon D, Davis RW, Brown PO (1995) Quantitative monitoring of gene-expression patterns with a complementary-DNA microarray. Science 270:467–470

Schuepp PH (1993) Leaf boundary layers, Tansley Review No. 59. New Phytologist 125:477–507

Serraj R, Sinclair TR (2002) Osmolyte accumulation: Can it really help increase crop yield under drought conditions? Plant Cell Environ 25:333–341

Shen X, Van Becelacre G, Kumar P, Davis RF, May OL, Chee P (2006) QTL mapping for resistance to root-knot nematodes in the M-120 RNR Upland cotton line (*Gossypium hirsutum* L.) of the Auburn 623 RNR source. Theor Appl Genet DOI 10.1007/s00122-006-0401-4

Showler AT (2002) Effects of water deficit stress, shade, weed competition, and kaolin particle film on selected foliar free amino acid accumulations in cotton, *Gossypium hirsutum* (L.). J Chem Ecol 28:631–651

Singh V, Pallaghy CK, Singh D (2006) Phosphorus nutrition and tolerance of cotton to water stress II. Water relations, free and bound water, and leaf expansion rate. Field Crops Res 96:199–206

Slafer GA (2003) Genetic basis of yield as viewed from a crop physiologist's perspective. Ann Appl Biol 142:117–128

Smirnoff N (1998) Plant resistance to environmental stress. Curr Opin Biotechnol 9:214–219

Sotirios KA, Argyrokastritis A, Loukas M, Eliopoulos E, Tsakas S, Kaltsikes PJ (2006) Isolation and characterization of stress related Heat shock protein calmodulin binding gene from cultivated cotton (Gossypium hirsutum L.). Euphytica 147:343–351

Stanhill G (1992) The limits of water-use efficiency in agriculture. First Volcani International Symposium, Bet-Dagan, pp 45–45

Stiller WN, Read JJ, Constable GA, Reid PE (2005) Selection for water use efficiency traits in a cotton breeding program: Cultivar differences. Crop Sci 45:1107–1113

Stiller WN, Reid PE, Constable GA (2004) Maturity and leaf shape as traits influencing cotton cultivar adaptation to dryland conditions. Agronomy J 96:656–664

Stuber CW, Polacco M, Lynn M (1999) Synergy of empirical breeding, marker-assisted selection, and genomics to increase crop yield potential. Crop Sci 39:1571–1583

Tangpremsri T, Fukai S, Fischer KS (1995) Growth and yield of sorghum lines extracted from a population for differences in osmotic adjustment. Aust J Agric Res 46:61–74

Tomkins JP, Peterson DG, Yang TJ, Main D, Wilkins TA, Paterson AH, Wing RA (2001) Development of genomic resources for cotton (*Gossypium hirsutum* L.): BAC library construction, preliminary STC analysis, and identification of clones associated with fiber development. Mol Breed 8:255–261

Tuinstra MR, Grote EM, Goldsbrough PB, Ejeta G (1997) Genetic analysis of post-flowering drought tolerance and components of grain development in *Sorghum bicolor* (L.) Moench. Mol Breed 3: 439–448

Ulloa M, Zeiger E, Lu Z, Cantrell RG, Percy RG (2000) QTL analysis of stomatal conductance and relationship to lint yield in an interspecific cotton. J Cotton Sci 4:10

Umezawa T, Fujita M, Fujita Y, Yamaguchi-Shinozaki K, Shinozaki K (2006) Engineering drought tolerance in plants: discovering and tailoring genes to unlock the future. Curr Opin Biotechnol 17:113–122

UNCTAD (2006) Cotton: Characteristics. United Nations Conference on Trade and Development http://r0.unctad.org/infocomm/anglais/cotton/sitemap.htm

Underhill PA, Jin L, Lin AA, Mehdi SQ, Jenkins T, Vollrath D, Davis RW, CavalliSforza LL, Oefner PJ (1997) Detection of numerous Y chromosome biallelic polymorphisms by denaturing high-performance liquid chromatography. Genome Res 7:996–1005

Voloudakis AE, Kosmas SA, Tsakas S, Eliopoulos E, Loukas M, Kosmidou K (2002) Expression of selected drought-related genes and physiological response of Greek cotton varieties. Funct Plant Biol 29:1237–1245

Wang J, Zhang H, Allen RD (1999) Overexpression of an *Arabidopsis* peroxisomal ascorbate peroxidase gene in tobacco increases protection against oxidative stress. Plant Cell Physiol 40:725–732

Ware JO (1951) Origin, rise, and development of American Upland cotton varieties and their status at present. University of Arkansas, College of Agriculture, Agric Experiment Sta, Fayetteville, AR

Wells R, Meredith WR (1986) Normal vs. okra leaf yield interactions in cotton. 2. Analysis of vegetative and reproductive growth. Crop Sci 26:223–228

Wendel JF (1989) New world tetraploid cottons contain old world cytoplasm. Proceedings of the National Academy of Sciences of the United States of America 86:4132–4136

Wendel JF, Brubaker CL, Percival AE (1992) Genetic diversity in *Gossypium hirsutum* and the origin of Upland cotton. Am J Bot 79:1291–1310

Wendel JF, Percy RG (1990) Allozyme diversity and introgression in the Galapagos Islands endemic *Gossypium darwinii* and its relationship to continental *Gossypium barbadense*. Biochem Syst Ecol 18:517–528

Westengen O, Huaman Z, Heun M (2005) Genetic diversity and geographic pattern in early South American cotton domestication. Theor Appl Genet 110:392–402

Wilkinson JR, Spradling KD, Yoder DW, Pirtle IL, Pirtle RA (2005) Molecular cloning and analysis of a cotton gene cluster of two genes and two pseudogenes for the PR5 protein osmotin. Physiol Mol Plant Path 67:68–82

Wilson ID, Barker GL, Edwards KJ (2003) Genotype to phenotype: A technological challenge. Ann Appl Biol 142:33–39

Wright RJ, Thaxton PM, El-Zik KH, Paterson AH (1999) Molecular mapping of genes affecting pubescence of cotton. J Hered 90:215–219

Wu CA, Yang GD, Meng QW, Zheng CC (2004) The cotton *GhNHX1* gene encoding a novel putative tonoplast Na+/H+ antiporter plays an important role in salt stress. Plant Cell Physiol 45:600–607

Yakir D, Deniro MJ, Ephrath JE (1990) Effects of water-stress on oxygen, hydrogen and carbon isotope ratios in 2 species of cotton plants. Plant CellEnviron 13:949–955

Yan JQ, He CX, Wang J, Mao ZH, Holaday SA, Allen RD, Zhang H (2004) Overexpression of the *Arabidopsis* 14-3-3 protein GF14 λ in cotton leads to a "Stay-Green" phenotype and improves stress tolerance under moderate drought conditions. Plant Cell Physiol 45:1007–014

Yan JQ, Wang J, Tissue D, Holaday AS, Allen R, Zhang H (2003) Photosynthesis and seed production under water-deficit conditions in transgenic tobacco plants that overexpress an *Arabidopsis* ascorbate peroxidase gene. Crop Sci 43:1477–1483

Yan JQ, Wang J, Zhang H (2002) An ankyrin repeat-containing protein plays a role in both disease resistance and antioxidation metabolism. Plant J 29:193–202

Zeng ZB, Kao CH, Basten CJ (1999) Estimating the genetic architecture of quantitative traits. Genet Res 74:279–289

Zhang H, Wang J, Nickel U, Allen RD, Goodman HM (1997) Cloning and expression of an *Arabidopsis* gene encoding a putative peroxisomal ascorbate peroxidase. Plant Mol Biol 34:967–971

Zhang JX, Nguyen HT, Blum A (1999) Genetic analysis of osmotic adjustment in crop plants. J Exp Bot 50:291–302

CHAPTER 32

RECENT ADVANCES IN MOLECULAR BREEDING OF FORAGE CROPS FOR IMPROVED DROUGHT AND SALT STRESS TOLERANCE

JI-YI ZHANG AND ZENG-YU WANG
Forage Improvement Division, The Samuel Roberts Noble Foundation, 2510 Sam Noble Parkway, Ardmore, OK 73401, USA
Email: zywang@noble.org

Abstract: Forages play a key role in ruminant livestock production and environmental protection. Because forage grasses and forage legumes often grow in marginal areas, stress tolerance is one of the most important traits in forage cultivar development. Conventional and genetic engineering approaches have been used to improve stress tolerance of forage grasses and legumes. This review summarizes recent advances in improving drought and salt tolerances of several major forage species

Keywords: forage grass, forage legume, drought tolerance, salt tolerance, genetic improvement

1. INTRODUCTION

Forages are normally referred to as plants and plant parts that are consumed by domestic livestock such as beef and dairy cattle, sheep, horses, and a wide range of other animals (Barnes and Baylor, 1995). The majority of the cultivated forages fit into two botanic families, Poaceae (Gramineae), the grasses, and Fabaceae (Leguminosae), the legumes (Nelson and Moser, 1995). The most intensively used forage grasses include tall fescue (*Festuca arundinacea* Schreb.), meadow fescue (*F. pratensis* Huds.), perennial ryegrass (*Lolium perenne* L.), Italian ryegrass (*L. multiflorum* Lam.), bermudagrass (*Cynodon dactylon* (L.) Pers.), bromegrass (*Bromus inermis* Leyss.) and orchardgrass (*Dactylis glomerata* L.). The most commonly used legume forages are alfalfa (*Medicago sativa* L.), white clover (*Trifolium repens* L.) and red clover (*T. pratense* L.). All these forage species exhibit gametophytic self-incompatibility and hence require cross-pollinated breeding systems.

Forages are the backbone of grassland agriculture system. Grasses and legumes play a key role in ruminant livestock production in South and North America,

M.A. Jenks et al. (eds.), Advances in Molecular Breeding Toward Drought and Salt Tolerant Crops, 797–817.

Australia, New Zealand and many EU countries. In the US alone, of the 938 million acres of land in farms, 48.6% or 456 million acres are classified as cropland pasture or grassland pasture (USDA, 2002). In Western Europe, 17% of the total land area consists of permanent grassland, and the UK grasslands currently comprise 64% of the agriculture land area (Wilkins and Humphreys, 2003).

In addition to serving as the major sources of feed nutrients for domestic and wild animals, forages contribute to human well-being through many other ways including: (1) protection and conservation of soil and water resources; (2) improvement of soil structure and fertility; (3) providing habitat for wildlife; (4) creation of recreation space for sport and leisure; (5) source of biomass for the productions of biofuels that has become attractive recently; (6) improvement and protection of the environment from pollution such as sediment, wind-blowing soil, municipal and farm wastes, and some toxic substances (Barnes and Baylor, 1995).

In the soil-plant-animal biological system, forages are in heavy demand from livestock. On the other hand, they bear severe growth limitations from soil and the environment because most of the forages are grown in marginal agricultural areas. An inevitable increase of world population during the next several decades will force the production of more food, meat and dairy products. We may face the fact of forcing grassland to far marginal areas that have even poorer soil and land management system featured with low water-holding capacity, infrequent irrigation, limited fertility or high salt content (Sanderson et al., 1997). Water resources available for irrigation are becoming scarce and this trend may increase drastically in the future with the likely scenarios of global warming (Breshears et al., 2005). For grassland agricultural systems with perennial forages and natural vegetations, the ability to survive periods of environmental constraint is an essential characteristic for success. To improve yield and sustainability in forage production, a survival strategy may be more important than a growth strategy, particularly in more severe or variable environments. This is characteristically different from many field agricultural systems such as annual crops, in which maximizing yield under optimum environmental conditions is a priority (Eagles et al., 1997).

2. STRESS RESPONSE AND PHYSIOLOGICAL STUDIES IN A FEW MAJOR FORAGE SPECIES

The fescues (genus *Festuca*) are cool-season C3 plants that are mainly used in temperate or cool climate regions. Among about 80 species in *Festuca*, tall fescue (*F. arundinacea* Schreb., $2n = 6x = 42$) is the most important forage species worldwide of the genus. In the US alone, tall fescue occupies approximately 12–14 million ha in pure stand and forms the forage basis for beef cow-calf production in the east-central and southeast US (Sleper and Buckner, 1995). Deep rooting accounts for a major share of the differences in drought tolerance between plants (Boyer, 1996). As a perennial grass species, tall fescue is better adapted to avoid drought than other cool-season grasses such as perennial ryegrass or Kentucky bluegrass (*Poa pratensis* L.) partially due to bigger root size (length or mass) and spatial distribution (Sheffer et al., 1987; Ervin and Koski, 1998). A link was found between drought tolerance of turf type tall fescue and major forage tall fescue cultivars with their enhanced deeper root growth under water limiting conditions (Carrow, 1996; Huang and Fry, 1998; Huang and Gao, 2000). This finding has been used in drought tolerance breeding programs by selecting low shoot-to-root ratios in turf type tall fescue populations with appropriate potting system (Bonos et al., 2004).

Summer dormancy is another useful strategy for tall fescue to survive the dry summer (Malinowski et al., 2005). Obligatory summer-dormancy (defined as plant dormancy in response to increased daylength and probably high temperature) has been characterized from some cool-season perennial grasses (Ofir and Kigel, 1999). By using a newly developed cultivar combined with novel endophyte, it was demonstrated that obligatory summer-dormant tall fescue had better drought tolerance in Texas Rolling Plains that were beyond the adaptation range of currently utilized semi-dormant tall fescue type and check cultivar (Malinowski et al., 2005). The mechanisms of obligatory summer-dormancy remain to be understood at the physiological, biochemical and molecular levels.

Alfalfa (*Medicago sativa* L., $2n = 4x = 32$, also called lucerne) is grown extensively throughout temperate and tropical regions for hay, silage and pasture with about 32 million hectares worldwide and 15 million hectares in North America (Volence et al., 2002). Alfalfa combines high biomass productivity, optimal nutritional profiles and adequate persistence, thus making it ideal for dairy cattle and other livestock (Brummer, 2004). In the US, alfalfa is the fourth most widely grown crop behind corn, wheat, and soybean. As a perennial forage crop, alfalfa is a fairly hardy species and has a relatively high level of drought tolerance compared with many food crops and other legume forages (Barnes and Sheaffer, 1995). When compared with other perennial forage legumes under field conditions in North Central US, average herbage yield of drought stressed alfalfa was 120% greater than yields of birdsfoot trefoil (*Lotus corniculatus* L.) and cicer milkvetch (*Astragalus cicer* L.), and 165% greater than red clover (*T. pratense* L.) (Peterson et al., 1992). The greater drought tolerance of alfalfa is partially due to its deeper roots and the ability to extract more available water in the root zone (Hall, 2001). Alfalfa becomes dormant during periods of cold or severe drought and may last for 1 to 2 years until the temperature or moisture available to resume growth (Hall et al., 1988; Barnes and Sheaffer, 1995).

The mechanisms controlling fall dormancy and winter hardiness in alfalfa have been investigated with biochemical and molecular approaches. The difference in the level of freezing tolerance between non-hardy and winter hardy alfalfa cultivars was found to be related to the capacity of the plants to accumulate raffinose and stachyose in their roots and crowns, other than the capacity to accumulate sucrose (Castonguay et al., 1995). The fall dormant alfalfa plants tended to have increased gene transcript and activity of galactinol synthase (GaS), one of the key enzymes in the raffinose family oligosaccharides (RFO), and start to accumulate raffinose and stachyose in crown earlier than non-dormant plants (Castonguay et al., 1995;

Castonguay and Nadeau, 1998; Cunningham et al., 2003). The correlation of fall dormancy and accumulation of RFO in critical organs of alfalfa plants revealed one more aspect to decipher the biochemical nature of alfalfa dormancy under drought and heat stress.

During drought or salt stress, plants induce processes that regulate osmotic adjustment to maintain sufficient cell turgor partially through accumulation of compatible solutes compromised of mainly nontoxic low molecule chemicals. Accumulation of proline upon dehydration due to water deficit, high salinity and low temperature has been widely reported in bacteria, algae and higher plants and the causal relationship between increased proline accumulation and plant tolerance of hyperosmotic stresses has been demonstrated (Hare et al., 1999). Proline content in alfalfa leaves and roots increased dramatically when plants were subjected to drought, no mater if the plants had been inoculated with mycorrhizal fungi or rhizobium bacteria or both (Goicoechea et al., 1998). However, it was found that proline content in alfalfa root did not show significant change during a 72 h treatment with 90 mM NaCl, while two Δ^1-pyrroline-5-carboxylate (*P5CS*) genes were transcriptionally induced to higher levels at all the time points assayed (Ginzberg et al., 1998a). As P5CS is recognized as the rate limited enzyme in proline biosynthesis, the weak link between transcript increase and proline accumulation in alfalfa remain to be understood. To characterize transcriptional regulation of the key proline cycle enzymes in alfalfa, a partial sequence of Δ^1-pyrroline-5-carboxylate dehydrogenase (*MsP5CDH*) gene and two proline dehydrogenase (*MsPDH*) genes that share a high nucleotide sequence homology and a similar exon/intron structure were identified and cloned. Estimation of transcript levels during salt stress and recovery revealed that proline accumulation during stress was linearly correlated with a strong decline of *MsPDH* transcript levels, while *MsP5CDH* steady-state transcript levels remained essentially unchanged. Salt-induced repression of *MsPDH1* promoter linked to the GUS reporter gene confirmed that the decline in *MsPDH* transcript level was due to less transcription initiation. Contrary to the salt-dependent repression, a rapid induction of *MsPDH* transcription occurred at a very early stage of the recovery process, independently of earlier salt treatments. Thus, two different regulatory modes of *MsPDH* expression exist, the repressing mode that quantifies salt concentration in an yet unknown mechanism and the 'rehydration'-enhancing mode that responds to stress relief with maximal induction of *MsPDH* transcription (Miller et al., 2005).

Besides proline, proline betaine and glycine betaine are among the major compatible solutes found in stressed alfalfa (Wood et al., 1991; Girousse et al., 1996; Ginzberg et al., 1998b). Alfalfa is one of the few plants that accumulates large amounts of both proline and proline betaine simultaneously (Trinchant et al., 2004). The combination might help to explain the relatively high overall osmotic tolerance of alfalfa. It was also demonstrated that proline betaine was efficiently catabolized through sequential demethylations via *N*-methylproline and proline. Salt stress was found to play a minor role on the biosynthesis rather it had stronger effect in reducing the turnover of proline betaine. In this way the proline betaine in shoots, roots, and nodules showed 10-, 4-, and 8 fold increase respectively (Trinchant et al., 2004). However, the mechanism of proline betaine biosynthesis and its turnover in alfalfa have not been elucidated.

Identification of signal transduction pathways in response to drought and salt stress in alfalfa has been a research topic for more than a decade. At least two mitogen-activated protein kinase (MAPKs) pathways have been found to play some roles in drought and salt stress signal cascades in alfalfa cells. MKK4 has been characterized as the MAP kinase that mediate drought and cold stress signals and plant response (Jonak et al., 1996). By means of MBP kinase assays of immunoprecipitations using antisera raised against kinase-specific peptides, it was found that kinase activity of MKK4 is activated rapidly and transiently by cold and drought stress, but not by heat or salt. At the same time, the transcription of *MKK4* gene is also induced by cold and drought stress, even though the steady state protein levels remained constant. Other MAPK members identified from alfalfa, MKK2 and MKK3, were not activated by these conditions (Jonak et al., 1996). As MKK4 activation in alfalfa appears to be associated with many different forms of stress, MKK4 was suggested to be renamed to SAM kinase (stress-activated MAP kinase) (Bogre et al., 1997). Another MAP kinase pathway was proposed, in which the 46 kDa kinase, SIMK, was identified and denoted as a salt induced MAP kinase (Munnik et al., 1999). It was not only activated by NaCl, but also by KCl and sorbitol, so it was believed to be an osmo-sensing protein. Compared with other MAPKs, this 46 kDa alfalfa SIMK was unique because it was constitutively localized in nuclei (Munnik et al., 1999). Its target molecules are most probably other nuclear proteins such as transcription factors that regulate the expression of unknown downstream genes of salt stress response pathway. It was also essential to root hair tip growth and the root hairs could be enhanced when it was overexpressed in tobacco plants (Šamaj et al., 2002). SIMKK was isolated and characterized as an upstream activator that interacted specifically with SIMK and phosphorylated both the threonine and tyrosine residues in its activation loop. Moreover, SIMKK enhanced the salt-induced activation of SIMK *in vivo* (Kiegerl et al., 2000). The activation of SIMK by SIMKK does not need stimulation by other upstream factors including MAPKKK, except for the requirement of salt stress (Cardinale et al., 2002). The proposed autoactive ability of SIMKK and SIMKK –like kinases and the activation enhancement of the downstream kinases make them useful candidates in salt stress tolerance improvement.

Formation of lipid peroxides either chemically or enzymatically is another integrated component of cellular damage of plants during drought and other stresses. A putative NADP-dependent aldose/adehyde reductase gene *MsALR*, whose transcription in somatic embryo was induced by osmotic (10% PEG treatment), heavy metal, oxidative, drought stress and ABA treatment, has been identified from alfalfa (Oberschall et al., 2000). The recombinant alfalfa enzyme is active on at least one known cytotoxic lipid peroxide degradation product, 4-hydroxynon-2-enal *in vitro*. Overexpressing this gene in tobacco plants provided considerable tolerance against oxidative damage and showed resistance to a long period of water deficiency and exhibited improved recovery after rehydration. These studies reveal a new and efficient detoxification pathway in alfalfa plants (Oberschall et al., 2000).

White clover (*Trifolium repens* L. $2n = 4X = 32$) is an allotetraploid species widely distributed in the world due to its wide range of climate adaptation (Pederson, 1995). But it is less tolerant of drought compared with other perennial temperate forage legumes because of its shallow root system and inability to effectively control transpiration (Hart, 1987; Annicchiarico and Piano, 2004). As white clover is used in systems for cattle or sheep grazing and is grown together with a companion grass, competition with associated grasses such as perennial ryegrasses, bermudagrass or tall fescue places white clover under additional water stress. Therefore, unlike many other crops, maximization of yield per se is not the main objective but rather the aim is to produce a balanced sward with a reliable, consistent white clover contribution.

The major feature of white clover is its stoloniferous habit. It spreads by growth of stolons with adventitious roots developing at the nodes. The persistence is largely dependent on the ability of vegetative stolons to survive variable periods of drought (Williams, 1987). So the development of a strong network of stolons is a prerequisite and stolon characters have been a major focus of breeding effects in this species (Sanderson et al., 2003). Biochemical studies indicated that when white clover was stressed with water deficit, the *de novo* amino acid synthesis including proline was increased in both leaves and roots (Lee et al., 2005). The phenomenon may serve as adaptive response during the first few days in drought stress, as the transient increase of amino acid concentration was followed by the decrease of protein synthesis that make the plants grow slower. The signal transduction of this early response has not received enough attention.

3. IMPROVEMENT OF STRESS TOLERANCE BY INTERGENERIC HYBRIDIZATION

Wide hybridization with relative species followed by chromosome and/or chromosome fragment introgression has been considered as an efficient way to transfer drought, salt and other stress tolerance gene (s) to the target species to widen the gene pool. Intergeneric hybrids between *Lolium* and *Festuca* species have received much attention by forage breeders.

Ryegrasses are considered the ideal grasses due to their rapid establishment, ability to withstand heavy grazing, good palatability and high nutritious value (Humphreys et al., 2003). However, their growth is restricted only to some European countries, some regions in Australia, New Zealand and Southeast US because they are not sufficiently robust to meet many of the environmental challenges in less favorite agriculture areas (Thomas et al., 2003). Among the genetically close relatives of ryegrasses are *Festuca* species that show better adaptation to abiotic and biotic stresses. Most of the species in this genus are more persistent due partially to their better developed root system (Sheffer et al., 1987; Ervin and Koski, 1998;

Humphreys et al., 1998). The close taxonomic relationship between *Lolium* and *Festuca* species makes it possible to hybridize between them and transfer genes through recombination of homoeologous chromosomes. This is the major reason why a man-made species *Festulolium* has been used in many grass breeding program worldwide (Dijkstra and Vos, 1975; Spangenberg et al., 1994; Casler et al., 2001; Kopecky et al., 2005; Yamada et al., 2005). Even though the amphidiploids are not widely used as forage crops due to low fertility and genetic instability (Canter et al., 1999), they serve as a potential resource to improve drought tolerance and other environmental stress tolerance of ryegrasses via gene introgression from fescues.

F. glaucescens is a tetraploid species with better drought resistance. The inter-species F_1 hybrid between *F. glaucescens* and a synthetic tetraploid Italian ryegrass was backcrossed with the diploid Italian ryegrass parent, and in BC_2 generation the predominate plants were diploid with introgressed *F. glaucescens* chromosome segments (Morgan et al., 2001). One line was selected, which could survive a combined severe drought and heat stress and contained one single short chromosome fragment from *F. glaucescens*. This transferred chromosome fragment located on the distal region of Italian ryegrass chromosome 3 viewed by GISH (Genomic in situ Hybridization) and *F. glaucescens*-specific molecular markers have been identified and used for molecular marker assisted selection in the breeding process (Humphreys et al., 2005).

Tall fescue is a valuable gene source of drought tolerance. In a back-crossing program involving *L. multiflorum* (the recurrent parent) and *F. arundinacea*, the diploid *L. multiflorum* phenotype was rapidly recovered with the inclusion of a small number of genes from the fescue parent. In field drought trials, it was found that 3% of the derivatives of these backcross populations were more drought resistant than the *L. multiflorum* parental populations and as drought resistant as *F. arundinacea*. After polycrossing of selected drought-resistant *Lolium*-like plants followed by one cycle of selection, the mean drought resistance of most progeny lines was significantly improved, in some cases to near that of *F. arundinacea* (Humphreys and Thomas, 1993). Genes for drought resistance transferred from *F. arundinacea* were mapped onto chromosome 2 in two *Lolium* genotypes. The two drought-resistant lines have the high water conductance of *Festuca* on their adaxial leaf surface and the low abaxial conductance of *Lolium* (Humphreys et al., 1997). The pentaploid hybrid ($2n = 5x = 35$) of autotetraploid *L. multiflorum* and *F. arundinacea* combines the high growth rate of *L. multiflorum* with the drought resistance and freezing-tolerance of *F. arundinacea*. To access different combinations of these characters, anther cultures were used to quickly select the plants with euploid chromosome numbers (14, 21, and 28) (Zare et al., 2002). Wide variation was found in plant height, leaf length, leaf width, tiller number and herbage dry matter among mature androgenic plants grown under field conditions. A number of lines have been identified that showed higher herbage dry matter under drought stress conditions (Zare et al., 2002).

White clover is a highly heterozygous outcrossing species with considerable variations available for the improvement of many traits, but this is not the case for some desirable attributes including drought tolerance (Brink and Pederson, 1998; Abberton and Marshall, 2005). Direct selection for drought tolerance has been carried out in the field and indirect methods have also been used, but success has been limited (Annicchiarico and Piano, 2004; Abberton and Marshall, 2005). In white clover drought tolerance improvement practice, introgression has also been used as a route to transfer the morphological or physiological traits from its related wide species that show more drought tolerance or have better persistence. Hybrids of white clover and related species Caucasian or Kura clover (*T. ambiguum* M. Bieb) and ball clover (*T. nigrescens* L. Viv.) have been developed to introgress key traits such as drought tolerance and grazing tolerance into the white clover gene pool. Caucasian or Kura clover (*T. ambiguum* M. Bieb) is a very persistent species with good drought tolerance due partially to its rhizomatous habit (Meredith et al., 1995). A range of backcross hybrids using white clover as recurrent parent have been generated (Anderson et al., 1991; Abberton et al., 1998). In the third generation of backcross, individual plants that were white clover but with more rhizomes as well as stolons were obtained and their drought tolerance were superior to the white clover parent (Abberton et al., 1998; Marshall et al., 2001).

4. IMPROVEMENT OF STRESS TOLERANCE BY GENETIC TRANSFORMATION

Genetic improvement of forages by conventional plant breeding is very slow. One reason is that most forage species are self-incompatible, which limits inbreeding to concentrate desired genes for use in rapid development of new cultivars. Genetic transformation allows the direct introduction of agronomic genes, thus offers new opportunities for molecular breeding of forages. Transformation systems for alfalfa and white clover have been well established, although genotype still plays an important role in *Agrobacterium* transformation. The production of transgenic grass plants has been more difficult than that of legumes. Transgenic grass plants were first obtained by biolistic transformation of embryogenic cell cultures. In recent years, protocols based on *Agrobacterium*-mediated transformation have been developed for major forage grasses including tall fescue, ryegrasses and bermudagrass (Wang and Ge, 2006). It is known that *Agrobacterium*-mediated transformation generally results in a lower copy number and an improved stability of gene expression than the free DNA delivery methods.

The responses of alfalfa to drought and salt stress signals have been extensively studied at the physiological and biochemical levels, however, limited information is available at the molecular level. The transfer and use of resources and information accumulated in the model legume *Medicago truncatula* is likely to translate the power of genomic and metabolic approaches into forage improvement. Two genes encoding the plant-specific AP2/ERF transcription factors have been isolated from this model species (Zhang et al., 2005; Zhang et al., 2006). They are designated as *WXP1* and *WXP2* respectively, with the name after wax production. Both *WXP1* and *WXP2* are distinctly different from the most studied genes in the AP2/ERF transcription factor family such as *AP2s*, *CBF/DREB1s*, *DREB2s*, *WIN1/SHN1* and *GL15*. Overexpression of *WXP1* under the control of the CaMV35S promoter led to significant increase in cuticular wax loading on leaves of transgenic alfalfa. It was revealed with electron microscopy scanning that wax crystals on the adaxial surface of newly expanded leaves accumulated earlier in transgenic plants than in control plants. The density of wax crystalline structures on both adaxial and abaxial surfaces of mature leaves was higher in transgenic than in control plants. The total leaf wax accumulation per surface area increased 29.6–37.7% in the transgenic lines, and the increase was mainly contributed by C30 primary alcohol. Transgenic leaves showed reduced water loss and chlorophyll leaching. Transgenic alfalfa plants with increased cuticular waxes showed enhanced drought tolerance demonstrated by delayed wilting after watering was ceased and quicker and better recovery when the dehydrated plants were re-watered (Zhang et al. 2005). Transgenic expression of either *WXP1* or *WXP2* in Arabidopsis led to significantly increase of cuticular wax deposition on leaves of 4-week-old and 6-week-old transgenic plants, even though differences in the accumulation of various wax components as well as their chain length distributions were found in the *WXP1* and *WXP2* plants. Analysis of fresh weigh loss from detached leaves revealed that the transgenic leaves tend to hold more water than the control. Under drought stress conditions, both *WXP1* and *WXP2* transgenic Arabidopsis plants showed significantly enhanced whole plant drought tolerance (Zhang et al. 2007). As *WXP1* is believed to be one of the useful candidate genes for improving plant drought and freezing tolerance, we have transferred this gene into a white clover under the control of a putative epidermal specific promoter. Preliminary results show that transgenic plants wilt later than empty vector controls growing in the same pot under drought conditions (Zhang and Wang, unpolished data).

The level of reactive oxygen intermediates/species (ROIs or ROS) in plant cells is alleviated by many environmental stresses (Mittler, 2002). There are at least two different mechanisms to regulate the intracellular concentration by scavenging the ROS, one for signal purposes which can modulate the low level of ROS, and another for detoxification of excess ROS during stress that is associated with oxidative damage at the cellular level (Mittler, 2002). Based on the later hypothesis, it appears to be a promising approach to obtain plants with diverse tolerance to abiotic stress by preventing oxidative stress or reducing the level of the reactive molecules (Allen, 1995). Superoxide dismutase (SOD) is a ubiquitous metal-containing enzyme of antioxidant system existing in many cellular compartments that can detoxify oxygen radicals to produce hydrogen peroxide and oxygen, in which the resulting hydrogen peroxide can be converted into O_2 or H_2O by other enzymes such as ascorbate peroxidase (APX), catalase (CAT), and glutathione peroxidase (GPX). When a Mn-SOD cDNA from *Nicotiana plumbaginifolia* was overexpressed in alfalfa genotype RA3, the transgenic plants tended to have reduced injury from water-deficit stress as determined by lower chlorophyll fluorescence, less electrolyte leakage, and better regrowth from crowns (McKersie et al., 1996). In a 3-year field trial with these transgenic plants, the survival rate and biomass production were significantly improved under drought stressed conditions. Under greenhouse conditions, transgenic alfalfa plants of the elite genotype N4 with the mitochondria Mn-SOD, the chloroplast Mn-SOD and the chloroplast Fe-SOD all showed 20% higher photosynthesis activity than wide type control plants at mild water stress (Rubio et al., 2002). The better performance was ascribed to a better stomatal conductance. However, pyramiding chloroplast targeted Mn-SOD and mitochondria targeted Mn-SOD resulted in decreased shoot and storage organ (crown + root) biomass when compared with transgenic lines harboring one of the genes or the wide type (Samis et al., 2002).

A common problem in irrigated agriculture is the gradual buildup of salts in the root zone, which can be detrimental to sustained crop production. Salt stress significantly limits productivity of alfalfa via its adverse effects on growth and symbiotic nitrogen-fixation capacity. As more than half of the alfalfa acreage in the US is irrigated with different water source, salt stress tolerance is required in newly developed cultivars that will be useful not only as a forage crop, but can be also used for bioremediation of salt-compromised land and as an efficient cover crop.

Salt stress can also alter gene expression via gene regulation and post-transcription modification. A zinc-finger family transcription factor, Alfin 1, is one that maintains normal root growth and development as well as modulates alfalfa tolerance through regulation of salt inducible gene expression in root. The cDNA sequence of *Alfin1* was isolated by differential screening of a cDNA library constructed with cells of a salt-tolerant alfalfa line (Winicov, 1993). It was predominantly expressed in root tissues and specifically induced by high salt. It was characterized as a putative transcription factor gene which had a Cys4 and His/Cys3 zinc finger motif in its deduced peptide (Bastola et al., 1998). The transcription activity and its possible regulation of salt stress response were confirmed *in vitro* that the recombinant Alfin1 could bind efficiently to adjacent G-rich triplet motifs in the promoter fragment of *MsPRP2* gene, which encoding a proline rich cell wall protein with a putative signal leading sequence (Bastola et al., 1998; Winicov, 2000). Alfin1 regulated the expression of *MsPRP2* gene in a root-specific manner in alfalfa (Winicov et al., 2004). Because the transgenic plants harboring antisense sequence of *Alfin 1* failed to establish and could not survive normal growth in the greenhouse, Alfin 1 is believed to be essential for root growth of alfalfa plant (Winicov and Bastola, 1999). Overexpression of *Alfin1* in alfalfa resulted in improved tolerance to salt stress of transgenic plants by enhancing root growth under normal and saline conditions. Calli expressing *Alfin 1* in the antisense orientation were more sensitive to NaCl inhibition (Winicov and Bastola, 1999).

Recent progress has been made in the identification and characterization of the mechanisms that allow plants to tolerate high salt concentrations. The identification of the different sodium transporters (in particular vacuolar and plasma membrane Na+/H+ antiporters) allows the engineering of crop plants with improved salt tolerance (Apse and Blumwald, 2002). The antiporters are prevalent membrane proteins present in bacteria, yeasts, animals, and plants. The Na^+/H^+ antiporter catalyzes the exchange of Na^+ and H^+ across the plasma membrane contributing to the regulation of internal pH, cell volume and sodium concentration. The vacuolar Na^+/H^+ antiporter can pump Na^+ from cytoplasm into vacuole, to maintain a higher K^+/Na^+ ratio in the cytoplasm than that in vacuoles, protecting cell from sodium toxicity. Recently a vacuolar Na^+/H^+ antiporter gene cloned from rice was overexpressed in perennial ryegrass by *Agrobacterium*-mediated transformation. The transgenic perennial ryegrass plants had dramatically improved salt-tolerance under 100–350 mmol/L NaCl treatment. After stress treatment for 10 weeks, while wild-type plants in pots were killed by watering with 350 mmol/l NaCl solution, transgenic plants survived the treatment (Wu et al., 2005). The leaves of transgenic plants accumulated higher concentrations of Na^+, K^+ and proline than those of the control plants (Wu et al., 2005).

5. MANIPULATION OF FRUCTAN BIOSYNTHESIS

Fructan metabolism has been studied intensively in tall fescue, perennial ryegrass and annual ryegrass due to their abundance and possible roles in osmotic adjustment and stress tolerance (Karsten and MacAdam, 2001; Ye et al., 2001; Chalmers et al., 2005). Fructans are a group of alternative carbohydrates found in bacteria, algae, and about 12–15% of the flowering plants including forage grasses (e.g. *Lolium* and *Festuca*) (Hendry and Wallace, 1993; Vijn and Smeekens, 1999; Cairns, 2003). They are linear or branched polyfructose molecules (< 50 for most of the plant fructans) produced by polymerization of glucose and fructose. They accumulate in growing and storage tissues of C3 or cool season grasses. In some plants, final fructan concentrations can be as high as 30% of the dry weight, with a gradient of accumulation between the apex and the base (Pollock and Cairns, 1991). Naturally they are predominantly stored in the vacuole of the cells in which they are synthesized from sucrose, lowering sucrose concentration in the cell and preventing sugar-induced feedback inhibition of photosynthesis.

In ryegrasses, fructans are found to be the major storage carbohydrates. They have been implicated in plant dehydration tolerance caused by drought and low temperature as the water-soluble carbohydrates can mediate osmotic adjustment (Hendry and Wallace, 1993; Pilon-Smits et al., 1995; Volaire et al., 1998; Hisano et al., 2004; Chalmers et al., 2005). In perennial ryegrass, fructan concentration was more affected by drought stress than other water soluble carbohydrates (Amiard et al., 2003). After eight-week-old perennial ryegrass plants were stopped watering for 14 days in the greenhouse, the most remarkable change was the two-fold increase in fructan concentration in elongating leaf base that was the most important survival organ. Fructans in leaf sheaths of the drought stressed plants were also dramatically increased. The increase was primarily due to the accumulation of high-degree of polymerization (DP) fructans (DP>8). Other water soluble carbohydrates such as sucrosyl-galactosides, raffinose and loliose did not show significant changes in leaf base, leaf sheaths and leaf blades. Thus, the role of fructans in drought tolerance of perennial ryegrass was assigned as protection of leaf base and leaf sheaths to better survive water deficit conditions (Amiard et al., 2003). Using a F_2 mapping population, Turner et al (2005) found that tiller bases of perennial ryegrass in the autumn had the highest water soluble carbohydrates (WSC) content, in which at least 74% being polymeric fructans. Compared with spring, the leaves in autumn also accumulated nearly 3-fold more polymeric fructans, which account for at least 59.5% of total WSC. These results indicated that perennial ryegrass uses mainly fructans as carbon source in autumn to maintain basal metabolism in period of adverse weather (Turner et al., 2006).

It has been shown that at least four enzymes are involved in fructan synthesis in higher plants, fructan:fructan 1-fructosyltransferase (1-FFT), Suc:Suc 1-fructosyltransferase (1-SST), fructan:fructan 6G-fructosyltransferase (6G-FFT), and Suc:fructan 6-fructosyltransferase (6-SFT) (Vijn and Smeekens, 1999). Based on the structure and composition of fructans found in perennial ryegrass, a hypothetical pathway of fructan biosynthesis was proposed that emphasized the metabolism in leaves (Pavis et al., 2001a). The accumulation and remobilization of fructans also need the concerted action of a number of specific enzymes such as fructosyltransferase (FT), fructan exohydrolases (FEH) and invertase (INV) that are tightly and coordinately regulated (Chalmers et al., 2005). Some of them have been isolated from perennial ryegrass. Characterization of the sequences from perennial ryegrass has shown they predominantly correspond to single copy genes, and in some cases map to genomic regions associated with phenotypic variation for carbohydrate content (Turner et al., 2006). Their transcript levels correlate with fructan metabolism enzyme activities in plant organs in which fructans typically accumulate (Pavis et al., 2001b). This opens up opportunities for the production of transgenic grasses with precisely altered endogenous fructan pools in grasses. Promoter sequences available for some of these genes will enable targeted and coordinated modification of fructan metabolism in transgenic grass plants. The knowledge obtained will ultimately benefit molecular breeding approaches for the development of cultivars with enhanced nutritive value and tolerance to abiotic stresses (Chalmers et al., 2005).

When either a 1-SST or a 6-SFT gene from wheat was transformed into perennial ryegrass under the control of the constitutive CaMV 35S promoter, selected overexpressors tended to accumulate 3- to 15-fold greater fructan than non-transgenic plants when sucrose content kept consistent (Hisano et al., 2004). Transgenic plants conferred enhanced freezing resistance while drought tolerance has not been assayed.

High concentrations of water-soluble carbohydrates (WSC) are accumulated in the growth zone of tall fescue leaf blades and stem base as the storage carbons for regrowth after defoliation and as a survival strategy after a period of adverse environmental stresses (Karsten and MacAdam, 2001). At normal growth conditions, about 80% of the WSC stored at the base of tall fescue leaves are fructans, which account for 40% of the dry matter in the leaf growth zone (Schnyder and Nelson, 1989). In response to drought stress, concentrations of fructan and other WSC decreases in tall fescue basal tissues, while sucrose and hexose concentration increases contributed by fructan hydrolysis, which decreases osmotic potential and improves the water status of plants (Spollen and Nelson, 1994). Sucrose hydrolysis in the basal 1.5 cm was largely due to fructosyltransferases, which had activities up to 10 times higher than in fully developed leaf tissue. Three fructosyltransferases that act both as (Suc):Suc 1-fructosyl transferase (1-SST) and as fructan:fructan 6^G-fructosyl transferase have been purified from the leaf growth zone of tall fescue (Luscher and Nelson, 1995). One cDNA sequence has been isolated from tall fescue, which corresponded to the predominant protein of one of the fructosyl transferases. The expression pattern of the corresponding mRNA in different zones of the growing leaves matched the 1-SST activity and fructan content. When the cDNA was transiently expressed in tobacco protoplasts, the corresponding enzyme preparations produced 1-kestose, showing that the cDNA encodes a 1-SST. When the cDNA was expressed in yeast, the recombinant protein had all the properties of known 1-SSTs, namely 1-kestose production, moderate nystose production, lack of 6-kestose production, and fructan exohydrolase activity with 1-kestose as the substrate (Luscher et al., 2000). Manipulation of fructan biosynthesis has not reported in tall fescue.

White clover does not accumulate fructan as a storage carbohydrate. By expressing the fructosyltransferase gene from bacterium *S. salivarius* under the control of the CaMV 35S promoter, the generated transgenic white clover plants accumulated fructan in leaves, petioles, stolons, flowers, and roots. Levels of fructan up to approximately 2% of dry weight were found in leaves. The fructan was of high molecular mass (>5000 kDa), typical of bacterial fructans. Fructosyltransferase enzyme activity in leaf extracts of the transformed plants appeared to be stable throughout leaf development. Most transformed lines appeared normal, flowered and produced seed, but the growth rate of some transformed lines decreased (Jenkins et al., 2002).

6. DROUGHT TOLERANCE VIA SYNBIOSIS WITH ENDOPHYTE

Evolution of a fungal-plant symbiosis around 400 million years ago may have been the key innovation that enabled plants to colonize the land (Remy et al., 1994). Related mycorrhizal fungi that grow in or on root continue to exist for more than 90% of land plants (Oldroyd et al., 2005). Besides, some grass endophytes live their entire life cycle within the aerial portion (shoot) of the host grass, forming nonpathogenic, systemic, and usually intercellular association (Malinowski and Belesky, 2000). Endophyte-infected grasses are better adapted than non-infected grasses to abiotic stresses, i.e., drought and marginal soil conditions due to direct changes affecting water status in shoots and indirect changes in root morphology and function (Malinowski and Belesky, 2000). These adaptations may arise from a chemical signaling system in the symbiotum. Apparently, drought signals sensed by roots can be received by endophyte and induce a range of responses in the host plants. Less is known about the chemistry of these signals at this time (Bultman, 2006). The possible mechanism of drought stress tolerance includes improved water uptake from the soil as of extensive root system (De Battista et al., 1990; Malinowski et al., 1999), better control of transpiration by rapid stomatal closure (Elmi and West, 1995), better water storage in tiller base by reduced leaf conductance (Elbersen and West, 1996), and enhanced drought tolerance by accumulating more compatible solutes including loline alkaloids (Richardson et al., 1992; Bacon, 1993; Bush et al., 1997).

Tall fescue plants on the pastures are commonly symbiotically infected with the fungal endophyte *Neotyphodium coenophialum* (Morgan-Jones and Gams) Glenn, Bacon and Hanlin (Bouton and Easton, 2005). Animals feed with endophyte-infected (E^+) tall fescue cultivars suffer from fescue toxicosis which causes poor weight gain and reproduction problems (Bouton et al., 1993; Sleper and West, 1996). Ergot alkaloids, especially ergovaline derived from the endophyte association are considered to be responsible for most animal problems (Lyons et al., 1986). However, introducing endophyte free tall fescue varieties has not been very successful because of their poor persistence once exposed to abiotic stresses. A very useful approach is to isolate naturally occurring, non-ergot-producing strains and re-infecting elite varieties. One such novel endophyte strain AR542 has been selected in New Zealand, which was used to re-infect tall fescue varieties Jesup and Georgia 5 in the US and commercialized by Pennington Seed, Inc. (Madison, GA) as the MaxQ technology (Bouton et al., 2002). More novel endophyte isolations have been characterized and are being used for re-infection of tall fescue and perennial ryegrass cultivars and breeding lines. Another promising approach to deal with this problem is genetic manipulation of *Neotyphodium* spp. endophytes to eliminate the toxin from the symbiosis (Panaccione et al., 2001).

Even though the symbiotic relationship between grass and its endophyte primarily affectthe growth of host plant under favorite and drought stresses conditions, maintenance of low shoot growth and enhanced root growth found in endophyte infected perennial ryegrass was believed one of the benefits to improve plant survival in dry areas (Cheplick, 2004; Hesse et al., 2005). It was also found that the effects of endophyte differed among host genotypes and associated with water supply (Cheplick and Cho, 2003; Lewis, 2004; Hesse et al., 2005). Further investigation indicated that endophyte colonization was a minor determinant of alkaloid levels, and that accumulation of the alkaloids relative to the endophyte mycelium is affected by plant genotype and tissue in a manner specific to each alkaloid (Spiering et al., 2005). Therefore, both host genotype and growth conditions need to be considered in characterizing the regulation of alkaloid levels in the grasses.

7. PERSPECTIVES

For perennial forage crops, good persistence, after years of growth with animal grazing and/or frequent cutting accompanied with unpredictable adverse weather periods, is highly required because full cultivation and re-seeding of permanent pasture is expensive and production during forage establishment period is normally very low (Wilkins and Humphreys, 2003). Maximum root depth should be combined with optimum leaf expansion and good control of water loss per unit leaf area via the cuticle and stomata (Pinheiro et al., 2005; Humphreys et al., 2006). Stress tolerance mechanisms that enable the plants to protect the water status of some critical tissues, such as stem base of grasses, crown of alfalfa and stolons of clover, should attract more attention in future research.

In forage crops, dormancy is an important developmental program allowing plants to withstand periods of extreme environmental conditions. But highly dormant genotypes tend to resume growth slowly after stresses are released. Precise control of dormancy, regarding time to build up and the intensity in the appropriate organs, may be a promising strategy for drought tolerance improvement (Mittler et al., 2001). Some issues related to this aspect are sensing the temperature change by root or root tips, signal transduction to shoot, deposition of C- and N- nutrients and drought/heat tolerance compatible solutes to root and/or crown for re-growth.

For most of the highly bred crops, many germplasms have been characterized with one or several specific traits that can be used to improve yield under drought. However, breeders have a number of good reasons not to introduce these so called unadapted parents into their breeding programs due to mainly the risk of rebuilding the harmonious combinations of these drought genes and their neighbor genes to act together with other good genes (Richards, 1996). Genetic transformation offers a solution to overcome this issue. Transgenic approaches are expected to complement or accelerate conventional breeding, since they offer the opportunity to generate unique genetic variation that is either absent or has very low heritability. Further understanding of the biochemical and molecular basis of plant stress response and tolerance remains a major challenge in plant biology research. The improvement of stress tolerance in forages should utilize the knowledge obtained from other plants, and at the same time, fully consider the unique aspect of forage crops regarding their growth and adaptation.

REFERENCES

Abberton, M.T. and Marshall, A.H., 2005, Progress in breeding perennial clovers for temperate agriculture, *J. Agri. Sci.* **143**: 117–135.

Abberton, M.T., Michaelson-Yeates, T.P.T., Marshall, A.H., Holdbrook-Smith, K. and Rhodes, I., 1998, Morphological characteristics of hybrids between white clover, *Trifolium repens* L., and Caucasian clover, *Trifolium ambiguum* M. Bieb, *Plant Breed.* **117**: 494–496.

Allen, R.D., 1995, Dissection of oxidative stress tolerance using transgenic plants, *Plant Physiol.* **107**: 1049–1054.

Amiard, V., Morvan-Bertrand, A., Billard, J.-P., Huault, C., Keller, F. and Prud'homme, M.-P., 2003, Fructans, but not the sucrosyl-galactosides, raffinose and loliose, are affected by drought stress in perennial ryegrass, *Plant Physiol.* **132**: 2218–2229.

Anderson, J.A., Taylor, N.L. and Williams, E.G., 1991, Cytology and fertility of the interspecific hybrid *Trifolium ambiguum* x *T. repens* backcross population, *Crop Sci.* **131**: 683–687.

Annicchiarico, P. and Piano, E., 2004, Indirect selection for root development of white clover and implications for drought tolerance, *J. Agron. Crop Sci.* **190**: 28–34.

Apse, M.P. and Blumwald, E., 2002, Engineering salt tolerance in plants, *Curr. Opin. Biotechnol.***13**: 146–150.

Bacon, C.W., 1993, Abiotic stress tolerance (moisture, nutrients) and photosynthesis in endophyte infevted tall fescue, *Agric., Ecosyst. Environ.***44**: 123–141.

Barnes, D.K. and Sheaffer, C.C., 1995, Alfalfa. In: *Forages. I. An Introduction to Grassland Agriculture*, R.F. Barnes, D.A. Miller, C.J. Nelson, eds, Iowa State University Press, Ames, Iowa, pp. 205–216.

Barnes, R.F. and Baylor, J.E., 1995, Forages in a changing world. In: *Forages: An Introduction to Grassland Agriculture*, R.F. Barnes, D.A. Miller, C.J. Nelson, eds, Iowa State University Press, Ames, pp. 3–13.

Bastola, D.R., Pethe, V.V. and Winicov, I., 1998, Alfin1, a novel zinc-finger protein in alfalfa roots that binds to promoter elements in the salt-inducible *MsPRP2* gene, *Plant Mol. Biol.* **38**: 1123–1135.

Bogre, L., Ligterink, W., Meskiene, I., Barker, P.J., Heberle-Bors, E., Huskisson, N.S. and Hirt, H., 1997, Wounding induces the rapid and transient activation of a specific map kinase pathway, *Plant Cell* **9**: 75–83.

Bonos, S.A., Rush, D., Hignight, K. and Meyer, W.A., 2004, Selection for deep root production in tall fescue and perennial ryegrass, *Crop Sci.* **44**: 1770–1775.

Bouton, J. and Easton, S., 2005, Endophytes in forage cultivars. In: *Neotyphodium in Cool-Season Grasses*, C.A. Roberts, C.P. West, D.E. Spiers, eds, Blackwell Publishing, Ames, IW

Bouton, J.H., Gates, R.N., Belesky, D.P. and Owsley, M., 1993, Yield and persistence of tall fescue in the southeastern coastal plain after removal of its endophyte, *Agron. J.***85**: 52–55.

Bouton, J.H., Latch, G.C.M., Hill, N.S., Hoveland, C.S., McCann, M.A., Watson, R.H., Parish, J.A., Hawkins, L.L. and Thompson, F.N., 2002, Reinfection of tall fescue cultivars with non-ergot alkaloid-producing endophytes, *Agron. J.***94**: 567–574.

Boyer, J.S., 1996, Advances in drought tolerance in plant, *Advances in Agron.* **56**: 187–218.

Breshears, D.D., Cobb, N.S., Rich, P.M., Price, K.P., Allen, C.D., Balice, R.G., Romme, W.H., Kastens, J.H., Floyd, M.L., Belnap, J., Anderson, J.J., Myers, O.B. and Meyer, C.W., 2005, Regional vegetation die-off in response to global-change-type drought, *PNAS* **102**: 15144–15148.

Brink, G. and Pederson, G., 1998, White clover response to a water-application gradient, *Crop Sci.* **38**: 771–775.

Brummer, E.C., 2004, Applying genomics to alfalfa breeding programs, *Crop Sci.* **44**: 1904–1907.

Bultman, T.L., 2006, Neotyphodium in cool-season grasses. *Crop Sci* 46: 493–494.

Bush, L.P., Wilkinson, H.H. and Schardl, C.L., 1997, Bioprotective alkaloids of grass-fungal endophyte symbioses, *Plant Physiol.* **114**: 1–7.

Cairns, A.J., 2003, Fructan biosynthesis in transgenic plants, *J. Exp. Bot.* **54**: 549–567.

Canter, P.H., Pašakinskiene, I., Jones, R.N. and Humphreys, M.W., 1999, Chromosome substitutions and recombination in the amphiploid *Lolium perenne—Festuca pratensis* cv Prior (2n=4x=28), *Theor. Appl. Genet.* **98**: 809–814.

Cardinale, F., Meskiene, I., Ouaked, F. and Hirt, H., 2002, Convergence and divergence of stress-induced mitogen-activated protein kinase signaling pathways at the level of two distinct mitogen-activated protein kinase kinases, *Plant Cell* **14**: 703–711.

Carrow, R.N., 1996, Drought resistance aspects of turfgrasses in the Southeast: root-shoot responses, *Crop Sci.* **36**: 687–694.

Casler, M.D., Pitts, P.G., Rose-Fricker, C., Bilkey, P.C. and Wipff, J.K., 2001, Registration of 'Spring Green' festulolium, *Crop Sci.* **41**: 1365–1366.

Castonguay, Y. and Nadeau, P., 1998, Enzymatic control of soluble carbohydrate accumulation in cold-acclimated crowns of alfalfa, *Crop Sci.* **38**: 1183–1189.

Castonguay, Y., Nadeau, P., Lechasseur, P. and Chouinard, L., 1995, Differential accumulation of carbohydrates in alfalfa cultivars of contrasting winterhardiness, *Crop Sci.* **35**: 509–516.

Chalmers, J., Lidgett, A., Cummings, N., Cao, Y., Forster, J. and Spangenberg, G., 2005, Molecular genetics of fructan metabolism in perennial ryegrass, *Plant Biotech. J.* **3**: 459–474.

Cheplick, G.P., 2004, Recovery from drought stress in *Lolium perenne* (Poaceae): are fungal endophytes detrimental? *Am. J. Bot.* **91**: 1960–1968.

Cheplick, G.P. and Cho, R., 2003, Interactive effects of fungal endophyte infection and host genotype on growth and storage in *Lolium perenne*, *New Phytol.* **158**: 183–191.

Cunningham, S.M., Nadeau, P., Castonguay, Y., Laberge, S. and Volenec, J.J., 2003, Raffinose and stachyose accumulation, galactinol synthase expression, and winter injury of contrasting alfalfa germplasms, *Crop Sci.* **43**: 562–570.

De Battista, J.P., Bouton, J.H., Bacon, C.W. and Siegel, M.R., 1990, Rhizome and herbage production of endophyte removed tall fescue clones and populations, *Agron. J.***82**: 51–54.

Dijkstra, J. and Vos, A.L.F.d., 1975, Meiotic doubling of chromosome number in Festulolium, *Euphytica* **24**: 743–749.

Eagles, C.F., Thomas, H., Volaire, F. and Howarth, C.J., 1997, Stress physiology and crop improvement. In: *Proceedings of the XVIII International Grassland Congress*, Christie B.R., ed, Canada, pp. 141–150

Elbersen, H.W. and West, C.P., 1996, Growth and water relations of field-grown tall fescue as influenced by drought and endophyte, *Grass and Forage Sci.* **51**: 333–342.

Elmi, A.A. and West, C.P., 1995, Endophyte infection effects on stomatal conductance, osmotic adjustment and drought recovery of tall fescue, *New Phytol.* **131**: 61–67.

Ervin, E.H. and Koski, A.J., 1998, Drought avoidance aspects and crop coefficients of Kentucky bluegrass and tall fescue turfs in the semiarid west, *Crop Sci.* **38**: 788–795.

Ginzberg, I., Stein, H., Kapulnik, Y., Szabados, L., Strizhov, N., Schell, J., Koncz, C. and Zilberstein, A., 1998a, Isolation and characterization of two different cDNAs of 1-pyrroline-5-carboxylate synthase in alfalfa, transcriptionally induced upon salt stress, *Plant Mol. Biol.* **38**: 755–764.

Ginzberg, I., Stein, H., Kapulnik, Y., Szabados, L., Strizhov, N., Schell, J., Koncz, C. and Zilberstein, A., 1998b, Isolation and characterization of two different cDNAs of Δ^1-pyrroline-5-carboxylate synthase in alfalfa, transcriptionally induced upon salt stress, *Plant Mol. Biol.* **38**: 755–764.

Girousse, C., Bournoville, R. and Bonnemain, J.L., 1996, Water deficit-induced changes in concentrations in proline and some other amino acids in the phloem sap of alfalfa, *Plant Physiol.* **111**: 109–113.

Goicoechea, N., Szalai, G., Antolín, M.C., Sánchez-Díaz, M. and Paldi, E., 1998, Influence of arbuscular mycorrhizae and *Rhizobium* on free polyamines and proline levels in water-stressed alfalfa, *J. Plant Physiol.* **153**: 706–711.

Hall, A.E., 2001, *Crop responses to environment.* CRC Press, Boca Raton, FL

Hall, M.H., Sheaffer, C.C. and Heichel, G.H., 1988, Partitioning and mobilization of photoassimilate by alfalfa subjected to water deficits, *Crop Sci.* **28**: 964–969.

Hare, P.D., Cress, W.A. and van Staden, J., 1999, Proline synthesis and degradation: a model system for elucidating stress-related signal transduction, *J. Exp. Bot.* **50**: 413–434.

Hart, A.L., 1987, Physiology. In: *Whiter Clover*, M.J. Baker, W.M. Williams, eds, CAB Int., Wallingford, Oxon, UK, pp. 125–151.

Hendry, G.A.F. and Wallace, R.K., 1993, The origin, distribution and evolutionaary significance of fructans. In: *Science and Technology of Fructans*, M. Suzuki, N.J. Chatterton, eds, CRC Press, Boca Raton, pp. 119–139.

Hesse, U., Schöberlein, W., Wittenmayer, L., Förster, K., Warnstorff, K., Diepenbrock, W. and Merbach, W., 2005, Influence of water supply and endophyte infection (*Neotyphodium* spp.) on vegetative and reproductive growth of two *Lolium perenne* L. genotypes, *Eur. J. Agron.* **22**: 45–54.

Hisano, H., Kanazawa, A., Kawakami, A., Yoshida, M., Shimamoto, Y. and Yamada, T., 2004, Transgenic perennial ryegrass plants expressing wheat fructosyltransferase genes accumulate increased amounts of fructan and acquire increased tolerance on a cellular level to freezing, *Plant Sci.* **167**: 861–868.

Huang, B. and Fry, J.D., 1998, Root anatomical, physiological, and morphological responses to drought stress for tall fescue cultivars, *Crop Sci.* **38**: 1017–1022.

Huang, B. and Gao, H., 2000, Root physiological characteristics associated with drought resistance in tall fescue cultivars, *Crop Sci.* **40**: 196–203.

Humphreys, J., Harper, J.A., Armstead, I.P. and Humphreys, M.W., 2005, Introgression-mapping of genes for drought resistance transferred from *Festuca arundinacea* var. *glaucescens* into *Lolium multiflorum*, *Theor. Appl. Genet.* **110**: 579–587.

Humphreys, M., Thomas, H.-M., Harper, J., Morgan, G., James, A., Ghamari-Zare, A. and Thomas, H., 1997, Dissecting drought- and cold-tolerance traits in the *Lolium-Festuca* complex by introgression mapping, *New Phytologist* **137**: 55–60.

Humphreys, M.W., Canter, P.J. and Thomas, H.M., 2003, Advances in introgression technologies for precision breeding within the *Lolium-Festuca* complex, *Annals of Applied Biol.* **143**: 1–10.

Humphreys, M.W., Pasakinskiene, I., James, A.R. and Thomas, H. 1998. Physically mapping quantitative traits for stress-resistance in the forage grasses, *J. Exp. Bot.* **49**: 1611–1618.

Humphreys, M.W. and Thomas, H., 1993, Improved drought resistance in introgression lines derived from *Lolium multiflorum* x *Festuca arundinacea* hybrids, *Plant Breed.* **111**: 155–161.

Humphreys, M.W., Yadav, R.S., Cairns, A.J., Turner, L.B., Humphreys, J. and Skot, L., 2006, A changing climate for grassland research, *New Phytol.* **169**: 9–26.

Jenkins, C.L.D., Snow, A.J., Simpson, R.J., Higgins, T.J., Jacques, N.A., Pritchard, J., Gibson, J. and Larkin, P.J., 2002, Fructan formation in transgenic white clover expressing a fructosyltransferase from *Streptococcus salivarius*, *Funct. Plant Biol.* **29**: 1287–1298.

Jonak, C., Kiegerl, S., Ligterink, W., Barker, P.J., Huskisson, N.S. and Hirt, H. 1996. Stress signaling in plants: A mitogen-activated protein kinase pathway is activated by cold and drought, *PNAS* **93**: 11274–11279.

Karsten, H.D. and MacAdam, J.W., 2001, Effect of drought on growth, carbohydrates, and soil water use by perennial ryegrass, tall fescue, and white clover, *Crop Sci.* **41**: 156–166.

Kiegerl, S., Cardinale, F., Siligan, C., Gross, A., Baudouin, E., Liwosz, A., Eklof, S., Till, S., Bogre, L., Hirt, H. and Meskiene, I., 2000, SIMKK, a mitogen-activated protein kinase (MAPK) kinase, is a specific activator of the salt stress-induced MAPK, SIMK, *Plant Cell* **12**: 2247–2258.

Kopecky, D., Lukaszewski, A.J. and Dolezel, J., 2005, Genomic constitution of *Festulolium* cultivars released in the Czech Republic, *Plant Breed.* **124**: 454–458.

Lee, B.-R., Jung, W.-J., Kim, K.-Y., Avice, J.-C., Ourry, A. and Kim, T.-H., 2005, Transient increase of de novo amino acid synthesis and its physiological significance in water-stressed white clover, *Funct. Plant Biol.* **32**: 831–838.

Lewis, G.C., 2004, Effects of biotic and abiotic stress on the growth of three genotypes of *Lolium perenne* with and without infection by the fungal endophyte *Neotyphodium lolii*, *Annals of Applied Biol.* **144**: 53–63.

Luscher, M., Hochstrasser, U., Vogel, G., Aeschbacher, R., Galati, V., Nelson, C.J., Boller, T. and Wiemken, A., 2000, Cloning and functional analysis of sucrose:sucrose 1-fructosyltransferase from tall fescue, *Plant Physiol.* **124**: 1217–1228.

Luscher, M. and Nelson, C.J., 1995, Fructosyltransferase activities in the leaf growth zone of tall fescue, *Plant Physiol.* **107**: 1419–1425.

Lyons, P.C., Plattner, R.D. and Bacon, C.W., 1986, Occurrence of peptide and clavine ergot alkaloids in tall fescue grass, *Science* **232**: 487–489.

Malinowski, D.P. and Belesky, D.P., 2000, Adaptations of endophyte-infected cool-season grasses to environmental stresses: Mechanisms of drought and mineral stress tolerance, *Crop Sci.***40**: 923–940.

Malinowski, D.P., Brauer, D.K. and Belesky, D.P., 1999, The endophyte *Neotyphodium coenophialum* affects root morphology of tall fescue grown under phosphorus deficiency, *J. Agron. Crop Sci.* **183**: 53–60.

Malinowski, D.P., Zuo, H., Kramp, B.A., Muir, J.P. and Pinchak, W.E., 2005, Obligatory summer-dormant cool-season perennial grasses for semiarid environments of the southern great plains, *Agron. J.* **97**: 147–154.

Marshall, A.H., Rascle, C., Abberton, M.T., Michaelson-Yeates, T.P.T. and Rhodes, I., 2001, Introgression as a route to improved drought tolerance in white clover (*Trifolium repens* L.), *J. Agron. Crop Sci.* **187**: 11–18.

McKersie, B.D., Bowley, S.R., Harjanto, E. and Leprince, O., 1996, Water-deficit tolerance and field performance of transgenic alfalfa overexpressing superoxide dismutase, *Plant Physiol.* **111**: 1177–1181.

Meredith, M.R., Michaelson-Yeates, T.P.T., Ougham, H.J. and Thomas, H., 1995, Trifolium ambiguum as a source of variation in the breeding of white clover, *Euphytica* **82**: 185–191.

Miller, G., Stein, H., Honig, A., Kapulnik, Y. and Zilberstein, A., 2005, Responsive modes of *Medicago sativa* proline dehydrogenase genes during salt stress and recovery dictate free proline accumulation, *Planta* **222**: 70–79.

Mittler, R., 2002, Oxidative stress, antioxidants and stress tolerance, *Trends in Plant Sci.* **7**: 405–410.

Mittler, R., Merquiol, E., Hallak Herr, E., Rachmilevitch, S., Kaplan, A. and Cohen, M., 2001, Living under a 'dormant' canopy: a molecular acclimation mechanism of the desert plant Retama raetam, *Plant J.* **25**: 407–416.

Morgan, W.G., King, I.P., Koch, S., Harper, J.A. and Thomas, H.M., 2001, Introgression of chromosomes of *Festuca arundinacea* var. *glaucescens* into *Lolium multiflorum* revealed by genomic *in situ* hybridisation (GISH), *Theor. Appl. Genet.* 103: 696–701.

Munnik, T., Ligterink, W., Meskiene, I., Calderini, O., Beyerly, J., Musgrave, A. and Hirt, H., 1999, Distinct osmo-sensing protein kinase pathways are involved in signalling moderate and severe hyper-osmotic stress, *Plant J.* **20**: 381–388.

Nelson, C.J. and Moser, E.L., 1995, Morphology and systematics. In: *Forages: An Introduction to Grassland Agriculture*, R.F. Barnes, D.A. Miller, C.J. Nelson, eds, Iowa State University Press, Ames, Iowa, pp. 15–30.

Oberschall, A., Deak, M., Torok, K., Sass, L., Vass, I., Kovacs, I., Feher, A., Dudits, D. and Horvath, G.V., 2000, A novel aldose/aldehyde reductase protects transgenic plants against lipid peroxidation under chemical and drought stresses, *Plant J.* **24**: 437–446.

Ofir, M. and Kigel, J., 1999, Photothermal control of the imposition of summer dormancy in *Poabulbosa*, a perennial grass geophyte, *Physiol. Planta.* **105**: 633–640.

Oldroyd, G.E.D., Harrison, M.J. and Udvardi, M., 2005, Peace Talks and Trade Deals. Keys to Long-Term Harmony in Legume-Microbe Symbioses, *Plant Physiol.* **137**: 1205–1210.

Panaccione, D.G., Johnson, R.D., Wang, J., Young, C.A., Damrongkool, P., Scott, B. and Schardl, C.L., 2001, Elimination of ergovaline from a grass-Neotyphodium endophyte symbiosis by genetic modification of the endophyte, *PNAS* **98**: 12820–12825.

Pavis, N., Boucaud, J. and Prud'homme, M.P., 2001a, Fructans and fructan-metabolizing enzymes in leaves of *Lolium perenne*, *New Phytol.* **150**: 97–109.

Pavis, N., Chatterton, N.J., Harrison, P.A., Baumgartner, S., Praznik, W., Boucaud, J. and Prud'homme, M.P., 2001b, Structure of fructans in roots and leaf tissues of *Lolium perenne*, *New Phytol.* **150**: 83–95.

Pederson, G.A., 1995, White clover and other perennial clovers. In: *Forages. I. An Introduction to Grassland Agriculture*, R.F. Barnes, D.A. Miller, C.J. Nelson, eds, Iowa State University Press, Ames, Iowa, pp. 227–236.

Peterson, P.R., Sheaffer, C.C. and Hall, M.H., 1992, Drought effects on perennial forage legume yield and quality, *Agron. J.***84**: 774–779.

Pilon-Smits, E.A.H., Ebskamp, M.J.M., Paul, M.J., Jeuken, M.J.W., Weisbeek, P.J. and Smeekens, S.C.M., 1995, Improved performance of transgenic fructan-accumulating tobacco under drought stress, *Plant Physiol.* **107**: 125–130.

Pinheiro, H.A., DaMatta, F.M., Chaves, A.R.M., Loureiro, M.E. and Ducatti, C., 2005, Drought tolerance is associated with rooting depth and stomatal control of water use in clones of *Coffea canephora*, *Ann. Bot.* **96**: 101–108.

Pollock, C.J. and Cairns, A.J., 1991, Fructan metabolism in grasses and cereals, *Annu. Rev. Plant Physiol. Plant Mol. Biol.* **42**: 77–101.

Remy, W., Taylor, T., Hass, H. and Kerp, H., 1994, Four hundred-million-year-old vesicular arbuscular mycorrhizae, *PNAS* **91**: 11841–11843.

Richards, R.A., 1996, Defining selection criteria to improve yield under drought, *Plant Growth Regul.* **20**: 157–166.

Richardson, M.D., Chapman, G.W., Jr., Hoveland, C.S. and Bacon, C.W., 1992, Sugar alkohols in endophyte infected tall fescue, *Crop Sci.* **32**: 1060–1061.

Rubio, M.C., Gonzalez, E.M., Minchin, F.R., Webb, K.J., Arrese-Igor, C., Ramos, J. and Becana, M., 2002, Effects of water stress on antioxidant enzymes of leaves and nodules of transgenic alfalfa overexpressing superoxide dismutases, *Physiol. Planta.* **115**: 531–540.

Šamaj, J., Ovecka, M., Hlavacka, A., Lecourieux, F., Meskiene, I., Lichtscheidl, I., Lenart, P., Salaj, J., Volkmann, D., Bögre, L., Baluška, F.e. and Hirt1, H., 2002, Involvement of the mitogen-activated protein kinase SIMK in regulation of root hair tip growth, *The EMBO J.* 21: 3296–3306.

Samis, K., Bowley, S. and McKersie, B., 2002, Pyramiding Mn-superoxide dismutase transgenes to improve persistence and biomass production in alfalfa, *J. Exp. Bot.* **53**: 1343–1350.

Sanderson, M.A., Byers, R.A., Skinner, R.H. and Elwinger, G.F., 2003, Growth and complexity of white clover stolons in response to biotic and abiotic stress, *Crop Sci.* **43**: 2197–2205.

Sanderson, M.A., Stair, D.W. and Hussey, M.A., 1997, Physiological and morphological response of perennial forages to stress, *Advances in Agron.* **59**: 171–224.

Schnyder, H. and Nelson, C.J., 1989, Growth rates and assimilate partitioning in the elongation zone of tall fescue leaf blades at high and low irradiance, *Plant Physiol.* **90**: 1201–1206.

Sheffer, K.M., Dunn, J.H. and Minner, D.D., 1987, Summer drought response and rooting depth of three cool-season turfgrasses, *HortScience* **22**: 296–297.

Sleper, D.A. and Buckner, R.C. 1995. The fescue. In: *Forages. I. An Introduction to Grassland Agriculture*, R.F. Barnes, D.A. Miller, C.J. Nelson, eds, Iowa State University Press, Ames, Iowa, pp. 345–371.

Sleper, D.A. and West, C.P. 1996. Tall fescue. In: *Cool-Season Forage Grasses*, L.E. Moser, D.R. Buxton, M.D. Casler, eds, American Society of Agronomy, Crop Science Society of America, Soil Science Society of America, Madison, WI, pp. 471–502.

Spangenberg, G., Vallés, M.P., Wang, Z.Y., Montavon, P., Nagel, J. and Potrykus, I., 1994, Asymmetric somatic hybridization between tall fescue (*Festuca arundinacea* Schreb.) and irradiated Italian ryegrass (*Lolium multiflorum* Lam.) protoplasts, *Theor. Appl. Genet.***88**: 509–519.

Spiering, M.J., Lane, G.A., Christensen, M.J. and Schmid, J., 2005, Distribution of the fungal endophyte *Neotyphodium lolii* is not a major determinant of the distribution of fungal alkaloids in *Lolium perenne* plants, *Phytochemistry* **66**: 195–202.

Spollen, W.G. and Nelson, C.J., 1994, Response of fructan to water deficit in growing leaves of tall fescue, *Plant Physiol.* **106**: 329–336.

Thomas, H.M., Morgan, W.G. and Humphreys, M.W., 2003, Designing grasses with a future - combining the attributes of *Lolium* and *Festuca*, *Euphytica* **133**: 19–26.

Trinchant, J.-C., Boscari, A., Spennato, G., Van de Sype, G. and Le Rudulier, D., 2004, Proline betaine accumulation and metabolism in alfalfa plants under sodium chloride stress: Exploring its compartmentalization in nodules, *Plant Physiol.* **135**: 1583–1594.

Turner, L.B., Cairns, A.J., Armstead, I.P., Ashton, J., Skot, K., Whittaker, D. and Humphreys, M.O., 2006, Dissecting the regulation of fructan metabolism in perennial ryegrass (*Lolium perenne*) with quantitative trait locus mapping, *New Phytol.* **169**: 45–58.

Vijn, I. and Smeekens, S., 1999, Fructan: more than a reserve carbohydrate? *Plant Physiol.* **120**: 351–360.

Volaire, F., Thomas, H. and Lelièvre, F., 1998, Survival and recovery of perennial forage grasses under prolonged Mediterranean drought:: growth, death, water relations and solute content in herbage and stubble, *New Phytol.* **140**: 439–449.

Volence, J.J., Cunningham, S.M., Haagenson, D.M., Berg, W.K., Joern, B.C. and Wiersma, D.W., 2002, Physiological genetics of alfalfa improvement: past failures, future prospects, *Field Crops Research* 75: 97–110.

Wang, Z.Y. and Ge, Y., 2006, Recent advances in genetic transformation of forage and turf grasses, *In Vitro Cell. Dev. Biol. Plant* **42**: 1–8.

Wilkins, P.W. and Humphreys, M.O., 2003, Progress in breeding perennial forage grasses for temperate agriculture, *J. Agri. Sci.* **140**: 129–150.

Williams, W.M., 1987, Genetics and breeding. In: *White Clover*, M.J. Baker, W.M. Williams, eds, CAB Int., Wallingford, Oxon, UK

Winicov, I., 1993, cDNA encoding putative zinc finger motifs from salt-tolerant alfalfa (*Medicago sativa* L.) cells, *Plant Physiol.* **102**: 681–682.

Winicov, I., 2000, *Alfin1* transcription factor overexpression enhances plant root growth under normal and saline conditions and improves salt tolerance in alfalfa, *Planta* **210**: 416–422.

Winicov, I. and Bastola, D.R., 1999, Transgenic overexpression of the transcription factor Alfin1 enhances expression of the endogenous msprp2 gene in alfalfa and improves salinity tolerance of the plants, *Plant Physiol.* **120**: 473–480.

Winicov, I., Valliyodan, B., Xue, L. and Hoober, J.K., 2004, The *MsPRP2* promoter enables strong heterologous gene expression in a root-specific manner and is enhanced by overexpression of *Alfin 1*, *Planta* **219**: 925–935.

Wood, K.V., Stringham, K.J., Smith, D.L., Volenec, J.J., Hendershot, K.L., Jackson, K.A., Rich, P.J., Yang, W.-J. and Rhodes, D., 1991, Betaines of alfalfa : Characterization by fast atom bombardment and desorption chemical Ionization mass spectrometry, *Plant Physiol.* **96**: 892–897.

Wu, Y.-Y., Chen, Q.-J., Chen, M., Chen, J. and Wang, X.-C., 2005, Salt-tolerant transgenic perennial ryegrass (*Lolium perenne* L.) obtained by *Agrobacterium tumefaciens*-mediated transformation of the vacuolar Na^+/H^+ antiporter gene, *Plant Sci.* **169**: 65–73.

Yamada, T., Forster, J.W., Humphreys, M.W. and Takamizo, T., 2005, Genetics and molecular breeding in *Lolium/Festuca* grass species complex, *Grassland Sci.* **51**: 89–106.

Ye, X.D., Wu, X.L., Zhao, H., Frehner, M., Noberger, J., Potrykus, I. and Spangenberg, G., 2001, Altered fructan accumulation in transgenic *Lolium multiflorum* plants expressing a *Bacillus subtilis sacB* gene, *Plant Cell Reports* **20**: 205–212.

Zare, A.G., Humphreys, M.W., Rogers, J.W., Mortimer, A.M. and Collin, H.A., 2002, Androgenesis in a *Lolium multiflorum* x *Festuca arundinacea* hybrid to generate genotypic variation for drought resistance, *Euphytica* **125**: 1–11.

Zhang, J.-Y., Broeckling, C.D., Blancaflor, E.B., Sledge, M.K., Sumner, L.W. and Wang, Z.-Y., 2005, Overexpression of WXP1, a putative Medicago truncatula AP2 domain-containing transcription factor gene, increases cuticular wax accumulation and enhances drought tolerance in transgenic alfalfa (*Medicago sativa*).*Plant J.* **42**: 689–707.

Zhang, J.-Y., Broeckling, C.D., Sumner, L.W. and Wang, Z.-Y., 2007, Heterologous expression of two medicago truncatula AP2 domain transcription factor genes, *WXP1* and *WXP2* , in arabidopsis led to increased leaf wax accumulation and improved drought tolerance, but differential response in freezing tolerance. *Plant Mol. Biol.* **64**: 265–278.

Advances in Molecular Breeding Toward Drought and Salt Tolerant Crops

Edited by

Matthew A. Jenks
Purdue University,
Horticulture Department,
West Lafayette, U.S.A.

Paul M. Hasegawa
Purdue University,
Horticulture Department,
West Lafayette, U.S.A.

and

S. Mohan Jain
University of Helsinki,
Department of Applied Biology,
Helsinki, Finland

Springer

A C.I.P. Catalogue record for this book is available from the Library of Congress.

ISBN 978-1-4020-5577-5 (HB)
ISBN 978-1-4020-5578-2 (e-book)

Published by Springer,
P.O. Box 17, 3300 AA Dordrecht, The Netherlands.

www.springer.com

Printed on acid-free paper

CONTENTS

PREFACE

Despite the intensive management practices available to modern agriculture, drought and salinity are still major constraints on crop production and food security. Yield losses to drought and salinity stress are commonly 80% and more, with actual losses dependant on the timing, intensity, and duration of the stress, and other location-specific environmental conditions. An appropriate yield capacity for sustainable food security could be greatly facilitated by improving crop plant productivity in drought and saline environments. A major constraint however to improving crop yield under these forms of abiotic stress is our lack of understanding of the complex physiological, biochemical, developmental, and genetic mechanisms that underlie this environmental stress tolerance, and the subsequent difficulty in combining favorable alleles to create improved high yielding genotypes. Furthermore, it appears certain that domestication has narrowed the genetic diversity within crops for stress tolerance, and thus limited options in traditional crop breeding. It's also likely that selection for high yield potential has negatively influenced crop plant responsiveness to environmental challenges. Consequently, traditional breeding strategies have made limited progress in enhancing harvest indices in environments plagued by drought and salinity stress.

Recent discoveries reveal a highly complex integration of response mechanisms involved in regulating plant adaptation to drought and salinity. For example, common genetic and biochemical networks and shared signal transduction pathways are a typical part of plant stress response. Emerging views of this interconnectedness in stress adaptation provides a platform for new thinking about crop improvement strategies. Merging recent discoveries in basic science with recent advances in molecular genetics and molecular breeding now offers new avenues for improving plant tolerance to drought and salinity. The application of new biotechnologies like gene and trait pyramiding, molecular-assisted selection, crop transformation, and mutation breeding may lead, not only to the development of new improved crop production systems, but also to advancing our fundamental understanding of drought and salt stress tolerance. This book will discuss a broad spectrum of reports and expertise regarding drought and salt tolerance determinants from the physiological, biochemical, developmental and genetic levels, and the new technologies now available to manipulate these determinants for germplasm improvement. Importantly, our new awareness of the remarkable complexity and interconnectedness of stress response mechanisms reveals a need to recognize that a more systems approach provides a more accurate means of integrating the traditionally diverse fields within plant stress studies, fields as diverse as water relations, biochemistry, genetics, and development. To emphasize the point, it is increasingly recognized that the relative importance to stress tolerance of individual determinants (whether genes, QTLs, or traits) often has more to do with genomic background and subsequent integration of diverse phenotypes than to the presence or absence of that single determinant.

This book will present a contemporary understanding of plant adaptation to drought and salinity that emphasizes fundamental physiological, biochemical, developmental and genetic mechanisms, as a prelude to thoughtful analyses of the integrated regulation of these determinants. The following section of this book will examine new strategies being employed to identify and then enhance these tolerance mechanisms. Later chapters are focused on efforts to discover existing genetic variation in crop germplasm and wild relatives, and manipulate genetic variation using mutation, transgenic, and molecular marker-assisted breeding approaches. This book seeks to integrate a broad cross-section of scientific knowledge and expertise about key determinants of drought and salt stress tolerance with modern crop improvement strategies. Information presented here will be especially useful to agronomists and horticulturists, crop breeders, molecular-geneticists, and biotechnologists, and serve as an important scholarly text for post-graduate students and researchers.

We, the editors, would like to thank the authors for their outstanding and timely work in producing such fine chapters. We would also like to thank Katie Vanvekoven for her clerical assistance, and Jacco Flipsen and Noeline Gibson of Springer for their advice and encouragement during the development of this important book.

Matthew A. Jenks, Paul M. Hasegawa, and S. Mohan Jain

CONTRIBUTORS

Rownak Afza Plant Breeding Unit, Joint FAO/IAEA Agriculture and Biotechnology Laboratory, International Atomic Energy Agency Laboratories, A-2444 Seibersdorf, Austria

Jose-Luis Araus CIMMYT (International Maize and Wheat Improvement Centre). Apdo. Postal 6-641, 06600 Mexico D.F. Mexico

Marianne Bänziger CIMMYT (International Maize and Wheat Improvement Centre). P.O. Box 1041, Village Market-00621, Nairobi, Kenya

Gozal Ben-Hayyim Institute of Plant Sciences, Agricultural Research Organization, The Volcani Center, Bet Dagan, Israel

D.G. Bonnett CSIRO Plant Industry, GPO Box 1600, Canberra, ACT 2601, Australia

Elizabeth A. Bray Department of Molecular Genetics and Cell Biology, University of Chicago, Chicago, IL 60637, USA

Myung-Ok Byun Department of Molecular Physiology and Biochemistry, National Institute of Agricultural Biotechnology, RDA, Suwon, 441-707, Korea

Laury Chaerle Unit Plant Hormone Signalling and Bio-imaging (HSB), Department of Molecular Genetics, Ghent University, K. L. Ledeganckstraat 35, B-9000 Gent, Belgium

Peng W. Chee Cotton Molecular Breeding Lab, NESPAL, University of Georgia, 2356 Rainwater Road, P.O. Box 748, Tifton, GA 31793

Viswanathan Chinnusamy Water Technology Centre, Indian Agricultural Research Institute, New Delhi, India

Cory A. Christensen Ceres, Inc., 1535 Rancho Conejo Blvd., Thousand Oaks, CA 91320

Grant R. Cramer MS 200, Department of Biochemistry & Molecular Biology, University of Nevada, Reno, NV 89557-0014, USA

John C. Cushman MS 200, Department of Biochemistry & Molecular Biology, University of Nevada, Reno, NV 89557-0014, USA

W.J. Davies Lancaster Environment Centre, Lancaster University, Bailrigg, Lancaster, LA1 4YQ UK

Laurent G. Deluc MS 200, Department of Biochemistry & Molecular Biology, University of Nevada, Reno, NV 89557-0014, USA

Kenneth A. Feldmann Ceres, Inc., 1535 Rancho Conejo Blvd., Thousand Oaks, CA 91320

Majid R. Foolad Department of Horticulture, The Pennsylvania State University, University Park, PA 16802, USA

Martin A. Fregene Centro Internacional de Agricultura Tropical (CIAT), Cali, Colombia

S.R. Grattan Department of Land, Air and Water Resources, University of California, One Shields Ave., Davis, CA 95616, USA

Glenn B. Gregorio International Rice Research Institute, Los Baños, Laguna, Philippines; Current address – African Rice Center (WARDA), c/o Internationa Institute of Tropicaal Agrioculture, Oyo Road, PMB 5320, Ibadan, Nigeria

Jérôme Grimplet MS 200, Department of Biochemistry & Molecular Biology, University of Nevada, Reno, NV 89557-0014, USA

P.A. Hollington CAZS Natural Resources, University of Wales Bangor, UK

Shri Mohan Jain Plant Breeding and Genetics Section, Joint FAO/IAEA Division, International Atomic Energy Agency, Vienna, Austria; Current address – Department of Applied Biology, University of Helsinki, Helsinki, Finland

Matthew A. Jenks Purdue University, Department of Horticulture and Landscape Architecture, Center for Plant Environmental Stress Physiology, West Lafayette, Indiana, 47907, USA

Jong-Myong Kim Plant Genomic Network Research Team, Plant Functional Genomics Research Group, RIKEN Plant Science Center (PSC), RIKEN Yokohama Institute, 1-7-22 Suehiro-cho, Tsurumi-ku, Yokohama 230-0045, Japan

Taiko Kim To Plant Genomic Network Research Team, Plant Functional Genomics Research Group, RIKEN Plant Science Center (PSC), RIKEN Yokohama Institute, 1-7-22 Suehiro-cho, Tsurumi-ku, Yokohama 230-0045, Japan

Karin Koehl Max Planck Institute of Molecular Plant Physiology, Am Muehlenberg 1, 14476 Potsdam-Golm, Germany

Joachim Kopka Max Planck Institute of Molecular Plant Physiology, Am Muehlenberg 1, 14476 Potsdam-Golm, Germany

Dylan K. Kosma Purdue University, Department of Horticulture and Landscape Architecture, Center for Plant Environmental Stress Physiology, West Lafayette, Indiana, 47907, USA

Hawk-Bin Kwon Division of Applied Biological Sciences, Sunmoon University, Asan, 336-708, Korea

Reg Lance Department of Agriculture and Food, Government of Western Australia, 3 Baron-Hay Court, South Perth WA6151, Australia

A. Läuchli Department of Land, Air and Water Resources, University of California, One Shields Ave., Davis, CA 95616, USA

Jeong-Dong Lee Delta Center, University of Missouri-Columbia, Missouri 65211, USA

Weiqi Li Kunming Institute of Botany, Chinese Academy of Sciences, Kunming 650204 and Department of Biology, Honghe University, Mengzi, Yunnan 661100, China

Zhi-Kang Li International Rice Research Institute, DAPO Box 7777, Metro Manila, Philippines; Institute of Crop Sciences/National Laboratory for Crop Gene Resources and Genetic Improvement, Chinese Academy of Agricultural Sciences

Chengdao Li Department of Agriculture and Food, Government of Western Australia, 3 Baron-Hay Court, South Perth WA6151, Australia

Edward L. Lubbers Cotton Molecular Breeding Lab, NESPAL, University of Georgia, 2356 Rainwater Road, P.O. Box 748, Tifton, GA 31793

Frans J.M. Maathuis Department of Biology, University of York, York YO10 5DD, UK

Akihiro Matsui Plant Genomic Network Research Team, Plant Functional Genomics Research Group, RIKEN Plant Science Center (PSC), RIKEN Yokohama Institute, 1-7-22 Suehiro-cho, Tsurumi-ku, Yokohama 230-0045, Japan

Chikelu Mba Plant Breeding Unit, Joint FAO/IAEA Agriculture and Biotechnology Laboratory, International Atomic Energy Agency Laboratories, A-2444 Seibersdorf, Austria

Girish Mishra Department of Biology, University of Missouri, St. Louis, MO 63121 and Donald Danforth Plant Science Center, St. Louis, MO 63132, USA

Gloria A. Moore Department of Horticultural Sciences, Plant Molecular and Cellular Biology Program, University of Florida, Gainesville, USA

Rana Munns CSIRO Plant Industry, GPO Box 1600, Canberra, ACT 2601, Australia

Henry T. Nguyen National Center for Soybean Biotechnology and Division of Plant Sciences

Eric S. Ober Rothamsted Research, Broom's Barn Research Station, Higham, Bury St Edmunds, IP28 6NP, UK

Andrew H. Paterson Department of Crop and Soil Science, Dept. Botany, and Dept. Genetics, University of Georgia, Athens, GA30602, USA

Md S. Pathan National Center for Soybean Biotechnology and Division of Plant Sciences

Ashwani Pareek Stress Physiology and Molecular Biology Laboratory, School of Life Sciences, Jawaharlal Nehru University, New Delhi 10067, India

Soo-Chul Park Department of Molecular Physiology and Biochemistry, National Institute of Agricultural Biotechnology, RDA, Suwon, 441-707, Korea

J.B. Passioura CSIRO Plant Industry, GPO Box 1600, Canberra, ACT 2601, Australia

R.A. Richards CSIRO Plant Industry, GPO Box 1600, Canberra, ACT 2601, Australia

Yehoshua Saranga Department of Field Crops, Vegetables and Genetics, Faculty of Agricultural, Food and Environmental Quality Sciences, The Hebrew University of Jerusalem, P.O. Box 12, Rehovot76100, Israel

Silvio Salvi Department of Agroenvironmental Sciences and Technology, Viale Fanin 44, 40127 Bologna, Italy

Motoaki Seki Plant Genomic Network Research Team, Plant Functional Genomics Research Group, RIKEN Plant Science Center (PSC), RIKEN Yokohama Institute, 1-7-22 Suehiro-cho, Tsurumi-ku, Yokohama 230-0045, Japan

Tim L. Setter Department of Crop and Soil Science, Cornell University, Ithaca, NY, USA

J. Grover Shannon Delta Center, University of Missouri-Columbia, Missouri 65211, USA

Robert E. Sharp Division of Plant Sciences, University of Missouri, Columbia, Missouri 65211, USA

Huazhong Shi Department of Chemistry and Biochemistry, Texas Tech University, Lubbock, TX 79409

Kazuo Shinozaki Plant Genomic Network Research Team, Plant Functional Genomics Research Group, RIKEN Plant Science Center (PSC), RIKEN Yokohama Institute, 1-7-22 Suehiro-cho, Tsurumi-ku, Yokohama 230-0045, Japan; Gene Discovery Research Team, RIKEN Plant Science Center (PSC), RIKEN Tsukuba Institute, 3-1-1 Koyadai, Tsukuba, Ibaraki 305-0074, Japan; CREST, Japan Science and Technology Corporation (JST), Japan

Sneh Lata Singla-Pareek International Centre for Genetic Engineering and Biotechnology, New Delhi 110067, India

Julissa Sosa Ceres, Inc. Thousand Oaks, CA 91320

Sudhir K Sopory International Centre for Genetic Engineering and Biotechnology, New Delhi 110067, India

W. Spielmeyer CSIRO Plant Industry, GPO Box 1600, Canberra, ACT 2601, Australia

Katherine A. Steele CAZS Natural Resources, University of Wales Bangor, UK

Roberto Tuberosa Department of Agroenvironmental Sciences and Technology, Viale Fanin 44, 40127 Bologna, Italy

Taishi Umezawa Gene Discovery Research Team, RIKEN Plant Science Center (PSC), RIKEN Tsukuba Institute, 3-1-1 Koyadai, Tsukuba, Ibaraki 305-0074, Japan

Dominique Van Der Straeten Unit Plant Hormone Signalling and Bio-imaging (HSB), Department of Molecular Genetics, Ghent University, K. L. Ledeganckstraat 35, B-9000 Gent, Belgium

Xuemin Wang Department of Biology, University of Missouri, St. Louis, MO 63121 and Donald Danforth Plant Science Center, St. Louis, MO 63132, USA

Zeng-Yu Wang Forage Improvement Division, The Samuel Roberts Noble Foundation, 2510 Sam Noble Parkway, Ardmore, OK 73401, USA

Liming Xiong Donald Danforth Plant Science Center, 975 N. Warson Road, St. Louis, MO 63132, USA

Jian-Long Xu Institute of Crop Sciences/National Laboratory for Crop Gene Resources and Genetic Improvement, Chinese Academy of Agricultural Sciences

Francisco Javier Zapata-Arias Plant Breeding Unit, Joint FAO/IAEA Agriculture and Biotechnology Laboratory, International Atomic Energy Agency Laboratories, A-2444 Seibersdorf, Austria; Current address – Bashati Vila, Apartment 5A, House 25, Road 68/A, Gulshan 1212, Dhaka Bangladesh

Wenhau Zhang College of Life Science, State Key Laboratory of Crop Genetics and Germplasm Enhancement, Nanjing Agricultural University, Nanijing 210095, P.R. China

Guoping Zhang Department of Agronomy, Zhejiang University, Hangzhou, 310029 China

Ji-Yi Zhang Forage Improvement Division, The Samuel Roberts Noble Foundation, 2510 Sam Noble Parkway, Ardmore, OK 73401, USA

Wenhau Zhang College of Life Science, State Key Laboratory of Crop Genetics and Germplasm Enhancement, Nanjing Agricultural University, Nanijing 210095, P.R. China

Fasong Zhou Ceres, Inc. Thousand Oaks, CA 91320

Jian-Kang Zhu Department of Botany & Plant Sciences, University of California-Riverside, Riverside, CA 92521, USA

Ellen Zuther Max Planck Institute of Molecular Plant Physiology, Am Muehlenberg 1, 14476 Potsdam-Golm, Germany

CHAPTER 1

PLANT GROWTH AND DEVELOPMENT UNDER SALINITY STRESS

A. LÄUCHLI AND S.R. GRATTAN
Department of Land, Air and Water Resources, University of California, One Shields Ave., Davis, CA 95616, USA

Abstract: Plant growth and development are adversely affected by salinity – a major environmental stress that limits agricultural production. This chapter provides an overview of the physiological mechanisms by which growth and development of crop plants are affected by salinity. The initial phase of growth reduction is due to an osmotic effect, is similar to the initial response to water stress and shows little genotypic differences. The second, slower effect is the result of salt toxicity in leaves. In the second phase a salt sensitive species or genotype differs from a more salt tolerant one by its inability to prevent salt accumulation in leaves to toxic levels. Most crop plants are salt tolerant at germination but salt sensitive during emergence and vegetative development. Root and shoot growth is inhibited by salinity; however, supplemental Ca partly alleviates the growth inhibition. The Ca effect appears related to the maintenance of plasma membrane selectivity for K over Na. Reproductive development is considered less sensitive to salt stress than vegetative growth, although in wheat salt stress can hasten reproductive growth, inhibit spike development and decrease the yield potential, whereas in the more salt sensitive rice, low yield is primarily associated with reduction in tillers, and by sterile spikelets in some cultivars.

Plants with improved salt tolerance must thrive under saline field conditions with numerous additional stresses. Salinity shows interactions with several stresses, among others with boron toxicity, but the mechanisms of salinity-boron interactions are still poorly known. To better understand crop tolerance under saline field conditions, future research should focus on tolerance of crops to a combination of stresses

Keywords: Vegetative growth, reproductive growth, development, salinity stress, boron, osmotic, ionic, crop

1. INTRODUCTION

In the preface to the 'Special Issue: Plants and salinity', Tim Flowers (2006) emphasized that "Salinity has been a threat to agriculture in some parts of the world for over 3000 years; in recent times, the threat has grown". As the world population continues to increase, more food needs to be grown to feed the people. This can

M.A. Jenks et al. (eds.), Advances in Molecular Breeding Toward Drought and Salt Tolerant Crops, 1–32.

be achieved by an increase in cultivated land and by an increase in crop productivity per area. The former has brought agriculture to marginal, salt-affected lands. Moreover, the salinity problem has been aggravated by the requirement of irrigation for crop production in arid and semiarid environments. It is estimated that at least 20% of all irrigated lands are salt-affected (Pitman and Läuchli, 2002). About 17% of the cultivated land is under irrigation; yet, irrigated agriculture contributes more than 30% of the total agricultural production (Hillel, 2000). The total global area of salt-affected soils has recently been estimated to be approximately 830 million hectares (Martinez-Beltran and Manzur, 2005). The different types of soil salinity that impact agricultural productivity, i.e. irrigation-induced salinity and 'transient' dry-land salinity have been characterized in detail by Rengasamy (2006), with special emphasis on Australia. Clearly, soil salinity is one of the major environmental stresses that limit agricultural productivity worldwide.

Population growth on the one hand and land degradation by salinization on the other have led plant scientists to the concept of developing salt-tolerant crops by genetic approaches (see recent reviews by Cuartero et al., 2006; Munns, 2005; Munns et al., 2006; Yamaguchi and Blumwald, 2005). However, the physiological, biochemical and molecular mechanisms of salt tolerance in plants are not yet sufficiently understood, and hence progress in developing salt tolerant crops has been slow. This chapter provides a brief overview of our present physiological knowledge of how growth and development of plants are affected by salinity. The focus is on annual crop species with special emphasis on cereals. Furthermore, crop growth and development under salinity stress will be discussed for both controlled and natural agricultural environments. The still poorly-understood relationship between sodium uptake and salt tolerance has been assessed in depth by Tester and Davenport (2003) but will only be covered briefly in this chapter. In the context of saline agricultural environments, soil salinity is often accompanied by additional abiotic and biotic stresses. For example, high boron concentrations often occur in saline environments. Therefore interactions between salinity and boron toxicity in crops are also examined. Our chapter does not focus on biochemical and molecular mechanisms of salt tolerance. For recent reviews that focus on these mechanisms, see for example Hasegawa et al. (2000), Zhu (2002), and Koiwa et al. (2006). Finally genomics-type technologies are beginning to enhance our understanding of how genes, proteins and metabolite profiles and their interactions and dynamic changes respond to salinity. For more information on these complex interactions see Bohnert et al. (2006).

2. SALINITY STRESS AND PLANT DEVELOPMENT

Salinity affects plants in different ways such as osmotic effects, specific-ion toxicity and/or nutritional disorders (Läuchli and Epstein, 1990). The extent by which one mechanism affects the plant over the others depends upon many factors including the species, genotype, plant age, ionic strength and composition of the salinizing solution, and the organ in question.

Plants undergo characteristic changes from the time salinity stress is imposed until they reach maturity (Munns, 2002a). This author describes these changes over different time scales in the plant's development. Moments after salinization, cells dehydrate and shrink, but regain their original volume hours later. Despite this recovery, cell elongation and to a lesser extent cell division, are reduced leading to lower rates of leaf and root growth. Over the next days, reductions in cell division and elongation translate into slower leaf appearance and size. Plants that are severely salt-stressed often develop visual injury due to excessive salt uptake. After weeks, lateral shoot development is affected and after months, clear differences in overall growth and injury are observed between salt-stressed plants and their non-stressed controls.

Understanding these temporal differences in response to salinity, Munns (2002a, 2005) developed the concept of the 'two-phase growth response to salinity' (Figure 1). The first phase of growth reduction happens quickly (within minutes) after exposure to salinity. This response is due to the osmotic changes outside the root causing changes in cell-water relations (osmotic effect). The osmotic effect initially reduces the ability of the plant to absorb water. This effect is similar to water stress and shows little genotypic differences. Several minutes after the initial decrease in leaf growth, there is a gradual recovery of the growth rate until a new steady state is reached, dependent upon the salt concentration outside the root (Munns, 2002a). The second much slower effect, taking days, weeks or months is the result of salt accumulation in leaves, leading to salt toxicity in the plant, primarily in the older leaves (i.e. salt-specific effect). This salt toxicity can result in the death of leaves and reduce the total photosynthetic leaf area. As a result, there is a reduction in the supply of photosynthate to the plant, affecting the overall carbon balance necessary to sustain growth (Munns, 2002a). Salt toxicity primarily occurs

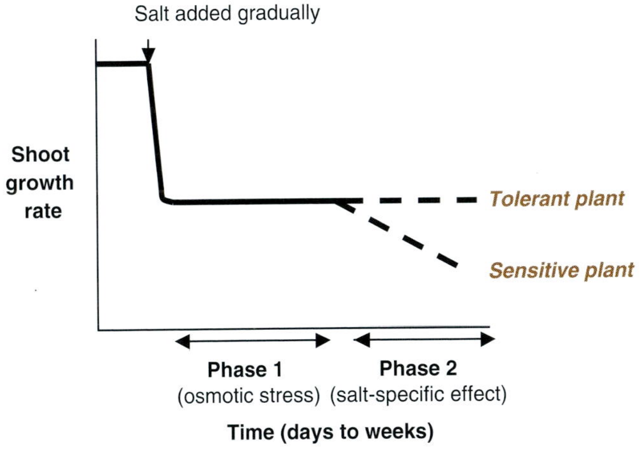

Figure 1. Schematic illustration of the two-phase growth response to salinity for genotypes that differ in the rate at which salt reaches toxic levels in leaves (Munns, 2005)

in the older leaves where Na and Cl build up in the transpiring leaves over a long period of time, resulting in high salt concentration and leaf death. Leaf injury and death is probably due to the high salt load in the leaf that exceeds the capacity of salt compartmentation in the vacuoles, causing salt to build up in the cytoplasm to toxic levels (Munns and Termaat, 1986; Munns 2002a; 2005; Munns et al, 2006). The rate at which leaves die and thus reduce the total photosynthetic leaf area determines the survival of the plant. If new leaves are produced at a rate greater than the rate at which old leaves die, there are enough photosynthesizing leaves for the plant to flower and produce seeds, although at reduced numbers. If, however, old leaves die faster than new leaves develop, the plant may not survive long enough to supply sufficient photosynthate to the reproductive organs and produce viable seeds. Based on this two-phase concept, the initial growth reduction for both salt sensitive and salt tolerant plants is caused by an osmotic effect of the salts in the medium outside the roots. In contrast, in the second phase, a salt-sensitive species or genotype differs from a more salt tolerant one by its inability to prevent salt from accumulating in transpiring leaves to toxic levels (Munns et al, 2006).

In light of the different mechanisms of plant response to salinity (Läuchli and Epstein, 1990) and characteristic sequential changes which the plants endure after being exposed to salinity (Munns, 2002a), are their specific developmental stages where the plants are more or less sensitive to salinity?

2.1. Salt Sensitivity in Relation to Developmental Growth Stage

It has long been recognized that a crop's sensitivity to salinity varies from one developmental growth stage to the next (Bernstein and Hayward, 1958). Although there are exceptions, the majority of the research indicates that most annual crops are tolerant at germination but are sensitive during emergence and early vegetative development (Läuchli and Epstein, 1990; Maas and Grattan, 1999). As plants mature, they become progressively more tolerant to salinity, particularly at later stages of development. While these statements are generally true (with the exception of perhaps a few crops), it is important to emphasize that the definition of salt tolerance is not the same for each growth stage. During germination and emergence, tolerance is based on percent survival, while during the later developmental stages, tolerance is usually based on relative growth reductions.

Salinity affects both vegetative and reproductive development which has profound implications depending on whether the harvested organ is a stem, leaf, root, shoot, fruit, fiber or grain. Salinity often reduces shoot growth more than root growth (Läuchli and Epstein, 1990) and can reduce the number of florets per ear, increase sterility and affect the time of flowering and maturity in both wheat (Maas and Poss, 1989a) and rice (Khatun et al. 1995). Since salt-tolerance from an agronomic or horticulturist perspective is based on the yield of the harvestable organ, relative to that in non-stressed environments, understanding how salinity affects vegetative and reproductive development is important for developing management strategies that can minimize stress at critical times.

2.1.1. *Germination and seedling emergence*

Although most plants are tolerant during germination, salinity stress delays this process even though there may be no difference in the percentage of germinated seeds from one treatment to another (Maas and Poss, 1989a). It is this observation that categorizes this developmental stage for most crops as 'salt tolerant'. For example, salinity up to 10 dS/m actually stimulated the germination of *Limonium perezii* seeds, a commercially grown ornamental flower, yet salinities above 6 dS/m reduced stem length, adversely affecting quality and marketability (Carter et. al., 2005). Even though salinity delays germination, higher salt concentrations will eventually reduce the percentage of germinated seeds (Kent and Läuchli, 1985; Badia and Meiri, 1994; Mauromicale and Licandro, 2002) (Figure 2). While most crops show enhanced tolerance to salinity during germination, this is not true for sugar beet, a crop categorized as salt tolerant which is somewhat sensitive to salinity at germination (Läuchli and Epstein, 1990). There are even differences in tolerance among cultivars (e.g. Ahmad et al., 2005; Bayuelo-Jimenez et al., 2002) and these differences do not necessarily correspond to seasonal tolerance, as shown for melon (Nerson and Paris, 1984), bean (Bayuelo-Jimenez et al., 2002) and rice (Heenan et al., 1988). On the other hand, salt tolerant barley varieties germinated faster and showed a much higher germination percentage than the more sensitive ones (Tajbakhsh et al., 2006). Regardless, salt tolerance screening at germination provides little basis for assessing crop salt tolerance.

The vast majority of these germination studies have been conducted in the laboratory using Petri-dish like containers with germination paper saturated with solutions that vary in salinity. While easy to observe germination, such artificial environments are uncharacteristic of field conditions (Esechie et. al., 2002). In

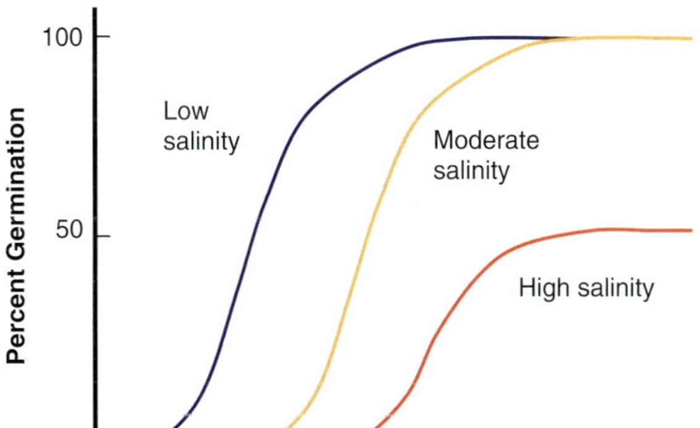

Figure 2. Generalized relationship between percent germination and time after water addition at low, moderate and high salinity. The germination rates and percentage of germinated seeds at a particular time various considerably among species and cultivars

addition, other variables such as seed viability, dormancy, seed coat pretreatment and permeability to water may complicate data interpretations and comparisons with other crops or to other developmental stages.

Unlike at germination, most crops are susceptible to salinity during emergence, which is based more on observation than quantitative research. Emergence studies have been conducted using different root media under various environmental conditions, making interpretation of the results and comparisons with other studies difficult, if not impossible. Moreover, most studies were conducted using NaCl as the sole salinizing salt, which is uncharacteristic of most salt-affected soils. When studies are conducted in mineral soils using NaCl solutions, sodicity (high sodium relative to calcium plus magnesium) can cause adverse effects on soil physical conditions, reducing oxygen diffusion rates and increasing soil strength (Grattan and Oster, 2003). This could inadvertently add unwanted stresses to the emerging seedlings.

Salinity delays emergence and if the stress is severe enough, stand establishment can be reduced (Maas and Grattan, 1999). Crop tolerance during this sensitive growth-stage differs considerably among crops and like germination, does not correlate well with crop tolerances based on yield-response functions. For example cotton, a crop known to be salt tolerant based on lint yields, is particularly prone to poor stands in fields that were previously irrigated with saline-sodic water (Grattan and Oster, 2003), despite the fact that salinity in the upper soil profile was less than the soil-salinity threshold[1] for cotton. In a related long-term field study, plant density of cotton was severely reduced by irrigation with saline drainage water of 4,500 mg/L TDS (EC~ 7ds/m) for three consecutive years (Goyal et al., 1999). These authors concluded that stand establishment was possibly the main reason for reduction in lint yield.

Under field conditions, germinated seedlings encounter a number of biotic and abiotic stresses. In addition to salinity, young seedlings near the soil surface are subjected to water stress (Katerji et al., 1994), fluctuating salinities due to capillary rise and evaporation (Pasternak et al., 1979), diurnal changes in soil temperature and surface crusts. Studies have shown that salinity is more detrimental to germination of seeds outside their optimal temperature range for germination (Vinizky and Ray, 1988). Also, because salinity delays germination and emergence, the young salt-stressed seedlings may be more susceptible to hypocotyl and cotyledon injury (Miyamoto et al., 1985; Esechie et al., 2002) or attack by pathogens. Although it is likely that this unavoidable combination of stresses that the emerging seedlings endure under field conditions can reduce the percentage of emerged seedlings, we are not aware of any in-depth evaluation on the tolerance of young seedlings under field or simulated field conditions. Such research would be valuable to better understand how crops respond to integrated biotic and abiotic stresses they encounter between germination and emergence.

---

[1] Yield threshold refers to the maximum soil salinity (expressed as the electrical conductivity of the saturated soil paste, ECe) that a plant can endure in the rootzone and still maintain optimal yield.

2.1.2. *Vegetative growth*

Most of the literature indicates that plants are particularly susceptible to salinity during the seedling and early vegetative growth stage as compared to germination. Examples are found in barley (Ayers et al., 1952), corn (Maas et al., 1983), cotton (Abul-Naas and Omran, 1974), cowpea (Maas and Poss, 1989b), melon (Botia et al., 2005), New Zealand spinach (Wilson et al., 2000), red orach (Wilson et al., 2000), rice (Pearson and Ayers, 1966), sorghum (Maas et al 1986), tomato (del Amor et al., 2001), and wheat (Maas and Poss, 1989a). In greenhouse experiments with corn and wheat, the total shoot biomass of salt-stressed plants relative to non-stressed plants was much lower than salinity's overall effect on relative grain yield (Maas et al, 1983; Maas and Poss, 1989a). Although it may not be true for most crops, some investigators found that salt tolerance among melon cultivars during early seedling growth correlated well with salt tolerance based on fruit yield at the end of the season (Nerson and Paris, 1984).

2.1.3. *Roots*

Why is early vegetative development so susceptible to salinity? It is well known that salinity with an adequate supply of calcium reduces shoot growth, particularly leaf area, more than root growth (Läuchli and Epstein, 1990). However, inadequate Ca supply under saline conditions can adversely affect membrane function and growth of the root within minutes (Epstein, 1961; Läuchli and Epstein, 1970; Cramer et al., 1988). When supplemental Ca was added to a salinized medium, cell elongation of cotton roots was favored at the expense of radial cell growth and cell production rates were maintained (Kurth et al., 1986). Additional studies with cotton roots revealed that supplemental Ca partly alleviated the inhibition of the elongation rate due to high salt in the medium but the shortening of the growth zone of the root caused by high salt stress was not restored by supplemental calcium (Zhong and Läuchli, 1993). High salt stress increased the deposition rate of Na in the growing region of the root and hence decreased the selectivity for K versus Na. The latter effect was partly mitigated by supplemental Ca, but only in the apical 2mm region (Zhong and Läuchli, 1994). The conclusion of these studies is that supplemental Ca alleviates the inhibitory effect of salt on cotton root growth by maintaining plasma membrane selectivity of K over Na (Zhong and Läuchli, 1994; reviews: Läuchli, 1990, 1999).

2.1.4. *Shoots*

Reduction in shoot growth due to salinity is commonly expressed by a reduced leaf area and stunted shoots (Läuchli and Epstein, 1990). Final leaf size depends on both cell division and cell elongation. Leaf initiation, which is governed by cell division, was shown to be unaffected by salt stress in sugar beet, but leaf extension was found to be a salt-sensitive process (Papp et al., 1983). Thus, cell division in leaves of sugar beet appears less salt sensitive than cell elongation. On the other hand, cell numbers in grass leaves were reduced by salinity (Munns and Termaat, 1986). As already described for roots the effect of salt stress on shoot growth in several species can also be partly alleviated by supplemental Ca (Läuchli and Epstein, 1990; Cramer, 2002). If, however, plants are exposed to high Na/Ca ratios, Ca-deficiency in the shoot can be induced, as for example demonstrated for developing corn leaves by Maas and Grieve (1987). The Ca status of the growing region of leaves is particularly sensitive to salt stress (Läuchli, 1990). This appears to be the consequence of inhibition by salt of symplastic xylem loading of Ca in the root (Lynch and Läuchli, 1985; Halperin et al, 1997), leading to reduced Ca status in growing region of leaves (Lynch et al., 1988; Lazof and Läuchli, 1991; Neves-Piestun and Bernstein, 2005; review: Lazof and Bernstein, 1999). The importance of supplemental Ca to alleviate salt stress effects in the shoot, as demonstrated originally by La Haye and Epstein (1971), has been clearly emphasized by Cramer (2002) and Munns (2002 b) who recommended adding at least 5–10 mM Ca to the medium for salinities of 100-150 mM NaCl, to counteract the inhibitory effect of high Na concentrations on growth.

As recently summarized in detail by Cramer (2002), many of the well known Na-Ca interactions in plants can be linked to Na-Ca interactions at the surface of the plasma membrane and subsequent Ca signaling events (Cramer et al., 1985). For a quantitative description of these Na-Ca interactions, ion activities instead of ion concentrations must be used (Cramer and Läuchli, 1986; Cramer et al., 1986; Yermiyahu et al., 1997; Kinraide, 1999). Ion activities usually are lower than their concentrations, particularly for Ca because of ion pair formation and precipitation as calcite (Cramer and Läuchli, 1986).

A detailed, quantitative study of the responses of leaf growth and development in sorghum to salt stress showed that the length of the growth zone was shortened by 20% under salt stress, and that salt stress also reduced the maximal relative elemental growth rate, particularly in the youngest region of the leaf (Bernstein et al, 1993a). Increasing the external Ca supply restored the length of the growing zone of the leaf and increased also the relative elemental growth rate (Bernstein et al., 1993b). This contrasts with the finding on roots where the shortening of the growing zone of cotton roots was not restored by supplemental Ca (Zhong and Läuchli, 1993). In barley leaves, salt stress did not affect the length of the elongation zone, but the Ca supply to the plant was not varied in this study (Fricke and Peters, 2002). Salt stress induced a dramatic decrease in Ca in the growing sorghum leaf which could be at least partly responsible for leaf growth inhibition (Bernstein et al., 1995). Sodium was preferentially accumulated in the basal part of the growing zone where growth was least affected by salt stress. Hence, it was concluded that high Na concentration in the salt-affected leaf tissue was not the primary cause for growth inhibition (Bernstein et al, 1995). Hu, Schmidhalter and coworkers (review by Hu et al, 2005a) conducted similar research on growing wheat leaves and also concluded that direct effects of Na and Cl toxicity on cell expansion and formation of the leaf cross-sectional area can be ruled out. However, one would need to know the cytoplasmic versus vacuolar Na concentrations in these tissues to draw more definitive conclusions. An additional important feature is that salinity has been demonstrated to reduce the area of proto-and metaxylem in growing leaves of sorghum (Baum et al, 2000) and wheat (Hu et al., 2005b) which may be responsible for decreased water deposition into the growing region of leaves. This could indirectly affect transport of Na and Cl and of nutrient ions to the growing leaves.

2.1.5. *Reproductive growth*

After the salt-sensitive early-vegetative growth stage, the bulk of the research suggests that most crops become progressively more tolerant as the plants grow older (Läuchli and Epstein, 1990; Maas and Grattan, 1999). There have been numerous studies characterizing crop response to salinity at various developmental growth stages. However, many of them did not evaluate plant response during the entire lifespan of the crop and those that did, most studies imposed salt stress at various times after emergence and continued the stress until harvest. The difficulty with the latter group of studies is that treatments give preferential favoritism to the later growth stages since the duration of salt-stress was less.

There were several studies, however, where the duration of salinity stress was held constant but the period of salt-stress imposition varied from one developmental stage to the next. These studies were conducted using re-circulating sand tanks where transient salinity conditions can readily be controlled. In experiments with wheat (Maas and Poss, 1989a), sorghum (Maas et al., 1986) and cowpea (Maas and Poss, 1989b), investigators found that these crops were most sensitive during vegetative and early reproductive stages, less sensitive during flowering and least sensitive during the seed filling stage. In all these studies, seed weight is the yield component of interest but similar conclusions regarding growth stage sensitivity were obtained with both determinate crops (the grain crops) and indeterminate (cowpea) crops.

Wheat and rice are not only two of the most important grain crops in the world but they have been the most intensively studied agronomic crops regarding salt sensitivity at different growth stages. Studies on these grain crops were conducted in the field, greenhouse and laboratory to better understand detailed changes in vegetative and reproductive developmental processes, as the plants endure various degrees of salt stress at different growth stages. Because of the extensive nature of the research on these crops, a summary of the key findings is presented below. Extensive research has also been conducted on the important horticultural crop tomato, but research on this crop will not be covered in a separate section of this chapter and the reader is referred to review articles by Cuartero et al. (2006) and Cuartero and Fernandez-Munoz (1999).

2.2. Wheat

It has long been known that salinity reduces the growth rate of the entire wheat plant and its specific organs, but it also affects plant development. The architecture of expanding wheat leaves from recently emerged seedlings subjected to 200 mM NaCl was greatly affected (Hu et al., 2005b). By close examination of the transverse section of leaf 4, investigators found that salinity reduced the cross sectional area, width and radii of epidermal and mesophyll cells along the leaf axis, indicating that adverse effects from salinity were occurring during leaf initiation.

The duration of plant development is also affected by salinity. The salt sensitivity of wheat at various growth stages was evaluated by Maas and Poss (1989a) by imposing salt stress [-0.05 to -1.25 MPa (1.4–28 dS/m)], using a combination of NaCl and $CaCl_2$ salts, either 10, 56, or 101 days after planting (referred to as vegetative and spikelet differentiation, reproductive, and maturation stages, respectively). At each developmental stage, the stress was imposed for a 45-day duration and then removed. Salt stress retarded leaf development and tillering but hastened plant maturity. When grain yield data were compared among treatments, 'Aldura' and the more tolerant variety 'Probred', became less sensitive to salinity the later plants were stressed, even though the duration of stress was held constant.

Salt stress, imposed while the shoot apex is in vegetative stage, can adversely affect spike development and decrease yields of wheat (Maas and Grieve, 1990). When wheat was salt-stressed during spike or panicle differentiation, reproductive development was stimulated but the number of spikelets was reduced. They found that salt stress accelerated the development of the shoot apex on the mainstem and decreased the number of spikelet primordia. The terminal spikelet stage occurred about two weeks earlier in salt-stressed wheat as compared to non-stressed controls. Anthesis also occurred earlier in salt-stressed plants but tillering was delayed several days. The investigators found that salt stress increased the phyllochron (the interval between appearance of successive leaves on the main stem based on thermal time) and reduced the number of leaves initiated on the main stem. Salt stress decreased the yield potential mostly by reducing the number of spike-bearing tillers. This conclusion was also reached by El-Hendawy et al. (2005) in a comprehensive evaluation of numerous wheat cultivars using cluster analysis. Therefore Maas and Grieve (1990) concluded that salinity stress needs to be avoided prior to and during spikelet development on all tiller spikes if full yield potential is to be achieved.

Grieve et al. (2001) conducted another salt stress release study on spring wheat where salinity was imposed and withdrawn, before or after, three growth stages; 1) late leaf primordial initiation, 2) double ridge stage, and 3) terminal spike formation. They found that grain yields were maximized when salt stress was delayed until after the terminal spike formation or by withdrawing stress at the late leaf primordial stage or double ridge stage. They found that short periods of salt stress during organogenesis have irreversible consequences on wheat growth and development.

In a more in-depth examination of semidwarf wheat varieties, Grieve et al (1993) used a three-piece linear-spline model and found that salinity decreased the rate of leaf primordium initiation but did not affect the duration of this phase. On the other hand, they found salinity reduced the duration of the spikelet primordium initiation phase, even though it had no effect on the rate of spikelet primordium initiation. This combination of effects resulted in less leaves and caused a reduction in the number of grain-bearing spikelets, severely affecting the yield potential of these wheat types.

Additional studies on wheat were conducted to examine salinity's effect on reproductive physiology. Khan and Abdullah (2003) found that pollen viability in two wheat cultivars differing in salinity tolerance was reduced 24–37%; depending upon cultivar. They also suggested that 80–90% of the carbon that fills wheat grains comes from current photosynthesis and not from stored vegetative carbon sources. While most of the carbon that is filling grains comes from active photosynthetic sources, the carbon is not distributed uniformly among tillers. Grieve et al (1992) analyzed the main spike yield components of salt-stressed wheat and found that grain yield from the main spike of two semidwarf Mexican wheat varieties increased up to 15% more in salt-stressed plants (-0.65 MPa OP) than non stressed plants. They found that decreases in kernel numbers per spike were offset by increases in kernel weight. Therefore moderately salt-stressed wheat plants distributed their carbohydrates preferentially towards the main stem tillers.

Other studies were directed towards ion relations in salt-stressed grain crops. Maas and Poss (1989a) found that K uptake was severely inhibited by salt stress imposed to wheat during the vegetative growth stage but not at later stages, even though the more tolerant variety 'Probred' accumulated less Na than the more sensitive "Aldura'. The effect of NaCl salinity on salt accumulation and reproductive development in the meristem of wheat and barley was studied by Munns and Rawson (1999). They selected two varieties of each species differing in salt tolerance to observe changes in the development of the apex as it changed from vegetative to reproductive growth. Apices were analyzed for ion contents when most of the spikelet primordia had been produced and the process of differentiation into floral organs had started. Potassium concentrations were unaffected by salinity (up to 175mM NaCl). In addition, they concluded that Na and Cl concentrations were too low to affect metabolism. Nevertheless, despite the small effect of salinity on apex ion relations, salinity still affected reproductive development; fewer spikelet primordia formed and the final spikelet numbers at ear emergence were reduced.

In summary, a mature wheat plant is a consequence of sequential developmental processes that are characterized by changes in shoot apex morphology. The yield components such as tillers per plant, number of spikelets per spike and individual grain weights, are developed sequentially as the crop develops. If salt stress is applied before and during the shoot apex transition from vegetative to reproductive stage, it can significantly affect vegetative and reproductive development. Salt stress can hasten reproductive development but also can adversely affect spike development and decrease the yield potential of wheat.

2.3. Rice

Although rice is one of the most important food crops in the world, both economically and nutritionally, it ranks among the most sensitive to salinity (Maas and Grattan, 1999). Not only is rice considerably less tolerant to salinity than wheat, but salinity affects its reproductive development quite differently.

Rice sensitivity to salinity varies considerably from one growth stage to the next. In terms of grain yield, rice is tolerant during germination (Heenan et al., 1988), sensitive to salinity during emergence and early seedling growth, becomes more tolerant later on in vegetative development, and then can becomes sensitive again during reproductive growth (Pearson and Bernstein, 1959; Flowers and Yeo, 1981; Khatun and Flowers, 1995; Abdullah et al., 2001). The vegetative shoot biomass of rice, on the other hand, is often affected much less than reproductive growth (except for young seedlings) (Khatun and Flowers, 1995; Munns et al., 2002). Field and greenhouse studies showed that salinity had a negative impact on stand establishment and adversely affected a number of yield components and even delayed heading (Grattan et al 2002). In one study, investigators found linear decreases in several yield components with increased salinity including the percent of sterile florets, tillers per plant and spikelets per panicle which translated into larger reductions in grain weight per plant at a given salinity (Zeng and Shannon, 2000 (Figure 3). However these investigators suggested that seedling emergence and early seedling growth stages were most sensitive to salinity, as was the 3-leaf panicle stage.

Being aware that rice response to salinity is a combination of the level of salinity, the duration of exposure and timing of exposure, Lee et al (2004) proposed a salt stress index that incorporates these factors. Using solution cultures, they found that the growth of rice was reduced over three times more with NaCl than synthetic sea water and that rice was two times more sensitive to salinity at the seedling stage than it was at the tillering stage. This not only implies that the tolerance of rice

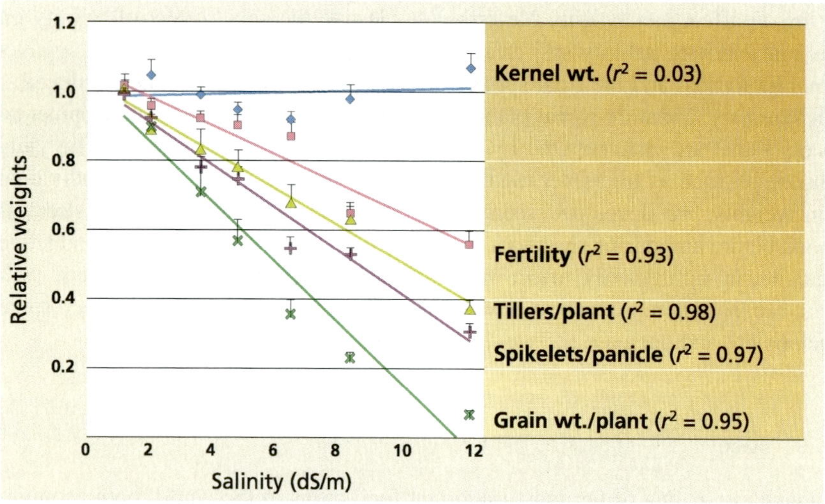

Figure 3. Relationship between salinity and various yield components of rice (*Oryza sativa* L. cv M-202). Fertility is inversely proportional to sterility. From Grattan et al., 2002 originally adapted from Zeng and Shannon, 2000

varies with stage of growth but it is strongly affected by the composition of the root media, particularly when NaCl is used as the sole salinizing salt.

Unlike what was observed on wheat, it has long been recognized that salinity can cause sterility in rice, particularly if imposed during pollination and fertilization (Pearson and Bernstein, 1959). However this effect has not been consistently observed. Akbar and Yabuno (1977) found that salinity caused panicle sterility in only some rice cultivars suggesting some genetic control (Khatun et al., 1995). Salinity's effect on rice resulted in delayed flowering, a decrease in the number of productive tillers and fertile florets per panicle and a reduction in individual grain weight (Khatun et. al, 1995; Lutts et al., 1995).

Zeng and Shannon (2000) examined salinity effects on seedling growth and yield components of rice. They found that seedling growth was adversely affected at salinity levels as low as 1.9 dS/m, but this effect did not translate into a reduction in grain yield. They also found that seedling survival was adversely affected at salinities >3.4 dS/m, confirming what has been known for decades that rice is extremely sensitive during early vegetative growth (Pearson and Ayers, 1966). Furthermore, individual seed size was not significantly affected by salinity but grain yield per plant was reduced primarily by a reduction in number of tillers per plant, number of spikelets per panicle, and the grain weight per panicle. Finally, they also found a substantial reduction in filled grains at 6 dS/m and higher suggesting that high salinity was causing some sterility.

Khatun and Flowers (1995) studied the effect of NaCl salinity on sterility and seed set in rice. Salinity increased the number of sterile florets and viability of pollen, becoming more pronounced with increased salinity. Seed set was reduced by 38% when female plants were grown in as low as 10 mM NaCl. When they compared crosses involving male and female parents grown at different salinities, effects on female plants dominated those on the pollinator plants.

The effects of 50 mM NaCl on floral characteristics, yield components and biochemical and physiological attributes of rice were studied to better understand the causes of sterility in rice (Abdullah et al., 2001). They concluded that sterility and reduction in seed set were primarily due to reduced translocation of soluble carbohydrates to primary and secondary spikelets, accumulation of more sodium and less potassium in all floral parts and inhibition of the specific activity of starch synthetase in developing rice grains, thus reducing seed set.

In summary, the reduction in number of spike-bearing tillers by salt stress during the vegetative and early reproductive development in most cereal crops appears to have a greater negative impact on grain yield than any other yield component. The time from planting to maturity in cereal crops typically decreases with increased salinity (Grieve et al (1993) but salinity has just the opposite effect on rice (Khatun et. al, 1995; Lutts et al., 1995). When salinity was applied to wheat from seedling emergence, it had a profound influence on reproductive development (Grieve et al, 1993). Leaf initiation rate decreased even though the time of flag leaf initiation was unchanged indicating salinity had no influence on the timing of the transition from vegetative to reproductive development, but greatly reduced the number of tillers and overall grain yield. Salt stress in rice can reduce seedling emergence and when imposed at early vegetative stages, it reduces tillers and grain-bearing panicles leading to low yields. However unlike wheat, certain rice cultivars can develop sterile spikelets, which appears to be genetically controlled, leading to further grain yield losses.

3. CELL ELONGATION AND CELL WALL PROCESSES UNDER SALINITY STRESS

Cell expansion is controlled by processes related to cellular water uptake and cell wall extension (Cramer and Bowman, 1993). Cell expansion is initiated by biochemical loosening of the cell wall under turgor pressure and uptake of water and solutes (Cosgrove, 1987, Hsiao et al., 1976; Boyer 1987). Although cell expansion is three-dimensional, it can be described in one dimensional space as a change in length (Nonami and Boyer, 1990). Quantitatively, cell growth can be described by the equation:

$$(1) \qquad \frac{1}{\nu}\frac{\delta \nu}{\delta t} = \mathrm{E}\Psi_p$$

where v is the cell volume, t is the time, E describes the cell wall yielding properties, and Ψ_p is the turgor pressure. More comprehensively, a quantitative description of growth should include mechanical and hydraulic components (Boyer, 1987):

$$(2) \qquad \frac{1}{\nu}\frac{\delta \nu}{\delta t} = \frac{mL}{m+L}(\Psi_0 - \Psi_S - Y)$$

where m, L, Ψ_o, Ψ_s and Y denote cell wall extensibility, hydraulic conductance, xylem water potential, cell osmotic potential and yield threshold, respectively. E in equation 1 and $mL/(m+L)$ are comparable; the expression Ψ_o - Ψ_s – Y denotes the driving force for cell growth (Cramer and Bowman, 1993). The yield threshold is the minimum turgor pressure (turgor threshold) at which cells expand. Thus, cell wall extensibility, hydraulic conductance, turgor and yield threshold are important components of these complex cell- growth processes and control the rate of leaf elongation. These growth parameters can be readily affected by salinity stress.

In an overview article, Hu and Schmidhalter (2004) concluded that the reduction of leaf elongation by salinity may either be related to decreases in cell wall extensibility or increases in yield threshold (see for example Cramer, 1991; Neumann, 1993). Other investigators focused on the response of Ψ_p to salinity; the result of these studies, however, were varied and not entirely conclusive (Cramer and Bowman, 1993). Whereas Thiel et al. (1988) found that Ψ_p in leaf epidermal cells was reduced by salinity, Yeo et al. (1991) determined that leaf elongation in rice declined after exposure to salinity, but no effect on Ψ_p in the growing zone was detected. In maize, leaf elongation was inhibited rapidly by salinity and then partially recovered to a new steady-state, while Ψ_p initially declined but then completely recovered to control values during the new but reduced steady-state elongation rate (Cramer and Bowman, 1991).

The role of turgor in the response of leaf elongation to salinity remains unclear. Cramer and Bowman (1993) considered the speed at which leaf elongation is reduced, suggesting a hydraulic signal might be occurring. In a more recent study (Cramer, 2003) the effects of salinity on leaf elongation rates of three grass species indicated that the inhibition of elongation was related either to the yield threshold or to hydraulic conductance or both, but cell-wall extensibility was not significantly affected by salinity. Other experiments showed variable effects of salinity on cell wall extensibility (Cramer and Bowman, 1993). Also, hydraulic conductance was not always reduced in salt-stressed plants (Cramer and Bowman, 1993). Focusing more specifically on cell wall properties, Cramer et al, (2001) argued that an increase in the yield threshold caused by salinity could be explained through an effect on the physical properties of the leaf cell walls. However, no changes in physical properties were detected in cell walls *in vitro*. Therefore the inhibition of cell elongation by salinity may not be related to a hardening of the physical structure of the cell walls. On the other hand, cell elongation has been considered to be stimulated by increased acidification of the cell wall (apoplast) space (see for example Hu and Schmidhalter, 2004) and hence salinity-induced inhibition of elongation growth would be related to a decrease in apoplastic acidification rate, as demonstrated for water stress-induced growth inhibition (e.g. Von Volkenburgh and Boyer, 1985). In contrast, Neves-Piestun and Bernstein (2001) did not find a significant effect of salinity on cell-wall acidification in maize leaves. This appears to be an important difference in the primary cause of inhibition of leaf elongation by water and salinity stress.

An early hypothesis proposed by Oertli (1968) stated that inhibition of leaf growth and leaf death by salinity could be caused by excessive salt accumulation in the apoplast of leaves, causing dehydration of leaf cells and loss of Ψ_p. Flowers et al. (1991) presented X-ray microanalysis data in support of the Oertli hypothesis. They found up to 600 mM Na in the leaf apoplast of rice plants that were subjected to 50 mM NaCl for a week. However, there is uncertainty whether the used technique would permit the required high spatial resolution for precise apoplastic ion localization. More recent studies using both an infiltration technique (Mühling and Läuchli, 2002a) and *in vitro* fluorescence imaging (Mühling and Läuchli, 2002b) showed that Na^+ concentrations in the leaf apoplast of maize and cotton remained too low to cause a decline in leaf growth under salinity stress. These results are not in support of Oertli's hypothesis. In contrast, solute concentrations in the leaf apoplast of the halophytic shrub *Sarcobatus vermiculatus*, obtained by the infiltration technique, reached values up to 230 mM Na^+ in plants subjected to 300 mM NaCl or higher (James et al., 2006). Thus, in halophytes, salt may accumulate in the leaf apoplast to quite high concentrations and then alter the water relations of the plant without causing salt toxicity in the leaves.

According to equation (2) cell elongation also depends on the cell osmotic potential, Ψ_s. In most plants, when leaf elongation partly recovered after the initial rapid drop in the elongation rate upon salinization of the medium, osmotic adjustment occurred with the solute content in the leaf cells becoming higher under saline than non-saline conditions. This adjustment occurs primarily by an increased accumulation of Na^+ and Cl- inside the cell but also by accumulation of organic solutes such as sugars (Munns et al., 2006). That significant solute accumulation, facilitating recovery of elongation growth, takes place in leaves following salinization has recently been demonstrated clearly by W. Fricke's group for developing barley leaves (Fricke and Peters, 2002; Fricke, 2004; Fricke et al., 2006). Specifically, Table 1 (see Fricke et al, 2006, Table 2) shows changes in osmolality in bulk leaf and epidermal cells of the growing barley leaf 3 after 20h exposure to 100 mM NaCl. Both tissues responded similarly to the salt stress. Osmolality increased by almost 200 mosmol kg^{-1} in the elongation zone Ez (10–30 mm from the point of leaf insertion), but only by about 75 mosmol kg^{-1} in the emerged, mature region of the leaf. The increase in osmolality, an indication of osmotic adjustment in the growing region of the leaf, was not caused by a decrease in water content but was due to a net increase in solute content (Fricke et al., 2004).

Sodium, K^+ and Cl- were the main inorganic solutes that contributed to osmotic adjustment, and with duration of the stress, Na^+ increasingly replaced K^+ as the main cation, particularly in the proximal region of the growth zone which has a high sink strength for solutes (See Table 1). This argues against ion toxicity due to Na^+ and Cl- controlling leaf growth. This general and important conclusion has now broad acceptance (e.g. Hu et al., 2005a; Fricke et al., 2006; Munns et al., 2006). Nevertheless, the positive contribution of Na^+ to osmotic adjustment in growing leaves under salinity stress can only occur if Na^+ is primarily compartmentalized in the vacuoles of leaf cells, and thus cytoplasmic toxicity would not be a potential problem. However, the volume of vacuoles in cells from young developing tissue is quite small.

Table 1. Osmolality (mosmol kg^{-1}) in bulk leaf and epidermal cell extracts of leaf 3 of barley, 20h following salinization of the root medium with 100 mM NaCl. Third leaves were analyzed within the elongation zone (EZ; 10–30mm from leaf insertion) and within the emerged part of the blade. Means and standard deviations (n=8) are shown. Reproduced from Fricke et al., 2006. J. Exp. Bot. 57: 1079–1095, Table 2) and permission of the Journal of Experimental Botany

Extract	Control (-NaCl)		Salt Treatment (100 mM NaCl)	
	Leaf-region		Leaf-region	
	EZ	Emerged Blade	EZ	Emerged Blade
Epidermal Cells	386 (22)	423(22)	578(34)	492(53)
Bulk Leaf	428(16)	451(15)	595(32)	530(28)

4. CROP GROWTH AND DEVELOPMENT IN SALINE AGRICULTURAL ENVIRONMENTS

Most research that studies the effect of salinity on crops has been conducted in controlled laboratory and greenhouse environments, allowing scientists to better understand detailed responses and determine possible mechanisms the plant uses to cope with this stress. However, such experimental conditions do not reflect the natural conditions the plant encounters in salt-affected areas. There are a number of additional abiotic and biotic stresses that plants may endure in the field such as extreme temperatures, water deficits, flooding, nutritional inadequacies, poor soil physical conditions, pathogens and pests (Mittler, 2006). Moreover, these stresses are not constant, but vary both spatially and temporally. Therefore geneticists must be aware that genetically-altered plants with higher salt tolerance must also thrive under field conditions with numerous additional interactive stresses for this improved plant to be commercially successful.

In the field, salt-affected crops must also contend with too much or too little water. Therefore, actual crop performance during the growing season is related to how the plant responds to both salinity and fluctuating soil water conditions, either excessive or deficit.

1. *Flooding.* The combined effects of salinity and flooding are common in saline areas, particularly where shallow saline-water tables exist or where soils are also sodic, reducing water infiltration and causing water to pond on the soil surface (Barrett-Lennard, 2003). In flooded or poorly-drained soils, diffusion of oxygen to roots is reduced, thereby limiting root respiration and plant growth (Sharpley et al. 1992). In addition, important nutrients such as nitrate, sulfate, iron and manganese can be chemically reduced, decreasing their availability to the plant (Kozlowski, 1997) and selective ion transport processes are disrupted (Drew et al., 1988). Such anaerobic conditions adversely affect crop growth and developmental processes, influence morphological and anatomical adaptations, and cause many physiological dysfunctions in the plant. When combined with salinity stress, Na and Cl concentrations increase in the shoot further decreasing plant growth and survival (Barrett-Lennard, 2003).

2. *Water deficit.* Plant stress from salinity and water deficit have much in common (Munns, 2002a), but how the plant responds to the combination of stresses remains unresolved (Meiri 1984; Homaee et al., 2002). Under field conditions, water deficit is practically unavoidable since the soil-water content varies temporally and spatially throughout the season. Therefore some degree of both stresses can be occurring at different times and places in the rootzone (Homaee et al., 2002). For example, stress from water deficit may predominate in the upper portion of the rootzone while salt stress may predominate in the lower portion. Clearly the combination of stresses is more damaging than either one alone, but quantifying the growth-limiting contribution of each is difficult and can vary depending upon environmental conditions.

Sixty years have passed since Wadleigh and Ayers (1945) first demonstrated that bean plants responded to the additive combination of matric[2] (i.e. related to water deficit) and osmotic (i.e. related to salinity) stresses. This controversial finding, however, does not imply that these stresses are additive in all situations (Shani and Dudley, 2001). For example, Meiri (1984) concluded that the matric potential preferentially affected the shoot growth of bean more than did the osmotic potential. Shalhevet and Hsiao (1986) also found that pepper and cotton were affected more by water stress than by salinity at equivalent reductions in soil-water potential. Although matric and osmotic components are additive, from a thermodynamic perspective, there are kinetic factors (*i.e.* water uptake and transpiration) that must be considered as well (Maas and Grattan, 1999). For example, plant response to these stresses under conditions of low evaporative demand is likely to be different than that observed under high evaporative demand since the matric and not the osmotic potential controls water flow to the roots from the surrounding soil. As the soil dries, the matric potential decreases, but increases the resistance of water flow to the roots in a non-linear fashion (Homaee, et al, 2002). On the other hand, increases in soil salinity, at a given water content, reduces the soil-water potential but does not reduce water flow to the root. Moreover the root cortical cells can osmotically adjust to some extent allowing water to readily move into the root. This is consistent with the observation of Shalhevet and Hsiao (1986) who observed much lower leaf water potentials in transpiring pepper and cotton leaves in water-stressed plants than those stressed by salinity at equivalent soil-water potentials. Furthermore, these investigators found that osmotic adjustment was incomplete in leaves of water-stressed plants as compared to salt-stressed plants. Lower leaf-turgor in water-stressed plants led to reduced transpiration, CO_2 assimilation rates and growth. The overall magnitude of the difference between matric and osmotic effects is likely related to differences in plant type, root-length density and evaporative demand.

More research is needed to assess the interactive effects of these stresses. A highly instrumented volumetric lysimeter system that characterizes osmotic and matric stresses continuously at various depths, such as the one described by Poss et al., (2004), could provide valuable insights into whether the plant responds equally to the combined stresses or whether one predominates over the other under certain environmental conditions. In addition, a newly introduced root water extraction model for non-uniform conditions of salinity stress and water stress (Homaee et al., 2002) may be appropriate to sort out individual contributions of combined stresses under variable evaporative demands.

3. *Plant pathogens.* Salinity can affect the soil microbe populations in the rhizospere and their interaction with roots. For example *Rhizobium* spp., which

---

[2] Matric potential is the reduction in the free energy of soil water due to water's attraction to the soil matrix. The matric potential is zero in saturated soils and becomes progressively more negative as the soil water content decreases.

are integral to legume production, seem more salt tolerant than their host plants, but evidence indicates that nodulation and N_2 fixation by some crops are impaired by salinity (Läuchli, 1984). Other investigators have suggested that mycorrhizal symbioses improve the ability of some crops to tolerate salt by improving phosphorus nutrition (Hirrel and Gerdemann, 1980, Ojala et al. 1983, Poss et al. 1985).

Salt-stressed plants may be predisposed to infection by soil pathogens. Salinity has been reported to increase the incidence of phytophthora root rot in chrysanthemum (MacDonald, 1982) and tomato (Snapp et al. 1991). The combined effects significantly reduced fruit size and yield of tomato (Snapp et al. 1991), but wetter soil under salt-stunted plants, due to less evapotranspiration than non-saline control plants, may contribute to increased susceptibility to fungal diseases. Research on salinity-pathogen interactions is rather limited despite its potential economic impact in salt-affected areas many of which are also prone to waterlogging. Therefore further research is warranted in this area.

4. *Climate.* It has long been known that climatic conditions have a profound influence on plant response to salinity. Crops are more sensitive to salinity in hot, dry climates than they are under cooler and more humid environments. The combined effects of salinity and conditions of high evaporative demand, whether caused by high temperature, low humidity, or increased wind are more stressful than salinity stress alone. Several crops including alfalfa, bean, beet, carrot, cotton, onion, squash, strawberry, clover, saltgrass, and tomato are more sensitive to salinity at higher temperatures than they are at lower temperatures (Ahi and Powers 1938, Magistad et al. 1943, Hoffman and Rawlins 1970). On the other hand, higher humidity allowed barley, bean, corn, cotton, onion, and radish to be more tolerant to salinity (Hoffman and Rawlins 1970, 1971, Hoffman et al. 1971, Nieman and Poulsen 1967). Because climate dramatically affects plant response to salinity, the time of year and location salt-tolerance experiments are conducted will likely affect the results.

5. *Soil physical conditions.* Poor soil physical conditions can also contribute additional stresses in salt-affected areas (Grattan and Oster, 2003). For example, soils with poor structure or impermeable layers could restrict root growth as well as influence water and salt distribution in the soil. Crusting at the soil surface acts as a physical barrier for emerging seedlings and can lead to poor stand establishment particularly if the young seedlings are already weakened by salt stress. Although there has been a considerable amount of research conducted on salinity and sodicity's effect on soil physical conditions, more research is needed to evaluate how these changes affect crop performance.

6. *Composition of the salinizing solution.* Agricultural soils around the world vary not only in salinity but also in the composition of salts in the soil (Tanji, 1990). The dominant cations in salinized soils are sodium (Na^+), calcium (Ca^{2+}) and

magnesium (Mg^{2+}) while the dominant anions are chloride (Cl-), sulfate (SO_4^{2-}) and bicarbonate (HCO_3^-). The ratios of these ions differ from one location to the next but the composition of the salts is usually characteristic of the geochemical characteristics of the area. Potassium (K^+) and carbonate (CO_3^{2-}) are usually very low in irrigation water and soil solutions since their concentration is controlled by pH and solid phase interactions.

The ratio of sodium to calcium varies dramatically in natural waters. Gibbs (1970) analyzed the chemical constituents of global waters and found an interesting relationship between salinity and the ratio of Na/(Ca + Na), expressed in mg/L. Only in very pristine (< 10 mg/L) or extremely saline (>10,000 mg/L) waters were the Na/(Ca + Na) ratio greater than 0.8. The ratio of Na/(Ca + Na) in most water sources used for agricultural production (100 – 1,200 mg/L) is between 0.05 and 0.6, indicating that calcium is an important salinizing constituent. However as salts in these waters become concentrated due to evapotranspiration and reuse, these ratios will begin to increase due to precipitation of calcite and other divalent ion minerals in comparison to sodium (Tanji, 1990).

Regardless of the fact that irrigation waters and agricultural soil solutions are comprised of multiple combinations of cations and anions, the vast majority of salinity experiments on plants use NaCl as the sole salinizing salt. Lazof and Bernstein (1999) discussed the shortcomings of research where not only NaCl was used as the sole salinizing salt, but those studies where non-saline control treatments contain unrealistic trace amounts of Na and Cl. These investigators emphasize that trace levels of NaCl in control treatments are problematic in light of observed stimulatory effects from small additions of NaCl up to 5mM in many glycophytes. Extremely high Na/Ca ratios, on the other hand, lead to nutritional disorders and secondary stresses due to adverse affects on soil conditions.

Not only is NaCl uncharacteristic of agriculturally saline environments, but experiments that use this as the sole salinizing salt create extreme ratios of Na/Ca, Na/K, Ca/Mg and Cl/NO_3 in the root media (Läuchli and Epstein, 1990). These extreme ratios can adversely affect the mineral-nutrient relations within the crop than would occur otherwise under normal saline environments (Grattan and Grieve, 1999). Nutrient imbalances in the crop may result from several factors including the effect of salinity on nutrient ion activity and availability, the uptake and/or distribution of a nutrient within the plant, and/or increasing the internal plant requirement for a nutrient element resulting from physiological inactivation.

The importance of calcium and its protective role for plants in saline environments has been known for a century (Kearney and Harter, 1907; LaHaye and Epstein, 1969). Calcium preserves the structural and functional integrity of cell walls and membranes and regulates ion transport and selectivity (Läuchli and Epstein, 1990; Cramer, 2002). Any changes in the cell Ca^{2+} homeostasis is suggested as a primary response to salinity stress as perceived by the root cells (Rengel, 1992). Sodium-induced Ca^{2+} deficiency has been observed by numerous investigators when the Na^+/Ca^{2+} ratio in the solution, at a given salinity level for a particular plant, increases above a critical level (Kopittke and Menzies, 2004).

Crops in the grass family such as barley, corn, rice, sorghum and wheat, are particularly prone to this effect and large differences have been observed among species and cultivars.

Calcium deficiency may be partly related to the effect of Na^+on Ca^{2+} distribution within the plant (Lazof and Bernstein, 1999). Some scientists have found Na^+ to restrict the radial movement of Ca^{2+} from the root epidermis to the root xylem vessels (Lynch and Läuchli, 1985) while others found it to inhibit Ca^{2+} transport to the shoot, particularly meristematic regions and developing leaves (Kent and Läuchli, 1985; Maas and Grieve 1987; Grieve and Maas 1988; Lazof and Läuchli, 1991). Salinity-induced Ca^{2+} deficiency has also been observed on crops from different families such as blossom-end rot in tomato and bell pepper and black heart in celery (Geraldson, 1957).

The effect of pH on Na induced Ca^{2+} deficiency was investigated because high pH is characteristic of many salt-affected agricultural soils. This alkaline pH may decrease the activity of Ca^{2+} and aggravate the condition (Kopittke and Menzies, 2004). Interestingly, pH changes in the alkaline range did not affect the critical calcium activity ratio (CAR), a value below which Ca^{2+} deficiency symptoms appear, for either mungbean or Rhodes grass in either soils or solution cultures.

Not all salt-affected soils are dominated by chloride salinity. Many salt-affected soils are sulfate dominated such as those found in the Canadian prairie (Curtain et al., 1993), California's San Joaquin Valley (Tanji, 1990), Egypt and India (Banuelos et al., 1993; Manchanda and Sharma, 1989). At moderate levels of salinity, sulfate was less deleterious to growth than was chloride salinity in alfalafa, pepper and sorghum (Rogers et al., 1998, Boursier and Läuchli, 1990 and Navarro et al., 2002). However at higher salinity levels, sorghum was more sensitive to sulfate salinity than it was to chloride salinity (Boursier and Läuchli, 1990).

7. *Interactions between salinity and boron toxicity.* Boron is essential for cell wall structure and plays an important role in membrane processes and metabolic pathways (Blevins and Lukaszewski, 1998; Läuchli, 2002; Brown et al., 2002). However, there is a small range where concentrations in the soil solution are optimal (Gupta et al., 1985). Above this range, boron becomes toxic and below it, boron is deficient. Toxicity can occur in crops when boron concentrations increase in young developing tissue or margins of mature leaves to lethal levels, but plant-tissue analyses can only be used as general guidelines for assessing the risk of B-toxicity (Nable et al., 1997).

It has been known for decades that boron mobility is affected by climatic conditions and that it varies among species (Eaton, 1944). He found that after boron enters the leaf, it remains immobile in most plant species while in others, particularly stone fruits, it can re-mobilize to fruits and other parts of the plant. More recent evidence has shown that boron can form complexes with polyols in some species allowing it to be phloem mobile (Brown and Shelp, 1997). In other species where these simple sugars exist in small amounts, boron remains immobile. This sheds light on why B-toxicity symptoms occur on margins of older leaves of some plants ('boron immobile') while toxicity symptoms appear on younger, developing tissue (i.e. tip die back) in others ('boron mobile'). In those plants where boron is phloem immobile, boron concentrations in growing tips and reproductive tissue is much lower than concentrations in mature leaves. In boron mobile plants, just the opposite is found.

High boron, like salinity, is an important abiotic stress that adversely affects sensitive crops in many arid and semi-arid climates. There are many agricultural areas around the world where both high salinity and high boron occur together (Tanji, 1990) or where both boron and salt concentrations in municipal wastewaters are high, potentially affecting the plant (Tsadilis, 1997). Despite the common occurrence of high boron and high salinity in many parts of the world, very little research has been done to study the interaction of the two (Grattan and Grieve, 1999; Ben-Gal and Shani, 2002).

The question has recently been raised, are the effects of salinity and boron on crops additive, synergistic, or antagonistic? From a review of the limited number of studies that addressed the combined effects of salinity and boron on the plant, it appears that the results are contradictory. In some cases salinity may enhance boron sensitivity while in others, it may reduce its' sensitivity or have no effect.

In sand-culture experiments conducted in a greenhouse, researchers found that wheat responded to boron in the soil solution independently of salinity (Bingham et al., 1987). They found that there was no salinity - B interaction with respect to leaf B concentration. Similarly, others have found that boron and salinity effects were independent of each other for corn, barley and alfalfa (Shani and Hanks, 1993 and Mikkelsen et al., 1988).

However in more recent studies, investigators found that salinity enhanced B-sensitivity in wheat (Grieve and Poss, 2000; Wimmer et al. 2001; Wimmer et al., 2003). Wheat, a boron immobile plant, is one of those crops that is tolerant to salinity relative to other crops but is relatively sensitive to B. Grieve and Poss (2000) found that salinity increased B accumulation in leaves and that boron concentrations increasing above 400 mg/kg dry wt were associated with more injury. However boron is not equally distributed in the plant. Wimmer et al., (2003) found that under saline conditions, total B concentration was reduced in the root, was unaffected in the basal portion of the leaf, and increased dramatically in the leaf tip. In a more recent study, Wimmer et al. (2005) found that in wheat, B-tolerance is multi-faceted and genotype specific. In one B tolerant genotype (GREEK) high B in the medium led to accelerated reproductive development and early maturation which indirectly kept B accumulation in the leaves to a low level.

More important than salinity's effect on boron distribution in wheat was it's effect on B-soluble fractions within the shoot. Wimmer et al. (2001, 2003) found that combined salt and boron stress significantly increased the B-soluble fractions in both inter- and intra-cellular portions of the basal leaf more than either stress did alone. They propose that the soluble-B fraction in cells is an indicator of B-toxicity.

In a greenhouse study using soil in pots, investigators found that salinity increased B sensitivity in tomato and cucumber (Alpaslan and Gunes, 2001). However they found that salinity reduced total B concentration in tomato but increased it in cucumber. Furthermore, these investigators found that NaCl increased membrane permeability but increasing B in the soil to toxic levels did not, except in the presence of salinity. These investigators did not examine soluble vs insoluble boron fractions as was done by Wimmer et al. (2003). Therefore it is unknown why tomato was more sensitive to boron in the presence of salinity when the total boron was reduced.

Other investigators found that salinity reduced boron's toxic effect. In one field study conducted in Northern Chile, a number of vegetable crop species and prickly pear cactus were irrigated with saline water (8.2 dS/m) containing a mixture of ions including 17 mg/L of boron (Ferreyra et al., 1997). Plant growth and crop yields of artichoke, asparagus, broad bean, red and sugar beets, Swiss chard, carrot, celery, a local variety of sweet corn, potato, prickly pear cactus, onion, shallot, spinach, were all greater than expected based on published salt and boron tolerance coefficients. These investigators found that salinity reduced leaf boron concentrations. Interactions likely occurred which increased the crop's tolerance for boron in the presence of saline conditions. The investigators suggested that a reduction in plant water uptake, due to higher salinity levels, would reduce the rate by which boron accumulates in the plant. This reduced rate would extend the time by which boron would reach damaging concentrations that would affect plant growth.

Others also found that salinity reduced leaf B concentration of chickpea (Yadav et al., 1989), wheat (Holloway and Alston, 1992), *Eucalyptus camaldulensis* (Poss et al., 1998), as well as reduced B uptake and accumulation in the stem of several *Prunus* rootstocks (El-Motaium et al., 1994), decreasing B-toxicity symptoms. In the latter study, the investigators found a negative relationship between B and SO_4^{2-} concentrations in tissue suggesting that SO_4^{2-} could be responsible for the salinity-induced reduction in tissue B. However, recent experiments with broccoli in greenhouse sand tank systems indicated that Cl salinity was equally effective as mixed sulfate-chloride salinity in reducing boron's detrimental effect (Grattan et al., 2004) even though the effect of the combined stresses was more detrimental than either one alone. Regardless of salt composition, they found that at low boron concentrations (< 1mg/L), salinity increased shoot boron concentration while at very high boron concentrations (24 mg/L), salinity reduced shoot boron concentration. These investigators also explored the hypothesis that boron is taken up passively via the transpirational stream. By measuring changes in the isotopic fractionation of water samples in these closed sand tank systems over time, they were able to separate transpiration from evapotranpiration and to make inferences regarding the passive uptake of boron in relation to the cumulated plant transpiration. They found that at low boron (< 1mg/L), total shoot B was higher than would be accounted for by simple transpirational mass flow (solution concentration x cumulative transpirational volume). One the other hand, at high boron concentrations (>14 mg/L), total shoot boron was substantially less than that predicted based on mass flow suggesting the plant is somehow able to regulate the accumulation of boron in the shoot, which is dependent upon the boron concentration of the external solution and not salinity. However, in cucumber, the $^{10}B/^{11}B$ ratio in the soil solution was equal to that in the plant tissue suggesting that the plant is unable to discriminate in uptake between the two isotopic species of boron (Grattan et al.,2005).

Another important finding in salinity-boron interactions is the influence pH has on this interaction (Grattan et al., 2005; 2006). Under slightly acidic conditions, boron in solution occurs as undissociated boric acid ($B(OH)_3$). In contrast, under slightly basic conditions, boron partly changes to borate ($B(OH)_4^-$). This change in the chemical speciation of boron under alkaline conditions may affect the mechanism and rate at which boron is transported through membranes. In a controlled sand tank experiment with cucumber, increased salinity, boron and pH (from 6 to 8) decreased fruit yield. Investigators did not find any significant salinity-boron interaction. However in slightly acidic conditions, regardless of salinity, increased boron was more detrimental than it was in slightly basic conditions. When the experiment was repeated with broccoli, these investigators found different results. They found that an increase in soil-water boron from 1 to 21 mg/L at pH 6, did not significantly reduce the head yields of broccoli at any salinity level. However at pH 8, as boron increased from 1 to 21 mg/L, head yields at both low (EC 2 dS/m) and high (EC 14 dS/m) were reduced by over 85%. Interestingly at moderate salinities (EC 5 to 11 dS/m), increased boron had very little detrimental effect.

Much has been learned over the past decade regarding salinity-boron interactions but many questions remain unresolved. More research is needed to (1) understand the relationship between visual injury symptoms, tissue boron concentrations, soluble boron fractions, the role pH plays and how these all interact affecting crop yield and (2) the influence of salinity on the soluble fractions of boron, boron mobility and distribution within the plant and how these relate to visual injury.

In summary, plants under field conditions often endure multiple stresses during their development. However, the vast majority of research has focused on individual stresses in the absence of others. Plant response to combined stresses can not be readily extrapolated based on their response to individual stresses (Mittler, 2006). Figure 4 shows potential interactions among several agricultural stress combinations. Some stress combinations show negative interactions while others exhibit positive interactions. For some stress combinations, there are no or unknown mode of interactions. Therefore Mittler (2006) suggested that tolerance to a combination of stresses should be the focus of future research, particularly those where the goal is to develop transgenic crops with enhanced tolerance to natural adverse field conditions.

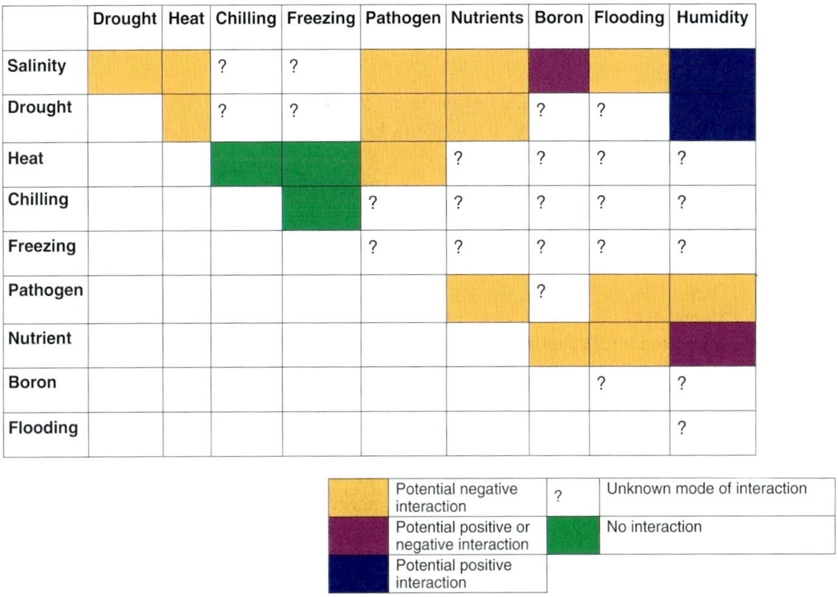

Figure 4. Agriculturally important environmental factors and their potential interactions. Modified from Mittler (2006). Boron-nutrient interaction source (Marschner, 1995)

REFERENCES

Abdullah, Z., M.A. Khan and T.J. Flowers. 2001. Causes of sterility in seed set of rice under salinity stress. J. Agron. Crop Sci. 187:25–32.

Abul-Naas, A.A. and M.S. Omran. 1974. Salt tolerance of seventeen cotton cultivars during germination and early seedling development. Acker Pflanzenbau 140:229–236.

Ahi, S.M., and W.L. Powers. 1938. Salt tolerance of plants at various temperatures. Plant Physiol. 13:767–789.

Ahmad, S., A. Wahid, E. Rasul and A. Wahid. 2005. Comparative morphological and physiological responses of green gram genotypes to salinity applied at different growth stages. Bot. Bull. Acad. Sin. 46:135–142.

Akbar, M. and T. Yabuno. 1977. Breeding for saline-resistant varieties of rice. IV. Inheritance of delayed-type panicle sterility induced by salinity. Japan J. Breed. 27: 237–240.

Alpaslan, M. and A. Gunes. 2001. Interactive effects of boron and salinity stress on the growth, membrane permeability and mineral composition of tomato and cucumber plants. Plant Soil 236: 123–128

Ayers, A.D., J.W. Brown and C.H. Wadleigh. 1952. Salt tolerance of barley and wheat in soil plots receiving several salinization regimes. Agron. J. 44:307–310.

Badia, D. and A. Meiri 1994. Tolerance of two tomato cultivars (Lycopersicum esculentum Mill) to soil salinity during emergence phase. Agr. Med 124:301–310

Banuelos, G.S., R. Mead, and G.J. Hoffman. 1993. Accumulation of selenium in wild mustard irrigated with agricultural effluent. Agric. Ecosyst. Environ. 43:119–126.

Barrett-Lennard, E.G. 2003. The interaction between waterlogging and salinity in higher plants: causes, consequences and implications. Plant Soil 253:35–54

Baum S.F., Tran P.N. and W.K. Silk. 2000. Effects of salinity on xylem structure and water use in growing leaves of sorghum. New Phytol. 146:119–127.

Bayuelo-Jimenez, J.S., R. Craig and J.P. Lynch. 2002. Salinity tolerance of Phaseolus species during germination and early seedling growth. Crop Sci. 42:1584–1594.

Ben-Gal, A. and U. Shani. 2002. Yield, transpiration and growth of tomatoes under combined excess boron and salinity stress. Plant Soil 247:211–221.

Bernstein, L. and H.E. Hayward. 1958. Physiology of salt tolerance. Ann. Rev. Plant Physiol. 9:25–46.

Bernstein N., Silk W.K. and A. Läuchli. 1993a. Growth and development of sorghum leaves under conditions of NaCl stress. Planta 191:433–439.

Bernstein N. Läuchli A. and W.K. Silk. 1993b. Kinematics and dynamics of sorghum (*Sorghum bicolor* L.) leaf development at various Na/Ca salinities. Plant Physiol. 103:1107–1114.

Bernstein N., Silk W.K. and A. Läuchli i. 1995. Growth and development of sorghum leaves under conditions of NaCl stress: possible role of some mineral elements in growth inhibition. Planta 196: 699–705.

Bingham, F.T., Strong, J. E., Rhoades, J. D. and Keren, R., 1987. Effects of salinity and varying boron concentrations on boron uptake and growth of wheat. Plant Soil 97: 345–351.

Blevins, D.G. and K.M. Lukaszewski. 1998. Boron in plant structure and function. Annu. Rev. Plant Physiol. Plant Mol. Biol. 49:501–523

Bohnert H.J., Gong Q., Li P. and S. Ma. 2006. Unraveling abiotic stress tolerance mechanisms – getting genomics going. Current Opinion in Plant Biology 9:180–188.

Botia, P., J.M. Navarro, A. Cerda and V. Martinez. 2005. Yield and fruit quality of two melon cultivars irrigated with saline water at different stages of development. Eur. J. Agron. 23:243–253

Boursier, P. and A. Läuchli. 1990. Growth responses and mineral nutrient relations of salt-stressed sorghum. Crop Sci. 30:1226–1233.

Boyer J.S. 1987. Hydraulics, wall extensibility and wall proteins. In: Physiology of Cell Expansion during Plant Growth, Proc. Second Annual Penn. State Symposium in Plant Physiology. Penn. State University, University Park, PA 16802, pp. 109–121.

Brown, P.H., N. Bellaloui, M.A. Wimmer, E.S. Bassil, J. Ruis, H. Hu, H. Pfeffer, F. Dannel, V. Römheld. 2002. Boron in plant biology. Plant boil. 4:211–229.

Brown, P.H and B.J. Shelp. 1997. Boron mobility in plants. Plant Soil:193:85–101

Carter, C.T., C.M. Grieve and J.P. Poss. 2005. Salinity effects on emergence, survival, and ion accumulation of *Limonium perezii*. J. Plant Nutr. 28:1243–1257.

Cosgrove D.J. 1987. Linkage of wall extension with water and solute uptake. In: Physiology of Cell Expansion during Plant Growth, Proc. Second Annual Penn. State Symposium in Plant Physiology., Penn. State University, University Park PA 16802, pp. 88–100.

Cramer G.R. 1991. Kinetics of maize leaf elongation. II. Responses of a sodium excluding cultivar and a Na including cultivar to varying Na/Ca salinity. J. Exp. Bot. 43:857–864.

Cramer, G.R. 2002. Sodium-calcium interactions under salinity stress. In: Salinity. Environment-Plants-Molecules.. A. Läuchli and U. Lüttge (Eds) . Kluwer Academic Publishers, Dordrecht, pp. 205–227

Cramer G.R. 2003. Differential effects of salinity on leaf elongation kinetics of three grass species. Plant Soil 253:233–244.

Cramer G.R. and D.C. Bowman. 1991. Kinetics of maize leaf elongation. I. Increased yield threshold limits short-term elongation rates after exposure to salinity. J. Exp. Bot. 42:1417–1426.

Cramer G.R. and D.C. Bowman. 1993. Cell elongation control under stress conditions. In: Handbook of Plant and Crop Stress, M Pessarakli (Ed.). Marcel Dekker, New York, pp. 303–319.

Cramer G.R., Epstein E. and A. Läuchli. 1988. Kinetics of root elongation of maize in response to short-term exposure to NaCl and elevated calcium concentration. J. Exp. Bot. 39:1513–1522.

Cramer G.R. and A. Läuchli. 1986. Ion activities in solution in relation to Na^+- Ca^{2+} interactions at the plasmalemma. J. Exp. Bot. 37:321–330.

Cramer G.R., Läuchli A. and E. Epstein. 1986. Effects of NaCl and $CaCl_2$ on ion activities in complex nutrient solutions and root growth of cotton. Plant Physiol. 81:792–797.

Cramer G.R., Läuchli A and V.S. Polito. 1985. Displacement of $Ca^{2+}$from the plasmalemma of root cells. A primary response to salt stress? Plant Physiol. 79:297–211.

Cramer G.R., Schmidt C.L. and C. Bidart. 2001. Analysis of cell wall hardening and cell wall enzymes of salt-stressed maize (*Zea mays*) leaves. Aust. J. Plant Physiol. 28:101–109.

Cuartero J. and R. Fernandez-Munoz. 1999. Tomato and salinity. Scientia Horticulturae 78:83–125.
Cuartero J., Bolarin M.C., Asins M.J. and V. Moreno. 2006. Increasing salt tolerance in the tomato. J. Exp. Bot. 57:1045–1058.
Curtain, D., H. Steppuhn and F. Selles. 1993. Plant responses to sulfate and chloride salinity: Growth and ionic relations. Soil Sci. Soc. J. 57:1304–1310.
del Amor, F.M., V. Martinez and A. Cerda. 2001. Salt tolerance of tomato plants as affected by stage of plant development. Hort. Sci 36:1260–1263.
Drew, M.C., J. Guenther, and A. Läuchli. 1988. The combined effects of salinity and root anoxia on growth and net Na^+ and K^+ accumulation in *Zea mays* grown in solution culture. Ann. Bot. 61:41–43
Eaton, F.M. 1944. Deficiency, toxicity, and accumulation of boron in plants. J. Agric. Res. 69:237–277.
El-Hendawy, S.E., Y. Hu, G.M. Yakout, A.M. Awad. S.E. Hafiz, and U. Schmidhalter. 2005. Evaluating salt tolerance of wheat genotypes using multiple parameters. Europ. J. Agron. 22:243–253
El-Motaium, R., Hu, H. and Brown, P. H., 1994. The relative tolerance of six Prunus rootstocks to boron and salinity. J. Amer. Soc. Hort. Sci 119: 1169–1175. UC Salinity-Drainage Task Force Annual report. Div. Agric. And Nat. Resources. University of California
Epstein E. 1961. The essential role of calcium in selective cation transport by plant cells. Plant Physiol. 36:437–444.
Esechie, H.A., A. Al-Saidi and S. Al-Khanjari. 2002. Effect of sodium chloride salinity on seedling emergence in chickpea. J. Agron. and Crop. Sci. 188:155–160
Ferreyra, R.E., A.U. Alijaro, R.S. Ruiz, L.P. Rojas and J.D. Oster. 1997. Behavior of 42 crop species grown in saline soils with high boron concentrations. Agric. Water Manag 32:111–124.
Flowers T. 2006. Preface. J. Exp. Bot. 57, p. iv.
Flowers T.J., Hajibagheri M.A. and A.R. Yeo. 1991. Ion accumulation in the cell walls of rice plants growing under saline conditions: evidence for the Oertli hypothesis. Plant Cell Environ. 14:319–325.
Flowers, T.J., and A.R. Yeo. 1981. Variability in the resistance of sodium chloride within rice (Oryza sativa L.) varieties. New Phytol. 88:363–373.
Fricke W. 2004. Rapid and tissue-specific accumulation of solutes in the growth zone of barley leaves in response to salinity. Planta 219:515–525.
Fricke W., Akhiyarova G., Wei W., Alexandersson E., Miller A., Kjellbom P.O., Richardson A., Wojciechowski T., Schreiber L., Veselov D., Kudoyarova G. and V. Volkov. 2006. The short-term growth response to salt of the developing barley leaf. J. Exp. Bot. 57:1079–1095.
Fricke W. and W.S. Peters. 2002. The biophysics of leaf growth in salt-stressed barley: a study at the cell level. Plant Physiol. 129:374–388.
Geraldson, C.M. 1957. Factors affecting calcium nutrition of celery, tomato, and pepper. Soil Sci. Soc. Am. Proc. 21:621–625.
Gibbs, R.J. 1970. Mechanisms controlling world water chemistry. Science 170:1088–1090.
Goyal S.S., Sharma S.K., Rains D.W. and A. Läuchli. 1999. Long-term reuse of drainage waters of varying salinities for crop irrigation in a cotton-safflower rotation system in the San Joaquin Valley of California – a nine year study. I. Cotton (*Gossypium hirsutum* L.). J. Crop Prod. 2, No. 2: 181–213.
Grattan, S.R. and C.M. Grieve. 1999. Salinity - Mineral nutrient relations in horticultural crops. Sci. Hort. 78:127–157.
Grattan, S.R., C.M. Grieve, J.P Poss, D. Suarez and A. Läuchli. 2005. Continued investigation into the interactions of saline drainage water on crop tolerance to boron. 2004–05 Technical Progress Report: UC Salinity/Drainage Research Program. DANR. University of California.
Grattan, S.R., C.M. Grieve, J.P Poss, D. Suarez, A. Läuchli and T. Smith. 2006. Continued investigation into the interactions of saline drainage water on crop tolerance to boron. 2005–06 Technical Progress Report: UC Salinity/Drainage Research Program. DANR. University of California.
Grattan, S.R., C. Grieve, J. Poss, D. Suarez and T. Smith. 2004. Does saline drainage water affect crop tolerance to boron?. 2003–04 Technical Progress Report: UC Salinity/Drainage Research Program. DANR. University of California. pp 19–32.
Grattan, S.R. and J.D. Oster. 2003. Use and reuse of saline-sodic waters for irrigation of crops. In: S.S. Goyal, S.K. Sharma and D.W. Rains (eds.), Crop Production in Saline Environments: Global and Integrative Perspectives. Haworth Press, New York. pp 131–162

Grattan, S.R., Shannon, M. C., Grieve, C. M., Poss, J. A., Suarez, D. L. and Francois, L. E. 1996. Interactive effects of salinity and boron on the performance and water use of eucalyptus. Acta Hort. 449:607–613

Grattan, S.R., L. Zeng, M.C. Shannon and S.R. Roberts, 2002. Rice is more sensitive to salinity than previously thought. Calif. Agric. 56:189–195.

Grieve, C.M. and J.P. Poss. 2000. Wheat response to interactive effects of boron and salinity. J. Plant Nutr. 23: 1217–1226.

Grieve, C.M, L.E. Francois and J.A. Poss. 2001. Salt stress during early seedling growth on phenology and yield of spring wheat. Cereal Res. Comm. ?? 167–174

Grieve, C.M., S.M. Lesch, L.E. Francois and E.V. Maas. 1992, Analysis of main-spike yield components in salt-stressed wheat. Crop Sci. 32:697–703

Grieve, C.M., S.M. Lesch, E.V. Maas, and L.E. Francois. 1993. Leaf and spikelet primordia initiation in salt-stressed wheat. Crop Sci. 22:1286–1294.

Grieve C.M., and E.V. Maas. 1988. Differential effects of sodium/calcium ratio on sorghum genotypes. Crop. Sci. 29:659–665.

Gupta, U.C., Jame, Y. W., Campbell, C. A., Leyshon, A. J. and Nicholaichuk, W., 1985. Boron toxicity and deficiency: A review. Can. J. Soil Sci. 65: 381–409.

Halperin S.J., Kochian L.V. and J. P. Lynch. 1997. Salinity stress inhibits calcium loading into the xylem of excised barley (*Hordeum vulgare*) roots. New. Phytol. 135:419–427.

Hasegawa P.M., Bressan R.A., Zhu J.-K. and H.J. Bohnert. 2000. Plant cellular and molecular responses to high salinity. Annu. Rev. Plant Physiol. Plant Mol. Biol. 51:463–499.

Heenan, D.P., L.G. Lewin and D.W. McCaffery. 1988. Salinity tolerance in rice varieties at different growth stages. Aust. J. Exp. Agric. 28:343–349.

Hillel D. 2000. Salinity Management for Sustainable Irrigation. The World Bank, Washington, D.C.

Hirrel, M.C. and J.W. Gerdemann. 1980. Improved growth of onion and bell pepper in saline soils by two vesicular-arbuscular mycorrhizal fungi. Soil Sci. Soc. Amer. J. 44:654–655.

Hoffman, G.J., and S.L. Rawlins. 1970. Design and performance of sunlit climate chambers. Trans. ASAE. 13:656–660.

Hoffman, G.J., and S.L. Rawlins. 1971. Growth and water potential of root crops as influenced by salinity and relative humidity. Agron. J. 63:877–880.

Hoffman, G.J., S.L. Rawlins, M.J. Garber, and E.M. Cullen. 1971. Water relations and growth of cotton as influenced by salinity and relative humidity. Agron. J. 63:822–826.

Holloway, R.E., and A.M. Alston. 1992. The effects of salt and boron on growth of wheat. Aust. J. Agric. Res., 43:987–1001.

Homaee, M., R.A. Feddes and C. Dirksen. 2002. A macroscopic water extraction model for nonuniform transient salinity and water stress. Soil Sci. Soc. Am. J. 66:1764–1772

Hsiao T.C., Acevedo E., Fereres E. and D.E. Henderson. 1976. Stress metabolism. Water stress, growth, and osmotic adjustment. Phil. Trans. R. Soc. Lond. (B), 273:470–500.

Hu, Y., J. Fromm and U. Schmidhalter. 2005a. Effect of salinity on tissue architecture in expanding wheat leaves. Planta 220:838–848.

Hu Y., Fricke W. and U. Schmidhalter. 2005b. Salinity and the growth of non-halophytic grass leaves: the role of mineral nutrient distribution. Funct. Plant Biol. 32:973–985.

Hu Y. and U. Schmidhalter. 2004. Limitation of salt stress to plant growth. In: Plant Toxicology (Hock, B.a nd E.F. Elstner (eds) Marcel Dekker, New York pp 191–224

James J.J., Alder N.N., Mühling K.H., Läuchli A.E., Shackel K.A., Donovan L.A. and J.H. Richards. 2006. High apoplastic solute concentrations in leaves alter water relations of the halophytic shrub, *Sarcobatus vermiculatus*.J. Exp. Bot. 57:139–147.

Katerji, N., J.W. van Horn, A. Hamdy, F. Karam and M. Mastrorilli. 1994. Effect of salinity on emergence and on water stress and early seedling growth of sunflower and maize. Agric. Water Mang. 26:81–91.

Kent, L.M. and A. Läuchli. 1985. Germination and seedling growth of cotton: salinity-calcium interactions. Plant Cell Envrion. 8:155–159

Kearney, T.H and L.L. Harter. 1907. The comparative tolerance of various plants for the salts common in alkali soils. U.S.D.A Plant industry bulletin 113:7–22

Khan, M.A. and Z. Abdullah. 2003. Reproductive physiology of two wheat cultivars differing in salinity tolerance under dense saline-sodic soil. Food, Agric. And Envrion. 1:185–189

Khatun, S., and T.J. Flowers. 1995. Effects of salinity on seed set in rice. Plant Cell Environ. 18:61–67.

Khatun, S., C.A. Rizzo and T.J. Flowers. 1995. Genotypic variation in the effect of salinity on fertility in rice. Plant Soil 173:239–250

Kinraide T.B. 1999. Interactions among Ca^{2+}, Na^{+} and K^{+} in salinity toxicity: quantitative resolution of multiple toxic and ameliorative effects. J. Exp. Bot. 50:1495–1505.

Koiwa H., Bressan R.A. and P.M. Hasegawa. 2006. Identification of plant stress-responsive determinants in arabidopsis by large-scale forward genetic screens. J. Exp. Bot. 57:1119–1128.

Kopittke, P.M. and N.W. Menzies. 2004. Effect of pH on Na induced calcium deficiency. Plant Soil 269:119–129

Kozlowski, T.T. 1997. Responses of woody plants to flooding and salinity. Tree Physiol. Monograph 1

Kurth E., Cramer G.R., Läuchli A. and E. Epstein. 1986. Effects of NaCl and $CaCl_2$ on cell enlargement and cell production in cotton roots. Plant Physiol. 82:1102–1106.

LaHaye, P.A. and E. Epstein. 1969. Salt toleration by plants: Enhancement with calcium. Science 166:395–396.

LaHaye P.A. and E. Epstein. 1971. Calcium and salt toleration by bean plants. Physiol. Plant. 25: 213–218.

Läuchli, A. 1984. Salt exclusion: An adaptation of legumes for crops and pastures under saline conditions. p.171–187. *In* R.C. Staples and G.H. Toenniessen (eds.) Salinity Tolerance in Plants. Strategies for Crop Improvement. John Wiley & Sons. New York.

Läuchli A. 1990. Calcium, salinity and the plasma membrane. In: Calcium in Plant Growth and Development, R.T. Leonard and P.K. Hepler (Eds.). The American Society of Plant Physiologists Symposium Series, Vol. 4, pp. 26–35.

Läuchli A. 1999. Salinity-potassium interactions in crop plants. In: Frontiers in Potassium Nutrition, D.M. Oosterhuis and G.A. Berkovitz (Eds.). Potash and Phosphate Institute, Norcross, Georgia, pp. 71–76.

Läuchli, A. 2002. Functions of boron in higher plants: Recent advances and open questions. Plant biol. 4:190–192.

Läuchli A. and E. Epstein. 1970. Transport of potassium and rubidium in plant roots. The significance of calcium. Plant Physiol. 45:639–641.

Läuchli, A. and E. Epstein. 1990. Plant responses to saline and sodic conditions. In K.K. Tanji (ed). Agricultural salinity assessment and management. ASCE manuals and reports on engineering practice No, 71. pp 113–137 ASCE New York

Lazof, D.B. and N. Bernstein. 1999. The NaCl induced inhibition of shoot growth: The case for disturbed nutrition with special consideration of calcium. Advances in Botanical Research 29:113–189.

Lazof, D.B. and A. Läuchli. 1991. The nutritional status of the apical meristem of Lactuca sativa as affected by NaCl salinization: an electron-probe microanalytic study. Planta 184:334–342

Lee, Y.S., S.R. Park, H.J. Park and Y.W. Kwon. 2004. Salt stress magnitude can be quantified by integrating salinity with respect to duration. Proceedings of 4th International Crop Sci Congress. Brisbane, Aust. 26 Sept- 1 Oct 2004 pp 1–5

Lutts, S., J.M. Kinet and J. Bouharmont. 1995. Changes in plant response to NaCl during development of rice (Oryza sative L.) varieties differing in salinity resistance. J. Exp Bot 46:1843–1852.

Lynch J. and A. Läuchli. 1985. Salt stress disturbs the calcium nutrition of barley (*Hordeum vulgare* L.). New Phytol. 99:345–354.

Lynch J., Thiel G. and A. Läuchli. 1988. Effects of salinity on the extensibility and Ca availability in the expanding region of growing barley leaves. Bot. Acta 101:355–361.

Maas E.V. and C.M. Grieve. 1987. Sodium-induced calcium deficiency in salt-stressed corn. Plant Cell Environ. 10:559–564.

Maas, E.V. and C.M. Grieve. 1990. Spike and leaf development in salt-stressed wheat. Crop Sci. 30:1309–1313.

Maas, E.V., G.J. Hoffman, G.D. Chaba, J.A. Poss and M.C. Shannon. 1983. Salt sensitivity of corn at various growth stages. Irrig. Sci. 4:45–57.

Maas, E.V. and J.A. Poss. 1989a. Salt sensitivity of wheat at different growth stages. Irrig. Sci. 10:29–40.

Maas, E.V. and J.A. Poss. 1989b. Salt sensitivity of cowpea at various growth stages. Irrig. Sci. 10: 313–320.

Maas, E.V., Poss, J.A., Hoffman, G.J. 1986. Salinity sensitivity of sorghum at three growth stages. Irrig. Sci. 7:1–11

Maas, E. V. and S. R. Grattan. 1999. Crop yields as affected by salinity. In R. W. Skaggs and J. van Schilfgaarde (eds) Agricultural Drainage. Agron. Monograph 38. ASA, CSSA, SSA, Madison, WI pp. 55–108.

MacDonald, J.D. 1982. Effect of salinity stress on development of *Phytophthora* root rot of chrysanthemum. Phytopath. 72:214–219.

Magistad, O.C., A.D. Ayers, C.H. Wadleigh, and H.F. Gauch. 1943. Effect of salt concentration, kind of salt, and climate on plant growth in sand cultures. Plant Physiol. 18:151–166.

Marschner, H. 1995. Mineral Nutrition of Higher Plants. Second edition. Academic Press, London, pp. 388–390

Martinez-Beltran J. and C.L. Manzur. 2005. Overview of salinity problems in the world and FAO strategies to address the problem. In: Proceedings of the International Salinity Forum, Riverside, California, April 2005, pp. 311–313.

Mauromicale, G. and P. Licandro. 2002. Salinity and temperature effects on germination, emergence and seedling growth of globe artichoke. Agronomie 22:443–450.

Meiri, A. 1984. Plant response to salinity: Experimental methodology and application to the field. p. 284–297. *In* I. Shainberg and J. Shalhevet (eds.) Soil Salinity Under Irrigation. Springer Verlag, New York.

Mikkelsen, R.L., B.H. Haghnia, A.L. Page and F.T. Bingham. 1988. The influence of selenium, salinity and boron on alfalfa tissue composition and yield. J. Environ. Qual:17:85–88

Mittler, R. 2006. Abiotic stress, the field environment and stress combination. Trends in Plant Sci 11:15–19

Miyamoto, S., K. Piela, and J. Patticrew. 1985 Salt effects on germination and seedling emergence of several vegetable crops and guayule. Irrig. Sci. 6:159–170.

Mühling K.H. and A. Läuchli. 2002a. Effect of salt stress on growth and cation compartmentation in leaves of two plant species differing in salt tolerance. J. Plant Physiol. 159:137–146.

Mühling K.H. and A. Läuchli. 2002b. Determination of apoplastic Na^+ in intact leaves of cotton by in vivo fluorescence ratio-imaging . Funct. Plant Biol. 29:1491–1499.

Munns, R. 2002a. Comparative physiology of salt and water stress. Plant Cell Environ. 25:239–250.

Munns R. 2002b. Salinity, growth and phytohormones. In: Salinity: Environment – Plants – Molecules, A. Läuchli and U. Lüttge (Eds.). Kluwer Academic Publishers, Dordrecht, pp. 271–290.

Munns R. 2005. Genes and salt tolerance: bringing them together. New Phytol. 167:645–663.

Munns R., James R.A. and A. Läuchli. 2006. Approaches to increasing the salt tolerance of wheat and other cereals. J. Exp. Bot. 57:1025–1043.

Munns R. and A. Termaat. 1986. Whole plant responses to salinity. Aust. J. Plant Physiol. 13:143–160.

Munns, R., S. Husain, A.R. Rivelli, R.A. James, A.G. Condon, M.P. Lindsay, E.S. Lagudah, D.P Schachtman and R.A. Hare. 2002. Avenues for increasing salt tolerance of crops, and the role of physiologically based selection traits. Plant Soil 247:93–105.

Munns, R. and H.M. Rawson. 1999. Effect of salinity on salt accumulation and reproductive development in the apical meristem of wheat and barley. Aust. J. Plant Physiol. 26:459–464

Nable, R.O., G.S. Bañuelos and J.G. Paull. 1997. Boron toxicity. Plant Soil, 193:181–198.

Navarro, J.M., C. Garrido, M. Carvajal and V. Martinez. 2002. Yield and fruit quality of pepper plants under sulfate and chloride salinity. J. Hort. Sci. and Biotech 77:52–57.

Nerson, H. and H.S. Paris. 1984. Effects of salinity on germination, seedling growth, and yield of melons. Irrig. Sci. 5:265–273

Neumann P.M. 1993. Rapid and reversible modifications of extension capacity of cell walls in elongating maize leaf tissues responding to root addition and removal of NaCl. Plant Cell Environ. 16:1107–1114,

Neves-Piestun B.G. and N. Bernstein. 2001. Salinity-induced inhibition of leaf elongation in maize is not mediated by changes in cell wall acidification capacity. Plant Physiol. 125:1419–1428.

Neves-Piestun B.G. and N. Bernstein. 2005. Salinity-induced changes in the nutritional status of expanding cells may impact leaf growth inhibition in maize. Funct. Plant Biol. 93:1610–1619.

Nieman, R.H., and L.L. Poulsen. 1967. Interactive effects of salinity and atmospheric humidity on the growth of bean and cotton plants. Bot. Gaz. 128:69–73.

Nonami H. and J.S. Boyer. 1990. Wall extensibility and cell hydraulic conductivity decrease in enlarging stem tissues at low water potentials. Plant Physiol. 93:1610–1619.

Oertli J.J. 1968. Extracellular salt accumulation, a possible mechanism of salt injury in plants. Agrochimica 12:461–469.

Ojala, J.C., W.M. Jarrell, J.A. Menge, and E.L.V. Johnson. 1983. Influence of mycorrhizal fungi on the mineral nutrition and yield of onion in saline soil. Agron J. 75:255–259.

Papp J.C., Ball M.C. and N. Terry. 1983. A comparative study of the effects of NaCl salinity on respiration, photosynthesis and leaf extension growth in *Beta vulgaris* (sugar beet). Plant Cell Environ 6:675–677.

Pasternak, D.M., M. Twersky and Y. de Malach. 1979. Salt resistance in agricultural crops. In: Stress physiology in crop plants (eds H. Mussell and R.C. Staples). Pp 127–142. Wiley, New York

Pearson, G.A. and A.D. Ayers. 1966. Relative salt tolerance of rice during germination and early seedling development. Soil Sci. 102:151–156.

Pearson, G.A. and L. Bernstein, 1959. Salinity effects at several growth stages of rice. Agron. J. 51:654:657.

Pitman MG. and A. Läuchli. 2002. Global impact of salinity and agricultural ecosystems. In: Salinity: Environment – Plants – Molecules, A. Läuchli and U. Lüttge (Eds.). Kluwer Academic Publishers, Dordrecht, pp. 3–20.

Poss, J.A., S.R. Grattan, and C. M. Grieve and M.C. Shannon. 1998. Characterization of leaf boron injury in salt-stressed *Eucalyptus* by Image Analysis. Plant Soil: 206: 237–245.

Poss, J.A., E. Pond, J.A. Menge, and W. M. Jarrell. 1985. Effect of salinity on mycorrhizal onion and tomato in soil with and without additional phosphate. Plant Soil 88:307–319

Poss, J.A., W.B Russell, P.J. Shouse, R.S. Austin, S.R. Grattan, C.M. Grieve, J.H. Lieth and L. Zeng. 2004. A volumetric lysimeter system (VLS): an alternative to weighing lysimeters for plant-water relations studies. Comp. Electronics Agric. 43:55–68.

Rogers, M.E., C.M. Grieve and M.C. Shannon. 1998. The response of lucerne (*Medicago sativa* L. to sodium sulfate and chloride salinity. Plant Soil 202:271–280.

Rengasamy P. 2006. World salinization with emphasis on Australia. J. Exp. Bot 57:1017–1023.

Rengel, Z. 1992. The role of calcium in salt toxicity. Plant Cell Environ. 15:625–632

Shalhevet, J. and T.C. Hsiao. 1986. Salinity and drought. Irrig. Sci. 7:249–264

Shani, U. and L.M. Dudley. 2001. Field studies of crop response to water and salt stress. Soil Sci. Soc. Am. J. 65:1522–1528

Shani, U. and R.J. Hanks. 1993. Model of integrated effects on boron, inert salt, and water flow on crop yield. Agron. J. 85:713–717.

Sharpley, A.N., J.J. Meisinger, J.F. Power, and D.L. Suarez. 1992. Root extraction of nutrients associated with long-term soil management. *In* J.L. Hatfield and B.A. Steward (eds.) Advances in Soil Science. Vol. 19. Springer-Verlag, New York.

Snapp, S.S., C. Shennan, and A.H.C. van Bruggen. 1991. Effects of salinity on severity of infection by *Phytophthora parasitica* Dast., ion concentrations and growth of tomato, *Lycopersicon esculentum* Mill. New Phytol. 119:275–284.

Tajbakhsh, M., M.X. Zhou, Z.H. Chen and N.J. Mendham. 2006. Physiological and cytological response of salt-tolerant and non-tolerant barley to salinity during germination and early growth. Aust. J. Exp. Agric. 46:555–562.

Tanji, K.K. 1990. (ed). Agricultural salinity assessment and management. ASCE manuals and reports on engineering practice No. 71. Am. Soc. Civil Eng., New York

Tester M. and R. Davenport. 2003. Na^+ tolerance and Na^+ transport in higher plants. Ann. Bot. 91:503–527.

Thiel G.H., Lynch J. and A. Läuchli. 1988. Short-term effects of salinity stress on the turgor and elongation of growing barley leaves. J. Plant Physiol. 132:38–44.

Tsadilas, C.D. 1997. Soil contamination with boron due to irrigation with treated municipal waste water. In Boron in Soils and Plants. R.W. Bell and B. Rerkasem (eds) Kluwer Academic Publishers, Dordrecht pp 265–270.

Van Volkenburgh E. and J.S. Boyer. 1985. Inhibitory effect of water deficit on maize leaf elongation. Plant Physiol. 77:190–194.

Vinizky, I and D.T. Ray. 1988. Germination of guar seed under salt and temperature stress. J. Am. Soc. Hort. Sci. 113:437–440.

Wadleigh, C.H. and A.D. Ayers. 1945. Growth and biochemical composition of bean plants as conditioned by soil moisture tension and salt concentration. Plant Physiol. 20: 106–132.

Wilson, C., S.M. Lesch and C.M. Grieve. 2000. Growth stage modulates salinity tolerance of New Zealand spinach (Tetragonia tetragonioides, Pall) and Red Orach (Atriplex hortensis L.) Annals Bot. 85:501–509.

Wimmer, M.A., K.H. Mühling, A. Läuchli, P.H. Brown and H.E. Goldbach. 2001. Interaction of salinity and boron toxicity in wheat (*Triticum asetivum* L.) In Plant nutrition - Food security and sustainability of agro-ecosystems. Kluwer Academic Publishers pp 426–427

Wimmer, M.A., K. H. Mühling, A. Läuchli, P.H. Brown and H.E. Goldbach. 2003. The interaction between salinity and boron toxicity affect subcellular distribution of in and proteins in wheat leaves. Plant Cell Envrion. 26:1267–1274

Wimmer M.A. Bassil E.S., Brown P.H. and A. Läuchli. 2005. Boron response in wheat is genotype-dependent and related to boron uptake, translocation, allocation, plant phenological development and growth rate. Funct. Plant Biol. 32:507–515.

Yadav, H. D., O. P.Yadav, O. P. Dhankar, and M. C. Oswal. 1989. Effect of chloride salinity and {\}boron on germination, growth, and mineral composition of chickpea (Cicer arietinum L.) Ann. Arid Zone 28:63–67.

Yamaguchi T. and E. Blumwald. 2005. Developing salt-tolerant crop plants: challenges and opportunities. Trends in Plant Science 10:615–620.

Yeo A.R., Lee K.S., Izard P., Boursier P. and T.J. Flowers. 1991. Short- and long-term effects of salinity on leaf growth in rice (*Oryza sativa* L.). J. Exp. Bot. 42:881–889.

Yermiyahu U., Nir S., Ben-Hayyim G., Kafkafi U. and T.B. Kinraide. 1997. Root elongation in saline solution related to calcium binding to root cell plasma membranes. Plant Soil 191:67–76.

Zeng, L. and M.C. Shannon. 2000. Salinity effects on seedling growth and yield components of rice. Crop Sci. 40:996–1003

Zhong H. and A. Läuchli. 1993. Spatial and temporal aspects of growth in the primary root of cotton seedlings: effects of NaCl and $CaCl_2$. J. Exp. Bot. 44:763–771.

Zhong H. and A. Läuchli. 1994. Spatial distribution of solutes, K, Na, and Ca and their deposition rates in the growth zone of primary cotton roots: effects of NaCl and $CaCl_2$. Planta 194:34–41.

Zhu J.-K. 2002. Salt and drought stress signal transduction in plants. Annu. Rev. Plant Biol. 53: 247–273.

CHAPTER 2

REGULATION OF ROOT GROWTH RESPONSES TO WATER DEFICIT

ERIC S. OBER[1] AND ROBERT E. SHARP[2]

[1]*Rothamsted Research, Broom's Barn Research Station, Higham, Bury St Edmunds, IP28 6NP, UK*
[2]*Division of Plant Sciences, University of Missouri, Columbia, Missouri 65211, USA*

Abstract: The growth and function of roots are essential for crop productivity under water-limiting conditions, but direct improvement of roots by plant breeding has been slow. One difficulty is the observation and quantitative measurement of root systems under conditions that are relevant to field environments. Another challenge is the identification of and selection for specific loci that could improve the acquisition of water from the soil profile. However, advances are being made in the understanding of root growth regulation and development. We review the evidence for the maintenance of root growth by ABA during water deficit, and the interactions with ethylene and other hormones. A biophysical model of cell expansion serves to focus discussion of topics relating to regulation of growth and development. The power of kinematic growth analysis is demonstrated by highlighting changes in growth regulatory processes and associated patterns of gene expression and protein composition that occur specifically in regions of the root where cell expansion is maintained under water deficit conditions. Growth is a complex process; new information adds further insight and further complexity to our understanding of how roots sense and respond to changes in environmental conditions. It is important to unravel these adaptive mechanisms so that it is clear how the manipulation of one process will affect the function of the whole plant, and so that the effect on final yield and water use can be predicted. This complexity makes simple linear models inadequate as explanatory tools, and a systems approach is needed to incorporate the weave of interacting networks of signaling and response pathways. The real challenge is to discover how root growth can be improved, to supply breeders with the practical tools to identify or introduce superior alleles in crop species, and ultimately to ensure that discoveries lead to improvements in productivity in the field

Keywords: Root growth, water deficit, ABA, ethylene, ROS, DELLA proteins, cell wall

1. ROOT GROWTH AND DROUGHT

Roots are essential to plant survival and play a critical role in determining the yield of crops. However, they are hidden from view, often deep in the soil, and this makes them difficult to study and easy to ignore. Historically, plant breeders have

M.A. Jenks et al. (eds.), Advances in Molecular Breeding Toward Drought and Salt Tolerant Crops, 33–53.

made selections for crop improvement based on visible traits, or at least traits that were easy to identify because of the need to rapidly assess large numbers of plants. It is not surprising, therefore, that root characters rarely feature in lists of active breeding targets. This may change with new discoveries and the development of innovative screening techniques.

One of the most critical challenges land plants face is drought, and under such circumstances the role of roots in acquiring soil water and nutrients is absolutely essential. Drought is the largest single factor limiting crop productivity worldwide (Boyer, 1982), yet it is unfortunate that many aspects of how roots grow, develop and respond to changing soil conditions are poorly understood. It is the growth of roots that determines root system architecture and exploration of the soil profile. Nearly all of this growth occurs within several mm of root apices. What happens in this relatively small mass of cells can make a huge impact on yield under sub-optimum conditions. This chapter focuses on what we have learned about one aspect of plant adaptation to drought: how root growth is maintained in drying soils. Much of what we know about root behavior comes from studies done on simple systems in controlled environments; many of the examples we will use here are from the maize (*Zea mays* L.) primary root system. In addition, we consider examples of root growth of mature plants in field conditions.

The maize seedling root system consists of the primary root and additional sets of seminal roots emanating from the mesocotyl-radicle junction (Cahn et al., 1989). In addition, nodal roots successively form on the plant stem, some of which emanate above the soil surface. In the root system of the mature plant, these seedling and nodal roots can produce up to 70 axile roots, and first- and second-order lateral roots can emerge from these (Hoppe et al., 1986). With respect to the root system of mature plants, the primary root is a minor constituent; however, it is a convenient system to study, and although root types can respond differently (Volkmar, 1997), most physiological and molecular features of the primary root may be generally applicable.

In several crop species including maize, the growth of roots and shoots is inhibited during water deficit, but roots continue growing at low water potentials (ψ_w) that are completely inhibitory to shoot growth (Spollen et al., 1993). This differential growth sensitivity may confer an advantage to plants in water-limited conditions by favoring the allocation of carbon below ground to permit greater exploration of soil while limiting the loss of water from shoot tissues. Understanding how growth is regulated in response to water deficit is necessary in order to find ways to improve crop productivity.

1.1. Hormonal Regulation of Root Growth

Roots growing slowly at low ψ_w synthesize and accumulate the plant hormone abscisic acid (ABA). ABA applied to well watered roots inhibits root growth (Sharp, 2002; Sharp et al., 1994). Therefore, it might seem logical to conclude that the endogenous production of ABA at low ψ_w causes the root growth inhibition.

However, it has been shown that this is incorrect. In the maize primary root ABA accumulation is required for the maintenance of growth. This was discovered by conducting a series of classic hormone response experiments, following the rules set out years ago by Jacobs (Jacobs, 1959). Briefly, ABA accumulation was blocked either by adding an inhibitor of carotenoid and therefore ABA synthesis (fluridone), or by using maize mutants deficient in ABA, such as *vp14*, which contains a lesion in 9-cis epoxycarotenoid dioxygenase, the rate-limiting step in ABA synthesis (Tan et al., 1997). Roots deficient in ABA showed severe growth inhibition at low ψ_w, and when exogenous ABA was added back to the tissues to restore the normal endogenous concentration of ABA, growth rates were restored (Saab et al., 1990; Sharp, 2002; Sharp et al., 1994). When ABA was added to roots such that the endogenous concentration increased above the normal level, growth was inhibited. In fact, a tissue concentration of ABA that is required for root growth at low ψ_w causes growth inhibition under well watered conditions, showing that the sensitivity to ABA also changes with tissue water status (Sharp, 2002; Sharp et al., 1994). Thus, exogenous ABA application leading to supranormal tissue ABA concentrations can cause artefactual responses and incorrect conclusions about the role of endogenous ABA. These cautions most probably also apply to the assessment of other hormones that may be involved in root growth regulation under water deficit conditions.

Further experimentation with the maize primary root system revealed that an important function of endogenous ABA is to keep ethylene production under control (Spollen et al., 2000). Under water deficit conditions, elevated levels of ABA in roots are sufficient to suppress excess ethylene production and, hence, further growth inhibition is prevented. Interestingly, ABA accumulation in shoot tissues at low ψ_w can be insufficient to prevent ethylene-induced growth inhibition (Sharp, 2002). This may be part of the mechanism of greater sensitivity of shoot than root growth to low ψ_w, although additional factors clearly modulate maize leaf growth under water deficit conditions, and some of these may be independent of ABA/ethylene interactions (Voisin et al., 2006). In the absence of stress, ABA may also help promote shoot growth via ethylene suppression. This was shown in *Arabidopsis* (LeNoble et al., 2004) and tomato (Sharp et al., 2000) by decreased growth of ABA deficient mutants when shoot water status was maintained by high humidity (to overcome effects of impaired stomatal functioning).

The effects of ethylene on growth are complex and variable. In many species there is a biphasic response of growth to ethylene, whereby very low concentrations of ethylene can stimulate growth while greater concentrations inhibit growth (Pierik et al., 2006). In addition, developmental stage can affect the growth response to ethylene. For instance, during the post-germination development of maize seedlings at low ψ_w, ethylene shifts from growth promotion of the mesocotyl to growth inhibition (Sharp, 2002; Sharp and LeNoble, 2002). In contrast to maize, ethylene stimulates root growth in deep water rice (Steffens et al., 2006), and ABA inhibits root growth by competitive inhibition of gibberellin (GA) stimulation of ethylene synthesis. The rice and maize systems are similar in that ABA blocks ethylene, but ethylene can promote or inhibit growth in the different systems.

Recent evidence suggests that ethylene may inhibit growth by stabilizing the activity of DELLA proteins, which restrict root growth (Achard et al., 2006). In this study, *Arabidopsis* seedlings growing on saline substrate showed increased ethylene production and decreased growth, but not in mutants deficient in ABA signaling (*abi1-1*) or lacking four of the five DELLA proteins. Under severe stress conditions, the disadvantage of decreased growth was offset by increased plant survival in the mutants compared with wild type. One of the major DELLA proteins (GAI) regulates GA action: GA stimulates growth by stimulating the breakdown of DELLA proteins. A generic view may be that ABA restricts ethylene, which permits GA breakdown of DELLAs, releasing the 'brakes' on growth. Conversely, if low ψ_w-induced ABA accumulation is prevented, increased ethylene may inhibit GA, which then protects DELLAs and their inhibition of growth. An exception may be in submerged tissues of plants adapted to aquatic environments where ethylene-GA interactions have the reverse effect, allowing ethylene to stimulate GA levels and DELLA breakdown.

It is not yet clear how ABA and the ethylene synthesis pathway interact. Other hormones also affect root growth, including auxin (Fuand Harberd, 2003; Rahman et al., 2001), brassinosteroids (Müssig et al., 2003), and cytokinins (Riefler et al., 2006). Multiple hormone pathways interact to affect growth in a complex manner. Hormones also affect root initiation, which can be separated from expansive growth *per se*. For instance, work with the *lrd2* mutant in *Arabidopsis* showed that both auxin and ABA mediate lateral root initiation, and that the suppression of lateral root initiation by ABA at low ψ_w does not involve changes in primary root growth rate (Deak and Malamy, 2005). (In soils, however, it should be noted that low ψ_w can lead to increased lateral root development [Ito et al., 2006]). In tomato, overexpression of the tonoplast H^+-pyrophosphatase led to increased root growth and drought tolerance, which may have been caused by increased apoplast acidification and polar auxin transport (Park et al., 2005). Auxin and ABA pathways may converge at ABI3 (de Smet et al., 2006), and ABA and ethylene pathways can intersect at ERF4 (Yang et al., 2005).

Part of the complexity in interpreting hormone-growth relationships is the need to separate cause and effect, and to consider environmental conditions that can alter hormone levels. For instance, a set of near-isogenic maize hybrids were developed that contrast for leaf ABA concentration, and QTLs linked to this trait were identified (Giuliani et al., 2005). However, the interesting result is that the main QTL appears to control root system architecture, and the effects on leaf ABA are probably pleiotropic. The proposed model suggests that the lines that accumulate greater levels of leaf ABA do so because root density is greater in the more superficial soil layers, which tend to dry out even under irrigated conditions. These roots in drying soil may synthesize ABA and transport it to leaves in the transpiration stream. A beneficial result is that the 'high ABA' allele is associated with larger root systems that reduce lodging. Breeders need this kind of genotypic and phenotypic information in order to manipulate root architecture.

1.2. Root Growth Biophysics

Roots, as all plant tissues, grow by production of new cells by the meristem, followed by expansion of these cells until growth ceases. Although cell division is vital, it is mostly cell expansion that drives roots through the soil matrix. Biophysical models of tissue growth provide a clear framework for the interpretation of growth responses. The Lockhart (Lockhart, 1965) equation, originally developed to explain the linear expansion of single cells, also has been applied to multicellular organ growth, such as the primary root. In simplified form:

$$(1) \qquad G = m(P - Y)$$

where G is expansive growth; m is cell wall extensibility; P is turgor; and Y is the minimum turgor threshold required to irreversibly extend the cell wall. This simplified form has been further elaborated, taking into account the driving forces and resistances to water flux through tissue (Passioura and Boyer, 2003), and soil penetration resistance (Bengough et al., 2006). It has been shown that m and Y are not merely physical constants describing the viscoelastic mechanical properties of the cell wall, but additionally comprise metabolic factors acting on wall polymer rheology (Passioura, 1994). This theoretical treatment of growth is important because it breaks down a complicated process into conceptually smaller components, each of which may be controlled by the expression of a gene or suite of genes that can be manipulated by plant breeding.

1.3. Root Growth Kinematics

In maize seedlings grown in darkness and near non-transpiring conditions at 29 °C, the primary root tip is advanced at about 3 mm h^{-1} in well watered vermiculite (an expanded phyllosilicate mineral that serves as a soil-like substrate) in which there is negligible penetration resistance (Sharp et al., 1988), or up to 4 mm h^{-1} in solution culture (Verslues et al., 1998). Roots are usually indeterminate structures: if optimum conditions could be maintained beyond the seedling stage, roots growing at this rate would reach a depth over 2 m in a month. It is not surprising, then, that the maize root systems excavated by Weaver (Weaver, 1926) in irrigated, deep prairie soils reached 2.5 m. In most cases, of course, environmental factors limit the full expression of the growth potential of the root system in the field.

In the seedling system, when the water content of the vermiculite decreased to a ψ_w of -1.6 MPa, root growth slowed to approximately 1 mm h^{-1} (Sharp et al., 1988). The rate of root growth is determined by the rate of cell production from the meristem, rates of cell expansion, and the final length of cells (or the time spent elongating). Following the first division in the meristem, cells at first expand anisotropically (Liang et al., 1997), then longitudinal expansion is favored over radial expansion. Each cell is displaced basipetally away from the root apex by further cell production and expansion. At a certain point cells cease and vacuole play an important role (Luu and Maurel, 2005). Aquaporins are also important in regulating water flux between the soil and the root. However, the quantity of water necessary for growth is small compared to the volume that is removed from the soil to meet the transpirational demand of the shoot. Water uptake in maize occurs 20–30 mm from the root tip (Frensch et al., 1996), but the bulk of water transport occurs in mature roots at least 30 cm from the root tip (Varney and Canny, 1993). Significant water movement depends on open metaxylem elements, which are slow to mature in maize (Wenzel et al., 1989). Aquaporins are a large gene family showing differential tissue expression patterns (Hachez et al., 2006). Aquaporins only passively facilitate the movement of water across membranes and cannot alter the magnitude or direction of the driving force. Thus, temporal downregulation of aquaporins could restrict the loss of water from root tissues into dry soil, but it remains to be shown that increased aquaporin activity can aid the acquisition of water from the soil (Vandeleur et al., 2005).

There is a large body of research on hydraulic conductivity of roots and root systems, but we only make a few observations here. It has been long noted that drought (Brown et al., 1987; Sharp and Davies, 1985) and ABA (Hose et al., 2000) can increase root hydraulic conductivity, possibly via modulation of aquaporin activity (Hartung et al., 2005), which can also be affected by reactive oxygen species (Henzler and Steudle, 2004). It is also well known that different root types and root ages within a root system have different conductivities (Pierret et al., 2006). For these reasons there is not necessarily a good correlation between root length density and soil water uptake; the mere presence of roots does not mean they are fully functional.

1.6. Changes in the Cell Wall

Compared with well watered roots, in maize primary roots growing under severe water deficit (−1.6 MPa), the cumulative effect of increased solute concentrations is a two-fold decrease in the total osmotic potential. Nevertheless, despite the increased driving force for water uptake, turgor pressure throughout the apical 10 mm of the maize primary root is 60% smaller than at high ψ_w (Spollen and Sharp, 1991). How do cells near the root apex maintain linear rates of expansion despite lower turgor? Equation 1 shows that in order for G to remain unchanged as P decreases, m must increase and/or Y must decrease. This means that cell walls in the apical few mm of the maize root tip must be more extensible at low ψ_w than at high ψ_w. Measurements of growing root cells showed that wall properties can compensate quickly to changes in turgor (Frensch, 1997). Biochemical evidence for this was provided by Wu et al. (1996) who showed that in this region acid-induced extensibility increased at low ψ_w. This fits with evidence that cell wall pH is significantly lower 0–3 mm from the root apex than in more basal regions (Fan and Neumann, 2004).

Plastic deformation of the cell wall requires that bonds between load-bearing structural elements must be broken, allowing some slip so that previously slack tethers take the strain (Passioura, 1994), fresh wall polymers are added (Proseus and Boyer, 2006), and then new bonds form to carry the tension in the wall produced by turgor pressure. Obviously this must happen in a controlled fashion, otherwise cells would explode with the internal force of hydrostatic turgor. The control occurs via activity of proteins acting on cellulosic, hemicellulosic and pectin components of the wall. Members of two protein families, xyloglucan endotransglycosylase (XET) and expansin, increase in activity specifically in the apical few mm of the elongation zone of maize primary roots at low compared to high ψ_w. XET activity at low ψ_w correlated with the spatial distribution of growth within the apical cm (Wu et al., 1994). The increase in XET activity at low ψ_w was dependent on ABA accumulation, and XET transcript levels also have been shown to be regulated by auxin in the *Arabidopsis* root tip (Osato et al., 2006). In contrast, effects of low ψ_w on expansin activity and transcript levels were not dependent on ABA in the maize primary root tip (Wu et al., 2001). Additional tests showed that tissues in the apical 5 mm of the maize primary root at low ψ_w exhibited increased susceptibility to expansin, whereas tissues in the zone of growth inhibition lose their sensitivity to expansin (Wu et al., 1996). The patterns of expansin activity, protein and transcript level are spatially and temporally complex (Wu et al., 2001). An important point to reinforce is that these discoveries would not have been made had whole roots been analyzed without regard to the spatial distribution of growth.

The synthesis of wall polymers and integration of new cell wall material is a complex process and is only partially understood. Construction of composite cell wall materials requires that deposition of cellulose microfibrils, which occurs outside the cell membrane, is coordinated with the synthesis of hemicellulose and pectin, which occurs in the Golgi apparatus. Furthermore, the rate of movement of these polymers from vesicles to the cell membrane depends on the supply of substrate for synthesis. A recently discovered family of wall-associated kinases, which are required for root growth, may play an important role in this regulation (Kohorn et al., 2006).

1.7. Functional Genomics and Proteomics

To discover further genes related to the pattern of growth within the root tip at high and low ψ_w, surveys of changes in the root transcriptome were undertaken (Figure 2; Poroyko et al., 2006). Results showed that EST profiles from distinct regions within the apical cm of the root tip, guided by the relative elongation rate distribution, showed populations of transcripts that were unique to accelerating, decelerating and non-growing cells. Other work also revealed a large number of specific genes up- or down-regulated by low ψ_w (Bassani et al., 2004). Cell wall proteomic studies of maize root tips (Zhu et al., 2006) are beginning to reveal information about the spatial and temporal regulation of proteins that may play a role in growth regulation. These approaches will supply a huge amount of new data, and an important challenge will be to narrow down large gene sets to a small number of key genes that can be studied in detail.

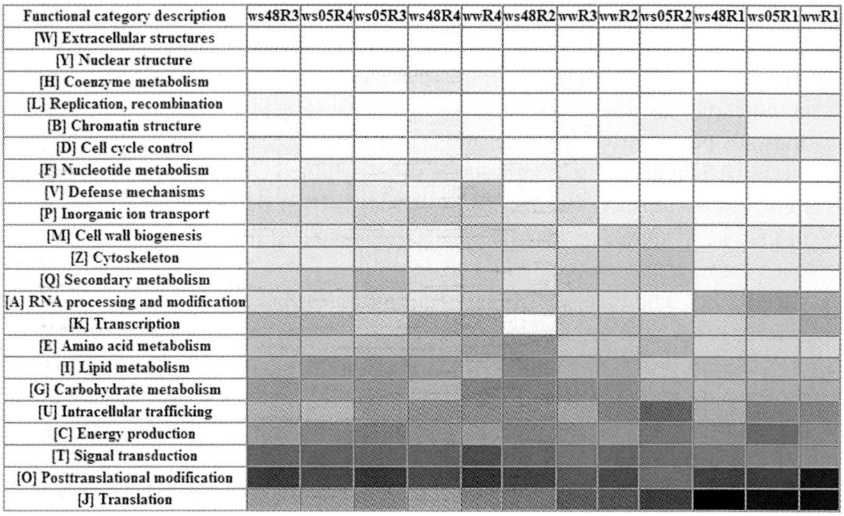

Figure 2. The relative abundance of transcripts, grouped according to functional category, expressed within the maize primary root tip at high (WW) and low (WS) ψ_w (Poroyko et al., 2006). The root tip was divided into four regions: 0–3 mm from the root apex-root cap junction (R1), 3–7 mm (R2), 7–12 mm (R3), 12–20 mm (R4). Refer to Figure 1 for the corresponding patterns in longitudinal growth rate. Roots were harvested at 5 and 48 h after transplanting seedlings to low ψ_w (−1.6 MPa). In the well watered treatment, roots were harvested at 5 and 48 h after transplanting and bulked together for analysis. Cells within the grid are shaded according to the proportional representation of each functional category within the unigene library for each experiment (columns). The scale ranges from 0 (white) to 12.6% (black)

1.8. The Role of Reactive Oxygen Species

In practically every study of stress-induced changes to the proteome there is a group of proteins classed under ‘oxidative stress’. Reactive oxygen species (ROS) such as superoxide ($O_2^{\cdot-}$) and hydroxyl radicals ($^{\cdot}OH$) accumulate under stress conditions and need to be kept under control to preserve the integrity of cellular macromolecules. The redox balance of cells is controlled by a series of enzymes and intermediate metabolites. Interestingly, ROS are not completely abolished, but also play important roles in signaling and growth regulation (Carol and Dolan, 2006). In growing cells, the controlled breakdown of cell wall polymers involves $^{\cdot}OH$ (Liszkay et al., 2004) and quenching ROS inhibits root growth (Demidchik et al., 2003). Overexpression in *Arabidopsis* of a peroxidase localized mainly in the root elongation zone stimulated root elongation (Passardi et al., 2006). Likewise, wall hardening via crosslinking polymers slows growth, and these reactions are controlled by ROS. For example, callose deposition and wall protein crosslinking via ROS production were induced by treating roots with ACC, the ethylene precursor, which reduced cell elongation (de Cnodder et al., 2005).

ABA also plays a role in regulating the balance between useful production and harmful over-production of apoplastic ROS. *Arabidopsis* mutants with defective NADPH oxidase could not generate the H_2O_2 required for ABA signaling during stomatal closure. However, root growth in the mutants was unaffected; only when exogenous ABA was added to well watered roots to cause root growth *inhibition* (see section 1.1) did it appear that ROS production mediated ABA (or induced ethylene?) effects on root growth (Kwak et al., 2003). In maize, ABA deficiency can cause uncontrolled ROS production and growth inhibition within the primary root apex (I-J Cho, M Sivaguru, RE Sharp, unpublished). Thus, too little or too much tissue ABA (in relation to normal physiological levels) can cause ROS-related growth inhibition. Another source of ROS is oxalate oxidase, a germin-type enzyme that releases H_2O_2 into the apoplast; other germin-type proteins show superoxide dismutase activity. Germins are a large and multi-functional family, and frequently appear in 'omic analyses of plants subjected to abiotic stresses (Bray, 2004). It is clear from current research that ROS play an important role in root growth regulation and the response to drought. However, finding a way to manipulate the control of cellular redox balance to favor root growth under dry conditions is a significant challenge.

1.9. Perception of Low ψ_w, Signal Transduction and Signal Maintenance

In the preceding sections we have described root growth responses to low ψ_w under steady-state conditions. An important question is how these responses are triggered as rhizosphere conditions begin to change. How do roots perceive a change in the surrounding ψ_w? One hypothesis is that the primary response occurs at the plasmamembrane-cell wall interface (Wojtaszek et al., 2005). Altered conformation of membrane-spanning proteins such as stretch-activated ion channels anchored to the cytoskeleton could affect ion fluxes and the electrochemical potential of the cell (Lew, 2004). This could then trigger a cascade of further events, perhaps including Ca^{2+} and pH, well-known intracellular signals (Gao et al., 2004).

Electrophysiological measurements showed that cortical cells within the elongation zone of maize primary roots undergo hyperpolarization (via activation of the plasmamembrane H^+-ATPase) in response to low ψ_w (Ober and Sharp, 2003). In another study, flux patterns of K^+, Cl^- and Na^+ at the root surface were transiently altered in response to low ψ_w (Shabala and Lew, 2002). A portion of the current in the elongation zone is also carried by Ca^{2+} (Kiegle et al., 2000), which is fundamental to growth in tip-growing cells such as pollen tubes and root hairs (Feijó et al., 2004). A hyperpolarized membrane potential would increase the driving force for increased K^+ uptake. This K^+ could contribute to short-term osmotic adjustment. However, under steady-state conditions during long-term exposure to low ψ_w, there is little change in tissue K^+ status (Sharp et al., 1990).

It is possible that short-term changes in cytoplasmic ion concentrations trigger the deposition of organic solutes, which accumulate during long-term exposure to low ψ_w, and as these compounds accumulate, ion concentrations return to normal levels. Accumulation of organic "compatible solutes" prevents deleterious effects of high ionic strength in the cytoplasm. Such a sequence of events in osmotic adjustment occurs in other organisms (e.g. bacteria; Yim and Villarejo, 1994), and most likely in plant roots as well.

In maize root tips, root cells hyperpolarized during the initial exposure to low ψ_w, but eventually returned to resting potentials near to but significantly more negative than those at high ψ_w (Ober and Sharp, 2003). In ABA-deficient roots, however, membrane potentials remained hyperpolarized specifically in the region in which cell growth is responsive to ABA. This could be an indication that in roots, as in stomatal guard cells, setpoints for ion homeostasis shift in response to low ψ_w, and maintenance of these setpoints may depend on ABA (MacRobbie, 2006).

It is important to note that experimental procedures used to investigate stress perception and signal transduction have a large influence on the results. The mode in which ψ_w is altered and the rate at which it is applied can produce different effects (Kacperska, 2004). For instance, roots subjected to a rapid decrease in ψ_w showed a depolarization, while slow imposition of low ψ_w caused a hyperpolarization (Ober and Sharp, 2003). Mannitol is a common osmotic agent, but is toxic to maize roots (Hohl and Schopfer, 1991); polyethylene glycol solutions with inadequate aeration can induce symptoms of hypoxia (Verslues et al., 1998). The challenge for investigators is to understand which set of conditions lead to accurate conclusions that can be applied outside the laboratory.

1.10. Cell Production

The emphasis in this chapter has been on root cell expansion, but growth also depends on cell production rate, which is a function of the rate of cell division and meristem size (Beemster et al., 2005). Within the meristem, under a range of environmental situations affecting supply of water and nutrients to the root, the duration of the cell cycle is uniform, but the proportion of cells that are dividing can change (Baskin, 2000). Thus, under most conditions, meristems are well protected and meristematic activity is robust and not easily perturbed. However, under severe water deficit, cell division rate can be inhibited (Saab et al., 1992; Sacks et al., 1997). Merely increasing the rate of cell division would not necessarily increase root growth as it could result in a large number of very small cells. However, overexpression of cyclin B genes resulted in increased growth without any effect on final cell size (Doerner et al., 1996; Lee et al., 2003). Also, the *CRL2* mutant in rice shows increased meristem size, cell flux and cortical cell length compared with wild type (Inukai et al., 2001). These results suggest that cell production could be manipulated by breeding to benefit root growth, although pleiotropic effects on whole plant function would have to be examined carefully.

1.11. Root Growth in the Real World

Roots of crop plants growing in the field encounter a range of situations that are rarely matched in controlled experimental conditions. Crops often have to face inhospitable subsoils: high penetration resistance, aluminum toxicity, pH extremes, and poor aeration are some of the problems roots face after they penetrate the surface layers (Passioura, 2006). In many cases, soils harden as they dry. In one study of wheat subjected to water deficit, soil penetration resistance inhibited yield more than soil water availability (Whalley et al., 2006).

The soil profile is often heterogeneous, with different soil textures and patches of water and nutrients (Hutchings and John, 2004). Roots do not grow in sterile media, but are surrounded by rhizosphere microflora (Watt et al., 2006) and many species are typically infected by mycorrhyzae. Neither is the rhizosphere a solitary root or neatly divided root system, but a complex weave of many roots, often clumped, and often of different species in weedy crops or intercropping systems. The highly variable and unpredictable nature of life in the field means that root systems must be equally flexible, as must be the models we employ to describe them.

Advantages to the plant provided by expression of 'phenotypic plasticity' in heterogeneous natural environments (Grime and Mackey, 2002) perhaps may be exploited for crop improvement. Examples of plasticity already mentioned are the ability of roots to grow thinner and longer, and changes in the number and length of lateral roots. Changes in root system architecture in response to P (Lynch et al., 2005) and N (Walch-Liu et al., 2006) status in local soil patches are well documented. The understanding of hydrotropism, the ability of roots to grow towards moist soil patches, is gaining ground with new studies utilizing mutants (Eapen et al., 2005; Tsuda et al., 2003), but the mechanisms remain unclear.

During water deficit as surface soil layers are depleted of moisture, root systems can proliferate deeper in the soil profile in permissive soils such that the density of roots is greater in stressed than non-stressed conditions (Klepper, 1990). Among numerous examples, this has been shown in maize roots growing in soil columns (Sharp and Davies, 1985), and in sugar beet (*Beta vulgaris* L.) roots in the field (Figure 3).

1.12. Selection Methods for Breeding for Improved Root Growth

One of the difficulties in breeding for complex quantitative traits such as root growth is the identification of a major character on which to base selections. One response is to make selections on phenotype without knowledge of what specific factors contribute to, say, greater root mass deep in the soil profile. With substantial effort and resources, this empirical approach can be used directly, or better, to identify QTLs conditioning these phenotypes. Markers linked to major QTLs can then be used for routine screening to select genotypes possessing superior alleles. This has been successful in rice (Steele et al., 2006). Another approach is to test the functional contribution of candidate genes, one by one, to determine which of the potential candidates has the greatest impact on the desired phenotypic trait. These efforts will be aided by innovations in laboratory and field techniques to observe and measure root systems. Rapid seedling screening techniques are essential when large numbers of genotypes need to be assessed (Bengough et al., 2004;

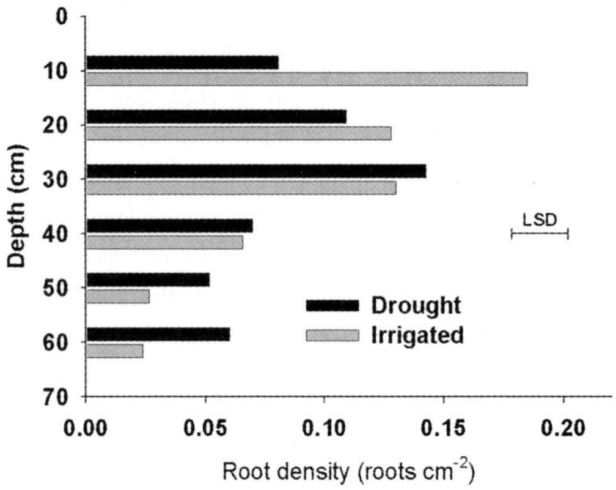

Figure 3. Plasticity in the proliferation of sugar beet roots within the soil profile under irrigated and droughted conditions in the field (CHG Smith, ES Ober, unpublished data). Note the significantly greater root density deep in the soil profile under droughted conditions. Root densities were determined using the trench profile face method, counting root contacts in 10 cm square grids. Bars indicate the mean of eight replicate plots. The LSD for the treatment x depth interaction term from ANOVA is shown

Kuchenbuch and Ingram, 2002). When the number of genotypes has been limited by this process, confirmation of genotypic rankings under field conditions can be considered. Most studies rely on the conventional techniques of quantifying root behavior using soil cores, but advances have been made in image analysis of samples (e.g. Vamerali et al., 2003). Another approach is to assess genotypic differences in rooting indirectly by measuring patterns of soil water depletion (Figure 4). Other useful field techniques that estimate different aspects of root systems are root pulling strength (Landi et al., 2002) and root electrical capacitance (van Beem et al., 1998).

1.13. Conclusions

Plants have evolved a highly complex regulation of root and shoot growth to achieve maximum fitness with the available resources. An important question to address is the extent to which plants have optimized this regulation, and what further advances are possible through genetic manipulation, whether via conventional plant breeding or transgenic technology. It is likely that benefits in one area may be balanced by increased costs in another. For instance, Passioura (Passioura, 1983) pointed out that "there is no point in a droughted crop investing a parcel of assimilate in its roots if the extra water thereby obtained does not allow the shoots to at least replenish the assimilate so spent". Despite the inevitable stress-related trade-offs within the plant (Weih, 2003), plant breeders have managed to find improvements in yield for water-limited environments, although it is slow and difficult work overcoming the conservative nature of plants geared towards survival. (We have seen, for instance, how DELLA proteins inhibit growth, but increase survival under

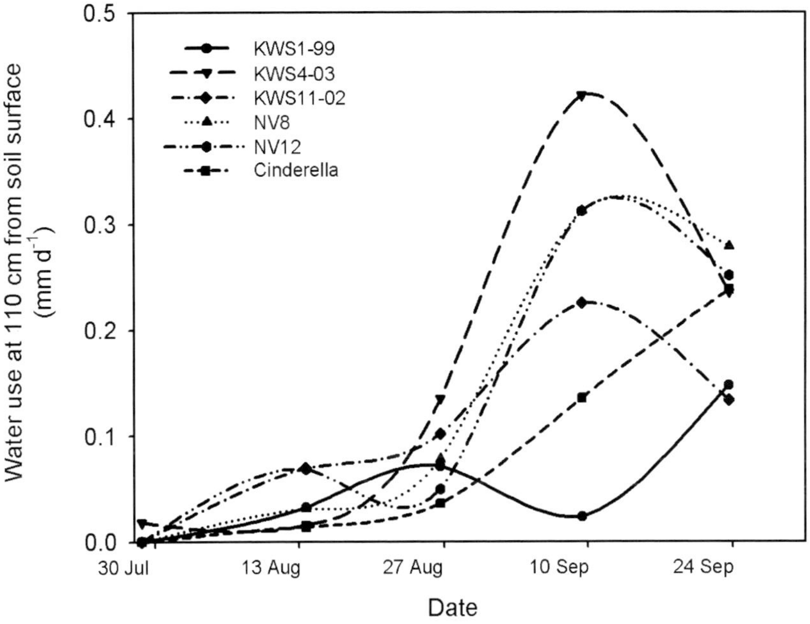

Figure 4. Patterns of soil water depletion during drought in five experimental sugar beet hybrids and one commercial cultivar (Cinderella). Differences between genotypes in summed water use from this layer were significant ($P<0.05$). Weekly changes in soil moisture content were made at 110 cm from the soil surface using a capacitance-type soil moisture probe (Ober et al., 2005). Smoothed lines connect mean values of four individual plots for each genotype

extreme conditions.) For a number of drought-related traits, there is substantial allelic variation among ecotypes within wild species (Mouchel et al., 2004), and within the pool of crop germplasm (Figure 4). These points provide encouragement for the process of crop improvement.

In this chapter we have reviewed a number of processes that are vital to root growth: cell wall modification, transport of organic solutes, ions and water, and control of oxidative stress. All of these were shown to be related to the spatial distribution of growth within the root tip, and most were regulated in some manner by ABA or ethylene. Factors that control hormonal synthesis or sensitivity, and the interactions between hormones are clearly important, but gross manipulation of hormone levels may produce confounding or undesirable side effects. Functional testing of candidate genes is necessary, but overexpression or silencing of genes must be done in a tissue-specific manner, using appropriate (e.g. stress-responsive) upstream elements. Growth kinematics is a useful tool to highlight which tissue regions should be targeted.

Functional genomic and proteomic studies are revealing numerous genes and proteins that are responsive to low ψ_w and spatially correlated with growth maintenance in the maize primary root. Taken together, so many interacting factors must coincide at the right time and place that the complexity of growth becomes impossible to portray in a simple linear fashion. The methodology of systems biology (Aderem, 2005) is required to describe the process of growth and to identify limiting factors for given situations. Also, neural network models can be applied to describe the multiple parameters controlling growth (Ushada and Murase, 2006). The tools of genomics and computational biology are developing at a fast pace, and will continue to aid discovery and identification of potential breeding targets.

A potential limiting factor for these new technologies is the quality of the data that are fed into the system. For improvement of root growth and function, this means providing accurate phenotypic data under relevant conditions. An important caveat for all genetic improvement projects is that at some point plants must be grown in the field to measure yield. The differences between conditions in controlled laboratory or glasshouse conditions and the field are often huge. Phenotypic expression can be radically altered by growing conditions resulting in large genotype x environment interactions. What were permissive conditions for expression of a trait in potting compost, for instance, may not exist in the field, making the trait irrelevant to plant breeders. The need for a multi-disciplinary approach emphasizes the importance of co-operation between breeders, physiologists, molecular biologists, agronomists, statisticians and crop modelers. At all levels, key components in this research are patience and persistence. Weaver (Weaver, 1926), who excavated by hand entire root systems of several crop species, noted

"There is no easy method of uncovering the root system, and unless one is willing to spend considerable time and energy, and exercise a great deal of patience, it is better not to begin. But once started, the work, although difficult, is very interesting and in fact even fascinating."

REFERENCES

Achard P, Cheng H, De Grauwe L, Decat J, Schoutteten H, Moritz T, Van Der Straeten D, Peng J, Harberd NP, 2006, Integration of plant responses to environmentally activated phytohormone signals. *Science 311: 91–94.*

Aderem A, 2005, Systems biology: its practice and challenges. *Cell 121: 511–513.*

Baskin TI, 2000, On the constancy of cell division rate in the root meristem. *Plant Molec Biol 43: 545–554.*

Bassani M, Neumann PM, Gepstein S, 2004, Differential expression profiles of growth-related genes in the elongation zone of maize primary roots. *Plant Molec Biol 56: 367–380.*

Beemster GTS, Mironov V, Inzé D, 2005, Tuning the cell-cycle engine for improved plant performance. *Curr. Op. Biotech. 16: 142–146.*

Bengough AG, Bransby MF, Hans J, McKenna SJ, Roberts TJ, Valentine TA, 2006, Root responses to soil physical conditions; growth dynamics from field to cell. *J.Exp. Bot. 57: 437–447.*

Bengough AG, Gordon DC, Al-Menaie H, Ellis RP, Allan D, Keith R, Thomas WTB, Forster BP, 2004, Gel observation chamber for rapid screening of root traits in cereal seedlings. *Plant Soil 262: 63–70.*

Boyer JS, 1982, Plant productivity and environment. *Science 218: 443–8.*

Bray EA, 2004, Genes commonly regulated by water-deficit stress in *Arabidopsis thaliana. J.Exp. Bot. 55: 2331–2341.*

Bret-Harte MS, Silk WK, 1994, Fluxes and deposition rates of solutes in growing roots of Zea mays. *J. Exp. Bot. 45: 1733–42.*

Brown KF, Messem AB, Dunham RJ, Biscoe PV, 1987, Effect of drought on growth and water use of sugar beet. *J. Agric. Sci. 109: 421–35.*

Cahn MD, Zobel RW, Bouldin DR, 1989, Relationship between root elongation rate and diameter and duration of growth of lateral roots of maize. *Plant Soil 119: 271–279.*

Carol RJ, Dolan L, 2006, The role of reactive oxygen species in cell growth: lessons from root hairs. *J. Exp. Bot. 57: 1829–1834.*

de Cnodder T, Vissenberg K, van der Straeten D, Verbelen JP, 2005, Regulation of cell length in the *Arabidopsis thaliana* root by the ethylene precursor 1-aminocyclopropane-1-carboxylic acid: a matter of apoplastic reactions. *New Phytol. 168: 541–550.*

de Smet I, Zhang H, Inzé D, Beeckman T, 2006, A novel role for abscisic acid emerges from underground. *Trends Plant Sci. doi:10.1016/j.tplants.2006.07.003.*

Deak KI, Malamy J, 2005, Osmotic regulation of root system architecture. *Plant J. 43: 17–28.*

Demidchik V, Shabala SN, Coutts KB, Tester MA, Davies JM, 2003, Free oxygen radicals regulate plasma membrane Ca^{2+}- and K^{+}-permeable channels in plant root cells. *J. Cell Sci. 116: 81–88.*

Doerner P, Jorgensen J-E, You R, Steppuhn J, Lamb C, 1996, Control of root growth and development by cyclin expression. *Nature 380: 520–23.*

Eapen D, Barroso ML, Ponce G, Campos ME, Cassab GI, 2005, Hydrotropism: root growth responses to water. *Trends Plant Sci. 10:44–51.*

Fan L, Neumann PM, 2004, The spatially variable inhibition by water deficit of maize root growth correlates with altered profiles of proton flux and cell wall pH. *Plant Physiol. 135: 1–10.*

Farrar JF, Minchin PEH, Thorpe MR, 1995, Carbon import into barley roots: effects of sugars and relation to cell expansion. *J. Exp. Bot. 46: 1859–1865.*

Feijó JA, Costa SS, Prado AM, Becker JD, Certal AC, 2004, Signalling by tips. *Curr. Op. Plant Biol. 7: 589–598.*

Frensch J, 1997, Primary responses of root and leaf elongation to water deficits in the atmosphere and soil solution. *J. Exp. Bot. 48: 985–999.*

Frensch J, Hsiao TC, 1995, Rapid response of the yield threshold and turgor regulation during adjustment of root growth to water stress in Zea mays. *Plant Physiol. 108: 303–312.*

Frensch J, Hsiao TC, Steudle E, 1996, Water and solute transport along developing maize roots. *Planta 198: 348–55.*

Fu X, Harberd NP, 2003, Auxin promotes *Arabidopsis* root growth by modulating gibberellin response. *Nature 421: 740–743.*

Gao D, Knight MR, Trewavas AJ, Sattelmacher B, Plieth C, 2004, Self-reporting Arabidopsis expressing pH and $[Ca^{2+}]$ indicators unveil ion dynamics in the cytoplasm and in the apoplast under abiotic stress. *Plant Physiol. 134: 898–908.*

Giuliani S, Sanguinetti MC, Tuberosa R, Bellotti M, Salvi S, Landi P, 2005, *Root-ABA1*, a major constitutive QTL, affects maize root architecture and leaf ABA concentration at different water regimes. *J. Exp. Bot. 56: 3061–3070.*

Gould N, Thorpe MR, Minchin PEH, Pritchard J, White PJ, 2004, Solute is imported to elongating root cells of barley as a pressure driven-flow of solution. *Funct. Plant Biol. 31: 391–397.*

Grime JP, Mackey JML, 2002, The role of plasticity in resource capture by plants. *Evol. Ecol. 16: 299–307.*

Hachez C, Moshelion M, Zelazny E, Cavez D, Chaumont F, 2006, Localization and quantification of plasma membrane aquaporin expression in maize primary root: a clue to understanding their role as cellular plumbers. *Plant Molec. Biol. 62: 305–323.*

Hartung W, Schraut D, Jiang F, 2005, Physiology of abscisic acid (ABA) in roots under stress–a review of the relationship between root ABA and radial water and ABA flows. *Aust. J. Agric. Res. 56: 1253–1259.*

Henzler T, Steudle E, 2004, Oxidative gating of water channels (aquaporins) in *Chara* by hydroxyl radicals. *Plant Cell and Environ. 27: 1184–1195.*

Hohl M, Schopfer P, 1991, Water relations of growing maize coleoptiles. Comparison between mannitol and polyethylene glycol 6000 as external osmotica for adjusting turgor pressure. *Plant Physiol. 95: 716–22.*

Hoppe DC, McCully ME, Wenzel CL, 1986, The nodal roots of Zea: their development in relation to structural features of the stem. *Can. J. Bot 64: 2524–37.*

Hose E, Steudle E, Hartung W, 2000, Abscisic acid and hydraulic conductivity of maize roots: a study using cell- and root-pressure probes. *Planta 211: 874–882.*

Hutchings MJ, John EA, 2004, The effects of environmental heterogeneity on root growth and root/shoot partitioning. *Ann. Bot. 94: 1–8.*

Inukai Y, Miwa M, Nagato Y, Kitano H, Yamauchi A, 2001, *RRL1*, *RRL2* and *CRL2* loci regulating root elongation in rice. *Breeding Sci. 51: 231–239.*

Ito K, Tanakamaru K, Morita S, Abe J, Inanaga S, 2006, Lateral root development, including responses to soil drying, of maize *(Zea mays)* and wheat (*Triticum aestivum*) seminal roots. *Physiol. Plant. 127: 260–267.*

Jacobs W.P., 1959, What substance normally controls a given biological process? I. Formulation of some rules. *Devel. Biol. 1: 527–533.*

Kacperska, A, 2004, Sensor types in signal transduction pathways in plant cells responding to abiotic stressors: do they depend on stress intensity? *Physiol. Plant. 122: 159–168.*

Kiegle E, Gilliham M, Haselhoff J, Tester M, 2000, Hyperpolarisation-activated calcium currents found only in cells from the elongation zone of Arabidopsis thaliana roots. *Plant J 21: 225–229.*

Klepper, B, 1990, Root growth and water uptake. *In, Irrigation of agricultural crops, Agronomy Monograph No. 30, ASA, CSSA, SSSA, pp. 281–322.*

Kohorn BD, Kobayashi M, Johansen S, Riese J, Huang L-F, Koch K, Fu S, Dotson A, Byers N, 2006, An Arabidopsis cell wall-associated kinase required for invertase activity and cell growth. *Plant J. 46: 307–316.*

Kuchenbuch RO, Ingram KT, 2002, Image analysis for non-destructive and non-invasive quantification of root growth and soil water content in rhizotrons. *J. of Plant Nutr. Soil Sci. 165: 573–581.*

Kwak JM, Mori IC, Pei Z-M, Leonhardt N, Torres MA, Dangl JL, Bloom RE, Bodde S, Jones JDG, Schroeder JI, 2003, NADPH oxidase *AtrbohD* and *AtrbohF* genes function in ROS-dependent ABA signaling in *Arabidopsis. EMBO J. 22: 2623–2633.*

Landi P, Sanguinetti MC, Darrah LL, Giuliani MM, Salvi S, Conti S, Tuberosa R, 2002, Detection of QTLs for vertical root pulling resistance in maize and overlap with QTLs for root traits in hydroponics and for grain yield under different water regimes. *Maydica 47: 233–243.*

Lee J, Das A, Yamaguchi M, Hashimoto J, Tsutsumi N, Uchimiya H, Umeda M, 2003, Cell cycle function of a rice B2-type cyclin interacting with a B-type cyclin-dependent kinase. *Plant J. 34: 417–425.*

LeNoble ME, Spollen WG, Sharp RE, 2004, Maintenance of shoot growth by endogenous ABA: genetic assessment of the involvement of ethylene suppression. *J. Exp. Bot. 55: 237–245.*

Lew, RR, 2004, Osmotic effects on the electrical properties of Arabidopsis root hair vacuoles in situ. *Plant Physiol. 134: 352–360.*

Liang BM, Sharp RE, Baskin TI, 1997, Regulation of growth anisotropy in well-watered and water-stressed maize roots. I. Spatial distribution of longitudinal, radial and tangential expansion rates. *Plant Physiol. 115: 101–111.*

Liszkay A, vander Zalm E, Schopfer P, 2004, Production of reactive oxygen intermediates ($O_2^{\cdot -}$, H_2O_2, and $^{\cdot}OH$) by maize roots and their role in wall loosening and elongation growth. *Plant Physiol. 136: 3114–3123.*

Lockhart, JA, 1965, An analysis of irreversible plant cell elongation. *J. Theor. Biol. 8: 264–275.*

Luu D-T, Maurel C, 2005, Aquaporins in a challenging environment: molecular gears for adjusting plant water status. *Plant Cell Environ. 28: 85–96.*

Lynch JP, Ho MD, 2005, Rhizoeconomics: carbon costs of phosphorous acquisition. *Plant Soil 269: 45–56.*

MacRobbie, EAC, 2006, Osmotic effects on vacuolar ion release in guard cells. *Proc. of the Natl. Acad. Sci. 103: 1135–1140.*

Mouchel CF, Briggs GC, Hardtke CD, 2004, Natural genetic variation in *Arabidopsis* identifies BREVIS RADIX, a novel regulator of cell proliferation and elongation in the root. *Genes Development 18: 700–714.*

Müssig C, Sin G-H, Altmann T, 2003, Brassinosteroids promote root growth in Arabidopsis. *Plant Physiol. 133: 1261–1271.*

Ober ES, Le Bloa M, Clark CJA, Royal A, Jaggard KW, Pidgeon JD, 2005, Evaluation of physiological traits as indirect selection criteria for drought tolerance in sugar beet. *Field Crops Res. 91:231–249.*

Ober ES, Sharp RE, 1994, Proline accumulation in maize (Zea mays L.) primary roots at low water potentials. I. Requirement for increased levels of abscisic acid. *Plant Physiol. 105: 981–987.*

Ober ES, Sharp RE, 2003, Electrophysiological responses of maize roots to low water potentials: relationship to growth and ABA accumulation. *J. Exp. Bot. 54: 813–824.*

Osato Y, Yokoyama R, Nishitani K, 2006, A principal role for AtXTH18 in *Arabidopsis thaliana* root growth: a functional analysis using RNAi plants. *J. Plant Res. 119: 153–162.*

Park S, Li J, Pittman JK, Berkowitz GA, Yang H, Undurraga S, Morris J, Hirschi KD, Gaxiola RA, 2005, Up-regulation of a H+-pyrophosphatase (H+-PPase) as a strategy to engineer drought-resistant crop plants. *Proc. Nat. Acad. Sci. 102:18830–18835.*

Passardi F, Tognolli M, De Meyer M, Penel C, Dunand C, 2006, Two cell wall associated peroxidases from *Arabidopsis* influence root elongation. *Planta 223: 965–974.*

Passioura JB, 1983, Roots and drought resistance. *Ag Water Manag. 7: 265–80.*

Passioura JB, 1994, The physical chemistry of the primary cell wall: implication for the control of expansion rate. *J. Exp. Bot. 45: 1675–82.*

Passioura JB, 2006, Increasing crop productivity when water is scarce–from breeding to field management. *Ag. Water Manag. 80: 176–196.*

Passioura JB, Boyer JS, 2003, Tissue stresses and resistance to water flow conspire to uncouple the water potential of the epidermis from that of the xylem in elongating plant stems. *Funct. Plant Biol. 30: 325–334.*

Peters, WS, 2004, Growth rate gradients and extracellular pH in roots: how to control an explosion. *New Phytol. 162: 571–574.*

Pierik R, Tholen D, Poorter H, Visser EJW, Voesenek LACJ, 2006, The Janus face of ethylene: growth inhibition and stimulation. *Trends Plant Sci. 11: 176–183.*

Pierret A, Doussan C, Pagès L, 2006, Spatio-temporal variations in axial conductance of primary and first-order lateral roots of a maize crop as predicted by a model of the hydraulic architecture of root systems. *Plant Soil 282: 117–126.*

Poroyko V, Spollen WG, Hejlek LG, Hernandez AG, LeNoble ME, Davis G, Nguyen HT, Springer GK, Sharp RE, Bohnert HJ, 2007, Comparing regional transcript profiles from maize primary roots under well-watered and low water potential conditions. *J. Exp. Bot. 58: 279–289.*

Proseus TE, Boyer JS, 2006, Periplasm turgor pressure controls wall deposition and assembly in growing *Chara corallina* cells. *Ann. Bot. 98: 93–105.*

Proseus TE, Zhu G-L, Boyer JS, 2000, Turgor, temperature and the growth of plant cells: using *Chara corallina* as a model system. *J. Exp. Bot. 51: 1481–1494.*

Rahman A, Amakawa T, Goto N, Tsurumi S, 2001, Auxin is a positive regulator for ethylene-mediated responses in the growth of Arabidopsis roots. *Plant Cell Physiol. 42: 301–307.*

Riefler M, Novak O, Strnad M, Schmülling T, 2006, *Arabidopsis* cytokinin receptor mutants reveal functions in shoot growth, leaf senescence, seed size, germination, root development, and cytokinin metabolism. *Plant Cell 18: 40–54.*

Saab IN, Sharp RE, Pritchard J, 1992, Effect of inhibition of abscisic acid accumulation on the spatial distribution of elongation in the primary root and mesocotyl of maize at low water potentials. *Plant Physiol. 99: 26–33.*

Saab IN, Sharp RE, Pritchard J, Voetberg GS, 1990, Increased endogenous abscisic acid maintains primary root growth and inhibits shoot growth of maize seedlings at low water potentials. *Plant Physiol. 93: 1329–36.*

Sacks M, Silk WK, Burman P, 1997, Effect of water stress on cortical cell division rates within the apical meristem of primary roots of maize. *Plant Physiol. 114:519–27.*

Shabala SN, Lew RR, 2002, Turgor regulation in osmotically stressed Arabidopsis epidermal root cells. Direct support for the role of inorganic ion uptake as revealed by concurrent flux and cell turgor measurements. *Plant Physiol. 129: 290–99.*

Sharp RE, 2002, Interaction with ethylene: changing views on the role of abscisic acid in root and shoot growth responses to water stress. *Plant, Cell Environ. 25: 211–222.*

Sharp RE, Davies WJ, 1985, Root growth and water uptake by maize plants in drying soil. *J. Exp. Bot. 36: 1441–56.*

Sharp RE, Hsiao TC, Silk WK, 1990, Growth of the maize primary root at low water potentials. II. Role of growth and deposition of hexose and potassium in osmotic adjustment. *Plant Physiol. 93: 1337–46.*

Sharp RE, LeNoble ME, 2002, ABA, ethylene and the control of shoot and root growth under water stress. *J. Exp. Bot. 53: 33–37.*

Sharp RE, LeNoble ME, Else MA, Thorne ET, Gherardi F, 2000, Endogenous ABA maintains shoot growth in tomato independently of effects on plant water balance: evidence for an interaction with ethylene. *J. Exp. Bot. 51: 1575–1584.*

Sharp RE, Silk WK, Hsiao TC, 1988, Growth of the maize primary root at low water potentials. I. Spatial distribution of expansive growth. *Plant Physiol. 87: 50–57.*

Sharp RE, Wu Y, Voetberg GS, Saab IN, LeNoble ME, 1994, Confirmation that abscisic acid accumulation is required for maize primary root elongation at low water potentials. *J Exp Bot 45: 743–51.*

Silk WK, 1984, Quantitative descriptions of development. *Annu. Rev. Plant Physiol. 35: 479–518.*

Spollen WG, LeNoble ME, Samuels TD, Bernstein N, Sharp RE, 2000, ABA accumulation maintains primary root elongation at low water potentials by restricting ethylene production. *Plant Physiol. 122: 967–976.*

Spollen WG, Sharp RE, 1991, Spatial distribution of turgor and root growth at low water potentials. *Plant Physiol. 96:438–43.*

Spollen WG, Sharp RE, Saab IN, Wu Y, 1993, Regulation of cell expansion in roots and shoots at low water potentials. *In, JAC Smith, H Griffiths, eds, Water Deficits. Plant Responses from the Cell to the Community, Bios Sci Publ, Oxford, pp 37–52.*

Steele KA, Price AH, Shashidhar HE, Witcombe JR, 2006, Marker-assisted selection to introgress rice QTLs controlling root traits into and Indian upland rice variety. *Theor. Appl. Genet. 112: 208–221.*

Steffens B, Wang J, Sauter M, 2006, Interactions between ethylene, gibberellin and abscisic acid regulate emergence and growth rate of adventitious roots in deepwater rice. *Planta 223: 604–612.*

Tan BC, Schwartz SH, Zeevaart JAD, McCarty DR, 1997, Genetic control of abscisic acid biosynthesis in maize. *Proc. Natl. Acad. Sci. 94: 12235–40.*

Tsuda S, Miyamoto N, Takahashi H, Ishihara K, Hirasawa T, 2003, Roots of *Pisum sativum* L. exhibit hydrotropism in response to a water potential gradient in vermiculite. *Ann. Bot. 92: 767–770.*

Ushada M, Murase H, 2006, Identification of a moss growth system using an artificial neural network model. *Biosystems Engineering 94: 179–189.*

Vamerali T, Guarise M, Ganis A, Bona S, Mosca G, 2003, Analysis of root images from auger sampling with a fast procedure: a case application to sugar beet. *Plant Soil 255: 387–397.*

van Beem J, Smith ME, Zobel RW, 1998, Estimating root mass in maize using a portable capacitance meter. *Agron. J. 90: 566–570.*

van der Weele, CM, Jiang, HS, Palaniappan, KK, Ivanov, VB, Palaniappan, K, Baskin TI, 2003, A new algorithm for computational image analysis of deformable motion at high spatial and temporal resolution applied to root growth. Roughly uniform elongation in the meristem and also, after an abrupt acceleration, in the elongation zone. . *Plant Physiol. 132: 1138–1148.*

Vandeleur R, Miemietz C, Tilbrook J, Tyerman SD, 2005, Roles of aquaporins in root responses to irrigation. *Plant Soil 274: 141–161.*

Varney GT, Canny MJ, 1993, Rates of water uptake into the mature root system of maize plants. *New Phytol. 123: 775–786.*

Verslues PE, Ober ES, Sharp RE, 1998, Oxygen relations and root growth at low water potentials. Impact of oxygen availability in polyethylene glycol solutions. *Plant Physiol. 116: 1403–1412.*

Verslues PE, Sharp RE, 1999, Proline accumulation in maize (*Zea mays* L.) primary roots at low water potentials. II. Metabolic source of increased proline deposition in the elongation zone. *Plant Physiol. 119: 1349–1360.*

Voetberg GS, Sharp RE, 1991, Growth of the maize primary root at low waterpotentials. III. Role of increased proline depositionin osmotic adjustment. *Plant Physiol 96: 1125–30.*

Voisin A-S, Reidy B, Parent B, Rolland G, Redondo E, Gerentes D, Tardieu F, Muller B, 2006, Are ABA, ethylene or their interaction involved in the response of leaf growth to soil water deficit? An analysis using naturally occurring variation or genetic transformation of ABA production in maize. *Plant, Cell and Environ. 29: 1829–1840.*

Volkmar, KM, 1997, Water stressed nodal roots of wheat: effects on leaf growth. *Aust J Plant Physiol 24: 49–56.*

Walch-Liu P, Ivanov II, Filleur S, Gan Y, Remans T, Forde BG, 2006, Nitrogen regulation of root branching. *Ann. Bot. 97: 875–881.*

Walter A, Spies H, Terjung S, Küsters R, Kirchgeßner N, Schurr U, 2002, Spatio-temporal dynamics of expansion growth in roots: automatic quantification of diurnal course and temperature response by digital image sequence processing. *J. Exp. Bot. 53: 689–698.*

Watt M, Silk WK, Passioura JB, 2006, Rates of root and organism growth, soil conditions, and temporal and spatial development of the rhizosphere. *Ann. Bot. 97: 839–855.*

Weaver, JE, 1926, Root habits of corn or maize. *Root development of field crops, McGraw Hill, New York, pp 180–191.*

Weih, M, 2003, Trade-offs in plants and the prospects for breeding using modern biotechnology. *New Phytol 158:1–9.*

Wenzel CL, McCully ME, Canny MJ, 1989, Development of water conducting capacity in the root systems of young plants of corn and some other C4grasses. *Plant Physiol. 89: 1094–1101.*

Whalley WR, Clark LJ, Gowing DJG, Cope RE, Lodge RE, Leeds-Harrison PB, 2006, Does soil strength play a role in wheat yield losses caused by soil drying? *Plant Soil 280: 279–290.*

Wojtaszek P, Anielska-Mazur A, Gabrys H, Baluska F, Volkmann D, 2005, Recruitment of myosin VIII towards plastid surfaces is root-cap specific and provides the evidence for actomyosin involvement in root osmosensing. *Funct. Plant Biol. 32: 721–736.*

Wu Y, Sharp RE, Durachko DM, Cosgrove DJ, 1996, Growth maintenance of the maize primary root at low water potentials involves increases in cell wall extension properties, expansin activity and wall susceptibility to expansins. *Plant Physiol 111: 765–772.*

Wu Y, Spollen WG, Sharp RE, Hetherington PR, Fry SC, 1994, Root growth maintenance at low water potentials. Increased activity of xyloglucan endotransglycosylase and its possible regulation by ABA. *Plant Physiol 106: 607–615.*

Wu Y, Thorne ET, Sharp RE, Cosgrove DJ, 2001, Modification of expansin transcript levels in the maize primary root at low water potentials. *Plant Physiol. 126: 1471–1479.*

Yang Z, Tian LN, Latoszek-Green M, Brown D, Wu KQ, 2005, Arabidopsis ERF4 is a transcriptional repressor capable of modulating ethylene and abscisic acid responses. *Plant Molec. Biol. 58: 585–596.*

Yim HH, Villarejo MR, 1994, Gene expression and osmoregulation in bacteria. *In, K. Strange, ed, Cellular and Molecular Physiology of Cell Volume Regulation, CRC Press, Boca Raton, Fla, pp 334–46.*

Zhu J, Chen S, Alvarez S, Asirvatham VS, Schachtman DP, Wu Y, Sharp RE, 2006, Cell wall proteome in the maize primary root elongation zone. I. Extraction and identification of water-soluble and lightly ionically bound proteins. *Plant Physiol. 140: 311–325.*

cms tall) could not exist. Soil provides most plants with a predictable supply of water (and nutrients) and some anchorage but the physical, biological and chemical properties of the rooting medium also mean that roots have had to evolve particular properties to ensure that much of the water in the soil within the potential rooting zone is made available to the plant.

Water availability from the soil becomes a particular issue if soil water is not replenished as it is used by the plant. Table 1 (modified from Robinson et al. 2003) lays out the basic design requirements for a root system faced with restricted availability of water. We focus here on desirable properties (for the point of view of sustained water uptake in drying substrate) of roots of crop plants where economic yield is an important issue. This is rather than focusing on survival of severe drought, which is an important component of drought resistance in wild plants but is largely irrelevant to the yielding of annual crops in particular. Here, yields are commonly restricted by soil moisture deficits well before the survival of the plant is at risk and therefore the mechanisms that contribute to the maintenance of yield are distinctly different from those that may contribute to plant survival of cellular desiccation. It is these mechanisms that can potentially be exploited in plant improvement programs for dryland agriculture. In the discussion that follows, we will use the framework laid out in Table 1. for an analysis of root properties that may be important for yield maintenance in situations where water supplies may be restricted.

Table 1. Summary of the design requirements of root systems of crop plants subjected to drought stress

1	Root growth and penetration of soil pores • Growth and turgor relations • Root proliferation • Root system topology • Impact of changes in root morphology and structure on the uptake of water and nutrients and yield.
2	Radial fluxes of water and ions into the root • Aquaporins • Water-proofing • Hydraulic lift • Impact of modified water and ion fluxes through roots
3	Root signals and the limitation of leaf growth and leaf functioning • Root signals and the limitation of leaf growth and leaf functioning • Abscisic acid synthesis, distribution and catabolism • Xylem sap pH • Ethylene and ACC
4	Signaling between substrate and roots • Rhizosphere micro-organisms
5	Resolution of design conflicts and behavior of roots of plants in communities

2. PENETRATION OF SOIL PORES

As water is lost from the leaves of a plant into comparatively dry air, resulting water potential gradients will pull water first from the xylem and then from the roots to replace that lost by transpiration. A reduction in root water potential will pull water into the root from the soil and cause water to move to the root through the soil, again down a gradient of decreasing water potential. As long as transpiration rates are not too high and plants are rooted in soil that is well charged with water, such movement can be comparatively rapid and plant water uptake can effectively keep pace with transpirational water loss. In conditions where transpiration rates are substantial, however, and particularly if soil water is not replenished, depletion zones of water (and nutrients) will develop around roots and in these regions the movement of water can be greatly slowed with a consequent significant restriction in the rate of uptake of water and nutrients by the plant. More drying of the bulk soil further from the root will further increase the resistance to water movement to the roots. Soil resistances to water movement are in series with radial root resistances and if the former are large (for example when the soil dries) then there is little benefit to be gained by engineering plants with low water uptake resistance per unit of root surface area (high hydraulic conductivity). Rather, water uptake can be sustained if root growth can be sustained as the substrate water potential declines, such that root tips grow into areas of soil where water contents are higher, soil resistances to water movement are consequently lower and therefore water availability is sustained.

The importance of capturing more of the water available in the soil is apparent from an analysis produced by Passioura (1977) showing crop yield (*Y*), particularly in 'water limited' crop production is a function of three variables:

(eqn 2) $Y = BWR \times W \times HI$

BWR is the biomass to water ratio, , *W* is the water available, and *HI* is the harvest index. The analysis is valuable for a variety of reasons, not least because it focuses attention on how to increase crop yield by increasing the water available to the crop, making more water available at key developmental periods so that a greater proportion of crop biomass is yield (*HI*) and how to improve the ratio of crop biomass produced to water lost (i.e increase water use efficiency). Root growth and functioning have important impacts on all of these variables.

2.1. Root Growth and Turgor Relations

Roots of most plants show reduced sensitivity to reductions in cellular water potential when compared to growth of shoot cells subjected to the same degree of dehydration, and the basis of this response if now comparatively well understood largely a result of an impressive body of work conducted by the Sharp laboratory using a model root system under growth conditions that can easily be replicated (the maize primary root system subjected to osmotic stresses imposed in a vermiculite growing medium) (see Sharp, this volume). Root growth maintenance at low substrate water potential requires turgor maintenance (e.g. Spollen and Sharp, 1991) and the capacity to loosen cell walls for irreversible extension despite the potential growth limitation imposed by a range of chemical inhibitors (e.g. LeNoble et al. 2004). The potential benefits to be obtained in terms of sustained water uptake and sustained plant growth in drying soil mean that increased understanding and modification of properties limiting root growth provides an attractive target for those interested in improving plants for dryland agriculture. The maize primary root system (Sharp, this volume) provides a wealth of information on the mechanisms behind root growth maintenance at low water potential and is already being exploited for plant improvement though functional genomics (Sharp et al. 2004). One issue with this system, however, is that as water potential is decreased in the vermiculite medium, root diameter decreases. Commonly, as soil water potential declines, roots become thicker, presumably to counteract increasing soil strength (not an issue with vermiculite). It will be important to address this as an issue in plant improvement programs as it is well known that plants can increase their potential water and nutrient uptake by producing a greater length of root from a given dry mass (i.e. by increasing SRL (Specific root length)), the common response to low water potential, while thicker roots, the common response to increasing soil strength, will be of benefit for penetration of drying soils.

In some crops, high yield under dryland conditions can be associated with deep root penetration (e.g. Mohamed et al. 2002) but as we emphasize below, this will not always be the case if increased root growth and root shoot ratio (a result of differential sensitivity of root and shoot growth to soil drying and common response to water scarcity) are achieved at the expense of economic yield, and this must be born in mind for plant improvement programs based on a modification of root properties .

2.2. Root Proliferation

In nearly all soils, water, nutrients, soil strength and other properties show considerable heterogeneity. Localised proliferation of roots is generally thought to be advantageous to plants in foraging for water and nutrients and involves the use of morphological plasticity in response to resource heterogeneity to selectively place resource-acquiring structures in favorable patches of habitat (Hutchings and John, 2003). Drew (1975) was among the first to describe root proliferation into patches of soil with high nutrient status, while Zhang and Forde (1998) have recently identified the gene in *Arabidopsis* responsible for the sensing of localized high concentrations of nitrate. Localised proliferation of parts of a root system into nutrient-rich patches may involve reduced growth of other parts of the same root in soil where nutrients are in short supply (Gersani and Sachs, 1992).

Increase in soil strength as soil dries can impact on root branching and generally more lateral roots per unit length of main root axis are found (Bengough, 2003). In some plants, however, total numbers of lateral roots may be decreased by higher mechanical impedance (Goss, 1977).

2.3. Root System Topology

The ecological literature (e.g. Kutschera (1960) shows us that the architecture of root systems is as varied as that of shoot systems. Perhaps as importantly, several authors have stressed the flexibility of architecture in response to changes in the local environment. Two extreme architectures that have been described (see e.g. Robinson et al. 2003) are the 'Herringbone' systems with a main axis and one or few developmental orders of laterals, and those systems with a 'dichotomous' architecture. Fitter et al. (1991) predicted that herringbone systems though more expensive to construct are more efficient at exploiting soil for water and nutrients. Importantly for crop improvement programs, Fitter (1987) showed that root systems of *Trifolium praetense* tended towards dichotomy when water was in ample supply but became more herringbone in structure as soil dried.

2.4. Impact of Changes in Root Morphology and Structure on the Uptake of Water and Nutrients and Yield

Local proliferation of roots can be shown to be advantageous in terms of growth of plants in soils with heterogeneous nutrient distributions. For example, Drew and Saker (1975) have shown that barley plants with only a few percent of roots in nutrient rich soil can achieve similar whole plant growth rates to plants with all of their roots exposed to high nutrient concentrations. Some of this apparent increase in uptake of nutrients from localized patches may be due to physiological adaptation of existing roots as well as to changes in root growth patterns (see below). Sharp and Davies (1979) have shown similarly that deeper rooting by only a few roots in maize plants can maintain substantial water uptake and vegetative growth as soil dries. Deeper rooting and sustained root growth late in the season in stay-green varieties can have beneficial effects on yield since water is then available during critical and sensitive phases of reproductive development (Borrell et al. 2001). The positive impact of this kind of response can be seen in equation 1 through an effect of extra water available on the *HI* component of yield.

It seems therefore that deeper rooting of individual plants or plants in competition for water with plants of other species can be beneficial in terms of extra water harvesting, particularly at critical stages of plant development. If this response can make what may be a relatively sustainable new source of water in the subsoil available to the plant then the effects of extra root production may be very positive. Proliferation in more superficial layers may increase water availability only rather temporarily, that is unless soil water is replenished. Such responses may be less obviously beneficial, particularly for plants in monoculture where extra investment in roots may yield little extra water for a crop of plants competing against themselves (see Passioura, 1977). Hutchings and John (2003) note that for mobile nutrients such as nitrogen there often seems to be little benefit to proliferation, unless plants are competing for nutrients. Under these circumstances proliferators can recover more N (or water) than competitors (Robinson et al. 1999) but the costs of this behaviour can be high and might outweigh benefits, particularly in a mono-culture. Once produced, roots have to be maintained and the allocation of extra carbohydrate to root systems, particularly during periods of grain filling can have adverse effects on the grain yield to water used ratio (equation 1 above).

One other consideration in analysis of benefits of extra resource investment in roots is the placement of roots in the soil. In compacted soil or in soil where mechanical impedance increases due to drying, roots will clump in channels or fissures. Clumping can also be produced by the production of many short laterals (see above) (Tardieu, 1988). Such clumping can enormously restrict the scavenging capacity of root systems for water and nutrients (Passioura, 1991).

There is now a considerable body of work which suggests that rather than engineering root properties to increase scavenging capacity for water in the soil, in environments where yielding is dependent upon stored water, there is some benefit to be taken from breeding crops with narrow xylem vessels which should increase the resistance to water flux and force plants to use water in the subsoil more slowly (Passioura 1972). In cereals, seminal roots grow deep into the subsoil and because crops in dryland environments can rely largely on subsoil water, seminal roots can be of crucial importance in determining water use patterns. If plants use the subsoil water too rapidly during the development of the vegetative plant, then too little will remain for the crucial period of development when grain is filling and HI will be reduced even if biomass production is substantial (see Equation 1 above). Use of subsoil water can be slowed if seminal roots have higher hydraulic resistances.

Wheat breeding in Australia has reduced the xylem vessel diameter of two commercial wheat varieties from 65μm to less than 55μm (Richards and Passioura, 1981a, 1981b). Selections with narrow vessels yielded 8% more than the controls in the driest environments, while yield differences in the wetter environments were largely non significant (Richards and Passioura, 1989).

Taken together these effects of structural and morphological variation of root systems show that an apparently simple target of 'keeping roots growing as soil dries' can have many and varied consequences, some of which will impact adversely on yielding. It is likely that selection for particular root traits will be beneficial in some soils and some environments with particular rainfall patterns, but not in others and that design requirements will vary from cropping region to cropping region.

3. RADIAL FLUXES OF WATER AND IONS INTO THE ROOT

Generally, the radial resistance to water movement into plants is much greater than the resistance to axial flow (in young maize roots 2 or 3X while in older roots the difference may be several hundred fold, Tyree, 2003). The resistances of the various components of the pathway have been debated over the years and there is still some controversy over their identity and magnitude. Steudle at al. (1993) have concluded that while the endodermis may be the major barrier to solute flow, this is not the case for water flow but this situation may vary with root age. In some plants (e.g. maize), apoplastic bypass to radial water flux can be important (Freundl et al. 1998) while in others (e.g. sunflower), the apoplastic pathway can be blocked due to lignification or suberisation. Steudle and Peterson (1998) have recently described a new model which helps our understanding of radial water flux. It is well known that root radial resistance to water uptake is apparently sensitive to the flux of water into the root with apparent resistance declining as the transpiration flux increases. Steudle and Peterson argue that this may be because of a change in the proportion of total water flux moving though different pathways into the root but their may also be changes in membrane properties to water flow (see Tyree, 2003).

3.1. Aquaporins

Channels in the membrane, analagous to those that are important for ion flux can influence the radial flow of water into roots. These are commonly referred to as aquaporins and Siefitz et al (2002) have demonstrated that these pores account for about half of the root conductance in tobacco. Activity of channels is under metabolic control (Tyerman et al. 2002; Maurel and Chrispeels, 2001). Steudle (2000) has suggested that this pathway can dominate water flux when movement is driven largely by osmotic gradients or when the apoplastic pathway becomes blocked, which can occur in response to some soil conditions. A variety of factors of soil and root factors will affect aquaporin activity, including pH, pCa and osmotic gradients. Clarkson et al. (2000) have shown how increased nitrogen availability increases the hydraulic conductivity of roots (Clarkson et al. 2000) and their may also be diurnal control of root hydraulic properties (see also Tsuda and Tyree, 2000).

3.2. Water Proofing

The role of stomata and cuticular development in the regulation of water loss is well known. However effective these mechanisms, they are of limited value if plants lose substantial quantities of water to the soil. Most species will show reduced root hydraulic conductivity as the soil dries and this change will restrict water loss from roots (e.g. Nobel and North, 1993; Nobel and Sanderson. 1984). As the soil dries, a vapour gap will develop between the root and the soil and this will also limit water loss. It appears that both of these changes are to some extent reversible when soil is rewetted and this property is referred to as rectifier-like activity. In some plants however water-proofed roots will not change their properties and increased water available in the soil can only be fully accessed if new roots are produced. Rectifier-like root properties may be of considerable significance for plants that grow in shallow soil which is prone to rapid and substantial variation in water status. Recent work suggests that aquaporins may help regulate water loss from very dry roots

3.3. Hydraulic Lift

In water-limited environments, one important feature of the survival of some plants is deep rooting (Canadell et al., 1996) with some woody plants rooting down to 10 metres or more. Under these conditions, hydraulic lift can be commonly observed. This is a passive mechanism where the water potential gradient transfers water through the root system, from deep wetter soil to shallower soil (Richards and Caldwell, 1987). With water moving in and out of roots on a daily basis, clearly this mechanism is not compatible with those mechanisms that contribute to water proofing of roots. During the night when transpiration rates are generally low, the mechanism can provide quite a lot of temporary stored water to the upper soil layers (more than the plant itself can store) During the following day, the roots of plants performing hydraulic lift (as well as of any neighbouring plants with shallower root systems) will extract this water from the soil and it can substantially increase plant transpiration in the following day and also contribute to enhanced carbon gain (Caldwell et al., 1998). A recent study by Kurz-Besson et al. (2006) shows that for cork oak trees in Portugal, hydraulic lift may provide between 17 and 81% of the water transpired.

3.4. Impact of Modified Water and Ion Fluxes Through Roots

The phenomenon of hydraulic lift described above is an excellent example of getting more water through the plant as a way of enhancing yield (see equation 1) when water is in short supply. Targets for a plant improvement programme might therefore include deeper rooting characteristics combined with shallow roots that do not show water proofing capacity.

Roots with high radial hydraulic conductivity can be beneficial for high biomass production where there is a lot of water available to the plant, or where water is regularly replenished by rainfall or by irrigation. Such plants may, however, have a tendency to use water too rapidly in water scarce situations, at least if water is required for reproductive development later in the season.

4. LONG-DISTANCE TRANSPORT BETWEEN ROOTS AND SHOOTS- ROOT SIGNALS

While the role of roots in scavenging for water and minerals in the soil is readily apparent and the contribution of water and ions as substrates for a variety of plant processes is well-discussed, roots also contribute other material to shoots in the form of signals, and the role of these signals in modifying plant growth and development is no less significant than that of the other root-sourced substrates. Roots signals have information content, for example allowing the plant to modify growth and functioning as a function of water and nutrient availability in the soil or soil mechanical impedance (see for example, Davies et al. 2002). Most importantly, canopy development and stomatal behavior can be restricted by root signaling, often in circumstances where the water relations of the shoots are not obviously changed by any modification of soil properties. Effects can be dramatic and over-coming or in certain circumstances enhancing the effects of root signals can be an appropriate target for a plant improvement program. Perhaps the most obvious target in this regard is a manipulation to allow sustained or even increased rates of canopy development as the soil dries. This can have a number of potential benefits. Firstly, when water is still in relatively abundant supply, the extreme sensitivity of root signaling (see e.g. Davies and Gowing, 2001) can limit leaf development, the consequent interception of solar radiation and the production of biomass. Suppressing root signaling that limits leaf growth or reducing the sensitivity of leaf growth to root signals can therefore allow the grower to produce more biomass in relatively moist soils. Another benefit can be achieved by intervening in the same way to allow the young crop to cover the soil more rapidly (Passioura, 2004). In Mediterranean-type climates where crops will largely grow on stored water derived from rains during the previous autumn, the soil surface can be wet in spring. Direct evaporation from the soil can be substantial (e.g. Leuning et al. 1994) and this loss of water will therefore be relatively unproductive (this water loss from the plant would generate extra carbon fixation and biomass production). van Herwaarden and Passioura (2001) have shown clearly how faster coverage of the soil by crops in these environments in the spring can greatly reduce seasonal evaporation from the soil and therefore increase water use productivity and impact positively on yield (equation 1).

Agriculture already uses an unsustainable 70% of the world's water supplies (Bacon, 2004). Reducing the use of water in agriculture can be achieved in a variety of ways but the use of deficit irrigation irrigation (DI) (the application of only a predetermined percentage of calculated potential plant water use) is an attractive means of saving water. Ideally the application of DI must be achieved without substantial yield penalty otherwise the yield/water use ratio (water use efficiency) will not be increased. We have already noted above how even mild soil drying will limit plant growth and development and so if plants are to be kept growing under a reduced supply of water then a plant improvement programme to suppress root signaling or the responses to root signaling will be needed.

4.1. Root Signals and the Limitation of Leaf Growth and Leaf Functioning

When the soil around the roots dries, dehydration of the root cortex will act to generate a number of chemical signals that will impact on plant growth and functioning. Extra synthesis of a number of growth regulators can positively inhibit leaf growth, while restricted synthesis of other regulators can act as a negative signal, with the lack of a promoter of leaf growth also restricting canopy development. The plant hormones abscisic acid (ABA) and ethylene will act in the first of the above categories, while reduction in the supply of cytokinins and in some cases ABA can act to restrict growth. Other plant growth regulators will also act as root signals, as will inorganic ions (see e.g. Roberts and Snowman,) and changes in pH of the xylem sap (Wilkinson and Davies, 2002). Not all hormonal root signals are synthesized in the root. Reductions in root turgor can act to more rapidly re-circulate hormones arriving in the phloem from the shoots and some hormonal signals may originate in the soil (see below). Some root-originated signals can act directly on the shoots and others act as part of a transduction chain to release or target shoot-sourced effectors.

4.2. Abscisic Acid Synthesis, Distribution and Catabolism

Of all of the so-called plant hormones, abscisic acid has received most attention as a compound mediating the effects of soil drying on plant growth and functioning. There is often a clear relationship between declining soil water availability and ABA content of the roots or ABA concentration in the xylem (Tardieu et al 1992). The extreme sensitivity of the stomatal response to ABA means that stomatal behavior can often be linked sensitively to changes in soil water availability (e.g. Zhang and Davies, 1990). One of the results of this is that as soil dries, sensitive responses of stomata can act to maintain shoot water status at a level comparable to that of the well watered plant. This turgor maintenance (or isohydric behaviour) can be important for plant development in drying soil.

ABA can act as a sensitive inhibitor of growth of shoots in drying soil (Bacon et al. 1998), but more recent work has suggested that this response is sensitive to the water status of the shoot, with ABA acting as an inhibitor of growth in plant parts where turgor is sustained but as turgor declines, this hormone is required to sustain some growth of both roots and shoots (see Sharp et al. 2004). This may be because ABA can suppress the run-away synthesis of ethylene, which itself acts as a growth inhibitor at low water potentials. The idea that the impact of a hormone can be either promotive or restrictive for growth, depending on the level of another variable is an intriguing one and suggests that manipulation of hormone action in plants can be a achieved by a variety of means other than the manipulation of hormone synthesis itself.

The root-sourced ABA signal can improve instantaneous water use efficiency (A/E) and in the longer term can modify a range of developmental variables which may be of adaptive significance under drought (see Trewavas and Jones, 1989). While quite subtle increases in ABA delivery to sites of action in the shoot can act to regulate gas exchange and growth (Jia and Davies, 2007), more substantial increases in hormone synthesis may be required to modify gene expression to affect development. ABA accumulation in developing reproductive structures can have deleterious effects on flowering and fruiting (Morgan, 1980) and there may be some advantage to be gained from manipulating plants to avoid such accumulations. We have shown recently (Jia et al. 2007) that ABA catabolism can be much more rapid than had previously been shown to be the case and there may be a case for targeting catabolism of this hormone in programs designed to increase yield under drought.

4.3. Xylem Sap pH

Wilkinson and Davies (1997) have shown that soil drying can act to alkalinize the xylem sap of some plants, and more recently Jia and Davies (2007) have shown that as had previously been hypothesized, these changes in the pH of sap delivered to leaves from roots in drying soil, are translated into changes in the pH of the apoplast of the leaves of these plants. Because ABA is a weak acid, the dissociated form arriving in leaves will partition according to pH gradients, tending to move to alkaline compartments. In the well-watered plant, these are the symplast of the leaf cells and the phloem. Alkalinisation of the apoplast as a result of soil drying (and other environmental changes – Jia and Davies, 2007) will result in more ABA residing for longer in the apoplast and therefore penetrating to the sites of action on the guard cells (which have only an apoplastic connection with the other cells in the leaf). Such changes in pH therefore have the effect of increasing the apparent stomatal response to a given delivery of ABA (i.e increasing the apparent sensitivity of stomata to the ABA signal). In many circumstances, an increase in xylem sap pH and an increase in ABA delivery occur together as an effect of soil drying and combine to generate a sensitive response to the change in soil conditions. The pH signal can be one of the most sensitive of all signals to a change in water availability in the soil (see e.g. Sobeih et al. 2004) and can occur without any extra ABA synthesis, purely by making more existing ABA available to sites of action in leaves. Changes in xylem sap and apoplastic pH are attributable to a range of changes in root, stem and leaf functioning (see e.g. Wilkinson and Davies, 2002) and some of these variables may be quite amenable, if not obvious, targets for the manipulation of stomatal behavior and water use efficiency.

4.4. Ethylene and ACC

Soil drying will increase concentrations of the ethylene precursor ACC (1-aminocyclopropane-carboxylic acid) both in the root and in the xylem (Gomez-Cadenas *et al.* 1996). Delivery of ACC to the shoot from the root system can account for shoot ethylene production (Else and Jackson 1998) and therefore can limit leaf growth under drought. The plant hormone ethylene can be involved in both the suppression of root growth during soil drying and the suppression of leaf growth via long-distance chemical signaling (Sharp et al. 2001). Drying of the soil around the roots of tomato plants can maintain leaf water potential at values equivalent to well-watered plants for up to 2 weeks (Sobeih *et al.* 2004), largely a function of partial stomatal closure following ABA/pH long distance signaling from roots in drying soil. Ethylene evolution of wild-type plants increases as soil dries but can be suppressed using transgenic ($ACO1_{AS}$) plants containing an antisense gene for one isoenzyme of ACC oxidase. Most importantly, $ACO1_{AS}$ plants show no inhibition of leaf growth when soil dries, even though both $ACO1_{AS}$ and WT plants show similar changes in other putative chemical inhibitors of leaf expansion (xylem sap pH and

ABA concentration). It seems likely that the enhanced ethylene evolution under PRD is responsible for leaf growth inhibition of WT plants. $ACO1_{AS}$ plants showed no leaf growth inhibition over a range of soil water contents which significantly restricted growth of WT plants.

5. SIGNALLING BETWEEN THE SUBSTRATE AND THE ROOT

We have described above a range of long-distance signaling pathways that may be manipulated to modify plant growth and functioning in drying soil. The emphasis has been upon changes in soil water availability impacting on uptake of inorganic ions from soil and the subsequent transport of these to the shoots through the xylem stream or on the impact of variation in root water status on the production and/or transport of hormonal signals. Of course the root will impact on the availability of inorganic ions for uptake, with one of the best examples of this is the acidification of the rhizosphere by roots which can increase the availability of ions. Exudation of organic acids (OAs) and phytases into the rhizosphere have been shown to greatly increase the availability of inorganic phosphate in soils, where the unavailability of this ion can often be greatly inhibiting to plant growth. In fact, exudation of malate and citrate from roots is thought to be the principle mechanism in alleviating Pi deficiency stress in plants (e.g. Ryan et al., 2001)). Secreted OAs mobilise bound and precipitated forms of Pi by anion exchange and may also enhance the activity of extracellular acid phytases which hydrolyse organic P in the rhizosphere. Transmembrane transporters probably exert primary control over OA secretion from higher plant roots (Ryan et al., 2001), although there is little information in crop systems which relates the presence of anion channels directly to Pi-induced OA efflux from roots. It seems likely that variation in OA efflux will impact on ionic signaling between roots and shoots via its impact on ion availability to the root surface.

Hormone fluxes from roots to shoots are comprised mainly of plant-sourced hormones but significant concentration of hormones can be found in the soil (e.g. Hartung et al., 1996). Presumably some of this hormone will originate from the roots while some may arise as a result of microbial activity in the rhizosphere. In addition to this some soil bacteria contain enzymes that will metabolise hormones as a carbon and nitrogen source. This is important, as Slovik has shown that low concentrations of ABA in the soil are important to sustain ABA accumulations in plants and to maintain root to shoot signaling in response to soil drying. ACC and ABA accumulated in the soil solution from whatever source can also be a source of signal for xylem transport as well as impacting on the signalling process through an equilibration effect on transport. As water moves into the root system along water potential gradients, some ACC and ABA molecules will be dragged along and these can be transferred into the xylem. The efficiency of radial ACC and ABA transport across the root is likely to vary between the different genotypes.

5.1. Rhizosphere Bacteria

Although some bacteria (containing ACC deaminase) can utilise ACC as a carbon and nitrogen source, bacterial ACC synthesis does not occur. Thus rhizobacteria utilising ACC must rely on efflux of ACC from plant roots or from fragments of plant material in the soil. This efflux may be considerable, as the soil solution of even well-watered plants contains appreciable amounts of ACC (0.23 μM - Else *et al.* 1995). Although no direct comparisons of ACC and ABA efflux have been made, plant roots appear to be more "leaky" with respect to ACC since the concentration of ACC in the soil solution (0.23 μM) of well-watered plants is 3 orders of magnitude greater than the concentration of ABA (0.67 nM). Edaphic conditions that stimulate root ACC synthesis such as soil drying and flooding are likely to increase root ACC efflux and soil ACC concentrations (by increasing root ACC concentrations and also increasing production in other plant material incorporated in the soil). Interestingly, rhizobacteria can decrease root ACC concentrations (Penrose *et al.* 2001) presumably by stimulating ACC efflux and there are now a few data indicating that rhizobacterial treatments can sustain growth in drying soil, presumably by reducing the accumulation of ethylene (Dodd et al. 2006). Alkaline soils stimulate efflux of weak acids (such as ABA and ACC) from roots according to the to the anion trap concept (Degenhardt *et al.* 2000) and modifying soil pH may also be a means of reducing the sensitivity of shoot growth and development to soil drying..

There is now good evidence that some soil bacteria will synthesise cytokinins (Arkipova et al. 2005) and there is interest in determining whether addition of these micro-organisms to the soil might prevent or slow the decline in cytokinin production by droughted plants and thereby act to maintain plant growth at low soil water potentials.

6. RESOLUTION OF DESIGN CONFLICTS AND BEHAVIOUR OF ROOTS OF PLANTS IN COMMUNITIES

We have argued above that selection for particular root traits in plant improvement programmes will be beneficial in some soils and some environments with particular rainfall patterns and at particular times in the development of the crop, but not in others. In other words, design requirements for root systems with respect to yield will vary from crop to crop, depending on the nature of the economic yield and whether the crop is determinate or indeterminate, from cropping region to cropping region where rainfall patterns differ and with developmental stage where relative impact of drought on vegetative development and reproductive development will vary. This should not be surprising because the same kinds of considerations are also important in selection for shoot traits that might impact positively on yielding in drought-prone environments (e.g. Condon et al. 2002).

Figure 1 shows a proposed ideotype for a very specific combination of crop and drought (grain crop growing largely on stored water), with inter-relationships shown between putative signalling capacities and vegetative and reproductive development.

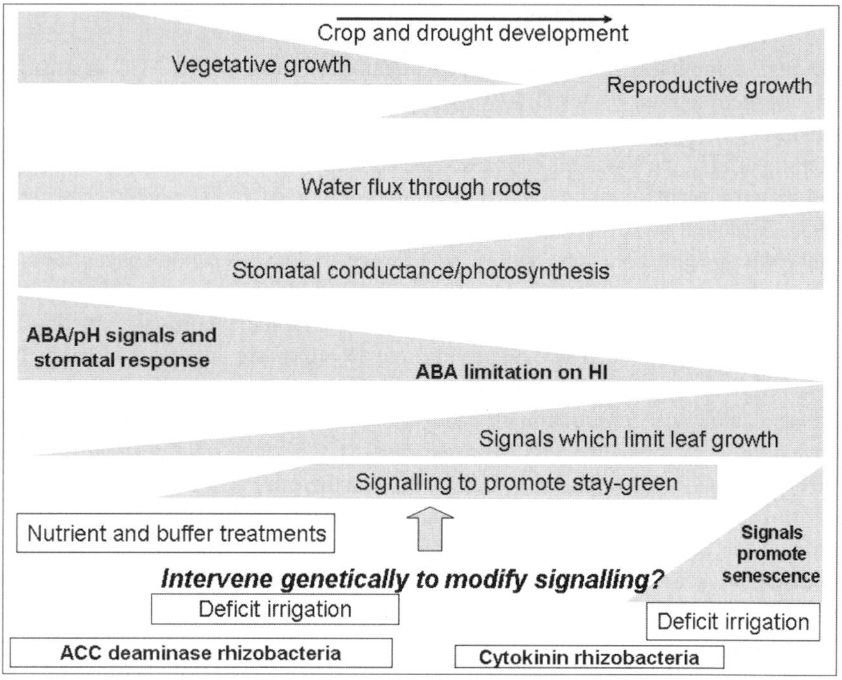

Figure 1. Proposed ideotype for a grain crop growing largely on stored water with inter-relationships shown between signaling capacities and vegetative and reproductive capacities

We also suggest application of particular management techniques and timing of possible genetic intervention (the use of a particular genotype with capacity for modified signaling capacity and response through the use of inducible promoters?). These interventions are proposed to modify signaling pathways and to enhance yield and efficiency of water use in dryland environments. The impacts on yield of the inter-relationships proposed in the diagram can be understood with reference to Figure 1.

It should be clear from Equation 1 that whatever the water availability throughout the growing season, there must be enough water available in the soil for the production and maturation of reproductive plant parts. In crops that produce grain yield towards the end of the season and are growing mainly on stored water, this can mean a requirement for judicious water use earlier in the season. This can be brought about by restricted root growth and/or restricted hydraulic conductivity of the root system. Water use can also be regulated by a sensitive system detecting soil drying, with this information passing to the shoots for effective control of water loss through stomatal regulation. Clearly the evolution of such a system (Cowan, 1988) allows water use (and plant development) to be linked to water availability with a fail-safe system operating to minimize the chances of damaging water deficits during reproductive development or even more catastrophic hydraulic failure at any point in the development of the crop (Sperry et al., 2002).

Root proliferation and effective scavenging for water can help to ensure that more water is available for reproductive development late in the season and these properties in combination with hydraulic lift can result in increased productivity simply by ensuring that more water goes through the plant (Equation 1). Restricted root signaling or low sensitivity of shoot growth to signaling early in the season will help plants cover the ground more rapidly, reduce water loss from the soil and thereby enable more water to move through the plant, thereby increasing carbon gain and water use efficiency. Restricted canopy growth once canopy closure has been reached (increased signaling?) can make more assimilates available for reproductive development, assuming that stomata are still partly open and photosynthesis is continuing (reduced stomatal signaling or reduced sensitivity of stomata to signals?) Root signals that induce stay-green characteristics and perhaps subsequently promote redistribution of assimilate and nitrogen to developing reproductive structures (Yang et al. 2001) can enhance harvest index and yield.

REFERENCES

Arkipova, T.N., Veselov, S.U., Melentiev, A.I., and Kudoyarova, G.R., 2005, *Bacillus subtilis* to produce cytokinins and to influence growth and endogenous hormone content of lettuce plants. *Plant and Soil* **272**: 201–209

Bacon, M.A., 2004, Water use efficiency in plant biology. In, *Water Use Efficiency in Plant Biology*. M.A. Bacon, ed., Blackwell, Oxford, pp 1–26.

Bengough, A.G., 2003, Root growth and function in relation to soil structure composition and strength. In, Root Ecology. H. de Kroon and EJW Visser, eds., Springer, pp 151–171.

Borrell, A., Hammer, G., and van Oosterom, E., 2001, Stay green: a consequence of the balance between supply and demand for nitrogen during grain filling. *Ann. Appl.Biol.* **138**: 91–95.

Caldwell, M. M., Dawson, T. E., and Richards, J. H., 1998, Hydraulic lift: consequences of water efflux from the roots of plants. *Oecologia* **113**: 151–161.

Canadell, J., Jackson, R. B., Ehleringer, J. R., Mooney, H. A., Sala, O. E., and Schulze, E- D., 1996, Maximum rooting depth of vegetation types at the global scale. *Oecologia* **108**: 13–18.

Clarkson, D.T., Carvajal, M., Henzler, T., Waterhouse, R.N., Smyth, A.M., Cooke, D.T., and Steudle, E., 2000, Root hydraulic conductance: diurnal aquaporin expression and the effects of nutrient stress. *J. Exp.Bot.* **51:** 61–70.

Condon, A.G., Richards, R.A., Rebetzke, G.J., and Farquhar, G.D. (2004) Breeding for high water use efficiency *J. Exp. Bot.***55**: 2447–2460.

Cowan, I.R., and Farquhar, G.D., 1977, Stomatal function in relation to leaf metabolism and environment. *Symp. Soc. Exp. Biol.* **31**: 471–505.

Davies, W.J., and Gowing, D.J.G., 1999, Plant responses to small perturbations in soil water status. In, *Physiological Plant Ecology* . M.C. Press et al., eds., Blackwell, Oxford, pp 67–90.

Davies WJ Wilkinson S., and Loveys BR, 2002, Stomatal control by chemical signalling and the exploitation of this mechanism to increase water use efficiency in agriculture. *New Phytol.* **153**: 449–460

Degenhardt, B., Gimmler, H., Hose, E., and Hartung, W., 2000, *Plant and Soil* **225**: 83–94.

Drew, M.C., 1975, Comparison of the effects of a localized supply of phosphate, nitrate, ammonium and potassium on the growth of the seminal root system ofand the shoot system of barley. *New Phytol.* **75**: 479–490.

Drew, M.C., and Saker, L.R., 1975, Nutrient supply and the growth of the seminal root system in barley. II Localised, compensatory increases in lateral root growth and rates of nitrate uptake when nitrate supply is restricted to only part of the root system. *J. Exp.Bot.* **26:** 79–90.

Fitter, A.H., 1987, An architectural approach to the comparative ecology of plant root systems. *New Phytol.* **106** (Suppl.): 61–77.

Fitter, A.H., Stickland, T.R., Harvey, M.L., and Wilson, G.W., 1991, Architectural analysis of plant root systems. I. Architectural correlates of exploitation efficiency. *New Phytol.* **118**: 375–382.

Freundl, E., Steudle, E., and Hartung, W., 1998, Water uptake by roots of maize and sunflower affects the radial transport of abscisic acid and the ABA concentration in the xylem. *Planta* **209**: 8–19.

Gersani, M., and Sachs, T., 1992, Developmental correlations between roots in heterogeneous environments. *Plant CellEnviron.*, **15**: 463–469.

Goss, M.J., 1977, Effects of mechanical impedance on root growth in barley. 1. Effects on the elongation and branching of seminal root axes. *J. Exp. Bot.*,**28**: 96–111.

Hartung, W., Sauter, A., Turner, N.C., Fillery, I.., and Heilmeier, H., 1996, Abscisic acid in soils: What is its function and which factors and mechanisms influence its concentration? *Plant and Soil* **184**:105–110.

Hutchings, M.J. and John, E.A., 2003, Distribution of roots in soil and root foraging activity. In, *Root Ecology*.H. de Kroon and EJW Visser, eds., Springer, Heidelberg, pp 33–60.

Jia, W. and Davies, W.J., 2007, Modification of leaf apoplastic pH in relation to stomatal sensitivity to root-sourced ABA signals. *Plant Physiol.*, in the press

Jia, W., Fan, Y., Ren, H., Davies, W.J., and Zhang, J., 2007, Dynamic analysis of ABA accumulation in relation to the rate of ABA catabolism in maize tissues under water deficit. *J. Exp. Bot.*, in the press.

Kutschera, L., 1960, Wurzelatlas mitteleuropaischer Ackerunkrauter und Kulturpflanzen. DLG Verlag, Frankfurt.

Kurz-Besson, C., Otieno, D., Lobo do Vale, R., Siegwolf, R., Schmidt, M., Herd, A., Nogueira, C., Soares David, T., Soares David, J., Tenhunen, J., Santos Pereira, J., and Chaves, M.M., 2006, Hydraulic lift in cork oak trees in a savannah-type Mediterranean ecosystem and its contribution to the local water balance. *Plant and Soil* **282**: 361–378.

LeNoble, M.E., Spollen, W.G., and Sharp, R.E., 2004, Maintenance of shoot growth by ABA: genetic assessment of the role of ethylene suppression. J. *Exp. Bot.*,**55**,:237–245.

Leuning, R., Condon, A.G., Dunin, F.X., Zegelin, S., and Denmead, O.T., 1994, Rainfall interception and evaporation from soil below a wheat canopy. *Agric. Forest Meteorol.*, **67**: 221–238.

Martre, P., North, G.B., and Nobel, P.S., 2001, Hydraulic conductance and mercury-sensitive water transport for roots of *Opuntia acanthocarpa* in relation to soil drying and rewetting.*Plant Physiol.*, **126**: 352–362.

Maurel, C., and Chrispeels, M.J., 2001, A molecular entry into plant water relations. *Plant Physiol.*, **125**: 135–138.

Mohamed, M.F., Keutgen, N., Tawfik, A.A., and Noga, G., 2002, Dehydration avoidance responses of tepary bean lines differing in drought resistance. *J. Plant Physiol.* **159**: 31–38.

Morgan, J.M., 1980, Possible role of abscisic acid in reducing seed set in water stressed wheat plants. *Nature* **289**: 655–657.

Nobel, P.S., and North, G. B., 1993, Rectifier-like behaviour of root-soil systems: new insights from desert succulents. In, *Water Deficits*, J.A.C. Smith & H Griffiths, eds., Bios Scientific, Oxford. pp 163–176

Nobel, P. S., and Sanderson, 1984, Rectifier like activities of roots of two desert succulents. J. Exp. Bot., **35**: 727–737.

Passioura, J.B., 1972, The effect of root geometry on the yield of wheat growing on stored water. *Aust. J. Agric. Res.* **23**: 745–752.

Passioura J B., 1977, Grain yield, harvest index, and water use of wheat. *Journal of the Aust.Inst. Agric. Sci.*, **43**: 117–121.

Passioura, J.B., 1981, Water collection by roots. In *Drought Resistance*, Aspinall, D. and Paleg, L. G., eds., Academic Press, New York, pp

Passioura, J.B., 1991, Soil structure and plant growth. Aust. J. Soil Res., **29**: 717–728.

Passioura, J.B., 2004, Water use efficiency in the farmers' fields. In, *Water Use Efficiency in Plant Biology*. M.A. Bacon, ed., Blackwell, Oxford. pp 302–321.

Penrose, D.M., Moffatt, B.A., and Glick, B.R., 2001, *Can. J. Microbiol.*, **47**: 77–80

Richards, J. M., and Caldwell, M. M., 1987, Hydraulic lift: substantial nocturnal water transport between layers by *Artemisia tridentada* roots. *Oecologia* **73**: 486–489.

Richards, R.A., and Passioura, J.B.,1981a, Seminal root morphology and water use of wheat I: Environmental effects. Crop *Sci.*, **21**: 249–252.

Richards, R.A., and Passioura, J.B., 1981b, Seminal root morphology and water use of wheat II: Genetic variation. *Crop Sci.* **21**: 253–255.

Richards, R.A., and Passioura, J.B., 1989, A breeding program to reduce the diameter of the major xylem vessel in the seminal roots of wheat and its effect on grain yield in rain-fed environments. *Aust. J. Agric. Res.* **40**: 943–950.

Robinson, D., Hodge, A. Griffith, B.S., and Fitter, A.H., 1999, Plant root proliferation in nitrogen rich patches confers competitive advantage. *Proc. Roy. Soc. Lond.* B. **266**: 431–435

Robinson, D., Hodge, A., and Fitter, A., 2003, Constraints on the form and function of root systems. In, *Root Ecology*. H. de Kroon and EJW Visser, esd.,. Springer, Heidelberg, pp 1–31.

Ryan, P.R., Delhaize, E., and Jones, D.R., 2001,. Function and mechanism of organic ion exudation from plant roots *Ann. Rev. Plant Physiol. Plant Mol. Biol.* **52**: 527–560.

Sharp, R.E., and Davies, W.J., 1979, Solute regulation and growth by roots and shoots of water-stressed maize plants. *Planta* **147**: 43–49.

Sharp, R.E., Poroyko, V., Hejlek, L.G., Spollen, W.G., Springer, G.K., Bohnert, H.J., and Nguyen, H., 2004a, Root Growth Maintenance during Water Deficits: Physiology to Functional Genomics. *J. Exp. Bot.* **55**: 2343–2352.

Sharp, R.E., Bonhert, H., and Nguyen, H., 2004, Root growth maintenance during water deficits: physiology to functional genomics *J. Exp. Bot.*,**55**: 2343–2351.

Siefritz, F., Tyree, M.T., Lovisolo, C., Schubert, A., and Kaldenhoff, R,. 2002, PIP1 plasma membrane aquaporins in tobacco: from cellular effects to function in plants. *Plant Cell* **14**: 869–876.

Sobeih, W., Dodd, I.C., Bacon, M.A., Grierson, D.C., and Davies, W.J., 2004, Long-distance signals regulating stomatal conductance and leaf growth in tomato (*Lycopersicon esculentum*) plants subjected to partial rootzone drying. *J. Exp.Bot.***55**: 2353–2364.

Sperry, J.S., Hacke, U.G., Oren, R., and Comstock, J.P., 2002, Water deficits and hydraulic limits to leaf water supply. *Plant Cell Env*. **25:** 251–263.

Spollen, W.G., and Sharp, R.E., 1991, Spatial distribution of turgor and root growth at low water potentials. *Plant Physiol.*, **96**: 438–443.

Steudle, E., 2000, Water uptake by roots: effects of water deficit. *J. Exp. Bot.*, **51**: 1531–1542.

Steudle, E., and Peterson, C.A., 1998, How does water get through roots? J. *Exp. Bot.***49**: 775–788.

Steudle, E., Murrmann, M., and Peterson, C.A., 1993, Transport of water and solutes across amize roots modified by puncturing the endodermis. Further evidence for the composite transport model of the root. *Plant Physiol.*, **103**: 335–349.

Tardieu, F., 1988, Analysis of the spatial variability of maize root density. II Distances between roots. *Plant and Soil* **107**: 267–272.

Tsuda, M., and Tyree, M.T., 2000, Plant hydraulic conductance measured by the high pressure flow meter in crop plants. J. Exp. Bot., **51:** 823–828.

Tyerman, S. D., Niemetz, C.M.., and Bramley, H., 2002, Plant aquaporins: multifunctional water and solute channels with expanding roles. *Plant Cell Env.* **25:** 173–194.

Tyree, M.T., 2003, Hydraulic properties of roots. In, *Root Ecology* H. de Kroon and EJW Visser., eds., Springer, Heidelberg, pp 125–150.

van Herwaarden, A.F., and Passioura, J.B., 2001, Improving estimates of water use efficiency in wheat. *Australian Grain* **11**: 3–5.

Wilkinson, S., and Davies, W.J., 1997, Xylem sap pH increase: A drought signal received at the apoplastic face of the guard cell which involves the suppression of saturable ABA uptake by the epidermal symplast. *Plant Physiol.*, **113:** 559–573

Wilkinson, S. and Davies, W.J., 2002, ABA-based chemical signalling: the co-ordination of responses to stress in plants. *Plant Cell Env.*, **25**: 195–210

Zhang, H., and Forde, B.G., 1998, An Arabidopsis MADS box gene that controls nutrient-induced changes in root architecture. *Science* **279**: 407–409.

Zhang, J., and Davies, W.J. 1990, Changes in the concentration of ABA in xylem sap as a function of changing soil water status can account for changes in leaf conductance and growth. *Plant Cell Environ.*, **13**: 271–285.

CHAPTER 4

REGULATING PLANT WATER STATUS BY STOMATAL CONTROL

LAURY CHAERLE AND DOMINIQUE VAN DER STRAETEN

Unit Plant Hormone Signalling and Bio-imaging (HSB), Department of Molecular Genetics, Ghent University, K. L. Ledeganckstraat 35, B-9000 Gent, Belgium

Abstract: The regulation of gas exchange at the leaf level is a key factor for plant survival under a fluctuating environment (Buckley, 2005). In this context, control of stomatal opening and closure is the evolutionary solution to balance water loss with CO_2 uptake and yield. A decrease in leaf/root water potential resulting from soil drought is typically accompanied by an elevated level of abscisic acid (ABA), which is well established as a stress hormone (Davies et al., 2005). ABA is a central component in drought-stress sensing leading to efficient stomatal control, thereby avoiding deleterious yield losses during stress conditions. Depending on the crop species, or its growing environment, different strategies for yield-optimization need to be chosen (Araus et al., 2002; Chaves and Oliveira, 2004). ABA effects are modulated by the levels of and sensitivity to other hormones, in an interdependent network. Unraveling the complex regulatory mechanisms of stomatal control between hormones, second messengers, ion channels and other classes of implicated proteins will lead to new insights in how to tailor plants to take maximum advantage of the available natural resources (Li et al., 2006). Possible strategies are either to trigger an earlier stress response without a negative impact on yield, or to attenuate the plant stress response so that assimilation will increase. These desired traits can be brought about by overexpressing or downregulating the expression of specific genes involved in the complex and possibly redundant signaling network of stomatal responses.

This chapter provides an overview of the mechanisms behind the changes in stomatal movements under water-limiting conditions, including hormonal regulation and developmental influences

Keywords: drought stress; screening; stomata; transpiration

1. INTRODUCTION

Aperture control of the microscopic pores at the leaf surface helps a plant to achieve growth, while avoiding dehydration (Buckley, 2005). Environmental parameters including air humidity, light intensity, temperature, air movement and concentration of atmospheric CO_2, but also endogenous hormonal and hydraulic signals regulate

M.A. Jenks et al. (eds.), Advances in Molecular Breeding Toward Drought and Salt Tolerant Crops, 73–90.

stomatal movements, and influence stomatal development and density (Hetherington and Woodward, 2003; Woodward et al., 2002). This multi-parameter control maximizes net photosynthesis, and allows the plant to effectively use the available water. Water use efficiency (WUE) is a parameter defined as the CO_2 assimilation per unit water transpired, which serves as a measure of plant yield (Condon et al., 2004). Control of stomatal aperture is a rapid adaptive response, while the effects on stomatal development (including density) are a longer-term response affecting newly emerging leaves. The stomatal control mechanism can however not be seen separately from the control of water transport at the root/soil interface or in the vascular system (Buckley, 2005; Davies et al., 2005; Jones, 1998). Therefore, despite the fact that WUE is most often determined at the leaf level, only whole crop WUE provides a correct picture of the whole plant system (Chaerle et al., 2005; Condon et al., 2004). An overview of the mechanisms exploited by plants to control stomatal aperture will be given, and the methods available to reveal these responses will be discussed.

2. STOMATAL CONTROL MECHANISMS

Drought stress is the major cause of stomatal closure. As outlined in the introduction, maintaining adequate photosynthesis and thus avoiding yield loss under adverse conditions is the primary route towards crop improvement. Multiple players in the complex regulatory system of leaf gas exchange have been identified and will be discussed in this section. An overview is given in Figure 1. As mentioned above, ABA has a central role in drought responses (Li et al., 2006); however, at least 4 independent drought signaling pathways exist, two of which are ABA-independent (Riera et al., 2005; Valliyodan and Nguyen, 2006). Stomatal perception of ABA induces a sequence of events initiated by a cytosolic pH increase, and followed by the accumulation of reactive oxygen species (ROS), nitric oxide (NO) synthesis, increase in the concentration of cytosolic calcium ions, synthesis of lipid-derived second messengers, activation of protein kinases and phosphatases, and finally, modulation of ion channel activity (both at the vacuolar and plasma membrane) (Garcia-Mata and Lamattina, 2003; Himmelbach et al., 2003). In addition, ABA induces a variety of transcription factors (TFs) that regulate the expression of stress-related genes; however, ABA-independent induction of TFs is an equally crucial component of stress tolerance (Riera et al., 2005). Stability and processing of mRNAs of ABA-responsive genes (Lee et al., 2006; Zhang et al., 2006) represents another level of regulation of stomatal opening (Riera et al., 2006; Verslues et al., 2006), but this is covered in another chapter of this book.

It is important to note that not all plant species follow the same strategy of stomatal control. In general, a division is made into two categories: isohydric and non-isohydric plants (Jones and Tardieu, 1998). Isohydric plants stabilize their leaf water contents by adjusting stomatal aperture; non-isohydric plants have a much slower stomatal reaction to drought stress.

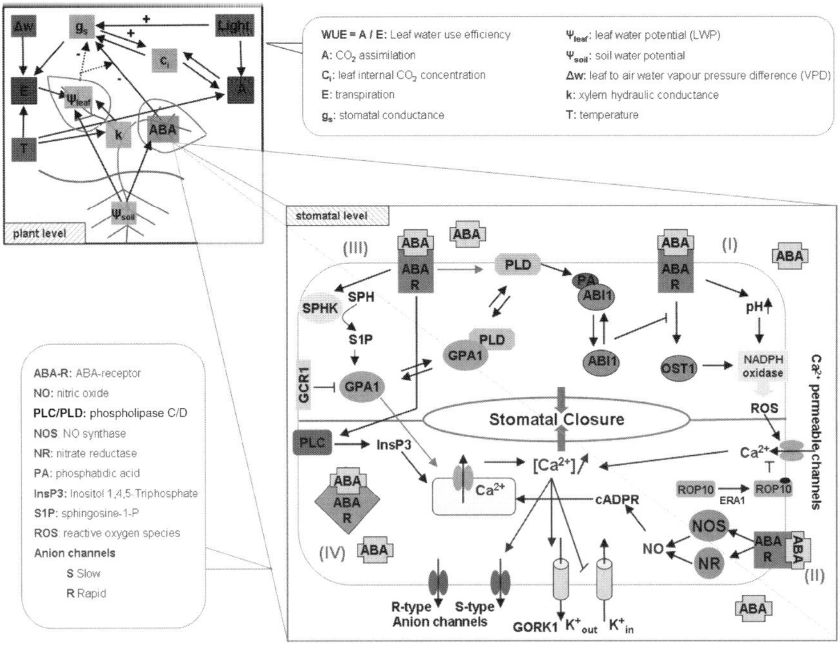

Figure 1. Schematic overview of stomatal aperture regulation

The top inset shows the response at the plant level to the varying environmental factors. The ABA response is worked out in detail in the main picture. ABA is either transported to the guard cells from the roots or vascular tissue, and binds to cell surface receptors (I-II-III). ABA synthesized in the guard cells is presumably recognized by an internal receptor (IV). Stomatal perception of ABA (I) leads to on accumulation of reactive oxygen species (ROS), through a cytosolic pH increase or mediated by the OST1 kinase. The ROS subsequently activate Ca2+ influx, leading to stomatal closure. The small G protein ROP10 is activated by the ERA farnesyltransferase and blocks Ca2+ influx. Stomatal perception of ABA (II) also results in the synthesis of NO, and the accumulation of cADPR (by Ca2+ activation of an ADP ribosyl cyclase). cADPR activates Ca2+ release from the vacuole, which amplifies the increases in cytoplasmic Ca2+, further promoting stomatal closure by modulation of the ion channels. Stomatal perception of ABA (III) liberates lipid-derived secondary messengers. Inositol-1,4,5- triphosphate (InsP3; derived from lipids through PLC activity), phosphatidic acid (PA; derived from ABA-activated PLD) and sphingosine-1-P (S1P). S1P induces stomatal closure in a process dependent on GPA1 (a Gα-subunit protein), whose function is inhibited by GCR1, a G protein coupled receptor-like protein. GPA1 interacts with PLD; free GPA1 inhibits stomatal opening. ABI1 is sequestered to the plasma membrane by PA, and thus cannot inhibit stomatal closure.

2.1. Signaling Factors (Secondary Messengers)

ABA-induced reactive oxygen species (ROS) production is catalyzed by two NADPH oxidases, encoded by AtrbohD and AtrbohF (Kwak et al., 2003). ROS production is induced through the action of the OST1 kinase (Mustilli et al., 2002) (also see below). A number of regulatory steps converge on the mobilization of $Ca^{2+}$from internal stores (Hetherington and Brownlee, 2004), followed by activation of ion channels. Lipid-derived signaling components such as phytosphingosine-1-phosphate (Coursol et al., 2005) and phosphatidic acid (PA) produced by phospholipase D (Bargmann and Munnik, 2006; Mishra et al., 2006) function in a signaling cascade that negatively regulates the ABI1 kinase activity. ABI1 (abscisic acid insensitive) promotes stomatal opening (Merlot et al., 2001). A second pathway leads to nitric oxide (NO) accumulation. NO is the first signaling intermediate in the reactions leading to changes in Ca^{2+} levels (Garcia-Mata and Lamattina, 2003; Li et al., 2006).

The most extensively studied ABA response mutants are the above mentioned *abi1* and the enhanced response to ABA (era) mutant, which is affected in a farnesyltransferase subunit (Pei et al., 1998). Farnesyltransferases posttranslationally modify proteins by farnesylation, resulting in membrane anchoring. Loss of the ERA1 gene not only leads to a lower transpiration level, but also to a reduction of growth (Pei et al., 1998). However, downregulating the expression of the farnesyltransferase gene proved to be a successful approach in engineering drought tolerance (Wang et al., 2005)

2.2. Kinase and Protease Regulatory Systems

Protein phosphatases of the 2C class, such as the ABI1 and 2 proteins, were identified as negative regulators of stomatal opening. The phylogeny of PP2C's revealed that ABI1, ABI2 and HAB are closely related (Saez et al., 2004). The loss of function mutant *hab1* (hypersensitive to ABA1) had a transpiration level similar to the wild type, likely due to the complementation of phosphatase activity by ABI1 and/or 2. However, overexpression led to a higher sensitivity to drought stress. Conversely, plants with knock-out mutations in both *HAB1* and *ABI1* had a high tolerance to drought stress (Saez et al., 2006).

The Snf1 (yeast sucrose non-fermenting)-related kinase 2 (SnRK2) OST1/ SNRK2E positively regulates ABA signal transduction (Belin et al., 2006). Loss of function mutants of *OST1* (open stomata) have an increased transpiration (Mustilli et al., 2002). In addition, over-expression of several other protein kinases (such as SRK2C (Umezawa et al., 2004) and NPK1 (Shou et al., 2004)) resulted in enhanced dehydration tolerance confirming their implication in the regulation of water usage.

2.3. Channels and Transporters

Stomatal guard cells are a model system to study the regulation of ion channels, central to the osmotic regulation of stomatal movements (Hetherington, 2001). It is therefore expected that modulation of ion channel activity will influence plant transpiration. The patch clamp technique in combination with a pharmacological approach is the method of choice to investigate activities of compounds that modulate ion channel activity. Such measurements at the single-cell level or at the microscopic scale (stomatal aperture determination) are complemented by measurements on the leaf or whole plant scale (transpiration).

Ion channels play a crucial role in changing the osmolyte content of guard cells and thus generate the driving force for turgor changes. Inward and outward movements of the key electrolytes K^+, Ca^{2+}, and Cl^- are regulated by separate ion-specific channels. Potassium selective channels cause the increase or decrease in guard cell osmotic potential that drives the change in cell volume and the subsequent pore aperture change. Five classes of K^+-channels have been identified in plants, all of which share homology with the "Shaker" type of K+-channels in animal systems. The inward rectifying channels cause an inward transport of K+ (KAT, AKT), while outward rectifying channels cause the transport of K+ out of cells. The guard cell outward rectifying K^+ channel GORK1 is the major voltage-gated potassium channel in the guard cell membrane. Not only the amplitude of the response (determining the degree of opening), but also the speed of stomatal movement in response to fluctuating water availability can be an important parameter in the optimization of plant yield (Hosy et al., 2003). By analogy with their importance as a target in animal and human physiology, signaling through plant ion channels could be modulated to fine-tune plant stress tolerance (Okuma and Murata, 2004).

Modulation of guard cell ion channel activities involves ABC-transporter proteins (multidrug resistance associated protein (MRP) type ATP binding cassette transporters). *MRP4* and *MRP5* knockout mutants (both T-DNA insertion mutants) were isolated to assess the effect on stomatal regulation of a defective ABC transporter (Klein et al., 2004; Klein et al., 2003). Absence of MRP4 expression causes stomata to be more opened both in light and in darkness as compared to wild type plants. A loss-of-function mutation of the *MRP5* gene resulted in reduced transpiration, but importantly also had a higher WUE compared to the wild type (as determined by continuous gas exchange measurements). Furthermore, *mrp5* plants are characterized by increased auxin levels (Gaedeke et al., 2001), pointing at the complex interplay of different regulatory mechanisms.

Efficient water use at the cellular level also involves activation of specific water channels. Water transport through the lipophilic cell membranes is facilitated by water-channels termed aquaporins, belonging to the class of plasma-membrane intrinsic proteins (PIPs) (Luu and Maurel, 2005). These channels are present throughout the plant, and their opening is induced by ABA (Ye and Steudle, 2006). Overexpression of a *Brassica napus* BnPIP in tobacco conferred increased drought tolerance, while expressing the antisense BnPIP had the opposite effect (Yu et al., 2005).

2.4. Transcription Factors

Transcription factors (TF) are grouped into classes depending on their conserved DNA-binding domain. The Apetala 2/ethylene-responsive element binding factor (AP2/ERF) class is one of the major TF families in plants (Shukla et al., 2006). A subfamily thereof, the dehydration response element binding protein /C-repeat binding factor (DREB/CBF) are implicated in ABA-independent regulation. DREB transcription factors activate DRE or C-repeat containing genes (Liu et al., 1998). Overexpression of the DREB/CBF transcription factor CBF4 resulted in drought stress tolerance (Haake et al., 2002). Likewise, a constitutively active form of DREB2A also leads to significant drought tolerance (Sakuma et al., 2006).

The bZIP family of TF is a second class of TF implicated in drought response. Within the bZIP family of TFs, several subfamilies were described (Bensmihen et al., 2005). One subgroup consists of TFs that bind to the conserved cis-acting sequences known as ABA-responsive elements (ABRE); these TFs are hence termed ABRE binding factors (ABFs). In rice, constitutively expressing CBF3/DREB1A or ABF3 did result in enhanced drought stress without growth penalty (Oh et al., 2005), in contrast to stunting seen in Arabidopsis (Liu et al., 1998).

A third important class is composed of the MYB-TF. One member of this transcription factor family, AtMYB60, an R2R3-MYB guard-cell specific TF, is down-regulated during drought stress, since knockout of AtMYB60 resulted in a constitutive reduction of stomatal opening, and consequently decreased wilting under water stress conditions (Cominelli et al., 2005). Another example of TF downregulation is provided by the disruption of the AP2-like ABA repressor 1 (ABR1) gene, leading to a higher level of ABA and thus drought resistance (Pandey et al., 2005). Importantly, the mutant was indistinguishable from the wild type under control conditions.

2.5. Metabolism

Stomata are characterized by a specialized physiology and an accordingly regulated metabolism (Outlaw, 2003). Malate is one of the solutes responsible for the turgor increase needed to open stomata and keep them open. Upregulating malate catabolism effectively reduced stomatal water loss and led to an increased WUE (Laporte et al., 2002). This was achieved in tobacco by expressing the maize NADP-malic enzyme (ME), which converts malate and NADP to pyruvate, NADPH, and CO_2.

Regulation of catabolic and anabolic enzymes also modulates ABA concentrations. Upon water deficit ABA is synthesized in roots and shoots and subsequently redistributed to the guard cells, where it triggers stomatal closure. After drought stress, ABA pools were detected in Arabidopsis shoot vasculature and in stomata, by using ABA-specific promotors coupled to the luciferase (LUC) reporter (Christmann et al., 2005). In Arabidopsis, expression of aldehyde oxidase 3 (AAO3), an enzyme involved in ABA synthesis, was revealed in root tips, root and shoot vasculature and in stomata (Koiwai et al., 2004). There is evidence for the existence of two pools of ABA, differing in their synthesis pathway and in their dynamics upon stress (Nambara and Marion-Poll, 2005; Seo and Koshiba, 2002). Foliar ABA was shown to be produced via the methyl-erythritol phosphate (MEP) pathway, but also the direct MEP-independent synthesis likely occurs in leaves (Nambara and Marion-Poll, 2005). This MEP-derived ABA pool was shown to regulate stomatal opening in response to rapid changes in water status. Inhibition of the MEP pathway resulted in an increase in leaf transpiration linked to a decrease in ABA-content (Barta and Loreto, 2006). However, the MEP-pathway derived ABA was shown not be involved in responses to high CO_2 or darkness. The regulation of ABA levels linked to the diurnal light/dark cycle was related to the cytochrome P450 mono-oxygenase enzyme (CytP450), that catabolises the endogenous guard cell ABA to 8'-hydroxy-ABA (Tallman, 2004). Manipulation of the expression of ABA 8'-hydroxylases might be the preferred strategy to modulate ABA levels and thus water usage, since overexpression of ABA-synthesizing genes induces an increased ABA catabolic activity that annihilates the desired ABA-effect (Yang and Zeevaart, 2006). ABA is also subject to inactivation by conjugation (glucosylation), which limits the timeframe in which an ABA signal exerts its effects (Priest et al., 2005). Agronomic use of glucosylation and hydroxylation resistant ABA-analogues with long lasting effect was proposed (Priest et al., 2005). A recent study provides indications for a role of *XERICO*, an Arabidopsis RING-H2 gene (really interesting new gene zinc finger protein), in ABA homeostasis. Constitutive overexpression of *XERICO* resulted in an accelerated response of ABA biosynthesis upon drought stress (Ko et al., 2006).

3. CROSSTALK OF STRESS PATHWAYS

Crosstalk and thus overlap between biotic and abiotic stress pathways is highly common; however plants have also evolved mechanisms that prioritize drought (or more general abiotic) stress responses to biotic responses (Fujita et al., 2006). As a consequence of this cross-talk, selection for drought resistance can have effects on the pathogen resistance traits of a crop (Timmusk and Wagner, 1999). Drought stress is the most prevalent cause of stomatal closure and subsequent leaf surface temperature increase. However, other stresses influencing the water status of plants can 'mimic' the drought response. Infections such as fungal and bacterial wilting diseases directly impinge on the water-use efficiency of plants, resulting in a decrease thereof (Guimaraes and Stotz, 2004). The toxin fusicoccin, commonly used to study stomatal responses of plant mutants, is released by these pathogens to divert plant resources to pathogen growth.

Wilting diseases typically block water transport in the plant leading to a higher leaf temperature (Pinter et al., 1979). In sunflower, the effect of a wilting disease (Verticillium) was found to resemble drought response (Sadras et al., 2000). In this non-isohydric plant, which by definition has a slow stomatal response to hydric stress (see above), drought (and wilting disease) can be quantified by a decrease in leaf area. Overexpression of the activated disease resistance 1 (*ADR1*) gene, which encodes a coiled-coil (CC)-nucleotide-binding site (NBS)-leucine-rich repeat (LRR) protein, confers in addition to broad-spectrum pathogen resistance also drought resistance, but also results in enhanced susceptibility to heat and salt stress (Chini et al., 2004). The *rcd1* (radical-induced cell death) mutant displays rapid programmed cell death upon ozon exposure, which is reminiscent of pathogen resistance by the hypersensitive response. In addition, *rcd1* has a higher transpiration rate than wild type, is less sensitive to ABA, ethylene and jasmonate, and is thus implicated in multiple hormone-and stress signaling pathways (Ahlfors et al., 2004).

RCD1 belongs to the (ADP-ribosyl)transferase domain-containing subfamily of the WWE protein-protein interaction domain protein family. An unexpected implication of disease resistance response to powdery mildew in barley was the loss of stomatal control due to epidermal cell death around the stomata through which pathogen ingress occurred (Prats et al., 2006). The lack of turgor pressure from the epidermal cells left the stomata continuously open, leaving the plants exposed to severe levels of drought stress.

As ABA, the plant hormone ethylene is often involved in stress responses (De Paepe and Van Der Straeten, 2005). Ethylene inhibits ABA-induced stomatal closure, and ethylene overproducing mutants have a higher transpiration rate (Tanaka et al., 2005). A decrease in ethylene sensitivity is one of the mechanisms by which overexpression of Hahb-4, an HD-Zip protein from *Helianthus annuus*, increases drought tolerance of Arabidopsis (Manavella et al., 2006). A similar effect was observed in maize ACC synthase (ZmACS6) loss-of-function mutants, which are affected in the first regulatory step of ethylene biosynthesis (Young et al., 2004).

Hormonal cross-talk with the ABA pathway in relation to stomatal regulation is not limited to ethylene. Levels of auxins and cytokinins, hormones known to promote stomatal opening (Tanaka et al., 2006), display pronounced diurnal patterns which follow reverse trends compared to the corresponding ABA levels (Novakova et al., 2005).

Understanding the integration of chemical, electrical and hydraulic signals as a response to (coinciding) stresses at the whole plant level is a challenge for the future (Brenner et al., 2006).

4. CUTICULAR AND STOMATAL TRANSPIRATION

Transpiration is determined by both regulation of stomatal aperture and stomatal density. The latter parameters and stomatal size, are largely determined by the developmental program, but are also influenced by hormonal signals (Bergmann, 2006; Chaerle et al., 2005). When grown at low humidity, plants adaptively increase cuticle wax load (Holroyd et al., 2002). In contrast, high humidity conditions result in a lower stomatal density (Bergmann, 2004).

Modification of the epidermal surface (wax load) affects the survival of plants under severe drought stress, when stomata are completely closed (Zhang et al., 2005). The *shn* (shine) gain-of-function mutant has an altered wax composition of the leaf cuticula, responsible for its shiny appearance (Aharoni et al., 2004). The *shn* leaf epidermis is more permeable, resulting in a higher cuticular transpiration, and is characterized by a lower stomatal density. The combined effect of these factors results in a drought-tolerant phenotype of the *shn* mutant. A single mutation can thus have multiple effects affecting leaf gas exchange. Another example of epidermal wax load modification resulting in increased drought tolerance is the overexpression of the ABA and drought-inducible AP2 transcription factor WXP1 (wax production) in alfalfa, leading to higher wax accumulation, with a minor growth retardation as a side-effect.

5. MONITORING OF DROUGHT STRESS RESPONSES

The stomatal pathway represents the major route for gas exchange, whereas the remaining part of the leaf surface (98 to 99,8% of its area) represents only a fraction of the total transpiration (10 to 100 times lower) since it is covered by a waxy cuticula (Nobel, 1991). The mechanisms described above have largely been discovered and characterized using techniques that reveal stomatal functioning (Merta et al., 2001) (see Table 1 for an overview).

5.1. Monitoring at the Lab Scale

An indication of modified water relations in a mutant plant is generally given by a wilty or withered phenotype (Kacira et al., 2002). Confirmation thereof is obtained by weight loss measurements, either using potted plants (with covered soil or substrate), detached leaves or shoots (Aharoni et al., 2004; Pandey and Assmann, 2004; Ruggiero et al., 2004). Integrative weight loss measurements over time, covering either complete dark or light periods allow to discriminate between stomatal and cuticular transpiration. However, this difference in transpiration between light and dark is more easily obtained by measuring changes in the humidity of air circulated over the leaf in a semi-closed measuring system. Direct assessment of leaf gas-exchange provides a real time, higher resolution measurement of the actual water loss (and CO_2 uptake) (Lasceve et al., 1997). Reduced transpiration can be monitored by porometric measurements, during which a small leaf region is enclosed in a cuvette for a measuring time of under 1 minute (Ahlfors et al., 2004). The use of multiple-cuvette systems enclosing leaves (or complete plants) yields time-courses allowing to compare the characteristics of different plants (Dodd et al., 2004). The high time-resolution also reveals differences in response to changing environmental factors, such as air humidity (Hosy et al., 2003).

Single cuvette portable systems are limited to short intermittent measurements on a batch of plants. This approach is labor intensive, and suffers from the lack

Table 1. Overview of the measuring techniques to reveal changes in stomatal control.
The time resolution of weight loss measurements is at the hour level in detached leaves due to accelerated water loss; for intact plants differences can be revealed with day resolution. Gas exchange measurements need an equilibration time for the air continuously circulated over a leaf enclosed in a measuring cuvette. Clamping of the leaf can influence leaf physiology, especially for longer time measurements (*)

Measuring technique	Measured parameter	Stomatal closure response	Destructive/ invasive	Time resolution
Weight loss	Amount of water evaporated	Decrease	– / –	Hours to days
Gas exchange	Change in water content of air	Decrease	– / *	Seconds to minutes
Thermography	Leaf temperature	Increase	– / –	Real-time
Carbon isotope discrimination	Discrimination of ^{13}C over ^{12}C	Decrease	+ / +	Integrative over growth period

of reproducible positioning of the measuring cuvette. Longer-term leaf clamping inevitably affects leaf physiology (e.g. by shading). Thermal imaging overcomes these limitations and monitors evaporation at the leaf surface non-invasively, in real time. Importantly measurements should be carried out in stable environmental conditions (Chaerle and Van Der Straeten, 2000). In addition, thermal imaging can also visualize the temperature of detached leaves, offering an alternative to integrative water measurement by weighing (Mustilli et al., 2002). Thermography has the additional benefit of visualizing heterogeneity in response of leaves. This might not be needed for field applications, where an average temperature measurement using a non-imaging infrared thermometer will be sufficient to monitor the temporal evolution of leaf temperature (under a developing stress). Light intensity, known to positively regulate stomatal aperture, was reported to influence drought stress detection by thermal imaging in Chrysanthemum (Blom-Zandstra and Metselaar, 2006). An approach to directly quantify stomatal conductance from thermal imaging data was recently proposed, and will provide the means to directly correlate temperature measurements to the above-described gas-exchange measurements (Leinonen et al., 2006). Furthermore, the reflectance of plant leaves also depends on water content. Changes in water status can be revealed by near infrared imaging, since in this spectral region, water absorbs part of the radiation (Peñuelas and Filella, 1998). This technique is mostly used in remote sensing applications, but has the potential to be used in a multi-sensor setup at the laboratory scale.

An integrative measurement of yield over a whole growing season can be obtained by the destructive carbon isotope discrimination technique at the end of the growth period (DELTA technique, see www.csiro.au). The heavier ^{13}C isotope containing CO_2 is discriminated against during fixation in the substomatal cavities. Upon stomatal closure, discrimination of the two isotopes becomes less likely, and values closer to zero are obtained for the delta (Δ) parameter. This parameter negatively correlates with transpiration efficiency and thus water use efficiency (WUE). The *ERECTA (ER)* gene, encoding a receptor-like kinase (RLK) from Arabidopsis was shown to confer increased WUE (Masle et al., 2005). Arabidopsis plants homozygous for *erecta* mutant alleles (Ler, Col*er105*, Col*er2*) had a higher Δ, higher transpiration and a higher stomatal density, compared to homozygous *ERECTA* plants, harboring the functional *ER* alleles. Using the delta screening approach, high yielding wheat cultivars were obtained (Condon et al., 2004). An important technical advance in the study of water relations is the possibility to measure water transport in-planta using magnetic resonance imaging (Windt et al., 2006). This allows to visualize changes in phloem and xylem transport, which also influence the water status of the plant.

5.2. Field Scale Monitoring

Thermal imaging can visualize early crop responses to water limitation from the plant level to the field scale. However, to be useful under field conditions at an early stage of plant development (before canopy closure), the image parts corresponding to leaf area need to be isolated selectively by dedicated software (Luquet et al., 2003).

To exploit thermography as a monitoring tool, water stress levels are expressed on a reference scale (0–1) by various Water Stress Indexes. The most basic approach is taking into account the temperature difference between canopy and air (T_{canopy} -T_{air}). Parameters based on these temperature measurements, such as Crop Water Stress Index (CWSI) and Water Deficit Index (WDI) are used to assess the water status of field plots (http://www.uswcl.ars.ag.gov/epd/remsen/irrweb/thindex.htm) (Jones, 2004a; Jones, 2004b). Leaf water potential (LWP) is determined by the osmotic status of the leaf, and can be measured on leaf discs using a vapor pressure osmometer (Verslues and Bray, 2004). Even though LWP is not always directly related to stomatal conductance (g_s), a correlation with the CWSI parameter was found (Cohen et al., 2005).

For early monitoring applications to be effective, it is important to take into account that non-isohydric plants do not display a change in stomatal conductance (and hence leaf temperature) upon early drought stress. As another complicating factor, crop water status at the field scale is characterized by spatial variability, due to soil characteristics, crop canopy variability, and inter-plant variability. This heterogeneity has to be discriminated against the effects of hydric stress. Approaches using normalized CWSI values that take into account crop canopy characteristics show great promise for making irrigation practices more efficient (Jones, 2004b).

Leaf temperature measurements are also amenable to simulation. The development of modeling approaches with virtual plants, allows to grow 'virtual crops' under different conditions and to assess their predicted responses (Tardieu, 2003). Especially in agronomically important crops, a longer generation time puts a limit on the development of new improved cultivars. Targeting the most promising approaches as revealed by the simulations allows to speed up crop breeding. Modeling specifically applied to the guard cell system regulatory network can also help predicting the effect of manipulations and guide the experimental approach (Li et al., 2006). Given the complexity of guard cell regulation, combining the available knowledge on interactions and regulations into a dynamic model can help to define missing links and to test new hypotheses. Predictive tools can therefore further advances to targeted improvement of water use.

5.3. Screening Applications

Using screening under controlled conditions, altered responses to drought stress among a batch of cultivars or within a mutated population can be pinpointed using thermal imaging. The isolation of the barley 'cool' mutant was the first example of a successful thermal screen (Raskin and Ladyman, 1988). Analogous screens have been carried out in Arabidopsis to isolate mutants with aberrant leaf temperature, shown to carry a mutation in kinases or phosphatases that regulate stomatal aperture (see above ABI and OST) (Merlot et al., 2002; Mustilli et al., 2002).

Using thermography to reveal stomatal responses upon a steep drop in air humidity, OST1 was subsequently revealed to be also implicated in the stomatal closure upon low air relative humidity (or low vapor pressure deficit VPD) (Xie et al., 2006).

As a consequence of a leaf temperature increase, assimilation can be directly affected. A limitation of photosynthesis is however predominantly caused by diffusion limitation (Flexas et al., 2006). Thus (drought) stress induced stomatal closure will limit crop yield. Therefore, screening for plants that have 'mild' reactions to a developing stress might be beneficial to increase yields.

The leaf temperature screening approach can also be carried out with infrared thermometers at the field plot scale. This technique was effectively used to screen Brassica genotypes for drought tolerance under decreasing soil moisture conditions (Singh et al., 1985).

6. ROUTES TO YIELD ENHANCEMENT

Constitutive expression of genes involved in the response to stress is of great benefit in applied research since it often results in strong phenotypes. However this approach mostly leads to a considerable growth penalty (Liu et al., 1998). The use of inducible promoter constructs can alleviate these adverse consequences (Chini et al., 2004; Umezawa et al., 2006). In some cases however, constitutive expression enhances yield significantly under stressed conditions without growth inhibition in optimal circumstances. As an example, overexpression of a NAC (NAM, ATAF, and CUC) transcription factor resulted in higher drought resistance both in the vegetative and in the reproductive stage of rice (Hu et al., 2006). Conversely, knockout of single genes in T-DNA insertion mutants (in general loss-of-function mutants) can remain without phenotype under normal growth conditions, yet confer a drought resistance phenotype, as exemplified by the *gcr1* mutant (G-Protein Coupled Receptor, GCR1) which is hypersensitive to ABA (Pandey and Assmann, 2004). GCR1 could thus be a key factor in engineering plant resistance to drought stress.

The use of modeling techniques together with the increasing genetic information available from whole genome sequencing efforts (achieved for Arabidopsis, Oryza, and Populus), micro-array gene-expression datasets and associated tools to extract signaling network information (Zimmermann et al., 2005), and expressed sequence tag (ESTs) databases (http://www.ncbi.nlm.nih.gov/dbEST/dbEST_summary.html) (Rudd, 2003) will provide the means needed to further increase crop yield in a world faced with an increased pressure on the available resources.

7. ACKNOWLEDGEMENT

L.C. is a post-doctoral fellow of the Research Foundation – Flanders.

REFERENCES

Aharoni, A., Dixit, S., Jetter, R., Thoenes, E., van Arkel, G., and Pereira, A., 2004, The SHINE Clade of AP2 Domain Transcription Factors Activates Wax Biosynthesis, Alters Cuticle Properties, and Confers Drought Tolerance when Overexpressed in Arabidopsis. *Plant Cell* **16**:2463–2480.

Ahlfors, R., Lang, S., Overmyer, K., Jaspers, P., Brosche, M., Tauriainen, A., Kollist, H., Tuominen, H., Belles-Boix, E., Piippo, M., Inze, D., Palva, E. T., and Kangasjarvi, J., 2004, Arabidopsis RADICAL-INDUCED CELL DEATH1 belongs to the WWE protein-protein interaction domain protein family and modulates abscisic acid, ethylene, and methyl jasmonate responses. *Plant Cell* **16**: 1925–1937.

Araus, J. L., Slafer, G. A., Reynolds, M. P., and Royo, C., 2002, Plant breeding and drought in C3 cereals: What should we breed for? *Ann. Bot.* **89**:925–940.

Bargmann, B. O. R., and Munnik, T., 2006, The role of phospholipase D in plant stress responses. *Current Opinion in Plant Biology* **9**:515–522.

Barta, C., and Loreto, F., 2006, The relationship between the Methyl-Erythritol Phosphate pathway leading to emission of volatile isoprenoids and abscisic acid content in leaves. *Plant Physiol.* **141**:1676–1683.

Belin, C., de Franco, P.-O., Bourbousse, C., Chaignepain, S., Schmitter, J.-M., Vavasseur, A., Giraudat, J., Barbier-Brygoo, H., and Thomine, S., 2006, Identification of features regulating OST1 kinase activity and OST1 function in guard cells. *Plant Physiol.* **141**:1316–1327.

Bensmihen, S., Giraudat, J., and Parcy, F., 2005, Characterization of three homologous basic leucine zipper transcription factors (bZIP) of the ABI5 family during Arabidopsis thaliana embryo maturation. *J. Exp. Bot.* **56**:597–603.

Bergmann, D., 2006, Stomatal development: from neighborly to global communication. *Current Opinion in Plant Biology* **9**:478–483.

Bergmann, D. C., 2004, Integrating signals in stomatal development. *Current Opinion in Plant Biology* **7**:26–32.

Blom-Zandstra, M., and Metselaar, K., 2006, Infrared thermometry for early detection of drought stress in Chrysanthemum. *Hortscience* **41**:136–142.

Brenner, E. D., Stahlberg, R., Mancuso, S., Vivanco, J., Baluska, F., and Van Volkenburgh, E., 2006, Plant neurobiology: an integrated view of plant signaling. *Trends in Plant Science* **11**:413–419.

Buckley, T. N., 2005, The control of stomata by water balance. *New Phytologist* **168**:275–292.

Chaerle, L., Saibo, N., and Van Der Straeten, D., 2005, Tuning the pores: towards engineering plants for improved water use efficiency. *Trends Biotechnol.* **23**:308–315.

Chaerle, L., and Van Der Straeten, D., 2000, Imaging techniques and the early detection of plant stress. *Trends Plant Sci.* **5**:495–501.

Chaves, M. M., and Oliveira, M. M., 2004, Mechanisms underlying plant resilience to water deficits: prospects for water-saving agriculture. *J. Exp. Bot.* **55**:2365–2384.

Chini, A., Grant, J. J., Seki, M., Shinozaki, K., and Loake, G. J., 2004, Drought tolerance established by enhanced expression of the CC-NBS-LRR gene, ADR1, requires salicylic acid, EDS1 and ABI1. *Plant J* **38**:810–822.

Christmann, A., Hoffmann, T., Teplova, I., Grill, E., and Muller, A., 2005, Generation of Active Pools of Abscisic Acid Revealed by In Vivo Imaging of Water-Stressed Arabidopsis. *Plant Physiol.* **137**:209–219.

Cohen, Y., Alchanatis, V., Meron, M., Saranga, Y., and Tsipris, J., 2005, Estimation of leaf water potential by thermal imagery and spatial analysis. *J. Exp. Bot.* **56**:1843–1852.

Cominelli, E., Galbiati, M., Vavasseur, A., Conti, L., Sala, T., Vuylsteke, M., Leonhardt, N., Dellaporta, S. L., and Tonelli, C., 2005, A Guard-Cell-Specific MYB Transcription Factor Regulates Stomatal Movements and Plant Drought Tolerance. *Current Biology* **15**:1196–1200.

Condon, A. G., Richards, R. A., Rebetzke, G. J., and Farquhar, G. D., 2004, Breeding for high water-use efficiency. *Journal of Experimental Botany* **55**:2447–2460.

Coursol, S., Le Stunff, H., Lynch, D. V., Gilroy, S., Assmann, S. M., and Spiegel, S., 2005, Arabidopsis sphingosine kinase and the effects of phytosphingosine-1-phosphate on stomatal aperture. *Plant Physiol.* **137**:724–737.

Davies, W., Kudoyarova, G., and Hartung, W., 2005, Long-distance ABA signaling and its relation to other signaling pathways in the detection of soil drying and the mediation of the plant's response to drought. *Journal of Plant Growth Regulation* **24**:285–295.

De Paepe, A., and Van Der Straeten, D., 2005, Ethylene biosynthesis and signaling: An overview. *Vitamins and Hormones-Advances in Research and Applications* **72**:399–430.

Dodd, A. N., Parkinson, K., and Webb, A. A. R., 2004, Independent circadian regulation of assimilation and stomatal conductance in the ztl-1 mutant of Arabidopsis. *New Phytologist* **162**:63–70.

Novakova, M., Motyka, V., Dobrev, P. I., Malbeck, J., Gaudinova, A., and Vankova, R., 2005, Diurnal variation of cytokinin, auxin and abscisic acid levels in tobacco leaves. *Journal of Experimental Botany* **56**:2877–2883.

Oh, S.-J., Song, S. I., Kim, Y. S., Jang, H.-J., Kim, S. Y., Kim, M., Kim, Y.-K., Nahm, B. H., and Kim, J.-K., 2005, Arabidopsis CBF3/DREB1A and ABF3 in Transgenic Rice Increased Tolerance to Abiotic Stress without Stunting Growth. *Plant Physiol.* **138**:341–351.

Okuma, E., and Murata, Y., 2004, Plant ion channels as potential targets of agro-chemicals. *Journal of Pesticide Science* **29**:304–307.

Outlaw, W. H., 2003, Integration of cellular and physiological functions of guard cells. *Critical Reviews in Plant Sciences* **22**:503–529.

Pandey, G. K., Grant, J. J., Cheong, Y. H., Kim, B. G., Li, L., and Luan, S., 2005, ABR1, an APETALA2-Domain Transcription Factor That Functions as a Repressor of ABA Response in Arabidopsis. *Plant Physiol.* **139**:1185–1193.

Pandey, S., and Assmann, S. M., 2004, The Arabidopsis Putative G Protein-Coupled Receptor GCR1 Interacts with the G Protein {alpha} Subunit GPA1 and Regulates Abscisic Acid Signaling. *Plant Cell* **16**:1616–1632.

Pei, Z. M., Ghassemian, M., Kwak, C. M., McCourt, P., and Schroeder, J. I., 1998, Role of farnesyltransferase in ABA regulation of guard cell anion channels and plant water loss. *Science* **282**:287–290.

Peñuelas, J., and Filella, I., 1998, Visible and near-infrared reflectance techniques for diagnosing plant physiological status. *Trends Plant Sci.* **3**:151–156.

Pinter, P. J., Stanghellini, M. E., Reginato, R. J., Idso, S. B., Jenkins, A. D., and Jackson, R. D., 1979, Remote detection of biological stresses in plants with infrared thermometry. *Science* **205**:585–586.

Prats, E., Gay, A. P., Mur, L. A. J., Thomas, B. J., and Carver, T. L. W., 2006, Stomatal lock-open, a consequence of epidermal cell death, follows transient suppression of stomatal opening in barley attacked by Blumeria graminis. *J. Exp. Bot.* **57**:2211–2226.

Priest, D. M., Jackson, R. G., Ashford, D. A., Abrams, S. R., and Bowles, D. J., 2005, The use of abscisic acid analogues to analyse the substrate selectivity of UGT71B6, a UDP-glycosyltransferase of Arabidopsis thaliana. *FEBS Letters* **579**:4454–4458.

Raskin, I., and Ladyman, J. A. R., 1988, Isolation and characterization of a barley mutant with abscisic-acid-insensitive stomata. *Planta* **173**:73–78.

Riera, M., Redko, Y., and Leung, J., 2006, Arabidopsis RNA-binding protein UBA2a relocalizes into nuclear speckles in response to abscisic acid. *FEBS Letters* **580**:4160–4165.

Riera, M., Valon, C., Fenzi, F., Giraudat, J., and Leung, J., 2005, The genetics of adaptive responses to drought stress: abscisic acid-dependent and abscisic acid-independent signalling components. *Physiologia Plantarum* **123**:111–119.

Rudd, S., 2003, Expressed sequence tags: alternative or complement to whole genome sequences? *Trends in Plant Science* **8**:321–329.

Ruggiero, B., Koiwa, H., Manabe, Y., Quist, T. M., Inan, G., Saccardo, F., Joly, R. J., Hasegawa, P. M., Bressan, R. A., and Maggio, A., 2004, Uncoupling the Effects of Abscisic Acid on Plant Growth and Water Relations. Analysis of sto1/nced3, an Abscisic Acid-Deficient but Salt Stress-Tolerant Mutant in Arabidopsis. *Plant Physiol.* **136**:3134–3147.

Sadras, V. O., Quiroz, F., Echarte, L., Escande, A., and Pereyra, V. R., 2000, Effect of Verticillium dahliae on photosynthesis, leaf expansion and senescence of field-grown sunflower. *Annals of Botany* **86**:1007–1015.

Saez, A., Apostolova, N., Gonzalez-Guzman, M., Gonzalez-Garcia, M. P., Nicolas, C., Lorenzo, O., and Rodriguez, P. L., 2004, Gain-of-function and loss-of-function phenotypes of the protein phosphatase 2C HAB1 reveal its role as a negative regulator of abscisic acid signalling. *The Plant Journal* **37**:354–369.

Saez, A., Robert, N., Maktabi, M. H., Schroeder, J. I., Serrano, R., and Rodriguez, P. L., 2006, Enhancement of abscisic acid sensitivity and reduction of water consumption in Arabidopsis by combined inactivation of the protein phosphatases type 2C ABI1 and HAB1. *Plant Physiol.* **141**:1389–1399.

Sakuma, Y., Maruyama, K., Osakabe, Y., Qin, F., Seki, M., Shinozaki, K., and Yamaguchi-Shinozaki, K., 2006, Functional analysis of an arabidopsis transcription factor, DREB2A, involved in drought-responsive gene expression. *Plant Cell* **18**:1292–1309.

Seo, M., and Koshiba, T., 2002, Complex regulation of ABA biosynthesis in plants. *Trends in Plant Science* **7**:41–48.

Shou, H. X., Bordallo, P., and Wang, K., 2004, Expression of the Nicotiana protein kinase (NPK1) enhanced drought tolerance in transgenic maize. *Journal of Experimental Botany* **55**:1013–1019.

Shukla, R. K., Raha, S., Tripathi, V., and Chattopadhyay, D., 2006, Expression of CAP2, an APETALA2-Family transcription factor from chickpea, enhances growth and tolerance to dehydration and salt stress in transgenic tobacco. *Plant Physiol.* **142**:113–123.

Singh, D. P., Singh, P., Kumar, A., and Sharma, H. C., 1985, Transpirational Cooling as a Screening Technique for Drought Tolerance in Oil Seed Brassicas. *Annals of Botany* **56**:815–820.

Tallman, G., 2004, Are diurnal patterns of stomatal movement the result of alternating metabolism of endogenous guard cell ABA and accumulation of ABA delivered to the apoplast around guard cells by transpiration? *J. Exp. Bot.* **55**:1963–1976.

Tanaka, Y., Sano, T., Tamaoki, M., Nakajima, N., Kondo, N., and Hasezawa, S., 2005, Ethylene inhibits abscisic acid-induced stomatal closure in Arabidopsis. *Plant Physiology* **138**:2337–2343.

Tanaka, Y., Sano, T., Tamaoki, M., Nakajima, N., Kondo, N., and Hasezawa, S., 2006, Cytokinin and auxin inhibit abscisic acid-induced stornatal closure by enhancing ethylene production in Arabidopsis. *Journal of Experimental Botany* **57**:2259–2266.

Tardieu, F., 2003, Virtual plants: modelling as a tool for the genomics of tolerance to water deficit. *Trends in Plant Science* **8**:9–14.

Timmusk, S., and Wagner, E. G. H., 1999, The plant-growth-promoting rhizobacterium Paenibacillus polymyxa induces changes in Arabidopsis thaliana gene expression: A possible connection between biotic and abiotic stress responses. *Molecular Plant-Microbe Interactions* **12**:951–959.

Umezawa, T., Fujita, M., Fujita, Y., Yamaguchi-Shinozaki, K., and Shinozaki, K., 2006, Engineering drought tolerance in plants: discovering and tailoring genes to unlock the future. *Current Opinion in Biotechnology* **17**:113–122.

Umezawa, T., Yoshida, R., Maruyama, K., Yamaguchi-Shinozaki, K., and Shinozaki, K., 2004, SRK2C, a SNF1-related protein kinase 2, improves drought tolerance by controlling stress-responsive gene expression in Arabidopsis thaliana. *Proceedings of the National Academy of Sciences of the United States of America* **101**:17306–17311.

Valliyodan, B., and Nguyen, H. T., 2006, Understanding regulatory networks and engineering for enhanced drought tolerance in plants. *Current Opinion in Plant Biology* **9**:189–195.

Verslues, P. E., and Bray, E. A., 2004, LWR1 and LWR2 are required for osmoregulation and osmotic adjustment in Arabidopsis. *Plant Physiol.* **136**:2831–2842.

Verslues, P. E., Guo, Y., Dong, C.-H., Ma, W., and Zhu, J.-K., 2006, Mutation of SAD2, an importin Beta-domain protein in Arabidopsis, alters abscisic acid sensitivity. *The Plant Journal* **47**:776–787.

Wang, Y., Ying, J., Kuzma, M., Chalifoux, M., Sample, A., McArthur, C., Uchacz, T., Sarvas, C., Wan, J., Dennis, D. T., McCourt, P., and Huang, Y., 2005, Molecular tailoring of farnesylation for plant drought tolerance and yield protection. *Plant Journal* **43**:413–424.

Windt, C. W., Vergeldt, F. J., De Jager, P. A., and Van As, H., 2006, MRI of long-distance water transport: a comparison of the phloem and xylem flow characteristics and dynamics in poplar, castor bean, tomato and tobacco. *Plant, Cell and Environment* **29**:1715–1729.

Woodward, F. I., Lake, J. A., and Quick, W. P., 2002, Stomatal development and CO2: ecological consequences. *New Phytologist* **153**:477–484.

Xie, X., Wang, Y., Williamson, L., Holroyd, G. H., Tagliavia, C., Murchie, E., Theobald, J., Knight, M. R., Davies, W. J., Leyser, H. M. O., and Hetherington, A. M., 2006, The Identification of Genes Involved in the Stomatal Response to Reduced Atmospheric Relative Humidity. *Current Biology* **16**:882–887.

Yang, S. H., and Zeevaart, J. A. D., 2006, Expression of ABA 8'-hydroxylases in relation to leaf water relations and seed development in bean. *The Plant Journal* **47**:675–686.

Ye, Q., and Steudle, E., 2006, Oxidative gating of water channels (aquaporins) in corn roots. *Plant, Cell and Environment* **29**:459–470.

Young, T. E., Meeley, R. B., and Gallie, D. R., 2004, ACC synthase expression regulates leaf performance and drought tolerance in maize. *Plant J* **40**:813–825.

Yu, Q., Hu, Y., Li, J., Wu, Q., and Lin, Z., 2005, Sense and antisense expression of plasma membrane aquaporin BnPIP1 from Brassica napus in tobacco and its effects on plant drought resistance. *Plant Science* **169**:647–656.

Zhang, B., Pan, X., Cobb, G. P., and Anderson, T. A., 2006, Plant microRNA: A small regulatory molecule with big impact. *Developmental Biology* **289**:3–16.

Zhang, J.-Y., Broeckling, C. D., Blancaflor, E. B., Sledge, M. K., Sumner, L. W., and Wang, Z.-Y., 2005, Overexpression of WXP1, a putative Medicago truncatula AP2 domain-containing transcription factor gene, increases cuticular wax accumulation and enhances drought tolerance in transgenic alfalfa (Medicago sativa). *The Plant Journal* **42**:689–707.

Zimmermann, P., Hennig, L., and Gruissem, W., 2005, Gene-expression analysis and network discovery using Genevestigator. *Trends in Plant Science* **10**:407–409.

CHAPTER 5

ECO-PHYSIOLOGICAL AND MOLECULAR-GENETIC DETERMINANTS OF PLANT CUTICLE FUNCTION IN DROUGHT AND SALT STRESS TOLERANCE

DYLAN K. KOSMA AND MATTHEW A. JENKS
Purdue University, Department of Horticulture and Landscape Architecture, Center for Plant Environmental Stress Physiology, West Lafayette, Indiana, 47907, USA

Abstract: A waxy cuticle covers the aerial organs of plants that functions to prevent uncontrolled water loss. The cuticle has often been considered a non-responsive adaptation that acts simply as a barrier to water loss, when in fact cuticle metabolism is quite responsive to environmental stresses. The responsiveness of the cuticle has been demonstrated by changes in cuticle chemistry and cuticle gene expression of drought and salt exposed plants. Alteration of cuticle traits through breeding and biotechnology approaches may prove useful in improving crops for drought and salt tolerance. However, work is still needed to lay the foundation for the use of cuticle genes and traits for agronomic purposes

Keywords: cuticle, wax, cutin, drought, salt, transpiration, water conservation, stomata, plant

1. INTRODUCTION

The plant cuticle is a hydrophobic coating that is composed of a cutin polyester membrane intermeshed and overlaid with free waxes that provides the outermost barrier over essentially all aerial plant organs, and whose foremost function across the plant kingdom is thought to be in plant water conservation (Goodwin & Jenks 2005). A dogma has persisted that a thick cuticle provides a more effective barrier to water loss, and thus can be categorized as a distinct xeromorphic trait. While cuticle does in fact have a significant influence on plant transpiration, recent studies challenge the simple notion of cuticle as an uncomplicated lipid coating whose barrier properties are determined by its thickness alone. This review describes recent advances in our understanding of plant cuticle function in plant drought and salt stress tolerance, and the underlying molecular genetic involvement in cuticle function and responsiveness to stress.

M.A. Jenks et al. (eds.), Advances in Molecular Breeding Toward Drought and Salt Tolerant Crops, 91–120.

2. CUTICLE ASSOCIATION WITH STRESS PHYSIOLOGY IN DROUGHT ADAPTED PLANTS

Xerophytic plants have extreme tolerance to arid environments and in general possess thicker cuticles than do mesophytes. Among mesophytic species, leaf cuticles range in thickness from less than 0.03 μm, as in *Arabidopsis thaliana* (L.) Heynh. (Franke et al. 2005), to seldom exceeding 0.2 μm to 0.3 μm in thickness (Jeffree 2006). A recent survey of 70 species from 21 genera of North American Cactaceae (a taxa of mainly xerophytic species) reported that 56 of these have cuticle thickness greater than 2 μm (Loza-Cornejo & Terrazas 2003), and a few possess cuticles well over 20 μm thick. The extremophile cactus *Ariocarpus fissuratus* (Engelman) K. Schuman. has a cuticle over 225 μm thick. It is typical that other xerophytic species, like *Hakea suaveolens* R. Br., *Clivia miniata*(Lindl.) Regal., and *Agave americana* L. (Jeffree, 2006), have cuticles over 4 μm. Likewise, many conifers have very thick cuticles, potentially associated with tolerance to dry sandy soils or desiccating alpine environments. For example, *Picea abies* (L.) Karst. has a needle cuticle 3.6 μm in thickness (Jeffree 2006) and *Pinus longaeva* D.K. Bailey needles have cuticles roughly 6 μm in thickness (Connor & Lanner 1991). A survey of many published articles however indicates that cuticle thickness does not correlate well with drought stress tolerance, cuticle permeability, or the degree of climatic dryness to which a species is adapted (Kamp 1930, Sitte & Rennier 1963, Radler 1965, Riederer & Schreiber 2001, Olyslaegers et al. 2002). There are many possible explanations for this. First, drought tolerance is a complex trait. Plants use many adaptations besides cuticle-based water conservation for avoiding tissue dehydration, such as; the formation of extensive root systems to more efficiently mine water, the ability to maintain extremely low osmotic potentials, and the ability to avoid drought by adaptive life cycles (Gibson 1996, Gutterman 2000). Thick cuticles likely have other functions in arid environments besides that of water barrier. Thick cuticles likely play important roles in preventing or ameliorating high temperature stress (Gibson 1998, Casado & Heredia 2001) and in the reduction of mutagenic ultraviolet radiation (Krauss et al. 1997, Holmes & Keiller 2002). In addition, cuticles can influence mechanical properties related to leaf/organ strength or toughness (Taylor 1971, Bargel et al. 2006), and thus could provide physical protection from damage by herbivorous pests (Potter & Kimmerer 1988, Gentry & Barbosa 2006). More extensive eco-physiological studies that correlate cuticle properties in xerophytes with specific requirements for survival and reproduction in arid environments would help explain the degree to which cuticle thickness has importance, as a permeability barrier, in mitigating high temperature stress, in UV protection, and in protection from herbivores.

Attempts to categorize cuticle water permeance values to life form and climate of origin have provided some insight on the eco-physiology of the cuticle (Riederer & Schreiber 2001). For example, deciduous plant species with mesomorphic leaves growing in temperate climates and evergreen epiphytic or climbing plants from tropical climates can be readily distinguished based on cuticle permeance to water values. In general, deciduous plant leaves have cuticles with high water permeance and tropical epiphytes have low permeability to water. However, as with cuticle thickness and permeance, relationships between life form/native-habitat and permeance do not always hold true. Cuticle permeance for leaves of some groupings (xerormorphic plants from Mediterranean climates) based on life form/habitat exhibit a broad range of permeance values that substantially overlap those groups defined as having high (deciduous plants) and low (tropical epiphytes) water permeance (Riederer & Schreiber 2001). Again, it is clear that cuticle thickness does not correlate well with the cuticle's permeability to water (Sitte & Rennier 1963, Radler 1965, Riederer & Schreiber 2001). As one case in point, fruits generally have among the thickest cuticles of any organs, but fruit cuticles also have some of the highest permeances to water (Schreiber & Riederer 1996, Riederer & Schreiber 2001). Rather than a simple homogenous lipid coating, several studies now demonstrate that the cuticle is structurally and chemically heterogeneous. Cuticle thickness and permeability varies over anticlinal and periclinal cell walls of the epidermis (Norris & Bukovac 1968, Norris 1974, Loza-Cornejo & Terrazas 2003) and even different epidermal cell types (guard cells, trichomes) have different cuticle properties, such as permeability, composition, and structure (Tanton & Crowdy 1972, Schlegel & Schönherr 2002, Schlegel et al. 2005). Other studies provide convincing evidence that the exact nanomolecular structure and packing arrangement of cutin and wax molecules within the cuticle membrane itself is a major determinant of cuticle permeability (Reynhardt & Riederer 1994, Riederer et al. 1995, Reynhardt 1997, Schreiber et al. 1997). More specifically, the size and placement of aliphatic wax constituents into impermeable crystalline regions within the cutin framework appear to play a major role in determining the number, size, and tortuosity of water diffusion pathways (Riederer 1991, Riederer & Schreiber 1995, Schreiber et al. 1997, Buchholz 2006, Burghardt & Riederer 2006). In this case, diffusion pathways are defined by amorphous inter-crystalline regions consisting of chain ends, polar functional groups, and possibly non-aliphatic (aromatic) wax compounds (Merk et al. 1998, Jenks 2002). The nature and exact nanomolecular structure within these intercrystalline regions however is still debatable. Transport kinetic studies have shown clearly that water, as a small, non-ionic, polar molecule, can diffuse through both polar pathways (reserved for the diffusion of ionic and small polar molecules) and lipophilic pathways reserved for lipophilic non-electrolytes; (Niederl et al. 1998, Schreiber et al. 2001, Schreiber 2005, 2006). New nanotechnological tools like, atomic force microscopy, Raman spectroscopic tools, nuclear magnetic resonance, and field emission scanning electron microscopy, hold much promise for future studies to understand cuticle structure, composition, and physical properties at a high spatial (nanometer) resolution. The utility of atomic force microscopy has been demonstrated, imaging the regeneration of wax crystals on the surface of plant cuticles (Koch et al. 2004), recrystallized, extracted wax (Koch et al. 2006), and isolated tomato cutin polymer (Beniitez et al. 2004ab, Benitez et al. 2004ba). The ability to obtain high-resolution images at higher magnifications than traditional scanning electron microscopes, and on unfixed and uncoated samples, promises to greatly expand our understanding of cuticle function.

2.1. A Comment on Methods used to Measure Cuticle Permeability

Plant cuticle permeability has been estimated using a variety of techniques. For example, chlorophyll extraction rate from plant tissues (with 80% ethanol) has frequently been used as an indicator of altered cuticle permeability (Lolle et al. 1998, Sieber et al. 2000, Chen et al. 2003, Aharoni et al. 2004, Schnurr et al. 2004, Zhang et al. 2005). There are still many questions regarding this method however, and Kerstiens et al. (2006) recently raised questions about the accuracy of using chlorophyll diffusion as an indication of cuticle permeability to water. Size selectivity of the diffusible compounds in polar and particularly lipophilic pathways may be an important issue (Buchholz et al. 1998, Schönherr & Schreiber 2004, Schreiber 2006) noting that the molecular size of water is much smaller than that of chlorophyll. Water loss decline curves of excised organs in darkness, termed minimum conductance (g_{min}), or mass change of whole growing plants in sealed pots, termed lowest conductance (g_{low}), have often been employed under the assumption that stomata are closed in darkness (Jenks et al. 1994, Chen et al. 2003, Aharoni et al. 2004, Chen et al. 2004, Xiao et al. 2004, Zhang et al. 2005). This assumption however may not be accurate for all plant species, as some plants do not appear to completely close their stomata at night and exhibit stomatal leakage (Kerstiens 1996, Burghardt & Riederer 2003). A new cuticle permeability assay employs a water-soluble dye, toluidine blue (TB), that preferentially binds to cell walls (Tanaka et al. 2004). Short duration submersion of plant organs in TB results in blue colored organs; organs having cuticles that are more permeable stain more intensely. Large scale screening with TB led to the discovery of a new set of cuticle permeability mutants, including new allelic members of previously characterized complementation groups (Tanaka et al. 2004). Other techniques used as indicators of cuticle permeability include plant response to herbicide sprays, wherein plants having cuticles that are more permeable exhibit earlier necrosis (Sieber et al. 2000, Chen et al. 2003). A more indirect means of assaying cuticle permeability has been to screen mutants for organ fusion (Lolle et al. 1998). To date, nearly all mutants having organ fusion that were examined had higher cuticle permeability to water than their respective wild type (Lolle et al. 1998, Tanaka & Machida 2006). Not all cutin mutants (e.g. *att1*) however, show organ fusion (Xiao et al. 2004). This may be due to differential diffusion of a morphogenic substance that causes fusion (Siegel & Verbeke 1989) or else a differential timing in the expression of the cuticle permeability phenotype in the respective mutant. To date organ fusion has been shown to occur primarily at the primordial organ development stage. When multiple assays of cuticle permeability have been employed (e.g. TB, chlorophyll leaching, organ or *in planta* water loss), they tend to give similar results. Organ fusion being the exception, as that a few cuticle mutants with altered permeability do not exhibit organ fusion (Lolle et al. 1997, Lolle et al. 1998, Sieber et al. 2000, Chen et al. 2003, Aharoni et al. 2004, Schnurr et al. 2004, Goodwin & Jenks 2005, Zhang et al. 2005, Tanaka & Machida 2006).

3. CUTICLE ASSOCIATION WITH STRESS PHYSIOLOGY IN SALINE ADAPTED PLANTS

Halophytes, like plants from arid climates, are adapted to water limiting environments created by the low osmotic potentials of saline soils or aerial salt sprays on tissue surfaces. Deposition of salty aerosol sprays from ocean wave action and plow thrown road salts has long been recognized as sources of salt damage (Bernstein 1975). The heavy cuticles commonly found on many seashore and salt marsh plants, such as *Ammophila arenaria* L., *Quercus obtusiloba* Michx., *Ilex opaca* Ait., and *Pinus thunbergii* Parl.(Harshberger 1909, Simini & Leone 1986), and salt tolerant plants, such as *Thellungiella halophilla* (C.A. Meyer) O.E. Schultz and *Thellungiella parvula* (Schrenk) Al-Shehbaz & OKane (Teusink et al. 2002) may not only provide an aerial barrier to water loss from plant tissues growing in desiccating saline soils, but the microrelief (microtopography) of the cuticle may also repel saline water droplets, and prevent the deposition of salt on the plant surface (Barthlott & Neinhuis 1997). For example, halophytic leaves like those of the salt spray zone and salt marsh ecotypes of halophyte *Agrostis stolonifera* L. exhibit lower wettability (higher advancing contact angles) and lower leaf sodium retention than those of inland ecotypes. The differences in wettability are likely attributed to differences in the physico-chemical and structural properties of waxes at the outermost surface (Ahmad & Wainwright 1976). Besides its role in the prevention of water loss, the cuticle may prevent the infiltration of toxic sodium ions into leaves. It is interesting to contemplate the function of cuticles as barriers to salt uptake based on recent studies that show that cuticles possess distinct polar pathways through which only charged molecules, like salt ions, can pass (Schreiber et al. 2001, Schreiber 2005, 2006). Potentially, future studies may show that plants adapted for the prevention of salt ion permeation into leaves have cuticles with unique polar pathways of diffusion.

Halophytic cuticles may reduce the uptake of salt from the rhizosphere into the plant transpiration stream by reducing overall transpiration rates. Growth of the halophyte *Suaeda maritima* (L.) Dumort. in elevated salt (0.34 M NaCl) concentrations led to pronounced changes in cuticle ultrastructure and wax crystal morphology. In salt grown *Suaeda maritima*, a 60% increase in cuticle thickness was accompanied by a 35% reduction in cuticular transpiration (g_{min}; (Hajibagheri et al. 1983). Whether this reduced total salt uptake in the transpiration stream by *Suaeda maritima* was not determined. These and other studies sugget that the cuticle plays a significant role in plant salt stress tolerance. Additional studies to elucidate the many possible mechanisms of cuticle function in plant salt tolerance are surely warranted.

4. MOLECULAR-GENETIC INVOLVEMENT IN DROUGHT AND SALT STRESS FUNCTIONS OF THE CUTICLE

Using mutagenesis and candidate gene testing strategies, a large collection of new plant genes directly linked to cuticle production has been identified, mostly in *Arabidopsis thaliana*, but also *Zea mays* L., *Medicago truncatula* Gaertn., and *Lycopersicon esculentum* Mill (Table 1). These mutants can be divided into three basic classes or types, 1) cuticular wax mutants, 2) cutin mutants and 3) mutants altered in both wax and cutin. Various means have been used to characterize the impact of mutations in these cuticle-associated genes on cuticle structure and composition, and overall plant physiology and growth. A series of assays have revealed that several mutants having altered cutin composition also possess greatly elevated transpiration rates (Goodwin & Jenks 2005). Surprisingly, mutants having only wax defects show very little or no change in transpiration. Since waxes are thought to be the hydrophobic cuticle component (whereas cutin is thought to be more hydrophilic), it is difficult to explain these results. Based on the recent model of cuticle structure wherein cutin provides a kind of framework that supports the packing of wax molecules into discreet crystalline and amorphous regions, one possible interpretation is that the framework or support function of cutin is very important in establishing cuticle permeability properties. Why doesn't the reduction in wax amount, observed on existing mutants, cause major changes in transpiration? Possibly, wax amounts are simply not reduced enough to begin to have an impact. In this scenario, it might be proposed that less wax is needed to form an effective cuticle water barrier than is normally produced on these plants. What the minimal wax load required to provide a normal water barrier is uncertain. A clue may come from transpiration studies of *Arabidopsis* wax mutants. *Arabidopsis* mutant *cer1* appears to have normal cutin, but has amongst the lowest leaf wax load of all wax mutants, 38% of wild-type wax amounts, respectively (Jenks et al. 1995). Curiously, despite demonstrating increased rates of excised stem water loss, *cer1* demonstrates only a minor increase in whole plant, night-time transpiration (g_{low}) rates (Goodwin and Jenks, 2005). A series of double *cer* mutants were recently created (Goodwin et al. 2005) with one goal being to lower the wax amounts further than identified wax mutants and test for effects on water loss. Surprisingly, of 14 new double *cer* mutants, only two had lower leaf wax than the *cer1* mutant; *cer1cer3* and *cer1cer4* had leaf wax quantities only 30% and 22% of wild type (Jenks, unpublished). Whether these double mutants have elevated transpiration has not been determined. Notwithstanding, the use of mutant and transgenic approaches to create extreme wax deficiencies could provide a powerful tool to answer these questions.

Additional clues to understanding the role of waxes in cuticle permeability came from a study by Vogg et al. (2004). Using both physical and genetic approaches to modify aliphatic epicuticular and intracuticular wax deposition of astomatous tomato fruit cuticles, they provided firm evidence that intracuticular waxes, and not epicuticular waxes provide the major permeability barrier function to the cuticle. Interestingly, intracuticular waxes in tomato include large amounts of triterpenoids, in addition to the typical aliphatics. Previous experimental and

Table 1. GENEVESTIGATOR data mining of water stress- and ABA-induced gene expression of cloned *Arabidopsis* cuticle genes and association of molecularly characterized cuticle genes with permeability defects as determined by chlorophyll leaching, whole plant or excised organ water loss curves, relative sensitivity to herbicides, organ fusion, or toluidine blue (TB) staining. Genes with a substantially increased abundance of transcript (ratio ≥ 1.4) are presented in bolded text. The ratio of treatment expression level to control expression level is given in a linear scale

Locus ID or Genbank #	Gene symbol	Predicted protein/function	10 μm ABA	Drought	300 mM mannitol	150 mM NaCl	Probeset	Associated permeability function	References
AT1G72970	*ACE/HTD*	Fatty Acid ω-alcohol dehydrogenase	1.178	0.811	0.222	0.605	262376_at	Yes	1,2, 3
AT1G36160	*ACC1/GK/PAS2*	Acetyl CoA-Carboxylase	**1.886**	0.967	1.239	1.051	263192_at	Yes	4,5,6,7
AT3G59420	*ACR4*	Receptor-like Protein Kinase	0.442	0.787	0.525	0.732	251521_at	Yes	8,9
AT1G62340	*ALE1*	Subtilisin-like Serine Protease	1.153	1.213	1.191	1.139	260630_at	Yes	10
AT1G64670	*BDG*	α/β-Hyrdrolase fold protein	1.141	0.884	0.697	0.864	261949_at	Yes	11
AT1G02205	*CER1*	Sterol Desaturase-like	**6.514**	1.012	**4.123**	**5.374**	264147_at	Yes	12,13,14,15,16
AT4G24510	*CER2*	Novel Regulatory Protein	**2.333**	0.805	**1.436**	1.179	254122_at	N.D.	13,17,18
AT4G33790	*CER4*	Fatty Acyl-CoA Reducatase	0.793	0.914	0.805	0.991	253309_at	Yes	13,19
AT1G51500	*CER5*	ABC Transporter	**1.991**	0.957	1.396	1.294	260490_at	Yes	15,16,20
AT1G68530	*CER6*	β-ketoacyl-CoA Synthase	**1.776**	1.240	**2.969**	**2.337**	260267_at	Yes	15,18,21,22,23,24
AT3G55360	*CER10*	Enoyl-Coa Reductase	1.106	0.840	0.637	0.869	251796_at	Yes	16,25
AT1G25450	*CER60*	β-ketoacyl-CoA Synthase	**1.875**	0.859	0.512	0.756	255732_at	N.D.	23,26
AT4G00360	*CYP86A2/ATT1*	Cytochrome P450-Dependent Monooyxgenase	**1.922**	1.212	**1.913**	**1.618**	255690_at*	Yes	13,27,28
AT2G45970	*CYP86A8/LCR*	Cytochrome P450-Dependent Monooyxgenase	1.294	1.031	0.712	0.897	266921_at	N.D.	27,29

(*Continued*)

Table 1. (Continued)

Locus ID or Genbank #	Gene symbol	Predicted protein/function	10 µm ABA	Drought	300 mM Mannitol	150 mM NaCl	Probeset	Associated permeability function	References
AT1G08510	*FATB1*	Acyl-Acyl Carrier Protein Thioesterase	1.273	0.978	1.300	**1.443**	261722_at	N.D.	30,31,32
AT2G26250	*FDH*	β-ketoacyl-CoA Synthase-like	1.234	0.872	0.975	1.084	267377_at	Yes	1,16,33,34,35,36,
AT2G46720	*HIC1*	β-ketoacyl-CoA Synthase	1.103	0.884	1.308	**1.400**	266319_s_at*	N.D.	37
AT1G01120	*KCS1*	β-ketoacyl-CoA Synthase	**2.748**	1.096	**1.469**	1.341	261570_at	Yes	38
AT1G49430	*LACS2*	Acyl-CoA Synthetase	**2.113**	0.827	0.868	0.759	262414_at	Yes	39
AT3G27670	*RST1*	Novel Protein	0.913	0.988	1.382	0.923	258238_at	No	40
AT5G57800	*WAX2/YRE/FLP1*	Sterol Desaturase/Dehydrogenase/ Reductase-like	**1.821**	1.065	1.152	1.194	247884_at	Yes	41,42,43
AT1G15360	*WIN1/SHN1*	AP2/EREBP Transcription Factor	0.582	1.018	1.196	1.345	262595_at	Yes	44, 45
AT4G14440	*N.A.*	β -Hydroxyacyl-CoA dehydratase	**1.551**	1.013	**1.460**	1.072	245612_at	N.D.	46
N.A.	*LeCER6*	β-ketoacyl-CoA synthase	N.D.	N.D.	N.D.	N.D.		Yes	47
U67422	*CR4*	Receptor-like Protein Kinase	N.D.	N.D.	N.D.	N.D.		Yes	48
AY505498	*GL1*	Desaturase/Receptor	N.D.	N.D.	N.D.	N.D.		No	49,50
X88779	*GL2*	Novel Regulatory Protein	N.D.	N.D.	N.D.	N.D.		N.D.	50,51
U89509	*GL8a*	β-ketoacyl-CoA Reductase	N.D.	N.D.	N.D.	N.D.		N.D.	50,52,53

AF527771	*GL8b*	β-ketoacyl-CoA Reductase	N.D.	N.D.	N.D.	N.D.		N.D.	53
AY714877	*GL15*	APETALA2-like Transcription Factor	N.D.	N.D.	N.D.	N.D.		Yes	50,54,55,56,57,58
N.A.	*WXP1*	AP2/EREBP Transcription Factor	N.D.	N.D.	N.D.	N.D.		Yes	59

Notes: * indicate ambiguous probe binding; N.A., not available; N.D., not determined; Experimental details available at: http://www.arabidopsis.org/info/expression/ATGenExpress.jsp

Cited Literature: 1. (Lolle et al. 1998), 2. (Krolikowski et al. 2003), 3. (Kurdyukov et al. 2006b), 4. (Baud et al. 2003), 5. (Baud et al. 2004), 6. (Bellec et al. 2002), 7. (Faure et al. 1998), 8. (Gifford et al. 2003), 9. (Tanaka et al. 2002), 10. (Tanaka et al. 2001), 11. (Kurdyukov et al. 2006a), 12. (Aarts et al. 1995), 13. (Goodwin & Jenks 2005), 14. (Hülskamp et al. 1995), 15. (Koornneef et al. 1989), 16. (Tanaka et al. 2004), 17. (Xia et al. 1996), 18. (Preuss et al. 1993), 19. (Rowland et al. 2006), 20. (Pighin et al. 2004), 21. (Fiebig et al. 2000), 22. (Hooker et al. 2002), 23. (Millar et al. 1999), 24. (Jenks et al., unpublished), 25. (Zheng et al. 2005), 26. (Trenkamp et al. 2004), 27. (Duan & Schuler 2005), 28. (Xiao et al. 2004), 29. (Wellesen et al. 2001), 30. (Bonaventure et al. 2003), 31. (Bonaventure et al. 2004a), 32. (Bonaventure et al. 2004b), 33. (Lolle et al. 1992), 34. (Lolle et al. 1997), 35. (Pruitt et al. 2000), 36. (Yephremov et al. 1999), 37. (Gray et al. 2000), 38. (Todd et al. 1999), 39. (Schnurr et al. 2004), 40. (Chen et al. 2005), 41. (Chen et al. 2003), 42. (Kurata et al. 2003), 43. (Ariizumi et al. 2003), 44. (Aharoni et al. 2004), 45. (Broun et al. 2004), 46. (Garcia et al. 2006), 47. (Vogg et al. 2004), 48. (Becraft et al. 1996), 49. (Sturaro et al. 2005), 50. (Beattie & Marcell 2002), 51. (Tacke et al. 1995), 52. (Xu et al. 1997), 53. (Dietrich et al. 2005), 54. (Lauter et al. 2005), 55. (Moose & Sisco 1996), 56. (Evans et al. 1994), 57. (Moose & Sisco 1994), 58. (Evans et al. 1994), 59. (Zhang et al. 2005) theoretical evidence indicates that terpenoids do not hinder water diffusion through the cuticle as well as long chain aliphatics do, perhaps due to poor packing in the cutin network and displacement of areas of tightly packed crystalline regions (Grncarevic & Radler 1967). This high terpenoid content in tomato cuticles may in part, explain why fruit cuticles, even though they are much thicker than most leaf cuticles, are nevertheless more permeable. More studies are needed to determine the role of specific aliphatics and terpenoids in determining cuticle permeability.

Studies using transgenic plants have provided further insight into the role of wax in cuticle function. Overexpression of the *Medicago truncatula* gene *WXP1* in *Medicago sativa* L., encoding a putative AP2/EREBP family transcription factor, caused a 37.7% increase in leaf cuticle wax deposition, primarily due to an increase in alcohols, that was associated with reductions in both water loss rate and chlorophyll efflux (Zhang et al. 2005). Surprisingly however, overexpression of *Arabidopsis WXP1* homolog *WIN1/SHN1*, in *Arabidopsis* led to a similar increase in *Arabidopsis* leaf wax accumulation, in this case primarily alkanes, but an increased rate of excised leaf water loss and chlorophyll leaching (Aharoni et al. 2004, Zhang et al. 2005). These contrasting results are quite peculiar since, intuitively increased amounts of hydrophobic wax in the *Arabidopsis* cuticle would be expected to reduce cuticle permeability by theoretically making more aliphatic crystalline wax regions. One possible explanation is that the increased wax displaced normal wax packing in the cutin framework of *Arabidopsis* differently than had occurred in *Medicago*, leading to an increase in the number and/or size of diffusion pathways. However, this is still quite speculative. The *lacs2* mutant (defective in the acyl-CoA synthase encoding *LACS2*) exhibits increased leaf wax amounts, especially alkanes, and like the *WIN1* overexpressor has increased chlorophyll efflux (Schnurr et al. 2004). By comparison, the *Arabidopsis* mutant *bdg* (defective in an α/β-hydrolase encoded by the *BDG* gene) exhibits large increases in the amount of alkane and aldehyde waxes, and this too is associated with increased chlorophyll efflux (Kurdyukov et al. 2006a). Once again, increased wax deposition leads to an unexpected increase in cuticle permeability. A mechanistic explanation for how transgenic overexpression of cuticle-associated genes that increase wax deposition but in one case increase leaf cuticle permeability, and in other cases decrease permeability, is still unavailable.

As mentioned above, changes in cutin structure or chemical composition cause a signficant change in permeability. The *att1*, *cer25*, *hth*, and *wax2* mutants all show reductions in the total amount of cutin monomers, a change in cutin monomer profiles, and a disrupted cuticle membrane (cutin) ultrastructure (Goodwin & Jenks 2005, Kurdyukov et al. 2006b). The *lacs2* mutant likewise shows a disruption in the cutin layer (Schnurr et al. 2004). All of these mutants show much higher leaf cuticle permeability than their respective isogenic wild-type parents (Lolle et al. 1997, Xiao et al. 2004, Goodwin & Jenks 2005). Notably however, *att1* and *wax2* have thicker cuticle membranes than wild type whereas *cer25* and *hth* have thinner cuticles, lending further support to arguments that cuticle thickness is not a primary determinant of cuticle permeability. Compared to all other cutin mutants, the *bdg* mutant is unique because it possesses more total cutin monomers than wild type and also a thicker cuticle membrane (Kurdyukov et al. 2006a). Cuticle permeability in *bdg* like other cutin mutants is greatly elevated. As such, elevation in cutin monomer deposition does not necessarily lead to reduced cuticle permeability as might be expected. It was postulated that the BDG protein plays a role in cross-linking cutin monomers. It is interesting to note that like the *bdg* mutant, *att1*, *hth*, and *wax2* have highly disorganized cuticle membrane ultrastructure leading to speculation that these too may be defective in cutin cross-linking. A new hypothesis can then be set forward here that cutin cross-linking may be a major determinant of a cuticle's permeability function. Recent characterization of *hth* cutin monomers reveals specific reductions in C_{16} mid-chain oxygenated hydroxyacids, C_{18} α/ω-dicarboxylic acids, and increased levels of precursor molecules C_{18} ω-hydroxy acids, all monomers with abundant hydroxy groups that should be important cross-linking sites. Potentially, a higher degree of cross linking among cutin monomers creates a denser or robust cutin scaffolding in which to pack wax molecules, thereby creating more, larger, or more dense crystalline regions. As well, linking hydroxyls into covalent, ester bonds precludes these polar groups from potential interactions with water and may cause reductions in polar pathways of diffusion in the cuticle membrane. Targeted studies are needed to determine whether more cutin cross-linking creates a less permeable cuticle.

5. CUTICLE FUNCTIONS ASSOCIATED WITH THE STOMATAL COMPLEX

New evidence suggests that the cuticle plays a major role in controlling stomatal transpiration. Microscopy studies of leaves and stems reveal that a cuticle membrane covers the entire surface of the substomatal chamber; the cavity below the stomatal pore made of inner walls of the guard cells and outer mesophyll cells (Osborn & Taylor 1990). In addition, the cuticle and cell wall forms a unique structure at the outer rim of the stomatal pore called the stomatal or cuticular ridge, a structure that forms an outer cavity above the pore called the stomatal ante-chamber (Zhao & Sack 1999, Jenks 2002). Many plants adapted to arid zones possess large and/or multiple rows of these ridges, and speculation has it that these ridges help seal the pore more tightly at night and during periods of high vapor pressure deficit or drought (Jenks 2002). If one erroneously assumes that differences in stomatal complex cuticle do not contribute to observed differences in water loss between cuticle mutants, then it might be postulated that water loss rates in light when the stomata are open, would be the same as observed night-time differences in transpiration rate (i.e. the amount of water loss from stomatal pores did not differ in these isogenic lines). However, transpiration studies of *Arabidopsis* wild type and the *wax2* demonstrate

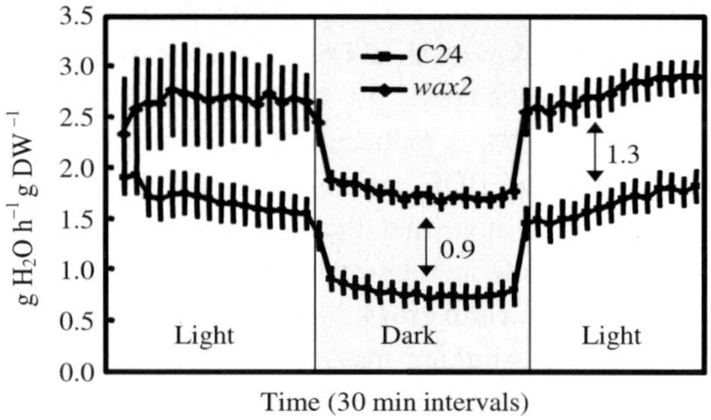

Figure 1. Dark and light transpiration of *Arabidopsis* wild type (C24) and *wax2* mutant revealing an unexpectedly higher transpiration differential in light than in darkness

that differences in water loss are 44% higher in lighted than in dark conditions (Figure 1). These differences appeared immediately upon lighting, and since a water filter was used to remove infrared heating at the plant surfaces, we assumed these differences were due to differences in water movement through stomatal pores, (i.e. more water vapor escaped the *wax2* stomata than those of wild type). Leaf areas and other aspects of leaf morphology were essentially identical, and an analysis of stomatal index revealed that the *wax2* mutant actually had slightly fewer stomata per unit area. The elevated water loss in *wax2* could not thus be attributed to an increase in the number of stomatal pores. Electron microscopy studies revealed, as expected, significant morphological alterations in the cuticles of the *wax2* mutant's stomatal complexes (Figure 2). Not only had the cuticle lining the substomatal chamber been disrupted, but also the size of the stomata's cuticular ridges on *wax2* was greatly reduced. Comparable results were found for *att1* and *cer25* (Xiao et al., 2004; Jenks, unpublished). Recent studies on polar pathways of cuticle transport suggest that guard cell and trichome cuticles may be more permeable to polar compounds (including water) as the cuticle of these epidermal cells demonstrate preferential precipitation of externally applied ionic salts of silver (Schlegel et al. 2005, Schreiber 2006, Schreiber et al. 2006). As well, hybrid *Populus* clones grown under water limiting conditions demonstrate increased cuticle deposition over leaf stomata (Pallardy & Kozlowski 1979). Likewise, wax deposits in the stomatal antechamber of *Picea sitchensis* (Bong.) Carr. were calculated to reduce the rate of leaf transpiration (Jeffree et al. 1971). Collectively, these results show that the cuticle plays a critical role in determining water loss through the stomatal pore. Future studies must now consider that transpiration rate measurements from stomatous organs can be impacted by the cuticle of the substomatal chamber and cuticle ridge, and that even a diminutive stomatal cuticle as in *Arabidopsis* will impact water loss through the stomatal pore. To what degree the separate stomatal ridges and substomatal cuticles contribute to differences in total water loss as observed is still unclear.

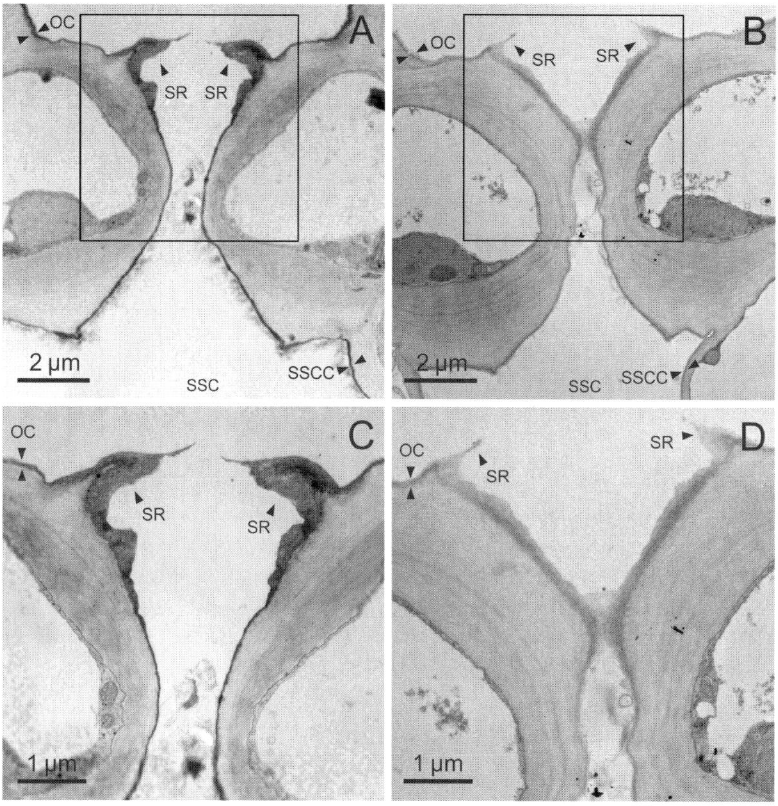

Figure 2. Stomatal complex of an internodal segment of *Arabidopsis thaliana* inflorescence stem of C24 wild type (A, C) and isogenic *wax2* (B, D) demonstrating alterations in the stomatal complex cuticle of *wax2*. Cuticle features are annotated as follows; outer cuticle (OC), stomal ridge (SR), substomatal chamber (SSC), substomatal chamber cuticle (SSCC). C and D are enlarged views of boxed areas from A and B

6. CUTICLE RESPONSE TO DROUGHT AND SALT

The cuticle has often been regarded, inaccurately, as a preformed, constitutive (i.e. non-responsive) morphological adaptation to water limited environments. In fact, cuticle wax metabolic pathways respond to osmotic stress in a very plastic manner, even in xerophytes, (Ahmad & Wainwright 1976, Hajibagheri et al. 1983). A typical cuticle response to water stress is an increase in cuticular wax quantity (Skoss 1955, Bondada et al. 1996, Jenks et al. 2001, Sanchez et al. 2001, Samdur et al. 2003, Cameron & Teece 2006, Kim et al. In preparation, Kim et al. In Press). In fact, an increase in leaf cuticular wax production by water stress exposure appears to be a near-universal response across the plant Kingdom, even in such ephemeral plants as *Arabidopsis* (Figure 3). In plants such as *Nicotiana glauca* Graham, the total leaf wax induction by drought treatments can exceed 150% (Cameron et al.,

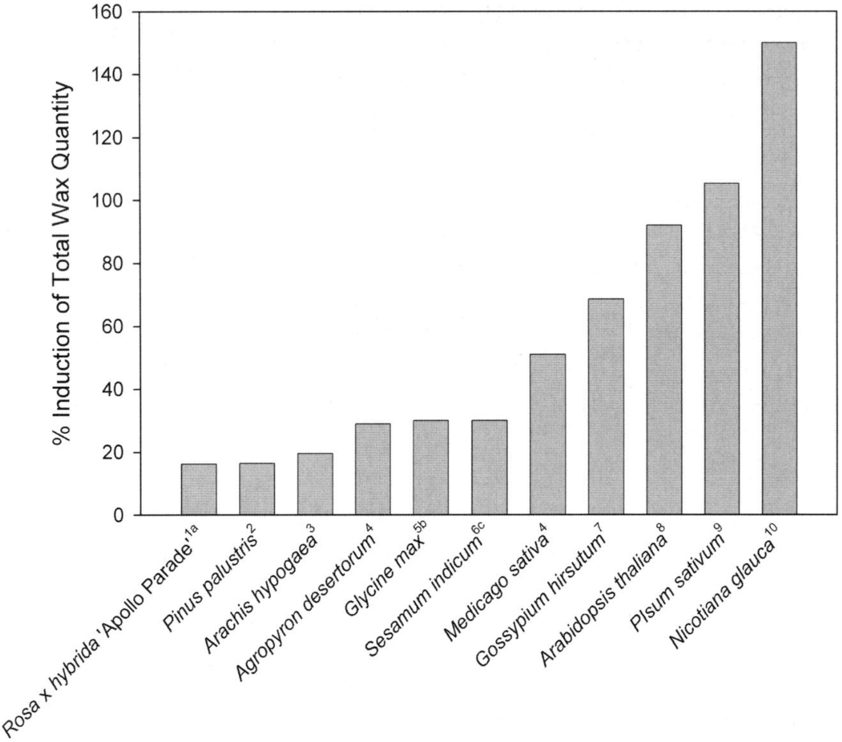

Figure 3. Percent induction (relative to non-treated controls) of total leaf cuticle wax quantity of several plant species resulting from drought treatment.
Cited Literature: 1. (Jenks et al. 2001), 2. (Prior et al. 1997), 3. (Samdur et al. 2003), 4. (Jefferson et al. 1989), 5. (Kim et al., in preparation), 6. (Kim et al. In Press), 4. (Jefferson et al. 1989), 7. (Bondada et al. 1996), 8. (Kosma et al., unpublished), 9. (Sanchez et al. 2001), 10. (Cameron & Teece 2006)
Notes: [a]maximum induction on most responsive cultivar; [b]mean of 17 cultivars; [c]mean of 18 cultivars visualize changes in signal intensity levels

2005). Cuticle induction can also arise from salt exposure (NaCl). When *Suaeda maritima* is grown in NaCl solution a 60% increase in cuticle thickness is observed, as well as thickening and increased density of epicuticular wax crystals (Hajibagheri et al. 1983). Cuticle thickening during growth in saline conditions is also observed in *Simmondsia chinensis* (Link) Schneider (Botti et al. 1998). Like wax, cutin monomer amounts on *Arabidopsis* leaves are also significantly increased by periodic salt treatment (Kosma & Jenks, unpublished results).

In the case of drought stress, induction of cuticle is observed in angiosperms, gymnosperms, xerophytes, and mesophytes, and is not limited to leaves alone but can also include stems and fruits (Skoss 1955, Bondada et al. 1996, Jenks et al. 2001, Sanchez et al. 2001, Samdur et al. 2003, Cameron & Teece 2006, Kim et al. In Press). In many of these studies, the induction occurs over a few days on preformed leaves, indicating that actual wax metabolic pathways have been induced. In other cases however, leaves formed during the drought may be smaller and, at least part of the measured increase in wax per leaf area may be due to shrinkage of the leaf and epidermal cell size (i.e. stress reduced surface area) rather than increased wax metabolism, per se. If leaf areas change noticeably, it may be best to represent total wax amount as a function of epidermal cell density (i.e. wax quantity per epidermal cell).

As it relates to plant growth in arid and saline environments, cuticle alterations are also elicited by non-osmotic stresses associated with dry climates like high temperature and intense of solar radiation (Skoss 1955, Steinmüller & Tevini 1985, Manetas et al. 1997, Gordon et al. 1998b). Skoss (1955) showed an increase in leaf wax weight of *Nicotiana glauca* with increasing temperature. He also showed that increasing temperature decreased the percentage of the total cuticle weight comprised by cutin. Light quantity and quality also have substantial impacts on cuticle anatomy and composition. The cuticles of sun leaves of *Quercus coccinea* Muenchh., *Quercus rubra* L. and *Quercus velutina* Lam. are nearly twice as thick as the cuticles of shade leaves (Ashton & Berlyn 1994). In *Quercus velutina* this holds true even for cuticle regions that cover the stomatal pore and extend into the substomatal chamber (Osborn & Taylor 1990). Ultraviolet-B (UV-B) light exposure (λ = 280-315 nm) causes differential increases of various wax components of seedlings in different *Picea* species (Gordon et al. 1998a). Enhanced levels of UV-B radiation in combination with water stress caused a two-fold increase in needle cuticle thickness of *Pinus pinea* L. Curiously, this increase was not evident in plants subjected solely to water stress or UV-B alone, indicating the potential for regulatory cross-talk between stress response pathways (Manetas et al. 1997). The induction of cuticle alterations by UV-B is not limited to conifers. Steinmüller and Tevini (1985) demonstrated that enhanced levels of UV-B stimulate a general increase in total wax amount (*ca.* 25%) on cucumber petioles (*Cucumis sativus* L. cv. Delikatess), barley leaves (*Hordeum vulgare* L. cv. Villa), and bean leaves (*Phaseolus vulgaris* L. cv. Favorit); in all three species an increased proportion of shorter carbon-chain length wax constituents explained the increase in total wax amount. In general, plants seem to respond to UV-B exposure with an increase in the proportion of short chained and in some cases branched aliphatics (Barnes et al. 1996). Curiously, UV induced changes in cuticle composition actually increase wettability of the leaf surface and cuticle permeability (Kerstiens 1994, Barnes et al. 1996). It is still unclear what biological advantages, if any, can be obtained by increasing cuticle wettability and permeability under high UV, or deposition of shorter chain wax components.

In addition to increasing total wax amount (Figure 3), drought and salt treatments differentially induce changes in the amounts of different wax constituent classes (e.g. alkanes, alcohols, aldehydes, etc.). When exposed to moderate drought stress, *Gossypium hirsutum* L. leaf alkane content increases from 11% to 66% of total waxes (Bondada et al. 1996) whereas *Sesamum indicum* L. plants show 30% and 13% increases in total leaf wax alkanes and aldehydes, respectively (Kim et al. In Press). In *Rosa* x *hybrida* prolonged drought stress causes moderate but significant increases in acids (C_{32}), aldehydes (C_{28} and C_{32}), and alkanes (C_{27}, C_{29}, C_{33}; (Jenks et al. 2001). In these three species, water stress causes an increased flux through the elongation and decarbonylation pathways of alkane synthesis specifically. This corresponds with previous studies that suggest that alkanes efficiently form crystalline regions theoretically most effective in limiting diffusion of water molecules (Reynhardt 1997, Jenks 2002). In general, longer chain-aliphatic waxes forming crystalline structures are attributed as being responsible for the barrier properties of the cuticle (Riederer & Schreiber 1995, Burghardt & Riederer 2006). Curiously, in insects it is generally agreed upon that warm, dry-climate inhabiting species, that exhibit the lowest rates of cuticular water loss, have cuticles containing longer chain-length alkanes (Lockey 1988, Gibbs 1998). The fact that many desert plants have greater long-chain ($>C_{31}$) alkane content supports a hypothesis that longer chain alkanes may contribute to reduced cuticle permeability (Wilkinson & Mayeux 1990, Stevens et al. 1994).

The cuticles of Graminaceous species may present a different strategy for responding to osmotic stress. Studies of *Avena sativa* L. and *Hordeum vulgare* L. have shown that imposed, periodic reductions in leaf water potential do not increase total cuticular wax quantities however, significant alterations in composition do occur (Larsson & Svenningsson 1986, Svenningsson & Liljenberg 1986, Svenningsson 1988). In some cultivars, lowering leaf water potential of *Avena sativa* L. seedlings lead to increases in the proportion of total epicuticular waxes comprised by fatty acids, alkanes, and primary alcohols. Interestingly, reducing leaf water potential increased the quantity of leaf intracuticular primary alcohols with a shift to shorter chain alcohols (C_{24} and C_{26}) and a reduction in longer chain alcohols (C_{28}) (Svenningsson & Liljenberg 1986). Similarly, several cultivars of *Hordeum vulgare* exhibit substantial alterations in the make-up of their leaf cuticular waxes when subjected to periodic reductions in leaf water potential manifested as a doubling in the percent of total wax comprised by esters and a reduction in the percent of wax made up of aldehydes and alcohols, also without a concurrent increase in total leaf wax amount (Larsson & Svenningsson 1986). These stress induced changes in wax composition of *Hordeum vulgare* were accompanied by shifts in the chain length distribution of wax constituent classes, with slight increases in longer chain alkanes (C_{31} and C_{33}) and esters (C_{48}) and reductions in alcohols (C_{26}) and fatty acids (C_{26}). Later research on multiple cultivars of *Hordeum vulgare* indicates that the observed increase in esters was largely attributed to an increase in the percent of total wax comprised, specifically, by epicuticular esters. It is interesting to note that Graminaceous species may exhibit different cuticle responses to decreased water potential. However, some caution should be used when interpreting the results of the aforementioned experiments pertaining to *Avena sativa* and *Hordeum vulgare*. Reductions in leaf water potential were accomplished by reducing root temperatures to 1.0°C for several hours, thus the changes in wax composition might actually reflect a cold-stress response. Nonetheless, consideration should be given to the notion that Graminaceous species may have developed unique mechanisms of response to drought and other osmostic stress.

On another note, studies of the impact of drought and salt stress on the cuticle of the stomatal complex have not been published. Based on the above discussions, it might be assumed that an acclimation treatment would lead to changes in structure and composition of cuticle in the substomatal chamber, and even change the size and functioning of the stomatal ridges. Further inquiry into stomatal cuticle response to stress, and its effect on stomatal water loss, should prove illuminating. Previous studies indicate that the role of induced cuticle synthesis is to reduce transpiration rate as a means to conserve water. Recent work in *Nicotiana glauca* showed that leaves of plants subjected to periodic drying had increased total wax quantity (1.5 to 2.5 fold) and exhibited a slower rate of water loss in the dark, suggestive of a negative relationship between total wax amount and water loss when stomata are closed (Cameron & Teece 2006). A similarly reduced transpiration rate after wax induction was evident in two *Rosa* cultivars (WIlliams et al. 1999, Jenks et al. 2001) and *Arabidopsis thaliana* (Kosma et al., unpublished). In the halophyte, *Suaeda maritima*, g_{min} declined in a step wise manner with increased sodium chloride concentration and cuticle thickness (Hajibagheri et al. 1983). Whether induced changes in cuticle permeability are responsible for reduced plant transpiration is still not verified. It must be considered that drought, salt, or other stress treatments can cause dramatic physiological changes other than changes in cuticle permeability that could impact plant transpiration rate measurements typically used in water relations studies. For example, residual (g_{min} or g_{low}) and day-time transpiration could be influenced by stomatal pores that close more fully after the stress, leaf cell adjustment to lower osmotic potentials, or even changes in stomatal index on leaves that develop during the stress. Notwithstanding, the very large induction in cuticle amount by these stress treatments indicates a stress tolerance function for cuticle, and a reduction in cuticle permeability specifically due to induced changes in cuticle appears likely. Future in-depth studies to link cuticle changes during drought and salt exposure to changes in cuticle permeability could shed much light on cuticle stress functions.

7. RESPONSE OF CUTICLE-ASSOCIATED GENES TO DROUGHT AND SALT

With the advent of genomics, an abundance of information about gene transcription profiles from stress and other treatments is now available on-line. Recent work has aimed at answering questions pertaining to developmental regulation of cuticle gene expression (Costaglioli et al. 2005, Suh et al. 2005). Data mining using the GENEVESTIGATOR meta-analysis tool (Zimmerman et al. 2004) can provide a unique look at the stress response of *Arabidopsis* genes associated with cuticle synthesis (Table 1). Using a gene expression ratio (treatment:control) of 1.4 as an arbitrary cutoff, it is observed that many cuticle-associated genes are induced by drought, salt (150 mM NaCl), low osmotic potential (300 mM mannitol), or the stress hormone abscisic acid (ABA; 10 μM). ABA induces many genes including, *ACC1*, *CER1*, *CER2*, *CER5*, *CER6*, *CER60*, *CYP86A2* (*ATT1*), *KCS1*, *LACS2*,

WAX2/YRE, and a gene encoding a β-hydroxyacyl-CoA dehydratase (At4g14440). At4g14440 is purported to be a component of the acyl-CoA elongase complex (Garcia et al. 2006). The gene with by far the highest induction by ABA treatment was *CER1* (ratio >6) however; *ACC1*, *CER2*, *KCS1*, and *LACS2* demonstrated significant upregulation (ratio >2) in response to ABA as well. *CER1*, *CER2*, *CER6*, *CYP86A2*, *KCS1*, and At4g14440 showed elevated transcript abundance to both osmotic stress and ABA treatment. The genes induced by salt, osmotic stress, and ABA, were *CER1*, *CER6*, and *CYP86A2*. A *CYP86A2* stress response cannot be absolutely determined from the GENEVESTIGATOR meta-analysis data as that a non-specific probe was used; the expression values given for *CYP86A2* may actually represent more than one gene (Table 1). Specific expression of *CYP86A2* has been analyzed and is described in the following paragraph. Surprisingly, no genes were significantly induced by drought when a 1.4 cutoff is used. This may be an artifact of the nature of the drought treatment, which involved the removal of entire plants from in vitro culture and exposure to a sterile air stream. A drought treatment as such may not accurately represent the gene response to an actual drought condition experienced by soil-grown plants in a field setting. It is surprising that drought stress caused no induction since many cuticle genes were induced by ABA and it is well know that ABA synthesis is induced by drought (Zhu 2002). ABA increased the transcript abundance of ten out of twenty-five genes. Osmotic and salt stress have less broad-based induction capacity, leading to increased accumulation of transcript of six and four out of twenty-five genes, respectively. The fact that many genes involved in elongation of aliphatic wax precursors (*ACC1*, *CER2*, *KCS1*, *LACS2*, and At4g14440) and synthesis of alkanes (*CER1*) were upregulated by osmotic stress and ABA raises the possibility that elongation and decarbonylation pathways in *Arabidopsis* may be primary metabolic targets for osmotic stress regulatory responses. Curiously, the recently characterized *CER4* gene was apparently repressed by ABA and osmotic stress and unaffected by salt. *CER4* is thought to be responsible for synthesis of long chain primary alcohols in the epidermis of *Arabidopsis* (Rowland et al. 2006). Combined, these results suggest an increased synthesis of cuticular alkanes as a primary stress response in *Arabidopsis*. Not surprising given the large increase of alkanes in leaf waxes of *Gossypium hirsutum* and *Sesamum indicum* plants exposed to water deficit (Bondada et al. 1996, Kim et al. In Press). Only three of twenty-five genes were repressed by ABA, two being regulatory in nature (*ACR4* and *WIN1/SHN1*). Interestingly, overexpression of *WIN1/SHN1* leads to an increase in cuticle permeability (Aharoni et al. 2004); hence downregulation under water-limiting conditions is logical. Nevertheless, it is difficult to read too much into these results since cuticle metabolism rate-limiting steps are unknown, and many metabolic and regulatory cuticle genes are yet to be discovered. It is apparent that ABA-dependent pathways are involved in the cuticle stress response; all genes induced by osmotic or salt stress are induced by ABA. Moreover, since ABA is a key regulator of diverse plant stress responses, these results suggest that the cuticle pathway may function in plant responses to many other kinds of stress besides drought and salt stress. Research is needed to explore the cuticle stress response network, as it would not only further our understanding of the genetic and physiological mechanisms involved in the cuticle's stress response, but it would also aid in the identification of candidate genes for crop improvement such as key regulatory and highly stress responsive cuticle genes.

In addition to discoveries from GENEVESTIGATOR, published studies of several cuticle associated genes also show that they are highly responsive to drought, salt, mannitol, or ABA (Hooker et al. 2002, Duan & Schuler 2005, Zhang et al. 2005). *WXP1,* a *WIN1* homolog, encodes an AP2/EREBP transcription factor from *Medicago truncatula,* is highly induced by cold stress and ABA, and to a lesser extent by water deprivation (Zhang et al. 2005). Curiously, *WIN1*, also an AP2/EREBP, shows a reduction in transcript abundance in response to ABA and drought stress (Table 1). Overexpression of *WXP1* in *Medicago sativa* L. causes changes in the expression of genes homologous to *Arabidopsis* cuticle genes; most notable are increases in *CER2* and *LCR* homologs, and decreases in *WAX2* and *CER1* homolog expression. In contrast, *WIN1* overexpression in *Arabidopsis* causes a significant increase in the accumulation of *CER1*, *CER2*, and *KCS1* transcripts (Broun et al. 2004). It is unclear why overexpression of *WIN1* homolog, *WXP1*, in *Medicago sativa* causes a decrease in transcript abundance of a putative alkane generating *CER1*-like gene. Notably, *WIN1* is only 29% identical to *WXP1* (Zhang et al. 2005) . The differential regulation of *WIN1* and *WXP1* expression to drought and osmotic related stress and the effects of overexpression on wax chemistry suggest that cuticle stress responses may be quite different in *Arabidopsis* and the *Medicago* species examined here. Interestingly, overexpression of *WIN1* in *Arabidopsis* and *WXP1* in *Medicago sativa* both resulted in increased leaf wax production but a more permeable cuticle in *Arabidpsis* and a less permeable cuticle in *Medicago sativa. Medicago sativa* leaves, unlike *Arabidopsis*, have a wax profile dominated by alcohols rather than alkanes, also indicative of different cuticle synthetic pathways in these species. The cuticle-associated gene *CER6* is highly induced by osmotic stress (polyethylene glycol), salt stress, and ABA (Hooker et al. 2002). In some cases *CER6* is induced to a higher degree than well-known stress responsive gene, *RD29A* (Shinozaki & Yamaguchi-Shinozaki 1997, Hooker et al. 2002). *CER6* is involved the elongation of very long chain ($>C_{24}$) fatty acids, precursors that would be necessary for increased synthesis of wax componenets like alkanes and aldehydes (Millar et al. 1999). Collectively, the metabolic role and high induction of *CER6* by osmotic stress and ABA are suggestive of a major stress response function.

Cutin genes and cutin synthesis may play an important role in ameliorating or signaling osmotic or other stresses. *CYP86A2* (*ATT1*) is a cytochrome P450-dependent monooxygenase involved in cutin synthesis that is associated with cuticle permeability (Table 1, Xiao 2004). *CYP86A2* is transiently induced to high levels by ABA, mannitol, and water deficit (Duan & Schuler 2005). *CYP86A8* (*LCR*) is also a P450-dependent monooxygenase involved in cutin synthesis (Wellesen et al. 2001), that is transiently induced by ABA but not by salt, drought, or mannitol treatment (Duan & Schuler 2005). The inducibility of cutin genes by osmotic stress and ABA brings to light interesting questions about the role of cutin and cutin genes in ameliorating water deficit. Little attention has been given to the role of the cutin polymer in the barrier properties of the cuticle. All cutin mutants examined exhibit increased cuticle permeability (Lolle et al. 1998, Chen et al. 2003, Schnurr et al. 2004, Xiao et al. 2004, Goodwin & Jenks 2005, Kurdyukov et al. 2006b). Although highly conjectural, increased synthesis of cutin monomers like dibasic acids and glycerol, which are thought to play a role in cross-linking ester chains, may lead to a less permeable cuticle. Notwithstanding, a high proportion of the cuticle-associated genes thus identified are significantly responsive to drought, salt, and related treatments and clear examples of gene interactions are evident. These findings increase the probability that future work with existing and yet to be identified genes will uncover a significant role for cuticle response in plant tolerance to drought and salt stress.

8. OPPORTUNITIES FOR CROP IMPROVEMENT

Traditional breeding strategies have focused on glaucousness (i.e. surface deposition of epicuticular wax crystals) as a target for selection, and research of this type has succeeded in associating glaucousness to drought resistance in a few crop plants (Richards et al. 1986, Blum 1988). Studies of near isogenic lines of several Gramineaceous species (*Triticum durum* Desf., *Triticum aestivum* L., and *Hordeum vulgare* L.) have shown that glaucousness is associated with increased water use efficiency, grain yield, straw biomass, and yield index, and at least part of this positive effect was thought due to the cooler canopy temperatures that glaucousness provided by the ability of glaucous wax coatings to reflect solar radiation, a phenomenon especially important under water-limited conditions (Richards et al. 1986, Febrero et al. 1998, Merah et al. 2000). Similar results with regard to yield, have been found in advanced inbred lines (F_8) differing in glaucousness; lines derived by single seed descent from a cross between *Triticum aestivum* varieties Seri and Baviacora (Monneveux et al. 2004).

Genetic studies have revealed some interesting facts about the existing genetic variation for glaucousness and wax quantity in cultivated varieties. In *Oryza sativa* L., the inheritance of leaf wax quantity is polygenic in nature (Haque et al. 1992). With the great amount of intraspecific variation in wax amount and composition found within many other plants, such as *Arabidopsis thaliana* (Rashotte & Feldmann 1998), *Zea mays* (Blaker et al. 1989) and *Picea pungens* Engelm. (Jenks, unpublished), it seems quite likely that the wax profiles of other species will be controlled by numerous genes with multiple alleles of varying dominance. For example, in breeding populations of *Triticum aestivum*, glaucousness is determined by two duplicated genes, *W* and *Iw*, with a copy of each found on chromosome 2B and 2D. *W* likely functions as a facilitator of wax production, whereas the *Iw* locus acts in the inhibition of wax production (Tsunewaki & Ebana 1999). Studies in *Musa* sp. have asserted that non-glaucousness is encoded by a dominant allele (*Wx*) but that the action of modifier genes with additive type action affect *Wx* expression leading to a glaucous phenotype (Ortiz et al. 1995). Further studies are needed to assess the potential of using genetic selection of glaucousness to improve the agronomic performance of important crops. Investigation into the genetic control of cuticle traits like wax and cutin composition and amounts, cuticle ultrastructure, cuticle permeability, and the development of molecular markers for use in molecular breeding, awaits a fuller elucidation of the physiological function of these specific cuticle characteristics, and gene control over them.

9. TRANSGENIC APPROACHES TO IMPROVE DROUGHT AND SALT TOLERANCE USING CUTICLE ASSOCIATED GENES

Many genes associated with cuticle production have been identified, but as mentioned above, there is still a great need to functionally characterize these genes, and the many yet to be discovered, before targeted genetic modifications can be effectively designed. A recently published screening technique for identifying plants with elevated epidermal permeability using toluidine blue stain has much promise to identify these new genes in a high-throughput manner (Tanaka et al. 2004). Other strategies that should be explored are the development and use of chemical based screens (via gas chromatography-mass spectrometry), such as those used by Rashotte et al. (2004) to find new wax mutants in *Arabidopsis*. Second site mutagenesis of wild-type lines carrying stress responsive *LUC* reporters of cuticle synthetic and regulatory promoters, an approach described by Koiwa et al. (2006) and Ishitani et al. (1997), could likewise do much to identify valuable cuticle genes, perhaps most useful being regulators of signal transduction pathways of the cuticle stress response.

Very little research has been done to explore the use of existing cuticle genes in crop improvement, except in the recent publication on *Medicago sativa* L. (Zhang et al. 2005). Heterologous expression of *WXP1* in *Medicago sativa* L. resulted in improved plant drought tolerance in greenhouse assays. Ectopic 35S-driven overexpression of *WXP1* in *Medicago sativa* was not associated with severe negative pleiotropic effects seen in *Arabidopsis* overexpressing other cuticle genes, like *CER6* and *WIN1/SHN1* (Hooker et al. 2002, Aharoni et al. 2004). The use of promoters from genes like *WAX2*, recently shown to have epidermal-specific expression (Nakayama et al. 2005), and *CER6* may prove useful in ameliorating difficulties associated with constitutive overexpression of cuticle-related genes (Hooker et al. 2002). Several cuticle genes (*CER1*, *CER6*, *CYP86A2*, *KCS1*, *LACS2*, etc.; Table 1) are known to be responsive to various forms of abiotic stress (Hooker et al. 2002, Duan & Schuler 2005). The use of epidermis-specific and/or highly stress responsive cuticle gene promoters driving the expression of cuticle genes that control cuticle permeability, transpiration, and water conservation, may prove to be effective strategies for the production of drought tolerant crop species without undesirable effects on other agronomic traits.

10. CONCLUSIONS

Many plants, especially xero- and halophytic species, possess unique characteristics like low cuticle permeability that contribute to their capacity to conserve water and survive and reproduce in naturally arid and saline habitats. Recent studies have begun to shed light on the physico-chemical bases for variation in cuticle permeability however, these studies are still at an early theoretical stage with the main research emphases revolving around ideas that intracuticular waxes and the cutin polyester interact at a nanomolecular scale to establish the cuticle's barrier properties. The plant cuticle metabolic pathway is now known to respond to osmotic stress signals, including salt, water deficit, and ABA. Despite this, it is still unknown what exact role cuticle induction has in providing drought and salt tolerance, even though reduced cuticle permeability and transpiration rate are postulated as major outcomes. The regulatory mechanisms controlling the genetic and metabolic networks involved in wax and cutin synthesis are far from characterized. It is hoped that newly discovered genes that function in cuticle permeability will be useful in scientific exploration of cuticle function, and for crop improvement. Notwithstanding, further fundamental studies of gene control over cuticle synthesis and cuticle permeability are expected to contribute substantially to the molecular toolbox of plant physiologists, plant breeders, and biotechnologists in the development of drought and salt tolerant crops.

REFERENCES

Aarts MGM, Keijzer CJ, Stiekema WJ, Pereira A (1995) Molecular characterization of the *CER1* gene of arabidopsis involved in epicuticular wax biosynthesis and pollen fertility. Plant Cell 7:2115–2127

Aharoni A, Dixit S, Jetter R, Thoenes E, van Arkel G, Pereira A (2004) The SHINE clade of AP2 domain transcription factors activates wax biosynthesis, alters cuticle properties, and confers drought tolerance when overexpressed in Arabidopsis. Plant Cell 16:2463–2480

Ahmad I, Wainwright S (1976) Ecotype differences in leaf surface properties of *Agrostis stolonifera* from salt marsh, spray zone and inland habitats. New Phytologist 76:361–366

Ariizumi T, Hatakeyama K, Hinata K, Sato S, Kato T, Tabata S, Toriyama K (2003) A novel male-sterile mutant of *Arabidopsis thaliana*, *faceless pollen-1*, produces pollen with a smooth surface and an acetolysis-sensitive exine. Plant Molecular Biology 53:107–116

Ashton P, Berlyn GP (1994) A comparison of leaf physiology and anatomy of *Quercus* (section *Erythrobalanus*-Fagaceae) species in different light environments. American Journal of Botany 81:589–597

Bargel H, Koch K, Cerman Z, Neinhuis C (2006) Evans Review No. 3: Structure–function relationships of the plant cuticle and cuticular waxes — a smart material? Functional Plant Biology 33:893–910

Barnes JD, Percy KE, Paul ND, Jones P, McLaughlin CK, Mullineaux PM, Creissen G, Wellburn AR (1996) The influence of UV-B radiation on the physicochemical nature of tobacco (*Nicotiana tabacum* L.) leaf surfaces. Journal of Experimental Botany 47:99–109

Barthlott W, Neinhuis C (1997) Purity of the sacred lotus, or escape from contamination in biological surfaces Planta, p 1–8

Baud S, Bellec Y, Miquel M, Bellini C, Caboche M, Lepiniec L, Faure J, Rochat C (2004) *gurke* and *pasticcino3* mutants affected in embryo development are impaired in acetyl-CoA carboxylase. European Molecular Biology Organization Reports 5:515–520

Baud S, Guyon V, Kronenberger J, Wuilleme S, Miquel M, Caboche M, Lepiniec L, Rochat C (2003) Multifunctional *ACETYL-CoA CARBOXYLASE1* is essential for very long chain fatty acid elongation and embryo development in Arabidopsis. The Plant Journal 33:75–86

Beattie GA, Marcell LM (2002) Effect of alterations in cuticular wax biosynthesis on the physicochemical properties and topography of maize leaf surfaces. Plant, Cell & Environment 25:1–16

Becraft PW, Stinard PS, McCarty DR (1996) *CRINKLY4*: A TNFR-like receptor kinase involved in maize epidermal differentiation. Science 273:1406–1409

Bellec Y, Harrar Y, Butaeye C, Darnet S, Bellini C, Faure J-D (2002) *Pasticcino2* is a protein tyrosine phosphatase-like involved in cell proliferation and differentiation in *Arabidopsis*. The Plant Journal 32:713–722

Beniitez JJ, Garcia-Segura R, Heredia A (2004a) Plant biopolyester cutin: a tough way to its chemical synthesis. Biochimica Et Biophysica Acta-General Subjects 1674:1–3

Benitez JJ, Matas AJ, Heredia A (2004b) Molecular characterization of the plant biopolyester cutin by AFM and spectroscopic techniques. Journal of Structural Biology 147:179–184

Bernstein L (1975) Effects of salinity and sodicity on plant growth. Annual Review of Phytopathology 13:295–312

Blaker T, Greyson R, Walden D (1989) Variation among inbred lines of maize for leaf surface wax composition. Crop Science 29:28–32

Blum A (1988) Plant breeding for stress environments, Vol. CRC press, Boca Raton, FL

Bonaventure G, Ba XM, Ohlrogge J, Pollard M (2004a) Metabolic responses to the reduction in palmitate caused by disruption of the *FATB* gene in Arabidopsis. Plant Physiology 135:1269–1279

Bonaventure G, Beisson F, Ohlrogge J, Pollard M (2004b) Analysis of the aliphatic monomer composition of polyesters associated with Arabidopsis epidermis: occurrence of octadeca-cis-6, cis-9-diene-1,18-dioate as the major component. The Plant Journal 40:920–930

Bonaventure G, Salas J, Pollard M, Ohlrogge J (2003) Disruption of the *FATB* gene in *Arabidopsis* demonstrates an essential role of saturated fatty acids in plant growth. Plant Cell 15:1020–1033

Bondada BR, Oosterhuis DM, Murphy JB, Kim KS (1996) Effect of water stress on the epicuticular wax composition and ultrastructure of cotton (*Gossypium hirsutum* L.) leaf, bract, and boll. Environmental and Experimental Botany 36:61-&

Botti C, Palzkill D, Munoz D, Prat L (1998) Morphological and anatomical characterization of six jojoba clones at saline and non-saline sites. Industrial Crops and Products 9:53–62

Broun P, Poindexter P, Osborne E, Jiang CZ, Riechmann JL (2004) *WIN1*, a transcriptional activator of epidermal wax accumulation in Arabidopsis. Proceedings of the National Academy of Sciences of the United States of America 101:4706–4711

Buchholz A (2006) Characterization of the diffusion of non-electrolytes across plant cuticles: properties of the lipophilic pathway. Journal of Experimental Botany 57:2501–2513

Buchholz A, Baur P, Schonherr J (1998) Differences among plant species in cuticular permeabilities and solute mobilities are not caused by differential size selectivities. Planta 206:322–328

Burghardt M, Riederer M (2003) Ecophysiological relevance of cuticular transpiration of deciduous and evergreen plants in relation to stomatal closure and leaf water potential. Journal of Experimental Botany 54:1941–1949

Burghardt M, Riederer M (2006) Cuticular transpiration. In: Riederer M, Müller C (eds) Biology of the plant cuticle. Blackwell Publishing, Oxford, p 292–311

Cameron KD, Teece MA (2006) Increased accumulation of cuticular wax and expression of lipid transfer protein in response to periodic drying events in leaves of tree tobacco. Plant Physiology 140:176–183

Casado CG, Heredia A (2001) Specific heat determination of plant barrier lipophilic components: biological implications. Biochimica Et Biophysica Acta-Biomembranes 1511:291–296

Chen G, Sagi M, Weining S, Krugman T, Fahima T, Korol A, Nevo E (2004) Wild barley *eibi1* mutation identifies a gene essential for leaf water conservation. Planta 219:684–693

Chen X, Goodwin SM, Liu X, Chen X, Bressan RA, Jenks MA (2005) Mutation of the *RESURRECTION1* locus of *Arabidopsis* reveals an association of cuticular wax with embryo development. Plant Physiology 139:909–919

Chen XB, Goodwin SM, Boroff VL, Liu XL, Jenks MA (2003) Cloning and characterization of the *WAX2* gene of Arabidopsis involved in cuticle membrane and wax production. Plant Cell 15:1170–1185

Connor KF, Lanner RM (1991) Cuticle thickness and chlorophyll content in bristlecone pine needles of various ages. Bulletin of the Torrey Botanical Club 118:184–187

Costaglioli P, Joubes J, Garcia C, Stef M, Arveiler B, Lessire R, Garbay B (2005) Profiling candidate genes involved in wax biosynthesis in *Arabidopsis thaliana* by microarray analysis. Biochimica et Biophysica Acta (BBA) – Molecular and Cell Biology of Lipids 1734:247–258

Dietrich CR, Perera MADN, D. Yandeau-Nelson M, Meeley RB, Nikolau BJ, Schnable PS (2005) Characterization of two *GL8* paralogs reveals that the 3-ketoacyl reductase component of fatty acid elongase is essential for maize (*Zea mays* L.) development. The Plant Journal 42:844–861

Duan H, Schuler M (2005) Differential expression and evolution of the *Arabidopsis* CYP86A subfamily. Plant Physiology 137:1067–1081

Evans M, Passas H, Poethig R (1994) Heterochronic effects of *glossy15* mutations on epidermal cell identity in maize. Development 120:1971–1981

Faure J, Vittorioso P, Santoni V, Fraisier V, Prinsen E, Barlier I, Van Onckelen H, Caboche M, Bellini C (1998) The *PASTICCINO* genes of *Arabidopsis thaliana* are involved in the control of cell division and differentiation. Development 125:909–918

Febrero A, Fernandez S, Molina-Cano J, Araus J (1998) Yield, carbon isotope discrimination, canopy reflectance and cuticular conductance of barley isolines of differing glaucousness. Journal of Experimental Botany 49:1575–1581

Fiebig A, Mayfield JA, Miley NL, Chau S, Fischer RL, Preuss D (2000) Alterations in *CER6*, a gene identical to *CUT1*, differentially affect long-chain lipid content on the surface of pollen and stems. Plant Cell 12:2001–2008

Franke R, Briesen I, Wojciechowski T, Faust A, Yephremov A, Nawrath C, Schreiber L (2005) Apoplastic polyesters in *Arabidopsis* surface tissues – A typical suberin and a particular cutin. Phytochemistry 66:2643–2658

Garcia C, Joubés J, Chevalier S, Laroche-Traineau L, Dieryck W, Lessire R (2006) At4g14440, the 3-hydroxyacyl-CoA dehydratase of the acyl-CoA elongase in *Arabidopsis*? 17th International Symposium on Plant Lipids, East Lansing, Michigan

Gentry G, Barbosa P (2006) Effects of leaf epicuticular wax on the movement, foraging behavior, and attack efficacy of *Diaeretiella rapae*. Entomologia Experimentalis et Applicata 121:115–122

Gibbs A (1998) Water-proofing properties of cuticular lipids. American Zoologist 38:471–482

Gibson A (1996) Special topics in water relations. In: Cloudsley-Thompson J (ed) Adapatations of desert organisms. Springer, Berlin, p 143–168

Gibson A (1998) Photosynthetic organs of desert plants: structural designs of nonsucculent desert plants cast doubt on the popular view that saving water is the key strategy. BioScience 48:911–920

Gifford ML, Dean S, Ingram GC (2003) The Arabidopsis *ACR4* gene plays a role in cell layer organisation during ovule integument and sepal margin development. Development 130:4249–4258

Goodwin SM, Jenks M (2005) Plant cuticle function as a barrier to water loss. In: Jenks M, Hasegawa PM (eds) Plant Abiotic Stress. Blackwell Publishing, p 14–36

Goodwin SM, Rashotte AM, Rahman M, Feldmann KA, Jenks MA (2005) Wax constituents on the inflorescence stems of double *eceriferum* mutants in Arabidopsis reveal complex gene interactions. Phytochemistry 66:771–780

Gordon DC, Percy KE, Riding RT (1998a) Effect of enhanced UV-B radiation on adaxial leaf surface micromorphology and epicuticular wax biosynthesis of sugar maple. Chemosphere 36:853–858

Gordon DC, Percy KE, Riding RT (1998b) Effects of u.v.-B radiation on epicuticular wax production and chemical composition of four *Picea* species. New Phytologist 138:441–449

Gray J, Holroyd G, van der Lee F, Bahrami A, Sijmons P, Woodward F, Schuch W, Hetherington A (2000) The *HIC* signalling pathway links CO_2 perception to stomatal development. Nature 408:713–716

Grncarevic M, Radler F (1967) The effect of wax components on cuticular transpiration-model experiments. Planta V75:23–27

Gutterman Y (2000) Environmental factors and survival strategies of annual plant species in the Negev Desert, Israel. Plant Species Biology 15:113–125

Hajibagheri M, Hall J, Flowers T (1983) The structure of the cuticle in relation to cuticular transpiration in leaves of the halophyte *Suaeda maritima* (L.) Dum. New Phytologist 94:125–131

Haque M, Mackill D, Ingram K (1992) Inheritance of leaf epicuticular wax content in rice. Crop Science 32:865–868

Harshberger J (1909) The comparative leaf structure of the strand plants of New Jersey. Proceedings of the American Philosophical Society 48:72–89

Holmes MG, Keiller DR (2002) Effects of pubescence and waxes on the reflectance of leaves in the ultraviolet and photosynthetic wavebands: a comparison of a range of species. Plant, Cell & Environment 25:85–93

Hooker TS, Millar AA, Kunst L (2002) Significance of the expression of the *CER6* condensing enzyme for cuticular wax production in Arabidopsis. Plant Physiology 129:1568–1580

Hülskamp M, Kopczak SD, Horejsi TF, Kihl BK, Pruitt RE (1995) Identification of genes required for pollen-stigma recognition in *Arabidopsis thaliana*. The Plant Journal 8:703–714

Ishitani M, Xiong L, Stevenson B, Zhu J-K (1997) Genetic analysis of osmotic and cold streess signal transduction in *Arabidopsis*: interactions and convergence of abscisic acid-dependent and abscisic acd-independent pathways. The Plant Cell 9:1935–1949

Jefferson P, Johnson D, Rumbaugh M, Asay K (1989) Water stress and genotypic effects on epicuticular wax production of alfalfa and crested wheatgrass in relation to yield and excised leaf water loss rate. Canadian Journal of Plant Science 69:481–490

Jeffree CE (2006) The fine structure of the plant cuticle. In: Riederer M, Müller C (eds) Biology of the Plant Cuticle. Blackwell Publishing Limited, Oxford, p 11–125

Jeffree CE, Johnson R, Jarvis P (1971) Epicuticular wax in the stomatal antechambers of Sitka spruce, and its effects on the diffusion of water vapour and carbon dioxide. Planta 98:1–10

Jenks M (2002) Critical issues with the plant cuticle's function in drought tolerance. In: Wood AJ (ed) Biochemical & Molecular Responses of Plants to the Environment. Research Signposts, Kerala, India, p 97–127

Jenks MA, Andersen L, Teusink RS, Williams MH (2001) Leaf cuticular waxes of potted rose cultivars as affected by plant development, drought and paclobutrazol treatments. Physiologia Plantarum 112:62–70

Jenks MA, Joly RJ, Peters PJ, Rich PJ, Axtell JD, Ashworth EN (1994) Chemically-induced cuticle mutation affecting epidermal conductance to water-vapor and disease susceptibility in *Sorghum bicolor* (L) Moench. Plant Physiology 105:1239–1245

Jenks MA, Tuttle HA, Eigenbrode SD, Feldmann KA (1995) Leaf epicuticular waxes of the *eceriferum* mutants in *Arabidopsis*. Plant Physiology 108:369–377

Kamp H (1930) Untersuchungen uber kutikularbau and kutikulare transpiration von blattern. Jarhrbucher fur Wissenschaftliche Botanik 72:403–465

Kerstiens G (1994) Air pollutants and plant cuticles: mechanisms of gas and water transport, and effects on water permeability. In: Percy KE, Cape JN, Jagels R, Simpson CJ (eds) Air pollutants and the leaf cuticles, Vol 36. Springer-Verlag, Berlin, p 39–55

Kerstiens G (1996) Diffusion of water vapour and gases across cuticles and through stomatal pores presumed closed. In: Kerstiens G (ed) Plant Cuticles: an integrated functional approach. BIOS Scientific Publishers Limited, Oxford, p 121–134

Kerstiens G, Schreiber L, Lendzian KJ (2006) Quantification of cuticular permeability in genetically modified plants. Journal of Experimental Botany 57:2547–2552

Kim K, Park S, Jenks M (In preparation) Influence of water deficit on leaf cuticular waxes of soybean.

Kim KS, Park SH, Jenks MA (In Press) Changes in leaf cuticular waxes of sesame (*Sesamum indicum* L.) plants exposed to water deficit. Journal of Plant Physiology In Press, Corrected Proof

Koch K, Barthlott W, Koch S, Hommes A, Wandelt K, Mamdouh W, De-Feyter S, Broekmann P (2006) Structural analysis of wheat wax (*Triticum aestivum*, c.v. 'Naturastar' L.): from the molecular level to three dimensional crystals. Planta V223:258–270

Koch K, Neinhuis C, Ensikat HJ, Barthlott W (2004) Self assembly of epicuticular waxes on living plant surfaces imaged by atomic force microscopy (AFM). Journal of Experimental Botany 55:711–718

Koiwa H, Bressan RA, Hasegawa PM (2006) Identification of plant stress-responsive determinants in *Arabidopsis* by large-scale forward genetic screens. Journal of Experimental Botany 57:1119–1128

Koornneef M, Hanhart CJ, Thiel F (1989) A genetic and phenotypic description of *eceriferum* (cer) mutants in *Arabidopsis thaliana*. Journal of Heredity 80:118–122

Krauss P, Markstadter C, Riederer M (1997) Attenuation of UV radiation by plant cuticles from woody species. Plant, Cell and Environment 20:1079–1085

Krolikowski KA, Victor JL, Wagler TN, Lolle SJ, Pruitt RE (2003) Isolation and characterization of the Arabidopsis organ fusion gene HOTHEAD. Plant Journal 35:501–511

Kurata T, Kawabata-Awai C, Sakuradani E, Shimizu S, Okada K, Wada T (2003) The *YORE-YORE* gene regulated multiple aspects of epidermal cell differentiation in *Arabidopsis*. The Plant Journal 36:55–66

Kurdyukov S, Faust A, Nawrath C, Bär S, Voisin D, Efremova N, Franke R, Schreiber L, Saedler H, Métreaux J-P, Yephremov A (2006a) The epidermis-specific extracellular *BODYGAURD* control cuticle development and morphogenesis in *Arabidopsis*. The Plant Cell 18:321–339

Kurdyukov S, Faust A, Trenkamp S, Bär S, Franke R, Efremova N, Tietjen K, Schreiber L, Saedler H, Yephremov A (2006b) Genetic and biochemical evidence for involvement of *HOTHEAD* in the biosynthesis of long-chain α-, ω-dicarboxylic fatty acids and formation of extracellular matrix. Planta 224:315–329

Larsson S, Svenningsson M (1986) Cuticular transpiration and epicuticular lipids of primary leaves of barley (Hordeum vulgare) doi:10.1111/j.1399–3054.1986.tb06589.x. Physiologia Plantarum 68:13–19

Lauter N, Kampani A, Carlson S, Goebel M, Moose SP (2005) microRNA172 down-regulates *GlOSSY15* to promote vegetative phase change in maize. Proceedings of the National Academy of Sciences 102:9412–9417

Lockey KH (1988) Lipids of the insect cuticle: origin, composition and function. Comparative Biochemistry and Physiology Part B: Biochemistry and Molecular Biology 89:595–645

Lolle SJ, Berlyn GP, Engstrom EM, Krolikowski KA, Reiter WD, Pruitt RE (1997) Developmental regulation of cell interactions in the *Arabidopsis fiddlehead-1* mutant: A role for the epidermal cell wall and cuticle. Developmental Biology 189:311–321

Lolle SJ, Cheung AY, Sussex IM (1992) *Fiddlehead*: an *Arabidopsis* mutant constitutively expressing an organ fushion program that involves interactions between epidermal cells. Developmental Biology 152:383–392

Lolle SJ, Hsu W, Pruitt RE (1998) Genetic analysis of organ fusion in *Arabidopsis thaliana*. Genetics 149:607–619

Loza-Cornejo S, Terrazas T (2003) Epidermal and hypodermal characteristics in North American Cactoidae (Cactaceae). Journal of Plant Research 116:27–35

Manetas Y, Petropoulou Y, Stamatakis K, Nikolopoulos D, Levizou E, Psaras G, Karabourniotis G (1997) Beneficial effects of enhanced UV-B radiation under field conditions: improvement of needle water relations and survival capacity of *Pinus pinea* L. seedlings during the dry Mediterranean summer. Plant Ecology 128:101–108

Merah O, Deleens E, Souyris I, Monneveux P (2000) Effect of Glaucousness on Carbon Isotope Discrimination and Grain Yield in Durum Wheat doi:10.1046/j.1439–037x.2000.00434.x. Journal of Agronomy and Crop Science 185:259–265

Merk S, Blume A, Riederer M (1998) Phase behaviour and crystallinity of plant cuticular waxes studied by Fourier transform infrared spectroscopy. Planta 204:44–53

Millar AA, Clemens S, Zachgo S, Giblin EM, Taylor DC, Kunst L (1999) *CUT1*, an *Arabidopsis* gene required for cuticular wax biosynthesis and pollen fertility, encodes a very-long-chain fatty acid condensing enzyme. Plant Cell 11:825–838

Monneveux P, Reynolds MP, Gonzalez-Santoyo H, Pena RJ, Mayr L, Zapata F (2004) Relationships between grain yield, flag leaf morphology, carbon isotope discrimination and ash content in irrigated wheat. Journal of Agronomy and Crop Science 190:395–401

Moose S, Sisco P (1996) *GLOSSY15*, an *APETALA2*-like gene from maize that regulates leaf epidermal cell identity. Genes Deve 10:3018–3027

Moose SP, Sisco PH (1994) Glossy15 Controls the Epidermal Juvenile-to-Adult Phase Transition in Maize 10.1105/tpc.6.10.1343. Plant Cell 6:1343–1355

Nakayama N, Arroyo JM, Simorowski J, May B, Martienssen R, Irish VF (2005) Gene Trap Lines Define Domains of Gene Regulation in Arabidopsis Petals and Stamens 10.1105/tpc.105.033985. Plant Cell 17:2486–2506

Niederl S, Kirsch T, Riederer M, Schreiber L (1998) Co-permeability of ^{3}H-labeled water and ^{14}C-labeld organic acids across isolated plant cuticles: investigating cuticular paths of diffusion and predicting cuticular transpiration. Plant Physiology 116:117–123

Norris RF (1974) Penetration of 2,4-D in relation to cuticle thickness. American Journal of Botany 61:74–79

Norris RF, Bukovac MJ (1968) Structure of the pear leaf cuticle with special reference to cuticular penetration. American Journal of Botany 55:975–983

Olyslaegers G, Nijs I, Roebben J, Kockelbergh F, Vanassche F, Laker M, Verbelen JP, Samson R, Lemeur R, Impens I (2002) Morphological and physiological indicators of tolerance to atmospheric stress in two sensitive and two tolerant tea clones in South Africa. Experimental Agriculture 38:397–410

Ortiz R, Vuylsteke, Ogburi N (1995) Inheritance of psuedostem waxiness in banana and plantain (*Musa* spp.). The Journal of Heredity 86:297–299

Osborn JM, Taylor TN (1990) Morphological and ultrastructural studies of plant cuticular membranes .1. sun and shade leaves of *Quercus velutina* (Fagaceae). Botanical Gazette 151:465–476

Pallardy S, Kozlowski T (1979) Cuticle development in the stomatal region of *Populus* clones. New Phytologist 85:363–368

Pighin JA, Zheng HQ, Balakshin LJ, Goodman IP, Western TL, Jetter R, Kunst L, Samuels AL (2004) Plant cuticular lipid export requires an ABC transporter. Science 306:702–704

Potter DA, Kimmerer TW (1988) Do holly leaf spines really deter herbivory? Oecologia V75:216–221

Preuss D, Lemieux B, Yen G, Davis R (1993) A conditional sterile mutation eliminates surface components from *Arabidopsis* pollen and disrupts cell signaling during fertilization. Genes & Development 7:974–985

Prior SA, Pritchard SG, Runion GB, Rogers HH, Mitchell RJ (1997) Influence of atmospheric CO_2 enrichment, soil N, and water stress on needle surface wax formatiion in *Pinus palustris* (Pinaceae). American Journal of Botany 84:1070–1077

Pruitt RE, Vielle-Calzada J-P, Ploense SE, Grossniklaus U, Lolle SJ (2000) *FIDDLEHEAD*, a gene required to suppress epidermal cell interactions in *Arabidopsis*, encodes a putative lipid biosynthetic enzyme. Proceedings of the National Academy of Sciences 97:1311–1316

Radler F (1965) Reduction of loss of moisture by the cuticle wax components of grapes. 207:1002–1003

Rashotte AM, Feldmann KA (1998) Correlations between epicuticular wax structures and chemical composition in *Arabidopsis thaliana*. International Journal of Plant Sciences 159:773–779

Reynhardt EC (1997) The role of hydrogen bonding in the cuticular wax of *Hordeum vulgare* L. European Biophysics Journal with Biophysics Letters 26:195–201

Reynhardt EC, Riederer M (1994) Structures and molecular dynamics of plant waxes: II. Cuticular waxes from leaves of *Fagus sylvatica* L. and *Hordeum vulgare* L. European Biophysics Journal with Biophysics Letters 23:59–70

Richards R, Rawson H, Johnson D (1986) Glaucousness in wheat: its development and effect on water-use efficiency, gas exchange and photosynthetic tissue temperatures. Australian Journal of Plant Physiology 13:468–473

Riederer M (1991) Cuticle as barrier between terrestrial plants and the atmosphere – significance of growth-structure for cuticular permeability. Naturwissenschaften 78:201–208

Riederer M, Burghardt M, Mayer S, Obermeier H, Schonherr J (1995) Sorption of monodisperse alcohol ethoxylates and their effects on the mobility of 2,4-D in isolated plant cuticles. Journal of Agricultural and Food Chemistry 43:1067–1075

Riederer M, Schreiber L (1995) Waxes: the transport barriers of plant cuticles. In: Hamilton RJ (ed) Waxes: Chemistry, Molecular Biology and Functions. The Oily Press, Dundee, p 131–156

Riederer M, Schreiber L (2001) Protecting against water loss: analysis of the barrier properties of plant cuticles. Journal of Experimental Botany 52:2023–2032

Rowland O, Zheng H, Hepworth SR, Lam P, Jetter R, Kunst L (2006) *CER4* encodes an alcohol-forming fatty acyl-Coenzyme A reductase involved in cuticular wax production in *Arabidopsis*. Plant Physiology 142:866–877

Samdur MY, Manivel P, Jain VB, Chikani BM, Gor HK, Desai S, Misra JB (2003) Genotypic differences and water-defecit induced enhancement in epicuticular wax load in peanut. Crop Science 43:1294–1299

Sanchez FJ, Manzanares M, de Andres EF, Tenorio JL, Ayerbe L (2001) Residual transpiration rate, epicuticular wax load and leaf colour of pea plants in drought conditions. Influence on harvest index and canopy temperature. European Journal of Agronomy 15:57–70

Schlegel TK, Schönherr J (2002) Selective permeability of cuticles over stomata and trichomes to calcium chloride. Acta Horticulturae 549:91–96

Schlegel TK, Schönherr J, Schreiber L (2005) Size selectivity of aqueous pores in stomatous cuticles of *Vicia faba*. Planta 221:648–665

Schnurr J, Shockey J, Browse J (2004) The acyl-CoA synthetase encoded by *LACS2* is essential for normal cuticle development in *Arabidopsis*. Plant Cell 16:629–642

Schönherr J, Schreiber L (2004) Size selectivity of aqueous pores in astomatous cuticular membranes isolated from *Populus canescens* (Aiton) Sm leaves. Planta 219:405–411

Schreiber L (2005) Polar paths of diffusion across plant cuticles: New evidence for an old hypothesis. Annals of Botany 95:1069–1073

Schreiber L (2006) Characterisation of polar paths of transport in plant cuticles. In: Riederer M (ed) Biology of the plant cuticle. Blackwell Publishing, Oxford, p 280–291

Schreiber L, Elshatshat S, Koch K, Lin J, Santrucek J (2006) AgCl precipitates in isolated cuticular membranes reduce rates of cuticular transpiration. Planta 223:283–290

Schreiber L, Riederer M (1996) Ecophysiology of cuticular transpiration: comparative investigation of cuticular water permeability of plant species from different habitats. Oecologia V107: 426–432

Schreiber L, Schorn K, Heimburg T (1997) ^{2}H NMR study of cuticular wax isolated from *Hordeum vulgare* L. leaves: identification of amorphous and crystalline wax phases. European Biophysics Journal with Biophysics Letters 26:371–380

Schreiber L, Skrabs M, Hartmann KD, Diamantopoulos P, Simanova E, Santrucek J (2001) Effect of humidity on cuticular water permeability of isolated cuticular membranes and leaf disks. Planta 214:274–282

Shinozaki K, Yamaguchi-Shinozaki K (1997) Gene expression and signal transduction in water-stress response. Plant Physiology 115:327–334

Sieber P, Schorderet M, Ryser U, Buchala A, Kolattukudy PE, Métreaux J-P, Nawrath C (2000) Transgenic *Arabidopsis* plants expressing a fungal cutinase show alterations in the structure and properties of the cuticle and postgenital organ fusion. The Plant Cell 12:721–737

Siegel B, Verbeke J (1989) Diffusible factors essential for epidermal cell redifferentiation in *Catharanthus roseus*. Science 244:580–582

Simini M, Leone IA (1986) Notes: The role of alkanes in epicuticular wax relative to tolerance of pine species to saline spray. Forest Science 32:487–492

Sitte P, Rennier R (1963) Untersuchungen an cuticularen Zellwandschichten. Planta 60:19–40

Skoss J (1955) Structure and composition of plant cuticle in relation to environmental factors and permeability. Botanical Gazette 117:55–72

Steinmüller D, Tevini M (1985) Action of ultraviolet radiation (UV-B) upon cuticular waxes in some crop plants. Planta 164:557–564

Stevens JF, Hart H, Bolck A, Swaving JH, Malingre TM (1994) Epicuticular wax composition of some European *Sedum* species. Phytochemistry 35:389–399

Sturaro M, Hartings H, Schmelzer E, Velasco R, Salamini F, Motto M (2005) Cloning and characterization of *GLOSSY1*, a maize gene involved in cuticle membrane and wax production. Plant Physiology 138:478–489

Suh MC, Samuels L, Jetter R, Kunst L, Pollard M, Ohlrogge J, Beisson F (2005) Cuticular lipid composition, surface structure and gene expression in *Arabidopsis* stem epidermis. Plant Physiology 139:1649–1645

Svenningsson M (1988) Epi- and intracuticular lipids and cuticular transpiration rates of primary leaves of eight barley (Hordeum vulgare) cultivars doi:10.1111/j.1399–3054.1988.tb05434.x. Physiologia Plantarum 73:512–517

Svenningsson M, Liljenberg C (1986) Changes in cuticular transpiration rate and cuticular lipids of oat (Avena sativa) seedlings induced by water stress doi:10.1111/j.1399–3054.1986.tb01224.x. Physiologia Plantarum 66:9–14

Tacke E, Korfhage C, Michel D, Maddaloni M, Motto M, Lanzini S, Salamini F, Doring H-P (1995) Transposon tagging of the maize *GlOSSY2* locus with the transposable element En/Spm. The Plant Journal 8:907–917

Tanaka H, Machida C (2006) The cuticle and cellular interactions. In: Riederer M (ed) Biology of the plant cuticle. Blackwell Publishing, Oxford, p 312–333

Tanaka H, Onouchi H, Kondo M, Hara-Nishimura I, Nishimura M, Machida C, Machida Y (2001) A subtilisin-like serine protease is required for epidermal surface formation in *Arabidopsis* embryos and juvenile plants. Development 128:4681–4689

Tanaka H, Watanabe M, Watanabe D, Tanaka T, Machida C, Machida Y (2002) *ACR4*, a putative receptor kinase gene of *Arabidopsis thaliana*, that is expressed in the outer cell layers of embryos and plants, is involved in proper embryogenesis. Plant and Cell Physiology 43:419–428

Tanaka T, Tanaka H, Machida C, Wantanabe M, Machida Y (2004) A new method for rapid visualization of defects in leaf cuticle reveals five intrinsic patterns of surface defects in *Arabidopsis*. The Plant Journal 37:139–146

Tanton TW, Crowdy SH (1972) Water pathways in higher plants: III. the transpiration stream within leaves. Journal of Experimental Botany 23:619–625

Taylor F (1971) Some aspects of the growth of mango (*Mangifera indica* L.). III. a mechanical analysis. New Phytologist 70:911–922

Teusink RS, Rahman M, Bressan RA, Jenks MA (2002) Cuticular waxes on *Arabidopsis thaliana* close relatives *Thellungiella halophila* and *Thellungiella parvula*. International Journal of Plant Sciences 163:309–315

Todd J, Post-Beittenmiller D, Jaworski JG (1999) *KCS1* encodes a fatty acid elongase 3-ketoacyl-CoA synthase affecting wax biosynthesis in *Arabidopsis thaliana*. Plant Journal 17:119–130

Trenkamp S, Martin W, Tietjen K (2004) Specific and differential inhibition of very-long-chain fatty acid elongases from *Arabidopsis thaliana* by different herbicides. Proceedings of the National Academy of Sciences 101:11903–11908

Tsunewaki K, Ebana K (1999) Production of near-isogenic lines of common wheat for glaucousness and genetic basis of this trait clarified by their use. Genes and Genetic Systems 74:33–41

Vogg G, Fischer S, Leide J, Emmanuel E, Jetter R, Levy AA, Riederer M (2004) Tomato fruit cuticular waxes and their effects on transpiration barrier properties: functional characterization of a mutant deficient in a very-long-chain fatty acid beta-ketoacyl-CoA synthase. Journal of Experimental Botany 55:1401–1410

Wellesen K, Durst F, Pinot F, Beneviste I, Nettesheim K, Wisman E, Steiner-Lange S, Saedler H, Yephremov A (2001) Functional analysis of the *LACERATA* gene of *Arabidopsis* provides evidence for different roles of fatty acid ω-hydroxylation in development. Proceedings of the National Academy of Sciences 98:9694–9699

Wilkinson R, Mayeux HS, Jr. (1990) Composition of epicuticular wax on *Opuntia engelmannii*. Botanical Gazette 151:342–347

WIlliams MH, Rosenqvist E, Bucchave M (1999) Response of potted miniature roses (Rosa x hybrida) to reduced water availability during production. The Journal of Horticultural Science and Biotechnology 74:301–308

Xia YJ, Nicolau BJ, Schnable PS (1996) Cloning and characterization of *CER2*, an Arabidopsis gene that affects cuticular wax accumulation. Plant Cell 8:1291–1304

Xiao FM, Goodwin SM, Xiao YM, Sun ZY, Baker D, Tang XY, Jenks MA, Zhou JM (2004) *Arabidopsis* CYP86A2 represses *Pseudomonas syringae* type III genes and is required for cuticle development. European Molecular Biology Organization Journal 23:2903–2913

Xu X, Dietrich CR, Delledonne M, Xia Y, Wen TJ, Robertson DS, Nikolau BJ, Schnable PS (1997) Sequence analysis of the cloned *GLOSSY8* gene of maize suggests that it may code for a β-ketoacyl reductase required for the biosynthesis of cuticular waxes. Plant Physiology 115:501–510

Yephremov A, Wisman E, Huijser P, Huijser C, Wellesen K, Saedler H (1999) Characterization of the *FIDDLEHEAD* gene of *Arabidopsis* reveals a link between adhesion response and cell differentiation in the epidermis. The Plant Cell 11:2187–2201

Zhang J, Broeckling CD, Blancaflor EB, Sledge MS, Sumner LW, Wang Z (2005) Overexpression of *WXP1*, a putative *Medicago trunculata* AP2 domain-containing transcription factor gene, increases cuticular wax accumulation and enhances drought tolerance in transgenic alfalfa (*Medicago sativa*). The Plant Journal 42:689–707

Zhao L, Sack F (1999) Ultrastructure of stomatal development in *Arabidopsis* (Brassicaceae) leaves. American Journal of Botany 86:929–939

Zheng HQ, Rowland O, Kunst L (2005) Disruptions of the Arabidopsis enoyl-CoA reductase gene reveal an essential role for very-long-chain fatty acid synthesis in cell expansion during plant morphogenesis. Plant Cell 17:1467–1481

Zhu J-K (2002) Salt and drought stress signal transduction in plants. Annual Review of Plant Biology 53:247–273

Zimmerman P, Hirsch-Hoffman M, Henning L, Gruissem W (2004) GENEVESTIGATOR: Arabidopsis microarray database and analysis toolbox. Plant Physiology 136:2621–2632

CHAPTER 6

MOLECULAR AND PHYSIOLOGICAL RESPONSES TO WATER-DEFICIT STRESS

ELIZABETH A. BRAY

Department of Molecular Genetics and Cell Biology, University of Chicago, Chicago, IL 60637, USA

Abstract: Soil-water-deficit stress causes many changes in the biology of the plant cell beginning with the perception of the stress followed by changes that promote the acclimation to the stress. The mechanism by which plant cells transduce the physical stress of loss of water to biochemical changes in the cell continues to elude plant biologists. Using modern techniques that allow measurements of thousands of changes in gene expression at one time, researchers have catalogued and are beginning to make progress in interpreting the function of the many changes in gene expression. Although, it still remains a challenge to understand the function and relevance of many of these responses. There are indications that laboratory stress conditions intended to mimic plant water-deficit stress do not cause a universal water stress response; only a small number of genes are commonly induced when plants are subjected to water-deficit stress in different laboratories. Researchers remain optimistic that lessons learned from the molecular response of Arabidopsis plants to stress can be used to improve crops for growth under non ideal field conditions and lessen the need for irrigation in areas of the world where water availabilty for agriculture is decreasing

Keywords: gene expression, microarray, soil water deficit, stress perception

1. INTRODUCTION

Plant water deficits caused by inadequate soil water content, especially during the growing season, may occur throughout the world triggering significant losses in crop productivity. Large areas of the world are prone to poor soil moisture conditions for plant growth and development (Figure 1). Significant problems are predicted for food production in the future due to the limited availability of fresh water suitable for agriculture (Jury and Vaux, 2005). The aridic or xeric soil moisture regimes, which generally can not support crop production without irrigation, are common throughout the world. Further obstacles for crop production must be considered in different regions of the world where varied combinations of stresses, depending upon such characteristics as soil type and temperature, can alter plant responses to the environment locally.

M.A. Jenks et al. (eds.), Advances in Molecular Breeding Toward Drought and Salt Tolerant Crops, 121–140.

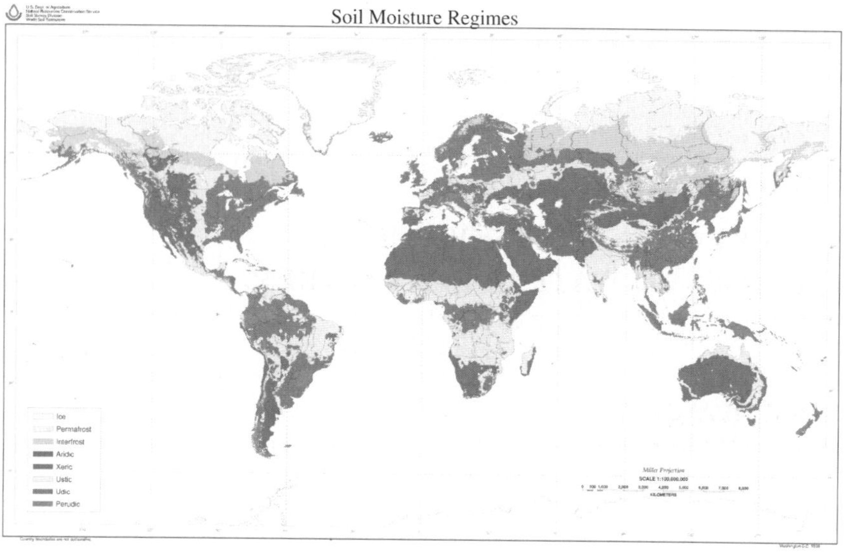

Figure 1. A map of the soil moisture regimes of the world. The aridic and xeric soil moisture regimes are the most limiting to plant growth and development. The aridic soil moisture regime does not have a period of water availability as long as 90 consecutive days when the soil temperature is above 8°C. The xeric soil moisture regime has a limited amount of water but it does not occur at optimum periods for plant growth. The ustic soil moisture regime has a limited amount of water available at a time when soil temperature is optimum for plant growth. (From the United States Department of Agriculture, Natural Resources Conservation Service: http://soils.usda.gov/use/worldsoils/mapindex/smr.html)

To develop improved genotypes for unfavorable growth conditions, it will require integrating our knowledge of plant response to water deficit at the physiological, cellular and molecular levels. Presently, studies to evaluate plant responses to water deficit have largely been done under controlled growth conditions. More studies are needed to determine if lessons learned in the greenhouse and growth chamber are applicable to field-grown plants.

2. RESISTANCE TO WATER-DEFICIT STRESS

Different species have different responses to soil water deficit owing to the responses programmed in the genome of each species. The genes that are expressed provide clues to the physiological and cellular responses that are required for maintenance of plant function in response to water-deficit stress.

Stress resistance, or the ability of a plant to survive periods of soil-water-deficit stress, can occur in plants that tolerate lowered cellular water content. Yet, resistance may also occur in situations where the cells of the plant are not subjected to decreased cellular water content (Figure 2). Thus a plant may be resistant to stress conditions by avoiding the soil water deficit. Desert ephemerals complete their life cycle prior to the development of soil water deficit avoiding decreased cellular

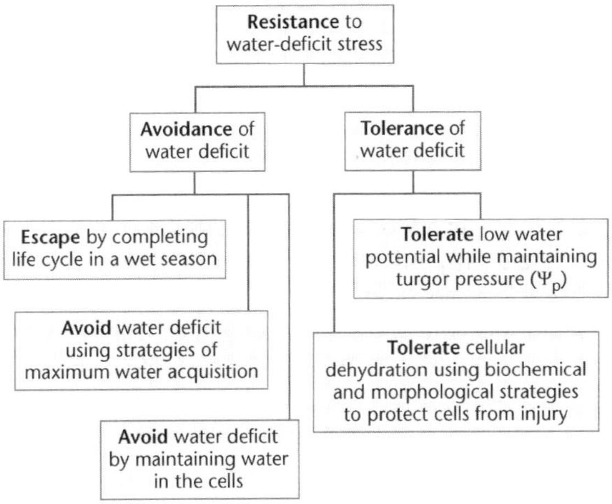

Figure 2. Plant resistance to water-deficit stress. (Originally published as Fig. 1 in Bray, 2001, DOI: 10.1038/npg.els.0001298)

water content. Avoidance of the cellular water deficit conditions may also occur in plants that have the ability to acquire more water through large root systems or the ability to prevent water loss through early closing of stomata. When osmotic adjustment occurs in cells, cellular water loss will also be avoided but the cells must be able to tolerate a higher cellular content of solutes and low cellular water potential. Plants with the ability to tolerate stress can withstand conditions of low cellular water content and low cellular water potential.

2.1. Adaptation or Acclimation

Many different species are adapted to dry environments, displaying characteristics that allow them to thrive under conditions of low water availability. For example, the morphology of succulents promotes storage of water and their physiology reduces loss of water since they fix CO_2 at night and keep their stomata closed during the day. Thus, adaptation is a function of the genomic make-up of the plant which is manifested in all aspects of plant, growth and development and can frequently be viewed in morphological characteristics. Adaptation to the environment is controlled by heritable differences that allow some species to be better able to function under soil water deficit due to their constitutive differences.

When challenged with soil water deficit, many species will respond to that lack of water in acclimation. Acclimation is an adjustment of physiology through inducible responses allowing plants to continue to grow or survive when the root media has a low water potential. For example, Arabidopsis seedlings acclimate to reduced water content of the media (Verslues and Bray, 2004). When Arabidopsis seedlings are subjected to low water potential in a PEG-conditioned agar media, acclimation occurs over a 96 h time period in several phases and patterns. Water relations, as measured by relative water content (RWC; Figure 3A) and leaf solute potential (Figure 3B), decrease in the first 24 h after seedlings are subjected to low water potential media. In the first phase of acclimation from 24 - 48 h, RWC increases and solute potential equilibrates with the water potential of the media. After 48 h, solute potential again drops, indicating that further acclimation occurs at the later time points. Other responses indicate that some aspects of acclimation occur in a different profile in time. Proline content increases steadily during the first 72 h of the stress period, after which there are only minimal increases in proline content (Figure 3C). When ABA is measured in seedlings, the ABA content is increased in the initial 12h of the stress decreasing to a content that is higher than the non-stressed steady-state level near 72 h (Figure 3D). Different measures of acclimation to stress indicate that each individual response occurs with its own timing.

Different species have different abilities to acclimate and can therefore withstand different degrees of stress. Thus adaptation may also be a result of the ability to acclimate.

When studying responses that are triggered by a change in the environment, our challenge remains to distinguish the responses that function to improve the plant's response to the environment compared to the responses that signify that the plant has been injured and is responding to that damage. Since many of the genes that are induced by water deficit do not have a known molecular function, we can not be certain that the function of a gene that is increased in response to stress promotes

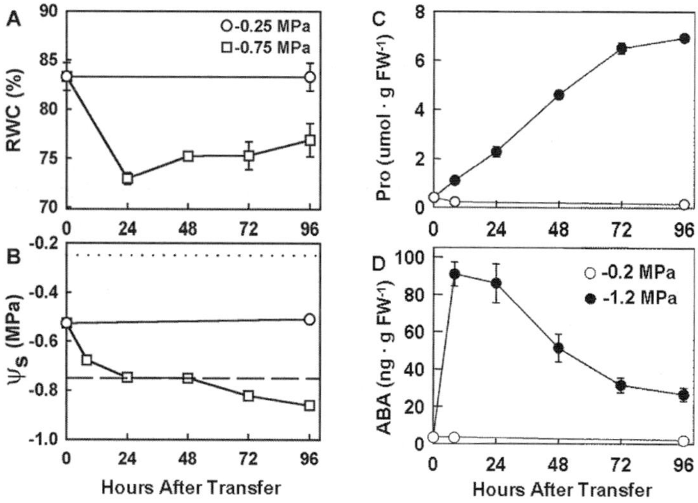

Figure 3. Acclimation in Arabidopsis seedlings. (A) Change in relative water content (RWC) in seedlings subjected to PEG-conditioned media at -0.75 MPa over a 96 h time course. (B) Solute potential of the seedlings at different time points after treatment as in A. (C) Proline content of seedlings subjected to PEG-conditioned media at -1.2 MPa. (D) Seedling ABA content at different timepoints after treatment as in C. (A and B originally published as Fig. 1 in Verslues and Bray, 2004. C and D adapted from Fig. 1 in Verslues and Bray, 2005.)

acclimation. In addition, we can not be certain of the impact of a change in gene expression of a gene with a known function, such as an enzymatic function, in its effect on the cellular homeostasis.

3. PERCEPTION OF WATER-DEFICIT STRESS

Among the many questions left to answer about the mechanisms of plant response to soil water deficit, the most intriguing may be, "By what means does the plant cell, and thus the whole plant, perceive a lack of soil water content?" Maintenance of cellular water relations is a critical component of life in all organisms. Cells must be able to regulate their water content as well as perceive alterations in cellular water relations. The ability of a cell to recognize loss of cellular water allows the cell to transduce a physical condition, such as loss of cell volume, into a biochemical response. This ability of physical forces to elicit biochemical responses is a critical event in the life of a cell throughout development as well as in response to water-deficit stress. Although this process, which may be referred to as mechanotransduction or osmosensing, has been studied in several model systems, the mechanisms in plants remain elusive.

Clues to the possible mechanisms that plant cells might use in the perception of water-deficit stress may come from mechanisms that are used in other organisms. Perhaps, the best studied model organisms are *E. coli* and yeasts.

3.1. Osmosensing Mechanisms in *E. coli* and Yeast

E. coli has two well-studied osmosensors. ProP is an osmosensor and an osmoprotectant transporter which is located in the cytoplasmic membrane. ProP promotes the transport of several osmolytes including proline and glycine betaine in response to increased external osmolarity (Racher et al., 1999; Figure 4). The transporter is activated under conditions that cause a decrease in cellular volume. ProP may act with ProQ to amplify the downstream responses. A second regulatory module, frequently called a two-component system, which propagates a His-Asp phosphorelay signal is also active in *E. coli* as an osmosensor (Cai and Inyoue, 2002). The first component is a histidine kinase called EnvZ, which has both kinase and phosphatase activity. EnvZ regulates the phosphorylation status of its response regulator, the second component of the two-component system, OmpR. OmpR is a transcription factor which in turn induces the expression of two outer membrane porins, OmpF and OmpC, which allow the diffusion of small molecular weight hydrophilic molecules.

The mechanisms that permit adaptation to the osmotic environment are perhaps best understood in yeast (*Saccharomyces cerevisiae*). Although this is not a simple linear regulatory process; there are a number of different processes that are coordinated which result in the osmotic stress response (Figure 5). A mitogen-activated protein (MAP) kinase cascade is initiated by the inactivation of the osmosensor Sln1 promoting glycerol production, the main osmolyte in yeast. When yeast cells

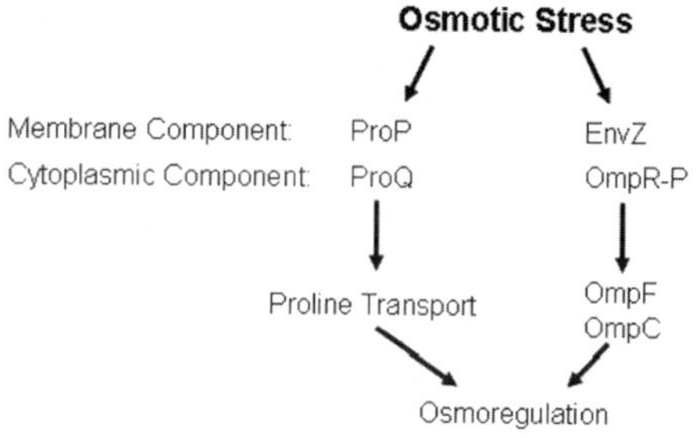

Figure 4. Diagram of two methods of osmosensing in *E. coli* that have been well-studied

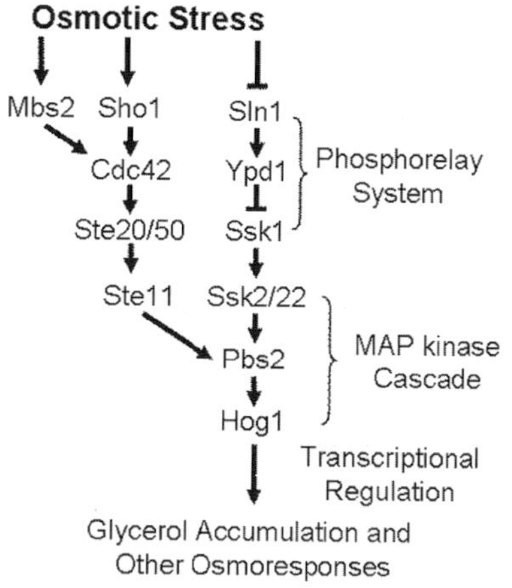

Figure 5. Diagram of the interrelated osmosensors that control glycerol accumulation and further downstream events in *Saccharomyces cerevisiae*

are not exposed to hyperosmotic stress, a phosphorelay similar to that found in *E. coli* is active from Sln1 through Ssk1. Concurrently, another osmosensor, Sho1, activates the Map kinase cascade through a second MAPKKK, Ste11 (Posas and Saito, 1997). At the same time, osmotic stress inactivates the glycerol channel, Fps1, preventing glycerol from being released from the cell. These three different pathways promote the increase in turgor pressure of the cell through an increase in glycerol content. Interestingly, both Sln1 and Fps1 can sense osmotic stress, through an unknown mechanism that possibly allows the cells to sense turgor pressure. Klipp et al. (2005) have modeled this response to osmotic stress, which provided insight into the importance of the rapid regulation of the glycerol channel in response to hyperosmotic stress. It remains possible that there are additional osmosensing mechanisms active in yeast cells.

Currently, there are two different theories on the mechanisms that control the ability of the cell to recognize water loss. One is that altered physical forces, especially to a membrane, change protein conformation causing changes to downstream processes that are directed by changes in gene expression. Another theory states that the change in pressure, and thus the change in water content of the cell, is sufficient to alter existing ligand concentration and activate receptors containing domains in the extracellular matrix (Tschumperlin et al., 2006). In normal human bronchial epithelial cells, compressive stress causes an increase in binding of epidermal growth factor family ligands to the epidermal growth factor receptor, due to increased local concentration of the ligand in stressed cells. Although, ligands are known in animal cells that can be involved in the response to loss of cellular water, analogous ligands have not been identified in plant cells. Yet, it is likely that in plant cells as well as in other eukaryotic cells, ligands would become concentrated as a result of cellular water loss (Hsiao, 1973).

3.2. Potential Mechanisms in Plants

Higher plants have many of the components that are involved in osmosensing in *E. coli* and yeast. Although, it has not been demonstrated that these act as osmosensors in plant cells. Presently, the molecule with the most potential to be an osmosensor in plants is ATHK1 identified in Arabidopsis (Urao et al., 1999). ATHK1 is a hybrid-type histidine kinase containing domains of both components of the typical two-component system, as well as predicted transmembrane domains. ATHK1 is able to complement yeast Sln1 mutants, indicating that ATHK1 may act as a histidine kinase in yeast cells. CRE1, which acts as a cytokinin receptor, is also a hybrid histidine kinase which can complement yeast Sln1 mutants in the presence of cytokinins (Inoue et al., 2001). The complementation assays indicate that the hybrid-type histidine kinases from plants are able to fulfill the role of these kinases in yeast cells, but it does not confirm that these proteins act as osmosensors in plant cells.

Reiser et al. (2003) present a paper in which they argue that both Sln1 derived from yeast and CRE1 from Arabidopsis are regulated by changes in cellular turgor pressure. Hyperosmotic stress, nystatin and cell wall removal all activate the Sln1 branch of the yeast osmoregulation pathway, but not the Sho1 branch. The authors argue that Sln1 senses the turgor pressure of the cell through pressure against the cell wall. CRE1 responded to the treatments in the same manner as Sln1. Although these experiments are not conclusive they do indicate that cell volume changes can be sensed by these proteins.

Given the importance of the ability to maintain cellular turgor and water relations properties, it is likely that higher plants have multiple osmosensing mechanisms which can be coordinated to fine-tune the response to the environment. In 1973, Dr. T.C. Hsiao proposed that turgor pressure may directly affect biochemical changes in the plant cell. Changes in molecular and ionic concentrations and spatial relations within the cell as a result of loss of cell volume may also be a means for a physical change to be transduced into changes in biochemical processes (Hsiao, 1973). Many signaling molecules are involved in osmotic sensing in plants, including abscisic acid, calcium, phospholipids and different types of kinases and phosphatases (Boudsocq and Laurière, 2005). More progress has been made in sorting out these signaling pathways than the initial events that trigger the cascades.

4. GENE EXPRESSION IN RESPONSE TO WATER-DEFICIT STRESS

Many different proteins are expressed in response to abiotic stresses such as water deficit. Perusing the types of genes that are induced by water-deficit conditions, it is easy to draw the conclusion that the expression of all classes of genes important in the biology of the plant is altered by stress. Indeed, there are genes involved in metabolism, cellular structure, signaling and regulation of cellular processes all induced by water-deficit stress. The extent of the changes in gene expression can be viewed using a two-dimensional gel comparison of proteins that accumulate in response to leaf water-deficit stress (Figure 6; after Bray, 1988). Many abundant proteins have altered patterns of expression in response to stress. At this time all of the proteins that are expressed have not been identified. The plant hormone

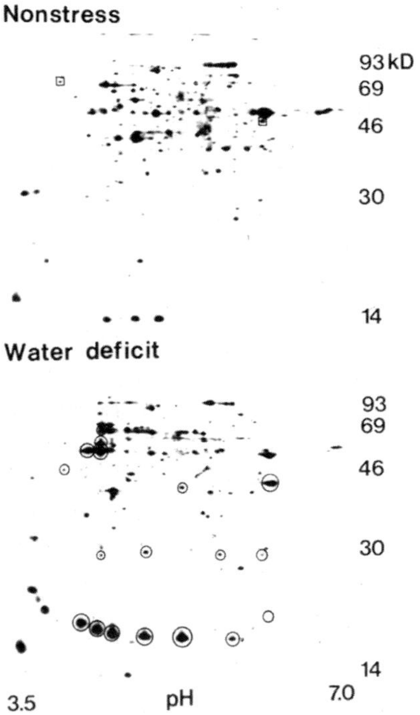

Figure 6. A comparison of proteins that accumulate in tomato leaflets after a period of nonstress or a period of leaf wilting. The most abundant proteins that are present in the water-deficit treatment and not in the nonstress treatment are circled. (After Bray, 1988)

ABA plays an important role in this regulation (Bray, 1988). The changes in gene expression are regulated by transcriptional mechanisms as well as by mechanisms that control the translation of specific mRNAs (Kawaguchi et al., 2004).

4.1. The First Glimpse at the Genes which are Induced by Water-Deficit Stress

Using cDNA library techniques that permitted differential screening for water-stress-induced genes, many of the major stress-induced genes were identified. Using RNA isolated from tomato leaves that had been wilted for 6 h, a cDNA library was constructed and screened to identify stress- and ABA-induced genes (Cohen and Bray, 1990). Two major types of stress-induced genes were identified using this approach: late embryogenesis abundant (LEA) and lipid-transfer proteins (LTP; Cohen et al., 1991; Plant et al., 1991).

LEA genes were originally identified as abundant transcripts that are expressed in the embryo in the late stages of seed development, during the period when seeds desiccate (Dure et al., 1989). These proteins are of unique character being extraordinarily hydrophilic and soluble at high temperatures. The expression during a developmental stage that requires survival of low cellular water content and expression in response to soil-water deficit has led many to propose that LEA proteins play an important role in cellular survival of low cellular water content possibly through a protective mechanism. There are several different classes of LEA genes based on the sequence of amino acids (Dure et al., 1989; Wise, 2003). Structural analyses indicate that at least two of the LEA groups are largely unstructured. Yet, protein of LEA14 (Pfam cluster PF03168) has a defined structure with its closest structural homolog being fibronectin type II (Singh et al., 2005). Given, the different amino acid sequences and structural forms, there are likely many different specific functions that these types of proteins perform. Recent evidence indicates that one of the roles of LEA proteins (group 1 and 3) may be to prevent protein aggregation under conditions of reduced cellular water content (Goyal et al., 2005).

LTP genes are induced under stress conditions. LTPs in plants were first studied for their ability to transfer lipids between membranes *in vitro* (Kader, 1996). These abundant proteins are small basic proteins which contain an internal hydrophobic pocket that can carry a lipid. However, it is unlikely that these proteins transfer lipids from membrane to membrane within the cell *in vivo*; LTPs, having signal peptides, enter the secretory pathway and are present in the cell wall rather than the cytoplasm. The true role of these proteins remains uncertain. In addition, LTPs have been identified as a major food allergen (van Ree, 2002).

The screening of a cDNA library constructed from mRNA isolated from Arabidopsis plants that were subjected to 10 h of rapid dehydration on filter paper led to the initial identification of 9 stress-induced genes, including members of LEA gene families and cysteine proteases (Yamaguchi-Shinozaki et al., 1992), and subsequently many more (Kiyosue et al., 1994).

4.2. Global Measures of Gene Expression—Microarrays

Microarrays facilitate the simultaneous measurement of numerous changes in gene expression, although a major challenge arises in cataloguing and interpreting the function of the hundreds to thousands of changes in gene expression that are identified. It is very difficult to ascertain the cellular or physiological relevance of the changes in gene expression that are altered under the relevant stress conditions. This is only partly because many of the induced or repressed genes do not have a known function. Using the model plant *Arabidopsis thaliana*, at least three microarray experiments with the goal to examine global changes in gene expression in response to plant water deficit stress have been published. More recently, the Arabidopsis Functional Genomics Network (AFGN), under the title of AtGenExpress, produced a set of microarray experiments using Affymetrix arrays ATH1 in which shoot and root gene expression were analyzed separately with two biological replicates. The "drought stress" treatment was a stream of dry air supplied to plants until they lost 10% of their fresh weight. After the treatment, plants were returned to the growth chamber and sampled for 24 h. An "osmotic stress" treatment was completed on plants continuously subjected to 300 mM mannitol. Whereas the experiments that have been completed thus far are a superb resource for studies on water-deficit stress, further microarray experiments are needed using plants that have been subjected to field conditions, or realistic controlled mimics of field conditions, to obtain a full appreciation of the changes in gene expression that occur under limiting water conditions in field-grown crop plants.

4.2.1. Genes commonly induced by water-deficit-stress treatments

A comparison of the *Arabidopsis thaliana* genes regulated by water-deficit stress among several different microarray experiments revealed that relatively few of the genes are commonly induced or repressed (Bray, 2004). The three experiments compared include: an oligonucleotide microarray containing approximately 7000 Arabidopsis genes used to analyze changes in gene expression of three-week-old plants to a 24 h desiccation treatment (Seki et al., 2002b). In a separate publication, gene expression in plants exposed to ABA treatment was examined (Seki et al., 2002a). Affymetrix arrays containing approx. 8000 Arabidopsis genes were used by Kreps et al. (2002) and Kawaguchi et al. (2004) to analyze changes in gene expression caused by two different methods of subjecting plants to cellular water deficit; four week-old plants grown in liquid culture were transferred to media containing 200 mM mannitol for 3 or 27 h (Kreps et al., 2002) or plants grown in soil were subjected to progressive loss of soil-water content and leaf samples were collected after 8 days when leaf RWC reached 65% (Kawaguchi et al., 2004).

In the three published experiments, there were 806 unique genes induced. Surprisingly, only 27 were induced in response to all three stress conditions (Figure 7A; Bray, 2004). Only 1.4% and 0.2% of the genes analyzed were commonly induced and repressed, respectively in the three experiments analyzed. Of the 27 commonly induced genes, all except At2g43570, an endochitinase isolog, were also

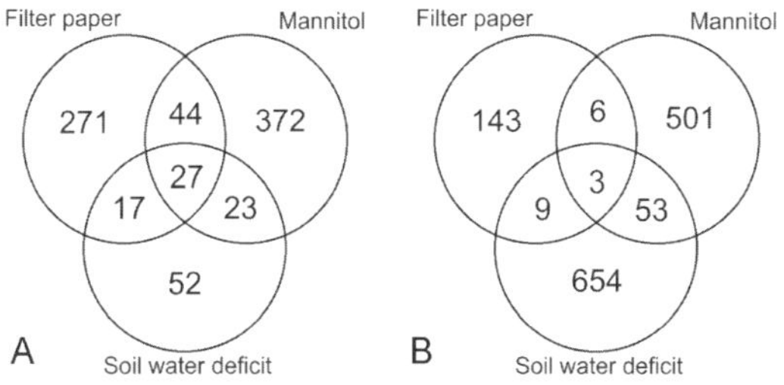

Figure 7. Genes commonly induced (A) or repressed (B) by water-deficit stress. (Originally published as Fig. 1 and 3 in Bray, 2004)

induced by ABA in a microarray experiment completed on ABA-treated plants (Seki et al., 2002a). Using the eFB Browser (http://bbc.botany.utoronto.ca/efp/cgi-bin/efpWeb.cgi), it was determined that the 27 genes were also induced in response to the osmotic stress conditions given in the AtGenExpress experiments. The 27 commonly induced genes were classified into six different categories: cellular metabolism, cellular transport, signal transduction, transcriptional regulation, hydrophilic proteins (LEAs), and unknown function (Bray, 2004).

Three genes were down regulated in all experiments (Figure 7B). These three genes were also down regulated in the AtGenExpress osmotic stress experiment. The three genes are all likely involved in the growth response through the repression of genes that increase cell wall extensibility. Two germin genes (AtGER1 and AtGER3; At1g72610, At5g20630) and an XTH (xyloglucan endotransglycosylase) are commonly down-regulated. Germin proteins are a member of the cupin superfamily, which have two histidine containing domains, yet the function is not clear. One possible function is in the alteration of cell wall properties that control growth. The xyloglucan endotransglucosylase/hydrolases (XTHs) are also involved in controlling cell wall extensibility through the cleavage and reformation of bonds between xyloglucan chains (Hyodo et al., 2003).

Why are there so few similarities in gene expression among these experiments? Since each laboratory experiment had individual characteristics including developmental stage of the plants, light, temperature, and degree and rate of stress, the lack of similarity may indicate exquisite control of gene expression in response to relatively small differences in the environment. An important difference between the experiments may be the rate of stress imposition and the timing of sampling. For example, in the progressive water loss experiment (Kawaguchi et al., 2004), plants had sufficient time to acclimate to the stress prior to sampling, whereas they were not likely to have acclimated in the more rapid rate of stress application in the other two experiments (Kreps et al., 2002, Seki et al, 2002b). Thus, in each case a set of changes in gene expression are sampled that are altered by the individual stress conditions. The genes that are identified as commonly induced are likely to be important under a broader range of conditions causing cellular water deficit.

4.3. What Can be Learned about the Physiologic/Metabolic State of the Cell from Microarray Experiments

The AtGenExpress data on drought and osmotic stress provide an excellent opportunity to study many changes in gene expression over a period of stress. Using the metabolism overview window on MapMan as a viewer to evaluate the stress response (http://gabi.rzpd.de/projects/MapMan/ data.shtml), it can be seen that the drought-stress treatment caused relatively few changes in expression of metabolic genes. In the shoot, there was mild down regulation of the light reactions after 6 h, and there were increases in genes involved in cell wall metabolism at the 30 min time point. In the root, the greatest changes were observed in the 1 h sample. As time progressed, changes in gene expression in comparison to the control were eliminated. This indicates that the plants recovered from the initial 10% loss of fresh weight after they were returned to the growth chamber in this experimental protocol. The osmotic stress treatment yielded many more changes in gene expression. The number of changes in the shoot increased throughout the 24 h time course, while the number of changes in the root appeared greatest at 12 h. After a 24 h treatment of 300 mM mannitol, there was a decrease in expression of genes in the shoot in the three categories of photosynthesis: light reactions, Calvin Cycle and photorespiration (Figure 8). The synthesis of tetrapyrroles, such as chlorophyll, would also be predicted to be reduced based on the decreased gene expression in this category. There was a general down regulation of genes involved in amino acid synthesis. The expression of enzymes involved in cell wall modification and degradation was also generally decreased. In categories that may serve to increase energy in the plant when photosynthesis is decreased, there were many genes that were up regulated. The expression of genes involved in mitochondrial electron transport, as well as sucrose degradation, was up regulated. Degradation of the amino acids and lipids through beta-oxidation and lipases also appears to be promoted based on the dramatic up regulation of genes involved in these processes. Increases were also evident for phenylpropanoid metabolism.

An analysis of the most abundant transcripts in the AtGenExpress osmotic stress data also provides some interesting insights into the physiology and cell biology of Arabidopsis plants as they are subjected to 300 mM mannitol. The assumption was made that the most abundant transcripts would highlight some of the major processes occurring in the cell at the time of stress and could be used to identify changes in response to the stress. The most abundant transcripts in roots and shoots were compared at two different time points, 3h and 24h, in the control and osmotic stress treated plants. The genes were ordered according to transcript abundance (hybridization value to the ATH1 microarray spot) and the top 250 genes were categorized into one of 13 different categories (Table 1).

In plants that were well watered, the category with the greatest number of genes in roots was protein synthesis, whereas in shoots the largest category was photosynthesis. After 24 h of stress in roots, there was a reduction in the number of genes involved in protein synthesis and genes of unknown function became the top

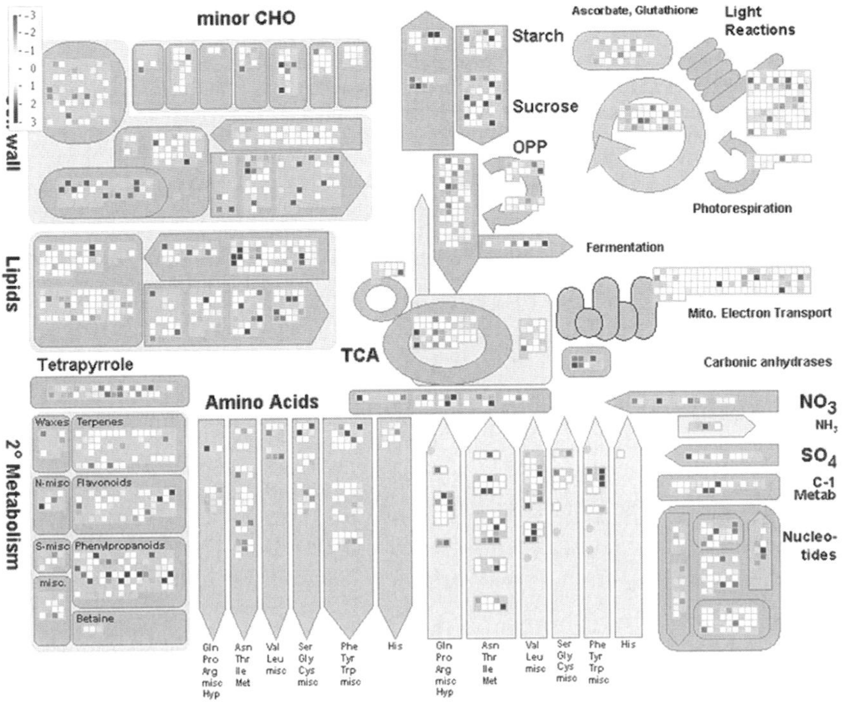

Figure 8. Metabolic overview viewer of MapMan for Arabidopsis shoots exposed to osmotic stress for 24 h collected for AtGenExpress. The squares outlined in black refer to genes that are up-regulated and those outlined in white indicate down-regulation. The intensity of the gray color indicates the increase or decrease in expression. (Captured from the output that can be viewed at http://gabi.rzpd.de/projects/MapMan/ data.shtml and then modified by outlining the squares to permit viewing in grayscale.)

category. There was an increase in the repair and degradation of proteins category owing to an increase in genes encoding chaperones and proteolytic enzymes.

The category of photosynthesis decreased after 24 h of osmotic stress, but not after 3 h of stress, in the shoots. There were also decreases in the carbohydrate metabolism and protein synthesis categories. In contrast, the carbohydrate metabolism category increased in roots. The categories that increased in response to the stress in shoots were repair and degradation of proteins, protection and unknown function. Many LEA genes and genes predicted to play a role in detoxification in the cell were present in the list of the 250 most abundant transcripts after 24 h of stress.

In response to stress, cellular resources were funneled away from building and maintaining the machinery required for protein synthesis and photosynthesis and into the production of genes involved in protection of the cell and repair and degradation of damaged proteins. In addition, the transcripts of more genes which are currently of unknown cellular function also became more abundant in roots and shoots as the stress conditions were sustained.

Table 1. A categorization of the function of the most abundant transcripts in Arabidopsis plants subjected to osmotic stress (AtGenExpress Osmotic Stress Experiment) for 3 and 24h. The percent of genes in each category is shown. Control and stress treatments and roots and shoots are considered separately. The energy category includes genes involved in electron transport and the protein maintenance category includes genes involved in protein repair as well as degradation

	3h				24h			
	Root		Shoot		Root		Shoot	
	control	stress	control	stress	control	stress	control	stress
	% of the Top 250 Genes							
Photosynthesis	0	0	25	23	0	1	21	16
Photorespiration	0	0	1	1	0	0	2	1
Energy	2	2	2	2	2	1	2	2
Carbohydrate metabolism	8	8	6	8	7	10	9	4
Other metabolism	7	6	8	5	7	8	9	7
Protein synthesis	30	23	14	7	33	10	13	1
Protein maintenance	6	8	4	6	5	10	4	10
Transport processes	4	3	3	4	4	3	4	4
Cellular structure	3	2	1	0	2	3	0	0
Signaling	2	2	1	2	3	2	2	3
Gene expression	2	1	1	1	2	3	2	2
Protection	7	10	6	10	8	10	5	12
Unknown function	12	20	16	17	11	25	16	26
No annotated gene	16	16	12	13	16	14	11	11

4.4. Translational Regulation of Gene Expression by Water-Deficit Stress

Although most studies on water-deficit-induced genes have been completed on regulation of gene expression by transcriptional and post-transcriptional mechanisms, other mechanisms of regulation of gene expression may occur. The production and accumulation of a gene product can also be regulated at the level of translation. In order to determine the extent of changes in gene expression that occur at the translational level, we used microarrays to compare the genes whose mRNAs are in the actively translated pools (polysomal complexes) compared to the nonactively translated pools (non-polysomal complexes) in well-watered and water-stressed Arabidopsis plants (Kawaguchi et al., 2004; Figure 9). The proportion of individual mRNA species in polysomal complexes in leaves of non-stressed and moderately dehydration-stressed Arabidopsis plants was used to estimate the extent of translational regulation in response to water-deficit stress. On average the proportion of each mRNA species associated with the polysome was reduced by the water deficit. Under non-stress conditions, an average of 82% of each mRNA species was isolated in the polysomal fraction with only a 72% average for the plants subjected to a progressive soil water deficit (Figure 9). On average there was a significant reduction in the proportion of mRNAs that were being translated in response to this stress. Yet, the translational status of individual genes varies considerably.

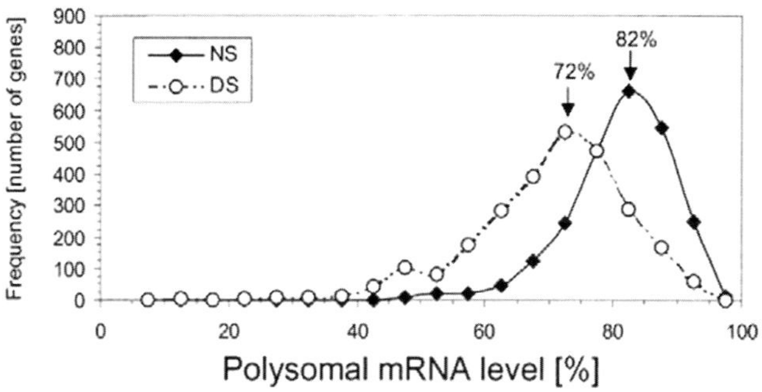

Figure 9. Reduction in polysomal mRNA level in response to moderate water-deficit stress. NS=non stress. DS=drought stress.(Originally published as Fig. 3B in Kawaguchi et al., 2004.)

Less than 1% of the 2136 genes analyzed by Kawaguchi et al. (2004) had an increase in translational status in response to progressive, soil-water-deficit stress (Table 2). This indicates that although translational regulation can be an important means of controlling the synthesis of gene products in response to water-deficit stress, it is not a major method on its own. However, when the translational status of mRNA species with increased abundance in response to water deficit were considered, it is noted that these genes are more likely to have an increase in translational status in response to water-deficit stress than the average. 15% of the dehydration-inducible mRNAs (2-fold or greater increase in abundance) had increased polysomal levels in response to water-deficit stress (Figure 10A). The translational status of 45% of the genes with increased mRNA abundance was maintained, compared to 28% maintenance of translational status when the entire set of genes was considered.

Table 2. Genes that are regulated by progressive water-deficit stress at the translational level in *Arabidopsis thaliana*

Locus ID	Annotation	Affy ID	p-value	Change in polysomal level [%]	mRNA Δ abundance
Metabolism					
At5g35790	glucose-6-phosphate dehydrogenase	16385_s_at	2.08E-03	16.3	0.45
Protection					
At2g21620	universal stress protein (RD2)	14697_g_at	6.31E-03	4.7	3.22
At3g49120	peroxidase, putative	14638_at	4.32E-02	6.5	5.13
At5g06760	late embryogenesis abundant-like	19152_at	5.03E-03	17.0	13.52

(*Continued*)

Table 2. (Continued)

Locus ID	Annotation	Affy ID	p-value	Change in polysomal level [%]	mRNA Δ abundance
Degrade and Repair Proteins					
At1g45145	thioredoxin, putative	13187_i_at	4.97E-02	2.9	3.28
At3g12580	heat shock protein 70	13284_at	2.16E-03	11.5	3.93
At4g39090	cysteine proteinase RD19A	14658_s_at	1.29E-02	6.1	2.57
Unknown Function					
At1g62510	similar to 14KD proline-rich protein DC2.15 precursor (sp|P14009)	18560_at	8.69E-05	3.6	5.67
At1g78850	curculin-like (mannose-binding) lectin family protein		1.24E-02	1.8	3.56
At1g78860	curculin-like (mannose-binding) lectin family protein		3.48E-04	2.0	4.08
At2g41100	calmodulin-like protein (TCH3)	19848_s_at	1.31E-02	5.3	1.59
At2g47770	TspO-MBR protein	14097_at	2.15E-05	22.4	25.16
At3g26740	light regulated protein, putative	16046_s_at	2.11E-03	7.1	2.36

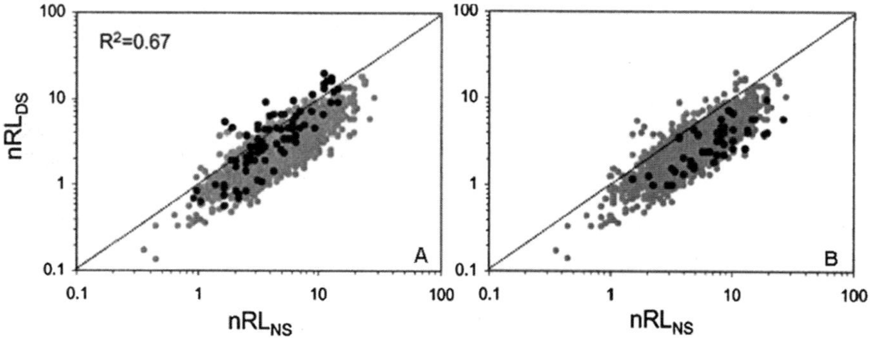

Figure 10. Translational regulation of mRNAs that (A) accumulate in response to water-deficit stress and (B) decrease in abundance. The normalized ratio of expression (nRL) in PS to NP were plotted for the nonstress (NS) or water-deficit (DS) treatments. (The gray points are all of the data and the black dots are the data points selected based on calculated mRNA abundance. (Originally published as Fig. 4A and C in Kawaguchi et al., 2004.)

Of the mRNAs which have decreased abundance in response to the stress, 92% of them have decreased translational status (Figure 10B). It is proposed that there is a connection between transcription and/or mRNA turnover and translation, as genes that have decreased transcription and or increased turnover in response to water deficit also have a decreased translational status.

Mechanisms that target specific genes for translation are not a solitary method of promoting cellular function in response to water-deficit stress. However, there is a tendency for genes with increased transcript abundance to maintain their translational status and for genes that have decreased abundance during stress to have decreased translational status.

Table 2 lists the 12 genes that are translationally regulated by water-deficit stress. One enzyme is included in the list. Three of the genes are in the protection category and three in the protein repair and degradation category: two of the categories that have many changes in gene expression and include many abundant mRNAs. All but two of the genes have a greater than 2-fold increase in mRNA abundance, highlighting the coordinate regulation by transcriptional and translational mechanisms.

5. CONCLUSION

Our understanding of the mechanisms that are involved in acclimation to water-deficit stress grow year-by-year. Yet, there is much to learn. It is important to use our advances in knowledge to develop better crops for food production in marginal lands. Genetic engineering remains a promising means to improve our crops. Thus far, the main use of transgenic plants has been to test the function of specific genes when they are mis-expressed. Few studies have been completed to determine the agricultural application of the over or under expression of a single gene or a suite of genes. Many genes have now been engineered into transgenic plants and tested under controlled growth conditions to determine if specific genes are involved in the stress response (Valliyodan and Nguyen, 2006). And, many different genes encoding transcription factors, specific enzymes, or protective proteins have been shown to function in the response to drought in the sense that the plants response to the stress is altered when the gene is mis-expressed. These clues to gene function continue to fuel our enthusiasm that this approach can lead to improved crops.

Is Arabidopsis a good model system for future agricultural studies? The answer to this question is yes and no. Arabidopsis acclimates to water-deficit stress, through an induction of many processes that have also been recorded in other plants. However, there are many other species that are more tolerant of water deficit, and thus should be studied to understand mechanisms that allow further acclimation or to identify the adaptive characteristics of a species.

Many attempts to improve stress resistance have centered on the thinking that improved resistance will come from increased expression of genes that protect the plant. This approach is likely to have a limit to its effectiveness, and other approaches will need to be added to our repertoire. An alternative may be to understand the mechanisms that cause the down regulation by stress of necessary processes for survival such as photosynthesis and protein synthesis. If the signaling pathway that directs the decrease in photosynthesis, decrease in protein synthesis and alterations in the cell wall could be blocked, plants may achieve an enhanced ability to function in the field during the stress. Gene expression associations in Arabidopsis can now be monitored, such as using the VxInsight tool to view global gene associations (http://www.arabidopsis.org/tools/bulk/microarray/analysis/index.jsp). For example, the two germin genes that are commonly down regulated by different water deficit conditions have strong expression associations with major processes that are also down regulated by stress. AtGER1 (At1g72610) has strong associations with genes involved in photosynthesis and AtGER3 (At5g20630) is associated with ribosomal genes.

ACKNOWLEDGEMENT

The Energy Biosciences Research Program at the Department of Energy has supported this work and future publications on two Arabidopsis mutants, *lwr1* and *lwr2*. I appreciate the University of California, Riverside and the many people there for their support during a large part of this work.

REFERENCES

Boudsocq, M., Laurière, C., 2005, Osmotic signaling in plants. Multiple pathways mediated by emerging kinase families. *Plant Physiol.* **138**:1185–1194.

Bray, E.A., 1988, Drought- and ABA-induced changes in polypeptide and mRNA accumulation in tomato leaves. *Plant Physiol.* **88**:1210–1214.

Bray, E.A., 2001, Plant Response to Water-deficit Stress. Encyclopedia of Life Sciences. John Wiley & Sons, Ltd., Chichester. http://www.els.net/ [DOI: 10.1038/npg.els.0001298]

Bray, E.A., 2004, Genes commonly regulated by water-deficit stress in *Arabidopsis thaliana*. *J. Exp. Bot.* **55**:2331–2341.

Cai, S.J. and Inouye, M., 2002, EnvZ-OmpR interaction and osmoregulation in *Escherichia coli*. *J. Biol. Chem.* **277**:24155–24161.

Cohen, A. and Bray, E.A., 1990, Characterization of three mRNAs that accumulate in wilted tomato leaves in response to elevated levels of endogenous abscisic acid. *Planta* **182**: 27–33.

Cohen, A., Plant, A.L., Moses, M.S., and Bray, E.A., 1991, Organ-specific and environmentally regulated expression of two abscisic acid induced genes of tomato: Nucleotide sequence and analysis of the corresponding cDNAs. *Plant Physiol.* **97**:1367–1374.

Dure, L. III, Crouch, M., Harada, J., Ho, T.-H.D., Mundy, J., Quatrano, R., Thomas, T., and Sung, Z.R., 1989, Common amino acid sequence domains among the LEA proteins of higher plants. *Plant Molec. Biol.*.**12**:475–486.

Goyal, K., Walton, L.J., and Tunnacliffe, A., 2005, LEA proteins prevent aggregation due to water stress. *Biochem J.* **388**:151–157.

Hasegawa, Y., Seki, M., Mochizuki, Y., Heida, N., Hirosawa, K., Okamoto, N., Sakurai, T., Satou, M., Akiyama, K., Iida, K., Lee, K., Kanaya, S., Demura, T., Shinozaki, K., Konagaya, A., Toyoda, T., 2006, A flexible representation of omic knowledge for thorough analysis of microarray data. *Plant Methods* **2**:5.

Hsiao, T.C., 1973, Plant response to water stress. *Annu. Rev. Plant Physiol.* **24**:519–570.

Hyodo, H., Yamakawa, S., Takeda, Y., Tsuduki, M., Yokota, A. Nishitani, K., Kohchi, T., 2003, Active gene expression of a xyloglucan endotransglucosylase/hydrolase gene, XTH9, in inflorescence apices is related to cell elongation in *Arabidopsis thaliana*. *Plant Molec. Biol.* **52**:473–482.

Inoue, T., Higuchi, M., Hashimoto, Y., Seki, M., Kobayashi, M., Kato, T., Tabata, S., Shinozaki, K., and Kakimoto, T., 2001, Identification of CRE1 as a cytokinin receptor from Arabidopsis. *Nature* **409**:1060–1063.

Jury, W.A., and Vaux Jr., H., 2005, The role of science in solving the world's emerging water problems. *PNAS* **102**:15715–15720.

Kader, J.-C., 1996, Lipid-transfer proteins in plants. *Ann. Rev. Plant Physiol. Plant Molec. Biol.* **47**:627–654.

Kawaguchi, R., Girke, T., Bray, E.A., Bailey-Serres, J.N., 2004, Differential mRNA translation contributes to gene regulation under non-stress and dehydration stress conditions in *Arabidopsis thaliana*. *Plant J.* **38**: 823–239.

Klipp, E., Nordlander, B., Kruger, R., Gennemark, P., and Hohmann, S., 2005, Integrative model of the response of yeast to osmotic shock. *Nat. Biotechnol.* **23**:975–82.

Kiyosue, T., Yamaguchi-Shinozaki, K., and Shinozaki, K., 1994, Cloning of cDNAs for genes that are early-responsive to dehydration stress (ERDs) in *Arabidopsis thaliana* L.: identification of three ERDs as HSP cognate genes. *Plant Molec. Biol.* **25**:791–798.

Kreps, J.A., Wu, Y., Chang, H.-S., Zhu, T., Wang, X., and Harper, J.F., 2002, Transcriptome changes for *Arabidopsis* in response to salt, osmotic and cold stress. *Plant Physiol.* **230:**2129–2141.

Plant, A.L., Cohen, A., Moses, M.S., and Bray, E.A., 1991, Nucleotide sequence and spatial expression pattern of a drought- and abscisic acid-induced gene of tomato. *Plant Physiol.* **97**:900–906.

Posas. F., and Saito, H., 1997, Osmotic activation of the HOG MAPK pathway via Ste11p MAPKKK: scaffold role of Pbs2p MAPKK. *Science* **276**:1702–1705.

Racher, K. I., Voegele, R. T., Marshall, E. V., Culham, D. E., Wood, J. M., Jung, H., Bacon, M., Cairns, M. T., Ferguson, S. M., Liang, W.-J., Henderson, P. J. F., White, G., and Hallett, F. R., 1999, Purification and reconstitution of an osmosensor: transporter ProP of *Escherichia coli* senses and responds to osmotic shifts. *Biochem.* **38:**1676–1684.

Reiser, V., Raitt, D.C., and Saito, H., 2003, Yeast osomosensor Sln1 and plant cytokinin receptor Cre1 respond to changes in turgor pressure. *J. Cell Biol.* **161**:1035–1040.

Seki, M., Ishida, J., Narusaka, M., et al. 2002a. Monitoring the expression pattern of around 7000 Arabidopsis genes under ABA treatments using a full-length cDNA microarray. *Funct. Int. Gen.* **2**:282–291.

Seki, M., Narusaka, M., Ishida, J., Nanjo, T., Fujita, M., Oono, Y., Kamiya, A., Nakajima, M., Enju, A., Sakurai, T., Satou, M., Akiyama, K., Taji, T., Yamaguchi-Shinozaki, K., Carninci, P., Kawai, J., Hayashizaki, Y., and Shinozaki, K., 2002b, Monitoring the expression profiles of 7000 Arabidopsis genes under drought, cold and high-salinity stresses using a full-length cDNA microarray. *Plant J.* **31**:279–292.

Singh, S., Cornilescu, C.C., Tyler, R.C., Cornilescu, G., Tonelli, M., Lee, M.S., and Markley, J.L., 2005, Solution structure of a late embryogenesis abundant protein (LEA14) from *Arabidopsis thaliana*, a cellular stress-related protein. *Prot. Sci.* **14**:2601–2609.

Tschumperlin, D.J., Dai, G., Maly, I.V., Kikuchi, T., Laiho, L.H., McVittie, A.K., Haley, K.J., Lilly, C.M., So, P.T.C., Lauffenburger, D.A., Kamm, R.D., and Drazen, J.M., 2004, Mechanotransduction through growth-factor shedding into the extracellular space. *Nature* **429**:83–86.

Urao, T., Yakubov, B., Satoh, R., Yamaguchi-Shinozaki, K., Seki, M., Hirayama, T., and Shinozaki, K., 1999, A transmembrane hybrid-type histidine kinase in Arabidopsis functions as an osmosensor. *Plant Cell* **11**: 1743–1754.

Valliyodan, B., and Nguyen, H.T., 2006, Understanding regulatory networks and engineering for enhanced drought tolerance in plants. *Curr Opin Plant Biol.* **9**:189–195.

Van Ree, R.. 2002, Clinical importance of non-specific lipid transfer proteins as food allergens. *Biochem Soc. Trans.* **30**:910–903.

Verslues, P.E. and Bray, E.A., 2004, *LWR1* and *LWR2* are required for osmoregulation and osmotic adjustment in *Arabidopsis thaliana. Plant Physiol.* **136**:2831–2842.

Verslues, P.E. and Bray, E.A., 2005, Role of abscisic acid acid (ABA) and *Arabidopsis thaliana* ABA-insensitive loci in low water potential-induced ABA and proline accumulation. *J. Exp. Bot.* **75**:201–212.

Wise, M., 2003, LEAping to conclusions: A computational reanalysis of late embryogenesis abundant proteins and their possible roles. *BMC Bioinform.* **4**:52.

Yamaguchi-Shinozaki, K., Koizumi, M., Urao, S., and Shinozaki, K., 1992, Molecular cloning and characterization of 9 cDNAs for genes that are responsive to desiccation in *Arabidopsis thaliana*: sequence analysis of one cDNA clone that encodes a putative transmembrane channel protein. *Plant Cell Physiol.* **33**: 217–224.

CHAPTER 7

INTEGRATION OF Ca^{2+} IN PLANT DROUGHT AND SALT STRESS SIGNAL TRANSDUCTION PATHWAYS

HUAZHONG SHI
Department of Chemistry and Biochemistry, Texas Tech University, Lubbock, TX 79409
E-mail: huazhong.shi@ttu.edu

Abstract: Plants cope with environmental changes by activating signal transduction cascades that control and coordinate the physiological and biochemical responses necessary for adaptation. Numerous signaling pathways that function as an integrated network have been implicated in plant abiotic stress response. Amongst them, calcium signaling was found to be incorporated in different signaling pathways during abiotic stress response, e.g. to heat, cold, drought, and salt. A well-recognized model of calcium signaling is that calcium signals characteristic of either elevation or oscillation of cytosolic Ca^{2+} is generated upon stimulation and then transduced through an array of Ca^{2+} activated proteins and downstream components, including calmodulins (CaMs) and CaM-binding proteins (CaMBPs), calcineurin B-like proteins (CBLs), Ca^{2+}-dependent protein kinases (CDPKs), Ca^{2+} and CaM-binding transcription factors, and other Ca^{2+}-binding proteins. Potential targeted effectors of calcium signaling include important enzymes/proteins involved in various cellular metabolism and physiological adjustment. This review begins with the generation of calcium signals followed by reviewing components decoding calcium signals. Implication of the signaling components in drought and salt stress response is emphasized and discussed

Keywords: calcium signaling, calcium sensor, abiotic stress, signal transduction

1. INTRODUCTION

Unlike animals, higher plants are sessile and therefore can not escape from unfavorable growth conditions such as drought and high salinity. Plants, including economically important crops frequently encounter these abiotic stresses, which represent major constraints on crop yield potential. Plants cope with environmental changes by activating signal transduction cascades that control and coordinate the physiological and biochemical responses necessary for adaptation. Abiotic stress signaling is extremely complex presumably because plants must have the capacity

M.A. Jenks et al. (eds.), Advances in Molecular Breeding Toward Drought and Salt Tolerant Crops, 141–182.

to tolerate several possible environmental extremes for their survival and reproduction. Numerous signaling pathways that function as an integrated network have been implicated in plant abiotic stress response. Amongst them, calcium signaling was found to be incorporated in different signaling pathways during abiotic stress response, e.g. heat, cold, drought, and salt.

Calcium is an essential micronutrient in plants. As a divalent cation, it has structural roles in the cell wall and membrane. Quantitatively it is most prominently present in the apoplast, the cell wall space where it fulfills at least two distinct functions: to cross-link pectin chains thereby contributing to their stability and mechanical properties. Calcium is essential for the integrity of the plasma membrane of the plant cells, specifically, the selectivity of the transport of the ions across the plasma membrane. Besides its nutritional role, calcium is also a well-known intracellular messenger mediating diverse responses of plants to both internal and external stimuli. This review focuses on the signaling role of calcium in plants. A well-recognized model of calcium signaling is that calcium signals characteristic of either elevation or oscillation of cytosolic Ca^{2+} is generated upon stimulation and then transduced through an array of Ca^{2+} activated proteins and downstream components, including calmodulins (CaMs) and CaM-binding proteins (CaMBPs), calcineurin B-like proteins (CBLs), Ca^{2+}-dependent protein kinases (CDPKs), Ca^{2+} and CaM-binding transcription factors, and other Ca^{2+}-binding proteins. Potential targeted effectors of calcium signaling include important enzymes/proteins involved in various cellular metabolism and physiological adjustment. This review will begin with the generation of calcium signals, followed by reviewing components decoding calcium signals. Implication of the signaling components in drought and salt stress response will be emphasized.

2. GENERATION OF CALCIUM SIGNALS

Although Ca^{2+} is an essential nutrient in plants, high concentration of Ca^{2+} in cytosol ($[Ca^{2+}]_{cyt}$) is toxic. The calcium concentration in soil is generally above 1.27 mM, while $[Ca^{2+}]_{cyt}$is around 0.15 μM. Plant cells maintain low cytosolic calcium by the functions of Ca^{2+}-ATPases and Ca^{2+}/H^+ antiporters (Sze et al., 2000; Hirsch, 2001). These enzymes transport cytosolic Ca^{2+} into either apoplast or the lumen of intracellular organelles, such as vacuole and endoplasmic reticulum (ER). Low concentration of cytosolic calcium facilitates rapid and dramatic increases in $[Ca^{2+}]_{cyt}$ during cell response to biotic and abiotic stress. Almost without exception abiotic stresses elicit a rise in the concentration of free calcium in the cytoplasm (White and Broadley, 2003). Repeated exposures of plants to NaCl treatments provoke prolonged alternations of both cytosolic and apoplastic Ca^{2+} concentrations (Gao et al., 2004).$[Ca^{2+}]_{cyt}$ elevation is also observed when plants are exposed to hypo-osmotic stress (Gao et al., 2004). Use of aequorin as a reporter has revealed that cold and wind can initiate specific Ca^{2+} signals that are spatially distinct in tobacco seedlings (van de Luit et al., 1999). Environmental stresses often cause the production of reactive oxygen intermediates (ROIs). There seems to be a relationship between the levels of ROIs and Ca^{2+} in plant cells. H_2O_2 triggers a biphasic Ca^{2+} elevation in tobacco cultured cells (Lecourieux et al., 2002). Rentel and Knight (2004) also observed a biphasic Ca^{2+} signature composed of two independent peaks. The magnitude of the Ca^{2+} signal correlates with the ROI levels. The elevation of $[Ca^{2+}]_{cyt}$ serves as a signal to mediate appropriate response of the cells to abiotic stress. The calcium signals are generated through the opening of Ca^{2+}-permeable channels that allow the downhill flow of Ca^{2+} from a compartment, such as apoplast, vacuole or ER, into cytoplasm. Conversely, upon cessation of the stress stimuli, the free calcium concentration in the cytoplasm reverts to its resting level by Ca^{2+}-ATPases and Ca^{2+}/H^+ antiporters. The interplay between influx through ion channels and efflux from pumps and carriers would be important to determine the form of a Ca^{2+} spike that is specific to both stimuli and signal decoders.

2.1. Characteristics of Calcium Signals

Although it is a common thought that elevating $[Ca^{2+}]_{cyt}$ is a primitive and universal response to stress, how calcium signals convey stimulus specificity to a variety of signaling pathways is still a debatable issue. The specificity of calcium-mediated signaling might be encoded by the spatial properties, and/or kinetics or magnitude of the $[Ca^{2+}]_{cyt}$ perturbation. The location of cytosolic Ca^{2+} elevation triggered by abiotic stress can be affected by the type, cellular location and abundance of Ca^{2+}-permeable channels. Since the diffusion of Ca^{2+} within the cytoplasm is slow (Clapham, 1995), transient Ca^{2+} influx to the cytosol from apoplast, ER or vacuole can make significant difference on the cytosolic location of Ca^{2+} elevation, thereby influence the spatial characteristics of $[Ca^{2+}]_{cyt}$. Different stimulus could activate distinct subcellularly localized Ca^{2+}-permeable channels, thus generate specific location of Ca^{2+} elevation with distinguishable spatial signal for the downstream Ca^{2+} signal transducing proteins. Pauly et al. (2001) showed that different cytosolic and nuclear calcium signals may be involved in discrimination between hyper- and hypo-osmotic treatments in tobacco suspension culture cells. In embryos of the multicellular alga *Fucus*, different spatial patterns of Ca^{2+} elevation were generated by different degrees of hypo-osmotic shock (Goddard et al., 2000). Spatial changes of $[Ca^{2+}]_{cyt}$ elevation was also observed in the guard cells after ABA treatment. The $[Ca^{2+}]_{cyt}$ elevation is first close to the plasma membrane and then adjacent to the vacuole (McAinsh et al., 1995; Allen et al., 1999). This spatial change is thought to be due to the sequential opening of hyperpolarization-activated Ca^{2+}-permeable channels at the plasma membrane and then second messenger activated Ca^{2+}-permeable channels in the tonoplast (Grabov and Blatt, 1998; Schroeder et al., 2001). The importance of cellular location of ion channels in determining stimulus specificity was evidenced by the observation that low temperature-induced stomatal closure involves primarily entry of Ca^{2+} across the plasma membrane, while intracellular mobilization appears to dominate if stomatal closure is initiated with ABA or mechanical stimulation (Wood et al., 2000). Spatial restricted Ca^{2+} elevation also appears to underlie the response of leguminous root hairs to Nod factors

(Hetherington and Brownlee, 2004). To sense and transduce the localized $[Ca^{2+}]_{cyt}$ elevation, the signal transducing proteins might be required to be either associated with the Ca^{2+} channel or recruited to membrane in close proximity to the channels. This notion is supported by the fact that many calcium-binding proteins including CDPKs and CBLs contain N-terminal myristoylation sites conferring membrane association. Another possibility defining specificity of calcium signaling is that elevation of $[Ca^{2+}]_{cyt}$ is necessary but insufficient to trigger specific response, rather, the combination of $[Ca^{2+}]_{cyt}$ elevation and the specific responding proteins decodes a specific stimulus and activate a corresponding signal transduction pathway. Plants possess numerous Ca^{2+}-binding proteins. Each or a group of these proteins could be specifically induced or activated by a specific stimulus.

The dynamic properties of the Ca^{2+} signal might determine the efficacy with which the response is elicited. In animal cells, stimulus-induced $[Ca^{2+}]_{cyt}$ oscillations and the mechanisms of generation and the potential for the information to be encoded in the frequency of $[Ca^{2+}]_{cyt}$ oscillation was largely studied (reviewed by Berridge et al., 2003).$[Ca^{2+}]_{cyt}$ oscillations and transients also occur in plant cells. This phenomena was observed in maize coleoptiles in response to IAA stimulus (Felle, 1988), in the shank of poppy pollen tubes during the self incompatibility response (Straatman et al., 2001; Rudd and Franklin-Tong, 2001), in the legume root hair in response to Nod factors (Hetherington and Brownlee, 2004), in the guard cells in response to different treatments (Schroeder et al., 2001), and in other type of cells (Evans et al., 2001). The frequency, period, and amplitude of $[Ca^{2+}]_{cyt}$ oscillations vary among different cell types and in response to different stimuli (Evans et al., 2001). Since the demonstration that stomatal guard cells could exhibit $[Ca^{2+}]_{cyt}$ oscillations (McAinsh et al., 1995), the guard cell has been used as a model in dissecting the functional significance of $[Ca^{2+}]_{cyt}$ oscillation. Ng et al. (2001) observed induced $[Ca^{2+}]_{cyt}$ oscillations in the guard cells by sphingosine-1-phosphate (S-1-P). The frequency and amplitude of $[Ca^{2+}]_{cyt}$ oscillations were S-1-P concentration dependent, which led to corresponding kinetics of stomatal closure. They concluded that drought-induced guard cell signal transduction involves S-1-P and $[Ca^{2+}]_{cyt}$ elevation. By combining cameleon (a ratiometic flurescent protein Ca^{2+} indication) technology with the use of Ca^{2+} homeostasis or signaling mutants, significant progress has been made in assigning specificity of Ca^{2+} signals in guard cells. Allen et al. (2000) showed that, in response to elevated external Ca^{2+} or oxidative stress, *det3* mutants defective of a vacuolar H^+-ATPase generate prolonged Ca^{2+} elevations, resulting in failure of stomatal closure, while wild type plants display repetitive transients of $[Ca^{2+}]_{cyt}$, leading to stomatal closure. In contrast, similar $[Ca^{2+}]_{cyt}$ oscillations and stomatal closure were observed in the guard cells of both *det3* and wild type in response to cold or ABA treatments. Moreover, Allen et al. (2001) found that, in guard cells of the ABA-insensitive mutant *gca2*, $[Ca^{2+}]_{cyt}$ oscillations induced by abscisic acid and extracellular calcium had increased frequencies and reduced transient duration, and steady-state stomatal closure was abolished. Experimentally imposing $[Ca^{2+}]_{cyt}$ oscillations with parameters that elicited closure in the wild type restored long-term closure in *gca2* stomata. Another study by use of *gca2* mutant (Pei et al., 2000) indicated that ABA-induced H_2O_2 production and the H_2O_2-activated Ca^{2+} channels leading to $[Ca^{2+}]_{cyt}$ elevation are important mechanisms for ABA-induced stomatal closing. Thus, Ca^{2+} signal is an essential component mediating drought induced ABA signaling cascade in guard cells that results in stomatal closure and reducing water loss.

2.2. Elements Encoding Calcium Signals

Elevation of $[Ca^{2+}]_{cyt}$ can arise via increased influx and/or decreased efflux. Ca^{2+}-permeable channels are the key entry points for Ca^{2+} into the cytosol (Sanders et al., 2002). These channels have been found in all plant membranes including plasma membrane, ER membrane, and vacuolar membrane. The primary roles of Ca^{2+}-permeable channels in the plasma membrane appear to be responsible for the generation of calcium signals, but they may also contribute to nutritional Ca^{2+} uptake (White, 2000). The presence of Ca^{2+}-permeable channels in diverse membrane is thought to contribute the specificity of calcium signals and to enable physiological flexibility. Although most calcium signals generally initiate from increased Ca^{2+} channel activity, the transport systems energizing calcium efflux from the cytosol provide critical functions in keeping low $[Ca^{2+}]_{cyt}$ to facilitate cytosolic Ca^{2+} perturbation in response to stimuli and in terminating a Ca^{2+} signal by restoring $[Ca^{2+}]_{cyt}$ to resting level. The efflux pathways might also help shape the dynamic form of a calcium spike and thereby help define the information encoded in the signals. The Ca^{2+} efflux systems include Ca^{2+} pumps and Ca^{2+}/H^+ antiporters. The following is to discuss each element encoding calcium signals and their implications in salt and drought stress.

2.2.1. Ca^{2+}-Permeable channels

In animal cells, specific $[Ca^{2+}]_{cyt}$ elevations have been assigned to specific Ca^{2+}-selective channels (Zou et al., 2002; Grimaldi et al., 2003). However, plants predominantly use Ca^{2+} permeable, rather than selective, channels. Ca^{2+}-permeable channels are present in the plasma membrane, tonoplast, ER and other endomembranes. These channels have been investigated with electrophysiological, biochemical and molecular approaches. At the molecular level the Ca^{2+}-permeable channels have been broadly classed as non-selective cation channels (NSCCs). NSCCs are a diverse group of ion channels characterized by their low discrimination between many essential and toxic cations. Members of this group are likely to function in low-affinity nutrient uptake, in distribution of cations within and between cells, and as plant Ca^{2+} channels. They are gated by diverse mechanisms, which can include voltage, cyclic nucleotides, glutamate, reactive oxygen species, and stretch. Accordingly, Ca^{2+}-permeable channels have been classified into depolarization-actived Ca^{2+} channels (DACCs), hyperpolarization-activated Ca^{2+} channels (HACCs), cyclic nucleotide gated channels (CNGCs), and glutamate receptor channels (GLR).

DACCs are Ca^{2+} permeable channels that are activated by membrane depolarization (White, 2000). Several types of DACCs have been observed in the plasma membrane of plant cells. All DACCs are permeable to both monovalent and divalent cations. Their activity appears to be controlled by microtubule cytoskeletal interactions and stabilized by the disruption of microtubules (Thion et al., 1998). Since plasma membrane depolarization is a frequently observed response to various biotic (Ehrhardt et al., 1992; Lhuissier et al., 2001) and abiotic (Okazaki et al., 2002) stresses that also elicit $[Ca^{2+}]_{cyt}$ elevations, it is deduced that DACCs might play a pivotal role at an early stage in transducing general stress-related signals by perception of a range of stimuli resulting in membrane depolarization. The molecular identities of DACCs have not been verified. An Arabidopsis homologue of the animal $\alpha 1$ subunit of voltage-dependent Ca^{2+} channels, AtTPC1, has been proposed to form a depolarization-activated channel (White et al., 2002). Predicted secondary structure of the AtTPC1 contains two shaker-like domains, each of which has six transmembrane spans forming a pore. A hydrophilic domain connecting these two shaker-like domains includes two EF hands, suggesting a calcium regulation of this channel. Overexpression of AtTPC1 in Arabidopsis enhanced an increase of $[Ca^{2+}]_{cyt}$ that was elicited by sugar-induced membrane depolarization, whereas antisense suppression of AtTPC1 also suppressed the $[Ca^{2+}]_{cyt}$elevation in response to sugar supply (Furuichi et al., 2001). Although firm conclusion regarding voltage gating awaits electrophysiological characterization, these results suggest that AtTPC1 is a likely DACC. The role of AtTPC1 in abiotic stress response is to be elucidated.

Patch clamp electrophysiological techniques have identified HACCs in different types of cells including stomatal guard cells (White and Broadly, 2003). Implications of HACCs in drought stress came from the studies in the guard cells in response to ABA. These channels activate at hyperpolarized membrane potentials (more negative than -100 mV) that is directly associated with $[Ca^{2+}]_{cyt}$ elevation in guard cells that follows ABA application (Grabov and Blatt, 1998; Pei et al., 1999). Drought stress increases ABA concentration in the guard cells which shifts the activation potential of HACCs to more positive voltage and the subsequent entry of Ca^{2+} not only depolarizes the plasma membrane but also initiates the $[Ca^{2+}]_{cyt}$-dependent events, including $[Ca^{2+}]_{cyt}$-dependent Ca^{2+} release from intracellular stores, that leads to stomatal closure (Blatt, 2000; White, 2000; Schroeder et al., 2001; White and Broadley, 2003). ABA-induced stomatal closure involves the production of reactive oxygen species (ROS) (Pei et al., 2000; Zhang et al., 2001). In Arabidopsis guard cells, H_2O_2 activates HACCs and thereby an increase in $[Ca^{2+}]_{cyt}$ (Pei et al., 2000). This process requires cytosolic NAD(P)H, suggesting that NADPH oxidases might be a component in the signaling chain of drought-ABA-H_2O_2-$[Ca^{2+}]_{cyt}$ elevation-stomatal closure (Murata et al., 2001). In Arabidopsis root hair, reactive oxygen species, especially hydroxyl radical, increase HACCs activity (Foreman et al., 2003). Molecular evidence has shown that the plasma membrane NADPH oxidase AtrbohC is required for the production of ROS and the generation of the root hair apical $[Ca^{2+}]_{cyt}$ gradient. The molecular identities of HACCs are not yet known.

Animal CNGCs were shown to play important roles in transduction of visual and olfactory stimuli via modulation of the membrane potential and Ca^{2+} signals in sensory cells. In addition, animal CNGCs are increasingly being implied in Ca^{2+} signaling in many other cell types (Kaupp and Seifert, 2002). Animal CNGCs can contain α and β subunits. The membrane-spaning, pore-forming α subunit forms a functional channel while the β subunit play a regulatory role (Kramer and Molokanova, 2001; Kaupp and Seifert, 2002). Animal cells encode only three α subunits and three β subunits. In Arabidopsis, CNGCs are encoded by no fewer than twenty genes (Talke et al., 2003). The twenty members of Arabidopsis CNGCs are structurally related to Shaker type K^+ channels with six putative transmembrane spans and a pore region between spans five and six. The C-terminus of Arabidopsis CNGCs contain overlapping CaM and cyclic nucleotide binding domains (Arazi et al., 2000; Kohler and Neuhaus, 2000). Binding CaM and cyclic nucleotides provides the potential for dual regulation by different signaling pathways and possible mechanisms for integrating these two signal molecules. Animal CNGCs have an N-terminal CaM binding domain and a C-terminal cyclic nucleotide binding domain. cNMP (cAMP or cGMP) binding to the CNGCs activates the channels, while CaM/Ca binding to the N-terminus of CNGCs alters the protein conformation, resulting in a decreased affinity of the channel for cNMP. Hence, cytosolic CaM modulates cNMP activation of CNGCs in animal cells (Kaupp and Seifert, 2002). Understanding of CNGC function in plants is fragmented. Evidence has been shown that CNGCs are involved in the pathogen response signaling pathways. An Arabidopsis mutant lacking ATCNGC2 shows a "defense no death" phenotype and fails to generate a typical hypersensitive response or program cell death (Clough et al., 2000). In addition, Ca^{2+} and monovalent cation homeostasis is affected in *cngc2* mutants (Chan et al., 2003). Some evidence suggests a role of CNGCs in heavy metal homeostasis. Arazi et al. (1999) found that overexpression of a tobacco CNGC, NtCBP4, led to hypersensitive to Pb^{2+}. In contrast, expression of a truncated version of NtCBP4 enhanced tolerance to Pb^{2+} and reduced uptake of this heavy metal (Sunkar et al., 2000). Arabidopsis CNGC10 was recently shown to be involved in K^+ uptake (Li et al., 2005; Borsics et al., 2006).

GLRs comprise another class of ion channel that might transport Ca^{2+} in a non-selective manner. The activation of GLRs by glutamate and other amino acids is a key event in animal cells in mediating fast chemical transmission and long-term synaptic potentiation (Dingledine et al., 1999). Arabidopsis possesses thirty members of GLRs (Davenport, 2001, Lacombe et al., 2001). Arabidopsis GLR structure is similar to that of animal glutamate receptors and is composed of four membrane-localized domains. Two glutamate binding domains are localized on the outside of the membrane; one is at the N-terminal and the other resides between membrane spans three and four (Sanders et al., 2002). Molecular identities of GLRs in plants have been lacking. Implication of these ion channels in salt and drought stress has not yet reported except the observation that glutamate stimulated unidirectional influx of Na^+ into intact roots by up to 25% (Demidchik et al., 2002).

2.2.2. Ca^{2+} pumps

Ca^{2+} pumps are active Ca^{2+} transporters directly energized by ATP hydrolysis (Ca^{2+}-ATPases) driving Ca^{2+} out of cytosol against the steep Ca^{2+} electrochemical gradient at the plasma membrane and across the endomembranes. Ca^{2+} pumps, together with Ca^{2+}/H^+ antiporters (see below) may be important in determining the peak amplitudes and duration of Ca^{2+} transients. Evidence to support the role of Ca^{2+} pumps in helping define the information encoded in the Ca^{2+} signals are still lacking, although the properties of Ca^{2+} pumps have been elucidated in some detail. Ca^{2+}-ATPases are high affinity Ca^{2+} transporters (Km=1-10μM, Evans and Williams, 1998). They can be grouped as Type IIA and Type IIB based on their protein sequence similarity to animal PMCA or SERCA, respectively (Sze et al., 2000; White and Broadley, 2003). Four Type IIA pumps have been identified in Arabidopsis (designated as AtECA1-4, Axelsen and Palmgren, 2001). Subcellular localization revealed that Type IIA pumps are present in the plasma membrane, toloplast, ER and Golgi apparatus (White and Broadley, 2003; Hetherington and Brownlee, 2004). Arabidopsis possesses at least ten genes encoding Type IIB Ca^{2+} pumps (AtACAs, Axelsen and Palmgren, 2001). Both plasma membrane and endomembrane localizatons of AtACAs have been found (Hetherington and Brownlee, 2004). Type IIA and IIB Ca^{2+}-ATPases have similar topology except that Type IIB has an N-terminal autoinhibitory domain but Type IIA lacks this domain. Both types of pumps contain ten transmembrane domains and a large cytoplasmic loop connecting membrane spans four and five. The large central cytoplasmic loop contains an ATP binding site and the aspartate residue that becomes phosphorylated during the reaction cycle (Sze et al., 2000). The N-terminal autoinhibitory domain in Type IIB pumps plays an important role in regulating pump activity. The autoinhibitory domain contains a binding site for Ca-CaM plus a serine phosphorylation site. Binding of CaM to the autoinhibitory site activates pump activity, while phosphorylation of this domain by CDPK inhibits pump activity (Sze et al., 2000).

Functional analysis of Ca^{2+} pumps has implicated their roles in salt stress response in both yeast cells and plants. PMR1, a yeast Ca^{2+}-ATPase in Golgi, controls salt tolerance by modulating the expression of the plasma membrane Na^+-ATPase PMR2 (Park et al., 2001). In yeast, activation of calcineurin is required for up-regulation of Na^+ efflux mechanism and thereby confers salt tolerance. Mutation in *PMR1* results in a maintained high cytosolic Ca^{2+} concentration, which causes continuous activation of calcineurin and increased expression of PMR2. Consequently, *pmr1* mutant cells are more salt tolerance (Park et al., 2001). This study suggests that modulation of cytosolic Ca^{2+} via Ca^{2+} pumps plays crucial role in salt stress response. The Arabidopsis *ACA4* gene encoding a vacuolar membrane Ca^{2+} pump can improve salt tolerance in yeast (Geisler et al., 2000). Although ACA4 expressing yeast cells displayed increased osmotic sensitivity at high external calcium, the ACA4 conferred significant osmotic stress tolerance to yeast cells at low external calcium. A more active N-terminal truncated form of ACA4 lacking CaM binding site specifically enhanced NaCl tolerance, whereas full-length ACA4 had less effect. The mechanism underlying salt tolerance conferred by ACA4 is likely due to the modulation of $[Ca^{2+}]_{cyt}$.

The involvement of Ca^{2+} pumps in salt stress response is also highlighted by the fact that the expression of many Ca^{2+}-ATPases is increased upon exposure to high salinity and some Ca^{2+}-ATPase genes are expressed only under stress conditions (Geisler et al., 2000; Garciadeblas et al., 2001). The expression of Type IIA Ca^{2+} ATPases from tomato (Wimmers et al., 1992) and tobacco (Perez-Prat et al., 1992) has been shown to be induced by NaCl. The plasma membrane Ca^{2+}-ATPase *SCA1* from soybean is highly and rapidly induced by NaCl stress but not by osmotic stress (Chung et al., 2000). The Arabidopsis *ACA4* gene is also up-regulated by NaCl treatment (Geisler et al., 2000). A survey of gene expression profiles of Ca^{2+}-ATPAase genes including both *AtECA*s and *AtACA*s in Arabidopsis based on the public available microarray data at AtGenExpress (http://jsp.weigelworld.org/expviz/expviz.jsp) revealed distinct basal and induced expression patterns of different Ca^{2+}-ATPase genes. The transcript level of Type IIA Ca^{2+}-ATPase *ECA2* (At4g00900) is increased 2- to 4-fold by NaCl treatments. *ECA4* (At1g07670) is slightly up-regulated by salt stress, while the expression of *ECA3* (At1g10130) is not affected by NaCl treatments. The expression of *AtACAs* is more dynamic. In spite of three *ACA* genes (*ACA4*: At2g41560, *ACA7*: At2g22950, *ACA9*: At3g21180) whose expression is not affected by cold, osmotic and salt stress, the expression of other *ACA* genes are significantly regulated by these stress treatments, especially by NaCl treatment. The expression of *ACA1* (At1g27770) is highly induced by NaCl in roots, while the induction of *ACA2* and *ACA10* expression is only moderate. The *ACA8* (At5g57110) is induced up to 10-fold by cold stress but not affected by salt stress, suggesting this Ca^{2+} pump might be important for calcium-mediated cold stress signaling. In contrast, the expression of *ACA11* (At3g57330) is down-regulated by cold stress for 24 hours. The *ACA12* (At3g63380) and *ACA13* (At3g22910) are two Type IIB Ca^{2+}-ATPase genes whose expression is strongly induced by both cold and salt stress. The expression of these two genes is induced up to 20-folds by cold stress. The expression level of *ACA13* is increased about 200-folds in root by NaCl treatment for 6 hours. Induced expression of Ca^{2+}-ATPase genes suggest that the Ca^{2+} pumps might be part of Ca^{2+}-dependent signal transduction pathway linked to abiotic stress including salt stress. The increase in cytosolic Ca^{2+} upon NaCl exposure is indeed an effector of salt tolerance. However, the elevated calcium level must be transitory, which could be adjusted by an increased capacity of Ca^{2+} pumps.

2.2.3. *Ca^{2+}/H^+ Antiporter*

Ca^{2+}/H^+ antiporters are low affinity Ca^{2+} transporters driven by the electrochemical gradients of H^+ to remove Ca^{2+} from the cytosol (Evans and Williams, 1998; Sanders et al., 2002). The Arabidopsis *CALCIUM EXCHANGER 1* (*CAX1*) was the first plant gene encoding an Ca^{2+}/H^+ antiporter to be cloned. This gene was identified by screening a cDNA library from Arabidopsis for clones able to complement a yeast mutant defective in vacuolar Ca^{2+} transporter (Hirschi et al.,

1996). CAX1 appears to have low Ca^{2+} affinity (Km is ~13 μM) and high Ca^{2+} transport capacity (Shigaki et al., 2001). CAX1 seems to be localized in the vacuolar membrane (Cheng et al., 2003). An N-terminal autoinhibitory domain has been determined to play a regulatory role in modulating CAX1 activity (Pittman and Hirschi, 2001). Loss of CAX1 function (Cheng et al., 2003) resulted in mass reduction of tonoplast Ca^{2+}/H^+ antiporter activity and, interestingly, increased activities of V-type H^+-ATPases and tonoplast Ca^{2+}-ATPases. Significantly, *cax1* mutants displayed increased expression of other Ca^{2+}/H^+ antiporters *CAX3* and *CAX4* and enhanced tolerance to Mn^{2+} and Mg^{2+}. The mutants exhibited altered plant development, perturbed hormone sensitivities, and altered expression of an auxin-regulated promoter-reporter gene fusion, which suggest a possible signaling role of CAX1 in stress and hormone response.

Arabidopsis possesses at least 12 genes encoding antiporters closely related to CAX1 (Maser et al., 2001), four (CAX1-4) of which have been characterized. CAX2 has a lower capacity for Ca^{2+} and appears to transport Mn^{2+} and Cd^{2+} in addition to Ca^{2+} (Pittman and Hirschi, 2003). CAX2 also contains an N-terminal autoinhibitory domain, suggesting that CAX1 and CAX2 may have shared regulatory features. Unlike *cax1*, *cax2* mutants displayed no discernable morphological phenotypes or alternations in Ca^{2+}/H^+ antiporter activities. However, *cax2* mutants exhibited a reduction in vacuolar Mn^{2+}/H^+antiport and reduced V-type H^+-ATPase activity. CAX3 was also biochemically and genetically characterized (Cheng et al., 2005). CAX3 is localized to the tonoplast. The expression of *CAX3* is predominate in roots, while *CAX1* is highly expressed in leaves. Knockout of *CAX3* were modestly sensitive to exogenous Ca^{2+} and also displayed reduction in vacuolar H^+-ATPase activity. Ionomic analysis of *cax1* and *cax3* single mutants and *cax1cax3* double mutant revealed synergistic function of CAX1 and CAX3 in plant growth and nutrition acquisition. CAX4 also appears to have vacuolar membrane localization (Cheng et al., 2002). The expression of *CAX4* is induced by Mn^{2+}, Na^+, and Ni^{2+} treatments.

In addition to the nutritional function, it has been proposed that CAXs may play a role in reducing cytosolic Ca^{2+} concentration to resting levels after a $[Ca^{2+}]_{cyt}$ elevation in response to external stimuli (Hirschi, 1999). Consistent with this notion is the observation that an Ca^{2+}/H^+ antiporter, but not the vacuolar Ca^{2+} pump, resets $[Ca^{2+}]_{cyt}$ in yeast following hypertonic shock (Denis and Cyert, 2002). In addition, Arabidopsis *det3* mutant with reduced tonoplast H^+-ATPase activity and presumably reduced Ca^{2+}/H^+ antiporter activity has a continuously high $[Ca2+]_{cyt}$ (Allen et al., 2000). Also support the signaling role of Ca^{2+}/H^+ antiporters is that plants treated with V-type ATPase inhibitor bafilomycin show greater $[Ca^{2+}]_{cyt}$ elevation in response to hyperosmotic shock (Takahashi et al., 1997). Signaling role of Ca^{2+}/H^+ antiporter was also evidenced at molecular level by the observation that mutations in *CAX1* increase *CBF/DREB1* expression and the cold-acclimation response in Arabidopsis (Catala et al., 2003). The expression of *CAX1* is induced in response to low temperature through an ABA-independent pathway. In fact, earlier study has identified a *RARE COLD INDUCIBLE* (*RCI*) gene named *RCI4* that is identical to *CAX1* (Jarillo et al., 1994; Capel et al., 1997; Llorente et al., 2002). The characterization of two T-DNA insertional mutants, *cax1-3* and *cax1-4*, demonstrated that mutations in *CAX1* gene do not affect the constitutive capacity to tolerate freezing temperature, dehydration, chilling, or high salt. Surprisingly, however, the *cax* mutants exhibit an increased ability to cold acclimation, which correlates with an enhanced expression of cold responsive transcription factor *CBF/DREB1* genes as well as the downstream targets in response to low temperature. These results indicate that CAX1 plays an essential role in the cold-acclimation response by controlling *CBF/DREB1* expression, likely by ensuring the proper control of Ca^{2+} homeostasis under low temperature condition. It seems that, following the $[Ca^{2+}]_{cyt}$ elevation elicited by environmental stimuli, subsequent reestablishment of $[Ca^{2+}]_{cyt}$ to resting level by Ca^{2+} pumps and Ca^{2+}/H^+ antiporters is a part of the generation of calcium signals. Implication of CAX1 in salt stress response stemmed from a study showing that the salt tolerance determinant SOS2, a Ser/Thr protein kinase can activate the Ca^{2+}/H^+ antiporter CAX1 to integrate calcium transport and salt tolerance (Cheng et al., 2004). SOS2 was able to interact with the N-terminus of CAX1 and activate CAX1 transport activity in a SOS3-independent manner in yeast cells. Moreover, the high level expression of a deregulated version of CAX1 *in planta* caused salt sensitivity. Thus, the vacuolar Ca^{2+}/H^+ antiporter CAX1 might be a component of salt stress signaling modulating $[Ca^{2+}]_{cyt}$ during calcium signal generation and subsequent signal transduction in response to salt stress.

3. DECODING CALCIUM SIGNALS

Once calcium signal is generated in response to internal or external stimuli, the signal is initially perceived by binding of Ca^{2+} to Ca^{2+} sensors. Most Ca^{2+} sensor proteins contain a common structural motif, the "EF hand". EF hand is a conserved helix-loop-helix structure that can bind a single Ca^{2+} ion (Kretsinger and Nockolds, 1973). Plants have huge number of EF-hand containing proteins. In Arabidopsis, approximately 250 genes encode EF-hand containing proteins (Day et al., 2002). Note that not all Ca^{2+}-binding proteins contain EF-hand and other protein motifs also confer Ca^{2+} binding ability. For example, the 70-amino acid annexin fold present in members of the membrane-associated annexin subfamily is a Ca^{2+}-binding motif (Delmer and Potikha, 1997). Another Ca^{2+}-binding motif termed the "C2 domain" is about 130-145 amino acids and is important for modulation of phospholipase D activity (Wang, 2005). Plant Ca^{2+} sensors include CaMs and CaM-like(CML) proteins, calcineurin B-like (CBL) proteins, Ca^{2+}-dependent protein kinases (CDPKs), and Ca^{2+}-binding proteins without EF-hand. The conformation change of Ca^{2+} sensors upon Ca^{2+} binding is essential for the calcium signal transduction. The Ca^{2+} sensors can be divided into two types, sensor relays and sensor responders (Sanders et al., 2002). Binding of Ca^{2+} onto a sensor relay results in its conformational change that is relayed to an interaction partner. The interaction partner then undergoes changes in structure or enzyme activity, which modulates the functions of the effectors. This type of sensor includes CaMs, CMLs, and CBLs. In contrast, sensor responders undergo a calcium-induced conformational change that alters their own activity or structure through intramolecular interactions. CDPKs are one of the sensor responders in plants. The following will discuss each type of Ca^{2+} sensors and their involvements in salt and drought stress response.

3.1. CaMs and CMLs

CaM is one of the best characterized Ca^{2+} responsive proteins in eukaryotic cells. Eukaryotic CaMs are well conserved proteins, reflecting their central role in eukaryotic biology (van Eldik and Watterson, 1998). CaMs are small acidic protein containing EF-hands. A CaM is arranged as two globular domains connected with a long flexible helix. Each globular domain contains a pair of intimately linked EF-hands (Bouche et al., 2005; McCormack et al., 2005). CaM is an unusual protein that has no catalytic activity of its own but activates numerous target proteins upon binding Ca^{2+}. Selectively binding of Ca^{2+} onto CaM with micromolar affinity results in the conformational change characteristic of exposing two hydrophobic surfaces surrounded by negative charges, one in each globular domain. The hydrophobic surfaces provide sites for interactions with target proteins with an affinity in the nanomolar range through non-specific van der Waals interactions (Crivici and Ikura, 1995). In this way, cytosolic Ca^{2+} perturbation perceived by CaM are transduced into altered target protein activity leading to subsequent cellular responses.

Plant CaMs are defined by their high sequence similarity, over 89% identity, to vertebrate CaMs and no predicted functional alteration caused by the amino acid variations (McCormack et al., 2005). By this definition, all CaMs in vertebrates are highly conserved, suggesting a structural requirement for CaMs to interact with diverse targets. CaM isoforms are encoded by small gene families in plants. In Arabidopsis, seven genes encode highly conserved or identical CaM proteins, named CaM1-7, having 148 amino acids in their mature forms (149 aa with the first Met that is cleaved following translation). These seven genes only encode four CaM isoforms. CaM2, 3, and 5 genes (At2g41110, At3g56800, and At2g27030, respectively) encode a single isoform and CaM1 and CaM4 genes (At5g37780 and At1g66410) encode another isoform. CaM6 (At5g21274) and CaM7 (At3g43810) are other two CaM isoforms in Arabidopsis. Amongst the CaM isoforms, only one to five amino acids substitutions are found. Comparing with CaM2/3/5, CaM7 has only one amino acid substitution, i.e. K^{127} to R^{127}, while CaM6 shows two amino acids substitutions of T^{118} to S^{118} and K^{127} to R^{127}. Therefore, CaM6 (S^{118}) only differs from CaM7 (T^{118}) on one amino acid. CaM1/4 has five amino acids differences from CaM2/3/5, which are D^{8} to E^{8}, R^{75} to K^{75}, D^{123} to E^{123}, K^{127} to R^{127}, and V^{145} to I^{145}. Although one to five amino acid substitutions exist, the function of these four CaM isoforms is very likely to be identical or very similar because the amino acid substitutions are from one to another with very similar properties. For example, aspartic acid (D) and glutamic acid (E) are structurally very similar and both have negative charges. Change from aspartic acid to glutamic acid is not likely to alter the protein functions. Similarly, the properties are very similar between the amino acids K and R, T and S, V and I. Substitution between these two amino acids is unlikely to change the protein functions. However, minor changes in CaMs, especially those that reside within Ca^{2+}-binding loop, might contribute to target specificity selection (Bhattacharya et al., 2004). It has been reported that different Arabidopsis CaM isoforms also differ in their affinity for the same protein. Liao et al. (1996) showed that the Arabidopsis CaM isoforms have different capacity to activate NAD kinase *in vitro*. It has also shown that the binding affinity of Arabidopsis CaM isoforms to kinesin-like motor protein is different (Reddy et al., 1999). CNGCs are ion channels with CaM binding motif. Different CaM isoforms from Arabidopsis possess distinct binding capacity to the CNGCs (Kohler and Neuhaus, 2000).

Multiple genes encoding identical or nearly identical CaM isoforms in plants indicate that the different CaM genes might have evolved distinct expression patterns, thus contribute to tissue-specific expression, developmental regulation and/or differential response to environmental stress. Indeed, expression studies show that the Arabidopsis CaM genes are differentially expressed in different tissues and circumstances (Perera and Zielinski, 1992; Gawienowski et al., 1993). The CaMs in Arabidopsis are expressed during all developmental stages (McCormack et al., 2005). Microarray results at AtGenExpress (http://jsp.weigelworld.org/expviz/expviz.jsp) revealed that, although Arabidopsis CaM transcripts are associated with all tissues and organs tested, the expression patterns are somewhat different. The Arabidopsis CaM gene expression is not significantly affected by different stress and hormone treatments tested in microarray analysis. However, some of the Arabidopsis CaMs were shown to be touch-inducible but at different levels and with different kinetics (Perera and Zielinski, 1992). CaM2 was also named as TCH1 because it is induced by touch stimulus. The soybean SCaM4 and ScaM5 are induced by a fungal elicitor (Heo et al., 1999). In a recent report three CaM genes from rice were shown to be differentially regulated by various stress signals (Phean-o-pas et al., 2005). *OsCaM1* mRNA levels is strongly increased in response to NaCl, mannitol and wounding treatments. In contrast, *OsCaM2* expression is relatively unchanged under these stress conditions. *OsCaM3* is also up-regulated by NaCl and wounding treatments, but more in a transient manner.

In addition to these highly conserved CaMs, plants also possess many CaM-like proteins (CMLs). McCormack and Braam (2003) defined the CMLs as proteins composed mostly of EF-hand Ca^{2+}-binding motif, having no other identifiable functional domains, and at least 16% identical with CaM, which excludes Ca^{2+}-binding proteins such as CDPKs and CBLs that contain additional functional domains. In Arabidopsis genome, fifty genes encode CMLs composed of CaM-like EF hand structures (McCormack and Braam, 2003; McCormack et al., 2005). All but one (CML1) of the CMLs have at least two EF hand-like motifs (McCormack and Braam, 2003). Thirty out of fifty CMLs have four predicted EF hands; one (CML12) has six EF hands. CMLs are more sequence diverse comparing with

CaMs and grouped into nine groups. In contrast to CaM genes, the fifty CML genes display distinct developmental regulation and abiotic stress response. Thirty-five CML genes in Arabidopsis have at least an average fivefold increase in expression by at least one of the stimuli tested based on Genevestigator online search (McCormack et al., 2005; http://www.genevestigator.ethz.ch). *CML37* (At5g42380), *CML39* (At1g76640) and *CML40* (At3g01830) are among the most strongly regulated genes. The expression levels of *CML37* and *CML39* are nearly 100- and 60-fold higher, respectively, in salt-stressed plants relative to the non-stresses control plants. Two CML genes are touch-inducible genes previously identified as TCH genes (CML12/TCH3, At2g41100; CML24/TCH2, At5g37770; Braam et al., 1997). Recently, a more detailed characterization of CML24/TCH2 has implicated its function in response to diverse stimuli (Delk et al. 2005). CML24 shares over 40% amino acid sequence identity with CaM, has four EF hands, and undergoes Ca^{2+}-dependent in hydrophobic interaction chromatography, indicating that CML24 binds Ca^{2+} and consequently undergoes conformational changes. *CML24* expression is up-regulated by touch, darkness, heat, cold, H_2O_2, ABA and IAA, suggesting its involvement in multiple stress and hormonal response. CML24-underexpressing transgenic plants are resistant to ABA inhibition of germination and seedling growth and have enhanced tolerance to ionic stress. These data suggest that *CML24* encodes a potential Ca^{2+} sensor that may function to enable response to ABA and presence of various salts. The Arabidopsis CML gene family also includes previously identified CaBP-22 (Ling and Zielinski, 1993), APC1 (Rozwadowski et al., 1999) and centrin-like protein (CNL20, At3g50360), suggesting diverse functions of CMLs in plant response to environmental, developmental and pathological challenges.

Use of bioinformatics tools and molecular approaches has enabled the identification of the cellular targets of CaM and CML in plants as well as in other organisms. The list of numerous CaM-binding proteins (CaMBPs) described by Snedden and Fromm (2001) reflects the functional diversity of plant CaMBPs and reveals an involvement in regulation of metabolism, cytoskeleton, ion transport, protein folding, transcription, protein phosphorylation and dephosphorylation, phospholipids metabolism, and unknown functions. Current findings support the idea that CaMs and CMLs are involved in response to environmental stimuli likely via activation/inactivation of specific CaMBPs. Plant CNGCs have diverse functions including control of ion homeostasis. Plant CNGCs contain a CaM binding domain overlapping with the cNMP binding motif (see the discussion in Ca^{2+}-permeable channels). Binding of Ca^{2+}/CaM inhibits cNMP-mediated channel activation (Hua et al., 2003), which might play roles during salt stress response. In Arabidopsis, *ACA4* encodes a vacuolar Ca^{2+}-ATPase and is up-regulated by salt stress (Geisler et al., 2000). ACA4 Ca^{2+} pump contains a CaM-binding domain within the N-terminal autoinhibitory domain. Binding of CaM to the N-terminal CaM binding site of ACA4 likely relieves autoinhibition, thus activating its Ca^{2+} pump activity. The regulation of ACA4 by CaM seems to play important role in salt tolerance (Geisler et al., 2000; also see the discussion in Ca^{2+} pumps). Another salt tolerance determinant *AtNHX1* encoding a vacuolar Na^+/H^+ antiporter also contains a CaM binding domain in its C-terminus (Yamaguchi et al., 2005). The expression of *AtNHX1* is up-regulated by salt stress in an ABA-dependent manner (Shi et al., 2002). Topological analysis revealed that the C-terminus of AtNHX1 resides in the vacuolar lumen and has regulatory role in antiporter cation selectivity (Yamaguchi et al., 2003). Yeast two-hybrid screening has identified a cDNA encoding a CML, named AtCaM15 (CML18, McCormack and Braam, 2003) that can interact with AtNHX1 C-terminus (Yamaguchi et al., 2005). The binding of AtCaM15 to AtNHX1 was shown to be Ca^{2+}- and pH-dependent. This interaction modifies the Na^+/K^+ selectivity of the antiporter, decreasing its Na^+/H^+ exchange activity (Yamaguchi et al., 2005). These studies suggest that CaMs and CMLs can perceive Ca^{2+} signals elicited by salt stress and then directly regulate ion transporters that are important for Na^+ and K^+ homeostasis and salt tolerance in plants. Two ABA-, cold- and osmotic stress-induced genes in bryophytes have been identified to encode CaMBPs sharing sequence similarity to mammalian inward rectifier potassium channels (Takezawa and Minami, 2004). These two putative membrane-bound transporter-like proteins, MCamb1 and MCamb2, have a central hydrophobic domain with two putative membrane spans and N- and C-terminal hydrophilic domains. CaM binds to MCamb1 and MCamb2 via an interaction with basic amphiphilic amino acids in the C-terminal domain. Expression of these two genes is dramatically increased following treatment with low temperature, hyperosmotic solutes, and ABA. It appears that one way of CaM participating in cellular signaling leading to enhancement of stress tolerance is through regulation of membrane transporters.

Some CaMBPs are induced by salt or osmotic stress treatments. Stress induction of above-mentioned MCamb genes from bryophytes is one example. In tobacco, a Ca^{2+}/CaM-binding kinase (NtCBK2) is induced by salt stress and GA treatment but not by cold or heat stress and other hormones (Hua et al., 2004). A recent identified CaMBP gene, *AtCaMBP25*, from Arabidopsis is highly induced by dehydration, low temperature and high salinity (Perruc et al., 2004). The 25 kDa AtCaMBP25 was identified by using a radiolabelled CaM probe to screen a cDNA library derived from Arabidopsis cell suspension cultures challenged with osmotic stress. AtCAMBP25 was shown to be nuclear localized and capable of binding to CaM in a Ca^{2+}-dependent manner. Overexpression of AtCaMBP25 enhances sensitivity to both salt and osmotic stress, while knockdown of this gene significantly increases tolerance to salt and osmotic stress. Thus, AtCaMBP25 functions as a negative regulator of osmotic stress tolerance in Arabidopsis. Some CaMBPs are induced not only by salt or osmotic stress but also by other stresses and different hormones. For example, members of the CAMTA (CaM-binding transcription activator) family in Arabidopsis, designated AtSR1-6, are rapidly and differentially induced by environmental signals such as temperature extremes, UV-B, salt, and wounding; hormones such as ethylene, ABA; and signal molecules such as methyl jasmonate, H_2O_2 and salicylic acid (Yang and Poovaiah, 2002). AtSR gene family in Arabidopsis encodes a family of CaM-binding/DNA-binding proteins that are located in nucleus and specifically recognize a novel 6-bp CGCG box found in the *cis*-elements of the gene promoters involved in ethylene, ABA, and light signaling. Another family gene in

Arabidopsis encodes a family of Ca^{2+}/CaM-binding proteins that are involved in transcriptional regulation of gene expression by interacting with fsh/Ring3 class transcription activators (Du and Poovaiah, 2004). This CaM-binding protein family was named as AtBT (Arabidopsis thaliana BTB and TAZ domain protein) family because it contains an N-terminal BTB domain and a C-terminal Zf-TAZ domain besides the C-terminal CaM-biding domain. Arabidopsis possesses five *AtBT* genes that exhibit varying responses to different stress stimuli, except that all five genes respond rapidly to H_2O_2 and salicylic acid treatments. *AtBT3* and *AtBT4* are induced by NaCl, whereas AtBT2 is down-regulated by NaCl. The expression of *AtBT1* and *AtBT5* are not affected by NaCl treatments. AtBT1 targets the nucleus and interacts with AtBET9 and AtBET10 that belong to the family of fsh/Ring3 class transcription regulators. AtBET10 also interacts with AtBT2 and AtBT4 and exhibits a transcriptional activation function in yeast cells. A recent study has identified another CaM-interacting transcription factor AtMYB2 that plays important role in salt tolerance in Arabidopsis (Yoo et al., 2005). AtMYB2 was isolated by using a salt-inducible CaM isoform GmCaM4 as a probe to screen a salt-treated Arabidopsis expression library. AtMYB2 contains a CaM binding motif located within the R2R3 DNA binding region. Interestingly, the DNA binding activity of AtMYB2 is enhanced by its interaction with the specific isoform GmCaM4, but inhibited by the interaction with a closely related CaM isoform GmCaM1, suggesting an antagonist regulation of AtMYB2 transcription activation activity by different CaM isoforms. Overexpression of GmCaM4 in Arabidopsis up-regulates the transcription of AtMYB2-regulated genes, including *RD22*, *P5CS1* and *ADH1*. The elevated transcription of *P5CS1* gene encoding a proline-synthesizing enzyme results in higher accumulation of proline in the overexpression transgenic plants. Overexpression of GmCaM4 in Arabidopsis confers salt tolerance in the transgenic plants. Thus, abiotic stress including salt stress elicits Ca^{2+} signal, which triggers the binding of Ca^{2+} onto CaM; conformational change of CaM upon Ca^{2+} binding promotes interaction of Ca^{2+}/CaM with its target proteins including transcription factors or CaM-binding proteins that can interact with transcription regulators; the interaction modulates the activities of transcription activators and consequently change gene expression. It seems that targeting transcription factors directly or indirectly via CaMBPs by CaM might be a fast way to alter gene expression pattern in response to stress signals.

CaM also targets metabolic or signaling enzymes. One of these enzymes called glyoxalase is on the list of identified CaM targets (Reddy et al., 2002; Reddy and Reddy, 2004). In animal systems, glyoxalase has been long known to be involved in various functions including cell division and proliferation, microtubule assembly, and protection against oxoaldehyde (Thornalley, 1990). Glyoxalase enzymes are important for the glutathione (GST)-based detoxification of methylglyoxal, which is formed primarily as a byproduct of carbohydrate and lipid metabolism. Glyoxalase I has been shown to be induced by salt and osmotic stress in tomato (Espartero et al., 1995). In Arabidopsis, glyoxalase I has also been found to be one of the genes induced by drought and cold stress (Seki et al., 2001). Overexpression of glyoxalase I or glyoxalase II alone in tobacco can improve salt tolerance of the transgenic plants (Veena et al., 1999; Singla-Pareek et al., 2003). Importantly, double transgenic plants overexpressing both glyoxalase proteins confers better response to salt stress than either of the single transgenic plants (Singla-Pareek et al., 2003). Ionic measurements revealed higher accumulation of Na^+ and K^+ in old leaves and negligible Na^+ accumulation in seeds of the transgenic lines as compared with the control plants. The redistribution of toxic Na^+ in the transgenic plants might contribute to salt tolerance.

Mitogen-activated protein kinase (MAPK) is one of the key molecules of signal transduction responding to various external stimuli in animals, plants and yeasts (MAPK group, 2002). In plants, the MAPKs investigated so far were mainly involved in stress responses (Jonak et al., 2002). The Arabidopsis MPK3, MPK4, and MPK6 are activated by a diverse set of stresses, including pathogens, osmotic, cold, and oxidative stress (Jonak et al., 2002). An upstream activator in Arabidopsis, MKK2 is specifically activated by cold and salt stress and directly targets MPK4 and MPK6. Overexpression of MKK2 results in constitutive MPK4 and MPK6 activity and confers freezing and salt tolerance, while null *mkk2* mutant plants are hypersensitive to cold and salt stress (Teige et al., 2004). Activated MAPKs by upstream kinases can be reversed by phosphotases, including dual-specificity MAPK phosphatases (MKPs). A study on a tobacco MKP, NtMKP1 has made a connection between Ca^{2+}/CaM signals and MAPK signaling pathway (Yamakawa et al., 2004). NtMKP1 interacts with CaMs and inactivates SIPK, an ortholog of MPK6 and WIPK, an ortholog of MPK3. Although whether NtMKP1 is induced or activated by salt or osmotic stress is not known, this study suggests that plant CaMs are involved in stress-activated MAPK cascades via MAPK phosphatases. The Arabidopsis MAPK phosphatase 1 (AtMKP1) has no predictable CaM-binding domain. In addition to the dual-specificity phosphatase domain (DSP), AtMKP1 contains a C-terminal domain named GEL (Gelsolin homology domain). The GEL domain seems to be capable of binding Ca^{2+} (Kolappan et al., 2003). Therefore, AtMKP1 could be directly regulated by Ca^{2+} perturbation in the cytosol. AtMKP1 interacts with MPK3, 4 and 6 and activates MPK6 in response to genotoxic stress. Consistently, *mkp1* mutants are hypersensitive to genotoxic stress treatments (UV-C and methyl methanesulphonate) (Ulm et al., 2001). Interestingly, microarray analysis in *mkp1* mutant revealed an increased mRNA level of a Na^+/H^+ exchanger belonging to a family of proteins involved in salt tolerance (Shi et al., 2000, 2005), suggesting that AtMKP1 might be involved in the response to salt stress in addition to genotoxic stress. Indeed, *mkp1* mutant plants exhibit elevated salt resistance in its early vegetative phase, indicating that AtMKP1 functions as a negative regulator of salt stress tolerance (Ulm et al., 2002).

3.2. CBLs and CBL-Interacting Protein Kinases (CIPKs)

CBLs are so called because of their sequence similarity with calcineurin B subunit in yeast and mammalian cells. Calcineurin is a calcium/calmodulin-dependent serine/threonine protein phosphatase (Rusnak and Mertz, 2000). It is a heterodimeric protein consisting of a catalytic subunit calcineurin A, and a regulatory calcium binding subunit, calcineurin B. Calcineurin belongs to a family of type 2B phosphatase. It is the only protein phosphatase dependent on Ca^{2+} and CaM for its activity thereby making it one of the most common intracellular transducer of Ca^{2+} signaling pathways. The active site of calcineurin is located on the A subunit, which catalyzes the removal of phosphate groups from its substrates. Calcineurin A consists of three conserved domains: the calcineurin B binding domain providing sites for hetero-dimerization, the CaM-binding domain, and the autoinhibitory domain. The autoinhibitory domain binds in the active site cleft in the absence of Ca^{2+}/CaM and inhibits the enzyme activity, which acts in concert with the CaM binding domain to confer CaM regulation. Three catalytic genes for calcineurin A subunit have been identified in vertebrate species. The calcineurin B subunit is also highly conserved, with mammalian calcineurin B showing 86% amino acid sequence identity with *Drosophila* calcineurin B and 54% identity with calcineurin B from yeast. Calcineurin B contains four Ca^{2+}-binding EF-hand motifs that bind four Ca^{2+} molecules with high affinity. Myristoylation of the N-terminus of the calcineurin B has been conserved throughout evolution from yeast to mammals, suggesting a crucial physiological role. Numerous functions have been identified for calcineurin in mammals, including induction of long-term potentiation and long-term depression, establishment of learning and memory, control of apoptosis, regulation of ion channels, and control of gene expression (for review, see Rusnak and Mertz, 2000). The role of mammalian calcineurin in regulation of gene expression has been well studied in the immune system. It was shown that binding of antigen to cell surface receptors of lymphocytes elicits Ca^{2+} entry and results in the activation of calcineurin, which then dephosphorylates transcription factors of the NF-AT family that are necessary for T-cell proliferation. Dephosphorylated NF-AT proteins translocate from the cytoplasm to the nucleus to bind the recognition sequence and promote the expression of the target genes (Shibasaki et al., 1996). In yeast, the function of calcineurin has been implicated in the recovery from pheromone-induced growth arrest, cell wall biosynthesis, cation homeostasis, and adaptation to salt stress. One downstream signaling component regulated by calcineurin in yeast is the zinc-finger type transcription factor Crz1P/Tcn1P (Cyert, 2003). Calcineurin dephosphorylates Crz1/Tcn1 in a manner analogous to calcineurin-dependent regulation of mammalian NF-AT transcription factors, resulting in the translocation of Crz1 to the nuleus, thereby regulating the expression of the vacuolar and secretory Ca^{2+} pumps *Pmc1P* and *Pmr1P*, one of the two genes encoding the β-1,3, glucan synthase *FKS2*, and the gene for the plasma membrane Na^+ pump *PMR2* (Stathopoulos and Cyert, 1997; Matheos et al., 1997). In budding yeast, null mutants lacking the calcineurin gene exhibit hypersensitivity to the monovalent cations Na^+ and Li^+, but not K^+, Ca^{2+} and Mg^{2+} (Nakamura et 1993; Mendoza et al., 1994). The role of calcineurin in salt tolerance is thought to be mediated by transcriptional and posttranslational mechanisms. Adaptation to high salt stress requires the presence of the plasma membrane Na^+ pump Pmr2p mediating Na^+ and Li^+ efflux. Yeast cells deficient in calcineurin accumulate Na^+ and Li^+ due to decreased expression of Pmr2p (Mendoza et al., 1994). In fact, DNA microarray has identified more than 160 genes, including *Pmr2p*, that are controlled by calcineurin/Crz1p signaling pathway (Yoshimoto et al., 2002). The list of calcineurin-dependent genes includes genes encoding proteins and enzymes for cell wall synthesis, ion transport, vesicle transport, lipid and steroid synthesis, protein degradation, and transcription, indicating variety of calcineurin functions in yeast.

By using two specific and potent calcineurin inhibitors, CsA and FK506, Luan et al. (1993) first demonstrated that a plant homolog of calcineurin is required for the Ca^{2+}-dependent inactivation of K^+ channels in guard cells. However, the molecular identity of calcineurin in plants has not yet been identified. In particular, the counterpart of mammalian and yeast calcineurin A has not been found in plants although plants have a superfamily of genes encoding protein phosphatase. Identical or nearly identical calcineurin B proteins are also not present in Arabidopsis and rice. The closest contenders are two EF-hand Ca^{2+}-binding proteins encoded by the CBL genes in Arabidopsis. The CBLs from Arabidopsis only share about 30% identity with calcineurin B from various organisms. The present study seems to support that plants, at least Arabidopsis, have no calcineurin and the CBL proteins activate protein kinases rather than protein phosphatase. Studies on CBLs and their interacting protein CIPKs have primarily implicated their function in abiotic stress response.

Role of CBLs in salt tolerance has been well documented by the studies of *SOS* genes, which led to the discovery of the SOS signaling pathway (Xiong et al., 2002; Shi et al., 2005). The tale of the SOS signaling pathway stems from the isolation of three *sos* (*salt overly sensitive*) mutants, i.e. *sos1*, *sos2* and *sos3*. By using root-bending assay, three complementary groups of Arabidopsis mutants showing hypersensitive to Na^+ were identified. Further characterization of these mutants indicated that mutations also cause hypersensitive of the mutants to Li^+, but not to K^+ and Cs^+ (Wu et al., 1996; Liu and Zhu, 1997; Zhu et al., 1998). Molecular identities of these three genes were uncovered through mapping-based cloning (Liu and Zhu, 1998; Liu et al., 2000; Shi et al., 2000). SOS3 is a member of CBLs in Arabidopsis. SOS3 encodes a Ca^{2+} sensor sharing about 30% identity and 50% similarity with calcineurin B subunit from various organisms and neuronal calcium sensor (NCS) from animals. The NCS proteins possess four EF-hand domains but only three or two bind Ca^{2+}. The NCS proteins are high affinity Ca^{2+}-binding proteins that act as Ca^{2+} sensor rather than Ca^{2+} buffer as they undergo conformational changes on Ca^{2+}-binding and regulate target proteins. NCS proteins are multifunctional and involved in the regulation of a wide range of neuronal functions including effects on receptors, ion channels, membrane traffic and cell survival (Burgoyne et al., 2004). The SOS3 protein is predicted to contain three typical EF-hand Ca^{2+}-binding motifs. Mutation in the SOS3 gene disrupts one of the EF-hand motifs and likely disables its Ca^{2+} binding (Liu and Zhu, 1998). The capability of SOS3 binding Ca^{2+} and reduced Ca^{2+}-binding ability of *sos3* mutant proteins was demonstrated by gel mobility shift and Ca^{2+} overlay assay (Ishitani et al., 2000). Like calcineurin B and NCS proteins, SOS3 protein also contains an N-terminal myristoylation site. The myristoylation and its functional significance has been experimentally examined (Ishitani et al., 2000). SOS3 protein was myrostoylated in an *in vitro* translation assay and the myristoylation requires the N-terminal myristoylation signature sequence. Although no significant difference in membrane association was observed between the myristoylated and nonmyristoylated SOS3, expression of the myristoylated but not the nonmyristoylated SOS3 can complement the salt hypersensitive phenotype of *sos3* mutant, suggesting that N-myristoylation is required for SOS3 function in plant salt tolerance.

A recent resolved crystal structure of SOS3 protein revealed that SOS3 is a two-domain structure connected by a short linker and each domain is formed by a pair of adjacent EF-hand motifs (Sanchez-Barrena et al., 2005). Each SOS3 protein can bind four Ca^{2+} molecules, which promotes dimerization of SOS3 proteins. SOS3 dimer displays a structure that exposes eight metal-binding sites, the N-terminal myristoylation sites, and the C-terminal end of the protein. This dimmer structure would increase the efficiency of N-terminal myristoylation and facilitate its C-terminal interaction with SOS2 with respect to the monomer. Ca^{2+} binding to SOS3 seems to be cooperative, as have been observed for other proteins of the EF-hand superfamily (Zhang et al., 1995; Finn et al., 1995). It appears that the binding of Ca^{2+} at sites EF3 and EF4 of SOS3 protein is responsible for the self-association of the macromolecule. The fact that a deletion of three amino acids at EF3 in *sos3* mutant produces a non-functional mutant protein suggests that Ca^{2+} binding at EF3 and probably dimerization are physiological relevant. Ca^{2+} binding and the subsequent dimmerization of SOS3 lead to a change in the global shape and surface properties of the protein that may be sufficient to transmit the Ca^{2+} signal elicited during salt stress onto its interaction partner SOS2.

SOS2 is a serine/threonine type protein kinase with an N-terminal catalytic domain similar to that of the yeast SNF1 kinase (Liu et al., 2000) and belongs to the CIPK family. Mutations in the N-terminal catalytic domain or C-terminal regulatory domain abolish the kinase function, suggesting that both domains are functional essential. SOS2 as the substrate of SOS3 Ca^{2+} sensor was determined in a yeast two-hybrid screening to identify SOS3-interacting proteins using SOS3 as bait (Halfter et al., 2000). SOS3 physically interacts with and activates SOS2 protein kinase in a Ca^{2+}-dependent manner. The interaction is mediated by the C-terminal regulatory domain of SOS2. Further analysis using yeast two-hybrid and *in vitro* binding assays revealed a 21-amino acid motif, designated FISL motif, in the regulatory domain of SOS2 that is necessary and sufficient for interaction with SOS3 (Guo et al., 2001). The C-terminal regulatory domain can interact with the N-terminal kinase domain, resulting in an autoinhibitory structure that blocks substrate access to the catalytic site of the kinase. Thus, the C-terminal regulatory domain functions as an autoinhibitory domain for SOS2 kinase activity. Removal of the regulatory domain of SOS2, including the FISL motif for SOS3 binding, creates a constitutively active SOS2 protein kinase, suggesting that the SOS3 interaction with SOS2 C terminal domain could relieve autoinhibition thereby activate SOS2 kinase. The essential role of the SOS2 C-terminal regulatory domain in plant salt tolerance was demonstrated *in planta* by expressing various activated forms of SOS2 in Arabidopsis and further evaluation of the salt tolerance of the transgenic plants (Guo et al., 2004). Some of the SOS2-interacting partners were identified by using yeast two-hybrid system. The protein phosphatase 2C ABI2 was found to be a SOS2-interacting protein (Ohta et al., 2003). A 37-amino acid domain, designated as the protein phosphatase interacting (PPI) motif, at SOS2 C-terminus downstream of and adjacent to the FISL motif was determined to be necessary and sufficient for interaction with ABI2. ABI2 is a well-known regulator of ABA signaling, which is important for plant tolerance to several abiotic stresses such as salt, drought, and freezing (Zhu, 2002). Interaction between SOS2 and ABI2 suggests a cross-talk between the ABA signaling and the SOS signaling pathways. Although the functional relationship between SOS2 and ABI2 remains to be elucidated, these two proteins, one kinase and one phosphatase, may control the phospharylation status of each other, or they may regulate the phosphorylation status of common protein substrates. SOS2 was proposed to be capable of phosphorylating ion transporters such as Ca^{2+}/H^+ antipoter CAX1 (see the discussion in 2.2.3. Ca^{2+}/H^+ antiporter), plasma membrane Na^+/H^+ antiporter SOS1 (Zhu, 2002; Shi et al., 2005) and vacuolar Na^+/H^+ antiporters (Qiu et al., 2004).

SOS1 is a plasma membrane Na^+/H^+ antiporter that transports Na^+ out of cells thereby reduces Na^+ accumulation in cytosol and confers plant salt tolerance (Shi et al., 2000; Shi et al., 2002; Shi et al., 2003). Mutations in *SOS1* gene render the mutant plants hypersensitive to salt stress, whereas overexpression of SOS1 improves plant salt tolerance. *In vitro* electrophysiological assays indicated that SOS1 antiport activity is controlled by SOS2 and SOS3 (Qiu et al., 2002). In highly purified plasma membrane vesicles, the Na^+/H^+ exchange activity is increased by *in vitro* addition of activated SOS2 proteins, suggesting SOS2 is required for SOS1 activation, presumably through phosphorylation. The plasma membrane vesicles from *sos2* and *sos3* mutant plants show significantly reduced Na^+/H^+ exchange activity, indicating SOS2 and SOS3 regulate SOS1 function. Reconstitution of the SOS components in yeast cells further supports the notion that SOS3/SOS2 protein complex phosphorylates and activates SOS1 (Quintero et al., 2002). In yeast cells, co-expression of SOS2 and SOS3 dramatically increases SOS1-dependent Na^+ tolerance. SOS3 was shown to be able to recruit SOS2 to the plasma membrane, where the SOS3/SOS2 complex phosphorylates SOS1. The ability of SOS2 phosphorylating SOS1 was also confirmed by the assay showing that the constitutively active form of SOS2 phosphorylates SOS1 *in vitro* in a SOS3-independent manner. Although more *in planta* evidence is still expected, both genetic and biochemical results obtained to date are in favor of the following model that links salt stress, Ca^{2+} signals, Ca^{2+} sensor, Ca^{2+} signal transducer, and final effectors. Under salt stress, cellular Ca^{2+} concentration increases, which leads to Ca^{2+} binding onto the SOS3 Ca^{2+} sensor proteins. Ca^{2+}-bound SOS3 interacts with the SOS2 protein and activates SOS2 kinase activity. SOS3 contains a myristoylation site at its N-terminus and the myristoylated SOS3 brings the complex onto the plasma membrane, providing the opportunity for SOS3 / SOS2 complex to interact with SOS1. SOS2 phosphorylates SOS1, which enhances SOS1 Na^+/H^+ exchange activity and promotes Na^+ efflux.

CBLs including SOS3 and CIPKs including SOS2 are encoded by a gene family, respectively, in Arabidopsis. The CBL/CIPK network is emerging as a new signaling system mediating a complex array of environmental stimuli. The CBL (also named SCaBPs, i.e. SOS3-like Ca^{2+} binding proteins) gene family consists of 10 members in both Arabidopsis and rice (Kolukisaoglu et al., 2004). The CBL family appears to be rather conserved at both DNA and protein levels. For example, almost all Arabidopsis CBL genes harbor six or seven introns and four introns are absolutely conserved in phase and position. The amino acid sequence identity of different CBLs ranges from 29% to 92% in Arabidopsis and from 40% to 92% in rice. Determination of the crystal structures of AtCBL2 and AtCBL4 (SOS3) revealed a similar overall structure of CBL proteins to that of calcineurin B and NCS featured with two globular domains separated by a short linker region (Nagae et al., 2003; Sanchez-Barrena et al., 2005). Each globular domain contains a pair of EF-hand motif. The EF-hand domain is a loop of 12 amino acids flanked by two α-helices, in which the amino acids at the positions 1(X), 3(Y), 5(Z), 7(-Y), 9(-X), and 12(-Z) are responsible for binding of the Ca^{2+} (Lewit-Bentley and Rety, 2000). Although the number of EF-hand as well as their spacing is absolutely conserved in all CBLs, it appears that each EF hand has different affinity to bind Ca^{2+}. This might be due to the difference in the EF-hand sequences. In fact, some of the EF hands differ significantly from the canonical EF-hand domain. For example, the conserved acidic residue E or D in the –Z position is replaced with A in the first EF hand of SCaBP2. In the second EF hand of SCaBP3, G is found in the –Z position. Crystal structure analysis also revealed that Ca^{2+} binding capacity between AtCBL4/SOS3 and AtCBL2 is different. Each AtCBL4/SOS3 protein molecule can bind four Ca^{2+} ions in a cooperative way, while AtCBL2 only binds two Ca^{2+} molecules. The two Ca^{2+} ions are coordinated in the first and fourth EF-hand motifs of AtCBL2, whereas the second and third EF-hand motifs are maintained in the open form by internal hydrogen bonding without coordination of Ca^{2+} ions. Difference in Ca^{2+} binding capacity and affinity amongst CBL proteins may result in different conformation of the proteins, which discriminates its interaction with the CIPKs to mediate transduction of specific Ca^{2+} signals elicited by distinct stresses.

Almost all known animal and yeast calcineurin and NCS proteins possess an N-terminal myristoylation signal sequence. N-myristoylation has been reported to promote protein-protein or protein-membrane association (Resh, 1999). However, only four of the Arabidopsis CBL proteins harbor a conserved myristoylation motif. Five of the rice CBLs have this motif. SOS3 is one of the AtCBLs containing N-terminal myristoylation site. Myristoylation is required for SOS3 function in plant (Ishitani et al., 2000). Moreover, AtCBL1, 5, and 9 contain N-terminal myristoylation site and appear to become myristoylated when translated *in vitro* in a rabbit reticulocyte lysate (Kolukisaoglu et al., 2004). Although N-terminal myristoylation seems to be required for protein functions in many systems, the exact role of myristoylation of CBLs remained a mystery. Since interaction of some of the CBLs with their target kinases is independent of myristoylation, it is possible that myristoylation promotes membrane association of the CBL-CIPK complex which modifies downstream membrane associated components.

One of the CBL family members, SOS3, has been extensively studied and shown to be essential for plant salt tolerance (see above discussion). Another member, CBL1 has also been shown to play important role in salt, drought and cold responses (Cheong et al., 2003; Albrecht et al., 2003). *CBL1* is highly induced by multiple stress signals, implicating its role in stress response pathways. Constitutive expression of CBL1 driven by the super promoter MAS enhances drought-induced gene expression but inhibits cold-induced gene expression in the transgenic plants. Consistently, the CBL1-overexpressing transgenic plants display enhanced tolerance to drought and salt stress but reduced tolerance to freezing. By contrast, disruption of *CBL1* gene results in enhanced cold induction and reduced drought induction of stress genes.*cbl1* null mutant plants display less tolerance to salt and drought but enhanced tolerance to freezing. These studies suggest that CBL1 functions as a positive regulator of salt and drought responses and a negative regulator of cold response in plants (Cheong et al., 2003). In another study, the AtCBL1/SCaBP5 was shown to be a negative regulator of ABA signaling (Guo et al., 2002). Arabidopsis mutants with silenced SCaBP5/AtCBL1 are hypersensitive to ABA in seed germination, seedling growth, stomatal closure, and display enhanced ABA-induced gene expression. Apart from SOS3 and AtCBL1, the biological function of other CBL members awaits to be elucidated.

CBLs function as a Ca^{2+} signal relay by specifically targeting a defined group of protein kinases designated as CBL-interacting protein kinases (CIPKs; Shi et al., 1999; Batistic and Kudla, 2004). CIPKs were also named as SOS2-like protein kinases (PKSes) because this protein family shares high sequence similarity with the salt tolerance determinamt SOS2. CIPKs are encoded by 25 genes in Arabidopsis and 30 genes in rice (Kolukisaoglu et al., 2004). Like SOS2, all CIPKs are comprised of a conserved N-terminal SNF1-type catalytic kinase domain with high sequence identity amongst the members and a C-terminal regulatory domain with higher variability. Like many other protein kinases, the catalytic domain of CIPKs contains an activation loop situated between the conserved Asp-Phe-Gly (DFG) and Ala-Pro-Glu (APE) motifs (Guo et al., 2001; Batistic and Kudla, 2004). Phosphorylation of this activation loop is often required for kinase activation by upstream protein kinases. A change from Thr to Asp in this loop, which mimic phosphorylation by upstream kinases, results in constitutively hyperactive and SOS3-independent SOS2 protein kinase (Guo et al., 2001; Gong et al., 2002a). Similar results were obtained for other two members of CIPKs, PKS6 and PKS11 (Gong et al., 2002b, 2002c). SOS2 could also be activated by the mutation of a Ser to Asp or a Tyr to Asp change within the activation loop. These results suggest the possibility that CIPKs may be activated via phosphorylation by a yet unknown upstream protein kinase.

The C-terminal regulatory domain of the CIPKs contains a conserved motif, named as FISL or NAF that is sufficient and required to mediate the interaction with CBL proteins (Guo et al., 2001; Albrecht et al., 2001). The 24-amino acid NAF domain in the CIPKs mediates interaction with all AtCBL proteins. The 21-amino acid FISL domain (24-amino acid in the NAF domain) in SOS2 is necessary and sufficient for interaction with SOS3. Interestingly, the FISL domain in SOS2 protein is also required but not sufficient for interaction with the SOS2 kinase domain. This interaction results in autoinhibition of SOS2 kinase activity (Guo et al., 2001). Since the FISL motif is well conserved in all CIPKs, it is conceivable that CIPK kinases may be maintained in an inactive state by autoinhibition but activated by interaction with CBLs triggered by environmental stimuli.

Another novel interaction module within the C-terminal regulatory domain of the CIPKs is the so-called PPI motif responsible for the interaction between CIPK proteins and the protein phosphatase 2C ABI1 and ABI2 (Ohta et al., 2003). The PPI motif is conserved in the CIPKs in Arabidopsis. Two amino acid residues, Arg and Phe, within this motif are highly conserved in the CIPK family and required for the interaction with the ABI phosphatases. Interestingly, CIPK proteins exhibit different interaction preference towards ABI1 and ABI2. SOS2/CIPK24, CIPK8/PKS11, CIPK14/PKS24, and CIPK15/PKS3 preferentially interact with ABI2, while CIPK20/PKS18 favors interaction with ABI1 (Ohta et al., 2003). Whether other CIPK members can interact with ABI proteins remains to be investigated. Moreover, biological significance of this molecular interaction requires further study.

Besides SOS2, several other CIPK proteins have been also implicated in stress response. The CIPK15/PKS3, through its interaction with the Ca^{2+} sensor AtCBL1/SCaBP5, acts as a global regulator of ABA signaling (Guo et al., 2002). RNAi mutants of both SCaBP5 and PKS3 are specifically hypersensitive to ABA in seed germination and seedling growth. ABA-regulated gene expression is also altered in these RNAi mutants. Moreover, PKS3 interacts with both SCaBP5 and ABI2/ABI1. In response to ABA treatment, the PKS3 activity is transiently decreased within the first 20 minutes, but recovered by 30 minutes of ABA treatment. Although dephosphorylation of PKS3 or phosphorylation of ABI2 was not experimentally observed, changes in the phosphorylation status of these two proteins during their interaction may serve as a signaling link between ABA, Ca^{2+} and downstream signaling components mediating ABA signal response. *PSK11/CIPK8* is preferentially expressed in roots in Arabidopsis. Transgenic plants expressing a constitutively hyperactive form of PKS11 (designated PKS11T161D) are more resistant to high concentrations of glucose, suggesting PKS11 might be involved in sugar signaling (Gong et al., 2002c). A recent study showed that CIPK23 is required for K^+ uptake under low-K^+ conditions (Xu et al., 2006; Li et al., 2006). Using a genetic screening in Arabidopsis, Xu et al. (2006) identified a low-K^+ sensitive (*lks*) mutant *lks1* showing leave chlorotic phenotype under low-K^+ growth condition. Mapping-based cloning has revealed that *LKS1* locus encodes the protein kinase CIPK23. Overexpression of *LKS1* significantly enhances K^+ uptake and tolerance to low K^+, while mutations in *LKS1* reduce K^+ uptake and cause leaf chlorosis and growth inhibition. CIPK23 interacts with six members of the CBL family, including CBL1, CBL2, CBL3, CBL5, CBL8, and CBL9. Among these six CBLs, CBL1 and CBL9 display strongest interaction with CIPK23. Interestingly, the *cbl1cbl9* double mutants show the same low-K^+ sensitive phenotype as *lks1* mutants, although *cbl1* and *cbl9* single mutants do not have the typical *lks* phenotype. These results indicate that CBL1 and CBL9 have overlapping functions and CIPK23 and CBL1/CBL9 function, through their interaction, in a linear signaling pathway to control high affinity K^+ uptake. Moreover, Xu et al. (2006) found that the K^+ channel AKT1 is a downstream target of CIPK23. CIPK23 interacts with and phosphorylates the cytosolic portion of plasma membrane-localized AKT1. Co-expression of CBL1 (or CBL9), CIPK23, and AKT1 in *Xenopus Oocytes* activates AKT1-mediated K^+ inward currents, while absence of either CIPK23 or CBL1 (or CBL9) results in complete inactivation of AKT1 activity. Consistently, *akt1* mutants display same low-K^+ sensitive phenotype as *lks1* does. A working model has been proposed to link the low-K^+ stress, Ca^{2+} signal, CBL/CIPK complex, protein phosphrylation, and K^+ transport in Arabidopsis (Xu et al., 2006). Low-K^+ stress signals may trigger the cytosolic Ca^{2+} signal and lead to activation of Ca^{2+} sensors CBL1 and CBL9. The CBL proteins interact with CIPK23 and recruit CIPK23 to the plasma membrane, where AKT1 is phosphorylated by CIPK23. As the result, AKT1 is activated for K^+ uptake under low K^+ conditions. CIPK14, another CIPK family member also designated as AtSR1, has been implicated in light response (Nozawa et al., 2001). Expression of CIPK14 gene is induced by light illumination. CIPK14 prominently interacts with AtCBL2, which is also up-regulated by illumination.

The CBL/CIPK network mediating Ca^{2+} signaling in Arabidopsis provides a combination of 10 sensor proteins and 25 interacting effector protein kinases. How CBL/CIPK combinations define the specificity of a Ca^{2+} signal elicited by a specific stress is still a question to be answered. Nonetheless, evidence suggests that the specificity may be contributed by the difference in gene expression, subcellular localization, Ca^{2+} binding affinity and capacity, preferential complex formation of CBLs and CIPKs. According to the AtGenExpress microarray data, CBL and CIPK genes exhibit differential expression developmentally and in response to abiotic stress. *AtCBL1* is induced by cold, osmotic, salt, drought, wounding stresses and cycloheximide treatment, but is down regulated by ABA treatment. *AtCBL2* and *AtCBL4/SOS3* are induced by cold but not osmotic, salt, drought and wounding. *AtCBL4/SOS3* but not *AtCBL2* is highly induced by ABA treatment. The expression of *AtCBL3*, *5*, *6*, *7*, *9*, and *10* seems not responding to abiotic stresses tested. *AtCBL5*, *7*, and *8* are expressed at rather low levels as compared with other *AtCBLs*. *AtCBL6*, *7*, *8*, and *9* are highly expressed in roots and pollen, while *AtCBL3* is expressed highly only in pollen and *AtCBL4* is expressed highly only in root. Expression of *AtCBL10* is very low in roots but high in leaves. *AtCBL5* is preferentially expressed in stem, senescing leaves and silique. *AtCBL9* expression seems highly induced by sulfate starvation. Gene expression of CIPKs also exhibits distinct patterns of developmental regulation and in response to abiotic stress and hormonal treatments. For example, *AtCIPK4*, *AtCIPK8*, *AtCIPK11*, *AtCIPK13*, *AtCIPK16*, *AtCIPK17*, *AtCIPK23*, and *AtCIPK25* are expressed highly in roots, while the expression levels of *AtCIPK9* and *AtCIPK20* are very low in roots but high in leaves. Some *AtCIPKs* (*AtCIPK5*, *14*, *17*, and *25*) are expressed predominantly in senescing leaves, and others (*AtCIPK11*, *18*, *19*, and *25*) high in flower and pollen. *AtCIPK* genes are also differentially regulated by abiotic stress and hormone treatments. Most of the AtCIPK genes (*AtCIPK1*, *2*, *4*, *5*, *6*, *7*, *9*, *10*, *11*, *12*, *14*, *19*, *22*, *25*) are induced by one or all of the three stress treatments (cold, osmotic, salt) at different extents. Among these abiotic stress-induced *AtCIPK* genes, some of them are also highly up-regulated by ABA treatments. *AtCIPK3* and *AtCIPK4* appear to be down-regulated by ABA. Interestingly, the expression of *AtCIPK24/SOS2* seems not to be regulated by abiotic stress and ABA, although its critical role in salt tolerance has been extensively documented. The expression of *AtCIPK23/LKS1* also appears not responding to cold, salt and osmotic stress. Consistent with its role in control K^+ uptake, *AtCIPK23/LKS1* is highly expressed in roots. Taken together, differential expression of CBLs and CIPKs may provide the possibilities for the CBL/CIPK network to act dynamically in response to environmental changes and developmental cues.

Cellular localization could potentially affect CBL-CIPK interaction and lead to modification of downstream components in different compartments. N-terminal miristoylation of CBLs could result in membrane association of these proteins. As discussed above, SOS3 and other three CBLs harbor N-terminal myristoylation signal sequence. SOS3 is indeed myristoylated and recruited to the plasma membrane, which is required for its *in planta* function. Thus, these four CBLs could discriminate the specificity of Ca^{2+} signals by recruiting CIPKs onto membranes to modify membrane-associated downstream proteins, e.g. transporters. Two CBLs in rice, OsCBL2 and OsCBL3 is localized to the toloplast of aleurone cell protein storage vacuoles and OsCBL4 is present on the plasma membrane (Hwang et al., 2005). OsCBL2 is up-regulated by gebberellin in the aleurone layer and plays an important role in promoting vacuolation of aleurone cell in response to gebberellin treatment. Although the interaction partner of OsCBL2 awaits to be identified, it is conceivable that the OsCBL2 interacting protein, possibly a CIPK, should be also localized in the toloplast. However, in contrast to the CBLs, known CIPKs do not have any recognizable localization signal or lipid-modification motif. The localization of CIPKs is presumably dependent on their interaction partner CBLs. Thus, different cellular localizations of CBLs could recruit CIPKs to distinct cellular compartments to fulfill diverse functions in response to different environmental stresses.

Another layer of control for CBLs to recognize a specific Ca^{2+} signal could be the affinity and capacity of Ca^{2+} binding. As discussed above, the EF-hand sequences are slightly different in different CBLs, which may result in different affinity as well as capacity for Ca^{2+} binding. Arabidopsis CBL protein family can be divided into three groups: group one with one canonical EF-hand (AtCBL6, 7, 8, 10), group two with two canonical EF-hands (AtCBL1, 9), and group three without any canonical EF-hand (AtCBL2, 3, 4, 5) (Kolukisaoglu et al., 2004). Experimental evidence has shown that AtCBL1, which has two canonical EF-hands, binds Ca^{2+} with relatively high affinity (Kudla et al., 1999), while AtCBL4/SOS3, which lacks canonical EF-hand, displays significantly lower affinity for Ca^{2+} binding (Ishitani et al., 2000). Crystal structure studies have indicated that AtCBL2 binds two Ca^{2+} ions (Nagae et al., 2003), while AtCBL4/SOS3 binds four Ca^{2+} molecules (Sanchez-Barrena et al., 2005). Differential Ca^{2+} binding of CBLs could result in distinct hydrophobic surface on the CBLs for interaction with CIPKs, which may cause preference of protein-protein interaction within the CBL/CIPK network. In fact, preferential interaction between CBLs and CIPKs has been well documented by several studies (Shi et al., 1999; Kim et al., 2000; Halfter et al., 2000; Albrecht et al., 2001; Guo et al., 2001; Guo et al., 2002; Kolukisaoglu et al., 2004). Preferential formation of CBL/CIPK complex might play an important role in generating required specificity in this signaling system.

3.3. Ca^{2+}-Dependent Protein Kinases (CDPKs)

CDPKs are a novel class of Ca^{2+} sensor, having both protein kinase domain and calmodulin-like domain harboring EF hands capable of binding Ca^{2+}. Thus, the CDPKs represent "sensor responders", in which Ca^{2+} binding to the C-terminal calmodulin-like domain induces conformational change that alters the kinase activity residing at the N-terminus of the CDPK. CDPKs are found in entire plant kingdom from green algae to angiosperms (Hrabak, 2000; Harmon et al., 2001). Besides plants, CDPKs are also present in certain protozoa. Notably, CDPKs have not been identified from the sequenced eukaryotic genomes of yeast, nematodes, fruitflies, and humans. Therefore, CDPKs are rather unique Ca^{2+} sensors in plants. Plant CDPKs are encoded by a large gene family. In Arabidopsis, 34 CDPK genes have been identified through sequence analysis, some of which have been studied at genetic and biochemical levels.

CDPKs typically contain four distinct domains: an N-terminal variable domain, a protein kinase domain, an autoinhibitory domain, and a calmodulin-like domain. The N-terminal variable domain, like its name indicated, represents the most diverse region amongst the CDPKs. The N-terminal variable domain of CDPKs in Arabidopsis is different not only in sequences, but also in length, ranging from 25 (AtCPK11) to 200 (AtCPK2) amino acids (Cheng et al., 2002). Although the function of the N-terminal variable domain of each CDPKs remains to be elucidated, sequence analysis and some experimental evidence suggest that this region might be important for subcellular localization of the proteins. 24 of the 34 Arabidopsis CDPKs have potential myristoylation sites at the beginning of the N-terminal variable domain. The Gly residue at the second position of the N-terminus is the target amino acid to be modified by covalent attachment of myristic acid. Like in CBLs, N-terminal myristoylation of CDPKs is though to promote membrane association of the proteins. Most Arabidopsis CDPKs are predicted to have an N-terminal myristoylation site. AtCPK2 has been shown experimentally to be myristoylated at the N-terminal Gly residue, and the first 10 amino acids are critical for localization to the ER membrane (Lu and Hrabak, 2002). Using GFP fusion, the subcellular targeting of nine Arabidopsis CDPKs was investigated (Dammann et al., 2003). Isoforms AtCPK3 and AtCPK4 showed a nuclear and cytosolic localization. A membrane association was observed for AtCPKs 1, 7, 8, 9, 16, 21, and 28, consistent with the presence of myristoylation sites in these CDPKs. All the membrane associated CDPKs except AtCPK1, which targeted to peroxisome, targeted extensively to the plasma membrane. Two CDPKs from other plant species have also been found to be membrane associated (Ellard-Ivey et al., 1999; Martin and Busconi, 2000). Myristoylation is often accompanied by an additional N-terminal modification, such as palmitoylation, to enhance protein-membrane interaction. All 24 Arabidopsis CDPKs harboring myristoylation sites also have at least one Cys residue at position 3, 4, or 5, a potential palmitoylation site (Milligan et al., 1995). The rice OsCPK2 has been shown to be N-terminal myristoylated and palmistoylated, which is important for CDPK-membrane association (Martin and Busconi, 2000). McCPK1, a salinity- and dehydration-responsive CDPK undergoes myristoylation but not palmistoylation *in vitro* (Chehab et al., 2004). Removal of the N-terminal myristate acceptor site partially reduces McCPK1 plasma membrane localization, while removal of the N-terminal domain completely abolishes plasma membrane localization, which suggests that myristoylation and possibly the N-terminal domain contribute to membrane association of the kinase (Chehab et al., 2004).

The kinase domain adjacent to the N-terminal variable domain is highly conserved amongst CDPKs. In Arabidopsis, the kinase domain, which contains all 12 subdomains highly conserved in eukaryotic Ser/Thr protein kinases, shares 44-95% identity and 60-98% similarity among all 34 CDPKs. The active site of the kinase domain is nearly identical in all CDPKs. All CDPKs that have been characterized in detail are activated by Ca^{2+} thereby provide a mechanism to decode Ca^{2+} signals. Next to the kinase domain is the autoinhibitory domain that functions as a pseudo-substrate (Harmon et al., 1994). CDPKs may be autoinhibited by autophosphorylation of the autoinhibitory domain in the absence of Ca^{2+}.

The camodulin-like domain resides at the C-terminus of the CDPKs. This domain contains EF hands capable of binding Ca^{2+}. The number of EF hands differs depending on the isoforms. Most Arabidopsis CDPKs harbor four EF hands, whereas a few of them have one, two or three (Cheng et al., 2002). Differences in numbers and positions of EF hands may contribute to the Ca^{2+} binding affinity and capacity, thereby provide specificity for CDPKs to recognize specific Ca^{2+} signals and downstream substrates. Ca^{2+} binding to the camodulin-like domain results in conformational change of the CDPK protein, thereby activates its kinase activity. Control of CDPK activity by Ca^{2+} fluctuation is largely through the interactions between the kinase, autoinhibitory, and camodulin-like domains. Current evidence supports the following model. Under the basal condition of low free Ca^{2+}, the autoinhibitory domain is bound by the kinase domain, keeping the kinase activity low. Upon binding Ca^{2+} via the EF hands in the camodulin like domain, CDPKs undergo conformational changes that release the pseudo-substrate (i.e. the autoinhibitory domain) from the catalytic site of the kinase domain, thus activating the CDPK proteins (Cheng et al., 2002).

A recent study on the crystal structure of the protein containing the autoinhibitory junction domain (J) and the camodulin-like domain (CaM-LD) of the Arabidopsis AtCPK1 suggested a more complex activation mechanism for CDPK (Chandran et al., 2005). The crystal structure reveals a symmetric dimmer of calcium-bound J-CaM-LD with domain-swap interactions, in which the J region of one promoter interacts extensively with the carboxy-terminal EF-hand domain of the partner promoter. However, as the J-CaM-LD is monomeric in solution, the activated monomer was modeled to account for the intramolecular recognition of the two domains. In CDPKs, the 31-residue J region joins the kinase domain and the CaM-LD. Vitart et al. (2000) found that the C-terminal part of the J region encompasses a pseudo-substrate autoinhibitor and a CaM-LD binding site. Similar to CaM, the CaM-LD in CDPKs consists of two structural domains termed the N- and C-lobes, each containing two EF hands for Ca^{2+} binding. Biophysical analyses of CDPK revealed that the Ca^{2+}-affinity of the C-lobe is significantly greater than that of the N-lobe (Christodoulou et al., 2004). Binding of the J region to the CaM-LD occurs extensively by interactions with residues in the C-lobe of the CaM-LD (Chandran et al., 2005). These findings suggest that there are differential roles of the two lobes in the activation of CDPKs, with the Ca^{2+}-loading of the N-lobe as the likely trigger for physiological CDPK activation. It was proposed that the regulatory mechanism of CDPK seems likely to involve some coupling of the interactions of the autoinhibitory sequence and the CaM-LD binding site in the J region and the N-lobe of the CaM-LD is a key component to this coupling (Chandran et al., 2005). It is possible that at basal levels of Ca^{2+} in the cell, the C-lobe of CDPK is likely to be constitutively Ca^{2+}-loaded, while the N-lobe remains unbound of Ca^{2+}, thereby serves as the Ca^{2+} sensor. Ca^{2+} binding in the N-lobe, which may alter the interaction between these two lobes, drives the process of CDPK activation.

Insights into the physiological roles of CDPKs have come from three types of study: identification of potential substrates, over-production of CDPKs *in planta*, and suppression of CDPKs *in planta*. These studies have shown that CDPKs are involved in many physiological processes, including hormone signaling, growth and development, metabolism, biotic and abiotic responses (Cheng et al., 2002). Some CDPK genes are regulated by abiotic stresses such as cold, salt and drought, suggesting a possible role for these CDPKs in abiotic stress response. For example, two CDPKs in alfalfa, *MsCK1* and *MsCK2* are differentially regulated by cold stress. *MsCK1* is induced but *MsCK2* is repressed by cold stress (Monroy and Dhindsa, 1995). The maize *ZmCPK1* is also transcriptionally up-regulated by cold stress (Berberich and Kusano, 1997). The rice CDPK gene *OsCPK7* displays transcriptional induction by salt stress (Saijo et al., 2000). Another rice CDPK gene, *OsCDPK13* is induced by cold stress but suppressed by salt and drought stresses (Abbasi et al., 2004). In Arabidopsis, both *AtCPK10* and *AtCPK11* are induced by dehydration and high concentrations of NaCl (Urao et al., 1994). The mung bean CDPK *VrCPK1* is also strongly induced by NaCl treatment (Botella et al., 1996). *McCDPK1* from common ice plant is also a salinity- and drought-induced CDPK (Patharkar and Cushman, 2000; Chehab et al., 2004).

Direct evidence establishing the role of CDPKs in ABA- and stress-responsive gene regulation came from Sheen's (1996) study using maize protoplasts expressing constitutive active CDPK proteins. In this study, a chimeric gene was generated by fusion of the barley ABA-responsive (HVA1) promoter to a reporter gene (GFP or LUC). The expression of the reporter gene was enhanced by exposure to cold, high salt, and ABA. The expression of the reporter gene was also substantially increased by the Ca^{2+} ionophore Ca^{2+}-ionomycin or Ca^{2+}-A23187, indicating that Ca^{2+} serves as a second messenger to mediate stress-responsive gene regulation. Moreover, expression of two constitutively activated CDPKs, AtCPK10 and AtCPK30 could activate reporter gene expression under the control of the HVA1 promoter. Thus, the CDPKs might sense Ca^{2+} elevation elicited by abiotic stress and transduce the stress signal onto downstream effectors through phosphorylation, which modulate stress-responsive gene expression.

A large number of CDPK substrates have been identified through various means. These protein substrates associated with CDPKs suggest a wide spectrum of CDPK functions (Cheng et al., 2002). The salt- and drought-inducible McCPK1 in common ice plant was shown to interact with the CDPK substrate protein 1 (CSP1), a pseudoresponse regulator (Patharkar and Cushman, 2000). CSP1 interacts with McCPK1 in a substrate-like fashion in both yeast two-hybrid assays and wheat germ interaction assays. Furthermore, McCPK1 is capable of phosphorylating CSP1 *in vitro* in a Ca^{2+}-dependent manner. Sequence analysis suggests that CSP1 is a novel member of a class of pseudoresponse regulator-like proteins that have a highly conserved helix-loop-helix DNA binding domain and a C-terminal activation domain. It appears that McCPK1 may regulate the function of CSP1 by reversible phosphorylation in plants. Another McCPK1-interacting protein identified through yeast two-hybrid screening is a novel coiled-coil protein named McCAP1 (CDPK adaptor protein 1) (Patharkar and Cushman, 2006). The McCPK1 has been shown to have a dynamic change in subcellular localization from the plasma membrane to the nucleus, ER, and actin microfilaments in response to reduction in humidity (Chehab et al., 2004). McCAP1 interacts with McCPK1 but appears not to be a substrate for McCPK1. It was speculated that McCAP1 might be important for the dynamic subcellular localization changes in response to low humidity stress. Using yeast two-hybrid system, the Arabidopsis stress-induced AtDi19 was identified as an AtCPK11-interacting protein (Milla et al., 2006). Besides interaction with AtCPK11, AtDi19 also interacts with AtCPK4, whereas other closely related CPKs from Arabidopsis interact weakly or do not interact with AtDi19. AtDi19 is phosphorylated by AtCPK11 in a Ca^{2+}-dependent manner. CDPKs are also implicated in ABA signaling pathways. Using yeast two-hybrid screening, Choi et al. (2005) identified AtCPK32 as an ABF4-interacting protein. ABF4 is a member of a subfamily of basic leucine zipper class transcription factors mediating ABA signaling control of gene expression. Transcription activity of ABF4 is induced by ABA treatment and the ABA-activated transcription is inhibited by protein kinase inhibitors, indicating the involvement of protein phosphorylation in this signaling event (Choi et al., 2000). AtCPK32 can interact and phosphorylate ABF4. Overexpression of AtCPK32 affects both ABA sensitivity and the expression of a number of ABF4-regulated genes. These results demonstrated that AtCPK32 is an ABA signaling component that regulates the ABA-responsive gene expression via ABF4 (Choi et al., 2005).

Important role of CDPK in salt and cold tolerance was evidenced by the fact that overexpression of a single CDPK confers enhanced tolerance to these stresses (Saijo et al., 2000). Rice *OsCDPK7* is induced by cold and salt stress. Overexpression of *OsCDPK7* enhances tolerance of the transgenic plants to both cold and salt stresses and the extent of tolerance is well correlated with the level of *OsCDPK7* expression in individual transgenic plants. Interestingly, transgenic plants overexpressing *OsCDPK7* display enhanced induction of some stress-responsive genes in response to salinity/drought, but not to cold. Thus, it was suggested that the downstream pathways leading to cold and salt/drought tolerance might be different from each other. It seems that OsCDPK7 acts as a branch point of cold and salt/drought stress signaling transduction. However, how the specificity of two different abiotic stress signals is maintained by a single CDPK remains elusive.

Little is known about specificity of CDPKs to sense and respond to different Ca^{2+} spikes elicited by distinct stresses. However, emerging theme is that the specificity could be provided by different subcellular localization, Ca^{2+} binding affinity and capacity, extent of enzyme activity, modification by other signaling pathways, and interaction with different downstream targets. There is evidence for cytosolic-, nuclear-, cytoskeleton-, plasma membrane-, and peroxisome-associated CDPKs. Different subcellular location of CDPKs could provide specificity spatially for Ca^{2+} signals. Sequence divergence is present in each EF hands in a given CDPK as well as different CDPK members. Such difference can influence Ca^{2+} activation thresholds, which may be used by CDPKs to discriminate specific Ca^{2+} spikes. This was evidenced by that three different CDPK isoforms display a different Ca^{2+} threshold for half-maximal activation, with the greatest difference between two isoforms being more than ten-fold (Lee et al., 1998). As discussed above, CDPKs can interact with a number of different protein substrates distinct in cellular functions. CDPKs could transduce specific Ca^{2+} signal to a unique downstream interaction partner to mediate stress response. Furthermore, some CDPKs have been shown to be activated by putative lipid messengers, interaction with 14-3-3 proteins, and phosphorylation by other kinases (Cheng et al., 2002; Sanders et al., 2002).

3.4. Other Ca^{2+} Binding Proteins

Besides Ca^{2+} sensor relays and responders discussed above, some other proteins also contain EF hands but do not fall into those protein families. The Arabidopsis respiratory burst oxidase homology proteins (AtRbohs) contain EF hands and belong to such proteins. Arabidopsis have ten *Atrboh* genes. All AtRobh proteins carry a presumably cytosolic 300-amino-acid N-terminal extension with two EF-hands that binds Ca^{2+} (Keller et al., 1998), which could account for the direct regulation of these oxidases by Ca^{2+}. In fact, plant Rboh proteins have been shown to be stimulated by Ca^{2+} (Sagi and Fluhr, 2001). Plant Rboh proteins are the source of reactive oxygen intermediates (ROI) produced following biotic and abiotic stresses. ROIs have broad role as signals that mediate plant response to environmental stimuli, developmental cues and programmed cell death in different cell types (Torres and Dangl, 2005).

ABI1 is another EF-hand containing protein that functions as a negative regulator of ABA response. ABI1 is a serine/threonine protein phosphatase 2C harboring a unique N-terminal extension containing an EF hand Ca^{2+} binding site. Muations in *ABI1* gene causes insensitive of the mutant plants to exogenous applied ABA in seed germination and seedling growth (Leung et al., 1994). Although no direct evidence showing Ca^{2+} binding and modulating ABI1 phosphatase activity, it was proposed that the ABI1 may function to integrate ABA and Ca^{2+} signals with phosphorylation-dependent response pathways. Interaction between ABI1 and CIPKs (see discussion in 3.2) might be a control node to modulate the dynamic change of phosphorylation status of downstream components in response to salt and ABA treatments.

Several proteins lacking EF hands are also capable of binding Ca^{2+}. These proteins include phospholipase D (PLD), annexin, calreticulin, calsequestrin and BiP (White and Broadley, 2003). It is beyond the scope of this chapter to review this type of proteins. Readers can refer to other reviews for the detailed functions of these proteins. Notably, PLD has been implicated in cellular response to ethylene and ABA, stomatal closure, pathogen response, ROS production and drought tolerance (Wang, 2005; Zhang et al., 2005).

ACKNOWLEDGEMENTS

Research in the H. Shi's laboratory is supported by grants from United States Department of Agriculture (2004-35100-14863 and 2006-34186-16976).

REFERENCES

Abbasi, F., Onodera, H., Toki, S., Tanaka, H. and Komatsu, S. (2004). OsCDPK13, a calcium-dependent protein kinase gene from rice, is induced by cold and gibberellin in rice leaf sheath. *Plant Mol. Biol.* **55,** 541–552.

Albrecht, V., Ritz, O., Linder, S., Harter, K. and Kudla, J. (2001). The NAF domain defines a novel protein-protein interaction module conserved in Ca^{2+}-regulated kinases. *EMBO J.* **20,** 1051–1063.

Albrecht, V., Weinl, S., Blazevic, D., D'Angelo, C., Batistic, O., Kolukisaoglu, U., Bock, R., Schulz, B., Harter, K. and Kudla, J. (2003). The calcium sensor CBL1 integrates plant responses to abiotic stresses. *Plant J.* **36,** 457–470.

Allen, G.J., Chu, S.P., Harrington, C.L., Schumacher, K., Hoffman, T., Tang, Y.Y., Grill, E., and Schroeder, J.I. (2001). A defined range of guard cell calcium oscillation parameters encodes stomatal movements. *Nature* **411,** 1053–1057.

Allen, G.J., Chu, S.P., Schumacher, K., Shimazaki, C.T., Vafeados, D., Kemper, A., Hawke, S.D., Tallman, G., Tsien, R.Y., Harper, J.F., Chory, J., and Schroeder, J.I. (2000). Alteration of stimulus-specific guard cell calcium oscillations and stomatal closing in *Arabidopsis det3* mutant. *Science* **289,** 2338–2342.

Allen, G.J., Kwak, J.M., Chu, S.P., Llopis, J., Tsien, R.Y., Harper, J.F., and Schroeder, J.I. (1999). Cameleon calcium indicator reports cytoplasmic calcium dynamics in *Arabidopsis* guard cells. *Plant J.* **19,** 735–747.

Arazi, T., Kaplan, B., and Fromm, H. (2000). A high-affinity calmodulin-binding site in a tobacco plasma-membrane channel protein coincides with a characteristic element of cyclic nucleotide-binding domains. *Plant Mol. Biol.* **42,** 591–601.

Arazi, T., Sunkar, R., Kaplan, B., and Fromm, H. (1999). A tobacco plasma membrane calmodulin-binding transporter confers Ni^{2+} tolerance and Pb^{2+} hypersensitivity in transgenic plants. *Plant J.* **20,** 171–182.

Axelsen, K.B. and Palmgren, M.G. (2001). Inventory of the superfamily of P-type ion pumps in Arabidopsis. *Plant Physiol.* **126,** 696–706.

Finn, B.E., Evenas, J., Drakenberg, T., Waltho, J.P., Thulin, E. and Forsen, S. (1995). Calcium-induced structural changes and domain autonomy in calmodulin, *Nature Struct. Biol.* **2,** 777–783.

Batistic, O. and Kudla, J. (2004). Integration and channeling of calcium signaling through the CBL calcium sensor/CIPK protein kinase network. *Planta.* **219,** 915–924.

Berberich, T. and Kusano, T. (1997). Cycloheximide induces a subset of low temperature-inducible genes in maize. *Mol. Gen. Genet.* **254,** 275–283.

Berridge, M.J., Bootman, M.D. and Roderick, H.L. (2003). Calcium signalling: dynamics, homeostasis and remodelling. *Nat. Rev. Mol. Cell Biol.* **4,** 517–529.

Bhattacharya, S., Bunick, C.G. and Chazin, W.J. (2004). Target selectivity in EF-hand calcium binding proteins. *Biochim. Biophys. Acta.* **1742,** 69–79.

Blatt, M.R. (2000). Ca^{2+} signalling and control of guard-cell volume in stomatal movements. *Curr. Opin. Plant Biol.* **3,** 196–204.

Borsics, T., Webb, D., Andeme-Ondzighi, C., Staehelin, L.A. and Christopher, D.A. (2006). The cyclic nucleotide-gated calmodulin-binding channel AtCNGC10 localizes to the plasma membrane and influences numerous growth responses and starch accumulation in *Arabidopsis thaliana*. *Planta* 2006 Aug 31 (Epub).

Botella, J.R., Arteca, J.M., Somodevilla, M. and Arteca, R.N. (1996). Calcium-dependent protein kinase gene expression in response to physical and chemical stimuli in mungbean (*Vigna radiata*). *Plant Mol. Biol.* **30,** 1129–1137.

Bouche, N., Yellin, A., Snedden, W.A. and Fromm, H. (2005). Plant-specific calmodulin-binding proteins. *Annu. Rev. Plant Biol.* **56,** 435–466.

Braam, J., Sistrunk, M.L., Polisensky, D.H., Xu, W., Purugganan. M.M., Antosiewicz, D.M., Campbell, P. and Johnson, K.A. (1997). Plant responses to environmental stress: regulation and functions of the *Arabidopsis TCH* genes. *Planta* **203**, S35–S41.

Burgoyne, R.D., O'Callaghan, D.W., Hasdemir, B., Haynes, L.P. and Tepikin, A.V. (2004). Neuronal Ca^{2+}-sensor proteins: multitalented regulators of neuronal function. *Trends Neurosci.* **27,** 203–209.

Capel, J., Jarillo, J.A., Salinas, J. and Martinez-Zapater, J.M. (1997). Two homologous low-temperature-inducible genes from Arabidopsis encode highly hydrophobic proteins. *Plant Physiol.* **115,** 569–576.

Catala, R., Santos, E., Alonso, J.M., Ecker, J.R., Martinez-Zapater. J.M. and Salinas, J. (2003), Mutations in the Ca^{2+}/H^+ transporter CAX1 increase CBF/DREB1 expression and the cold-acclimation response in Arabidopsis. *Plant Cell* **15,** 2940–2951.

Chan, C.W., Schorrak, L.M., Smith, R.K. Jr., Bent, A.F. and Sussman, M.R. (2003). A cyclic nucleotide-gated ion channel, CNGC2, is crucial for plant development and adaptation to calcium stress. *Plant Physiol.* **132,** 728–731.

Chandran, V., Stollar, E.J., Lindorff-Larsen, K., Harper, J.F., Chazin, W.J., Dobson, C.M., Luisi, B.F. and Christodoulou, J. (2005). Structure of the regulatory apparatus of a calcium-dependent protein kinase (CDPK): a novel mode of calmodulin-target recognition. *J. Mol. Biol.* **357,** 400–410.

Chehab, E.W., Patharkar, O.R., Hegeman, A.D., Taybi, T. and Cushman, J.C. (2004). Autophosphorylation and subcellular localization dynamics of a salt- and water deficit-induced calcium-dependent protein kinase from ice plant. *Plant Physiol.* **135,** 1430–1446.

Cheng, N.H., Pitman, J.K., Barkla, B.J., Shigaki, T. and Hirschi, K.D. (2003). The Arabidopsis *cax1* mutant exhibits impaired ion homeostasis, development, and hormonal responses and reveals interplay among vacuolar transporters. *Plant Cell* **15**, 347–364.

Cheng, N.H, Pittman, J.K., Shigaki, T. and Hirschi, K.D. (2002). Characterization of CAX4, an Arabidopsis H^+/cation antiporter. *Plant Physiol.* **128,** 1245–1254.

Cheng, N.H., Pittman, J.K., Shigaki, T., Lachmansingh, J., LeClere, S., Lahner, B., Salt, D.E. and Hirschi, K.D. (2005). Functional association of Arabidopsis CAX1 and CAX3 is required for normal growth and ion homeostasis. *Plant Physiol.* **138,** 2048–2060.

Cheng, N.H., Pittman, J.K., Zhu, J.K. and Hirschi, K.D. (2004). The protein kinase SOS2 activates the Arabidopsis H^+/Ca^{2+} antiporter CAX1 to integrate calcium transport and salt tolerance. *J. Biol. Chem.* **279,** 2922–2926.

Cheng, S., Willmann, M.R., Chen, H. and Sheen, J. (2002), Calcium signaling through protein kinases: the Arabidopsis calcium-dependent protein kinase gene family. *Plant Physiol.* **129,** 469–485.

Cheong, Y.H., Kim, K.N., Pandey, G.K., Gupta, R., Grant, J.J. and Luan, S. (2003). CBL1, a calcium sensor that differentially regulates salt, drought, and cold responses in *Arabidopsis. Plant Cell* **15,** 1833–1845.

Choi, H., Hong, J., Ha, J., Kang, J. and Kim, S.Y. (2000). ABFs, a family of ABA-responsive element binding factors. *J. Biol. Chem.* **275,** 1723–1730.

Choi, H.I., Park, H.J., Park, J.H., Kim, S., Im, M.Y., Seo, H.H., Kim, Y.W., Hwang, I. and Kim, S.Y. (2005). Arabidopsis calcium-dependent protein kinase AtCPK32 interacts with ABF4, a transcriptional regulator of abscisic acid-responsive gene expression, and modulates its activity. *Plant Physiol.* **139,** 1750–1761.

Chung, W.S., Lee, S.H., Kim, J.C., Heo, W.D., Kim, M.C., Park. C.Y., Park, H.C., Lim, C.O., Kim, W.B., Harper, J.F. and Cho, M.J. (2000). Identification of a calmodulin-regulated soybean Ca^{2+}-ATPase (SCA1) that is located in the plasma membrane. *Plant Cell* **12,** 1393–1407.

Clapham, D.E. (1995). Calcium signaling. *Cell* **80,** 259–268.

Clough, S.J., Fengler, K.A., Yu, I.C., Lippok, B., Smith, R.K., and Bent, A.F. (2000). The *Arabidopsis dnd1* "defense, no death" gene encodes a mutated cyclic nucleotide-gated ion channel. *Proc. Natl. Acad. Sci. USA* **97,** 9323–9328.

Crivici, A. and Ikura, M. (1995). Molecular and structural basis of target recognition by calmodulin. *Annu. Rev. Biophys. Biomol. Struct.* **24,** 85–116.

Cyert, M.S. (2003). Calcineurin signaling in *Saccharomyces cerevisiae*: how yeast go crazy in response to stress. *Biochem. Biophys. Res. Commun.* **311,** 1143–1150.

Dammann, C., Ichida, A., Hong, B., Romanowsky, S.M., Hrabak, E.M., Harmon, A.C., Pickard, B.G. and Harper, J.F. (2003). Subcellular targeting of nine calcium-dependent protein kinase isoforms from Arabidopsis. *Plant Physiol.* **132,** 1840–1848.

Day, I., Reddy, V.S., Shad Ali, G. and Reddy, A.S.N. (2002). Analysis of EF-hand-containing proteins in Arabidopsis. *Genome Biol.* **3,** 1–24.

Delk, N.A., Johnson, K.A., Chowdhury, N.I. and Braam, J. (2005). CML24, regulated in expression by diverse stimuli, encodes a potential Ca^{2+} sensor that functions in responses to abscisic acid, daylength, and ion stress. *Plant Physiol.* **139,** 240–253.

Delmer, D.P. and Potikha, T.S. (1997). Structure and functions of annexins in plants. *Cell. Mol. Life Sci.* **53,** 546–553.

Demidchik, V., Davenport, R.N., and Tester, M. (2002). Non-selective cation channels in plants. *Annu. Rev. Plant Mol. Biol. Plant Physiol.* **53,** 67–107.

Denis, V. and Cyert, M.S. (2002). Internal Ca^{2+} release in yeast is triggered by hypertonic shock and mediated by a TRP channel homologue. *J.Cell Biol.* **156,** 29–34.

Dingledine, R., Borges, K., Bowie, D. and Traynelis, S.F. (1999). The glutamate receptor ion channels. *Pharmacol Rev.* **51,** 7–61.

Du, L. and Poovaiah, B.W. (2004). A novel family of Ca^{2+}/calmodulin-binding proteins involved in transcriptional regulation: interaction with fsh/Ring3 class transcription activators. *Plant Mol. Biol.* **54,** 549–569.

Ehrhardt, D.W., Atkinson, E.M. and Long, S.R. (1992). Depolarization of alfalfa root hair membrane potential by Rhizobium meliloti Nod factors. *Science* **256,** 998–1000.

Ellard-Ivey, M., Hopkins, R.B., White, T. and Lomax, T.L. (1999). Cloning, expression and N-terminal myristoylation of CpCPK1, a calcium-dependent protein kinase from zucchini (*Cucurbita pepo* L.). *Plant Mol. Biol.* **39,** 199–208.

Espartero, J., Sanchez-Aguayo, I. and Pardo, J.M. (1995). Molecular characterization of glyoxalase-I from a higher plant; upregulation by stress. *Plant Mol. Biol.* **29,** 1223–1233.

Evans, D.E, and Williams, L.E. (1998). P-type calcium ATPases in higher plants – biochemical, molecular and functional properties. *Biochim. Biophy. Acta.* **1376**, 1–25.

Evans, N.H., McAinsh, M.R., and Hetherington, A.M. (2001). Calcium oscillations in higher plants. *Curr. Opin. Plant Biol.* **4,** 415–420.

Felle, H. (1988). Auxin causes oscillations of cytosolic free calcium and pH in *Zea mays* coleoptiles. *Planta* **174**, 495–499.

Foreman, J., Demidchik, V., Bothwell, J.H.F., Mylona, P., Miedema, H., Torres, M.A., Linstead, P., Costa, S., Brownlee, C., Jones, J.D.G. *et al.* (2003). Reactive oxygen species produced by NADPH oxidase regulate plant cell growth. *Nature* **422**, 442–446.

Furuichi, T., Cunningham, K.W., and Muto, S. (2001). A putative two-pore channel AtTPC1 mediates Ca^{2+} flux in Arabidopsis leaf cells. *Plant Cell Physiol.* **42,** 900–905.

Gao, D., Knight, M.R., Trewavas, A.J., Sattelmacher, B. and Plieth, C. (2004). Self-reporting Arabidopsis expressing pH and [Ca^{2+}] indicators unveil ion dynamics in the cytoplasm and in the apoplast under abiotic stress. *Plant Physiol.* **134,** 898–908.

Garciadeblas, B., Benito, B. and Rodríguez-Navarro, A. (2001). Plant cells express several stress calcium ATPases but apparently no sodium ATPase. *Plant Soil* **235**, 181–192.

Gawienowski, M.C., Szymanski, D., Perera, I.Y. and Zielinski, R.E. (1993). Calmodulin isoforms in *Arabidopsis* encoded by multiple divergent mRNAs. *Plant Mol. Biol.* **22**, 215–225.

Geisler, M., Frangne, N., Gomes, E., Martinoia, E., and Palmgren, M.G. (2000). The ACA4 gene of Arabidopsis encodes a vacuolar membrane calcium pump that improves salt tolerance in yeast. *Plant Physiol.* **124,** 1814–1827.

Goddard, H., Manison, N.F.H., Tomos, D., and Brownlee, C. (2000). Elemental propagation of calcium signals in response-specific patterns determined by environmental stimulus strength. *Proc. Natl. Acad. Sci. USA* **97,** 1932–1937.

Gong, D., Gong, Z., Guo, Y., Chen, X. and Zhu, J.K. (2002c). Biochemical and functional characterization of PKS11, a novel *Arabidopsis* protein kinase. *J. Biol. Chem.* **277,** 28340–28350.

Gong, D., Gong, Z., Guo, Y. and Zhu, J.K. (2002b). Expression, activation, and biochemical properties of a novel *Arabidopsis* protein kinase. *Plant Physiol.* **129,** 225–234.

Gong, D., Guo, Y., Jagendorf, A.T. and Zhu, J.K. (2002a). Biochemical characterization of the *Arabidopsis* protein kinase SOS2 that functions in salt tolerance. *Plant Physiol.* **130,** 256–264.

Grabov, A., Blatt, M.R. (1998). Membrane voltage initiates Ca^{2+} waves and potentiates Ca^{2+} increases with abscisic acid in stomatal guard cells. *Proc. Natl. Acad. Sci. USA* **95**, 4778–4783.

Grabov, A., and Blatt, M.R. (1998). Membrane voltage initiates Ca^{2+} waves and potentiates Ca^{2+} increases with abscisic acid in stomatal guard cells. *Proc. Natl. Acad. Sci. USA* **95,** 4778–4783.

Grimaldi, M., Maratos, M. and Verma, A. (2003). Transient receptor potential channel activation causes a novel form of $[Ca^{2+}]_I$ oscillations and is not involved in capacitative Ca^{2+} entry in glial cells. *J Neurosci.* **23,** 4737–4745.

Guo, Y., Qiu, Q., Quintero, F.J., Pardo, J.M., Ohta, M., Zhang, C., Schumaker, K.S. and Zhu, J.K. (2004). Transgenic evaluation of activated mutant alleles of SOS2 reveals a critical requirement for its kinase activity and C-terminal regulatory domain for salt tolerance in *Arabidopsis thaliana*. *Plant Cell* **16,** 435–449.

Guo, Y., Halfter, U., Ishitani, M., and Zhu, J.K. (2001). Molecular characterization of functional domains in the protein kinase SOS2 that is required for plant salt tolerance. *Plant Cell* **13,** 1383–1400.

Guo, Y., Xiong, L., Song, C.-P., Gong, D., Halfter, U., and Zhu, J.K. (2002). A calcium sensor and its interacting protein kinase are global regulators of abscisic acid signaling in *Arabidopsis*. *Dev. Cell* **3,** 233–244.

Halfter, U., Ishitani, M. and Zhu, J.K. (2000). The *Arabidopsis* SOS2 protein kinase physically interacts with and is activated by the calcium-binding protein SOS3. *Proc. Natl. Acad. Sci. USA* **97,** 3735–3740.

Harmon, A.C., Gribskov, M., Gubrium, E. and Harper, J.F. (2001). The CDPK superfamily of protein kinases. *New Phytol.* **151,** 175–183.

Harmon, A.C., Yoo, B-C. and McCaffery, C. (1994). Pseudosubstrate inhibition of CDPK, a protein kinase with a calmodulin-like domain. *Biochemistry* **33,** 7278–7287.

Heo, W.D., Lee, S.H., Kim, M.C., Kim, J.C., Chung, W.S., Chun, H.J., Lee, K.J., Park, C.Y., Park, H.C., Choi, J.Y. and Cho, M.J. (1999). Involvement of specific calmodulin isoforms in salicylic acid-independent activation of plant disease resistance responses. *Proc. Natl. Acad. Sci. USA* **96**, 766–771.

Hetherington, A.M. and Brownlee, C. (2004). The generation of Ca^{2+} signals in plants. *Annu. Rev. Plant Biol.* **55,** 401–427.

Hirschi, K. (2001). Vacuolar H^+/Ca^{2+} transport: Who's directing the traffic? *Trends Plant Sci.* **6,** 100–104.

Hirschi, K.D. (1999). Expression of Arabidopsis CAX1 in tobacco: Altered calcium homeostasis and increased stress sensitivity. *Plant Cell* **11,** 2113–2122.

Hirschi, K.D., Zhen, R.G., Cunningham, K.W., Rea, P.A., and Fink, G.R. (1996). CAX1, an H^+/Ca^{2+} antiporter from Arabidopsis. *Proc. Natl. Acad. Sci. USA* **93,** 8782–8786.

Hrabak, E.M. (2000). Calcium-dependent protein kinases and their relatives. *Adv. Bot. Res.* **32,** 185–223.

Hua, B.G., Mercier, R.W., Zielinski, R.E., Berkowitz, G.A. (2003). Functional interaction of calmodulin with a plant cyclic nucleotide gated cation channel. *Plant Physiol. Biochem.* **41,** 945–954.

Hua, W., Li, R.J., Wang, L. and Lu, Y.T. (2004). A tobacco calmodulin-binding protein kinase (NtCBK2) induced by high-salt/GA treatment and its expression during floral development and embryogenesis. *Plant Sci.* **166,** 1253–1259.

Hwang, Y.S., Bethke, P.C., Cheong, Y.H., Chang, H.S., Zhu, T. and Jones, R.L. (2005). A gibberellin-regulated calcineurin B in rice localizes to the tonoplast and is implicated in vacuole function. *Plant Physiol.* **138,** 1347–1358.

Ishitani, M., Liu, J., Halfter, U., Kim, C.-S., Shi, W., and Zhu, J.K. (2000). SOS3 function in plant salt tolerance requires N-myristoylation and calcium-binding. *Plant Cell* **12,** 1667–1677.

Christodoulou, J., Malmendal, A., Harper, J.F. and Chazin, W.J. (2004). Evidence for differing roles for each lobe of the calmodulin-like domain in a calcium-dependent protein kinase, *J. Biol. Chem.* **279,** 29092–29100.

Jarillo, J.A., Capel, J., Leyva, A., Martinez-Zapater, J.M. and Salinas, J. (1994). Two related low-temperature-inducible genes of Arabidopsis encode proteins showing high homology to 14-3-3 proteins, a family of putative kinase regulators. *Plant Mol. Biol.* **25,** 693–704.

Jonak, C., Okresz L., Bogre L. and Hirt, H. (2002). complexity, cross talk and integration of plant MAP kinase signaling. *Curr. Opin. Plant Biol.* **5,** 415–424.

Kaupp, U.B. and Seifert, R. (2002). Cyclic nucleotide-gated ion channels. *Physiol. Rev.* **82,** 769–824.

Keller, T., Damude, H.G., Werner, D., Doerner, P., Dixon, R.A. and Lamb, C. (1998). A plant homolog of the neutrophil NADPH oxidase gp91phox subunit gene encodes a plasma membrane protein with Ca^{2+} binding motifs. *Plant Cell* **10,** 255–266.

Kim, K.N., Cheong, Y.H., Gupta, R. and Luan, S. (2000). Interaction specificity of *Arabidopsis* calcineurin B-like calcium sensors and their target kinases. *Plant Physiol.* **124,** 1844–1853.

Kohler, C., and Neuhaus, G. (2000). Characterisation of calmodulin binding to cyclic nucleotide-gated ion channels from *Arabidopsis thaliana*. *FEBS Lett.* **471,** 133–136.

Kolappan, S., Gooch, J.T., Weeds, A.G. and McLaughlin, P.J. (2003). Gelsolin domains 4-6 in active, actin-free conformation identifies sites of regulatory calcium ions. *J. Mol. Biol.* **329,** 85–92.

Kolukisaoglu, U., Weinl, S., Blazevic, D., Batistic, O. and Kudla, J. (2004). Calcium sensors and their interacting protein kinases: genomics of the Arabidopsis and rice CBL-CIPK signaling networks. *Plant Physiol.* **134**, 43–58.

Kramer, R.H. and Molokanova, E. (2001). Modulation of cyclic-nucleotide-gated channels and regulation of vertebrate phototransduction. *J. Exp. Biol.* **204**, 2921–2931.

Kretsinger, R.H. and Nockolds, C.E. (1973). Carp muscle calcium-binding protein. II. Structure determination and general description. *J. Biol. Chem.* **248,** 3313–3326.

Kudla, J., Xu, Q., Harter, K., Gruissem, W. and Luan, S. (1999). Genes for calcineurin B-like proteins in *Arabidopsis* are differentially regulated by stress signals. *Proc. Natl. Acad. Sci. USA* **96,** 4718–4723.

Lacombe, B., et al. (2001). The identity of plant glutamate receptors. *Science* **292,** 1486–1487.

Lecourieux, D., Mazars, C., Pauly, N., Ranjeva, R. and Pugin, A. (2002). Analysis and effects of cytosolic free calcium increases in response to elicitors in *Nicotiana plumbaginifolia* cells. *Plant Cell* **14,** 2627–2641.

Lee, J-Y., Yoo, B-C. and Harmon, A.C. (1998). Kinetic and calcium-binding properties of three calcium-dependent protein kinase isoenzymes from soybean. *Biochemistry* **37,** 6801–6809.

Leung, J., Bouvier-Durand, M., Morris, P.C., Guerrier, D., Chefdor, F. and Giraudat, J. (1994). Arabidopsis ABA response gene ABI1: features of a calcium-modulated protein phosphatase. *Science* **264,** 1448–1452.

Lhuissier, F.G.P., De Ruijter, N.C.A., Sieberer, B.J., Esseling, J.J. and Emons, A.M.C. (2001). Time course of cell biological events evoked in legume root hairs by *Rhizobium* Nod factors: state of the art. *Ann. Bot.* **87**, 289–302.

Li, L., Kim, B.G., Cheong, Y.H., Pandey, G.K. and Luan, S. (2006). A Ca^{2+} signaling pathway regulates a K(+) channel for low-K response in Arabidopsis. *Proc. Natl. Acad. Sci. USA* **103,** 12625–12630.

Li, X-L., Borsics, T., Harrington, H.M. and Christopher, D.A. (2005). *Arabidopsis* AtCNGC10 rescues potassium channel mutants of *E. coli*, yeast and *Arabidopsis* and is regulated by calcium/calmodulin and cyclic GMP in *E. coli*. *Funct. Plant Biol.* **32,** 643–653.

Liao, B., Gawienowski, M.C. and Zielinski, R.E. (1996). Differential stimulation of NAD kinase and binding of peptide substrates by wild-type and mutant plant calmodulin isoforms. *Archives of Biochemistry and Biphysics* **327**, 53–60.

Ling, V. and Zielinski, R.E. (1993). Isolation of an *Arabidopsis* cDNA sequence encoding a 22 kDa calcium-binding protein (CaBP-22) related to calmodulin. *Plant Mol. Biol.***22**, 207–214.

Liu, J. and Zhu, J.-K. (1997). Proline accumulation and salt-stress-induced gene expression in a salt-hypersensitive mutant of Arabidopsis. *Plant Physiol.* **114,** 591–596.

Liu, J. and Zhu, J.-K. (1998). A calcium sensor homolog required for plant salt tolerance. *Science* **280,** 1943–1945.

Liu, J., Ishitani, M., Halfter, U., Kim, C.-S. and Zhu, J.K. (2000). The *Arabidopsis thaliana SOS2* gene encodes a protein kinase that is required for salt tolerance. *Proc. Natl. Acad. Sci. USA* **97,** 3730–3734.

Llorente, F., Lopez-Cobollo, R.M., Catala, R., Martinez-Zapater, J.M. and Salinas, J. (2002). A novel cold-inducible gene from Arabidopsis, RCI3, encodes a peroxidase that constitutes a component for stress tolerance. *Plant J.* **32,** 13–24.

Lu, S.X. and Hrabak, E.M. (2002). An Arabidopsis calcium-dependent protein kinase is associated with the endoplasmic reticulum. *Plant Physiol.* **128,** 1008–1021.

Luan, S., Li, W., Rusnak, F., Assmann, S.M. and Schreiber, S.L. (1993). Immunosuppressants implicate protein phosphatase regulation of K^+ channels in guard cells. *Proc. Natl. Acad. Sci. USA* **90,** 2202–2206.

Zhang, M., Tanaka, T. and Ikura, M. (1995). Calcium-induced conformational transition revealed by the solution structure of apo calmodulin, *Nature Struct. Biol.* **2,** 758–767.

MAPK Group (Ichimura et al.,) (2002). Mitogen-activated protein kinase cascades in plants: a new nomenclature. *Trends Plant Sci.* **7,** 301–308.

Martin, M. and Busconi, L. (2000). Membrane localization of a rice calcium-dependent protein kinase (CDPK) is mediated by myristoylation and palmitoylation. *Plant J.* **24,** 429–435.

Maser, P., et al. (2001). Phylogenetic relationships within cation transporter families of Arabidopsis. *Plant Physiol.* **126,** 1646–1667.

Matheos, D.P., Kingsbury, T.J., Ahsan, U.S. and Cunningham, K.W. (1997). Tcn1p/Crz1p, a calcineurin-dependent transcription factor that differentially regulates gene expression in *Saccharomyces cerevisiae*. *Genes Dev.* **11,** 3445–3458.

McAinsh, M.R., Webb, A.A.R., Taylor, J.E., and Hetherington, A.M. (1995). Stimulus-induced oscillations in guard cell cytoplasmic free calcium. *Plant Cell* **7,** 1207–1219.

McCormack, E. and Braam, J. (2003). Calmodulins and related potential calcium sensors in Arabidopsis. *New Phytol.* **159,** 585–598.

McCormack, E., Tsai, Y.C. and Braam, J. (2005). Handling calcium signaling: Arabidopsis CaMs and CMLs. *Trends Plant Sci.* **10,** 383–389.

Mendoza, I., Rubio, F., Rodriguez-Navarro, A. and Pardo, J.M. (1994). The protein phosphatase calcineurin is essential for NaCl tolerance of *Saccharomyces cerevisiae. J. Biol. Chem.* **269,** 8792–8796.

Milla, M.A., Townsend, J., Chang, I.F. and Cushman, J.C. (2006). The Arabidopsis AtDi19 gene family encodes a novel type of Cys2/His2 zinc-finger protein implicated in ABA-independent dehydration, high-salinity stress and light signaling pathways. *Plant Mol. Biol.* **61,** 13–30.

Milligan, G., Parenti, M. and Magee, A.I. (1995). The dynamic role of palmitoylation in signal transduction. *Trends Biol. Sci.* **20,** 181–186.

Monroy, A.F. and Dhindsa, R.S. (1995). Low-temperature signal transduction: induction of cold acclimation-specific genes of alfalfa by calcium at 25 C. *Plant Cell* **7,** 321–331.

Murata, Y., Pei, Z.M., Mori, I.C., and Schroeder, J. (2001). Abscisic acid activation of plasma membrane Ca^{2+} channels in guard cells requires cytosolic NAD(P)H and is differentially disrupted upstream and downstream of reactive oxygen species production in *abi1–1* and *abi2–1* protein phosphatase 2C mutants. *Plant Cell* **13,** 2513–2523.

Nagae, M., Nozawa, A., Koizumi, N., Sano, H., Hashimoto, H., Sato, M. and Shimizu, T. (2003). The crystal structure of the novel calcium-binding protein AtCBL2 from *Arabidopsis thaliana. J. Biol. Chem.* **278,** 42240–42246.

Nakamura, T., Liu, Y., Hirata, D., Namba, H., Harada, S., Hirokawa, T. and Miyakawa, T. (1993). Protein phosphatase type 2B (calcineurin)-mediated, FK506-sensitive regulation of intracellular ions in yeast is an important determinant for adaptation to high salt stress conditions. *EMBO J.* **12,** 4063–4071.

Ng, C.K.Y., Carr, K., McAinsh, M.R., Powell, B., and Hetherington, A.M. (2001). Drought-induced guard cell signal transduction involves sphingosine-1-phosphate. *Nature* **410,** 596–598.

Nozawa, A., Koizumi, N. and Sano, H. (2001). An *Arabidopsis* SNF1-related protein kinase, AtSR1, interacts with a calcium-binding protein, AtCBL2, of which transcripts respond to light. *Plant Cell Physiol.* 42:976–981.

Ohta, M., Guo, Y., Halfter, U. and Zhu, J.K. (2003). A novel domain in the protein kinase SOS2 mediates interaction with the protein phosphatase 2C ABI2. *Proc. Natl. Acad. Sci. USA* **100,** 11771–11776.

Okazaki, Y., Ishigami, M. and Iwasaki, N. (2002). Temporal relationship between cytosolic free Ca^{2+} and membrane potential during hypotonic turgor regulation in a brackish water Charophyte *Lamprothamnium succinctum. Plant Cell Physiol.* **43,** 1027–1035.

Park, S.Y., Seo, S.B., Lee, S.J., Na, J.G. and Kim, Y.J. (2001). Mutation in PMR1, a Ca^{2+}-ATPase in Golgi, confers salt tolerance in *Saccharomyces cerevisiae* by inducing expression of PMR2, an Na^{+}-ATPase in plasma membrane. *J. Biol. Chem.* **276,** 28694–28699.

Patharkar, O.R. and Cushman, J.C. (2000) A stress-inducible calcium-dependent protein kinase from *Mesembryanthemum crystallinum* phosphorylates a two-component pseudo-response regulator. *Plant J.* **24,** 679–691.

Patharkar, O.R. and Cushman, J.C. (2006). A novel coiled-coil protein co-localizes and interacts with a calcium-dependent protein kinase in the common ice plant during low-humidity stress. *Planta* Jun 14 (Epub).

Pauly, N., Knight, M.R., Thuleau, P., Graziana, A., Muto, S., Ranjeva, R., and Mazars, C. (2001). The nucleus together with the cytosol generates patterns of specific cellular calcium signatures in tobacco suspension culture cells. *Cell Calcium* **30,** 413–421.

Pei, Z.M., Murata, Y., Benning, G., Thomine, S., Klusener, B., Allen, G.J., Grill, E., and Schroeder, J.I. (2000). Calcium channels activated by hydrogen peroxide mediate abscisic acid signalling in guard cells. *Nature* **406,** 731–734.

Pei, Z.M., Ward, J.M., and Schroeder, J.I. (1999). Magnesium sensitizes slow vacuolar channels to physiological cytosolic calcium and inhibits fast vacuolar channels in fava bean guard cell vacuoles. *Plant Physiol.* **121,** 977–986.

Perera, I.Y. and Zielinski, R.E. (1992). Structure and expression of the Arabidopsis CaM-3 calmodulin gene. *Plant Mol. Biol.* **19,** 649–664.

Perez-Prat, E., Narasimhan, M.L., Binzel, M.L., Botella, M.A., Chen, Z., Valpuesta, V., Bressan, R.A. and Hasegawa, P.M. (1992). Induction of a putative Ca-ATPase mRNA in NaCl-adapted cells. *Plant Physiol.* **100,** 1471–1478.

Perruc, E., Charpenteau, M., Ramirez, B.C., Jauneau, A., Galaud, J.P. et al. (2004). A novel calmodulin-binding protein functions as a negative regulator of osmotic stress tolerance in *Arabidopsis thaliana* seedlings. *Plant J.* **38,** 410–420.

Phean-O-Pas, S., Punteeranurak, P. and Buaboocha, T. (2005). Calcium signaling-mediated and differential induction of calmodulin gene expression by stress in *Oryza sativa* L. *J. Biochem. Mol. Biol.* **38,** 432–439.

Pittman, J.K. and Hirschi, K.D. (2003). Don't shoot the (second) messenger: endomembrane transporters and binding proteins modulate cytosolic Ca^{2+} levels. *Curr. Opin. Plant Biol.* **6,** 257–262.

Pittman, J.K., and Hirschi, K.D. (2001). Regulation of CAX1, an Arabidopsis Ca^{2+}/H^+ antiporter. Identification of an N-terminal autoinhibitory domain. *Plant Physiol.* **127,** 1020–1029.

Qiu, Q.S., Guo, Y., Quintero, F.J., Pardo, J.M., Schumaker, K.S. and Zhu, J.K. (2004). Regulation of vacuolar Na^+/H^+ exchange in *Arabidopsis thaliana* by the salt-overly-sensitive (SOS) pathway. *J. Biol. Chem.* **279,** 207–215.

Qiu, Q.S., Guo, Y., Dietrich, M.A., Schumaker, K.S. and Zhu J.K. (2002). Regulation of SOS1, a plasma membrane Na^+/H^+ exchanger in *Arabidopsis thaliana*, by SOS2 and SOS3. *Proc. Natl. Acad. Sci. USA* **99,** 8436–8441.

Quintero, F.J., Ohta, M., Shi, H., Zhu, J.K. and Pardo, J.M. (2002). Reconstitution in yeast of the Arabidopsis SOS signaling pathway for Na^+ homeostasis. *Proc. Natl. Acad. Sci. USA* **99,** 9061–9066.

Reddy, V.S., Ali, G.S. and Reddy, A.S.N. (2002). Genes encoding calmodulin-binding proteins in the Arabidopsis genome. *J. Biol. Chem.* **277,** 9840–9852.

Reddy, V.S. and Reddy, A.S.N. (2004). Proteomics of calcium-signaling components in plants. *Phytochemistry* **65,** 1745–1776.

Reddy, V.S., Safadi, F., Zielinski, R.E. and Reddy, A.S.N. (1999). Interaction of a kinesin-like protein with calmodulin isoforms from *Arabidopsis*. *J. Biol. Chem.* **274**, 31727–31733.

Rentel, M.C. and Knight, M.R. (2004). Oxidative stress-induced calcium signaling in Arabidopsis. *Plant Physiol.* **135,** 1471–1479.

Resh, M.D. (1999). Fatty acylation of proteins: new insights into membrane targeting of myristoylated and palmitoylated proteins. *Biochim. Biophys. Acta* **1451,** 1–16.

Rozwadowski, K., Zhao, R., Jackman, L., Huebert, T., Burkhart, W.E., Hemmingsen, S.M., Greenwood, J. and Rothstein, S.J. (1999). Characterization and immunolocalization of a cytosolic calcium-binding protein from Brassica napus and Arabidopsis pollen. *Plant Physiol.* **120,** 787–798.

Rudd, J.J., and Franklin-Tong, V.E. (2001). Unravelling response-specificity in Ca^{2+} signalling pathways in plant cells. *New Phytol.* **151,** 7–33.

Rusnak, F. and Mertz, P. (2000). Calcineurin: form and function. *Physiol. Rev.* **80,** 1483–1521.

Sagi, M. and Fluhr, R. (2001). Superoxide production by plant homologues of the gp91(phox) NADPH oxidase. Modulation of activity by calcium and by tobacco mosaic virus infection. *Plant Physiol.* **126,** 1281–1290.

Saijo, Y., Hata, S., Kyozuka, J., Shimamoto, K. and Izui, K. (2000). Over-expression of a single Ca^{2+}-dependent protein kinase confers both cold and salt/drought tolerance on rice plants. *Plant J.* **23,** 319–327.

Sanchez-Barrena, M.J., Martinez-Ripoll, M., Zhu, J.K. and Albert, A. (2005). The structure of the *Arabidopsis thaliana* SOS3: molecular mechanism of sensing calcium for salt stress response. *J. Mol. Biol.* **345,** 1253–1264.

Sanders, D., Pelloux, J., Brownlee, C. and Harper, J.F. (2002). Calcium at the crossroads of signaling. *Plant Cell* **14(Suppl),** S401–S17.

Schroeder, J.I., Allen, G.J., Hugouvieux, V., Kwak, J., and Waner, D. (2001). Guard cell signal transduction. *Annu. Rev. Plant Physiol. Plant Mol. Biol.* **52,** 627–658.

Seki, M., Narusaka, M., Abe, H., Kasuga, M., Yamaguchi-Shinozaki, K., Carninci, P., Hayashizaki, Y. and Shinozaki, K. (2001). Monitoring the expression pattern of 1300 Arabidopsis genes under drought and cold stresses by using a full-length cDNA microarray. *Plant Cell.* **13,** 61–72.

Sheen, J. (1996). Ca^{2+}-dependent protein kinases and stress signal transduction in plants. *Science* **274,** 1900–1902.

Shi, H., Bressan, R., Hasegawa, P.M. and Zhu, J.K. (2005). Sodium. In: *Plant Nutritional Genomics* (Eds Broadley M, White P), Blackwell Publishing, London, pp 127–149.

Shi, H., Ishitani, M., Kim, C. and Zhu, J.K. (2000). The *Arabidopsis thaliana* salt tolerance gene *SOS1* encodes a putative Na^+/H^+ antiporter. *Proc. Natl. Acad. Sci. USA* **97,** 6896–6901.

Shi, H. and Zhu, J.K. (2002). Regulation of the vacuolar Na^+/H^+ antiporter gene *AtNHX1* expression by salt stress and abscisic acid. *Plant Mol Biol.* **50,** 543–550.

Shi, J., Kim, K.N., Ritz, O., Albrecht, V., Gupta, R., Harter, K., Luan, S. and Kudla, J. (1999). Novel protein kinases associated with calcineurin B-like calcium sensors in *Arabidopsis. Plant Cell* **11,** 2393–2405.

Shi, H., Lee, B., Wu, S.-J., and Zhu, J.-K. (2003). Overexpression of a plasma membrane Na^+/H^+ antiporter improves salt tolerance in Arabidopsis. *Nature Biotech.* **21,** 81–85.

Shi, H., Quintero, F.J., Pardo, J.M. and Zhu, J.-K. (2002). The putative plasma membrane Na^+/H^+ antiporter SOS1 controls long-distance Na^+ transport in plants. *Plant Cell* **14,** 465–477.

Shibasaki, F., Price, E.R., Milan, D. and McKeon, F. (1996). Role of kinases and the phosphatase calcineurin in the nuclear shuttling of transcription factor NF-AT4. *Nature* **382,** 370–373.

Shigaki, T., Cheng, N.H., Pittman, J.K., and Hirschi, K. (2001). Structural determinants of Ca^{2+} transport in the Arabidopsis H^+/Ca^{2+} antiporter CAX1. *J. Biol. Chem.* **276,** 43152–43159.

Singla-Pareek, S.L., Reddy, M.K. and Sopory, S.K. (2003). Genetic engineering of the glyoxalase pathway in tobacco leads to enhanced salinity tolerance. *Proc. Natl. Acad. Sci. USA* **100,** 14672–14677.

Snedden, W.A. and Fromm, H. (2001). Calmodulin as a versatile calcium signal transducer in plants. *New Phytol.* **151,** 35–66.

Stathopoulos, A.M. and Cyert, M.S. (1997). Calcineurin acts through the CRZ1/TCN1-encoded transcription factor to regulate gene expression in yeast. *Genes Dev.* **11,** 3432–3444.

Straatman, K.R., Dove, S.K., Holdaway-Clarke, T., Hepler, P.K., Kunkel, J.G. and Franklin-Tong, V.E. (2001). Calcium signalling in pollen of *Papaver rhoeas* undergoing the self-incompatibility SI response. *Sexual Plant Reproduction* **14**, 105–110.

Sunkar, R., Kaplan, B., Bouche, N., Arazi, T., Dolev, D., Talke, I.N., Maathuis, F.J.M., Sanders, D., Bouchez, D., and Fromm, H. (2000). Expression of a truncated tobacco NtCBP4 channel in transgenic plants and disruption of the homologous Arabidopsis CNGC1 gene confer Pb^{2+} tolerance. *Plant J.* **24,** 533–542.

Sze, H., Liang, F., Hwang, I., Curran, A.C., and Harper, J.F. (2000). Diversity and regulation of plant Ca^{2+} pumps: Insights from expression in yeast. *Annu. Rev. Plant Physiol. Plant Mol. Biol.* **51,** 433–462.

Takahashi, K., Isobe, M., Knight, M.R., Trewavas, A.J. and Muto, S. (1997). Hypoosmotic shock induces increases in cytosolic Ca^{2+} in tobacco suspension-culture cells. *Plant Physiol.* **113**, 587–594.

Takezawa, D. and Minami, A. (2004). Calmodulin-binding proteins in bryophytes: identification of abscisic acid-, cold-, and osmotic stress-induced genes encoding novel membrane-bound transporter-like proteins. *Biochem. Biophys. Res. Commun.* **317,** 428–436.

Talke, I.N., Blaudez, D., Maathuis, F.J. and Sanders, D. (2003). CNGCs: prime targets of plant cyclic nucleotide signalling? *Trends Plant Sci.* **8,** 286–293.

Teige, M., Scheikl, E., Eulgem, T., Doczi, R., Ichimura, K., Shinozaki, K., Dangl, J.L. and Hirt, H. (2004). The MKK2 pathway mediates cold and salt stress signaling in Arabidopsis. *Mol. Cell* **15,** 141–152.

Thion, L., Mazars, C., Nacry, P., Bouchez, D., Moreau, M., Ranjeva, R. and Thuleau, P. (1998). Plasma membrane depolarization-activated calcium channels, stimulated by microtubule-depolymerizing drugs in wild-type *Arabidopsis thaliana* protoplasts, display constitutively large activities and a longer half-life in *ton2* mutant cells affected in the organization of cortical microtubules. *Plant J.* **13,** 603–610.

Thornalley, P.J. (1990). The glyoxalase system: new developments towards functional characterization of a metabolic pathway fundamental to biological life. *Biochem. J.* **269,** 1–11.

Torres, M.A. and Dangl, J.L. (2005). Functions of the respiratory burst oxidase in biotic interactions, abiotic stress and development. *Curr. Opin. Plant Biol.* **8,** 397–403.

Ulm, R., Ichimura, K., Mizoguchi, T., Peck, S.C., Zhu, T., Wang, X., Shinozaki, K. and Paszkowski, J. (2002). Distinct regulation of salinity and genotoxic stress responses by Arabidopsis MAP kinase phosphatase 1. *EMBO J.* **21,** 6483–6493.

Ulm, R., Revenkova, E., di Sansebastiano, G.P., Bechtold, N. and Paszkowski, J. (2001). Mitogen-activated protein kinase phosphatase is required for genotoxic stress relief in Arabidopsis. *Genes Dev.* **15,** 699–709.

Urao, T., Katagiri, T., Mizoguchi, T., Yamaguchi-Shinozaki, K., Hayashida, N. and Shinozaki, K, (1994). Two genes that encode Ca^{2+}-dependent protein kinases are induced by drought and high-salt stresses in *Arabidopsis thaliana. Mol. Gen. Genet.* **244,** 331–340.

Vitart, V., Christodoulou, J., Huang, J.F., Chazin, W.J. and Harper, J.F. (2000) Intramolecular activation of a Ca^{2+}-dependent protein kinase is disrupted by insertions in the tether that connects the calmodulin-like domain to the kinase. *Biochemistry* **39**, 4004–4011.

van der Luit, A.H., Olivari, C., Haley, A., Knight, M.R., and Trewavas, A.J. (1999). Distinct calcium signaling pathways regulate calmodulin gene expression in tobacco. *Plant Physiol.* **121,** 705–714.

Van Eldik, L.J. and Watterson, M.D. (1998). Calmodulin and calcium signal transduction: an introduction. In: van Eldik, L.J., Watterson, M.D., eds. Calmodulin and signal transduction. New York, USA: Academic Press, 1–15.

Veena, Reddy, V.S. and Sopory, S.K. (1999). Glyoxalase I from Brassica juncea: molecular cloning, regulation and its over-expression confer tolerance in transgenic tobacco under stress. *Plant J.* **17,** 385–395.

Wang, X. (2005). Regulatory functions of phospholipase D and phosphatidic acid in plant growth, development, and stress responses. *Plant Physiol.* **139,** 566–573.

White, P.J., Broadley, M.R. (2003). Calcium in plants. *Ann. Bot.* **92,** 487–511.

White, P.J., Bowen, H.C., Demidchik, V., Nichols, C. and Davies, J.M. (2002). Genes for calcium-permeable channels in the plasma membrane of plant root cells. *Biochim. Biophys. Acta* **1564,** 299–309.

White, P.J. (2000). Calcium channels in higher plants. *Biochim. Biophys. Acta* **1465,** 171–189.

Wimmers, L.E., Ewing, N.N. and Bennett, A.B. (1992). Higher plant Ca^{2+}-ATPase: primary structure and regulation of mRNA abundance by salt. *Proc. Natl. Acad. Sci. USA* **89,** 9205–9209.

Wood, N.T., Allan, A.C., Haley, A., Viry-Moussaid, M., and Trewavas, A.J. (2000). The characterization of differential calcium signalling in tobacco guard cells. *Plant J.* **24,** 335–344.

Wu, S.-J., Lei, D., and Zhu, J.-K. (1996). *SOS1*, a genetic locus essential for salt tolerance and potassium acquisition. *Plant Cell* **8,** 617–627.

Xiong, L., Schumaker, K.S. and Zhu, J.-K. (2002). Cell signaling for cold, drought, and salt stresses. *Plant Cell* **14(suppl),** S165–183.

Xu, J., Li, H.D., Chen, L.Q., Wang, Y., Liu, L.L., He, L. and Wu, W.H. (2006). A protein kinase, interacting with two calcineurin B-like proteins, regulates K^+ transporter AKT1 in Arabidopsis. *Cell* **125,** 1347–1360.

Yamaguchi, T., Aharon, G.S., Sottosanto, J.B. and Blumwald, E. (2005). Vacuolar Na^+/H^+ antiporter cation selectivity is regulated by calmodulin from within the vacuole in a Ca^{2+}- and pH-dependent manner. *Proc. Natl. Acad. Sci. USA* **102,** 16107–16112.

Yamaguchi, T., Apse, M.P., Shi, H. and Blumwald, E. (2003). Topological analysis of a plant vacuolar Na^+/H^+ antiporter reveals a luminal C terminus that regulates antiporter cation selectivity. *Proc. Natl. Acad. Sci. USA* **100,** 12510–12515.

Yamakawa, H., Katou, S., Seo, S., Mitsuhara, I., Kamada, H. and Ohashi, Y. (2004). Plant MAPK phosphatase interacts with calmodulins. *J. Biol. Chem.* **279,** 928–936.

Yang, T. and Poovaiah, B.W. (2002). A calmodulin-binding/CGCG box DNA-binding protein family involved in multiple signaling pathways in plants. *J. Biol. Chem.* **277,** 45049–45058.

Yoo, J.H., Park, C.Y., Kim, J.C., Heo, W.D., Cheong, M.S. et al. (2005). Direct interaction of a divergent CaM isoform and the transcription factor, MYB2, enhances salt tolerance in Arabidopsis. *J. Biol. Chem.* **280,** 3697–3706.

Yoshimoto, H., Saltsman, K., Gasch, A.P., Li, H.X., Ogawa, N., Botstein, D., Brown, P.O. and Cyert, M.S. (2002). Genome-wide analysis of gene expression regulated by the calcineurin/Crz1p signaling pathway in *Saccharomyces cerevisiae*. *J. Biol. Chem.* **277,** 31079–31088.

Zhang, W., Yu, L., Zhang, Y. and Wang, X. (2005). Phospholipase D in the signaling networks of plant response to abscisic acid and reactive oxygen species. *Biochim. Biophys. Acta* **1736,** 1–9.

Zhang, X., Zhang, L., Dong, F.C., Gao, J.F., Galbraith, D.W. and Song, C.P. (2001). Hydrogen peroxide is involved in abscisic acid-induced stomatal closure in *Vicia faba*. *Plant Physiol.* **126,** 1438–1448.

Zhu J.K. (2002). Salt and drought stress signal transduction in plants. *Annu. Rev. Plant Biol.* **53,** 247–273.

Zhu, J.K., Liu, J. and Xiong, L. (1998). Genetic analysis of salt tolerance in *Arabidopsis thaliana*: evidence of a critical role for potassium nutrition. *Plant Cell* **10,** 1181–1192.

Zou, H., Lifshitz, L.M., Tuft, R.A., Fogarty, K.E. and Singer, J.J. (2002). Visualization of Ca^{2+} entry through single stretch-activated cation channels. *Proc. Natl. Acad. Sci. USA* **99,** 6404–6409.

CHAPTER 8

PHOSPHOLIPID SIGNALING IN PLANT RESPONSE TO DROUGHT AND SALT STRESS

XUEMIN WANG[1], WENHAU ZHANG[2], WEIQI LI[3], AND GIRISH MISHRA[1]

[1]*Department of Biology, University of Missouri, St. Louis, MO 63121 and Donald Danforth Plant Science Center, St. Louis, MO 63132, USA*

[2]*College of Life Science, State Key Laboratory of Crop Genetics and Germplasm Enhancement, Nanjing Agricultural University, Nanijing 210095, P.R. China*

[3]*Kunming Institute of Botany, Chinese Academy of Sciences, Kunming 650204 and Department of Biology, Honghe University, Mengzi, Yunnan 661100, China*

Abstract: Many stresses trigger transient increases in minor phospholipids, such as phosphatidic acid (PA) and phosphoinositides (PIs), in plants. Such changes are early events in signaling plant stress response. Lipid mediators affect cellular functions through direct interaction with proteins and/or structural effects on cell membranes. The identified lipid targets in plants include protein phosphatases, kinases, and proteins involved in membrane trafficking and cytoskeleton. The effect of lipids on signaling, intracellular trafficking, and cytoskeletal organization plays important roles in plant coping with drought and salinity

Keywords: lipid signaling; phospholipases; phosphoinositides; drought; salt; stress response; osmotic stress; abscisic acid; phosphatidic acid

1. INTRODUCTION

As the physical barrier that separates the interior of a cell from its surroundings, cell membranes play a pivotal role in plant responses to environment stresses. Membranes are the initial site of cellular perception of stress cues. Many subsequent steps in signaling cascades, such as activation of effector proteins, generation of second messengers, and alteration in cellular metabolism, are often associated with membranes. While proteins have been the focus of most research on membrane-associated signaling events, recent advances have made it evident that membrane lipids and their derivatives are important players in the signaling network of plant responses to stress, including drought and salinity.

M.A. Jenks et al. (eds.), Advances in Molecular Breeding Toward Drought and Salt Tolerant Crops, 183–192.

Membrane lipids give rise to various signaling messengers, such as phosphatidic acid (PA), diacylglycerol (DAG), DAG-pyrophosphate (DAG-PP), lysophospholipids, free fatty acids (FFAs), oxylipins, phosphoinositides, and inositol polyphosphates. The production of these mediators is regulated by different families of enzymes, particularly phospholipases, lipid kinases, and/or phosphatases (Wang, 2004). Great strides have been made recently to understand the role of lipid signaling in different plant processes. Several recent reviews have dealt with signaling aspects of various lipids and enzymes in plants (Chapman, 2004; Ryu, 2004; Testerink and Munnik, 2005; van Schooten et al, 2006; Wang, 2004; Wang et al., 2006; Zhang et al., 2005). Here, we will focus on the involvement of phospholipid-mediated signaling in plant response to drought and salt stress.

2. PHOSPHATIDIC ACID AS A MESSENGER IN OSMOTIC STRESS

PA has been identified as a new class of lipid mediators regulating numerous cellular processes, including signal transduction, cytoskeletal rearrangement, secretion, endo/exocytosis,and oxidative burst. PA is a minor membrane lipid, constituting less than 1% of total phospholipids in Arabidopsis leaves (Zhang et al., 2004; Wang et al., 2006). Cellular levels of PA in plants change rapidly under various conditions, including abiotic and biotic stresses as well as during plant growth and development (Testerink and Munnik, 2005; Wang et al., 2006). In particular, PA is produced under different forms of osmotic stress, such as dehydration, drought, salinity, freezing, and treatment with the stress hormone abscisic acid (ABA).

Cellular PA may be produced by multiple enzymes (Wang et al., 2006; Figure 1). Available data indicate that signaling PA is generated by two principal routes in plants. One is the phospholipase D (PLD)-catalyzed hydrolysis of common membrane phospholipids to produce directly PA. Another is phospholipase C (PLC) hydrolysis of phosphatidylinositol (4, 5) bisphosphate PI(4, 5)P_2 followed by phosphorylation of DAG by DAG kinase (DGK). PLD, PLC, and DGK each consist of multiple enzymes in plants. For example, Arabidopsis has 12 PLD genes that are grouped into six types, PLDα, β, γ, δ, ε, and ζ. Different type of PLDs display distinguishable properties, such as their requirements for Ca^{2+}, phosphoinositides (PI), and/or free fatty acids, their lipid vesicle composition, substrate preferences, subcellular location, and patterns of gene expression. These differences play an important role in regulating the spatial and temporal production of PA and also indicate distinguishable functions among different PLDs. It has been shown that Arabidopsis PLDα1 is responsible for ABA-induced PA (Zhang et al., 2004), whereas PLDδ is involved in the dehydration-induced PA formation (Katagiri et al. 2001). PLD and PA have been suggested to affect osmotic-stress induced production of proline (Thiery et al., 2004). 1-Butanol, an inhibitor of PA production by PLD, reduces NaCl- induced H^+-ATPase activation, whereas applied PA stimulated H^+-ATPase activity (Zhang et al., 2006). These results point to a role of PLD and PA in salt stress response.

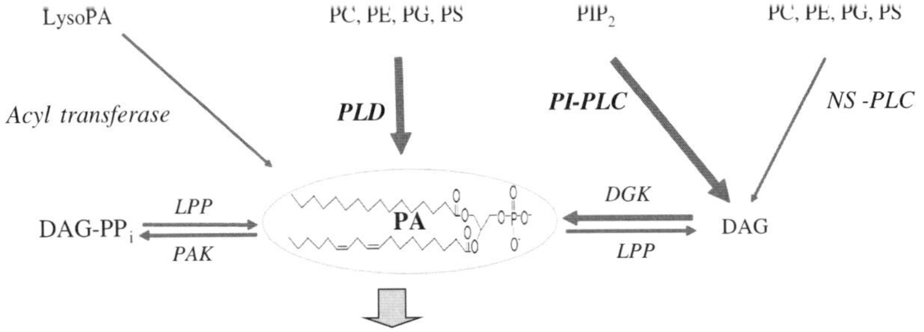

PA-binding protein	Protein function	Effect of PA binding
ABI1	PP2C and a negative regulator of ABA response	Inhibit ABI1 function and promote stomatal closure
AtPDK1	3-Phopshoinositide-dependent kinase 1	Activate AGC kinase and promote root growth
AtCP	Heterodimeric actin capping protein	Inhibit capping and stimulate actin polymerization
mTOR	Mammalian target of rapamycin	Activate TOR signaling and promote cell survival
Opi1p	Yeast soluble transcriptional repressor	Regulate Opi1p location between ER and nucleus and the hemeostasis of glycerolipid synthesis

Figure 1. PA production and selected PA effects (Wang et al., 2006). Enzymatic reactions leading to the PA production (*upper*) and removal (*lower*). DGK, diacylglycerol kinases; DAG-PP_i, diacylglycerol pyrophosphate; LPP, lipid phosphate phosphatase;LysoPA, lysophosphatidic acid; PAK, phosphatidic acid kinase; PI-PLC, phosphoinostide-specific PLC; NS-PLC, non-specific PLC. PC, phosphatidylcholine; PE, phosphatidylethanolamine; PG, phosphatidylglycerol; PS, phosphatidylserine. PA binds to different types of proteins and the specific examples including proteins from plants, animals, and yeast

Arabidopsis has muliple PI-PLCs that hydrolyze PI(4, 5)P_2 to generates DAG and inositol 1, 4, 5-trisphosphate (IP_3). DAG serves as a potent activator of protein kinase C in animal cells, but its target is unclear in plants. Under salt and hyperosmotic conditions, PLC-produced DAG is phosphorylated to PA by DGK. In several cases, it has been found that the PLD and PLC/DGK reactions are activated differentially in response to different stimuli (den Hartog et al., 2003; Zonia and Munnik, 2004).

The increase in PA can impact cell function in different ways:i) PA can act as a messenger by its interaction with specific target proteins. PA has been found to tether target proteins to membranes and/or modulate the catalytic activity of its targets (Fan et al., 2001; Anthony et al., 2004; Zhang et al., 2004; Huang et al., 2006). ii) PA may alter membrane structure, promoting membrane fusion and interaction of certain soluble proteins with membranes (Kooijman et al., 2005; Wang et al., 2006). iii) PA may be converted to other signaling molecules, such as DAG, lysoPA, DAG-PP_i, and free fatty acids, or may be involved in membrane lipid metabolism (Testerink and Munnik, 2005; Wang et al., 2006; Figure 1). A number of PA-binding proteins have been identified in plants, animals, and yeast, which include protein kinases (Anthony et al., 2004; Fang et al., 2001), protein phosphatases (Zhang et al., 2004), transcriptional factors (Loewen et al., 2004), and proteins involved in vesicular trafficking and cytoskeletal dynamics (Huang et al., 2006; Testerink et al., 2004).

Potentially relevant to drought and salt stress is the PA binding to ABI1 protein phosphatase 2C (PP2C) (Zhang et al., 2004), to 3-phosphoinositide-dependent kinase 1 (AtPDK1) (Anthony et al., 2004), and to the mammalian target of rapamycin (mTOR), a phosphatidylinositol-3 kinase like protein kinase (Fang et al., 2001). The PA-ABI1 interaction is required for ABA promotion of stomatal closure (Mishra et al., 2006), whereas the PA-AtPDK1 binding promotes root growth (Anthony et al., 2004). The TOR kinase pathway mediates translational regulation of cell growth and proliferation in animal cells. The TOR pathway in plants, animals, and yeast are affected by various adverse conditions including osmotic stress (Mahfouz et al., 2006). The TOR target ribosomal S6 kinase 1 (S6K1) is also regulated by AtPDK1. Thus, it would be of interest to test whether the TOR signaling pathway is a PA target in plant osmotic stress response.

3. PHOSPHATIDIC ACID AND PHOSPHOLIPASE D IN STOMATAL CLOSURE AND WATER LOSS

One documented function of PA is to mediate stomatal response to ABA. ABA is a phytohormone involved in diverse plant processes, including stomatal closure. ABA causes stomatal closure by affecting two separable processes:it promotes the closing of opened stomata and inhibits the opening of closed stomata. Recent results show that stomatal responses to ABA are regulated by a bifurcating pathway that includes PLDα1, PA, ABI1, and Gα (Mishra et al., 2006; Figure 2). To promote closure of open stomata, PLDα1 produces PA that binds to the ABI1 PP2C (Zhang et al., 2004). ABI1 is a negative regulator of ABA response, and the PA-ABI1 interaction is necessary to remove the ABI1 inhibition of the ABA promotion of stomatal closure. PA regulates ABI1 function by inhibiting its phosphatase activity and by sequestering it in the plasma membrane. The membrane tethering by PA decreases ABI1's translocation from the cytosol to the nucleus and promotes ABA signaling (Zhang et al., 2004).

To inhibit opening of closed stomata, PLDα1 modulates the function of Gα (only canonical Gα subunit of a heterotrimeric G protein in Arabidopsis) through multiple interactions (Figure 2). PLDα1 activates the intrinsic GTPase activity that converts active $G\alpha\text{-}_{GTP}$ to inactive $G\alpha\text{-}_{GDP}$ (Zhao and Wang 2004). In turn, $G\alpha\text{-}_{GDP}$ binds to PLDα1 and decreases its activity. Weakening the $G\alpha\text{-}_{GDP}$ and PLDα1 interaction renders Arabidopsis plants hypersensitive to ABA because both the Gα and PLDα1 functions are less inhibited by the subdued interaction between PLDα1 and Gα (Mishra et al., 2006). On the other hand, both PLDα1 and Gα are positive regulators in ABA inhibition of stomatal opening. The positive role of Gα may result from the exchange of GTP with GDP; the binding of GTP to Gα ($G\alpha\text{-}_{GTP}$) dissociates Gα from PLDα1, thus removing the inhibition of PLDα1 activity. PA resulting

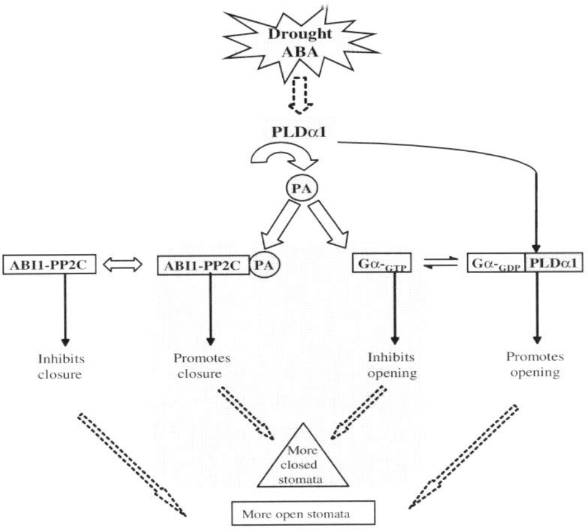

Figure 2. A bifurcating pathway by which PLD and PA in signaling ABA response in guard cells (Mishra et al., 2006). PLDα1-derived PA binds to ABI1, and this interaction tethers ABI1 to the plasma membrane and also decreases the PP2C activity. Thus, PA promotes ABA response by suppressing the negative effect of ABI1. In addition, PLDα1 binds to Gα, the α subunit of heterotrimeric G proteins, and this interaction regulates reciprocally the activity of PLDα1 and Gα, and thus the production of PA. Note that this model is not comprehensive and only includes some of the signaling components implicated in the ABA signaling cascades

from PLDα1 activity promotes inhibition of stomatal opening (Figure 3). Thus, the PLDα1- Gα interaction regulates mutually the activity of both proteins.

The PA/PLD regulation of stomatal closure affects plant water loss. Terrestrial plants lose water primarily via stomata. During drought, ABA levels in plants increase and promotes stomatal closure. This change is crucial to maintaining a hydration status in leaves and permitting plant survival. In Arabidopsis, the stomata of PLDα1-deficient plants fail to close in response to ABA (Sang et al., 2001; Mishra et al., 2006). *PLDα1*-deficient plants exhibit a higher rate of transpirational water loss than wild-type plants, whereas overexpression of *PLDα1* reduces transpirational water loss in tobacco by rendering the plants more sensitive to ABA (Sang et al., 2001). These insights into the pathways regulating stomatal function may be used to produce plants with enhanced water-usage efficiency and drought tolerance.

4. PHOSPHOINOSITIDES IN OSMOTIC STRESS

Phosphoinositides (PIs) are phosphorylated phosphatidylinositols. They include three monophosphorylated PI3P, PI4P, and PI5P; three bisphosphorylated PI(4, 5)P_2, PI(3, 4)P_2, and PI(3, 5)P_2; and one trisphosphorylated PI(3, 4, 5)P_3 (Figure 3). Except for PI(3, 4, 5)P_3, the occurrence of all other PIs have been reported in plants. Several PIs, such as PI5P, PI(3, 5)P_2, and PI(4, 5)P_2, are elevated in plants under hyperosmotic stress (Meijer and Munnik 2003; Zonia and Munnik, 2004).

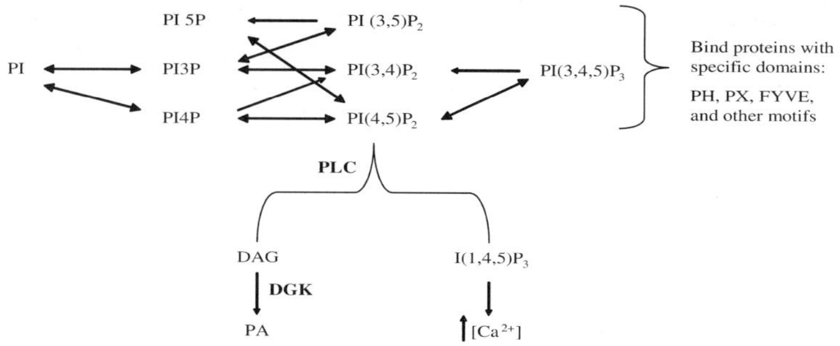

Figure 3. Phosphoinositide metabolism and selected functions. The phosphorylation and dephosphorylation of PIs are catalyzed by specific PI kinases and phosphatases. The arrow points to known direction of PI metabolism, but some of the reactions are yet to be demonstrated in plants. One function of PIs is to interact with proteins with specific domains such as PH, PX, and FYVE. In addition, PI(4, 5)P_2 is a substrate of PLC that produces IP_3 and DAG

A study on Sshlp, a soybean Sec14-like, phosphatidylinositol transfer proteins (PITP), has shed light on the early events of osmosensory signaling and PI synthesis in plants (Monks et al., 2001). Under hyperosmotic stress, Sshlp kinases, SPK1 and/or SPK2, are activated and rapidly phosphorylate Sshlp. This modification decreases membrane association of Sshlp. Sshlp enhances the activities of plant PI 3-kinase and PI 4-kinase, suggesting that Sshlp's function in cellular signaling is to alter the plant's capacity to synthesize PIs during hyperosmotic stress (Figure 4).

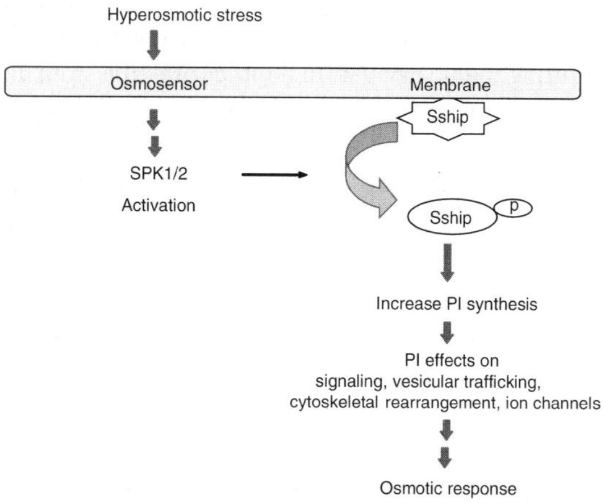

Figure 4. A hypothetic model of hyperosmotic stress-induced production of phosphoinositides and functions (Monks et al., 2001). Hyperosmotic stress triggers activation of the kinases SPK 1 and 2 that phosphorylate the Sec14-like Sshlp. Activated Sshlp stimulates PI 3-kinase– and PI 4-kinase to increase the production of PIs. PIs modulate cellular processes and participate in osmotic stress responses

Recently, dysfunctions of specific PITPs have been linked to plant root hair growth and stress response (Bohme et al., 2004; Vincent et al., 2005). The Arabidopsis Sec14p-nodulin domain proteins AtSfh1p regulate intracellular and plasma membrane PI polarity directing membrane trafficking, Ca^{2+} signaling, and cytoskeleton functions to the growing root hair apex (Vincent et al., 2005). It has been proposed that Sec14p-nodulin domain proteins represent a family of regulators of polarized membrane growth in plants (Vincent et al., 2005). In yeast, PLD activity is required for suppression of PITP defect (Xie et al., 1998), indicating that crosstalk exists among PITPs, PLDs, PIs, and PA in cell regulation.

PIs are important signaling molecules that regulate actin organization, membrane trafficking, endo/exocytosis, and ion channels. For example, PIs are required for activating plant shaker-type K^+ channels (Liu et al., 2005) and for normal stomatal movement (Jung et al., 2002). PIs affect the location and function of proteins with specific PI-binding domains, such as pleckstrin homology (PH), phox homology (PX), FYVE finger, and other motifs. In plants, several proteins have shown to interact with PIs, and these include certain PLDs (Pappan et al., 1997; Zheng et al., 2002) and Patellin1, a novel Sec14-like protein, localized to the cell plate (Peterman et al., 2004). PI(4, 5)P_2 is a required activator of PLDβ, γ, and ζ, and also stimulates PLDα1 activity (Pappan et al., 2004; Qin and Wang 2002). PLDζs contain two PI-binding domains, PH and PX, whereas of PLDβ has been shown to bind to PIs through a polybasic residue-motif (Zheng et al., 2002). The role of PLD in osmotic stress-induced production of PA and ABA signaling is described earlier. Thus, it is possible that increases in PIs under hyperosmotic stress contribute to the activation of PLD and PA production in the stress response.

In addition to be a mediator on its own, PI(4, 5)P_2 is the substrate of PLC that produces DAG and IP_3. While osmotic stress-induced DAG has been shown to be rapidly converted to PA, IP_3 accumulates in salt, cold, and osmotically stressed plants (Smoleńska and Kacperska 1996; Takahashi et al., 2001; Xiong et al., 2001). IP_3 promotes an increase in Ca^{2+} in guard cells and stomatal closure, and suppression of a PLC reduces ABA-promoted closure of stomata (Hunt et al., 2003).

The signaling process of PIs and the derivative inositol polyphosphates is terminated through the action of PI phosphatases and inositol polyphosphate phosphatases. Perturbation of these phosphatases by either overexpression or ablation affects the expression of stress-responsive genes under salt, drought, and ABA treatment and alters plants stress tolerance. At5PTase1 is up-regulated in response to ABA and has bee suggested to act as a signal terminator of ABA signaling (Burnette et al., 2003). FRY1 encodes an inositol polyphosphate 1 phosphatase, and fry1 mutant plants had elevated levels of IP_3 after ABA treatment. The mutant plants are compromised in tolerance to freezing, drought, and salt stresses (Xiong et al., 2001). Knockout of the suppressor of actin mutation (SAC) domain phosphatase results in elevated levels of PI(4, 5)P_2 and IP_3 as compared to wild-type plants under unstressed conditions (Williams et al., 2005). The *sac9* mutants display constitutive stress response, including closed stomata, anthocyanin accumulation, overexpressing stress-induced genes, and accumulating reactive-oxygen species. The results indicate that cellular levels of PIs and inositol polyphosphates are tightly regulated to achieve optimal plant performance under stress.

5. PERSPECTIVES

Drought and salinity are the two crucial environmental factors that limit plant growth, productivity, and geographic distribution. Lipid-mediated signaling plays important roles in plant responses to these stresses. Information is growing rapidly on the production of potential lipid mediators under different growth conditions. However, the current knowledge is still rather limited on the exact role of specific lipid signaling reactions in plant adaptation to drought and salt stress.

Genetic manipulation of enzymes that produce lipid mediators provides valuable insights into the function of specific lipid signaling reaction. The distinguishable phenotypes resulting from the loss of different PLDs indicate that some PLDs have unique functions in plant response to different forms of osmotic stresses. Such distinctions may be related to the location and timing of PA production regulated by different PLDs. Spatial and temporal regulation is important to all signaling events, but it is particularly critical to intracellular lipid messengers because of their limited mobility in the cell. Similarly, distinguishable functions are expected to occur for other phospholipid-signaling enzyme families, such as PLCs, DGKs, PI kinases, and phosphatases. Therefore, it is important to identify specific gene and enzymes involved when the roles of given type of lipid signaling enzymes in a specific physiological response are addressed. Such information will also be necessary for potential biotechnological manipulation of lipid signaling pathways to improve plant stress tolerance.

ACKNOWLEDGMENTS

We thank Eric Weldon for critical reading of the manuscript. Work from authors' laboratory was supported by grants from the National Science Foundation (IBN-0454866) and the U.S. Department of Agriculture (2005-35318).

REFERENCES

Anthony, R. G., Henriques, R., Helfer, A., Meszaros, T., Rios, G., Testerink, C., Munnik, T., Deak, M., Koncz, C., and Bogre, L, 2004, A protein kinase target of a PDK1 signalling pathway is involved in root hair growth in Arabidopsis, *EMBO J.***23:** 572–581.

Bohme, K., Li, Y., Charlot, F., Grierson, C., Marrocco, K., Okada, K., Laloue, M., and Nogue, F., 2004, The Arabidopsis COW1gene encodes a phophatidylinositol transfer protein essential for root hair tip growth, *Plant J.***40**: 686–698.

Burnette, R. N., Gunesekera, B. M., and Gillaspy, G. E., 2003, An Arabidopsis inositol 5-phosphatase gain-of-function alters abscisic acid signaling. *Plant Physiol.***132:** 1011–1019.

Chapman, K. D., 2004, Occurrence, metabolism, and prospective functions of N-acylethanolamines in plants. *Prog. Lipid Res.* **43**: 302–327.

den Hartog, M., Verhoef, N., and Munnik, T., 2003, Nod factor and elicitors activate different phospholipid signaling pathways in suspension-cultured alfalfa cells. *Plant Physiol.* **132**: 311–317.

Fang, Y., Vilella-Bach, M., Bachmann, R., Flanigan, A., and Chen, J., 2001, Phosphatidic acid-mediated mitogenic activation of mTOR signaling. *Science*.**294**: 1942–1945.

Huang, S., Gao, L., Blanchoin, L., and Staiger C. J., 2006, Heterdimeric capping protein from arbidopsis is regulated by phosphatidic acid. *Mol. Biol. Cell*.**4**: 1946–1958.

Hunt, L., Mills, L. N., Pical, C., Leckie, C. P., Aitken, F. L., Kopka, J., Mueller-Roeber, B., McAinsh, M. R., Hetherington, A. M., and Gray J. E., 2003, Phospholipase C is required for the control of stomatal aperture by ABA. *Plant J*.**34:** 47–55.

Jung, J. Y., Kim Y. W., Kwak, J. M., Hwang, J. U., Young, J., Schroeder, J. I., Hwang. I., and Lee, Y., 2002, Phosphatidylinositol 3- and 4-phosphate are required for normal stomatal movements. *Plant Cell*. **14:** 2399–2412.

Katagiri T, Takahashi S, Shinozaki K (2001) Involvement of a novel *Arabidopsis* hospholipase D, AtPLDδ, in dehydration-inducible accumulation of phosphatidic acid in stress signalling. *Plant J.* **26**: 595–605.

Kooijman, E. E., Chupin, V., Fuller, N. L., Kozlov, M. M., de Kruijff, B., Burger, K. N., and Rand, P. R., 2005, Spontaneous curvature of phosphatidic acid and lysophosphatidic acid. *Biochemistry*. **44**:2097–2102.

Li, W., Li, M., Zhang, W., Welti, R., and Wang, X., 2004, The plasma membrane-bound phospholipase Dδ enhances freezing tolerance in Arabidopsis thaliana. *Nature Biotechnol*.**22**: 427–433.

Liu, K., Li, L., and Luan, S., 2005, An essential function of phosphatidylinositol phosphates in activation of plant shaker-type K+ channels. *Plant J.* **42:** 433–443.

Loewen, C.J., Gaspar, M. L., Jesch, S. A., Delon, C., Ktistakis, N. T., Henry, S. A., and Levine, T. P., 2004, Phospholipid metabolism regulated by a transcription factor sensing phosphatidic acid. *Science*.**304**: 1644–1647.

Mahfouz, M. M., Kim, S., Delauney, A. J., and Verma, D. P., 2005, Arabidopsis TARGET OF RAPAMYCIN interacts with RAPTOR, which regulates the activity of S6 kinase in response to osmotic stress signals. *Plant Cell*.**18**:477–490.

Meijer, H. J., and Munnik, T., 2003, Phospholipid-based signaling in plants. *Annu. Rev. Plant Biol*.**54**: 265–306.

Mishra, G., Zhang, W., Deng, F., Zhao, J., and Wang X., 2006, A bifurcating pathway directs abscisic acid effects on stomatal closure and opening in Arabidopsis. *Science*.**312:** 264–266.

Monks, D. E., Aghoram, K., Courtney, P. D., DeWald, D. B., Dewey, R. E., 2001, Hyperosmotic stress induces the rapid phosphorylation of a soybean phosphatidylinositol transfer protein homolog through activation of the protein kinases SPK1 and SPK2. *Plant Cell*. **13**: 1205–1219.

Pappan K, Zheng L, Krishnamoorthi R, Wang X., 2004, Evidence for and characterization of Ca^{2+} binding to the catalytic region of Arabidopsis thaliana phospholipase Dβ. *J. Biol. Chem*.**279**: 47833–47833.

Peterman, T. K., Ohol, Y. M., McReynolds, L. J., and Luna, E. J., 2004, Patellin 1, a novel Sec14-like protein, localizes to the cell plate and binds phosphoinositides. *Plant Physiol.* **136**: 3080–3094.

Ryu, S. B., 2004. Phospholipid-derived signaling mediated by phospholipase A in plants. *Trends Plant Sci.* **9**: 229–235.

Sang, Y., Zheng, S., Li, W., Huang, B., and Wang, X., 2001, Regulation of plant water loss by manipulating the expression of phospholipase Dα. *Plant J.* **28**: 135–144.

Smoleńska, G., and Kacperska, A., 1996, Inositol 1, 4, 5-trisphosphate formation in leaves of winter oilseed rape plants in response to freezing, tissue water potential and abscisic acid. *Physiol Plant.* **96**: 692–698.

Takahashi, S., Katagiri, T., Hirayama, T., Yamaguchi-Shinozaki, K., and Shinzaki, K., 2001, Hyperosmotic stress induces a rapid and transient increase in inositol 1, 4, 5-trisphosphate independent of abscisic acid in Arabidopsis cell culture. *Plant Cell Physiol.* **42**: 214–222.

Testerink, C., Dekker, H. L., Lim, Z. Y., Johns, M. K., Holmes, A. B., Koster, C.G., Ktistakis, N.T., and Munnik, T., 2004, Isolation and identification of phosphatidic acid targets from plants. *Plant J.* **39**:527–536.

Testerink, C., and Munnik, T. 2005. Phosphatidic acid: a multifunctional stress signaling lipid in plants. *Trend Plant Sci.* **10**: 368–375.

Thiery L, Leprince AS, Lefebvre D, Ghars MA, Debarbieux E, Savoure A (2004) Phospholipase D is a negative regulator of proline biosynthesis in Arabidopsis thaliana. *J. Biol. Chem.*279: 14812–14818.

van Shooten, B., Testerink, C., and Munnik, T., 2006. Signalling diacylglycerol pyrophosphate, a new phosphatidic acid metabolite. *Biochim. Biophys. Acta.* **1761**: 151–159.

Vincent, P., Chua, M., Nogue, F., Fairbrother, A., Mekeel, H., Xu, Y., Allen, N., Bibikoba, T. N., Gilroy, S., and Bankaitis, V. A., 2005, A Sec14p-nodulin domain phosphatidylinositol transfer protein polarizes membrane growth of Arabidopsis thaliana root hairs. *J. Cell Biol.* **168**: 801–812.

Wang, X., 2004, Lipid signaling. *Curr. Opin. Plant Biol.* **7**: 329–336.

Wang, X, Devaiah, S.D., Zhang, W., and Welti, R. 2006, Signaling functions of phosphatidic acid. *Prog. Lipid Research* **45**: 250–278.

Williams, M.E., Torabinejad, J, Cohick, E., Parker, K., Drake, E. J., Thompson, J. E., Hortter, M., and Dewald, D. B., 2005, Mutations in the Arabidopsis phosphoinositide phosphatase gene SAC9 lead to overaccumulation of PtdIns(4, 5)P2 and constitutive expression of the stress-response pathway. *Plant Physiol.* **138**: 686–700.

Xie, Z., Fang, M., Rivas, M. P., Faulkner, A. J., Sternweis, P. C., Engebrecht, J. A., and Bankaitis, V. A., 1998, Phospholipase D activity is required for suppression of yeast phosphatidylinositol transfer protein defects. *Proc. Natl. Acad. Sci. USA.* **95**:12346–12351.

Xiong L., Lee B., Ishitani M., Lee H., Zhang C., and Zhu J-K., 2001, FIERY1 encoding an inositol polyphosphate 1-phosphatase is a negative regulator of abscisic acid and stress signaling in Arabidopsis. *Genes Dev.* **15**:1971–1984.

Zhang, W., Qin, C., Zhao, J., and Wang, X., 2004, Phospholipase Dalpha1-derived phosphatidic acid interacts with ABI1 phosphatase 2C and regulates abscisic acid signaling. *Proc. Natl. Acad. Sci. USA.* **101**: 9508–9513.

Zhang Y., Wang L., Liu Y., Zhang Q., Wei Q., and Zhang W., 2006, Nitric oxide enhances salt tolerance in maize seedlings through increasing activities of proton-pump and Na^+/H^+ antiport in the tonoplast. *Planta*, Epub ahead of print

Zhang, W., Yu, L., Zhang, Y., and Wang, X., 2005, Phospholipase D in the signaling network of plant responses to abscisic acid and reactive oxygen species, *Biochim. Biophys. Acta* **1736**:1–9.

Zhao, J., and Wang, X., 2004, Arabidopsis phospholipase Dα1 interacts with the heterotrimeric G-protein a subunit through a motif analogous to the DRY motif in G-protein-coupled receptors. *J. Biol. Chem.* **279**: 1794–1800.

Zheng, L., Shan, J., Krishnamoorthi, R., and Wang, X., 2002, Activation of plant phospholipase Dβ by phosphatidylinositol 4, 5-bisphosphate: characterization of binding site and mode of action. *Biochemistry.* **41**: 4546–4553.

Zonia, L., and Munnik, T., 2004, Osmotically induced cell swelling versus cell shrinking elicits specific changes in phospholipid signals in tobacco pollen tubes. *Plant Physiol.* **134**: 813–823.

CHAPTER 9

ABSCISIC ACID IN PLANT RESPONSE AND ADAPTATION TO DROUGHT AND SALT STRESS

LIMING XIONG

Donald Danforth Plant Science Center, 975 N. Warson Road, St. Louis, MO 63132, USA

Abstract: The plant stress hormone abscisic acid (ABA) plays several critical roles in plant response to stress and stress tolerance. ABA is well studied for its roles in the activation of stress-responsive genes and the regulation of guard cell movement. More recently, ABA has also been demonstrated to regulate root adaptation to drought stress. To date, limited success has been achieved in regulating plant ABA action for increasing plant drought tolerance. Revealing the mechanisms of ABA action in stress adaptation will further help the development of hardy crop plants

Keywords: ABA; drought stress; drought tolerance; salt stress; root development

1. INTRODUCTION

Abscisic acid (ABA) is one of the five classic growth regulators that play critical roles in plant growth and development. The chemical identity of this sesquiterpene (C_{15}) was established in the 1960's through the landmark work of several research groups while identifying the compounds responsible for leaf abscission and bud dormancy (reviewed in Zeevaart and Creelman, 1988). It is now known that ABA does not play major roles in these processes for which it was named. ABA does have essential roles in seed maturation and dormancy. However, the critical roles it plays during plant response to environmental stress won its reputation as a 'stress hormone' and as such ABA is central to any discussion of plant adaptation to adverse environmental conditions. This chapter will focus on the role of ABA in plant response and adaptation to drought and salt stresses. Other aspects of ABA biology such as ABA biosynthesis and ABA signal transduction will be discussed only in the context of their involvement in stress response. ABA biosynthesis (Schwartz et al., 2003; Nambara and Marion-Poll, 2005; Taylor et al., 2005) and ABA signal transduction (Leung and Giraudat, 1998; Finkelstein and Rock, 2002; Himmelbach et al., 2003; Christmann et al., 2003; Xie et al., 2006) are discussed in more detail in these recent reviews.

M.A. Jenks et al. (eds.), Advances in Molecular Breeding Toward Drought and Salt Tolerant Crops, 193–221.

2. MAJOR PHYSIOLOGICAL PROCESSES REGULATED BY ABA

The function of ABA in various physiological processes was mainly inferred from studies using exogenous ABA as well as plant mutants defective either in ABA biosynthesis or responses. Other approaches such as examining the spatial and temporal localization of ABA were sometimes also used to speculate about cellular processes that may be regulated by ABA. These approaches may not be able to uncover all functions of ABA. Certain cellular activities may only need such a low level of ABA that even ABA deficient or insensitive mutants can satisfy the needs since all these mutants are leaky (i.e., they either still produce a limited amount of ABA or show some response to ABA). Furthermore, cellular machinery tends to be more sensitive to ABA in case of ABA deficiency (e.g., Xiong et al., 2001). The negative impacts of ABA deficiency in these biosynthetic mutants might have been significantly alleviated. It is thus likely that there are still undiscovered cellular processes that require ABA or that ABA may be essential for plant growth and development. In the latter scenario, complete lack of ABA or ABA signaling (e.g., loss of all ABA receptors) would be lethal. This perhaps is one of the reasons why the ABA receptor(s) was not identified in genetic screens. The significantly reduced vigor of all ABA deficient mutants under normal unstressed conditions also implies that ABA is required for certain cellular processes that may not be related to stress adaptation. Here we will briefly discuss some of the physiological processes that are regulated by ABA.

ABA is widely known for its involvement in seed maturation and dormancy. Seed maturation requires the synthesis of storage proteins and the preparation for desiccation tolerance so that the embryo will remain viable under extreme dehydration conditions as seen with dry seeds (McCarty, 1995; Finkelstein and Rock, 2002; Bentsink and Koornneerf, 2002). ABA activates genes involved in both processes, although seed maturation may require an even lower level of ABA than do dormancy initiation and maintenance. The essentiality of ABA in seed development is witnessed by the observation that once ABA level was suppressed immunologically the embryos will not be able to acquire desiccation tolerance and thus are not viable when desiccated (Phillips et al., 1997). On the other hand, the role of ABA in seed dormancy can readily be seen by the pre-mature (viviparous) germination of maize ABA deficient mutants and by the inhibition of exogenous ABA on seed germination.

Another developmental process that is potentially affected by ABA is the vegetative to reproductive phase transition. It is well known that plants under moderate drought stress will flower earlier (a 'drought escape' strategy). However, the relationship of flowering time with drought susceptibility is complex. When using the carbon isotope discrimination ratio ($\delta^{13}C$) as an index of water use efficiency (WUE), it was found that $\delta^{13}C$ is positively correlated with flowering time in several plant species (Araus et al., 2002; Farquhar et al., 1989). However, it is suspicious that late-flowering ecotypes really have a high WUE. Generally, plants (and in particular ecotypes within a plant species, such as Arabidopsis) that are late flowering usually grow more slowly than rapidly flowering plants.

Slow-growing plants consume less water at a given period of time. Since drought stress is well known to promote flowering, one may expect that ABA would promote flowering too. However, Arabidopsis ABA-deficient mutant *aba1* and insensitive mutant *abi1* (Landsberg background) tend to flower earlier under short-day conditions (Martinez-Zapater et al., 1994) while the *aba1/los6* mutant (C24 background) flowered earlier both under long-day and short-day conditions (Xiong et al., 2002). On the other hand, ABA-hypersensitive mutants are either early flowering (*sad1; abh1-2*, C24 ecotype; *chl1*, L. Xiong, unpublished) or late flowering (*hyl1*, Lu and Fedoroff, 2000; *era1*, *fry1*, L. Xiong, unpublished). It is not known whether or not the flowering time changes in these mutants have to do with altered ABA responses. Recently, it was reported that prolonged exogenous ABA treatment in Arabidopsis up-regulates the flowering-suppressing gene *FLC* and delays flowering (Razem et al., 2006). Yet, short-term ABA treatments do not affect *FLC* gene expression or flowering time in Arabidopsis (H. Chen and L. Xiong, unpublished). Nonetheless, the facts that the flowering time regulator FCA can bind to exogenous ABA (Razem et al., 2006) and that ABA is detected in floral primordia (e.g. Peng et al., 2006) suggest that ABA may affect floral development and flowering time regulation.

In vegetative tissues, ABA plays a role in several developmental processes. It was noted that ABA may affect root hair cell patterning through the regulation of *GLABRA (GL) 2* and other cell patterning genes (van Hengel et al., 2004). The role of ABA in root hair patterning and root hair growth may relate to the observation that drought stress and ABA treatments result in abnormal root hair development (Schnall and Quatrano, 1992) (see below). While studying plant response to drought stress, we found that drought stress inhibits the elongation of lateral roots both in artificial growth media and in soil. ABA has similar effects in inhibiting lateral root elongation both in Arabidopsis and in crop plants (Xiong et al. 2006; L. Xiong, unpublished). Observations on osmotic stress and ABA inhibition of lateral root growth were also reported under different experimental conditions (De Smet et al., 2003; Deak and Malamy, 2005). Thus, it is likely that this rhizogenesis process may represent an adaptive response to drought stress, as was demonstrated in recent genetic studies (Xiong et al., 2006; see below).

ABA has also been implicated in plant response to wounding, ozone, light, and pathogens. This may partly result from the fact that these various stress-signaling pathways may share some common components. Some of these stresses may also give rise to osmotic or desiccation stress that indirectly activates the ABA signaling pathways. For example, the hypersensitive response of pathogenesis in plant cells could result in a significant osmotic stress (Wright and Beattie, 2004). On the other hand, the antagonistic role of ABA on disease resistance may have to do with ABA suppression of salicylic acid, jasmonic acid and ethylene signaling (Mauch-Mani and Mauch, 2005).

Perhaps the best known and also the most studied process that ABA is involved in is plant response to abiotic stress such as drought, salt, and cold stress. These different stresses do share some common features. For example, they all induce dehydration stress to the plant cells. Accordingly, these abiotic stresses all activate ABA biosynthesis to various extents and induce a common set of stress responsive genes.

3. REGULATION OF CELLULAR ABA LEVELS AND THE EXPRESSION OF STRESS-RESPONSIVE GENES

The magnitude of cellular response to ABA is determined by ABA level and ABA sensitivity, and both aspects may involve complex signal transduction processes. Cellular ABA levels are dynamically regulated through biosynthesis/degradation, conjugation/de-conjugation, compartmentalization, and transport. Signal transduction from ABA perception to gene activation involves complex regulatory circuits (network) and multiple components. Here we briefly introduce some of the research on the regulation of ABA levels and ABA activated gene expression. ABA signal transduction will not be discussed here.

3.1. ABA Biosynthesis and Catabolism

In higher plants, ABA is mainly produced by the cleavage of a C_{40} carotenoid precursor that initially occurs in the plastids but the final steps of the pathway occur in the cytosol. This 'indirect' pathway is now well understood and the major enzymes catalyzing these reactions were identified (Schwartz et al., 2003; Nambara and Marion-Poll, 2005). Several of the enzymes in this pathway are encoded by single copy genes in the Arabidopsis genome (e.g., *ZEP*, *ABA3* and perhaps *ABA2*). Surprisingly, null mutations in these single copy genes do not completely block ABA biosynthesis, implying that there are either certain shortcuts in the major pathway that can circumvent these steps or minor pathways that can still produce a limited amount of ABA.

ABA biosynthesis is regulated both by internal developmental cues as well as by environmental stresses. The regulation of the biosynthetic pathway largely occurs at the level of transcriptional regulation of the biosynthetic genes, although regulation at other levels is also possible (Xiong and Zhu, 2003). Furthermore, ABA can self-regulate its own biosynthesis in that all of the ABA biosynthetic genes can be up-regulated to various extents by ABA (Xiong and Zhu, 2003).

The temporal and spatial regulation of ABA biosynthesis constitutes another layer of control of ABA biosynthesis. The tissue-specific regulation of ABA biosynthesis is particularly interesting since it may reveal the sites of ABA action, although ABA and perhaps its precursors as well can be transported over long distance. Among the known ABA biosynthetic genes, several of them appear to be expressed ubiquitously in those major tissues examined (Xiong and Zhu, 2003). However, the short-chain dehydrogenase/reductase gene *ABA2* (González-Guzmán et al., 2002; Cheng et al., 2002) is mainly expressed in vascular tissues of roots, stems, and leaves (Cheng et al., 2002), suggesting that these places are probably the key places of ABA synthesis. Similar expression in vascular tissues and guard cells was also reported for the ABA aldehyde oxidase 3 (*AAO3*) gene and protein (Koiwai et al., 2004). Using ABA-inducible promoters of the *RD29B* and *AtHB6* genes to drive the expression of the firefly luciferase reporter gene, Christmann et al. (2005) found that these promoters are mainly activated at vascular tissues and guard cells upon drought stress treatment, implying that these sites might be the sites of ABA biosynthesis or ABA response. To determine the actual sites of ABA synthesis and accumulation, more direct approaches to detect ABA *in situ* need to be developed.

ABA can be conjugated with glucose to produce ABA-glucose ester (ABA-GE). This conjugate may not have biological activities similar to ABA. Glycosylation of ABA could thus serve as a regulatory process to inactivate ABA. Nevertheless, ABA can be released from ABA-GE by the hydrolytic enzyme β-glucosidase. These extracellular enzymes in leaves were shown to be able to release ABA from ABA-GE transported from xylem sap (Dietz et al., 2000). The ratio of ABA-GE to free ABA in xylem sap is low, yet ABA-GE can account for the majority of total ABA in older leaves (Weiler, 1980). Like free ABA, the level of conjugated ABA also increases in response to drought and salt stress (see Sauter et al., 2002 and references therein). Thus, conjugated ABA is still originated from de novo biosynthesis and its level may be kept in balance with free ABA.

The significant amount of conjugated ABA accumulated under drought stress may serve important roles in guard cell regulation and stress adaptation. To understand the role of these conjugates in drought stress adaptation requires the identification of the specific glucosidase(s) that hydrolyze these conjugates. Genetic approaches to knock out or over-express these genes can then be used to ascertain the contribution of ABA-GE in stress response and stress tolerance. However, genes encoding ABA-GE glucosidases have not been identified. Plant genomes encode a family of β-glucosidase genes (about 50 in the Arabidopsis genome). It would be interesting to identify the particular glucosidases that regulate ABA levels during drought stress. Proteomics analysis of drought-treated plants often found that the level of β-glucosidase proteins increased (e.g., Riccardi et al., 1998). More alkaline pH in the xylem sap under drought stress may also increase the activity of these enzymes (Sauter et al., 2002). These drought and ABA up- regulated glucosidases are likely candidates for ABA-GE glucosidases.

Upon stress-relief, ABA level can quickly decrease to pre-stress level. Catabolism of ABA is enhanced by both stress-relief signal and by ABA itself. The major enzyme responsible for the early step of the catabolism is ABA 8-hydroxylase (e.g., Krochko et al., 1998). Genes encoding these enzymes were identified (Kushiro et al., 2004; Saito et al., 2004). Regulation of these genes may alter ABA levels and plant drought response.

3.2. Gene Gegulation by ABA

In the decade between late 80's to early 90's, the discovery that stress and ABA can up-regulate gene expression inspired a great interest in the isolation and characterization of these stress and ABA-responsive genes and their promoter *cis* elements as well as transcription factors responsible for their activation (e.g., Skriver and Mundy, 1990; Finkelstein and Rock, 2002; Busk and Pages, 1998). To date, genome-wide microarray analysis has made possible to identify all the genes regulated by ABA. Genes regulated by ABA amounted to several hundreds to a few thousands in Arabidopsis dependent on the particular treatment conditions and definition of regulation. Many of these genes are also regulated by drought, salt and low temperature stresses. Thus, relatively few genes are specifically regulated by ABA, implying a common role of ABA in plant stress response. In Arabidopsis, up to 30 percent of the genes can be regulated by abiotic stress (Kreps et al., 2002). The large number of genes up-regulated under stress suggests that a significant re-programming of cellular activities occurs when plants encounter the stress challenge.

ABA-responsive genes fall into a diverse range of functional groups that are categorized in different ways (Ramanjulu and Bartels, 2002; Leung and Giraudat, 1998; Finkelstein and Rock, 2002, Bray, 1997). These may include genes encoding enzymes that function in the production of compatible solutes, antioxidants, and genes encoding peptides with unclear functions as well as genes encoding signal transduction components. Among them, two related groups of ABA regulated genes encoding peptides of unclear functions received intensive research. These are the *COR/RD/LT* group stress-responsive genes and the late embryogenesis abundant (LEA) genes. Although they are expressed at different developmental stages and in different tissues, they share many commonalities and have similar functions: to preserve the cells from dehydration damages. These genes may also be activated by similar mechanisms.

Many of these genes share similar regulatory elements in the promoter regions. Early work identified the ABA-responsive element, ABRE, in the EM genes (encoding LEA proteins) in wheat and rice as a core sequence containing ACGT. This *cis* element is responsible for ABA up-regulation of these EM genes (Marcottee et al., 1989; Guiltinan et al., 1990; Mundy et al., 1990). In addition, coupling elements work together with ABRE in conferring ABA response in these ABA responsive promoters (Shen and Ho, 1995). Other *cis* elements such as MYC and MYB binding sequences are found in other ABA-responsive genes (Shinozaki et al., 2003). In vegetative tissues several basic domain/Leu zipper (bZip) transcription factors (AREB/ABF) that bind to ABRE were later isolated and they confer ABA induction of many stress-responsive genes (reviewed in Finkelstein and Rock, 2002; Rock, 2000; Xie et al., 2005). In contrast, the AP2/ERF transcription factors DREB/CBF that bind to the DRE/CRT element are responsible for dehydration-induced gene expression (Shinozaki et al., 2003). Both the ABRE and DRE/CRT *cis* elements exist in the promoters of many stress- and ABA-responsive genes. Although DRE and ABRE elements have different core sequences and their own binding factors, both pathways collaboratively activate gene expression. Furthermore, stress induction of certain stress-responsive genes often requires ABA (Xiong et al., 2001; Kizis and Pages, 2002; Narusaka et al., 2003). Although the modes of action for many of the ABA and stress-responsive gene products are unknown, overexpression of these ABA-activated genes or their upstream transcription factor genes was found to confer increased tolerance of the transgenic plants to drought and other abiotic stresses (reviewed in Seki et al. 2003, Bajaj et al., 1999; Chinusamy et al., 2005, Vinocur and Altman, 2005; Umezawa et al., 2006). These experiments demonstrated that these ABA responsive genes do play important roles in plant stress tolerance.

4. ROLE OF ABA IN DROUGHT STRESS RESPONSE AND ADAPTATION

Drought stress is often caused by prolonged water shortage in the soil that could not meet plant transpiration demand. Different plant species may adopt different strategies to deal with drought stress. Researchers sometimes divide plant drought adaptation into several categories (Levitt, 1980): 'drought escape' (shortening life cycle by flowering earlier), 'drought avoidance' (growing deeper roots, depositing leaf wax, and closing stomata), and 'drought tolerance' (production of osmolytes, antioxidants, and other stress-relieving agents). These terms are sometimes confusing and here we loosely define drought tolerance as the ability of plants to withstand water deficit while maintaining appropriate physiological activities.

For a given plant species, one can image that plants will have three ways to deal with drought challenges:to reduce water consumption, to increase water uptake, and to mitigate the negative impacts of water deficit. These tasks are accomplished in several ways. First, guard cell stomatal pores are closed upon drought stress and thus the transpirational water loss is minimized. This is a relatively quick response. Second, an array of stress-responsive genes is activated. The products of these genes function directly or indirectly in drought tolerance (see Section 3.2). Third, in a longer term there are certain developmental changes that may make the plants more adaptive to drought stress. These changes may occur in root development, phase transition, wax deposition, guard cell patterning and perhaps leaf morphology (for newly emerged leaves). ABA, whose level is up-regulated by drought stress (Section 3.1), is either required or is involved in all these processes.

4.1. Guard Cell Regulation

Water potential-driven influx of water in guard cells swells these cells and opens the stomatal pore. Likewise, efflux of water from guard cells shrinks the cells and closes the stomatal pore. Water potentials in these cells in turn are regulated by ion and solute fluxes through respective channels and transporters. The activities of these channels and transporters are regulated by a number of factors such as light, CO_2, and ABA (Schroeder et al., 2001; Luan, 2002; Fan et al., 2004; Hetherington and Woodward, 2003). Thus, the opening and closing of stomata are also controlled by light, CO_2, drought and salt stress. Among them, light and CO_2 regulation of stomata may be independent of ABA. Here we confine our discussion to guard cell regulation by ABA.

ABA regulation of guard cell ion channels is the major basis of ABA regulation of stomatal closing. The dogma of this regulation is that ABA induces a transient increase in cytosolic Ca^{2+} that in turn inhibits plasma membrane proton pumps and inward K^+ channels and also activates anion channels that lead to the release of anions from the guard cells. Anion efflux-induced depolarization activates outward K^+ channels and leads to K^+ efflux as well (reviewed in Schroeder et al., 2001; Fan et al., 2004; Pei et al., 2006). Reduced osmolarity in guard cells thus leads to water efflux and stomata closure. Although the molecular identities of these channels involved in guard cell movement have not been identified, several transporters were shown to affect stomata response to ABA and drought stress (Hosy et al., 2003; Klein et al., 2003). In these processes, ABA induced transient increases of Ca^{2+} was suggested as an early event that regulates subsequent gating of other channels. Since ABA itself has not been suggested to bind to any ion channels and regulate their activities, ABA regulation of ion channels will involve certain intermediate molecules and additional signal transduction processes.

By using various electrophysical, fluorescence imaging, pharmacological, biochemical, molecular, and genetic approaches, several intermediate molecules or second messengers that induce internal Ca^{2+} release in animal cells are also found to mediate ABA-induced Ca^{2+}release in isolated guard cell protoplasts, membrane vesicles, or guard cells. These include inositol phosphates [inositol 1, 4, 5-trisphosphate (IP_3) and inositol 1, 2, 3, 4, 5, 6-hexaphosphate (IP_6)], phosphatidic acid (PA), cADPR (cyclic adenosine 5'-diphosphoribose), NAADP (nicotinic acid adenine dinucleotide phosphate), NO (nitric oxide), H_2O_2, and sphingosphine 1-phosphate (Schroeder 2001; Pei, 2006; Fan et al., 2004). Many of the studies were based on analysis of the currents from unknown channels in isolated protoplasts or membrane vesicles that were treated with exogenous compounds; it is not entirely clear whether all these molecules are the endogenous second messengers in living guard cells (Levchenko et al., 2005). However, the fact that mutations in some of the enzymes involved in the generation of these second messengers were found to affect ABA regulation of stomatal movement and plant drought tolerance suggests that they may indeed play roles in ABA and drought stress responses in planta (e.g., Staxen et al., 1999; Kwak et al., 2003; Zhang et al., 2004). In particular, stomatal regulation requires phosphatidic acid (PA) produced by PLDα1 and PA could bind to ABI1 as well as the heterotrimeric G protein α subunit GPA1 to induce stomatal closure and to inhibit opening in response to ABA (Mishra et al., 2006). Thus, PA may provide a link between ABI1 and GPA1 that are previously demonstrated to regulate stomatal response to ABA (Murata et al., 2001; Wang et al., 2001).

ABA regulation of stomatal closure may also have Ca^{2+}-independent routes, although these routes are not well characterized. Using florescence dyes and in vivo imaging techniques, Levchenko et al. (2005) reported that cytosolic ABA activation of guard cell anion channels does not involve ABA-induced Ca^{2+}transients, although a basal level of $Ca^{2+}$is required. Likewise, the intermediate molecules of ABA action such as IP_3, IP_6, NAADP, and cADPR were not able to mimic ABA in the activation of the anion channels. It is unclear whether the anion channels responsible for the currents observed in this study are the same kind of channels referred in previous studies that require transient Ca^{2+} to activate. In previous patch-clamping studies of guard cells, ABA was shown to be able to activate anion channels that contribute to the closure of stomatal guard cells. The signaling pathway for this activation is unclear, but it appears that components requiring farnesylation may negatively regulate the process since in *era1* mutant, which has a mutation in the β subunit of a farnesyl transferase, the activation of S-type anion channels by ABA was enhanced. This facilitates the closure of stomata and leads to increased drought tolerance of the *era1* mutant (Pei et al., 1998).

Stomatal regulation by ABA occurs fairly quickly (with only a few minutes' lag period for channel regulation by exogenous ABA). Thus, protein posttranslational modification is likely the major mode of action for ABA signaling in stomatal regulation. Several protein phosphatases (such as the 2C type and 2A type) and kinases are involved in ABA regulation of guard cells (Finkelstein and Rock, 2002; Xie et al., 2006). ABI1 and ABI2 are well-studied phosphatases affecting guard cell ABA response. The *abi1*-1 mutation impairs ABA-induced K^+ currents (Armstrong et al., 1995). It was shown that ABI1 was impaired in ABA-induced production of reactive oxygen species (ROS) where ABI2 was impaired in steps downstream of ROS production but before the activation of anion channels (Murata et al., 2001). Similar to *abi1*-1, mutations in the OST1 kinase also impair ABA-induced ROS production (Mustilli et al., 2002). Accordingly, guard cells of the *ost1/snrk2* mutants are insensitive to ABA and the mutants exhibited a wilty phenotype (Yoshida et al., 2002 and Mustilli et al., 2002). Interestingly, OST1 was found to interact with ABI1 and the activation of OST1 by ABA was impaired in the dominant *abi1*-1 mutant (Yoshida et al., 2006). These observations suggest that this pair of phosphoproteins may constitute an auto-regulated module in mediating ABA-induced stomata movement.

The control of ion channel gating is a late step of guard cell ABA response. There are other events preceding the gating of these channels. Regulation of membrane proteins including ion channels and transporters often requires vesicle trafficking, targeting and fusion. The swelling and shrinking of guard cells as well as the dynamics of channels and transporters may evoke significant vesicle trafficking. Several components potentially involved in vesicle trafficking have been showed to affect guard cell ABA responses. These include for example, small GTPase (e.g., AtRac1, see Lemichez et al., 2001; Rop10, see Zhang et al., 2001) and SNARE proteins (e.g., OSM1/SYP61 see Zhu et al., 2002; NtSyr1, see Leyman et al., 1999). Some of these GTPases requires prenylation in order to target to the plasma membrane. They are likely the targets of the farnesyl transferases that include the β subunit ERA1 (enhanced response to ABA 1) (Culter et al., 1996). The *osm1/syp61* (osmotic stress-sensitive mutant 1/syntaxin 61) mutant exhibited reduced sensitivity to ABA in guard cell response (but not in seed germination) and increased sensitivity to salt and osmotic stresses (Zhu et al., 2002). In another study, the tobacco syntaxin protein AtSyp1 was found to mediate ABA-induced Cl^- flux in oocytes (Leyman et al., 1999). ABA treatment inactivates the Rho-like small GTPase AtRAC1/ARAC3/Rop6 and results in the disruption of actin fiber in guard cells. These events precede the stomatal closure induced by ABA. In the ABA-insensitive mutant *abi1*-1, ABA was unable to inactivate AtRAC1 and also failed to reorganize the actin cytoskeleton (Lemichez et al., 2001).

In addition to channel regulation, transcription regulation of genes may also contribute to stomatal regulation. Recently, it was found that a R2R3 type Myb transcription factor AtMYB60 affects stomata regulation (Cominelli et al., 2005). The *myb60* mutant showed constitutively a smaller stomata aperture. Interestingly, the *MYB60* gene was down-regulated by ABA and drought stress, suggesting that regulation of this gene may contribute to ABA and drought-induced stomatal closure. However, in the *myb60* mutant, only a limited number of genes were moderately down regulated. It is unclear how this transcription factor would affect stomata movement.

4.2. ABA Regulation of Drought-Responsive Genes in Drought Tolerance

Drought stress induces the expression of a large number of stress-inducible genes. Many of these genes are also up-regulated by ABA (Section 3.2). The products of these genes may contribute to much of the so-call 'drought tolerance' that emphasizes the ability of cells to tolerate the stress. Drought stress creates several challenges to plant cells. First, it causes an increased production of reactive oxygen species that could be detrimental to cellular membranes and other macromolecules. Second, some proteins may undergo misfolding, aggregating, and denaturation. Third, the low water potential in soil requires the plant cells to lower water potential as well in order to retain and uptake water. Many of the genes that are up-regulated by drought and ABA encode proteins presumably with these functions. For example, some of the drought/ABA up-regulated genes encode enzymes that function in the biosynthesis of compatible solutes (e.g., proline, sugars) that could lower the water potential and facilitate water uptake and retention. Others encode enzymes that can directly detoxify reactive oxygen species. These enzymes include glutathione peroxidase (GPX) (Rodriguez Milla et al., 2003), glutathione S-transferases (GST) (Moon, 2003), superoxide dismutase (SOD), catalase (CAT) (Guan, et al., 2000; Pei et al., 2000), ascorbate peroxidase (APX) and glutathione reductase (GR) (Jiang and Zhang, 2003). Certain stress-responsive genes encode polypeptides that may help to restore the nature structures of abnormal proteins. For proteins that could not be repaired, ABA up-regulated genes that encode various components in the proteolysis pathway (Hoth et al., 2002) will promote the degradation of these unfold proteins to avoid the negative effects of their accumulation on cellular activities.

Although drought alone can activate these stress-responsive genes, ABA can synergistically enhance their expression. This was demonstrated by both exogenous applied ABA and ABA deficient or insensitive mutants. Two possibilities may account for this synergy. One is that the signaling pathways for ABA and drought stress may act in parallel but may also interact with one another. Another possibility is that the transcription factors that respectively bind to the ABRE and the DRE/CRT elements in the promoters of stress responsive genes may cooperate in gene activation. In any event, the outcome of this interaction results in even higher expression of stress responsive genes that are an advantage to the plants under stress.

4.3. Roots Signal Drought to the Shoot

It was thought that the ability of plants to sense water deficit in soil may have to do with ABA production in the roots and its translocation to the leaves where it serves as a signal to close stomata. This has been a subject receiving intensive study and debate (Wilkins and Davies, 2002). Other hormones (auxin, cytokinine, and ethylene), metabolites, various cations and anions, reactive oxygen species as well as pH sometimes are also associated with drought-induced stomatal closure (e.g., Goodger et al., 2005).

4.4. Root Adaptation to Drought Stress

It has been well documented that the growth of roots is generally less inhibited by drought stress relative to that of the shoot (Hsiao and Xu, 2000; Wu and Cosgrove, 2000; Serraj and Sinclair, 2002; Sharp et al., 2004). Thus, plants growing under lower water potential conditions usually have a higher ratio of root to shoot mass (Fisher and Turner, 1978). Here we will confine our discussion to root development under drought stress.

While root growth is more adapted to drought stress than that of the shoot, it is not clear whether root development also has some adaptation to drought stress. Many studies have been conducted to explore the root growth and development of crop plants under drought stress. Due to the genetic complexity of crop cultivars, however, most of these studies did not provide a clear clue about root development under drought stress (see below). In Arabidopsis, several earlier reports had described roots' responses to drought stress. It was reported that in response to drought root hairs become 'bulbous' and 'shortening' (Schnall and Quatrano, 1992) or 'short, tuberized, hairless roots' form (Vartanian et al., 1994). Interestingly, ABA was shown to have similar effects on roots in inducing these alterations. However, it is not known whether these responses give any advantage to plants under drought stress. Thus, whether these responses are adaptive is unclear.

Assuming that roots may play a critical role in drought tolerance, researchers had tried to link root development with drought tolerance in crop plants such as rice, maize, soybean, sorghum, barley, and coffee tree (e.g., Champoux et al., 1995; Nguyen et al., 1997; Maggio et al., 2001; Xu et al., 2001; Lafitte and Courtois, 2002; Sharp et al., 2004; Pinheiro et al., 2005). Many of these studies were intended to identify root traits that could be used in breeding for drought tolerance. In these studies root characteristics (such as dry mass, thickness, length, etc) between drought tolerant and drought sensitive cultivars were compared and their correlation to drought tolerance was inferred. The conclusions of these studies vary and are often contradictory (Price et al., 2002). Apparently, these correlative studies may have had difficulty in ascertaining whether the differences in root systems or architecture are responsible for or linked to drought tolerance, because the genetic backgrounds of the tolerant and the sensitive strains often are different or unclear. In this regard, quantitative trait locus (QTL) analysis using progenies derived from crosses between drought tolerant and drought sensitive lines may yield more reliable information regarding whether a particular trait has anything to do with the tolerance.

Despite some conflicting results in different QTL analyses, several studies did suggest a connection between root characteristics and drought tolerance (Nguyen et al., 1997; Zhang et al., 2001; Yue et al., 2005). For instance, using a double haploid mapping population derived from a cross between a strain of upland rice (drought tolerant) and a strain of low land rice (drought sensitive), Mu et al. (2003) reported that drought tolerance correlates with longer maximum root length and fewer root numbers. However, with multiple QTLs controlling overall drought tolerance, analysis of variance alone will be difficult to determine the actual contribution of the root traits to overall drought tolerance. In some analyses, the QTL effects were considered pleiotropic rather than direct linkage (Giuliani et al., 2005). Furthermore, it is still not easy to identify genes underlying the QTL loci in crop plants. Therefore, direct genetic study of the relationship between root traits and drought tolerance in a model plant species would be desirable in order to clarify the role of roots in drought adaptation and to reveal novel drought tolerance mechanisms.

To isolate drought tolerance determinants, we searched for possible responses of roots to drought stress in Arabidopsis. It was noticed that under osmotic stress the Arabidopsis root system underwent a characteristic change that had not clearly been described in the literature before we started our work. Whereas the control plants developed a number of lateral roots, those subjected to osmotic stress (by supplementing nutrient media with 50 mM or 75 mM mannitol) failed to develop or were delayed in lateral root development. Importantly, this inhibition of lateral root growth by drought stress was also observed in seedlings growing in soil and was observed in several crop plants (Xiong et al., 2006; L. Xiong, unpublished).

Like many other drought responses, drought inhibition of lateral root development is also partly mediated by ABA. Although exogenous ABA at lower concentrations (0.1 to 1.0 μM) has little effect on the growth of the primary roots (sometimes primary root elongation is stimulated by lower concentrations of ABA, e.g., Xiong et al., 2001), it clearly inhibits lateral root elongation. ABA deficient mutants (*aba1*, *aba2*, and *aba3*) generally tend to have more lateral roots under non-stressful conditions. On agar plates supplemented with mannitol, the magnitude of inhibition of lateral root elongation was reduced in *aba* mutants compared to the wild type, although the mutants still responded to the treatment in reducing lateral root elongation. These data suggest that the inhibition of lateral root elongation by mannitol is partly mediated by ABA (Xiong et al., 2006). Further evidence to support ABA's role in mediating inhibition of lateral root development is that the inhibition of lateral root elongation by osmotic stress and ABA is significantly compromised in the ABA-insensitive mutant *abi1*-1 (Xiong et al., 2006). During the course of our study, reports on the influence of osmotic stress and ABA on Arabidopsis root development were recently published (De Smet et al., 2003; Deak and Malamy, 2005). These authors also found that ABA and osmotic stress inhibit lateral root development, although the experimental conditions used in these studies are different from ours. In fact, osmotic stress or drought stress inhibition of lateral root growth was also documented in early reports (e.g., van der Weele et al., 2000), although its significance was previously unclear. Our study and those of others thus demonstrate that osmotic stress and drought stress can regulate lateral root development. With these findings, we further hypothesized and subsequently confirmed that the characteristic inhibition of lateral root development by drought/osmotic stress may represent an adaptive response to drought and therefore it is a typical 'drought rhizogenesis' process. This drought rhizogenesis process might be related to the one reported in a previous study on morphological changes of roots observed with soil-grown Arabidopsis (Vartanian et al., 1994). In our follow-up studies, the *dig* (*d*rought-*i*nduced rhizo*g*enesis) mutants defective in drought rhizogenesis were isolated. It was found that those *dig* mutants that exhibit a hypersensitive response to drought or ABA in drought rhizogenesis are more tolerant to drought stress and the insensitive ones are drought sensitive (Xiong et al., 2006). Our genetic studies thus demonstrate that drought rhizogenesis response is closely linked to whole plant drought tolerance and is an adaptive response to drought stress.

Now that drought rhizogenesis has been established as an adaptive response, what would its benefits be to the plants? Under drought or any other abiotic stresses, there is a significant decrease in photosynthesis and, consequently, a reduction in the amount of metabolites and energy. It is imperative for the plants to use this reduced amount of resources to their maximum advantage – usually to survive stresses. Apparently, under drought stress conditions, an urgent need of the plants would be to increase the uptake of water, which is usually more available deep down in the soil. Restriction of the horizontal proliferation of lateral roots in the top soil and allocation of more resources to the growth of primary roots certainly would offer an advantage to the plants by expanding their domains of water supply. Thus, the adaptive response of roots to water deficit by means of drought rhizogenesis is in sharp contrast to their response to nutrient deficiency. Under nutrient starvation conditions, increased proliferation of lateral roots are commonly observed, which may help the plants to increase their exploitation of the topsoil where bioavailable nutrients are more enriched relative to the subsoil.

The regulation of lateral root development by ABA under drought stress may result from the interplay of drought and ABA with other hormones such as auxin, ethylene, gibberillic acid and cytokinins, yet current study on this topic is very limited.

4.5. Hydrotropism, Hydraulic Conductivity and Water Uptake

Plant roots have the ability to grow toward soil patches with more available water or grow away from dry soil regions. This hydrotropic response may be important for plants to find water resources. Using existing mutants defective in auxin and ABA biosynthesis or signaling, it was reported that hydrotropism requires ABA since seedlings of ABA deficient mutant *aba1*-1 and ABA-insensitive mutant *abi2*-1 are less responsive to hydrotropic stimuli (Takahashi et al., 2002). On the other hand, auxin insensitive mutants *axr1*-3 and *axr2*-1 showed enhanced hydrotropism (Takahashi et al., 2002). However, another group reported that ABA deficient mutants and ABA insensitive mutants were not defective in hydrotropism (Eapen et al., 2003). Recently, Arabidopsis hydrotropic mutants were isolated (Kobayashi et al., 2003; Eapen et al., 2003). Among the ahydrotropic mutants, some exhibited normal gravitropic responses whereas others were impaired in gravitropism (Kobayashi et al., 2003). In a separate screen, one ahydrotropic mutant *ahr1* showed enhanced gravitropic response. Thus, hydrotropism differs from gravitropism but both responses may interact. Furthermore, the perception of gravity and water availability might share similar mechanisms and auxin may have been recruited in the perception and response to water availability. Future identification of the genes required for hydrotropism may help to reveal the mechanisms of hydrotropism.

Once roots reach the water source, the ability to absorb water depends on the driving force (created by water potential difference across plasma membrane) and the resistance of plant cells to water passage. ABA was shown to transiently activate the expression of certain water channel (aquoporin) genes in a number of plant species (reviewed in Javot and Maurel, 2002). Several experiments showed that exogenous ABA could increase root water conductivity Lp (e.g., Ludewig et al., 1988; Zhang et al., 1995; Quintero et al., 1999; Hose et al., 2000) but how this occurs is unclear. In addition to up-regulation of the aquoporin expression and activity, presumably, ABA may alter root structure and/or decrease water potential and thus would facilitate water uptake. ABA up-regulated gene products may have a significant effect on lowering water potentials. When wheat roots were treated with ABA, there was a significant increase in osmolarity and turgor pressure, although levels of cations were not changed (Jones et al., 1987). This suggests that increased non-ionic solutes after ABA treatments are responsible for the increased osmolarity.

5. ABA IN PLANT RESPONSE TO SALT STRESS

Relative to its well-described functions in drought stress response, less is known about the role of ABA in plant salt stress response. Like drought stress, salt stress also imposes osmotic stress to plant cells and results in the accumulation of toxic compounds such as reactive oxygen species. Therefore, combating osmotic stress and detoxifying toxic compounds are also important for salt tolerance. In these aspects, ABA may have similar functions in plant salt tolerance as in drought tolerance. Comparison of salt tolerance between plant species or genotypes differing in ABA responses may help to reveal the role of ABA in salt tolerance. Variations in salt tolerance among Arabidopsis ecotypes have been investigated (e.g., Quesada et al., 2002) yet it is unclear whether these differences have anything to do with ABA biosynthesis and responses. In their comparison of the expression profiles of the salt tolerant *Thellungiella halophila* with *Arabidopsis thaliana*, Gong et al. (2005) and Taji et al. (2004) found that the expression level of ABA biosynthetic and ABA responsive genes was higher in *Thellungiella*. Nonetheless, using ABA deficient or ABA response mutants may better address the potential role of ABA in salt tolerance.

Seed germination is highly sensitive to environmental conditions such as water availability, temperature, and salinity. It is known that salt stress enhances ABA biosynthesis which in turn inhibits germination. ABA deficient mutants are thus less inhibited by salt stress during germination. Accordingly, some mutants isolated for their tolerance to salt stress are found to be allelic to ABA deficient or ABA insensitive mutants (e.g., Quesada et al., 2000; Ruggiero et al., 2004). At the vegetative stage, however, ABA deficient mutants are more sensitive to salt stress partly because these mutants are impaired in the activation of stress-responsive genes (Xiong et al., 2001; 2002). Likewise, it is expected that ABA insensitive mutants would be salt sensitive too if ABA signaling is critical to salt tolerance. Nonetheless, the salt sensitivity of adult ABA response mutants was not well studied. Ohta et al. (2004) reported that seedlings of *abi1*-1 and *abi2-1* are more tolerant to salt stress. A similar observation was also reported by Achard et al. (2006). Overall, the effects of mutations in ABA signaling components on plant salt stress response are less obvious than on drought stress response. Measurement of several growth parameters of adult Arabidopsis wild type and *abi1*-1 and *abi2*-1 plants growing under salt stress also did not reveal any differences among these genotypes (Cramer, 2002). Perhaps this has to do with the different impacts of drought stress and salt stress on plants.

While drought stress imposes ‘physical’ water deficit stress to the plant cells, high salinity in soil creates a ‘physiological’ water deficit stress. In fact, high external Na^+ often is much more detrimental to glycophytes than the resulting lower water potential. The mechanisms that condition ionic stress tolerance are therefore of predominate importance for plant tolerance to high salinity. Consequently, limiting the uptake, reducing root to shoot transport, increasing the exclusion and compartmentalization of Na^+ are the methods of choice that glycophytic plants may use in dealing with salt stress. This was clearly demonstrated in the genetic study of salt tolerance in Arabidopsis. The most salt sensitive mutants (in terms of primary root elongation and seedling growth) in Arabidopsis are those that are impaired in maintaining ion homeostasis (Zhu, 2000). The importance of regulating Na^+ transporters in salt tolerance was also witnessed in other plants including rice (e.g., Ren et al., 2005). ABA, whose level increases upon salt stress through transcriptional up-regulation of its biosynthetic genes (Xiong and Zhu, 2003), may directly or indirectly modulate ion homeostasis during salt stress.

ABA regulates the expression of some of the transporters involved in salt uptake and compartmentalization. For example, the vacuole localized Na^+/H^+antiporter genes are up-regulated by ABA (Yokoi, et al., 2002; Shi and Zhu, 2002). In addition, protein modification is critical to the function of these transporters. ABA may regulate transporter activities through posttranslational modification of the transporters or their regulators, yet the mechanisms involved in the regulation are unclear at this time. The protein kinase SOS2 can directly phosphorylate and regulate the plasma membrane-localized Na^+/H^+ antiporter SOS1 (Qiu et al., 2002; Quintero et al., 2002) in excluding Na^+ from the cytosol. Interestingly, SOS2 could interact with ABI2, a 2C type phosphatase that negatively regulates ABA signaling (Ohta et al., 2003). Other 2C type phosphatases, for example PP2CA, which may act as a negative regulator of several ABA responses (Kuhn et al., 2006; Yoshida et al., 2006), also interact with K^+ channels (Chérel et al., 2002; Vranova et al., 2001). While K^+ channels in guard cells play critical roles in stomatal opening and closing (Section 4.1), disturbed K^+ homeostasis in roots and other tissues and cell types may contribute to salt sensitivity (e.g., Zhu et al., 1998; Rus et al., 2004). It is thus likely that ABA may play a role in regulating ion transporter activities under salt stress. Nonetheless, this regulation may not be as evident as the regulation of ion channels and transporters in guard cells under drought stress (Section 4.1), although salt stress also quickly induces stomata closure.

While about one quarter of salt stress-regulated genes are specifically regulated by salt stress (Ma et al., 2006), most others can also be regulated by drought and cold stress as well. Salt stress induction of at least a subset of these genes is ABA-dependent. Early studies suggested that the transcript levels of stress-responsive genes are lower in both *aba* and *abi* mutants when treated with salt or osmotic stress, although controversy also existed in the literature (see references cited in Xiong et al., 2001). Since the *LOS5* and *LOS6* genes are required for salt induction of stress responsive genes and they encode the ABA biosynthetic enzymes ABA3 (molybdenum cofactor sulfurase) and ABA1 (zeaxanthin epoxidase), respectively, this demonstrates that ABA is indeed required for stress induction of these common stress responsive genes. However, it should be noted that there are other stress-responsive genes whose expression may be independent of ABA.

6. ABA INTERACTION WITH OTHER HORMONES IN PLANT STRESS RESPONSE

The growth and development of plants, like that of animals, is regulated by a diverse set of growth hormones. Plants are distinct from animals in that they cannot move and therefore they have evolved more robust mechanisms to deal with adverse environmental conditions. Stress responses in plants evoke a wide array of genes and intensive signaling pathways. Under abiotic stress, the stress hormone ABA works together with other phytohormones to adjust growth and development programs so that the plants may be better adapted to the adverse conditions. Under these conditions, plant growth will generally be slowed down as a result of reduced synthesis and signaling of growth promoting hormones (e.g., auxin, gibberrelin, and cytokinine) and increased synthesis of growth inhibition hormones (ABA and ethylene). Among these hormones, the interaction of ethylene and ABA in abiotic stress response was the most studied.

Under drought and salt stress, ethylene production increases (McMichae et al., 1972; Apelbaum and Yang, 1981) because of the activation of the biosynthetic genes and enzymes (e.g., Liu and Zhang, 2004). Increased accumulation of ethylene under abiotic stress may inhibit plant growth. It was thought that ABA may restrict the production of ethylene and thus could promote growth under abiotic stresses (Sharp, 2002). On the other hand, ethylene may promote ABA biosynthesis under drought stress. It was suggested that ethylene-induced ABA production may contribute to the inhibition of auxin-related herbicides on plant growth. This is because high levels of IAA or synthetic auxin (such as 2, 4-D) induce ethylene production (e.g., Burg and Burg, 1965), which in turn promotes ABA biosynthesis (Hansen and Grossmann, 2000). However, it is unclear whether indigenous auxin could induce ABA biosynthesis via enhanced ethylene production. The auxin maxima in plant tissues do not appear to completely overlap with the sites of ABA biosynthesis. Furthermore, auxin biosynthesis and signaling may be generally compromised under drought stress, although detailed experimental evidence in this aspect is lacking. Thus, auxin may not play a major role in the interaction of ABA and ethylene under drought stress.

Enhanced production of ethylene may be beneficial for plants under abiotic stress. For example, senescence of old leaves induced by ethylene will remobilize nutrients from these leaves and reduce transpirational water loss of the plant as a whole. When the ethylene biosynthetic gene ACS (1-aminocyclopropane-1-carboxylic acid synthase) was knocked out in maize, the leaves of the mutant plants exhibited delayed senescence and had high levels of chlorophyll and high CO_2 fixation rate. However, these leaves also had a high transpiration rate (Young et al., 2004). So it is expected that these mutant plants will delay leaf senescence (a 'Stay-Green' trait) but may use up the available water in soil more quickly. Thus, 'Stay Green' is beneficial for the plants only when water will be available imminently. Perhaps because of these reasons, the soybean stay-green mutations actually increased drought susceptibility (Luquez and Guiarmet, 2002) although it was previously reported that the Stay-Green trait in sorghum and rice might contribute to increased biomass production under drought stress (e.g., Borrell et al., 2000).

In addition to the biosynthesis, ethylene responsiveness may be regulated by drought or salt stress too. The ethylene receptor ETR1 was down-regulated by salt stress at transcription and protein levels (Zhao and Schaller, 2004). This would result in increased sensitivity to ethylene (ethylene receptors negatively regulate the downstream signaling) and perhaps would offer some advantages to plants under stress. When the ethylene receptor NTHK1 was overexpressed in tobacco (so that the downstream ethylene signaling pathway would be suppressed), the transgenic plants were found to have a higher Na^+/K^+ ratio and were more sensitive to salt stress (Cao et al., 2006). Similarly, the ethylene insensitive mutant *ein3* was more sensitive to salt stress whereas the constitutive ethylene response mutant *ctr1* was salt-tolerant during the early seedling stage (Achard et al., 2006). Interaction between ABA and ethylene under abiotic stress is also suggested by the fact that certain transcription factors responsible for the activation of ABA/stress-responsive genes and ethylene-responsive genes are of a similar class and may be subject to similar regulations. The ethylene responsive factor binding protein (ERF/EREBP) and the CBF/DREB class of transcription factors may cross-activate stress responsive genes (e.g., Fujimoto et al., 2000). Some ERF proteins act as transcription repressors regulating ethylene and ABA responses (Yang et al., 2005; Song et al., 2005). Accordingly, regulating these ERF transcriptional regulators may result in altered drought and salt stress sensitivity (e.g., Yang and Wu, 2005; Song et al., 2005; Zhang et al., 2005).

The antagonism between ethylene and ABA was also found in other stress response processes. Ethylene inhibits ABA-induced stomatal closure and reduces the induction of the ABA-induced gene *Rab18* (Tanaka et al., 2005). Yet, how this was achieved is unclear. It was noted that the *ABI1* and *ABI2* genes were highly up-regulated by ethylene (De Paepe et al., 2004). These negative regulators of ABA signaling may thus reduce ABA responses. In previous studies, it was found that ethylene and ABA play antagonizing roles in controlling seed germination (Beaudoin et al., 2000; Ghassemian et al., 2000). Ethylene and ABA may also interact in regulating drought rhizogenesis (Section 4.4). The ethylene insensitive mutant *ein2* is hypersensitive to ABA in drought rhizogenesis, implying a role of ethylene in regulating cell fate and cell elongation in response to abiotic stress (L. Xiong, unpublished data). In addition, *ein2* also exhibited reduced lateral root development in the absence of exogenous ABA.

Phytohormones may act together with or independently of ABA in regulating stomata movement (reviewed in Dodd, 2003). Exogenous IAA can stimulate stomatal opening and suppress the inhibition of ABA on the opening. Auxin-induced stomatal opening may result from auxin-induced ethylene production since inhibition of ethylene biosynthesis can inhibit IAA-induced stomatal opening (Merritt et al., 2001). Exogenous ABA alkalinizes the cytosol whereas IAA can acidify the cytosol (Gehring et al., 1990). The pH changes induced by IAA or ABA may regulate ion channels and ion flux and thus stomata movement. Evidence has suggested that cytosolic alkalization occurs before ABA-induced ROS production and stomata closure (Suhita et al., 2004). In some of these studies on hormonal regulation of stomata movement, hormone biosynthesis or response mutants were not always used and epidermal strips were the major materials used to observe stomatal regulation. Further investigation of the role of non-ABA hormones in regulating stomata movement is needed. Because of the significant accumulation of ABA within or in the vicinity of the guard cells during drought stress, however, ABA should be the prevailing hormone in regulating stomata under drought stress.

In summary, increased functions of growth inhibitory hormones and simultaneously reduced functions of growth promoting hormones contribute to the retarded plant growth under drought or salt stress. This may represent an adaptive strategy for the plants to survive the stress at the expense of their growth. This strategy may work well for a population as a whole to survive a drought spell: retarded plants will use less water resource and will be able to recover growth when the next rainfall comes. To genetically abrogate this response may temporarily enhance plant growth but may not result in increased productivity of a population if water will not be available at later stages of plant development.

7. HARNESS ABA AND ITS SIGNALING FOR THE REGULATION OF PLANT DROUGHT AND STRESS TOLERANCE

Since ABA plays such important roles in drought and salt adaptation, it is possible that by manipulating ABA levels or ABA sensitivity one may be able to obtain stress tolerant crop plants. Before the isolation of ABA biosynthetic genes, crop breeders had tried to use ABA levels as a trait in breeding for drought resistant crops. Theoretically, a higher ABA level under drought stress may confer increased drought tolerance. However, it should be born in mind that ABA biosynthesis and catabolism are drought-stress regulated. Thus, drought sensitive plants may in fact experience a higher degree of stress and thus may lead to increased production of ABA (Xiong and Zhu, 2003). Therefore, the causal relation between internal ABA levels and drought susceptibility may be complex. Accordingly, both positive (Samet et al., 1980; Henson et al., 1981) and negative (Ilahi and Dorffling, 1982; Quarrie and Jones, 1979; Wang and Huang, 2003) correlations between leaf ABA content and drought tolerance were found (reviewed in Quarrie, 1991). Since the accumulation of ABA in leaves does not necessarily correlate with stomatal closure even within the same plant (Jia and Zhang 1999), variations in leaf ABA levels may not be able to account for differences in drought tolerance among cultivars of diverse genetic backgrounds. Thus, the correlation between ABA and drought tolerance among different plant species or cultivars within species is pleiotropic at the best (Giuliani et al., 2005). Without some certainty as to the contribution of ABA levels to drought tolerance, breeding ABA levels using conventional approaches for enhancing drought tolerance may not work well. Pertinent to breeding ABA levels, there are also other breeding programs using various traits that appear to correlate with drought resistance of crop plants. Considering the genetic diversity of many crop plants and the complex mechanisms of drought tolerance, it may be advisable to use yield performance as the sole trait in the breeding for drought tolerance.

With the identification of the ABA biosynthetic genes, it becomes feasible to use these genes in enhancing ABA production and, potentially, plant stress tolerance. To maximize the possibility of bursting ABA production, it is ideal to enhance the expression of the rate-limiting enzyme in the biosynthetic pathway. Researchers have suggested that the cleavage step is rate-limiting (Schwatz and Zeevaart, 2003). Therefore, initial efforts to regulate ABA biosynthesis were mainly focused on the *NCED3* gene. It was reported that overexpression of this gene in Arabidopsis and in tobacco resulted in increased ABA production and enhanced drought and salt tolerance (Iuchi et al., 2001; Qin and Zeevaart, 2002). On the other hand, it was proposed that ABA biosynthetic genes might be subject to self-regulation (Xiong and Zhu, 2003). In this scenario, up-regulation of any of the biosynthetic genes would result in increased ABA biosynthesis to various extents. In consistence with this idea, overexpressing the *ZEP* gene, which is the most abundantly expressed among all these known ABA biosynthetic genes, and *ABA3/LOS5* either resulted in enhanced ABA responses in stress gene induction or increased drought tolerance (Xiong et al., 2002; unpublished). Similarly, regulation of *ZEP* in tobacco also results in increased ABA accumulation and enhanced seed dormancy (Frey et al., 1999). In addition to the regulation of biosynthesis, reduced ABA catabolism and conjugation will also result in enhanced ABA accumulation and potentially will increase drought tolerance of the plants (Section 3.1).

Another approach to enhance drought tolerance is to increase the sensitivity of plant cells to ABA. Since Arabidopsis mutants with altered sensitivity to ABA are available (Finkelstein and Rock, 2002), regulating the expression level of these genes may confer altered drought tolerance in transgenic plants. One example is that suppression of the *ERA1* gene in canola resulted in increased sensitivity to ABA and increased drought tolerance and better yield under mild drought stress (Wang et al., 2005). Because the pathway for ABA signal transduction involves far more components than those in the ABA biosynthesis pathway, there would be many opportunities to regulate ABA responses to control stomata and whole plant response to drought and salt stress. Many successful laboratory studies were reported in enhancing drought or salt tolerance by expressing these signal transduction components. These components include putative receptors/sensors, G-protein subunits, second messenger producers, protein kinases (including Ca^{2+}-dependent protein kinases, MAP kinase components), and transcription factors. In other experiments, stress-inducible genes are directly regulated. Research on the enhancement of stress tolerance by regulating stress responsive genes and stress signaling components has been intensively reviewed elsewhere (Bajaj et al., 1999; Chinusamy et al., 2005; Vinocur and Altman, 2005; Umezawa et al., 2006).

Despite many successful examples in enhancing stress tolerance using transgenic techniques, currently there has been no field application of these techniques. In fact, very few field trials of these transgenic plants (Wang et al., 2005) were conducted. In field conditions, the effectiveness of the transgenic plants in improving drought tolerance may also vary considerably (Bahieldin et al., 2005). A concern about public acceptance of genetically modified crops may not be the only reason for the lack of field application of the above laboratory research. Current transgenic techniques can also be improved. For example, overexpression of stress responsive genes often impairs plant growth under normal conditions. This negative effect could be reduced by using inducible promoters. However, certain stress inducible promoters (such as the *RD29A* promoter) are also regulated by other environmental factors (such as light, circadian rhythm, and mechanical stress). In the field condition, the transgene may still be turned on even if there is no drought stress. Therefore, one important task in the future is to develop artificial promoters that can be specifically turned on by drought or other designated stresses. Furthermore, the transgenes should be expressed in the specific tissue or cell types (e.g., guard cells) where they are supposed to function. In these ways, the negative effect of their genetic manipulations may be minimized.

ACKNOWLEDGEMENTS

The author would like to thank Christine Ehret for critical reading of the manuscript. Work in the author's laboratory has been supported by the National Science Foundation and the United States Department of Agriculture.

REFERENCES

Achard, P., Cheng, H., De Grauwe, L., Decat, J., Schoutteten, H., Moritz, T., Van Der Straeten, D., Peng, J., and Harberd, N. P., 2006. Integration of plant responses to environmentally activated phytohormonal signals. Science. **311**: 91–94.

Apelbaum, A., and Yang, S. F. 1981. Biosynthesis of stress ethylene induced by water deficit. Plant Physiol. **68**: 594–596.

Araus, J. L., Slafer, G. A., Reynolds, M. P., and Royo, C. 2002. Plant breeding and drought in C_3 cereals: what should we breed for? Ann Bot. **89**: 925–40.

Armstrong, F., Leung, J., Grabov, A., Brearley, J., Giraudat, J., and Blatt, M. R. 1995. Sensitivity to abscisic acid of guard-cell K^+ channels is suppressed by *abi-1*, a mutant Arabidopsis gene encoding a putative protein phosphatase. Proc. Natl. Acad. Sci. USA. **92**: 9520–9524.

Bahieldin, A., Mahfouz, H. T., Eissa, H. F., Saleh, O. M., Ramadan, A. M., Ahmed, I. A., Dyer, W. E., El-Itriby, H. A., and Madkour, M. A. 2005. Field evaluation of transgenic wheat plants stably expressing HVA1 gene for drought tolerance. Physiol. Plant. **123**: 421–427.

Bajaj S., Targolli, J., Liu, L. F., Ho, T. H. D., and Wu, R. 1999. Transgenic approaches to increase dehydration-stress tolerance in plants. Mol. Breeding **5**: 493–503.

Beaudoin, N., Serizet, C., Gosti, F., and Giraudat, J. 2000. Interactions between abscisic acid and ethylene signaling cascades. Plant Cell. **12**: 1103–1115.

Bentsink, L., and Koornneerf, M. 2002. Seed dormancy and germination. In C. R. Somerville, E. M. Meyerowitz, eds, The Arabidopsis Book. American Society of Plant Biologists, Rockville, MD

Borrell, A. K., Hammer, G. L., and Henzell, R. G. 2000. Does maintaining green leaf area in sorghum improve yield under drought? II. Dry matter production and yield. Crop Sci. **40**: 1037–1048.

Bray, E. A. 1997. Plant responses to water deficit. Trends Plant Sci. **2**: 48–54.

Burg, S. P., and Burg, E. A. 1965. The interaction between auxin and ethylene and its role in plant growth. Proc. Natl. Acad. Sci. USA. **55**: 262–269.

Busk, K., and Pages, M. 1998. Regulation of abscisic acid-induced transcription. Plant Mol. Biol. **37**: 425–435.

Cao, W. H., Liu, J., Zhou, Q. Y., Cao, Y. R., Zheng, S. F., Du, B. X., Zhang, J. S., and Chen, S. Y. 2006. Expression of tobacco ethylene receptor NTHK1 alters plant responses to salt stress. Plant Cell Environ. **29**: 1210–1219.

Champoux, M. C., Wang, G., Sarkarung, S., Mackill, D. J., O'Toole, J. C., Huang, N., and McCouch, S. R. 1995. Locating genes associated with root morphology and drought avoidance in rice via linkage to molecular markers. Theor. Appl. Genet. **90**: 969–981.

Cheng, W. H., Endo, A., Zhou, L., Penney, J., Chen, H. C., Arroyo, A., Leon, P., Nambara, E., Asami, T., Seo, M., Koshiba, T., and Sheen, J. 2002. A unique short-chain dehydrogenase/reductase in Arabidopsis glucose signaling and abscisic acid biosynthesis and functions. Plant Cell **14**: 2723–2743.

Chérel, I., Michard, E., Platet, N., Mouline, K., Alcon, C., Sentenac, H., and Thibaud, J. B. 2002. Physical and Functional Interaction of the Arabidopsis K^+ Channel AKT2 and Phosphatase AtPP2CA. Plant Cell **14**: 1133–1146.

Chinusamy, V., Xiong L., and Zhu, J. K. 2005. Use of Genetic Engineering and Molecular Biology Approaches for Crop Improvement for Stress Environments. In: Abiotic Stresses: Plant Resistance Through Breeding and Molecular Approaches. Edited by Ashraf, M., Haworth Press, pp. 47–107.

Christmann, A., Grill, E., and Meinhard, M. 2003. Abscisic acid signaling. Topics Cur. Genet. **4**: 39–71.

Christmann, A., Hoffmann, T., Teplova, I., Grill, E., and Muller, A. 2005. Generation of active pools of abscisic acid revealed by in vivo imaging of water-stressed Arabidopsis. Plant Physiol. **137**: 209–219.

Cominelli, E., Galbiati, M., Vavasseur, A., Conti, L., Sala, T., Vuylsteke, M., Leonhardt, N., Dellaporta, S. L., and Tonelli, C. 2005. A guard-cell-specific MYB transcription factor regulates stomatal movements and plant drought tolerance. Curr Biol. **15**: 1196–1200.

Cramer, G. R. 2002. Response of abscisic acid mutants of Arabidopsis to salinity. Funct. Plant Biol. **29**: 561–567.

Cutler, S., Ghassemian, M., Bonetta, D., Cooney, S., and McCourt, P. 1996. A protein farnesyl transferase involved in abscisic acid signal transduction in Arabidopsis. Science. **273**: 1239–1241.

Deak, K. I., and Malamy, J. 2005. Osmotic regulation of root system architecture. Plant J **43**: 17–28

De Paepe, A., Vuylsteke, M., Van Hummelen, P., Zabeau, M., and van der Straeten, D. 2004. Transcriptional profiling by cDNA-AFLP and microarray analysis reveals novel insights into the early response to ethylene in Arabidopsis. Plant J. **39**: 537–559.

De Smet, I., Signora, L., Beeckman, T., Inze, D., Foyer, C. H., and Zhang, H. 2003. An abscisic acid-sensitive checkpoint in lateral root development of Arabidopsis. Plant J. **33**: 543–555.

Dietz, K. J., Sauter, A., Wichert, K., Messdaghi, D., and Hartung, W. 2000. Extracellular β-glucosidase activity in barley involved in the hydrolysis of ABA glucose conjugate in leaves. J. Exp. Bot. **51**: 937–944.

Dodd, I. C. 2003. Hormonal interactions and stomatal responses. J. Plant Growth Reg. **22**: 32–46.

Eapen, D., Barroso, M. L., Campos, M. E., Ponce, G., Corkidi, G., Dubrovsky, J. G., and Cassab, G. I. 2003. A no hydrotropic response root mutant that responds positively to gravitropism in Arabidopsis. Plant Physiol. **131**: 536–546.

Fan, L. M., Zhao, Z., and Assmann, S. M. 2004. Guard cells: a dynamic signaling model. Curr. Opin. Plant Biol. **7**: 537–546.

Farquhar, G. D., Ehleringer, J. R., and Hubick, K. T. 1989. Carbon isotope discrimination and photosynthesis. Ann. Rev. Plant Physiol. Plant Mol. Biol. **40**: 503–537.

Finkelstein, R. R., and Rock, C. D. 2002. Abscisic acid biosynthesis and response. In C. R. Somerville, E. M. Meyerowitz, eds, The Arabidopsis Book. American Society of Plant Biologists, Rockville, MD

Fisher, R. A., and Turner, N. C. 1978. Plant productivity in the arid and semiarid zones. Ann. Rev. Plant Physiol. **29**: 277–317.

Frey, A., Audran, C., Marin, E., Sotta, B., and Marion-Poll, A. 1999. Engineering seed dormancy by the modification of zeaxanthin epoxidase gene expression. Plant Mol Biol. **39**: 1267–1274.

Fujimoto, S. Y., Ohta, M., Usui A., Shinshi, H., and Ohme-Takagi M. 2000. Arabidopsis ethylene-responsive element binding factors act as transcriptional activators or repressors of GCC box–mediated gene expression. Plant Cell **12**: 393–404.

Gehring, G. A., Irving, H. R., and Parish, R. W. 1990. Effects of auxin and abscisic acid on cytosolic calcium and pH in plant cells. Proc. Natl. Acad. Sci. USA **87**: 9645–9649.

Ghassemian, M., Nambara, E., Cutler, S., Kawaide, H., Kamiya, Y., and McCourt, P. 2000. Regulation of abscisic acid signaling by the ethylene response pathway in Arabidopsis. Plant Cell. **12**: 1117–1126.

Giuliani, S., Sanguineti, M. C., Tuberosa, R., Bellotti, M., Salvi, S., and Landi, P. 2005. Root-ABA1, a major constitutive QTL, affects maize root architecture and leaf ABA concentration at different water regimes. J. Exp. Bot. **56**: 3061–3070.

Gong, Q., Li, P., Ma, S., Rupassara, S. I., and Bohnert, H. J. 2005. Salinity stress adaptation competence in the extremophile *Thellungiella halophila* in comparison with its relative *Arabidopsis thaliana*. Plant J. **44**: 826–839.

González-Guzmán, M., Apostolova, N., Belles, J. M., Barrero, J. M., Piqueras, P., Ponce, M. R., Micol, J. L., Serrano, R., and Rodriguez, P. L. 2002. The short-chain alcohol dehydrogenase ABA2 catalyzes the conversion of xanthoxin to abscisic aldehyde. Plant Cell **14**:1833–1846.

Goodger, J. Q., Sharp, R. E., Marsh, E. L., and Schachtman, D. P. 2005. Relationships between xylem sap constituents and leaf conductance of well-watered and water-stressed maize across three xylem sap sampling techniques. J. Exp. Bot. 56: 2389–2400.

Guan, L. M., Zhao, J. and Scadalios, J. G. 2000. Cis-elements and trans-factors that regulate expression of the maize Cat1 antioxidant gene in response to ABA and osmotic stress: H2O2 is the likely intermediary signaling molecule for the response. Plant J. **22**: 87–95

Guiltinan, M. J., Marcotte, W. R. Jr, and Quatrano, R. S. 1990. A plant leucine zipper protein that recognizes an abscisic acid response element. Science. **250**: 267–271.

Hansen, H., and Grossmann, K. 2000. Auxin-Induced Ethylene Triggers Abscisic Acid Biosynthesis and Growth Inhibition. Plant Physiol. **124**: 1437–1448

van Hengel, A.. J., Barber, C., and Roberts, K. 2004. The expression patterns of arabinogalactan-protein *AtAGP30* and *GLABRA2* reveal a role for abscisic acid in the early stages of root epidermal patterning. Plant J. **39**: 70–83.

Henson, I. E., Mahalakshmi, V., Bidinger, F. R., and Alagarswamy, G. 1981. Genotypic variation in pearl millet (*Pennisetum americanum* (L.) Leeke), in the ability to accumulate abscisic acid in response to water stress. J. Exp. Bot. **32**: 899–910;

Hetherington, A. M., and Woodward, F. I. 2003. The role of stomata in sensing and driving environmental change. Nature **424**: 901–908.

Himmelbach, A., Yang Y., and Grill, E. (2003). Relay and control of abscisic acid signaling. Curr. Opinion Plant Biol. **6**: 470–479.

Hose, E., Steudle, E., and Hartung, W. 2000. Abscisic acid and hydraulic conductivity of maize roots: a study using cell- and root pressure probes. Planta **211**: 874–882.

Hosy, E., Vavasseur, A., Mouline, K., Dreyer, I., Gaymard, F., Poree, F., Boucherez, J., Lebaudy, A., Bouchez, D., Very, A. A., Simonneau, T., Thibaud, J. B., and Sentenac, H. 2003. The Arabidopsis outward K^+ channel GORK is involved in regulation of stomatal movements and plant transpiration. Proc. Natl. Acad. Sci. USA. **100**: 5549–5554.

Hoth, S., Morgante, M., Sanchez, J. P., Hanafey, M. K., Tingey, S. V., and Chua, N. H. 2002. Genome-wide gene expression profiling in *Arabidopsis thaliana* reveals new targets of abscisic acid and largely impaired gene regulation in the *abi1*-1 mutant. J. Cell Sci. **15**: 115: 4891–900.

Hsiao, T. C., and Xu, L. K. 2000. Sensitivity of growth of roots versus leaves to water stress: biophysical analysis and relation to water transport. J Exp. Bot. **51**: 1595–1616.

Ilahi, I., and Dörffling, K. 1982. Changes in abscisic acid and proline levels in maize varieties of different drought resistance. Physiol. Plant. **55**: 129–135.

Iuchi, S., Kobayashi, M., Taji, T., Naramoto, M., Seki, M., Kato, T., Tabata, S., Kakubari, Y., Yamaguchi-Shinozaki, K., and Shinozaki, K. 2001. Regulation of drought tolerance by gene manipulation of 9-cis-epoxycarotenoid dioxygenase, a key enzyme in abscisic acid biosynthesis in Arabidopsis. Plant J. **27**: 325–333.

Javot, H., and Maurel, C. 2002. The role of aquaporins in root water uptake. Ann. Bot. 90:.301–313.

Jia, W., and Zhang, J. 1999. Stomatal closure is induced rather by prevailing xylem abscisic acid than by accumulated amount of xylem-derived abscisic acid. Physiol. Plant. **106**: 268–275.

Jiang, M., and Zhang, J. 2002. Water stress-induced abscisic acid accumulation triggers the increased generation of reactive oxygen species and up-regulates the activities of antioxidant enzymes in maize leaves. J Exp. Bot. **53**: 2401–2410.

Jones, H., Leigh, R. A., Tomos, A. D. & Jones, R. G. W. 1987. The effect of abscisic acid on cell turgor pressures, solute content and growth of wheat roots. Planta **170**: 257 262.

Kizis, D., and Pages, M. 2002. Maize DRE-binding proteins DBF1 and DBF2 are involved in *rab17* regulation through the drought-responsive element in an ABA-dependent pathway. Plant J. **30**: 679–689.

Klein, M., Perfus-Barbeoch, L., Frelet, A., Gaedeke, N., Reinhardt, D., Mueller-Roeber, B., Martinoia, E., and Forestier, C. 2003. The plant multidrug resistance ABC transporter AtMRP5 is involved in guard cell hormonal signalling and water use. Plant J. **33**: 119–129.

Kobayashi, A., Kakimoto, Y., Fujii, N., and Takahashi, H. 2003. Physiological and genetic characterization of hydrotropic mutants of *Arabidopsis thaliana*. Biol. Sci. Space. **17**: 243–244.

Koiwai, H., Nakaminami, K., Seo, M., Mitsuhashi, W., Toyomasu, T., and Koshiba, T. 2004. Tissue-specific localization of an abscisic acid biosynthetic enzyme, AAO3, in Arabidopsis. Plant Physiol. **134**: 1697–1707.

Kreps, J. A., Wu, Y., Chang, H. S., Zhu, T., Wang, X., and Harper, J. F. 2002. Transcriptome changes for Arabidopsis in response to salt, osmotic, and cold stress. Plant Physiol. 130: 2129–2141.

Krochko, J. E., Abrams, G. D., Loewen, M. K., Abrams, S. R., and Cutler, A. J. 1998. (+)-Abscisic acid 8'-hydroxylase is a cytochrome P450 monooxygenase. Plant Physiol. **118**: 849–860.

Kuhn, J. M., Boisson-Dernier, A., Dizon, M. B., Maktabi, M. H., and Schroeder, J. I. 2006. The protein phosphatase AtPP2CA negatively regulates abscisic acid signal transduction in Arabidopsis, and effects of *abh1* on AtPP2CA mRNA. Plant Physiol. **140**: 127–139.

Kushiro, T., Okamoto, M., Nakabayashi, K., Yamagishi, K., Kitamura, S., Asami, T., Hirai, N., Koshiba, T., Kamiya, Y., and Nambara, E. 2004. The Arabidopsis cytochrome P450 CYP707A encodes ABA 8'-hydroxylases: key enzymes in ABA catabolism. EMBO J. **23**: 1647–1656.

Kwak, J. M., Mori, I. C., Pei, Z. M., Leonhardt, N., Torres, M. A., Dangl, J. L., Bloom, R. E., Bodde, S., Jones, J. D., and Schroeder, J. I. 2003. NADPH oxidase AtrbohD and AtrbohF genes function in ROS-dependent ABA signaling in Arabidopsis. EMBO J. **22**: 2623–2633.

Lafitte, R. H., and Courtois, B. 2002. Interpreting cultivar × environment interactions for yield in upland rice: assigning value to drought adaptive traits. Crop Sci **42**:1409–1420.

Lemichez, E., Wu, Y., Sanchez, J. P., Mettouchi, A., Mathur, J., and Chua, N. H. 2001. Inactivation of AtRac1 by abscisic acid is essential for stomatal closure. Genes. Dev. **15**: 1808–1816.

Leung, J., and Giraudat, J. 1998. Abscisic acid signal transduction. Ann. Rev. Plant Physiol. Plant Mol. Biol. **49**:199–222.

Levchenko, V., Konard, K. R., Dietrich, P., Roelfsema, R. G., and Hedrick, R. (2005). Cytosolic abscisic acid activates guard cell anion channels without preceding Ca^{2+} signals. Proc. Natl. Acad. Sci. USA **102**: 4203–4208

Levitt, J. 1980. Responses of plants to environmental stress. Volume II. Water, radiation, salt and other stresses. Academic Press, New York.

Leyman, B., Geelen, D., Quintero, F. J., Blatt, M. R. 1999. A tobacco syntaxin with a role in hormonal control of guard cell ion channels. Science. **283**: 537–540.

Liu, Y. and Zhang, S. 2004. Phosphorylation of 1-aminocyclopropane-1-carboxylic acid synthase by MPK6, a stress-responsive mitogen-activated protein kinase, induces thylene biosynthesis in Arabidopsis. Plant Cell **16**: 3386–3399.

Lu, C., and Fedoroff, N. 2000. A mutation in the Arabidopsis HYL1 gene encoding a dsRNA binding protein affects responses to abscisic acid, auxin, and cytokinin. Plant Cell 12: 2351–2365.

Ludewig, M., Dorffling, K., and Siefert, H. 1988. Abscisic acid and water transport in sunflowers. Planta **175**: 325–333.

Luquez, V. M., and Guiarmet, J. J. 2002. The stay green mutations d1 and d2 increase water stress susceptibility in soybeans. J. Exp. Bot. **53**: 1421–1428.

Ma, S., Gong, Q., and Bohnert, H. 2006. Dissecting salt stress pathways. J. Exp. Bot. **57**: 1097–1107.

Maggio, A., Hasegawa, P. M., Bressan, R. A., Consiglio, M. F., and Joly, R. J. 2001. Unravelling the functional relationship between root anatomy and stress tolerance. Aust J Plant Physiol. 28: 999–1004.

Marcotte, W. R. Jr., Russell, S. H., and Quatrano, R. S. 1989. Abscisic acid-responsive sequences from the em gene of wheat. Plant Cell. 1: 969–976.

Martinez-Zapater, J. M., Coupland G, Dean, C., and Koornneef, M. 1994. The Transition to Flowering in Arabidopsis. In Arabidopsis, E. M. Meyerowitz, and C. R. Somerville, eds (Plainview, New York: Cold Spring Harbor Laboratory Press), pp. 403–433.

Mauch-Mani, B., and Mauch, F. 2005. The role of abscisic acid in plant-pathogen interactions. Curr. Opinon Plant Biol. **8**:409–414

McCarty, D. R. 1995. Genetic control and integration of maturation and germination pathways in seed development. Ann. Rev. Plant Physiol. **46**: 71–93.

McMichael, B. L., Jordan, W. R. and Powell, R. D. 1972. An effect of water stress on ethylene production by intact cotton petioles. Plant Physiol. **49**: 658–660.

Merritt, F., Kemper, A., and Tallman, G. 2001. Inhibitors of ethylene synthesis inhibit auxin-induced stomatal opening in epidermis detached from leaves of *Vicia faba* L. Plant Cell Physiol **42**: 223–230.

Mishra, G., Zhang, W., Deng, F., Zhao, J., and Wang, X. 2006. A bifurcating pathway directs abscisic acid effects on stomatoal closure and opening in Arabidopsis. Science **312**: 264–266.

Moons, A. 2003. *Osgstu3* and *osgtu4*, encoding tau class glutathione S-transferases, are heavy metal- and hypoxic stress-induced and differentially salt stress-responsive in rice roots. FEBS Lett. **553**: 427–432.

Mundy, J., Yamaguchi-Shinozaki, K., and Chua, N. H. 1990. Nuclear proteins bind conserved elements in the abscisic acid-responsive promoter of a rice *rab* gene. Proc. Natl. Acad. Sci. USA. **87**:1406–1410.

Murata, Y., Pei, Z. M., Mori, I. C., and Schroeder, J. 2001. Abscisic acid activation of plasma membrane Ca^{2+}channels in guard cells requires cytosolic NAD(P)H and is differentially disrupted upstream and downstream of reactive oxygen species production in *abi1*-1 and *abi2*-1 protein phosphatase 2C mutants. Plant Cell. 13: 2513–2523.

Mustilli, A. C., Merlot, S., Vavasseur, A., Fenzi, F., and Giraudat, J. 2002. Arabidopsis OST1 protein kinase mediates the regulation of stomatal aperture by abscisic acid and acts upstream of reactive oxygen species production. Plant Cell. **14**: 3089–3099.

Nambara, E., and Marion-Poll, A. 2005. Abscisic acid biosynthesis and catabolism. Annu Rev Plant Biol. **56**: 165–185.

Narusaka, Y., Nakashima, K., Shinwari, Z. K., Sakuma, Y., Furihata, T., Abe, H., Narusaka, M., Shinozaki, K., and Yamaguchi-Shinozaki, K. 2003. Interaction between two *cis*-acting elements, ABRE and DRE, in ABA-dependent expression of Arabidopsis rd29A gene in response to dehydration and high-salinity stresses. Plant J. **34**: 137–148.

Nguyen, H. T., Babu, R. C., and Blum, A. 1997. Breeding for drought resistance in rice: physiology and molecular genetics considerations. Crop Sci. **37**:1426–1434.

Ohta, M., Guo, Y., Halfter, U., and Zhu, J. K. 2003. A novel domain in the protein kinase SOS2 mediates interaction with the protein phosphatase 2C ABI2. Proc. Natl. Acad. Sci. USA **100**: 11771–11776.

Osakabe, Y., Maruyama, K., Seki, M., Satou, M., Shinozaki, K., and Yamaguchi-Shinozaki, K. 2005. Leucine-rich repeat receptor-like kinase1 is a key membrane-bound regulator of abscisic acid early signaling in Arabidopsis. Plant Cell **17**: 1105–19.

Pei, Z. M., Ghassemian, M., Kwak, C. M., McCourt, P., and Schroeder, J. I. 1998. Role of farnesyltransferase in ABA regulation of guard cell anion channels and plant water loss. Science **282**: 287–290.

Pei, Z. M., and Kuchitsu, K. 2005. Early ABA signaling events in guard cells. J. Plant Growth Reg. **24**: 296–307.

Pei, Z. M., Murata, Y., Benning, G., Thomine, S., Klusener, B., Allen, G. J., Grill, E. and Schroeder, J. I. 2000. Calcium channels activated by hydrogen peroxide mediate abscisic acid signalling in guard cells. Nature **406**: 731–734

Peng, Y. B., Zou, C., Wang, D. H., Gong, H. Q., Xu, Z. H., and Bai, S. N. 2006. Preferential localization of abscisic acid in primordial and nursing cells of reproductive organs of Arabidopsis and cucumber. New Phytol. 170: 459–466.

Phillips, J., Artsaenko, O., Fiedler, U., Horstmann, C., Mock, H. P., Muntz, K., Conrad, U. 1997. Seed-specific immunomodulation of abscisic acid activity induces a developmental switch. EMBO J. **16**: 4489–4496.

Pinheiro, H. A., Damatta, F. M., Chaves, A. R., Loureiro, M. E., and Ducatti, C. 2005. Drought tolerance is associated with rooting depth and stomatal control of water use in clones of *Coffea canephora*. Ann. Bot. **96**: 101–108.

Qin, X., and Zeevaart, J. A. 2002. Overexpression of a 9-cis-epoxycarotenoid dioxygenase gene in *Nicotiana plumbaginifolia* increases abscisic acid and phaseic acid levels and enhances drought tolerance. Plant Physiol. **128**: 544–551.

Qiu, Q. S., Guo, Y., Dietrich, M. A., Schumaker, K. S., and Zhu, J. K. 2002. Regulation of SOS1, a plasma membrane Na^+/H^+ exchanger in Arabidopsis thaliana, by SOS2 and SOS3. Proc. Natl. Acad. Sci. USA. **99**: 8436–8441.

Quarrie, S. A. 1991. Implications of genetic differences in ABA accumulation for crop production. In: Davies, W. J., Jones, H. G. eds. Abscisic acid: physiology and biochemistry. Oxford, UK: Bios Scientific Publishers Ltd. 227–243.

Quarrie S. A., and Jones, H. G. 1979. Genotypic variation in leaf water potential, stomatal conductance and abscisic acid concentration in spring wheat subjected to artificial drought stress. Ann. Bot. **44**: 323–332.

Quesada, V., Ponce, M. R., and Micol, J. L. (2000). Genetic analysis of salt-tolerant mutants in *Arabidopsis thaliana*. Genetics **154**: 421–436

Quesada, V., Garcia-Martinez, S., Piqueras, P., Ponce, M. R., and Micol, J. L. 2002. Genetic architecture of NaCl tolerance in Arabidopsis. Plant Physiol. **130**: 951–963.

Quintero, J. M., Fournier, J. M., Benlloch, M. 1999. Water transport in sunflower root systems: effects of ABA, Ca^{2+} status and $HgCl_2$. J. Exp. Bot. **50**: 1607–1612.

Quintero, F. J., Ohta, M., Shi, H., Zhu, J. K., and Pardo, J. M. 2002. Reconstitution in yeast of the Arabidopsis SOS signaling pathway for Na+ homeostasis. Proc. Natl. Acad. Sci. USA. **99**: 9061–9066.

Raamanjulu, S., and Bartels, D. 2002. Drought- and desciccation-induced modulation of gene expression in plants. Plant Cell Environ. **25**: 141–151.

Razem, F. A., El-Kereamy, A., Abrams, S. R., and Hill, R. D. 2006. The RNA-binding protein FCA is an abscisic acid receptor. Nature **439**: 290–294.

Ren, Z. H., Gao, J. P., Li, L. G., Cai, X. L., Huang, W., Chao, D. Y., Zhu, M. Z., Wang, Z. Y., Luan, S., and Lin, H. X. 2005. A rice quantitative trait locus for salt tolerance encodes a sodium transporter. Nat. Genet. **37**: 1141–1146.

Riccardi, F., Gazeau, P., de Vienne, D., and Zivy, M. 1998. Protein changes in response to progressive water deficit in maize. Quantitative variation and polypeptide identification. Plant Physiol. **117**: 1253–1263.

Rock, C. D. and Zeevaart, J. A. D. (1991). The aba mutant of Arabidopsis thaliana is impaired in epoxy-carotenoid biosynthesis. Proc. Natl. Acad. Sci. USA. **88**: 7496–7499.

Rodriguez, Milla, M. A., Maurer, A., Rodriguez Huete, A., and Gustafson, J. P. 2003. Glutathione peroxidase genes in Arabidopsis are ubiquitous and regulated by abiotic stresses through diverse signaling pathways. Plant J. **36**: 602–615.

Ruggiero, B., Koiwa, H., Manabe, Y., Quist, T. M., Inan, G., Saccardo, F., Joly, R. J., Hasegawa, P. M., Bressan, R. A., and Maggio, A. 2004. Uncoupling the effects of abscisic acid on plant growth and water relations. Analysis of *sto1/nced3*, an abscisic acid-deficient but salt stress-tolerant mutant in Arabidopsis. Plant Physiol. 136: 3134–3147.

Rus, A., Lee, B. H., Munoz-Mayor, A., Sharkhuu, A., Miura, K., Zhu, J. K., Bressan, R. A., Hasegawa, P. M. 2004. AtHKT1 facilitates Na^+ homeostasis and K^+ nutrition in planta. Plant Physiol. **136**: 2500–2511.

Saito, S., Hirai, N., Matsumoto, C., Ohigashi, H., Ohta, D., Sakata, K., Mizutani, M. 2004. Arabidopsis CYP707As encode (+)-abscisic acid 8'-hydroxylase, a key enzyme in the oxidative catabolism of abscisic acid. Plant Physiol. **134**: 1439–1449.

Samet, J. S., and Sinclair, T. R. 1980. Leaf Senescence and Abscisic Acid in Leaves of Field-grown Soybean. Plant Physiol. 66: 1164–1168.

Sauter, A., Dietz, K. J., and Hartung, W. 2002. A possible stress physiological role of abscisic acid conjugates in root-to-shoot signaling. Plant Cell Environ. **25**: 223–228

Schnall, J. A., and Quatrano, R. S. 1992. Abscisic acid elicits the water-stress response in root hairs of Arabidopsis. Plant Physiol. **100**: 216–218

Schwartz, S. H., Qin, X., and Zeevaart, J. A. 2003. Elucidation of the indirect pathway of abscisic acid biosynthesis by mutants, genes, and enzymes. Plant Physiol. **131**:1591–601.

Seki, M., Kamei, A., Yamaguchi-Shinozaki, K., and Shinozaki, K. 2003. Molecular responses to drought, salinity and frost: common and different paths for plant protection. Curr. Opin. Biotechnol. **14**: 194–199.

Serraj, R., and Sinclair, T. R. 2002. Osmolyte accumulation: can it really help increase crop yield under drought conditions? Plant Cell Environ. **25**: 333–341.

Sharp, R. E. 2002. Interaction with ethylene: changing views on the role of abscisic acid in root and shoot growth responses to water stress. Plant Cell Environ. **25**: 211–222. -

Sharp, R. E., Poroyko, V., Hejlek, L. G., Spollen, W. G., Springer, G. K., Bohnert, H. J., and Nguyen, H. T. 2004. Root growth maintenance during water deficits: physiology to functional genomics. J. Exp. Bot. **55**: 2343–2351.

Shen, Q., and Ho, T.-H. D. (1995). Functional dissection of an abscisic acid (ABA)-inducible gene reveals two independent ABA responsive complexes each containing a G-box and a novel cis acting element. Plant Cell **7**: 295–307.

Shi, H., and Zhu, J. K. 2002. Regulation of expression of the vacuolar Na^+/H^+ antiporter gene AtNHX1 by salt stress and abscisic acid. Plant Mol. Biol. **50**: 543–550.

Shinozaki, K., Yamaguchi-Shinozaki, K., and Seki, M. 2003. Regulatory network of gene expression in the drought and cold stress responses. Curr. Opin. Plant Biol. **6**: 410–417.

Skriver, K., and Mundy, J. 1990. Gene expression in response to abscisic acid and osmotic stress. Plant Cell. **2**: 503–512.

Song, C. P., Agarwal, M., Ohta, M., Guo, Y., Halfter, U., Wang, P., and Zhu, J. K. 2005. Role of an Arabidopsis AP2/EREBP-type transcriptional repressor in abscisic acid and drought stress responses. Plant Cell. **17**: 2384–2396.

Staxen, I., Pical, C., Montgomery, L. T., Gray, J. E., Hetherington, A. M., and McAinsh, M. R. 1999. Abscisic acid induces oscillations in guard-cell cytosolic free calcium that involve phosphoinositide-specific phospholipase C. Proc. Natl. Acad. Sci. USA. **96**: 1779–1784

Suhita, D., Raghavendra, A. S., Kwak, J. M., and Vavasseur, A. 2004. Cytoplasmic alkalization precedes reactive oxygen species production during methyl jasmonate- and abscisic acid-induced stomatal closure. Plant Physiol. **134**:1536–1545.

Taji, T., Seki, M., Satou, M., Sakurai, T., Kobayashi, M., Ishiyama, K., Narusaka, Y., Narusaka, M., Zhu, J. K. and Shinozaki, K. 2004. Comparative genomics in salt tolerance between Arabidopsis and Arabidopsis-related halophyte salt cress using Arabidopsis microarray. Plant Physiol. **135**: 1697–1709.

Takahashi, N., Goto, N., Okada, K., and Takahashi, H. 2002. Hydrotropism in abscisic acid, wavy, and gravitropic mutants of *Arabidopsis thaliana*. Planta. **216**: 203–211.

Tanaka, Y., Sano, T., Tamaoki, M., Nakajima, N., Kondo, N., and Hasezawa, S. 2005. Ethylene inhibits abscisic acid-induced stomatal closure in Arabidopsis. Plant Physiol. **138**: 2337–2343.

Taylor, I. B., Sonneveld, T., Bugg, T. D. H., and Thompson, A. L. 2005. Regulation and manipulation of the biosynthesis of abscisic acid, including the supply of xanthophylls precursors. J. Plant Growth Reg. **24**: 253–273.

Umezawa, T., Fujita, M., Fujita, Y., Yamaguchi-Shinozaki, K., and Shinozaki, K. 2006. Engineering drought tolerance in plants: discovering and tailoring genes to unlock the future. Curr. Opin. Biotechnol. **17**: 113–122.

Vartanian, N., Marcotte, L., and Ciraudat, J. 1994. Drought rhizogenesis in *Arabidopsis thaliana*, differential responses of hormonal mutants. Plant Physiol. **104**: 761–767.

Vinocur, B., and Altman, A. 2005. Recent advances in engineering plant tolerance to abiotic stress: achievements and limitations. Curr. Opin. Biotechnol. **16**: 123–132.

Vranova, E., Tähtiharju, S., Sriprang, R., Willekens, H., Heino, P., Palva, E. T., Inzé, D., and Van Camp, W. V. 2001. The AKT3 potassium channel protein interacts with the AtPP2CA protein phosphatase 2C. J. Exp. Bot. **52**: 181–182.

Wang, X. Q., Ullah, H., Jones, A. M., and Assmann, S. M. 2001. G protein regulation of ion channels and abscisic acid signaling in Arabidopsis guard cells. Science. **292**: 2070–2072.

Wang, Z., and Huang, B. 2003. Genotypic variation in abscisic acid accumulation, water relations, and gas exchange fro Kentucky bluegrass exposed to drought stress. J. Am. Soc. Hort. Sci. **128**: 349–355.

Wang, Y., Ying, J., Kuzma, M., Chalifoux, M., Sample, A., McArthur, C., Uchacz, T., Sarvas, C., Wan, J., Dennis, D. T., McCourt, P., and Huang, Y. 2005. Molecular tailoring of farnesylation for plant drought tolerance and yield protection. Plant J. **43**: 413–424.

van der Weele, C. M., Spollen, W. G., Sharp, R. E., and Baskin, T. I. 2000. Growth of *Arabidopsis thaliana* seedlings under water deficit studied by control of water potential in nutrient-agar media. J. Exp. Bot. **51**: 1555–1562.

Weiler, E. W. 1980. Radioimmunoassays for the differential and direct analysis of free and conjugated abscisic acid in plant extracts. Planta **148**: 26–36.

Wilkinson, S., and Davies, W. J. 2002. ABA-based chemical signaling: the co-ordination of responses to stress in plants. Plant Cell Environ. **25**: 195–210.

Wright, C. A., and Beattie, G. A. 2004. *Pseudomonas syringae* pv. tomato cells encounter inhibitory levels of water stress during the hypersensitive response of *Arabidopsis thaliana*. Proc. Natl. Acad. Sci. USA. **101**: 3269–3274.

Wu, Y., and Cosgrove, D. J. 2000. Adaptation of roots to low water potentials by changes in cell wall extensibility and cell wall proteins. J. Exp. Bot. **51**:1543–1553.

Xie, Z., Ruas, P., Shen, Q. J. 2006. Regulatory networks of the phytohormone abscisic acid. Vitamins and Hormones **72**: 235–269.

Xiong, L., Ishitani, M., Lee, H., and Zhu, J. K. 2001. The Arabidopsis *LOS5/ABA3* encodes a molybdenum cofactor sulfurase and modulates cold stress- and osmotic stress-responsive gene expression. Plant Cell **13**: 2063–2083.

Xiong, L., Lee, H., Ishitani, M., and Zhu, J. K. 2002. Regulation of osmotic stress-responsive gene expression by the *LOS6/ABA1* locus in Arabidopsis. J. Biol. Chem. **277**: 8588–8596.

Xiong, L., Wang, R., Mao, G., and Koczan, J. (2006). Identification of drought tolerance determinants by genetic analysis of root response to drought stress and abscisic acid. Plant Physiol. **142**: 1065–1074.

Xiong, L., and Zhu, J. K. 2003. Regulation of abscisic acid biosynthesis. Plant Physiol. **133**: 29–36.

Xu, J. C., Li, J. Z., Zheng, X. W., Zou, L. X., and Zhu, L. H. 2001. QTL mapping of the root traits in rice seedling. Acta Genetica Sinica **28**: 433–438.

Yang, Z., Tian, L., Latoszek-Green, M., Brown, D., and Wu, K. 2005. Arabidopsis ERF4 is a transcriptional repressor capable of modulating ethylene and abscisic acid responses. Plant Mol. Biol. **58**: 585–596.

Yokoi, S., Quintero, F. J., Cubero, B., Ruiz, M. T., Bressan, R. A., Hasegawa, P. M., and Pardo, J. M. 2002. Differential expression and function of Arabidopsis thaliana NHX Na^+/H^+ antiporters in the salt stress response. Plant J. **20**: 529–539.

Yoshida, R., Hobo, T., Ichimura, K., Mizoguchi, T., Takahashi, F., Aronso, J., Ecker, J. R., and Shinozaki, K. 2002. ABA-activated SnRK2 protein kinase is required for dehydration stress signaling in Arabidopsis. Plant Cell Physiol. **43**: 1473–1483.

Yoshida, R., Umezawa, T., Mizoguchi, T., Takahashi, S., Takahashi, F., and Shinozaki, K. 2006. The regulatory domain of SRK2E/OST1/SnRK2.6 interacts with ABI1 and integrates abscisic acid (ABA) and osmotic stress signals controlling stomatal closure in Arabidopsis. J. Biol. Chem. **281**: 5310–5318.

Yoshida, T., Nishimura, N., Kitahata, N., Kuromori, T., Ito, T., Asami, T., and Shinozaki, K., and Hirayama, T. 2006. *ABA-hypersensitive germination 3* encodes a protein phosphatase 2C (AtPP2CA) that strongly regulates abscisic acid signaling during germination among Arabidopsis protein phosphatase 2Cs. Plant Physiol. **140**: 115–126.

Young, T. E., Meeley, R. B., and Gallie, D. R. 2004. ACC synthase expression regulates leaf performance and drought tolerance in maize. Plant J. **40**: 813–825.

Zeevaart, J. A. D. and Creelman, R. A. 1988. Metabolism and physiology of abscisic acid. Ann. Rev. Plant Physiol. **39**: 439–473.

Zhang, J. H., Zhang, X., and Liang, J. 1995. Exudation rate and hydraulic conductivity of maize roots are enhanced by soil drying and abscisic acid treatment. New Phytol. **131**: 329–336.

Zhang, W., Qin, C., Zhao, J., and Wang, X. (2004). Phospholipase Dα1-derived phosphatidic acid interacts with ABI1 phosphatase 2C and regulates abscisic acid signaling. Proc. Natl. Acad. Sci. USA **101**: 9508–9513.

Zhang, X., Zhang, Z., Chen, J., Chen, Q., Wang, X. C., and Huang, R. 2005. Expressing TERF1 in tobacco enhances drought tolerance and abscisic acid sensitivity during seedling development. Planta. **222**: 494–501.

Zhao, X. C., and Schaller, G. E. 2004. Effect of salt and osmotic stress upon expression of the ethylene receptor ETR1 in Arabidopsis thaliana. FEBS Let. **562**: 189–192.

Zhu, J., Gong, Z., Zhang, C., Song, C. P., Damsz, B., Inan, G., Koiwa, H., Zhu, J. K., Hasegawa, P. M., and Bressan, R. A. 2002. OSM1/SYP61: a syntaxin protein in Arabidopsis controls abscisic acid-mediated and non-abscisic acid-mediated responses to abiotic stress. Plant Cell. 14: 3009–3028.
Zhu, J. K. 2001. Genetic analysis of plant salt tolerance using Arabidopsis. Plant Physiol. **124**: 941–948.
Zhu, J. K., Liu, J., and Xiong, L. 1998. Genetic analysis of salt tolerance in arabidopsis. Evidence for a critical role of potassium nutrition. Plant Cell. **10**: 1181–1191.

CHAPTER 10

SMALL RNAs: BIG ROLE IN ABIOTIC STRESS TOLERANCE OF PLANTS

VISWANATHAN CHINNUSAMY[1], JIANJUN ZHU[2], TAO ZHOU[2] AND JIAN-KANG ZHU[3]

[1]*Water Technology Centre, Indian Agricultural Research Institute, New Delhi, India*

[2]*Department of Horticulture and Landscape Architecture, Purdue University, West Lafayette, Indiana*

[3]*Department of Botany & Plant Sciences, University of California-Riverside, Riverside, CA 92521, USA*

Abstract: Understanding gene regulation mechanisms is important for genetic improvement of abiotic stress resistance of crops. In response to developmental and environmental cues, plants employ a plethora of gene regulation mechanisms, one of which is posttranscriptional regulation of gene expression by non-protein coding small RNAs. Samll RNAs, namely, microRNAs (miRNAs) and short interfering RNAs (siRNAs), are ~20 to 24-nucleotide single stranded RNAs. miRNAs are synthesized from MIR gene transcripts, while siRNAs are synthesized from dsRNA formed by transcripts of heterochromatin DNA repeats, mRNAs encoded by natural cis-antisense gene pairs and miRNA directed cleavage of ssRNA/mRNA. Small RNAs regulate the expression of complementary/partially complementary genes by directing mRNA cleavage, translational repression, chromatin remodeling and DNA methylation. Several stress responsive small RNAs have been identified in plants and their role in oxidative stress tolerance, osmolyte accumulation/osmoprotection and nutrient starvation response have been established. Under abiotic stresses, stress-upregulated miRNAs may down-regulate their target genes, which are likely negative regulators of stress tolerance, while stress down regulated miRNAs may result in accumulation of their target gene mRNAs, which may positively regulate stress tolerance. Overexpression of miRNA-resistant target genes will help overcome post-transcriptional gene silencing, and thus may lead to better expression of engineered trait in transgenic plants. Understanding the roles of small RNAs in transcriptome homeostasis, cellular tolerance, phenological and developmental plasticity of plants under abiotic stress and recovery will help genetic engineering of abiotic stress resistance in crop plants

Keywords: microRNAs, short interfering RNAs, osmoprotection, osmoregulation, osmotic stress management, ubiquitination, sugar sensing

M.A. Jenks et al. (eds.), Advances in Molecular Breeding Toward Drought and Salt Tolerant Crops, 223–260.

1. INTRODUCTION

Abiotic stresses such as drought, excess salt, extremes of temperatures (low or high) and mineral nutrient deficiencies impose severe production constrains on food, fodder, fiber and fuel production. Drought is a major abiotic stress that affects agriculture in 45% of the world geographical area (Bot et al. 2000). Food security of many countries depends upon irrigated agriculture. However, the available amount of water for irrigation is increasingly getting scarce worldwide. Use of underground water for irrigation (around 10% of total irrigation) led to unsustainable decline in underground water tables (Somerville and Briscoe 2001). About 20% of irrigated agricultural land and 2% of dryland agriculture are affected by salinity (Yeo et al. 1999). Temperature stresses often affect crop production in irrigated agriculture. Further, temperatures lower or higher than the physiological optimum exacerbate drought and salinity stresses. Global climate changes suggest a future increase in the frequency of drought and extreme-temperature stresses in many parts of the world (IPCC 2001). Growing population demands more production from the shrinking agricultural land. This can be achieved by increasing yield under irrigated agriculture and enhancing the productivity of stress affected lands. As yield levels already reached a plateau in irrigated agriculture, the productivity of abiotic stress affected area needs to be increased to meet the growing global demand. Therefore, the greatest challenge for the coming decades will be increasing crop production from abiotic stress affected lands. Genetic enhancement of abiotic stress tolerance of crops is one of the important strategies to enhance productivity of crops in these areas. Research conducted during the past four decades has led to the understanding of major morphological and physiological mechanisms of drought and salt tolerance. With the advent of molecular biology, efforts have been made to unravel the molecular basis of these morphological and physiological mechanisms of abiotic stress tolerance (Reviews, Zhu, 2002; Chinnusamy et. al. 2004, 2005 and 2006; Yamaguchi and Blumwald 2005; Bohnert et al. 2006; Umezawa et al. 2006; Yamaguchi-Shinozaki and Shinozaki 2006).

The identification of microRNAs (miRNAs) as a regulator of the timing of *C. elegans* heterolarval development revealed a new mechanism of gene repression (Lee et al. 1993). Studies on the molecular basis of posttranscriptional gene silencing led to the discovery of small interfering RNAs (siRNAs) (Hamilton and Baulcombe 1999). Followed by these pioneering studies, research during the past decade identified ~21 to 24 nt small RNAs as ubiquitous repressors of gene expression in animals and plants. Small RNAs, which do not code for proteins but negatively regulate gene expression, are classified into two types based on their biogenesis: 1) microRNAs (miRNAs), and 2) short interfering RNAs (siRNAs). Small RNAs play a pivotal role in gene regulation in plants. This chapter will cover briefly the biogenesis of small RNAs and small RNA mediated gene regulation in abiotic stress tolerance of plants.

2. BIOGENESIS AND ACTION OF SMALL RNAS

2.1. Biogenesis of miRNAs

miRNAs are synthesized from single stranded primary miRNA (pri-miRNA) transcripts, which are transcribed from miRNA genes (*MIR* genes) by RNA polymerase II. The structure of pri-miRNA is similar to that of protein coding mRNAs, i.e., it has 5' 7mG cap and 3' poly-A tail. The pri-miRNA transcript forms one or more stem-loop secondary structures (imperfectly paired hairpin) of 60–300 nucleotides. In plants, the hairpin structure of pri-miRNA is cleaved by a Ribonuclease III-like enzyme called Dicer-like 1 (DCL1) protein to produce miRNA-miRNA* duplex in the nucleus. DCL1 protein interacts with a nuclear dsRNA-binding protein, HYPONASTIC LEAVES 1 (HYL1) (Kurihara et al. 2006) for recognition and accurate cleavage of pri-miRNA into miRNA-miRNA* duplex (stem region of hairpin) of ~21 nt in length. A nuclear methyltransferase protein HUA ENHANCER 1 (HEN1) methylates the 3-terminal nucleotides on their 2'-OH group in miRNA-miRNA* duplex and this methylation may help stabilize miRNAs. Thus generated miRNA-miRNA* duplex is exported from nucleus into cytoplasm by HASTY (HST), a member of the importin β family of nucleocytoplasmic transporters. The miRNA-miRNA* duplex is then unwound into a single stranded mature miRNA by an unknown helicase. The miRNA (~21 nt length) then enters the RNA induced silencing complex (Figure 1; Bartel 2004; Kidner and Martienssen 2005; Jones-Rhoades et al. 2006; Mallory and Vaucheret 2006).

2.2. Biogenesis of siRNAs

Short interfering RNAs (siRNAs) are synthesized from long double-stranded RNAs (dsRNAs) of exogenous or endogenous origin. Viral- or cellular-encoded RNA-dependent RNA polymerases (RDRs) generate dsRNAs in viral infected cells of plants. In case of transgene silencing, plant encoded RDRs generate dsRNAs from RNAs produced by the transgene. The endogenous sources of dsRNAs are 1) miRNA directed cleavage products of non-coding single stranded RNAs (ssRNAs)/mRNA, which are then converted into dsRNAs by RDRs, 2) dsRNAs formed from the mRNAs encoded by natural *cis*-antisense gene pair (Borsani et al. 2005), and 3) dsRNAs generated from heterochromatin and DNA repeats (Mallory and Vaucheret, 2006). The siRNAs produced by miRNA directed cleavage of ssRNAs/transgene mRNAs are referred to as trans-acting siRNAs (ta-siRNAs), while the siRNAs derived from dsRNAs formed from the mRNAs encoded by natural cis-antisense gene pair are called natural antisense gene generated siRNAs (nat-siRNAs). RDRs and DCL-like proteins process the dsRNAs formed from different sources. *Arabidopsis* genome encodes four DCL-like proteins and six RDR proteins. Biogenesis of different classes of siRNAs is carried out by specific RDR-DCL protein combinations (Jones-Rhoades et al. 2006; Mallory and Vaucheret 2006).

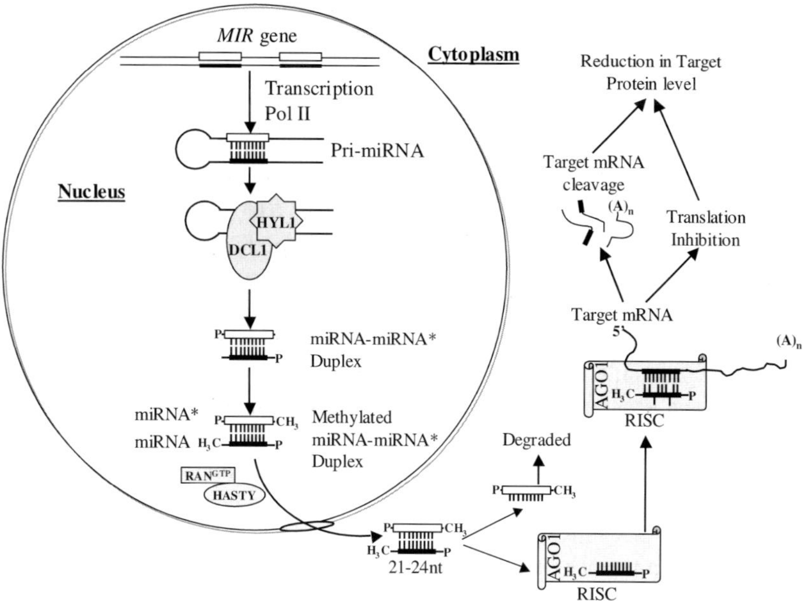

Figure 1. miRNA Biogenesis and Post-Transcriptional RNA Silencing in Plants. Pri-miRNA is transcribed from miRNA gene (*MIR* gene) by RNA polymerase II. The pri-miRNA transcript forms one or more stem-loop structure, which is cleaved by Ribonuclease III-like enzyme called Dicer-like 1 (DCL1) protein. DCL1 protein interacts with a nuclear dsRNA-binding protein, HYPONASTIC LEAVES 1 (HYL1) for recognition and accurate cleavage of pri-miRNA into miRNA-miRNA* duplex (~21 bp), which is then methylated by a nuclear methyltransferase, HUA ENHANCER 1 (HEN1) at the 3-terminal nucleotides on their 2'-OH group in miRNA-miRNA* duplex. miRNA-miRNA* duplex is exported to cytosol by HASTY (HST), a member of the importin β family of nucleocytoplasmic transporters. The miRNA-miRNA* duplex is unwound into single stranded mature miRNA by an unknown helicase. The miRNA (~21 nucleotide length) then enters the RNA induced silencing complex (RISC) containing Argonaute (AGO) family protein, AGO1. RISC then cleaves the target mRNA or inhibits translation and thus reduces the protein level of target gene

Biogenesis of ta-siRNAs involves miRNA directed cleavage of target mRNAs. These cleavage products are recognized by *SUPPRESSOR OF GENE SILENCING3* (*SGS3,* At5g23570), a coiled-coil protein with zinc finger domain, followed by synthesis of complementary RNA strand by RDR6 (At3g49500 = SUPPRESSOR OF GENE SILENCING2, SILENCING DEFECTIVE1). These dsRNAs are then cleaved into 21 nt siRNA duplex by DCL4 to produce ta-siRNAs (Figure 2; Peragine et al. 2004; Gasciolli et al. 2005; Vazquez et al. 2004; Xie et al. 2005; Yoshikawa et al. 2005; Ronemus et al. 2006).

Biogenesis of nat-siRNAs begins with the formation of dsRNAs by the transcripts of natural *cis*-antisense gene pairs. These dsRNAs are processed by DCL2, RDR6, SGS3, and a plant-specific RNA polymerase, NRPD1A to generate a 24-nt nat-siRNA, which then direct biogenesis of 21-nt nat-siRNAs by DCL1 (Figure 3; Borsani et al. 2005).

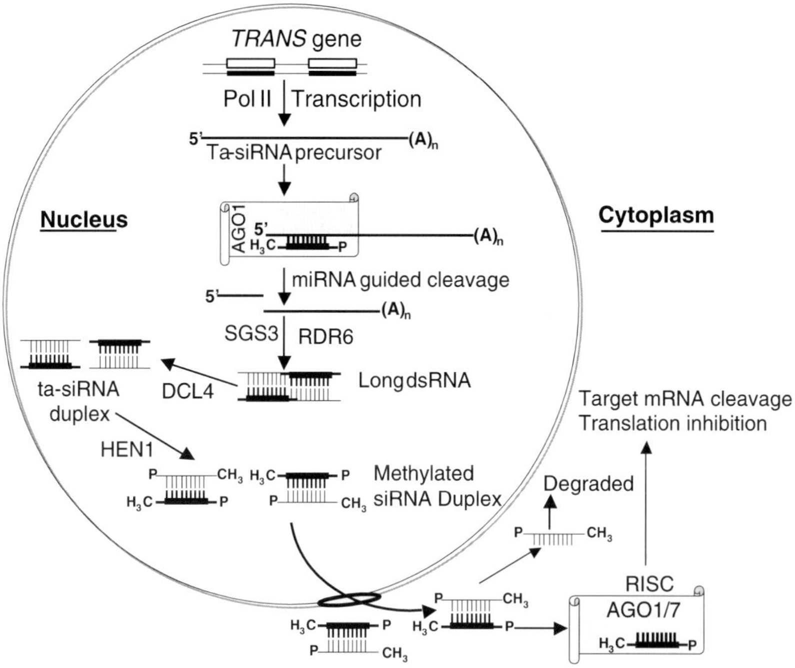

Figure 2. ta-siRNA Biogenesis and Post-Transcriptional RNA Silencing in Plants. Transcription of transgene or viral infection leads to formation pri-siRNA transcripts, which are cleaved by miRNA guided cleavage by RISC complex. The cleaved mRNA is then converted into dsRNAs by RNA-dependent RNA polymerases (RDRs). The long dsRNA is again cleaved by DCL4 ta-siRNA duplex. ta-siRNA duplex is methylated at 3' end by HEN1 and exported to cytoplasm. In cytoplasm, appropriate strand of ta-siRNA enters RISC containing AGO1 or AGO7. RISC then cleaves the target mRNA or inhibit translation and thus reduces the protein level of target gene

The third type of siRNAs (24-nt) is generated by DCL3, RDR2 and NRPD1A by processing RNAs from transposons, 5S rRNA genes, endogenous genes with direct invert repeats and transgenes with direct repeats (Chan et al. 2004; Xie et al. 2004; Zilberman et al. 2004; Mallory and Vaucheret, 2006).

2.3. Mechanism of Action of Small RNAs

The modes of action of gene silencing directed by miRNAs and siRNAs can be grouped into the following three categories: 1) Cleavage of target mRNA by complementary miRNA or siRNA, 2) miRNA or siRNA mediated translational repression and 3) transcriptional silencing mediated mainly by heterochromatic siRNAs and in some cases by miRNA (Fig. 1–3; Bartel 2004; Bao et al. 2004; Chan et al. 2005). Small RNAs are incorporated into RNA-induced silencing complex (RISC) or RNAi-induced transcriptional silencing (RITS) complex, which contain Argonaute (AGO) family proteins. AGO proteins contain two conserved domains, namely the

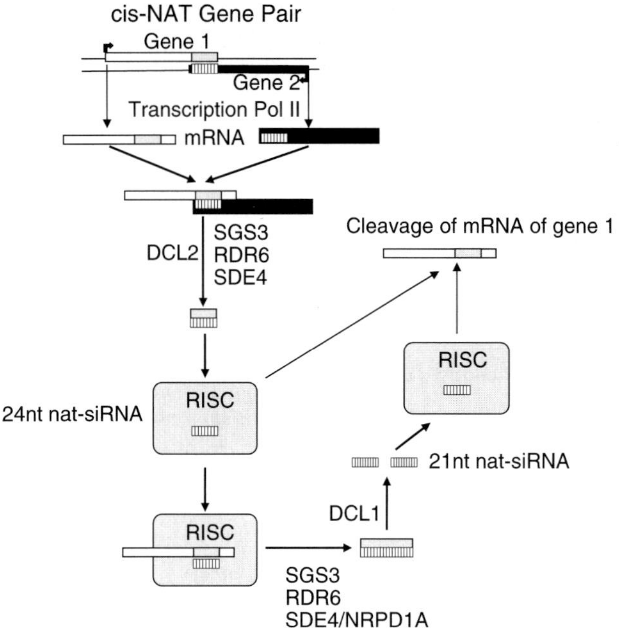

Figure 3. nat-siRNA Biogenesis and Post-Transcriptional RNA Silencing in Plants. In overlapping genes on opposite strands of DNA, (*cis*-NAT gene pair), when both the genes are transcribed, transcribed mRNAs form dsRNA in the overlapping region. This dsRNA undergoes a siRNA biogenesis pathway requiring DCL2, RDR6, SGS3, and NRPD1A to produce a 24-nt nat-siRNA. The 24-nt nat-siRNA guides the cleavage of the target gene transcript to further produce a 21-nt nat-siRNAs by DCL1. These nat-siRNAs guide cleavage of target gene mRNAs, and thus reduce the mRNA level of target genes

PAZ (an RNA-binding domain) and PIWI (similar to RNase H enzyme) domains. *Arabidopsis* genome encodes ten Argonaute proteins (Jones-Rhoades et al.2006).

The mature miRNA-miRNA* duplex is incorporated into AGO protein of RNA-induced silencing complex (RISC) and the miRNA* is degraded. The RISC represses the target gene expression mainly by mRNA cleavage and in some cases by translational repression in plants. miRNA in the AGO1 complex pairs with complementaty mRNA of the target gene and guides the cleavage of mRNA by AGO1. In the case of ta-siRNAs and nat-siRNAs, one strand of siRNA is incorporated into RISC containing AGO1 or AGO7, which then cleaves complementary mRNAs. A miRNA cannot regulate its own expression (MIR gene) as it is identical to it, while siRNAs can act in both *cis* and *trans*. In *cis*, siRNA derived from the negative strand of RNA duplex can guide degradation of RNA from which the siRNA was generated, and in trans by cleavage of any other RNA with substantial complementarity to the positive and negative strands of siRNAs (Bartel 2004; Mallory and Vaucheret 2006). When complementarity between small RNAs and protein-coding mRNAs is less but suitable constellation of complementarity exists, the translation of mRNAs will be repressed (Aukerman and Sakai 2003; Bartel 2004).

The 24-nt siRNAs derived from transposons and repetitive sequences are loaded into a putative RNA-induced transcriptional silencing (RITS) complex containing AGO4 in plants. The siRNA-RITS mediates transcriptional silencing of the loci from which the siRNAs are derived. Transcriptional silencing is imposed by chromatin remodeling and DNA methylation (Chan et al. 2004; Lippman et al. 2004; Jones-Rhoades et al. 2006; Mallory and Vaucheret 2006). Further, miRNAs can also guide DNA methylation, and thus may participate in transcriptional gene silencing (Bao et al. 2004). Thus, there is an inverse correlation between the expression of small RNAs and their target genes.

3. SMALL RNA REGULATED GENE EXPRESSION UNDER ABIOTIC STRESSES

Plants have evolved some stress-specific tolerance mechanisms and some common tolerance mechanisms to resist all abiotic stresses. Drought tolerance mechanisms of plants can be grouped into the following three mechanisms: 1) avoidance of cellular water deficit, 2) tolerance to low tissue water potential, and 3) phenotypic/developmental plasticity. Plants can avoid cellular water deficit by uptake of water from deep soil, which is controlled by root growth under drought, deep root system, change in hydraulic conductivity, aquaporin expression, etc., and by minimization of cuticular and stomatal (transpirational) water loss from aerial parts of plants. Plants regulate the rate of transpiration by adjusting stomatal number, distribution, sunken stomata, stomatal hairs, and control of stomatal aperture, opening and closing. Plants minimize water loss by closing their stomata under brief exposure to drought, while under long term drought, plants employ non-stomatal mechanisms of water loss minimization such as deposition of epicuticular wax in the epidermal cells of stems and leaves, radiation reflectance (leaf pubescence, leaf color), reduction in transpiring surface i.e., reduction in leaf area, leaf rolling (monocots), folding (dicots) and senescence. Osmotic adjustment or active solute (organic and inorganic) accumulation under osmotic stresses helps plant to overcome osmotic stresses. Tolerance to low tissue water potential is brought about by mechanisms that allow cells to function under low tissue water potential. These mechanisms include protection of cellular machinery by chaperone proteins, late embryogenesis abundant type proteins, osmoprotectants, change in plasma membrane and organellar membrane composition, etc. Detoxification of reactive oxygen species by antioxidant defense is an important tolerance mechanism under abiotic stresses. Under salinity stress, plants employ mechanisms to maintain cellular ion homeostasis (SOS pathway) and osmotic homeostasis, and damage control/repair mechanisms (Zhu 2003; Chinnusamy et al. 2005). Oxidative stress management and osmotic adjustment/osmoprotection are common mechanism of abiotic stress tolerance. Molecular genetic evidences for small RNA mediated regulation of oxidative stress management and osmolyte accumulation/osmoprotection have been established recently (Borsani et al. 2005; Sunkar et al. 2006). Further, some stress regulated small RNAs have been identified (Jones-Rhoades and Bartel 2004; Sunkar and Zhu 2004) and their target genes have been predicted. The potential regulation of stress responses by small RNAs and their target genes is discussed below.

3.1. Abiotic Stress Regulated Small RNAs

Cloning of small RNAs from stressed *Arabidopsis* plants led to the identification of some stress responsive miRNAs and siRNAs (Sunkar and Zhu 2004; Borsani et al. 2005). Construction and analysis of a library of small RNAs from *Arabidopsis* seedlings treated with ABA, dehydration, salinity, or cold stresses led to the identification of 34 new miRNAs comprising 15 families. Expression analysis of these miRNAs under abiotic stresses revealed that expression of some of these miRNAs is indeed regulated by abiotic stresses in *Arabidopsis*. These miRNAs and their target genes are listed in Table 1. miR393 is strongly upregulated by ABA, cold, dehydration, and salt stress, while miR397b and miR402 are slightly upregulated by all these stresses. Stress specific regulation of miRNAs was also found as in the case of miR319c, which is upregulated by cold but not dehydration, salt, or ABA. Further, all of these stresses down-regulate miR389a.1 (Sunkar and Zhu 2004). Many genes involved in stress tolerance of plants are either upregulated or down regulated, depends upon their role under stress. Genes involved in primary metabolism are, in general, down regulated by abiotic stresses. Under abiotic stresses, stress-upregulated miRNAs may down-regulate their target genes, which are likely negative regulators of stress tolerance, while down regulation of miRNAs under stress may result in accumulation of their target gene mRNAs, which are positive regulators of stress tolerance (Sunkar and Zhu 2004).

3.2. Target Genes of Stress Regulated Small RNAs

Predicted target genes for the stress-regulated miRNAs are involved in various cellular functions (Table 2). Stress-induced or upregulated miRNAs are expected to target negative regulators of stress responses (for example, repressors of stress responsive genes) and genes involved in plant processes that are inhibited by stresses (e.g., cell division and expansion). On the other hand, under non-stress conditions positive regulators and/or stress upregulated genes may be the targets of stress down-regulated miRNAs, and thus, down regulation of these miRNAs leads to upregulation of their target genes under stress. Table 2 shows the expression pattern of target genes of stress-regulated miRNAs.

The target genes of abiotic stress regulated miRNAs encode proteins for ubiquitin mediated proteolysis (E3 ubiquitin ligase SCF complex F-box protein), sugar sensing (At4g03190, identical to GRR1), cell wall metabolisms (laccases), cytochrome P450 and proteins with unknown functions. Interestingly these target genes also show stress responsive expression patterns (Table 2).

Table 1. Abiotic stress responsive miRNAs and their target genes in *Arabidopsis* (Source: Sunkar and Zhu 2004)

Arabidopsis miRNA	Expression under abiotic stress	Predicted target genes and their functions
miR393	Strongly upregulated by cold, dehydration, NaCl, and ABA	At3g26810(2), putative E3 ubiquitin ligase SCF complex F-box protein At1g12820(2) At3g62980(2) At3g26830(2), putative cytochrome P450 At4g03190(3)
miR397b	Slightly upregulated by cold, dehydration, NaCl, and ABA	At3g60250(2) At2g29130(3) At2g38080(3)
miR402	Slightly upregulated by cold, dehydration, NaCl, and ABA	At4g34060(1)
miR319c	Upregulated by cold but not dehydration, NaCl, or ABA	
miR389a.1	Down-regulated by cold, dehydration, NaCl, and ABA	At5g18040(1), At5g18065(1), At4g29760(2), At1g51670(2), At4g29770(3)

In addition, Sunkar and Zhu (2004) have also identified 102 endogenous small RNAs in *Arabidopsis*.

3.2.1. Ubiquitination

Regulated protein degradation mediated by the ubiquitin/26S proteosome contributes significantly to developmental processes such as embryogenesis, hormone signaling, and senescence. About 5% (1400 genes) of the *Arabidopsis* genome encodes components of the ubiquitin/26 proteasome (Ub/26S) pathway, of which ~90% of the genes encode subunits of the E3 ubiquitin ligases. Ubiquitin protein is attached to a substrate through the action of three enzymes: the ubiquitin activating enzyme (E1), ubiquitin conjugating enzyme (E2), and ubiquitin protein ligase (E3). E3 ubiquitin ligases confer substrate specificity to the pathway (Moon et al. 2004). In addition to the developmental processes, protein degradation plays a very crucial role in various abiotic stress responses of plants. Most of the abiotic stresses induce senescence, and often senescing older leaves supply the nutrients, mainly the N required by young leaves and reproductive parts. Leaf senescence is also used as a strategy to reduce water loss under drought stress. Heat stress induces protein denaturation and the denatured proteins are either repaired or undergo ubiquitin-mediated proteolysis. In fact, heat stress has been shown to increase conjugated ubiquitin in plants (Ferguson et al. 1990; Ortiz and Cardemil 2001). Further, ubiquitination also regulate gene expression under cold stress. During cold acclimation, Inducer of CBF expression 1 (ICE1) protein is activated and ICE1 induces the expression of C-repeat (CRT)-binding factors (*CBFs*) and other transcription factors leading to the transcription of downstream effector genes

Table 2. Regulation of target genes of stress-responsive miRNA by ABA and abiotic stresses in Arabidopsis

Arabidopsis miRNA	Target genes and their functions	Fold change in expression of target genes under abiotic stresses (Genevestigator Response viewer, https://www.genevestigator.ethz.ch)					
		Drought	Salt	Cold	Heat	Oxidative	ABA
miR393	At3g26810(2), putative E3 ubiquitin ligase SCF complex F-box protein	0.82	0.59	0.78	0.61	0.80	0.66
	At1g12820(2), putative E3 ubiquitin ligase SCF complex F-box protein, induced by phosphate starvation (Martin et al. 2000)	1.02	0.76	0.65	1.02	1.05	0.85
	At3g62980(2), putative E3 ubiquitin ligase SCF complex F-box protein = transport inhibitor response 1 = TIR1	1.26	1.07	0.76	0.90	1.24	1.12
	At3g26830(2), putative cytochrome P450, indole phytoalexin biosynthesis	1.08	4.85	1.57	0.42	1.73	0.90
	At4g03190(3), F-box family protein (FBL18), almost identical to GRR1-like protein 1 (GRH1=GRR1)	1.28	1.01	0.71	0.93	1.24	0.64
miR397b	At3g60250(2) Regulatory (beta) subunit (CKB3)of the protein kinase CK2. Involved in regulation of the circadian clock in *Arabidopsis*	1.06	0.88	1.04	1.52	1.09	0.87
	At2g29130(3), putative laccase	0.66	0.80	0.63	0.67	0.75	0.45
	At2g38080(3), putative laccase, cell wall biosynthesis.	1.24	1.26	1.04	0.88	1.15	1.02
miR402	At4g34060(1) ROS1, excision DNA repair family protein						
miR389a.1	At5g18040(1) expressed protein	1.03	1.47	1.04	4.13	1.21	1.02
	At5g18065(1) expressed protein						
	At4g29760(2) expressed protein						
	At1g51670(2) expressed protein	0.72	1.8	1.11	1.29	0.97	0.58
	At4g29770(3) expressed protein	0.9	0.79	0.9	12.42	1.05	0.83

and cold acclimation (Chinnusamy et al. 2003 & 2006; Lee et al. 2005). The HOS1 (high expression of osmotically responsive genes), an E3 ubiquitin ligase protein negatively regulates cold responses through the ubiquitination of ICE1. After 12 hours of cold stress, degradation of ICE1 is mediated by HOS1 (Dong et al. 2006). Thus ubiquitination-mediated proteolysis is an important process of abiotic stress tolerance. The miR393 target genes encoding E3 ubiquitin ligases, cytochrome P450 and F-box family protein (Table 2). One of the target genes At3g26810 encoding E3 ubiquitin ligase showed down regulation under abiotic stresses namely drought, salt, cold, heat and oxidative stresses and ABA treatment in *Arabidopsis* (Table 2, microarray data from response viewer of Genevestigator, https://www.genevestigator. ethz.ch). Thus, upregulation of miR393 by ABA, cold, dehydration, and salt stresses (Sunkar and Zhu 2004) may down regulate its target genes encoding E3 ubiquitin ligases under these stresses. The target proteins for At3g26810 and At1g12820 mediated ubiquitination may accumulate under these stresses. Another target gene At3g62980, an auxin receptor encoding E3 ubiquitin ligase (transport inhibitor response 1 = TIR1) showed down regulation under cold stress, suggesting that miR393 is one possible regulator of At3g62980 expression under cold stress (Table 2). These results suggest that these E3 ubiquitin ligases may be involved in proteolysis of transcriptional and other regulators of stress tolerance. The stress upregulation of miR393 suggests that a down-regulation of target TIR1 (At3g62980) mRNA or its translational repression. Auxin is involved in many plant processes including cell elongation and thus growth. TIR1 is a positive regulator of auxin signaling by promoting the degradation of Aux/IAA proteins (Dharmasiri and Estelle, 2002). The miRNA inhibition of TIR1 would down-regulate auxin signaling and may contribute to the inhibition of plant growth under stress. However, a role of these E3 ubiquitin ligases in abiotic stress responses is yet to be defined experimentally.

3.2.2. Sugar sensing

In plants, energy and carbon requirement for various processes of growth and development is met through sugars. Sugars synthesized in photosynthesis are transported to sink tissues, and channeled to respiration or converted into storage compounds (lipids, starch, protein, sucrose, fructans). Hence, it is expected that the metabolic processes involved are dependent upon the concentration of sugars. Low sugar levels in source tissue increase source activities such as photosynthesis, nutrient mobilization, and export, while high sugar levels in sink tissues stimulate growth and storage. Sugar accumulation in source tissues down-regulates photosynthesis and ensures sugar homeostasis. Abiotic stresses influence the sugar status of cells due to a decrease in photosynthesis, change in metabolism, and demand for high maintenance respiration. Plants use sugar status as a signal to modulate source-sink activity, and thus, growth and development in response to various environmental cues. The plant stress hormone, ABA, also plays a very crucial role in cellular sugar budget mediated regulation of plant growth and development as evident from the glucose/sugar insensitivity of ABA deficient and ABA insensitive mutants (Rolland et al. 2006). Sunkar and Zhu (2004) have shown the upregulation of miR393 by ABA, cold, dehydration and salt stresses. One of the miR393 target genes, At4g03190, encodes an F-box protein that shows similarity to Glucose Repression Resistance 1 (GRR1), a yeast protein involved in glucose repression. Genevestigator response viewer data suggest that At4g03190 is down-regulated in response to cold and ABA treatments (Table 2). In yeast, GRR1 is involved in glucose suppression. In plants, the GRR1 homolog, At4g03190, may also play a similar role, and may be important in sugar sensing. Investigations into the role of miR393 and its target At4g03190 may shed light on sugar sensing in plants under stress.

3.2.3. Oxidative stress management

Reactive oxygen species (ROS) namely, superoxide radicals ($O_2^{\cdot -}$), hydrogen peroxide (H_2O_2) and hydroxyl radicals ($OH^{\cdot}$) are produced in aerobic cellular processes such as mitochondrial and chloroplast electron transport, or oxidation of glycolate (photorespiration), xanthine, and glucose. The antioxidants (ascorbate, glutathione, α-tocopherol and carotenoids) and ROS detoxifying enzymes (superoxide dismutase, catalases, and enzymes of the ascorbate- glutathione cycle) detoxify ROS so as to prevent the increase of ROS to toxic levels. Abiotic stresses cause impairment of photosynthesis and other metabolic processes, and thus, cause production of toxic levels of ROS, which then cause oxidative damage to membrane lipids, proteins and nucleic acids. Therefore, ROS detoxification is crucial for abiotic stress tolerance of plants. Transgenic studies have shown that abiotic stress tolerance of plants can be improved through overexpression of genes encoding ROS detoxifying enzymes (Alscher et al. 2002; Mittler 2002; Foyer and Noctor 2005).

Abiotic stresses impair photosynthesis and respiration in plants and lead to production of superoxide radical. Superoxide dismutases (SODs) form the first line of defense against superoxide radicals by rapidly converting superoxide to hydrogen peroxide (H2O2). Depending upon the metal co-factor required, the SODs are classified into three groups: iron SOD (FeSOD), manganese SOD (MnSOD), and copper-zinc SOD (Cu/Zn-SOD), which are localized in different cellular compartments. Overexpression of SODs in transgenic plants resulted in enhanced stress tolerance (Alscher et al. 2002). In *Arabidopsis*, two Cu-Zn SOD genes namely CSD1 (encodes the cytosolic form) and CSD2 (encodes the chloroplastic form) have been identified as targets of miR398 (Jones-Rhoades and Bartel 2004; Sunkar and Zhu 2004). Hence, the role of miR398 in regulation of CSD1 and CSD2 gene expression was examined by Sunkar et al. (2006). Expression pattern of miR398 and its target genes showed an inverse correlation in various developmental stages and under oxidative stresses. Expression of miR398 is downregulated by oxidative stress (high light, heavy metals and methyl viologen), while *CSD1* and *CSD2* transcript accumulation was enhanced under oxidative stress conditions. miR398 promoter::GUS reporter also supported that miR398 expression is repressed by oxidative stress. A transient coexpression assay with miR398 with its target genes in *Nicotiana* showed that miR398 directs the degradation of *CSD1* and *CSD2* mRNAs *in vivo*. Nuclear-run-on assays showed that oxidative stress induced upregulation of *CSD1* and *CSD2* is not due to increased transcription, supporting that the upregulation due to decreased miR398-guided posttranscriptional regulation (Sunkar et al. 2006).

Several attempts have been made to improve plant stress tolerance by overproduction of Cu/Zn-SODs in transgenic plants and in some cases minimal or no increase in stress tolerance was observed in some of the studies (Tepperman and Dunsmuir 1990; Pitcher et al. 1991; Payton et al. 1997). This may be due to miR398-guided degradation of *CSD* mRNAs in these transgenic plants. This problem can be avoided if *CSD* gene is modified to destroy miR398 recognition site. This hypothesis was tested in transgenic *Arabidopsis* overexpressing normal CSD2 gene or miR398-resistant form of *CSD2* gene, *mCSD2* (miR398 target site mutated without modifying amino acid sequence). Northern blot analysis revealed that transgenic plants overexpressing *mCSD2* accumulated two-fold higher transcript levels as compared with transgenic plants overexpressing wild-type *CSD2* gene. The performance of transgenic plants was compared with wild-type plants under various oxidative stresses. The *mCSD2* transgenic plants showed highest oxidative stress tolerance followed by *CSD2* transgenics. The *mCSD2* transgenics showed higher chlorophyll content, PSII yield, germination and biomass, and lesser membrane damage as compared with *CSD2* transgenics and wild-type plants under oxidative stress conditions (Sunkar et al. 2006). miR398 and its target sites on *CSD1* and *CSD2* mRNAs are conserved in dicotyledonous and monocotyledonous plants (Bonnet et al. 2004; Jones-Rhoades and Bartel 2004; Sunkar and Zhu 2004; Lu et al. 2005; Sunkar et al. 2005), and hence use of miR398-resistant CSDs and similar strategies with other miRNA-targeted stress tolerance effector genes offers an improved means to engineer crop plants with enhanced stress tolerance (Sunkar et al. 2006).

3.2.4. Osmoregulation and osmoprotection

Osmotic stress is caused by soil water deficit, salinity and temperature extremes. Decrease in soil water availability under drought or decrease in water potential of soil solution under salinity cause osmotic stress, which leads to decreased water uptake and loss of turgor. Low temperature stress is often accompanied by osmotic stress as low temperature may limit water uptake by roots, while freezing induced ice formation in the apoplast results in movement of water from cytoplasm to apoplast, and thus severe cellular dehydration. Osmotic adjustment, one of the vital cellular responses to osmotic stress conserved in both halophytic and glycophytic plants, refers to the lowering of osmotic potential due to net accumulation of compatible solutes in response to osmotic stresses. Osmotic adjustment helps to maintain cell turgor at low water potentials. Further, solute accumulation may also help lower the freezing point of cells and thus enhance freezing tolerance. In addition to their role in osmotic adjustment, compatible solutes such as proline, betaine, polyols, sugar alcohols and soluble sugars, protect plants from stress by detoxification of radical oxygen species, and stabilization of proteins and membranes. Several transgenic plants engineered to over-produce osmolytes/ osmoprotectants showed enhanced abiotic stress tolerance (Chinnusamy et al. 2005).

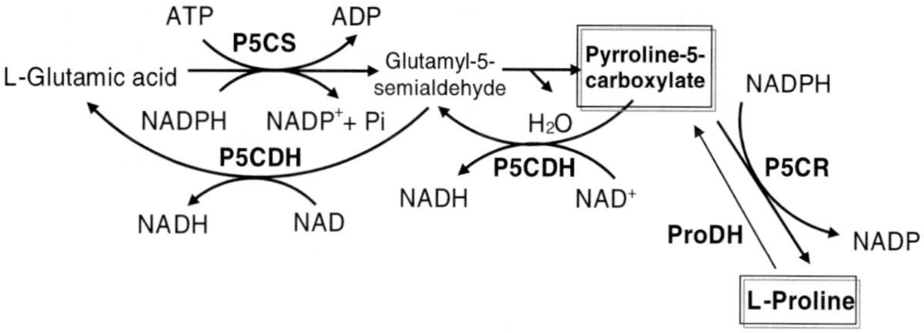

Figure 4. Proline metabolism in plants. Proline is synthesized in the cytosol from L-Glutamic acid via Glutamyl-5-semialdehyde (GSA), which spontaneously cyclizes to Δ^1-pyrroline-5-carboxylate (P5C), and then P5C is reduced to proline. Proline degradation is catalyzed in mitochondria by proline dehydrogenase (ProDH) and P5C-dehydrogenase (P5CDH). P5CS: Δ^1Pyrroline-5 carboxylate synthetase; P5CR: Δ^1Pyrroline-5 carboxylate reductase

Proline plays a key role in osmoprotection, osmotic adjustment and redox regulation across diverse organisms. Plants accumulate free proline in response to the abiotic stresses and transgenic plants engineered to over-produce proline showed enhanced abiotic stress tolerance (Kavi Kishor et al. 1995; Hong et al. 2000). Proline is synthesized from L-glutamic acid (Figure 4). Proline is catabolized to L-glutamic acid in mitochondria by Pro-dehydrogenase (ProDH) and P5C-dehydrogenase (P5CDH) during stress recovery to supply nitrogen and energy. The *AtProDH* gene is repressed by dehydration but induced by both the application of exogenous proline and rehydration, indicating an important role of proline catabolism in stress tolerance of plants. *Arabidopsis* transgenic plants engineered with an antisense *AtProDH* showed enhanced accumulation of proline and tolerance to freezing and high salinity stresses (Nanjo et al. 1999). *p5cdh* knockout mutant of *Arabidopsis* accumulate Δ^1-pyroline-5-carboxylate (P5C). Accumulation of P5C leads to apoptosis, with callose deposition, reactive oxygen species production, and DNA laddering, involving a salicylic acid signal transduction pathway (Deuschle et al., 2004). Further, P5CDH is also induced by pathogen at infection sites and leads to ROS accumulation and programmed cell death (Ayliffe et al. 2002). Thus, the proline and P5C metabolism plays a crucial role in abiotic and biotic stress response of plants and other organisms.

Recent research findings from our lab revealed a new mechanism of *P5CDH* regulation mediated by endogenous siRNAs under salt stress (Borsani et al. 2005). Genome analyses have revealed thousands of genes in convergent overlapping pairs that can generate complementary transcripts (Boi et al. 2004, Jen et al. 2005; Wang et al. 2005b), and expression profiling approaches showed widespread antisense transcription throughout genomes in both animals and plants (Brenner et al. 2000; Yamada et al. 2003; Bertone et al. 2004). Natural antisense transcripts (NAT) show sequence complementarity in partial or whole sequence of their transcripts. In overlapping genes on opposite strands of DNA, *cis*-NATs are transcribed from the same genomic loci as their sense transcripts but from the opposite strands. In

Arabidopsis, one such *cis*-NAT gene pair identified was P5CDH (At5g62530) and SRO5 (At5g62520). Sunkar and Zhu (2004) have cloned miRNAs and putative siRNAs from *Arabidopsis* treated with various abiotic stresses. One putative siRNA (clone # P96-F02) was found to match with the overlapping region between the 3' end of P5CDH and the 3'UTR of SRO5. This siRNA of 21-nt matched with SRO5 3'end and was complementary to the overlapping region of P5CDH. When this 21-nt siRNA was used as probe, a 24-nt siRNA (24-nt SRO5-P5CDH nat-siRNA) was detected in only NaCl-treated *Arabidopsis* plants. Northern blot analyses revealed that *SRO5* was induced by NaCl, while the constitutively expressed *P5CDH* was reduced by NaCl treatment. Salt stress induced SRO5 mRNA complements with the P5CDH mRNA to produce a dsRNA, which is processed by a siRNA biogenesis pathway requiring DCL2, RDR6, SGS3, and NRPD1A to produce a 24-nt SRO5-P5CDH nat-siRNA. The 24-nt nat-siRNA guides the cleavage of the P5CDH transcript to further produce a 21-nt P5CDH nat-siRNAs by DCL1. These nat-siRNAs all guide cleavage of *P5CDH* mRNAs, leading to proline accumulation (Borsani et al. 2005).

Downregulation of *P5CDH* also causes P5C mediated ROS accumulation. SRO5 protein is targeted to mitochondria, the site of proline catabolism (Borsani et al. 2005). SRO5 is similar to RADICLE INDUCED CELL DEATH 1 (RCD1), which prevents ROS-induced cell death, as *rcd1* plants are hypersensitive to ROS induced cell death (Ahlfors et al. 2004). Salt treatment causes accumulation of H_2O_2, and both salt and H_2O_2 induce the expression of *SRO5*. *sro5* mutant also showed hypersensitivity to ROS (H_2O_2). These evidences suggest that ROS detoxification under salt stress is mediated by SRO5 protein (Borsani et al. 2005). Thus, the SRO5-P5CDH nat-siRNAs together with the P5CDH and SRO5 proteins forms an important regulatory loop controlling proline and ROS production and stress tolerance (Borsani et al. 2005).

4. GROWTH AND DEVELOPMENT UNDER ABIOTIC STRESSES

Crop yield under abiotic stresses depends not only on the mere survival of plants under stress conditions but also on the phenological and developmental plasticity of plants. For example, grain yield of cereal crops is a function of total biomass accumulated and the fraction of biomass partitioned for grain development (harvest index). Biomass accumulation depends upon the leaf area (leaf area index and leaf area duration), photosynthetic rate and respiration rate (both growth and maintenance respiration). Harvest index depends upon the total spikes per unit area, spikelets per spike, florets per spikelet, spikelet fertility, rate and durations of grain filling. Under abiotic stress conditions, plants adjust the durations of phenological phases, and the rate of developmental processes, which modify biomass and harvest index. Phenotypic plasticity is a very important mechanism of abiotic stress tolerance of plants. Change in the duration of various phenological phases (vegetative phase, days to flowering, grain development duration, etc) helps plants to avoid critical growth phases under stress conditions. In plants, each phenological/developmental phase is programmed in growing degree days or thermal time. Hence, temperature extremes significantly affect the duration of phenological phases in plants. High temperature as well as drought-high temperature stress combinations reduces durations of various phenological phases. Tolerant genotypes often enhance their growth rate in order to compensate for the reduction in phenological durations. Reproductive development appears to be the phase of crop development that is the most susceptible to abiotic stresses, as any damage at this stage is irrecoverable. Drought stress reduces days to flowering in wheat, while it delays flowering in rice. In maize, drought stress increases anthesis to silking interval. Reduction in reproductive organ number and size helps plants to use the available resources efficiently so that some viable healthy seeds are produced. However, to date the molecular basis of phenological and developmental plasticity under abiotic stress is poorly understood. Understanding of the regulation of genes that control growth and development under abiotic stresses warrants immediate attention.

Mutations in genes encoding proteins involved in miRNA biogenesis and action exemplified the pivotal roles of miRNAs in plant development. The role of miRNAs in various developmental processes, such as leaf and root development, phase transitions, floral identity and flowering under non-stress environments is well established (Mallory and Vaucheret 2006; Jones-Rhoades et al. 2006). Table 3 shows target genes of miRNAs in *Arabidopsis* and their expression pattern under drought, salinity, cold, heat and oxidative stresses and ABA treatment in plants. Many of the miRNA target genes, involved in regulation of growth and development, show stress-regulated expression as revealed from Genevestigator (Zimmermann et. al. 2004) Response Viewer data (Table 3).

Roots play a pivotal role in maximization of water and nutrient use efficiency as 1) roots take up water and nutrient as well as sense water and nutrient status of the soil and signal the shoots for optimization of plant growth, 2) root-based hormonal (ABA) signal is an important determinat of stomatal responses, which determine the rate of transpiration and carbon fixation 3) biochemical carboxylation capacity is associated with nutrient acquisition by roots and 4) transpirational cooling ability of plants depends upon efficient function of root water uptake system.

Under conditions of drought, roots can adapt to continue growth for water and nutrient acquisition from deep soil layers. Some of the miRNAs and their target genes are involved in the regulation of root growth. The plant hormone auxin is a major regulator of root growth. Overexpression of miR160, which targets auxin response factors (ARFs) resulted in agravitrophic roots with disorganized root caps and increased lateral rooting, while overexpression of miR160-resistant ARF16 resulted in reduced lateral roots and reduced fertility (Wang et al 2005a). In contrast, transgenic *Arabidopsis* overexpressing miR164, which targets NAC transcription factors, exhibited reduced lateral rooting, whereas overexpression of miR164-resistant *NAC1* resulted in increased number of lateral roots (Guo et al. 2005).

Leaf development is also regulated by miRNAs. Transgenic plants overexpressing of miR159-resistant *MYB33* exhibited reduction in size, petiole length,

Table 3. Predicted/validated target genes of miRNAs in *Arabidopsis* and their expression under abiotic stress in plants

miRNA Gene	Target genes	Target gene functions	Fold change in expression of target genes under abiotic stresses (https://www.genevestigator.ethz.ch)					
			Drought	Salt	Cold	Heat	Oxidative	ABA
MIR156 Controls Flowering time, apical dominance	At1g27360 (1)	SPL11, squamosa promoter-binding protein-like 11 TF	1.01	0.97	0.81	1.09	1.21	1.03
	At1g27370 (1)	SPL10	1.07	0.92	0.99	1.14	1.07	0.87
	At1g53160 (2)	SPL4	0.57	0.87	0.76	0.68	0.57	1.1
	At1g69170 (1)	APL4 (SPL6),	1.07	1.41	0.75	1.21	1.22	0.92
	At2g33810 (1.5)	SPL3, binds to AP1 promoter, involved in regulation of flowering	1.01	0.77	0.59	0.94	1.07	0.75
	At2g42200 (1)	SPL9	0.77	1.18	1.48	0.73	0.86	2.12
	At3g15270 (3)	SPL5	0.85	1.47	0.44	0.70	0.63	1.03
	At3g57920 (1)	SPL	0.86	1.09	0.65	0.85	0.84	1.35
	At5g43270 (1)	SPL2	1.09	1.19	0.72	1.09	1.03	0.78
	At5g50570 (1)	SPL	0.71	1.45	0.70	1.00	0.88	0.86
	At5g50670 (1)	SPL	0.71	1.45	0.70	1.00	0.88	0.86
MIR159 Controls flowering time, male sterility	At2g26950 (1.5)	MYB104	0.71	1.23	1.33	1.23	0.67	0.8
	At2g26960 (2.5)	MYB81	0.76	0.71	0.76	0.9	0.9	1.35
	At2g32460 (1.5)	MYB101	0.61	0.89	1.16	0.96	0.49	0.72
	At3g11440 (1.5)	MYB65, R2R3-MYB gene, Double mutants with MYB33 are male sterile, showing defects in pollen development and anther development.	1.06	1.15	0.99	1.14	1.01	0.98

(*Continued*)

Table 3. (Continued)

miRNA Gene	Target genes	Target gene functions	Fold change in expression of target genes under abiotic stresses (https://www.genevestigator.ethz.ch)					
			Drought	Salt	Cold	Heat	Oxidative	ABA
	At3g60460 (1.5)	R2R3 MYB TF, DUO POLLEN 1, required for male gamete formation	–	–	–	–	–	–
	At4g26930 (2.5)	MYB97	0.94	1.14	0.74	1.22	0.76	0.93
	At5g06100 (1.5)	MYB33	1.04	0.99	1.21	1.08	1.04	0.83
	At5g55020 (2)	MYB120	0.41	0.83	0.99	0.96	0.43	0.63
	At1g30210 (2.5)	TCP24, Leaf development	0.69	0.65	1.01	0.71	0.73	0.93
	At1g53230 (3)	TCP3, flower and leaf development	0.74	0.59	0.92	0.49	0.75	0.74
	At2g31070 (2.5)	TCP10	1.21	0.61	0.58	0.57	0.58	0.84
	At3g15030 (2.5)	TCP4, Leaf development	0.64	0.34	1.63	0.5	0.76	1.09
	At4g18390 (2.5)	TCP2	0.74	1.17	3.05	0.77	0.8	1.53
MIR160 Controls root growth	**At1g77850** (0.5)	ARF17, AUXIN RESPONSE FACTOR 17	0.81	0.77	2.62	0.95	0.80	0.8
	At2g28350 (1)	ARF10	1.35	0.99	1.0	1.32	1.53	0.81
	At4g30080 (1.5)	ARF16, root cap cell differentiation	1.03	1.31	0.91	0.84	1.06	0.94

MIR161	**At1g06580**	Pentatricopeptide (PPR) repeat-containing protein	0.96	0.50	0.31	1.77	1.46	0.42
MIR162	**At1g01040** (2)	DCL1	1.12	0.71	1.08	1.31	1.13	0.85
MIR163	**At1g66690**	S-adenosyl-L-methionine:carboxyl methyltransferase family protein	1.76	5.01	9.12	1.13	1.84	1.42
	At1g66700	S-adenosyl-L-methionine:carboxyl methyltransferase family protein	–	–	–	–	–	–
	At1g66720	S-adenosyl-L-methionine:carboxyl methyltransferase family protein	0.86	1.10	0.97	1.41	0.85	1.04
	At3g44860	S-adenosyl-L-methionine:carboxyl methyltransferase family protein, catalytic specificity towards farnesoic acid	0.51	5.91	0.30	0.60	0.29	1.52
MIR164 Regulates leaf development, floral identity, root branching	**At1g56010** (1)	NAC1, downregulates auxin signals and promotes lateral root emergence	0.98	0.98	0.82	0.92	1.20	1.21
	At3g15170 (1)	CUC1	0.62	1.24	1.43	0.76	0.86	0.97

(Continued)

Table 3. (*Continued*)

miRNA Gene	Target genes	Target gene functions	Fold change in expression of target genes under abiotic stresses (https://www.genevestigator.ethz.ch)					
			Drought	Salt	Cold	Heat	Oxidative	ABA
	At5g07680 (1.5)	Similar to CUC2	0.94	2.30	0.55	1.59	1.34	4.12
	At5g39610 (2)	NAC6	1.81	5.67	1.48	1.57	1.71	2.59
	At5g53950 (1)	CUC2	0.88	1.04	1.23	1.01	0.86	1.12
	At5g61430 (1.5)	Similar to CUC2	1.21	1.47	1.07	1.98	1.42	1.86
MIR165 Regulates Leaf development, leaf polarity, auxin response, female sterility	**At1g30490** (1.5)	PHV, homeobox-leucine zipper transcription factor (HB-9); Dominant PHV mutations cause transformation of abaxial leaf fates into adaxial leaf fates	1.14	0.94	1.09	1.21	0.98	0.87
	At5g60690 (1.5)	REV, regulates meristem initiation at lateral positions. a member of a small homeodomain-leucine zipper family	1.27	0.88	0.89	0.99	1.12	0.67

	At2g34710 (1.5)	PHB, ATHB-14, meristem initiation, primary shoot apical meristem specification	1.47	0.80	0.92	0.99	1.25	0.66
	At4g32880 (1.5)	ATHB8, involved in meristem initiation, primary shoot apical meristem specification, xylem histogenesis	1.46	1.06	0.97	1.00	1.16	0.76
MIR166 (miRNA166 differs from miRNA165 by single nucleotide)	**At1g52150** (1.5)	ATHB15, Critical for vascular development and negatively regulates vascular cell differentiation	1.45	0.89	1.02	1.13	1.39	0.90
MIR167	**At1g30330** (2)	ARF6, auxin-responsive factor, Acts redundantly with ARF8 to control stamen elongation and flower maturation.	0.90	0.64	1.47	0.88	0.83	0.73
	At5g37020 (2)	ARF8, acts redundantly with ARF6 to control stamen elongation & flower maturation; negatively regulates fruit initiation	1.03	0.76	1.15	0.80	0.90	0.91
MIR168	**At1g48410** (2.5)	AGO, argonaute protein	1.13	1.27	1.15	1.19	1.17	1.12
MIR169	**At1g17590** (1.5)	CCAAT-binding transcription factor	1.12	0.85	0.53	1.14	1.12	0.53

(*Continued*)

Table 3. (Continued)

miRNA Gene	Target genes	Target gene functions	Fold change in expression of target genes under abiotic stresses (https://www.genevestigator.ethz.ch)					
			Drought	Salt	Cold	Heat	Oxidative	ABA
	At1g54160 (2)	CCAAT-binding transcription factor (CBF-B/NF-YA) family protein	1.24	6.35	0.9	1.09	1.14	8.92
	At1g72830 (1.5)	CCAAT-binding transcription factor	1.47	1.41	0.61	1.12	1.39	1.15
	At3g05690 (1.5)	CCAAT-binding transcription factor	1.38	1.31	0.78	1.51	1.69	0.97
	At3g20910 (2)	CCAAT-binding transcription factor	1.08	1.19	1.17	0.88	0.88	1.17
	At5g06510 (1.5)	CCAAT-binding transcription factor	1.53	2.54	1.02	1.26	1.90	0.97
	At5g12840 (1.5)	CCAAT-binding transcription factor	0.95	1.80	1.24	1.03	0.95	2.87
MIR171	At2g45160 (2)	Scarecrow-like (SCL) transcription factor	0.98	0.79	1.74	0.74	0.94	0.66
	At4g00150 (2)	Scarecrow-like transcription factor 6 (SCL6)	1.13	1.03	1.04	0.8	1.04	1.06
	At3g60630 (2)	Scarecrow transcription factor family protein	1.01	0.66	1.35	0.81	0.95	0.79
MIR172 Regulates floral organ identity, leaf development	**At2g28550** (1.5)	*TARGET OF EAT1* (*TOE1*), AP2 domain-containing transcription factor RAP2.7	0.80	0.79	3.21	0.85	0.88	1.15
	At2g39250 (1)	AP2-type TF that can repress flowering	0.66	0.95	1.27	0.75	0.73	3.04

	At4g36920 (0.5)	APETALA 2, floral homeotic gene	1.05	0.88	1.55	0.93	0.99	0.74
	At5g60120 (0.5)	TOE2, AP2 –type TF	0.95	1.02	1.14	1.15	0.87	1.00
	At5g67180 (1.5)	TOE3, AP2 –type TF	1.20	1.41	1.18	0.86	1.11	2.84
MIR390	At1g03475	AtCPO-I, coproporphyrinogen III oxidase	0.73	0.53	0.88	0.86	0.81	0.88
	At1g30040	Gibberellin 2-oxidase, AtGA2OX2; expression is responsive to cytokinin	1.21	3.65	0.86	1.08	1.19	0.94
	At1g56130	Leucine-rich repeat family protein/protein kinase	–	–	–	–	–	–
	At1g60720	Expressed protein	0.92	1.42	0.89	1.10	0.89	0.73
	At2g39480 (3)	ABC transporter family protein	0.97	1.30	0.93	0.92	1.01	0.86
	At3g49250 (3)	Expressed protein	1.11	1.15	0.84	1.10	1.01	0.81
	At4g03205	Coproporphyrinogen III oxidase	1.00	0.88	0.48	0.72	0.89	0.90
	At4g08850	Leucine-rich repeat family protein kinase	0.88	2.55	0.83	0.69	0.97	0.51
	At4g13430	Aconitase family protein	1.13	0.69	0.98	0.78	0.91	0.81
	At4g17450	Copia-like retrotransposon family						

(Continued)

Table 3. (Continued)

miRNA Gene	Target genes	Target gene functions	Fold change in expression of target genes under abiotic stresses (https://www.genevestigator.ethz.ch)					
			Drought	Salt	Cold	Heat	Oxidative	ABA
	At5g07280	EXCESS MICROSPORO-CYTES1, Leucine-rich repeat receptor protein kinase involved in tapetal cell fate specification	0.94	1.22	1.10	1.03	0.84	1.14
	At5g14210	Leucine-rich repeat transmembrane protein kinase	1.01	0.82	0.92	0.93	1.05	0.73
	At5g53470	Acyl-CoA binding protein	1.02	0.79	1.18	0.74	0.87	1.03
	At5g14050 (3)	Transducin family protein/WD-40 repeat family protein	1.10	0.90	1.50	1.05	1.12	0.65
MIR394	**At1g27340** (1)	F-box family protein	1.01	0.81	0.97	0.68	1.00	0.84
	At1g61520	PSI type III chlorophyll a/b-binding protein	0.75	0.70	0.84	0.83	0.72	0.74
	At1g10200	Transcription factor LIM	0.87	0.71	1.25	0.92	0.93	0.76
	At4g24860	AAA-type ATPase family protein	0.90	1.23	1.37	1.33	0.87	0.65
MIR395	At2g28780	Expressed protein	1.08	0.44	1.24	0.81	1.66	0.46
	At3g49430	Pre-mRNA splicing factor	0.89	0.85	1.01	0.94	0.96	0.81

	At3g22890 (1.5)	APS1, ATP sulfurylase, the first enzyme in the sulfate assimilation pathway	1.20	0.99	0.77	0.68	1.11	0.81
	At4g14680 (1.5)	APS3, ATP-sulfurylase 3	0.92	2.38	0.57	0.69	1.01	0.57
	At5g43780 (0.5)	APS4, ATP-sulfurylase 4	1.25	1.28	1.03	1.39	1.12	1.18
	At5g10180	AST68, a low-affinity sulfate transporter expressed in the root cap and central cylinder, where it is induced by sulfur starvation	1.31	0.91	0.85	0.69	1.17	0.89
MIR396	**At2g22840** (3)	GRL1, GROWTH-REGULATING FACTOR 1, Mutants result in smaller leaves indicating the role of the gene in leaf development. Expressed in root, shoot and flower	1.05	0.68	2.49	0.99	0.94	1.12
	At4g37740 (3)	GRL2 TF, Mutants result in smaller leaves indicating the role of the gene in leaf development	1.28	1.05	1.09	1.09	1.15	0.71

(Continued)

Table 3. (Continued)

miRNA Gene	Target genes	Target gene functions	Fold change in expression of target genes under abiotic stresses (https://www.genevestigator.ethz.ch)					
			Drought	Salt	Cold	Heat	Oxidative	ABA
	At2g36400 (3)	GRL3 TF, Mutants result in smaller leaves indicating the role of the gene in leaf development	1.27	0.90	1.01	1.19	1.08	0.64
	At3g52910 (3)	GRL4 TF, Involved in leaf development	0.96	0.89	1.01	0.97	0.93	0.91
	At5g53660 (3)	GRL7 TF, Involved in leaf development	1.11	0.93	1.09	1.06	0.88	0.92
	At4g24150 (3)	GRL8 TF, Involved in leaf development	0.79	0.87	0.92	2.21	1.26	1.11
	At2g45480 (3)	GRL9 TF, Involved in leaf development	–	–	–	–	–	–
	At2g40760 (2.5)	Rhodanese-like domain-containing protein	1.06	0.74	0.84	1.22	1.19	0.89
	At4g27180 (3)	ATK2, kinesin-like protein B, microtubule-based movement	1.36	0.94	0.82	0.98	1.24	0.84
MIR397A	**At2g29130** (2)	putative laccase	0.66	0.80	0.63	0.67	0.75	0.47
	At2g38080 (2)	Putative laccase	1.24	1.26	1.04	0.88	1.15	1.02
MIR399	At3g54700 (2)	Phosphate transporter	–	–	–	–	–	–

	At1g28540	Expressed protein	0.96	1.17	1.24	1.01	0.92	1.06
	At1g75980	Expressed protein	1.18	1.01	0.91	2.23	1.08	0.82
	At2g31700	Hypothetical protein	–	–	–	–	–	–
	At2g33770 = PHO2	ubiquitin-conjugating E2 enzyme	1.25	0.74	0.80	0.83	1.16	1.03
	At3g16760	tetratricopeptide repeat (TPR)-containing protein	0.75	0.83	0.83	0.89	0.80	1.13
	At4g09730	DEAD/DEAH box helicase	0.77	0.51	1.28	1.22	0.90	0.69
	At5g05250	Expressed protein	1.42	1.20	0.17	0.76	0.70	0.59
	At4g00170 (3)	Vesicle-associated membrane family protein	1.03	1.09	1.00	1.02	1.05	0.86
MIR400	At1g06580 (0)	Pentatricopeptide (PPR) repeat-containing protein	0.96	0.50	0.31	1.77	1.46	0.42
	At1g62720 (1)	Pentatricopeptide (PPR) repeat-containing protein	1.07	1.88	1.38	1.20	0.85	1.18
	At1g62670 (1)	Pentatricopeptide (PPR) repeat-containing protein	0.96	0.71	1.00	1.16	1.14	0.94
	At1g62590 (2)	Pentatricopeptide (PPR) repeat-containing protein	1.08	0.83	0.71	1.33	1.13	0.82

(Continued)

Table 3. (Continued)

miRNA Gene	Target genes	Target gene functions	Fold change in expression of target genes under abiotic stresses (https://www.genevestigator.ethz.ch)					
			Drought	Salt	Cold	Heat	Oxidative	ABA
	At1g22960 (2)	Pentatricopeptide (PPR) repeat-containing protein	1.08	1.36	0.80	1.26	1.08	0.77
	At1g20300 (2)	Pentatricopeptide (PPR) repeat-containing protein	1.27	0.96	1.01	1.23	1.20	0.89
	At2g31400 (2)	Pentatricopeptide (PPR) repeat-containing protein	0.88	0.73	0.99	0.82	0.94	1.04
	At5g39710 (2)	Pentatricopeptide (PPR) repeat-containing protein	1.02	1.17	1.18	1.12	0.97	0.79
	At3g16010 (3)	Pentatricopeptide (PPR) repeat-containing protein	1.06	0.85	0.54	1.22	0.99	0.86
	At4g19440 (3)	Pentatricopeptide (PPR) repeat-containing protein	1.01	1.14	0.98	1.39	1.01	0.81
MIR401	At2g06095 (1)	Expressed protein	0.62	0.77	1.08	1.62	0.62	0.47
	At3g42350 (2)	Hypothetical protein	–	–	–	–	–	–
	At2g13270 (3)	Expressed protein	–	–	–	–	–	–
MIR402	At4g34060 (1)	Similar to HhH-GPD base excision DNA repair family protein (ROS1)	–	–	–	–	–	–
MIR403	At3g07780 (5)	Expressed protein	0.84	0.95	0.78	1.19	0.95	0.89
	At1g31280	Argonaute 2	1.03	1.96	1.51	1.34	1.4	1.00
MIR404	At1g07650 (1)	Leucine-rich repeat transmembrane protein kinase	0.73	0.85	1.00	0.70	0.74	1.02
MIR405	At5g48280 (3)	Hypothetical protein	–	–	–	–	–	–
MIR406	At2g42510 (3)	spliceosome protein-related	–	–	–	–	–	–

	At1g54380 (3)	spliceosome protein-related	1.01	0.78	1.04	1.25	1.15	0.77
MIR407a	At1g17770 (3)	SET domain-containing protein (SUVH7),						
	At3g13100 (3)	ABC transporter family protein	1.11	3.11	1.71	1.14	1.25	1.27
	At5g49870 (3)	jacalin lectin family protein, similar to myrosinase-binding protein homolog	1.19	0.35	0.79	0.54	0.99	0.87
	At1g12010 (3)	Sequence is similar to 1-amino-cyclopropane-1-carboxylic acid oxidase	1.54	0.46	0.65	2.11	2.22	1.26
MIR407b	At1g54870 (3)	Short-chain dehydro-genase/reductase (SDR) family protein	0.89	2.73	1.13	1.23	0.77	1.89
MIR408	At2g47020 (0)	Peptide chain release factor	1.04	1.1	0.86	0.99	1.05	0.8
	At2g02850 (2)	Plantacyanin one of blue copper proteins. Involved in anther development and pollination.	1.25	1.0	1.57	1.1	1.22	0.89
MIR414	At1g20920	DEAD box RNA helicase	1.17	0.94	1.39	2.15	1.15	0.87
	At1g15670	Kelch repeat-containing F-box family protein	0.86	0.97	0.47	0.94	0.92	0.61
	At2g44130	Kelch repeat-containing F-box family protein	1.73	1.46	0.52	1.19	1.56	1.17

(Continued)

Table 3. (*Continued*)

miRNA Gene	Target genes	Target gene functions	Fold change in expression of target genes under abiotic stresses (https://www.genevestigator.ethz.ch)					
			Drought	Salt	Cold	Heat	Oxidative	ABA
	At3g42670	SNF2 domain-containing protein/helicase domain-containing protein, RNA-directed DNA methylation.	1.27	0.75	0.91	1.23	1.22	1.02
	At4g26110	Nucleosome assembly protein (NAP)	1.09	0.71	0.9	1.2	1.25	0.93
MIR419	At2g01830	Histidine kinase (AHK4), involved in phosphate-starvation, sugar, and cytokinin signaling	1.21	1.1	0.75	0.59	1.07	0.78
MIR447	**<u>At5g60760</u>**	2-phosphoglycerate kinase	1.21	1.22	1.17	1.09	1.45	0.43

Source for miRNA target genes: Sunkar and Zhu, 2004; Jones-Rhoades et al. 2004 & 2006; Adai et al. 2005
Note: miRNAs and their target genes listed in Table 2 are not included here. Validated target genes are in bold and underlined letters. Values in parenthesis indicate no. of mismatches between miRNA and their targets in complementary base-pair. Expression fold values are calculated from mean of 4-12 experiments with various durations of stresses, and hence indicate only mean expression over time.

apical dominance and fertility, and leaves with round shape (Millar and Gubler 2005). Transgenic *Arabidopsis* overexpressing miR160-resistant *ARF17* showed extra cotyledons, leaf symmetry and leaf defects, extra petals and reduced fertility (Mallory et al. 2005). Similarly, overexpression of miR164-resistant CUC1, miR164-resistant CUC2, miR165/166-resistant PHB and miR165/166-resistant REV in transgenic *Arabidopsis* resulted in leaf polarity defects (Laufs et al. 2004; Mallory et al. 2004a & b).

Reproductive phase is very sensitive to abiotic stresses. Hence, plants modulate flowering time and flower number (sink size) in response to abiotic stresses. One of the target genes of stress-regulated miR397b is At3g60250, which encodes the regulatory (beta) subunit (CKB3) of the protein kinase CK2. Transgenic plants overexpressing CKB3, display increased CK2 activity and shorter periods of rhythmic expression of circadian clock-associated 1 (CCA1) and LHY, and reduced phytochrome induction of an *LHCb* gene. Further, *CKB3* overexpressing plants flowered earlier under both long and short photoperiods (Sugano et al. 1999). Stress upregulation of miR397b may result in downregulation of *CKB3*, which in turn may delay flowering in adverse abiotic stress conditions. Similarly transgenic overexpression of some miRNAs resulted in alteration of flowering time. Overexpression of miR156, which targets the SPL family transcription factors, showed enhanced leaf initiation, decreased apical dominance and delayed flowering time (Schwab et al. 2005). Similarly transgenic *Arabidopsis* plants overexpressing gibberellin-regulated miR159 and miR319 showed a delay in flowering time (Palatnik et al. 2003; Achard et al. 2004). In contrast, overexpression of miR172, which targets AP2-type transcription factors, resulted in early flowering (Chen 2004). In addition to the time of flowering, fertility is highly affected by abiotic stresses. Transgenic overexpression of miR159 and miR166 resulted in enhanced male sterility and female sterility, respectively, under non-stress conditions (Achard et al. 2004; Williams et al. 2005). It is interesting to note that many of the miRNA target genes involved in plant growth and development and hormone signaling are regulated by abiotic stresses and ABA (Table 3). Studies on the expression of the miRNAs and their target genes (for root development, leaf development, flowering time and fertility) under abiotic stresses will help further understanding of these processes under abiotic stress environments.

5. MANAGEMENT OF MINERAL NUTRIENT DEFICIENCY

Mineral nutrient deficiency is one of the primary reasons for low crop yield, as resource poor farmers do not apply adequate fertilizers. Current evidences show that miRNAs regulate the acquisition and use of mineral nutrients such phosphorous and sulphur. Among the major mineral nutrients, use of nitrogenous fertilizers is more common than the use of phosphate and potash fertilizers. Even when phosphate fertilizers are applied to crops, most of the phosphate is fixed (>80%) in compounds such as phytate that are not readily available for crops. Further, as phosphate is present in readily available form ($H_2PO_4^-$) in soil in a narrow pH range (pH 6-7), phosphorous deficiency is very common in acidic, alkaline and calcareous soils. Phosphate diffusion rate is very slow and requires adequate soil moisture to keep adequate phosphate concentration at the root zone and hence phosphate availability is reduced under drought conditions. Thus, phosphorous is a limiting factor of crop productivity in many parts of the world. Phosphorus is a constituent of ATP, nucleic acids and phospholipids, and also participates in energy transfer, activation/inactivation of protein, and carbon and amino acid metabolic processes. Plants take up phosphorous as $H_2PO_4^-$ through Pi transporters. Plants employ strategies such as enhanced root growth, root hairs, activation of high affinity Pi transporters, exudation of phosphate solublizing organic acids, phosphatases and nucleases to release Pi from organic sources, and symbiotic associations with arbuscular mycorrhizal fungi (Raghothama 1999).

In silico analyses led to the identification of miR399 target genes such as At3g54700, a phosphate transporter (Jones-Rhoades and Bartel, 2004) and At2g33770, a putative ubiquitin conjugating (UBC) enzyme (Sunkar and Zhu 2004). Expression of miR399 is induced both in roots and shoots under Pi deficiency (Fujii et al. 2005; Bari et al. 2006; Chiou et al. 2006) and reversal to Pi rich conditions rapidly decreases miR399 levels (Bari et al. 2006). Expression of miR399 correlated with down regulation of its target gene UBC24 under Pi deficiency (Fujii et al. 2005; Aung et al. 2006; Bari et al. 2006; Chiou et al. 2006). Overexpressing *MIR399* in transgenic plants led to the down regulation of UBC transcript levels and accumulation of Pi in shoots to toxic levels (Fujii et al. 2005; Aung et al. 2006; Bari et al. 2006; Chiou et al. 2006). Overexpression of miR399 also impaired Pi remobilization from old leaves to developing young tissues (Chiou et al. 2006). miRNA398-mediated down regulation of UBC mediates Pi starvation response such as primary root elongation and induction of high affinity Pi transporter, AtPT1, as evident from the transgenic plants overexpressing *MIR399* and 5'UTR minus UBC (Δ*UTR-UBC*, deletion of miR399 target region), respectively (Fujii et al. 2005). Pi overaccumulator (*pho2*) mutation is caused by a single nucleotide change resulting in early termination within the *UBC24* gene and *UBC24* T-DNA knockout mutants also showed Pi hyper-accumulation and Pi toxicity similar to miR399 overexpressing plants (Aung et al. 2006). Micrografting of *pho2* root genotype also resulted in leaf Pi accumulation (Bari et al. 2006). Co-localization of miR399 and UBC24 in the vascular cylinder suggests their function in Pi translocation and remobilization (Aung et al. 2006). Identification of putative *PHO*2 orthologs containing five miR399-binding sites in their 5'-UTR in other higher plants, and Pi-dependent miR399 expression in rice suggest a conserved miR399-mediated regulatory mechanism of Pi deficiency response in plants (Bari et al. 2006). These evidences show that miR399 regulates Pi homeostasis through UBC/PHO2 proteolysis pathway. Identification of the targets of UBC/PHO2 pathway will shed more light on understanding Pi deficiency response of plants.

Sulfate is an essential nutrient for growth and development of plants. Plants adapt to sulfate deficiency by modulating expression of genes encoding sulfate transporters and enzymes of sulfate metabolism (Hawkesford and De Kok 2006).

miRNAs also play crucial role in sulfate acquisition and use in plants. *In silico* identification of miR395 targets such as At5g10180 (a low-affinity sulfate transporter) and ATP sulfurylases (APS1, 3 & 4, catalyzes the first and rate limiting step in the sulfate assimilation pathway) (Sunkar and Zhu, 2004; Jones-Rhoades et al. 2004 & 2006). Expression of miR395 is induced with a concomitant down regulation of ATP sulfurylases (APS) transcripts under sulfate deficiency stress (Jones-Rhoades and Bartel 2004). Moreover, oxidative, pathogen infection and heavy metal stresses appear to increase the sulfur demand, as these stresses enhance the expression of sulfate transporters and enzymes of the assimilatory pathway (Hawkesford and De Kok 2006). The target genes of miR395 also show differential regulation in response to various abiotic stresses (Table 3). These results suggest a crucial role of miR395 in regulation of sulfate uptake and homeostasis in plants.

6. CONCLUSIONS AND PERSPECTIVES

Plants employ complex mechanisms of gene regulation in response to developmental and environmental cues. Small RNAs, namely miRNAs and siRNAs, have emerged as a novel regulatory mechanism of gene expression, as expression levels of small RNAs have inverse relationship with the transcript levels of their target genes. Post-transcriptional gene repression is mediated by small RNA-guided cleavage of target mRNAs and translational repression, while transcriptional silencing is mediated mainly by heterochromatic siRNAs and in some cases by miRNAs. Abiotic stresses induce upregulation and downregulation of numerous genes. It is expected that small RNAs may play a significant role in stress regulated gene expression, as, the *Arabidopsis hen1-1* and *dcl1-9* mutants that are impaired in the production of miRNAs are hypersensitive to abiotic stresses (Zhu JK, unpublished data). Further, microarray analysis of *ago1* and *dcl1* mutants revealed impaired expression of stress responsive genes for oxidative stress management, heat shock proteins, aquaporins, LEA-type proteins, fatty acid desaturases, etc (Ronemus et al. 2006).

Recently some stress-regulated miRNAs have been identified. Oxidative stress mediated downregulation of miR398 gene results in posttranscriptional induction of Cu-Zn SOD and thus ROS management (Sunkar et al. 2006). Osmoprotection and ROS management under salinity stress is regulated by nat-siRNAs in plants. Transcription of antisense overlapping gene pair of Δ^1-pyrroline-5-carboxylate dehydrogenase (P5CDH) and the salt-stress induced gene SRO5, generates dsRNA, which is processed into a 24-nt siRNA by DCL2, RDR6, SGS3, and NRPD1A. Initial cleavage of the P5CDH transcript guided by the 24-nt siRNA results in generation of 21-nt siRNAs by DCL1 and further cleavage of P5CDH transcripts. Proline accumulation increases as P5CDH is downregulated, and thus accumulated proline contributes to osmoprotection/osmotic adjustment. In addition, stress induced SRO5 protein mediates ROS detoxification (Borsani et al. 2005).

The functional relationship between cis-natural antisense gene pair P5CDH-SRO5 and their regulation by nat-siRNAs suggest that similar functional relationships may be applied to other cis-antisense gene pairs. About 2000 genes in convergent overlapping pairs in *Arabidopsis* are regulated by various environmental or hormonal stimuli (Girke T and Zhu JK, unpublished data) and thus these may be regulated by nat-siRNAs similar to the P5CDH-SRO5 gene pair.

Several stress regulated transcription factors and effector genes have been identified. It is not known whether small RNAs play a role in homeostasis of stress responsive transcriptome. Abiotic stresses cause significant changes in growth and development of plants. Although, several small RNAs that regulate plant growth and development under non-stress environments are known, their role under abiotic stress environments is yet to be established. Studies on stress regulated small RNAs will help in understanding the phenological and developmental plasticity of plants under abiotic stress environments.

Small RNAs are being identified in a wide range of plants, and many conserved small RNAs have been identified. Further understanding of posttranscriptional gene regulation under abiotic stress is important for better genetic engineering of stress tolerance in plants. Stress regulated manipulation of miRNA genes and miRNA-resistant target genes will help overcome post-transcriptional gene silencing, and thus better expression of engineered traits in transgenic plants. Understanding the role of small RNAs and their target genes in abiotic stress responses of plants will be imperative for better understanding of abiotic stress tolerance of plants, and will further enhance our knowledge of transcriptome homeostasis in abiotic stress response of plants.

ACKNOWLEDGEMENTS

Work done in our lab was supported by National Institute of Health grants R01GM59138 and R01GM0707501, and National Science Foundation grant IBN-0212346.

REFERENCES

Achard P, Herr A, Baulcombe DC, Harberd NP. (2004). Modulation of floral development by a gibberellin-regulated microRNA. *Development* 131: 3357–65.

Adai A, Johnson C, Mlotshwa S, Archer-Evans S, Manocha V, Vance V, Sundaresan V. (2005). Computational prediction of miRNAs in *Arabidopsis thaliana*. *Genome Res.* 15: 78–91.

Ahlfors R, Lang S, Overmyer K, Jaspers P, Brosche M, Taurianinen A, Kollist H, Tuominen H, Belles-Boix E, Piippo M, Inze D, Palva ET, Kangasjarvi J. (2004). *Arabidopsis* radical-induced cell death1 belongs to the WWE protein-protein interaction domain protein family and modulates abscisic acid, ethylene, and methyl jasmonate responses. *Plant Cell* 16: 1925–1937.

Alscher RG, Erturk N, Heath LS. (2002). Role of superoxide dismutases in controlling oxidative stress in plants. *J Exp Bot.* 53: 1331–1341.

Aukerman MJ, Sakai H. (2003). Regulation of flowering time and floral organ identity by a microRNA and its APETALA2-like target genes. *Plant Cell* 15: 2730–2741.

Aung K, Lin SI, Wu CC, Huang YT, Su Cl, Chiou TJ. (2006). *pho2*, a phosphate overaccumulator, is caused by a nonsense mutation in a microRNA399 target gene. *Plant Physiol.* 141: 1000–1011.

Ayliffe MA, Roberts JK, Mitchell HJ, Zhang R, Lawrence GJ, Ellis JG, Pryor TJ. (2002). A plant gene up-regulated at rust infection sites. *Plant Physiol.* 129: 169–180.

Bao N, Lye KW, Barton MK. (2004). MicroRNA binding sites in *Arabidopsis* class III HD-ZIP mRNAs are required for methylation of the template chromosome. *Dev Cell* 7: 653–662.

Bari R, Datt Pant B, Stitt M, Scheible WR. (20060. PHO2, microRNA399, and PHR1 define a phosphate-signaling pathway in plants. *Plant Physiol.* 141: 988–999.

Bartel DP. (2004). MicroRNAs: genomics, biogenesis, mechanism, and function. *Cell* 116: 281–297.

Bertone P, Stolc V, Royce TE, Rozowsky JS, Urban AE, Zhu X, Rinn JL, Tongprasit W, Samanta M, Weissman S. Gerstein M, Snyder M. (2004). Global identification of human transcribed sequences with genome tiling arrays. *Science* 306: 2242–2246.

Bohnert HJ, Gong Q, Li P, Ma S. 2006. Unraveling abiotic stress tolerance mechanisms - Getting genomics going. *Curr Opin Plant Biol.* 9: 180–188

Boi S, Solda G, Tenchini ML. (2004). Shedding light on the dark side of the genome: overlapping genes in higher eukaryotes. *Curr Genomics* 5, 509–524.

Bonnet E, Wuyts J, Rouzé P, Van de Peer Y. (2004). Detection of 91 potential conserved plant microRNAs in *Arabidopsis thaliana* and *Oryza sativa* identifies important target genes. *Poc Natl Acad Sci USA.* 101: 11511–11516

Borsani O, Zhu J, Verslues PE, Sunkar R, Zhu JK. (2005). Endogenous siRNAs derived from a pair of natural *cis*-antisense transcripts regulate salt tolerance in *Arabidopsis. Cell* 123: 1279–1291.

Bot AJ, Nachtergaele FO, Young A. (2000). Land resource potential and constraints at regional and country levels. *In*: *World Soil Resources Reports 90*, Land and Water Development Division, FAO, Rome, 2000. pp: 17–24

Brenner S, Johnson M, Bridgham J, Golda G, Lloyd DH, Johnson D, Luo SJ, McCurdy S, Foy M, Ewan M, Roth R, George D, Eletr S, Albrecht G, Vermaas E, Williams SR, Moon K, Burcham T, Pallas M, DuBridge RB, Kirchner J, Fearon K, Mao J, Corcoran K. (2000). Gene expression analysis by massively parallel signature sequencing (MPSS) on microbead arrays. *Nat Biotechnol.* 18: 630–634.

Chan SW, Henderson IR, and Jacobsen SE. (2005). Gardening the genome: DNA methylation in *Arabidopsis thaliana. Nat Rev Genet.* 6: 351–360.

Chan SW, Zilberman D, Xie Z, Johansen LK, Carrington JC, Jacobsen SE. (2004). RNA silencing genes control *de novo* DNA methylation. *Science* 303: 1336

Chen X. (2004). A microRNA as a translational repressor of *APETALA2* in *Arabidopsis* flower development. *Science* 303: 2022–25

Chinnusamy V, Jagendorf A, Zhu JK. (2005). Understanding and improving salt tolerance in plants. *Crop Sci.* 45: 437–448.

Chinnusamy V, Ohta M, Kanrar S, Lee B.-h, Hong X, Agarwal M, Zhu JK. (2003). ICE1, a regulator of cold induced transcriptome and freezing tolerance in *Arabidopsis. Genes Dev.* 17: 1043–1054.

Chinnusamy V, Schumaker K, Zhu JK. (2004). Molecular genetic perspectives on cross-talk and specificity in abiotic stress signalling in plants. *J Exp Bot.* 55: 225–236.

Chinnusamy V, Zhu J, Zhu JK (2006) Gene regulation during cold acclimation in plants. *Physiol Plant.* 126: 52–61.

Chiou TJ, Aung K, Lin SI, Wu CC, Chiang SF, Su CL. (2006). Regulation of phosphate homeostasis by MicroRNA in Arabidopsis. *Plant Cell* 18: 412–21

Dharmasiri S, Estelle M. (2002). The role of regulated protein degradation in auxin response. *Plant Mol Biol.* 9: 401–409.

Deuschle K, Funck D, Forlani G, Stransky H, Biehl A, Leister D, van der Graaff E, Kunze R, Frommer WB. (2004). The Role of {delta}1-pyrroline-5-carboxylate dehydrogenase in proline degradation. *Plant Cell* 16: 3413–3425.

Dong CH, Agarwal M, Zhang Y, Xie Q, Zhu JK. (2006). The negative regulator of plant cold responses, HOS1, is a RING E3 ligase that mediates the ubiquitination and degradation of ICE1. *Proc Nat Acad Sci USA* 103: 8281–8286.

Ferguson DL, Guikema JA, Paulsen GM. (1990). Ubiquitin pool modulation and protein degradation in wheat roots during high temperature stress. *Plant Physiol.* 92: 740–746.

Foyer CH, Noctor G. (2005). Redox homeostasis and antioxidant signaling: A metabolic interface between stress perception and physiological responses. *Plant Cell* 17: 1866–1875.

Fujii H, Chiou TJ, Lin SI, Aung K, Zhu JK. (2005). A miRNA involved in phosphate-starvation response in *Arabidopsis. Curr Biol.* 15: 2038–43.

Gasciolli V, Mallory AC, Bartel DP, Vaucheret H. (2005). Partially redundant functions of *Arabidopsis* DICER-like enzymes and a role for DCL4 in producing trans-acting siRNAs. *Curr Biol.* 15: 1494–1500.

Guo HS, Xie Q, Fei JF, Chua NH. (2005). MicroRNA directs mRNA cleavage of the transcription factor *NAC1* to downregulate auxin signals for *Arabidopsis* lateral root development. *Plant Cell* 17: 1376–86

Hamilton AJ, Baulcombe DC. (1999). A novel species of small antisense RNA in posttranscriptional gene silencing in plants. *Science* 286: 950–952.

Hawkesford MJ and De Kok LJ. 2006. Managing sulphur metabolism in plants. *Plant, Cell Environ.* 29: 382–395

Hong Z, Lakkineni K, Zhang Z, Verma DPS. (2000). Removal of feedback inhibition of Δ^1-pyrroline-5-carboxylate synthetase results in increased proline accumulation and protection of plants from osmotic stress. *Plant Physiol.* 122:1129–1136.

IPCC (2001). *Climate change 2001*: the scientific basis. Contribution of working group I to the third assessment report of the Intergovernmental Panel on Climate Change (IPCC). Houghton JT, Ding Y, Griggs DJ, Noguer M, van der Linden PJ, Xiaosu D, eds. Cambridge, UK: Cambridge University Press.

Jen CH, Michalopoulos I, Westhead DR, Meyer P. (2005). Natural antisense transcripts with coding capacity in *Arabidopsis* may have a regulatory role that is not linked to double-stranded RNA degradation. *Genome Biol.* 6: R51.

Jones-Rhoades MW, Bartel DP, Bartel B. (2006). MicroRNAs and their regulatory roles in plants. *Annu Rev Plant Biol.* 57: 19–53

Jones-Rhoades, MW, Bartel DP. (2004). Computational identification of plant miRNAs and their targets, including a stress-induced miRNA. *Mol Cell* 14: 787–799

Kavi Kishor PB, Hong Z, Miao G, Hu C, Verma DPS (1995) Overexpression of ^{1}-pyrroline-5-carboxylate synthetase increases proline overproduction and confers osmotolerance in transgenic plants. *Plant Physiol.* 108:1387–1394

Kidner CA, Martienssen RA. (2005). The developmental role of microRNA in plants. *Curr Opin Plant Biol.* 8:38–44.

Kurihara Y, Takashi Y, Watanabe Y. (2006). The interaction between DCL1 and HYL1 is important for efficient and precise processing of pri-miRNA in plant microRNA biogenesis. *RNA* 12: 206–212.

Laufs P, Peaucelle A, Morin H, Traas J. (2004). MicroRNA regulation of the CUC genes is required for boundary size control in *Arabidopsis* meristems. *Development* 131: 4311–22.

Lee B-h, Henderson DA, Zhu JK. (2005). The Arabidopsis cold-responsive transcriptome and its regulation by ICE1. *Plant Cell* 17: 3155–3175.

Lee RC, Feinbaum RL, Ambros V. (1993). The *C. elegans* heterochronic gene *lin-4* encodes small RNAs with antisense complementarity to *lin-14*. *Cell 75*: 843–854.

Lippman Z, Gendrel AV, Black M, Vaughn MW, Dedhia N, McCombie WR, Lavine K, Mittal V, May B, Kasschau KD, Carrington JC, Doerge RW, Colot V, Martienssen R. (2004). Role of transposable elements in heterochromatin and epigenetic control. *Nature* 430: 471–476

Lu S, Sun YH, Shi R, Clark C, Li L, Chiang VL. (2005). Novel and mechanical stress-responsive microRNAs in *Populus trichocarpa* that are absent from *Arabidopsis. Plant Cell* 17: 2186–2203.

Mallory AC, Reinhart BJ, Jones-Rhoades MW, Tang G, Zamore PD, Barton MK, Bartel DP (2004a). MicroRNA control of PHABULOSA in leaf development: importance of pairing to the microRNA 5' region. *EMBO J.* 23: 3356–64

Mallory AC, Dugas DV, Bartel DP, Bartel B. (2004b). MicroRNA regulation of NAC-domain targets is required for proper formation and separation of adjacent embryonic, vegetative, and floral organs. *Curr Biol.* 14:1035–46.

Mallory AC, Bartel DP, Bartel B. (2005). MicroRNA-directed regulation of *Arabidopsis AUXIN RESPONSE FACTOR17* is essential for proper development and modulates expression of early auxin response genes. *Plant Cell* 17: 1360–1375.

Mallory AC, Vaucheret H. (2006). Functions of microRNAs and related small RNAs in plants. *Nature Genet.* 38: S31–S36.

Millar AA, Gubler F. (2005). The *Arabidopsis GAMYB-Like* genes, *MYB33* and *MYB65*, are microRNA-regulated genes that redundantly facilitate anther development. *Plant Cell* 17: 705–721.

Mittler R. (2002). Oxidative stress, antioxidants and stress tolerance. *Trends Plant Sci.* 7: 405–410.

Moon J, Parry G, Estelle M. 2004. The ubiquitin-proteasome pathway and plant development. *Plant Cell* 16: 3181–95.

Nanjo, T., Kobayashi, M., Yoshiba, Y., Sanada, Y., Wada, K., Tsukaya, H., Kakubari, Y., Yamaguchi-Shinozaki, K., and Shinozaki, K. (1999). Biological functions of proline in morphogenesis and osmotolerance revealed in antisense transgenic *Arabidopsis thaliana*. *Plant J.* 18: 185–193.

Ortiz C, Cardemil L (2001) Heat-shock responses in two leguminous plants: a comparative study. *J Exp Bot.* 52: 1711–9.

Palatnik JF, Allen E, Wu X, Schommer C, Schwab R, Carrington JC, Weigel D. (2003) Control of leafmorphogenesis by microRNAs. *Nature* 425: 257–263.

Payton P, Allen RD, Trolinder N, Holaday AS.(1997). Over-expression of chloroplast-targeted Mn superoxide dismutase in cotton (*Gossypium hirsutum* L., cv. Coker 312) does not alter the reduction of photosynthesis after short exposures to low temperature and high light intensity. *Photosynth Res.* 52: 233-244

Peragine A, Yoshikawa M, Wu G, Albrecht HL, and Poethig RS. (2004). SGS3 and SGS2/SDE1/RDR6 are required for juvenile development and the production of trans-acting siRNAs in *Arabidopsis*. *Genes Dev.* 18: 2368–2379.

Pitcher LH, Brennan E, Hurley A, Dunsmuir P, Tepperman JM, Zilinskas BA.(1991). Overproduction of petunia chloroplastic copper/zinc superoxide dismutase does not confer ozone tolerance in transgenic tobacco. *Plant Physiol.* 97: 452–455

Raghothama KG. (1999). Phosphate acquisition. *Annu Rev Plant Physiol Plant Mol Biol.* 50: 665–693.

Rolland F, Baena-Gonzalez E, Sheen J. (2006) Sugar sensing and signaling in plants: Conserved and novel mechanisms. *Annu Rev Plant Biol.* 57: 675–709

Ronemus M, Vaughn MW, Martienssen RA. (2006). MicroRNA-targeted and small interfering RNA–mediated mRNA degradation is regulated by argonaute, dicer, and RNA-dependent RNA polymerase in *Arabidopsis*. *Plant Cell* 18: 1559–1574

Schwab R, Palatnik JF, Riester M, Schommer C, Schmid M, Weigel D. (2005). Specific effects of microRNAs on the plant transcriptome. *Dev Cell* 8: 517–527

Somerville CR, Briscoe J (2001). Genetic engineering and water. *Science* 292: 2217

Sugano S, Andronis C, Ong MS, Green RM, Tobin EM. (1999). The protein kinase CK2 is involved in regulation of circadian rhythms in *Arabidopsis*. *Poc Natl Acad Sci USA*. 96: 12362-12366.

Sunkar R, Girke T, Jain PK, Zhu JK. (2005). Cloning and characterization of microRNAs from rice. *Plant Cell* 17:1397–411.

Sunkar R, Kapoor A, Zhu JK. (2006) Posttranscriptional induction of two Cu/Zn superoxide dismutase genes in *Arabidopsis* is mediated by downregulation of miR398 and important for oxidative stress tolerance. *Plant Cell* 18: 2051–2065

Sunkar R, Zhu JK. (2004). Novel and stress-regulated microRNAs and other small RNAs from Arabidopsis. *Plant Cell* 16: 2001–2019.

Tepperman JM, Dunsmuir P. (1990). Transformed plants with elevated levels of chloroplastic SOD are not more resistant to superoxide toxicity. *Plant Mol Biol.* 14: 501–11.

Umezawa T, Fujita M, Fujita Y, Yamaguchi-Shinozaki K, Shinozaki K. 2006.Engineering drought tolerance in plants: discovering and tailoring genes to unlock the future. *Curr Opin Biotech.* 17: 113–22.

Vazquez F, Vaucheret H, Rajagopalan R, Lepers C, Gasciolli V, Mallory AC, Hilbert JL, Bartel DP, Crete P. (2004). Endogenous trans-acting siRNAs regulate the accumulation of Arabidopsis mRNAs. *Mol Cell* 16: 69–79.

Wang JW, Wang LJ, Mao YB, Cai WJ, Xue HW, Chen XY. (2005a). Control of root cap formation by microRNA-targeted auxin response factors in *Arabidopsis*. *Plant Cell* 17:2204–16

Wang XJ, Gaasterland T, Chua NH. (2005b). Genome-wide prediction and identification of cis-natural antisense transcripts in *Arabidopsis thaliana*. *Genome Biol.* 6: R30.

Williams L, Grigg SP, Xie M, Christensen S, Fletcher JC. (2005). Regulation of *Arabidopsis* shoot apical meristem and lateral organ formation by microRNA *miR166g* and its *AtHDZIP* target genes. *Development* 132: 3657–3668

Xie Z, Johansen LK, Gustafson AM, Kasschau KD, Lellis AD, Zilberman D, Jacobsen SE, Carrington JC. (2004). Genetic and functional diversification of small RNA pathways in plants. *PLoS Biol.* 2: 642–652.

Xie ZX, Allen E, Wilken A, Carrington JC. (2005). DICER-LIKE 4 functions in trans-acting small interfering RNA biogenesis and vegetative phase change in *Arabidopsis thaliana. Proc Natl Acad Sci USA* 102: 12984–12989.

Yamada K, Lim J, Dale JM, Chen HM, Shinn P, Palm CJ, Southwick AM, Wu AC, Kim C, Nguyen M, Pham P, Cheuk R, Karlin-Newmann G, Liu SX, Lam B, Sakano H, Wu T, Yu G, Miranda M, Quach HL, Tripp M, Chang CH, Lee JM, Toriumi M, Chan MM, Tang CC, Onodera CS, Deng JM, Akiyama K, Ansari Y, Arakawa T, Banh J, Banno F, Bowser L, Brooks S, Carninci P, Chao Q, Choy N, Enju A, Goldsmith AD, Gurjal M, Hansen NF, Hayashizaki Y, Johnson-Hopson C, Hsuan VW, Iida K, Karnes M, Khan S, Koesema E, Ishida J, Jiang PX, Jones T, Kawai J, Kamiya A, Meyers C, Nakajima M, Narusaka M, Seki M, Sakurai T, Satou M, Tamse R, Vaysberg M, Wallender EK, Wong C, Yamamura Y, Yuan S, Shinozaki K, Davis RW, Theologis A, Ecker JR. (2003). Empirical analysis of transcriptional activity in the *Arabidopsis* genome. *Science* 302: 842–846.

Yamaguchi T, Blumwald E. (2005). Developing salt-tolerant crop plants: challenges and opportunities. *Trends Plant Sci.* 10: 615–620.

Yamaguchi-Shinozaki K, Shinozaki K. (2006). Transcriptional regulatory networks in cellular responses and tolerance to dehydration and cold stresses. *Annu Rev Plant Biol.* 57: 781–803.

Yeo AR, Flowers SA, Rao G, Welfare K, Senanayake N and Flowers TJ. 1999. Silicon reduces sodium uptake in rice (*Oryza sativa* L.) in saline conditions and this is accounted for by a reduction in the transpirational bypass flow. *Plant Cell Environ.* 22: 559–565.

Yoshikawa M, Peragine A, Park MY, and Poethig RS (2005). A pathway for the biogenesis of trans-acting siRNAs in *Arabidopsis*. *Genes Dev.* 19: 2164–2175.

Zhu JK. (2002). Salt and drought stress signal transduction in plants.*Annu Rev Plant Biol.* 53: 247–73.

Zhu JK. (2003). Regulation of ion homeostasis under salt stress. *Curr Opin Plant Biol.* 6: 441–445.

Zilberman D, Cao X, Johansen LK, Xie Z, Carrington JC, Jacobsen SE. (2004). Role of *Arabidopsis* ARGONAUTE4 in RNA-directed DNA methylation triggered by inverted repeats. *Curr Biol.* 14: 1214–20.

Zimmermann P, Hirsch-Hoffmann M, Hennig L, Gruissem W. (2004). GENEVESTIGATOR. *Arabidopsis* microarray database and analysis toolbox. *Plant Physiol.* 136: 2621–2632

CHAPTER 11

TRANSCRIPTOME ANALYSIS OF PLANT DROUGHT AND SALT STRESS RESPONSE

MOTOAKI SEKI[1,*], TAISHI UMEZAWA[2], JONG-MYONG KIM[1], AKIHIRO MATSUI[1], TAIKO KIM TO[1]
AND KAZUO SHINOZAKI[1,2,3]

[1]*Plant Genomic Network Research Team, Plant Functional Genomics Research Group, RIKEN Plant Science Center (PSC), RIKEN Yokohama Institute, 1-7-22 Suehiro-cho, Tsurumi-ku, Yokohama 230-0045, Japan*
[2]*Gene Discovery Research Team, RIKEN Plant Science Center (PSC), RIKEN Tsukuba Institute, 3-1-1 Koyadai, Tsukuba, Ibaraki 305-0074, Japan*
[3]*CREST, Japan Science and Technology Corporation (JST), Japan*

Abstract: Plants must adapt to drought and high-salinity stresses in order to survive. Molecular and genomic studies have shown that many genes with various functions are induced by drought and high-salinity stresses, and that the various signaling factors are involved in the stress responses. The development of microarray-based expression profiling methods, together with the availability of genomic and/or cDNA sequence data, and gene-knock-out mutants, has allowed significant progress in the characterization of the plant stress response. Recent studies also revealed that small RNAs, RNA processing and chromatin regulation are involved in the abiotic stress responses. In this review, we highlight recent progress in the research on the transcriptome for the response to plant drought and salt stress

Keywords: drought stress, high-salinity stress, transcriptome, transcription factors, small RNAs, RNA processing, chromatin remodeling

1. INTRODUCTION

Plant growth is greatly affected by environmental abiotic stresses, such as drought and high salinity. Plants must adapt to these stresses in order to survive. These stresses induce various biochemical and physiological responses in plants. Several

**Corresponding author:* Dr. Motoaki Seki, Plant Genomic Network Research Team, Plant Functional Genomics Research Group, RIKEN Plant Science Center (PSC), RIKEN Yokohama Institute, 1-7-22 Suehiro-cho, Tsurumi-ku, Yokohama 230-0045, Japan Tel: +81-45-503-9625; Fax: +81-45-503-9586; e-mail: mseki@psc.riken.jp

M.A. Jenks et al. (eds.), Advances in Molecular Breeding Toward Drought and Salt Tolerant Crops, 261–283.

hundred genes that respond to these stresses at the transcriptional level have been identified (Figure 1, Shinozaki and Yamaguchi-Shinozaki 2000; Kreps et al., 2002; Seki et al. 2002b, 2003; Xiong et al., 2002; Zhu 2002; Shinozaki et al. 2003; Lee et al., 2005). It is important to analyze the functions of stress-inducible genes not only to understand the molecular mechanisms of stress tolerance and the responses of higher plants but also to improve the stress tolerance of crops by gene manipulation. Stress-inducible genes have been used to improve the stress tolerance of transgenic plants (Thomashow 1999; Hasegawa et al. 2000; Shinozaki and Yamaguchi-Shinozaki 2000; Zhang 2003; Umezawa et al., 2006).

Molecular and genetic analyses have recently revealed that newly identified small RNAs, RNA processing and chromatin regulation have functions in the drought and salt stress responses (Figure 1). Several small RNAs are regulated by the abiotic stresses (Sunkar and Zhu, 2004; Borsani et al., 2005). Several genes involved in RNA processing (Lu and Fedoroff, 2000; Hugouvieux et al., 2001; Xiong et al., 2001; Gong et al., 2002; Koiwa et al., 2002; Xiong et al., 2002; Papp et al., 2004; Nishimura et al., 2005) and chromatin regulation (Sridha and Wu, 2006) have been identified as components of the drought and salt stress signal transduction.

In this review, we highlight recent progress on research on the transcriptome for the response to drought and salt stresses.

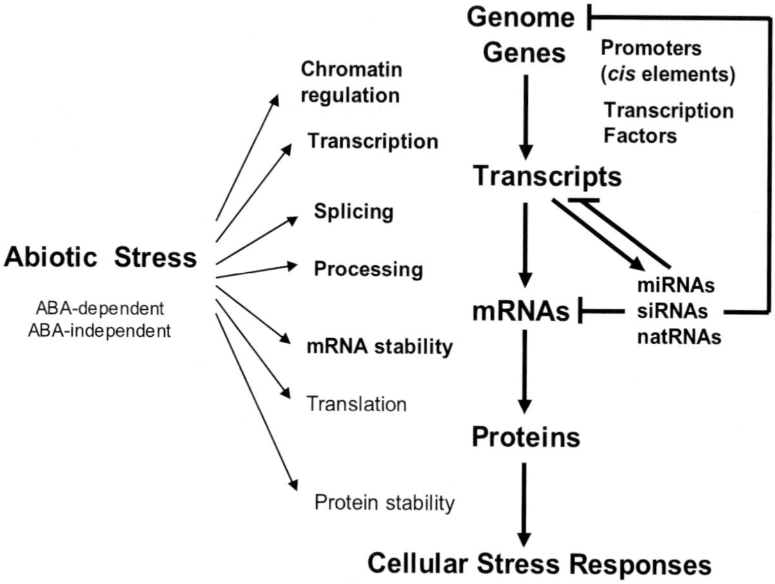

Figure 1. New concept of transcriptional regulatory networks in abiotic stress responses

2. DNA MICROARRAYS ARE AN EXCELLENT TOOL FOR IDENTIFYING GENES REGULATED BY VARIOUS STRESSES

Microarray technology is a powerful tool for the systematic analysis of expression profiles of large numbers of genes (Richmond and Somerville 2000; Seki et al. 2004). Microarray technology has been applied to the analysis of expression profiles in response to abiotic stresses, such as drought, cold and high-salinity by several groups (Kawasaki et al. 2001; Seki et al. 2001, 2002b; Chen et al. 2002; Fowler and Thomashow 2002; Kreps et al. 2002; Lee et al., 2005). Several hundred genes that respond to drought, cold and high-salinity stresses have been identified at the transcriptional level by microarray technology in *Arabidopsis* (Fowler and Thomashow 2002; Kreps et al. 2002; Seki et al., 2002b; Lee et al., 2005). Stress-responsive genes have also been identified in other plant species, such as an *Arabidopsis*-related halophyte, *Thellungiella halophila* (Inan et al., 2004; Taji et al., 2004; Gong et al., 2005; Wong et al., 2006), barley (Oztur et al., 2002), maize (Wang et al., 2003; Yu and Setter, 2003), rice (Kawasaki et al., 2001; Rabbani et al., 2003; Lan et al., 2005), wheat (Gulick et al., 2005), hot pepper (Hwang et al., 2005), pine (Watkinson et al., 2003), poplar (Gu et al., 2004; Brosche et al., 2005), potato (Rensink et al., 2005), and sorghum (Buchanan et al., 2005).

The products of the drought-, high-salinity- or cold-stress-inducible gene products can be classified into 2 groups (Shinozaki and Yamaguchi-Shinozaki 1999, 2000; Seki et al. 2002b, 2003). The first group includes functional proteins that probably function in stress tolerance. The second group contains regulatory proteins, that is, protein factors involved in further regulation of signal transduction and gene expression that probably function in the stress response (Shinozaki and Yamaguchi-Shinozaki, 2000; Seki et al. 2002b, 2003), such as various transcription factors, protein kinases, protein phosphatases, enzymes involved in phospholipid metabolism, F-box proteins, and other signaling molecules, such as calmodulin-binding protein(Seki et al. 2002b). Several reviews on the transcriptome in abiotic stress conditions in higher plants have been published recently (Bray 2002; Ramanjulu and Bartels 2002; Hazen et al., 2003; Seki et al., 2003, 2004, 2005; Yamaguchi-Shinozaki and Shinozaki, 2005, 2006).

3. MANY STRESS-INDUCIBLE TRANSCRIPTION FACTOR GENES HAVE BEEN IDENTIFIED BY TRANSCRIPTOME ANALYSIS

Transcription factors play important roles in response of the plant to environmental stresses and its development. Transcription factors are sequence-specific DNA-binding proteins that are capable of activating and/or repressing transcription. The *Arabidopsis* genome encodes more than 1,500 transcription factors (Riechmann et al., 2000) and a number of transcription factor families have been implicated in plant stress responses (Shinozaki et al. 2003; Yamaguchi-Shinozaki and Shinozaki, 2005, 2006). Microarray analyses have also revealed many stress-inducible transcription factor genes and demonstrated overlap among various stress- or hormone-signaling pathways (Seki et al., 2001, 2002b, 2002c, 2004; Chen et al., 2002; Cheong et al., 2002; Narusaka et al. 2003; Shinozaki et al. 2003).

Recently, Iida et al. (2005) identified ca. 2,000 *Arabidopsis* transcription factor proteins by PSI-BLAST and InterProScan. The list of the transcription factors is available from RARTF (RIKEN *Arabidopsis* Transcription Factor database, http://rarge.gsc.riken.jp/rartf/). We prepared a new custom-made oligomicroarray containing the 2,000 transcription factor genes and identified 73 and 93 transcription factor genes that are induced by drought and high-salinity, respectively (Seki et al., unpublished results). The major transcription factor families that are induced by the abiotic stresses are ERF/AP2, bZIP, NAC, MYB, bHLH, Cys2Cys2 zinc-finger, Cys2His2 zinc-finger, WRKY, HB and HSF (Seki et al., unpublished results). These transcription factors probably function in stress-inducible gene expression, although most of their target genes have not yet been identified.

DNA microarrays are useful for identifying the target genes of the stress-related transcription factors. Target genes of the following stress-related transcription factors have been studied by microarray analyses (Table 1): ERF/AP2 TF family, such as DREB1A/CBF3 (Seki et al., 2001, Fowler and Thomashow, 2002, Maruyama et al., 2004), DREB1C/CBF2 (Vogel et al., 2005), rice DREB1/CBF homolog, OsDREB1A (Dubouzet et al., 2003; Ito et al., 2006) and *Brassica* DREB1/CBF homologs, BNCBF5 and 17 (Savitch et al., 2005), DREB2A (Sakuma et al., 2006); bZIP TF family, such as AREB1 (Fujita et al., 2005; Furihata et al., 2006); MYB TF family, such as AtMYB2 (Abe et al., 2003), AtMYB60 (Cominelli et al., 2005) and HOS10 (Zhu et al., 2005); bHLH TF family, such as AtMYC2 (Abe et al., 2003) and ICE1 (Lee et al., 2005); NAC TF family, such as RD26/ANAC072 (Fujita et al., 2004, Tran et al., 2004), ANAC019 (Tran et al., 2004) and ANAC055 (Tran et al., 2004); C2H2-Type Zinc Finger TF family, such as ZAT12 (Davletova et al., 2005); Homeodomain TF family, such as HOS9 (Zhu et al., 2004). The roles of the transcription factors in the abiotic stress signaling and the expression profiling results are summarized in recent reviews (Shinozaki et al., 2003, Bartels and Sunkar, 2005; Seki et al., 2005; Yamaguchi-Shinozaki and Shinozaki, 2005, 2006; Umezawa et al., 2006). The information of the target genes is useful for understanding the transcriptional regulatory networks in cellular responses to the abiotic stresses.

4. MICROARRAY ANALYSIS OF THE MUTANTS OR TRANSGENIC PLANTS RELATIVE TO STRESS-RELATED SIGNALING FACTORS PROVIDES A BROAD PICTURE OF THE TRANSCRIPTOME REGULATED BY THE SIGNALING FACTOR

Molecular and genetic studies have shown that various signal transduction systems function in abiotic stress responses, involving protein phosphorylation and/or dephosphorylation, phospholipid signaling, calcium signaling, protein degradation

Table 1. Microarray analysis on the Arabidopsis signaling factor genes involved in the abiotic stress responses

Functional Categories	Genes	AGI code	Coded protein	Mutant or Transgenic Plants	Phenotypes of mutants or transgenic plants	Microarrays used	Expression profiling results	References
Transcription Factors								
	DREB1A/CBF3	At4g25480	ERF/AP2 transcription factor	Transgenic Plants overexpressing DREB1A/CBF3		1.3K RAFL cDNA microarray	upregulation of 12 genes	Seki et al. (2001) Plant Cell 13:61–72
				Transgenic Plants overexpressing DREB1A/CBF3		8K Affymetrix GeneChip	upregulation of 41 genes, downregulation of 27 genes	Fowler and Thomashow (2002) Plant Cell 14:1675–1690
				Transgenic Plants overexpressing DREB1A/CBF3		7K RAFL cDNA microarray, 8K Affymetrix GeneChip	upregulation of 38 genes	Maruyama et al.(2004) Plant J. 38:982–993
	DREB1B/CBF1	At4g25490	ERF/AP2 transcription factor	Transgenic Plants overexpressing DREB1B/CBF1		8K Affymetrix GeneChip	upregulation of 41 genes, downregulation of 27 genes	Fowler and Thomashow (2002) Plant Cell 14:1675–1690
	DREB1C/CBF2	At4g25470	ERF/AP2 transcription factor	Transgenic Plants overexpressing DREB1C/CBF2		8K Affymetrix GeneChip	upregulation of 41 genes, downregulation of 27 genes	Fowler and Thomashow (2002) Plant Cell 14:1675–1690
				Transgenic Plants overexpressing DREB1C/CBF2		24K Affymetrix ATH1 GeneChip	upregulation of 151 genes,downregulation of 43 genes	Vogel et al. (2005) Plant J. 41: 195–211
	DREB2A	At5g05410	ERF/AP2 transcription factor	Transgenic Plants overexpressing DREB2A (active form with internal deletion)	increased tolerance to drought stress	7K RAFL cDNA microarray	upregulation of 17 genes	Sakuma et al. (2006) Plant Cell 18:1292–1309
	AREB1/ABF2	At1g45249	bZIP transcription factor	Transgenic Plants overexpressing AREB1/ABF2 (active form with internal deletion)	increased tolerance to drought stress, hypersensitive to ABA	Agilent 22K oligo microarray	upregulation of 8 genes	Fujita et al. (2005) Plant Cell 17:3470–3488

(Continued)

Table 1. (Continued)

Functional Categories	Genes	AGI code	Coded protein	Mutant or Transgenic Plants	Phenotypes of mutants or transgenic plants	Microarrays used	Expression profiling results	References
				Transgenic Plants overexpressing AREB1/ABF2 (phosphorylated active form)	increased tolerance to dehydration	Agilent 22K oligo microarray	upregulation of 26 genes	Furihata et al. (2006) PNAS 103:1988–1993
	AtMYB2	At2g47190	MYB transcription factor	Transgenic Plants overexpressing AtMYB2 and AtMYC2	increased tolerance to osmotic stress, hypersensitive to ABA	7K RAFL cDNA microarray	upregulation of 32 genes	Abe et al. (2003) Plant Cell 15:63–78
	AtMYB60	At1g08810	MYB transcription factor	T-DNA knock-out line	increased tolerance to drought stress	7K RAFL cDNA microarray	upregulation of 6 genes, downregulation of 30 genes	Cominelli et al. (2005) Curr Biol. 15:1196–1200
	HOS10	At1g35515	MYB transcription factor	T-DNA knock-out line	hypersensitive to freezing and salt stress	24K Affymetrix ATH1 GeneChip	upregulation of 6 genes, downregulation of 6 genes	Zhu et al. (2005) PNAS 102:9966–9971
	AtMYC2/JIN1 (JASMONATE INSEN-SITIVE 1)	At1g32640	bHLH transcription factor	Transgenic Plants overexpressing AtMYB2 and AtMYC2	increased tolerance to osmotic stress, hypersensitive to ABA	7K RAFL cDNA microarray	upregulation of 32 genes	Abe et al. (2003) Plant Cell 15:63–78
	ICE1 (Inducer of CBF Expression1)	At3g26744	bHLH transcription factor	EMS-mutagenized T-DNA knock-out line	decreased cold acclimation	24K Affymetrix ATH1 GeneChip	up- or down-regulation of 369 genes	Chinnusamy et al. (2003) Genes & Dev. 17:1043-1054; Lee et al. (2005) Plant Cell 17:3155–3175
	ZAT12/RHL41	At5g59820	C2H2-Type zinc finger transcription factor	Transgenic Plants overexpressing ZAT12/RHL41	increased freezing tolerance	24K Affymetrix ATH1 GeneChip	upregulation of 47 genes, downregulation of 158 genes	Vogel et al. (2005) Plant J. 41:195–211
	ANAC019	At1g52890	NAC transcription factor	Transgenic Plants overexpressing ANAC019	increased tolerance to drought stress	7K RAFL cDNA microarray	upregulation of 17 genes	Tran et al. (2004) Plant Cell 16:2481–2498

	ANAC055	At3g15500	NAC transcription factor	Transgenic Plants overexpressing ANAC055	increased tolerance to drought stress	7K RAFL cDNA microarray	upregulation of 14 genes	Tran et al. (2004) Plant Cell 16:2481–2498
	ANAC072/ RD26	At4g27410	NAC transcription factor	Transgenic Plants overexpressing RD26/ANAC072	increased tolerance to drought stress	7K RAFL cDNA microarray	upregulation of 28 genes	Tran et al. (2004) Plant Cell 16:2481–2498
				Transgenic Plants overexpressing RD26/ANAC072	hypersensitive to ABA	Agilent 22K oligo microarray	upregulation of 20 genes	Fujita et al. (2004) Plant J. 39:863–876
				Transgenic Plants repressing RD26/ANAC072	insensitive to ABA	Agilent 22K oligo microarray	downregulation of 15 genes	Fujita et al. (2004) Plant J. 39:863–876
	HOS9	At2g01500	Homeo-domain transcription factor	EMS-mutagenized T-DNA knock-out line	sensitive to freezing stress	24K Affymetrix ATH1 GeneChip	upregulation of 140 genes	Zhu et al. (2004) PNAS 101:9873–9878
Protein kinases								
	SRK2C/ SnRK2.8	At1g78290	SNF1-related protein kinase 2	Transgenic Plants overexpressing SRK2C	increased tolerance to drought stress	Agilent 22K oligo microarray	upregulation of 18 genes, downregulation of 14 genes	Umezawa et al. (2004) PNAS 101:17306–17311
	SOS2/ CIPK24/ SnRK3.11	At5g35410	SNF1-related protein kinase 3	EMS-mutagenized mutant	hypersenstive to salt stress	Agilent 22K oligo microarray	upregulation of 27 genes, downregulation of 3 genes	Kamei et al. (2005) Plant Cell Environ 28:1267–1275
	AtMKK2	At4g29810	MAP kinase kinase 2	Transgenic Plants overexpressing MKK2	increased tolerance to freezing and salt stress	24K Affymetrix ATH1 GeneChip	upregulation of 127 genes,downregulation of 25 genes	Teige et al. (2004) Mol. Cell 15:141–152
	RPK1	At1g69270	LRR receptor-like protein kinase	Transgenic Plants overexpressing chimeric RPK1 LRR and BRI1 kinase domain	increased sensitivity to ABA	Agilent 22K oligo microarray	upregulation of 27 genes, downregulation of 19 genes	Osakabe et al. (2005) Plant Cell 17:1105–1119
				T-DNA knock-out line	insensitive to ABA	Agilent 22K oligo microarray	downregulation of 39 genes	Osakabe et al. (2005) Plant Cell 17:1105–1119

(*Continued*)

Table 1. (Continued)

Functional Categories	Genes	AGI code	Coded protein	Mutant or Transgenic Plants	Phenotypes of mutants or transgenic plants	Microarrays used	Expression profiling results	References
Others								
	MKP1	At3g55270	MAP kinase phosphatase	T-DNA knock-out line	hypersensitive to genotoxic stress, increased tolerance to salt stress	8K Affymetrix GeneChip	upregulation of 21 genes	Ulm et al. (2002) EMBO J. 21:6483–6493
	SOS3/CBL4	At5g24270	Protein with the calcineurin B subunit	EMS-mutagenized plant	hypersenstive to salt stress	Agilent 22K oligo microarray	upregulation of 4 genes, downregulation of 1 gene	Kamei et al. (2005) Plant Cell Environ 28:1267–1275
	ADR1	At1g33560	disease resistance protein (CC-NBS-LRR class)	Transgenic Plants overexpressing ADR1	increased tolerance to drought stress	7K RAFL cDNA microarray	upregulation of 20 genes	Chini et al. (2004) Plant J. 34: 810–822
	SLH1	At5g45260	disease resistance protein (TIR-NBS-LRR-WRKY class)	Ds knock-out line	sensitive to low humidity	7K RAFL cDNA microarray	upregulation of several genes	Noutoshi et al. (2005) Plant J. 43:873–888
	ABH1/CBP80	At2g13540	mRNA cap binding protein	T-DNA knock-out line	ABA hypersensitive stomatal closing and reduced wilting during drought	8K Affymetrix GeneChip	upregulation of 13 genes, downregulation of 18 genes	Hugouvieux et al. (2001) Cell 106: 477–487
	AHG2	At1g55870	poly(A)-specific ribonuclease	EMS-mutagenized plant	ABA-hypersensitive germination	7K RAFL cDNA microarray	upregulation of 19 genes	Nishimura et al.(2005) Plant J. 44:972–984

and so on (Bartels and Sunkar, 2005; Boudsocq and Lauriere, 2005; Mahajan and Tuteja, 2005; Vinocur and Altman, 2005). Although these complex signaling processes are not yet fully understood, several genes encoding the signaling factors involved in the abiotic stress responses have been identified (Shinozaki et al., 2003; Chinnusamy et al., 2004; Bartels and Sunkar, 2005; Umezawa et al., 2006).

Microarray analysis is useful for studying the function of the stress-related signaling factors. The transcriptome regulated by the following signaling factors involved in the abiotic stress responses has been studied by microarray analyses (Table 1): a SNF1-related protein kinase 2 (SnRK2), SRK2C (Umezawa et al., 2004); mitogen-activated protein kinase (MAPK) kinases, AtMKK2 (Teige et al., 2004) and MKK3 (Takahashi et al., unpublished results); a MAP kinase phosphatase, MKP1 (Ulm et al., 2002); a SnRK3 protein kinase, SOS2 (Kamei et al., 2005); a Ca^{2+}-binding protein with EF-hands, SOS3 (Kamei et al., 2005); a coiled-coil (CC)-nucleotide-binding site (NBS)-leucine-rich repeat (LRR) protein, activated disease resistance 1 (ADR1) (Chini et al., 2004); a LRR receptor-like kinase, RPK1 (Osakabe et al., 2005); a resistance (R)-like protein consisting of a domain with a Toll and interleukin-1 receptor homology (TIR), a nucleotide-binding domain (NB), LRR and a WRKY domain, sensitive to low humidity 1 (SLH1) (Noutoshi et al., 2005). The information on the transcriptome helps us understand the regulatory network of the abiotic stress responses: specificity and cross-talk. The roles of the signaling molecules in the abiotic stress signaling and the expression profiling results are summarized in recent reviews (Bartels and Sunkar, 2005; Boudsocq and Lauriere, 2005, Umezawa et al., 2006).

5. RECENT MOLECULAR AND GENETIC ANALYSES REVEALED THAT THE SMALL RNAS, RNA PROCESSING AND CHROMATIN REGULATION HAVE FUNCTIONS IN THE ABIOTIC STRESS RESPONSES

Recently, accumulating evidence indicates that small RNAs, RNA processing and chromatin regulation are involved in the abiotic stress responses and tolerance. We summarize the recent progress on the small RNAs, RNA processing and chromatin regulation in the abiotic stress responses and tolerance in the following.

6. SEVERAL SMALL RNAS ARE INVOLVED IN THE ABIOTIC STRESS RESPONSES

Small RNAs (21- to 25-nt), such as microRNAs (miRNAs) and small interfering RNAs (siRNAs) function in silencing genes by multiple mechanisms and are present in both plants and animals (Carrington and Ambros, 2003). miRNAs are generated by endonucleolytic processing by the enzyme Dicer from hairpin-structured single-stranded precursor RNAs that are transcribed from endogenous nonprotein-coding genes (Bartel, 2004). siRNAs are also produced by a Dicer, but differ from miRNAs in that they are generated from double-stranded RNAs (dsRNAs) as a result of antisense transcription or due to the activity of cellular RNA-dependent RNA polymerases (RdRPs) (Baulcombe, 2004). Recent studies indicated that several miRNAs and siRNAs are involved in the abiotic stress responses (Sunkar and Zhu, 2004; Borsani et al., 2005).

Sunkar and Zhu (2004) reported that several miRNAs are responsive to abiotic stresses. miR393 which targets the F-box protein TIR1 (Dharmasiri et al., 2005; Kepinski and Leyser, 2005), encoding an auxin receptor, is strongly upregulated by drought, high-salinity, cold and ABA treatments, suggesting that the upregulation of miR393 contributes to the inhibition of plant growth under stress condition (Sunkar and Zhu, 2004). Recently, Navarro et al. (2006) showed that a flagellin-derived peptide triggered induction of the miR393. Repression of auxin signaling by overexpression of miR393a increased the resistance to bacterium *Pseudomonas syringae* (Navarro et al., 2006). Sunkar and Zhu (2004) also reported slight upregulation of miR397b, miR402 and miR319c, and downregulation of miR389a.1 by abiotic stress treatments.

Borsani et al. (2005) found that the antisense overlapping gene pair of Δ^1-pyrroline-5-carboxylate dehydrogenase (P5CDH), a stress-related gene, and SRO5, a gene of unknown function, generates two types of siRNAs, 24-nt siRNA and 21-nt siRNAs. The expression of SRO5 is induced by salt and H_2O_2 treatments and this induction is required for generation of the siRNAs to cleave the p5CDH transcript. When the SRO5 expression is induced by salt treatment, a 24-nt siRNA is formed by a biogenesis pathway that is dependent on DCL2, RDR6, SGS3, and NRPD1A. Initial cleavage of the P5CDH transcript guided by the 24-nt siRNA establishes a phase for the subsequent generation of 21-nt siRNAs by DCL1 and further cleavage of P5CDH transcripts. *p5cdh* knock-out mutants are more tolerant to high-salinity stress due to the higher proline accumulation. There is substantially more accumulation of reactive oxygen species (ROS) in salt-stressed *sro5* knock-out mutants and the *sro5* mutants are more sensitive to high-salinity and H_2O_2-mediated oxidative stresses. Because salt treatment causes oxidative stress, these results suggest that the salt-stress induction of SRO5 and SRO5-P5CDH nat-siRNA formation might be mediated by increased ROS under high-salinity stress (Borsani et al., 2005).

7. RNA PROCESSING EVENTS ARE LINKED TO THE RESPONSES TO ABA AND DROUGHT STRESS

Several genes involved in RNA processing have been identified as components of ABA or drought signal transduction (Table 2).

The hyponastic leaves 1 (*hyl1*) mutant, which exhibits ABA hypersensitivity in seed germination and root elongation, was identified in a screen for ABA-hypersensitive *Arabidopsis* transposon insertion lines (Lu and Fedoroff, 2000). HYL1 encodes a nuclear-localized double-stranded RNA (dsRNA) binding protein. The *hyl1* mutant is also hypersensitive to glucose, NaCl and osmotic stress (Han et al., 2004). The ABA-hypersensitivity of the *hyl1* mutant is correlated with

Table 2. RNA processing-related Arabidopsis genes involved in the abiotic stress responses

Gene	AGI code	Coded protein	Mutant or Transgenic Plants	Phenotypes of mutants or transgenic plants	References
HYL1	At1g09700	nuclear-localized double-stranded RNA (dsRNA) binding protein	*Ds* knock-out line	hypersensitive to ABA, glucose, NaCl and osmotic stress	Lu and Fedoroff (2000) Plant Cell 12: 2351-2366 Lu et al. (2002) PNAS 99: 15812-15817
ABH1/CBP80	At2g13540	mRNA cap binding protein	T-DNA knock-out line	ABA hypersensitive inhibition of seed germination. ABA hypersensitive stomatal closing and reduced wilting during drought	Hugouvieux et al. (2001) Cell 106: 477-487
CBP20	At5g44200	mRNA cap binding protein	T-DNA knock-out line	hypersensitive to ABA during germination. Reduction of stomatal conductance, increased tolerance to drought stress	Papp et al. (2004) Plant Mol. Biol. 55: 679-686
AHG2	At1g55870	poly(A)-specific ribonuclease (AtPARN)	EMS-mutagenized plants	hypersensitive to ABA in seed germination and post-germination growth	Nishimura et al. (2005) Plant J. 44: 972-984
SAD1	At3g10730	Sm-like snRNP protein	EMS-mutagenized T-DNA line	hypersensitive to drought and ABA in seed germination and root growth	Xiong et al. (2001) Dev. Cell 1: 771-781
FRY2/CPL1	At4g21670	transcriptional repressor with two	EMS-mutagenized T-DNA line	increased tolerance to salt stress and ABA during seed germination.	Xiong et al. (2002) PNAS 99: 10899-10904

(*Continued*)

Table 2. (*Continued*)

Gene	AGI code	Coded protein	Mutant or Transgenic Plants	Phenotypes of mutants or transgenic plants	References
		double-stranded RNA-binding domains and a region homologous to the catalytic domain of RNA polymerase II C-terminal domain phosphatases	T-DNA knock-out line	hypersensitive to freezing stress at the seedling stage	Koiwa et al. (2002) PNAS 99: 10893-10898 Koiwa et al. (2004) PNAS 101: 14539-14544
CPL3	At2g33540	transcriptional repressor with a region homologous to the catalytic domain of RNA polymerase II C-terminal domain phosphatases	T-DNA knock-out line	hyperexpression of RD29A-LUC in response to ABA	Koiwa et al. (2002) PNAS 99: 10893-10898
LOS4	At3g53110	nuclear localized RNA helicase	EMS-mutagenized T-DNA line	reduced expression of DREB1/CBF and its target genes in response chilling sensitive and defective in cold acclimation.	Gong et al. (2002) PNAS 99: 11507-11512
FCA	At4g16280	RNA-binding protein, ABA receptor	EMS-mutagenized plants	late flowering	Macknight et al. (2002) Plant Cell 14: 877-888 Quesada et al. (2006) EMBO J. 22: 3142-3152

accumulation of ABI5, a key player in ABA-triggered postgermination growth arrest, at a lower ABA concentration in *hyl1* mutants than in wild-type seedlings (Lu et al., 2002). The DNA microarray experiments showed that the *ANP1* and the *AtMPK3* genes, encoding components of the stress-activated MAP kinase cascade, and the *ABI3* and *ABI5* transcription factor genes are overexpressed in *hyl1* seedlings when compared to wild-type seedlings (Lu et al., 2002). HYL1 plays a role in microRNA-mediated gene regulation (Han et al., 2004). Recent studies showed that HYL1 performs miRNA processing in collaboration with DCL1 (Hiraguri et al., 2005; Kurihara et al., 2006).

Hugouvieux et al. (2001) isolated a recessive ABA hypersensitive *Arabidopsis* mutant, *abh1*. *abh1* mutant exhibits ABA hypersensitivity during seed germination and stomatal closure and reduced wilting during drought stress. *ABH1* encodes a functional mRNA cap binding protein (CBP80), a protein comprising the eukaryotic nuclear cap-binding complex (CBC). DNA chip experiments showed that 18 genes including *RD20*, *KIN2* and *COR15b* had reduced the transcript levels 3-fold in the *abh1* mutant, and 7 of these genes are ABA-regulated in the wild-type plant. These expression profiling results are consistent with the fact that *abh1* plants showed ABA-hypersensitive stomatal closing and reduced wilting during drought. Hugouvieux et al. (2001) also showed ABA-hypersensitive cytosolic calcium increases in *abh1* guard cells.

A mutation in the mRNA cap binding protein (CBP20), another subunit comprising the eukaryotic nuclear cap-binding complex (CBC), also confers hypersensitivity to ABA during germination, significant reduction of stomatal conductance and greatly enhanced drought tolerance (Papp et al., 2004). The phenotype is very similar to that of the *abh1* mutant, suggesting that both gene products have the same function in responses to ABA and drought.

The ABA-hypersensitive germination2 (*ahg2*) mutant, which exhibits ABA hypersensitivity in seed germination and post-germination growth, was identified (Nishimura et al., 2005). AHG2 encodes a poly(A)-specific ribonuclease (AtPARN) that is presumed to function in mRNA degradation. Expression of *AHG2* gene was induced by treatment with ABA, high-salinity and osmotic stress. Microarray experiments showed increased expression of the ABA-, salicylic acid- and stress-inducible genes in untreated *ahg2* plants, suggesting that the *ahg2* mutation affects various stress responses as well as ABA responses.

The *Arabidopsis sad1* (supersensitive to ABA and drought) mutant was identified as an ABA-induced bioluminescence activation mutant in a screen of transgenic *Arabidopsis* plants harboring the stress-responsive *RD29A* promoter fused to a luciferase reporter (Xiong et al., 2001). *Sad1* mutant shows hypersensitivity to drought and ABA in seed germination and root growth.

The expression of the stress-responsive genes, such as *RD29A*, was increased in the *sad1* mutant. *SAD1* encodes a polypeptide similar to multifunctional Sm-like snRNP proteins that are required for mRNA splicing, export, and degradation.

The *Arabidopsis fry2* (fiery2)/*cpl1* (C-terminal domain phosphatase-like 1) mutant was also identified as a stress-induced bioluminescence activation mutant in a screen of transgenic *Arabidopsis* plants harboring the *RD29A-LUC* cassette (Koiwa et al., 2002; Xiong et al., 2002). *fry2/cpl1* mutants show increased tolerance to salt stress and to ABA during seed germination, and hypersensitiveness to freezing damage at the seedling stage (Xiong et al., 2002). FRY2/CPL1 encodes a novel transcriptional repressor harboring two double-stranded RNA-binding domains and a region homologous to the catalytic domain of RNA polymerase II C-terminal domain phosphatases that regulate gene transcription. These results indicate that FRY2/CPL1 is a negative regulator of stress gene transcription. CPL1 is a phosphatase that specifically dephosphorylate Ser-5 of the C-terminal domain (CTD) of RNA polymerase II, which consists of tandem repeats of a $Y^1S^2P^3T^4S^5P^6S^7$ heptapeptide (Koiwa et al., 2004). Koiwa et al. (2002) also reported a T-DNA insertion in the *CPL3* gene, another C-terminal domain phosphatase homolog in *Arabidopsis*, causes hyperresponsiveness to ABA.

The *Arabidopsis los4* mutant was also identified as a mutant with deregulated expression of the *RD29A-LUC* reporter gene in a screen of transgenic *Arabidopsis* plants harboring the *RD29A-LUC* cassette (Gong et al., 2002). *Los4* mutants show a reduced expression of DREB1/CBF and its target genes, such as RD29A, in response to cold, but not to ABA or salinity. The *los4* mutants are also sensitive to chilling and defective in cold acclimation. LOS4 encodes a nuclear localized RNA helicase.

Expression of a dominant negative form of the ABA-activated protein kinase (AAPK), a SnRK2 protein kinase identified from *Vicia faba* in guard cells prevented activation by ABA of anion channels and stomatal closure, implicating AAPK in rapid ABA signaling events in the guard cells (Li et al., 2000). The AAPK-interacting protein 1 (AKIP1) with sequence homology to heterogeneous nuclear RNA-binding protein A/B (hnRNP A/B) is phosphorylated by AAPK (Li et al., 2002). hnRNPs are involved in alternative pre-mRNA splicing, 3' end processing and mRNA export. AKIP1 binds to a dehydrin mRNA after phosphorylation by AAPK (Li et al., 2002).

Recently, an ABA receptor, FCA was identified (Razem et al., 2006).

FCA is a nuclear RNA-binding protein that promotes flowering by preventing the accumulation of mRNA encoding FLC, a MADS box transcription factor that is a potent repressor of the floral transition (Simpson, 2004). To function FCA requires a second protein, the RNA 3'-end processing factor FY, which binds to its tryptophan-tryptophan (WW) protein interaction domain (Simpson et al., 2003). FCA autoregulates its expression by promoting premature cleavage and polyadenylation in intron 3 of its own precursor mRNA (pre-mRNA) (Macknight et al., 2002; Quesada et al., 2003). Razem et al. (2006) showed that FCA binds ABA with high affinity in an interaction that is stereospecific. Binding of ABA to FCA abolishes the interaction of FCA with FY, leading to an increase in full-length FCA transcripts and a delay in flowering through increased FLC activity.

A number of alternative splicing events in response to environmental changes and ABA application have been reported, although the biological significance of the alternatively spliced transcripts produced is unknown. Xu et al. (2004)

showed that the alternative splicing of the *Arabidopsis* PIMT2 gene encoding protein-L-isoaspartate methyltransferase is regulated by ABA. The ABA application enhanced the accumulation of the PIMT2 transcript accompanied by the predominant production of the shorter PIMT2 transcript form (Xu et al., 2004). Xue and Loveridge (2004) showed that three transcript forms of a dehydration-, salt- or ABA-responsive ERF/AP2 transcription factor, *HvDRF1* from barley showing sequence similarity with *Arabidopsis* DREB2A were produced through alternative splicing and that two of them encoded ERF/AP2 transcriptional activators. This alternative splicing pattern was also observed in a wheat homolog gene, *TaDRF1* (Xue and Loveridge, 2004). Alternative splicing is also induced in the SOS4 gene (Shi et al., 2002) encoding a pyridoxal kinase under salt, ABA and cold treatments. Genome-wide analysis of *Arabidopsis* full-length cDNAs also indicated that alternative splicing profiles changed by cold stress (Iida et al., 2004).

8. CHROMATIN REGULATION IS INVOLVED IN THE ABIOTIC STRESS RESPONSES

Acetylation and deacetylation of histone are promoter-dependent, locus-specific and genetically reversible, which provides a general mechanism for reversible gene regulation responsive to development and environmental changes (Tian et al., 2005). Histone deacetylase and linker histone genes have been found to be involved in the response of the plant to abiotic stresses.

Overexpression of an *Arabidopsis* histone deacetylase homolog, *AtHD2C*, in transgenic *Arabidopsis* plants resulted in ABA insensitivity and enhanced tolerance to salt and drought stresses (Sridha and Wu, 2006). The expression of several ABA-responsive genes, such as *RD29B* and *RAB18*, was upregulated in the overexpressor of *AtHD2C*. These results indicate that *AtHD2C* can modulate the responses to ABA, drought and salinity stresses.

An *Arabidopsis* AP2/ERF-type transcription factor, AtERF7, interacts with the *Arabidopsis* homolog of a human global transcriptional corepressor, AtSin3, which in turn may interact with HDA19/AtRPD3A, a histone deacetylase (Song et al., 2005). AtSin3 and HDA19 enhance the transcriptional repression activity of AtERF7 (Song et al., 2005). Overexpression of AtERF7 in transgenic Arabidopsis plants reduced ABA responses in guard cells and decreased drought tolerance, whereas reduction in AtERF7 expression caused ABA hypersensitivity in guard cells, seed germination, and seedling growth (Song et al., 2005).

Recent expression profiling studies indicated that a number of genes involved in chromatin remodeling and post transcriptional regulation were also upregulated by cold stress (Lee et al., 2005), suggesting their involvement in cold-responsive gene regulation. Expression of a drought and ABA-inducible gene, *his1-3*, encoding a linker histone H1-3 protein, is upregulated in the plants overexpressing an activated form of AREB1 (Fujita et al., 2005). The transgenic plants overexpressing an activated form of AREB1 showed ABA hypersensitivity and enhanced drought tolerance (Fujita et al., 2005). These results suggest that histone H1-3 plays a role in drought-responsive gene expression.

9. CONCLUSIONS AND PERSPECTIVES

The microarray-based expression profiling method is useful for analyzing the expression pattern of plant genes under various stress treatments, and for identifying target genes of the stress-related signaling factors. By the expression profiling approach, many stress-inducible genes have been identified. Functional analysis of these stress-inducible genes has provided more information on the signal transduction in these stress responses.

Large sets of *Arabidopsis* microarray data, such as the expression dataset provided through the AtGenExpress Consortium (http://www.arabidopsis.org/info/expression/ATGenExpress.jsp) and NASCArrays (http://affymetrix.arabidopsis.info.) (Craigon et al., 2004) generated by the Nottingham *Arabidopsis* Stock Center (NASC)'s transcriptomics service, are available. Large sets of plant microarray data are also available from GEO (http://www.ncbi.nlm.nih.gov/geo) (Edgar et al., 2002) and ArrayExpress (http://www.ebi.ac.uk/arrayexpress) (Brazma et al., 2003). Analysis tools for the *Arabidopsis* expression data, such as Genevestigator (http://www.genevestigator.ethz.ch/) (Zimmermann et al., 2004), are also available. The Genevestigator is a user-friendly web-based tool for a large *Arabidopsis* Affymetrix GeneChip data and provides categorized quantitative information about elements, such as genes, contained in large microarray database (Zimmermann et al., 2004). Availability of large sets of the plant microarray data and user-friendly analytical tools should aid the functional analysis of the stress-related genes and our better understanding the transcriptional regulatory networks in the abiotic stress responses.

By genetic approaches and biochemical analyses of signal transduction and stress tolerance of drought and salt stress, many mutants on the signal transduction and stress tolerance of these stresses have been identified (Xiong et al. 2002; Zhu 2002; Bartels and Sunkar, 2005; Mahajan and Tuteja, 2005). Reverse genetic approaches, such as transgenic analyses, have become useful for studying the function of the signaling components (Hasegawa et al. 2000; Xiong et al. 2002; Bartels and Sunkar, 2005; Umezawa et al., 2006; Yamaguchi-Shinozaki and Shinozaki, 2006). The availability of the full-length cDNAs (Seki et al., 2002a; Yamada et al., 2003) and gene-knock-out mutants (Alonso et al., 2003; Kuromori et al., 2004) will greatly facilitate the functional analysis of the signaling components.

Whole genome tiling array studies (Yamada et al., 2003; Stolc et al., 2005) will also become powerful tools for identification of stress-regulated small RNAs or non-coding RNAs, and for analysis of alternative splicing and chromatin remodeling. New sequencing technologies using massively parallel signature sequencing (MPSS) (Lu et al., 2005) and a pyrosequencing-based method (Margulies et al., 2005) developed by 454 Life Sciences will enable the identification of large numbers of the small RNAs involved in the abiotic stress responses. Comparative genomics studies using bioinformatic approaches will also lead to an improved understanding of function and biological significance of the small RNAs.

ACKNOWLEDGEMENTS

This work was supported in part by a grant for Genome Research from RIKEN, the Program for Promotion of Basic Research Activities for Innovative Biosciences, the Special Coordination Fund of the Science and Technology Agency, and a Grant-in-Aid from the Ministry of Education, Culture, Sports, Science and Technology of Japan (MECSST) to K.S. It was also supported in part by a Grant-in-Aid for Scientific Research on Priority Areas "Systems Genomics" from MECSST and the President Discretionary Fund from RIKEN to M.S.

REFERENCES

Abe, H., Urao, T., Ito, T., Seki, M., Shinozaki, K. and Yamaguchi-Shinozaki, K. (2003) *Arabidopsis* AtMYC2 (bHLH) and AtMYB2 (MYB) function as transcriptional activators in abscisic acid signaling, *Plant Cell*, **15**: 63–78.

Alonso, J.M., Stepanova, A.N., Leisse, T.J., Kim, C.J., Chen, H., Shinn, P., Stevenson, D.K., Zimmerman, J., Barajas, P., Cheuk, R., Gadrinab, C. et al. (2003) Genome-wide insertional mutagenesis of Arabidopsis thaliana, *Science* **301**: 653–657.

Bartel, D.P. (2004) MicroRNAs: genomics, biogenesis, mechanism and function, *Cell* **116**: 281–297.

Bartels and Sunkar (2005) Drought and salt tolerance in plants, *Critical Reviews in Plant Sciences* **24**:23–58.

Baulcombe, D. (2004) RNA silencing in plants, *Nature* **431**: 356–363.

Borsani, O., Zhu, J., Verslues, P.E., Sunkar, R. and Zhum J.K. (2005) Endogenous siRNAs derived from a pair of natural cis-antisense transcripts regulate salt tolerance in Arabidopsis, *Cell* **123:** 1279–1291.

Boudsocq, M. and Lauriere, C. (2005) Osmotic signaling in plants. Multiple pathways mediated by emerging kinase families, *Plant Physiol.* **138**: 1185–1194.

Bray, E.A. (2002) Classification of genes differentially expressed during water-deficit stress in Arabidopsis thaliana: an analysis using microarray and differential expression data, *Ann. Bot.* **89**: 803–811.

Brazma, A., Parkinson, H., Sarkans, U., Shojatalab, M., Vilo, J., Abeygunawardena, N., Holloway, E., Kapushesky, M., Kemmeren, P., Lara, G.G., Oezcimen, A., Rocca-Serra, P. and Sansone, S.A. (2003) ArrayExpress-a public repository for microarray gene expression data at the EBI, *Nucleic Acids Res.* **31**: 68–71.

Brosche, M., Vinocur, B., Alatalo, E.R., Lamminmaki, A., Teichmann, T., Ottow, E.A., Djilianov, D., Afif, D., Bogeat-Triboulot, M.B., Altman, A., Polle, A., Dreyer, E., Rudd, S., Paulin, L., Auvinen, P. and Kangasjarvi, J. (2005) Gene expression and metabolite profiling of *Populus euphratica* growing in the Negev desert, *Genome Biol.* **6**: R101.

Buchanan, C.D., Lim, S., Salzman, R.A., Kagiampakis, I., Morishige, D.T., Weers, B.D., Klein, R.R., Pratt, L.H., Cordonnier-Pratt, M.M., Klein, P.E. and Mullet, J.E. (2005) *Sorghum bicolor*'s transcriptome response to dehydration, high salinity and ABA, *Plant Mol. Biol.* **58**:699–720.

Carrington, J.C. and Ambros, V. (2003) Role of microRNAs in plant and animal development, *Science* **301**: 336–338.

Chen, W., Provart, N.J., Glazebrook, J., Katagiri, F., Chang, H.S., Eulgem, T., Mauch, F., Luan, S., Zou, G., Whitham, S.A., Budworth, P.R., Tao, Y., Xie, Z., Chen, X., Lam, S., Kreps, J.A., Harper, J.F., Si-Ammour, A., Mauch-Mani, B., Heinlein, M., Kobayashi, K., Hohn, T., Dangl, J.L., Wang, X. and Zhu, T. (2002) Expression profile matrix of *Arabidopsis* transcription factor genes suggests their putative functions in response to environmental stresses, *Plant Cell* **14**: 559–574.

Cheong, Y.H., Chang, H.S., Gupta, R., Wang, X., Zhu, T. and Luan, S. (2002) Transcriptional profiling reveals novel interactions between wounding, pathogen, abiotic stress, and hormonal responses in *Arabidopsis*, *Plant Physiol.* **129**: 661–677.

Chini, A., Grant, J., Seki, M., Shinozaki, K. and Loake, G. (2004) Drought tolerance established by enhanced expression of the CC-NBS-LRR gene, ADR1, requires salicylic acid, EDS1 and ABI1. *Plant J* **38**:810–822.

Chinnusamy, V., Ohta, M., Kannar, S., Lee, B., Hong, Z., Agarwal, A. and Zhu, J.K. (2003) ICE1: a regulator of cold-induced transcriptome and freezing tolerance in *Arabidopsis*, *Genes Dev.* **17**: 1043–1054.

Chinnusamy, V., Schumaker, K. and Zhu, J.K. (2004) Molecular genetic perspectives on cross-talk and specificity in abiotic stress signaling in plants, *J. Exp. Bot.* **55**: 225–236.

Cominelli, E., Galbiati, M., Vavasseur, A., Conti, L., Sala, T., Vuylsteke, M., Leonhardt, N., Dellaporta, S. and Tonelli, C. (2005) A guard-cell-specific MYB transcription factor regulates stomatal movements and plant drought tolerance, *Curr. Biol.* **15**: 1196–1200.

Craigon, D.J., James, N., Okyere, J., Higgins, J., Jotham, J. and May, S. (2004) NASCArrays: a repository for microarray data generated by NASC's transcriptomics service, *Nucleic Acids Res.* **32**: D575–D577.

Davletova, S., Schlauch, K., Coutu, J. and Mittler, R (2005) The Zinc-finger protein Zat12 plays a central role in reactive oxygen and abiotic stress signaling in *Arabidopsis*, *Plant Physiol.* **139**:847–856.

Dharmasiri, N., Dharmasiri, S. and Estelle, M. (2005) The F-box protein TIR1 is an auxin receptor, *Nature* **435**:441–445.

Dubouzet, J.G., Sakuma, Y., Ito, Y., Kasuga, M., Dubouzet, E.G., Miura, S., Seki, M., Shinozaki, K. and Yamaguchi-Shinozaki, K. (2003) OsDREB genes in rice, Oryza sativa L., encode transcription activators that function in drought-, high-salt- and cold-responsive gene expression, *Plant J* **33**: 751–763.

Edgar, R., Domrachev, M. and Lash, A.E. (2002) Gene expression omnibus: NCBI gene expression and hybridization array data repository, *Nucleic Acids Res.* **30**: 207–210.

Fowler, S. and Thomashow, M.F. (2002) *Arabidopsis* transcriptome profiling indicates that multiple regulatory pathways are activated during cold acclimation in addition to the CBF cold response pathway, *Plant Cell* **14**:1675–1690.

Fujita, M., Fujita, Y., Maruyama, K., Seki, M., Hiratsu, K., Ohme-Takagi, M., Tran, L.S.P., Yamaguchi-Shinozaki, K. and Shinozaki, K. (2004) A dehydration-induced NAC protein, RD26 is involved in ABA-dependent stress signaling pathway. *Plant J.* **39**:863–876.

Fujita, Y., Fujita, M., Sato, R., Maruyama, K., Parvez, M.M., Seki, M., Hiratsu, K., Ohme-Takagi, M., Shinozaki, K. and Yamaguchi-Shinozaki, K. (2005) AREB1 is a transcription activator of novel ABRE-dependent ABA signaling that enhances drought stress tolerance in *Arabidopsis*, *Plant Cell* **17**: 3470–3488.

Furihata, T., Maruyama, K., Fujita, Y., Umezawa, T., Yoshida, R., Shinozaki, K. and Yamaguchi-Shinozaki, K. (2006) Abscisic acid-dependent multisite phosphorylation regulates the activity of a transcription activator AREB1, *Proc. Natl. Acad. Sci. USA* **103**: 1988–1993.

Gong, Z., Lee, H., Xiong, L., Jagendorf, A., Stevenson, B. and Zhu, J.K. (2002) RNA helicase-like protein as an early regulator of transcription factors for plant chilling and freezing tolerance, *Proc. Natl. Acad. Sci. USA* , **99**:11507–11512.

Gong, Q., Li, P., Ma, S., Rupassara, S.I. and Bohnert, H.J. (2005) Salinity stress adaptation competence in the extremophile *Thellungiella halophila* in comparison with its relative *Arabidopsis thaliana*, *Plant J.* **44**:826–839.

Gu, R., Fonseca, S., Puskas, L.G., JR, L.H., Zvara, A., Dudits, D. and Pais, M. (2004) Transcript identification and profiling during salt stress and recovery of *Populus euphratica*, *Tree Physiol.* **24**:265–276.

Gulick, P.J., Drouin, S., Yu, Z., Danyluk, J., Poisson, G., Monroy, A.F. and Sarhan, F. (2005) Transcriptome comparison of winter and spring wheat responding to low temperature, *Genome* **48**: 913–923.

Han, M.H., Goud, S., Song, L. and Fedoroff, N. (2004) The *Arabidopsis* double-stranded RNA-binding protein HYL1 plays a role in microRNA-mediated gene regulation, *Proc. Natl. Acad. Sci. USA* **101**:1093–1098.

Hasegawa, P.M., Bressan, R.A., Zhu, J.K. and Bohnert, H.J. (2000) Plant cellular and molecular responses to high salinity, *Annu. Rev. Plant Physiol. Plant Mol. Biol.* **51**: 463–499.

Hazen, S.P., Wu, Y. and Kreps, J.A. (2003) Gene expression profiling of plant responses to abiotic stress, *Funct. Integr. Genomics* **3**: 105–111.

Hiraguri, A., Itoh, R., Kondo, N., Nomura, Y., Aizawa, D., Murai, Y., Koiwa, H., Seki, M., Shinozaki, K. and Fukuhara, T. (2005) Specific interactions between Dicer-like proteins and HYL1/DRB-family dsRNA-binding proteins in *Arabidopsis thaliana. Plant Mol. Biol.* **57**:173–188.

Hugouvieux, V., Kwak, J.M. and Schroeder, J.I. (2001) An mRNA cap binding protein, ABH1, modulates early abscisic acid signal transduction in *Arabidopsis, Cell* **106**: 477–487.

Hwang, E.W., Kim, K.A., Park, S.C., Jeong, M.J., Byun, M.O. and Kwon, H.B. (2005) Expression profiles of hot pepper (*Capsicum annuum*) genes under cold stress conditions, *J. Biosci.* **30**: 657–667.

Iida, K., Seki, M., Sakurai, T., Satou, M., Akiyama, K., Toyoda, T., Konagaya, A. and Shinozaki, K. (2004) Genome-wide analysis of alternative pre-mRNA splicing in Arabidopsis thaliana based on full-length cDNA sequences, *Nucleic Acids Res.* **32**: 5096–5103.

Iida, K., Seki, M., Sakurai, T., Satou, M., Akiyama, K., Toyoda, T., Konagaya, A. and Shinozaki, K. (2005) RARTF: database and tools for complete sets of *Arabidopsis* transcription factors. *DNA Res.***12**:247–256.

Inan, G., Zhang, Q., Li, P., Wang, Z., Cao, Z., Zhang, H., Zhang, C., Quist, T.M., Goodwin, S.M., Zhu, J., Shi, H., Damsz, B., Charbaji, T., Gong, Q., Ma, S., Fredricksen, M., Galbraith, D.W., Jenks, M.A., Rhodes, D., Hasegawa, P.M., Bohnert, H.J., Joly, R.J., Bressan, R.A. and Zhu, J.K. (2004) Salt Cress. A halophyte and cryophyte Arabidopsis relative model system and its applicability to molecular genetic analyses of growth and development of extremophiles, *Plant Physiol.* **135**: 1718–1737.

Ito, Y., Katsura, K., Maruyama, K., Taji, T., Kobayashi, M., Seki, M., Shinozaki, K., and Yamaguchi-Shinozaki, K. (2006) Functional analysis of rice DREB1/CBF-type transcription factors involved in cold-responsive gene expression in transgenic rice. *Plant Cell Physiol.* **47**:141–153.

Kamei, A., Seki, M, Umezawa, T., Ishida, J., Satou, M., Akiyama, K., Zhu, J.K. and Shinozaki, K. (2005) Analysis of gene expression profiles in *Arabidopsis* salt overly sensitive mutants, *sos2* and *sos3* mutants. *Plant Cell and Environ.* **28**:1267–1275.

Kawasaki, S., Borchert, C., Deyholos, M., Wang, H., Brazille, S., Kawai, K., Galbraith, D. and Bohnert, H. (2001) Gene expression profiles during the initial phase of salt stress in rice, *Plant Cell* **13**: 889–905.

Kepinski, S. and Leyser, O. (2005) The Arabidopsis F-box protein TIR1 is an auxin receptor, Nature 435: 446–451.

Koiwa, H., Barb, A.W., Xiong, L., Li, F., McCully, M.G., Lee, B.H., Sokolchik, I., Zhu, J., Gong, Z., Reddy, M., Sharkhuu, A., Manabe, Y., Yokoi, S., Zhu, J.K., Bressan, R.A. and Hasegawa, P.M. (2002) C-terminal domain phosphatase-like family members (AtCPLs) differentially regulate *Arabidopsis thaliana* abiotic stress signaling, growth, and development, *Proc. Natl. Acad. Sci. USA* **99**:10893–10898.

Koiwa, H., Hausmann, S., Bang, W.Y., Ueda, A., Kondo, N., Hiraguri, A., Fukuhara, T., Bahk, J.D., Yun, D.J., Bressan, R.A., Hasegawa, P.M. and Shuman, S. (2004) *Arabidopsis* C-terminal domain phosphatase-like 1 and 2 are essential Ser-5-specific C-terminal domain phosphatases, *Proc. Natl. Acad. Sci. USA* **101**:14539–14544.

Kreps, J.A., Wu, Y., Chang, H.S., Zhu, T., Wang, X. and Harper, J.F. (2002) Transcriptome changes for Arabidopsis in response to salt, osmotic, and cold stress, *Plant Physiol.* **130**: 2129–2141.

Kurihara, Y., Yuasa, T. and Watanabe, Y. (2006) The interaction between DCL1 and HYL1 is important for efficient and precise processing of pri-miRNA in plant microRNA biogenesis, *RNA* **12**:206–212.

Kuromori, T., Hirayama, T., Kiyosue, Y., Takabe, H., Mizukado, S., Sakurai, T., Akiyama, K., Kamiya, A., Ito, T. and Shinozaki, K. (2004) A collection of 11 800 single-copy Ds transposon insertion lines in *Arabidopsis. Plant J.* **37**: 897–905.

Lan, L., Li, M., Lai, Y., Xu, W., Kong, Z., Ying, K., Han, B. and Xue, Y. (2005) Microarray analysis reveals similarities and variations in genetic programs controlling pollination/fertilization and stress responses in rice (*Oryza sativa* L.), *Plant Mol. Biol.* **59**: 151–164.

Lee, B.H., Henderson, D.A. and Zhu, J.K. (2005) The Arabidopsis cold-responsive transcriptome and its regulation by ICE1, *Plant Cell* **17**: 3155–3175.

Li, J., Wang, X.Q., Watson, M.B. and Assmann, S.M. (2000) Regulation of abscisic acid-induced stomatal closure and anion channels by guard cell AAPK kinase, *Science* **287**: 300–303.

Li, J., Kinoshita, T., Pandey, S., Ng, C.K.Y., Gygi, S.P., Shimazaki, K. and Assmann, S.M. (2002) Modulation of an RNA-binding protein by abscisic-acid-activated protein kinase, *Nature* **418**: 793–797.

Lu, C. and Fedoroff, N. (2000) A mutation in the *Arabidopsis* HYL1 gene encoding a dsRNA-binding protein affects responses to abscisic acid, auxin, and cytokinin, *Plant Cell* **12**:2351–2366.

Lu, C., Han, M.H., Guevara-Garcia, A. and Fedoroff, N.V. (2002) Mitogen-activated protein kinase signaling in postgermination arrest of development by abscisic acid, *Proc. Natl. Acad. Sci. USA* **99**:15812–15817.

Lu, C., Tej, S.S., Luo, S., Haudenschild, C.D., Meyers, B.C. and Green, P.J. (2005) Elucidation of the small RNA component of the transcriptome, *Science* **309**: 1567–1569.

Macknight, R., Duroux, M., Laurie, R., Dijkwel, P., Simpson, G. and Dean, C. (2002) Functional significance of the alternative transcript processing of the *Arabidopsis* floral promoter FCA, *Plant Cell* **14**: 877–888.

Mahajan, S. and Tuteja, N. (2005) Cold, salinity and drought stresses: an overview, *Archives of Biochemistry and Biophysics* **444**: 139–158.

Margulies, M., Egholm, M., Altman, W.E., Attiya, S., Bader, J.S., Bemben, L.A., Berka, J., Braverman, M.S., Chen, Y.J., Chen, Z., Dewell, S.B., Du, L., Fierro, J.M. et al. (2005) Genome sequencing in microfabricated high-density picolitre reactors, *Nature* **437:** 376–380.

Maruyama, K., Sakuma, Y., Kasuga, M., Ito, Y., Seki, M., Goda, H., Shimada, Y., Yoshida, S., Shinozaki, K. and Yamaguchi-Shinozaki, K. (2004) Identification of cold-inducible downstream genes of the Arabidopsis DREB1A/CBF3 transcriptional factor using two microarray systems, *Plant J.* **38**:982–993.

Narusaka, Y., Narusaka, M., Seki, M., Ishida, J., Nakashima, M., Kamiya, A., Enju, A., Sakurai, T., Satoh, M., Kobayashi, M., Tosa, Y., Park, P. and Shinozaki, K. (2003) The cDNA microarray analysis using an *Arabidopsis pad3* mutant reveals the expression profiles and classification of genes induced by *Alternaria brassicicola* attack, *Plant Cell Physiol* **44**: 377–387.

Navarro, L., Dunoyer, P., Jay, F., Arnold, B., Dharmasiri, N., Estelle, M., Voinnet, O. and Jones, J.D.G. (2006) A plant miRNA contributes to antibacterial resistance by repression by repressing auxin signaling, *Science* **312**: 436–439.

Nishimura, N., Kitahata, N., Seki, M., Narusaka, Y., Narusaka, M., Kuromori, T., Asami, T., Shinozaki, K. and Hirayama, T. (2005) Analysis of *ABA Hypersensitive Germination2* revealed the pivotal functions of PARN in stress response in *Arabidopsis*, *Plant J.* **44**:972–984.

Osakabe, Y., Maruyama, K., Seki, M., Satou, M., Shinozaki, K. and Yamaguchi-Shinozaki, K. (2005) An LRR receptor kinase, RPK1, is a key membrane-bound regulator of abscisic acid early signaling in *Arabidopsis*, *Plant Cell* **17**:1105–1119.

Oztur, Z.N., Talame, V., Deyholos, M., Michalowski, C.B., Galbraith, D.W., Gozukirmizi, N., Tuberosa, R. and Bohnert, H.J. (2002) Monitoring large-scale changes in transcript abundance in drought- and salt-stressed barley, *Plant Mol. Biol.* **48**: 551–573.

Papp, I., Mur, L.A., Dalmadi, A., Dulai, S. and Koncz, C. (2004) A mutation in the Cap Binding Protein 20 gene confers drought tolerance to *Arabidopsis*, *Plant Mol. Biol.* **55**:679–686.

Quesada, V., Macknight, R., Dean, C. and Simpson, G.G. (2003) Autoregulation of FCA pre-mRNA processing controls *Arabidopsis* flowering time, *EMBO J.* **22**: 3142–3152.

Rabbani, M.A., Maruyama, K., Abe, H., Khan, M.A., Katsura, K., Ito, Y., Yoshiwara, K., Seki, M., Shinozaki, K. and Yamaguchi-Shinozaki, K. (2003) Monitoring expression profiles of rice (*Oryza sativa* L.) genes under cold, drought and high-salinity stresses, and ABA application using both cDNA microarray and RNA gel blot analyses, *Plant Physiol.* **133**: 1755–1767.

Ramanjulu, S. and Bartels, D. (2002) Drought- and desiccation-induced modulation of gene expression in plants, *Plant Cell Environ.* **25**: 141–151.

Razem, F.A., El-Kereamy, A., Abrams, S.R. and Hill, R.D. (2006) The RNA-binding protein FCA is an abscisic acid receptor, *Nature* **439**: 290–294.

Rensink, W.A., Lobst, S., Hart, A., Stegalkina, S., Liu, J. and Buell, C.R. (2005) Gene expression profiling of potato responses to cold, heat, and salt stress, *Funct. Integr. Genomics*, **5**: 201–207.

Richmond, T. and Somerville, S. (2000) Chasing the dream: plant EST microarrays, *Curr. Opin. Plant Biol.* **3**: 108–116.

Riechmann, J.L., Heard, J., Martin, G., Reuber, L., Jiang, C.Z., Keddie, J., Adam, L., Pineda, O., Ratcliffe, O.J., Samaha, R.R., Creelman, R., Pilgrim, M., Broun, P., Zhang, J.Z., Ghandehari, D., Sherman, B.K. and Yu, G.L. (2000) *Arabidopsis* transcription factors: genome-wide comparative analysis among eukaryotes, *Science* **290**: 2105–2110.

Sakuma, Y., Maruyama, K., Osakabe, Y., Feng, Q., Seki, M., Shinozaki K. and Yamaguchi-Shinozaki, K. (2006) Functional analysis of an *Arabidopsis* transcription factor, DREB2A, involved in drought-responsive gene expression. *Plant Cell* **18**: 1292–1309.

Savitch, L.V., Allard, G., Seki, M., Robert, L.S., Tinker, N.A., Huner, N.P.A., Shinozaki, K. and Singh, J. (2005) The effects of overexpression of two *Brassica* CBF/DREB1-like transcription factors on phytosynthetic capacity and freezing tolerance in *Brassica napus*, *Plant Cell Physiol.* **46**: 1525–1539.

Seki, M., Narusaka, M., Abe, H., Kasuga, M., Yamaguchi-Shinozaki, K., Carninci, P., Hayashizaki, Y. and Shinozaki, K. (2001) Monitoring the expression pattern of 1300 *Arabidopsis* genes under drought and cold stresses using a full-length cDNA microarray, *Plant Cell* **13**: 61–72.

Seki, M., Narusaka, M., Kamiya, A., Ishida, J., Satou, M., Sakurai, T., Nakajima, M., Enju, A., Akiyama, K., Oono, Y., Muramatsu, M., Hayashizaki, Y., Kawai, J., Carninci, P., Itoh, M., Ishii, Y., Arakawa, T., Shibata, K., Shinagawa, A. and Shinozaki, K. (2002a) Functional annotation of a full-length *Arabidopsis* cDNA collection, *Science* **296**: 141–145.

Seki, M., Narusaka, M., Ishida, J., Nanjo, T., Fujita, M., Oono, Y., Kamiya, A., Nakajima, M., Enju, A., Sakurai, T., Satou, M., Akiyama, K., Taji, T., Yamaguchi-Shinozaki, K., Carninci, P., Kawai, J., Hayashizaki, Y. and Shinozaki, K. (2002b) Monitoring the expression profiles of 7000 *Arabidopsis* genes under drought, cold, and high-salinity stresses using a full-length cDNA microarray, *Plant J.* **31**: 279–292.

Seki, M., Ishida, J., Narusaka, M., Fujita, M., Nanjo, T., Umezawa, T., Kamiya, A., Nakajima, M., Enju, A., Sakurai, T., Satou, M., Akiyama, K., Yamaguchi-Shinozaki, K., Carninci, P., Kawai, J., Hayashizaki, Y. and Shinozaki, K. (2002c) Monitoring the expression pattern of ca. 7000 Arabidopsis genes under ABA treatments using a full-length cDNA microarray, Funct. Integr. Genomics **2**: 282–291.

Seki, M., Kamei, A., Satou, M., Sakurai, T., Fujita, M., Oono, Y., Yamaguchi-Shinozaki, K. and Shinozaki, K. (2003) Transcriptome Analysis in Abiotic Stress Conditions in Higher Plants, *Topics Curr. Genet.* **4**: 271–295.

Seki, M., Satou, M., Sakurai, T., Akiyama, K., Iida, K., Ishida, J., Nakajima, M., Enju, A., Narusaka, M., Fujita, M., Oono, Y., Kamei, A/, Yamaguchi-Shinozaki. K. and Shinozaki, K. (2004) RIKEN Arabidopsis full-length (RAFL) cDNA and its applications for expression profiling under abiotic stress conditions, *J. Exp. Bot.* **55**: 213–223.

Seki, M., Ishida, J., Nakajima, M., Enju, A., Iida, K., Satou, M., Fujita, M., Narusaka, Y., Narusaka, M., Sakurai, T., Akiyama, K., Oono, Y., Kamei, A., Umezawa, T., Mizukado, S., Maruyama, K., Yamaguchi-Shinozaki, K. and Shinozaki, K. (2005). Genomic analysis of stress response. In "Plant Abiotic Stress (Edited by Drs. M.A. Jenks and P.M. Hasegawa)", pp. 248–265. Blackwell Publishing Ltd., Sheffield, UK

Shi, H., Xiong, L., Stevenson, B., Lu, T. and Zhu, J.K. (2002) The Arabidopsis *salt overly sensitive 4* mutants uncover a critical role for vitamin B6 in plant salt tolerance, *Plant Cell* **14**: 575–588.

Shinozaki, K. and Yamaguchi-Shinozaki, K. (2000) Molecular responses to dehydration and low temperature: Differences and cross-talk between two stress signaling pathways, *Curr. Opin. Plant Biol.* **3**: 217–223.

Shinozaki, K., Yamaguchi-Shinozaki, K. and Seki, M. (2003) Regulatory network of gene expression in the drought and cold stress responses, *Curr. Opin. Plant Biol.* **6**: 410–417.

CHAPTER 12

COMPARATIVE METABOLOME ANALYSIS OF THE SALT RESPONSE IN BREEDING CULTIVARS OF RICE

ELLEN ZUTHER, KARIN KOEHL AND JOACHIM KOPKA

Max Planck Institute of Molecular Plant Physiology, Am Muehlenberg 1, 14476 Potsdam-Golm, Germany

E-mail: Kopka@mpimp-golm.mpg.de

Abstract: Metabolomics aims for comprehensive analysis of the metabolic complement. The metabolic phenotype is typically described by changes in metabolic pool sizes. Today investigations are technologically limited to a few hundred metabolites. Metabolomics studies are typically restricted to a single analytical technology, such as GC-TOF-MS which will be the focus technology of this chapter. Two strategies for data analysis are applied. Metabolite fingerprinting investigates all analytical signals. Metabolite profiling considers only information which represents known metabolites. In the last 8–10 years functional metabolome analysis has passed from concept discussion, method development and feasibility assessment into a phase of method automation and increased scope of applications for enhanced hypothesis generation. It is, however, still an early time for lessons to be learned from high-throughput metabolome analyses. This chapter attempts to exemplify the potential of metabolome analysis for the screening of genetic diversity selected by breeding. This diversity is a widely recognized but also a hard to investigate biological resource. In land races, selection has lead to successful adaptation, for example towards environmental stress tolerance. However, the underlying genomic changes remain elusive. Metabolic phenotyping analysis may circumvent the problem by identifying metabolic markers for a targeted selection. Ultimately metabolic profiling may allow an initial functional insight into metabolic modes of tolerance acquisition without prior knowledge of genomic modifications

Keywords: Metabolome, metabolite profiling, GC-TOF-MS, salinity, rice

1. INTRODUCTION

1.1. Introduction to Metabolome Fingerprinting and Profiling

At the onset of the post-genomic era the concept of metabolome analysis has sparked a new interest in plant metabolism (Trethewey et al. 1999, Fiehn et al. 2000, Fiehn 2002, Sumner et al. 2003, Bino et al. 2004, Fernie et al. 2004, Jenkins et al. 2004,

M.A. Jenks et al. (eds.), Advances in Molecular Breeding Toward Drought and Salt Tolerant Crops, 285–315.

Trethewey 2004). Plant metabolism might be viewed as one source of signals which regulate transcriptional, translational and post-translational processes in plants. On the other hand, metabolism and the subcellular network of dynamic changes in metabolite pool sizes or fluxes may be the ultimate force that delivers the phenotype of an organism (e.g. Roessner-Tunali et al. 2004, Sauer 2004, Carrari et al. 2006, Ratcliffe and Shachar-Hill 2006). In this chapter we intend to discuss the potential new contribution that metabolite fingerprinting and profiling may add to our understanding of molecular stress physiology in plants and thus to improved molecular breeding strategies towards stress tolerant crops.

We will focus on GC-MS technology (Kopka 2006a). This technology was one of the first analytical technologies utilized for functional genomics analyses and molecular physiological studies in plants (Fiehn et al. 2000, Roessner et al. 2000, Roessner et al. 2001a, Roessner et al. 2001b, Roessner et al. 2002). It may still be the most comprehensive and robust metabolomic technology of today. GC-MS profiling is now joined by an emerging multitude of mutually complementing analytical technologies which each open up broad but nevertheless limited windows into metabolism. GC-MS profiling technology was among the first metabolomic technologies applied to rice (e.g. GC-MS, Frenzel et al. 2002; ^{13}C-NMR, Fan et al. 2003; CE-MS, Sato et al. 2004; LC-UV and LC-MS, Li et al. 2004, Morino et al. 2005).

Detailed method descriptions for the GC-MS profiling technology (e.g. Gullberg et al. 2004, Jonsson et al. 2004, Jonsson et al. 2005, Erban et al. 2007) and recent status evaluations are available (Kopka et al. 2004, Kopka 2006b). Here an exemplary study utilizing GC-MS based profiling to reinvestigate adaptation to salinity in *Oryza sativa* L. is presented. Two general types of data mining will be applied to GC-MS analyses of leaf and root samples: (1) fingerprinting analysis, which is defined as the comprehensive "non-biased" analysis of all signals obtained through one analytical technology, and (2) profiling analysis, which utilizes only the subset of identified analytical signals, which represent known metabolites (Fiehn 2002). It is self-evident that only profiling may lead to functional insights into metabolic phenotypes or metabolic modes of stress tolerance. On the other hand profiling is biased towards those metabolic components from a study that are identified at present, for example those that have been made publicly available in dedicated electronic libraries (Halket et al. 1999, Wagner et al. 2003, Halket et al. 2005, Kopka et al. 2005, Schauer et al. 2005). As a consequence important but previously not observed marker metabolites may only be discovered by comprehensive fingerprinting analysis. "Non-biased" fingerprinting, however, is still subject to the limitations of the applied chemical-analytical technology. Truly comprehensive metabolome analyses are currently not possible. An enormous technological effort is required for the combination of analytical technologies for increased metabolome coverage.

Metabolite profiling broadens the scope of metabolite coverage compared to traditional methods. Metabolite profiling by GC-MS allows simultaneous analysis of approximately 50 – 150 known metabolites from the same small amount of sample. A similar analysis when attempted with separate targeted analytical methods is technically more demanding and requires much higher amounts of sample material.

The technological leap in obtaining multi-parallel information does not necessarily come at the cost of quantitative accuracy. If required, GC-MS profiling experiments are perfectly suited for exact quantification provided calibration experiments are performed (e.g. Roessner et al. 2001a, Roessner-Tunali et al. 2003). For most purposes relative quantification of changes in metabolite pool sizes are sufficient (e.g. Cook et al. 2004, Kaplan et al. 2004). Respective calculation of response ratios and normalization procedures for GC-MS profiling studies were described previously (Kopka et al. 2006b).

In conclusion metabolite profiling simplifies comparative metabolic analyses and strongly reduces the analytical bias of metabolic assessments. But admittedly the metabolic window is still rather narrow.

1.2. Present Knowledge on Metabolic Responses to Salt Stress in Rice

Breeding for salinity tolerance, an important goal for the rice crop, has been limited by the complex and polygenic nature of the trait (Yeo and Flowers 1983, Jain et al. 2003). As salinity stress is of marginal importance to the natural ecological range of the species *Oryza sativa L.* (Ponnameruma 1984, Yeo et al. 1990), surveys, for example on traditional Indian cultivars (Yeo et al. 1990) failed to reveal differential tolerance strategies as a basis for combinatorial breeding. Highly tolerant cultivars, such as Pokkali, seem to exploit vigorous vegetative growth and are consequently found among traditional long-straw cultivars which are of limited use for high yield breeding (Yeo et al. 1990). However, strong heterotic effects were observed when crossing moderately tolerant cultivars, a finding which is indicative of the importance of subtle combinatorial effects for improving salt tolerance (Gregorio et al. 2002).

The mechanistic basis of salt tolerance is delineated mainly to the maintenance of low internal Na^+ levels, the sequestration of Na^+ away from growing leaves, the Na^+ tolerance within tissues and to the homeostasis of essential ions, namely K^+, Ca^{2+} and P_i (Yeo and Flowers 1982, Yeo et al. 1990, Zhu et al. 2001, Hien et al. 2003, Babu et al. 2005, Sohn et al. 2005). So far Na^+, K^+, Ca^{2+} and P_i ion accumulation in the absence of salinity stress did not appear to yield unambiguous selection markers for the prediction of salt tolerance (Zhu et al. 2001). Thus there still is a demand for markers, which would allow enhanced selection of parent lines and high-throughput screening of breeding populations. As far as investigations of halophytes indicate, these markers may be found among metabolic adaptations which serve osmotic adjustment and osmoprotection (Bohnert et al. 1995). In the following we shortly summarize the findings concerning major metabolic compound classes.

1.2.1. Proline and amino acids

Accumulation of amino acids and polyamines under salt stress may be interpreted as a sink-reaction to trap excess ammonium (Bouchereau et al. 1999, Yamamoto et al. 2004) but also as an osmoprotectant adaptation. The function of a compatible osmolyte is traditionally assigned to proline. Proline levels increase upon salt treatment, but increased proline concentration is also discussed to represent a symptom of salt stress injury (Lutts et al. 1999). In rice, proline, but also arginine, leucine, alanine, valine and glutamine have been reported to accumulate under salt stress in both shoots and roots (Dubey and Rani 1989, Alpaslan et al. 1999, Hien et al. 2003, Babu et al. 2005). Higher proline accumulation in both organs has been observed in tolerant rice cultivars (Igarashi et al. 1997, Babu et al. 2005), but proline treatment proved to be neutral with respect to salt tolerance (Garcia et al. 1997). Activation of glutamate conversion to proline through overexpression of delta(1)-pyrroline-5-carboxylate synthetase (P5CS; EC 1.5.1.12) appeared to enhance salt tolerance in rice (Zhu et al. 1998, Anoop and Gupta 2003). In the tolerant cultivar Dee-gee-woo-gen (DGWG), P5CS expression increased more under salt stress then in the sensitive IR28 (Igarashi et al. 1997). In contrast, Hien et al. (2003) found that proline accumulation did not result from P5CS regulation. Proline synthesis may be fuelled through other channels. Candidate pathways are the putrescine catabolism via diamine oxidase (DAO; EC 1.4.3.6) (Hien et al. 2003, Bouchereau et al. 1999) and lysine degradation via the bifunctional enzyme, lysine 2-oxoglutarate reductase/saccharopine dehydrogenase (LKR; EC1.5.1.8/ SDH; EC 1.5.1.9). The LKR gene was found to be one of the induced transcripts in the tolerant rice cultivar DGWG (Shiozaki et al. 2005). In addition, enhanced glutamate biosynthesis could also be transmitted to the proline pool through the urea cycle and ornithine-5-aminotransferase activity (OAT; EC 2.6.1.13) (Lutts et al. 1999).

Hydroxyproline-containing diketopiperazines (HPCDs), including naturally occurring L-hydroxyprolyl-L-proline anhydride (D-104) and L-hydroxyprolyl-L-leucine anhydride (D-301), increased stress resistance in rice significantly, when the seeds were treated with these compounds during the initial germination period (Ienaga et al. 1990). Based on structure-activity analysis of six chemical derivatives, D-104 and D-301 were discussed to represent novel phytohormone candidates.

1.2.2. Polyamines

Polyamines are ubiquitous aliphatic amines. Putrescine, spermidine and spermine, for example, are soluble metabolites, which can be conjugated to phenolic compounds or bound to macromolecules (Martin-Tanguy 1997, Bouchereau et al. 1999). Besides a role in regulating cell proliferation and differentiation, polyamines appear to be involved in plant responses to environmental stresses, such as nutrient deficiency, low and high temperature stress, salinity and osmotic stress, hypoxia and oxidative stress (Chattopadhyay and Ghosh 1998, Bouchereau et al. 1999). It is unknown, how polyamines contribute to stress tolerance. Interpretations range from ion balancing (Young and Galston 1984) to regulation of structure, function and synthesis of macromolecules (Jacob and Stetler 1989). Even an unspecific detrimental metabolic reaction which might cause growth retardation and reduce viability was suggested (Slocum et al. 1984).

Results on the role of polyamines in salt acclimation of rice appear to be contradictory. Polyamine concentrations typically increase in response to salt treatment

(Basu et al. 1988, Katiyar and Dubey 1990, Kakkar et al. 2000). But other investigations report a decrease in polyamine concentrations upon salt stress (Prakash and Prathapasenan 1988).

Some studies report increased putrescine levels in sensitive cultivars (Katiyar and Dubey 1990). Other reports comparing sensitive and tolerant cultivars show changes in the ratios among higher polyamines and putrescine and differences between root and shoot metabolism (Kakkar et al. 2000, Krishnamurthy and Bhagwat 1989, Maiale et al. 2004), for example a strong putrescine accumulation in roots of a tolerant compared to a sensitive cultivar and an opposite trend within shoots (Lefevre et al. 2001).

Analyses of enzyme activities and transcription paint an equally conflicting picture. Key steps in polyamine synthesis are catalysed by arginine decarboxylase (ADC, EC 4.1.1.19), the committing step in polyamine synthesis, and S-adenosyl-L-methionine decarboxylase. The salt tolerant cultivar Pokkali had an increased activity and transcript level of ADC. In comparison to Pokkali the sensitive cultivar M-1-48 showed reduced ADC activity and transcription upon prolonged salt acclimation (Chattopadhyay et al. 1997). In the salt tolerant cultivar Giza 181 A, however, ADC, S-adenosyl-L-methionine decarboxylase, and spermidine synthase activities were reduced and polyamine oxidase activity was not detectable (Maiale et al. 2004). Expression or proteome profiling indicate induction of S-adenosyl-L-methionine decarboxylase (Kawasaki et al. 2001, Rabbani et al. 2003, Shiozaki et al. 2005) and S-adenosyl-L-methionine synthetase (Shiozaki et al. 2005, Wu et al. 2005, Yan et al. 2005).

In transgenic approaches expression of the S-adenosyl-L-methionine decarboxylase 1 gene was shown to convey salt tolerance in rice seedlings (Li and Chen 2000). Expression of either ADC (Roy and Wu 2001, Capell et al. 2004) or S-adenosyl-L-methionine decarboxylase (Roy and Wu 2002) enhanced NaCl or osmotic tolerance and caused higher spermidine and spermine concentrations.

1.2.3. Betaines and quaternary ammonium compounds (QAC)

Betaines and other QACs accumulate in many salt tolerant species, including *Poaceae*, namely barley and wheat (Ishitani et al. 1993), but their presence in rice is debated. Betaines and QAC in their function as compatible solutes may represent marker molecules for salt tolerance. However, only few authors (Krishnamurthy et al. 1988, Kishitani et al. 2000) report the presence of glycinebetaine and a total QAC fraction in rice shoot and roots. Other studies on rice and related species using either traditional chemical analysis or mass spectrometry suggest that glycinebetaine is not accumulated in rice (Rathinasabapathi et al. 1993, Takabe et al. 1998). Activity of betaine aldehyde dehydrogenase has been detected in rice leaves (Nakamura et al. 1997) and overexpression of transgenic choline oxidase seems to be beneficial for salt tolerance in rice (Mohanty et al. 2002). So far the role of betaines and QAC for salt tolerance of rice remains elusive.

1.2.4. Sugars

Total soluble sugars typically accumulate under salt stress, especially in monocot species (Kinzel 1982, Bohnert et al. 1995). Reducing and non-reducing sugars, namely sucrose, fructans and trehalose, appear to have opposing effects: salt tolerance of wheat was correlated to accumulation of sucrose and fructans, whereas accumulation of reducing sugars seemed to exert detrimental effects (Kerepesi and Galiba 2000, Kafi et al. 2003).

Fructans have not yet been reported in rice and trehalose appeared to fall below detection limits in rice and other *Poaceae* (Penna 2003). Trehalose acts as a highly effective osmoprotectant in resurrection plants (Penna 2003) and thus may move into the focus of investigations. Rabbani et al. (2003) and Chao et al. (2005) both demonstrated for the rice crop increased expression of the trehalose-6-phosphate phosphatase gene in response to salinity stress and discovered differential regulation in a salt-tolerant compared to a sensitive cultivar. In detail, eight genes which were related to trehalose biosynthesis appeared to be induced in the salt tolerant cultivar Nona Bokra (Chao et al. 2005). Genetic modification for enhanced trehalose biosynthesis in rice demonstrated the potential for an improved performance under salt stress (Garg et al. 2002, Jang et al. 2003).

A possible role of sucrose was also implied in recent investigations. Modification of sucrose transport under salt stress was implied by the discovery of two salt responsive clones with high homology to sucrose transporters (Shiozaki et al. 2005). In addition the sensitive cultivar Giza 35 showed low sugar levels under control conditions which increased upon salt stress. In contrast the tolerant genotype Giza 159 had constitutively high sucrose levels (Rathert 1983, Zhou et al. 2004). Sucrose synthase (SuSy) may be one point of regulation. Salt responsiveness of SuSy gene expression (Wu et al. 2005) and a differential transcriptional regulation of SuSy in a salt tolerant compared to a salt-responsive rice cultivar (Chao et al. 2005) were both recently found.

1.2.5. Polyols

Polyols are thought to have functions as osmoprotectants and appear to contribute to osmotic adjustment under salt stress (Bohnert et al. 1995). Polyols comprise straight-chain metabolites, such as mannitol and sorbitol, and cyclic substances, for example *myo*-inositol and methylated inositol derivatives like ononitol or pinitol. Enzymes of the inositol synthesis and methylation pathway, namely inositol monophosphatase (INO1) and inositol-O-methyltransferase (IMT1), are under tight stress regulation in highly salt tolerant ice plants (Bohnert et al. 1995). Inositol is in addition a building block for stress related conjugates, for example galactinol, which is required for raffinose, stachyose or verbascose biosynthesis.

Information on the role of polyols in rice and Poaceae in general is scarce, for example arabitol and mannitol accumulation was linked to pathogen resistance (Yakubov and Chkanikov 1994). Furthermore Chao et al. (2005) proposed that reduced expression of two GDP-mannose dehydrogenase genes in response to 140 mM NaCl treatment might contribute to mannitol accumulation in the salt-tolerant cultivar

Nona Bokra, whereas a mannitol transporter exhibited increased gene expression in the sensitive cultivar IR29 (Walia et al. 2005). Inositol metabolism has been linked to salt tolerance through the discovery of a conserved L-*myo*-Inositol-1-phosphate synthase in *Porteresia coarctata*, a salt tolerant relative of *Oryza sativa* (Majee et al. 2004) The same enzyme exhibited increased expression upon salt stress in rice (Shiozaki et al. 2005). Finally galactinol synthase, an enzyme which is thought to be involved in temperature stress responses, was also discovered to be an adaptive feature of salinity stress in rice. This enzyme was induced upon salinity stress in the tolerant rice cultivar FL478 but not in the sensitive cultivar IR29 (Walia et al. 2005).

1.2.6. Organic acids

Organic acid biosynthesis, specifically organic acid exchange with the rhizosphere, has been associated with nutrient deficiency, particularly phosphorous, exposure to toxic cations, such as Al^{3+}, and anoxia (Lopez-Bucio et al. 2000, Ryan et al. 2001). Under salt stress organic acids are thought to compensate for charge imbalance in those plants that either take up less Cl^- than Na^+ ions, for example *Chenopodiaceae*, or translocate more Cl^- than Na^+ ions to the shoots, such as the *Poaceae* (Kinzel 1982). Data from field surveys and cultivation experiments indicate that organic acid concentrations decrease in glycophytes under salt stress (Kinzel 1982). In rice seedlings, salt tolerance has been linked to increased activities of malate-dehydrogenase and isocitrate-dehydrogenase (Ritambhara and Dubey 1995). The authors interpret these findings by assuming inhibition of the pentose phosphate pathway under salt stress, which might be compensated by NADPH generation through cytosolic isocitrate-dehydrogenase activity.

1.2.7. Metabolic coverage of GC-TOF-MS metabolite profiling

Most of the metabolites mentioned above can be monitored by GC-TOF-MS metabolite profiling (Kopka et al. 2005, Schauer et al. 2005, and refer to the GMD web resource (http://csbdb.mpimp-golm.mpg.de/gmd.html). A few general restrictions apply. Quaternary ammonium salts, such as betaines, have low volatility. Sugars larger than trisaccharides, for example fructans or stachyose, or polyphosphates, such as phytic acid, exhibit boiling points which are beyond the upper temperature limit of routine GC. Due to the employed chemical derivatization phosphates which are bound to the reduced carbon atom of sugars are lost. Guanidino- ($-NH-CNH-NH_2$) and ureido- ($-NH-CO-NH_2$) moieties are instable. Typically arginine and citrulline are converted into ornithine and agmatine into putrescine (for further details refer to Steinhauser and Kopka 2007).

1.3. Lessons from transcriptome and proteome analysis applied to metabolite profiling

Metabolomic studies are at an early stage in rice (see paragraph 1.1). None has directed attention to metabolic stress adaptation. In contrast expression profiling studies (Chao et al. 2005, Kawasaki et al. 2001, Rabbani et al. 2003,

Walia et al. 2005, Ueda et al. 2006), screenings for salt-induced ESTs (Sahi et al. 2003, Shiozaki et al. 2005, Wu et al. 2005) and several proteome profiles have been performed (Salekdeh et al. 2002, Abbasi and Komatsu 2004, Kim et al. 2005, Yan et al. 2005, Parker et al. 2006). Thus we were able to draw a few general consequences for metabolite profiling. (1) Leaf and root need to be analysed in parallel. Both organs may exhibit different adaptations as was implied by comparative expression analysis of rice shoot and root (Chao et al. 2005) and differential quantitative trait loci (QTL) for ion accumulation (Koyama et al. 2001) and transport (Lin et al. 2004). (2) Salt stress may fundamentally affect plant physiology, for example photosynthetic carbon dioxide assimilation (Chao et al. 2005, Kim et al. 2005), photorespiration (Kim et al. 2005) as well as carbohydrate-, nitrogen- and energy-metabolism (Yan et al. 2005). (3) Salt acclimation in rice may have a long-term impact on the proteome and presumably on the metabolome, which may become apparent only after seven days exposure to salt (Parker et al. 2006). (4) Adaptations may be constitutive or stress-inducible (Salekdeh et al. 2002, Chao et al. 2005).

1.4. The Challenge of Utilizing Selected Diversity

Targeted genetic modifications, such as systematic genomic knock-out projects, gain of function populations or defined mutant populations were the first obvious choice for the exploitation of the new GC-MS metabolite profiling technology. These sources of genetic diversity allow analysis of genes with known function and classification of non-characterized orphan genes by similarity of metabolic phenotypes. The nature of multigenic quantitative traits has been addressed for example by GC-MS based investigations of introgression populations (Schauer et al. 2006). Because of the high costs and time required for breeding introgression lines, the vast resource of genetic diversity, which is generated by either natural selection of wild crop ancestors or breeding, is difficult to utilize.

Large germ-line collections of land races and new or traditional cultivars are available through national projects, such as the Vietnamese source of our rice cultivars (see below), or through international efforts, for example those of the International Rice Research Institute (IRRI). These collections are extensively characterized for physiological phenotype information, but lack a detailed systems characterization at genomic, transcriptomic, proteomic and metabolomic level. A detailed systems characterization is prerequisite to unravel the nature of multi-genic traits and complex, possibly synergistic adaptations to environmental stresses. Metabolite profiling is one feasible step towards this aim, because GC-MS analyses have comparatively low costs associated. Metabolic phenotyping for patterns beneficial or linked to the breeding goal may prove to be the first choice of pre-selecting and screening genetic diversity generated by either natural or human selection.

In this sense we performed a feasibility study with a Vietnamese collection of rice cultivars. These cultivars comprised 14 breeding lines and land races from different geographic regions of Vietnam. For the traditional landraces, the genetic pedigree is not available. The 14 lines had been previously characterised as differentially salt tolerant (Figure 1). We added one additional cultivar, IR57311, and introduced an out-group of three cultivars of subspecies *japonica*, namely Nipponbare, Taipai and Zhong hua (Figure 1). These four cultivars were not chosen for their salt tolerance, but especially cv. Nipponbare is known to be salt sensitive. *Oryza sativa subspec. japonica* cv. Nipponbare is here used as a reference cultivar to compare this study to subsequent investigations. Metabolite profiling experiments typically use reference genotypes under control conditions to facilitate comparison between experiments. The resulting cultivar population was reassessed for the tolerance trait in a controlled environment study. This reinvestigation was accompanied in parallel by GC-TOF-MS based metabolic classification.

In conclusion, our investigation is in contrast to the mainstream of current metabolomics investigations which tackle the functional genomics aspect of molecular plant physiology. Our challenge is the gain of functional and physiological insight into metabolic patterns without knowledge of the genetic disposition of the underlying crop diversity. While insight into gene function will be almost impossible at this stage of investigation, we can expect the discovery of potential metabolic markers or even metabolic modes of stress tolerance. We intend to demonstrate that this approach is in essence functional and not purely descriptive.

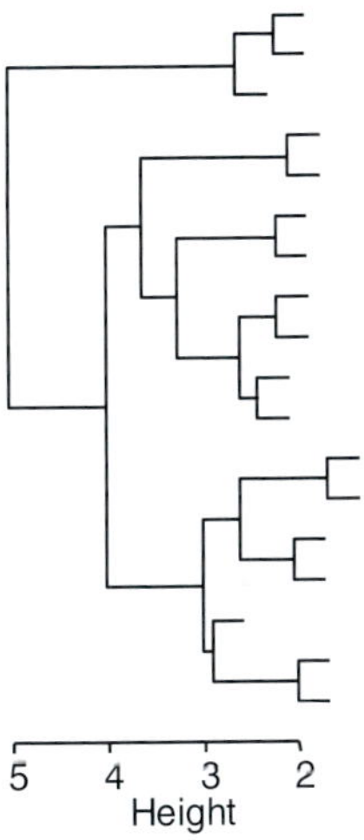

Cultivar	Code	SSI	Class
Oryza sativa subsp. japonica (Tai Pai)	51	1.16	S
Oryza sativa subsp. japonica (Nipponbare)	50	1.00	S
Oryza sativa subsp. japonica (Zhong Hua)	53	0.70	S
Oryza sativa subsp. indica (Doc Phung)	8	0.29	IS
Oryza sativa subsp. indica (Cha Va)	28	0.26	IT
Oryza sativa subsp. indica (DR2)	2	0.44	S
Oryza sativa subsp. indica (C71)	5	0.22	IT
Oryza sativa subsp. indica (IR57311)	52	0.19	T
Oryza sativa subsp. indica (C70)	4	0.18	T
Oryza sativa subsp. indica (Doc Do)	7	0.25	IT
Oryza sativa subsp. indica (CR203)	1	0.11	T
Oryza sativa subsp. indica (Nuoc Man 1)	25	0.34	IS
Oryza sativa subsp. indica (Nuoc Man)	21	0.32	IS
Oryza sativa subsp. indica (Lua Man)	22	0.48	S
Oryza sativa subsp. indica (Cuom)	14	0.27	IS
Oryza sativa subsp. indica (Nep Man)	23	0.15	T
Oryza sativa subsp. indica (Cham Bien)	27	0.22	IT
Oryza sativa subsp. indica (Cham)	26	0.10	T

Figure 1. Cultivar code, average salt sensitivity index (SSI), salt tolerance ranking (Class) and metabolic hierarchical cluster analysis (HCA). HCA demonstrates the degree of overall similarity between metabolite phenotypes of the *Oryza sativa* L. cultivars from the present study. Metabolite profiles were cultivar averages of 4 - 5 plants each. Organ as well as NaCl dose dependent profiles were concatenated prior to agglomerative HCA. HCA was performed using complete linkage of Euclidian distances. Note that tolerant cultivars occur in two of the three major branches

2. ASSESSMENT OF SALT TOLERANCE

2.1. Experimental

Salt water frequently floods rice cultivation areas in the Vietnamese costal plains or tidal river banks. Salinity influences yield when it occurs during two critical growth stages, namely after transplanting of young plantlets and during seed setting of the mature crop. We concentrated on the first, the three-leaf stage. For the characterisation of salt tolerance in 18 cultivars (see above) we performed 3 independent experiments in environmentally controlled growth chambers. After germination (10 days at 28 °C), plants were grown on hydroponic medium (Yang et al. 1994). Environmental conditions were: 12 h day length at 600 $\mu E\ m^{-2}\ s^{-1}$, temperature 26 °C in the light and 22 °C at night, relative humidity 75% in the light and 70% at night. After a growth period of 14 days, coincident with the three-leaf stage, the medium was changed to fresh medium with 0 (control), 50 or 100 mM NaCl. Each salt level was replicated in five blocks. In each block all cultivars were randomized. After 14 days of acclimation, a representative root sample and the youngest fully expanded leaf were harvested from each plantlet for subsequent GC-MS metabolite profiling. Samples were stored at -80 °C. Randomized GC-MS analysis of 240 plantlets was performed in two separate batches with 4-5 replicate plants per cultivar and condition. Automated, timed, in-line derivatization was applied prior to GC-TOF-MS. One representative GC-TOF-MS data set was used for the following comparative study. In addition to the GC-TOF-MS metabolite phenotype, photosynthetic yield and growth parameters, such as tiller number, leaf and root length, fresh and dry weight of shoot and root were measured (e.g. Figure 2) and the plants scored according to IRRI scheme (http://www.knowledgebank.irri.org/ses/).

2.2. Cultivar Classification

The re-assessment of salinity tolerance was performed using a modified salinity susceptibility index (SSI; Fischer and Maurer, 1978) and quantile ranking to create four tolerance classes, namely tolerant (T), intermediate tolerant (IT), intermediate sensitive (IS) and sensitive (S). SSI allows the comparison of salinity tolerance relative to a reference cultivar, here the salt-sensitive cultivar Nipponbare (code 50). We show one exemplary SSI analysis using the photosynthetic yield parameter 6 days after acclimation to 100 mM NaCl (Figure 3). With exception of the outgroup of sensitive *japonica* cultivars, the assignment of cultivars to a tolerance class was affected by the experiment and by the morphological response parameter, on which the respective assignment was based. We concluded that the Vietnamese collection represented a gradient of increasingly salt tolerant cultivars rather than two distinct classes of tolerant and non-tolerant cultivars.

We supported this conclusion by comparing averaged root and shoot fingerprints of each of the four generated tolerance classes. Averages were calculated from all constituent fingerprints at the level of reproducibly occurring fragment masses (see

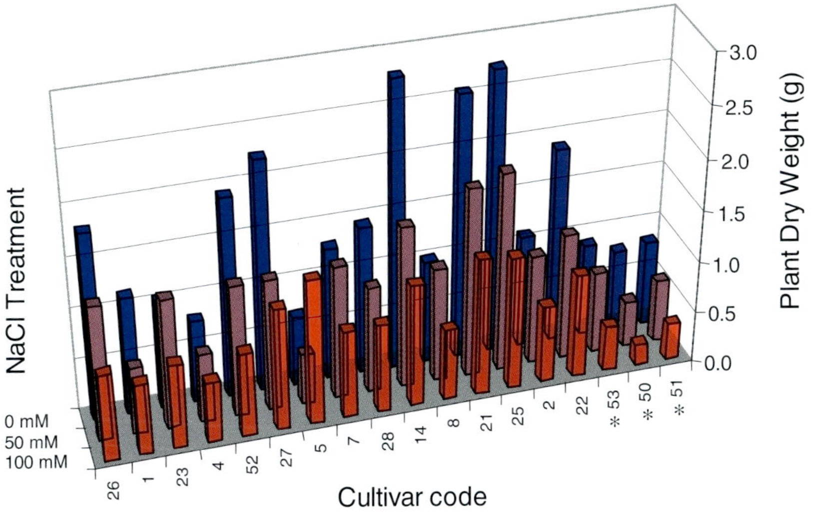

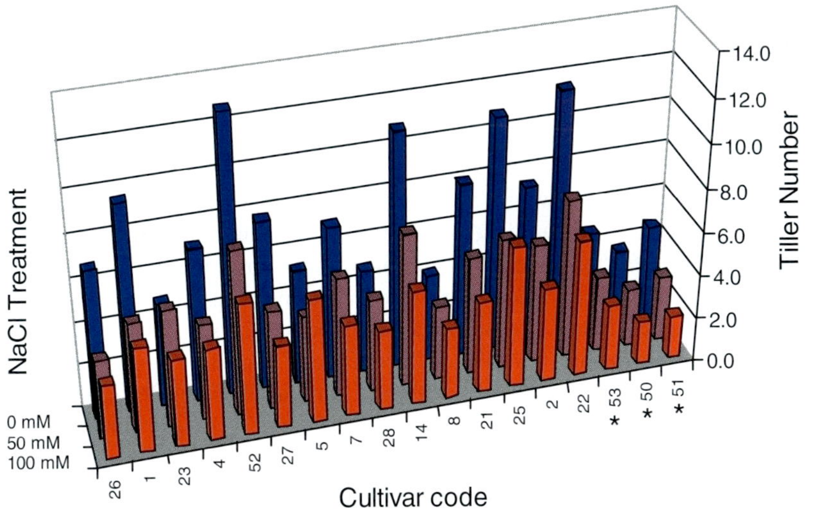

Figure 2. Changes of plant dry weight (A) and tiller number (B) in response to NaCl acclimation. The sequence of cultivars is as shown in Figure 3. Members of subspecies *japonica* are indicated by asterisk

paragraph 3.2; e.g. Figure 7). We applied hierarchical cluster analysis (HCA) to compare the metabolic changes during salinity acclimation in these classes. HCA demonstrated that the salt acclimation was the main determinant factor of metabolic phenotypes (Figure 4). Root fingerprints exhibited higher variance compared to leaf. Both organs exhibited clear changes in response to salt. The clustering pattern indicated dose dependent changes in salt sensitive and tolerant classes. The tolerant

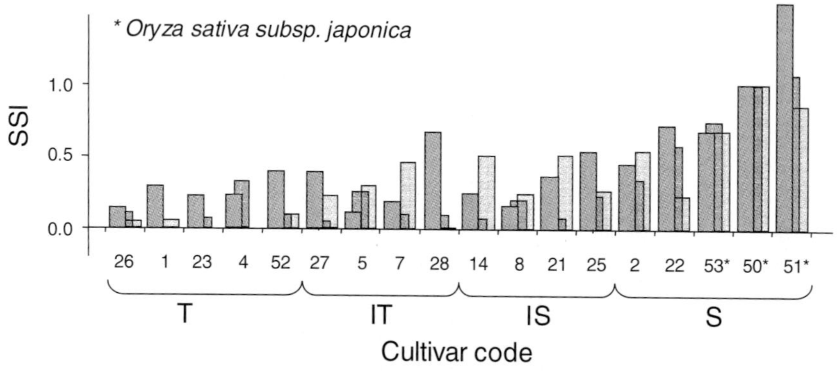

Figure 3. Salt susceptibility index (SSI) based on photosynthetic yield of the *Oryza sativa* L. cultivars from the present study (cultivar code see Figure 1). Photosynthesis was measured as yield = (F'm-F'v)/F'm with the pulse-amplitude-modulation (PAM) fluorimetric method using a portable fluorimeter (PAM-2000) from WALZ, Germany. Leaves were adapted at 300 μE $m^{-2}s^{-1}$ light during growth to record the Ft-level before a single saturating flash was applied to reach Fm (maximal fluorescence). Salt acclimation was started at the three-leaf seedling stage. Measurements were performed 6 days after acclimation to 100 mM NaCl. Phenotype stability was assessed in three independent experiments as represented by three stacked bars. SSI was calculated according to SSI = $(1-(X_k/X_t)) * (1-(R_k/R_t))^{-1}$, where X_k represents the photosynthetic yield of each cultivar under control conditions and X_t the measurement after salt acclimation. R represents the behavior of the reference cultivar Nipponbare, code 50. Classification was according to quantile ranking of the averaged SSI of three independent experiments: tolerant (T), intermediate tolerant (IT), intermediate sensitive (IS) and sensitive (S)

and intermediate tolerant classes (T, IT) exhibited a high similarity. The S-class and to a minor extend the IS-class exhibited an increasing dose dependent metabolic differentiation: in roots S- and IS-classes were increasingly dissimilar compared to both T- and IT-classes; in leaf S- and IS-classes associated at 50 mM NaCl with the metabolic phenotypes of the 100 mM dose, whereas T- and IT-classes at 50 mM NaCl were more similar to the control treatment. Similar HCA results, specifically a distinctive S-class but rather non-distinctive IS-, IT- and T-classes, were obtained with SSI classifications based on morphological data.

We furthermore performed a global cultivar specific HCA using concatenated profiles which comprised all fingerprint information of both organs at the three treatment levels (Figure 1). This analysis supported the overall difference of the *japonica* out-group compared to the *indica* cultivars and interestingly the presence of two major metabolic phenotypes among the remaining cultivars. Cultivar IR 57311, code 52, exhibited metabolic similarity to C70, code 4, from Vietnam and was equally tolerant to salt stress. Most importantly, sensitive and tolerant cultivars were found among both major metabolic phenotypes.

We furthermore investigated the possible presence of differential metabolite phenotypes under control conditions which may be indicative of salt tolerance or sensitivity. These metabolic features will in the following be called a metabolic predisposition. Separate HCA of leaf and root profiles without salt treatment clearly supported the

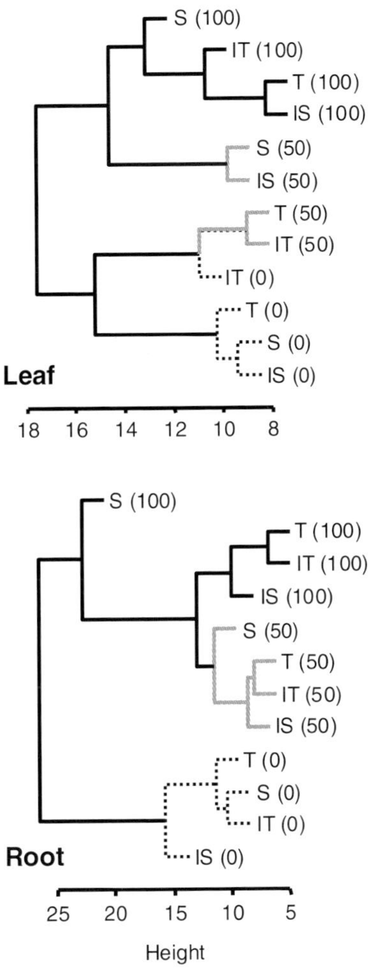

Figure 4. HCA analyses of salinity tolerance classes (see Figure 3). The metabolic phenotype was assessed separately using fingerprints of leaf and root organs. Response ratios of all mass fragments were averaged per cultivar class. HCA was performed using complete linkage of Euclidian distances. Class identity, salt dose (round brackets), and a variance scale are indicated

presence of two metabolic phenotypes with more tolerant cultivars (Figure 5). The metabolic classification was less clear in the leaf compared to the root and the association of some sensitive and intermediate cultivars changed from root to leaf.

In conclusion we classified rice cultivars into two major metabolic phenotypic groups. We made the unexpected discovery that enhanced salt tolerance may be acquired through two types of metabolic acclimation. We demonstrated a predisposition under control conditions which can be traced more clearly in the metabolome of roots compared to leaves.

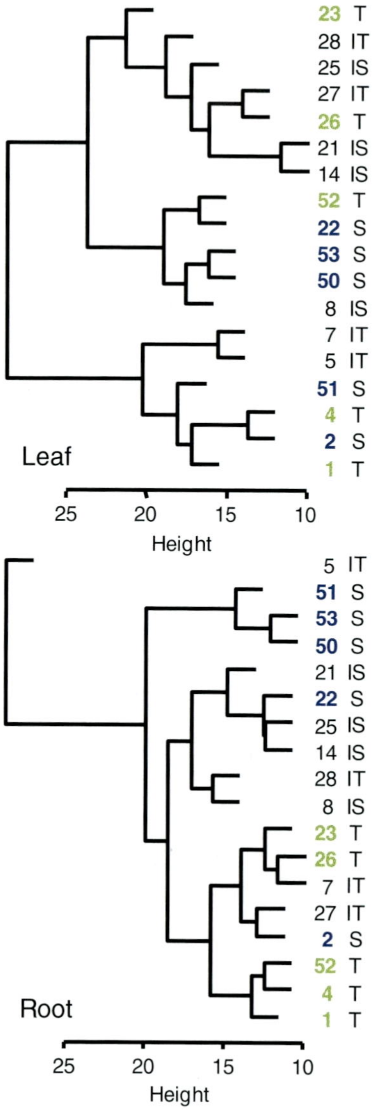

Figure 5. HCA analysis of leaf and root profiles from control treatments. HCA was as described in the legend to Figure 4. Response ratios of all mass fragments were averaged per cultivar

3. ANALYSIS OF THE METABOLIC PHENOTYPE

3.1. Independent Component Analysis Reveals Metabolites Relevant for Salt Acclimation

Analysis of the metabolite phenotype requires a tool box of statistical algorithms and visualizations. In the following we will show an independent component analysis (ICA; Stone 2002, Scholz et al. 2004) applied to GC-TOF-MS fingerprints of root samples from single plants. ICA is a variant of the principal component analysis (PCA) which is widely applied in metabolite profiling studies (e.g. Roessner et al. 2001a, Scholz et al. 2005, Trygg et al. 2006). PCA is a technology for sample classification which reduces multivariate data sets to few variables comprising the major variances in the data set. Thus PCA analyses of metabolite profiles make the assumption that high variance indicates high relevance. ICA allows in addition the non-supervised search for best bimodal sample partitions. The comparison of ICA and PCA applied to the root GC-TOF-MS fingerprinting data illustrates the enhanced discovery potential of ICA (Figure 6). ICA is well suited for the confirmation of known experimental sample classes but allows also the discovery of unexpected classes or trends among samples. Each independent component (IC) encodes a single partition among samples. A so-called loadings analysis (e.g. Figure 7) unravels those fingerprinting signals which are most relevant for the distinction of the embedded sample partitions. For these reasons ICA is most appropriate for analyses of data sets which a priori lack clear sample classifications and to test for unforeseen sample classes or outlier profiles.

ICA clearly supported HCA analyses and demonstrated that the salt treatment induced the major variance within the data sets of leaves (data not shown) and roots (Figure 6). Each organ had one IC, namely $IC1_{root}$ and $IC1_{leaf}$, which described the partition between control and salt treatment. The data set of roots exhibited in addition $IC7_{root}$ which demonstrated an unexpected sample partition among the plantlets which were treated with the highest salt dose. This finding would have passed unnoticed without ICA. No sample partition was detected which separated the more tolerant from the more salt sensitive cultivars. But in both, $IC1_{root}$ and $IC1_{leaf}$, individual plantlets of the sensitive cultivars were shifted towards the general salt acclimated state, some even overlapped. We therefore concluded that metabolic markers for salt sensitivity may be represented as quantitative changes of metabolite pools rather than as "black-and-white" absence or presence of marker metabolites. A subsequent ICA loadings analysis (Figure 7) may demonstrate the transition in data mining from fingerprinting to profiling analysis.

3.2. From Fingerprints to Profiles

GC-TOF-MS fingerprints comprise thousands of mass fragments at hundreds of retention time index (RI) windows (Figure 7). For example the present experiment yielded 18.158 mass fragments from root and 12.079 from leaf which occurred repeatedly within identical RI windows of replicate GC-TOF-MS profiles. So-called mass spectral tags (MSTs), defined as mass spectra with linked RI information (Kopka 2006a, Kopka 2006b), are reconstituted from co-eluting sets of mass fragments which exhibit the same quantitative changes. These MSTs can be identified by comparison to mass spectral and RI libraries of pure authentic metabolites (Kopka et al. 2005, Schauer et al. 2005). Mass spectral and RI comparison are supported by automated matching tools. Metabolite identification must however be checked by subsequent manual validation. Table 1 shows an overview of identified

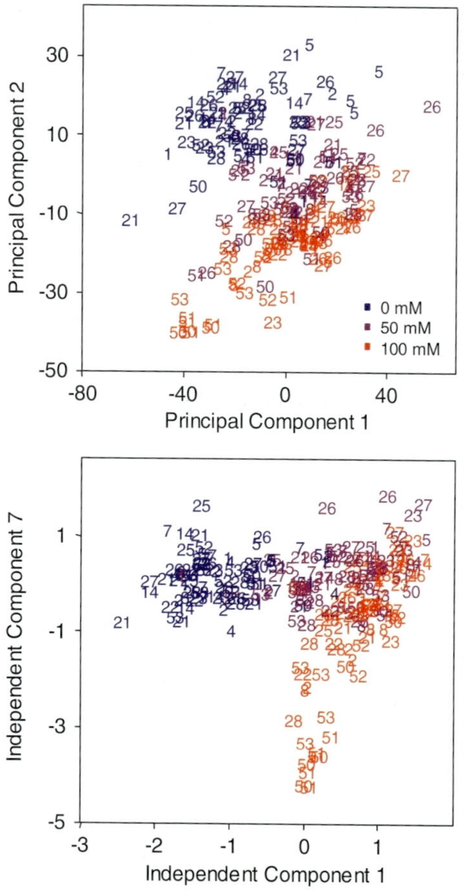

Figure 6. Principal component analysis (PCA; top) and an independent component analysis (ICA; bottom) of GC-TOF-MS metabolite fingerprints from root samples. Numbers indicate individual plants of the respective cultivar code (refer to Figure 1). Salt doses are color-coded. Two bimodal sample distributions were detected. Independent component 1 distinguishes control and salt acclimated samples

metabolite classes from our experiment. An exemplary detailed list of the class of organic acids is presented in Table 2.

Loadings analysis of $IC1_{root}$ and $IC1_{leaf}$ demonstrated that three metabolites were among the most relevant for salt acclimation in both root and leaf metabolism: citric acid, malic acid and quinic acid (e.g. Figure 7). In addition we found that shikimic acid, oxalic acid and proline contributed substantially to salt induced changes in leaf metabolism, whereas root metabolism exhibited additional major changes of fructose, mannitol, putrescine and four MSTs, which represented currently non-identifed metabolites (Figure 7). The loadings analysis of $IC7_{root}$ showed sucrose as the most influential metabolite for this component. Sucrose levels were low in those plantlets which differed from the majority of salt acclimated plants.

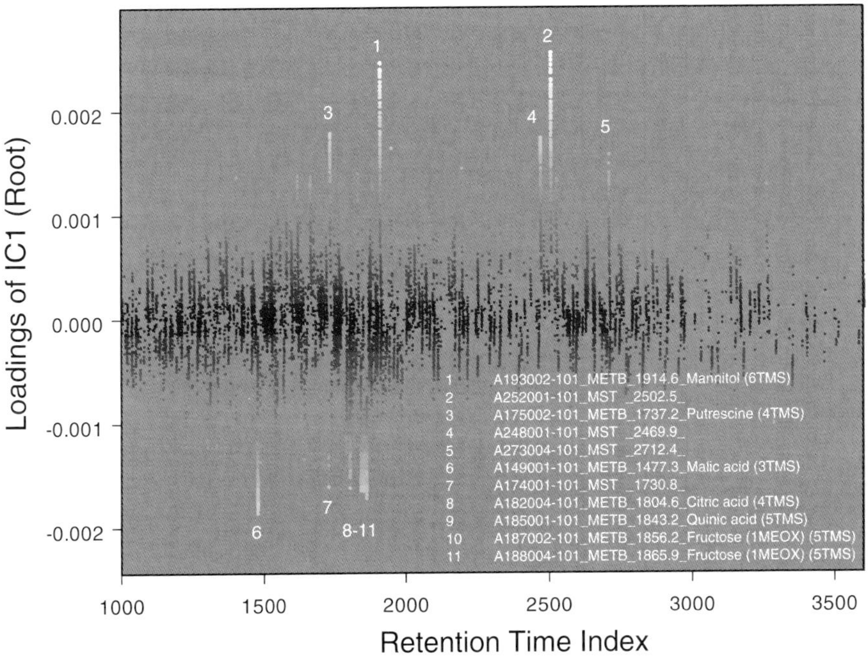

Figure 7. Loadings analysis of independent component 1 (Figure 7) revealing the metabolites which are most relevant for salt acclimation. Each vertex represents a mass fragment at a gas chromatographic retention time index. High loadings amplitude in white color code indicates relevancy. Numbers indicate identified metabolites (METB) and jet non-identified mass spectral tags (MST) which are each represented by multiple co-eluting mass fragments. The number code, for example 175002 allows queries at GMD (http://csbdb.mpimp-golm.mpg.de/gmd.html) for the respective mass spectral data and supplementary information. Note that GC-MS inherent chemical derivatization may generate multiple analytes, which represent the same metabolite, for example fructose

In addition respective plantlets were small, exhibited low photosynthetic yield and were individuals of sensitive cultivars. We concluded that these plants were beyond salt acclimation and had entered toxicity induced carbohydrate starvation which became apparent first in the root through strongly reduced carbohydrate levels.

A detailed analysis of the most variable metabolites demonstrated that fructose and organic acids, such as malic, citric, shikimic (data not shown) and quinic (data not shown) acid, were in general depleted upon salt acclimation (Figure 8). This depletion was observed at 50 as well as at 100 mM NaCl and exhibited a dose dependency in both leaf and root. Only oxalic acid was strongly reduced in leaves but not in roots. Interestingly the leaf pools of most Vietnamese cultivars exhibited a stronger acid depletion than the respective root pools. An even more extreme behavior of the acid metabolites was found in the out-group of subspecies *japonica* cultivars. The 3 chosen cultivars exhibited an enhanced depletion of these

Table 2. Summary of an automated inventory of substances detected by GC-TOF-MS profiling of polar extracts from *Oryza sativa* L. leaf and root samples. 494 substances were identified and among those 156 metabolites were represented. Because of the chemical derivatization which is required for GC analysis of polar compounds, metabolites typically form 1 or 2 analytes. Lipid and fatty acid components, which were present at trace levels in polar extracts, result from incomplete partitioning into a chloroform fraction, which was discarded for the present study. These residual lipid components, internal standards and reagent artifacts are disregarded for subsequent analysis

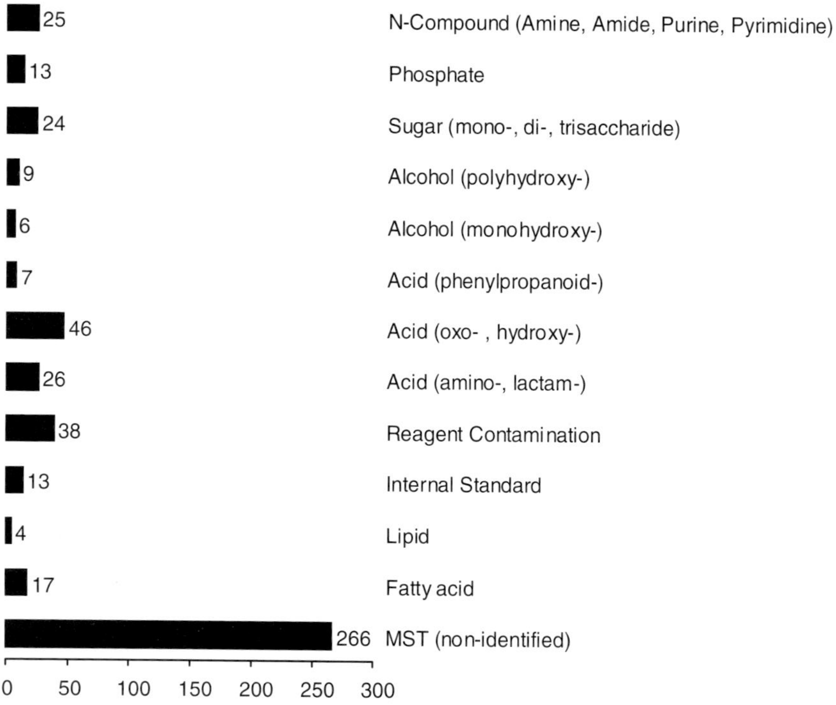

metabolites from the root pools. A typical marker metabolite for salt acclimation, proline, was confirmed to accumulate in leaf and for most cultivars also in roots (Figure 8). Accumulation was highly variable between cultivars. Putrescine and mannitol levels increased upon salt acclimation. Both exhibited stronger changes in root compared to leaf. The three tolerant lines, code 1, 4 and 52, exhibited high similarity of fructose, citric, malic, shikimic and quinic acid changes compared to the more sensitive lines of the same HCA branch. But the remaining tolerant cultivars, code 23 and 26, members of the alternate HCA branch, were clearly different with respect to these metabolites.

In conclusion ICA yielded insight into the set of metabolites which appeared to be most relevant - according to variance - for salt acclimation in rice. Among the most variable metabolites no single marker was found which would indicate an adaptation towards enhanced salt tolerance. Combining ICA and HCA results we concluded

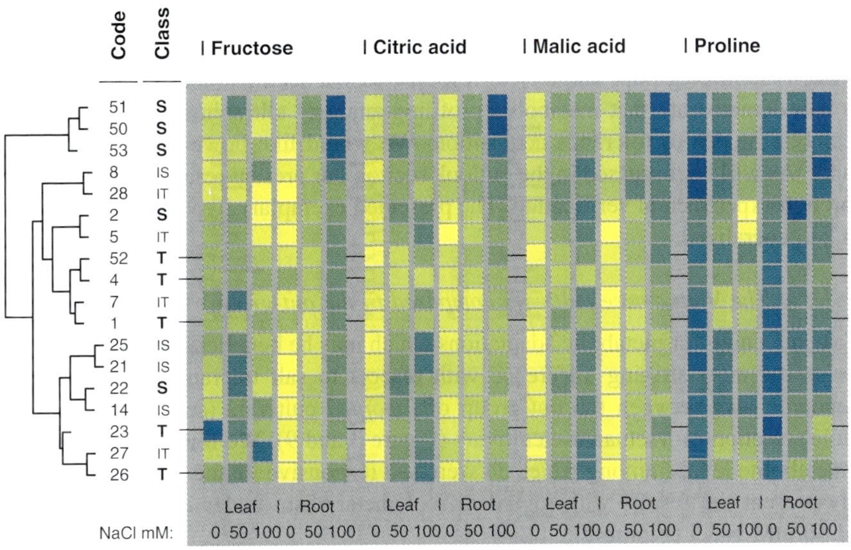

Figure 8. Overview of the changes of four metabolite pools in leaves and roots of *Oryza sativa L.* Fructose, citric acid, malic acid and proline are shown. Increases relative to the average pool size are color coded yellow, decreases are indicated blue

that patterns of multiple pool size changes rather than a single marker molecule may indicate successful adaptations towards enhanced salinity tolerance. But we currently lack bioinformatics protocols to automate a systematic screening of our data sets for combinatorial metabolic patterns and thereby reduce the complexity of information for hypothesis generation.

3.3. Comparative Analysis of Metabolic Phenotypes

In the following we will redirect the data mining process to the finding that tolerant cultivars exhibited a common, however bipartite, metabolic phenotype in roots under control conditions (Figure 5). In addition we address the metabolic difference of the sensitive out-group cultivars to the Vietnamese selection under these conditions. This approach reduces the complexity of information to be considered and tests again for the presence of common markers among the tolerant or sensitive cultivars of our study. Instead of the most variable metabolites we now considered all metabolites including those exhibiting minor variances. This analysis appeared to be most promising among the multiple options of data mining, because it was based on clear separations of tolerant and sensitive cultivars (Figure 5), (1) the out-group of sensitive cultivars, code 50, 51, and 53, formed a separate class, (2) a HCA branch was present which comprised only sensitive and intermediate Vietnamese cultivars, (3) only a single sensitive cultivar, code 2, was classified within the branch of tolerant cultivars. In addition the analysis of the metabolic phenotype under control

The concentration of orthophosphate was higher indicating a possible interaction between phosphate accumulation and acid depletion (Lopez-Bucio et al. 2000, Ryan et al. 2001, Vance et al. 2003).

3.3.2. The tolerant Vietnamese cultivars compared to the sensitive and intermediate Vietnamese cultivars

The Vietnamese cultivars appeared to form a gradient of increasing salt tolerance. However two distinct metabolic phenotypic adaptations were expected from the initial analyses of the metabolite phenotype. We separately screened the two groups of tolerant cultivars, code 1, 4, 52, and code 23, 26, respectively, under control conditions for significant changes in root metabolism compared to the sensitive and intermediate Vietnamese cultivars. No single marker metabolite enabling unequivocal distinction of the tolerant cultivars from the remaining Vietnamese cultivars was found (Figure 9). Some potential common marker metabolites, such as low P_i, 3-PGA, glycine or phenylalanine content, were shared by other Vietnamese cultivars. The most striking differences between the two tolerant metabolic classes were the different levels of fructose under control condition and the drastic increase of amino acids, for example glycine, phenylalanine, glutamate and glutamine (Figure 9), in cultivars 23 and 26 upon salt acclimation.

4. OUTLOOK AND TENTATIVE LESSONS FROM METABOLOME PROFILING

The presented preliminary study on metabolic modes of acquiring salt tolerance in rice demonstrated the feasibility of our approach. Previous and sometimes conflicting studies on the importance of single or a small number of metabolites for salt tolerance can now be combined into a more comprehensive picture (e.g. Figure 8 and 9). The necessity for further experimentation is clearly indicated. (1) More and previously characterized cultivars with increased salt tolerance need to be investigated, ideally sensitive parent cultivars and tolerant progeny of both rice subspecies, namely *japonica* and *indica*. Characterised recombinant inbred populations would be ideal to obtain in addition gene function information. (2) Time course analyses of salt acclimation would yield information on transient metabolic changes and establish a sequence of metabolic events. (3) Ultimately, a transfer from lab to field will be required. It is evident that even though metabolite profiling may be comparatively low cost, only some of these tasks can be performed in an academic laboratory. Some of the above experiments, specifically dealing with the transfer to the field, are best performed in a combined effort of academic and commercial partners.

Two tentative lessons may be drawn from our investigation. (1) Single unambiguous marker molecules may be found. But because of the high pathway connectivity these single markers will be rare in primary metabolism and perhaps more frequent in secondary metabolism. Instead of single markers we found patterns of simultaneous metabolic changes which we hypothesise to reflect metabolic modules that might be under common regulation. Today however we still lack systematic metabolite profiling studies for comparison with other environmental stresses or with targeted genetic modifications, which would allow the discovery of unique or ubiquitous metabolic patterns involved in stress response. For example, depletion of TCA cycle intermediates may be common for processes which have a high demand for energy. The redistribution or exclusion of Na^+ and Cl^- ions within and from the plant requires fuelling of transport mechanisms. TCA cycle intermediates may also be lost through other processes, for example acid secretion or sequestration of nitrogen into amino acid pools. And finally acid metabolism might interact with the availability of internal phosphate as previously discussed (Lopez-Bucio et al. 2000, Ryan et al. 2001, Vance et al. 2003).

(2) As stated above successful metabolic modes of adaptation may be combinatorial, i.e. involve patterns of changes in many metabolite levels. In addition alternate metabolic adaptations may achieve the same purpose for one particular type of stress. If only single metabolites were monitored results could be conflicting, because a metabolite might be required to be at low levels in one successful metabolic context and high in the alternate adaptation.

In conclusion metabolomics today is well established to generate novel hypotheses and insight into successful adaptations. In order to support hypothesis generation we need to test and develop bioinformatics tools which automatically pick up conditional patterns of metabolic changes. For improved interpretation of these changes we require a detailed comparative description of metabolic responses under a multitude of defined conditions and developmental stages. An integrative systems-wide picture of changes needs to be compiled which considers changes at transcriptome, proteome, ionome and metabolome levels (e.g. Urbanczyk-Wochniak et al. 2003, Colebatch et al. 2004, Weckwerth et al. 2004, Denby and Gehring 2005, Oksman-Caldentey et al. 2005). Building the required "systems phenotyping factories" for plants is certainly worth the effort.

ACKNOWLEDGEMENTS

We would like to thank M. R. Siahpoosh, A. Erban, Dr. D. Sanchez and Dr. D. Hincha, Max Planck Institute of Molecular Plant Physiology, Potsdam-Golm, Germany for critical contributions to our discussions on this manuscript. Our thanks extend to Prof. Dr. Le Tran Binh, Institute of Biotechnology (IBT), Hanoi, Vietnam, for sharing his expertise in rice cultivation and contributing most of the rice cultivars. The *japonica* cultivars were contributed by the IRRI and Prof. Dr. Xue Hongwei, Shanghai Institute of Plant Physiology and Ecology. This work was generously supported by the Max Planck Society, and the German Bundesministerium für Bildung und Forschung (BMBF), grant PTJ-BIO/0312854.

REFERENCES

Abbasi FM, Komatsu S (2004) A proteomic approach to analyze salt-responsive proteins in rice leaf sheath. Proteomics 4:2072–2081.

Alpaslan M, Gunes A, Taban S (1999) Salinity resistance of certain rice (*Oryza sativa* L.) cultivars. Turkish J Biol 23:499–506.

Anoop N, Gupta AK (2003) Transgenic *indica* rice cv IR-50 over-expressing *Vigna aconitifolia* DELTA-1-pyrroline-5-carboxylate synthetase cDNA shows tolerance to high salt. J Plant Biochem Biotechnol 12:109–116.

Babu CR, Vijayalakshmi C, Mohandass S (2005) Evaluation of rice (*Oryza sativa* L.) genotypes for salt tolerance. J Food Agricult Environ 3:190–194.

Basu R, Maitra N, Ghosh B (1988) Salinity results in polyamine accumulation in early rice *Oryza sativa* L. seedlings. Austral J Plant Physiol 15:777–786.

Bino RJ, Hall RD, Fiehn O, Kopka J, Saito K, Draper J, Nikolau BJ, Mendes P, Roessner-Tunali U, Beale MH, Trethewey RN, Lange BM, Wurtele ES, Sumner LW (2004) Potential of metabolomics as a functional genomics tool. Trends Plant Sci 9:418–425.

Bohnert HJ, Nelson DE, Jensen RG (1995) Adaptations to environmental stresses. Plant Cell 7:1099–1111.

Bouchereau A, Aziz A, Larher F, Martin-Tanguy J (1999) Polyamines and environmental challenges: Recent development. Plant Sci 140:103–125.

Capell T, Bassie L, Christou P (2004) Modulation of the polyamine biosynthetic pathway in transgenic rice confers tolerance to drought stress. Proc Natl Acad Sci USA 101:9909–9914.

Carrari F, Schauer N, Willmitzer L, Fernie AR (2006) Systems biology: a renaissance of the top-down approach. In: Nagata T, Lörz H, Widholm JM (eds) Biotechnology in agriculture and forestry Vol 57: Saito K, Dixon RA, Willmitzer L (eds) Plant metabolomics. Springer-Verlag, Berlin Heidelberg New York, pp 185–198.

Chao DY, Luo YH, Shi M, Luo D, Lin HX (2005) Salt-responsive genes in rice revealed by cDNA microarray analysis. Cell Res 15:796–810.

Chattopadhyay MK, Gupta S, Sengupta DN, Ghosh B (1997) Expression of arginine decarboxylase in seedlings of *indica* rice (*Oryza sativa* L.) cultivars as affected by salinity stress. Plant Mol Biol 34:477–483.

Chattopadhyay MK, Ghosh B (1998) Molecular analysis of polyamine biosynthesis in higher plants. Current Sci 74:517–522.

Colebatch G, Desbrosses G, Ott T, Krusell L, Kloska S, Kopka J, Udvardi MK (2004) Global changes in transcription orchestrate metabolic differentiation during symbiotic nitrogen fixation in *Lotus japonicus*. Plant J 39:487–512.

Cook D, Fowler S, Fiehn O, Thomashow MF (2004) A prominent role for the CBF cold response pathway in configuring the low-temperature metabolome of *Arabidopsis*. Proc Natl Acad Sci USA 101:15243–15248.

Denby K, Gehring C (2005) Engineering drought and salinity tolerance in plants: lessons from genome-wide expression profiling in *Arabidopsis*. Trends Biotech 23:547–552.

Dubey RS, Rani M (1989) Salinity induced accumulation of free amino acids in germinating rice seeds differing in salt tolerance. J Agronomy Crop Sci 163:236–247.

Erban A, Schauer N, Fernie AR, Kopka J (2007) Non-supervised construction and application of mass spectral and retention time index libraries from time-of-flight GC-MS metabolite profiles. In: Weckwerth W (ed) Metabolomics: methods and protocols. Humana Press, Totowa, pp 19–38.

Fan TWM, Lane AN, Higashi RA (2003) In vivo and in vitro metabolomic analysis of anaerobic rice coleoptiles revealed unexpected pathways. Russian J Plant Physiol 50:787–793.

Fernie AR, Trethewey RN, Krotzky AJ, Willmitzer L (2004) Metabolite profiling: from diagnostics to systems biology. Nat Rev Mol Cell Biol 5:763–769.

Fiehn O (2002) Metabolomics - the link between genotypes and phenotypes. Plant Mol Biol 48:155–171.

Fiehn O, Kopka J, Dörmann P, Altmann T, Trethewey RN, Willmitzer L (2000) Metabolite profiling for plant functional genomics. Nat Biotechnol 18:1157–1161.

Fischer RA, Maurer R (1978) Drought resistance in spring wheat cultivars. 1. Grain-yield responses. Australian J Agricultural Res 29:897–912.

Frenzel T, Miller A, Engel KH (2002) Metabolite profiling - A fractionation method for analysis of major and minor compounds in rice grains. Cereal Chem 79:215–221.

Garcia AB, Engler JD, Iyer S, Gerats T, VanMontagu M, Caplan AB (1997) Effects of osmoprotectants upon NaCl stress in rice. Plant Physiol 115:159–169.

Garg AK, Kim JK, Owens TG, Ranwala AP, Choi YD, Kochian LV, Wu RJ (2002) Trehalose accumulation in rice plants confers high tolerance levels to different abiotic stresses. Proc Natl Acad Sci USA 99:15898–15903.

Gregorio GB, Senadhira D, Mendoza RD, Manigbas NL, Roxas JP, Guerta CQ (2002) Progress in breeding for salinity tolerance and associated abiotic stresses in rice. Field Crops Res 76:91–101.

Gullberg J, Jonsson P, Nordström A, Sjöström M, Moritz T (2004) Design of experiments: an efficient strategy to identify factors influencing extraction and derivatization of *Arabidopsis thaliana* samples in metabolomic studies with gas chromatography/mass spectrometry. Anal Biochem 331:283–295.

Halket JM, Przyborowska AM, Stein SE, Mallard WG, Down S, Chalmers RA (1999) Deconvolution gas chromatography mass spectrometry of urinary organic acids - potential for pattern recognition and automated identification of metabolic disorders. Rapid Commun Mass Spectrom 13:279–284.

Halket JM, Waterman D, Przyborowska AM, Patel RKP, Fraser PD, Bramley PM (2005) Chemical derivatization and mass spectral libraries in metabolic profling by GC/MS and LC/MS/MS. J Exp Bot 56:219–243.

Hien DT, Jacobs M, Angenon G, Hermans C, Thu TT, Van Son L, Roosens NH (2003) Proline accumulation and DELTA-1-pyrroline-5-carboxylate synthetase gene properties in three rice cultivars differing in salinity and drought tolerance. Plant Sci 165:1059–1068.

Ienaga K, Nakamura K, Kurohashi M, Nakanishi T, Ichii T (1990) Simple peptides. 5-Hydroxyproline-containing diketopiperazines inducing drought resistance in rice. Phytochem 29:35–39.

Igarashi Y, Yoshiba Y, Sanada Y, Yamaguchi-Shinozaki K, Wada K, Shinozaki K (1997) Characterization of the gene for DELTA-1-pyrroline-5-carboxylate synthetase and correlation between the expression of the gene and salt tolerance in *Oryza sativa* L. Plant Mol Biol 33:857–865.

Ishitani M, Arakawa K, Mizuno K, Kishitani S, Takabe T (1993) Betaine aldehyde dehydrogenase in the *Gramineae*: levels in leaves of both betaine-accumulating and nonaccumulating cereal plants. Plant Cell Physiol 34:493–495.

Jacob ST, Stetler DA (1989) Polyamines and RNA synthesis. In: Bachrach U, Heimer QM (eds.) The physiology of polyamines Vol. 1. CRC Press, Boca Raton, FL, pp 133–140.

Jain RK, Saini N, Jain S, Singh R (2003) Molecular strategies for developing salt tolerant crops. Indian J Biotechnol 2:121–137.

Jang IC, Oh SJ, Seo JS, Choi WB, Song SI, Kim CH, Kim YS, Seo HS, Choi YD, Nahm BH (2003) Expression of a bifunctional fusion of the *Escherichia coli* genes for trehalose-6-phosphate synthase and trehalose-6-phosphate phosphatase in transgenic rice plants increases trehalose accumulation and abiotic stress tolerance without stunting growth. Plant Physiol 131:516–524.

Jenkins H, Hardy N, Beckmann M, Draper J, Smith AR, Taylor J, Fiehn O, Goodacre R, Bino RJ, Hall RD, Kopka J, Lane GA, Lange BM, Liu JR, Mendes P, Nikolau BJ, Oliver SG, Paton NW, Rhee S, Roessner-Tunali U, Saito K, Smedsgaard J, Sumner LW, Wang T, Walsh S, Wurtele ES, Kell DB (2004) A proposed framework for the description of plant metabolomics experiments and their results. Nature Biotechnol 22:1601–1606.

Jonsson P, Gullberg J, Nordström A, Kusano M, Kowalczyk M, Sjöström M, Moritz T (2004) A strategy for identifying differences in large series of metabolomic samples analyzed by GC/MS. Anal Chem 76:1738–1745

Jonsson P, Johansson AI, Gullberg J, Trygg J, Jiye A, Grung B, Marklund S, Sjöström M, Antti H, Moritz T (2005) High-throughput data analysis for detecting and identifying differences between samples in GC/MS-based metabolomic analyses. Anal Chem 77:5635–5642.

Kafi M, Stewart WS, Borland AM (2003) Carbohydrate and proline contents in leaves, roots, and apices of salt-tolerant and salt-sensitive wheat cultivars. Russian J Plant Physiol 50:155–162.

Kaplan F, Kopka J, Haskell DW, Zhao W, Schiller KC, Gatzke N, Sung DY, Guy CL (2004) Exploring the temperature-stress metabolome of *Arabidopsis*. Plant Physiol 136:4159–4168.

Kakkar RK, Bhaduri S, Rai VK, Kumar S (2000) Amelioration of NaCl stress by arginine in rice seedlings: Changes in endogenous polyamines. Biol Plantarum 43:419–422.

Katiyar S, Dubey RS (1990) Changes in polyamines titer in rice seedlings following sodium chloride salinity stress. J Agron Crop Sci 165:19–27.

Kawasaki S, Borchert C, Deyholos M, Wang H, Brazille S, Kawai K, Galbraith D, Bohnert HJ (2001) Gene expression profiles during the initial phase of salt stress in rice. Plant Cell 13:889–905.

Kerepesi I, Galiba G (2000) Osmotic and salt stress-induced alteration in soluble carbohydrate content in wheat seedlings. Crop Sci 40:482–487.

Kim D-W, Rakwal R, Agrawal GK, Jung Y-H, Shibato J, Jwa N-S, Iwahashi Y, Iwahasi H, Kim DH, Shim I-S, Usui K (2005) A hydroponic rice seedling culture model system for investigating proteome of salt stress in rice leaf. Electrophoresis 26:4521–4539.

Kinzel H (1982) Pflanzenökologie und Mineralstoffwechsel. Ulmer, Stuttgart.

Kishitani S, Takanami T, Suzuki M, Oikawa M, Yokoi S, Ishitani M, Alvarez-Nakase AM, Takabe T (2000) Compatibility of glycinebetaine in rice plants: evaluation using transgenic rice plants with a gene for peroxisomal betaine aldehyde dehydrogenase from barley. Plant Cell Environ 23:107–114.

Kopka J (2006a) Current challenges and developments in GC-MS based metabolite profiling technology. J Biotechnol 124:312–322.

Kopka J (2006b) Gas chromatography mass spectrometry. In: Nagata T, Lörz H, Widholm JM (eds) Biotechnology in agriculture and forestry Vol 57: Saito K, Dixon RA, Willmitzer L (eds) Plant metabolomics. Springer-Verlag, Berlin Heidelberg New York, pp 3–20.

Kopka J, Fernie AF, Weckwerth W, Gibon Y, Stitt M (2004) Metabolite profiling in plant biology: platforms and destinations. Genome Biol 5:109–117.

Kopka J, Schauer N, Krueger S, Birkemeyer C, Usadel B, Bergmüller E, Dörmann P, Gibon Y, Stitt M, Willmitzer L, Fernie AR, Steinhauser D (2005) GMD@CSBDB: The Golm metabolome database. Bioinformatics 21:1635–1638.

Koyama ML, Levesley A, Koebner RMD, Flowers TJ, Yeo AR (2001) Quantitative trait loci for component physiological traits determining salt tolerance in rice. Plant Physiol 125:406–422.

Krishnamurthy R, Anbazhagan M, Bhagwat KA (1988) Evaluation of quaternary ammonium compounds as index of salt tolerance using rice cultivars. Indian J Experiment Biol 26:137–139.

Krishnamurthy R, Bhagwat KA (1989) Polyamines as modulators of salt tolerance in rice cultivars. Plant Physiol 91:500–504.

Lefevre I, Gratia E, Lutts S (2001) Discrimination between the ionic and osmotic components of salt stress in relation to free polyamine level in rice (*Oryza sativa*). Plant Sci 161:943–952.

Li YG, Zhang F, Wang ZT, Hu ZB (2004) Identification and chemical profiling of monacolins in red yeast rice using high-performance liquid chromatography with photodiode array detector and mass spectrometry. J Pharmaceut Biomed Anal 35:1101–1112.

Li ZY, Chen SY (2000) Differential accumulation of the S-adenosylmethionine decarboxylase transcript in rice seedlings in response to salt and drought stresses. Theoret Appl Genet 100:782–788.

Lin HX, Zhu MZ, Yano M, Gao JP, Liang ZW, Su WA, Hu XH, Ren ZH, Chao DY (2004) QTLs for Na^+ and K^+ uptake of the shoots and roots controlling rice salt tolerance. Theoret Appl Genet 108:253–260.

Lopez-Bucio J, Nieto-Jacobo MF, Ramirez-Rodriguez V, Herrera-Estrella L (2000) Organic acid metabolism in plants: from adaptive physiology to transgenic varieties for cultivation in extreme soils. Plant Sci 160:1–13.

Lutts S, Majerus V, Kinet JM (1999) NaCl effects on proline metabolism in rice (*Oryza sativa*) seedlings. Physiol Plantarum 105:450–458.

Maiale S, Sanchez DH, Guirado A, Vidal A, Ruiz OA (2004) Spermine accumulation under salt stress. J Plant Physiol 161:35–42.

Majee M, Maitra S, Dastidar KG, Pattnaik S, Chatterjee A, Hait NC, Das KP, Majumder AL (2004) A novel salt-tolerant L-*myo*-inositol-1-phosphate synthase from *Porteresia coarctata* (Roxb.) Tateoka, a halophytic wild rice - Molecular cloning, bacterial overexpression, characterization, and functional introgression into tobacco-conferring salt tolerance phenotype. J Biol Chem 279:28539–28552.

Martin-Tanguy J (1997) Conjugated polyamines and reproductive development: Biochemical, molecular and physiological approaches. Physiol Plantarum 100:675–688.

Mohanty A, Kathuria H, Ferjani A, Sakamoto A, Mohanty P, Murata N, Tyagi AK (2002) Transgenics of an elite *indica* rice variety Pusa Basmati 1 harbouring the *codA* gene are highly tolerant to salt stress. Theoret Appl Genet 106:51–57.

Morino K, Matsuda F, Miyazawa H, Sukegawa A, Miyagawa H, Wakasa K (2005) Metabolic profiling of tryptophan-overproducing rice calli that express a feedback-insensitive alpha subunit of anthranilate synthase. Plant Cell Physiol 46:514–521.

Nakamura T, Yokota S, Muramoto Y, Tsutsui K, Oguri Y, Fukui K, Takabe T (1997) Expression of a betaine aldehyde dehydrogenase gene in rice, a glycinebetaine nonaccumulator, and possible localization of its protein in peroxisomes. Plant J Cell Mol Biol 11:1115–1120.

Oksman-Caldentey K-M, Saito K (2005) Integrating genomics and metabolomics for engineering plant metabolic pathways. Curr Opin Biotech 16:174–179.

Parker R, Flowers TJ, Moore AL, Harpham NVJ (2006) An accurate and reproducible method for proteome profiling of the effects of salt stress in the rice leaf lamina. J Exp Bot 57:1109–1118.

Penna S (2003) Building stress tolerance through over-producing trehalose in transgenic plants. Trends Plant Sci 8:355–357.

Ponnameruma F (1984) Role of cultivar tolerance in increasing rice production on saline lands. In: Staples R, Toenniessen G (eds) Salinity tolerance in plants - strategies for crop improvement. Wiley, New York, pp 255 - 271.

Prakash L, Prathapasenan G (1988) Effect of NaCl salinity and putrescine on shoot growth, tissue ion concentration and yield of rice (*Oryza sativa*). J Agron Crop Sci 160:325–334.

Rabbani MA, Maryuama K, Abe H, Khan A, Katsura K, Ito Y, Yoshiwara K, Seki M, Shinozaki K, Yamaguchi-Shinozaki K (2003) Monitoring expression profiles of rice genes under cold, drought, and high-salinity stresses and abscisic acid application using cDNA microarray and RNA gel-blot analyses. Plant Physiol 133:1755–1767.

Ratcliffe RG, Shachar-Hill Y (2006) Measuring multiple fluxes through plant metabolic networks. Plant J 45:490–511.

Rathert G (1983) Carbohydrate status in response to ion regulation of two rice varieties (*Oryza sativa*) grown in saline medium. J Plant Nutrit 6:817–830.

Rathinasabapathi B, Gage DA, Mackill DJ, Hanson AD (1993) Cultivated and wild rices do not accumulate glycinebetaine due to deficiencies in two biosynthetic steps. Crop Sci 33:534–538.

Ritambhara GK, Dubey RS (1995) Influence of NaCl salinity on the behaviours of malate, isocitrate and glucose 6-phosphate dehydrogenases in growing rice seedlings in relation to salt tolerance. Indian J Plant Physiol 38:48–53.

Roessner U, Luedemann A, Brust D, Fiehn O, Linke T, Willmitzer L, Fernie AR (2001a) Metabolic profiling allows comprehensive phenotyping of genetically or environmentally modified plant systems. Plant Cell 13:11–29.

Roessner U, Wagner C, Kopka J, Trethewey RN, Willmitzer L (2000) Simultaneous analysis of metabolites in potato tuber by gas chromatography-mass spectrometry. Plant J 23:131–142.

Roessner U, Willmitzer L, Fernie AR (2001b) High-resolution metabolic phenotyping of genetically and environmentally diverse plant systems - identification of phenocopies. Plant Physiol 127:749–764.

Roessner U, Willmitzer L, Fernie AR (2002) Metabolic profiling and biochemical phenotyping of plant systems. Plant Cell Rep 21:189–196

Roessner-Tunali U, Hegemann B, Lytovchenko A, Carrari F, Bruedigam C, Granot D, Fernie AR (2003) Metabolic profiling of transgenic tomato plants overexpressing hexokinase reveals that the influence of hexose phosphorylation diminishes during fruit development. Plant Physiol 133:84–99.

Roessner-Tunali U, Lui J, Leisse A, Balbo I, Perez-Melis A, Willmitzer L, Fernie AR (2004) Flux analysis of organic and amino acid metabolism in potato tubers by gas chromatography-mass spectrometry following incubation in ^{13}C labelled isotopes. Plant J 39:668–679.

Roy M, Wu R (2001) Arginine decarboxylase transgene expression and analysis of environmental stress tolerance in transgenic rice. Plant Sci 160:869–875.

Roy M, Wu R (2002) Overexpression of S-adenosylmethionine decarboxylase gene in rice increases polyamine level and enhances sodium chloride-stress tolerance. Plant Sci 163:987–992.

Ryan PR, Delhaize E, Jones DL (2001) Function and mechanism of organic anion exudation from plant roots. Annu Rev Plant Phys Plant Mol Biol 52:527–560.

Sahi C, Agarwal M, Reddy MK, Sopory SK, Grover A (2003) Isolation and expression analysis of salt stress-associated ESTs from contrasting rice cultivars using a PCR-based subtraction method. Theor Appl Genet 106:620–628.

Salekdeh GH, Siopongco J, Wade LJ, Ghareyazie B, Bennett J (2002) A proteomic approach to analyzing drought- and salt-responsiveness in rice. Field Crops Res 76:199–219.

Sato S, Soga T, Nishioka T, Tomita M (2004) Simultaneous determination of the main metabolites in rice leaves using capillary electrophoresis mass spectrometry and capillary electrophoresis diode array detection. Plant J 40:151–163.

Sauer U (2004) High-throughput phenomics: experimental methods for mapping fluxomes. Curr Opin Biotechnol 15:58–63.

Schauer N, Semel Y, Roessner U, Gur A, Balbo I, Carrari F, Pleban T, Perez-Melis A, Bruedigam C, Kopka J, Willmitzer L, Zamir D, Fernie AR (2006) Comprehensive metabolic profiling and phenotyping of interspecific introgression lines for tomato improvement. Nat Biotech 24:447–454.

Schauer N, Steinhauser D, Strelkov S, Schomburg D, Allison G, Moritz T, Lundgren K, Roessner-Tunali U, Forbes MG, Willmitzer L, Fernie AR, Kopka J (2005) GC-MS libraries for the rapid identification of metabolites in complex biological samples. FEBS Lett 579:1332–1337.

Scholz M, Gatzek S, Sterling A, Fiehn O, Selbig J (2004) Metabolite fingerprinting: detecting biological features by independent component analysis. Bioinformatics 20:2447–2454.

Scholz M, Kaplan F, Guy CL, Kopka J, Selbig J (2005) Non-linear PCA: a missing data approach. Bioinformatics 21:3887–3895.

Shiozaki N, Yamada EM, Yoshiba EY (2005) Analysis of salt-stress-inducible ESTs isolated by PCR-subtraction in salt-tolerant rice. Theor Appl Genet 110:1177–1186.

Slocum RD, Kaur-Sawhney R, Galston AW (1984) The physiology and biochemistry of polyamines in plants. Arch Biochem Biophys 235:283–303.

Sohn YG, Lee BH, Kang KY, Lee JJ (2005) Effects of NaCl stress on germination, antioxidant responses, and proline content in two rice cultivars. J Plant Biol 48:201–208.

Steinhauser D and Kopka J (2007) Methods, applications and concepts of metabolite profiling: primary metabolism. In: Fernie AR, Baginsky S (eds) Plant systems biology. Birkhäuser, in press

Stone JV (2002) Independent component analysis: an introduction. Trends Cognitive Sci 6:59–64.

Sumner LW, Mendes P, Dixon RA (2003) Plant metabolomics: large-scale phytochemistry in the functional genomics era. Phytochem 62:817–836.

Takabe T, Nakamura T, Nomura M, Hayashi Y, Ishitani M, Muramoto Y, Tanaka A (1998) Glycine-betaine and the genetic engineering of salinity tolerance in plants. In: Stress responses of photosynthetic organisms: molecular mechanisms and molecular regulations. Amsterdam, New York, Elsevier, pp 115–131.

Trethewey RN (2004) Metabolite profiling as an aid to metabolic engineering in plants. Curr Opin Plant Biol 7:196–201.

Trethewey RN, Krotzky AJ, Willmitzer L (1999) Metabolic profiling: a Rosetta stone for genomics? Curr Opin Plant Biol 2:83–85.

Trygg J, Gullberg J, Johansson AI, Jonsson P, Moritz T (2006) Chemometrics in metabolomics – an introduction. In: Nagata T, Lörz H, Widholm JM (eds) Biotechnology in agriculture and forestry Vol 57: Saito K, Dixon RA, Willmitzer L (eds) Plant metabolomics. Springer-Verlag, Berlin Heidelberg New York, pp 117–128.

Ueda A, Kathiresan A, Bennet J, Takabe T (2006) Comparative transcriptome analysis of barley and rice under salt stress. Theor Appl Genet 112:1286–1294.

Urbanczyk-Wochniak E, Luedemann A, Kopka J, Selbig J, Roessner-Tunali U, Willmitzer L, Fernie AR (2003) Parallel analysis of transcript and metabolic profiles: a new approach in systems biology. EMBO Reports 4:989–993.

Vance CP, Uhde-Stone C, Allan DL (2003) Phosphorus acquisition and use: critical adaptations by plants for securing a non renewable resource. New Phytologist 157:423–447.

Wagner C, Sefkow M, Kopka J (2003) Construction and application of a mass spectral and retention time index database generated from plant GC/EI-TOF-MS metabolite profiles. Phytochem 62:887–900.

Walia H, Wilson C, Condamine P, Liu X, Ismail AM, Zeng L, Wanamaker SI, Mandal J, Xu J, Cui X, Close TJ (2005) Comparative transcriptional profiling of two contrasting rice genotypes under salinity stress during the vegetative growth stage. Plant Physiol 139:822–835.

Weckwerth W, Wenzel K, Fiehn O (2004) Process for the integrated extraction, identification and quantification of metabolites, proteins and RNA to reveal their co-regulation in biochemical networks. Proteomics 4:78–83.

Wu Y, Wang Q, Ma Y, Chu C (2005) Isolation and expression analysis of salt up-regulated ESTs in upland rice using PCR-based subtractive suppression hybridization method. Plant Sci 168:847–853.

Yakubov BA, Chkanikov DI (1994) Use of polyol quantitative analysis for the estimation of rice varieties resistance to pyriculariosis. Mikologiya i Fitopatologiya 28:65–69.

Yamamoto A, Shim IS, Fujihara S, Yoneyama T, Usui K (2004) Effect of difference in nitrogen media on salt-stress response and contents of nitrogen compounds in rice seedlings. Soil Sci Plant Nutrit 50:85–93.

Yan S, Tang Z, Su W, Sun W (2005) Proteomic analysis of salt stress-responsive proteins in rice root. Proteomics 5:235–244.

Yang X, Romheld V, Marschner H (1994) Effect of bicarbonate on root-growth and accumulation of organic-acids in Zn-inefficient and Zn-efficient rice cultivars (*Oryza sativa* L). Plant Soil 164:1–7.

Yeo AR, Flowers TJ (1982) Accumulation and localization of sodium ions within the shoots of rice *Oryza sativa* varieties differing in salinity resistance. Physiol Plantarum 56:343–348.

Yeo AR, Flowers TJ (1983) Varietal differences in the toxicity of sodium ions in rice *Oryza sativa* leaves. Physiol Plantarum 59:189–195.

Yeo AR, Yeo ME, Flowers SA, Flowers TJ (1990) Screening of rice (*Oryza sativa* L.) genotypes for physiological characters contributing to salinity resistance, and their relationship to overall performance. Theoret Appl Genet 79:377–384.

Young ND, Galston A.W. (1984) Physiological control of arginine decarboxylase activity in K-deficient oat shoots. Plant Physiol 76:331–335.

Zhou W, Sun QJ, Zhang CF, Yuan YZ, Ji Z, Lu BB (2004) Effect of salt stress on ammonium assimilation enzymes of the roots of rice (*Oryza sativa*) cultivars differing in salinity resistance. Acta Botanica Sinica 46:921–927.

Zhu BC, Su J, Chang M, Verma DPS, Fan YL, Wu R (1998) Overexpression of a delta(1)-pyrroline-5-carboxylate synthetase gene and analysis of tolerance to water- and salt-stress in transgenic rice. Plant Sci 139:41–48.

Zhu GY, Kinet JM, Lutts S (2001) Characterization of rice (*Oryza sativa* L.) F3 populations selected for salt resistance. I. Physiological behaviour during vegetative growth. Euphytica: Netherlands J Plant Breeding 121:251–263.

CHAPTER 13

ROOT SIGNALING IN RESPONSE TO DROUGHT AND SALINITY

FRANS J.M. MAATHUIS
Department of Biology, University of York, York YO10 5DD, UK
E-mail: fjm3@york.ac.uk

Abstract: Plant responses to water and salinity stress have been studied extensively. Perception of and adaptation to salt and drought stress take place with time scales that vary from seconds to months. Several important signaling intermediates have been identified that contribute to this process including the second messengers Ca^{2+}, IP_3 and cGMP, hormones such as ABA, regulatory proteins such as kinases and phosphatases, and many specific transcription factors. Extensive available data allow us to build up a simplified chronological record which indicates that initial osmosensing may rely on the action of specific receptor kinases and/or mechanosensitive ion channels. Second messengers are responsible for the subsequent signal transduction to the nucleus where transcription factors, for example of the DREB family, induce gene transcription. The upregulation of hormone synthesis, particularly of ABA, instigates a cascade of responses including altered transcription of many genes. Ultimately, these signaling events lead to changed activity of target proteins such as those involved in compatible osmolyte synthesis and the transport of water and ions

Keywords: Root signaling, drought, salinity, Ca^{2+}, IP_3, cGMP, ABA, kinases, phosphatases, transcription factors

1. INTRODUCTION

Like any living organism, plants must respond to endogenous and external stimuli, a process that crucially depends on information processing. This will typically include the perception of signals, transduction through several intermediate signaling compounds and ultimately a targeted response, for example, alteration of gene transcription.

The overall process of perception, transduction and responses, can take place at the (sub)cellular level, in specific tissues, or in a whole organism. It can occur at a time scale of milliseconds, hours or months. It can be based on a single signaling pathway but more frequently it is underpinned by multiple pathways that interact in antagonistic or synergistic ways.

M.A. Jenks et al. (eds.), Advances in Molecular Breeding Toward Drought and Salt Tolerant Crops, 317–331.

The response of plants to water and salinity stress has been studied extensively at various levels, during different developmental stages and in a range of species. Using several approaches, a number of specific signaling intermediates has been identified of which several are dealt with in more details elsewhere in this volume. For example, the second messenger Ca^{2+} has been shown to be involved in the transduction of many environmental signals, including drought and salinity. Phospholipid based signaling is another generic mechanism found in many eukaryotes and in plants the phospholipid inositol-3-phosphate in particular is implicated in osmotic stress (Drobak and Watkins, 1999). Other second messengers such as the cyclic nucleotide cGMP have also been shown to impact on salt and drought perception (Donaldson et al., 2004). Many transcription factors have been identified that are important in modulating expression of appropriate genes in response to drought/salinity stress (e.g. Uno et al., 2000) and it is clear that phosphorylation/dephosphorylation of specific proteins is also essential (e.g. Teige et al., 2004).

With such diversity in participating components, plant tissues and timescales, unravelling signaling pathways is no mean feat. Many of the current insights have come from cellular studies, the use of forward genetics and application of findings obtained with non-plant organisms in particular yeast. The advent and wholesale application of microarray technology, allowing researchers to study transcriptional regulation of many or all genes in a plant or in a plant organ, potentially opened up avenues to study signaling in a far more comprehensive way and also to assess whether separate pathways interacted. Such studies have made important contributions to our understanding of how plants respond to drought/salinity stress. However, a number of possible caveats has to be kept in mind: The rationale behind most microarray experiments is the identification of genes, and by implication proteins, that are involved in specific conditions, treatments, genotypes etc. The underlying assumption in such studies is that changes in gene expression *per se* may reflect functional properties of the encoded protein. For example, up-regulation of genes after a dehydration experiment may suggest that the encoded proteins are involved in a plant's response or recovery. However, many signaling compounds will not fit into this interpretation for obvious reasons. A rapid increase in cytoplasmic Ca^{2+} concentration for example does not involve any transcriptional change nor does the phosphorylation of proteins. Indeed, any rapid signal, i.e. less than a few minutes, is very unlikely to be based in transcriptional changes.

Thus, by their nature transcriptomics data are more likely to reveal signaling targets in the form of gene regulation rather than signals themselves which will often have to be derived indirectly through interpretation of the transcript changes. But, since the protein targets are crucial to generate real physiological responses, transcriptomics studies are potentially of great importance. In this chapter we will try to integrate the current understanding of how drought and salinity invoke signals in plant roots. Though many data derive from transcriptomics studies we will also include results from other sources.

Before we look into specific signals and pathways, it is useful to reiterate some of the major events that take place in plants when faced with drought or salinity stress. Drought and salinity stress lead to water deficits, ionic toxicity, imbalances in ion homeostasis and the occurrence of oxidative stress. General responses to such detrimental conditions include reduced growth (Munns, 2002) and this is reflected in a downregulation of genes that encode proteins involved in cell wall expansion, protein synthesis in the form of ribosomal components, and DNA synthesis. Photosynthesis is also inhibited and genes encoding specific components of photosystem I and II are concomitantly downregulated. Dehydration and a perturbed ion balance require transcriptional regulation of aquaporins and ion transport functions to accommodate changes in water status and ionic content. Osmotic imbalance leads to the activation of biochemical pathways for example through the upregulation of transcripts that are responsible for the biosynthesis of compatible solutes (Lefevre et al., 2001). Protein ubiquitination becomes more prevalent indicating generally increased turnover of proteins and a reshaping of the proteome to adapt to new conditions. The generation of reactive oxygen species (ROS) during drought and salinity stress requires upregulation of detoxifying systems such as SOD and catalase enzymes and biosynthesis of ROS scavengers. Bioenergetics are modulated to serve changing energy demands for example in the form of augmented H^+-ATPase activity to drive the increased demand for extrusion and compartmentation of Na^+ ions.

Several of these responses, e.g. reduced growth and detoxification of ROS, can be observed in response to many abiotic stresses and are thus not salt/drought specific. Type and intensity of responses will also depend on the severity of the stress, the tissue that is studied, and will have divergent time components. Where salinity is concerned, the initial effects are osmotic causing an almost instantaneous reduction in growth particularly in shoot tissue. The build up of ions in the cell to toxic levels will take considerably longer and affect root tissues first.

Such disparate responses will require many signaling processes taking place at different times and with different durations. A generic signaling pathway (Figure 1) consists of some means to perceive a stimulus which ultimately leads to an adjustment of the cellular proteome to create an adequate response. Perception of external stimuli typically occurs through the action of a specific receptor in the plasma membrane. Subsequently, perception is translated into a cellular signal often involving secondary messengers. Intermediate steps that lead to the ultimate goal of transcriptional regulation generally include protein and hormone based alteration of transcription factors that promote transcription of genes dedicated to the desired response.

This is clearly a gross oversimplification and many of the shown steps in this linear scheme will themselves include transcriptional regulation, posttranslational changes in proteins, relocalisation of proteins and the interaction of hormones. Furthermore, linear schemes would crucially depend on ‘every link in the chain’ and biological systems generally have built-in redundancy in the form of signaling networks rather than one way streets. We will assess the available data regarding

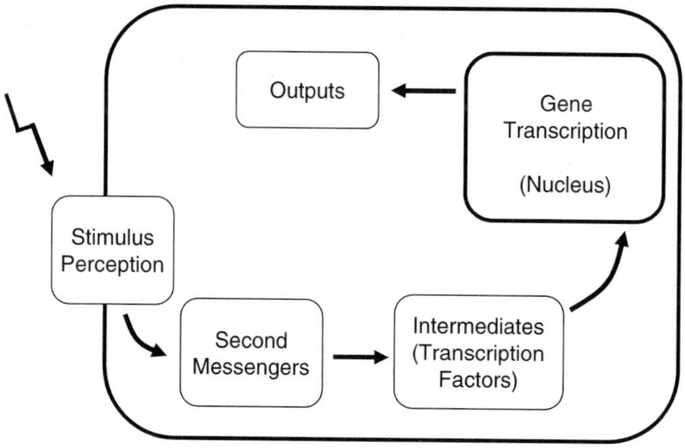

Figure 1. Generic signal transduction scheme. Perception of external stimuli such as drought and salinity is typically mediated by sensors and receptors in the plasma membrane, e.g. receptor kinases and mechanosensitive ion chanels. Stimulus perception is subsequently translated into a secondary messenger signal such as a raise in cytoplasmic Ca^{2+}. Through mostly unknown intermediary steps that may include phosphorylation relays, hormones, and transcription factors, signals are transduced to the nucleus where specific genes are transcribed. Transcription-induced alteration of the cellular proteome generates outputs in the form of physiological responses

signaling events in a chronological order to construct an idea of how these would function in plant roots faced with salinity or drought stress.

2. INITIAL SIGNALING EVENTS

2.1. Perception of Drought and Salinity Stress

The onset of both salinity and drought lowers the external water potential and causes an almost immediate reduction of water delivery to plant tissues and a reduction in growth that can be observed within seconds (Munns, 2001). How plants sense the loss of cellular water is not clear but this likely depends on membrane receptors that report the changed physical forces on membranes and cell walls that follow changes in turgor. In yeast, the HOG (high osmolarity glycerol) pathway is the major response to osmotic stress and it is initiated by histidine kinases Sln1 or Sho1 (Reiser et al., 2003). There is good evidence that the Sln1 kinase has periplasmic domains that record changes in turgor by measuring the distance between membrane and cell wall. Activation of Sln1 results in a MAP kinase based phosphorelay that culminates in the transcription of genes for glycerol synthesis.

Similar mechanisms may exist in plants where several kinases have been described with putative functions in osmosensing. One candidate is the Arabidopsis histidine kinase AtHK1 which has structural similarity to the yeast Sln1 (Urao et al., 1999) and is associated with a phosphorelay cascade consisting of histidine containing phosphotransfer proteins. In Arabidopsis, AtHK1 transcript is particularly pronounced in root tissue and rapidly accumulates during osmotic stress (Urao et al., 1999) as was also observed in poplar (Chefdor et al., 2006).

A second plausible mechanism could be based on mechanosensitive ion channels. Such channels are gated open in response to membrane stretching and consequently form excellent osmosensors. The fairly non-selective properties of these transporters means that channel opening would cause large membrane depolarisations and increased cytoplasmic Ca^{2+} levels, either of which is a potent signal to report changes in environmental osmolarity.

Unlike drought stress, salinity stress leads to ion influx in addition to water deficits. Toxicity of ions such as Cl^- and in particular Na^+ will not occur instantaneously but only after substantial cytoplasmic accumulation. Nevertheless, plants may perceive actual changes in ionic concentration long before they become detrimental. In mammalian cells, specific Na^+ selective ion channels function as sensors to regulate body fluid Na^+ levels (e.g. Watanabe et al., 2006) presumably through a gating mechanism that depends on the Na^+ concentration itself. In plants, similar mechanisms have been postulated for regulation of K^+ homeostasis via 'K^+ sensitive' potassium channels. However, no Na^+ selective channels have been found in plants, and it is therefore difficult to envisage how plants would directly perceive Na^+ levels. In addition, it is unclear whether this would entail sensing of the apoplastic, cytoplasmic or vacuolar concentrations, or a mixture of these.

2.2. The Role of Secondary Messengers in Early Events

The translation of the primary perception event into a cellular signal often involves changes in the levels of non proteinaceous second messengers. In drought/salinity signaling three groups of secondary messengers have been shown to play a role: cytoplasmic calcium (Ca^{2+}_{cyt}), the cyclic nucleotide 3',5'-cyclic GMP (cGMP), and phospholipid based compounds such as inositol phosphate (IP_3).

Imposition of osmotic or salinity stress evokes a rapid (within seconds) transient elevation of Ca^{2+}_{cyt} which appears to be derived from both extra- and intracellular stores and has a signal intensity that correlates with the degree of the applied stress conditions (Knight et al., 1997; Lynch et al., 1989). How the Ca^{2+} signal is generated in plants remains to be resolved but hypothetically, plasma membrane stretch activated ion channels could function as osmosensors allowing initial Ca^{2+} entry which subsequently, through Ca^{2+} induced Ca^{2+} release, may affect internal stores.

A key question is whether the observed Ca^{2+} transient reports changes in osmotic conditions, ionic conditions, or both. Although not observed in earlier studies (Knight et al., 1997) more recent evidence suggests that the imposition of ionic stress in the form of NaCl generates a different Ca^{2+} signal than application of equiosmolar amounts of sorbitol, especially at moderate concentrations (Donaldson et al., 2004). The notion that increased levels of ions *per se* generate Ca^{2+} signals within seconds would imply that ion levels are perceived in the apoplast since it is very unlikely that cytoplasmic or vacuolar levels will change significantly during such a brief period. The mechanism for this is unknown and it also remains to be tested whether such a response is specific for Na^+ and/or Cl^- and thus constitutes a genuine salinity stress signal rather than a generic response to variations in ionic strength such as changes in surface charge or membrane depolarisation. Similar experiments in yeast showed that rapid Ca^{2+} transients are entirely due to osmotic effects and did not vary for different salts or between ionic and non-ionic osmotica (Matsumoto et al., 2002; Denis and Cyert, 2002).

For several plant species inhibitory effects of cGMP on Na^+ fluxes have been described (Maathuis and Sanders, 2001; Rubio et al., 2003; Essah et al., 2003). On the basis of these observations it was postulated that a cyclic nucleotide based signaling event might be part of the early response of plants to salinity stress (Maathuis and Sanders, 2001) as was previously shown for *Dictyostelium* cells (Kuwayama et al., 1996). This hypothesis was recently confirmed by showing that osmotic and ionic stress induce a sustained increase in cellular cGMP within seconds (Donaldson et al., 2004). The rise in cellular cGMP did not appear to rely on either Ca^{2+} or the presence of phospholipids (Donaldson et al., 2004). cGMP is synthesised from GTP by a guanylyl cyclase which in animal cells is often stimulated through a receptor coupled G-protein. Whether a similar mechanism occurs in plants remains to be seen.

Osmotic stress activates phospholipase C (PLC) which leads to the formation of diacylglycerol (DAG) and inositol(1,4,5)triphosphate (IP_3). DAG is involved in the generation of ROS whereas the main target of IP_3 are IP_3-gated ion channels found in endomembranes that are capable of conducting Ca^{2+} (Alexandre et al., 1990). Thus, formation of IP_3 is one of the predominant mechanisms to release Ca^{2+} from internal stores. The sequence and nature of events upstream of PLC are unknown but in kidney cells PLC is believed to be activated by a membrane bound tyrosine receptor kinase that senses osmotic stress (e.g. Zhuang et al., 2000) which is analogous to the above discussed kinase-based phosphorelay that initiates osmoperception in yeast and possibly in plants.

Other phospholipases such as PLD are also activated after osmotic stress, and in many cases the activity of enzymes like PLC and PLD produces IP_3 signals that are transient and peak 1.5 to 2.0 minutes after the onset of stress (Heilmann et al., 1999).

Although it is very possible that parallel signaling pathways are initiated in response to osmotic/salt stress it likely that these are not linear and that components such as Ca^{2+}, cGMP and phospholipids interact at many stages: When cGMP and Ca^{2+} signals were studied it was found that the Ca^{2+} signal in response to ionic stress depended on a preceding rise in the concentration of cytoplasmic cGMP (Donaldson et al., 2004) whereas the signal that reports osmotic stress may be independent of cGMP. Mechanisms to convert the cGMP signal into elevation of Ca^{2+}_{cyt} are available in the form of plasma membrane and endomembrane cyclic nucleotide gated ion channels (Talke et al., 2003) that are activated by cGMP and can mediate Ca^{2+} flux.

Similarly, although the minute timescale of IP_3 production rules out phospholipid-based signaling as a source for the very rapid cGMP and Ca^{2+} elevations, it may well participate in the later phase of the Ca^{2+} transient since it often takes minutes for cytoplasmic Ca^{2+} concentrations to return to their control level (Donaldson et al., 2004; Matsumotu et al., 2002).

3. TRANSDUCTION OF INITIAL EVENTS

Perception of osmotic stress appears to occur within seconds and possibly involves the action of receptor kinases and mechanosensitive channels. Perception is rapidly translated into changes in secondary messenger levels perhaps relying on intermediates such as phosphorelays, G-proteins, cyclases and lipases.

The secondary messengers may constitute signals that are sustained for many minutes: The rapid Ca^{2+} transient induced by osmotic/salt stress is often followed by a slower phase that can last many minutes, cGMP showed an increase that lasted at least 15 minutes in Arabidopsis (Donaldson et al., 2004) and similarly IP_3 has occasionally been reported to increase for more than 30 minutes (DeWald et al., 2001). This makes these intermediates prime candidates to direct the early perception events towards cellular responses which often require modulation of gene transcription. The earliest changes that can be recorded in gene transcript level occur 5 to 10 min after stress imposition. Transcripts that are modulated this rapidly typically encode proteins that are themselves involved in signal transduction and include kinases, phosphatases and many transcription factors.

One particular class of transcription factor consists of DREB (dehydration response element binding) type proteins that have been shown to be induced within 10 min after the onset of drought stress (Dubouzet et al., 2003). DREB proteins show considerable sequence homology to ERF/AP2 (ethylene response factor) type transcription factors that are involved in a large range of cellular functions and bind to specific DNA sequences of target transcripts. The target sequence TACCGACAT is called the dehydration responsive element (DRE) and is found in the promoter region of many salt and drought inducible genes (Shinozaki et al., 2003). In both rice and Arabidopsis, the DREB2 group is specifically induced by salt/drought stress whereas other members of the DREB family respond to other abiotic stresses such as cold. DREB targets that contain the DRE cis element include many proteins that are well known for their upregulation after salt/drought stress such as members of the RD (response to dehydration), COR (cold induced), KIN (stress induced) and ERD (early response to dehydration) gene families. The exact functions of the latter groups are often unclear but some or many may act as chaperone (e.g. Vinocur and Almann, 2005) to prevent protein damage during stress conditions and as such form a first line of defence in protection of the proteome from the adverse effects of drought or salinity.

Other types of transcription factor that are induced during the early stages of salt/drought stress, belong to the MYB and NAC families and several of these have been shown to play important roles in the response to abiotic stress (Chen et al., 2002; He et al., 2005).

4. THE ROLE OF HORMONES

A further group of root genes whose transcripts start to increase during the first approximately 10–30 minutes, are those involved in the biosynthesis of drought stress related hormones such as jasmonic acid (JA) and in particular abscisic acid (ABA). ABA is important in maintaining water homeostasis and dehydration response. The biosynthesis of ABA must start within minutes of stress perception since levels of the hormone have been found to be significantly increased after 10-20 min (Fricke et al., 2004). Biosynthesis of ABA requires rapid upregulation of 9-*cis*-epoxycarotenoid dioxygenase (NCED3), zeaxanthin epoxidase (ZEP) and aldehyde oxidase 3 (AAO3) (Qin and Zeevaart, 1999). Although ABA may have direct effects in regulating protein activity its main target is probably the concerted modulation of gene transcription to evoke physiological responses to abiotic stress (Figure 2). Many of the steps in ABA induced transcript regulation are unknown but there is some evidence that a leucine rich repeat receptor kinase such as RPK1 (Osakabe et al., 2005) functions in the perception of raised ABA levels. RPK1 type kinases could be responsible for the activation of ABA induced transcription factors via phosphorylation. A number of transcription factors has been described that are activated by ABA: bZIP type transcription factors of the AREB (ABA-responsive element binding) family are induced after drought, NaCl or ABA treatment (Uno et al., 2000). AREBs recognise specific cis-elements in target genes (ABREs or ABA responsive elements) with a consensus sequence of ACGTGGC. Other classes of ABA dependent transcription factors include RD26 a NAC transcription factor (Fujita et al., 2004) and members of the Zn-finger and MYB/MYC families (Abe et al., 2003).

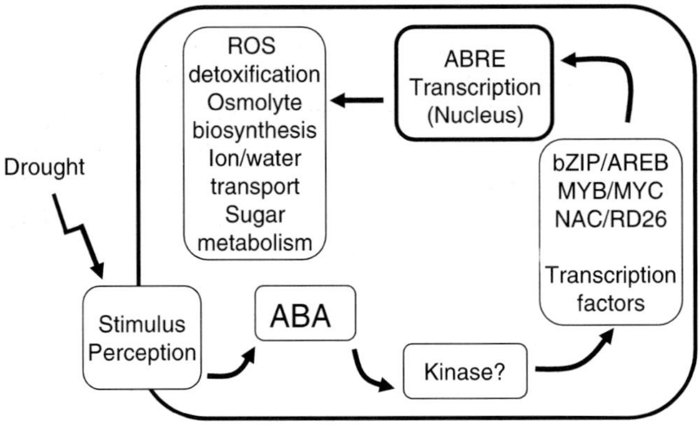

Figure 2. The central role of ABA in drought response. Although the initial perception of drought and osmotic stress in plants has not yet been elucidated, a rapid synthesis of the hormone abscisic acid (ABA) is always observed. ABA, possibly via kinase intermediates, activates transcription factors of several different classes which recognise the ABA responsive element (ABRE) motif in upstream regions of many genes. Increased transcription results in increased protein activity of cellular processes such as reactive oxygen (ROS) detoxification, the synthesis of compatible osmolytes, and readjustment of ion and water homeostasis and sugar metabolism

The number of genes affected directly and indirectly by ABA is at least in the hundreds and, over a longer timescale, possibly thousands and includes genes from many functional categories. The largest group consists of genes involved in metabolism (e.g. Wang et al., 2006) and signifies major changes in sugar metabolism and photosynthesis, and the biosynthesis of compatible osmolytes. In particular, ABA has been shown to mediate detoxification of reactive oxygen species through induction of superoxide dismutases, catalases and peroxidases (Bellaire et al., 2000). All these processes occur in response to perturbation of water and ion homeostasis. A further important down stream target of ABA is the transport of amino acids, monovalent cations and water. Transcriptomics studies have shown that ABA affects transcription of a large number of root membrane proteins such as H^+-ATPases, sugar transporters, ion channels and aquaporins (e.g. https://www.genevestigator.ethz.ch/).

5. MEDIUM TERM RESPONSES

Hormones such as ABA will remain elevated for hours and thus provide a long lasting signal to keep cellular metabolism and physiology geared toward appropriate responses to drought and salinity stress. ABA sensitive bZIP and MYC/MYB transcription factors are directly linked to induction of stress sensitive genes such as rd22 and rd29B, possibly through interaction of the protein kinase SRK2E (Yoshida et al., 2002). In turn rd22 and rd29B are believed to function as intermediates for many of the downstream targets of ABA that show alteration in activity in the time course of hours such as catalase expression, osmolyte biosynthesis and regulation of ion transporters and aquaporins. Similarly, the previously mentioned family of DREB transcription factors contains several members that are upregulated early on after the onset of stress, e.g. DREB2A transcript levels can be seen to increase within 10 min, but levels continue to increase for at least 24 h (Dubouzet et al., 2003; Chen et al., 2002). Other isoforms like DREB1A are switched on after approx. 40 min and show levels that gradually decrease after a maximum that occurs around 5 h. DREB1A is responsible for the transcriptional regulation of around 30 genes (Maruyama et al., 2004) that include other transcription factors, a sugar transporter, and a large number of LEA-type stress response genes.

6. OUTPUTS

A sustained stress in the form of salinity or drought will require short term and long term responses and signaling pathways. The pathways described so far ensure that after a couple of hours, most plants will largely have recovered from the initial osmotic effects. Typically, growth will have resumed by then, albeit at a slower rate than before, and the cellular osmolarity will have been adjusted.

The targets and outputs of signaling events comprise responses at the cellular, organ and whole plant level. These outputs may include significant alterations in biochemistry, partitioning of nutrients, photosynthesis, gene transcription etc. and it is beyond the scope of this chapter to deal with all of them. As an example we will therefore look at ion homeostasis which forms an important output of salt/drought induced signaling pathways.

6.1. Regulation of Transport

In Arabidopsis root tissue, Na^+ levels increase to over 30 mM on a FW basis after plants have been exposed to 80 mM NaCl for 2 hours (Maathuis et al., 2003). This is the net result of a number of processes that includes Na^+ uptake, Na^+ efflux to the apoplast and vacuole and translocation to shoot tissue. The increase in Na^+ tissue level is likely to affect transport and compartmentation of other ions in particular K^+, Ca^{2+} and NO_3-. During the same period Cl^- ions will be taken up and compartmentalised and water homeostasis needs to be reset.

These processes require the concerted activity of membrane transporters that are responsible for the movement of specific ions and water molecules across membranes and are therefore important outputs for signaling pathways. However, some transporters, particularly those involved with Na^+ and Ca^{2+}, may also participate in sensing and signaling functions. Some of the major classes of transporter that are affected during salt and water stress are primary pumps such as the tonoplast H^+ ATPase and pyrophosphatase and Ca^{2+} pumps, proteins involved in uptake of Na^+ (non selective ion channels, HKT transporters), Na^+ extrusion (NHX type and CHX type antiporters), and K^+ transporters. In addition, a large number of aquaporins has been shown to be regulated during salt and drought stress.

For a few of these transporters more information is available regarding their regulation. The use of forward genetics identified a particular Na^+:H^+ antiporter (SOS1) that is involved in extrusion of Na^+ from the cytoplasm, an important aspect of salinity tolerance. SOS1 is expressed in many tissues but particularly in the root epidermis and around the vascular tissue. SOS1 transcript is elevated after salt stress and its activity is directly increased after phosphorylation by the kinase SOS2 (Chinusamy et al., 2004). SOS2 itself associates with SOS3 a Ca^{2+}-binding protein (Liu and Zhu, 1997) similar to calcineurin type phosphatases directly sensitive to levels of cytoplasmic Ca^{2+}. Thus a linear pathway can be envisaged (Figure 4) that starts with a salt induced Ca^{2+} transient and terminates in the increased activity of an antiporter that limits cytoplasmic Na^+ accumulation. In addition, SOS2 may also affect the activity of other transporters such as HKT1 which is involved in Na^+ uptake (Laurie et al., 2001; Rus et al., 2001) and NHX1 which is responsible for extrusion of Na^+ into the vacuole (Apse et al., 1999). NHX1 transcription and activity is also under control of an ABA dependent pathway that possibly includes the ABA dependent phosphatase ABI1 and MYC/MYB transcription factors (Yokoi et al. 2002; Chinusamy et al., 2004).

A cyclic nucleotide-based pathway (Figure 3) may affect relevant transport functions in several ways: the quick rise in cGMP (Donaldson et al., 2004) can directly deactivate non selective ion channels that are responsible for Na^+ influx by binding to the channel protein (Maathuis and Sanders, 2001). In this way,

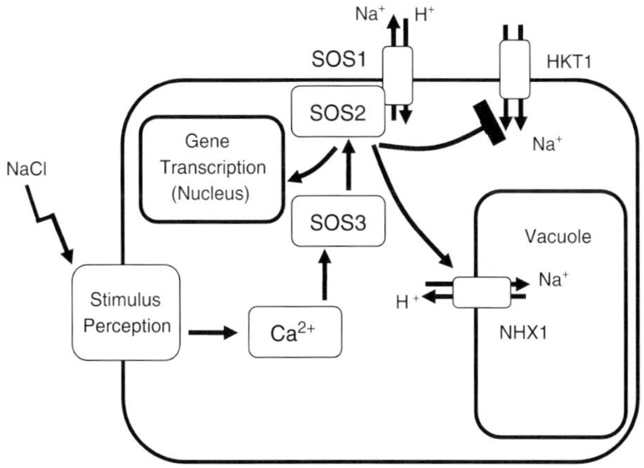

Figure 3. The SOS pathway is pivotal in response to salinity stress. The ionic stress induced Ca^{2+} signal is transduced by the Ca^{2+} binding SOS3, and through association with the SOS2 kinase to target proteins. A main target for SOS2 is activation of SOS1, the plasma membrane located Na^+:H^+ antiporter. SOS2 can also affect the activity of other Na^+ extruding systems such as the vacuolar Na^+:H^+ antiporter NHX1. In addition, SOS2 may limit Na^+ influx through HKT1 type mechanisms and affect transcription of further components involved in ion homeostasis

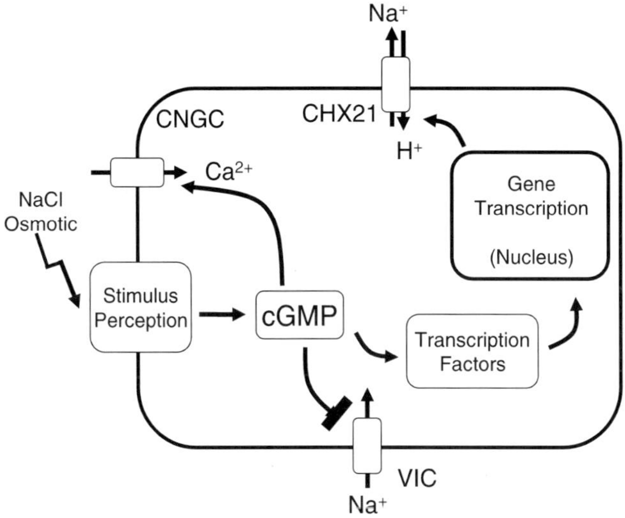

Figure 4. The second messenger cGMP signals drought and ionic stress. Exposure of plants to ionic and osmotic stress leads to a rapid and sustained increase in cellular levels of the second messenger cGMP. cGMP reduces the open probability of some voltage independent non selective cation channels (VICs) that conduct Na^+ and thus reduces Na^+ influx. cGMP can activate cyclic nucleotide gated channels (CNGCs) generating a pathway for the influx of external Ca^{2+}. In addition, cGMP affects transcription of many genes including cation proton antiport mechanisms such as CHX21

Na^+ influx is reduced (Maathuis and Sanders, 2001, Essah et al., 2003; Rubio et al., 2003). However, cGMP may also affect transporter activity through modulation of gene transcription. An example of the latter is the upregulation of the cation:proton antiporter CHX21 by cGMP (Maathuis, 2006). CHX21 probably functions as a Na^+:H^+ antiport involved in root Na^+ xylem loading (Hall et al., 2006). In addition, large gene families of cyclic nucleotide gated channels have been found in plants (Talke et al., 2001) of which several isoforms are involved in cation transport (Li et al., 2005; Gobert et al., 2006)

SKOR, an outward rectifying channel involved in the long distance transport of K^+ from root to shoot, is transcriptionally downregulated during drought stress via the action of ABA (Gaymard et al., 1998). This ostensibly occurs to retain K^+ in the root and thus decrease the root water potential to combat dehydration.

7. CONCLUDING REMARKS

Drought and salinity stress are major constraints on agriculture and likely to increase their impact in the future due to population growth and climate change. Roots are the plant organs that initially perceive these stresses and it is now clear that a large number of signaling responses ensues this perception. The immediate effects of both salinity and drought are osmotic and as such likely to be relayed by sensors such as kinases with periplasmic domains or mechanosensitive ion channels. Secondary messengers such as Ca^{2+}, cyclic nucleotides and phospholipids play intermediary functions in the transduction of stress stimuli to the nucleus through the action of hormones and transcription factors where regulation of gene transcription ensures that appropriate responses are evoked.

Although in particular transcriptomics studies have greatly contributed to the identification of putative signaling elements there are two major issues that need resolving: (i) very specific questions such as the exact mechanism of osmotic stress perception and sensing of ion levels such as Na^+ and Cl^- and (ii) integrating the data into comprehensive models that encompass the general scheme outlined above.

As far as (i) is concerned, the use of forward and reverse genetics is likely to give more definitive answers in the near future. Progress for the second point will heavily rely on the application of 'omics' analyses in combination with extensive cell biological research and ultimately will require systems biology based approaches.

REFERENCES

Abe, H., Urao, T., Ito, T., Seki, M., Shinozaki, K., and Yamaguchi-Shinozaki, K. (2003) Arabidopsis AtMYC2 (bHLH) and AtMYB2 (MYB) function as transcriptional activators in abscisic acid signaling. Plant Cell 15, 63–78.

Alexandre, J., Lassalles, J.P., and Kado, R.T. (1990) Opening of Ca^{++} channels in isolated red beet root vacuole membrane by inositol 1,4,5-triphosphate. Nature 343, 567–570.

Apse, M.P., Aharon, G.S., Snedden, W.A. and Blumwald, E. (1999) Salt tolerance conferred by overexpression of a vacuolar Na^+/H^+ antiport in *Arabidopsis*.. *Science*, 285, 1256–1258.

Bellaire, B.A., Carmody, J., Braud, J., Gossett, D.R., Banks, S.W., Lucas, M.C., and Fowler, T.E. (2000) Involvement of abscisic acid-dependent and - Independent pathways in the upregulation of antioxidant enzyme activity during NaCl stress in cotton callus tissue. Free Radical Research 33, 531–545.

Chefdor, F., Benedetti, H., Depierreux, C., Delmotte, F., Morabito, D., and Carpin, S. (2006) Osmotic stress sensing in Populus: Components identification of a phosphorelay system. FEBS letters 580, 77–81.

Chen, W.Q., Provart, N.J., Glazebrook, J., Katagiri, F., Chang, H.S., Eulgem, T., Mauch, F., Luan, S., Zou, G.Z., Whitham, S.A., Budworth, P.R., Tao, Y., Xie, Z.Y., Chen, X, Lam, S., Kreps, J.A., Harper, J.F., Si-Ammour, A., Mauch-Mani, B., Heinlein, M., Kobayashi, K., Hohn, T., Dangl, J.L., Wang, X and Zhu, T. (2002) Expression profile matrix of *Arabidopsis* transcription factor genes suggests their putative functions in response to environmental stresses. *Plant Cell*, 14, 559–574.

Chinnusamy, V., Schumaker, K., and Zhu, J.K. (2004) Molecular genetic perspectives on cross-talk and specificity in abiotic stress signaling in plants. Journal Experimental Botany 55, 225–236.

Denis, V. and Cyert, M.S. (2002) Internal Ca^{2+} release in yeast is triggered by hypertonic shock and mediated by a TRP channel homologue. Journal Cell Biology 156, 29–34.

DeWald, D.B., Torabinejad, J., Jones, C.A., Shope, J.C., Cangelosi, A.R., Thompson, J.E., Prestwich, G.D., and Hama, H. (2001) Rapid accumulation of phosphatidylinositol 4,5-bisphosphate and inositol 1,4,5-trisphosphate correlates with calcium mobilization in salt-stressed Arabidopsis. Plant Physiology 126, 759–769.

Donaldson, L., Ludidi, N., Knight, M.R., Gehring, C., and Denby, K. (2004) Salt and osmotic stress cause rapid increases in Arabidopsis thaliana cGMP levels. FEBS letters 569, 317–320.

Drobak, B.K. and Watkins, P.A.C. (2000) Inositol(1,4,5)trisphosphate production in plant cells: an early response to salinity and hyperosmotic stress. FEBS letters 481, 240–244.

Dubouzet, J.G., Sakuma, Y., Ito, Y., Kasuga, M., Dubouzet, E.G., Miura, S., Seki, M., Shinozaki, K., and Yamaguchi-Shinozaki, K. (2003) OsDREB genes in rice, Oryza sativa L., encode transcription activators that function in drought-, high-salt- and cold-responsive gene expression. Plant Journal 33, 751–763.

Essah, P.A., Davenport, R., and Tester, M. (2003) Sodium influx and accumulation in Arabidopsis. Plant Physiology 133, 307–318.

Fricke, W., Akhiyarova, G., Veselov, D., and Kudoyarova, G. (2004) Rapid and tissue-specific changes in ABA and in growth rate in response to salinity in barley leaves. Journal Experimental Botany 55, 1115–1123.

Fujita, M., Fujita, Y., Maruyama, K., Seki, M., Hiratsu, K., Ohme-Takagi, M., Tran, L.S.P., Yamaguchi-Shinozaki, K., and Shinozaki, K. (2004) A dehydration-induced NAC protein, RD26, is involved in a novel ABA-dependent stress-signaling pathway. Plant Journal 39, 863–876.

Gaymard, F., Pilot, G., Lacombe, B., Bouchez, D., Bruneau, D., Boucherez, J., Michaux-Ferriere, M., Thibaud, J., and Sentenac, H. (1998) Identification and disruption of a plant Shaker-like outward channel involved in K^+ release into the xylem sap. Cell 94, 647–655.

Gobert, A., Park, G., Amtmann, A., Sanders, D., and Maathuis, F.J.M. (2006) Arabidopsis thaliana Cyclic Nucleotide Gated Channel 3 forms a non-selective ion transporter involved in germination and cation transport. Journal Experimental Botany 57, 791–800.

Hall, D., Evans, A.R., Newbury, H.J., and Pritchard, J. (2006) Functional analysis of CHX21: a putative sodium transporter in Arabidopsis. Journal Experimental Botany 57:1201–1210.

He, X.J., Mu, R.L., Cao, W.H., Zhang, Z.G., Zhang, J.S., and Chen, S.Y. (2005) AtNAC2, a transcription factor downstream of ethylene and auxin signaling pathways, is involved in salt stress response and lateral root development. Plant Journal 44, 903–916.

Heilmann, I., Perera, I.Y., Gross, W., and Boss, W.F. (1999) Changes in phosphoinositide metabolism with days in culture affect signal transduction pathways in Galdieria sulphuraria. Plant Physiology 119, 1331–1339.

Knight, H., Trewavas, A.J., and Knight, M.R. (1997) Calcium signaling in arabidopsis thaliana responding to drought and salinity. Plant Journal 12, 1067–1078.

Kuwayama, H., Ecke, M., Gerisch, G., and VanHaastert, P.J.M.(1996) Protection against osmotic stress by cGMP-mediated myosin phosphorylation. Science 271, 207–209.

Laurie, S., Feeney, K.A., Maathuis, F.J.M., Heard, P.J., Brown, S.J., and Leigh, R.A. (2002) A role for HKT1 in sodium uptake by wheat roots. Plant Journal 32, 139–149.

Lefevre, I., Gratia, E., and Lutts, S. (2001) Discrimination between the ionic and osmotic components of salt stress in relation to free polyamine level in rice (Oryza sativa). Plant Science 161, 943–952.

Li, X.L., Borsics, T., Harrington, H.M., and Christopher, D.A. (2005) Arabidopsis AtCNGC10 rescues potassium channel mutants of E- coli, yeast and Arabidopsis and is regulated by calcium/calmodulin and cyclic GMP in E-coli. Functional Plant Biology 32, 643–653.

Liu, J.P. and Zhu, J.K. (1997) An *Arabidopsis* mutant that requires increased calcium for potassium nutrition and salt tolerance. Proceedings National Academy Science (USA) 94, 14960–14971.

Lynch, J. and Laeuchli, A. (1988) Salinity affects intracellular calcium in corn root protoplasts. Plant Physiology 87, 351–356.

Maathuis, F.J.M., Filatov, V, Herzyk, P., Krijger, G.C., Axelsen, K.B., Chen, S.X., Green, B.J., Li, Y., Madagan, K.L., Sanchez-Fernandez, R., Forde, B.G., Palmgren, M.G., Rea, P.A., Williams, L.E., Sanders, D. and Amtmann, A. (2003) Transcriptome analysis of root transporters reveals participation of multiple gene families in the response to cation stress. *Plant Journal*, 35, 675–692.

Maathuis, F.J.M. (2005) cGMP modulates gene transcription and cation transport in Arabidopsis roots. Plant Journal 45, 700–711.

Maathuis, F.J.M. and Sanders, D. (2001) Sodium uptake in *Arabidopsis thaliana* roots is regulated by cyclic nucleotides. Plant Physiology 127, 1617–1625.

Maruyama, K., Sakuma, Y., Kasuga, M., Ito, Y., Seki, M., Goda, H., Shimada, Y., Yoshida, S., Shinozaki, K., and Yamaguchi-Shinozaki, K. (2004) Identification of cold-inducible downstream genes of the Arabidopsis DREB1A/CBF3 transcriptional factor using two microarray systems. Plant Journal 38, 982–993.

Matsumoto, T.K., Ellsmore, A.J., Cessna, S.G., Low, P.S., Pardo, J.M., Bressan, R.A., and Hasegawa, P.M. (2002) An osmotically induced cytosolic Ca^{2+} transient activates calcineurin signaling to mediate ion homeostasis and salt tolerance of Saccharomyces cerevisiae. Journal Biological Chemistry 277, 33075–33080.

Munns, R. (2002) Comparative physiology of salt and water stress. Plant Cell Environment 25, 239–250.

Osakabe, Y., Maruyama, K., Seki, M., Satou, M., Shinozaki, K., and Yamaguchi-Shinozaki, K. (2005) Leucine-rich repeat receptor-like kinase1 is a key membrane-bound regulator of abscisic acid early signaling in Arabidopsis. Plant Cell 17, 1105–1119.

Qin, X.Q. and Zeevaart, J.A.D. (1999) The 9-cis-epoxycarotenoid cleavage reaction is the key regulatory step of abscisic acid biosynthesis in water-stressed bean. Proceedings National Academy Sciences (USA) 96, 15354–15361.

Reiser, V., Raitt, D., and Saito, H. (2003) Yeast osmosensor Sln1 and plant cytokinin receptor Cre1 respond to changes in turgor pressure. Yeast 20, S169.

Rubio, F., Flores, P., Navarro, J.M., and Martinez, V. (2003) Effects of Ca^{2+}, K^+ and cGMP on Na^+ uptake in pepper plants. Plant Science 165, 1043–1049.

Rus, A., Yokoi, S., Sharkhuu, A., Reddy, M., Lee, B.H., Matsumoto, T.K., Koiwa, H., Zhu, J.K., Bressan, R.A., and Hasegawa, P.M. (2001) AtHKT1 is a salt tolerance determinant that controls Na^+ entry into plant roots. Proceedings National Academy Sciences (USA) 98, 14150–14155.

Shinozaki, K., Yamaguchi-Shinozaki, K., and Seki, M. (2003) Regulatory network of gene expression in the drought and cold stress responses. Current Opinion Plant Biology 6, 410–417.

Talke, I.N., Blaudez, D., Maathuis, F.J.M. and Sanders, D. (2003) CNGCs: prime targets of plant cyclic nucleotide signaling? *Trends in Plant Science*, 8, 286–293.

Teige, M., Scheikl, E., Eulgem, T., Doczi, F., Ichimura, K., Shinozaki, K., Dangl, J.L., and Hirt, H. (2004) The MKK2 pathway mediates cold and salt stress signaling in Arabidopsis. Molecular Cell 15, 141–152.

Uno, Y., Furihata, T., Abe, H., Yoshida, R., Shinozaki, K., and Yamaguch-Shinozaki, K. (2000) *Arabidopsis* basic leucine zipper transcription factors involved in an abscisic acid-dependent signal transduction pathway under drought and high-salinity conditions. Proceedings National Academy Science (USA) 23, 11637.

Urao, T., Yakubov, B., Satoh, R., Yamaguchi Shinozaki, K., Seki, M., Hirayama, T., and Shinozaki, K. (1999) A transmembrane hybrid-type histidine kinase in arabidopsis functions as an osmosensor. Plant Cell 11, 1743–1754.

Vinocur, B. and Altman, A. (2005) Recent advances in engineering plant tolerance to abiotic stress: achievements and limitations. Current Opinion Biotechnology 16, 123–132.

Wang, P.C., Du, Y.Y., An, G.Y., Zhou, Y., Miao, C., and Song, C.P. (2006) Analysis of global expression profiles of Arabidopsis genes under abscisic acid and H2O2 applications. Journal Integrative Plant Biology 48, 62–74.

Watanabe, E., Hiyama, T.Y., Shimizu, H., Kodama, R., Hayashi, N., Miyata, S., Yanagawa, Y., Obata, K., and Noda, M. (2006) Sodium-level-sensitive sodium channel Na-x is expressed in glial laminate processes in the sensory circumventricular organs. American Journal Physiology-Regulatory Integrative Comparative Physiology 290, R568-R576.

Yokoi, S., Quintero, F.J., Cubero, B., Ruiz, M.T., Bressan, R.A., Hasegawa, P.M., and Pardo, J.M. (2002) Differential expression and function of Arabidopsis thaliana NHX Na^+/H^+ antiporters in the salt stress response. Plant Journal 30, 529–539.

Yoshida, R., Hobo, T., Ichimura, K., Mizoguchi, T., Takahashi, F., Aronso, J., Ecker, J.R., and Shinozaki, K. (2002) ABA-activated SnRK2 protein kinase is required for dehydration stress signaling in Arabidopsis. Plant Cell Physiology 43, 1473–1483.

Zhuang, S.G., Hirai, S.I., and Ohno, S. (2000) Hyperosmolality induces activation of cPKC and nPKC, a requirement for ERK1/2 activation in NIH/3T3 cells. American Journal Physiology-Cell Physiology 278, C102–C109.

CHAPTER 14

BIOTECHNOLOGY APPROACHES TO ENGINEERING DROUGHT TOLERANT CROPS

CORY A. CHRISTENSEN AND KENNETH A. FELDMANN

Ceres, Inc., 1535 Rancho Conejo Blvd., Thousand Oaks, CA 91320

Abstract: In the last decade, the sequencing of several plant genomes has greatly amplified the number of genes being evaluated for their ability to confer stress tolerance. Over 50 genes have been reported to confer drought tolerance when overexpressed and the number of field trials for transgenic drought tolerant crops is on the rise. Nevertheless, no transgenic drought tolerant crop has yet been commercialized. In this chapter, we examine the approaches being taken by academic labs and the agricultural biotechnology industry to identify and evaluate candidate genes. We address criteria used for selecting candidate genes, developing high-throughput phenotyping platforms and applying drought stress in the lab. In addition, we highlight promising genes that are at more advanced stages of evaluation

Keywords: drought tolerance; desiccation; arabidopsis; overexpression; abiotic stress

1. INTRODUCTION

In the late 1990's, as the first complete plant genome (*Arabidopsis thaliana*) was sequenced, it was possible to envision the systematic determination of the function of all the genes within a plant species and functional genomics projects were initiated in the public and private sectors. While the initial approaches taken differed by some degree (Riechmann et al., 2000; Boyes et al., 2001; Kjemtrup et al., 2003; Alexandrov et al., 2006), the fundamental element of the strategy was the same: overexpress genes in plants and screen for agronomically valuable phenotypes.

In the drought literature, approximately 50 unique genes involved in plant water relations have been described (Umezawa et al., 2006). The number of field trials for transgenic plants involving drought tolerance is on the rise (Figure 1). These field trials comprise 17 institutions and 13 crop species. Nevertheless, drought tolerance

M.A. Jenks et al. (eds.), Advances in Molecular Breeding Toward Drought and Salt Tolerant Crops, 333–357.

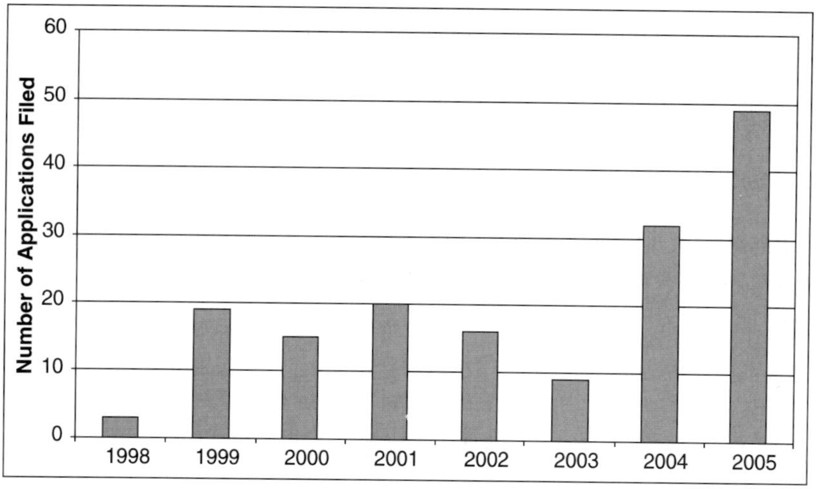

Figure 1. United States field trial applications for transgenic plants involving drought tolerance (data from Information Systems for Biotechnology (http://www.nbiap.vt.edu/))

resulting from genetic engineering has not been commercialized in any major crop to date (Information Systems for Biotechnology (http://www.nbiap.vt.edu/)).

It has become clear that in order to successfully deploy such technology in the field, molecular biologists need to work closely with agronomists, physiologists, and plant breeders to understand the required traits under the specific cropping system of interest. The genes that have been shown to affect water relations or desiccation tolerance in the lab should be considered a valuable tool kit for the development of these traits. In this review, we focus on describing approaches used to identify drought tolerance genes and highlight promising genes for engineering drought tolerance in crops.

2. INTRODUCTION TO DROUGHT STRESS

2.1. Drought Effects on Plants

Detailed discussions of the effects of drought at the physiological, cellular, and molecular level have been provided earlier in this volume. In this section, we provide a brief synopsis as a foundation for the discussion of engineering drought-tolerant crops. Two fundamental facts confront researchers seeking to improve crop performance under drought stress. First, as soil dries, transpiration decreases and, second, transpiration is highly correlated with plant productivity (in terms of photosynthetic rate; Bacon, 2004). Under moderate stress, reduced stomatal conductance results in decreased intercellular CO_2 concentrations which limit photosynthesis (reviewed in Chaves and Oliveira, 2004).

Table 1 Breakdown of United States field trial applications by institution and crop species (data from Information Systems for Biotechnology (http://www.nbiap.vt.edu/))

Institution	Organism	Applications
Monsanto	Corn	95
Syngenta	Corn	13
Rutgers U	Creeping bentgrass	11
Rutgers U	Kentucky bluegrass	10
Rutgers U	Bermudagrass	10
Biogemma	Corn	8
Monsanto	Soybean	6
Oklahoma State U	Wheat	5
Monsanto	Cotton	5
Montana State U	Wheat	5
Rutgers U	Perennial ryegrass	3
Kansas State U	Creeping bentgrass	3
BASF	Corn	3
RiceTec, Inc.	Rice	2
U of California/Davis	Tomato	2
U of California/Davis	Persimmon	2
North Carolina State U	Festuca arundinacea	2
ARS	Cotton	1
Stine Biotechnology	Corn	1
Texas Tech U	Cotton	1
U of Arizona	Guayule	1
Kansas State U	Wheat	1
DeKalb	Corn	1
Bowdoin C	Cotton	1

Drought stress has been divided into three stages. In stage I, soil moisture is sufficient to maintain maximal stomatal conductance. Stage II begins as soil dries and stomatal conductance declines to balance the uptake of water from the soil and transpiration. Beginning in this stage, processes contributing to crop yield, such as photosynthesis and leaf growth, are inhibited. Stage III is reached when stomatal conductance is at a minimum and uptake of water from the soil is unable to meet the transpirational demand (Serraj and Sinclair, 2002). For an excellent review on the types of damage caused by drought and plant responses for coping with them see Chaves and Oliveira, 2004.

Drought tolerance is a non-specific phrase used to cover any type of enhancement in plant productivity in response to limited water availability. More specifically, plant responses to drought can be divided into three mechanisms: drought escape, desiccation postponement, and desiccation tolerance. To date, the most important contribution to drought tolerance in crops has come from breeding for altered phenological traits such as early flowering and maturity (Araus et al., 2002). Such traits constitute an escape from drought conditions rather than tolerance *per se*.

During stage II drought, desiccation can be postponed by preserving more water in the soil (water conservation) or increasing the fraction of soil water extractable by the plant (enhanced water uptake). Desiccation postponement is usually achieved by structural modification such as increased cuticle thickness, deeper roots, reduced leaf area or reduced stomatal conductance (Jones, 2004). Limiting stomatal conductance can result in enhanced water use efficiency (WUE). WUE is a special type of desiccation postponement and can be defined on the whole-plant level as the ratio of biomass or harvestable yield produced to the amount of water transpired (Bacon, 2004). Given the strong dependence of yield on transpiration, any trait resulting in decreased water use, if expressed constitutively, will likely result in decreased productivity under non-drought stress conditions.

At stage III drought the issue is survival, not productivity. Desiccation tolerance is achieved through changes in biochemical composition that protect macromolecules and membranes or that maintain turgor by osmotic adjustment. Engineering for desiccation tolerance may be less important in industrialized agriculture because the impact on yield under conditions where desiccation tolerance comes into play would be so severe as to render harvesting the crop unprofitable (Serraj and Sinclair, 2002). Furthermore, a negative correlation between yield and tolerance to extreme drought conditions has been observed, suggesting a fundamental trade-off that may be difficult to overcome by genetic engineering (Mitra, 2001). However, under subsistence farming or for perennial crops, the benefits provided by desiccation tolerance may be valuable.

The identification of 50 genes that confer drought tolerance (Umezawa et al., 2006) bodes well for the prospects of genetically engineering drought tolerant crops. However, in this context it is important to make a distinction between affecting plant water relations and the kind of drought tolerance that will be valuable for the agricultural industry. For example, genes that confer desiccation tolerance in the lab may not be useful for enhancing yield in the field. Nevertheless, many laboratory studies address only desiccation recovery not increased biomass or seed yield under water limitation. From an industry perspective, the research focus should be on crop productivity rather than desiccation tolerance *per se*. That is not to say that genes discovered in this manner will not be agronomically useful, but successful deployment in the field will require more detailed physiological studies than are typically reported (Chaves and Oliveira, 2004).

3. IDENTIFYING GENES THAT CONFER DROUGHT TOLERANCE

3.1. Arabidopsis as a Model System for Gene Function Discovery

As discussed above, the sequencing of the arabidopsis genome spawned a high-throughput gene function discovery industry. The suitability of arabidopsis as a model organism for laboratory research has been well established (Somerville and Koornneef, 2002). However, whether genetic traits discovered in arabidopsis are valuable to the agricultural industry will depend on how frequently these traits translate into crops of interest such as corn.

Research on the CBF/DREB family of transcription factors has shown that drought and cold responsive gene regulons are conserved between widely divergent species such as arabidopsis and rice (Zhang et al., 2004a). In addition, structural proteins such as ion channels are also conserved (Zhang et al., 2004a). In these cases, genes discovered in arabidopsis confer the same traits when transferred to other species. However, this is not always the case. The tomato genome contains orthologs of arabidopsis *CBF1*, *2*, and *3*, but the tomato CBF regulon appears to be smaller and constitutive overexpression of arabidopsis *CBF3* or tomato *CBF1* did not confer enhanced stress tolerance (Zhang et al., 2004b). Nevertheless, the approach of engineering stress tolerance in crops through first understanding the molecular basis of tolerance in arabidopsis has been validated.

Several features of arabidopsis make it amenable to high-throughput pipeline development: ease of transformation, small size, and short life cycle. There are a number of approaches that can be taken to gene function discovery in arabidopsis. In most cases, the agbiotech industry has chosen to generate large populations of lines containing a single misexpressed transgene. In this section considerations that were important in the construction of these populations will be discussed.

Classical or forward genetics studies begin with a mutant phenotype and work towards gene identification. Genome sequencing has made reverse genetics approaches that begin with a gene and work towards a gene function more feasible. Discovery of gene function in these studies depends on altering gene activity. Down-regulation of gene activity can be accomplished by a number of methods including: chemical or high-energy (x-ray, neutrons, gamma-ray) mutagensis, insertional mutagenesis (T-DNA or transposon), anti-sense expression or RNA interference (RNAi). Of these methods, insertional mutagenesis has been the most practical approach to high-throughput down-regulation in plants. Large collections of mutagenized populations of arabidopsis are publicly available and insertions have been mapped to over 75% of the approximately 30,000 genes in arabidopsis (Winkler et al., 1998; Krysan et al., 1999; Parinov and Sundaresan, 2000; Alonso et al., 2003). RNAi has been used in *Caenorhabditis elegans* to knockout the function of approximately 86% of its annotated genes (Kamath et al., 2003), but has not been used in large-scale programs in plants except for Arabidopsis (Hilson et al., 2003). RNAi constructs will have limitations in terms of the same construct being useful in heterologous species.

An alternative and complementary strategy to down-regulation is overexpression. This can be accomplished by activation tagging. In one approach T-DNA or transposon insertional mutagenesis is carried out with constructs that contain multiple copies of a transcriptional enhancer, such as those derived from the cauliflower mosaic virus 35S promoter (35S), near the border sequence. Constitutive or altered expression may occur if the insert lands in the vicinity of a gene. Several activation tagging populations have been created and screened in arabidopsis (Weigel et al., 2000; Marsch-Martinez et al., 2002; Schneider et al., 2005).

The most common approach to gene activity alteration in industry is directed high throughput gene overexpression. This is usually accomplished by transforming plants with constructs where expression of the gene of interest is driven by a highly and broadly expressing promoter. Typically the 35S promoter is used, but tissue-specific, inducible, and developmentally regulated promoters are also used. Overexpression is preferred over down-regulation because transferring a gain-of-function phenotype from a model organism to a crop species is much more tractable than down-regulating gene function in crop plants with more complex genomes.

Neither down-regulation nor overexpression is sufficient to discover the function of all genes within a species. Indeed, a combination of these and other functional genomics approaches will be required (Zhang, 2003; Gutterson and Zhang, 2004). In many cases, down-regulation of a gene does not result in a phenotype due to functional redundancy (Bouche and Bouchez, 2001). Alternatively, the function of genes with post-transcriptional regulation may be discovered only by down-regulation and not by overexpression (Vinocur and Altman, 2005). Nevertheless, the primary approach being taken by industry in the first wave of functional genomics projects has been overexpression.

3.2. Which Genes Should be Evaluated?

Tens of thousands of unique genes have been identified in plants. The question is "which genes are going to have a higher than average probability of conferring drought tolerance when overexpressed?" Of the drought tolerance genes reported, many were selected for analysis because of their increased expression under drought conditions (see example in Stockinger et al., 1997). The underlying assumption in this approach is that plants possess the functional proteins required for stress tolerance and that genes encoding these proteins are upregulated during stress. The paradox is that overexpression of these genes can confer enhanced stress tolerance when the plant is already expressing these genes in response to stress.

An illustration of one reason why this works can be found in the overexpression of the SRK2C kinase from arabidopsis (Umezawa et al., 2004; Denby and Gehring, 2005). Since *SRK2C* is post-translationally regulated, 35S-driven overexpression does not induce the negative impact on growth seen for other genes such as *DREB1A* (Kasuga et al., 1999). However, overexpression does result in increased kinase activity upon stress induction relative to controls. Thus, overexpression appears to have primed the plant for an increased magnitude of signaling in response to stress.

Global gene transcription profiling (gTxP) experiments can be useful in identifying both the functional and regulatory genes involved in plant stress response. However, not all of these genes may be good targets for transgenic intervention. As discussed in Denby and Gehring (2005), overexpression of regulatory genes may elicit too broad a response resulting in growth retardation, overexpression of a single functional gene may be unlikely to confer stress tolerance, and tissue and temporal specificity of gene regulation may be essential for conferring stress tolerance. In addition, a yeast study on salt stress-induced genes has shown that very few of the up-regulated genes are essential for stress tolerance (Giaever et al., 2002). However, it should be noted that this study looked at stress sensitivity after down-regulation of target gene activity rather than overexpression. It is possible that overexpression of some of these genes could confer marginal improvements in stress tolerance.

Additional evidence in support of using stress-inducible genes as overexpression targets comes from a study of the halophyte *Thellungiella halophila*. This species, which is closely related to arabidopsis, appears to constitutively express many genes that are stress-inducible in arabidopsis. These results suggest that the basis of salt tolerance in *T. halophila* may be differential regulation of common stress response genes (Inan et al., 2004; Taji et al., 2004; Denby and Gehring, 2005).

While gTxP is a valuable tool for enriching candidate populations with stress-related genes, there are other approaches that have also proven to be valuable. These include selecting targets with homology to known stress tolerance genes within the species (paralogs) or across species (orthologs), selecting genes from stress tolerance pathways outside the plant kingdom (e.g., TPS1 from yeast (Romero et al., 1997)), and random selection to allow for serendipity.

3.3. Developing a High-Throughput Functional Genomics Pipeline

As discussed above, gene overexpression is the most common tool employed in agricultural biotechnology for functional genomics analysis. Typical projects executed in the last decade have examined from thousands of genes to over ten thousand genes. At peak throughput, hundreds of genes may be processed in a single week. The scale of these projects requires the adoption of high-throughput (HTP) strategies. A key component of any HTP strategy is compartmentalization of the process. HTP pipelines will typically have teams of researchers dedicated to sequencing, cloning, transformation, and phenotypic analysis. These research pipelines are supported by databases that track the progress of genes and record phenotype information. For an excellent discussion of the considerations that go into developing and supporting a HTP phenotyping pipeline, see Boyes et al. (2001) and Kjemtrup et al. (2003). Another key element for enhancing efficiency is automation. Recently, Granier et al. (2006) described a drought assay that would be quite impractical if not for automated pot weighing and watering.

At Ceres, we have generated a misexpression population in arabidopsis representing over 15,000 unique genes. For each of these genes, we typically collect seeds from five independent transformation events. To facilitate rapid screening for genes that confer desired traits, we create superpools containing seeds from five events each for 100 genes.

Superpools are screened under various abiotic stress conditions including drought, heat, cold, low nitrogen, salt, low light and others. Individual seedlings (candidates) that appear to be performing qualitatively better under these conditions are selected for further analysis. The transgenes being overexpressed in the candidate plants are determined by PCR and sequencing. Once the transgene is identified, the original T_2 seeds are retrieved from the five transformation events and tested in quantitative assays for stress tolerance. Lines that are positive are carried forward to the T_3 generation for analysis. If overexpression of a transgene confers stress tolerance in multiple events and multiple generations, it is considered a Lead and moved forward for additional analysis and testing in target crops.

3.4. Application of Drought Stress in the Lab

A variety of methods for imposing drought stress have been applied in laboratory settings including: 1) osmotically adjusting growth media, 2) detaching leaves and 3) withholding water from soil-grown plants. Osmoticum supplements such as polyethylene glycol (van der Weele et al., 2000), mannitol, or sorbitol (Dejardin et al., 1999) are commonly added to a nutrient media such as MS salts to adjust the osmotic potential of the media and mimic dry soil conditions. These surrogate assays for drought can be performed in agar-solidified media or hydroponically. Plants grown under these conditions exhibit typical drought responses including the expression of stress response genes (Kreps et al., 2002) and altered growth (van der Weele et al., 2000). One advantage of using osmoticum supplements is the uniformity and reproducibility of osmotic stress as well as increased throughput attained by processing seedlings on plates.

Another type of surrogate screen for drought tolerance involves exposing plants to exogenous abscisic acid (ABA). Plants can be exposed to high concentrations of abscisic acid (ABA) and assayed for resistance (Lopez-Molina and Chua, 2000) or exposed to low concentrations of ABA and assayed for sensitivity (Nishimura et al., 2004; Villalobos et al., 2004). Both of these approaches have uncovered genes involved in ABA signaling. Crop species may respond 'conservatively' to drought by closing stomata and minimizing water use or 'pessimistically' by maintaining maximal transpiration until they wilt (Jones, 2004). ABA resistance may be valuable for engineering a more pessimistic response, while ABA hypersensitivity may be valuable for engineering a more conservative response (Kang et al., 2002).

An extreme assay to measure water relations involves detaching leaves (or uprooting whole plants) and measuring water loss rates as the plant material dries (Romero et al., 1997; Qin and Zeevaart, 2002). Since the majority of water loss occurs through the stomates, this assay can serve as a surrogate for stomatal conductance measurements. However, it may be the least like field conditions of any assay type.

Soil drought experiments in the lab have been done at varying levels of sophistication ranging from simply withholding water until plants wilt and then re-watering to carefully controlled long-term studies where precise soil moisture levels are maintained for prolonged periods by monitoring and irrigating with a robot (Granier et al., 2006). Soil assays can be divided into two categories, acute and chronic.

Acute drought assays typically involve the complete withholding of water until control plants wilt followed by re-watering and recovery (see (Fujita et al., 2005;

Sakuma et al., 2006) for recent examples). The measure of drought tolerance in these studies is often given as percent survival. If re-watering is timed precisely for the point at which the controls have wilted beyond recovery, but the experimental plants have not, these studies will tend to amplify the apparent tolerance because they measure recovery not desiccation postponement. Slight differences in the degree of wilting can translate into large differences in survival. Thus, for an accurate assessment of a plant's ability to recover from dehydration stress, it is critical to ensure that experimentals and controls are stressed to the same level of soil moisture before re-watering. A number of variables impact soil moisture including: soil volume, plant biomass and environmental conditions. Furthermore, developmental differences can result in differential water use and result in false-positives. For example, plants with a late-flowering phenotype will wilt later than controls because they are not using as much water as the controls that have already flowered. The acute drought approach has been successful at identifying genes involved in plant water relations. However, this approach assays for tolerance to extremely dry conditions that may be relevant for plant survival, but less relevant for crop productivity under field conditions.

Chronic drought assays involving long term water deprivation or cycles of water stress and recovery coupled with practical measurements of biomass and yield are more likely to uncover transgenes that will be valuable for engineering drought tolerant crops. In practice, chronic stress experiments are often designed to give an approximately 50% reduction in the parameter being measured relative to non-stressed controls (Bruce et al., 2002). At Ceres, we have developed an arabidopsis assay to measure plant biomass and yield under chronically sub-optimal soil moisture. Soil moisture levels are maintained above the wilting-point but below that required for optimal plant growth by deficit irrigation. Deficit irrigation is implemented after seedling establishment and maintained through plant maturity and senescence. Under this watering regime, rosette area, plant height, shoot biomass, and seed yield are reduced by 25–50% relative to well-watered controls.

A similar assay platform, "Phenopsis," with robotic automation of deficit irrigation has recently been reported (Granier et al., 2006). Phenopsis was designed to control for variability in developmental stage and uniformity of stress conditions. A key feature of this approach is robotic weighing of pots to determine soil moisture levels and precise deficit irrigation to maintain soil moisture within 0.05 g H_20 g^{-1} dry soil per sample. Several ecotypes of arabidopsis were analyzed and An-1 was found to have enhanced WUE. Of note is the observation that An-1 has intrinsically lower water use and reduced biomass (Granier et al. 2006). If fixed amounts of water had been added to each pot, An-1 would have performed better simply because of increased soil moisture and the enhanced WUE would not have been discovered.

3.5. Drought Tolerance Traits to be Measured

When designing drought assays for the lab or the field, a key consideration is the traits that are to be measured. Harvestable yield is ultimately the trait of interest and value for row crops. However, yield, is characterized by low heritability and high gene by environment interactions. Thus, secondary traits are typically chosen as surrogates for yield. In principle, secondary traits should have high heritability, correlate with yield, be practical to measure, be integrative over the plant life cycle and not confer any negative effects on yield under non-stressed conditions. In breeding programs, it is particularly relevant to ask what plant characteristics are limiting yield under drought stress and to approach the problem with a firm understanding of the ecophysiology of the target crop (Araus et al., 2002; Bruce et al., 2002).

When dealing with model plants (e.g., arabidopsis) in the laboratory environment such questions are still important. However, selections based on narrowly defined physiological traits such as stomatal conductance or specific structural traits such as rooting depth may unnecessarily limit the scope of the project. Furthermore, such traits may negatively impact yield under well-watered conditions (Bruce et al., 2002). More emphasis should be placed on creating drought stress conditions that are relevant to the field and then selecting traits empirically for yield or biomass enhancement. In this respect, model plant research projects should be thought of in terms of developing a genetic toolbox. Particular tools (promoters and coding sequences) can subsequently be deployed to engineer drought tolerance according to the specific cropping system and environmental conditions. Indeed, the drought tolerance literature is replete with examples of traits that are thought to be valuable under some conditions, but not others. Nearly any trait has both positive and negative attributes depending on the application. Figure 2 illustrates some of the traits that are potential targets for bioengineering.

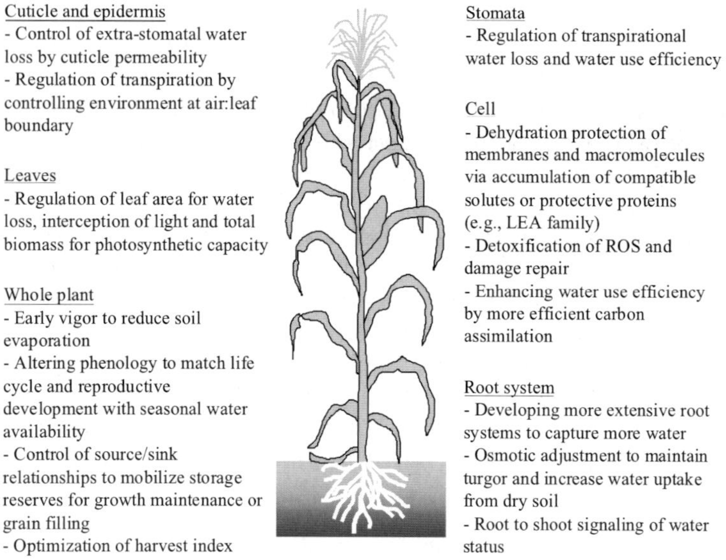

Figure 2. Mechanisms of drought tolerance that are potential targets for bioengineering of drought tolerant crops

4. GENES THAT HAVE BEEN IDENTIFIED

Several recent reviews have provided comprehensive lists of the genes shown to confer drought tolerance when mis-expressed (Rontein et al., 2002; Wang et al., 2003; Zhang et al., 2004a; Umezawa et al., 2006). These proteins have a variety of functions including: transcription regulation, signal transduction, desiccation protection, synthesis of osmolytes, ROS-scavenging enzymes, and ABA biosynthesis and signal transduction. In addition, genes involved in epidermal patterning (Masle et al., 2005; Shpak et al., 2005) and cuticle biogenesis (Aharoni et al., 2004; Zhang et al., 2005) have also been linked to drought tolerance.

In most cases, the evidence indicating that these genes confer drought tolerance is based on laboratory-grown plants under artificial drought conditions. Thus, a key question is whether expression of these genes in target crops species will confer increased biomass or yield under natural drought conditions in the field. Translatability of drought tolerance from model species in the lab to crop species in the field is dependent on the ability to transfer the mechanism of tolerance (e.g., a transcription factor would need to activate the same targets in the crop plant as the model plant) and on whether that mechanism of tolerance will actually lead to increased biomass or yield under drought.

Rather than replicate the long list of genes that have been implicated in drought tolerance, we will focus on several case studies of promising genes. This will be done with the aim of illustrating how successfully transgenic approaches are being applied and to give some indication of where we stand today relative to the ultimate goal of engineering drought tolerant crops. For the reasons discussed above, preference in presenting case studies will be given to those involving field trials or chronic drought conditions and measuring biomass or yield. These studies will be presented in the context of specific trait areas of focus.

4.1. ABA Signaling and Stomatal Conductance

Many transgenic approaches to engineering drought tolerance have focused on components of the ABA signaling pathway including ABA biosynthesis. ABA signaling affects the expression of stress-responsive genes and results in decreased stomatal conductance through direct action on guard cell anion channels (Leung and Giraudat, 1998; Himmelbach et al., 2003; Levchenko et al., 2005). Reduced transpirational water loss results in delayed wilting in short-term desiccation and recovery assays. However, simply reducing stomatal conductance may not be a practical approach to confer enhanced biomass and yield under water-limiting conditions in the field owing to the tight linkage of stomatal conductance and photosynthetic productivity. The efficacy of this approach will depend on whether stomatal conductance in a given species is optimized for maximal WUE (Chaves and Oliveira, 2004). Nevertheless, the benefits of ABA-induced gene expression can be achieved without affecting stomatal conductance (Fujita et al., 2005). In this section, we will focus on studies that employ modulation of ABA signaling or alternative approaches to optimizing stomatal conductance to achieve practical drought tolerance.

One straightforward way to increase ABA signaling is to increase the biosynthesis of ABA. In this regard, a number of studies have dealt with overexpression of 9-cis-epoxycarotenoid dioxygenase (NCED) a critical enzyme in the ABA biosynthesis pathway (Taylor et al., 2000). NCED overexpression has been carried out in tomato (Thompson et al., 2000), arabidopsis (Iuchi et al., 2001), and tobacco (Qin and Zeevaart, 2002). In each case, transgenic plants exhibited decreased transpirational water loss and reduced wilting in dry-down assays. Enhanced biomass or yield under water-limiting conditions was not tested. As reviewed in Bruce et al. (2002), ABA accumulation may be valuable for enhancing plant survival, but often results in reduced productivity. However, Qin and Zeevaart (2002) report that tobacco plants over-expressing NCED showed normal plant stature under well-watered conditions suggesting that enhanced WUE could be achieved by altering ABA levels.

Several components of the ABA signal transduction pathway are targets for engineering drought tolerance including *ERA1* (Wang et al., 2005b). *ERA1* encodes the β-subunit of arabidopsis farnesyltransferase and *era1* loss-of-function mutants show enhanced ABA sensitivity suggesting that a negative regulator of ABA signaling requires farnesylation to function (Cutler et al., 1996). However, loss-of-function mutations in *ERA1* result in pleiotropic phenotypes that negatively impact yield potential (Yalovsky et al., 2000; Ziegelhoffer et al., 2000). To circumvent the pleiotropy, Wang et al. (2005b) expressed anti-sense BnFTB (the putative *Brassica napus* ortholog of arabidopsis *ERA1*) under the control of a drought-inducible promoter, rd29A, in *B. napus*. In the lab, transgenic plants exhibited reduced stomatal conductance and transpiration under water limiting conditions that induced expression of the antisense transgene relative to controls. In addition, transgenic plants and non-transgenic controls were examined in field trials over three growing seasons in areas of Canada that require irrigation to supplement minimal summer rainfall. In each trial, transgenic plants showed higher seed yields than controls under water-limited conditions. Under well-watered conditions, transgenics performed similarly to controls and showed no impact on key agronomic parameters including canola seed quality (Wang et al., 2005b). Thus, modulation of ABA-signaling, if done in an inducible manner rather than a constitutive manner, can be an effective approach.

Plants have developed elaborate mechanisms for regulating transpirational water loss and carbon assimilation (see discussion of optimization theory on p.10, Bacon, 2004). A fundamental assumption made when approaching the engineering of drought tolerance is that these mechanisms are inefficient and can be improved by genetic modification. Inefficiencies could include imprecise regulation of stomatal conductance or inefficient enzymatic steps in photosynthesis. Alternatively, plant adaptations to drought could be in conflict with agronomic objectives. For example, a plant response may favor long-term survival over short term productivity.

A recent paper dealing with the overexpression of NADP-malic enzyme (ME) in tobacco illustrates that stomatal conductance in tobacco may not be optimized for maximal WUE. In transgenic plants, overexpression of ME reduced stomatal conductance while increasing the amount of fresh weight gained per unit water used (Laporte et al., 2002). As the authors point out, the extent to which stomatal conductance can be reduced while still maintaining high WUE needs to be determined empirically in the field for the target crop species. Because the fine-tuning of stomatal conductance may differ across species or germplasm, this approach may not work in every situation.

4.2. Signal Transduction and Gene Regulation

In the past several years, much has been learned about the transcription factors (TFs) that regulate drought-responsive gene expression (Wang et al., 2003; Zhang et al., 2004a; Vinocur and Altman, 2005; Umezawa et al., 2006). These TFs come from several families including DREB/CBF, AP2/ERF, bZIP, MYB/MYC, Cys2His2-type zinc-finger, and NAC (Umezawa et al., 2006). Comparatively less is known about the upstream signal transduction cascade that activates these TFs. Nevertheless, misexpression of several signal transduction components has been shown to confer drought tolerance (Umezawa et al., 2006). In this section, we will examine both TFs and signal transduction components that have been tested in advanced drought studies.

Mitogen-activated protein kinase (MAPK) cascades have been targets of interest because of their activation by H_2O_2 and role in regulating stress-responsive gene expression (Kovtun et al., 2000). Recently NPK1, a tobacco MAPKKK, overexpressed by a modified 35S promoter in maize was tested in a chronic drought assay. The soil moisture level of pot-grown maize plants was maintained at either 25% or 100% of holding capacity for drought-treated and well-watered controls, respectively. Plant performance measurements included apparent photosynthetic rate, timing of maturation, leaf number, kernel number and kernel weight. In all cases, NPK1-expressing plants performed better than controls under drought without any significant performance reductions under well-watered conditions (Shou et al., 2004). Earlier studies involving overexpression of NPK1 in tobacco showed tolerance to multiple stresses (freezing, heat and salt) in plate-based assays (Kovtun et al., 2000).

Perhaps the most studied gene target in this category is DREB1A, a member of the CBF/DREB family of TFs that have been shown to activate drought-tolerance genes. This gene and its paralogs were discovered in arabidopsis (Stockinger et al., 1997; Liu et al., 1998) and have been misexpressed in arabidopsis, maize, tomato, tobacco, wheat and rice illustrating the translatability of the arabidopsis phenotype to these other dicot and monocot species (Qin et al., 2004; Umezawa et al., 2006). In early experiments, constitutive overexpression of these genes caused stunted growth phenotypes which not only made evaluation of stress tolerance challenging, but precluded any practical application of the technology for crops (Liu et al., 1998). Misexpression under the control of drought-inducible promoters, such as rd29A, has resulted in plants that are morphologically normal, but show enhanced stress tolerance (Kasuga et al., 1999).

This result raises an interesting question: why should drought-inducible expression of a transgene confer stress tolerance when the transgene itself is already induced by drought endogenously? Perhaps the explanation lies in the fact that expression of *rd29A* is regulated by *DREB1A* in wildtype (Seki et al., 2001). Thus, driving *DREB1A* with the *rd29A* promoter sets up a positive feedback loop that may reinforce and amplify drought-inducible gene expression. This hypothesis is supported by the observation that under non-stressed conditions, limited transgene expression was observed while strong transgene expression was observed under stressed conditions (Kasuga et al., 2004). However, a comparison between endogenous tobacco *DREB1A* transcription and *rd29A* driven *DREB1A* transcription was not made.

One of Monsanto's most advanced lead genes for drought tolerance is also a TF belonging to the NF-YB class of CCAAT-binding factors (United States Patent Application Publication No. US 2005/0086718 A1). Transgenic maize expressing an arabidopsis NF-YB showed reduced wilting and enhanced yield in field trials under dry conditions and reduced oxidative stress damage and increased chlorophyll content under normal conditions. This gene may confer tolerance by enhancing chloroplast function (Heard, J. et al., Abstract L8.02, Interdrought II, Rome, September 2005).

4.3. Osmolytes, Compatible Solutes, Protective Proteins

Trehalose, mannitol, fructans, glycine betaine, and proline are among the metabolites that are considered compatible solutes with the capacity to protect cells from dehydration damage. An abundance of research has been done to engineer drought tolerance through increased compatible solute accumulation (Rontein et al., 2002; Serraj and Sinclair, 2002). These efforts have proceeded in spite of the lack of a firm understanding of their mode of action (Serraj and Sinclair, 2002; Ramachandra Reddy et al., 2004). The idea that these solutes are conferring stress tolerance through osmotic adjustment and enhanced water uptake is controversial (Serraj and Sinclair, 2002). More likely roles for compatible solutes in stress tolerance include stabilizing biological and macromolecular structures, scavenging reactive oxygen species, or modifying carbon metabolism. Indeed, when compatible solutes do accumulate to the levels required for osmotic adjustment, plant growth is typically stunted (Maggio et al., 2002). In this section, we will examine three case studies for trehalose, mannitol, and HVA1 a LEA family protein from Barley.

Trehalose, a non-reducing disaccharide of glucose, has been a focus for engineering drought tolerance owing to its accumulation under stress in a wide variety of species from yeast, bacteria, and invertebrates (Crowe et al., 1992). It does not accumulate to detectable levels in most plants with the exception of desiccation tolerant species such as *Selaginella lepidophylla*. However, many plant genomes contain the necessary enzymes for trehalose synthesis (Goddijn and van Dun, 1999). Trehalose is synthesized in two steps. Trehalose-6-phosphate synthase (TPS) catalyzes the combination of UDP-glucose and glucose-6-phosphate to make trehalose-6-phosphate, which is then converted to trehalose by the action of trehalose-6-phosphate phosphatase (TPP). Recently, Garg et al. (2002) have transformed an indica rice variety with a fusion gene derived from the *E. coli otsA* (TPS) and *otsB* (TPP) coding regions driven by either an ABA-inducible promoter or a light regulated promoter. Transgenic lines were subjected to two cycles of drying and recovery to impose drought stress in the lab and tolerance was measured by visible symptoms (leaf rolling and wilting) as well as measurements of photosynthetic activity (quantum yield of PSII photochemistry and Fv/Fm). Both the ABA-inducible and light-inducible transgenic constructs conferred significant decreases in visible stress symptoms and maintenance of photosynthetic activity (Garg et al., 2002). Previous studies involving overexpression of TPS by the 35S promoter in tobacco resulted in stunted plants (Romero et al., 1997; Pilon-Smits et al., 1998). Such pleiotropic and negative phenotypes were not observed in the study by Garg and colleagues possibly due to the use of inducible promoters, a TPSP fusion gene, or reduced toxicity of trehalose in monocots (Jang et al., 2003). These possibilities illustrate that one size may not fit all in transgenic intervention.

Mannitol is another widely studied compatible solute which is present in many plant species and increases in response to water stress (Patonnier et al., 1999). Mannitol is produced by the action of mannitol-1-phosphate dehydrogenase which converts fructose-6-phosphate to mannitol-1-phosphate. Mannitol-1-phosphate is converted to mannitol by nonspecific phosphatases. In the second case study, misexpression of *E. coli mtl1D* driven by the maize *ubi1* promoter in wheat has been shown to confer enhanced drought tolerance (Abebe et al., 2003). Drought stress was applied by deficit irrigation (one-third the amount of well-watered controls) over a 30 day period. Mtl1D-expressing plants tolerated water deficits better than non-transgenic controls as evidenced by greater shoot biomass. However, mannitol accumulation was slight and no difference in osmotic potential was observed between transgenics and controls strongly suggesting that mannitol is not conferring tolerance through osmotic adjustment. Rather, Abebe et al. suggest that mannitol is functioning as a scavenger of hydroxyl ions or by stabilizing macromolecular structures.

HVA1 is a member of the group 3 late embryogenesis abundant (LEA) genes from barley that is up-regulated in response to stress (Hong et al., 1992; Sutton et al., 1992). In the final case study, overexpression of *HVA1*, under the control of the maize *ubi1* promoter, in wheat, resulted in increased root and shoot biomass and seed weight under moderate water deficits relative to non-transgenic and non-expressing controls. Under non-stressed conditions no differences were observed between transgenics and controls (Sivamani et al., 2000). In a more detailed physiological analysis, some of the same researchers evaluated *HVA1* overexpression in rice. In a prolonged dry-down experiment, *HVA1* overexpressed by the *Act1* promoter resulted in longer maintenance of leaf relative water content, less reduction in biomass and reduced membrane leakage. No difference in osmotic adjustment was observed causing the authors to speculate that HVA1 functions through membrane protection (Chandra Babu et al., 2004). However, since soil moisture was not measured or controlled in the study, an indirect desiccation postponement mode of action (through decreased transpiration) cannot be ruled out. Furthermore, since membrane leakage was measured when controls were at ~51% and transgenics were at ~92% little can be concluded about the role of HVA1 in protecting membranes against desiccation. Further studies will be needed to determine the precise physiological basis of the apparent drought tolerance in *HVA1*-expressing lines.

4.4. Oxidative Stress

Reactive oxygen species (ROS) accumulate in response to multiple stresses and damage cellular components through oxidation. To detoxify excess ROS, plants use scavenging enzymes such as superoxide dismutase (SOD), ascorbate peroxidase (APX) and catalase (CAT) together with antioxidants such as ascorbic acid and glutathione (Mittler, 2002). Transgenic approaches to enhance the capacity for detoxifying ROS have yielded plants with increased stress tolerance.

A manganese-SOD from *Nicotiana plumbaginifolia* was over-expressed in alfalfa using the 35S promoter and targeted to either the mitochondrion or chloroplast. In controlled dry-down experiments in the lab, SOD-expressing plants exhibited reduced damage to PSII, reduced membrane damage and (for the mitochondria-targeted SOD only) increased shoot regrowth after defoliation. In three years of field trials, transgenic plants consistently yielded more (up to double) biomass than controls (McKersie et al., 1996). Subsequently, these authors evaluated transgenic plants containing both the mitochondria- and chloroplast-targeted transgenes and found that "pyramiding" of these two constructs did not result in synergistically increased biomass. In fact, biomass yields were reduced when both genes were expressed (Samis et al., 2002). SOD continues to be pursued as a transgenic intervention point (e.g., (McKersie et al., 1999; McKersie et al., 2000; Samis et al., 2002; Wang et al., 2005a)).

Lipid peroxidation is one type of cellular damage inflicted by ROS. Members of the aldo-keto reductase superfamily can detoxify degradation products of lipid peroxides (Vander Jagt et al., 1992). A member of this family from alfalfa, aldose/aldehyde reductase (MsALR), has been shown to reduce oxidative damage and confer drought stress tolerance when over-expressed in tobacco. Transgenic plants subjected to a desiccation and recovery assay in the lab showed increased photosynthetic efficiency (Fv/Fm) and reduced accumulation of lipid peroxide degradation products (Oberschall et al., 2000).

4.5. Plant Morphology, Development and Structure

To date, the most important trait in the generation of drought tolerant crops is altered phenology. For example, breeding programs have selected for early maturing varieties in regions with late season drought (Araus et al., 2002). A number of additional developmental or structural traits may be useful for drought tolerance (depending on the crop and environmental conditions) including: shoot and root morphology, cuticle thickness, water transporters (i.e., aquaporins), stay-green and so forth (Boyer, 1992; Araus et al., 2002; Chaves and Oliveira, 2004). These traits mainly result in desiccation avoidance rather than desiccation tolerance. Efforts to engineer drought tolerance by modifying or incorporating each of these traits have been made, yet this trait category is often overlooked in reviews of recent progress. Some traits that have a significant impact on crop production under drought such as the anthesis-silking interval in maize (Bruce et al., 2002) are not studied in model dicot species due to their distinct anatomy and life history.

One of the more tractable traits in this area is cuticle thickness and composition because a number of single genes that affect this trait have been identified. The cuticle is comprised of cutin polymers and waxes which cover aerial plant organs and form a barrier to extra-stomatal transpiration (Kunst and Samuels, 2003). Arabidopsis has proven to be a valuable system for studying cuticle biogenesis. A large number of loss-of-function mutations have been identified that result in reduced cuticles and the functions of these genes in cutin and cuticular wax synthesis and deposition are being elucidated (Jenks et al., 1996; Chen et al., 2003; Goodwin et al., 2005). Wax-deficient mutants have also been identified in maize, sorghum, barley and rape (Kunst and Samuels, 2003). Three members of the AP2/EREBP family in arabidopsis comprise the SHINE clade which is involved in cuticle biogenesis. Overexpression of these SHINE TFs results in an apparent drought avoidance phenotype in desiccation and recovery assays. In this case, avoidance likely results from the observed decrease in stomatal density because *SHN1* overexpression actually results in increased cuticle permeability (based on chlorophyll leaching) and increased water loss rates in detached leaves (Aharoni et al., 2004). The contradiction apparent in these results was not investigated further.

Another AP2 domain-containing TF from *Medicago truncatula*, WXP1, also regulates cuticular wax deposition, but does not appear to be orthologous to members of the SHINE clade in arabidopsis. In contrast to the SHINE clade results, overexpression of WXP1 in *M. truncatula* resulted in increased wax deposition on leaves, decreased water loss rates in detached leaves and decreased cuticle permeability (based on chlorophyll leaching), while also showing delayed wilting in desiccation and recovery assays (Zhang et al., 2005). However, the reduced shoot biomass and delayed flowering that were also observed in the overexpression lines may also have contributed to the delay in wilting by reducing total plant transpiration. This possibility was not addressed in the study.

Aquaporins are transmembrane water channels that facilitate symplastic movement of water (Johansson et al., 2000). Overexpression of aquaporins in tobacco has been shown to increase transpiration rates, which enhances growth when water is not limiting, but results in earlier wilting under water limitation (Aharon et al., 2003). This result is consistent with the observation that most arabidopsis aquaporins are down-regulated at the mRNA and protein level in response to prolonged drought stress (Alexandersson et al., 2005). However, overexpression of two rice aquaporins in arabidopsis has been reported to enhance growth on high osmotic media (Guo et al., 2006) and other TxP experiments show more complex up- and down-regulation of aquaporin genes (Jang et al., 2004). Thus, the utility of aquaporins for biotechnology solutions to drought tolerance is still unresolved.

Additional possible transgenic intervention points include the *ERECTA* gene for stomatal patterning. ERECTA impacts WUE as measured by carbon isotope discrimination (Masle et al., 2005; Shpak et al., 2005), but long-term plant performance under water limiting conditions for ERECTA misexpression lines has not been reported.

5. FUTURE PERSPECTIVES

Some of the experimental results that were presented in this chapter were obtained over ten years ago. This raises the question: why have no transgene-based technologies addressing drought tolerance been commercialized? One contributing factor to the delay is the development cycle for transgenic crops (McElroy, 2004) which can take over a decade to complete. The increasing number of field trials (Figure 1) is a good indicator that lead genes are advancing through product development programs and that we may soon see transgenic drought tolerance on the market.

Another possible contributing factor is the failure of many molecular and genetic studies to address whether laboratory results translate into *bona fide* drought tolerance under field conditions. This disconnect was discussed recently at the Interdrought II conference (Rome, September 2005). Quoting from the conference conclusions and recommendations:

> "There is an explosive growth of information in genomics with a proportionally minute rate of application of this information to problem solving in farming under water-limited conditions … Any research that claims to impact plant production under water-limited conditions must address crop yield or its major components in the research plan and the research report … We urge that the measurement of plant water status in a comprehensive and instructive manner is required in order to justify statements about drought stress, water deficit, drought adaptation, etc."

The Interdrought II report further stresses the need for molecular biologists to work closely with physiologists, plant breeders and agronomists in the evaluation of transgenic technology. Most reports of drought tolerance are based on delayed wilting in laboratory desiccation and recovery assays. There are numerous ways to reduce total plant transpiration, thus improving performance in this type of assay. It is comparatively more difficult to engineer an optimal plant response to water deficit that achieves yield protection under drought without negatively impacting yield under favorable conditions. Researchers seeking to impact drought tolerance in crops through transgenic intervention should utilize assays that more adequately measure plant performance under field-relevant conditions.

One difficulty that confronts researchers in all disciplines is the trade-off between drought tolerance and productivity under favorable conditions which many traits exhibit. A negative correlation between drought resistance and productivity has been observed in breeding programs (reviewed in Chaves and Oliveira, 2004). In this context, we should note that traits that are adaptive in an ecological context are not necessarily important in an agricultural context (Wang et al., 2003). For example, one might consider it self-evident that deeper or more extensive root systems would confer an advantage under drought by enhancing the ability to capture scarce water resources. However, in a long-term breeding study of maize, recombinant inbred lines exhibiting more vigorous early root development consistently showed reduced grain yield under both well-watered and drought stress conditions (Bruce et al., 2002). Thus, the potential for increased water acquisition failed to offset the costs of investment in a more extensive root system.

Current biotechnology approaches to stress tolerance are still in their first phase. The lead genes identified constitute a valuable tool kit. Successful deployment of these tools in the field will require sophisticated approaches that control the magnitude, tissue-specificity and timing of gene expression. We have already seen that using stress-inducible promoters can mitigate the negative phenotypes associated with constitutive overexpression (Kasuga et al., 2004). The identification and development of a set of promoters capable of driving expression in tissue-specific, inducible and/or developmentally regulated manner will likely prove integral to engineering drought tolerant crops.

The impact of desiccation on a plant cell is wide ranging and multiple cellular factors and pathways are involved in coping with this stress. Yet, most research to date has involved misexpression of single genes. When that single gene is a TF, an entire set of genes (regulon) can be activated. That is a primary advantage of using TFs in biotechnology approaches. Of course, affecting the regulation of many genes can also be a disadvantage and can lead to unwanted negative pleiotropic effects. When metabolic pathways are targeted, single gene approaches must confront the tendency of cell systems to restore homeostasis. Multiple steps in the biosynthetic pathway may have to be engineered in order to more precisely control metabolic flux (Rontein et al., 2002; Vinocur and Altman, 2005). For functional proteins such as LEA proteins or SOD that mitigate specific types of cellular damage, single gene approaches have shown improvement in stress tolerance. Even greater levels of stress tolerance may be achieved by creating plants in which multiple transgenes from different functional categories are misexpressed.

Most laboratory studies deal with stress conditions in isolation. For example, drought treatment is distinct from heat treatment. However, it is rarely the case that crop plants are dealing with a single stress in the field. Periods of drought are often associated with increased heat and this combination of stresses can impose conflicting physiological demands on the plant (Mittler, 2006). Increased leaf transpiration is an effective way to dissipate excess heat, but is not an appropriate response under drought stress. Thus, an additional challenge for biotechnology is to develop combination stress conditions in the laboratory screening environment that more faithfully represent actual field conditions.

REFERENCES

Abebe, T., Guenzi, A.C., Martin, B., and Cushman, J.C. (2003). Tolerance of mannitol-accumulating transgenic wheat to water stress and salinity. Plant Physiol **131,** 1748–1755.

Aharon, R., Shahak, Y., Wininger, S., Bendov, R., Kapulnik, Y., and Galili, G. (2003). Overexpression of a plasma membrane aquaporin in transgenic tobacco improves plant vigor under favorable growth conditions but not under drought or salt stress. Plant Cell **15,** 439–447.

Aharoni, A., Dixit, S., Jetter, R., Thoenes, E., van Arkel, G., and Pereira, A. (2004). The SHINE clade of AP2 domain transcription factors activates wax biosynthesis, alters cuticle properties, and confers drought tolerance when overexpressed in Arabidopsis. Plant Cell **16,** 2463–2480.

Alexandersson, E., Fraysse, L., Sjovall-Larsen, S., Gustavsson, S., Fellert, M., Karlsson, M., Johanson, U., and Kjellbom, P. (2005). Whole gene family expression and drought stress regulation of aquaporins. Plant Mol Biol **59,** 469–484.

Alexandrov, N.N., Troukhan, M.E., Brover, V.V., Tatarinova, T., Flavell, R.B., and Feldmann, K.A. (2006). Features of Arabidopsis genes and genome discovered using full-length cDNAs. Plant Mol Biol **60,** 69–85.

Alonso, J.M., Stepanova, A.N., Leisse, T.J., Kim, C.J., Chen, H., Shinn, P., Stevenson, D.K., Zimmerman, J., Barajas, P., Cheuk, R., Gadrinab, C., Heller, C., Jeske, A., Koesema, E., Meyers, C.C., Parker, H., Prednis, L., Ansari, Y., Choy, N., Deen, H., Geralt, M., Hazari, N., Hom, E., Karnes, M., Mulholland, C., Ndubaku, R., Schmidt, I., Guzman, P., Aguilar-Henonin, L., Schmid, M., Weigel, D., Carter, D.E., Marchand, T., Risseeuw, E., Brogden, D., Zeko, A., Crosby, W.L., Berry, C.C., and Ecker, J.R. (2003). Genome-wide insertional mutagenesis of Arabidopsis thaliana. Science **301,** 653–657.

Araus, J.L., Slafer, G.A., Reynolds, M.P., and Royo, C. (2002). Plant breeding and drought in C3 cereals: what should we breed for? Ann Bot (Lond) **89 Spec No,** 925–940.

Bacon, M.A. (2004). Water use efficiency in plant biology. In Water use efficiency in plant biology, M.A. Bacon, ed (Boca Raton: CRC Press LLC), pp. 1–22.

Bouche, N., and Bouchez, D. (2001). Arabidopsis gene knockout: phenotypes wanted. Curr Opin Plant Biol **4,** 111–117.

Boyer, J.S. (1992). Mechanisms for obtaining water use efficiency and drought resistance. In Plant breeding in the 1990s, H.T. Stalker and J.P. Murphy, eds (Wallingford, UK: CAB International), pp. 181–200.

Boyes, D.C., Zayed, A.M., Ascenzi, R., McCaskill, A.J., Hoffman, N.E., Davis, K.R., and Gorlach, J. (2001). Growth stage-based phenotypic analysis of Arabidopsis: a model for high throughput functional genomics in plants. Plant Cell **13,** 1499–1510.

Bruce, W.B., Edmeades, G.O., and Barker, T.C. (2002). Molecular and physiological approaches to maize improvement for drought tolerance. J Exp Bot **53,** 13–25.

Chandra Babu, R., Zhang, J., Blum, A., David Ho, T.-H., Wu, R., and Nguyen, H.T. (2004). HVA1, a LEA gene from barley confers dehydration tolerance in transgenic rice (Oryza sativa L.) via cell membrane protection. Plant Sci **166,** 855–862.

Chaves, M.M., and Oliveira, M.M. (2004). Mechanisms underlying plant resilience to water deficits: prospects for water-saving agriculture. J Exp Bot **55,** 2365–2384.

Chen, X., Goodwin, S.M., Boroff, V.L., Liu, X., and Jenks, M.A. (2003). Cloning and characterization of the WAX2 gene of Arabidopsis involved in cuticle membrane and wax production. Plant Cell **15,** 1170–1185.

Crowe, J.H., Hoekstra, F.A., and Crowe, L.M. (1992). Anhydrobiosis. Annu Rev Physiol **54,** 579–599.

Cutler, S., Ghassemian, M., Bonetta, D., Cooney, S., and McCourt, P. (1996). A protein farnesyl transferase involved in abscisic acid signal transduction in Arabidopsis. Science **273,** 1239–1241.

Dejardin, A., Sokolov, L.N., and Kleczkowski, L.A. (1999). Sugar/osmoticum levels modulate differential abscisic acid-independent expression of two stress-responsive sucrose synthase genes in Arabidopsis. Biochem J **344 Pt 2,** 503–509.

Denby, K., and Gehring, C. (2005). Engineering drought and salinity tolerance in plants: lessons from genome-wide expression profiling in Arabidopsis. Trends Biotechnol **23,** 547–552.

Fujita, Y., Fujita, M., Satoh, R., Maruyama, K., Parvez, M.M., Seki, M., Hiratsu, K., Ohme-Takagi, M., Shinozaki, K., and Yamaguchi-Shinozaki, K. (2005). AREB1 Is a Transcription Activator of Novel ABRE-Dependent ABA Signaling That Enhances Drought Stress Tolerance in Arabidopsis. Plant Cell **17,** 3470–3488.

Garg, A.K., Kim, J.K., Owens, T.G., Ranwala, A.P., Choi, Y.D., Kochian, L.V., and Wu, R.J. (2002). Trehalose accumulation in rice plants confers high tolerance levels to different abiotic stresses. Proc Natl Acad Sci U S A **99,** 15898–15903.

Giaever, G., Chu, A.M., Ni, L., Connelly, C., Riles, L., Veronneau, S., Dow, S., Lucau-Danila, A., Anderson, K., Andre, B., Arkin, A.P., Astromoff, A., El-Bakkoury, M., Bangham, R., Benito, R., Brachat, S., Campanaro, S., Curtiss, M., Davis, K., Deutschbauer, A., Entian, K.D., Flaherty, P., Foury, F., Garfinkel, D.J., Gerstein, M., Gotte, D., Guldener, U., Hegemann, J.H., Hempel, S., Herman, Z., Jaramillo, D.F., Kelly, D.E., Kelly, S.L., Kotter, P., LaBonte, D., Lamb, D.C., Lan, N., Liang, H., Liao, H., Liu, L., Luo, C., Lussier, M., Mao, R., Menard, P., Ooi, S.L., Revuelta, J.L., Roberts, C.J., Rose, M., Ross-Macdonald, P., Scherens, B., Schimmack, G., Shafer, B., Shoemaker, D.D., Sookhai-Mahadeo, S., Storms, R.K., Strathern, J.N., Valle, G., Voet, M., Volckaert, G., Wang, C.Y., Ward, T.R., Wilhelmy, J., Winzeler, E.A., Yang, Y., Yen, G., Youngman, E., Yu, K., Bussey, H., Boeke, J.D., Snyder, M., Philippsen, P., Davis, R.W., and Johnston, M. (2002). Functional profiling of the Saccharomyces cerevisiae genome. Nature **418,** 387–391.

Goddijn, O.J., and van Dun, K. (1999). Trehalose metabolism in plants. Trends Plant Sci **4,** 315–319.

Goodwin, S.M., Rashotte, A.M., Rahman, M., Feldmann, K.A., and Jenks, M.A. (2005). Wax constituents on the inflorescence stems of double eceriferum mutants in Arabidopsis reveal complex gene interactions. Phytochemistry **66,** 771–780.

Granier, C., Aguirrezabal, L., Chenu, K., Cookson, S.J., Dauzat, M., Hamard, P., Thioux, J.J., Rolland, G., Bouchier-Combaud, S., Lebaudy, A., Muller, B., Simonneau, T., and Tardieu, F. (2006). PHENOPSIS, an automated platform for reproducible phenotyping of plant responses to soil water deficit in Arabidopsis thaliana permitted the identification of an accession with low sensitivity to soil water deficit. New Phytol **169,** 623–635.

Guo, L., Wang, Z.Y., Lin, H., Cui, W.E., Chen, J., Liu, M., Chen, Z.L., Qu, L.J., and Gu, H. (2006). Expression and functional analysis of the rice plasma-membrane intrinsic protein gene family. Cell Res **16,** 277–286.

Gutterson, N., and Zhang, J.Z. (2004). Genomics applications to biotech traits: a revolution in progress? Curr Opin Plant Biol **7,** 226–230.

Himmelbach, A., Yang, Y., and Grill, E. (2003). Relay and control of abscisic acid signaling. Curr Opin Plant Biol **6,** 470–479.

Hong, B., Barg, R., and Ho, T.H. (1992). Developmental and organ-specific expression of an ABA- and stress-induced protein in barley. Plant Mol Biol **18,** 663–674.

Inan, G., Zhang, Q., Li, P., Wang, Z., Cao, Z., Zhang, H., Zhang, C., Quist, T.M., Goodwin, S.M., Zhu, J., Shi, H., Damsz, B., Charbaji, T., Gong, Q., Ma, S., Fredricksen, M., Galbraith, D.W., Jenks, M.A., Rhodes, D., Hasegawa, P.M., Bohnert, H.J., Joly, R.J., Bressan, R.A., and Zhu, J.K. (2004). Salt cress. A halophyte and cryophyte Arabidopsis relative model system and its applicability to molecular genetic analyses of growth and development of extremophiles. Plant Physiol **135,** 1718–1737.

Iuchi, S., Kobayashi, M., Taji, T., Naramoto, M., Seki, M., Kato, T., Tabata, S., Kakubari, Y., Yamaguchi-Shinozaki, K., and Shinozaki, K. (2001). Regulation of drought tolerance by gene manipulation of 9-cis-epoxycarotenoid dioxygenase, a key enzyme in abscisic acid biosynthesis in Arabidopsis. Plant J **27,** 325–333.

Jang, I.C., Oh, S.J., Seo, J.S., Choi, W.B., Song, S.I., Kim, C.H., Kim, Y.S., Seo, H.S., Choi, Y.D., Nahm, B.H., and Kim, J.K. (2003). Expression of a bifunctional fusion of the Escherichia coli genes for trehalose-6-phosphate synthase and trehalose-6-phosphate phosphatase in transgenic rice plants increases trehalose accumulation and abiotic stress tolerance without stunting growth. Plant Physiol **131,** 516–524.

Jang, J.Y., Kim, D.G., Kim, Y.O., Kim, J.S., and Kang, H. (2004). An expression analysis of a gene family encoding plasma membrane aquaporins in response to abiotic stresses in Arabidopsis thaliana. Plant Mol Biol **54,** 713–725.

Jenks, M.A., Rashotte, A.M., Tuttle, H.A., and Feldmann, K.A. (1996). Mutants in Arabidopsis thaliana Altered in Epicuticular Wax and Leaf Morphology. Plant Physiol **110,** 377–385.

Johansson, I., Karlsson, M., Johanson, U., Larsson, C., and Kjellbom, P. (2000). The role of aquaporins in cellular and whole plant water balance. Biochim Biophys Acta **1465,** 324–342.

Jones, H.G. (2004). What is water use efficiency? In Water use efficiency in plant biology, M.A. Bacon, ed (Boca Raton: CRC Press LLC), pp. 27–41.

Kamath, R.S., Fraser, A.G., Dong, Y., Poulin, G., Durbin, R., Gotta, M., Kanapin, A., Le Bot, N., Moreno, S., Sohrmann, M., Welchman, D.P., Zipperlen, P., and Ahringer, J. (2003). Systematic functional analysis of the Caenorhabditis elegans genome using RNAi. Nature **421,** 231–237.

Kang, J.Y., Choi, H.I., Im, M.Y., and Kim, S.Y. (2002). Arabidopsis basic leucine zipper proteins that mediate stress-responsive abscisic acid signaling. Plant Cell **14,** 343–357.

Kasuga, M., Miura, S., Shinozaki, K., and Yamaguchi-Shinozaki, K. (2004). A combination of the Arabidopsis DREB1A gene and stress-inducible rd29A promoter improved drought- and low-temperature stress tolerance in tobacco by gene transfer. Plant Cell Physiol **45,** 346–350.

Kasuga, M., Liu, Q., Miura, S., Yamaguchi-Shinozaki, K., and Shinozaki, K. (1999). Improving plant drought, salt, and freezing tolerance by gene transfer of a single stress-inducible transcription factor. Nat Biotechnol **17,** 287–291.

Kjemtrup, S., Boyes, D.C., Christensen, C., McCaskill, A.J., Hylton, M., and Davis, K. (2003). Growth stage-based phenotypic profiling of plants. Methods Mol Biol **236,** 427–442.

Kovtun, Y., Chiu, W.L., Tena, G., and Sheen, J. (2000). Functional analysis of oxidative stress-activated mitogen-activated protein kinase cascade in plants. Proc Natl Acad Sci U S A **97,** 2940–2945.

Kreps, J.A., Wu, Y., Chang, H.S., Zhu, T., Wang, X., and Harper, J.F. (2002). Transcriptome changes for Arabidopsis in response to salt, osmotic, and cold stress. Plant Physiol **130,** 2129–2141.

Krysan, P.J., Young, J.C., and Sussman, M.R. (1999). T-DNA as an insertional mutagen in Arabidopsis. Plant Cell **11,** 2283–2290.

Kunst, L., and Samuels, A.L. (2003). Biosynthesis and secretion of plant cuticular wax. Prog Lipid Res **42,** 51–80.

Laporte, M.M., Shen, B., and Tarczynski, M.C. (2002). Engineering for drought avoidance: expression of maize NADP-malic enzyme in tobacco results in altered stomatal function. J Exp Bot **53,** 699–705.

Leung, J., and Giraudat, J. (1998). Abscisic Acid Signal Transduction. Annu Rev Plant Physiol Plant Mol Biol **49,** 199–222.

Levchenko, V., Konrad, K.R., Dietrich, P., Roelfsema, M.R., and Hedrich, R. (2005). Cytosolic abscisic acid activates guard cell anion channels without preceding Ca2+ signals. Proc Natl Acad Sci U S A **102,** 4203–4208.

Liu, Q., Kasuga, M., Sakuma, Y., Abe, H., Miura, S., Yamaguchi-Shinozaki, K., and Shinozaki, K. (1998). Two transcription factors, DREB1 and DREB2, with an EREBP/AP2 DNA binding domain separate two cellular signal transduction pathways in drought- and low-temperature-responsive gene expression, respectively, in Arabidopsis. Plant Cell **10,** 1391–1406.

Lopez-Molina, L., and Chua, N.H. (2000). A null mutation in a bZIP factor confers ABA-insensitivity in Arabidopsis thaliana. Plant Cell Physiol **41,** 541–547.

Maggio, A., Miyazaki, S., Veronese, P., Fujita, T., Ibeas, J.I., Damsz, B., Narasimhan, M.L., Hasegawa, P.M., Joly, R.J., and Bressan, R.A. (2002). Does proline accumulation play an active role in stress-induced growth reduction? Plant J **31,** 699–712.

Marsch-Martinez, N., Greco, R., Van Arkel, G., Herrera-Estrella, L., and Pereira, A. (2002). Activation tagging using the En-I maize transposon system in Arabidopsis. Plant Physiol **129,** 1544–1556.

Masle, J., Gilmore, S.R., and Farquhar, G.D. (2005). The ERECTA gene regulates plant transpiration efficiency in Arabidopsis. Nature **436,** 866–870.

McElroy, D. (2004). Valuing the product development cycle in agricultural biotechnology–what's in a name? Nat Biotechnol **22,** 817–822.

McKersie, B.D., Bowley, S.R., and Jones, K.S. (1999). Winter survival of transgenic alfalfa overexpressing superoxide dismutase. Plant Physiol **119,** 839–848.

McKersie, B.D., Bowley, S.R., Harjanto, E., and Leprince, O. (1996). Water-Deficit Tolerance and Field Performance of Transgenic Alfalfa Overexpressing Superoxide Dismutase. Plant Physiol **111,** 1177–1181.

McKersie, B.D., Murnaghan, J., Jones, K.S., and Bowley, S.R. (2000). Iron-superoxide dismutase expression in transgenic alfalfa increases winter survival without a detectable increase in photosynthetic oxidative stress tolerance. Plant Physiol **122,** 1427–1437.

Mitra, J. (2001). Genetics and genetic improvement of drought resistance in crop plants. Current Science **80,** 758–763.

Mittler, R. (2002). Oxidative stress, antioxidants and stress tolerance. Trends Plant Sci **7,** 405–410.
Mittler, R. (2006). Abiotic stress, the field environment and stress combination. Trends Plant Sci **11,** 15–19.
Nishimura, N., Yoshida, T., Murayama, M., Asami, T., Shinozaki, K., and Hirayama, T. (2004). Isolation and characterization of novel mutants affecting the abscisic acid sensitivity of Arabidopsis germination and seedling growth. Plant Cell Physiol **45,** 1485–1499.
Oberschall, A., Deak, M., Torok, K., Sass, L., Vass, I., Kovacs, I., Feher, A., Dudits, D., and Horvath, G.V. (2000). A novel aldose/aldehyde reductase protects transgenic plants against lipid peroxidation under chemical and drought stresses. Plant J **24,** 437–446.
Parinov, S., and Sundaresan, V. (2000). Functional genomics in Arabidopsis: large-scale insertional mutagenesis complements the genome sequencing project. Curr Opin Biotechnol **11,** 157–161.
Patonnier, M.P., Peltier, J.P., and Marigo, G. (1999). Drought-induced increase in xylem malate and mannitol concentration and closure of *Fraxinus excelsior* L. stomata. J Exp Bot **50,** 1223–1229.
Pilon-Smits, E.A.H., Terry, N., Sears, T., Kim, H., Zayed, A., Hwang, S.B., van Dun, K., Voogd, E., Verwoerd, T.C., Krutwagen, R.W., and Goodijn, O.J.M. (1998). Trehalose-producing transgenic tobacco plants show improved growth performance under drought stress. J Plant Physiol **152,** 525–532.
Qin, F., Sakuma, Y., Li, J., Liu, Q., Li, Y.Q., Shinozaki, K., and Yamaguchi-Shinozaki, K. (2004). Cloning and functional analysis of a novel DREB1/CBF transcription factor involved in cold-responsive gene expression in Zea mays L. Plant Cell Physiol **45,** 1042–1052.
Qin, X., and Zeevaart, J.A. (2002). Overexpression of a 9-cis-epoxycarotenoid dioxygenase gene in Nicotiana plumbaginifolia increases abscisic acid and phaseic acid levels and enhances drought tolerance. Plant Physiol **128,** 544–551.
Ramachandra Reddy, A., Chaitanya, K.V., and Vivekanandan, M. (2004). Drought-induced responses of photosynthesis and antioxidant metabolism in higher plants. J Plant Physiol **161,** 1189–1202.
Riechmann, J.L., Heard, J., Martin, G., Reuber, L., Jiang, C., Keddie, J., Adam, L., Pineda, O., Ratcliffe, O.J., Samaha, R.R., Creelman, R., Pilgrim, M., Broun, P., Zhang, J.Z., Ghandehari, D., Sherman, B.K., and Yu, G. (2000). Arabidopsis transcription factors: genome-wide comparative analysis among eukaryotes. Science **290,** 2105–2110.
Romero, C., Belles, J.M., Vaya, J.L., Serrano, R., and Culianez-Macia, F.A. (1997). Expression of the yeast trehalose-6-phosphate synthase gene in transgenic tobacco plants: Pleiotropic phenotypes include drought tolerance. Planta **201,** 293–297.
Rontein, D., Basset, G., and Hanson, A.D. (2002). Metabolic engineering of osmoprotectant accumulation in plants. Metab Eng **4,** 49–56.
Sakuma, Y., Maruyama, K., Osakabe, Y., Qin, F., Seki, M., Shinozaki, K., and Yamaguchi-Shinozaki, K. (2006). Functional Analysis of an Arabidopsis Transcription Factor, DREB2A, Involved in Drought-Responsive Gene Expression. Plant Cell **18,** 1292–1309.
Samis, K., Bowley, S., and McKersie, B. (2002). Pyramiding Mn-superoxide dismutase transgenes to improve persistence and biomass production in alfalfa. J Exp Bot **53,** 1343–1350.
Schneider, A., Kirch, T., Gigolashvili, T., Mock, H.P., Sonnewald, U., Simon, R., Flugge, U.I., and Werr, W. (2005). A transposon-based activation-tagging population in Arabidopsis thaliana (TAMARA) and its application in the identification of dominant developmental and metabolic mutations. FEBS Lett **579,** 4622–4628.
Seki, M., Narusaka, M., Abe, H., Kasuga, M., Yamaguchi-Shinozaki, K., Carninci, P., Hayashizaki, Y., and Shinozaki, K. (2001). Monitoring the expression pattern of 1300 Arabidopsis genes under drought and cold stresses by using a full-length cDNA microarray. Plant Cell **13,** 61–72.
Serraj, R., and Sinclair, T.R. (2002). Osmolyte accumulation: can it really help increase crop yield under drought conditions? Plant Cell Environ **25,** 333–341.
Shou, H., Bordallo, P., and Wang, K. (2004). Expression of the Nicotiana protein kinase (NPK1) enhanced drought tolerance in transgenic maize. J Exp Bot **55,** 1013–1019.
Shpak, E.D., McAbee, J.M., Pillitteri, L.J., and Torii, K.U. (2005). Stomatal patterning and differentiation by synergistic interactions of receptor kinases. Science **309,** 290–293.

Sivamani, E., Bahieldin, A., Wraith, J.M., Al-Niemi, T., Dyer, W.E., Ho, T.D., and Qu, R. (2000). Improved biomass productivity and water use efficiency under water deficit conditions in transgenic wheat constitutively expressing the barley HVA1 gene. Plant Science **155,** 1–9.

Somerville, C., and Koornneef, M. (2002). A fortunate choice: the history of Arabidopsis as a model plant. Nat Rev Genet **3,** 883–889.

Stockinger, E.J., Gilmour, S.J., and Thomashow, M.F. (1997). Arabidopsis thaliana CBF1 encodes an AP2 domain-containing transcriptional activator that binds to the C-repeat/DRE, a cis-acting DNA regulatory element that stimulates transcription in response to low temperature and water deficit. Proc Natl Acad Sci U S A **94,** 1035–1040.

Sutton, F., Ding, X., and Kenefick, D.G. (1992). Group 3 LEA Gene HVA1 Regulation by Cold Acclimation and Deacclimation in Two Barley Cultivars with Varying Freeze Resistance. Plant Physiol **99,** 338–340.

Taji, T., Seki, M., Satou, M., Sakurai, T., Kobayashi, M., Ishiyama, K., Narusaka, Y., Narusaka, M., Zhu, J.K., and Shinozaki, K. (2004). Comparative genomics in salt tolerance between Arabidopsis and aRabidopsis-related halophyte salt cress using Arabidopsis microarray. Plant Physiol **135,** 1697–1709.

Taylor, I.B., Burbidge, A., and Thompson, A.J. (2000). Control of abscisic acid synthesis. J Exp Bot **51,** 1563–1574.

Thompson, A.J., Jackson, A.C., Symonds, R.C., Mulholland, B.J., Dadswell, A.R., Blake, P.S., Burbidge, A., and Taylor, I.B. (2000). Ectopic expression of a tomato 9-cis-epoxycarotenoid dioxygenase gene causes over-production of abscisic acid. Plant J **23,** 363–374.

Umezawa, T., Yoshida, R., Maruyama, K., Yamaguchi-Shinozaki, K., and Shinozaki, K. (2004). SRK2C, a SNF1-related protein kinase 2, improves drought tolerance by controlling stress-responsive gene expression in Arabidopsis thaliana. Proc Natl Acad Sci U S A **101,** 17306–17311.

Umezawa, T., Fujita, M., Fujita, Y., Yamaguchi-Shinozaki, K., and Shinozaki, K. (2006). Engineering drought tolerance in plants: discovering and tailoring genes to unlock the future. Curr Opin Biotechnol **17,** 1–10.

van der Weele, C.M., Spollen, W.G., Sharp, R.E., and Baskin, T.I. (2000). Growth of Arabidopsis thaliana seedlings under water deficit studied by control of water potential in nutrient-agar media. J Exp Bot **51,** 1555–1562.

Vander Jagt, D.L., Robinson, B., Taylor, K.K., and Hunsaker, L.A. (1992). Reduction of trioses by NADPH-dependent aldo-keto reductases. Aldose reductase, methylglyoxal, and diabetic complications. J Biol Chem **267,** 4364–4369.

Villalobos, M.A., Bartels, D., and Iturriaga, G. (2004). Stress tolerance and glucose insensitive phenotypes in Arabidopsis overexpressing the CpMYB10 transcription factor gene. Plant Physiol **135,** 309–324.

Vinocur, B., and Altman, A. (2005). Recent advances in engineering plant tolerance to abiotic stress: achievements and limitations. Curr Opin Biotechnol **16,** 123–132.

Wang, F.Z., Wang, Q.B., Kwon, S.Y., Kwak, S.S., and Su, W.A. (2005a). Enhanced drought tolerance of transgenic rice plants expressing a pea manganese superoxide dismutase. J Plant Physiol **162,** 465–472.

Wang, W., Vinocur, B., and Altman, A. (2003). Plant responses to drought, salinity and extreme temperatures: towards genetic engineering for stress tolerance. Planta **218,** 1–14.

Wang, Y., Ying, J., Kuzma, M., Chalifoux, M., Sample, A., McArthur, C., Uchacz, T., Sarvas, C., Wan, J., Dennis, D.T., McCourt, P., and Huang, Y. (2005b). Molecular tailoring of farnesylation for plant drought tolerance and yield protection. Plant J **43,** 413–424.

Weigel, D., Ahn, J.H., Blazquez, M.A., Borevitz, J.O., Christensen, S.K., Fankhauser, C., Ferrandiz, C., Kardailsky, I., Malancharuvil, E.J., Neff, M.M., Nguyen, J.T., Sato, S., Wang, Z.Y., Xia, Y., Dixon, R.A., Harrison, M.J., Lamb, C.J., Yanofsky, M.F., and Chory, J. (2000). Activation tagging in Arabidopsis. Plant Physiol **122,** 1003–1013.

Winkler, R.G., Frank, M.R., Galbraith, D.W., Feyereisen, R., and Feldmann, K.A. (1998). Systematic reverse genetics of transfer-DNA-tagged lines of Arabidopsis. Isolation of mutations in the cytochrome p450 gene superfamily. Plant Physiol **118,** 743–750.

Yalovsky, S., Kulukian, A., Rodriguez-Concepcion, M., Young, C.A., and Gruissem, W. (2000). Functional requirement of plant farnesyltransferase during development in Arabidopsis. Plant Cell **12,** 1267–1278.

Zhang, J.-Y., Broeckling, C.D., Blancaflor, E.B., Sledge, M.K., Sumner, L.W., and Wang, Z.-Y. (2005). Overexpression of WXP1, a putative Medicago truncatula AP2 domain-containing transcription factor gene, increases cuticular wax accumulation and enhances drought tolerance in transgenic alfalfa (Medicago sativa). Plant J **42,** 689–707.

Zhang, J.Z. (2003). Overexpression analysis of plant transcription factors. Curr Opin Plant Biol **6,** 430–440.

Zhang, J.Z., Creelman, R.A., and Zhu, J.K. (2004a). From laboratory to field. Using information from Arabidopsis to engineer salt, cold, and drought tolerance in crops. Plant Physiol **135,** 615–621.

Zhang, X., Fowler, S.G., Cheng, H., Lou, Y., Rhee, S.Y., Stockinger, E.J., and Thomashow, M.F. (2004b). Freezing-sensitive tomato has a functional CBF cold response pathway, but a CBF regulon that differs from that of freezing-tolerant Arabidopsis. Plant J **39,** 905–919.

Ziegelhoffer, E.C., Medrano, L.J., and Meyerowitz, E.M. (2000). Cloning of the Arabidopsis WIGGUM gene identifies a role for farnesylation in meristem development. Proc Natl Acad Sci U S A **97,** 7633–7638.

CHAPTER 15

HIGH THROUGHPUT APPROACHES FOR THE IDENTIFICATION OF SALT TOLERANCE GENES IN PLANTS

FASONG ZHOU, JULISSA SOSA AND KENNETH A. FELDMANN

Ceres, Inc. Thousand Oaks, CA 91320

Abstract: Salt tolerance in plants is a complex trait, which involves multiple genes participating in a myriad of processes that limit uptake, promote efflux, enhance vacuolar storage of Na^+ and recycle Na^+ from shoots to roots. In addition, the suppression of high Na^+-triggered oxidative stress and cell death also increases salt tolerance. A number of salt tolerance genes have been identified and characterized using arabidopsis and rice as model plants. Mutant screens have been frequently utilized and most genes identified with this approach are overly-sensitive to salt stress, *i.e. sos* genes, implying that positive gene function is required for salt tolerance. To identify genes that positively contribute to salt tolerance and are more easily transferred to crops, Ceres has developed a large population of arabidopsis transgenic lines overexpressing genes from several species and used a seed pooling strategy to screen these for enhanced salt tolerance. Thus far, we have identified 10 genes that when overexpressed result in increased salt tolerance. The encoded proteins are related to calmodulin, calmodulin-binding, zinc-finger, putative cyclases, stress-related and novel proteins. These genes may be involved in the regulation of *AtNHX1* and/or *SOS* genes, as well as the suppression of high Na^+-triggered generation of reactive oxygen species and cell death. We discuss methods, such as stacking genes that provide different mechanisms for salt tolerance or using salt inducible promoters, to develop super-tolerant cultivars of crops to be grown in high salinity soils

Keywords: arabidopsis, salt tolerance, high throughput screen, misexpression

1. INTRODUCTION

Soil salinity is a serious problem in agriculture; about 20% of the arable land in the world is affected by high salinity (Xiong 2002). Two major factors have led to the salinization of arable land. First, vast tracts of potentially productive soil along coasts have become saline because of salt water incursion. Secondly, agricultural practices have contributed to the salinization of large areas of farmland

M.A. Jenks et al. (eds.), *Advances in Molecular Breeding Toward Drought and Salt Tolerant Crops*, 359–379.

(Rengasamy 2006). For example, in one of the most productive irrigated farming areas in the world, California's San Joaquin Valley, a comprehensive analysis and modeling of the cumulative change in salt storage over 57 years showed a net salt increase of 8 to 10 million tons per year since 1940 in the 1,400 km^2 simulation area. However, in the past 20 years, salinity prevention measures have dramatically reduced the net deposition of salt from the irrigation water (Schoups, Hopmans et al. 2005). Most crop plants are hypersensitive to salt stress. High levels of sodium concentration in the soil cause stunted growth and in more severe conditions plants are unable to survive. Thus, soil salinity not only reduces the area of arable land available for crop production, but also dramatically reduces yields in affected area.

To meet the demands for food and consumable agricultural products of an ever increasing world population, efficient usage of salinity-affected land appears to be an unavoidable choice. Unlike the conceivable costly engineering approaches of leaching the salt from saline soils, developing salt tolerant cultivars that are able to produce relatively good yields in the presence of high salt seems to be a more practical solution to the problem.

2. PHYSIOLOGICAL RESPONSES TO SALT STRESS

High levels of sodium in soils affect plant growth through four distinct mechanisms. First, high levels cause osmotic stress. Second, they inhibit the uptake of K^+, a major nutrient for plants. Third, Na^+ itself is toxic to cytosolic enzymes at high concentrations. Finally, high levels of sodium trigger oxidative stress and cell death (Xiong 2002). In order to survive high salt stress conditions, plants have evolved sophisticated mechanisms to defend themselves against stresses. Salt tolerance in plants is a complex trait which requires the orchestrated expression and functioning of genes participating in a myriad of processes relating to sodium influx (A, Figure 1), efflux (B), storage (C), recycling (D) and the suppression of high Na^+-triggered oxidative stress and cell death (E). Limiting the absorption of Na^+ from saline soil would be the first line of defense in protecting a plant from high Na^+ toxicity. The comparative analysis of ion homeostasis in salt cress (*Thellungiella halophila*) and arabidopsis under the same level of salt stress revealed that the limitation of sodium influx was the main mechanism of salt tolerance and lower net Na^+ accumulation in the salt tolerant plant species (Wang, Davenport et al. 2006). However, because of the similar ionic properties of K^+ and Na^+, plants are usually unable to efficiently block the entry of Na^+ into the cells. Many ion channels are permeable to Na^+ and one of them, HKT1, has been reported to be expressed in roots and likely to be responsible for the uptake of Na^+(Rus, Yokoi et al. 2001). The negative regulation of AtHKT1 activity by Ca^{++}-mediated signaling has been suggested to reduce salinity-caused damage to plants (Xiong 2002). Once Na^+ enters the cell, a mechanism to pump it out appears to be critical for salt tolerance. A H^+/Na^+ antiporter, SOS1, has been identified and been implicated to play an important role for the extrusion of Na^+ as evidenced by the knockout mutant *sos1*, which is more sensitive to salinity stress (Shi, Ishitani et al. 2000). Sodium storage

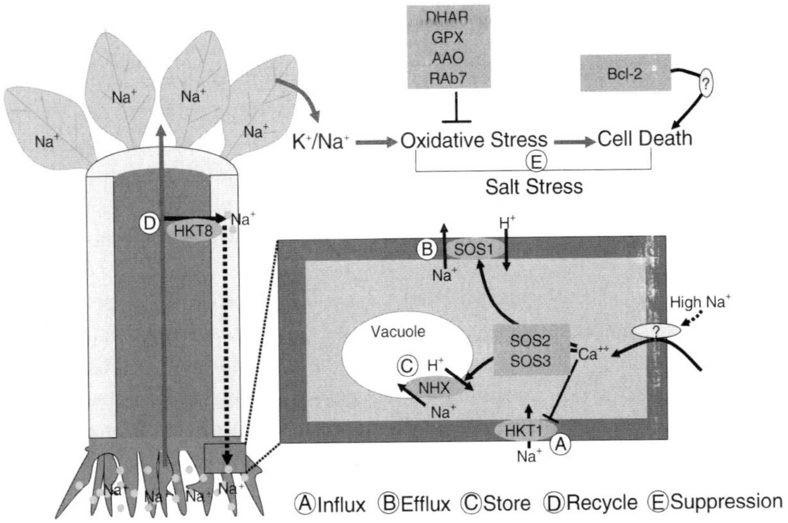

Figure 1. Mapping salt tolerance genes in the stress response network. Five major physiological processes critical for salt tolerance are highlighted as A, B, C, D, and E. Ca^{++}-mediated signals appear to regulate multiple responses. Representative genes involved in salt tolerance are depicted

in the vacuole is also a mechanism to avoid high Na^+ toxicity. Overexpression of a vacuole membrane anchored H^+/Na^+ antiporter, *AtNHX1*, has been reported to confer salt tolerance in various plant species (Apse, Aharon et al. 1999; Zhang and Blumwald 2001; Zhang, Hodson et al. 2001; He, Yan et al. 2005).

Once Na^+ enters a plant, it can be transported to the shoot via transpiration. The accumulation of high Na^+ in shoots lowers the K^+/Na^+ ratio and causes the imbalance of ion-homeostasis, triggering excessive production of reactive oxygen species (ROS) ultimately resulting in cell death. The toxicity of high levels of Na^+ in aerial tissues in the end manifests itself as necrotic lesions, stunted growth and wilting. The higher K^+/Na^+ ratio in salt tolerant plants and the characterization of *sas2* (sodium over-accumulation in shoots) mutants led to the speculation that Na^+ recycling from shoots to roots also plays a role in salt tolerance (Berthomieu, Conejero et al. 2003). Recently, a Na^+-selective transporter, *SKC1*, has been identified in rice to be partially responsible for salt tolerance in the resistant variety, Nona Bokro, and is suggested to play a role in recycling the Na^+ from shoots to roots (Ren, Gao et al. 2005).

Besides direct toxicity effects, high levels of Na^+ cause secondary damage to plants by promoting ROS generation (Chaparzadeh, D'Amico et al. 2004) and cell death (Huh, Damsz et al. 2002). Genes that suppress oxidative stress or inhibit cell death may potentially increase salt tolerance. The overexpression of the vesicle trafficking protein, Rab7, and several antioxidant regulators, such as dehydroascorbate reductase (DHAR), glutathione peroxidase (GPX) and ascorbate oxidase (AAO), have provided supportive evidence to the notion that suppression of oxidative stress enhances salt tolerance (Roxas, Smith et al. 1997; Roxas, Lodhi et al. 2000; Kwon, Choi et al. 2003; Yamamoto, Bhuiyan et al. 2005; Ushimaru 2006). Examples of cell death suppression to enhance salt tolerance can be found in the literature. The expression of a human anti-apoptotic protein, Bcl-2, in a transgenic yeast strain increased salt tolerance (Huh, Damsz et al. 2002). In contrast, the loss-of-function mutant, *shmt1* (disrupted in serine hydroxymethyltransferase involved in the photorespiratory pathway) displayed hypersensitivity to salt stress, showed greater accumulation of ROS and promoted cell death (Moreno, Martin et al. 2005).

The coordination of salt tolerance-related physiological processes appears to be critical for a plant to be able to adapt to high salinity conditions. How a plant senses high Na^+ concentrations in the soil and initiates signal transduction to activate a set of stress-responsive genes for salt tolerance has yet to be elucidated. Nevertheless, several genes that regulate signal transduction have been identified. For example, SOS3, a Ca^{++}-binding protein, appears to be involved in the regulation of SOS1, AtNHX1 and AtHKT1 activities. It likely senses high Na^+-induced Ca^{++} signaling and transmits it to downstream signaling components via interaction with SOS2, a Ser/Thr protein kinase (Xiong 2002). Loss-of-function mutants, *sos2* and *sos3*, both displayed hypersensitivity to salt stress. In addition, MAPK-mediated signaling cascades have also been suggested to play an important role in stress responses by sensing ROS and phosphorylation of regulatory proteins. Transgenic tobacco plants that constitutively expressed an active MAPKKK, *NPK1*, displayed better tolerance to salt and temperature shocks than wild-type plants (Kovtun, Chiu et al. 2000). Similarly, transgenic arabidopsis that overexpressed a MAP kinase kinase, *MKK2*, showed tolerance to salt and cold stresses (Teige, Scheikl et al. 2004). Finding critical components of signal transduction for salt tolerance may be the key for the genetic improvement of salt tolerance in crops and is a significant challenge for plant biologists.

3. QUANTITATIVE NATURE OF SALT TOLERANCE

The quantitative nature of salt tolerance has its roots in the physiological processes that involve multiple genes, each with a small and unknown effect. Quesada and colleagues investigated 102 arabidopsis wild-type accessions and 100 recombinant inbred lines (RILs) derived from a cross between the Col-4 and Landsberg *erecta* (L*er*-0) for variation in germination rate and growth rate under salt conditions (Quesada, Garcia-Martinez et al. 2002). Their results showed quantitative variation and the polygenic nature of salt tolerance during seed germination and vegetative growth. Three major QTLs were identified to be associated with salt tolerance at the germination stage; they are positioned on chromosomes 2 (19.1 cM) and 4 (48.7 and 72.2 cM). Three other QTLs, related to salt tolerance assessed for growth rate, are located on chromosomes 4 (54.5) and 5 (37.8 and 96.9 cM). No clear correlation between the tolerances of these two developmental stages was observed in either of the populations of natural accessions or RILs.

Even in salt cress, a halophyte that shows extreme tolerance to high salinity, salt tolerance seems to be inherited as a quantitative trait although no direct evidence could be obtained from a genetic experiment because of sexual barriers between halophyte and glycophyte species. Differential gene expression profiling between arabidopsis and salt cress, two plant species showing opposite responses to salt stress, in the presence and absence of high salt stress (150 or 250 mM NaCl), showed that 51 genes are highly expressed in salt cress but not in arabidopsis (3 fold difference cutoff used) without salt stress, and that 128 genes are steadily up-regulated in salt cress while down-regulated in arabidopsis in response to salt stress. Many genes showed differential down-regulation as well under salt stress. However, they are less convincing because arabidopsis oligo-microarray chips were used in the study and the sequence diversity between the two species might have caused low levels of hybridization signal in salt cress (Gong, Li et al. 2005). Morphological and biochemical comparison of the two species also suggests that multiple morphological features and physiological processes likely contribute to better salt tolerance in salt cress. The major site of Na^+ accumulation is in old leaves, followed by young leaves and taproots. Salt cress usually possesses more mature leaves and larger taproots than arabidopsis. The higher levels of activity of the tonoplast H^+-ATPase in salt-stressed leaves and roots of salt cress appear to contribute to better tolerance to salt by sequestration of Na^+ ions in vacuoles (Vera-Estrella, Barkla et al. 2005). All these quantitative variations may collectively determine the different levels of salt tolerance between arabidopsis and salt cress.

Molecular isolation of salt tolerance-associated QTLs would enhance our understanding of the quantitative nature of salt tolerance in plants. Molecular isolation of a major effective gene located in a QTL is currently feasible with the aid of high density genetic maps and completely or partially sequenced genomes of several plant species. The successful cloning of a QTL that controls fruit size in tomato was an early successful example of map-based QTL (Frary, Nesbitt et al. 2000). Recently, a major QTL, *SKC1*, responsible for salt tolerance in rice has been molecularly isolated by map-based cloning from an *indica* variety, Nona Bokra (Ren, Gao et al. 2005). *SKC1* encodes a Na^+-selective transporter that belongs to the *HKT* family. The SKC1 protein, encoded by a resistance allele, showed higher activity than the susceptible allele from the variety, Koshihikari. However, the difference does not account for all the phenotypic variations associated with the higher K^+/Na^+ ratio in shoots, and the higher K^+/Na^+ ratio itself does not completely account for the enhanced salt tolerance in Nona Bokro. It was previously shown that 8 QTLs were associated with salt tolerance in Nona Bokra and of them two major QTLs were responsible for the variations of K^+/Na^+ ratio between salt tolerant and susceptible varieties (Lin, Zhu et al. 2004). These examples demonstrate that with the assistance of available molecular genetic tools, it is feasible to identify and isolate major factors of a complex trait. However, it is still a big challenge to determine the minor, but equally important, factors without molecular manipulation of their function, *e.g.* either up- or down-regulation of gene expression to amplify their functional performance. Understanding of the molecular and physiological processes that are involved in salt tolerance will give us clues in the selection of candidate genes for molecular engineering of crop plants.

4. GENETIC MANIPULATION OF CROP PLANTS FOR SALT TOLERANCE

Manipulation of crop genotypes for better performance has been accomplished conventionally for centuries through selection and hybridization-breeding. In the past century, hybridization-breeding programs have achieved a great deal in terms of increasing crop yields. Breeders introduced many valuable genes from land races or closely related wild species into commercial cultivars through genetic crossing, which subsequently greatly improved crop resistance to biotic and abiotic stresses and enhanced the efficiency of fertilizer usage. For example, wheat yields were significantly increased because of the strong lodging resistance and the improved harvest index in the semi-dwarf wheat varieties developed through hybridization-breeding. This achievement has been widely recognized as the "green revolution". The wheat dwarf gene that led to the "green revolution" has been cloned and subsequently used to improve other crop species. Transgenic expression of the dwarf gene (*Rht-B1/Rht-D1*) in rice also reduced the height of stems (Peng, Richards et al. 1999).

Conventional breeding is still an effective strategy for new cultivar development, but has its limitations. A major gene-controlled trait can be readily transferred from a valuable germplasm into a related crop species through hybridization-breeding. However, for the improvement of quantitative traits, hybridization-breeding is less effective. Because of the genetic segregation of quantitative trait loci (QTL), trait transfer is usually incomplete, and the selection process is both time-consuming and labor-intensive. Furthermore, hybridization breeding cannot transfer traits between species when sexual barriers exist. The best germplasm for salt tolerance, found in halophytes, cannot be naturally crossed with today's crops. Attempts to improve the salt tolerance of crops via conventional hybridization-breeding have met with very limited success due to the complexity of inheritance and physiology of the tolerant trait (Flowers 2004).

The recent advances in biotechnology have provided us with opportunities to transfer useful genes between species. In the past decade, significant progress has been made using molecular approaches to manipulate crop plants. Through the introduction and expression of genes in plants, researchers are now poised to provide the world community with crop species tailored to grow more efficiently and with increased yields despite suboptimal geographic and/or climatic environments. These new approaches have the additional advantage of not being limited to one plant species, but instead being applicable to multiple plant species through genetic transformation. A gene can be functional across species, and even in different kingdoms of organisms. For instance, when a yeast protein (YCF1) was expressed in Arabidopsis, salt tolerance was significantly enhanced in the transgenic plants.

YCF1 is a member of the ATP-binding cassette (ABC) transporter family associated with multi-drug resistance. YCF1-mediated salt tolerance in arabidopsis is likely due to the sequestration of salt into vacuoles (Koh, Song et al. 2005). More examples of transgene-mediated salt tolerance listed in Table 1 show that a gene from one species can be used in another to improve the performance of transgenic plants under salt stress.

The identification of genes that confer salt tolerance is a prerequisite to the dramatic improvement of crop plants on saline soils. In the past several years, a number of genes have been identified and tested in various species and some of them have been shown to be effective in conferring salt tolerance to crop plants (Table 1). For example, overexpression of the arabidopsis tonoplast membrane H^+/Na^+ antiporter, AtNHX1, not only enhanced salt tolerance in arabidopsis (Apse, Aharon et al. 1999), but also in *Brassica napus* (Zhang, Hodson et al. 2001), tomato (Zhang and Blumwald 2001) and cotton (He, Yan et al. 2005). Similarly, overexpression of the gene, *SOS1*, encoding a plasma membrane anchored H^+/Na^+ protein also enhanced salt tolerance in arabidopsis (Shi, Lee et al. 2003). Despite these advances, there has not been a release of salt tolerant cultivars of commercial crops that is able to produce a relatively good yield on high salinity soils. This is probably due to the quantitative nature of salt tolerance, *i.e.* many genes each with a small effect that are involved in the myriad processes that suppress high Na^+-triggered stresses. The number of available genes that can be used to manipulate salt tolerance thus becomes a bottleneck for the development of salt tolerant cultivars.

5. HIGH THROUGHPUT SCREENS TO IDENTIFY SALT TOLERANCE GENES

5.1. Mutant Screens

Typically, variants of the wild-type are identified by screening mutagenized populations. Mutations can be generated by insertion mutagens (T-DNA or transposons) irradiation or chemical mutagens. Theoretically, every gene in the genome can be mutated and the resulting homozygous plants characterized. Mutant screens are a great approach for the identification of genes with loss-of-function or change-of-function phenotypes. As such, a species with strong salt tolerance, a halophyte, may be a better choice for mutagenesis followed by screens for sensitive mutants. However, most mutant screens for salt sensitivity have been performed in arabidopsis, a salt sensitive species, because of its tractability to molecular genetic approaches. The available genetic and genomics tools in arabidopsis greatly enhance our capability to do map-based cloning. High density oligo-microarray chips can even make gene identification possible (Mockler, Chan et al. 2005).

One of the earliest screens for salt tolerance was performed by Saleki and colleagues (Saleki, Young et al. 1993). They screened EMS-mutagenized Columbia for mutants that were tolerant to high salt stress at the germination stage. Three mutants, RS17, RS19 and RS20, were identified and characterized, but none of

Table 1. Genes reported to be involved in salt tolerance

Gene name and product	Gene source	Tested species	Gene manipulation	Salt tolerance	Reference
Reduced absorption of Na^+					
AtHKT1 (high affinity K^+ transporter)	Arabidopsis	Arabidopsis	T-DNA KO	$+^*$	(Rus, et al. 2001; Rus, et al. 2004)
HKT1 (high affinity K^+ transporter)	Wheat	Wheat	Antisense	+	(Laurie, et al. 2002)
Pumping out Na^+					
SOS1 (plasma membrane Na^+/H^+ antiporter)	Arabidopsis	Arabidopsis	Overexpression	+	(Shi, Lee et al. 2003)
Enhanced vacuole storage of Na^+					
YCF1 (ABC transporter)	Yeast	Arabidopsis	Overexpression	+	(Koh, et al. 2005)
SsVP (Vacuolar H^+-pyrophosphatase)	*Suaeda salsa*	Arabidopsis	Overexpression	+	(Guo, et al. 2006)
AtNHX1 (vacuolar Na^+/H^+ antiporter)	Arabidopsis	Arabidopsis *B. napus,* Cotton Tomato	Overexpression	+	(Apse, et al. 1999; Zhang and Blumwald 2001; Zhang, et al. 2001; He, et al. 2005)
GhNHX1 (vacuolar Na^+/H^+ antiporter)	Cotton	Tobacco	Overexpression	+	(Wu, et al. 2004)
NHX (vacuolar Na^+/H^+ antiporter)	*A. gmelini*	Rice	Overexpression	+	(Ohta, et al. 2002)
ACA4 (Vacuolar Ca^{2+}-ATPase)	Arabidopsis	Yeast	Overexpression	+	(Geisler, et al. 2000)
Na^+ recycling from shoots to roots					
SKC1 (Na^+-transporter)	Rice	Rice	Map-based cloning	+	(Ren, et al. 2005)
AtHKT1 (high affinity K^+ transporter)	Arabidopsis	Arabidopsis	Mutations	$-^{**}$	(Maser, et al. 2002; Berthomieu, et al. 2003)

Suppress oxidative stress and enhance salt tolerance					
DHAR (Dehydroascorbate reductase)	Rice Human	Arabidopsis Tobacco	Overexpression	+	(Ushimaru 2006) (Kwon, et al. 2003)
GmTP55 (Antiquitin-like protein)	Soybean	Arabidopsis Tobacco	Overexpression	+	(Rodrigues, et al. 2006)
AAO (ascorbate oxidase)	Tobacco Arabidopsis	Tobacco Arabidopsis	Antisense or T-DNA insertion	+	(Yamamoto, et al. 2005)
AtRab7 (vesicle trafficking protein)	Arabidopsis	Arabidopsis	Overexpression	+	(Mazel, et al. 2004)
GPX (glutathione peroxidase)	Tobacco	Tobacco	Overexpression	+	(Roxas, et al. 1997; Roxas, et al. 2000; Yoshimura, et al. 2004)
Suppression of cell death					
shmt1 (Serine hydroxymethyltransferase)	Arabidopsis	Arabidopsis	Mutant	–	(Moreno, et al. 2005)
Bcl-2 (Cell death suppressor)	Human	Yeast	Overexpression	+	(Huh, et al. 2002)
Transcription regulation					
CBF3	Arabidopsis	Rice	Overexpression	+	(Oh, et al. 2005)
JERF3	Tomato	Tobacco	Overexpression	+	(Wang, et al. 2004)
OsDREB1A	Rice	Arabidopsis	Overexpression	+	(Dubouzet, et al. 2003)
SNAC1	Rice	Rice	Overexpression	+	(Hu, et al. 2006)
STO	Arabidopsis	Arabidopsis	Overexpression	+	(Nagaoka and Takano 2003)
ABF2	Arabidopsis	Arabidopsis	Overexpression	+	(Kim 2003)
Tsi1	Tobacco	Tobacco	Overexpression	+	(Park, et al. 2001)
Signal transduction					
MKK2 (Mitogen-activated protein kinase kinase 2	Arabidopsis	Arabidopsis	Overexpression	+	(Teige, et al. 2004)
AtCaMBP25 (Calmodulin binding protein)	Arabidopsis	Arabidopsis	Antisense expression	+	(Perruc, et al. 2004)
NPK1 (MAPKKK)	Tobacco	Arabidopsis	Constitutively activation	+	(Kovtun, et al. 2000)

(*Continued*)

Table 1. (*Continued*)

Gene name and product	Gene source	Tested species	Gene manipulation	Salt tolerance	Reference
BvCKA2 (protein kinase CK2)	Sugar beet	Yeast	Overexpression	+	(Kanhonou, et al. 2001)
SOS2 (Ser/Thr kinase)	Arabidopsis	Arabidopsis	Mutation	–	(Liu, et al. 2000)
SOS3 (Ca^{++}-sensor)	Arabidopsis	Arabidopsis	Mutation	–	(Liu and Zhu 1998)
Unknown Mechanisms					
PcSrp (Serine-rich-protein)	*Porteresia coarctata*	Yeast Finger Millet	Overexpression	+	(Mahalakshmi, et al. 2006)
XTH	Hot pepper	Arabidopsis	Overexpression	+	(Cho, et al. 2006)
PDH45 (DNA helicase)	Pea	Tobacco	Overexpression	+	(Sanan-Mishra, et al. 2005)
OsISAP1 (Zinc-finger protein)	Rice	Tobacco	Overexpression	+	(Mukhopadhyay, et al. 2004)
BADH (betaine aldehyde dehydrogenase)	*Daucus carota*	*Daucus carota*	Overexpression	+	(Kumar, et al. 2004)
ALDH (Aldehyde dehydrogenase)	Arabidopsis	Arabidopsis	Overexpression	+	(Sunkar, et al. 2003)
BvelF1A (Translation initiation factor)	Sugar beet	Yeast	Overexpression	+	(Rausell, et al. 2003)
TPSP (Trehalose production)	*E. coli*	Rice	Overexpression	+	(Garg, et al. 2002)
CDH (Choline dehydrogenase)	*E. coli*	Tobacco	Overexpression	+	(Saijo, et al. 2000)

* Reduced accumulation of Na^{+} in roots

** Increased Na^{+} accumulation in shoots

them showed enhanced salt tolerance in later stages of development as compared to wild-type. A few similar direct mutant screens performed in arabidopsis later also turned out to have achieved limited success (Werner 1995; Quesada, Ponce et al. 2000). This is likely because loss-of-function mutations usually cause even greater sensitivity to salinity stress. Thus, salt tolerance mutants cannot be easily detected in direct screens at the vegetative growth stage. To overcome this problem, Zhu and colleagues engineered a transgenic arabidopsis line by using luciferase as a reporter to visualize salt stress responses (Chinnusamy, Stevenson et al. 2002). A luciferase gene was fused with a stress-responsive promoter, *RD29A*, and an arabidopsis transgenic line with a single copy of the transgene was mutagenized. Under salt stress, the mutants which showed either enhanced or reduced expression of the reporter gene, as an indication of increased salt tolerance or over-sensitivity to salt, were selected. In these screens, several salt-overly-sensitive mutants were isolated (Zhu, Liu et al. 1998). Their function in salt tolerance was validated by molecular complementation. The transgenic arabidopsis plants that constitutively express *SOS1* under the 35S promoter displayed enhanced salt tolerance (Shi, Lee et al. 2003).

An EMS-induced mutant screen system has also been developed using salt cress (Bressan, Zhang et al. 2001). The ability of salt cress to withstand salinity stress is much better than arabidopsis as salt cress can tolerate salinity shock up to 500 mM NaCl, equivalent to sea water. Salinity sensitive mutants, due to the loss-of-function of salt tolerance genes, would be potentially identified under high salt stress conditions. However, the cloning of the mutated genes will be challenging because of the lack of genome sequence and genetic markers.

5.2. A High Throughput Misexpression Approach

Most of the genes known to be important in salt tolerance display a positive function, *i.e.* overexpression of the gene enhances salt tolerance. Several such examples include *AtNHX1, SOS1, CBF1*, *DREB1* and *Tsi1* (Table 1). As such, a high throughput screen of a population of lines overexpressing thousands of unique genes in model plants such as arabidopsis or rice appears to offer huge potential for the identification of salt tolerance genes. The selection of candidate genes that are putatively functional in salt tolerance when overexpressed is the first step in this approach if there is a need to prioritize genes for cloning.

A large number of salinity-induced genes have been identified by comparing the gene expression profiles in salt-treated and -untreated plants. Various experiments such as mRNA differential display (Park, Park et al. 2001; Shiozaki, Yamada et al. 2005), cDNA-AFLP (Chen, Ma et al. 2003) and cDNA or oligonucleotide microarray-hybridization (Kreps, Wu et al. 2002; Gu, Fonseca et al. 2004; Taji, Seki et al. 2004; Gong, Li et al. 2005; Walia, Wilson et al. 2005) have been described for identifying salt stress-induced genes. Bohnert and colleagues (Gong, Li et al. 2005) recently reported a group of genes that are highly expressed in salt cress, but not in arabidopsis after salt treatment. Functional characterization of this group of salt responsive genes with a strong constitutive or stress-induced promoter will likely shed light on the mechanisms of salt tolerance in halophytes.

The number of genes induced by salinity can be enormous due to the secondary effects of salt stress. A large fraction of salt stress-induced genes might be effectors that are activated by oxidative stress or cell death; they may not directly contribute to salt stress tolerance. There is no reliable way to predict causal genes from secondary effectors. The number of candidate genes that need to be tested can be huge. Therefore, an efficient transformation system is required to generate a population of overexpression lines which contains all of the novel genes from a single species as well as genes from other species that appear to be important in stress tolerance. Arabidopsis is one of the easiest plant species to transform with agrobacterium. This along with its short life cycle, small stature and sensitivity to salt stress make it nearly ideal for salt tolerance screens via overexpression of single genes. At Ceres, we have already overexpressed more than 15,000 unique, mostly arabidopsis, genes. This collection of arabidopsis transgenic lines has been extensively screened for various phenotypes such as high yield, flowering time, high or low temperature, and nitrogen, light, drought, salt tolerances, etc.

Several hundred transgenic seeds can be sown on one agar plate (150 mm in diameter) containing sodium chloride. In two to three weeks, the stress tolerance phenotypes are obvious to the unaided eye, *i.e.* wild-type plants show necrotic lesions and stunted growth and die, while the tolerant transgenic plants display continued growth with fewer necrotic lesions. Although most of the tolerance phenotypes observed in transgenic plants potentially result from the overexpression of a transgene, T-DNA insertion-caused knockouts of endogenous genes (easy to identify), overexpression-caused co-suppression of a homolog and expression of small RNA-caused interference may contribute to salt tolerance as well. Therefore, the term, misexpression (ME), is assigned to this high throughput genetic engineering pipeline.

6. DEVELOMENT OF MISEXPRESSION RESOURCES FOR HIGH THROUGHPUT SCREENS

6.1. Gene Selection and Vector Construction

Molecular isolation of full-length genes for misexpression vector construction is the first step in this pipeline. Full-length cDNA cloning is our first choice for isolation of candidate genes to be used in misexpression vectors, since the synthesis of cDNAs does not require genomic information and can be done in any species. Production of high quality cDNA libraries is the key for the isolation of full-length cDNAs from species where genomic sequence is not yet known. A few processes, such as extraction of undegraded mRNA, complete reverse transcription, highly efficient ligation and transformation efficiency determine the quality of a cDNA library. Several biotech companies such as Promega, Invitrogen, and Vitrotechlabs etc., provide ingredients, or services for full-length cDNA library construction. For the identification of cDNAs in a targeted gene approach, PCR with gene specific primers, colony hybridization using a gene specific probe, or high throughput sequencing can be performed. At Ceres, full-length cDNA libraries of arabidopsis and other plant species have been extensively sequenced and tens of thousands of full-length cDNAs constitute the rich gene resources for construction of misexpression vectors. These sequences along with the complete genome sequence of several plant species provides a plentiful set of genes for the overexpression pipeline.

In recent years, microRNAs have been shown to play important roles in plant growth and development and stress tolerance (Borsani, Zhu et al. 2005). A survey of the arabidopsis genome and ESTs has yielded 117 miRNAs. In rice, 175 miRNAs have been identified (Zhang, Pan et al. 2006). Overexpression of these miRNAs and characterization of their phenotypes would shed light on their biological functions. At Ceres, large numbers of vectors overexpressing full-length cDNAs, genomic CDSs or miRNAs from arabidopsis, rice, wheat, corn, soybean, or oilseed rape, etc., have been constructed and transformed into arabidopsis for various biological function screens and product development.

6.2. Plant Transformation and Screen Development

For high throughput functional screens, wild-type arabidopsis (Wassilewskija) plants were transformed with agrobacteria containing single misexpression constructs. Five or six independent transformants were selected for each construct in the T_1 generation, and the seeds were harvested and stored individually. An aliquot of seeds from the individual transgenic events of the same construct were pooled to make a masterpool for each line (construct). An equal portion of seeds from 100 masterpools was mixed to make a superpool. The superpool seeds were then bulked under ideal growth conditions to generate an essentially unlimited supply of T_3 seeds for each superpool for various screens.

A plate screen was developed to identify genes involved in salt tolerance. A dose response test showed that 150 mM NaCl was optimal for superpool screens. 2,500 seeds from each superpool (this number represents a >95% probability of screening every genotype in each superpool (500 events)) were sown on MS agar plates containing 150 mM NaCl. After stratification at 4°C for 3 days, the plates were placed to a growth chamber at 22°C, 16:8 hour light:dark cycle, 70% humidity and light intensity of $\sim$100 μEinsteins. Two-week-old seedlings were screened for salt tolerance along with non-transgenic seedlings as controls. Two formats were employed for seed plating, manual plating or robotic (COPAS robot, Union Biometrica), and each showed distinctive advantages. Manual plating allows more seeds to be sown on a single plate. The disadvantage is the uneven plant density and generation of more frequent false-positive candidates. A secondary screen is necessary to confirm the selected candidates from this screening method. The robot can sow seeds that are evenly spaced on the plates. The candidates selected from these plates are more reliable. However, it takes much more time to plate the

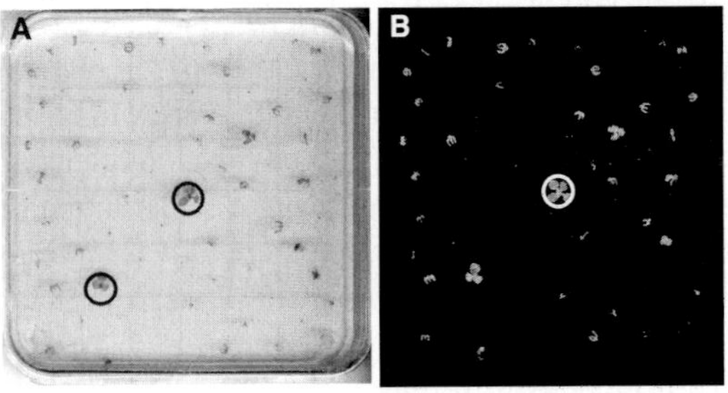

Figure 2. A representative plate of a superpool screen for salt tolerance. Transgenic superpool seeds were sterilized and sowed by a robot on a plate containing agar-solidified MS medium supplemented with 150 mM NaCl. The 2 week-old seedlings were scanned with an EPSON color scanner (A) and chlorophyll fluorescence imager (B). The circled plants were marked as positive candidates

large number of superpool seeds. To further ensure that high quality candidates are selected, we use both a visual screen and a chlorophyll fluorescence screen. In the visual screen, the plants that are larger than controls and show continuous growth beyond two weeks are marked as candidates. In the chlorophyll fluorescence screen, red fluorescence, an indication of higher photosynthetic activity, is used as a marker for salt tolerance. Candidates are selected if they show a larger size and higher photosynthetic activity (Figure 2). To further eliminate false-positive candidates, the selected plants are transferred to soil for two weeks and treated with 200 mM NaCl solution for 48 hours. In this way, additional false-positive plants are eliminated from the candidate population. Leaf tissue from the remaining tolerant candidates is harvested to isolate DNA for PCR amplification of the transgene and identification of the ME line. Thus far, we have screened and analyzed 104 superpools consisting of >10,000 ME lines. Salt tolerant candidate plants were identified for ~300 unique ME lines.

7. IDENTIFICATION OF GENES THAT ENHANCE SALT TOLERANCE VIA MISEXPRESSION

7.1. Identification of Misexpression Lines with Increased Salt Tolerance

In superpool screens, candidates are selected based on individual plant performance because the salt tolerance phenotype could result from various effects, such as transgene expression, T-DNA insertion, seed quality or random variation caused by environmental conditions. To distinguish the salt tolerance phenotypes caused by transgene expression from those resulting from other effects is a critical process in this screen. Two measures are performed in this process; first, we look for multiple independent transgenic events that show similar phenotypes, and secondly, we compare transgenic events with wild-type based on population performance. A prevalidation assay was developed to further eliminate false-positives. In the prevalidation assay, the five or six independent transgenic events of a candidate ME line are compared to wild-type Ws on the MS agar plates containing 150 mM NaCl based on population performances (Figure 3). ME lines that have two or more independent transgenic events performing significantly better than the non-transgenic control and segregating in a dominant ratio for salt tolerance, are selected for further characterization. More than 30 ME lines passed the prevalidation assay and were subjected to further confirmation of the transgene function in salt tolerance.

7.2. Confirmation of the Function of the Transgene in salt Tolerance

To confirm that a transgene enhanced salt tolerance, a validation assay was developed to investigate the association of salt tolerance phenotypes with the corresponding transgene. For each positive ME line, transgenic plants were compared with the internal non-transgenic segregants, as well as external wild-type controls on the same salt plate (Figure 4). Segregating populations of T_2 and T_3 seeds of a positive ME line were assayed to verify the function of the transgene in salt tolerance. For each event by generation, two independent replicates (36 seedlings each) were analyzed. Seedling area and photosynthetic efficiency (Fv/Fm) were measured on each plant at 1–2 weeks after seed sowing. Salt growth index (SGI) = seedling area (cm^2) x photosynthesis efficiency (Fv/Fm) was calculated to reflect the growth rate under high salt stress. After the analysis of salt tolerance phenotypes, the seedlings were recovered on normal growth media and the transgenic status of each plant was determined. The statistical analysis of SGI variations between transgenic plants and the internal non-transgenic segregants, as well as the wild-type controls was performed to investigate the significance of transgene-mediated salt tolerance.

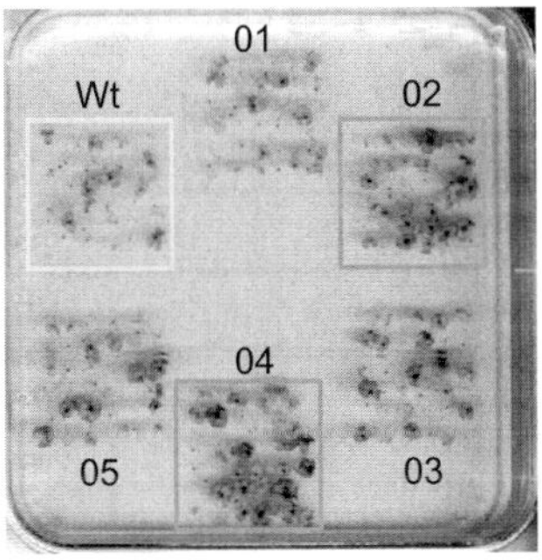

Figure 3. Prevalidation of a misexpression line for salt tolerance. Five independent transgenic events of a misexpression line and the corresponding Ws wild-type control (yellow-squared) were grown on agar-solidified MS medium supplemented with 150 mM NaCl. The green-squared events were selected as positive candidates. The image was taken 2 weeks after sowing seeds

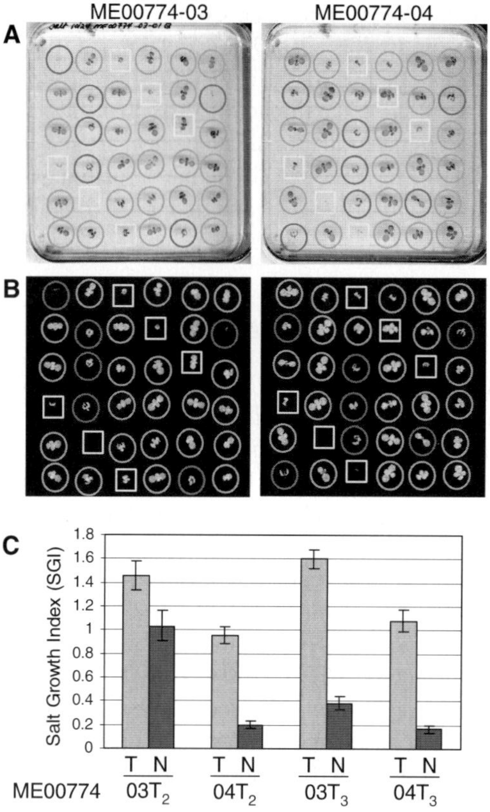

Figure 4. Association between transgene expression and salt tolerance. The plants were grown on MS agar medium containing 150 mM salt for 2 weeks and scanned using an EPSON color scanner (A) and chlorophyll fluorescence imager (B). The squared plants represent wild-type controls; the red-circled plants are non-transgenic controls, and the green-circled plants are transgenics. The salt growth index (SGI) = seedling area X photosynthesis efficiency (Fv/Fm) was calculated for each plant. The bar represents the average value +/- standard error of SGI for transgenic plants (T) or pooled non-transgenic plants (N) that include 6 wt Ws controls (C). 03 and 04 represent the third and 4th events of ME00774. Two plates were used as independent replicates for each event per generation

Several genes that enhance salt tolerance when overexpressed were confirmed with validation assays (Table 2). The transgene-encoded proteins belong to calmodulin, calmodulin-binding, Zinc-finger, putative cyclase, stress-related protein families and unknown functions. The novel Ca^{++}-signaling components identified in this misexpression screen further demonstrated that Ca^{++}-mediated signaling pathways likely coordinate the physiological processes leading to enhanced salt tolerance potentially through the regulation of *AtNHX1* and *SOS1* genes as suggested by Zhu and colleagues (Xiong 2002). The zinc finger protein identified in this screen is identical to ZAT12 that was shown before to suppress oxidative stress and to

Table 2. Ceres-identified salt tolerance genes (*CST*)

Misexpression line	Gene name	Putative gene function	Gene source organism
ME03807	*CST* 1	Cyclase	Arabidopsis
ME02064	*CST* 2	Calmodulin binding protein	*Zea mays*
ME02907	*CST* 3	Calmodulin	Arabidopsis
ME04074	*CST* 4	Shikimate kinase precursor	Arabidopsis
ME00774	*CST* 5	Universal stress protein family	Arabidopsis
ME01142	*CST* 6 (ZAT12)	Zinc finger protein transcription factor	Arabidopsis
ME01468	*CST* 7	Similar to an oxygen evolving complex in rice	Arabidopsis
ME00199	*CST* 8	Steroid sulfotransferase	Arabidopsis
ME09814	*CST* 9	Novel protein	*Brassica napus*
ME09090	*CST* 10	Novel protein	Arabidopsis

enhance salt tolerance when overexpressed (Davletova, Schlauch et al. 2005). The molecular mechanisms of each gene in regulation of salt tolerance need to be further investigated.

8. CONCLUSIONS

The high throughput screen of misexpression lines has turned out to be efficient for the identification of genes with quantitative effects on salt tolerance. In addition to genes shown in table 2, we have more candidates awaiting validation. Similarly, many more superpools, containing thousands of additional unique genes, need be screened. Characterization of these genes and their interactions with the known signaling components as listed in Table 1 will hopefully help us to understand the mode of action of each gene in salt tolerance. Based on their distinctive roles in salt tolerance (as illustrated in Figure 1), representative genes important for each process will be selected for stacking. By pyramiding these genes in a single genetic background, we expect to develop a super-tolerant transgenic plant that can be grown under high salt stress condition. The knowledge gained from investigations of model plant species promises to speed up the process of engineering crop cultivars that are suitable for high saline soils.

REFERENCES

Apse, M. P., G. S. Aharon, et al. (1999). "Salt tolerance conferred by overexpression of a vacuolar Na+/H+ antiport in Arabidopsis." *Science* **285**(5431): 1256–8.

Berthomieu, P., G. Conejero, et al. (2003). "Functional analysis of AtHKT1 in Arabidopsis shows that Na(+) recirculation by the phloem is crucial for salt tolerance." *Embo J* **22**(9): 2004–14.

Borsani, O., J. Zhu, et al. (2005). "Endogenous siRNAs derived from a pair of natural cis-antisense transcripts regulate salt tolerance in Arabidopsis." *Cell* **123**(7): 1279–91.

Bressan, R. A., C. Zhang, et al. (2001). "Learning from the Arabidopsis experience. The next gene search paradigm." *Plant Physiol* **127**(4): 1354–60.

Chaparzadeh, N., M. L. D'Amico, et al. (2004). "Antioxidative responses of Calendula officinalis under salinity conditions." *Plant Physiol Biochem* **42**(9): 695–701.

Chen, G. P., W. S. Ma, et al. (2003). "[Isolation and characterization of SIR73 gene fragment in salt resistant mutant of wheat involved in salt stress]." *Yi Chuan* **25**(2): 173–6.

Chinnusamy, V., B. Stevenson, et al. (2002). "Screening for gene regulation mutants by bioluminescence imaging." *Sci STKE* **2002**(140): PL10.

Cho, S. K., J. E. Kim, et al. (2006). "Constitutive expression of abiotic stress-inducible hot pepper CaXTH3, which encodes a xyloglucan endotransglucosylase/hydrolase homolog, improves drought and salt tolerance in transgenic Arabidopsis plants." *FEBS Lett* **580**(13): 3136–44.

Davletova, S., K. Schlauch, et al. (2005). "The zinc-finger protein Zat12 plays a central role in reactive oxygen and abiotic stress signaling in Arabidopsis." *Plant Physiol* **139**(2): 847–56.

Dubouzet, J. G., Y. Sakuma, et al. (2003). "OsDREB genes in rice, Oryza sativa L., encode transcription activators that function in drought-, high-salt- and cold-responsive gene expression." *Plant J* **33**(4): 751–63.

Flowers, T. J. (2004). "Improving crop salt tolerance." *J Exp Bot* **55**(396): 307–19.

Frary, A., T. C. Nesbitt, et al. (2000). "fw2.2: a quantitative trait locus key to the evolution of tomato fruit size." *Science* **289**(5476): 85–8.

Garg, A. K., J. K. Kim, et al. (2002). "Trehalose accumulation in rice plants confers high tolerance levels to different abiotic stresses." *Proc Natl Acad Sci U S A* **99**(25): 15898–903.

Geisler, M., N. Frangne, et al. (2000). "The ACA4 gene of Arabidopsis encodes a vacuolar membrane calcium pump that improves salt tolerance in yeast." *Plant Physiol* **124**(4): 1814–27.

Gong, Q., P. Li, et al. (2005). "Salinity stress adaptation competence in the extremophile Thellungiella halophila in comparison with its relative Arabidopsis thaliana." *Plant J* **44**(5): 826–39.

Gu, R., S. Fonseca, et al. (2004). "Transcript identification and profiling during salt stress and recovery of Populus euphratica." *Tree Physiol* **24**(3): 265–76.

Guo, S., H. Yin, et al. (2006). "Molecular cloning and characterization of a vacuolar H+ -pyrophosphatase gene, SsVP, from the halophyte Suaeda salsa and its overexpression increases salt and drought tolerance of Arabidopsis." *Plant Mol Biol* **60**(1): 41–50.

He, C., J. Yan, et al. (2005). "Expression of an Arabidopsis vacuolar sodium/proton antiporter gene in cotton improves photosynthetic performance under salt conditions and increases fiber yield in the field." *Plant Cell Physiol* **46**(11): 1848–54.

Hu, H., M. Dai, et al. (2006). "Overexpressing a NAM, ATAF, and CUC (NAC) transcription factor enhances drought resistance and salt tolerance in rice." *Proc Natl Acad Sci USA*.

Huh, G. H., B. Damsz, et al. (2002). "Salt causes ion disequilibrium-induced programmed cell death in yeast and plants." *Plant J* **29**(5): 649–59.

Kanhonou, R., R. Serrano, et al. (2001). "A catalytic subunit of the sugar beet protein kinase CK2 is induced by salt stress and increases NaCl tolerance in Saccharomyces cerevisiae." *Plant Mol Biol* **47**(5): 571–9.

Kim, Y. S., KS. Zhu, JK (2003). "EMS mutagenesis of Arabidopsis." *Methods in Molecular Biology* **323**.

Koh, E. J., W. Y. Song, et al. (2005). "Expression of yeast cadmium factor 1 (YCF1) confers salt tolerance to Arabidopsis thaliana." *Plant Science* **170**(3): 534–541.

Kovtun, Y., W. L. Chiu, et al. (2000). "Functional analysis of oxidative stress-activated mitogen-activated protein kinase cascade in plants." *Proc Natl Acad Sci USA* **97**(6): 2940–5.

Kreps, J. A., Y. Wu, et al. (2002). "Transcriptome changes for Arabidopsis in response to salt, osmotic, and cold stress." *Plant Physiol* **130**(4): 2129–41.

Kumar, S., A. Dhingra, et al. (2004). "Plastid-expressed betaine aldehyde dehydrogenase gene in carrot cultured cells, roots, and leaves confers enhanced salt tolerance." *Plant Physiol* **136**(1): 2843–54.

Kwon, S. Y., S. M. Choi, et al. (2003). "Enhanced stress-tolerance of transgenic tobacco plants expressing a human dehydroascorbate reductase gene." *J Plant Physiol* **160**(4): 347–53.

Laurie, S., K. A. Feeney, et al. (2002). "A role for HKT1 in sodium uptake by wheat roots." *Plant J* **32**(2): 139–49.

Lin, H. X., M. Z. Zhu, et al. (2004). "QTLs for Na+ and K+ uptake of the shoots and roots controlling rice salt tolerance." *Theor Appl Genet* **108**(2): 253–60.

Liu, J., M. Ishitani, et al. (2000). "The Arabidopsis thaliana SOS2 gene encodes a protein kinase that is required for salt tolerance." *Proc Natl Acad Sci USA* **97**(7): 3730–4.

Liu, J. and J. K. Zhu (1998). "A calcium sensor homolog required for plant salt tolerance." *Science* **280**(5371): 1943–5.

Mahalakshmi, S., G. S. Christopher, et al. (2006). "Isolation of a cDNA clone (PcSrp) encoding serine-rich-protein from Porteresia coarctata T. and its expression in yeast and finger millet (Eleusine coracana L.) affording salt tolerance." *Planta*: 1–13.

Maser, P., B. Eckelman, et al. (2002). "Altered shoot/root Na+ distribution and bifurcating salt sensitivity in Arabidopsis by genetic disruption of the Na+ transporter AtHKT1." *FEBS Lett* **531**(2): 157–61.

Mazel, A., Y. Leshem, et al. (2004). "Induction of salt and osmotic stress tolerance by overexpression of an intracellular vesicle trafficking protein AtRab7 (AtRabG3e)." *Plant Physiol* **134**(1): 118–28.

Mockler, T. C., S. Chan, et al. (2005). "Applications of DNA tiling arrays for whole-genome analysis." *Genomics* **85**(1): 1–15.

Moreno, J. I., R. Martin, et al. (2005). "Arabidopsis SHMT1, a serine hydroxymethyltransferase that functions in the photorespiratory pathway influences resistance to biotic and abiotic stress." *Plant J* **41**(3): 451–63.

Mukhopadhyay, A., S. Vij, et al. (2004). "Overexpression of a zinc-finger protein gene from rice confers tolerance to cold, dehydration, and salt stress in transgenic tobacco." *Proc Natl Acad Sci USA* **101**(16): 6309–14.

Nagaoka, S. and T. Takano (2003). "Salt tolerance-related protein STO binds to a Myb transcription factor homologue and confers salt tolerance in Arabidopsis." *J Exp Bot* **54**(391): 2231–7.

Oh, S. J., S. I. Song, et al. (2005). "Arabidopsis CBF3/DREB1A and ABF3 in transgenic rice increased tolerance to abiotic stress without stunting growth." *Plant Physiol* **138**(1): 341–51.

Ohta, M., Y. Hayashi, et al. (2002). "Introduction of a Na+/H+ antiporter gene from Atriplex gmelini confers salt tolerance to rice." *FEBS Lett* **532**(3): 279–82.

Park, J. M., C. J. Park, et al. (2001). "Overexpression of the tobacco Tsi1 gene encoding an EREBP/AP2-type transcription factor enhances resistance against pathogen attack and osmotic stress in tobacco." *Plant Cell* **13**(5): 1035–46.

Peng, J., D. E. Richards, et al. (1999). "'Green revolution' genes encode mutant gibberellin response modulators." *Nature* **400**(6741): 256–61.

Perruc, E., M. Charpenteau, et al. (2004). "A novel calmodulin-binding protein functions as a negative regulator of osmotic stress tolerance in Arabidopsis thaliana seedlings." *Plant J* **38**(3): 410–20.

Quesada, V., S. Garcia-Martinez, et al. (2002). "Genetic architecture of NaCl tolerance in Arabidopsis." *Plant Physiol* **130**(2): 951–63.

Quesada, V., M. R. Ponce, et al. (2000). "Genetic analysis of salt-tolerant mutants in Arabidopsis thaliana." *Genetics* **154**(1): 421–36.

Rausell, A., R. Kanhonou, et al. (2003). "The translation initiation factor eIF1A is an important determinant in the tolerance to NaCl stress in yeast and plants." *Plant J* **34**(3): 257–67.

Ren, Z. H., J. P. Gao, et al. (2005). "A rice quantitative trait locus for salt tolerance encodes a sodium transporter." *Nat Genet* **37**(10): 1141–6.

Rengasamy, P. (2006). "World salinization with emphasis on Australia." *J Exp Bot* **57**(5): 1017–23.

Rodrigues, S. M., M. O. Andrade, et al. (2006). "Arabidopsis and tobacco plants ectopically expressing the soybean antiquitin-like ALDH7 gene display enhanced tolerance to drought, salinity, and oxidative stress." *J Exp Bot*.

Roxas, V. P., S. A. Lodhi, et al. (2000). "Stress tolerance in transgenic tobacco seedlings that overexpress glutathione S-transferase/glutathione peroxidase." *Plant Cell Physiol* **41**(11): 1229–34.

Roxas, V. P., R. K. Smith, Jr., et al. (1997). "Overexpression of glutathione S-transferase/glutathione peroxidase enhances the growth of transgenic tobacco seedlings during stress." *Nat Biotechnol* **15**(10): 988–91.

Rus, A., B. H. Lee, et al. (2004). "AtHKT1 facilitates Na+ homeostasis and K+ nutrition in planta." *Plant Physiol* **136**(1): 2500–11.

Rus, A., S. Yokoi, et al. (2001). "AtHKT1 is a salt tolerance determinant that controls Na(+) entry into plant roots." *Proc Natl Acad Sci USA* **98**(24): 14150–5.

Saijo, Y., S. Hata, et al. (2000). "Over-expression of a single Ca2+-dependent protein kinase confers both cold and salt/drought tolerance on rice plants." *Plant J* **23**(3): 319–27.

Saleki, R., P. G. Young, et al. (1993). "Mutants of Arabidopsis thaliana Capable of Germination under Saline Conditions." *Plant Physiol* **101**(3): 839–845.

Sanan-Mishra, N., X. H. Pham, et al. (2005). "Pea DNA helicase 45 overexpression in tobacco confers high salinity tolerance without affecting yield." *Proc Natl Acad Sci U S A* **102**(2): 509–14.

Schoups, G., J. W. Hopmans, et al. (2005). "Sustainability of irrigated agriculture in the San Joaquin Valley, California." *Proc Natl Acad Sci USA* **102**(43): 15352–6.

Shi, H., M. Ishitani, et al. (2000). "The Arabidopsis thaliana salt tolerance gene SOS1 encodes a putative Na+/H+ antiporter." *Proc Natl Acad Sci USA* **97**(12): 6896–901.

Shi, H., B. H. Lee, et al. (2003). "Overexpression of a plasma membrane Na+/H+ antiporter gene improves salt tolerance in Arabidopsis thaliana." *Nat Biotechnol* **21**(1): 81–5.

Shiozaki, N., M. Yamada, et al. (2005). "Analysis of salt-stress-inducible ESTs isolated by PCR-subtraction in salt-tolerant rice." *Theor Appl Genet* **110**(7): 1177–86.

Sunkar, R., D. Bartels, et al. (2003). "Overexpression of a stress-inducible aldehyde dehydrogenase gene from Arabidopsis thaliana in transgenic plants improves stress tolerance." *Plant J* **35**(4): 452–64.

Taji, T., M. Seki, et al. (2004). "Comparative genomics in salt tolerance between Arabidopsis and a Rabidopsis-related halophyte salt cress using Arabidopsis microarray." *Plant Physiol* **135**(3): 1697–709.

Teige, M., E. Scheikl, et al. (2004). "The MKK2 pathway mediates cold and salt stress signaling in Arabidopsis." *Mol Cell* **15**(1): 141–52.

Ushimaru, T. N., T. Fujioka, Y. etc. (2006). "Transgenic Arabidopsis plants expressing the rice dehydroascorbate reductase gene are resistant to salt stress." *J Plant Physiol.*

Vera-Estrella, R., B. J. Barkla, et al. (2005). "Salt stress in Thellungiella halophila activates Na+ transport mechanisms required for salinity tolerance." *Plant Physiol* **139**(3): 1507–17.

Walia, H., C. Wilson, et al. (2005). "Comparative transcriptional profiling of two contrasting rice genotypes under salinity stress during the vegetative growth stage." *Plant Physiol* **139**(2): 822–35.

Wang, B., R. J. Davenport, et al. (2006). "Low unidirectional sodium influx into root cells restricts net sodium accumulation in Thellungiella halophila, a salt-tolerant relative of Arabidopsis thaliana." *J Exp Bot* **57**(5): 1161–70.

Wang, H., Z. Huang, et al. (2004). "Ectopic overexpression of tomato JERF3 in tobacco activates downstream gene expression and enhances salt tolerance." *Plant Mol Biol* **55**(2): 183–92.

Werner, J. F., RR. (1995). "Arabidopsis mutants with reduced response to NaCl and osmotic stress. Physiol. Plant. 93, 659–666." *Physiol. Plant.* **93**: 659–666.

Wu, C. A., G. D. Yang, et al. (2004). "The cotton GhNHX1 gene encoding a novel putative tonoplast Na(+)/H(+) antiporter plays an important role in salt stress." *Plant Cell Physiol* **45**(5): 600–7.

Xiong, L. Z., JK. (2002). "Salt Tolerance." *The Arabidopsis Book* **24**(1, DOI: 10.1199/tab.0048): 1–22.

Yamamoto, A., M. N. Bhuiyan, et al. (2005). "Suppressed expression of the apoplastic ascorbate oxidase gene increases salt tolerance in tobacco and Arabidopsis plants." *J Exp Bot* **56**(417): 1785–96.

Yoshimura, K., K. Miyao, et al. (2004). "Enhancement of stress tolerance in transgenic tobacco plants overexpressing Chlamydomonas glutathione peroxidase in chloroplasts or cytosol." *Plant J* **37**(1): 21–33.

Zhang, B., X. Pan, et al. (2006). "Conservation and divergence of plant microRNA genes." *Plant J* **46**(2): 243–59.

Zhang, H. X. and E. Blumwald (2001). "Transgenic salt tolerant tomato plants accumulate salt in foliage but not in fruit." *Nat Biotechnol* **19**(8): 765–8.
Zhang, H. X., J. N. Hodson, et al. (2001). "Engineering salt-tolerant Brassica plants: characterization of yield and seed oil quality in transgenic plants with increased vacuolar sodium accumulation." *Proc Natl Acad Sci USA* **98**(22): 12832–6.
Zhu, J. K., J. Liu, et al. (1998). "Genetic analysis of salt tolerance in arabidopsis. Evidence for a critical role of potassium nutrition." *Plant Cell* **10**(7): 1181–91.

CHAPTER 16

DISSECTING QTLS FOR TOLERANCE TO DROUGHT AND SALINITY

ROBERTO TUBEROSA* AND SILVIO SALVI

Department of Agroenvironmental Sciences and Technology, Viale Fanin 44, 40127 Bologna, Italy

E-mail: roberto.tuberosa@unibo.it

Abstract: Compared to conventional breeding approaches, the dissection of the genetic basis of quantitative traits into their single components (i.e. Quantitative Trait Loci: QTLs) provides a more direct access to valuable allelic diversity at the loci governing the adaptive response to drought and salinity. Genomics and post-genomics platforms offer unprecedented opportunities to map, clone and manipulate the suite of QTLs affecting tolerance to drought and salinity in model species and crops. New high-throughput platforms capable of reducing the cost of molecular profiling coupled with a rapidly expanding amount of sequence information will streamline QTL dissection and the identification of superior alleles to enhance tolerance to drought and/or salinity. Therefore, it is expected that yield improvement under drought and/or saline conditions will increasingly benefit from the manipulation of QTLs through marker-assisted selection and, following QTL cloning, genetic engineering. Approaches based on the screening of wild relatives will unveil new allelic variants lost during domestication and early selection. Allele mining (e.g. association mapping, TILLING) in germplasm and mutant collections coupled with marker-assisted backcrossing and/or genetic engineering will further expand the possibilities to improve elite materials. QTL-based modelling approaches will contribute to better understand 'Genotype x Environment' interactions and to single out the most promising genotypes based upon the available QTL information. The impact of QTL-based approaches on the release of improved cultivars more resilient to drought and salinity will depend on their successful integration with conventional breeding methodologies and a thorough understanding of the biochemical and physiological processes limiting yield under such adverse conditions.

Keywords: candidate gene, drought, genomics, metabolome, proteome, QTL, salinity, TILLING, transcriptome

1. INTRODUCTION

In the past decade, remarkable advances have been achieved in unveiling the molecular complexity of the adaptive response of plants to drought and salinity

**Corresponding Author*: Fax +39-051-2096241

M.A. Jenks et al. (eds.), Advances in Molecular Breeding Toward Drought and Salt Tolerant Crops, 381–411.

(Blum et al., 1996; Shinozaki and Yamaguchi-Shinozaki, 1996 and 1997; Zhu et al., 1997; Hasegawa et al., 2000; Coraggio and Tuberosa, 2004; Bohnert et al., 2006). This progress has mainly been possible through studies of Arabidopsis mutants at loci inherited in a Mendelian fashion (Werner and Finkelstein, 1995; Abe et al., 1997; Stockinger et al., 1997; Kasuga et al., 1999; Mizoguchi et al., 2000; Quesada et al., 2000; Zhu, 2001; Qiu et al., 2004; Koiwa et al., 2006) rather than through the analysis of naturally occurring allelic variation, which is what plant breeders usually contend with in order to improve crop performance. The quantitative inheritance of tolerance to drought and/or salinity and its low heritability have hindered a more complete understanding of the genetic and physiological bases of yield in crops exposed to such adverse conditions (Blum, 1988; Boyer, 1996; Nguyen et al., 1997; Zhu et al., 1997; Passioura, 2002 and 2007; Tuberosa et al., 2002b; Munns et al., 2006; Richards, 2006). Further complexity is contributed by the concomitance of other abiotic stresses (e.g. heat, Al toxicity, etc.) that can amplify the negative effects of drought and salinity. This notwithstanding, conventional breeding has allowed for a steady, albeit slow increase in yield under conditions of limited water availability (Blum, 1988; Tollenaar and Wu, 1999; Trethowan et al., 2002; Duvick, 2005) or excessive soil salinity (Munns and Richards, 2007). To overcome the low response to a direct selection for yield in crops exposed to drought and/or salinity, substantial efforts have targeted the manipulation of morpho-physiological and biochemical traits with heritability higher than that of yield itself (Ludlow and Muchow, 1990; Blum, 1996; Turner, 1997; Araus et al., 2002; Slafer, 2003; Colmer et al., 2005; Munns et al., 2006). Nonetheless, this indirect selection strategy has been successful only in a few cases, a notable example being represented by the release of the drought-tolerant wheat cultivars Drysdale and Rees (www.csiro.au/csiro), which have been selected for the unique dryland conditions of eastern Australia using carbon isotope discrimination, a molecular signature which provides an indirect measure of water-use efficiency (Farquhar et al., 1989; Condon et al., 2002; Richards et al., 2002).

The dissection of the genetic basis of tolerance to drought and salinity has greatly improved following the introduction of the molecular platforms that enable us to identify the quantitative trait loci (QTLs) governing the relevant genetic variation in cultivated and wild germaplsm (Tanksley, 1993; Ribaut and Hoisington, 1998; Flowers et al., 2000; Koyama et al., 2001; Munns, 2005; Varshney et al., 2005; Tuberosa and Salvi, 2006; Passioura et al., 2007). Therefore, it is expected that improving yield under drought and/or saline conditions will increasingly benefit from marker-assisted selection of target QTLs and, following their cloning, genetic engineering. During the past decade, a growing number of studies have strived to map QTLs for yield in plants exposed to water deficit and/or salinity (for an updated list, see also other chapters in this volume and http://www.plantstress.com). New dimensions for deciphering the role and function of genes governing the response to drought and/or salinity have been added by bioinformatics (Bray, 2002) and the information generated by sequencing and post-genomics platforms (Tuberosa et al., 2002a; Hilson et al., 2004 Hirai et al., 2005; Sakurai and Shibata, 2006).

As a consequence, the list of cloned QTLs has expanded rapidly in the past few years (Salvi and Tuberosa, 2005; Price, 2006). Nonetheless, the number of cloned QTLs relevant for improving tolerance to drought and/or salinity remains disappointingly low, mainly due to the difficulties in (i) identifying major QTLs amenable to cloning and (ii) accurately characterizing the phenotypic effects of QTLs in the field, particularly when drought and salinity conditions prevail. This chapter presents a few selected examples in dissecting QTLs for tolerance of plants to drought and/or salinity and critically reviews the merits and pitfalls of a number of genomic platforms and approaches that enable us to map and clone QTLs.

2. THE QTL APPROACH: WHERE MENDELIAN AND QUANTITATIVE GENETICS MERGE

The goal of the QTL approach is to dissect the complex inheritance of quantitative traits into 'Mendelian-like' factors amenable to marker-assisted selection (MAS) and, eventually, their cloning. QTL dissection is usually an unbiased approach, one in which plants lead us to identify the relevant chromosome regions and the sequences whose functional polymorphisms regulate the observed phenotypic variability. A number of authoritative reviews have addressed the detailed analysis of the basic methods and approaches required to identify QTLs (Tanksley, 1993; Zeng, 1994; Lee, 1995; Utz and Melchinger, 1996; Korol et al., 2001). Here, we summarize some basic concepts on a number of issues relevant for the success of QTL mapping and cloning.

At its simplest, QTL mapping entails the phenotypic evaluation and the profiling with molecular markers (e.g. RFLPs, SSRs, AFLPs, SNPs, etc.) of a mapping population (usually comprising ca. 100–200 related families derived from a biparental cross) coupled with a statistical analysis to test the level of significance in the differences between the phenotypic values of the parental marker alleles, averaged across all individuals of the mapping population. The availability for a particular species of maps obtained with different crosses and sharing common polymorphisms allows for the construction of a reference map that enables us to more effectively compare the position of QTLs and mutants mapped in different studies, an important prerequisite for the identification of major QTLs (Tuberosa et al., 2002b; Zheng et al., 2003; Sawkins et al., 2004) and candidate genes. Importantly, Robertson (1985) postulated that a mutant phenotype at a particular locus could be caused by a variant allele with a much more drastic effect on the phenotype in comparison to that of the naturally-occurring QTL alleles at the same locus. Robertson's hypothesis was first validated in maize for a plant height QTL that colocalized with the mutant *dwarf3* (Touzet et al., 1995). More recently, a mutant at a locus (*ERECTA*) influencing drought tolerance in Arabidopsis (Masle et al., 2005) has been shown to colocalize with a QTL for the same trait. These results indicate that no real boundary exists between Mendelian and quantitative genetics, and that loci will be classified in one of the two categories based on the magnitude and heritability of the additive effect of the parental alleles segregating in the mapping population. Hence, the information provided by mutants is of great relevance and value for QTL studies, particularly due to the availability of methods (e.g. Ecotilling; Comai et al., 2004) that allow us to gauge the effect of allelic variation at candidate loci on the phenotypic variability of target traits (Buckler and Thornsberry, 2002).

2.1. From QTLs to Genes

QTL cloning represents an essential entry point towards a more effective exploitation of sequence variability at selected target loci and to unlock the allelic diversity present in germplasm collections (Tanksley and McCouch, 1997). Positional cloning and association mapping are the two approaches that have been most frequently applied for QTL dissection. Both approaches exploit linkage disequilibrium (LD) to identify the most promising candidate sequence for the subsequent validation phase. The large number of drought-/salt-induced genes described so far (Ozturk et al., 2002; Hazen et al., 2003; Bohnert et al., 2006) is a valuable source for the construction of functional maps that can facilitate the identification of candidate genes for QTLs. Specific efforts toward enriching linkage maps with function-specific genes have been undertaken and are being utilized for QTL analysis (Davis et al., 1999; Andersen and Lubberstedt, 2003; Birnbaum et al., 2003; Gorantla et al., 2005; Gupta and Rustgi, 2004; Qi et al., 2004; Musket et al., 2005).

2.1.1. Positional cloning

The first prerequisite for the positional cloning of a major QTL is the production of a large segregating population in a nearly isogenic background where only the target QTL segregates. The large number of plants (ca. 2,000 or preferably more) in the segregating population allows for the recovery of a sufficiently high number of recombination events in the target region, an essential feature for achieving an adequate level of map resolution. Additional markers in the target region can be added through bulk segregant analysis (Salvi et al., 2002), comparative mapping based on synteny with model species and also among crops (Bennetzen and Ma, 2003; Sorrels et al., 2003) and/or microarray analysis (Giuliani et al., 2005a). After completing the fine mapping, the markers more tightly linked to the QTL are anchored to a chromosome physical map. When the sequence of the entire genome is available (e.g. Arabidopsis, rice, etc.) the anchoring can be extended to the whole genome, thus strongly facilitating QTL mapping (e.g. any monomorphic probe can be mapped directly onto the map) and cloning (Borewitz and Chory, 2004; Yazaki et al., 2004). When the genome sequence is unavailable, genomic libraries (e.g. BAC clones) are screened. Polymorphic genes or genomic sequences that cosegregate with the QTL are then functionally tested with a number of different approaches (e.g. genetic engineering, identification of knockouts, association mapping, etc.). Other procedures such as RNAi (Waterhouse and Helliwell, 2003) and TILLING (Targeted Induced Local Lesions IN Genomes; McCallum et al., 2000;

Stemple, 2004) allow for a genome wide functional screening and validation applicable to almost any species.

2.1.2. *Association mapping*

Differently from positional cloning, QTL discovery through association mapping (Flint-Garcia et al., 2003) is based on the molecular and phenotypic characterization of unrelated accessions. The analysis evaluates the difference in allele frequency in case-control samples or, preferably when dealing with complex traits, the change in the mean of the investigated traits caused by allele substitution. The interest in association mapping is due to the possibility of performing QTL analysis and cloning without the time-consuming production of large experimental populations (Buckler and Thornsberry, 2002; Morgante and Salamini, 2003). The applicability of association mapping is influenced by the level of LD, availability and cost of molecular markers, and the presence of population structure among the investigated accessions (Pritchard et al., 2000; Remington et al., 2001; Flint-Garcia et al., 2003). Populations characterized by high LD (> 1 cM) are more suitable for a genome-wide search of genes/QTLs, particularly when the panel of accessions has been profiled at a rather limited number of loci (Maccaferri et al., 2005 and 2006; see also http://www.distagenomics.unibo.it/iduwue/index.html for further details). Conversely, the validation of the role of a candidate gene requires the utilization of panels with much lower LD (< 0.01 cM), hence a much higher level of genetic resolution. In barley, association mapping has been applied to identify chromosome regions influencing tolerance to salt stress (Pakniyat et al., 1997). A large-scale association mapping effort to identify drought-related QTLs is presently underway in several major crops through the Generation Challenge Program on biodiversity (http://www.generationcp.org). Because the candidate gene approach relies on the information available on open reading frames (ORFs), the effectiveness of its application is reduced when the QTL is caused by a polymorphism at a distant (> 5 kb), cis-acting, non-coding sequence, as recently shown in maize for *Vgt1*, a QTL for flowering time (Salvi et al., 2007). Therefore, as recognized by Rafalski and Morgante (2004), the identification of regulatory regions often quite distant from the effector genes indicates that the selection of a candidate sequence to be tested for association mapping with a phenotype is by no means a challenging undertaking if the genomic scan aims to be comprehensive.

2.1.3. *The candidate gene approach*

The candidate gene approach relies on prior information about the role and function of a particular coding sequence and seeks evidence to validate its causal role in determining the variability among plants for the target trait (Pflieger et al., 2001). Although the candidate gene approach can be deployed also with no prior knowledge about QTLs for the target trait (for an example see Yamasaki et al., 2005), its application usually involves genes mapped within the support interval of QTLs (Salvi and Tuberosa, 2005) for which a plausible cause-effect relationship can be hypothesized between the target trait and the function of the candidate gene. The role of a candidate gene can be validated through forward (e.g. genetic engineering, association mapping, etc.) or reverse genetics (e.g. screening of knockout mutants, TILLING, RNA interference, etc.) approaches. Therefore, the candidate gene approach bypasses the tedious procedures of positional cloning. The identification of suitable candidate genes and the elucidation of their function can be facilitated by combining different approaches and high-throughput platforms applied to the target crop and/or to model species (Markandeya et al., 2005; Bohnert et al., 2006). From a technical standpoint, it should be noted that the combination of the 'omics' platforms (e.g. transcriptomics, proteomics, metabolomics, etc.) with laser capture microdissection provides unprecedented levels of functional resolution at the tissue level, down to a single-cell layer (Nakazono et al., 2003). In maize, a combination of laser capture microdissection and subsequent microarray analyses applied to the root pericycle of wild-type and *rum1* mutant allowed Woll et al. (2005) to identify 19 genes involved in signal transduction, transcription and the cell cycle that are active before lateral root initiation. These findings will contribute to the identification of the developmental checkpoints involved in lateral root formation downstream of *rum1*, thus providing relevant clues to unravel the functional basis of root growth plasticity, an important factor for the adaptive response of plants to drought.

Among the different platforms available for the mass-scale profiling of the transcriptome, microarrays have been frequently utilized to elucidate the changes in gene expression elicited by exposure to drought and/or salinity (Ozturk et al., 2002; Zinselmeier et al., 2002; Hazen et al., 2003; Oono et al. 2003; Seki et al., 2003; Yu and Setter, 2003; Kawaguchi et al., 2004; Schnable et al., 2004; Giuliani et al., 2005a; Rensink, 2005). Transcriptome analysis has revealed that changes in expression of a number of genes are regulated by both drought and salinity (Zhu et al., 1997; Chen et al., 2002; Ozturk et al., 2002). Because the functional basis of a number of cloned plant QTLs relates to differences in the level of expression (Salvi and Tuberosa, 2005; Salvi et al., 2007), QTL cloning may in some cases be facilitated through a direct profiling approach applied to suitable genetic materials (Wayne and McIntyre, 2002; Hazen et al., 2003). In this context, an interesting application of transcriptome analysis is the identification of the so-called eQTLs, i.e. QTLs able to influence the level of expression (hence the 'e') of a particular gene. In this case, the application of QTL analysis to the level of gene expression of each progeny of a mapping population will identify eQTLs influencing the observed variability in mRNA level of the profiled genes. Circumstantial evidence regarding the importance of each ORF in governing variability for yield under conditions of drought/salinity can be obtained by comparing the map position of QTLs for yield with the map position of the ORFs and the corresponding eQTLs. In plants, eQTLs were first identified in maize (Schadt et al., 2003). The eQTLs were found to map both at the gene loci for which expression was analysed (indicating allelic differences at *cis* regulatory regions) and at different chromosome locations (indicating allelic differences for *trans*-acting regulatory factors). Due to the high cost for profiling RNA samples of an entire mapping population, transcriptome profiling based on microarrays is better suited for studies involving a limited number of samples extracted from congenic strains differing at key genomic regions (e.g. NILs) and/or bulked RNA samples obtained from the tails of mapping populations. More recently, cDNA-AFLPs have been used as an alternative to microarrays to identify eQTLs in Arabidopsis (Vuylsteke etal., 2006). As compared to microarrays, the cDNA-AFLP approach (i) has a relatively low start-up cost and requires no prior sequence information (Breyne et al., 2003), (ii) avoids bias for abundant transcripts, and (iii) can distinguish the expression of highly homologous genes (Breyne and Zabeau, 2001; Breyne et al., 2003). However, distinct drawbacks of a cDNA-AFLP platform are the limited coverage of the transcriptome and the identification of differential genes, a procedure which requires purification and sequencing of individual AFLP fragments. The studies conducted so far in animal species and humans have demonstrated the feasibility of the eQTL approach. However, the limited statistical power of such studies due to the small population size exacerbates the problem of detection of false-positive eQTLs, which requires caution in interpreting the results. For a more robust implementation of eQTL analysis, Haley and de Koning (2006) have proposed to combine expression studies with the fine mapping of functional trait loci. Selective transcriptional profiling based on available information on individual quantitative traits and marker data has been advocated as a means to select a subset of individuals for optimizing the effectiveness and accuracy of RNA profiling (Nettleton and Wang, 2006).

More in general, the interpretation of the results obtained from profiling experiments carried out under controlled conditions should take into due consideration the conditions utilized to expose the plant to drought and/or salinity. In several studies aimed at identifying genes differentially expressed under drought, plants were exposed to severe stress intensity in a very short time, commonly only a few hours (Ozturk et al., 2002; reviewed in Hazen et al., 2003). These experimental conditions will be more damaging as compared to similar levels of water deficit that plant tissues may experience in the field, where dehydration unfolds over a prolonged period of time (commonly days or weeks), thus allowing for a more proper activation of the molecular mechanisms leading to those beneficial adaptive responses (e.g. osmotic adjustment, early flowering, thickening of leaf cuticles, etc.) allowing the plant to partially counteract the negative effects of drought and/or salinity. In barley, the changes in gene expression observed in excised leaves following a rapidly-induced dehydration have shown a low correlation (from 0.19 to 0.41) with the changes attained under a slower dehydration regime (in pots) which mimicked more closely field conditions (Talamè et al., 2007). Therefore, molecular results obtained under artificially-induced conditions of water or salt stress should be dealt with caution and duly validated prior to their utilization in a more applicative context. Another factor that could greatly reduce the effectiveness of profiling experiments to capture the key events triggering important adaptive responses is the timing of the sampling in relation to the dynamics of the stressing event(s). Indeed, Boyer and Westgate (2004) have indicated that the results obtained with microarrays on the role of invertase activities in ovary abortion in drought-stressed maize (Zinselmeier et al., 2002) are difficult to reconcile with the pivotal role played by assimilate supply in preventing maize ovaries from aborting under drought conditions (Zinselmeier et al., 1999; McLaughlin and Boyer, 2004).

Additional leads to the changes in cellular metabolism in response to drought and/or salinity can be acquired through profiling of the proteome (Hochholdinger et al., 2005 and 2006; Wen et al., 2005; Sauer et al., 2006) and metabolome (Fiehn, 2002; Steuer et al., 2003; Morgenthal et al., 2006) which, as compared to the transcriptome, are functionally 'closer' to the phenotype and thus can also account for post-transcriptional regulation. Nonetheless, it should be appreciated that proteomics and metabolomics as currently performed report changes for only a limited portion of the genome; additionally, proteomics is often unable to detect the changes in gene products (e.g. transcription factors) that, despite their low level, are more likely to play an important role in adaptation to adverse conditions and as such, underline QTLs. In maize, proteome profiling is in progress to ascertain the role of cell wall proteins (CWPs) in the elongation of the primary root (Zhu et al., 2006). Although many of the CWPs identified in this study have previously been shown to be involved in cell wall metabolism and cell elongation, a number of CWPs (e.g. endo-1, 3;1, 4-β-D-glucanase and α-L-arabinofuranosidase) had not been described in previous cell wall proteomic studies. In rice, more than 2,000 proteins were detected reproducibly in drought-stressed and well-watered leaves (Salekdeh et al., 2002). Among the 1,000 proteins that were reliably quantified, 42 changed significantly in abundance and/or position, one of which involved an actin depolymerization factor (ADF) accumulated in drought-stressed leaves. *In silico* work has indicated that the ADF gene family includes 11 paralogues (*OsADF1-11*) spread over seven chromosomes; additionally, a more detailed analysis has indicated an interaction of OsADF5 with the actin cytoskeleton (Liu et al., 2005).

Profiling the proteome of a mapping population offers the opportunity to identify QTLs influencing protein quantity (PQLs, Protein Quantity Loci; de Vienne et al., 1999; Zivy and de Vienne, 2000; Consoli et al., 2002). Co-localization of a PQL with its protein-coding locus would indicate that allelic differences at that locus influence the expression of the protein, whereas co-localization between a PQL and a QTL for a different trait would suggest an association between the candidate protein and trait variation (de Vienne et al., 1999; Pelleschi et al., 1999). In maize, Jeanneau et al. (2002a) have shown that under conditions of mild water stress the *Asr1* gene, a putative transcription factor, co-localizes with a PQL for its protein (ASR1) and a QTL for anthesis-silking interval (ASI) and leaf senescence. Based on these findings, it was hypothesized that the *Asr1* polymorphism is responsible for the presence or absence of the ASR1 protein, which would also affect ASI and leaf senescence; the validity of this hypothesis was confirmed through genetic engineering (Jeanneau et al., 2002b).

Metabolome profiling aims at the identification and quantification of metabolites in a biological sample in order to provide a more comprehensive view of the functional characteristics under investigation (Fell, 2001; Morgenthal et al., 2006; Sakurai and Shibata, 2006). With the present technology, up to ca. 2000 different metabolites can be profiled in a single sample (Fiehn, 2002). Metabolome profiling applied to a mapping population can be used to identify QTLs regulating the level of a particular metabolite and verify its coincidence with QTLs for yield and/or genes involved in metabolic pathways. In maize, QTLs for invertase activity have been identified in a population subjected to drought stress (Pelleschi et al., 1999). The number of QTLs for invertase activity detected under drought (nine in total) was more than twice the number detected under well-watered conditions (four in total), an indirect indication of the important role of this enzyme under drought conditions. One QTL common to both treatments was located near *Ivr2*, an invertase-encoding gene on chromosome bin 5.03. Drought produced an early stimulation of acid-soluble invertase activity in adult leaves, whereas the activity of the cell wall invertase was found to be unaffected. These studies imply invertase activity as an important limiting factor for grain yield in maize exposed to drought during the reproductive phase (McLaughlin and Boyer, 2004; Boyer and Westgate, 2004). More recently, Pelleschi et al. (2006) reported co-location between the activities of three enzymes (invertase, sucrose-P synthase and ADP-glucose pyrophosphorylase) implied in sucrose and starch metabolism and a corresponding structural gene, which can thus be considered as a candidate gene for explaining part of the variability in enzyme activity. These results clearly indicate that carbohydrate metabolism provides valuable leads for understanding and improving maize responses to water stress.

3. TARGETING QTLS FOR TOLERANCE TO DROUGHT AND SALINITY

Presently, the number of drought- and salinity-related QTLs with an additive effect sufficiently large to allow for a positional cloning approach is very limited. Clearly, such major QTLs are, and will remain, the exception rather than the rule. The accurate characterization and validation of a QTL requires its isogenization, i.e. the production of congenic strains differing only for a small chromosome region flanking the target QTL. In allogamous and highly heterotic species like maize the consistency of a QTL effect and its breeding value should be further investigated through the evaluation of testcrosses obtained by crossing different tester lines with pairs of near isogenic lines (NILs) contrasted for parental alleles at the QTL region (Landi et al., 2007). A distinct advantage related to the availability of NILs at different QTLs, is the possibility to test in a more systematic and accurate way for possible epistatic interactions. Although the derivation of NILs and other congenic strains does not lead to short-term applications, it is an essential step towards the 'Mendelization' of single QTLs and their positional cloning. A number of NILs have been obtained for QTLs of traits relevant for drought tolerance (Tuinstra et al., 1998; Shen et al., 2001; Price et al., 2002; Sanchez et al., 2002; Landi et al., 2005) or salinity (Ren et al., 2005; Huang et al., 2006). A more systematic and high-throughput approach to generate a series of NILs covering the whole genome, irrespectively from the investigated trait, is provided by the construction of a series of lines, each one carrying a small portion (usually ca. 15–30 cM) of a donor genome in an otherwise common genetic background. An important example of the effectiveness of this approach for gene/QTL discovery and cloning has already been provided in tomato (Zamir, 2001). A similar effort recently completed in maize has allowed for the identification of chromosome regions influencing root traits at the seedling stage (Salvi et al., 2005). In order to gain a more complete understanding of complex polygenic phenotypes in rice, Li Zhikang et al. (2005) have developed over 20,000 introgression lines (ILs) in three elite rice genetic backgrounds for a wide range of complex traits, including tolerance to drought and salinity. Together, these ILs contain a significant portion of loci affecting the selected complex phenotypes at which allelic diversity exists in the primary gene pool of rice. Complementary to the genome-wide knock-out mutants, this IL collection opens a new way for highly efficient QTL discovery, candidate gene identification and QTL cloning of specific phenotypes.

3.1. Prioritizing the Choice of Target Traits and QTLs for Marker-Assisted Selection and Cloning

A number of reviews have analysed and discussed the mechanisms and the traits underlying tolerance to drought (Blum, 1996; Richards, 1996 and 2000; Turner, 1997; Passioura, 2002; Richards et al., 2002) and salinity (Zhu, 2001; Tester and Davenport, 2003; Munns, 2005; Bohnert et al., 2006). The identification of QTLs for the morpho-physiological traits that more strongly influence yield under drought and/or saline conditions relies on the accurate phenotyping under the appropriate field conditions in the target environment. This issue is particularly crucial for the identification and characterization of QTLs for traits categorized as adaptive (e.g. accumulation of osmolytes or other metabolites in response to cellular dehydration) as compared to constitutive traits (e.g. root elongation rate). Indeed, one of the major difficulties in enhancing drought/salinity tolerance through MAS relates to the high QTL x Environment interaction shown by the majority of QTLs in trials conducted under varying water/salinity regimes and/or during different seasons (Ribaut et al., 2002). A major advantage in targeting constitutive traits is that their phenotyping does not require conditions of programmed stress as carefully controlled as for drought-/salinity-adaptive traits; additionally, ranking of the genotypes of a mapping population scored for a constitutive trait under different water regimes and/or salinity levels will be less affected by environmental factors. A critical step from an applicative perspective (e.g. MAS), is to verify to what extent the effect of the beneficial QTL allele is consistent in the genetic backgrounds to be improved. One reason for the limited applicative results contributed so far by the QTL approach for improving tolerance to drought and/or salinity is due to the fact that the parental lines of the crosses evaluated for QTL discovery have often been chosen based on their morpho-physiological attributes rather than their agronomic value. While this approach facilitates the identification of major QTLs, it does not guarantee any real progress when MAS is applied to introgress the desirable QTL alleles in the elite germplasm routinely deployed by breeders, also because such alleles, or those with even more beneficial effects, could already be prevalent or even fixed in elite accessions.

3.1.1. *QTLs for tolerance to drought*

For categorizing the mechanisms conferring tolerance to drought, we have adopted the nomenclature followed by Ludlow and Muchow (1990) which distinguishes traits that allow the plant to escape drought (e.g. early flowering in crops grown in Mediterranean-like environments) from the traits that influence resistance to drought, with the latter ones further categorized in terms of dehydration avoidance and dehydration tolerance. Dehydration avoidance depends on maintenance of turgor through an increase in water uptake (e.g. deeper roots) and/or reduction in water loss (e.g. increased leaf waxiness), while dehydration tolerance involves biochemical mechanisms (e.g. accumulation of compatible solutes to preserve membrane integrity) that allow the cell to tolerate the negative effect caused by cellular dehydration.

The limited success in improving drought resistance through molecular approaches is primarily related to the difficulty in identifying the key physiological determinants of yield under varying drought conditions (Blum, 1988; Ludlow and Muchow, 1990; Boyer, 1996; Turner, 1997; Passioura, 2002; Nguyen and Blum, 2004; Tuberosa, 2004). As an example, a greater capacity of root meristems to adjust osmotically at a given water potential will positively impact root mass and final yield only if deeper roots allow for the extraction of additional moisture from the soil. However, when deeper soil layers do not provide additional moisture, a condition quite common in many drought-prone environments (e.g. the Mediterranean basin), growing larger/deeper roots will not provide any clear advantage and might even influence negatively final yield due to an excessive partitioning of photosynthates to the root and the high metabolic cost required for sustaining root growth and functions.

A trait that has been extensively investigated as an indirect measure of drought tolerance is the capacity to accumulate abscisic acid (ABA). This phytoregulator is accumulated in response to dehydration and regulates the adaptive response of the plant to a decreased moisture (Quarrie, 1991; Sharp et al., 2004). One problem in the interpretation of the results of field studies investigating QTLs for ABA is due to the confounding effect of water status on (i) the value of the investigated morpho-physiological traits and (ii) the interpretation of their association with yield. Variation in ABA, as well as other metabolic traits influenced by drought, has both an environmental and genetic component and thus interpreting the results of a QTL analysis for ABA accumulation disregarding the water status of the plant does not allow one to appreciate to what extent a high ABA concentration is due to a constitutively higher capacity of a genotype to accumulate ABA at a given water status or to lower water content, hence a higher level of water stress possibly due to a weaker root. Monitoring the water status of the vast number of plants typically included in a mapping population is a rather daunting task, particularly when the water status of the plant changes rapidly as a result of the fluctuations in the evapotranspirative demand during the day. For traits highly influenced by water status, a more accurate evaluation of a set of genotypes can be obtained under controlled conditions which allow for better control of daily fluctuations in the water status of the plants. This condition can be achieved by exposing plants to a given concentration of osmolytes (e.g. mannitol solution). This approach has been recently adopted to investigate the capacity to accumulate ABA in maize exposed to a water deficit (Sanguineti et al., 2006). In particular, a historical series of hybrids representing the last six decades of breeding were grown in hydroponics and were exposed to polyethylene glycol (PEG) in order to simulate water deficit. Interestingly, although maize breeders have never selected for the capacity to accumulate ABA, a highly significant, linear decrease (ca. -30%) in ABA accumulation was recorded; additionally, this decrease was significantly associated with the linear increase observed in grain yield. Based on these results, Sanguineti et al. (2006) suggested that the decreased capacity to accumulate ABA might be related to a negative effect of ABA on reproductive fertility, a trait that maize breeders have traditionally selected for (Duvick, 2005), particularly under drought conditions. In maize, Landi et al. (2005) derived pairs of backcross-derived, near-isogenic lines (BDLs) differing for the parental alleles at a major QTL (*root-ABA1*) on bin 2.04 that affects the concentration of ABA in the leaf (L-ABA; Tuberosa et al., 1998), root architecture and other drought-related traits (Giuliani et al., 2005b). A field evaluation conducted under well-watered and water-stressed conditions during two consecutive seasons indicated that each pair of *root-ABA1* BDLs differed significantly and markedly for L-ABA (Landi et al., 2005). More recently, the evaluation of testcrosses with the BDLs has shown a highly significant effect of *root-ABA1*on several agronomic traits, including grain yield (Landi et al., 2007). Furthermore, the BDLs and derived near-isogenic hybrids showed significant differences for root lodging, root mass and brace root angle (Giuliani et al., 2005a; Landi et al., 2005), thus supporting the hypothesis that some of the QTLs for L-ABA could derive from primary QTL effects on root architecture as previously postulated by Tuberosa et al. (1998). Interestingly, the experimental evidence gathered on this QTL from the cross Polji17 x F-2 also suggested that the effect of this QTL on L-ABA might be consequent to a primary effect on root architecture (Lebreton et al., 1995). The positional cloning of this QTL is presently underway in our laboratory.

In rice, roots have often been targeted for identifying the corresponding QTLs and to evaluate their effects on grain yield under different water regimes (Champoux et al., 1995; Nguyen et al., 1997; Zheng et al., 2000; Price et al., 2002; Courtois et al., 2003; Li Zichao et al., 2005). These studies have highlighted a rather complex picture and in some cases have allowed for the production of NILs (Shen et al., 2001; Steele et al., 2006) suitable for attempting the cloning of the corresponding QTLs. This notwithstanding, to our best knowledge none of the isogenized QTLs has so far been cloned.

3.1.2. QTLs for salinity tolerance

Salinity tolerance is governed by a suite of genes acting at different hierarchical levels of functional and morphological complexity that influence uptake of Na^+ and K^+ by the roots, Na^+ transport to the shoot and its cellular compartmentation as well as the ionic and osmotic balance of cells (Tester and Davenport, 2003; Munns, 2005). Accumulation of Na^+ in the cytoplasm disrupts metabolic processes and reduces growth. Maintaining low levels of cytoplasmic Na^+requires the coordinate regulation of transport proteins on numerous cellular membranes. As an example, the extensive work carried out in Arabidopsis has been instrumental in elucidating the molecular events involved in the adaptive response to salinity (Zhu et al., 1997; Zhu, 2001) and has highlighted a number of interesting candidates for naturally occurring variation in salt tolerance. Qiu et al. (2004) have linked components of the *Salt-Overly-Sensitive pathway (SOS1–3)* to salt tolerance and demonstrated that the activity of the plasma membrane Na^+/H^+ exchanger (SOS1) is regulated by SOS2 (a protein kinase) and SOS3 (a calcium-binding protein). Additionally, their work demonstrated that (i) the tonoplast Na^+/H^+ exchanger in *Arabidopsis* is a target of the SOS regulatory pathway and (ii) the regulation of the exchangers in the tonoplast and plasma membrane may be coordinated. In rice, Martinez-Atienza et al. (2006) have identified a plasma membrane Na^+/H^+ exchanger that is the functional homologue of the *Arabidopsis thaliana* SOS1 protein. The rice transporter, denoted by OsSOS1, demonstrated a capacity for Na^+/H^+ exchange in plasma membrane vesicles of yeast cells and reduced their net cellular Na^+ content. Additionally, OsSOS1 suppressed the salt sensitivity of a *sos1–1* mutant of Arabidopsis. These results represent the first molecular and biochemical characterization of a Na^+ efflux protein from monocots and demonstrate that the SOS salt tolerance pathway operates in cereals and evidence a high degree of structural conservation among the SOS proteins from dicots and monocots.

Notwithstanding the large body of information on the role of mutant loci in the response to salinity, little evidence exists on the role of naturally occurring variation at such loci in governing QTLs for salt tolerance in crops. Two notable examples recently reported in rice (Ren et al., 2005) and in durum wheat (Huang et al., 2006) are analyzed in more detail. In a rice mapping population derived from the cross between Nona Bokra, a salt-tolerant *indica* variety, and Koshihikari, a susceptible *japonica* variety, the *SKC1* QTL accounted for ca. 40% of the phenotypic variation in shoot K^+ accumulation under salt stress conditions (Ren et al., 2005). Earlier findings indicated that K^+ homeostasis is important in salt tolerance. To understand the molecular basis of this QTL, the *SKC1* gene was isolated by positional cloning. For this purpose, a high-resolution map with 2973 BC_3F_2 plants confined *SKC1* to a 7.4 kb region which harboured only one predicted ORF coding for a putative K-Na symporter considered to be a plausible candidate. To validate its role, a 4219 bp Nona Bokra fragment containing the *SKC1* promoter region and the entire ORF was transferred into Zonghua 11, a *japonica* cv. that contains the Koshihikari allele. This experiment fully confirmed the role of *SKC1* in maintaining K^+ homeostasis. Database searches showed a high similarity between SKC1 and the HKT-type transporters found in plants, indicating that *SKC1* represents a novel member of the extended HKT family known to affect Na^+ unloading from xylem in roots and sheaths. Some members of the HKT family encode for high affinity K^+ transporter and function as Na^+ transporters in Arabidopsis and rice (Ren et al., 2005;

Rodriguez-Navarro and Rubio, 2006). HKT transporters also appear to be important for the control of Na^+ transport in durum and bread wheat (Laurie et al., 2002; James et al., 2006). A homology search identified *OsHKT8* as the ORF more closely related to *SKC1*. The functional difference at *SKC1* between the Nona Bokra and the Koshihikari alleles appeared to be related to four amino acid changes. Accordingly, the *SKC1* expression pattern did not differ significantly between Koshihikari and its near isogenic counterpart, further supporting the functional role of the four amino acid changes in determining alleles functionality. More detailed analyses revealed that *SKC1* is preferentially expressed in the parenchyma cells surrounding the xylem vessels and showed that SKC1 protein functions as a Na^+-selective transporter, suggesting its involvement in regulating K^+/Na^+ homeostasis under salt stress through the regulation of loading or unloading of xylem vessels. Under salinity, the allele contributed by Nona Broka allele was associated to a higher K^+ and lower Na^+ than Koshihikari. Additional studies conducted using *Xenopus* oocytes demonstrated that the *SKC1* locus encodes a functional Na^+ selective transporter. The analysis of the sequence data revealed seven members of the HKT family with one of them (*SKC1*/*OsHKT8*) playing a key role in providing tolerance to salinity (Ren et al., 2005).

In hexaploid wheat (AABBDD genome) and other Triticeae, Na^+ exclusion is one of the major mechanisms conferring salt tolerance (Huang et al., 2006). As compared to bread wheat, durum wheat (AABB genome) has a higher rate of Na^+ transport to the shoot and is characterized by a lower K^+/Na^+ ratio in leaves. The higher K^+/Na^+ ratio and salt tolerance in bread wheat is related to the *KNa1* locus which maps in the distal portion of chromosome 4DL. Trials conducted in the greenhouse with durum landraces characterized by different degrees of Na^+ accumulation have indicated that at moderate salinity levels, the introduction of Na^+ exclusion trait can boost yield by ca. 20% (Husain et al., 2004). In durum wheat, Munns et al. (2000) have identified Line 149 as a novel source of Na^+ exclusion with low Na^+ concentrations and with a K^+/Na^+ ratio similar to those found in bread wheat. Line 149 was derived from a cross between accession C68–101 of diploid wheat (*Triticum monococcum*; AA genome) and the durum cv. Marrocos. Genetic studies conducted with a population derived from the cross between Line 149 x Tamaroi revealed that two major loci controlled Na^+ accumulation in leaf blades (Munns et al., 2003). One locus, named *Nax1*, accounted for 38% of the phenotypic variation for Na^+ concentration and mapped to the long arm of chromosome 2A (Lindsay et al., 2004). At this locus, *T. monococcum* contributed the allele for low Na^+ concentration in Line 149 (James et al., 2006). More detailed studies showed that net xylem loading and leaf sheath sequestration in Line 149 interacted to control leaf blade Na^+ concentration (Davenport et al., 2005) and that the major effect of the *Nax1* allele contributed by *T. monococum* was to enhance Na^+ removal from the xylem in the roots and in the leaf sheath, hence reducing Na^+ concentrations in the leaf blade and consequently, the negative effects of salinity on leaf photosynthesis and senescence (James et al., 2006). More recently, the identification in rice of a feasible candidate in the region syntenic to *Nax1* has been accomplished by Huang et al. (2006) based on a wheat functional map comprising 8200 ESTs assigned to chromosome bins through deletion stocks (Qi et al., 2004) and rice genome sequence information. The analysis of syntenic relationship between wheat chromosome 2AL and rice chromosome 4L coupled with mapping data showed cosegregation of *Nax1* with marker *OsHKT7*, a gene belonging to the HKT family. Hybridization with wheat EST corresponding to *OsHKT7* revealed the presence of one polymorphic band (*TmHKT-A1*) cosegregating with *Nax1* and a monomorphic band (*TmHKT-A2*) while amino acid sequence analysis revealed the presence of a filter Ser in P-loop A in the HKT7 transporter, suggesting its function as a Na^+ transporter (Maser et al., 2002). These results suggested the presence in wheat of a putative Na^+transporter closely related to *OsHKT7* as the most likely candidate gene for *Nax1*. This hypothesis was further substantiated through the analysis of RT-PCR expression profiles which showed strikingly different levels of expression of *TmHKT-A2* in roots and leaf sheaths, but not in leaf blades of Line 149 and Tamroi, in accordance with their susceptibility to salinity and the role of *Nax1* in decreasing the Na^+ concentration in blades by retaining Na^+ in the sheaths (James et al., 2006). Conversely, no cDNA product was detected corresponding to *TmHKT7-A1*. The removal of Na^+ from the xylem resulted in a nearly 4-fold difference in blade Na^+ concentration between Line 149 and Tamaroi. In summary, the study of Huang et al. (2006) clearly shows the usefulness of using the rice genome sequence to identify suitable candidates for QTLs in cereals.

3.2. Searching for Alleles for Tolerance to Drought and Salinity in Wild Germplasm

The domestication process has inevitably caused a severe bottleneck in the genetic variability present in the wild relatives of modern crops. Therefore, it is likely that during domestication a number of potentially favourable alleles present in the genetic pool of wild germplasm (Colmer et al., 2006) were lost to early farmers and to modern breeding. This condition is more likely to occur at the loci that are more closely linked to the loci controlling the traits (e.g. ear shattering in cereals) that have played a major role in domestication. This limitation can be partially overcome through the application of advanced backcross quantitative trait locus analysis (AB-QTL), an approach that enables us to identify and exploit valuable QTL alleles present exclusively in wild germplasm (Tanksley and Nelson, 1996; Grandillo and Tanskley, 2005). The AB-QTL approach relies on the evaluation of backcross (BC) families derived from a cross between an elite variety used as recurrent parent and a donor accession, usually a wild species that is sexually-compatible with the crop. Usually, a cycle of selection is carried out in BC_1 before proceeding to QTL mapping in the BC_2 generation. The validity of AB-QTL has already been tested in different crops (Tanksley et al., 1996; Moncada et al., 2001; Grandillo and Tanksley, 2005). An example is offered by wild barley (*Hordeum spontaneum*), a valuable source of alleles for improving tolerance to abiotic stresses

(Forster et al., 2000 and 2004; Baum et al., 2003). An *H. vulgare* x *H. spontaneum* backcross population was evaluated under rainfed conditions in three Mediterranean countries in order to identify agronomically valuable alleles contributed by the wild parent (Talamè et al., 2004). In particular, among the 81 putative QTLs that influenced heading date, plant height, ear length, ear extrusion, grain yield and/or grain weight, in 43 cases (53%) the alleles increasing traits' value were contributed by *H. spontaneum*. As to grain yield, *H. spontaneum* contributed the agronomically favourable allele at six of the 17 QTLs that influenced this trait. Therefore, although the majority (65%) of the favourable QTL alleles was contributed by *H. vulgare*, a sizeable number of such alleles were from the wild parent. These results are encouraging as to the possibility of using AB-QTL as a germplasm enhancement strategy for identifying wild progenitor alleles capable of improving yield of the related crop cultivated under arid conditions. This approach may be particularly valuable for the identification of beneficial wild alleles improving survival at an early growth stage to conditions of severe drought and/or salinity. Ideally, the introgression of such beneficial alleles should bear no negative consequences under more favourable conditions (Johnson et al., 2000).

Wild relatives can also be deployed to identify novel alleles for agronomically relevant traits by focusing on loci targeted by selection during both domestication and modern breeding (Yamasaki et al., 2005). To this end, the comparative analysis of a large-scale screening of the allelic diversity of elite accessions, landraces and the undomesticated wild relative of a particular crop allows for the identification of loci devoid of genetic variation within the elite germplasm as a result of domestication and subsequent man-made selection. In this case, the assumption is that the observed loss of genetic diversity observed from the wild parent to the cultivated crop pinpoints the strong selection occurred at loci controlling traits of agronomic importance, including those relevant for adaptation to abiotic stress. Therefore, both this 'diversity screen' approach and the AB-QTL approach provide the distinct advantage of identifying agronomically valuable loci which would otherwise go undetected due to a lack of allelic diversity in the genetic pool presently cultivated. Additionally, the diversity screen approach allows for the identification of candidate genes of potential agronomic importance even without prior knowledge of gene function and the phenotype of interest. This notwithstanding, Yamasaki et al. (2005) recognized that the applicability of the diversity screen approach is limited by a number of factors, most notably that some of the identified genes may only be hitchhiking with neighbouring selected genes. The validity of diversity screening is being tested in maize (Yamasaki et al., 2005), a species particularly suited for this approach due to the extensive allelic richness of teosinte, the wild progenitor of maize, and the landraces presently cultivated.

3.3. Arabidopsis as a Model

Despite the extensive genetic information and materials available for Arabidopsis and its value as a model species (Alonso-Blanco and Koornneef, 2000; Zhang et al.,

2005; Maggio et al., 2006), limited work has been carried out toward the cloning of QTLs imparting tolerance to drought and/or salinity in this model species. Due to the relative ease to carry out extensive phenotyping in Arabidopsis, an area worthy of future exploration relates to the identification of QTLs controlling root architecture and its plasticity, both of which play an important role in the adaptive response to drought. The power of applying QTL analysis on root traits in Arabidopsis was recently shown by the identification of a naturally-occurring allele at a new type of transcription factor regulating root development (Mouchel et al., 2004 and 2006) and by the association of the activity of a sucrose-splitting enzyme and a QTL for root elongation (Sergeeva et al., 2006).

An additional area worthy of exploration relates to the mechanisms regulating the level of gene expression. Also in this case, Arabidopsis has provided useful insights. Although several microRNAs (miRNAs) have been shown to play a role in plant development, for the first time a reduced expression of a miRNA in Arabidopsis has been shown to influence the root phenotype (Guo et al., 2005). *Arabidopsis thaliana* miR164 was predicted to target five NAM/ATAF/CUC (NAC) domain-encoding mRNAs, including NAC1, which transduces auxin signals for lateral root emergence. Cleavage of endogenous and transgenic NAC1 mRNA by miR164 was shown to be blocked by NAC1 mutations that disrupt base pairing with miR164. Compared with wild-type plants, Arabidopsis mir164 mutants expressed less miR164 and more NAC1 mRNA and produced more lateral roots. The results of this landmark study indicate that auxin induction of miR164 provides a homeostatic mechanism to clear NAC1 mRNA to down-regulate auxin signals and provide an example of the value of using Arabidopsis as a model for elucidating the molecular mechanisms regulating root growth. Further insights on the role of auxins on root growth were provided by the study of Okushima et al. (2005): their results suggest that the ARF7 (Auxin Response Factor 7) and ARF19 proteins play essential roles in auxin-mediated growth of lateral roots by regulating both unique and partially overlapping sets of target genes.

The screening of nine Arabidopsis accessions grown under rigorously-controlled conditions revealed that one accession was unaffected by water deficit in terms of root growth (Granier et al., 2006). A mapping population including this accession as one of the parents will facilitate the identification of QTLs modulating the response of roots to a decreasing soil moisture. Fitz Gerald et al. (2006) examined the root systems of the closely related Arabidopsis ecotypes Landsberg *erecta* (L*er*) and Columbia (Col) grown under mild osmotic stress conditions and found that L*er* initiates more lateral root primordia, has an overall larger root system and shows a decreased sensitivity to osmotica than Col. To understand the genetic basis for these differences, QTLs for root architecture and size under mild osmotic stress were mapped in a L*er* x Col recombinant inbred population. Two major QTLs were identified and confirmed in NILs. The NILs also allowed for the identification of distinct physiological roles for the gene(s) at each locus. This study provides useful insights to dissect the molecular basis for naturally occurring differences in developmental plasticity of the root system in Arabidposis.

The importance of root architecture in the adaptive response to drought was further investigated by focusing on the inhibition of lateral root development as an adaptive response to drought. Xiong et al. (2006) showed that this response is partly mediated by the sensitivity to ABA and devised genetic screens to isolate *drought inhibition of lateral root growth* (*dig*) mutants with altered responses to drought or ABA in lateral root development. Characterization of these *dig* mutants revealed an altered drought stress tolerance, indicating that the response of lateral roots to drought stress is linked to drought adaptation. In Arabidopsis, ABA has also been shown to play a central role in mediating the regulatory effects of nitrate on root branching (Signora et al., 2001), thus adding evidence to its role in regulating plasticity of lateral root growth. Several QTLs for lateral root number (LRN) and density (LRD) and for total length of the lateral root system (LRL) and for primary root length (PRL), were identified in the Bay-0 x Shahdara population (Loudet et al., 2005). The results showed that variation in the extent of the lateral root system depends mainly on the growth of the existing lateral roots rather than in a change in LRN. Additionally, factors controlling lateral root growth showed no major effect on primary root growth. One QTL for PRL was isogenized and its effects further confirmed, thus making this QTL a good candidate for further fine-mapping and cloning.

Positional cloning has pinpointed the role of the *ERECTA* gene in Arabidopsis in the regulation of plant transpiration efficiency, i.e. the ratio between photosynthesis and transpiration rates. The evaluation of a RIL population derived from the cross of two ecotypes differing for carbon-isotope discrimination (Δ), an indirect measure of water-use efficiency, revealed a major QTL on chromosome 2 which, depending on growth conditions, accounted for up to 64% of the total phenotypic variation in Δ. This QTL (referred to as transpiration efficiency 1: *TE1*) spanned a region of ca. 37 genes and peaked on the *ERECTA* gene, a putative leucine-rich-repeat receptor-like kinase known to affect inflorescence development. Subsequent experiments demonstrated a role for the *ERECTA* gene in controlling leaf photosynthetic capacity and balancing biochemical and stomatal limitations on photosynthesis. Additionally, it was shown that the function of *ERECTA* varies according to the genetic background and is modulated by other polymorphisms between the parental ecotypes. The main mechanisms by which *ERECTA* controls Δ include effects on stomatal density, epidermal cell expansion, mesophyll cell proliferation and cell-cell contact. *ERECTA* is the first gene to be shown to act on the coordination between transpiration and photosynthesis, and, as such, to be identified as a transpiration efficiency gene, as opposed to simply a gene regulating stomatal density or photosynthesis. The magnitude of changes observed in null mutants suggests that *ERECTA* might act as a master gene. *ERECTA* homologues have been identified in several species and would represent an interesting target for an association study in crops. Phylogenetic analysis has pinpointed that *ERECTA* has evolved during or before early Angiosperms evolution, hence underlining its likely role on plant fitness under the selective pressure of water-limited conditions. From an applicative standpoint, it is worth noting that genetic engineering of *ERECTA* improved transpiration efficiency without detectable penalty in growth, leading Masle et al. (2005) to suggest the potential value of manipulating *ERECTA* as a path for improving crop performance under dry conditions through a decreased stomatal conductance on one hand, and removal of stomatal limitations to improve yield potential under well-watered conditions on the other.

Columbia and Landsberg *erecta* were also used by Quesada et al. (2002) as parental lines of a recombinant inbred population to identify QTLs conferring tolerance to salinity. A total of 11 QTLs were found to contribute to natural variation in Na^{+} tolerance in Arabidopsis, six at the germination and five at the vegetative growth stages, respectively. At least five of these QTLs are likely to represent loci not yet described by their relationship with salt stress. It should be noted that differences in salt tolerance between Arabidopsis ecotypes are small, thus limiting a more comprehensive discovery of QTLs imparting salt tolerance. However, several close relatives of Arabidopsis are extremely salt tolerant and could thus represent an interesting source for salt tolerance studies through QTL or mutational analyses (Inan et al., 2004). As an example, salt cress (*Thellungiellahalophila*) can survive seawater-level salinity and complete its life cycle in the presence of 300 mM NaCl. This small annual crucifer has a small genome (ca. 2-fold bigger than Arabidopsis) with high sequence identity (ca. 92%) with Arabidopsis, and can be genetically transformed by the simple floral dip procedure (Inan et al., 2004). Analysis of salt cress ESTs provides evidence for the presence of paralogs, missing in the Arabidopsis genome, and for genes with abiotic stress-relevant functions. Hybridization of salt cress RNA targets to an Arabidopsis whole-genome oligonucleotide array has shown that commonly stress-associated transcripts are expressed at a noticeably higher level in unstressed salt cress plants and are induced rapidly under stress (Taji et al., 2004).

The growing interest in QTL mapping and cloning in Arabidopsis and other related model species will provide additional insights into the genetic basis of adaptation to drought and salinity. To what extent this knowledge will have an impact on the release of better performing crops will depend on our capacity to identify crops' orthologs to Arabidopsis at the target QTLs and to single out the most agronomically valuable alleles at these loci.

4. CONCLUDING REMARKS

QTL-based approaches will impact plant breeding for drought and salt resistance in two different ways. First, it will be possible to identify relevant QTL alleles and pyramid them in correct combinations through MAS. Second, it is under the QTL cloning framework that the molecular basis of natural adaptation to hostile environments will be dissected and understood, providing leads for more accurately tailoring through genetic engineering the morphology of crops, their physiology and metabolism in order to face environmental stresses.

The effectiveness of MAS for producing drought-/salinity-tolerant cvs. will depend on the identification of the relevant QTL alleles and their pyramiding in the correct combinations. This approach could be regarded as an evolution of the so-called 'ideotype' breeding, the main difference being that we now rely on more powerful tools for dissecting the genetic basis of the phenotype and for piecing back together the best QTL alleles, in a sort of molecular jigsaw puzzle. This new concept of 'breeding by design' (Peleman and van der Voort, 2003), although already applicable from a purely technical standpoint in a number of major crops, in the case of drought and salinity tolerance it is still far from being routinely applicable, in view of our incomplete understanding of the molecular basis of drought and salinity tolerance and, most importantly, the difficulty in predicting the phenotypic value of a new genotype tailored through MAS. In this context, QTL-based modelling holds promise to allow for a more effective design of ideotypes on the basis of QTLs for parameters of response curves to varying levels of an environmental factor (Cooper et al., 2002; Hammer et al., 2006). Crop modelling can potentially be a powerful tool to resolve Genotype x Environment (G x E) interactions as well as the genetic basis of traits' plasticity (Chapman et al., 2003; Reymond et al., 2004). An example on how an ecophysiological model and QTL analysis can be integrated to investigate the genetic basis of leaf growth in response to drought has been provided by Reymond et al. (2003), who have identified QTLs for leaf elongation rate in maize as a function of water vapor pressure difference, soil water status and meristem temperature. Therefore, crop modelling based on QTLs for the response to drought and salinity might help us to more appropriately address and resolve G x E interactions and to identify the relevant genetic determinants (Yin et al., 2003 and 2004).

From a functional standpoint, we foresee that growing attention will be devoted to investigate the genetic basis of sensitivity to growth regulators as a means to reduce reproductive failure, probably the most important factor in curtailing yield of crops exposed to drought during the reproductive phase (Duvick, 2005; Campos et al., 2006), and more in general, to optimize growth response to drought and salinity. Comparative analysis of mutants hypersensitive to osmotic stress in tomato indicates that appropriate ABA perception and signalling is essential for developing appropriate osmotic tolerance (Borsani et al., 2002). In rice, Chen et al. (2006) have shown that sensitivity to ABA regulates lateral root growth and have identified genetic variability for this trait that plays a primary role in adaptation to drought. QTLs for root architecture are likely to receive increasing interest, in view of the difficulty of dissecting the genetic basis of root growth using conventional approaches. Additional traits suitable for QTL dissection are osmotic adjustment, relocation of stem reserves and stay green. QTLs for these traits have already been described (reviewed in Tuberosa and Salvi, 2004). In terms of experimental materials utilized for QTL discovery and cloning, a growing attention will be devoted to the exploitation of (i) progenies derived from multiparental crosses and (ii) adequately large mini-core collections of germplasm accessions with varying levels of LD. In the mapping populations so far utilized for QTL discovery, most QTLs go undetected due to the small size of the population and the presence of functionally-monomoprhic alleles. Evaluating multiparental crosses and mini-core collections will increase the chances of identifying functional variability at QTLs and to select the most beneficial allele in terms of agronomic performance. Additionally, high-throughput proteome and metabolome profiling will expand the ability to identify the causative mechanisms contributing to adaptive responses to drought and salinity, or simply to increase yield potential *per se*. Nevertheless, the deluge of information originated by the 'omics' platforms will not automatically translate into knowledge on how to improve tolerance to drought and salinity. This translation will be facilitated by bridging more systematically the different platforms and approaches applied for the dissection of QTLs in model and crop species. A systems biology-like approach (Stephanopoulos et al., 2004) will be increasingly instrumental in optimizing the accurate integration and exploitation of all this information in breeding terms.

Although it is not possible to predict to what extent QTl-based approaches will eventually affect conventional breeding practices, we remain confident that future progress toward the release of cultivars more resilient to drought and salinity will be accelerated through a more systematic discovery of the function of the QTLs governing the naturally occurring variation relevant for yield under water-limited and saline conditions. On a realistic ending note, the successful exploitation of QTL dissection and its applications to enhance yield under drought and salinity conditions will depend on their successful integration with conventional breeding methodologies and a thorough understanding of the biochemical and physiological processes limiting yield under such adverse conditions.

REFERENCES

Abe, H., Yamaguchi-Shinozaki, K., Urao, T., Iwasaki, T., Hosakawa, D., and Shinozaki, K., 1997, Role of Arabidopsis MYC and MYB homologs in drought- and abscisic acid-regulated gene expression, *Plant Cell* **9**:1859–1868.

Alonso-Blanco, C., and Koornneef, M., 2000, Naturally occurring variation in *Arabidopsis*: an under-exploited resource for plant genetics. *Trends Plant Sci.* **5**:22–29.

Andersen, J. R., and Lubberstedt, T., 2003, Functional markers in plants, *Trends Plant Sci.* **8**:554–560.

Araus, J. L., Slafer, G. A., Reynolds, M. P., and Royo, C., 2002, Plant breeding and drought in C_3 cereals: what should we breed for? *Ann. Bot.* **89**:925–940.

Baum, M., Grando, S., Backes, G., Jahoor, A., Sabbagh, A., and Ceccarelli, S., 2003, QTLs for agronomic traits in the Mediterranean environment identified in recombinant inbred lines of the cross 'Arta' × *H. spontaneum* 41–1, *Theor. Appl. Genet.* **107**:1215–1225.

Bennetzen, J. L., and Ma, J., 2003, The genetic colinearity of rice and other cereals on the basis of genomic sequence analysis, *Curr. Opin. Plant Biol.* **6**:128–133.

Birnbaum, K., Shasha, D. E., Wang, J. Y., Jung, J. W., Lambert, G. M., Galbraith, D. W., and Benfey, P. N., 2003, A gene expression map of the *Arabidopsis* root, *Science* **302**:1956–1960.

Blum, A., 1988, Breeding for stress environments, CRC Press, Boca Raton.

Blum, A., 1996, Crop responses to drought and the interpretation of adaptation, *Plant Growth Regul.* **20**:135–148.

Blum, A., Munns, R., Passioura, J. B., and Turner, N. C., 1996, Genetically engineered plants resistant to soil drying and salt stress: How to interpret osmotic relations? *Plant Physiol.* **110**:1051–1053.

Bohnert, H. J., Gong, Q., Li, P., and Ma, S., 2006, Unraveling abiotic stress tolerance mechanisms -getting genomics going, *Curr. Opin. Plant Biol.* **9**:180–188.

Borewitz, J. O., and Chory, J., 2004, Genomics tools for QTL analysis and gene discovery, *Curr. Opin. Plant Biol.* **7**:132–136.

Borsani, O., Cuartero, J., Valpuesta, V., and Botella, M. A., 2002, Tomato *tos1* mutation identifies a gene essential for osmotic tolerance and abscisic acid sensitivity, *Plant J.* 32:905–914.

Boyer, J. S., 1996, Advances in drought tolerance in plants, *Adv. Agron.* **56**:187–218.

Boyer, J. S., and Westgate, M. E., 2004, Grain yields with limited water, *J. Exp. Bot.* **55**:2385–2394.

Bray, E. A., 2002, Abscisic acid regulation of gene expression during water-deficit stress in the era of the *Arabidopsis* genome, *Plant Cell Environ.* **25**:153–161.

Breyne, P., and Zabeau, M. 2001 Genome-wide expression analysis of plant cell cycle modulated genes, *Curr. Opin. Plant Biol.* **4**:136–142.

Breyne, P., Dreesen, R., Cannoot, B., Rombaut, D., Vandepoele, K., Rombauts, S., Vanderhaeghen, R., Inze, D., and Zabeau, M., 2003, Quantitative cDNA-AFLP analysis for genome-wide expression studies, *Mol. Genet. Genomics* **269**:173–179.

Buckler, E. S. I., and Thornsberry, J. M., 2002, Plant molecular diversity and applications to genomics, *Curr. Opin. Plant Biol.* **5**:107–111.

Campos, H., Cooper, M., Edmeades, G. O., Loffler, C., Schussler, J. R., and Ibanez, M., 2006, Changes in drought tolerance in maize associated with fifty years of breeding for yield in the U.S. Corn Belt, *Maydica* **51**:369–381.

Champoux, M. C., Wang, G., Sarkarung, S., Mackill, D. J., O'Toole, J. C., Huang, N., and McCouch, S. R., 1995, Locating genes associated with root morphology and drought avoidance in rice via linkage to molecular markers, *Theor. Appl. Genet.* **90**:969–981.

Chapman, S., Cooper, M., Podlich, D., and Hammer, G., 2003, Evaluating plant breeding strategies by simulating gene action and dryland environment effects, *Agronomy J.* **95**:99–113.

Chen, C. W., Yang, Y. W., Lur, H. S., Tsai, Y. G., and Chang, M. C., 2006, A novel function of abscisic acid in the regulation of rice (*Oryza sativa* L.) root growth and development, *Plant Cell Physiol.* **47**:1–13.

Chen, W., Provart, N. J., Glazebrook, J., Katagiri, F., Chang, H., Eulgem, T., Mauch, F., Luan, S., Zou, G., Whitham, S. A., Budworth, P. R., Tao, Y., Xie, Z., Chen, X., Lam, S., Kreps, J. A., Harper, J.F., Si-Ammour, A., Mauch-Mani, B., Heinlein, M., Kobayashi, K., Hohn, T., Dangl, J. L., Wang, X., and Zhu, T., 2002, Expression profile matrix of Arabidopsis transcription factor genes suggests their putative functions in response to environmental stresses, *Plant Cell* **14**:559–574.

Colmer, T. D., Flowers, T. J., and Munns, R., 2006, Use of wide crosses and wild relatives to improve salt tolerance of wheat, *J. Exp. Bot.* **57**:1059–1078.

Colmer, T. D., Munns, R., and Flowers, T. J., 2005, Improving salt tolerance of wheat and barley: future prospects, *Aust. J. Exp. Ag.* **45**:1425–1443.

Comai, L., Young, K., Till, B. J., Reynolds, S. H., Greene, E. A., Codomo, C. A., Enns, L. C., Johnson, J. E., Burtner, C., Odden, A. R., and Henikoff, S., 2004, Efficient discovery of DNA polymorphisms in natural populations by Ecotilling, *Plant J.* **37**:778–786.

Condon, A. G., Richards, R. A., Rebetzke, G. J., and Farquhar, G. D., 2002, Improving intrinsic water-use efficiency and crop yield, *Crop Sci.*, **42**:122–131.

Consoli, L., Lefévre, C. L., Zivy, M., de Vienne, D., and Damerval, C., 2002, QTL analysis of proteome and transcriptome variations for dissecting the genetic architecture of complex traits in maize, *Plant Mol. Biol.* **48**:575–581.

Cooper, M., Chapman, S. C., Podlich, D. W., and Hammer, G. L., 2002, The GP problem: quantifying gene-to-phenotype relationships, *In Silico Biol.* **2**:151–164.

Coraggio, I., and Tuberosa, R., 2004, Improving crops' tolerance to abiotic stresses, in: *Handbook of Plant Biotechnology*, P. Christou, and H. Klee, eds., John Wiley & Sons, Ltd., Chichester, pp. 413–468.

Courtois, B., Shen, L., Petalcorin, W., Carandang, S., Mauleon, R., and Li, Z., 2003, Locating QTLs controlling constitutive root traits in the rice population IAC 165 × Co39, *Euphytica* **134**:335–345.

Davenport, R., James, R. A., Zakrisson-Plogander, A., Tester, M., and Munns, R., 2005, Control of sodium transport in durum wheat, *Plant Physiol.* **137**:807–818.

Davis, G. L., McMullen, M. D., Baysdorfer, C., Musket, T., Grant, D., Staebell, M., Xu, G., Polacco, M., Koster, L., Melia-Hancock, S., Houchins, K., Chao, S., and Coe, E. H. J., 1999, A maize map standard with sequenced core markers, grass genome reference points and 932 expressed sequence tagged sites (ESTs) in a 1736 locus map, *Genetics* **152**:1137–1172.

de Vienne, D., Leonardi, A., Damerval, C., and Zivy, M., 1999, Genetics of proteome variation for QTL characterization: application to drought-stress responses in maize, *J. Exp. Bot.* **50**:303–309.

Duvick, D. N., 2005, Contribution of breeding to yield advances in maize, in: *Advances in Agronomy*, vol. 86, D. N. Sparks, ed., Academic Press, San Diego, pp. 83–145.

Farquhar, G. D., Ehleringer, J. R., and Hubick, K. T., 1989, Carbon isotope discrimination and photosynthesis, *Ann. Rev. Plant Physiol. Plant Mol. Biol.* **40**:503–537.

Fell, D. A., 2001, Beyond genomics, *Trends Genet.* **17**:680–682.

Fiehn, O., 2002, Metabolomics - the link between genotypes and phenotypes, *Plant Mol. Biol.* **48**: 155–171.

Fitz Gerald, J. N., Lehti-Shiu, M. D., Ingram, P. A., Deak, K. I., Biesiada, T., and Malamy, J. E., 2006, Identification of quantitative trait loci that regulate Arabidopsis root system size and plasticity, *Genetics* **172**:485–498.

Flint-Garcia, S. A., Thornsberry, J. M., and Buckler, E. S. IV, 2003, Structure of linkage disequilibrium in plants, *Ann. Rev. Plant Biol.* **54**:357–374.

Flowers, T. J., Koyama, M. L., Flowers, S. A., Sudhakar, C., Singh, K. P., and Yeo, A. R., 2000, QTL: their place in engineering salt tolerance of rice to salinity, *J. Exp. Bot.* **51**:99–106.

Forster, B. P., Ellis, R. P., Moir, J., Talamè, V., Sanguineti, M. C., Tuberosa, R., This, D., Teulat-Merah, B., Ahmed, I. A., Mariy, S. A., Bahri, H., El Quahabi, M., Zoumarou-Wallis, N., El-Fellah, M., and Ben Salem, M., 2004, Genotype and phenotype associations with drought tolerance in barley tested in North Africa, *Ann. Appl. Biol.* **144**:157–168.

Forster, B. P., Ellis, R. P., Thomas, W. T. B., Newton, A. C., Tuberosa, R., This, D., El Enein, R. A., Bahri, M. H., and Ben Salem, M., 2000, The development and application of molecular markers for abiotic stress tolerance in barley, *J. Exp. Bot.* **51**:19–27.

Giuliani, S., Clarke, J., Kreps, J. A., Sanguineti, M. C., Salvi, S., Landi, P., Zhu, T., and Tuberosa, R., 2005a, Microarray analysis of backcrossed-derived lines differing for *root-ABA1*, a major QTL controlling root characteristics and ABA concentration in maize, in: *Proceedings of an International Congress "In the Wake of the Double Helix: From the Green Revolution to the Gene Revolution", 27–31 May 2003, Bologna, Italy*, R. Tuberosa, R. L. Phillips and M. Gale, eds., Avenue media, Bologna, pp. 463–490.

Giuliani, S., Sanguineti, M. C., Tuberosa, R., Bellotti, M., Salvi, S., and Landi, P., 2005b, *Root-ABA1*, a major constitutive QTL, affects maize root architecture and leaf ABA concentration at different water regimes, *J. Exp. Bot.* **56**:3061–3070.

Gorantla, M., Babu, P. R., Lachagari, V. B. R., Feltus, F. A., Paterson, A. H., and Reddy, A. R., 2005, Functional genomics of drought stress response in rice: transcript mapping of annotated unigenes of an indica rice (*Oryza sativa* L. cv. Nagina 22), *Curr. Sci.* **89**:496–514.

Grandillo, S., and Tanksley, S. D., 2005, Advanced backcross QTL analysis: results and perspectives, in: *Proceedings of an International Congress "In the Wake of the Double Helix: From the Green Revolution to the Gene Revolution", 27–31 May 2003, Bologna, Italy*, R. Tuberosa, R. L. Phillips and M. Gale, eds., Avenue media, Bologna, pp. 115–132.

Granier, C., Aguirrezabal, L., Chenu, K., Cookson, S. J., Dauzat, M., Hamard, P., Thioux, J. J., Rolland, G., Bouchier-Combaud, S., Lebaudy, A., Muller, B., Simonneau, T., and Tardieu, F., 2006, PHENOPSIS, an automated platform for reproducible phenotyping of plant responses to soil water deficit in *Arabidopsis thaliana* permitted the identification of an accession with low sensitivity to soil water deficit, *New Phytol.* **169**:623–635.

Guo, H. S., Xie, Q., Fei, J. F., and Chua, N. H., 2005, MicroRNA directs mRNA cleavage of the transcription factor NAC1 to downregulate auxin signals for *Arabidopsis* lateral root development, *Plant Cell* **17**:1376–1386.

Gupta, P. K., and Rustgi, S., 2004, Molecular markers from the transcribed/expressed region of the genome in higher plants, *Funct. Integr. Genomics* **4**:139–162.

Haley, C., and de Koning, D. J., 2006, Genetical genomics in livestock: potentials and pitfalls, *Anim. Genet.* **37**:10–12.

Hammer, G., Cooper, M., Tardieu, F., Welch, S., Walsh, B., van Eeuwijk, F., Chapman, S., and Podlich, D., 2006, Models for navigating biological complexity in breeding improved crop plants, *Trends Plant Sci.* **11**:587–593.

Hasegawa, P. M., Bressan, R. A., Zhu, J. K., and Bohnert, H. J., 2000, Plant cellular and molecular responses to high salinity, *Ann. Rev. Plant Physiol. Plant Mol. Biol.* **51**:463–499.

Hazen, S. P., Wu, Y. J., and Kreps, J. A., 2003, Gene expression profiling of plant responses to abiotic stress, *Funct. Integr. Genomics* **3**:105–111.

Hilson, P., Allemeersch, J., Altmann, T., Aubourg, S., Avon, A., Beynon, J., Bhalerao, R. P., Bitton, F., Caboche, M., Cannoot, B., Chardakov, V., Cognet-Holliger, C., Colot, V., Crowe, M., Darimont, C., Durinck, S., Eickhoff, H., Falcon de Longevialle, A., Farmer, E. E., Grant, M., Kuiper, M. T. R., Lehrach, H., Léon, C., Leyva, A., Lundeberg, J., Lurin, C., Moreau, Y., Nietfeld, W., Paz-Ares, J., Reymond, P., Rouzé, P., Sandberg, G., Segura, M. D., Serizet, C., Tabrett, A., Taconnat, L., Thareau, V., Van Hummelen, P., Vercruysse, S., Vuylsteke, M., Weingartner, M., Weisbeek, P. J., Wirta, V., Wittink, F. R. A., Zabeau M., and Small I., 2004, Versatile gene-specific sequence tags for *Arabidopsis* functional genomics: transcript profiling and reverse, *Genome Res.* **14**:2176–2189.

Hirai, M. Y., Klein, M., Fujikawa, Y., Yano, M., Goodenowe, D. B., Yamazaki, Y., Kanaya, S., Nakamura, Y., Kitayama, M., Suzuki, H., Sakurai, N., Shibata, D., Tokuhisa, J., Reichelt, M., Gershenzon, J., Papenbrock, J., and Saito, K., 2005, Elucidation of gene-to-gene and metabolite-to-gene networks in arabidopsis by integration of metabolomics and transcriptomics, *J. Biol. Chem.* **280**:25590–25595.

Hochholdinger, F., Sauer, M., Dembinsky, D., Hoecker, N., Muthreich, N., Saleem, M., and Yan, L., 2006, Proteomic dissection of plant development, *Proteomics* **6**:4076–4083.

Hochholdinger, F., Woll, K., Guo, L., and Schnable, P. S., 2005, The accumulation of abundant soluble proteins changes early in the development of the primary roots of maize (*Zea mays* L.), *Proteomics* **5**:4885–4893.

Huang, S., Spielmeyer, W., Lagudah, E. S., James, R. A., Platten, J. D., Dennis, E. S., and Munns, R., 2006, A sodium transporter (HKT7) is a candidate for *Nax1*, a gene for salt tolerance in durum wheat, *Plant Physiol.* **142**:1–11.

Husain, S., von Caemmerer, S., and Munns, R., 2004, Control of salt transport from roots to shoots of wheat in saline soil, *Funct. Plant Biol.* **31**:1115–1126.

Inan, G., Zhang, Q., Li, P., Wang, Z., Cao, Z., Zhang, H., Zhang, C., Quist, T. M., Goodwin, S. M., Zhu, J., Shi, H., Damsz, B., Charbaji, T., Gong, Q., Ma S., Fredricksen, M., Galbraith, D. W., Jenks, M. A., Rhodes, D., Hasegawa, P. M., Bohnert, H. J., Joly, R. J., Bressan, R. A., and Zhu, J.-K., 2004, Salt cress. A halophyte and cryophyte Arabidopsis relative model system and its applicability to molecular genetic analyses of growth and development of extremophiles, Plant Physiol. **135**:1718–1737.

James, R. A., Davenport, R., and Munns, R., 2006, Physiological characterisation of two genes for Na^+ exclusion in wheat: *Nax1* and *Nax2*, *Plant Physiol.* **142**:1537–1547.

Jeanneau, M., Gerentesb, D., Foueillassarc, X., Zivyd, M., Vidala, J., Toppand, A., and Perez, P., 2002a, Improvement of drought tolerance in maize: towards the functional validation of the *Zm-Asr1* gene and increase of water use efficiency by over-expressing C4PEPC, *Biochimie* **84**:1127–1135.

Jeanneau, M., Vidal, J., Gousset-Dupont, A., Lebouteiller, B., Hodges, M., Gerentes, D., and Perez, P., 2002b, Manipulating PEPC levels in plants, *J. Exp. Bot.* **53**:1837–1845.

Johnson, W. C., Jackson, L. E., Ochoa, O., Peleman, J., St Clair, D. A., Michelmore, R. W., and van Wijk, R., 2000, Lettuce, a shallow-rooted crop, and *Lactuca serriola*, its wild progenitor, differ at QTL determining root architecture and deep soil water exploitation, *Theor. Appl. Genet.* **101**:1066–1073.

Kasuga, M., Liu, Q., Miura, S., Yamaguchi-Shinozaki, K., and Shinozaki, K., 1999, Improving plant drought, salt, and freezing tolerance by gene transfer of a single stress-inducible transcription factor, *Nat. Biotechnol.* **17**:287–291.

Kawaguchi, R., Girke, T., Bray, E. A., and Bailey-Serres, J., 2004, Differential mRNA translation contributes to gene regulation under non-stress and dehydration stress conditions in *Arabidopsis thaliana*, *Plant J.* **38**:823–839.

Koiwa, H., Bressan, R. A., and Hasegawa, P. M., 2006, Identification of plant stress-responsive determinants in Arabidopsis by large-scale forward genetic screens, *J. Exp. Bot.* **57**:1119–1128.

Korol, A. B., Ronin, Y. I., Itskovich, A. M., Peng, J., and Nevo, E., 2001, Enhanced efficiency of quantitative trait loci mapping analysis based on multivariate complexes of quantitative traits, *Genetics* **157**:1789–1803.

Koyama, M. L., Levesley, A., Koebner, R. M. D., Flowers, T. J., and Yeo A. R., 2001, Quantitative trait loci for competent physiological traits determining salt tolerance in rice, *Plant Physiol.* **125**:406–422.

Landi, P., Sanguineti, M. C., Liu, C., Li, Y., Wang, T. Y., Giuliani, S., Bellotti, M., Salvi, S., and Tuberosa, R., 2007, *Root-ABA1* QTL affects root lodging, grain yield and other agronomic traits in maize grown under well-watered and water-stressed conditions, *J. Exp. Bot.* **58**:319–326.

Landi, P., Sanguineti, M. C., Salvi, S., Giuliani, S., Bellotti, M., Maccaferri, M., Conti, S., and Tuberosa, R., 2005, Validation and characterization of a major QTL affecting leaf ABA concentration in maize, *Mol. Breed.* **15**:291–303.

Laurie, S., Feeney, K. A., Maathuis, F. J., Heard, P. J., Brown, S. J., and Leigh, R. A., 2002, A role for HKT1 in sodium uptake by wheat roots, *Plant J.* **32**:139–149.

Lebreton, C., Lazic-Jancic, V., Steed, A., Pekic, S., and Quarrie, S.A., 1995, Identification of QTL for drought responses in maize and their use in testing causal relationships between traits, *J. Exp. Bot.* **46**:853–865.

Lee, M., 1995, DNA markers and plant breeding programs, *Adv. Agron.* **55**:265–344.

Li, Zichao, Mu, P., Li, C., Zhang, H., Li, Z., Gao, Y., and Wang, X., 2005, QTL mapping of root traits in a doubled haploid population from a cross between upland and lowland japonica rice in three environments, *Theor. Appl. Genet.* **110**:1244–1252.

Li, Zhikang, Fu, B.-Y., Gao, Y.-M., Xu, J.-L., Ali, J., Lafitte, H. R., Jiang, Y.-Z., Rey, J. D., Vijayakumar, C. H. M., Maghirang, R., Zheng, T.-Q., and Zhu, L.-H., 2005, Genome-wide introgression lines and their use in genetic and molecular dissection of complex phenotypes in rice (*Oryza sativa* L.), *Plant Mol. Biol.* **59**:33–52.

Lindsay, M. P., Lagudah, E. S., Hare, R. A., and Munns, R., 2004, A locus for sodium exclusion (*Nax1*), a trait for salt tolerance, mapped in durum wheat, *Funct. Plant Biol.* **31**:1105–1114.

Liu, J., Raveendran, M., Mushtaq, R., Ji, X., Yang, X., Bruskiewich, R., Katiyar, S., Cheng, S., Lafitte, R., and Bennett, J., 2005, Proteomic analysis of drought-responsiveness in rice: OsADF5, in: *Proceedings of an International Congress "In the Wake of the Double Helix: From the Green Revolution to the Gene Revolution", 27–31 May 2003, Bologna, Italy*, R. Tuberosa, R. L. Phillips and M. Gale, eds., Avenue media, Bologna, pp. 491–505.

Loudet, O., Gaudon, V., Trubuil, A., and Daniel-Vedele, F., 2005, Quantitative trait loci controlling root growth and architecture in *Arabidopsis thaliana* confirmed by heterogeneous inbred family, *Theor. Appl. Genet.* **110**:742–753.

Ludlow, M. M., and Muchow, R. C., 1990, A critical evaluation of traits for improving crop yields in water-limited environments, *Adv. Agron.* **43**:107–153.

Maccaferri, M., Sanguineti, M. C., and Tuberosa, R., 2005, Analysis of linkage disequilibrium in a collection of elite durum wheat genotypes, *Mol. Breed.* **15**:271–289.

Maccaferri, M., Sanguineti, M. C., Natoli, V., Araus-Ortega, J. L., Ben Salem, M., Bort, J., Chenenaoui, S., Deambrogio, E., Garcia del Moral, L., De Montis, A., El-Ahmed, A., Maalouf, F., Machlab, H., Moragues, M., Motawaj, J., Nachit, M., Nserallh, N., Ouabbou, H., Royo, C., and Tuberosa, R., 2006, A panel of elite accessions of durum wheat (*Triticum durum* Desf.) suitable for association mapping studies, *Plant Gen. Res.* **4**:79–85.

Maggio, A., Zhu, J.-K., Hasegawa, P. M., and Bressan, R. A., 2006, Osmogenetics: Aristotle to Arabidopsis, *Plant Cell* **18**:1542–1557.

Markandeya, G., Babu, P. R., Lachagari, V. B. R., Feltus, F. A., Paterson, A. H., and Reddy, A.R., 2005, Functional genomics of drought stress response in rice: transcript mapping of annotated unigenes of an indica rice (*Oryza sativa* L. cv. Nagina 22), *Curr. Sci.* **89**:496–514.

Martínez-Atienza, J., Jiang, X., Garciadeblas, B., Mendoza, I., Zhu, J.-K., Pardo, J. M., and Quinterno, F. J., 2006, Conservation of the SOS salt tolerance pathway in rice, *Plant Physiol.* **143**:1001–1012. First published on December 8, 2006; 10.1104/pp.106.092635.

Maser, P., Eckelman, B., Vaidyanathan, R., Horie, T., Fairbairn, D. J., Kubo, M., Yamagami, M., Yamaguchi, K., Nishimura, M., Uozumi, N., Robertson, W., Sussman, M. R., and Schroeder, J. I., 2002, Altered shoot/root Na^+ distribution and bifurcating salt sensitivity in Arabidopsis by genetic disruption of the Na^+ transporter AtHKT1, *FEBS Lett.* **531**:157–161.

Masle, J., Gilmore, S. R., and Farquhar, G. D., 2005, The ERECTA gene regulates plant transpiration efficiency in Arabidopsis, *Nature* **436**:866–870.

McCallum, C. M., Comai, L., Greene, E. A., and Henikoff, S., 2000, Targeted screening for induced mutations, *Nat. Biotechnol.* **18**:455–457.

McLaughlin, J. E., and Boyer, J. S., 2004, Sugar-responsive gene expression, invertase activity, and senescence aborting maize ovaries at low water potentials. *Ann. Bot.* **94**:675–689.

Mizoguchi, T., Ichimura, K., Yoshida, R., and Shinozaki, K., 2000, MAP Kinase cascade in Arabidopsis: their role in stress and hormone response, *Results Probl. Cell Differ.* **27**:29–38.

Moncada, P., Martinez, C., Borrero, J. M. C., Gauch, H. J., Guimaraes, E., Tohme, J., McCouch, S., 2001, Quantitative trait loci for yield and yield components in an *Oryza sativa* × *Oryza rufipogon* BC_2F_2 population evaluated in an upland environment, *Theor. Appl. Genet.* **102**:41–52.

Morgante, M., and Salamini, F., 2003, From plant genomics to breeding practice, *Curr. Opin. Biotech.* **14**:214–219.

Morgenthal, K., Weckwerth, W., and Steuer, R., 2006, Metabolomic networks in plants: transitions from pattern recognition to biological interpretation, *BioSystems* **83**:108–117.

Mouchel, C. F., Briggs, G. C., and Hardtke, C. S., 2004, Natural genetic variation in Arabidopsis identifies *BREVIS RADIX*, a novel regulator of cell proliferation and elongation in the root, *Genes Dev.* **18**:700–714.

Mouchel, C. F., Osmont, K. S., and Hardtke, C. S., 2006, BRX mediates feedback between brassinosteroid levels and auxin signalling in root growth, *Nature* **43**:458–461.

Munns, R. 2005, Genes and salt tolerance: bringing them together, *New Phytol.* **167**:645–663.

Munns, R., and Richards, R. A. 2007, Recent advances in breeding wheat for drought and salt stresses, in: *Advances in molecular-breeding toward drought and salt tolerant crops*, M. A. Jenks, P. M. Hasegawa, and S. Mohan Jain, eds., Springer, pp. 565–586.

Munns, R., Hare, R. A., James, R. A., and Rebetzke, G. J., 2000, Genetic variation for improving the salt tolerance of durum wheat, *Aust. J. Ag. Res.* **51**:69–74.

Munns, R., James, R. A., and Läuchli, A., 2006, Approaches to increasing the salt tolerance of wheat and other cereals, *J. Exp. Bot.* **57**:1025–1043.

Munns, R., Rebetzke, G. J., Husain, S., James, R. A., and Hare, R. A., 2003, Genetic control of sodium exclusion in durum wheat, *Aust. J. Ag. Res.* **54**:627–635.

Musket, T., Davis, D., Gerau, M., Schroeder, S., Sanchez-Villeda, H., Spollen, W., Springer, G., Poroyko, V., Bohnert, H., Hejlek, L., Sharp, R., Nguyen, H., and Davis, G., 2005, A maize root transcriptome map, *47*th *Annual Maize Meeting*, Lake Geneva, Wisconsin, USA, pp. 106.

Nakazono, M., Qiu, F., Borsuk, L. A., and Schnable, P. S., 2003, Laser-capture microdissection, a tool for the global analysis of gene expression in specific plant cell types: identification of genes expressed differentially in epidermal cells or vascular tissues of maize, *Plant Cell* **15**:583–596.

Nettleton, D., and Wang, D., 2006, Selective transcriptional profiling for trait-based eQTL mapping, *Anim. Genet.* **37**:13–17.

Nguyen, H. T., and Blum, A., 2004, *Physiology and Biotechnology Integration for Plant Breeding*, Marcel Dekker, Inc., New York.

Nguyen, H. T., Babu, R. C., and Blum, A., 1997, Breeding for drought resistance in rice: physiology and molecular genetics considerations, *Crop Sci.* **37**:1426–1434.

Okushima, Y., Overvoorde, P. J., Arima, K., Alonso, J. M., Chan, A., Chang, C., Ecker, J. R., Hughes, B., Lui, A., Nguyen, D., Onodera, C., Quach, H., Smith, A., Yu, G., and Theologis, A., 2005, Functional genomic analysis of the AUXIN RESPONSE FACTOR gene family members in *Arabidopsis thaliana*: unique and overlapping functions of ARF7 and ARF19, *Plant Cell* **17**:444–463.

Oono, Y., Seki, M., Nanjo, T., Narusaka, M., Fujita, M., Satoh, R., Satou, M., Sakurai, T., Ishida, J., Akiyama, K., Lida, K., Maruyama, K., Satoh, S., Yamaguchi-Shinozaki, K., and Shinozaki, K.,

2003, Monitoring expression profiles of Arabidopsis gene expression during rehydration process after dehydration using ca. 7000 full-length cDNA microarray, *Plant J.* **34**:868–887.

Ozturk, Z. N., Talamè, V., Deyholos, M., Michalowski, C. B., Galbraith, D. W., Gozukirmizi, N., Tuberosa, R., and Bohnert, H. J., 2002, Monitoring large-scale changes in transcript abundance in drought- and salt-stressed barley, *Plant Mol. Biol.* **48**:551–573.

Pakniyat, H., Powell, W., Baird, E., Handley, L. L., Robinson, D., Scrimgeour, C. M., Nevo, E., Hackett, C. A., Caligari, P. D. S., and Forster, B. P., 1997, AFLP variation in wild barley (*Hordeum spontaneum* C. Koch) with reference to salt tolerance and associated ecogeography, *Genome* **40**: 332–341.

Passioura, J. B., 2002, Environmental biology and crop improvement, *Funct. Plant Biol.* **29**:537–546.

Passioura, J. B., 2007, The drought environment: physical, biological and agricultural perspectives, *J. Exp. Bot.* **58**:113–118.

Passioura, J. B., Spielmeyer, W. I., and Bonnett, D. G., 2007, Requirements for success in marker-assisted breeding for drought-prone environments in: *Advances in molecular-breeding toward drought and salt tolerant crops*, M. A. Jenks, P. M. Hasegawa, and S. Mohan Jain, eds., Springer, pp. 479–500.

Peleman, J. D., and Van der Voort, J. R., 2003, Breeding by design, *Trends Plant Sci.* **8**:330–334.

Pelleschi, S., Guy, S., Kim, J. Y., Pointe, C., Mahe, A., Barthes, L., Leonardi, A., and Prioul, J. L., 1999, *Ivr2*, a candidate gene for a QTL of vacuolar invertase activity in maize leaves. Gene-specific expression under water stress, *Plant Mol. Biol.* **39**:373–380.

Pelleschi, S., Leonardi, A., Rocher, J. P., Cornic, G., de Vienne, D., Thévenot, C., and Prioul, J. L., 2006, Analysis of the relationships between growth, photosynthesis and carbohydrate metabolism using quantitative trait loci (QTLs) in young maize plants subjected to water deprivation, *Mol. Breed.* **17**:21–39.

Pflieger, S., Lefebvre, V., and Causse, M. 2001, The candidate gene approach in plant genetics: a review, *Mol. Breed.* **7**:275–291.

Price, A. H., 2006, Believe it or not, QTLs are accurate! *Trends Plant Sci.* **11**:213–216.

Price, A., Steele, K., Gorham, J., Bridges, J., Moore, B., Evans, J., Richardson, P., and Jones, R., 2002, Upland rice grown in soil-filled chambers and exposed to contrasting water-deficit regimes I. Root distribution, water use and plant water status, *Field Crop Res.* **76**:11–24.

Pritchard, J. K., Stephens, M., Rosenberg, N. A., and Donnelly, P., 2000, Association mapping in structured populations, *Am. J. Hum. Genet.* **67**:170–181.

Qi, L. L., Echalier, B., Chao, S., Lazo, G. R., Butler, G. E., Anderson, O. D., Akhunov, E. D., Dvoák, J., Linkiewicz, A. M., Ratnasiri, A., Dubcovsky, J., Bermudez-Kandianis, C. E., Greene, R. A., Kantety, R., La Rota, C. M., Munkvold, J. D., Sorrells, S. F., Sorrells, M. E., Dilbirligi, M., Sidhu, D., Erayman, M., Randhawa, H. S., Sandhu, D., Bondareva, S. N. K., Gill, S., Mahmoud, A. A., Ma, X.-F., Miftahudin, Gustafson, J. P., Conley, E. J., Nduati, V., Gonzalez-Hernandez, J. L., Anderson, J. A., Peng, J. H., Lapitan, N. L. V., Hossain, K. G., Kalavacharla, V., Kianian, S. F., Pathan, M. S., Zhang, D. S., Nguyen, H. T., Choi, D.-W., Fenton, R. D., Close, T. J., McGuire, P. E., Qualset, C. O., and Gill B. S., 2004, A chromosome bin map of 16,000 expressed sequence tag loci and distribution of genes among the three genomes of polyploid wheat, *Genetics* **168**:701–712.

Qiu, Q.-S., Guo, Y., Quintero, F. J., Pardo, J. M., Schumaker, K. S., and Zhu, J.-K., 2004, Regulation of vacuolar Na^+/H^+ exchange in *Arabidopsis thaliana* by the salt-overly-sensitive (SOS) pathway, *J. Biol. Chem.* **279**:207–215.

Quarrie, S. A., 1991, Implications of genetic differences in ABA accumulation for crop production, in: *Abscisic acid: physiology and biochemistry*, W. J. Davies and H. G. Jones, eds., Bios Scientific Publishers, Oxford, pp. 227–243.

Quesada, V., Garcia-Martinez, S., Piqueras, P., Ponce, M. R., and Micol, J. L., 2002, Genetic architecture of NaCl tolerance in Arabidopsis, *Plant Physiol.* **130**:951–963.

Quesada, V., Ponce, M.R., and Micol, J.L., 2000, Genetic analysis of salt-tolerant mutants in *Arabidopsis thaliana*, *Genetics* **154**:421–436.

Rafalski, A., and Morgante, M., 2004, Corn and humans: recombination and linkage disequilibrium in two genomes of similar size, *Trends Genet.* **20**:103–111.

Remington, D. L., Thornsberry, J. M., Matsuoka, Y., Wilson, L. M., Whitt, S. R., Doebley, J., Kresovich, S., Goodman, M. M., Buckler, E. S. IV, 2001, Structure of linkage disequilibrium and phenotypic associations in the maize genome, *P. Natl. Acad. Sci. USA* **98**:11479–11484.

Ren, Z.-H., Gao, J.-P., Li, L.-G., Cai, X.-L., Huang, W., Chao, D.-Y., Zhu, M.-Z., Wang, Z.-Y., Luan, S., and Lin, H.-X., 2005, A rice quantitative trait locus for salt tolerance encodes a sodium transporter, *Nat. Genet.* **37**:1141–1146.

Rensink, W. A., 2005, Microarray expression profiling resources for plant genomics, *Trends Plant Sci.* **10**:603–609.

Reymond, M., Muller, B., and Tardieu, F., 2004, Dealing with the genotypexenvironment interaction via a modelling approach: a comparison of QTLs of maize leaf length or width with QTLs of model parameters, *J. Exp. Bot.* **55**:2461–2472.

Reymond, M., Muller, B., Leonardi, A., Charcosset, A., and Tardieu, F., 2003, Combining quantitative trait loci analysis and an ecophysiological model to analyze the genetic variability of the responses of maize leaf growth to temperature and water deficit, *Plant Physiol.* **131**:664–675.

Ribaut, J. M., and Hoisington, D., 1998, Marker-assisted selection: new tools and strategies, *Trends Plant Sci.* **3**:236–239.

Ribaut, J. M., Banziger, M., Betran, J., Jiang, C., Edmeades, G. O., Dreher, K., and Hoisington, D., 2002, Use of molecular markers in plant breeding: drought tolerance improvement in tropical maize, in: *Quantitative Genetics, Genomics, and Plant Breeding*, M. S. Kang, ed., CABI Publishing, Wallingford, pp. 85–99.

Richards, R. A., 1996, Defining selection criteria to improve yield under drought, *Plant Growth Regul.* **20**:157–166.

Richards, R. A., 2000, Selectable traits to increase crop photosynthesis and yield of grain crops, *J. Exp. Bot.* **51**:447–458.

Richards, R. A., 2006, Physiological traits used in the breeding of new cultivars for water-scarce environments., *Agr. Water Manage.* **80**:197–211.

Richards, R. A., Rebetzke, G. J., Condon, A. G., and van Herwaarden, A. F., 2002, Breeding opportunities for increasing the efficiency of water use and crop yield in temperate cereals, *Crop Sci.* **42**:111–121.

Robertson, D. S., 1985, A possible technique for isolating genic DNA for quantitative traits in plant, *J. Theor. Biol.* **117**:1–10.

Rodryguez-Navarro, A., and Rubio, F., 2006, High-affinity potassium and sodium transport systems in plants, *J. Exp. Bot.* **57**:1149–1160.

Sakurai, N., and Shibata, D., 2006, KaPPA-View for integrating quantitative transcriptomic and metabolomic data on plant metabolic pathway maps, *J. Pestic. Sci.* **31**:293–295.

Salekdeh, G. H., Siopongco, J., Wade, L. J., Ghareyazie, B., and Bennett, J., 2002, A proteomic approach to analyzing drought- and salt-responsiveness in rice, *Field Crops Res.* **76**:199–219.

Salvi, S., and Tuberosa, R., 2005, To clone or not to clone plant QTLs: present and future challenges, *Trends Plant Sci.* **10**:297–304.

Salvi, S., Costrini, P., Reynard, J. S., Zhao, Q., and Tuberosa, R., 2005, Mapping major QTLs for seminal root traits and flowering time using an introgression library in maize, Volume of Abstracts of *Plant GEMs 4*, Amsterdam, The Netherlands, p. 148.

Salvi, S., Sponza, G., Morgante, M., Fengler, K., Meeley, R., Ananiev, E., Svitashev, S., Bruggemann, E., Niu, X., Li, B., Hainey, C. F., Rafalski, A., Tingey, S. V., Tomes, D., Miao, G.-H., Phillips, R. L., and Tuberosa. R., 2007, Conserved non-coding genomic sequences controlling flowering time differences in maize, *P. Natl. Acad. Sci* 104:11376–11381.

Salvi, S., Tuberosa, R., Chiapparino, E., Maccarerri, M., Veillet, S., Van Beuningen, L., Isaac, P., Edwards, K., Phillips, R. L., 2002, Toward positional cloning of *Vgt1*, a QTL controlling the transition from the vegetative to the reproductive phase in maize, *Plant Mol. Biol.* **48**:601–613.

Sanchez, A. C., Subudhi, P. K., Rosenow, D. T., and Nguyen, H. T., 2002, Mapping QTLs associated with drought resistance in sorghum (*Sorghum bicolor* L. Moench), *Plant Mol. Biol.* **48**:713–726.

Sanguineti, M. C., Duvick, D. N., Smith, S., Landi, P., and Tuberosa, R., 2006, Effects of long-term selection on seedling traits and ABA accumulation in commercial maize hybrids, *Maydica* **51**: 329–338.

Sauer, M., Jakob, A., Nordheim, A., and Hochholdinger, F., 2006, Proteomic analysis of shoot-borne root initiation in maize (*Zea mays* L.), *Proteomics* **6**:2530–2541.

Sawkins, M. C., Farmer, A. D., Hoisington, D., Sullivan, J., Tolopko, A., Jiang, Z., and Ribaut, J. M., 2004, Comparative map and trait viewer (CMTV): an integrated bioinformatic tool to construct consensus maps and compare QTL and functional genomics data across genomes and experiments, *Plant Mol. Biol.* **56**:465–480.

Schadt, E. E., Monks, S.A., Drake, T. A., Lusis, A. J., Che, N., Colinayo, V., Ruff, T. G., Milligan, S. B., Lamb, J. R., Cavet, G., Linsley, P. S., Mao, M., Stoughton, R. B., and Friend, S.H., 2003, Genetics of gene expression surveyed in maize, mouse and man, *Nature* **422**:297–302.

Schnable, P. S., Hochholdinger, F., and Nakazono, M., 2004, Global expression profiling applied to plant development, *Curr. Opin. Plant Biol.* **7**:50–56.

Seki, M., Kamei, A., Yamaguchi-Shinozaki, K., and Shinozaki, K., 2003, Molecular responses to drought, salinity and frost: common and different paths for plant protection, *Curr. Opin. Biotech.* **14**:194–199.

Sergeeva, L. I., Keurentjes, J. J., Bentsink, L., Vonk, J., van der Plas, L. H., Koornneef, M., and Vreugdenhil, D., 2006, Vacuolar invertase regulates elongation of *Arabidopsis thaliana* roots as revealed by QTL and mutant analysis, *P. Natl. Acad. Sci. USA* **103**:2994–2999.

Sharp, R. E., Poroyko, V., Hejlek, L. G., Spollen, W. G., Springer, G. K., Bohnert, H. J., and Nguyen, H.T., 2004, Root growth maintenance during water deficits: physiology to functional genomics, *J. Exp. Bot.* **55**:2343–2351.

Shen, L., Courtois, B., McNally, K. L., Robin, S., and Li, Z., 2001, Evaluation of near-isogenic lines of rice introgressed with QTLs for root depth through marker-aided selection, *Theor. Appl. Genet.* **103**:75–83.

Shinozaki, K., and Yamaguchi-Shinozaki, K., 1996, Molecular responses to drought and cold stress, *Curr. Opin. Biotech.* **7**:161–167.

Shinozaki, K., and Yamaguchi-Shinozaki, K., 1997, Gene expression and signal transduction in water-stress response, *Plant Physiol.* **115**:327–334.

Signora, L., De Smet, I., Foyer, C. H., and Zhang, H., 2001, ABA plays a central role in mediating the regulatory effects of nitrate on root branching in Arabidopsis, *Plant J.* **28**:655–662.

Slafer, G. A., 2003, Genetic basis of yield as viewed from a crop physiologist's perspective, *Ann. Appl. Biol.* **142**:117–128.

Sorrells, M. E., La Rota, M., Bermudez-Kandianis, C. E., Greene, R. A., Kantety, R., Munkvold, J. D., Miftahudin, Mahmoud, A., Ma, X., Gustafson, P. J., Qi, L. L., Echalier, B., Gill, B. S., Matthews, D. E., Lazo, G. R., Chao, S., Anderson, O. D., Edwards, H., Linkiewicz, A. M., Dubcovsky, J., Akhunov, E. D., Dvorak, J., Zhang, D., Nguyen, H. T., Peng, J., Lapitan, N. L. V., Gonzalez-Hernandez, J. L., Anderson, J. A., Hossain, K., Kalavacharla, V., Kianian, S. F., Choi, D. W., Close, T. J., Dilbirligi, M., Gill, K. S., Steber, C., Walker-Simmons, M. K., McGuire, P. E., and Qualset, C. O., 2003, Comparative DNA sequence analysis of wheat and rice genomes, *Genome Res.* **13**:1818–1827.

Steele, K. A., Price, A. H., Shashidha, H. E., and Witcombe, J. R., 2006, Marker-assisted selection to introgress rice QTLs controlling root traits into an Indian upland rice variety, *Theor. Appl. Genet.* **112**:208–221.

Stemple, D. L., 2004, TILLING: a high-throughput harvest for functional genomics, *Nat. Rev. Genet.* **5**:145–150.

Stephanopoulos, G., Alper, H., and Moxley, J., 2004, Exploiting biological complexity for strain improvement through systems biology, *Nat. Biotechnol.* **22**:1261–1267.

Steuer, R., Kurths, J., Fiehn, O., and Weckwerth, W., 2003, Observing and interpreting correlations in metabolomic networks, *Bioinformatics* **19**:1019–1026.

Stockinger, E. J., Gilmour, S.J., and Thomashow, M.F., 1997, *Arabidopsis thaliana* CBF1 encodes an AP2 domain containing transcriptional activator that binds to the C-repeat/DRE, a cis-acting DNA regulatory element that stimulates transcription in response to low temperature and water deficit, *P. Natl. Acad. Sci. USA* **94**:1035–1040.

Taji, T., Seki, M., Satou, M., Sakurai, T., Kobayashi, M., Ishiyama, K., Narusaka, Y., Narusaka, M., Zhu, J. K., and Shinozaki, K., 2004, Comparative genomics in salt tolerance between Arabidopsis and arabidopsis-related halophyte salt cress using Arabidopsis microarray, *Plant Physiol.* **135**:1697–709.

Talamè V., Ozturk, N. Z., Bohnert, H. J., and Tuberosa, R., 2007, Barley transcript profiles under dehydration shock and drought stress treatments: a comparative analysis, *J. Exp. Bot.* **58**:229–240.

Talamè, V., Sanguineti, M. C., Chiapparino, E., Bahri, H., Ben Salem, M., Forster, B. P., Ellis, R. P., Rhouma, S., Zoumarou, W., Waugh, R., and Tuberosa, R., 2004, Identification of *Hordeum spontaneum* QTL alleles improving field performance of barley grown under rainfed conditions, *Ann. Appl. Biol.* **144**:309–319.

Tanksley, S. D., 1993, Mapping polygenes, *Ann. Rev. Genet.* **27**:205–233.

Tanksley, S. D., and McCouch, S. R., 1997, Seed banks and molecular maps: unlocking genetic potential from the wild, *Science* **277**:1063–1066.

Tanksley, S., and Nelson, J., 1996, Advanced backcross QTL analysis: a method for the simultaneous discovery and transfer of valuable QTLs from unadapted germplasm into elite breeding lines, *Theor. Appl. Genet.* **92**:191–203.

Tanksley, S., Grandillo, S., Fulton, T., Zamir, D., Eshed, T., Petiard, V., Lopez, J., and Beck, B.T., 1996, Advanced backcross QTL analysis in a cross between an elite processing line of tomato and its wild relative L. *pimpinellifolium*, *Theor. Appl. Genet.* **92**:213–224.

Tester, M., and Davenport, R., 2003, Na^+ tolerance and Na^+ transport in higher plants, *Ann. Bot.* **91**:503–527.

Tollenaar, M., and Wu, J., 1999, Yield improvement in temperate maize is attributable to greater stress tolerance, *Crop Sci.* **39**:1597–1604.

Touzet, P., Winkler, R. G., and Helentjaris, T., 1995, Combined genetic and physiological analysis of a locus contributing to quantitative variation, *Theor. Appl. Genet.* **91**:200–205.

Trethowan, R. M., van Ginkel, M., and Rajaram, S., 2002, Progress in breeding for yield and adaptation in global drought affected environments, *Crop Sci.* **42**:1441–1446.

Tuberosa, R., and Salvi, S., 2004, QTLs and genes for tolerance to abiotic stress in cereals, in: *Cereal Genomics*, R. Varshney and P. K. Gupta, eds., Kluwer Academic Publishers, Dordrecht, pp. 253–315.

Tuberosa, R., and Salvi, S., 2006, Genomics approaches to improve drought tolerance in crops, *Trends Plant Sci.* **11**:415–412.

Tuberosa, R., Gill, B. S., and Quarrie, S. A., 2002a, Cereal genomics: ushering in a brave new world, *Plant Mol. Biol.* **48**:445–449.

Tuberosa, R., Salvi, S., Sanguineti, M. C., Landi, P., Maccaferri, M., and Conti, S., 2002b, Mapping QTLs regulating morpho-physiological traits and yield: case studies, shortcomings and perspectives in drought-stressed maize, *Ann. Bot.* **89**: 941–963.

Tuberosa, R., Sanguineti, M. C., Landi P., Salvi S., and Conti S., 1998, RFLP mapping of quantitative trait loci controlling abscisic acid concentration in leaves of drought-stressed maize (*Zea mays* L.), *Theor. Appl. Genet.* **97**:744–755.

Tuinstra, M. R., Ejeta, G., and Goldsbrough, P., 1998, Evaluation of near-isogenic sorghum lines contrasting for QTL markers associated with drought tolerance, *Crop Sci.* **38**:835–842.

Turner, N. C., 1997, Further progress in crop water relations, *Adv. Agron.* **528**:293–338.

Utz, H. F., and Melchinger, A. E., 1996, PLABQTL: a program for composite interval mapping of QTL, *J. Agric. Genomics* **2**:1–6.

Varshney, R .K., Graner, A., and Sorrells, M. E., 2005, Genomics-assisted breeding for crop improvement, *Trends Plant Sci.* **10**:621–30.

Vuylsteke, M., van den Daele, H., Vercauteren, A., Zabeau, M., and Kuiper, M., 2006, Genetic dissection of transcriptional regulation by cDNA-AFLP, *Plant J.* **45**:439–446.

Waterhouse, P. M., and Helliwell, C. A., 2003, Exploring plant genomes by RNA-induced gene silencing, *Nat. Rev. Genet.* **4**:29–38.

Wayne, M. L., and McIntyre, L., 2002, Combining mapping and arraying: an approach to candidate gene identification, *P. Natl. Acad. Sci. USA* **99**:14903–14906.

Wen, T. J., Hochholdinger, F., Sauer, M., Bruce, W., and Schnable, P. S., 2005, The *roothairless1* gene of maize encodes a homolog of sec3, which is involved in polar exocytosis, *Plant Physiol.* **138**:1637–1643.

Werner, J., and Finkelstein, R. R., 1995, *Arabidopsis* mutants with reduced response to NaCl and osmotic stress, *Physiol. Plant.* **93**:659–666.

Woll, K., Borsuk, L. A., Stransky, H., Nettleton, D., Schnable, P. S., and Hochholdinger, F., 2005, Isolation, characterization, and pericycle-specific transcriptome analyses of the novel maize lateral and seminal root initiation mutant rum1, *Plant Physiol.* **139**:1255–1267.

Xiong, L., Wang, R.-G., Mao, G., and Koczan, J. M., 2006, Identification of drought tolerance determinants by genetic analysis of root response to drought stress and abscisic acid, *Plant Physiol.* **142**:1065–1074.

Yamasaki, M., Tenaillon, M. I., Bi, I. V., Schroeder, S. G., Sanchez-Villeda, H., Doebley, J. F., Gaut, B. S., and McMullen, M. D., 2005, A large-scale screen for artificial selection in maize identifies candidate agronomic loci for domestication and crop improvement, *Plant Cell* **17**:2859–2872.

Yazaki, J., Kojima, K., Suzuki, K., Kishimoto, N., and Kikuchi, S., 2004, The Rice PIPELINE: a unification tool for plant functional genomics, *Nucleic Acids Res.* **32**:383–387.

Yin, X. Y., Stam, P., Kropff, M. J., and Schapendonk, A. H. C. M., 2003, Crop modeling, QTL mapping, and their complementary role in plant breeding, *Agron. J.* **95**:90–98.

Yin, X., Struik, P. C., and Kropff, M. J., 2004, Role of crop physiology in predicting gene-to-phenotype relationships, *Trends Plant Sci.* **9**:426–432.

Yu, L. X., and Setter, T. L., 2003, Comparative transcriptional profiling of placenta and endosperm in developing maize kernels in response to water deficit, *Plant Physiol.* **131**:568–582.

Zamir, D., 2001, Improving plant breeding with exotic genetic libraries, *Nat. Rev. Genet.* **2**:983–989.

Zeng, Z. B., 1994, Precision mapping of quantitative trait loci, *Genetics* **136**:1457–1468.

Zhang, W. X., Ruan, J. H., Ho, T. H. D., You, Y. S., Yu, T. T., and Quatrano, R. S., 2005, Cis-regulatory element based targeted gene finding: genome-wide identification of abscisic acid- and abiotic stress-responsive genes in *Arabidopsis thaliana*, *Bioinformatics* **21**:3074–3081.

Zheng, B. S., Yang, L., Zhang, W. P., Mao, C. Z., Wu, Y. R., Yi, K. K., Liu, F. Y., Wu, P., 2003, Mapping QTLs and candidate genes for rice root traits under different water-supply conditions and comparative analysis across three populations, *Theor. Appl. Genet.* **107**:1505–1515.

Zheng, H. G., Babu, R. C., Pathan, M. S., Ali, L., Huang, N., Courtois, B., and Nguyen, H.T., 2000, Quantitative trait loci for root-penetration ability and root thickness in rice: comparison of genetic backgrounds, *Genome* **43**:53–61.

Zhu, J. K., 2001, Plant salt tolerance, *Trends Plant Sci.* **6**:66–71.

Zhu, J. K., Hasegawa, P., and Bressan, R., 1997, Molecular aspects of osmotic stress in plants, *Crit. Rev. Plant Sci.* **16**:253–277.

Zhu, J., Chen, S., Alvarez, S., Asirvatham, V. S., Schachtman, D. P., Wu, Y., and Sharp, R. E., 2006, Cell wall proteome in the maize primary root elongation zone. I. Extraction and identification of water-soluble and lightly ionically bound proteins, *Plant Physiol.* **140**:311–325.

Zinselmeier, C., Jeong, B. R., and Boyer, J. S., 1999, Starch and the control of kernel number in maize at low water potentials, *Plant Physiol.* **121**:25–36.

Zinselmeier, C., Sun, Y., Helentjaris, T., Beatty, M., Yang, S., Smith, H., and Habben, J., 2002, The use of gene expression profiling to dissect the stress sensitivity of reproductive development in maize, *Field Crop Res.* **75**:111–121.

Zivy, M., and de Vienne, D., 2000, Proteomics: a link between genomics, genetics and physiology, *Plant Mol. Biol.* **44**:575–580.

CHAPTER 17

INDUCED MUTATIONS FOR ENHANCING SALINITY TOLERANCE IN RICE

CHIKELU MBA*[,1], ROWNAK AFZA[1], SHRI MOHAN JAIN[2,3], GLENN B. GREGORIO[4,5] AND FRANCISCO JAVIER ZAPATA-ARIAS[1,6]

[1] *Plant Breeding Unit, Joint FAO/IAEA Agriculture and Biotechnology Laboratory, International Atomic Energy Agency Laboratories, A-2444 Seibersdorf, Austria*

[2] *Plant Breeding and Genetics Section, Joint FAO/IAEA Division, International Atomic Energy Agency, Vienna, Austria*

[3] *Current address – Department of Applied Biology, University of Helsinki, Helsinki, Finland*

[4] *International Rice Research Institute, Los Baños, Laguna, Philippines*

[5] *Current address – African Rice Center (WARDA), c/o Internationa Institute of Tropicaal Agrioculture, Oyo Road, PMB 5320, Ibadan, Nigeria*

[6] *Current address – Bashati Vila, Apartment 5A, House 25, Road 68/A, Gulshan 1212, Dhaka Bangladesh*

**E-mail: c.mba@iaea.org*

Abstract: Salt accumulation in soil surfaces, known as soil salinity, could lead to the impairment of plant growth and development and is manifested mostly under irrigated and dryland agriculture. Excess salts in the soil affects plants through osmotic stress; accumulation to toxic levels within the cells; and through the interference with the uptake of mineral nutrients. Rice productivity in several parts of the world is therefore severely limited by salinity on account of the prevalence of irrigation in rice farming. Tolerance to salt toxicity in plants is a genetic and physiologically complex trait. Halophytes (salt tolerant plants) are different from the salt-sensitive glycophytes in terms of peculiarities in their anatomy, ability to sequester otherwise toxic ions, and other physiologic processes. It is logical therefore to infer complexity also at the genetic level on account of the several pathways involved in these mechanisms. These complexities have confounded genetic improvement strategies for salinity tolerance in plants resulting in a paucity of saline tolerant plants, with only about 30 officially released saline tolerant crop varieties world-wide. Only one saline tolerant rice variety, Bicol, has been officially released to farmers. We review strategies being currently employed in the development of saline tolerant rice varieties. These include conventional plant breeding which is hampered by the lack of suitable genetic variation for this trait; the modest progress made through doubled haploidy; and the reliance on somaclonal variation, an unsustainably unpredictable strategy. This review also posits that while genetic transformation has led to the modification of certain physiological indices implicated in salinity tolerance in rice, in isolation, these modifications have not been translated to improved yield under salt stress. A more recently adopted strategy, induced mutagenesis, has led to some

M.A. Jenks et al. (eds.), Advances in Molecular Breeding Toward Drought and Salt Tolerant Crops, 413–454.

promising results. We argue that the production of induced rice mutants holds the greatest promise of these strategies for mitigating the scourge of soil salinity considering the relative ease with which other traits in this crop have been modified using this methodology. The underlying principles of induced mutagenesis; the modes of action of different mutagenic agents; and procedures for the rapid production and detection of mutants are also summarised. In order to enhance efficiency in the production, detection and incorporation of induced mutants into crop improvement programmes, we suggest the coupling of *in vitro* (such as doubled haploidy and cell suspension cultures) and molecular genetic techniques to this methodology. It is posited also that the efficiency of this process can be greatly enhanced by marker-aided selection while high throughput reverse genetics strategies could lead to the rapid detection of mutation events in target genes. It is concluded that with the plethora of genomics resources available for rice, the use of induced mutations for improving salinity tolerance (and other traits) would rely significantly on the concerted application of efficiency enhancing *in vitro* techniques and functional genomics strategies (including reverse genetics)

Keywords: Induced mutation, in vitro breeding techniques, salinity, rice

1. SOIL SALINITY AND IMPACT ON AGRICULTURE

Salinity refers to the increase in the soil surface of dissolved salts, mostly sodium chloride or common table salt (with sodium and chloride ions), but calcium, magnesium, sulphates and bicarbonates are also implicated in soil contamination by salts. These salts occur in excess at the surface of the soil with negative impacts on the ability of plants (crops, pasture and other vegetation) to grow, develop or produce expected yields. Salinity is measured in electrical conductivity (EC) and expressed in deciSiemens per meter (dS/m or dS m^{-1}) or previously as the equivalent millimhos per centimetre (mmhos/cm or mmhos cm^{-1}). There are 3 main categories of salt affected lands, the saline, sodic (alkali), and saline-sodic soils. Though they all have debilitating effects on plant productivity, the differences between these forms of manifestation are found in the amounts and kinds of salt present. The classification into these 3 categories is based on total soluble salts (estimated by EC); soil pH; and exchangeable sodium percentage or sodium adsorption ratio (SAR), which is the ratio of sodium to calcium and magnesium (Lamond and Whitney, 1992; Flowers and Flowers, 2005). Table 1 summarises the characteristics of different soil types on the bases of these characteristics.

Sodicity affects plant growth through its effects on the soil characteristics, mostly the clogging of soil pores by clay particles leading to surface crusting, reduced water infiltration and low aeration of the soil profile. These result in less available water for the plant's use. On the other hand, salinity, through the impacts of excess of Na^+ and Cl^- ions, affects plants through the toxicity of absorbed ions, effects on osmotic potential and interference in the absorption of other mineral elements (Sairam and Tyagi, 2004).

Table 1. Characteristics of saline and sodic soils

Parameter	Sodic Soils	Saline Soils	Saline-Sodic Soils
Prevailing excess ions	Sodium, carbonate/bicarbonate	Sodium, chloride and sulphate	Variable
pH	High (8.5 to 10.8)	Less than 8.5	Less than 8.5
Sodium absorption ratio	Higher than 15	Lower than 15	Higher than 15
Electrical conductivities	Less than 4.0	Higher than 4.0	Less than 4.0
Total soluble salts	low	high	high
Soil physical structure	Poor. Soil particles tend to disperse; soil is sticky when wet; and nearly impermeable	Normal (good structure and permeability)	Normal

2. SCOPE OF THE PROBLEM

Oldeman et al. (1991) had estimated that of the 230 million ha. of irrigated land, 45 million ha were salt-affected soils (19.5 percent) and that of the almost 1,500 million ha of dryland agriculture, 32 million were salt-affected soils (2.1 percent). The estimate of Flowers (2003) puts the figure of salt polluted soils at 900×10^6 ha. The Food and Agriculture Organization of the United Nations (FAO), Rome estimated more recently that based on the FAO/UNESCO Soil Map of the World, the total area of saline soils (arable and non-arable lands) is 397 million hectares (ha) and that of sodic soils is 434 million ha (http://www.fao.org/AG/AGL/agll/spush/intro.htm). All estimates indicate a disturbing trend for this major constraint to irrigated and dry land agriculture. The global distribution of these salt-polluted soils is summarised in Table 2.

Table 2. Regional distribution of salt-affected soils in million ha

Regions	Total area	Saline soils	%	Sodic soils	%
Africa	1899.1	38.7	2.0	33.5	1.8
Asia and the Pacific and Australia	3107.2	195.1	6.3	248.6	8.0
Europe	2010.8	6.7	0.3	72.7	3.6
Latin America	2038.6	60.5	3.0	50.9	2.5
Near East	1801.9	91.5	5.1	14.1	0.8
North America	1923.7	4.6	0.2	14.5	0.8
Total	**12781.3**	**397.1**	**3.1%**	**434.3**	**3.4%**

Source: (Reproduced with permission from FAO from http://www.fao.org/ag/agl/agll/spush/table1.htm)

At high concentrations of salt in the soil, plant growth and development (including yield) are reduced through 3 main ways (Flowers and Flowers, 2005):

a. the hydrophilic nature of salt enables it to attract water to itself thereby effectively limiting the water available for plants to absorb, leading to osmotic stress and eventually dehydration of the plants and possibly death;
b. the plants also absorb the salt and since they have no mechanism for excluding or expelling salt, chloride and sodium ions accumulate to toxic levels in their cells, eventually leading to death in adverse situations; and
c. interference by these excess salts of the uptake of essential nutrients

The effects of salinity are manifested in 2 main forms in agriculture, dryland salinity and irrigated land salinity. In both cases, excess salt brought up to the surface of the soil by rising water tables left after the water evaporates. Salinisation in drylands is brought about by the rise to the surface of the soil of ground water on account of recharge (addition of water to ground water) being more than discharge (water that is used up from the ground water). This imbalance leads to the rise of the water table. As the water table rises, salts that are held in the soil profile are dissolved and carried along with the rising water table to the soil surface. Eventually, the water evaporates leaving the salts behind at the soil surface. This scenario prevails in the dry land salinity situation of Australia where 5.7 million hectares were considered in 2000 to be salinized or at risk of salinization (http://audit.ea.gov.au/ANRA/docs/fast_facts/fast_facts_21.html). This report also predicted that this value would rise to over 17 million hectares in 2050, a very worrying prediction considering that these mapped areas are in the Western Australia wheat belt region. The situation is similar in irrigated lands as the water table also rises except that it becomes even more critical when the water used for irrigation is saline. Problems of salinity in the soil are greatest in arid regions where there are high rates of evapotranspiration and where irrigated agriculture is practised. Gleick (1993) included most of the US southwest, as well as large areas in Africa, Australia, Spain, Chile, the Middle East, and Asia amongst the most seriously affected areas. About one third of the world's irrigated land, especially in arid and semi-arid areas, has been damaged by salinity. Salt build-up in soils and groundwater is a global problem that affects 20 to 30 percent of the world's 260 million hectares (about 642 million acres) of irrigated land, thus limiting world global food production. The management of salinized soils costs billions of dollars annually through additional investments in order to mitigate the reduced productivity of the land, water, agricultural inputs, genetic resources and manpower inputs. Additionally, salt can be carried in the rain, blown in from the sea or produced as rock dissolves to form soil.

Soil salinity, the principal soil chemical stress in tidal wetlands (Quijano-Guerta and Kirk, 2002), is usually associated with poor irrigation practices or insufficient irrigation water, alkaline soils in inland areas, increase in the level of saline groundwater, and intrusion of saline seawater in coastal areas. The severity of the impact of salinity on rice agriculture is enhanced by the fact that salinity-related stress is usually accompanied by other stress inducing factors such as mineral deficiencies and toxicities. Gregorio et al. (2002) listed alkalinity, phosphorus (P) and zinc

(Zn) deficiencies and boron (Bo) toxicity as usually occurring in tandem with soil salinity that has been brought about by poor quality groundwater or irrigation. On the other hand, when soil salinity occurs as a result of tidal intrusion, according to the same authors, the problem becomes even complicated as the following stress factors must also be reckoned with:

- Soil acidity where the associated high levels of P, Zn, Iron (Fe), and Aluminium (Al) are toxic;
- Acid sulphate conditions characterised by deficiencies in P and Zn and toxicities in Al and sulphides;
- Peatiness that is characterised by P and Zn deficiencies but with an accompanying Fe and organic acid toxicities;

Most plants, and basically all crops, are sensitive to high concentrations of salt in the soil. For reasons of brevity, the scope of this review will be restricted to salinity in rice agriculture.

3. IMPACT OF SALINITY ON RICE AGRICULTURE

Rice is the world's most important food crop and occupies over 148 million hectares. As a major staple crop especially in Asia, rice represents about half of the calorie intake of 3 billion people. Saline-affected soils estimated at least at 900×10^6 ha globally (Flowers 2003) poses a significant threat to agriculture (Flowers and Yeo, 1995; Munns, 2002), a threat that is even more pronounced for rice agronomy, on account of the correlation between salinity and wetlands where a majority of rice is grown. Rice is a crop that usually relies on some form of irrigation. Salinity therefore remains the most widespread soil problem in rice growing countries, drastically limiting rice production especially where irrigation is mandatory. In extreme cases, salt pollution has completely prevented growing rice over large land areas around the world. The surface evaporation of soil water, that may initially have had low concentrations of salt, and the direct intrusion of sea water are the main causes of elevated levels of salt in the soil (Quijano-Guerta and Kirk, 2002).

It is estimated that crop productivity especially that of irrigated rice, has been grievously compromised in over 400 million hectares of land globally with about 54 million ha in South and Southeast Asia alone. Indeed, 40% of irrigated land is salinized globally. This accounts for drastic yield losses and hence loss of income for rice farmers in such places as the Mekong Delta of Vietnam; India; Pakistan; Bangladesh; northeast Thailand; etc. In northern Africa also, say in Egypt, over 25% of rice-cultivated area in the northern part of the Nile Delta is affected by salinity (El-Bably, 2002; Gregorio et al. 2002). These dire figures arise because though some rice genotypes have some level of tolerance to salinity, none has been identified with tolerance to salinity all through the growth and development cycle, and most crop species are generally intolerant of one-third of the concentration of salts found in seawater (Flowers, 2004). In addition to the above mentioned factors, rice agronomy on many salt polluted soils is also impacted by submergence and drought.

From the foregoing it becomes obvious that the development of concerted strategies to mitigate the effects of these confounding factors must be taken into account in any crop improvement strategy aimed at developing saline tolerant rice varieties. This is in part because of the complex nature of the ecology of the tidal wetland as well as the need to develop tolerance to a wide range of stress factors (Quijano-Guerta and Kirk, 2002). Simply, the breeder must device a mechanism for pyramiding the tolerance to above factors in addition to those controlling yield and biotic stress resistance into the germplasm, a formidable challenge considering the sheer complexity of introgressing genes controlling such disparate factors into a single germplasm. This is compounded by the paucity of information on the genetics of the mechanisms of inheritance of these traits.

4. MECHANISM OF SALINITY TOLERANCE

It is commonly accepted that salt tolerance is complex genetically and physiologically (Flowers, 2004; Sairam and Tyagi, 2004) with halophytes (salt tolerant plants) and less tolerant plants (glycophytes, salt sensitive plants) displaying a wide range of adaptations under saline conditions. Also, the review of Flowers and Flowers (2005) had in support of Flowers et al. (1977) suggested that the key to understanding this phenomenon and hence the possible genetic manipulation of salt tolerance in crops might lie in the elucidation of both the genetic and physiological mechanisms of salt tolerance exhibited in halophytes, especially those of the families Chenopodiaceae and Poaceae. The logic is that the understanding of the mechanism of tolerance in halophytes would aid the 'design' of other plants to 'mimic' this mechanism. Mansour et al. (2003) posited that plant salt tolerance operates at the cellular level with ion sequestration in vacuoles or ion exclusion at plasma membranes being the most compelling scenarios for the manifestation of salinity tolerance in plants. The injury suffered by plants under toxic salt levels result therefore from both ionic and osmotic damages on account of lowered water potential (Kefu et al. 2003). We present below a brief overview of the physiological and genetic mechanisms that have been reported for plants' response to salt pollution.

5. PHYSIOLOGY OF SALT TRANSPORT AND ACCUMULATION IN PLANTS

Compartmentalization would seem to be the halophytes' mechanism for simultaneously using the same salts (that could potentially build up to toxic levels) for maintaining the vital function of osmotic pressure within the cells as well as for metabolic processes. The latter takes place in the cytoplasm while osmotic balance is maintained through the accumulation of the necessary salts in the vacuoles (Flowers et al. 1986). The mechanism for the uptake of sodium (or chloride) ions by plant cells is not yet fully understood though similar information is available for potassium (Flowers and Flowers, 2005). Ordinarily ion transport is 'screened' at the root endodermal level, but any breakdown in this process, such as injuries to the endodermis, could lead to a by pass of this protective mechanism and hence significant quantities of ions, such as sodium and chloride, reaching the shoots as could be the case for paddy rice in high concentrations of salt (Flowers and Flowers, 2005; Yeo et al. 1987). Over all, the conclusion seems to be that halophytes are able to survive under otherwise toxic saline levels through a balancing of the requirements of salt for maintaining osmotic balance and those for essential metabolic processes necessary for growth and development (Flowers, 2005; Flowers and Yeo, 1986). Such 'balancing' may be found in features that impose reduced transpiration (and hence less salts reaching the shoots) including reduced leaf surface area and the more specialised salt glands (Thomson et al. 1988) in the leaves that excrete excess salts.

To be able to thrive in an environment with excess salts, Flowers and Flowers (2005) concluded that the plant's response would depend on all or a combination of the following:

A. Morphology of the plant being that salt tolerant halophytes are distinct from the salt-sensitive glycophytes in terms of their anatomy and other adaptive features;
B. Compartmentalization of the salts required for different purpose (osmotic pressure versus vital metabolic processes);
C. Regulation of transpiration in order to reduce the transport of toxic ions to the shoots;
D. Mechanisms for the control of ion movement;
E. Characteristics of the cell membranes that facilitate the screening of solutes entering the cell;
F. Mechanisms for tolerating high sodium to potassium ratios in the cytoplasm; and
G. The specialised salt glands for excreting excess salts.

6. GENETIC BASES FOR SALT TOLERANCE

Considering the above rather divergent factors implicated in the physiological mechanisms for salt tolerance, it would be plausible to expect the implication of several genes and gene interactions in salt tolerance. According to Flowers (2004), there is sufficient evidence to support the conclusion that salinity tolerance in plants is a multigenic trait with research findings on the physiology of salt tolerance suggesting that this is constituted by sub-traits, the inheritance of any of which may in turn be governed by any number of genes. The same author (Flowers, 2004) had also highlighted earlier works that implicated heterosis and dominance factors in the inheritance of salt tolerance in tomato, pigeon pea (and its wild relative, *Atylosia albicans*) and sorghum. In rice specifically, sterility under saline conditions had been shown to be under the control of 3 genes with additive and dominance gene actions and with high heritability also implicated (Flowers, 2004; Moeljopawiro and Ikehashi, 1981; Subbarao et al. 1990; Azhar and McNeilly, 1988). In their review of this subject, Flowers and Flowers (2005) listed the identified quantitative trait loci (QTLs) associated with salinity and ion transport for yield and morphological characters in tomato, citrus, rice, Arabidopsis and barley. As is usual with QTLs, these were specific to populations and environments thereby limiting the utility of the markers that could be derived from them across varying populations.

7. CURRENT STATUS OF RICE GERMPLASM IMPROVEMENT FOR SALINITY TOLERANCE

Globally, there may be no more than 33 officially registered and released saline tolerant crop varieties (Flowers, 2005). Only one salt tolerant rice variety, Bicol, has been released for cultivation to farmers (Gregorio et al. 2002). The limited progress in enhancing saline tolerance in rice therefore reflects the general status of slow progress in conventional crop breeding for this trait. Flowers (2004) as mentioned in earlier parts of this review had attributed this state of affairs to the complexity of the trait being that salt tolerance is complex genetically and physiologically. With this perspective in mind, Flowers and Yeo (1995) had earlier suggested the following 5 ways as holding promise for developing saline tolerant crops:

- Development of halophytes as alternative crops;
- Use of interspecific hybridizations to enhance the salinity tolerance ofavailable crops;
- Use of the existing variation in crop germplasm;
- Generation of desirable variation in existing crop germplasm through recurrent selection, induced mutations or *in vitro* techniques; and
- Breeding for yield rather than tolerance to salt; and
- More recently (Bohnert and Jensen, 1996; Flowers and Yeo, 1996; Flowers, 2004), genetic transformation has also been recognized as a potent tool for addressing this problem.

The six suggestions except for the first one dealing with the development of halophytes as alternative crops would hold true for rice and have indeed been pursued by scientists with modest progress over the years. While the rest of this article will be devoted to a review of the progress made through the application of induced mutagenesis to mitigate this problem, we present below a succinct overview of achievements made in addressing salinity in rice through conventional breeding, *in vitro* techniques (doubled haploidy), genetic engineering and mutagenesis.

7.1. Conventional Breeding

Major breeding efforts at International Agricultural Research Centres (IARCs) such the International Rice Research Institute (IRRI), Manila, Philippines and the West African Rice Development Authority (WARDA) Bouake, Côte D'Ivoire and National Agricultural Research Systems (NARS) initiated in the 1960s had resulted in the release to farmers of improved varieties with higher yield, better quality traits and increased tolerance to various biotic and abiotic stresses. This has resulted in a general increase in rice productivity especially in Asia and South America. This impetus was driven largely by the introduction into breeding programmes of the spontaneous mutant, Dee-Geo-Woo-Gen, which possessed a dwarfing gene. The result was dwarf, photo-insensitive and upright-effective rice plant types which were additionally amenable to improved agronomic practices. However, the efforts invested in the development of saline tolerant varieties have not been equally successful.

IRRI for instance has been conducting extensive breeding programmes for the development of saline tolerant rice varieties (Gregorio et al. 2002; Quijano-Guerta and Kirk, 2002; Senadhira et al. 2002) but only modest achievements have been made so far. Gregorio et al. (2002) listed the following as progress made in this regard:

- The identification and improvement of donor germplasm as well as the study of the mode of inheritance of several soil stress factors in these germplasm have led to the development of a number of advanced lines. However, these lines that were developed using traditional salt tolerant parents such as Nona Bokra, Pokkali, SR26B and Kalarata did not replicate the level of saline tolerance found in the parents.
- Achievement of a fair understanding of the physiological mechanisms for some of these stress factors. To this end, recombinant inbred line (RIL) mapping populations for several traits were developed and genotyped; major genes and quantitative traits loci (QTLs) for these traits mapped; while activities were ongoing for the development of tools for marker-aided selection for these traits; and
- The development of rapid and reliable screening techniques for elongation ability and tolerance to salt; submergence; Fe and Al toxicity; and P use efficiency; and

The above are hardly sufficient to mitigate this global problem and the slow pace of achieving progress has been attributed by the same authors (Gregorio et al. 2002) to the following:

- Limited knowledge on the mechanisms for tolerance to these debilitating factors in rice;
- The involvement of (and probably interaction between) several complex mechanisms;
- Inadequate (high throughput) screening techniques;
- Low selection efficiency; and
- Limited understanding of the stress and environmental interactions in play.

7.2. *In vitro* Techniques for Enhancing Saline Tolerance in Rice

In vitro techniques, especially doubled haploidy, have been used extensively in attempts to enhance salinity tolerance. The strategy usually involves the irradiation of the anthers followed by ploidy doubling and regeneration of plantlets. Alternatively, seeds could be irradiated and doubled haploidy applied to the anthers harvested from plants of the first mutant generation (M_1). Doubled haploid (DH) protocols such as anther/microspore culture provide ideal systems for applying mutagenic treatments. This process induces mutations that can be immediately fixed (made homozygous) by the doubled haploid process. The potential contribution of these DH lines as sources of salt resistance genes for use in hybridisation programmes seems immense as initial results seem to suggest. However, the identification of somaclonal variants following doubled haploidy was the strategy that has contributed most to the development of salt tolerant rice varieties as the two instances below indicate.

1. Development of salt-tolerant rice cultivar through indica/indica anther culture

Senadhira et al. (2002) and Gregorio et al. (2002) reported the production of 79 dihaploid rice lines through the anther culture (AC) of the progeny of two indica rice breeding lines (IR5657-33-2 X IR4630-22-2-5-1-3). The aim of this cross was to combine the high yield of the former with the salinity tolerance of the latter. Following the evaluation of these AC-derived plants for salinity tolerance and other agronomic traits in the greenhouse and field (especially in salt polluted soils), eight lines with desirable levels of yield, salinity tolerance, early maturity, good plant architecture and resistance to pests and diseases were identified within 3 years. Gregorio et al. (2002) reported that these lines have been used as donor lines for breeding programmes in Bangladesh, Dominican Republic, Egypt, Mexico, Myanmar, Philippines and Thailand. The most striking result from this was the official registration and release in the Philippines in 1995 of one of these lines, IR51500-AC11-1 as PSBRc50 or the so-called 'Bicol' for cultivation in salt affected areas in the country. Since then, Bicol has become a donor parent of choice in rice genetic improvement programmes with enhanced salinity tolerance as a breeding objective with a scheme drawn up by Senadhira et al. (2002) for the integration of AC in rice improvement programmes.

The significance of this finding is that the use of the AC strategy has the potentials for significantly reducing the length of the breeding cycle through the timely availability of homozygous diploid lines (Senadhira et al. 2002). This was also a landmark achievement in that while AC production in the japonica rice types had become quite widespread, indica rice types had hitherto remained largely recalcitrant for this. In this study under review, this recalcitrance was still observed as evidenced in the over 30% sterility observed. Even amongst the fertile ones, as much as 25% had to be further discarded on account of undesirable plant architecture.

2. Somaclonal variation for inducing the semi-dwarf characteristic in the salt tolerant landrace Pokalli

Afza and co-workers (unpublished data) of the Plant Breeding Unit of the Joint FAO/IAEA Agriculture and Biotechnology Laboratory, Agency's Laboratories, Seibersdorf, Austria had reported the development of a semi-dwarf variant of the landrace Pokalli that maintained the high level of tolerance to salt in this landrace with otherwise undesirable agronomic characteristics. Genetic variation for salt tolerance is rare in contemporary semi-dwarf (*sd-1*) rice cultivars and breeding lines. Some landraces however exhibit greater tolerance to salt, but are agronomically unacceptable because of their tall stature. One such landrace, Pokalli was subjected to anther culture in an attempt to induce gametoclonal variation for agronomically important traits such as plant height and photoperiodic insensitivity. Over 100 green plants were regenerated from 2,000 cultured anthers. Among theses, two DH lines exhibited a semi-dwarf stature. Mutations, induced during the culture procedure, in the semi-dwarfing gene *sd-1* were confirmed by PCR using locus specific primers. The semi-dwarf DH lines were multiplied and checked for response to salinity in hydroponics system (10 dS m^{-1}) and field tests at IRRI, Philippines. These semi-dwarf lines will be evaluated for salt tolerance in field conditions in other parts of the world, e.g. Asia, Africa and Latin America.

Similarly, Gregorio et al. (2002) reported work carried out at IRRI, the Philippines, in which Pokkali and other traditionally used salt tolerance donor parents were subjected to cell culture with the aim of inducing and identifying somaclonal variants. One such variant, TCCP 266-2-49-B-B-3 again displayed the semi-dwarf characteristic as well as other superiority in quality traits such as the possession of white pericarp and improved cooking quality, traits that are absent in Pokkali. This genotype has subsequently replaced Pokkali as the donor parent of choice for introgressing salinity tolerance.

7.3. Genetic Transformation for Salinity Tolerance in Rice

Transgenic strategies have been severally adopted for the improvement of indices related to salinity tolerance in plants. Examples include rice (Hu et al. 2006); tomato (Xhang and Blumwald 2001; Jia et al. 2002); tobacco (Pandey et al. 2002; Veena et al. 1999; Singla-Pareek et al. 2003); and *Sorghum bicolor* (Sanan-Mishra et al. 2002). Flowers (2004) conducted an extensive survey of reports on the genetic transformation of plants including alfalfa, tomato, tobacco and rice, for enhanced salinity tolerance and concluded that there had not been conclusive evidence that the overall yield of the transformants had been improved vis-à-vis those of the wild types under saline conditions. Most of the reports while indicating the over- or down-regulation of genes that had been implicated in the expression of traits related to salt tolerance failed to actually relate this to yield in valid experiments. In tomato, for instance, such a transformant while showing superior biomass over the wildtype under saline and normal conditions, yielded far less (only 50%) than the wildtype in the absence of salt. In the more recent work involving the overexpression of stress responsive genes in rice (Hu et al. 2006), no data was presented on seed set in the studies relating to salinity tolerance. Flowers (2004) also reviewed the work of Garg et al. (2002) involving the transformation of rice to overexpress genes that led to the synthesis of trehalose. The transformants accumulated less sodium ions in the shoots and grew better than the control wildtypes under saline conditions. This seemed to contradict earlier instances where the exogenous application of trehalose had been shown to have deleterious effect on plant growth. Flowers (2004) concluded that the interesting feature of this finding was that the synthesis of trehalose was under the influence of a stress-inducible promoter, thereby opening up the possibility that the use of stress-inducible promoters in genetic transformation may eliminate the stunting that is usually observed in non-stressed plants in the presence of this substance. It is probable therefore that this path could be taken in concert with targeting other mechanisms that operate in halophytes including the regulation of transpiration, synthesis of compatible solutes, and the ability to function with low cytoplasmic potassium concentrations (Flowers, 2004).

The above imply therefore that the routine use of genetic transformation to effect salinity tolerance in rice (and other plants) while remaining promising is far from certain as the appropriate combinations of target genes and promoters will need to be determined and evaluated.

7.4. Mutation Induction in Rice for Salinity Tolerance

There are no reported instances of induced mutant rice varieties with enhanced tolerance to saline soils that have been officially released to farmers for cultivation. However, there are promising mutant lines that could be entered into national varietal registration and release processes soon. One example is the Pakistani induced rice mutant, Shu-92, which has yielded significantly more than the standard salt tolerant checks, Nona Bokra and Pokkali, by 40% and 49% margins, respectively in multi-locational trials in Pakistan (Balooch et al. 2003). The collaborative efforts between the IAEA and IRRI in the use of induced mutagenesis for rice genetic improvement has focussed on generating salt tolerant variants from IR29 which is a commercial glutinous rice cultivar developed at IRRI and improving the plant architecture, agronomic traits and grain quality attributes of a salt tolerant wild rice species, Pokkali. IR29 has low amylose content, an average plant height of 88cm and matures at118 days after sowing. It grows on irrigated land and is highly susceptible to salinity, a character that has made it one of the most commonly used rice varieties in studying the genetics of salinity tolerance (Gregorio and Senadhira 1993; Lee et al. 1996; Gregorio et al. 2002). Promising stable mutants that meet above requirements have been developed (Afza *et al.* unpublished data) and are undergoing further field evaluations and molecular genetic characterization. Under green house conditions, putative mutants have displayed superior plant vigour; of acceptable plant heights; had more productive tillers; and higher grain yield per plant than the wild types. The development of useful salt-tolerant mutants from IR29 that retains the above-listed attributes will be a major contribution to addressing the scourge of salinity in rice agriculture in SE Asia as this variety is well-adapted and widely cultivated by farmers in this region.

8. THE IMPACT OF MUTATION-ASSISTED CROP IMPROVEMENT IN RICE GENETIC IMPROVEMENT

Globally, more than 2,250 officially released mutant crop varieties from 175 plant species are being cultivated by farmers in 59 countries of Africa, Asia, Australia, Europe, South America and North America (Maluszynski et al. 2000; FAO/IAEA

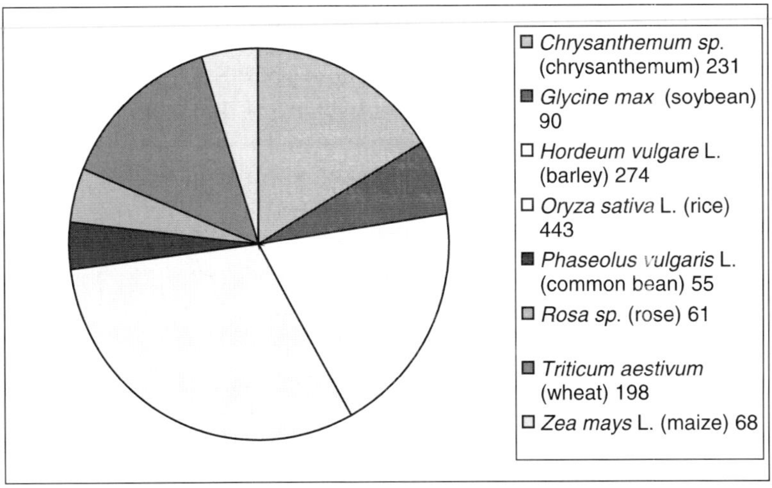

Figure 1. Distribution of officially released crop mutants amongst the top 8 crops with the most number of mutants. Data culled from http://www-mvd.iaea.org/MVD/Default.htm

Mutant Varieties Database (http://www-mvd.iaea.org/MVD/Default.htm; Chopra, Jain, 2005). Out of these, 440 are rice mutants, thereby making rice the crop plant with the greatest number of mutant varieties officially released to farmers (Figure 1). Table 3 shows the countries where these mutants have been released to farmers for cultivation while Table 4 summarises the economic impact of rice mutants in the top rice growing countries. These mutants possess superior attributes in such agronomic

Table 3. Distribution of officially released rice mutants by countries

Country	No. of released rice mutants
China	191
Japan	46
India	40
Brazil	27
Côte d'Ivoire, Guyana	26
USA	23
Vietnam	18
Indonesia, Pakistan, Russia	6
Bangladesh, France	5
Cameroon, Philippines, Thailand	4
Burkina Faso, Hungary, Iraq, Nigeria	3
Costa Rica, Italy, Korea, Myanmar	2
Isle of Man, Madagascar, Portugal	1

Source: (Culled from: Maluszynski et al. 2000; FAO/IAEA Mutant Varieties Database (http:// www-mvd.iaea.org/MVD/Default.htm).

Table 4. Economic impact of mutation-derived rice varieties in some countries

Mutant variety and year of release in brackets	Trait	Mutagen/ cross with mutant	Country	Value or area	Basis of value assessment
RD6 (1977) RD15 (1978)	Glutinous Earliness	Gamma rays	Thailand	US$ 16.9 billion	Total crop value at farm gate for the period 1989–98
Reimei and its sd1 progenies-18 varieties; (starting 1966)	Shortness	Gamma rays	Japan	US$ 937 million	Total value per year (1997)
Zhefu 802 (1981)	Earliness	Gamma rays	China	10.6 million ha	Cumulative planted area between 1986–1994
Amaroo (1987)	Semi-dwarfness	Cross (sd_1 from Calrose 76- gamma ray induced)	Australia	60–70% of rice growing area	Current annual planted area
Shwewartun (1975)	Grain quality	Gamma rays	Myanmar	800,000 ha	Total planted area in 1993
Camago 8 (1996)	Blast resistance	Gamma rays	Costa Rica	30% of rice growing area in the country	Current annual planted area
TNDB100 (1997) THDB (1999)	Semi-dwarfness	Gamma rays	Vietnam	220,000 ha	Total planted area in 1999
Pusa-NR-162(1988) Pusa-NR-381 (1989)	Earliness Blast resistance	Crosses with mutants	India	US$1,748 million	Annual crop value

Source: Adapted from Ahloowalia et al. (2004) and Maluszynski et al. (2000).

traits as reduced plant height (defence against lodging), tolerance to abiotic stresses, pests and diseases resistance, increased tillering and several grain quality traits.

9. CURRENT STATUS OF THE USE OF INDUCED MUTATIONS FOR THE GENETIC IMPROVEMENT OF RICE IN ASIA

Countries in Asia have overwhelmingly adopted the use of induced mutagenesis for improving their most important staple crop, rice, as evidenced in the overwhelming majority of officially released rice mutants being found in this continent. Many NARS rice programmes in Asia have integrated induced mutagenesis as a permanent component of their rice breeding activities with many superior rice varieties developed and officially released to farmers in these countries. Table 5 shows the highlights of these activities in China, Japan, Pakistan, India, Indonesia, Vietnam, and Malaysia. The induced rice mutants possess agronomic and grain quality traits that make them usually the most preferred varieties in the market. Dwarf plant architecture which confers lodging resistance is a commonly occurring plant architecture modification for these mutants. Grain quality traits include low contents of phytic acid; low amylase; large embryos; photoperiod insensitivity; and resistance to biotic and abiotic stress factors.

10. MAKING THE CASE FOR THE USE OF INDUCED MUTAGENESIS TO ADDRESS SALINITY IN RICE AGRICULTURE

With above successes in the modulation of agronomic and grain quality traits in rice through induced mutations, it is envisaged that the achievement of salinity tolerance in this crop will also be relatively easily achieved through the same mechanism – induced mutations. We therefore present in the following sections information that would guide investigators starting off on this route. The aim of presenting some of these aspects of induced mutagenesis in crop improvement is to posit the argument that induced mutagenesis would be a method of choice for addressing the seeming intractable problem of salinity in rice agriculture. By reviewing the results of the hugely successful efforts at developing induced rice mutants, we make the case that induced mutagenesis holds a lot of promise for this crop. We review below therefore the validated methodologies for producing mutants in rice (and other crops) in order to provide a guide for scientists who want to adopt this strategy.

11. AN OVERVIEW OF THE STRATEGIES FOR INDUCED MUTAGENESIS IN CROP IMPROVEMENT

Crop genetic improvement is dependent on the availability of useful and exploitable genetic variation within the genepool, usually germplasm of the crop, which is available to the plant breeder. Natural sources of genetic variation include

Table 5. Status of some recent applications of induced mutagenesis in rice breeding in some Asian countries

Country	Highlights and references
China	77 new mutant rice varieties were officially released between 1991 and 2004. The novel traits include: • Leaf colour (green-revertible albinism and yellow colouration) being used as a marker for tracing the introgression of mutated parts of the genome in breeding programmes • Contents of low phytic acid, an anti-nutritional factor that reduces the bioavailability of proteins, mineral elements and carbohydrate in the diet. …… • Possession of giant embryo and thus enhanced levels of gamma amino butyric acid (GABA), a compound that is important for the stabilization of blood pressure and reduction of blood lipid levels in humans. • High resistant starch (RS) types which is important in the management of human metabolic disorders such as diabetes and hyperlipidemia as RS is not digested in the small intestine and thus decreases the postprandial glucose and insulin responses. The products of its digestion in the in the colon, short-chain fatty acids, help prevent several diseases of the colon (Chen et al. 2006). Functional genomics resource and molecular breeding • Over 700 characterized morphological and physiological rice mutant lines are being maintained at the Chinese National Rice Research Institute (CNRRI) and are being used to study the gene actions involved in the inheritance of the mutated traits • The brittle culm gene (*bc-1*) has been cloned while the genome location of a number of other genes have been mapped • The genetic analysis of the *Lrd-* series of genes implicated in resistance have indicated control by a single recessive gene in some as well as multi-locus action in others (Zhu et al. 2006).
India	• Development of semi-dwarf, non-lodging, disease resistant induced mutant of basmati rice that retains the grain quality that is sought after in the export market (Patnaik et al. 2006).
Indonesia	• A total of 14 officially released mutant rice varieties with superiority in terms of disease and pests resistance, semi-dwarfness, grain quality and early maturity (Ismachin and Sobrizal 2006).
Japan	• Mutant varieties with resistance to lodging; giant embryos (containing more plant oils); low amylose content; low protein contents (for special dietary needs) (Amano 2006).
Malaysia	• Superior mutants with enhanced dwarf plant stature; earliness; photoperiod insensitivity; grain and quality traits; and resistance to pests and diseases (Mohamad et al. 2006).
Myanmar	• Mutants with superior grain quality traits and resistance to pests and diseases (New 2006).

Pakistan	• Most significant highlights include the development of basmati mutants (aromatic) with dwarf stature; early maturity; superior grain quality; tolerance to cold and high altitudes;Non-aromatic salt tolerant mutant with fine grain quality and high yields (Cheema 2006; Balooch et al. 2006).
Sri Lanka	• Advanced mutant line with lodging resistance and preferred gain traits developed from a well-adapted variety that is tolerant to iron toxicity (Bentota 2006).
Viet Nam	• High yielding mutant rice varieties with good grain quality, earliness, photoperiod insensitivity, tolerance to lodging, abiotic stresses (acidity and drought), and resistance to pests and diseases have been released to farmers (Tran et al. 2006; Do et al. 2006).

spontaneous mutations and hybridisations between wild and closely related species. In seeking to mimic this natural phenomenon, scientists have over the past 70 years been artificially inducing changes to the genetic make-up of organisms and thereby generating variations in putative parental materials which are in turn incorporated into breeding programmes. Plant breeders engaged in the development of new superior varieties exploit such variations when they are useful. In rare cases, the mutants possess traits of agronomic or economic importance to such an extent that they require little or indeed no further manipulation before being released to farmers. Most of the time however, the mutants are just "raw materials" (pre-breeding material) that must be included in a normal varietal development mechanism. This would normally involve controlled crosses with otherwise well established varieties which lack the desirable trait identified in the mutant, followed by several cycles of field evaluation.

12. MUTAGENIC AGENTS

Mutations are induced in plants through the exposure of their propagules, such as seeds and other meristematic regions, to both physical and chemical agents with mutagenic properties. For over 70 years, both chemical and physical mutagens have been deliberately used to induce mutations and more recently, transfer DNA (T-DNA) insertional mutagenesis is also being applied especially in reverse genetics strategies for the development of mutant lines for specific genes. The excellent review by Montelone (1998) and that of Kodym and Afza (2003) on commonly used mutagens are summarised in Tables 6 and 7 for chemical and physical mutagens, respectively.

12.1. Chemical Mutagens

The first reported case for the mutagenic activity of a chemical was the rather inauspicious demonstration by Charlotte Auerbach in 1942 that nitrogen mustard

Table 6. Classification of some commonly used chemical mutagens

Type of chemical agent	Mode of action	Examples of chemical agents
Base analogs	Structurally resemble purines and pyrimidines and cause mutations through incorporation into DNA in place of the normal bases during DNA replication. They cause transitions (purine to purine or pyrimidine to pyrimidine); and tautomerization (existing in two forms between which they interconvert e.g. guanine can exist in keto or enol forms)	• **Bromouracil** (BU), a synthetic compound that resembles thymine (has Br atom instead of methyl group) and incorporates into DNA, pairing with adenine (A) just like thymine (T). Displays high propensity to tautomerization to the enol form (BU*) • **2-aminopurine (2AP),** an A analog which pairs with T or (less frequently and efficiently) with C. It also causes A:T to G:C or G:C to A:T transitions.
Chemical agents that modify the structure and pairing properties of bases	They act through deamination, the replacement of cytosine by uracil which can pair with A and thus from subsequent cycles of replication lead to transitions (where C is replaced by T, and G is replaced by A on the other strand of DNA). They react with bases and add methyl or ethyl groups and depending on the affected atom, the alkylated base may then degrade to yield a baseless site, which is mutagenic and recombinogenic, or mispair to result in mutations upon DNA replication	• **nitrous acid,** causes C to uracil (U), methylcytosine (meC) to T, and A to hypoxanthine deaminations. Hypoxanthine in DNA pairs with C and therefore subsequently causes transitions. • **Alkylating agents e.g.** ethylethane sulfonate, ethylmethane sulfonate – EMS, hydroxylamine (NH_2OH), **N**-methyl-N'-nitro-N-nitrosoguanidine (MNNG).
Intercalating agents	These are flat, multiple ring molecules that by interacting with bases of DNA insert between them. This insertion causes a "stretching" of the DNA duplex and the DNA polymerase in turn recognises this stretch as an additional base andinserts an extra base opposite at this stretched (intercalated) molecule. The result is that intercalating agents cause frameshifts i.e. an alteration of the reading frame: since codons are groups of three nucleotides.	• acridine orange, proflavin, ethidium bromide
Agents altering DNA structure	This encompasses a wide range of chemical agents with differing activities	• Large molecules which bind to bases in DNA and cause them to be noncoding, referred to as "bulky" lesions (e.g. **NAAAF**) . They block transcription and DNA replication. • Agents that cause intra- and inter-strand crosslinks (e.g. **Psoralens)** • Chemical agents that cause DNA strand breaks (e.g. **peroxides**)

Table 7. Classification of some commonly used physical mutagens

Type of Agent	Mode of Action	Examples
Electromagnetic spectrum (EM) which includes visible light and other forms of radiation	Causes covalent bonding between adjacent pyrimidines leading to pyrimidine dimers. These dimers are bulky and can block transcription and DNA replication. They can also stimulate mutations and chromosome rearrangements	ultra violet (UV) radiation is the biologically important one as its wavelength is preferentially absorbed by DNA and amino acids with important biological and genetic implications
Ionizing radiation	They produce reactive ions (charged particles) when they react with biological systems, especially in the presence of water producing reactive oxygen species – oxygen; superoxide anion; peroxide; hydroxyl radical; and hydroxyl ion.. Their effects on DNA either through the free radicals produced or through direct action include: • Single or double strand breaks • Damage to or loss of bases • Crosslinking of DNA to itself or to proteins	X-, Gamma- and cosmic rays; particle radiation such as fast and thermalneutrons; and alpha- and beta-particles

(component of poisonous mustard gas used in World Wars I and II) could cause mutations in cells. Since then of course, many more such agents have been discovered and reported. The most commonly used chemical agents for inducing mutations in plants are presented in Table 6. They range from base analogs; agents that modify the pairing properties of bases; intercalating agents; to chemicals that directly modify DNA structure. Their effects include deamination, transitions, insertions, all resulting ultimately in frameshifts. In extreme cases, DNA strand breaks and the stoppage of transcription and replication also occur.

12.2. Physical Mutagens

The pioneering discoveries by Roentgen of X-rays in 1895 (http://physics.nobel.brainparad.com/wilhelm_conrad_rontgen.html); Becquerel's discovery of radioactivity in 1896 (http://nobelprize.org/nobel_prizes/physics/laureates/1903/becquerel-bio.html); and Marie and Pierre Curie's of radioactive elements in 1898 (http://nobelprize.org/nobel_prizes/physics/articles/curie/) led eventually to the report in the 1920's of the effects of radiation on genes. Table 7 summarises the most commonly used physical mutagens by their mode of action and with examples also provided. They induce changes through the production of dimmers and reactive ions. Their effects range from point mutations to gross chromosomal damages.

12.3. T-DNA Insertional Mutagenesis

Transfer DNA (T-DNA) is the DNA segment of the tumour-inducing plasmid that is present in the pathogenic bacterium, *Agrobacterium tumefaciens that* is transferred to plant cells and inserted into the plant's DNA as part of the infection process. It is commonly used as a vector for transferring foreign genes into the genome during the production of transgenic plants. This property is exploited by randomly inserting T-DNA or transposons into the genome of a plant. Depending on where the T-DNA has been inserted in the gene and other factors such as the redundancy of the gene, the gene action is usually disrupted. Since the sequence of the foreign DNA (T-DNA or transposable element) is known, appropriate polymerase chain reaction (PCR) primers and sequencing of the amplification product can be used to identify the genome locations where the T-DNA has been successfully inserted (McKinney et al. 1995; Krysan et al. 1996; Krysan et al. 1999; Sessions et al. 2004; Chopra, 2005; Shikazono et al. 2005) and confirmed by hybridisation (Sessions et al. 2004). The ease with which T-DNA insertion mutants are produced implies that it is fast becoming a method of choice for developing functional genomics resources (Ryu et al. 2004) e.g. the over 188,000 developed for rice (Jong-Seong et al. 2000; An et al. 2005); and over 380,000 for Arabidopsis (Krysan et al, 1999; An et al. 2005). The latter authors (Krysan et al. 1999) also proposed an elegant procedure for proper identification of the mutants and mutation events and also for ensuring a correlation of gene disruption to the phenotypic effects observed.

In spite of the increasing interest in the use of this genetic transformation – induced mutagenesis interface for basic research in functional genetics, the routine application in crop improvement strategies is still far fetched on account of its laborious nature and the concerns surrounding the deployment of genetically modified organisms (GMOs) in food and agriculture.

13. RADIOSENSITIVITY

The efficient use of induced mutations in crop improvement is usually dependent on the exposure of the propagules to appropriate levels and doses of the mutagen usually necessitating prior establishment before proceeding with the actual mutation induction. This process of determination of the optimal doses of the mutagens is referred to as radiosensitivity tests. Radiosensitivity, coined from radiation sensitivity, is a relative measure that provides an indication of the magnitude of observed effects on an irradiated material as a result if its exposure to radiation (Van

Harten, 1998; Kodym and Afza, 2003). Wide variations in radiosensitivity have been observed within genera, species and even cultivars. These differences have been ascribed to differences in interphase chromosome volume (ICV), a property determined by nuclear volume and chromosome number, content of nucleic acids, ploidy level and in the chromosome size (Evans and Sparrow, 1961). It is therefore recommended to establish this value before carrying out elaborate irradiation of plant propagules.

In the absence of data on radiosensitivity estimates for a particular material, a pilot study is usually carried out. This is achieved through the irradiation of a sample of the propagule over a wide range of doses and collecting data after germination or regeneration of the seeds or vegetative propagules, respectively. For gamma irradiation, a range of 0–800 at 200 Gy intervals; and for fast neutrons, a range of 0–40 at 10 Gy intervals, would be suitable for a pilot survey. For this, 50 seeds per treatment are usually sufficient but 4 to 5 replications using smaller samples of 10 seeds could also be done (Hermelin, 1997; Konzak et al. 1967). A non-irradiated control is always included as an internal reference and evaluated along with the irradiated samples. It is assumed that lethality would be 100% at high doses but usually, the dose of the mutagen leading to death of 50% of the test materials, the so-called LD_{50}, is recommended. In practice however, this is determined by the percentage reduction in or effect on certain morphological traits. Also, other LD values such as $LD_{30,}$ i.e. 30% percentage damage or reduction, taking the non-irradiated control as the reference point, could also be used by the experienced investigator who desires to vary the stringency conditions of an experiment.

Figure 2 shows the estimation of radiosensitivity for a rice genotype, at the Seibersdorf laboratories of the Plant Breeding Unit of the Joint FAO/IAEA Agriculture and Biotechnology Laboratory in Austria. The reaction to variations in irradiation dosage by botanical seeds of this material was estimated as percentage departure of the seedling height of irradiated materials from the values of the non-irradiated control. The point corresponding to 50% damage, i.e. 50% mark on the X-axis (read off from the line of best fit) was considered as the LD_{50}. For this genotype and using this parameter, seedling height, this point falls at about 300Gy. This could also be calculated more precisely using the linear regression equation. In practise however, irradiation for generating mutants in crop improvement programmes is carried out over a range of plus/minus 5 of this determined optimal dose.

Protocols for the use of physical mutagens, gamma and fast neutron irradiations, for mutagenising seeds have been described by Kodym and Afza (2003) while more detailed information is provided on the background and experimental design of radiosensitivity tests using these agents is in preparation (Kodym et al., pers. comm.). Also, the radiosensitivity of numerous crop species to gamma rays and fast neutron radiation was tested and compiled by Brunner (1977; 1985). Suggested procedures for the generation of mutants using both physical and chemical mutagens are given in Boxes 1 and 2, respectively.

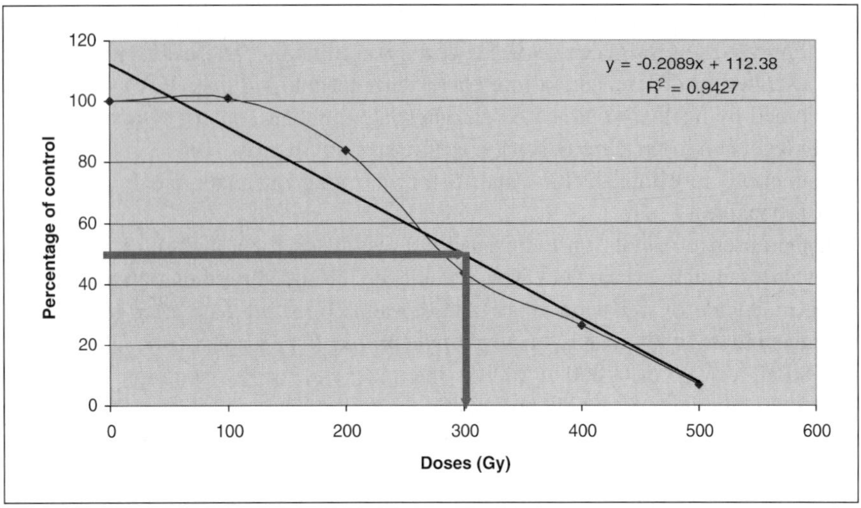

Figure 2. Graph of percentage height reduction of putative mutant seedling against irradiation doses for a rice genotype. The LD_{50} corresponds to 300Gy

Box 1. Determination of optimal irradiation dose (radiosensitivity tests) for rice.

Procedures

1. Select dry, clean, disease-free dormant seeds with uniform size to achieve a higher degree of uniformity. Only viable seeds with high germination rate should be used.
2. Choose a wide range of doses from 100Gy to 600Gy with a minimum range of 50Gy.
3. Use 40–50seeds per treatment at 3 replications.
4. Pack the seeds according to the treatment dose in a water permeable bag.
5. Label each bag appropriately (e.g. indicating the variety name, date and the treatment to be carried out). Non irradiated, i.e. control should be included.
6. Keep the seeds for a period of 5–7 days in a vacuum desiccator over a 60% (vol.) solution of glycerol at room temperature. This is to achieve an equilibration to seed moisture contents of 12–14%, the most suitable condition for efficient mutation induction
7. Collect the precise dosimetry data for the gamma sources and determine the treatment exposure times using following equation: Exposure time in seconds = Desired doses/dose rate. Or 7. Hand over the seeds to the GAMMA source operator for irradiation.

(*Continued*)

8. Sow the seeds in rows in increasing order of doses in trays (which can each accommodate a minimum of 7 rows) in the glass house. Seed sowing should be done in a way that would allow for periodically taking measurements and also for good visual assessments.
9. Collect the following data:
   Germination rate
   Seedling height
   Survival rate
   Chlorophyll mutation
   Number of tillers
   Seed setting
   Sterility test in M_2 generation
10. Plot a graph of the absorbed doses against percentage changes in the mutants from the values of the non irradiated control for each parameter. Percentage plant height reduction is a good parameter for determining damage due to mutagenic treatment. The LD_{50}, i.e. the dosage corresponding to 50% damage (reduction in height) is an appropriate dose for irradiation. This can be read off from the line of best fit or more precisely calculated using the straight line equation.

Box 2. Procedures for optimally inducing mutations in rice using a chemical mutagen, EMS

Seed selection:

1. Select disease-free seeds with uniform size and shape. Viable seeds with high germination rates must be used.
2. Choose a wide range of the concentration of the chemical for use in the induction treatments with varying time durations in order to determine the optimum concentration – by - time for mutation induction. A minimum of three hours duration is usually required for EMS treatments.
3. Use 40– 50 seeds per treatments with a minimum of three replications.
4. Put the seeds in a polyethylene mesh bag, fold the bag and hold in place with plastic clip.
5. Label the bag appropriately (e.g. indicating the variety name, date and the treatment to be carried out).
6. A control, i.e. not treated with EMS should be included.

Pre-soaking:

7. Place the bags (each containing separate seeds of each genotype – by – mutagen type – by – duration treatment combination) in a beaker and keep

(*Continued*)

under running water at least for 20–24 hours. Pre-soaking helps facilitate total uptake of the mutagen in the embryo tissue.

8. Remove the bags and dry off the excess water.
9. Put the bags in the treatment solution (see below) according to the label of Step 5 above.

Preparation of solution:

EMS solution

10. Just before use, prepare the EMS solution according to the desired concentration in an airtight bottle. To prepare 0.5% EMS with 2% DMSO (one of the carrier agents), for example, add 0.5 ml of EMS and 2ml of DMSO to a bottle and use distilled water to make up the volume to 100ml.
11. Shake the solution vigorously to mix properly.

Treatment:

12. Put the bags containing the seeds in the respective bottles according to the treatment combination and place the bottle in gyratory shaker at a rate of 30rpm to allow all the seeds to be exposed homogeneously to the EMS solution

Post-treatment washing and drying

13. Remove the bags with the seeds after the treatment and wash the seeds three to four times in distilled water. The water used at this step must be handled as a toxic liquid waste.
14. Put the seeds again under running water to get rid of any bit of EMS adhering to the seeds.
15. Shake the bag to get rid of excess water and place them on blotting paper for drying.

Sowing and data collection

16. Sow the seeds in rows in a tray which can accommodate all the treatments.
17. Sowing should be done in order of increasing concentration for visualization/documentation and should permit the taking of measurements at intervals.
18. Collect the following data:
    Germination rate
    Seedling height
    Survival rate
    Chlorophyll mutation
    Number of tillers (e.g. for rice)
    Seed setting
    Sterility test in M_2 generation

(*Continued*)

19. Plot a graph of the absorbed doses against percentage changes in the mutants from the values of the non irradiated control for each parameter. Percentage plant height reduction is a good parameter for determining damage due to mutagenic treatment.

Notes

- EMS is carcinogenic.
- It should be labelled properly as "carcinogen" when used.
- Wear gloves during the above procedures.
- Use disposable pipette and balloon for pipetting.
- Work should be done in confined area and under hood. If hood is not available, use a face mask.

Dispose of the unused EMS solution and all the liquid waste by adding the 4% NaOH and excess water.

14. DETECTION OF USEFUL MUTANTS

The ability to detect mutant variants is at the core of any induced mutagenesis endeavour (both for crop improvement and functional genomics). Mutations alter the DNA and when such alterations affect gene function, changes to the phenotype are observed on account of the altered expression of the gene in question. In general, the detection of phenotypic changes due to mutations would depend on whether mutations occur in translated or untranslated segments of the gene, with the latter having less or usually nil effect on protein function and hence phenotypic manifestation. In translated regions of the gene, other than duplications, the most common type of mutation is a base substitution (Robert, 2003) and this holds true for both spontaneous and induced mutations brought about by both chemical and physical mutagens. Point mutations, involving one base, could be in the form of a base substitution or either of deletion or insertion of a base that would lead to a frameshift mutation. In the case of substitution, the resulting mutation could be 'samesense' or wildtype where the base substitution mutation causes the same amino acid as is found in the wild-type to be inserted. This would usually not be detectable as no gene function is affected. In instances where the substitution leads to the insertion of a different amino acid, this is referred to as 'missense' mutation, and the resulting phenotype could display a wide range of difference from the wild-type, starting from normally active proteins, through inactive proteins, less active proteins, to unstable proteins. The effects of missense mutations are therefore readily detected in the mutants. Base substitutions could also lead the creation of a premature stop codon resulting in a truncated and usually non-functional protein product. This is referred to as 'nonsense' mutation and relatively easily detectable on account of the altered protein function.

When the mutation event leads to the insertion or deletion of a nucleotide at one point, the overall number of nucleotides changes and this leads to a rearrangement of the codons with profound effects on a protein product, including both loss of function and instability. This again is easily detectable in the phenotype. Mutations could also involve small deletions in which case the protein function would be significantly affected, usually resulting in total loss of function.

Mutations in untranslated regions are largely of less importance but could still lead to alterations in protein function if for instance this is in a transcribed region by affecting recognition signals. On the other hand, mutations in regions that are neither transcribed nor translated might affect either transcriptional start or stop signals and thus the regulation of the region in question. It is also possible that the effect could involve the "structural" regions of the DNA, which would then impact upon gene expression albeit indirectly.

Large deletions, inversions and duplications usually have profound effects on gene expression with deletions and inversions typically eliminating the activity of the affected gene. Duplication could also be manifested in gene function when the copy number is critical. Also, mutations involving insertions, insertion sequences, and transposons usually eliminate the gene function through the destruction of whatever information is encoded in that region.

Taking the foregoing into consideration therefore, the success of any induced mutagenesis work would depend on careful planning which would include a clear understanding of how to efficiently detect the mutation events either through phenotypic characterization or by the use of molecular tools to query the target regions. We shall review some of the methodologies in the following sections with particular reference to detecting salt tolerant rice mutants. The isolation or identification of putative mutants requires an efficient screening method that must take into account the requirement for efficiently handling large mutant populations. The screening technique must be reliable and able to evaluate large amounts of mutated material.

15. A MODIFIED HYDROPONICS SCREENING METHODOLOGY FOR THE RAPID DETECTION OF SALT TOLERANT RICE MUTANTS

Salinity screening under field conditions is most of the time inaccurate and difficult due to large environmental effects associated with this trait. A hydroponics system that was amenable to the controlled environment of the greenhouse was developed at IRRI, Philippines for selecting salt tolerant rice genotypes at seedling stage (Gregorio et al. 1997). This was modified at the greenhouse of the Plant Breeding Unit FAO/IAEA Laboratory in Seibersdorf, Austria (situated at 48:13:00 N 16:22:00 E) in order to adapt this screening methodology to handling large mutated plant populations and to enhance efficiency (Afza et al. 1999).

The set up developed at IRRI included styrofoam floats in the size of 36.5 x 26.5cm having 100 holes 10 x 10 and a nylon net bottom, placed on top of a rectangular 18l plastic tray. However, the set up with styrofoam is washable only for a few times, is not very stable and can screen only a limited number of seedlings. After 2–3 times of use, the seedling float gets contaminated with algae and needs to be replaced. This modified version included polyvinylchloride (PVC) plates in the size of 36.5 x 26.5 cm having 100 holes 10 x 10 and a nylon net bottom, placed on top of a rectangular

18l plastic tray. Another variant has a PVC plate measuring 56 x 36 cm with 24 holes (6x7cm) and placed on top of a 25l plastic tray. This set up could accommodate up to 720 seedlings per tray thus making this screening technique adequate and appropriate for selection of mutated seedlings for salinity tolerance (Afza et al. 1999).

Seeds that had been pre-germinated for two days are sown 1 in each hole in the PVC seedling plate that has been fitted as a lid over the tray filled with distilled water such that the bottom of the plate makes contact with the distilled water. The set up is left to stand for a further 3 days and then salinity stress is introduced by adding sodium chloride (NaCl) to the nutrient solution described by Yoshida et al. (1976) up to the desired EC. The desired pH is checked daily and maintained at 5.5 by adding 1N of NaOH or HCl. Scores of salinity injury based on visual symptoms (1- tolerant and 9-sensititive) at seedling stage, according to the modified Standard Evaluation System of IRRI (Gregorio et al. 1997), are taken at 5, 7, 9 and 12 days after salinization. It is recommended that this set-up be replicated 3 times using ten seedlings of each variety per replication for each assay. A case study for using this set up to establish a proof of concept for detecting salt-tolerant rice mutants is given in Box 3.

Box 3. Establishing the hydroponics system for screening for salt tolerant rice variants

Nine different rice varieties of known levels of tolerance (Susceptible types: IR29, PP2462–11, Taipei 309 and Wagwag; moderate types: IR58430-6B-14-1-2, IR51500-AC-11-1 and IR63731-1-1-4-3-2; and tolerant types: Nona Bokra and Pokkali) were tested at five different salinity levels, EC 2, 6, 8, 10 and 12 dS/m. The aim was to determine the optimum salinity level and best time for scoring injury symptoms. Salinity injury was scored four times: 5, 7, 9 and 12 days after salinization. The screening was done in the greenhouse with day/night temperatures 30 / 20 ^{0}C and relative humidity of at least 50% during the day. The experiments were replicated three times using ten seedlings of each variety per replication.

Salinity of EC 6 dS/m showed very low salt injury even after 12 days. The varietal differences were observed 9 days after salinization at EC 8 dS/m. The sensitive check IR 29 had an average score of 6.6 that could be classified as moderate. At EC 10 dS/m and 12 days after salinization, a clear distinction of the three categories (tolerant, moderate and susceptible) of the varieties tested was established. Most varieties were severely injured at EC 12 dS/m within 10 days of salinity stress, except the two tolerant checks Pokkali and Nona Bokra. After 20 days at this salinity level, all varieties died, including the tolerant varieties. Thus EC 12 dS/m was too high for isolating the moderately tolerant lines. The conclusion therefore was that the optimum salinity level for screening rice at the green house of FAO/IAEA Laboratory in Seibersdorf, Austria should be EC 10 dS/m and visual symptoms scoring be done starting 12 days after salinization.

16. *IN VITRO* SELECTION FOR SALT TOLERANCE IN RICE

Selection at the cellular level, i.e. under *in vitro* conditions offers yet another possibility for integrating efficiency enhancing methodologies besides the traditional methods in plant germplasm selection, evaluation and characterization (Croughan et al. 1978; Heszsky et al. 1991). For a trait like salinity for instance, plant breeders usually do not have precise tools to collect data that would lead to valid inferences under field conditions. Also, available *in vivo* methodologies are time consuming and a scaling up to an *in vitro* platform promises savings in time, equipment, cost and space. Additionally, i*n vitro* selection enables the investigator to control the experimental environment more precisely. The strategy therefore would be to use *in vitro* techniques to rapidly get rid of non-promising materials and then validate the results under field conditions.

The detection of salt tolerant variants of rice through the culturing of their calli under high saline condition has been severally reported (Li and Heszky, 1986; Reddy and Vaidyanath, 1986; Binh and Heszky, 1990; Binh et al. 1992; and Wincov, 1976; Lee et al. 2003). The methodology has also bee used for the selection for other traits in rice such as disease resistance (Nakajima, 1991) and in other crops like tomato and tobacco (Nakajima, 1991). This methodology can also be adapted to induced mutagenesis in rice facilitated by DH. Afza et al. (2006) reported the identification of salt tolerant wheat mutants in China through this method. The spikes of the F_1 plants with pollen at the uni-nucleate stage were irradiated with gamma rays and as an initial screening method, the calli that developed were cultured on media containing an excess of salt. The second step was to screen the seedlings of the DH lines regenerated from these calli in saline hydroponics tests of seedlings. Ultimately, the selected putative salt tolerant DH lines were multiplied and evaluated in saline field trials in China.

16.1. The Use of Molecular Genetic Markers for Identifying Putative Mutants and Tracking the Inheritance of the Mutated Segment of the Genome

The use of molecular genetic markers to track the inheritance of genes influencing a trait of interest in a genetic improvement program is referred to as marker-assisted (or aided) selection, MAS. Normally MAS is used to circumvent the confounding effects of environmental influences in varietal evaluations. This could lead to economy of time, space and savings in cost in a breeding scheme for a trait that can only be measured at crop maturity. The use of MAS is predicated on the fact that the selection for a marker flanking a gene of interest is in effect a selection for the presence (or absence) of the desired allele (variant) of the gene in the progeny being evaluated. Therefore, a molecular genetic marker associated with a gene of interest can be used to probe for such a gene once DNA can be extracted from a progeny being evaluated. It is for this reason that molecular markers are used to track the mutated segments of genomes. A marker that distinguishes a parent from the mutant, identified either from an appropriate segregating population or from polymorphisms in the gene(s) of interest, can therefore be used in early screening of progeny. Current efforts are therefore geared at generating the appropriate rice segregating populations for salinity tolerance through crosses between the tolerant and susceptible variants (Afza, et al., pers. com.). This proposed strategy will involve the use of bulk segregant analyses to identify the discriminating markers and subsequently use them in screening appropriately designed mutant populations.

16.2. Reverse Genetics Strategies

A major drawback to the routine application of induced mutagenesis to both crop improvement and genomics studies (through forward and reverse genetics strategies, respectively) remains the drudgery of producing, handling and assaying the requisite large populations of mutant stocks. The use of large populations of starting materials is imperative, as the outcome of any particular mutation induction is never known until after the assays. Therefore, large population sizes must be created in order to have a fair chance of detecting desirable mutations. This is expensive, laborious and time consuming. Recent advances in genomics, especially the ever increasing volume of publicly available genomics resources, imply that a high throughput platform such as Targeting Induced Local Lesions IN Genomes - TILLING - (McCallum et al. 2000; 2000a) will make the rapid evaluation of mutant stocks for specific genomic sequence alteration more practicable. TILLING is a reverse genetics strategy that through the provision of allelic series of point mutations in genes of interest allows for a rapid low-cost discovery of induced point mutations in mutated populations. In very simple terms, this involves the pooling of DNA and the use of appropriate polymerase chain reaction (PCR) primers to identify point mutations in genes of interest whose target regions are being amplified. This strategy employs a mismatch-specific endonuclease to detect natural or induced DNA polymorphisms in PCR. (McCallum et al. 2000a; Till et al. 2003; Till et al. 2004; Colbert et al. 2001; Henikoff and Comai, 2003; Gilchrist and Haughn, 2005; Greene et al. 2003).

Its advantages over other reverse genetic strategies include its versatility (applicable to virtually any organism); adaptability to high throughput; independence of genome size, reproductive system or generation time (Gilchrist and Haughn, 2005). TILLING employs no transgenic methods and in addition to functional genomics also holds promise for the genetic improvement of agricultural crops via induced mutagenesis (Henikoff et al. 2004). This has been successfully demonstrated for maize (Till et al. 2004); and for wheat it has been used in identifying over 200 alleles of the waxy starch genes (Slade et al. 2004). The United States Department of Agriculture (USDA) is also employing this strategy in soybean improvement. (http://www.ars.usda.gov/is/AR/archive/jul05/genes0705.htm)

This platform, on account of its high throughput nature and with continuing innovations to reduce its costs, promises to be a method of choice for rapidly and efficiently identifying point mutations (Comai and Henikoff, 2006). The review by

Table 8. Examples of some major current TILLING Projects

Plant	Institutional Affiliations and lead scientists	Features
Arabidopsis	Collaborative endeavour between the Comai Laboratory at the University of Washington and the HenikoffLaboratory at the Fred Hutchinson Cancer Research Centre, Seattle, WA, USA	Partial cost recovery service that provides requestors with mutations in genes of interest. The results (sequences, accession numbers of mutations and seed stocks) are ultimately deposited in the public domain. • Capacity building in this and advocacy for this technology through training programmes and workshops and other support mechanisms for other laboratories setting up the facility • Development of supporting web-based freeware, CODDLE for primer design; and PARSESNP (Taylor and Greene, 2003) for graphical simulation of the effects of the mutation; SIFT (Ng and Henikoff, 2003) for predicting whether amino acid substitution affects protein function; GelBuddy for analysis of electrophoresis gel image (Zerr and Henikoff, 2005).
Lotus japonicus	Perry and group at the John Innes Centre, UK (Perry et al. 2003)	This was aimed at identifying useful alleles of genes in this legume that currently has no insertional mutagenesis tool.
Maize	Purdue University & Seattle (FHCRC and University of Washington)	This initiative which includes the collaborating laboratories of the ATP is also a partial cost recovery service provider for mutations in maize genes of interest.

Source: Adapted from Gilchrist and Haughn (2005)

Gilchrist and Haughn (2005) listed the available TILLING projects and these are included in the summary presented in Table 8.

16.3. Facilitating Induced Mutagenesis Through *in vitro* Techniques

Plant tissue culture involves the growing of plant cells, tissues or organs (referred to as ex-plants) such as apical meristems, axillary buds, micro-cuttings and micro-plants that have been harvested from the mother plant, on artificial growth media under controlled and aseptic conditions. *In vitro* techniques involving somatic embryogenesis, organogenesis and anther culture have all been applied extensively in seed propagated crops such as rice, in order to speed up and increase efficiency in crop improvement schemes. Being that induced mutagenesis requires the handling of large mutant populations and efficiently evaluating such populations, *in vitro* techniques, especially doubled haploidy and cell suspension cultures, therefore hold promise for significantly increasing efficiency in the generation and screening of mutant populations. The rapid path for attaining homozygousity which *in vitro* methods permit is another reason for advocating their integration into induced crop mutagenesis. Additionally, this method also helps to substantially reduce the chances of chimeras, for instance.

17. DOUBLED HAPLOIDY

Kasha and Mauluszynski (2003); Atanassov et al. (1995); Jain et al. (1996); Zapata-Arias et al. (1995) referred to haploidy as the possession of the gametophytic (i.e. half of the) normal number of chromosomes by plants (sporophytes). In a haploid state, genes exist in only one allele implying therefore that even the recessive forms of a gene, as most mutations usually are, will be expressed and hence observed phenotypically. Doubling the chromosome number of a haploid plant therefore leads to the production of a completely homozygous individual, a prerequisite for the start-off of most crop improvement strategies. The production of homozygous lines however usually requires several generations of selfing and selection. The doubling of the chromosomes in the gametophyte to achieve the diploid state, referred to as doubled haploidy (DH), therefore leads to significant savings in time and resources required to produce a homozygous line (Brown and Thorpe, 1995). The authors (Brown and Thorpe, 1995) also estimated that doubled haploidy technique had been successfully applied in 171 plant species, a figure that had risen to 250 by 2003 (Maluszynski et al., 2003a). The main methods that had been used for generating DHs have been severally reviewed (Brown and Thorpe, 1995) and they include the following procedures:

- Androgenesis i.e. the culture of excised anthers and pollen;
- Gynogenesis i.e. the culturing of excised ovaries and ovules;
- The culture of embryo (the *bulbosum* technique, named after *Hordeum bulbosum*, in which this technique involving embryo rescue after endosperm failure was first established); and
- Parthenogenesis involving the harvesting of chimeric haploid sectors from embryos that have been induced even though the nuclei have not fused to produce a diploid zygote through semigamy, pseudogamy or apogamy.

Information is now readily available on efficient methodologies for the routine production of DH from several plant species. Maluszynski et al. (2003) documented protocols for DH production for 33 plant species from barley, wheat, maize, rice, triticale, rye, oat, durum wheat, Timothy, ryegrass and other grasses, rapeseed, broccoli, brassicas, potato, tobacco, linseed/flax, sugar beet, asparagus, onion, apple, aspen, cork oak, and citrus. Additionally, references were provided for protocols for doubled haploidy in 226 additional plant species, providing therefore a compendium on this subject covering at least 250 plant species (Maluszynski et al., 2003a). Kasha et al. (2005) provided a detailed and comprehensive review of the status of doubled haploid production in plants.

Doubled haploidy therefore holds a lot of promise for enhancing the efficiency of induced mutations as the gametophytes could be induced to mutate and the diploid status subsequently restored through fairly simple laboratory procedures such as colchicine treatment. Alternatively, anthers harvested from the first mutant generation (M_1) could also be used to initiate DH lines thereby eliminating the risk of damage to anthers by irradiation or ether chemical mutagens. This strategy is being used currently in induced mutagenesis in rice such as the report of Lee et al (2003) on the production of stable DH-derived mutant japonica rice lines. At the diploid stage, the mutations that have been fixed can be easily reviewed as all the alleles would be homozygous. Over 20 rice varieties have been developed with the intervention of DH (Zapata-Arias, 2003). Box 4 shows the protocol being used at the Plant Breeding Unit of Agriculture and Biotechnology Laboratory of the Agency's Laboratories, Seibersdorf for the integration of DH (anther culture) in induced mutagenesis in rice.

Box 4: Protocol for rice anther culture

1. Panicle collection and cold treatment:

- Select primary tillers having an auricle distance of around 8-12cm from flag leaf to that of the next leaf (subtending leaf) and with anthers containing pollen at the mid- to late uninucleate stage.
- Harvest the panicles early in the morning, wash the panicles in tap water and wrap in moistened paper towel.
- Keep in plastic bag or in glass tubes inside incubators at 8°C for 7–10 days .

2. Panicle sterilization

- Surface sterilize the panicles with 70% ethanol for 30 seconds in large test tubes.
- Immerse the panicles for 20 minutes in 20% commercial bleach (5.25% NaOCl)
- Rinse three times with sterile distilled water.
- Cut the panicles in three parts – bottom, middle and the top of the panicles
- Select the florets having yellow-green, or anther length of half of the size of the florets. This should give the optimum stages for anthers culture, i.e. pollen stage of mid- to late uninucleate. Keep the florets in a petri dish.
- To detach anther, cut the florets at the lower part.

(Continued)

- Release the anthers, hold each floret with a sterile forceps and tap lightly at the edge of vessels or petri dish containing the callus induction medium to release the anthers from the florets into the medium.
- Callus induction medium (N-6 basal medium, 0.5mg/l 2.4-D, 6% sucrose, 0.8% agarose, pH 5.6–5.8)
- Incubate in the dark at 25±2°C
- Calli are formed in about 30–45 days after anther plating. Record calli induction efficiency.

3. Plant regeneration:

- Transfer calli of 1–3mm size to plant regeneration medium (Murashige and Skoog, 1962).
- Incubate culture at 12-hour photoperiod, about around 1.000 lux ($66\mu Em^{-2}s^{-1}$) light intensity and temperature of 25±2°C.

4. Rooting medium:

- Transfer regenerated plants to rooting medium to MS basal medium without growth regulators.
- Collect data on green and albino plants

5. Acclimatization:

- Wash the roots with tap water to remove the agar and transfer the plants with well developed roots to pots and let grow in the greenhouse
- Regenerated plants are prone to dehydration. To avoid this, it is advisable to cut the tips of the leaves and keep the plants in an atmosphere with a high humidity
- Identify the haploid plants and treat them with colchicines in order to double the chromosome number. In rice, haploid plants are usually bushy, with a lot of tillers but displaying no fertile seeds. In practice, the bushy plant architecture alone is used for identifying the haploids.

18. CELL SUSPENSION CULTURES

Plant cell suspension culture refers to the growth in liquid medium of individual cells that have been isolated from induced callus in tissue explants. Through the exploitation of totipotency, plantlets are fairly easily regenerated from a single to few individual cells from the suspension cultures through the modification of *in vitro* culture media, implying that the occurrence of chimeras, sectoral differences – mutated and non-mutated sectors in same plant – is greatly minimized if not eliminated completely. This is because a plantlet originates from, at most, a few number of cells. Compared with irradiating a nodal segment or some other regenerative plant part made up of many cells, cell suspension cultures, from the consideration of the need to avoid the investment of extra resources in dissociating chimeras, especially in vegetatively propagated plants, would therefore be the ideal starting material for induced mutagenesis. Therefore, cell suspensions, especially embryogenic suspensions, would so long as the protocols for regenerating plantlets have been established, be the most efficient strategy for producing mutants with minimal chimeras.

From the foregoing, it follows therefore that *in vitro* techniques would lead to greater efficiency in the production and multiplication of mutants. This is because relative to *in vivo* conditions, significantly larger numbers of regenerative plant parts can be induced to mutate per unit time and space thereby rapidly producing large numbers of mutants, an imperative for the use of induced mutations to generate utilisable genetic variation for crop improvement. The advantages for the coupling of *in vitro* techniques to induced mutations therefore include the massive generation of mutants from homozygous backgrounds; handling of large numbers of mutants in space- and time-efficient manners; adoption of rapid screening methods; production of progenies from plants with low fecundity; and the rapid cloning and distribution of disease-free planting materials of putative mutants. Another important advantage is that *in vitro* conditions permit several cycles of regeneration, a necessity for the dissociation of chimeras (a major drawback to induced mutagenesis), within a relatively short period of time.

19. SUMMARY AND CONCLUSIONS

Salinity, the presence of elevated levels of salt in the soil, is a major problem of rice agriculture with a prognosis that forebodes even worse scenarios than the current significant problems it poses for agriculture along the coasts of the Indian Ocean, the Mediterranean Sea and the dry regions of the Americas, Australia, Middle East and North Africa. While an understanding of its impact on plants, including the mechanisms for attaining toxicity are well-understood, scientists are still accumulating information that hopefully will lead to a clear understanding of and hence manipulation of the mechanisms for tolerance to salinity. Both the physiological and genetic bases for tolerance to salts in the soil seem to be very complicated, explaining the rather frustratingly slow pace for developing tolerant plant varieties. The problems are even compounded for a crop like rice with the dependence on irrigated agriculture for paddy rice cultivation. Strategies for developing saline tolerant rice varieties have included conventional breeding methods; *in vitro* techniques; genetic transformation; and induced mutagenesis. The employment of conventional breeding methods has been hampered by the dearth of appropriate germplasm (with a minimum of linked deleterious attributes) for use in introgressing this trait into well-adapted genotypes. Genetic transformation, in spite of its promises, has also not been effective as it has only been possible to engineer a few genes out of the plethora that obviously are implicated in the expression of several genes involved at different stages of the plant's development. So far, the up- or down-regulation of genes, usually one at a time, affecting a certain pathway of import in salinity has not led to the ultimate goal of enhanced grain yield under saline conditions. With regards to *in vitro* techniques, some success has been achieved through anther culture, for instance, but until the mechanism for enhancing saline tolerance through this method is better understood, it might well be ascribed to a chance occurrence, somaclonal variation being one of the plausible explanations, and hence not extendable as a verified strategy. Induced mutagenesis seems to be the method of choice as it holds the promise for affecting several genes at the same time and is a technology that is not susceptible to the concerns being expressed in relation to genetic transformation.

The grave impact of salinity on rice agriculture is recognised by many organizations at both the national and international levels. In countries where salinity severely constrains crop productivity like Australia and in the Middle East, there are robust concerted efforts to mitigate this scourge. It is for this reason that in SE Asia with the use of sea water for irrigating rice paddies such as in Viet Nam, the development of saline tolerant rice varieties has been of vital importance. The result is the development of a number of advanced lines if saline tolerant rice mutants for this country. The IAEA and IRRI have over the past 7 years collaborated in the use of induced mutagenesis to broaden the genetic base of rice germplasm with the aim of identifying and incorporating tolerant mutant variants into breeding schemes. The resulting saline tolerant variants include the IAEA1, IAEA2, IAEA3 and IAEA4 that are being further evaluated in other agroecological zones outside of the test sites in Philippines. Segregating populations are also being developed from these materials with the aim of using them in classical genetic studies aimed at better understanding the mechanisms for inheriting saline tolerance as well as developing molecular genetic markers that can be used in tracking the inheritance of the mutated segments of the genome. Also, the somaclonal variant developed from Pokkali, TCCP 266-2-49-B-B-3, has on account of the induced semi-dwarf plant architecture and superior grain qualities replaced the parent (Pokkali) as the donor parent for salinity tolerance in IRRI breeding schemes.

FAO is also investing considerable efforts in advocacy issues to highlight the impact of salinity on agriculture in general hence the several meetings and conferences that it has organized; publications produced; partnerships fostered with other international organizations such as with UNESCO. Through its joint programme with the IAEA, both organizations' strategies to address salinity have, in addition to laboratory activities in Seibersdorf mentioned in earlier sections, included the empanelling of Coordinated Research Projects (CRP). One such ongoing initiative is the CRP on "*Identification and pyramiding of mutated genes: novel approaches for improving crop tolerance to salinity and drought*". This CRP, coordinated by the Joint FAO/IAEA Programme, has participating institutes from Australia, Bulgaria, China, Cuba, Egypt, Ghana, India, Indonesia, Israel, Italy, Pakistan, Thailand, Tunisia, Turkey, USA and Viet Nam. IRRI is also participating as an international organization. The principal aim is "To generate genetic variability and to use existing mutated and naturally tolerant germplasm of crop plant genetic resources to identify genes controlling various traits contributing to tolerance to drought and salinity in defined environments and so gain a better understanding of the physiological and molecular basis of plant tolerance to drought and salinity."

After a review of the current strategies being deployed in the development of saline tolerant rice varieties which include conventional breeding; genetic transformation; *in vitro* techniques; and induced mutagenesis, it is the position of this article that the use of induced mutagenesis facilitated by enabling biotechnologies (such as doubled haploidy and molecular biology techniques) might hold the greatest promise for rapidly developing saline tolerant rice germplasm. Additionally, while the selection for indices implicated in the physiological response to salinity may aid the garnering of empirical information on the subject, it would seem that for practical purposes, selection should be for grain yield under saline stress. This is because so far, selection for these indices has not translated to actual grain yield just as the engineering of a few isolated pathways has so far not led to development of transgenic materials with enhanced yield under saline environments. Induced mutagenesis in order to be meaningful also will be facilitated by modern biotechnologies that enhance efficiency.

- Doubled haploidy will lead to the rapid generation of homozygous materials obviating the need for several cycles of selfing in order to recover mutants in homozygous recessive forms.
- Cell suspension cultures will increase efficiency of the production of mutants through the elimination or reduction of chimeras as the mutants will be arising from one or a few cells. Along with this however will be a requirement to develop reproducible regeneration protocols for most rice genotypes.
- Molecular techniques including the use of MAS, high throughput reverse genetics strategies such as TILLING also hold a lot of promise. Whole genome scan methodologies such as Diversity Array Technology (DArT) also hold promise for rapidly identifying and eliminating uninteresting mutants.
- Efficient screening methodologies are also a prerequisite for efficiency in the over all process. The modified hydroponics system for screening for response to salt toxicity while producing reliable data still relies on growing the whole plant and taking agronomic data through the different stages of its growth and development. Space requirement in the greenhouses is also an issue. For tens of thousands of putative mutants, this is usually a daunting task to accomplish on account of the sheer drudgery. To this extent, the development of reliable *in vitro* screening methods is imperative.

20. FUTURE PERSPECTIVES

20.1. Functional Genomics

Future endeavours in the genetic improvement of rice will be relying heavily on the immense genomics information that has become publicly available with the sequencing of the rice genome. Efforts are now geared towards harnessing this information especially with the aim of determining the functionsof the different parts of the rice genome. Induced mutant rice genetic stocks, available at IRRI, remain vital in achieving this. The organization of these mutants into mutation grids, i.e. a phenotypically and genotypically characterised mutant population with sufficient density of mutations as to include a mutation in every gene, is paramount. Such a population, almost 50,000 rice mutants (by 2002) already exists in IRRI.

With the appropriate density of mutations, new developments in high throughput reverse genetic strategies such as TILLING, could be exploited to achieve a rapid identification of genes implicated in salinity tolerance and a clear understanding of their functions. The achievement of a routine application of such a strategy would facilitate the development and deployment of saline tolerant rice mutants as the large prerequisite mutant populations could then be easily queried for desired genome alterations at very early seeding stages.

20.2. Concerted Application of Biotechnologies (DH, Cell Suspension Cultures, etc.) in Support of Induced Mutagenesis

A major hindrance to the adoption of the induced mutations strategies in crop improvement is the low level of efficiency, in terms of quality and quantity, for the production of the induced mutant populations. One aspect of this problem, the quality of the mutants, relates to the inherent recessive nature of non-lethal mutations. In a heterozygous background therefore, phenotypic manifestations of mutations are practically impossible to detect in the early progenies necessitating several cycles of selfing. *In vitro* techniques such as the induction of doubled haploidy (DH) using mutagenised sex cells in anther or ovule cultures circumvent this bottleneck. Again, even when the protocols have been established such as anther culture in rice, the use of this technique in induced mutagenesis is not common.

Related to the problem of the production and subsequent detection of mutation events in homozygous genetic backgrounds is that of the confounding effects of chimeras. Though more pronounced in vegetatively propagated crops, it is also problematic in seed propagated plants such as rice and could be circumvented through the use of cell suspension cultures as starting materials for inducing mutations. The use of this strategy however requires the establishment of efficient regeneration protocols such as for somatic embryogenesis. Such protocols, both for creating cell suspension cultures and for regeneration of plantlets from the cultures are therefore clearly required.

REFERENCES

Afza, R; Zapata-Arias, F.J; Zwiletitsch, F; Berthold, G; Gregorio G. 1999 Modification of a rapid screening method for rice mutants to NaCl tolerance using liquid nutrient culture. Mutation Breeding News Letter No.144. International Atomic Energy Agency, Vienna, Austria.

Afza, R; Jain, SM; Shu, Q; M. Guzmann, M; Zapata, FJ; Tumimbang, E; Greogorio, G; Mba, C. 2006. Doubled haploidy and induced mutation in breeding for salt tolerance in rice and wheat. Book of Abstracts, The International Conference on "Haploids in Higher Plants III', Vienna, Austria. February 12–15, 2006.

Ahloowalia, BS; Maluszynski, M; Nichterlein, K. 2004. Global impact of mutation-derived varieties. Euphytica 135: 187–204.

Amano, E. 2006. Use of Induced Mutants in Rice Breeding in Japan. Plant Mutation Reports 1:21–24. International Atomic Energy Agency, Vienna, Austria.

An, G; Lee, S; Kim, S-H; Kim, S-R. 2005. Molecular Genetics Using T-DNA in Rice. Plant Cell Physio. 46(1); 14–22.

Aneeta; Sanan-Mishra, N; Tuteja, N; Kumar Sopory, S. 2002. Salinity- and ABA-induced up-regulation and light-mediated modulation of mRNA encoding glycine-rich RNA-binding protein from Sorghum bicolor. Biochem Biophys Res Commun. 296(5):1063–8.

Atanassov, A; Zagorska, N; Boyadijiev, P; Djilianov, D. 1995. *In vitro* production of haploid plants. World Journal of Microbiology and Biotechnology. 11:400–408.

Azhar, FM; McNeilly, T. 1988. The genetic basis for salt tolerance in *Sorghum bicolor* (L) Moench seedlings. Plant Breeding 101:114–121.

Balooch, AW; Soomro, AM; Naqvi, MH; Bughio, HR; Bughio, MS. 2006. Sustainable Enhancement of Rice (*Oryza sativa* L.) Production Through the Use of Mutation Breeding. Plant Mutation Reports 1:40–42. International Atomic Energy Agency, Vienna, Austria.

Balooch, AW; Soomro, AM; Javed, MA; Bughio, H-ur-R; Alam, SM; Bughio, MS; Mohammed, T; Mastoi, N-ur-N. 2003. Induction of Salt Tolerance in Rice Through Mutation Breeding. Asian Journal of Plant Sciences 2(3): 273–276.

Bentota, AP. 2006. Mutation Improvement of Rice Variety Bw-267–3 for Red Pericarp Grains and Lodging Resistance. Plant Mutation Reports 1:42–43. International Atomic Energy Agency, Vienna, Austria.

Binh, DQ; Heszsky, LE. 1990. Restoration of the regeneration potential of long term culture in rice (*Oryza sativa* L) by salt pretreatment. J. Plant Physiol. 136:336–340.

Binh, DQ; Heszsky, LE; Gyulai, G; Csillag, A. 1992; Plant regeneration of NaCl-pretreated cells from long-term suspension culture of rice (*Oryza sativa* L.) in high saline conditions. Plant Cell, Tissue and organ Culture **29**:75–82.

Bohnert, HJ; Jensen, RG. 1996. Metabolic engineering for increased salt tolerance – the next step. Australian Journal of Plant Physiology 23: 661–666.

Brown, DCW; Thorpe, TA. 1995. Crop improvement through tissue culture. World Journal of Microbiology and Biotechnology. 11:409–415.

Cheema, AA. 2006. Mutation Breeding for Rice Improvement in Pakistan: Achievements and Impact. Plant Mutation Reports 1:36–39. International Atomic Energy Agency, Vienna, Austria.

Chen, X; Liu, X; Wu, D; Shu, QY. 2006. Recent Progress of Rice Mutation Breeding and Germplasm Enhancement in China. Plant Mutation Reports 1:4–6. International Atomic Energy Agency, Vienna, Austria.

Chopra, VL. 2005. Mutagenesis: Investigating the process and processing the outcome for crop improvement. Current Science. 89(2): 353–359.

Colbert, T; Till, BJ; Tompa, R; Reynolds, S; Steine, MN; Yeung, AT; McCallum, CM; Comai, L; Henikoff, S. 2001. High-throughput screening for induced point mutations. Plant Physiology. 126(2): 480–484.

Comai, L; Henikoff, S. 2006. TILLING: Practical single-nucleotide mutation discovery. Plant J. 45: 684–94.

Croughan, TP; Stavarek, SJ; Rains, DW, 1978. Selection of a NaCl tolerant line of Cultured Alfalfa, Crop Science, 18:959–963.

Do, KT; Dao, MS; Hung, PQ; Nguyen, TC. Rice Mutation Improvement for Short Duration, High Yield and Tolerance to Adverse Conditions in Mekong Delta of Viet Nam. Plant Mutation Reports 1:49–51. International Atomic Energy Agency, Vienna, Austria.

Don Palmer, CE; Keller, WA. 2005. Overview of Haploidy. In: Kasha, KJ; Keller, WA; Palmer CE (eds.). 2005. Haploids in Crop Improvement II. Biotechnology in Agriculture and Forestry Series. Springer-Verlag Berlin and Heidelberg GmbH & Co., Germany. Pp 3 – 7.

El-Bably, AZ. 2002. Advanced and integrated approaches for crop tolerance to poor quality irrigation water in Egypt. In Zdruli P., Steduto P., Kapur S. (eds.). 7th international meeting on soils with Mediterranean type of climate (selected papers). Bari : CIHEAM-IAMB, p. 363–378 (Options Méditerranéennes : Série A. Séminaires Méditerranéens ; n. 50). 7. International Meeting on: Soils with Mediterranean Type of Climate, 2001/09/23–28, Valenzano (Italy)

Evans, HJ; Sparrow, AH. 1961. Nuclear factors effecting radiosensitivity II. Dependence on nuclear and chromosome structure and organization. Brokkhaven Symp. In Biol. Vol 14, 101 – 127.

Fast Facts 21. Dryland Salinity in Australia - key findings. http://audit.ea.gov.au/ANRA/docs/fast_facts/fast_facts_21.html. ISBN 0 642 371 091 December 2000.

Flowers, T; Troke, P; Yeo, A 1977. The mechanism of salt tolerance in halophytes. Annu. Rev. Plant Physiol. 28: 89–121.

Flowers, TJ; Hajibagheri, MA; Clipson, NJW. 1986. Halophytes. Q. Rev. Biol. 61: 313–337.

Flowers, TJ; Yeo, AR. 1996. Metabolic engineering for increased salt tolerance – the next step. Australian Journal of Plant Physiology 23: 666–667.

Flowers, TJ. 2004. Improving crop salt tolerance. J. Exp. Bot. 55 (396): 307–319.

Flowers, TJ; Flowers SA. 2005. Why does salinity pose such a difficult problem for plant breeders? Agricultural Water Management 78: 15–24.

Garg, A; Kim, JK; Owens, TG; Ranwala, AP; Choi, YDC; Kochian, LV; Wu, RJ. 2002. Trehalose accumulation in rice plants confers high tolerance levels to different abiotic stresses. Proceedings of the National Academy of Sciences, USA. 99: 15898–15903.

Gilchrist, EJ; Haughn, GW. 2005. TILLING moves beyond functional genomics into crop improvement. Curr Opin Plant Biol. 8(2):211–5.

Gleick, P.H. (Editor). 1993. Water in crisis. Oxford University Press, New York, NY, 473 pp.

Greene, EA; Codomo, CA; Taylor, NE; Henikoff, JG; Till, BJ; Reynolds, SH; Enns, LC; Burtner, C; Johnson, JE; Odden, AR; Comai, L; Henikoff S. 2003. Spectrum of chemically induced mutations from a large-scale reverse-genetic screen in Arabidopsis. Genetics 164:731–740.

Gregorio, GB; Senadhira, D. 1993. Genetic analysis of salinity tolerance in rice. Theor. Appl. Gent. 86: 333–338.

Gregorio, GB; Senadhira, D; Mendoza, RD; Manigbas, NL; Roxas, JP; Guerta, CQ. 2002. Progress in breeding for salinity tolerance and associated abiotic stresses in rice. Field Crops Research 76: 91–101.

Henikoff, S; Comai, L. 2003. Single-nucleotide mutations for plant functional genomics. Ann Rev Plant Biol 54:375–401

Henikoff, S; Till, BJ; Comai, L. 2004. TILLING. Traditional Mutagenesis Meets Functional Genomics. Plant Physiol. 135:1–7.

Hermelin, T. 1997. SOP's for radiation services for the induction of mutation in plant breeding. Plant Breeding Unit and Plant Genetics Section. FAO/IAEA Internal report.

Heszsky, LE; Nam, LS; Kiss, E; Simon Kiss, I; Lokos, K; Binh, DQ. 1991. *in vitro* studies on rice in Hungary. In Bajaj YPS(ed) Biotechnology in Agriculture and Forestry, Vol-14, pp 619–641, Rice Springer Verlag, Berlin-Heidelberg- New-York.

http://www.ars.usda.gov/is/AR/archive/jul05/genes0705.htm.

http://physics.nobel.brainparad.com/wilhelm_conrad_rontgen.html

http://nobelprize.org/nobel_prizes/physics/laureates/1903/becquerel-bio.html

http://nobelprize.org/nobel_prizes/physics/articles/curie/

Hu, H; Dai, M; Yao, J; Xiao, B; Li, X; Zhang, Q; Xiong, L. 2006. Overexpressing a NAM, ATAF, and CUC (NAC) transcription factor enhances drought resistance and salt tolerance in rice. PNAS 103(35):12987–12992.

Ismachin, M; Sobrizal. 2006. A Significant Contribution of Mutation Techniques to Rice Breeding in Indonesia. Plant Mutation Reports 1:18–21. International Atomic Energy Agency, Vienna, Austria.

Jain SM; Sopory SK; Veilleux RE. (Eds.). 1996. In vitro haploid production in higher plants. Vol. 1–5. Kluwer. The Netherlands

Jain, SM. 2005. Major mutation-assisted plant breeding programs supported by FAO/IAEA. Plant Cell Tissue and Organ Culture 82:113–123.

Jeon, J-S; Lee, S; Jung, K-H; Jun, S-H; Jeong, D-H; Lee, J; Kim, C; Jang, S; Lee, S; Yang, K; Nam, J; An, K; Han, M-J; Sung, R-J; Choi, H-S; Yu, J-H; Choi, J-H; Cho, S-Y; Cha, S-S; Kim, S-I; An, G. 2000. T-DNA insertional mutagenesis for functional genomics in rice. The Plant Journal. 22(6):561–570.

Jia, G-X; Zhu, Z-Q; Chang, F-Q; Li, Y-X. 2002. Transformation of tomato with the BADH gene from Atriplex improves salt tolerance. Plant Cell Reports 21(2): 141–146.

Kasha, KJ; Maluszynski, MM. 2003. Production of doubled haploids in crop plants. An Introduction. In: Maluszynski, M; Kasha, KJ; Forster, BP; Szarejko, I (eds.). Doubled Haploid Production in Crop Plants: A Manual. Kluwer Academic Publishers, Dordrecht, The Netherlands / Boston, USA / London, UK. Pp 1–4. 428pp.

Kasha, KJ; Keller, WA; Palmer, CE (Eds.). 2005. Haploids in Crop Improvement II. Biotechnology in Agriculture and Forestry Series. Springer-Verlag Berlin and Heidelberg GmbH & Co., Germany.300pp.

Kefu, Z; Hai, F; San, Z; Jie, S. 2003. Study on the salt and drought tolerance of *Suaeda salsa* and *Kalanchoe claigremontiana* under iso-osmotic salt and water stress. Plant Sci 165: 837–844.

Kodym, A; Afza, R. 2003. Physical and Chemical Mutagenesis. In: Erich Grotewold (ed.). Plant Functional Genomics. Humana Press, Totowa, New Jersey, USA. pp189–204.

Konzak, C.F., Mikaelsen, K., Sigurbjörnsson, B. Burtscher, A. 1967. Recommended standard procedures for irradiating, cultivating and measuring cereal seeds to determine the effects of neutron irradiation in the neutron-seed-irradiation program. In: Neutron irradiation of seeds (Technical Reports Series, No.76), IAEA, Vienna, 103–107.

Krysan, P; Young, JC; Sussman, MR. 1999. T-DNA as an Insertional Mutagen in Arabidopsis. The Plant Cell 11: 2283–2290.

Krysan, PJ; Young, JC; Tax, F; Sussman, MR. 1996. Identification of transferred DNA insertions within *Arabidopsis* genes involved in signal transduction and ion transport. Proc. Natl. Acad. Sci. USA. 93:8145–8150.

Lamond, RE; Whitney, DA. 1992. Management of Saline and Sodic Soils.MF-1022. Kansas State University, Cooperative Extension Service, Manhattan, Kansas. 4pp.

Lee, SY; Cheong, JI; Kim, TS. 2003. Production of doubled haploids through anther culture of M_1 rice plants derived from mutagenized fertilized egg cells. Plant Cell Reports 22(3): 218–223.

Lee, KS. 1995. Variability and genetics of salt tolerance in japonica rice (*Oryza sativa* L.). Ph.D Thesis. University of the Philippines, Los Baños, Philippines.

Lee, KS; Senadhira, D; Gregorio, GB. 1996. Genetic analysis of salinity tolerance in japonica rice. SABRAO J. 28(2): 7–13.

Lee, SY; Lee, JH; Kwon, TO. 2003. Selection of salt-tolerant doubled haploids in rice anther culture, Plant Cell. Tiss. Org. Cult. 74(2): 143–149

Li, SN; Heszsky, LE. 1986. Testing of salt tolerance and regeneration in callus (n, 2n) of rice. In: Horn, W, Jensen, JC. Odenbach, W & Schieder, JO (eds): Genetic manipulation in Plant Breeding. Pp 617–619 Walter de Gruyter and Co, Berlin-New York.

Maluszynski, M; Nichterlein, K; van Zanten, L; Ahloowalia, BS. 2000. Officially released mutant varieties – the FAO/IAEA database Mutation Breeding Reviews. The Joint FAO/IAEA Programme of Nuclear Techniques in Food and Agriculture, International Atomic Energy Agency, Vienna, Austria. Pp 88.

Maluszynski, M; Kasha, KJ; Forster, BP; Szarejko, I. 2003. (eds.). Doubled Haploid Production in Crop Plants: A Manual. Kluwer Academic Publishers, Dordrecht / Boston, USA / London, UK. 428pp.

Maluszynski, M; Kasha, KJ; Szarejko, I. 2003. Published doubled haploid protocols in plant species. In: Maluszynski, M; Kasha, KJ; Forster, BP; Szarejko, I. 2003. (eds.). Doubled Haploid Production in Crop Plants: A Manual. Kluwer Academic Publishers, Dordrecht The Netherlands / Boston, USA / London, UK. Pp 309–335.

Mansour, MMF; Salama, KHA; Al-Mutawa, MM. 2003. Transport proteins and SALT tolerance in plants. Plant Science 164: 891–900.

McCallum, CM; Comai, L; Greene, EA; Henikoff, S. 2000. Targeted screening for induced mutations. Nature Biotechnology. 18(4): 455–7.

McCallum CM; Comai, L; Greene, EA; Henikoff, S. 2000. Targeting Induced Local Lesions in Genomes (TILLING) for plant functional genomics. Plant Physiol. 123(2): 439–442.

McKinney, EC; Ali, N; Traut, A; Feldmann, KA; Belostotsky, DA; McDowell, JM; Meagher, RB. 1995. Sequence-based identification of T-DNA insertion mutation in *Arabidopsis:* Actin mutants act2–1 and act4–1. Plant J. 8:613–622.

Mishra, B; Akbar, M; Seshu, DV. 1990. Genetic studies on salinity tolerance in rice towards better productivity in salt-affected soils. In: Proceedings of the Papers Presented at the Rice Research Seminar, July 12, 1990. IRRI, Los Baños, Philippines

Moeljopawiro, S; Ikehashi H. 1981. Inheritance of salt resistance in rice. Euphytica 30: 291–300.
Mohamad, O; Mohd. Nazir, B; Alias I; Azlan, S; Abdul Rahim, H; Abdullah, MZ; Othman, O; Hadzim, K; Saad, A; Habibuddin, H; Golam F. 2006. Development of Improved Rice Varieties Through the Use of Induced Mutations in Malaysia. Plant Mutation Reports 1:27–34. International Atomic Energy Agency, Vienna, Austria.
Montelone, BA. 1998. Mutation, Mutagens, and DNA Repair. http://www-personal.k-state.edu/~bethmont/mutdes.html. Division of Biology, Kansas State University, USA.
Murashige, T; Skoog, F. 1965. A revised medium for rapid growth and bioassay with tobacco tissue culture. Physiol. Plant., 18: 473–497.
Nakajima, K. 1991. Biotechnology for crop improvement and production in Japan. Paper presented at the Regional Expert Consultation on the Role of Biotechnology in Crop Production, FAO Regional Office for Asia and the Pacific, Bangkok, June 18–21, 1991. pp21.
New, KT. 2006. Rice Mutation Breeding for Varietal Improvement in Myanmar. Plant Mutation Reports 1:34–36. International Atomic Energy Agency, Vienna, Austria
Ng, PC; Henikoff, S. 2003. SIFT: predicting amino acid changes that affect protein function. Nucl. Acids Res., 31:3812–3814.
Oldeman, LR; Hakkeling, RTA; Sombroek, WG. 1991. *World Map of the Status of Human-Induced Soil Degradation*: An Explanatory Note, Second revised version. Wageningen: International Soil and Reference Center, The Netherlands, and *Nairobi, Kenya: International Soil Reference and Information Centre/United Nations Environment Programme.*
Patnaik, D; Chaudhary, D; Rao, GJN. 2006. Genetic Improvement of Long Grain Aromatic Rices through Mutation Approach. Plant Mutation Reports 1:11–16. International Atomic Energy Agency, Vienna, Austria.
Pandey, GK; Reddy, VS; Reddy, MK; Deswal, R; Bhattacharya, A; Sopory, SK. 2002. Transgenic tobacco expressing *Entamoeba histolytica* calcium binding protein exhibits enhanced growth and tolerance to salt stress. Plant Sci. 162: 41–47
Perry, JA; Wang, TL; Welham, TJ; Gardner, S; Pike, JM; Yoshida, S; Parniske, M. 2003. A TILLING reverse genetics tool and a web-accessible collection of mutants of the legume Lotus japonica. Plant Physiol. 131: 866–71.
Quijano-Guerta, C; Kirk, GJD. 2002. Tolerance of rice germplasm to salinity and other soil chemical stresses in tidal wetlands. Field Crops Research 76: 111–21.
Reddy, PJ; Vaidyanath, K. 1986. *In vitro* characterization of salt stress effects and the selection of salt tolerant plants in rice (*Oryza sativa* L.) Theor.Appl.Genet.71. 757–760.
Roberts, G. 2003. Effects of mutations. http://www.bact.wisc.edu/Bact370/effectsofmut.html.
Ryu, C-H; You, J-H; Kang, H-G; Hur, J; Kim, Y-H; Han, M-J; An, K; Chung, B-C; Lee, C-H; An, G. 2004 Generation of T-DNA gene tagging lines with a bidirectional gene trap vector and the establishment of an insertion-site database. Plant Mol Biol 54: 489–502.
Sairam, RK; Tyagi, A. 2004. Physiology and molecular biology of salinity stress tolerance in plants. Current Science 86(3): 407–421.
Senadhira, D; Zapata-Arias, FJ; Gregorio, GB; Alejar, MS; De la Cruz, HC; Padolina, TF; Galvez, AM. 2002. Development of the first salt-tolerant rice cultivar through indica/indica anther culture. Field Crops Research 76:103–110.
Sessions, A; Burke, E; Presting, G; Aux, G; McElver, J; Patton, D; Dietrich, B; Ho, P; Bacwaden, J; Ko, C; Clarke, JD; Cotton, D; Bullis, D; Snell, J; Miguel, T; Hutchison, D; Kimmerly, B; Mitzel, T; Katagiri, F; Glazebrook, J; Law, M; Goff, SA. 2002. A high-throughput Arabidopsis reverse genetics system.The Plant Cell 2985–2994
Shikazono, N; Suzuki, C; Kitamura, S; Watanabe, H; Tano, S; Tanaka, A. 2005. Analysis of mutations induced by carbon ions in *Arabidopsis thaliana.* J. Exp. Bot. 56 (412): 587–596.
Singla-Pareek, SL; Reddy, MK; Sopory, SK. 2003. Genetic engineering of the glyoxalase pathway in tobacco leads to enhanced salinity tolerance. PNAS 100(25): 14672–14677.
Slade, AJ; Fuerstenberg, SI; Loeffler, D; Steine, MN; Facciotti, D. 2004. A reverse genetic, nontransgenic approach to wheat crop improvement by TILLING. Nat. Biotech. On-line version doi:10.1038/nbt1043.

Subbarao, GV; Johansen, C; Kumar Rao, JVDK; Jana, MK. 1990. Salinity tolerance in F1 hybrids of pigeonpea and a tolerant wild relative. Crop Science 30: 785–788.

Taylor, NE; Greene, EA. 2003. PARSESNP: A tool for the analysis of nucleotide polymorphisms. Nucl. Acids Res. 31:3808–3811.

Thomson, WW; Faraday, CD; Cross JW. 1988. Salt glands. In: Baker, DA; Hall JL (Eds.). Solute Transport in Plant Cells and Tissues. Longman Scientific and Technical, Harlow, Essex, England pp. 498–537

Till, BJ; Reynolds, SH; Greene, EA; Codomo, CA; Enns, LC; Johnson, JE; Burtner, C; Odden, AR; Young, K; Taylor, NE; Henikoff, JG; Comai, L; Henikoff, S. 2003. Large-scale discovery of induced point mutations with high-throughput TILLING. Genome Res. 13(3):524–530.

Till, BJ; Reynolds, SH; Weil, C; Springer, N; Burtner, C; Young, K; Bowers, E; Codomo, CA; Enns, LC; Odden, AR; Greene, EA; Comai, L; Henikoff, S. 2004. Discovery of induced point mutations in maize genes by TILLING. BMC Plant Biol. 28; 4:12.

Till, BJ; Burtner, C; Comai, L; Henikoff, S. 2004. Mismatch cleavage by single-strand specific nucleases. Nucleic Acids Res. 32(8):2632–2641.

Till, BJ; Colbert, T; Tompa, R; Enns, L; Codomo, C; Johnson, J; Reynolds, SH; Henikoff, JG; Greene, EA; Steine, MN; Comai, L; Henikoff, S. 2003. High-throughput TILLING for functional genomics, in Plant Functional Genomics: Methods and Protocols, ed. Grotewald, E. Humana Press, 236:205–220.

Tran, DQ; Dao, TTB; Nguyen, HD; Lam, QD; Bui, HT; Nguyen, VB; Nguyen, VX; Le, VN; Do, HA; Phan, P. 2006. Rice Mutation Breeding in Institute of Agricultural Genetics, Viet Nam. Plant Mutation Reports 1:47–49. International Atomic Energy Agency, Vienna, Austria.

Van Harten, AM. 1998. Mutation Breeding. Theory and practical Applications. Cambridge, U.K.: New York, Cambridge University Press, 111–127.

Veena; Reddy, SV; Sopory, SK. 1999. Glyoxalase I from Brassica juncea: molecular cloning, regulation and its over-expression confer tolerance in transgenic tobacco under stress. The Plant J. 17(4): 385–395.

Wincov, I. 1996. Characterization of rice (*Oryza sativa* L.) plants regenerated from salt tolerant cell-lines. Plant Science 113 105–111;

Yeo, AR; Yeo, ME; Flowers TJ. 1987. The contribution of an apoplastic pathway to sodium uptake in rice roots in saline conditions. J. Exp. Bot. 38: 1141–1153.

Yoshida, S; Forno, DA; Cock, JH; Gomez, KA. 1976. Laboratory manual for physiological Studies of rice. IRRI, Las Banos, Laguna, Philippines. Pp 83.

Zapata-Arias, FJ; Torrizo, LB; Ando, A. 1995. Current developments in biotechnology for genetic improvement: the case of rice (*Oryza saliva* L.). World Journal of Microbiology and Biotechnology. 11:393–399.

Zapata-Arias, FJ. 2003. Laboratory protocol for anther culture technique in rice. In: Maluszynski, M; Kasha, KJ; Forster, BP; Szarejko, I. 2003. (Eds.). Doubled Haploid Production in Crop Plants: A Manual. Kluwer Academic Publishers, Dordrecht, The Netherlands / Boston, USA / London, UK. Pp

Zerr, T; Henikoff, S. 2005. Automated band mapping in electrophoretic gel images using background information. Nucleic Acids Res. 33(9):2806–2812.

Zhang, H-X; Blumwald, E. 2001. Transgenic salt-tolerant tomato plants accumulate salt in foliage but not in fruit. Nature biotech. 19: 765–768.

Zhu, XD; Chen, HQ; Shan, JX. 2006. Nuclear Techniques for Rice Improvement and Mutant Induction in China National Rice Research Institute. Plant Mutation Reports 1:7–10. International Atomic Energy Agency, Vienna, Austria.

CHAPTER 18

PARTICIPATORY BREEDING FOR DROUGHT AND SALT TOLERANT CROPS

P.A. HOLLINGTON AND KATHERINE A. STEELE

CAZS Natural Resources, University of Wales Bangor, UK

Abstract: Although enormous effort has been put into conventional breeding programmes for both drought and salt tolerance, there has been little progress in producing varieties that are adopted by farmers in their fields. This is largely due to the lack of consideration given to the specific needs of farmers in droughted and salt-affected environments, in particular in terms of non-yield post-harvest traits. We discuss with examples the advantages and disadvantages of participatory variety selection (PVS) and participatory plant breeding (PPB) in their various forms, as well as the use of the term client-oriented breeding to describe the process of involving the end-users of the breeding programme. Methods for the analysis of participatory trials, and practical considerations for their management, are presented. We also show how both participatory and molecular approaches can be combined into an integrated, client-oriented breeding programme

Keywords: participatory variety selection, participatory plant breeding, client-oriented breeding, PVS, PPB, COB

1. INTRODUCTION

Drought and salinity both cause severe reductions in crop yields, with consequent increases in poverty, in many areas of the world, and their importance is likely to increase in response to the effects of global change (Yeo, 1999; Mannion, 1995; Hillel and Rozenweig, 2002) and increased competition for water (e.g. Hobbs and Gupta, 2003). Traditional plant breeding has had limited effect in improving crop production in areas subject to these stresses (e.g. Araus et al., 2002), and we argue that increased farmer involvement in the whole breeding process would lead to better targeted varieties suited in particular to the needs of poor farmers in drought and salt-affected regions. We concentrate on cereals, as that is the area of our expertise.

Arid and semiarid regions cover around a quarter of the world's land area, and are inhabited by around one sixth of the population (WRI, 2000). Water is limited in most of the countries around the eastern and southern Mediterranean (Parry et al., 2005), and

M.A. Jenks et al. (eds.), Advances in Molecular Breeding Toward Drought and Salt Tolerant Crops, 455–478.

in many other parts of Africa and Asia (FAO, 2003). With climate change, droughts will become both more frequent and more severe (Fischer et al., 2002), and by 2025 one third of the population of the developing world will face severe water shortages (Seckler et al., 1998). Although rainfed farming accounts for around 80% of agricultural land (Rockström et al., 2003), it has low yields and accounts for only about 60% of world crop production (FAO, 2002). Little extra land is available to increase the production area (Young, 1999), increasing pressure to raise yields per unit area of soil and unit volume of water. Plant breeding to increase crop production under drought may be the greatest challenge facing agricultural science (Reynolds et al., 2005).

Salinity, and related problems such as sodicity and waterlogging, has long been a major constraint on crop production, particularly in irrigated systems (Szabolcs, 1994; Shannon, 1997). Around 800 M ha are affected by salinity or sodicity worldwide (FAO, 2005). Most is due to natural causes (primary salinity), but about 2% of dryland agriculture is affected by secondary salinity (resulting from human activities), as is 20% of irrigated land, which produces about one third of the world's food (Munns, 2005). In the developing world, wheat yields can be reduced by 65% (Quayyum and Malik, 1988), leading to increased poverty and reliance on imports. The problem will worsen as population growth forces more land under irrigation, and climate change and water shortages make it essential to exploit marginal lands and water.

2. PROBLEMS WITH CONVENTIONAL BREEDING

2.1. Lack of Success of Conventional Methods in Breeding

In breeding for low potential environments subject to abiotic stress such as drought and salinity, one of the main problems is that selection efficiency decreases as the difference between the environment being selected in and the target environment increases (Ceccarelli and Grando, 1999), due to high genotype x environment (G x E) interactions (Ceccarelli, 1994, Ceccarelli et al., 1994). Thus genotypes selected on research stations under high input conditions do not in general do well in low potential environments such as marginal farms in dry areas (Ceccarelli et al., 1998). Despite this, selection for high yield potential has often led to high production in a range of environments (Slafer et al., 1999; Richards, 2000; Araus et al., 2002; Richards et al., 2002; Tambussi et al., 2005). There is an argument and evidence that selection in high-potential environments does lead to higher yields in poorer ones (Richards, 2000; Araus et al, 2002), but only where those are characterised by frequent mild or moderate stresses, and a distinction must be drawn between dry and very dry environments. In the latter (yield potential < 1 t ha^{-1}), breeding barley for survival (tolerance to severe stress) rather than production has been successful (Ceccarelli and Grando, 1996) by using locally-adapted germplasm (Ceccarelli et al., 1998). For barley, high yield potential might be an advantage under moderate stress, and drought tolerance only so in extreme conditions.

2.1.1. *Drought*

Drought tolerance is a complex, multigenic trait (Bennett, 2003), and plants have developed a range of strategies to balance the need for growth and reproduction on the one hand with stress resistance on the other. This makes it hard to identify strategies and selection criteria that contribute to high and stable yields under drought. Several traits increase wheat yields in Mediterranean-type environments (Turner, 2004a). The most important is early phenology and better matching of growth to the rainfall pattern (Perry and D'Antuono, 1989; Turner, 2003). However, progress in increasing crop yields under drought conditions has been much slower than for temperate crops (Turner, 2004a; 2004b). One explanation is that, due to G x E, indirect selection for traits based on deep understanding of the physiological and molecular processes of the crop and its response to stress would be more effective than selection for yield (Araus et al., 2002).

However, the use of physiological traits, except for carbon isotope discrimination, has had little benefit (Araus et al., 2003; Slafer et al., 2005), and attempts to improve crops using marker-assisted selection (MAS) for yield under drought have yet to realise their potential (Parry et al., 2005; Araus et al., 2003). Genetic transformation has focussed largely on cellular processes (Araus et al., 2003), while the important processes involved in crop response to drought are at higher levels and so largely multigenic (Richards, 1996). Transformation has also largely focused on traits related to survival under drought rather than on stress avoidance and higher yield potential, as have many other molecular efforts (Passioura, 2002).

2.1.2. *Salinity*

Many efforts have been made to breed salt-tolerant crops, but with little progress in producing varieties accepted and used by farmers (Flowers, 2004; Gregorio and Cabuslay, 2005).) Fewer than 30 salt-tolerant cultivars had been released by 1995 (Flowers and Yeo, 1995) and in the following nine years only three more were registered, and one patented (Flowers, 2004). Although transgenics with some salt tolerance have been produced, none have been field-tested and few of the claims made to success seem valid due to flawed testing programmes, e.g. experiments under conditions of zero transpiration, or failure to compare the wild-type with transformants under appropriate salinity levels (Flowers, 2004).

The multigenic nature of salt tolerance, and rapid spatial and temporal changes in field salinity (Richards, 1983) make reliable screening difficult. In some crops this is compounded by additional stresses associated with salinity: mineral deficiencies and toxicities, submergence, deep water and drought may be important. These vary with time, so cultivar adaptability depends upon long term tolerance (Gregorio et al., 2002). The dynamism affects the level of salinity at which to screen. Large areas of a field may be of low or moderate salinity, with only small areas of high salinity. Richards (1992) concluded that selection at low salinity was preferable, as most yield was from the non-saline areas. However, he drew his conclusions from work on drying saline fields inappropriate to the irrigated conditions of many developing countries, and did not consider salinity/waterlogging interactions. Isla et al. (2003)

3.1.1. Stages in a PVS programme

A successful programme of PVS has 4 phases:
1. Identification of farmers' requirements
2. Search for suitable material for testing
3. Experimentation with farmers
4. Dissemination of farmer-preferred material

The first phase, identification, allows farmers to be given appropriate material to test, and is carried out using several methods, including participatory rural appraisal (PRA), crop walks at harvest, and pre-selection by farmers of material in multi-entry trials, either on research stations or on farms, and including local landraces, recommended cultivars and advanced breeding lines. This allows:
- identification of the best performing lines;
- comparison of recommended germplasm with local material;
- evaluation of the extent of diversity; and
- assessment of the degree of agreement between farmers' names for landraces and their phenotypes

The search phase looks for cultivars that meet farmers' needs in terms of, in particular, maturity, height, grain quality and agro-environmental niche. They are selected from those released nationally and regionally (including old varieties), and from advanced pre-release material.

The third phase, experiments on farmers' fields, includes a number of techniques (Table 1), with the extent of participation varying from very little, basically "on-station" trials moved to the farm (valuable to broaden the range of soils, pests and diseases encountered and so encourage interactions with farmers), to almost total, with very little input from researchers. Such informal research and development (IRD) is very cost-effective, particularly for organisations such as NGOs that might wish to provide farmers with acceptable improved genotypes but that have only limited resources.

The degree of farmer participation, and the resources required, can vary between these extremes. Greater scientist input is usually needed when farmers are asked to grow more than one new cultivar, as assistance is needed with the design, and assessment of yield and other traits from replicated trials may need considerable scientific input. Many IRD trials have only one new genotype alongside the farmer's usual variety, and involve the collection of simple data which might only be perceptions rather than measured yields. In an IRD programme for rice in Nepal (albeit in a high-potential environment), farmers were offered small amounts of seed of different varieties to grow without any researcher intervention (Joshi et al., 1997). In a region where 90% of the rice grown was of one variety, within two years over 35% of households chose to grow the new genotype again.

An important part of the evaluation phase is post-harvest participatory assessment of organoleptic qualities such as aroma, taste, grain consistency, cooking quality and taste, rarely assessed in traditional plant breeding but important for poor farmers. By doing this, expensive field evaluations of agronomically suitable lines that will be rejected by consumers can be avoided, and such evaluations have

Table 1. Methods of varietal selection with varying degrees of farmer participation

Methods in increasing order of farmer-participation	Evaluation includes	Example institutions
1. Researcher-managed and evaluated on-station trials; farmers may visit station to identify farmer-acceptable material	Yield data; possibly farmer evaluation	Research
2. Researcher-managed on-farm trials, replicated design; farmers may be involved in evaluation	Yield data; possibly farmer evaluation	Research
3. Farmer-managed, replicated design, on-farm trials, with scientists' supervision; several entries per farmer	Yield data; farmers' perceptions	Research
4. Farmer-managed, unreplicated design, on-farm trials; one cultivar per farmer; replication across farmers	Yield data; farmers' perceptions	Research, Extension, Non-governmental organisation (NGO)
5. Trials as in 4	Farmers' perceptions only	NGO, Extension, Research
6. Farmer-managed trials; no formal design either within a farm or across farmers	Informal, anecdotal	NGO, Extension, Research

Source: Witcombe et al., 1996

been developed for rice (e.g. Virk et al., 2003). Consumers in one region may well have different preferences to those in another for these traits (Stirling and Witcombe, 2004).

The final stage of PVS is further dissemination of the farmer-preferred cultivars, and developing linkages between plant breeding organisations and those responsible for seed production is important. Whether or not the cultivar identified from PVS has been officially released in the area is critical (Witcombe et al., 1996). If not, it is not usually possible for it to be recommended by extension services, or for its multiplication by the public-sector, so it is important to include released cultivars in a PVS programme. If a variety has not been released, a time-consuming release process is needed, often requiring data unavailable from PVS. We therefore feel it is important for varietal release committees to consider data on farmer perceptions and farmer demands, rather than relying solely on yield data.

PVS, using farmers' knowledge to identify useful characteristics in a variety, can provide valuable information to breeding programmes. It can identify general adaptive traits for particular environments or cropping systems, and identify specific traits wanted by farmers in particular areas.

3.1.2. *Examples of PVS in drought areas*

PVS was carried out on a range of crops in rainfed areas of western India (Joshi and Witcombe, 1996) to identify and overcome the constraints that caused farmers to continue growing landraces of rice and chickpea. Farmer-acceptable varieties were not present among the released cultivars for the area, which generally had at least one undesirable trait, for example slow maturity, short straw, or the perceived need for additional unaffordable inputs, but was identified from varieties released elsewhere. The PVS trials identified the rice variety Kalinga III as outyielding the landraces and meeting farmers' requirements in almost all respects, confirmed by seed sales after the trials, in which demand exceeded supply, despite large areas being sown by farmers who had saved seed. Kalinga III had been released in several states in India, but not those in which the PVS was carried out, i.e. its domain was much greater than its area of release. Three chickpea cultivars were also preferred by farmers that were recommended only in other areas.

Mulatu and Belete (2001) carried out PVS in lowland sorghum in Ethiopia, to provide farmers with an alternative to their usual landraces, and to identify their needs for future breeding programmes. Following identification, varieties were searched for and seed sourced. Eight varieties were initially tested on-station. In the following two years, farmers evaluated these against local cultivars, but selected only three that had been selected by the breeders. These were rapidly introduced into the local system through farmer-to-farmer seed exchange. An unexpected outcome was to disprove the generally-held view that farmers in the region would only grow long-duration varieties.

Baidu-Forson (1997) used on-station PVS to identify farmer-preferred traits in pearl millet in western Niger. High yield was regarded as much less important than high numbers of productive tillers, large grains, tall stature and short duration. Women farmers were less likely to reject a variety because of agronomic traits than were men, and there was no evidence of the preferred traits differing between locations.

Advanced rice lines were tested by villagers in India (Maurya et al., 1988), and superior genetic material preferred by farmers was identified. Participatory selection is also being carried out by IRRI and partners for rice in drought-affected environments in India (IRRI, 2006a) and Thailand (IRRI, 2006b).

3.2. Participatory Plant Breeding

Participatory plant breeding, where farmers contribute to selection from segregating material, is a logical extension of PVS, creating new variability but consuming more resources (Witcombe et al., 1996). It should be used when PVS fails to identify suitable cultivars, and can use cultivars identified through PVS as the parents for crosses. Again, PPB can have varying degrees of participation. These range from growing all generations on-station, with farmers involved only at the pre-release or even post-release stages (targeting wide adaptation, although early generations may all be grown at a single location with multi-location testing later), to training expert farmers to make crosses and carry out the selection, either with or without assistance from breeders, when the material produced will be specific to farmers' requirements. Witcombe et al. (2006) noted that if participation was to increase the efficiency of the breeding programme, rather than for reasons of equity or empowerment, then it was optional whether or not farmers themselves made selections in the segregating generations.

The most common situation is where the breeder gives relatively early generation material to farmers, leaving selection to them, and at the later stages breeders monitor diversity on-farm and identify material for conventional trials. Such a system is easy to run over a large number of locations. Although not using breeders in the intermediate generations, it relies on them in the early and late stages. Breeders are also vital to disseminate the material into the official release and seed systems – sometimes this can be accelerated if scientists run a breeding scheme in parallel with the farmers, where lines popular with farmers are entered into formal trials and purified for certification, as done by Sthapit et al. (1996) for rice in Nepal.

PPB can be either collaborative, when farmers grow a bulk in their own fields and select from it, or consultative, when breeders consult farmers for their opinions on, for example, the material grown by the breeders on- station (Witcombe et al., 2006). Collaborative breeding allows cost-effective replication of selections, by giving seed of a particular bulk to many farmers, allowing replication across physical environments (farmers' fields) and also across farmers who may have different selection strategies and select on different traits. PPB is much faster than conventional breeding, due to the reduced time between making the cross and the uptake of the variety.

3.2.1. *Examples of PPB*

Examples of truly collaborative participatory breeding are rare. Thakur (1995) screened F_3 rainfed rice with farmers, although subsequent selection and generation advance was by researchers. In Ethiopia, farmers were involved in enhancing sorghum landraces through mass selection, in cooperation with scientists (Worede and Mekbib, 1993). Work in the Philippines has involved farmers in selecting rice progeny (Salazar, 1992).

Sthapit et al. (1996) conducted PPB wth rice in high-altitude areas of Nepal. Farmer participation began at the F_5 stage, and continued over two seasons. Farmers were enthusiastic, and made successful selections in the segregating material. There were large differences in farmers' preferences between bulks, and the most preferred were rapidly adopted. The best variety did well in formal trials in Nepal, and was much better than varieties produced by centralised breeding.

PPB has been carried out in maize in rainfed areas of western India (Witcombe et al., 2003), following the failure of an earlier PVS programme. This led to the development of several varieties that performed well both in on-station and on-farm trials. One was officially released for cultivation in hill areas of Gujarat that outyielded the local check in on-station trials, and was one week earlier to silk. In the farmers' trials, where average yields were lower, its yield advantage was almost 30%, and it was felt to have better grain quality than the local landraces. The returns from PPB were higher than from conventional breeding, as it was cheaper, and the benefits to farmers were realised earlier. The same conclusion was reached by Virk et al. (2003), who give examples of the use of both collaborative and consultative breeding for rainfed rice in hill areas of India.

Barley breeders at the International Centre for Agricultural Research in Dry Areas (ICARDA) have developed highly successful participatory programmes for marginal, rainfed environments in West Asia and North Africa. In trials in Syria, Morocco and Tunisia, selection criteria used by farmers included grain filling and straw traits, as well as yield (Ceccarelli et al., 2001). Breeders put a high priority on disease resistance, but this was not considered important by farmers. Farmer selection in a range of generations was effective, when carried out on-farm and also on the research station, and decentralised, participatory selection could be the best method for identifying those lines that would do best in farmers' fields. Genotypes were identified that were adapted to a range of biophysical and socioeconomic environments, that would also increase productivity, stability and biodiversity, and that were more environmentally friendly (Ceccarelli et al., 2003). The programme later was extended from barley to lentil (Ceccarelli et al., 2002), and to other countries in the region (Kafawin et al., 2005).

3.3. Mother and Baby Trials

One of the most common designs for a participatory trial is the Mother and Baby protocol, originally developed by Snapp (1999) and summarised in Table 2 (which includes also a comparison with IRD trials), although many modifications to the concept have been made for different situations. They are usually single-replicate, multi-entry trials (mother trials) and single intervention trials of a new entry versus a local check (baby trials). Mother trials are more effective than repli-

Table 2. Summary of differences between Mother, Baby and IRD trials

Mother	Baby	IRD
Few trials	Many trials	More trials than Baby trials
Researcher-designed and supervised	Simple-design, farmer supervised	No design
All entries, single replicate, small plots	One or two entries, single replicate, large plots	One entry only
Yield recorded	Yield not recorded	Yield not recorded
Farmer perceptions usually measured by matrix ranking	Farmer perceptions usually measured household-level questionnaire (HLQ)	Farmer perceptions measured informally (by anecdote)
Farmer management, but more weeding if needed	Farmer management	Farmer management
Farmer can receive compensation for resources used	Farmers bear cost and risk, but get free seed	Farmer has free seed and benefit
Repeated on-station as replicated complete block design or similar	Not repeated on research station	Not repeated on research station

Source: Witcombe, 2002

cated on-station trials as they sample more environments (Johnson et al., 1992), and Baby trials allow the cost-effective use of many replicates (Witcombe et al., 2005).

Mother trials generally consist of a single block of a randomised complete block design. Each trial consists of several germplasm lines, selected to meet the identified needs of farmers. Mother trials facilitate direct comparison of all entries and produce statistically analysable yield data as they are replicated across farmers' fields. In general, one Mother trial should be carried out per village, if possible sited prominently, for example near a road or path that many other farmers use, to give the opportunity for them to see and discuss the varieties. Baby trials are of one, or sometimes two, of the lines from the Mother trial grown by individual farmers who will compare them to a local check variety, and are far more numerous than Mother trials. Baby trials give statistically analysable data on farmers' perceptions and acceptance of new germplasm. Very often it unnecessary to record yield data, and simple comparisons (better, the same or worse) with the standard variety suffice. Several examples of Mother and Baby trial designs, as well as references to some of the statistical considerations, are given in Witcombe (2002) and Virk and Witcombe (2002).

3.4. Practical Considerations: Selection of Farmers and Villages

The essential component of successful participatory breeding, as for any participatory work, is the cooperation of farmers or other end users. Within the work our group and colleagues have carried out in Asia this has generally been enthusiastic, and it is unlikely to be a constraint in most programmes. However, there is a need for careful consideration of a number of factors when selecting the participating villages and farmers, but this is no different to the selection of testing sites in conventional breeding programmes. Farmers and villages must be relevant to the area and socio-economic environment (recommendation domain) for which the new material is intended. Farms must be carefully selected to ensure that the conclusions will apply to the appropriate group of farmers. However, some consideration must be given as to how long a village is associated with a research station, as villages continually used for research work may become less representative of the region over the years. Similarly, selecting only "good" farmers restricts the recommendation domain – the justification that they set an example is not valid for research work, only for demonstrations, although it may be useful for such farmers to be used for the "Mother" trials in a "Mother and baby" system.

Preliminary surveys can group farms according to socio-economic or environmental conditions, and decisions must be made on whether the results will apply to all groups or to a subset. Following this, a random sample, which may need to be stratified, should be taken of the relevant farms or groups of farms, using enough to have a reasonable estimate of between-farm variability. Multistage sampling using villages as primary units and farms as secondary units is effective. The sample must be large enough for the analysis to be valid when the farms are split into groups (soil type, owner/tenant, access to credit, for example).

3.5. Analysis of Participatory Trials

Participatory trials have been considered difficult to analyse, for a number of reasons, although robust methods have now been developed (see examples in Bellon and Reeves, 2002). Data range from "consensus"-type data from farmer meetings and focus-group discussions (FGDs), through questionnaires, to plot-level yield and other data recorded by the farmer or the researcher. Particularly important are plot level data such as the degree and time of waterlogging or pest damage, or at a farm level such as rainfall or soil type, or the wealth category of the farmers, which help to explain the other data. Some data (for example sowing date, weeding dates, and other management practices) could be at either level.

Very often, researchers collect too much "yield-type" data, and the collection of more concomitant information would help understand the causes of variability. Often much data is collected and never analysed, so care is needed to select only useful, or potentially useful, concomitant data. In general, what is not controlled experimentally should be measured at both plot and farm levels, if it is of direct interest, or if it might help explain some of the data variation. It may be unnecessary to record yields formally, simply counting bundles or piles and a rough estimate of harvest index, may suffice. It is important to remember that participatory trials are not participatory unless records are kept of farmers' contributions – these may be recorded in several ways. It is good practice for the researcher to measure the cropped area in baby trials, and to obtain yield data from the farmer to ensure accurate comparisons.

Very often, the data is irregular, with different experimental treatments and many missing observations, so common methods of statistical analysis cannot be used. Coe (2002a and 2002b) provides an excellent summary of the problems and ways around them. A framework for combining quantitative and qualitative data is given by Marsland et al., (2000). On-farm trials are more variable than on-station trials, with less consistency of management and more soil differences, and the variability increases with the degree of participation. This is often useful to gain further insights into the responses of new varieties to different environments.

FGDs after a farm walk give reliable estimates of farmer perceptions. The same comments on quality traits tend to recur across FGDs, farmers and villages, e.g. perceptions on cooking quality, milling quality, ease of threshing etc, which are important traits that are hard to measure without farmer participation. Simple scoring can be used, for example whether the new variety is better, the same or worse than the local. This type of data needs to be converted to percentages and transformed, and should be analysed using villages as replicates.

The data may need to be split into subsets, e.g. groups of similar farms, and if necessary particular plots or particular farms omitted, and it is important to pay attention to comments about individual plots, particularly where data is either zero or missing. For example, if the plot was eaten by animals, or damaged by a storm, a missing value should be recorded, but a zero yield for some other reason should be recorded as zero. The "comment" facility in Microsoft Excel is useful to record such information.

In many cases, the data will be a combination of one or more types, with the added complication that it may be collected at a number of levels, from the plant or plot through to the community or beyond. Analysis of such data, with different levels of variation, can be carried out using mixed methods, and can be accomplished easily in standard packages such as GENSTAT or SAS, even when there is a high degree of imbalance in the structure. For a brief introduction to these methods, see Allan and Rowlands (2001).

It is vital to feed data and results back to the farmers, for both scientific and ethical reasons. Farmers are supposed to be the beneficiaries of the work, and can benefit much earlier if they get results straight away. It is also a common courtesy, as they have made the work possible, and may well have devoted time and resources to it. Finally, they can provide a great deal of information to assist in the analysis and interpretation of the data.

3.5.1. Quantitative data

The analysis of quantitative data, e.g. yield, will vary according to the type of trial. For Mother trials the ideal method is to use the location as the replicate, and analyse conventionally by ANOVA, or use regression-based methods to assess G x E interactions. With Baby trials the best way is to use paired *t*-tests or 2-way ANOVA to compare the new varieties with the control. Which method is chosen depends upon the researcher's particular interests. In order to test for variability between or within fields, then 2-way ANOVA is appropriate as it provides a separate sum of squares to test variation among fields. The paired *t*-test assumes that the pairs of observations on a farm are related. Non-parametric methods, which make no assumptions about the distribution of the population, can also be used, for example Friedman's method for randomised blocks (an alternative to 2-way ANOVA), and Wilcoxon's signed ranks test (instead of the paired *t*-test).

3.5.2. Qualitative data

To analyse discrete scores recorded in mother trials (e.g. the impressions of the farmers for yield or taste), a matrix ranking analysis should be used, either through 2-way ANOVA, or by using Friedman's method with ranks, which gives a χ^2 for each comparison. Such analysis is commonly used in IRD trials. Analysis of perception data from Baby trials can use either a *Z*-test to compare percentages, or a χ^2 test. More detail of all these analyses is given in Virk and Witcombe (2002).

Household Level Questionnaires (HLQs) record the proportion of farmers holding a particular perception. Such data can be summarised in 2-way or *n*-way tables by farm type, and often show excellent agreement between farmers in different villages. If there is enough data models can be fitted to explore how the responses vary across different farm types. Adoption data is the best measure of overall perceptions, as it expresses the complex trade-off of the various traits as seen by farmers (Joshi and Witcombe, 2002). If they adopt a variety with known weaknesses, it is because these are outweighed by its strengths. Such data becomes reliable three seasons after first introduction, adoption earlier than this is still experimentation.

Finally, the analyses should be combined, using the results from the interviews to understand the yield variation. If there is enough within-site replication and detailed yield responses have been measured, then separate within-site analyses can be carried out, followed by combined analysis. This will usually only be the case with researcher-designed and managed trials. Where there is no replication within farms, then treatment contrasts at the farm level can be used to interpret farm x treatment interactions, or alternatively the information gleaned from the other strata can be used.

4. ADVANTAGES OF PARTICIPATORY BREEDING

Table 3 summarises the differences between participatory and conventional plant breeding. Apart from the participation of farmers, and the differences associated with this, a fundamental difference is in the number of lines grown in the intermediate (F_4 and F_5) stages, which is much higher in conventional breeding programmes.

Participatory breeding makes a virtue of the limited numbers of lines grown by each farmer, and the likelihood of success is increased as at least one parent of any cross should be well-adapted to the local environment. G x E interactions are greatly reduced, as selection is always in the target environment, and their impact is likely to be less, as the local parental material is already adapted to local climatic variation. Due to the small number of crosses, large F_2 and F_3 populations can be grown to increase the possibility of identifying transgressive segregants giving rise to desirable F_4 and F_5 progeny. Further consideration of population size and numbers of crosses is given in Witcombe and Virk (2001).

The advantages apply to any decentralised breeding, whether or not it includes farmer participation. Farmer inclusion as an integral part of the process reduces the demand on research station land and eliminates the need for breeders to carry out single-plant selection over many generations. It also ensures that all traits which are relevant to farmers, including post-harvest traits, are evaluated, and is particularly efficient when these are traits which are difficult to evaluate in the laboratory (taste, aroma, for example). Although PVS trials in Bangladesh had identified widely-adapted rice varieties, the initial involvement in goal-setting by Nepalese farmers had led to the creation of suitable material, and no amount of PVS could compensate for a lack of suitable material to choose from (Joshi et al., 2006).

By giving seed directly to farmers both PVS and PPB allow farmers across all wealth categories to adopt new varieties rapidly. Farmers grow and assess the varieties on their own farms, using their own management, and so can choose varieties specifically adapted to their own conditions and needs. Both methods lead to increased local and regional varietal diversity, reducing the vulnerability of crops to attack by pests and diseases (Joshi and Witcombe, 2003).

Table 3. Comparison of participatory and conventional plant breeding in self-pollinated crops*

Participatory plant breeding	Conventional breeding
Parents, crosses and early generation population size	
Landraces and locally-adapted cultivars often used as parents	Elite cultivars often used as parents
A few carefully-chosen crosses	Often many crosses
Very large early-generation populations from each cross	Small or medium-sized early-generation populations
Methods	
Farmer-unacceptable material identified early; very suitable when quality traits are important	Farmer-unacceptable cultivars can be released; more suitable when high yield is the most important criterion
Bulk pedigree breeding. Number of lines is limited by number of farmers; requires less resources but may be less efficient	Pedigree breeding often preferred; may give greater genetic advance than bulk pedigree breeding
Increased intra-cultivar diversity, but causes seed certification difficulties; a pure-line cultivar or more uniform mixture can be produced using few extra resources	Produces uniform pure-line cultivar; procedures for testing release, and certification of cultivar already in place
Adaptation	
Most suitable for producing cultivars with specific adaptation to marginal environments	Most suitable for producing cultivars with wide adaptation for non-stressed, high-input environments
Replication across locations in early generations simpler, ensuring reasonably wide adaptation	More difficult to replicate early generations across location
Genotype x location (GxL) interaction between farmers' fields and research station eliminated	GxL large because multi-location trials are conducted at higher input levels than those used by resource-poor farmers
Farmer awareness, adoption and risk	
Farmers' awareness of cultivar choice raised	No participatory involvement of farmers
Early farmer-to-farmer spread of material occurs	Little and late farmer involvement; several years can elapse between finished product and exposure to farmers
Farmer exposed to risk of poor material (but farmers are very good at managing risk and quantities are small)	Farmers exposed to risk due to genetic vulnerability caused by widely adapted cultivars being grown over large areas
Material can be given to farmers only after disease screening; plants exposed to farmer-relevant races of pathogens	Disease screening often done on multi-locational basis to breed for broad range of host-plant resistance

Source: Witcombe et al, 1996
*Many of these comparisons also apply to open-pollinated crops, in particular those regarding farmer awareness, adoption and risk.

Participatory programmes can promote the development of informal seed dissemination networks, which can rapidly increase the uptake rate of new varieties (Joshi and Witcombe, 2002; Virk et al., 2005), and the whole process is much faster than conventional breeding (Morris et al., 1994; Joshi et al., 2006), and shows an increased rate of return on investment (Pandey and Rajatasereekul, 1999).

Mangione et al. (2006) found no difference between the cost of the participatory and non-participatory barley-breeding programmes at ICARDA, and concluded that this was a result of the decentralised, participatory programme reaching the same level of development of the breeding material 3 years earlier than the centralised, non-participatory scheme. For the same costs, PPB generated more information due to the greater number of trials at each site, so improving selection efficiency and providing a tool to optimise numbers of sites and farmers per site.

5. COMBINING MOLECULAR AND PARTICIPATORY TECHNIQUES

Molecular DNA analysis techniques such as quantitative trait loci (QTL) mapping have great potential for crop improvement, but QTL mapping requires analysis both of molecular markers and of measurable, pre-defined phenotypic traits in experimental populations, and is difficult for complex traits such as yield, particularly when evaluation has to be in the field. In dry and saline environments, uncontrolled environmental variation and high G x E interactions increase the problems, so many efforts to use MAS to select for QTLs controlling traits associated with drought or salt tolerance, or yield under these conditions, have been severely constrained (Reyna and Sneller, 2001; Quarrie et al., 2005). MAS can be augmented with PPB to improve its application in more marginal environments.

There are several possible approaches. Markers can be used to evaluate the diversity between and within different varieties (Mkumbira et al., 2003) to identify genetically diverse or uniform varieties for PVS. It is often the case that identical varieties or landraces have different names given them by farmers, and molecular markers can identify them as the same genotype (Bajracharya et al., 2006), which could save time if they are to be used as parents in crosses.

PPB has been combined with MAS (Steele et al., 2007) to introgress QTL from the tropical *japonica* rice Azucena into Kalinga III. Although preferred by farmers for its early maturity, drought resistance, yield and quality, PVS showed that Kalinga III was prone to lodging under better conditions and had a poor root system: better roots could improve its terminal drought resistance. Marker-assisted backcrossing (MABC) was used to introgress a QTL on chromosome 9 that significantly increased root length in Kalinga III (Steele et al., 2006), as well as three other root QTL and one for aroma. Near isogenic lines (NILs) were developed, containing around 15% of Azucena alleles, and four were selected from on-station trials for early maturity, long, thick stems, lodging resistance, grain quality and yield using consultative PPB in which farmers visited the research station at the flowering stage.

Trials of these lines were carried out as replicated on-station trials and with farmers in ten villages. The participatory trials were Mother and Baby trials, and evaluations included a range of field and post-harvest traits. In both sets of trials, which were all carried out under drought conditions, the NILs outperformed Kalinga III for grain and straw yield. Although the QTL on chromosome 9 had no detectable effect on these traits, a combination of several root QTLs did have a significant effect. The Azucena alleles did not need to be at QTLs to improve the performance of Kalinga III, as a line with no root QTL out-performed that variety. The work showed that "wide" crosses combined with backcross breeding, MAS and a client-oriented approach, could play a major role in variety improvement.

5.1. Marker Evaluated Selection

Steele et al. (2004) tested a novel method to evaluate the effects of selection in rice on marker frequency – marker evaluated selection (MES), using the results of PPB in India (Virk et al., 2003) and Nepal (Witcombe et al., 2001). Instead of selecting for a pre-defined trait, they selected for a wide range of agronomic characteristics that determined adaptation to a particular environment. Selection from bulk populations of an upland (Kalinga III) x lowland (IR64) cross, was made by working closely with farmers in several ecosystems in India and Nepal, replicating selections within ecosystems. Representative samples of promising bulks were genotyped and evaluated to test for allele frequencies with the aim of identifying QTLs controlling agronomic traits. Farmer selection was either collaborative, where farmers selected within bulks in their own fields grown under traditional management, or consultative, where they selected within populations (usually bulk lines) grown on research stations under fertility and moisture conditions close to those of the farmers' fields. Unselected F_2 plants were grown in the greenhouse and genotyped to give a baseline allele frequency against which those in the selected bulks were tested.

Genomic regions from Kalinga III were strongly selected by farmers in upland environments, and regions from IR64 in the lowland ones, although there were exceptions where the upland parent contributed positively to adaptation to the lowland environment and *vice versa*. MES is potentially more powerful than QTL analysis when used to evaluate the genotypes selected from a wide cross, as it can measure the results of selection along he entire genome for every considered trait. It could then be applied in a second breeding cycle using MAS to develop ideotypes with all the loci contributing to adaptation to a particular agroecosystem.

5.2. Incorporation of Participation in an Integrated Breeding Programme for Salinity Tolerance

Bennett and Khush (2003) developed an integrated programme to develop salt-tolerant crops, based on a detailed physiological, biochemical and molecular understanding of the impact of salt on growth and reproduction. They argued that the G x E for individual mechanisms should be simpler to understand than for tolerance as a whole. Multiple donors selected on the their capacity to contribute useful alleles conferring specific mechanisms of tolerance, which may correspond to major known genes, finely-mapped major QTLs or transgenes transferred by recombinant DNA technology are used. Salt-tolerance is initially assembled by gene pyramiding using robust molecular tools during pre-breeding, and is later backcrossed into elite lines by DNA-assisted backcrossing. The stages are:

1. Salinity appraisal,
2. Mechanism discovery,
3. Gene and allele discovery,
4. Pre-breeding for salt tolerance,
5. Molecular breeding, and
6. Participatory evaluation

Farmers are involved at the beginning and end of the process. Salinity appraisal provides key information on the target environment, including the intensity and variability of the stress, the ions involved, rainfall, the growth stages most affected, the affordable management options, availability of supplementary irrigation, and the additional traits that farmers consider important. This gives information on how much of the problem is genetic, how much is management and how much is due to rainfall. Farmers are also vital in the final stage, participatory evaluation, ideally using trials that are simple in design, simple to report and simple to aggregate, so they can reach as many farming communities as possible for any given level of financial and human input. Due to the great diversity of saline sites and the strong G x E interactions of tolerance, both PVS and PPB would be useful to aid the adoption of successful varieties and identify those for which iteration of the second and third phases would be desirable.

6. CONCLUSIONS

The methods outlined above have improved upon more conventional varietal testing:

1. In many breeding programmes fewer resources are allocated to testing more advanced lines (Witcombe *et al*, 1998). PVS redresses this balance by allocating more resources to these important lines.
2. PVS allows farmers to evaluate varieties for all traits and make trade-offs between traits e.g., grain yield v fodder yield, maturity, and grain quality.
3. PVS, PPB and COB test varieties under realistic management.
4. PVS, PPB and COB test varieties across more physical niches as trials are replicated in more locations.
5. PVS, PPB and COB test varieties across social niches where food preferences might vary.
6. All sectors of agricultural society can input into and benefit from plant breeding programmes. Men and women may have different priorities for desired traits, so both must be active in the selection of seeds to be saved for the next season's crop, and other marginalised groups.
7. Seed from participatory trials rapidly enters the informal seed system, leading to accelerated uptake of improved material.

Variants of these methods have been used to great effect in many crops. Recent work combining molecular and participatory approaches has shown potential, in particular in improving MAS for quantitative traits in breeding for marginal environments. The benefits of farmer participation are not universal, and although breeding programmes should always have some client-orientation, intensive collaboration with farmers is not always needed (Witcombe et al., 2005). There is increasing realisation that defining breeding programmes as "participatory" or as "formal", "conventional" or "classical" develops a them-and-us mentality between breeders, and that it is better to describe the programme according to the degree of client-orientation (Biggs and Gauchan, 2001; Witcombe et al., 2005).

REFERENCES

Abdus Salam, Hollington, P. A., Gorham, J., Wyn Jones, R. G., and Gliddon, C., 1999. Physiological genetics of salt tolerance in wheat (*Triticum aestivum* L.): Performance of wheat varieties, inbred lines and reciprocal F1 hybrids under saline conditions. *J. Agron. Crop Sci.* **183**: 145–156.

Allan, E. and Rowlands, J., 2001. *Mixed models and multilevel data structures in agriculture*. Statistical Services Centre, University of Reading, UK.

Araus, J L., Bort, J., Steduto, P., Villegas, D., and Royo, C., 2003. Breeding cereals for Mediterranean conditions: Ecophysiological clues for biotechnology application. *Ann. Appl. Biol.* **142**: 129–141

Araus, J. L., Slafer, G. A., Reynolds, M. P., and Royo, C., 2002. Plant breeding and drought in C3 cereals : what to breed for. *Ann. Bot.* **89**: 925–940.

Baidu-Forson, J., 1997. On-station participatory varietal evaluation: a strategy for client-oriented breeding. *Exp. Agric.* **33**: 43–50.

Bajracharya, J., Steele, K. A., Jarvis, D. I., Sthapit, B. R., and Witcombe, J. R., 2006. Rice landrace diversity in Nepal: variability of agro-morphological traits and SSR markers in landraces from a high-altitude site. *Fld. Crops Res.* **95**: 327–335

Barrett-Lennard, E. G., 2003. The interaction between waterlogging and salinity in higher plants: causes, consequences and implications. *Plant Soil* **253**: 35–54

Barrett-Lennard, E. G., van Ratingen, P., and Mathie, M. H., 1999. The developing pattern of damage in wheat (*Triticum aestivum* L.) due to the combined stresses of salinity and hypoxia: experiments under controlled conditions suggest a methodology for plant selection. *Aust. J. Agric. Res.* **50**: 129–136.

Bellon, M. R. and Reeves, J., (eds.) 2002. *Quantitative Analysis of Data from Participatory Methods in Plant Breeding*. Mexico, D.F. CIMMYT.

Bennett, J., 2003. Opportunities for increasing water productivity of CGIAR crops through plant breeding and molecular biology, in *Water Productivity in Agriculture: Limits and Opportunities for Improvement*, J.W. Kijne, R. Barker and D. Molden, eds., Wallingford, CABI International, pp 103–126

Bennett, J. and Khush, G. S., 2003. Enhancing salt tolerance in crops through molecular breeding: a new strategy. *J. Crop Prod.* **7**: 11–65

Biggs, S. and Gauchan, D., 2001. Resource-poor farmer participation in research: a synthesis of experiences from nine national agricultural research systems. *Special series on the Organisation and Management of On-Farm Client-Oriented Research (OFCOR). OFCOR-Comparative Study Paper 3.* The Hague: International Service for National Agricultural Research (ISNAR).

Ceccarelli, S., 1994. Specific adaptation and breeding for marginal conditions. *Euphytica* **77**: 205–219

Ceccarelli, S. and Grando, S., 1996. Drought as a challenge for the plant breeder. *Plant Growth Reg.* **20**: 149–155.

Ceccarelli, S. and Grando, S., 1999. Decentralised participatory plant breeding. *ILEIA Newsletter* **15**: 3/4 , 36–37.

Ceccarelli, S., Erskine, W., Hamblin, J., and Grando, S., 1994. Genotype by environment interaction and international breeding programmes. *Exp. Agric.* **30**: 177–187

Ceccarelli, S., Grando, S., Bailey, E., Amri, A., El-Felah, M., Nassif, F., Rezgui, S., and Yahyaoui, A., 2001. Farmer participation in barley breeding in Syria, Morocco and Tunisia. *Euphytica* **122**: 521–536

Ceccarelli, S, Grando, S., and Impiglia, A., 1998. Choice of selection strategy in breeding barley for stress environments. *Euphytica* **103**: 307–318.

Ceccarelli, S., Grando, S., Martini, M., and Lutf, A., 2002. Participatory barley and lentil breeding in Yemen. *ICARDA Caravan* **16**: 18–19

Ceccarelli, S., Grando, S., Singh, M., Michael, M., Shikho, A., Al-Issa, M., Al-Saleh, A., Kaleonjy, G., Al-Ghanem, S. M., Al-Hasan, A. L., Dalla, H., Basha, S., and Basha, T., 2003. A methodological study on participatory barley breeding. II. Response to selection. *Euphytica* **133**: 185–200.

Ceccarelli, S., Grando, S., Tutwiler, R., Baha, J., Martini, A. M., Salahieh, H., Goodchild, A., and Michael, M., 2000. A methodological study on participatory barley breeding. I. Selection phase. *Euphytica* **111**: 91–104.

Chambers, R., 1989. Institutions and practical change. Reversals, institutions and change, in *Farmer First,* R. Chambers, A. Pacey and L. A. Thrupp, eds., Overseas Development Institute, London, pp 181–195

Coe, R., 2002a. Analyzing data from participatory on-farm trials, in *Quantitative Analysis of Data from Participatory Methods in Plant Breeding*, M. R. Bellon and J. Reeves, eds., CIMMYT, Mexico, DF, pp 18–35

Coe, R., 2002b. Analyzing ranking and rating data from participatory on-farm trials, in *Quantitative Analysis of Data from Participatory Methods in Plant Breeding*, M. R. Bellon and J. Reeves, eds., CIMMYT, Mexico, DF, pp 44–65

Evans, L. T. 1998. *Feeding the Ten Billion: Plants and Population Growth*, Cambridge University Press, Cambridge.

FAO. Food and Agriculture Organisation of the United Nations. 2002. *Agriculture: Towards 2015 / 2030.* Interim report, Economic and Social Department. FAO, Rome, Italy (3 August 2006) http://www.fao.org/es/ESD/AT2050web.pdf

FAO. Food and Agriculture Organisation of the United Nations. 2003. *Agriculture, food and water. A contribution to the World Water Development Report.* Rome, Italy. FAO Land and Water Development Division (31 August, 2006): ftp://ftp.fao.org/agl/aglw/docs/agricfoodwater.pdf

Fischer, G., Sham, M., and van Veltuizen, H., 2002. *Climate Change and Agricultural Variability.* Laxemburg, Austria: IIASA.

Flowers, T. J., 2004. Improving crop salt tolerance. *J. Exp. Bot.* **55**: 307–319

Flowers, T. J., and Yeo, A. R., 1995. Breeding for salinity resistance in crop plants: where next? *Aust. J. Plant Physiol.* **22**: 875–884

Gregorio, G. B., and Cabuslay, G. S., 2005 Breeding for abiotic stress tolerance in rice, in *Abiotic Stresses: Plant Resistance through Breeding and Molecular Approaches,* M. Ashraf and P. J. C. Harris, eds., Food Products Press, The Haworth Press Inc, pp. 513–544

Gregorio, G. B., Senadhira, D., Mendoza, R. D., Manigbas, N. L., Roxas, J. P. and Guerta, C. Q. 2002. Progress in breeding for salinity tolerance and associated abiotic stresses in rice. *Fld. Crops Res.* **76**: 91–101

Hillel, D., and Rosenzweig, C., 2002. Desertification in relation to climate variability and change. *Adv. Agron.* **77**: 1–38

Hobbs, P .R., and Gupta, R. K., 2003. Rice-wheat cropping systems in the Indo-Gangetic plains: issues of water productivity in relation to new resource-conserving technologies, in *Water Productivity in Agriculture: Limits and Opportunities for Improvement*, J. W. Kijne, R. Barker, and D. Molden, eds., CABI International, Wallingford, UK. ISBN 0 85199 669 8, pp. 239–253

Hollington, P. A., 2000. Technological breakthroughs in screening/breeding wheat varieties for salt tolerance. Invited paper presented at National Conference on *Salinity Management in Agriculture,* S. K. Gupta, S. K. Sharma and N. K. Tyagi, eds., CSSRI Karnal, India, December 2 - 5, 1998. Karnal, India: Central Soil Salinity Research Institute, pp 273–289.

Hollington, P. A., Royo, A., Miller, T. E, Quarrie, S. A., Mahmood, A., and Aragüés, R., 1994. The use of doubled haploid breeding techniques to develop wheat varieties for saline areas. *Proc 3rd Congr, Eur Soc Agron*, pp 156–157

IRRI 2006a (July 10, 2006): http://www.irri.org/cure/workplans%5CWG1_plan05.htm

IRRI 2006b (July 10, 2006); http://www.irri.org/cure/workplans%5CWG5_plan05.htm

Isla, R., Aragüés, R., and Royo, A. (2003.) Spatial variability of salt-affected soils in the middle Ebro valley (Spain) and implications in plant breeding for increased productivity. *Euphytica* **134**: 325–334

Johnson, J. J., Alldredge, J. R., Ullrich, S. E., and Dangi, O., 1992. Replacement of replications with additional locations for grain sorghum cultivar evaluation. *Crop Sci.*.**32**: 43–46.

Joshi, A., and Witcombe, J. R., 1996. Farmer participatory crop improvement II: Participatory varietal selection in India. *Exp. Agric.* **32**: 461–477

Joshi, K. D., and Witcombe, J. R., 2002. Participatory varietal selection in rice in Nepal - a comparison of two methods assessed by varietal adoption. *Euphytica* **127**: 445–458

Joshi, K. D., and Witcombe, J. R., 2003. The impact of participatory plant breeding on landrace diversity: a case study for high-altitude rice in Nepal. *Euphytica* 134: 117–125

Joshi, K. D., Musa, A. M., Johansen, C., Gyawali, S., Harris, D., and Witcombe, J. R., 2006. Highly client-oriented breeding, using local preferences and selection, produces widely adapted rice varieties. *Fld. Crops Res.* (in press)

Joshi, K. D., Subedi, M., Rana, R. B., Kadayat, K. B., and Sthapit, B. R., 1997. Enhancing on-farm varietal diversity through participatory varietal selection: a case study for *Chaite* rice in Nepal. *Exp. Agric.* **33**: 335–344

Kafawin, O., Saoub, H., Ceccarelli, S., Shakhatreh, Y., Yasin, A., Grando, S. , Bwaliez, A. R., and Khazaleh, A., 2005 Participatory barley breeding for improving production in stress environments. *Dirasat. Agricultural Sciences* **32:** 57–63.

Maas, E. V., 1990. Crop salt tolerance. In: Tanji, K.K. (ed) *Agricultural Salinity Assessment and Management*. ACSE Manuals and reports on engineering practice No. 71. New York, ASCE. ISBN 0-87262-762-4, pp. 262–304

Mangione, D., Senni, S., Puccioni, M., Grando, S., and Ceccarelli, S., 2006. The cost of participatory plant breeding. *Euphytica*. In press.

Mannion, A. M., 1995. Biotechnology and environmental quality. *Progr. Phys. Geog.* **19**: 192–215.

Marsland, N., Wilson, I., Abeyasakera, S., and Kleih, U., 2000 *A methodological framework for combining quantitative and qualitative survey methods*. Statistical Services Centre, University of Reading, UK.

Maurya, D. M., Bottrall, A., and Farrington, J., 1988. Improved livelihoods, genetic diversity and farmers' participation: a strategy for rice breeding in rainfed areas of India. *Exp. Agric.* **24**: 311–320.

Mkumbira, J., Chiwona-Karltun, L., Lagercrantz, U., Mahungo, N. M., Saka, J., Mhone, A., Bokanga, M., Brimer, L., Gullberg, U., and Rosling, H., 2003. Classification of cassava into "bitter" and "cool" in Malawi: frm farmers' perception to characterisation by molecular markers. *Euphytica* **132**: 7–22.

Morris, M. L., Dubin, H. J., and Pokherel, T., 1994. Returns to wheat breeding research in Nepal. *Agric Econ.* **10**: 269–282

Mulatu, E., and Belete, K., 2001. Participatory varietal selection in lowland sorghum in eastern Ethiopia: impact on adoption and genetic diversity. *Exp. Agric.* **37**: 211–229.

Munns, R. 2005. Genes and salt tolerance: bringing them together. *New Phytol.* **167**: 645–663

Pandey, S. and Rajatasereekul, S.. 1999. Economics of plant breeding: the value of shorter breeding cycles for rice in Northeast of Thailand. *Fld. Crops Res.* **64**: 187–197.

Parry, M. A. J., Flexas, J., and Medrano, H., 2005. Prospects for crop production under drought: research priorities and future directions. *Ann. Appl. Biol.* **147**: 211–226

Passioura, J., 2002. Environmental biology and crop improvement. *Funct. Plant. Biol.* **29**: 537–546.

Perry, M. W., and D'Antuono, M. F., 1989. Yield improvement and associated characteristics of some Australian spring wheat cultivars introduced between 1860 and 1982. *Aust. J. Agric. Res.* **40**: 457–472.

Quarrie, S. A., and Mahmood, A., 1993. Improving salt tolerance in hexaploid wheat. *Annual report 1992, AFRC Institute of Plant Science Research Cambridge Laboratory John Innes Institute Nitrogen Fixation Laboratory and Sainsbury Laboratory*, p 4.

Quarrie, S. A., Steed, A., Calestani, C., Semikhodskii, A., Lebreton, C., Chinoy, C., Steele, N., Pljevljakusic, D., Waterman, E., Weyen, J., Schondelmaier, J., Habash, D. Z., Farmer, P., Saker, L., Clarkson, D. T., Abugalieva, A., Yessimbekova, M., Turuspekov, Y., Abugalieva, S., Tuberosa, R., Sanguineti, M.-C., Hollington, P. A., Aragüés, R., Royo, A., and Dodig, D., 2005. A high density genetic map of hexaploid wheat (*Triticum aestivum* L.) from the cross Chinese Spring x SQ1 and its use to compare QTLs for grain yield across a range of environments. *Theor. Appl. Gen.* **110**: 865–880

Quayyum, M. A., and Malik, M. D., 1988. Farm production losses in salt-affected soils. *Proc. 1st Nat. Congr. Soil Sci.* Lahore, Pakistan, October 1985, pp 356–364

Qureshi, R. H., Rashid, A., and Ahmad, N., 1990. A procedure for quick screening of wheat cultivars for salt tolerance, in *Genetic Aspects of Plant Mineral Nutrition,* N. el Bassam, M. Dambroth, and B. C. Loughman, eds., Kluwer, pp 315–324

Reyna, N., and Sneller, C. H., 2001 Evaluation of marker-assisted introgression of yield QTL alleles into adapted soybean. *Crop Sci.* **41**: 1317–1321

Reynolds, M. P., Mujeeb-Kazi, A., and Sawkins, M., 2005. Prospects for utilising plant-adaptive mechanisms to improve wheat and other crops in drought- and salinity-prone environments. *Ann. Appl. Biol.* **146**; 239–259.

Richards, R. A., 1983. Should selection for yield in saline regions be made on saline or non-saline soils? *Euphytica* **32**: 431–438

Richards, R. A., 1992. Increasing salinity tolerance of grain crops: Is it worthwhile? *Plant Soil* **146**: 89–98.

Richards, R. A., 1996. Defining selection criteria to improve yield under drought. *Plant Growth Reg.* **20**: 57–166

Richards, R. A., 2000. Selectable traits to increase crop photosynthesis and yield of grain crops. *J. Exp. Bot.* **51**: 447–458.

Richards, R. A., Rebetzke, G. J., Condon, A., and van Herwaarden, A.F., 2002. Breeding opportunities for increasing the efficiency of water use and crop yield in temperate cereals. *Crop Sci.* **42**: 111–121.

Rockström, J., Barron, J., and Fox, P., 2003. Water productivity in rain-fed agriculture: challenges and opportunities for smallholder farmers in drought-prone tropical agroecosystems, in *Water Productivity in Agriculture: Limits and Opportunities for Improvement*, J. W. Kijne, R. Barker, and D. Molden, eds., CABI International, Wallingford, UK. ISBN 0 85199 669 8, pp 145–162

Salazar, R., 1992. MASIPAG: alternative community rice breeding in the Philippines. *Approp. Technol.* **18**: 20–21

Seckler, D., Molden, D., and Barker, R., 1998. Water scarcity in the twenty-first century. *IWMI Water Brief* 1. Colombo, Sri Lanka, IWMI

Shannon, M. C., 1997. Adaptation of plants to salinity. *Adv. Agron.* **60**: 75–120

Shen, L., Courtois, B., McNally, K. L., Robin, S., and Li, Z., 2001. Evaluation of near-isogenic lines of rice introgressed with QTLs for root depth through marker-aided selection. *Theor. Appl. Gen.* **103**: 427–437

Slafer, G. A., Araus, J. L., and Richards, R. A., 1999. Promising traits for future breeding to increase wheat yield, in *Wheat: Ecology and Physiology of Yield Determination*, E. H. Satorre and G. A. Slafer, eds., New York: Food Product Press, pp. 379–415.

Snapp, S., 1999. Mother and baby trials: a novel trial design being tried out in Malawi. In: TARGET. *The Newsletter of the Soil Fertility Research Network for Maize-Based Cropping Systems in Malawi and Zimbabwe*. Jan. 1999 issue. CIMMYT, Zimbabwe

Srivastava, J. P., and Jana, S., 1984. Screening wheat and barley germplasm for salt tolerance, in *Salinity Tolerance in Plants – Strategies for Crop Improvement*, R. C. Staples and G. H. Toenniessen, eds., New York, Wiley., pp. 273–283

Steele, K. A., Edwards, G., Zhu, J., and Witcombe, J. R., 2004. Marker-evaluated selection in rice: shifts in allele frequency among bulks selected in contrasting agricultural environments identify genomic regions of importance to rice adaptation and breeding. *Theor. Appl. Genet.* **109**: 1247–1260

Steele, K. A., Price, A. H., Shashidar, H. E., and Witcombe, J. R., 2006. Marker-assisted selecton to introgress rice QTLs controlling root traits into an Indian upland rice variety. *Fld. Crops Res.* **112**: 28–221

Steele, K. A., Virk, D. S., Kumar, R., Prasad, S. C., and Witcombe, J. R., Field evaluation of upland rice lines selected for QTLs controlling root traits. *Fld. Crops Res.* **101**:180–186.

Sthapit, B. R., Joshi, K. D., and Witcombe, J. R., 1996. Farmer participatory crop improvement. III. Participatory plant breeding, a case study of rice in Nepal. *Exp. Agric.* **32**: 487–504

Stirling, C. M., and Witcombe, J. R., 2004. *Farmers and Plant Breeders In Partnership.* 2nd edition. Dept. for International Development (DFID) Plant Sciences Research Programme (PSP) and Centre for Arid Zone Studies (CAZS). Bangor, UK: CAZS, University of Wales Bangor.

Szabolcs, I., 1994. Soils and salinization, in *Handbook of Plant and Crop Stress*, M. Pessarakli, M., ed., New York, USA: Marcel Dekker, pp. 3–11

Tambussi, E. A., Nogués, S., Ferrio, P., Voltas, J., and Araus, J.L., 2005. Does higher yield improve barley performance in Mediterranean conditions? A case study. *Fld. Crops Res.* **91**: 149–160.

Thakur, R., 1995. Prioritization and development of breeding strategies for rainfed lowlands: a critical appraisal. In: Proc. IRRI Conference, 1995: *Fragile Lives in Fragile Ecosystems.* Los Baños, Philippines, IRRI, Pp 817–824.

Turner, N. C., 2003. Drought resistance: a comparison of two research frameworks, in *Management of agricultural drought: agronomic and genetic options.* N. P. Saxena, ed., Enfield, Science Publishers, pp. 89–102.

Turner, N. C., 2004a. Sustainable production of crops and pastures under drought in a Mediterranean environment. *Ann. Appl. Biol.* **144**: 139–147.

Turner, N. C., 2004b. Agronomic options for improving rainfall-use efficiency of crops in dryland farming systems. *J. Exp. Bot.* **55**: 2413–2425

Virk, D. S., and Witcombe, J. R., 2002. An introduction to data management and analysis for participatory varietal selection trials, in *Breeding Rainfed Rice for Drought-Prone Environments: Integrating Conventional and Participatory Plant Breeding in South and Southeast Asia. Proc. DFID Plant Sci. Res. Prog. / IRRI Conf, 12–15 March 2002, IRRI, Los Baños, Laguna, Philippines*, J. R. Witcombe, L. B. Parr and G. N. Atlin, eds., Bangor and Manila: Department for International Development (DFID) Plant Sciences Research Programme, Centre for Arid Zone Studies (CAZS) and International Rice Research Institute (IRRI), pp. 69–72.

Virk, D. S., Chakraborty, M., Ghosh, J., Prasad, S. C., and Witcombe, J. R., 2005. Collaborative and consultative participatory plant breeding of rice for the rainfed uplands of eastern India. *Euphytica* **132**: 95–108.

Virk, D. S., Packwood, A. J. P., and Witcombe, J. R., 1996. Plant breeding, varietal testing and popularisation, and research linkages. Paper presented at *ODA / ICAR workshop on Reorganising Research for Rainfed Farming.* CRIDA, Hyderabad, 11–15 September, 1995.

Virk, D. S., Singh, D. N., Prasad, S. C., Gangwal, J. S., and Witcombe, J. R., 2003. Collaborative and consultative participatory plant breeding of rice for the rainfed uplands of eastern India. *Euphytica* **132**: 95–108.

Witcombe, J. R., 2002a A mother and baby trial system, in *Breeding Rainfed Rice for Drought-Prone Environments: Integrating Conventional and Participatory Plant Breeding in South and Southeast Asia. Proc. DFID Plant Sci. Res. Prog. / IRRI Conf, 12–15 March 2002, IRRI, Los Baños, Laguna, Philippines*, J. R. Witcombe, L. B. Parr and G. N. Atlin, eds., Bangor and Manila: Department for International Development (DFID) Plant Sciences Research Programme, Centre for Arid Zone Studies (CAZS) and International Rice Research Institute (IRRI), pp. 79–89.

Witcombe, J. R., and Virk, D. S., 2001. Number of crosses and population size for participatory and classical plant breeding. *Euphytica* **122**: 451–462

Witcombe, J. R., Gyawali, S., Sunwar, S., Sthapit, B. R., and Joshi, K. D., 2006. Participatory plant breeding is better described as highly client-oriented plant breeding. II. Optional farmer collaboration in the segregating generations. *Exp. Agric.* **42**: 79–90

Witcombe, J. R., Joshi, A., and Goyal, S. N., 2003. Participatory plant breeding in maize: A case study from Gujarat, India. *Euphytica* **130**: 413–422

Witcombe, J. R., Joshi, A., Joshi, K. D., and Sthapit, B. R., 1996. Farmer participatory crop improvement. I. Varietal selection and breeding methods and their impact on biodiversity. *Exp. Agric.* **32**: 445–460

Witcombe, J. R., Joshi, K. D., Gyawali, S., Musa, A. M., Johansen, C., Virk, D. S., and Sthapit, B R., 2005. Participatory plant breeding is better described as highly client-oriented plant breeding. I. Four indicators of client-orientation in plant breeding. *Exp. Agric.* **41**: 299–319

Witcombe, J. R, Packwood, A. J. P., Raj, A. G. B., and Virk, D. S., 1998. The extent and rate of adoption of modern cultivars in India. In *Seeds of choice: Making the most of New Varieties for Small Farmer*

53–58 (Eds J. R. Witcombe, D. S. Virk and J. Farrington). New Delhi: Oxford IBH, and London: Intermediate Technology Publications.

Witcombe, J. R., Subedi, M., and Joshi, K. D., 2001. Towards a practical participatory plant-breeding strategy in predominantly self-pollinated crops, in *An Exchange of Experiences from South and Southeast Asia: Proc. Int. Symp. on Participatory Plant Breeding and Participatory Plant Genetic Resources Enhancement*, CGIAR Programme for Participatory Research and Gender Analysis (PGRA), Cali, Colombia, pp 243–248

Worede. M., and Mekbib, H., 1993. Linking genetic resource conservation to farmers in Ethiopia, in *Cultivating knowledge: genetic diversity, farmer experimentation and crop research,* W. de Boef, K. Amanor and K. Wellard, eds., London: Intermediate Technology Publications, pp 78–84.

WRI. 2000. World Resources Institute. World resources: People and ecosystems, the fraying web of life. Oxford: Elsevier Science.

Yeo, A. R., 1999. Predicting the interaction between the effects of salinity and climate change on crop plants. *Sci. Hort.* **78**: 159–174

Yeo, A. R., Yeo, M. E., and Flowers, T. J., 1988. Selection of lines with high and low sodium transport from within varieties of an inbreeding species, rice (*Oryza sativa* L.). *New Phytol.* **110**: 13–19

Young, A., 1999. Is there really spare land? A critique of estimates of available cultivable land in developing countries. *Env, Dev and Sust* **1**: 3–18.

CHAPTER 19

REQUIREMENTS FOR SUCCESS IN MARKER-ASSISTED BREEDING FOR DROUGHT-PRONE ENVIRONMENTS

J.B. PASSIOURA, W. SPIELMEYER, AND D.G. BONNETT

CSIRO Plant Industry, GPO Box 1600, Canberra, ACT 2601, Australia

Abstract: The challenge that commercial breeders have in improving the yield of crops in drought-prone environments is to produce cultivars that capture more of the water supply for use in transpiration; exchange transpired water for CO_2 more effectively in producing biomass; and convert more of the biomass into grain. Many traits affect these requirements, and assume greater or lesser importance depending on the severity and timing of a drought. Although hundreds of QTLs have been found that are associated with improved yield during drought, few have been converted into markers useful to breeders owing to the difficulty of understanding the phenotype in realistic environments. The few markers that are in use target disease resistance and morphological or ecophysiological traits known to affect the water economy of a crop. Faster progress in developing markers useful to breeders will come with the evidently increasing interaction between breeding programs and marker laboratories

Keywords: Marker assisted breeding, breeders markers, water limited environments, transpiration efficiency, harvest index

1. INTRODUCTION

Commercial breeding programs need to produce varieties that are attractive to farmers. A successful new variety must meet minimal standards for a range of important traits that include grain quality, disease resistance, appropriate phenology for the target environment, and ability to yield well across a range of seasons. As well as meeting these minimal standards, it must also be demonstrably better than existing varieties in at least one of these traits. Ability to yield well during drought, while desirable, is difficult to select for and tends to be a bonus. It is difficult to select for because the timing and severity of episodes of drought in the field are so variable.

To a breeder or an agronomist the term "drought" usually means that the yield of a crop in a given season is substantially limited by the water supply. The term also has a wide range of other meanings, the diversity of which arises mostly from the

M.A. Jenks et al. (eds.), Advances in Molecular Breeding Toward Drought and Salt Tolerant Crops, 479–500.

time scale of interest of whoever is using it (Passioura, 2006). To a plant scientist working in a laboratory it may mean suddenly depriving a plant of water so that it suffers substantial water stress in a few days, possibly even a few hours. To an insurer it is a statistical idea, say, the driest decile of growing seasons.

The terms "drought tolerance" or "drought resistance" likewise have diverse meanings – ranging from the ability of a plant to survive severe rapid dehydration, as in many laboratory explorations in this area, to the ability of crops growing in the field to achieve a reasonable yield despite a poor water supply. The latter meaning is often used by plant breeders in relation to their breeding lines, and is sometimes used to denote a low coefficient of variation in yield in response to a wide range of water supply. Along with a low coefficient of variation, though, comes the breeders' sardonic idea of "yield resistance", that is, the given line is unable to respond to good seasons, and is therefore of little use.

In the agronomic world, "drought tolerance" is giving way to the notions of "water-use efficiency" or "water productivity". The latter are quantifiable, with units of amount of crop yield per volume of water supplied or used, say, kg ha^{-1} mm^{-1}, though sometimes expressed as kg m^{-3} (with a value one tenth of the former), especially in an irrigation area. While "water productivity" encompasses many aspects of the physiology, biochemistry and molecular biology of water stress in plants, it also transcends these aspects, and concentrates rather on water as a limiting resource and what are the requirements for making best use of that resource in building the yield of a water-limited crop. Because it is quantifiable it enables progress to be more easily charted.

There are three avenues for converting a limiting water supply into grain yield. We can ensure (a) that as much as possible of that water is used by the plant, rather than lost to say, weeds, drainage, run-off, or direct evaporation from the soil; (b) that the trade of water for carbon dioxide by the leaves is most effectively converted into dry matter; and (c) that as much of that dry matter as possible is converted into grain, i.e. that we achieve a large harvest index.

Our aims in this paper are to explore how molecular and other markers may be used to select plants in segregating populations, using the focus provided by dissecting water productivity into these three components, and how best to take promising lines through to incorporation into commercial breeding programs. We concentrate on water-limited dryland wheat as our main example.

2. WHAT DRIVES COMMERCIAL BREEDING PROGRAMS?

In the semi-arid environments where potential yield is strongly limited by water supply, there is a wide-spread view that water is indeed the predominant limitation to crop growth and yield. There are however many other problems facing farmers that are just as difficult to deal with as inadequate precipitation, such as weeds, pests and disease, poor nutrition, frost, heat, and even waterlogging in wet years. Any of these will reduce water productivity. The result is that farmers' yields are in practice often well below what would be expected if water were the main limitation, as we expand on later. Genetic or agronomic improvements that help deal with any of these will help improve water productivity. Beyond yield, the quality of the grain produced is a major determinant of its value and returns to farmers. Thus it is necessary to explore the ecophysiological and agronomic background to water productivity, while also considering the need to produce grain with reasonable market value, as a guide to identifying markers that would be of most use in a breeding program aiming to produce varieties optimized for these environments.

Resistance to major diseases, acceptable grain quality, and greater productivity are the main targets of commercial breeding programs. Underlying the productivity are the well-established requirements of appropriate flowering time for the target environment and semi-dwarf habit. A new variety must possess a demonstrably better combination of characters and meet minimum acceptable standards for all important traits in its target production zone(s) or it will have little chance of being accepted in the market place.

Most of the effort of commercial breeding programs must be directed towards developing lines with these superior combinations that can be released as varieties, thus imposing limitations on the types of crosses and selection methodologies that can be used and the progress likely to be made for any particular trait in a single breeding cycle. In this context a breeding cycle is the period between making a simple or complex cross to when an inbred line is produced that can be either released as a variety or used as a parent in a further breeding cycle. Considering that the time from crossing to varietal release is commonly 10 years or greater, the imperative to focus on the most commercially important traits becomes clear. Most programs will also have a smaller effort directed at developing germplasm aiming to achieve greater advances across a narrower spectrum of traits that will not have all the attributes needed in a new variety but will be useful in subsequent crosses directed at varietal release. It is here that targeted pre-breeding efforts producing lines with novel traits, or stronger expression of existing desirable traits in adapted backgrounds, may usefully complement efforts of commercial breeding programs, thereby allowing more resources to be directed at crosses with a greater chance of commercial success.

Molecular markers are improving the ability of plant breeders to efficiently combine greater numbers of desirable alleles in a single breeding cycle and have been rapidly adopted in recent years by wheat breeding programs in Australia (Wilson et al., 2006), Canada (De Pauw et al., 2005), the USA (http://maswheat.ucdavis.edu), CIMMYT (William et al., 2006) and no doubt by others including the private sector. Initial successes have been with simply inherited, although sometimes difficult-to-phenotype, traits such as resistance to cereal cyst nematode (Ogbonnaya et al., 2001), rust resistance (Spielmeyer et al., 2003; Lagudah et al., 2006) and tolerance to high levels of boron (Jefferies et al., 2000). Markers for important components of end-use quality such as glutenin and puroin-doline alleles, together with a better understanding of how alleles at different loci interact to affect dough strength, have led to adoption of these markers more recently (Eagles et al., 2005). Given the time frames involved, it is not surprising that only a few varieties have been released using marker-assisted selection although the

Table 1. Traits that may influence water productivity of wheat crops and for which markers are being used in breeding programs

Trait	Significance	Reference
Rust resistance genes	Maintains leaf area under disease pressure	Seah et al. 2000; Spielmeyer et al. 2003; Mago et al. 2005; Lagudah et al. 2006
Aluminum tolerance gene *ALMT-1*	Competent root system	Sasaki et al. 2004
Boron tolerance	Competent root system	Jefferies et al. (2000)
Root lesion nematode	Competent root system	Williams et al. (2002)
Resistance to cereal cyst nematode (CCN)	Reduces nematode numbers to improve root health of following crop	Ogbannaya et al (2001)
Rht-B1b, *Rht-D1b* dwarfing genes	Remove to avoid short coleoptiles	Ellis et al. (2002)
Alternative dwarfing genes	Semi-dwarfing phenotype without reducing coleoptile length	Ellis et al.(2005)
Tiller inhibiting gene (*tin*)	Inhibits excess production of tillers	Spielmeyer and Richards (2004)

number is expected to grow substantially in coming years. Current examples include the Australian wheats "MacKellar" (virus tolerance) and "Young" (rust resistance), and the Canadian wheats "Lillian" and "Somerset" (high protein content).

A number of the traits for which markers are currently being used directly improve the ability of crops to use available water by retaining photosynthetic area in the presence of foliar disease, or by maintaining a healthy root system able to access more available water in the presence of root diseases or toxic levels in the soil of sodium, boron or aluminum (see Table 1). Marker assisted selection (MAS) for these traits can also contribute indirectly to improving yield performance of new varieties by more effectively removing undesirable genotypes in the early stages of breeding. As a result, a greater proportion of lines progressing to yield trials will have acceptable levels of disease resistance, and greater emphasis can be placed on selection for improved grain quality and yield. Markers for aspects of grain quality will contribute to improving yield in the same way, even though the trait(s) themselves may have no direct effect on yield.

Increasing research effort is being directed towards identifying physiological and morphological traits that will help improve water productivity in dryland cereal crops in addition to traits conferring disease resistance and tolerance to other abiotic stresses such as boron or aluminum toxicity. A number of potentially useful traits have been identified within the ecophysiological and agronomic context outlined below.

3. ECOPHYSIOLOGY AND AGRONOMY OF WATER-LIMITED YIELD

A crop's yield is the culmination and integration of processes occurring over the whole of its growing season. Many of these processes are especially influential in affecting yields in water-limited environments. Their relative importance depends on both the amount and the timing of the availability of water. Yield is particularly vulnerable to insufficient water at certain times during a crop's life, most notably at sowing and at flowering. However, it also depends strongly on how effectively the crop uses water throughout the growing season – in trading carbon dioxide for transpired water most effectively, and on achieving an adequate balance between water used during vegetative growth and that used during grain filling, so that a good harvest index is attained. The following summarizes the main processes involved, some of which, though not yet all, can be modified genetically with the help of markers.

3.1. Floral and Vegetative Development

Matching the crop's phenology to its environment is the primary requirement for getting good water productivity. There is an optimal time for sowing, weather permitting, in a given environment, and an optimal time for flowering. Because time of flowering is so important and easy to select for in a given environment, it ranks, along with disease resistance and grain quality, as one of the breeders' primary selection criteria. The reason that time of flowering is so important is that it balances, particularly in winter cereals in mediterranean environments, the risk of frost damage if flowering is too early with the risk of inadequate water supply during grain filling as the crop matures into hot weather, both of which result in a low harvest index (Richards et al., 2002).

The trajectory of the development of leaf area between sowing and flowering influences the loss of water by direct evaporation from the soil surface (a major loss of water, often exceeding 50% of the rainfall, in many environments), competition with weeds, and the development of sufficient (but not excessive) biomass by flowering for the plant to be able to set an adequate number of seeds (Fischer, 1979). This trajectory depends critically on adequate nutrition, timeliness of sowing, control of weeds and diseases of roots, and ability of the seedlings to establish well. The latter can be especially important with semi-dwarf varieties of winter cereals whose coleoptiles are also dwarfed and are unable to reach the soil surface if the seeds are sown deeper than about 50 mm, which can easily happen if a farmer is obliged to sow when conditions are not ideal.

It is notable that these developmental issues, while they greatly influence water productivity, are typically unrelated to the biochemistry or physiology of water stress. They deal predominantly with the lifestyle of the crop, for example, whether it behaves conservatively or opportunistically in relation to using up soil water.

3.2. Transpiration Efficiency - Trading Water for Carbon

The transpiration efficiency of leaves, i.e. the ratio of carbon fixed to the amount of water transpired, depends on both evaporative demand by the environment and the CO_2 concentration within the leaves (Tanner and Sinclair, 1983; Condon et al., 2002; Kemanian et al., 2005). While it is in the first instance an instantaneous process producing photosynthate, the integration of this process through time, coupled with the many processes involved in converting photosynthate into new tissues and organs, conspire to generate a reasonably constant amount of above-ground biomass in relation to a given water supply during a whole growing season in a given environment. For example, in a mediterranean environment, with spring wheats growing during the autumn to spring growing season, the water productivity of above-ground biomass is about 55 kg ha^{-1} per mm of transpired water (French and Schultz, 1984; Condon et al., 2002). This is not to say that there is no useful genetic variation in this – there is, but it is only about 10%, whereas variation in the proportion of water lost to productive use, or the variation in harvest index, can vary several fold.

The concentration of CO_2 within a leaf has implications not only for transpiration efficiency but also for the discrimination that occurs during photosynthesis against the heavy stable isotope of carbon, ^{13}C. In general, for a given evaporative demand and stomatal conductance, the lower is the concentration of CO_2 within a leaf, the greater is the transpiration efficiency and the less discrimination there is against ^{13}C. These two relationships together provide an effective tool, based on isotopic analysis of plant tissue, for estimating average internal CO_2 concentration within leaves, and thence the intrinsic transpiration efficiency (Farquhar and Richards, 1984).

In practice, the isotopic signature varies somewhat during the season and with the tissue being measured (Condon and Richards, 1993). Nevertheless, breeding lines in wheat selected for intrinsically higher transpiration efficiency yielded better than those not so selected (Rebetzke et al., 2002). This trait, which tends to result in a more sparing use of water during vegetative growth, and thence better availability of water during flowering and grainfilling, has greater impact when rainfall is low, but does not reduce yield when rainfall is high. These lines have culminated in the release of two varieties, "Drysdale" and "Rees" which yield well in dry years compared to the parents from which they were derived (Richards, 2006) .

3.3. Harvest Index – Optimising the Proportion of Biomass in the Grain

In winter rainfall environments, crops that flower too early may not have built enough biomass to set and fill a large number of seeds (Fischer, 1979). They may also be prone to frost damage at flowering. Those that flower too late may fail to fill their grain adequately because they have too little water left in the soil and may be exposed to the heat and aridity of late spring and early summer (Richards, 1991) which reduces post-flowering photosynthesis and gives too little time to transfer to the grain any carbohydrate stored before flowering (Blum, 1998). These points are illustrated in Figure 1.

Given the variability among seasons, the optimal flowering time is necessarily an average. As an average there may be little room for further improvement, though breeders have been producing slower maturing varieties of wheat that can be sown

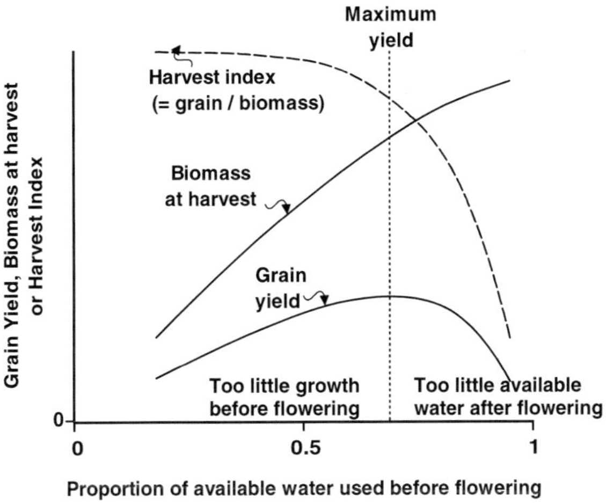

Figure 1. Schematic graph of grain yield of wheat, biomass at harvest, and harvest index, in relation to the proportion of the available water supply used before flowering. The scale of the y-axis is arbitrary, though the maximal harvest index is typically 0.5. (From Passioura, 2002)

earlier in the season while still flowering at the optimal time (Anderson et al., 1996) thereby giving farmers more flexibility in sowing time to cover different seasonal opportunities.

Water deficits during specific stages of floral development can severely damage seed set, through pollen sterility or abortion of embryos, or can prematurely end grain filling (Passioura, 2006). Low water potentials during pollen-mother-cell meiosis can induce severe pollen sterility and thence low yields in the cereals even though subsequent conditions might be good. Low water potentials around the time of anthesis are especially damaging in rice and maize. In rice, panicles may fail to emerge fully, spikelets lose water readily, lemma and palea may die, and anthers may fail to dehisce (Saini and Westgate, 2000). Maize is prone to severe embryo abortion (Boyer and Westgate, 2004) and to a mismatch in the timing of anthesis and silking, such that silking is delayed until after the pollen has been shed, leading to lack of fertilization (Bolaños and Edmeades, 1993). Breeding has successfully dealt with this latter problem (Ribaut et al., 2004).

These various effects of water deficits on fertility can lead to severe, sometimes complete, loss of yield in droughted grain crops. While total loss is rare, it is likely that drought-induced infertility can unnecessarily reduce yields in seasons in which there is a reasonable water supply but in which severe transient water deficits occur at these especially sensitive times.

Even if floral fertility and seed set are adequate, crops that suffer water deficits during grain filling may have poor harvest indices. Excessive vegetative growth, often stimulated by excessive nitrogen supply, can worsen such water deficits by using too much water before flowering (Figure. 1). The result is that the crop senesces prematurely and its yield falls (van Herwaarden et al., 1998). Grain filling is inhibited both by too little concurrent photosynthesis or by too little remobilization of carbohydrates stored during vegetative growth, which could otherwise be an important contributor to water-limited yield (Blum, 1998).

3.4. Integrating Over the Whole Season

The above discussion serves as a checklist of the various processes that contribute to the yield of a water-limited cereal crop, and the various problems other than water supply *per se* that can arise.

Many things can go wrong during the season-long growth of a crop: poor or untimely establishment, weeds, pests, diseases, poor nutrition, frost, heat, and even waterlogging in wet years. The upshot is that, in practice, yields are often well-below the water-limited potential. Figure 2 shows a compilation of the yields of several hundred wheat crops in relation to their water use, across four major environments (Sadras and Angus, 2006). The solid line, of slope 22 kg ha^{-1} mm^{-1}, marks the water-limited potential yield, which is approximately similar across these environments. This yield is consistent with the current practical limit of about 55 kg ha^{-1} mm^{-1} for above-ground biomass production (see section 3.2), coupled with a practical maximal harvest index of about 0.40 which is typical of crops that achieve their water-limited potential.

Some of this shortfall would have been due to the water supply being poorly distributed, perhaps delaying sowing to well past the optimal time, or inducing

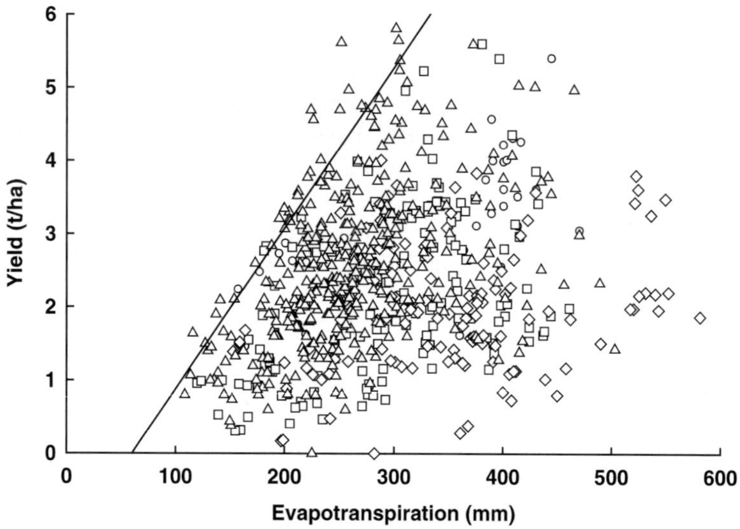

Figure 2. Scatter plot of grain yield and seasonal evapotranspiration from 691 published experiments in four mega-environments: ○, China Loess Plateau; □, Mediterranean basin; ◇, North American Great Plains; Δ, South Eastern Australia. The line, of slope 22 kg $ha^{-1}mm^{-1}$, uses the French and Schultz (1984) frontier concept. Adapted, with permission, from Sadras and Angus (2006)

infertility through severe water stress at flowering. Nevertheless, most of the yields well below the line are likely to be due to agronomic problems (French and Schultz, 1984), though several of these can be ameliorated through breeding – for example, the risk of poor establishment of wheat because of seeds being sown too deeply can be reduced by using lines that have long coleoptiles despite being otherwise dwarfed, as described in section 5.

It is notable that the region empty of data in the top left of Figure 2 is beyond well-understood agronomic, ecophysiological, and biochemical limits. Although unusually favorable seasons (for example, having moderate temperatures and high radiation during flowering and grainfilling) could occasionally result in yields substantially above the line, there is little chance of consistently getting such yields. This is not to say that there is no hope of producing varieties capable of frequently exceeding, say, 25 kg ha^{-1} per mm of transpired water, for novel science always produces surprises, but the chances look slim.

4. SCALING UP FROM LABORATORY TO FIELD

The previous section deals predominantly with ecophysiological processes known to influence crop yield in farmers' fields. Most of these processes act slowly, with time scales of weeks to months. In parallel with this research, which is carried out predominantly in the field, there is much research done in laboratory or glasshouse under the general banner of "drought tolerance".

Most of this latter research, well reviewed by Chaves et al. (2003), deals with events and processes that occur at short time scales, ranging from minutes for the stomatal control that so influences the exchange of carbon dioxide for water vapour by the leaves, through hours for the expression of genes in response to sudden severe water deficits, to days for the partitioning of assimilate among various organs that influences the trajectory of leaf area development. Although many of these fast processes form the basis for the slower processes whose connection with crop yield is reasonably clear, relating them directly to yield is difficult. The connections can be subtle, and may be unrelated to plant water relations. For example, few laboratory scientists would become aware of the importance of coleoptile length, outlined in section 5, without first being apprised of the operational problem in the field, yet it is a trait that could markedly improve crop yields if sowing is hampered by early drought (Rebetzke et al., 2007a).

It is notable that there are more than 2700 patents or patent applications returned by CAMBIA's Patent Lens database http://www.cambia.org patentlens/simple.cgi when queried with a full-text search using: (drought near/2 tolerant or drought near/2 tolerance) and "plant breeding". Random sampling of these patents suggests that a large proportion of them involve expression of genes in response to water stress that have no obvious connection with the processes outlined in section 3 that are known to influence the performance of water-limited crops. Few seem to have stimulated the development of selectable markers that could be used in commercial breeding programs.

The assays for drought tolerance described in these patents typically involve growing plants in small pots for a few days, then depriving them of water until they have wilted severely, then seeing how well they recover after rewatering. Assays such as this essentially explore desiccation tolerance, and could be useful for improving the robustness of perennial pasture plants. However, with annual crop plants, desiccation tolerance is unlikely to have much effect in the field, for droughts that are severe enough to kill crops are not commonly relieved by good rains during the particular growing season, and thus represent too small a target to interest commercial breeders.

This is not to say that all processes with time scales of hours or a few days have no practical significance. Important processes that occur this fast include effects of water deficits (Saini and Westgate, 2000) or of frost on floral fertility. Indeed, greater tolerance of freezing, perhaps enabled by CBF transcription factors (Miller et al., 2006) would enable breeders of crops with a winter-spring growing season to aim for earlier flowering, which would then give the crops a longer period of grain-filling in mild conditions before the heat and aridity of late spring and summer, thereby avoiding some of the effects of the late drought.

Although much laboratory research dealing with how plants respond to drought seems to be of limited use to commercial breeders and is therefore largely aspirational in relation to producing drought tolerant crops, there are signs of closer interactions developing between laboratory and field scientists which offer promise of improving the effectiveness of scaling up from laboratory to field.

5. MARKERS FOR TUNING THE CROP TO ITS ENVIRONMENT

Several of the issues covered in the previous sections in relation to improving water productivity of crops are being tackled in commercial breeding programs. A few are amenable to marker-assisted selection, as summarised in Table 1. Of particular importance are markers for traits that influence the establishment and development of the crop, the competence of the crop's root system, and the ability to maintain functional leaf area under disease pressure.

5.1. Sowing and Establishment of the Crop

While the optimal time of flowering in a given environment is well understood and paramount in breeders' minds, the time of sowing is more flexible. Unreliable weather at optimal sowing time means that farmers often have to sow when they can and to do so as fast as they are able. In mechanised agriculture sowing equipment that is 10 m or more wide, traveling at 10 km per hour, makes it hard to consistently sow seeds at the right depth. In drier regions that experience substantial pre-sowing rains, farmers may have to sow deeply to place seed near moisture for germination to occur. Standard semidwarf wheats containing the widely used dwarfing genes *Rht-B1b* and *Rht-D1b* have short coleoptiles unable to extend more than about 50 mm (Whan 1976; Allan et al., 1980). Wheats with long coleoptiles emerge with higher frequency than those with short coleoptiles especially when sown deep or where stubble has been retained (Rebetzke et al., 2005, 2007a). Alternative dwarfing genes are available that reduce plant height but not coleoptile length (Ellis et al., 2004). These have been mapped and linked microsatellite markers identified (Ellis et al., 2005). "Perfect" markers have also been developed from the coding sequences of the *Rht-B1b*/*Rht-D1b* genes that are specific for the mutations responsible for the height reduction (Peng et al., 1999; Ellis et al., 2002). These markers are now being used to select against *Rht-B1b*/*Rht-D1b* genes thus making it easier to identify plants with the desirable alternative dwarfing genes. Substituting the latter will produce wheats with longer coleoptiles by eliminating the negative effects of *Rht-B1b*/*Rht-D1b* genes and thereby uncover considerable residual genetic variation in coleoptile length, still common among modern varieties (Rebetzke et al., 2007b).

In some water-limited environments rapid development of leaf area after the emergence of seedlings may be desirable where crops are relying predominantly on in-season rainfall for their water supply. The advantages are twofold. Less water is lost by direct evaporation from the soil, and the crops compete better with weeds. As part of germplasm enhancement activities in CSIRO Plant Industry, breeding lines such as "Vigour18" have been developed which produce long coleoptiles and large early leaf area. (Richards and Lukacs, 2002). Quantitative trait analysis identified a QTL on chromosome 6A in "Vigour18" that accounted for much of the phenotypic variation for both early leaf area and coleoptile length in an experimental segregating population (Spielmeyer et al., 2007). The same QTL region was associated with greater plant height at maturity suggesting that genes located in this region may have pleiotropic effects on early as well as late developmental stages. Markers linked to the QTL were associated with greater leaf area and longer coleoptiles in breeding populations that segregated for these traits. Marker-assisted selection is now being used to enrich for the "vigour" allele in early generation breeding lines to complement existing phenotype-based selection methodologies based on measuring leaf width of seedlings grown in soil under natural conditions.

Crops that develop leaf area quickly may be in danger of becoming too leafy. Most current wheat varieties produce more tillers than they can sustain to grain maturity (van Herwaarden et al., 1998). Particularly in water-limited environments, many tillers die before flowering and are a waste of resource that could otherwise be used productively. Yield trials in Australia have shown that reduced tillering wheats produced larger kernels, therefore reducing the potential for shriveled grain (Duggan et al., 2005). The low tillering phenotype has also been associated with greater root biomass (Richards et al., 2006). A recessive gene that inhibits excessive tillering (tin) has been identified and mapped to chromosome 1AS (Richards, 1988). A tightly linked microsatellite marker (gwm136) amplifies a unique sized PCR product which is present in donor lines carrying the tiller inhibition gene (Spielmeyer and Richards, 2004). The phenotype of wheat lines carrying the tin gene can range from monoculm (complete absence of tillers) to an intermediate oligoculm type depending on genetic background and a wide range of environmental effects that influence tiller number in wheat. Marker-assisted selection for this recessive gene the leaf area QTL on chromosome 6A also accounted for much of the phenotypic variation for greater early leaf area in a breeding population coming from a cross between the same donor line but different adapted backgrounds. The marker was therefore useful in selecting lines with greater average leaf area in adapted germplasm.

The more valuable QTLs however are those that are expressed in different genetic backgrounds and occur in unrelated donor lines expressing the same trait. To use the previous example again, markers from the chromosome 6A region were associated with long coleoptiles in both the breeding line "Vigour 18" and an unrelated donor line HMIOS. These results demonstrated that (a) the QTL region on chromosome 6A contains gene(s) for both longer coleoptiles and greater early leaf area, (b) the QTL contributes to the phenotypic variation in adapted germplasm, and (c) two unrelated donor lines carry alleles for long coleoptiles probably corresponding to the same QTL. Markers from the 6A QTL region are being used to develop adapted germplasm with longer coleoptiles and increased vigour (Bonnett, unpublished). With greater emphasis on QTL validation in the future, we anticipate that breeders will also use more QTL markers in their programs.

An alternative approach, that may allow breeders to develop and use more markers for complex traits such as yield, is to identify and track QTLs in breeding populations by association mapping. Differences in the level of linkage disequilibrium between species (Gupta et al., 2005) and, particularly for wheat, the need to recover existing combinations of alleles for grain quality, will influence the number and value of the associations identified. Given the likelihood that the effects of QTLs for complex traits will change over time through fixation of important regions and differing interactions with new alleles at other loci, a continual reassessment of the value of QTLs in breeding populations may be needed in parallel with their use in selection (Podlich et al., 2005). This will require integration of good multi-environment yield data and efficient whole-genome fingerprinting techniques that can be applied to the large numbers of lines making up the yield trials of commercial breeding programs. Association mapping approaches may not give good clues to the underlying mechanisms responsible for the yield effects of QTLs, but given that conventional breeding has achieved yield gains despite ignorance of contributory mechanism(s), this need not be an impediment to their use. Further, QTLs for yield may help identify candidate regions for further study.

6.3. Mapping the Pathway from Marker Laboratory to the Field

Further success in developing and applying markers in commercial breeding will depend on strong linkages between laboratory and field scientists. Germplasm-enhancement projects and breeding programs must jointly define the most suitable targets and donor lines. Similarly, every marker-development project should have a clear path of delivery to the germplasm-enhancement or breeding program. This path needs to be well mapped from an early stage to ensure that markers are being developed for traits that are of interest to breeders. In particular, the development of breeders' markers for QTLs that account for only a small part of the variation for a trait with complex inheritance will benefit from close linkages between people who are developing markers, crop physiologists, and agronomists. For example, a QTL for early vigour was identified by gaining a better understanding of components that contribute to the overall phenotype which the breeder might refer to as "rapid production of early, above ground biomass". Prior to any marker work, the width of seedling leaves was identified as the most heritable component that was correlated with leaf area and biomass (Rebetzke and Richards, 1999). Leaf width and not biomass was used to phenotype the mapping family resulting in the identification of a QTL for leaf width. The same QTL was not identified in the mapping family using biomass measurements, a trait with lower heritability than leaf width.

Marker projects that are integrated with germplasm enhancement or breeding programs have a much greater chance of delivering markers useful for breeders than those not so integrated. Identifying breeders' markers for QTLs poses particular challenges. Effectively phenotyping and identifying more heritable components will greatly increase the likelihood of detecting useful QTLs. This approach necessitates strong linkages between breeding, crop physiology, and molecular genetics. In many instances these linkages have been formed by a new generation of molecular breeders who are skilled both in conventional breeding and in developing markers. With the growing number of useful markers, breeders also need to consider how best to make use of them. Apart from the cost of carrying out the marker assays, the cost/benefit analysis needs to include the timing of marker use, and the population structure and size needed to recover target alleles. These issues are discussed in more detail in the following section.

7. CONVERTING ELITE GERMPLASM INTO COMMERCIALLY COMPETITIVE VARIETIES

Plant breeding is an iterative process in which the gains made in one breeding cycle are built on in subsequent cycles. Because many commercially directed crosses fail to produce their desired outcome, breeders must make multiple crosses to spread risk and ensure a steady supply of improved varieties. A commercially directed cross will usually involve a small number of parents that, between them, carry all of the attributes needed in a variety. Often these parents will have a high coancestry and many attributes in common while differing for a small number of commercially important traits. If parents have a lower coancestry and differ for a greater number of traits, or even if they are phenotypically similar but the genetic control of common traits differs, it may be difficult to derive a commercially attractive combination from a simple biparental cross. In such crosses prohibitively large population sizes may be needed to combine desirable alleles across large numbers of loci even with the most efficient strategy (Table 2). In such cases, or where one parent contributes only a small number of desirable attributes and the other contributes many more, one or more backcrosses may be necessary to recover a commercially viable line.

Table 2. Population sizes needed to recover a target genotype (P=0.05) in biparental crosses segregating at different numbers of unlinked loci with different selection strategies

Loci	A. Select homozygotes in F_2	B. Select only among inbreds	C. F_2 enrichment followed by selection among derived inbreds	D. F_2 enrichment at 50% of loci followed by selection among derived inbreds
2	46	10	11	17
4	765	46	26	55
6	12269	190	59	165
8	196327	765	127	480
10	3141251	3066	274	1385
12	5.0×10^7	12269	589	3998
14	8.0×10^8	49081	1276	11561
16	1.3×10^{10}	196327	2778	33516
18	2.1×10^{11}	785312	6081	97421
20	3.3×10^{12}	3141251	13380	283915

In spite of these constraints, breeding programs have successfully improved water-limited yields over an extended period. New approaches will therefore need to offer tangible benefits to be adopted (Vandeleur and Gill, 2004; Perry and D'Antuono, 1991; O'Brien, 1982)

Increasing uptake of marker technology and identification of ever more marker-trait associations provides breeders with a greater range of potential crossing and selection strategies to recover a commercially attractive line. The opportunity to combine a greater number of desirable alleles in a single breeding cycle is much greater than in the past. In almost all cases, however, marker-assisted selection will need to be used in combination with phenotypic selection for important variation for which molecular markers are not available. The most appropriate selection strategy will vary depending on: the relative commercial importance of traits; their genetic control and response to phenotypic selection; the number of important regions for which markers are available; and the relative costs of phenotypic versus MAS and of MAS versus the cost of additional cycles of inbreeding or doubled haploid production. Although it appears that there are many different options to combine multiple genomic regions, it is possible to distil some general principles (Table 2).

One of the often cited benefits of molecular markers (e.g. Koebner and Summers, 2003) is the possibility of identifying homozygotes for multiple desirable alleles in the earliest generation possible. While this may initially seem attractive, it requires large population sizes even with relatively small numbers of unlinked polymorphic loci (Table 2, strategy A). Delaying selection until populations are more inbred requires substantially smaller population sizes (Table 2, strategy B). Unless the cost of producing inbred lines is high or there is a considerable benefit in isolating a target individual more quickly, inbreeding prior to selecting target homozygotes is likely to be preferable. Even just one generation of inbreeding can greatly reduce population sizes. For example, in a simple cross between two inbred parents that are polymorphic at 4 important loci, an F_2 population of 765 is needed to be 95% confident of recovering at least one homozygous line carrying the desirable alleles at these loci. Delaying selection by only one generation reduces the population size by over 80% to only 150 individuals in the F_3 (cf. 46 in complete inbreds). This phenomenon is now commonly exploited in commercial breeding through production of inbred lines by single seed descent or doubled haploid techniques to combine relatively small numbers of desirable alleles.

In crosses with greater numbers of segregating loci, an F_2 enrichment strategy may be desirable (Bonnett et al., 2005). This approach removes homozygotes for undesirable alleles yet retains both heterozygotes and homozygotes for desirable alleles. Population sizes required to recover desirable homozygotes among inbred lines derived from the selected F_2s are substantially reduced using this strategy compared with selecting only among inbred lines (Table 2, strategy C vs. strategy B). Markers can play an important role in F_2 enrichment, but phenotypic screens may do as well for a more limited number of traits. These idealized scenarios illustrate that, even with completely linked markers for all the important alleles segregating in a cross, when to use markers in the breeding program is an important factor in successfully constructing new allelic combinations.

In real-world breeding populations, it is rare that the number and location of all important polymorphic loci is known and even rarer that markers are available for all of them. Phenotypic selection must then be used to identify favorable combinations of the untagged alleles controlling commercially important traits. Phenotypic selection is usually applied over successive generations, beginning with the most heritable traits requiring the smallest seed quantities, followed by those with lower heritability requiring higher levels of inbreeding and greater seed quantities (e.g. yield and end-product quality). Even with highly heritable traits it is usually possible to select simultaneously for only a small number in early generations. In such cases however, partial F_2 enrichment is still often a useful strategy (Table 2, strategy D). This strategy, used routinely in CSIRO's germplasm development program, involves enriching the marker alleles in F_2 while maintaining populations large enough to prevent loss of untagged alleles before selecting phenotypically among more inbred material. While this requires larger populations than if markers were available for all loci (Table 2, strategy C), the populations are substantially smaller than needed if none of the loci are enriched (Table 2, strategy B). This strategy of partial F_2 enrichment has proved to be effective in assembling combinations of marker alleles linked to dwarfing genes, grain quality, and root traits, with alleles for important unmarked traits for water productivity such as long coleoptiles (Bonnett, unpublished). This approach should be applicable in a wide range of scenarios where retention of unselected variation until later generations is desirable.

Further increases in required population sizes beyond those described above and shown in Table 2 are necessary if there is recombination between target alleles and the marker(s) used to select for them and if target alleles from different parents are linked in repulsion. These are common "real world" complexities in marker-assisted plant breeding. More sophisticated modeling approaches to determine minimum population sizes in these scenarios are now being developed (Wang et al., 2007).

Limits to how many alleles can be combined in a single breeding cycle mean that breeders wont put effort into selecting for a marker-linked trait unless it is of sufficient value to add to the suite of traits they must select for. Most markers currently used in commercial breeding are for traits of sufficient value that they were selected phenotypically before the marker was developed (e.g. boron tolerance, cereal cyst nematode resistance). Markers for glutenins and puroindoline alleles are now being adopted due both to the development of markers and a growing understanding of the contribution of different combinations of alleles to grain quality (Eagles et al., 2005). Novel traits will be more attractive to commercial breeders if research has determined their value, if they are available in "breeder friendly" backgrounds (i.e. related to parents used in commercial breeding programs, especially if they are under complex genetic control), and if they can be selected using simple, preferably marker-based, selection tools that integrate easily with the routine operations of breeding programs. The reduced tillering gene appears to meet these criteria and "breeder friendly" germplasm will soon be transferred to commercial breeding programs.

Work on the transpiration efficiency (TE) trait is a good example of the steps required for development of a variety incorporating a novel, polygenically inherited trait. Initially, the effect of transpiration efficiency on yield was determined by transferring the trait from an agronomically inferior donor line with high expression of the trait into a modern variety, Hartog, and testing high and low TE backcross lines in multi-environment trials. Selection was based on a physiological marker, carbon isotope discrimination (Rebetzke et al., 2002). As well as establishing a useful increase in water-limited yields, backcrossing TE into the "Hartog" background at CSIRO allowed collaborators in commercial breeding programs to select varieties "Drysdale" and "Rees" from among these lines (Richards, 2006). Unfortunately, measuring carbon isotope discrimination is expensive to apply on a large scale and adds complexity to the routine operations of breeding programs because it is different to any of the routine screens currently used. As a result, selection for TE is not being actively applied in commercial wheat breeding and awaits the development of molecular markers or other simpler and cheaper screening methods. In spite of this, "Drysdale" and "Rees" are being used as parents in commercial breeding and should allow development of improved "Hartog-like" varieties with high TE through backcrossing, even without targeted screening for TE. This illustrates the value of incorporating novel polygenically inherited traits into well-adapted backgrounds, even in the absence of simple selection screens.

Work on traits such as alternative dwarfing genes and early vigour has concentrated, from the earliest stages, on establishing trait value, on availability in "breeder friendly" backgrounds, and on developing molecular markers to select for the major genetic determinants of the trait.

8. CONCLUDING COMMENTS

The yield of a crop in a water-limited environment is the culmination of many processes occurring at various stages of the crop's life. It reflects the ability of the crop to deal with many deleterious impacts of biotic and abiotic stresses, which, in addition to inadequate water, may include weeds, pests and diseases, poor nutrition, frost, or heat. These other stresses often interact with water stress, sometimes negatively, sometimes positively, depending on the season – for example, mild nitrogen starvation can improve water-limited yield if it results in a crop conserving soil water during vegetative growth so that there is more for it to use during grain filling.

Thus there are many facets to the problem of improving water-limited yield. The commercial breeder has to consider many important traits, and is acutely aware that these must be present in superior combinations in any variety that can be released. Molecular markers offer a powerful tool for helping select these superior combinations, and their use has become routine in many breeding programs. Future advances in marker technologies will reduce the cost per assay thereby allowing not only more marker assays for single gene traits and QTLs but also providing the tools for implementing strategies for selecting genome wide markers.

To make best use of these technologies, future marker development will require ever closer links between the molecular laboratory and the breeder. Frequent dialogues will ensure that markers are developed for important traits where the breeder sees a clear need for marker-assisted selection and is prepared to spend resources to fund such marker screening in the program. It is important to recognize that the best marker developed in an experimental population may lack sufficient polymorphism across the range of genetic backgrounds in the program. Additional resources may be required to develop "breeders markers" which can distinguish the donor allele from the recurrent parent alleles and are in a format amenable to high throughput screening.

The importance of understanding the phenotype can not be overemphasized. Especially when dissecting complex traits into genetic parameters through QTL analysis, it is often useful to dissect the phenotype into parameters which are more heritable and can be measured under controlled conditions. This strategy is only successful if the relevance of these parameters on the trait is understood when scaling up to the field.

REFERENCES

Allan, R.E., 1980, Influence of Semidwarfism and Genetic Background on Stand Establishment of Wheat. *Crop Sci* **20**:634–638

Anderson, W.K., Heinrich, A. and Abbotts, R. 1996, Long-season wheats extend sowing opportunities in the central wheat belt of Western Australia. *Australian Journal of Experimental Agriculture* **36**: 203–208.

Blum, A. 1998, Improving wheat grain filling under stress by stem reserve mobilization. *Euphytica* **100**: 77–83.

Bonnett, D.G., Rebetzke, G.J, Spielmeyer, W., 2005, Strategies for efficient implementation of molecular markers in wheat breeding. *Mol. Breed* **15**:75–85

Bolaños, J. and Edmeades, G.O., 1993, Eight cycles of selection for drought tolerance in lowland tropical maize. II. Responses in reproductive behavior. *Field Crops Research* **31**: 253–268.

Boyer, J.S., and Westgate, M.E., 2004, Grain yields with limited water. *Journal of Experimental Botany* **55**: 2385–2394.

Chaves, M.M., Maroco, J.P., and Pereira, J.S., 2003. Understanding plant responses to drought - from genes to the whole plant. *Functional Plant Biology* **30**: 239–264.

Condon, A.G., Richards, R.A., Rebetzke, G.J., and Farquhar, G.D., 2002, Improving intrinsic water-use efficiency. *Crop Science* **42**: 122–131.

DePauw, R.M., Townley-Smith, T.F., Humphreys, G., Knox, R.E., Clarke, F.R. and Clarke, J.M.2005, Lillian hard red spring wheat. *Can. J. Plant Sci.* **85**: 397 – 401

Duggan, B.L., Richards, R.A., and van Herwaarden, A.F., 2005, Agronomic evaluation of a tiller inhibition gene (tin) in wheat. II. Growth and partitioning of assimilate. *Aust J Agric Res* **56**:179–186

Ellis, M., Rebetzke, G., Azanza, F., Richards, R.A., and Spielmeyer, W., 2005, Molecular mapping of alternative dwarfing genes in wheat. *Theor.Appl. Genet..* **111**: 423–430

Ellis, M., Rebetzke, G., Chandler, P., Bonnett, D.G., Spielmeyer, W., and Richards, R.A., 2004, The effect of different height reducing genes on the early growth of wheat. *Funct. Plant Biol.***31**:583–589

Ellis, M.H., Spielmeyer, W., Rebetzke, G.J., and Richards, R.A., 2002,"Perfect" markers for the Rht-B1b and Rht-D1b dwarfing genes in wheat. *Theor. Appl. Gen.* **105**:1038–1042.

Fischer, R.A., 1979, Growth and water limitation to dryland wheat yield in Australia: a physiological framework. Journal of the *Aust. Inst. Agric. Sci.* **45**: 83–94.

French, R.J., and Schultz, J.E., 1984, Water use efficiency of wheat in a mediterranean-type environment. I. The relation between yield, water use and climate. *Australian Journal of Agricultural Research*, **35**: 743–764.

Gupta, P.K., Rustgi, S., Kulwal, P.L., 2005. Linkage disequilibrium and association studies in higher plants: present status and future prospects. *Plant Molecular Biology* **57**: 461–485.

Jefferies, S.P., Pallotta, M.A., Paull, J.G., Karakousis, A., Kretschmer, J.M., Manning, S., Islam, A.K.M.R., Langridge, P., Chalmers, K.J., 2000, Mapping and validation of chromosome regions conferring boron toxicity tolerance in wheat (*Triticum aestivum*). *Theor. Appl. Genet.* **101**:767–777

Koebner, R.M.D., and Summers, R.W., 2003,21st Century Wheat Breeding: Plot Selection or Plate Detection? *Trends in Biotechnology* **21**: 59–63.

Lagudah, E.S., McFadden, H., Singh, R.P., Huerta-Espino, J., Spielmeyer, W., 2006, Molecular characterization of the *Lr34/Yr18* slow rusting resistance gene region in wheat. *Theor.Appl. Genet.* In Press.

Mago, R., Bariana, H.S., Dundas, I.S., Spielmeyer, W., Lawrence, G.J., Pryor, A.J., Ellis, J.G., 2005, Development of PCR markers for the selection of wheat stem rust resistance genes *Sr24* and *Sr26* in diverse wheat germplasm. *Theor. Appl. Genet.* **111**: 496–504

Miller, A.K., Galiba, G., Dubcovsky, J.2006, A cluster of 11 CBF transcription factors is located at the frost tolerance locus fr-a(m)2 in Triticum monococcum. *Molecular Genetics and Genomics* **275**: 193–203.

Ogbonnaya, F.C., Subrahmanyam, N.C., Moullet, O., de Majnik, J., Eagles, H.A., Brown, J.S., Eastwood, R.F., Kollmorgen, J., Appels, R., Lagudah, E.S., 2001, Diagnostic DNA markers for cereal cyst nematode resistance in bread wheat. *Aust J Agric Res* **52**: 1367–1374.

Passioura JB2002, Environmental biology and crop improvement. Functional Plant Biology, 29, 537–546.

Passioura, J.B., 2006, Increasing crop productivity when water is scarce – from breeding to field management. *Agricultural Water Management* **80**:176–196.

Peng, J., Richards, D.E., Hartley, N.M., Murphy, G.P, Devos, K.M., Flintham, J.E., Beales, J., Fish, L.J., Worland, A.J., Pelica, F., Duralalagaraja Sudhakar, Christou, P., Snape, J.W., Gale, M.D., Harberd, N.P., 1999,'Green revolution' genes encode mutant gibberellin response modulators. *Nature* **400**:256–261.

Podlich, D.W., Winkler, C.R., Cooper, M., 2004, Mapping as you go: An effective approach for marker-assisted selection of complex traits. *Crop Sci* **44**:1560–1571.

Rebetzke, G.J., Condon, A.G., Richards, R.A., Farquhar, G.D., 2002, Selection for reduced carbon isotope discrimination increases aerial biomass and grain yield of rainfed bread wheat. *Crop Science*. **42**: 739–745.

Rebetzke, G.J., Richards, R.A., 1999, Genetic improvement of early vigour in wheat. *Aust J Agric Res* **50**:291–301

Rebetzke, G.J., Bruce, S.E., Kirkegaard, J.A., 2005, Longer coleoptiles improve emergence through crop residues to increase seedling number and biomass in wheat. *Plant and Soil* **272**: 87–100.

Rebetzke, G.J., Richards, R.A., Fettell, N.A., Long, M., Condon, A.G., Forrester, R.I., Botwright, T.L., 2007a. Genotypic increases in coleoptile length improves stand establishment, vigour and grain yield of deep-sown wheat. *Field Crops Research*. **100**: 10–23.

Rebetzke, G.J., Ellis, M.H., Bonnett, D.G., Richards, R.A. 2007b. Molecular mapping of genes for coleoptile growth in bread wheat (*Triticum aestivum* L.). *Theoretical and Applied Genetics* **114**: 1173–1183.

Rengasamy, P., Chittleborough, D., Helyar, K., 2003, Root-Zone Constraints and Plant-Based Solutions for Dryland Salinity. *Plant and Soil* **257**: 249–260.

Ribaut, J.-M., Bänziger, M., Setter, T., Edmeades, G., and Hoisington, D., 2004, Genetic dissection of drought tolerance in maize: A case study. In H. Nguyen and A. Blum (eds.) *Physiology and Biotechnology Integration for Plant Breeding*. Marcel Dekker, Inc. New York

Richards, R.A., 1988, A Tiller Inhibitor Gene in Wheat and its Effect on Plant Growth. *Aust J Agric Res* **39**: 749–757

Richards, R.A., 1991, Crop improvement for temperate Australia - future opportunities. *Field Crops Research*, **26**: 141–169.

Richards, R.A., 2006, Physiological traits used in the breeding of new cultivars for water-scarce environments. *Agricultural Water Management* **80**: 197–211.

Richards, R.A., Lukacs, Z., 2002, Seedling vigour in wheat-sources of variation for genetic and agronomic improvement. *Aust J Agric Res* **53**:41–50

Richards, R.A., Rebetzke, G.J., Condon, A.G., and van Herwaarden, A.F., 2002, Breeding opportunities for increasing the efficiency of water use and crop yield in temperate cereals. *Crop Science* **42**: 111–121.

Saini, H.S., and Westgate, M.E., 2000, Reproductive development in grain crops during drought. *Advances in Agronomy* **68**:59–96.

Sasaki, T., Yamamoto, Y., Ezaki, B., Katsuhara, M., Ahn, S.J., Ryan, P., Delhaize, E., Matsumoto, H.2004, A wheat gene encoding an aluminum-activated malate transporter. *Plant J.* **37**:645–653

Seah, S., Spielmeyer, W., Jahier, J., Sivasithamparam, K., and E.S. Lagudah, 2000, Families of resistance gene sequences within an introgressed chromosomal segment derived from T. ventricosum in wheat, which confer resistance to nematode and rust pathogens. *Molecular Plant-Microbe Interactions* **13**: 334–341.

Spielmeyer, W., Sharp, P., Lagudah, E.S., 2003, Identification of markers linked to broad-spectrum stem rust resistance gene *Sr2* in wheat. *Crop Science* **43**: 333–336.

Spielmeyer, W., Richards, R.A., 2004, Comparative mapping of wheat chromosome 1AS carrying tiller inhibition gene with corresponding rice chromosome 5. *Theoretical and Applied Genetics* **109**: 1303–1310.

Spielmeyer, W., Hyles, H., Paul, J., Azanza, F., Bonnett, D., Ellis, M., Moore, C., Richards, R.A. (2007) A QTL on chromosome 6A in bread wheat (Triticum aestivum) is associated with longer coleoptiles, greater seedling vigour and final plant height. *Theoretical and Applied Genetics* (In press).

Tanner, C.B., and Sinclair, T.R., 1983, Efficient water use in crop production: research or re-search? In *Limitations to Efficient Water Use in Crop Production* (eds. HM Taylor, WR Jordan and TR Sinclair) American Society of Agronomy, Wisconsin, USA, pp. 1–27.

Vandeleur, R.K., Gill, G.S., 2004, The impact of plant breeding on the grain yield and competitive ability of wheat in Australia. *Aust J Agric Res* **55**:855–861 van Herwaarden, A.F., Angus, J.F., Richards, R.A., and Farquhar, G.D., 1998,"Haying-off", the negative grain yield response of dryland wheat to nitrogen fertiliser. II. Carbohydrate and protein dynamics. *Australian Journal of Agricultural Research* **49**:1083–1093.

Wang, J., Chapman, S.C., Bonnett, D.G., Rebetzke, G.J., Crouch, J., 2007. Application of population genetic theory and simulation models to efficiently pyramid alleles via marker-assisted selection. *Crop Science* **47**: 582–588.

Watt, M., Kirkegaard, .JA., Rebetzke, G.J., 2005, A wheat genotype developed for rapid leaf growth copes well with the physical and biological constraints of unploughed soil. *Func. Plant Biol.* **32**: 695–706.

Whan, B.R., 1976, The association between coleoptile length and culm length in semidwarf and standard wheats. *J Aust Inst Agric Sci* **42**:194–196.

Williams, K.J., Taylor, S.P., Bogacki, P., Pallotta, M., Bariana, H.S., Wallwork, H., 2002, Mapping of the root lesion nematode (*Pratylenchus neglectus*) resistance gene Rlnn1 in wheat. *Theor. Appl. Genet.* **104**:874–879.

William, H.M., 2006,. Wheat molecular breeding- does it offer any hope? *In* Reynolds MP, Godinez D. "*Challenges to International Wheat Breeding*" March 20–24th, 2006 Cd. Obregon, Mexico. CIMMYT, Mexico, D.F.

Wilson, R., McLean, R., Barclay, I., Cakir, M., Appels, R., Devlin, G., Li, D.2006, Marker implementation in the Department of Agriculture, Western Australia Wheat Breeding Program. *Proceedings 13th Australasian Plant Breeding Conference*. Christchurch, New Zealand 18–21 April.

CHAPTER 20

TRANSGENIC PLANTS FOR DRY AND SALINE ENVIRONMENTS

SNEH LATA SINGLA-PAREEK[1], ASHWANI PAREEK[2] AND SUDHIR K SOPORY[1]

[1]*International Centre for Genetic Engineering and Biotechnology, New Delhi 110067, India*
[2]*Stress Physiology and Molecular Biology Laboratory, School of Life Sciences, Jawaharlal Nehru University, New Delhi 10067, India*
E-mail: sopory@icgeb.res.in

Abstract: In the past decade, the scientific community has witnessed a major leap in our understanding about how plant perceives and respond to abiotic stresses. Various candidates participating in this coordinated and orchestrated relays have been identified and their molecular mechanisms of operation have also been worked out. This analysis has clearly established a complex network of cellular machinery operative in plants under such conditions. Tools of functional genomics have been utilized to decipher the contributions of several of these individual components towards the complex stress response. Some of these studies have also been extended beyond model plants, and crop systems such as rice have been utilized to document the usefulness of some of these strategies towards genetic modifications of crop plants which are better adapted towards unfavorable environmental conditions. It is heartening to see the extension of few efforts beyond laboratory to field level testing. Indeed, a few of selected candidate genes have also passed these field level tests. However, it is also true that drought/salinity tolerant transgenic crop plants are yet away from the reach of farmers. A conscious deliberate and strategic action plan along with the right choice of battery of genes is required to achieve this important goal

Keywords: Transgenic plants, dry and saline environments, signaling, transcription factors, osmolytes, ROS, membrane transport

1. INTRODUCTION

In the present era of 'omics', where information about genes and their regulation patterns is just 'pouring-in' from many laboratories, the usefulness of the strategy based on testing the role of an individual gene through genetic engineering is not limited. Keeping in mind, the large yield gap in agricultural productivity caused due to environmental stresses, there have been considerable efforts to develop

M.A. Jenks et al. (eds.), Advances in Molecular Breeding Toward Drought and Salt Tolerant Crops, 501–530.

stress tolerant plants via genetic engineering. The success has been forthcoming in engineering plants for improved tolerance towards biotic stresses, however, abiotic stress tolerant transgenic plants neither have been tested on large scale under field conditions nor have these plants found their way in breeding programs yet. This is despite sustained efforts having been carried out in many laboratories targeting towards manipulating genes belonging to diverse categories. Recently, a number of excellent reviews have been compiled on elucidation of the stress tolerance mechanisms and use of transgenic technology in agriculture for developing abiotic stress tolerant crop plants (Singla-Pareek et al., 2001; Grover et al., 2003; Sreenivasulu et al., 2004; Vinocur and Altman, 2005; Bajaj and Mohanty, 2005). It follows that, in addition to using the known potential genes for developing stress tolerant transgenic plants, detailed analysis towards understanding of abiotic stress tolerance mechanisms using molecular biology and genomic approaches are required as well (Bohnert et al., 2006; Grennan, 2006). Also, one may have to adopt a specific unique strategy for each kind of abiotic stress, as is being done for developing saline tolerant plants (Yamaguchi and Blumwald, 2005; Cuartero et al., 2006) and drought tolerant plants (Umezawa et al., 2006).

In the present compilation, we have made an attempt to highlight different strategies which the researchers are employing worldwide with respect to the selection of genes and their use in transformation for developing drought and saline tolerant transgenic plants. In this category, extensive work has been carried out for genetic manipulation of important components pertaining to transcription factors, reactive oxygen species (ROS) scavenging, osmolytes, ion homeostasis mechanisms etc. There have been several compilations where related aspects have been presented to the scientific community (Zhang et al., 2004; Vinocur and Altman, 2005; Yamaguchi and Blumwald, 2005; Mittler, 2006), the present one is an updated version in the context of recent interesting scientific findings. For brevity sake, we have presented only the representative reports of each category, without any intention to omit any other similar report by other group(s). We have also tried to bring in some of the recent reports which have raised hopes towards the possibilities of raising plants having better tolerance for both biotic and abiotic stresses. At the end, we discuss the lessons learnt from the work done so far and present our perspectives which we think may be important in tailoring stress tolerant crop plants in the future.

2. TRANSGENIC PLANTS OVERXPRESSING SIGNAL COMPONENTS AND TRANSCRIPTION FACTORS

Understanding how plant perceives the stress and how the signal is transduced downstream to bring about a specific effect has been an active area of research for several decades. Owing to impressive progress made in molecular biology techniques and the classical techniques of mutagenesis, it has been possible to make attempts to decipher the minute gears of stress response in plants (Valliyodan and Nguyen, 2006). We now know that signaling pathways exist which are dependent or independent on ABA. Similarly, Ca^{++} plays an important role in the cascading phenomenon. In the present section, we discuss the reports where genetic modifications have been attempted for signal components and transcription factors.

2.1. Calcium and Signaling Components

Calcium is known to be an important signal transducer for many stress responsive genes. One important calcium binding protein that modulates the activity of many other proteins in the pathway is calmodulin (CaM). In plants, many isoforms of CaM are known (Zielinski, 1998). In a recent study, one of the isoform of CaM was found to bind to a transcription factor MYB2 and was reported to enhance its DNA binding activity. Overexpression of this isoform of CaM in Arabidopsis upregulated the transcription of MYB2 regulated genes including P5CS1 which is known to confer salt tolerance by overproducing proline (Yoo et al., 2005). The mechanism by which calcium regulates sodium homeostasis via SOS pathway has been well elucidated from work of the Zhu's group (Zhu, 2002). One of the first components of this pathway is calcineurin like proteins which in turn activate CIP kinases (CIPK). The activated kinase can regulate the expression of a plasma membrane Na^{+} antiporter (SOS1) to efflux sodium out of the system (Guo et al., 2004). It has been further shown that overexpression of SOS1 can lead to enhanced tolerance of plants to NaCl stress (Shi et al., 2003). Earlier, overexpression of yeast calcineurin was also found to confer stress tolerance (Marin-Manzano et al., 2004). Recently, overexpression of mouse calcineurin gene in rice resulted in its higher salt tolerance and less sodium was found to be accumulated in roots. The transgenic plants also showed higher expression of a group 2 LEA protein (Ma et al., 2005). A rice gene encoding a calcium-dependent protein kinase (CDPK) was overexpressed in rice which showed tolerance to salt and drought stress (Saijo et al., 2000). In a recent study, Jin et al. (2005) cloned a MAP kinase gene (EhHOG) from a fungus, *Eurotium herbariorum* that grows in Dead Sea in Israel. Further, the EhHOG was able to complement hog1 mutant of *S. pombe* which could be grown under high osmotic stress. It has been suggested that such genes may prove to be very useful for developing crop plants for saline areas. However, suitability of EhHOG for this purpose remains to be tested. Recently, a rice GTPase OsRacB has also been overexpressed in rice and tobacco (Luo et al., 2006). The overexpressing plants grew much better than control under salinity stress, while the antisense plants did not show any change in response towards stress treatment indicating that OsRacB is only an accessory factor in plant stress tolerance.

2.2. Transcription Factors

Genomic analysis in Arabidopsis (Arabidopsis Genome Initiative, 2000), rice (Goff et al., 2002; Yu et al., 2002; IRGSP, 2005) and other plants has revealed that a large number of genes are upregulated in response to various stresses. Some of these genes are common while others may be unique to a specific stress. Classically, two pathways have been implicated in the induction of these genes - an ABA independent and an ABA dependent pathway. In addition, calcium is also known to affect the induction of stress responsive genes. Several transcription factors have been implicated in these pathways. CBFs (DREBs) are the transcription factors that bind to CRT/DRE cis elements in the stress induced promoters and ABFs bind to ABRE cis elements (Yamaguchi-Shinozaki and Shinozaki, 2006). The former usually fall in ABA –independent pathway and the latter in the ABA dependent pathway. In addition to these two classes, other transcription factors have also been shown to play an important role in stress responses. DREB1A and DREB2A cDNA were isolated first from Arabidopsis (Stockinger et al., 1997; Liu et al., 1998), followed by their isolation from a wide variety of plants (Agarwal et al., 2006). However, while expression of DREB1 genes has been investigated extensively in several crop species, DBRE2 is explored in a limited species.

Studies have shown that overexpression of CBF3 (DREB1) confers stress tolerance, however, enhanced expression under the constitutive promoter resulted in dwarfing of the phenotype in Arabidopsis (Kasuga et al., 1999). This phenotype could be corrected if the expression is regulated via a stress inducible promoter, like rd29 (Oh et al., 2005). It was thus shown that use of rd29A promoter with DREB1A conferred both drought and low-temperature stress tolerance in tobacco (Kasuga et al., 2004). However, it has also been shown recently that Arabidopsis CBF3 and ABF3, when overexpressed in rice, increased tolerance to salinity and drought without any penalty on plant growth (Oh et al., 2005). Unlike transgenic Arabidopsis, the rice transgenic plants did not show much tolerance towards cold stress. This could be due to differences in the fine regulation of transcript accumulation in rice and Arabidopsis. In another study, when ABF2 was overexpressed, the resultant transgenic plants showed tolerance to multiple stresses and altered sensitivity to ABA (Kim et al., 2004). ABF2 overexpression also promoted glucose-mediated inhibition of seedling development. This data together with an analysis of other ABF mutants showed that ABF3 and 4 have a specific role in stress response and ABF2 is also required for glucose response.

A gene encoding homeobox - leucine zipper protein named Hahb-4 was cloned from sunflower and found to be upregulated in response to drought conditions and to ABA. When overexpressed in Arabidopsis, the transgenic plants showed shorter stems and internodes and more compact inflorescence. However, these plants were more tolerant to water stress conditions and produced same seed weight under both, non-stress and stress conditions as compared to wild type plants under normal conditions (Dezar et al., 2005). A novel jasmonate (JA) and ethylene responsive factor, JERF3, which acts as a transcription activator in yeast, was found to be induced in response to ethylene, JA, cold, salt and ABA in tomato (Wang et al., 2004). This factor was found to bind to GCC box, cis element that responds to ethylene and JA and also to DRE element which responds to drought, salinity and cold. Overexpression of JERF3 in tobacco was found to result in the induction of pathogenesis related genes and the plants showed enhanced tolerance towards salinity stress. A novel class of transcription factors called NAC (NAM, ATAF1,2, CUC2) are involved in many diverse plant functions (Olsen et al., 2005). It was recently shown that AtNAC2 might be involved in salinity stress tolerance and responds to auxin and ethylene signaling pathways in addition to ABA signaling. Overexpression of AtNAC2 resulted in the promotion of lateral root development and one of the genes that showed upregulation was found to be glyoxalase I (He et al., 2005). Earlier Fujita et al. (2004) had shown that dehydration induced protein - RD26 is a NAC transcription factor and was shown to transactivate glyoxalase I promoter. Our group has shown that overexpression of glyoxalase I and II can confer salinity stress tolerance (Veena et al., 1999; Singla-Pareek et al., 2003). Taking a clue from the above findings, we feel that manipulation of NAC transcription may turn out to be another important strategy for developing stress tolerant transgenic plants.

Recently, a RING zinc-finger protein has been overexpressed in Arabidopsis (Ko et al., 2006). These proteins have been reported to be having important regulatory roles in the development of a variety of organisms. The protein is only 162 amino acids long with an N-terminal trans-membrane domain and a RING-H2 zinc finger motif located at C-terminus. Microarray analysis has shown that the expression of many of the genes involved in the biosynthesis of plant hormones (e.g. ethylene, brassinosteroid, gibberellic acid) were significantly changed in these transgenic plants.

As mentioned above, under signaling section, CaM overexpression led to an increase in the MYB regulated gene expression. Malik and Wu (2005) overexpressed AtMYB2 in japonica rice under the control of ABA inducible promoter. The transgenic plants showed tolerance against salt stress and exhibited higher biomass together with decreased leakage of ions. Zhang et al. (2005) recently found a novel AP2 domain containing transcription factor from Medicago trunculata. It is one of the longest peptides of all the known AP2/ERF transcription factor family. It was found to be regulated by drought, cold and ABA and involved in the activation of pathway leading to wax production. The gene when overexpressed in M. sativa led to a significant enhancement in the production of wax crystals on the adaxial side of young leaves. The transgenic plants were also found to be drought tolerant as seen by delayed wilting and faster recovery. Similarly, another study involved expression of ERF/AP2-type transcription factor (CaPF1) from Capsicum annuum in pine calli and was reported to counteract the inhibitory effects of salt stress on adventitious shoot formation (Tang et al., 2006a).

3. DEVELOPING TRANSGENIC PLANTS THAT SCAVENGE ROS AND MAINTAIN REDOX STATE

Reactive oxygen species (ROS) were earlier believed to be the toxic by-products of aerobic metabolism. Several antioxidants and antioxidative enzymes of this pathway have been discovered and analyzed in past few years (Mittler, 2002; Foyer and Noctor, 2005; Ogawa, 2005). Recent studies have clearly established the important role of ROS as signaling intermediates in processes such as growth, development and response to biotic and abiotic stresses and programmed cell death (Van Breusegem and Dat, 2006; Gapper and Dolan, 2006; Pitzschke and Hirt, 2006). In this section, we bring out the representative reports where these antioxidative enzymes have been tested directly or indirectly as suitable candidate genes for raising plants with improved tolerance towards abiotic stresses.

3.1. ROS Enzymes

Superoxide dismutase (SOD) is a critical component of the ROS scavenging system in plant cells. Overexpression of Mn-SOD improved drought tolerance in transgenic rice plants (Wang et al., 2005). Similarly, ascorbate peroxidase (APX) cDNA from Arabidopsis was fused to the chloroplast transit peptide of GR and overexpressed in the chloroplasts. The resulting transgenic plants showed tolerance towards salinity as well as water stress by reducing the toxicity caused by the production of H_2O_2 under stress (Badawi et al., 2004). In a recent study, transgenic potato plants overexpressing both SOD and APX showed enhanced tolerance against oxidative stress and high temperature (Tang et al., 2006b). In another report, a Chlamydomonas glutathione peroxidase (GPX) was overexpressed in tobacco either in cytosol or in chloroplast. The transgenic plants showed decreased MDA production under stress and were tolerant to salinity (upto 250 mM NaCl) as well as chilling stress (Yoshimura et al., 2004). It follows that these plants had developed capacity to remove unsaturated fatty acid hydroperoxides generated under stress and thus, were able to maintain membrane integrity.

3.2. GSH

It has been established in literature that GSH plays an important role in antioxidative defense system in plants. Further, an increase in glutathione synthesis as well as GSH/GSSG redox state has been shown to be related to stress tolerance (Tausz et al., 2004). Koh et al. (2006) tested the function of yeast cadmium factor1 (YCF1) in transgenic Arabidopsis. This factor is known to sequester glutathione chelates of heavy metals into vacuoles. In addition to improved tolerance towards heavy metals and xenobiotics, the transgenic plants also showed tolerance towards higher concentrations of NaCl. This tolerance was lost if the plants were treated with BSO, an inhibitor of gamma glutamylcysteine synthase thus showing an important role of glutathione biosynthesis and maintenance of GSH levels in salinity tolerance as well. This is similar to the work of Singla-Pareek et al. (2003, 2006), where it was shown that overexpression of glyoxalases lead to tolerance towards both salinity and heavy metal stress by maintaining GSH homeostasis (Yadav et al., 2005).

In examining the role of ascorbate oxidase (AO) in salt stress tolerance, Yamamoto et al. (2005) examined transgenic tobacco plants expressing AO in either sense or antisense orientation. They also carried out investigations on Arabidopsis AO mutants. Under salt stress conditions, the antisense plants (having only 0.2 fold enzyme activity compared to non-transgenic plants) showed higher seed germination, increased photosynthesis and better seed yields. The phenotype of the

Arabidopsis mutant was similar to the antisense plants. It seems that a decrease in the AO levels leads to low level of H_2O_2 accumulation under salt stress.

4. TRANSGENIC PLANTS WITH HIGHER LEVELS OF OSMOLYTES/COMPATIBLE SOLUTES

Osmotic stresses such as salinity or drought results in cellular dehydration, and plants try to survive under limited availability of water by means of a physiological process of osmo-adaptation that consists of intracellular accumulation of compatible solutes or membrane stabilization. Sugars, polyols, amino acids and diverse substances are synthesized and accumulated intracellularly to counterbalance the osmotic pressure of the environment, and maintain cell turgor. Osmoprotectants such as proline, glycine betaine, mannitol, pinitol, ononitol and trehalose are the common candidates for this purpose of osmo-adjustment in a range of organisms. We present in the following section, some of the representative reports – demonstrating the 'proof of principle', where transgenic plants have been generated with overexpression of one or the other gene of the biosynthesis pathway resulting in hyper accumulation of the respective osmolyte. It is to be mentioned here that, in most of these cases, the source of such genes has been lower organisms (see Table 1).

4.1. Trehalose

Trehalose, a disaccharide, is found in various organisms - mostly in bacteria, algae, fungi, yeast and insects. Trehalose is known to affect sugar metabolism and acts as an osmoprotectant. Its role in stress response was indicated from a study of desiccation tolerant lower plant, Selaginella lepidophylla which accumulated 12% of its dry weight under stress as trehalose (Goddijn and vanDun, 1999). The gene encoding trehalose -6-phosphate synthase (TPS1)- a key enzyme for trehalose biosynthesis, was engineered into tobacco. The transgenic plants exhibited drought tolerance as detached leaves from transformants lost water slowly as compared to control plants (Romero et al., 1997). Cortina and Culianez-Macia (2005) showed that overexpression of yeast TPS1 under the control of CaMV 35S promoter in tomato resulted in enhanced tolerance to drought, salt and oxidative stress. The transgenic plants showed higher chlorophyll and starch content but had thick shoots, erected branches and somewhat aberrant root development. However, this did not affect the overall productivity. The TPS1 gene was also constitutively expressed in potato and the resulting transgenic plants showed increased drought tolerance (Yeo et al., 2000).

Introduction of a gene encoding bifunctional fusion (TPSP) of TPS and T-6-P phosphatase (TPP) from E. coli was expressed in rice under the control of ubiquitin promoter. The trehalose levels were found to increase and the transgenic plants resulted in an increased tolerance to drought, salt and cold without having any growth inhibition (Jang et al., 2003). Earlier, Garg et al. (2002) had shown that

Table 1. Genes used from non-plant sources to transform plants for abiotic stress tolerance

Gene	Source	Recipient	Tolerance	Reference
CodA	A. globiformis	Rice, B. juncea, Tobacco	s	Mohanty et al 2002 Prasad et al., 2000 Lilius et al., 1996
OtsA, B	E. coli	Tobacco, Rice	d, s	Pilon-Smits et al.,1998
TPS/TPP	Yeast E. coli	Rice	d, s	Romero et al., 1997 Garg et al., 2002 Jang et al., 2003
HAL1	Yeast	Tomato, Arabidopsis, Watermelon	s	Gisbert et al., 2000 Yang et al., 2001 Ellul et al., 2003
Invertase	Yeast	Tobacco	s	Fukushima et al., 2001
CaN	Yeast	Tobacco	s	Pardo et al., 1998
Gst/Gpx	E.coli	Tobacco	s	Roxas et al., 1997
EctA, B, C	Halomonas elongata	Tobacco	osmotic	Nakayama et al., 2000
Coq2	Yeast	Tobacco	s, methylviologen	Ohara et al., 2004
MtlD	E. coli	Tobacco, Arabidopsis, Populus	s, Ox.	Tarczynski et al., 1993 Thomas et al., 1995 Shen et al., 1997 Hu et al., 2005
YCF1	Yeast	Arabidopsis	s	Koh et al., 2006
Calmodulin	Bovine	Tobacco	s	Olsson et al., 2004
SAM	Human	Tobacco	s	Waie et al., 2003
CaBP	E. histolytica	Tobacco	s	Pandey et al., 2002
GPX	Chlamydomonas	Tobacco	s, d	Yoshimura et al., 2004
Dehydro ascorbate reductase	Human	Tobacco	s, d	Kwon et al., 2003
AHS(MET25)	Yeast	Tobacco	Ox.	Matiyahu et al., 2006
Calcineurin	Mouse	Rice	S	Ma et al., 2005

overexpression of trehalose biosynthetic genes from E. coli into pusa basmati rice resulted in increased amounts of trehalose and sustainable plant growth under salt and drought conditions.

4.2. Glycine Betaine

Glycine betaine is accumulated in the cells of a number of halophytes and bacteria as an adaptive response to saline and water stress conditions. The bacterial choline oxidase gene (codA) isolated from Arthrobacter globiformis, converts choline into glycine betaine. Overexpression of codA gene has been shown to confer stress tolerance in various plants species such as Arabidopsis, Brassica and rice (Hayashi et al., 1997; Sakamoto et al., 1998; Prasad et al., 2000; Mohanty et al., 2002). This was shown to be due to higher accumulation of glycine betaine. In many of these studies the tolerance was seen at the seed germination and vegetative growth phase. In another study, Sulpice et al. (2003) showed that salt shock to non-transformed plants induced abortion of flower buds by blocking the development of anthers, pistils and petals. In codA overexpressing Arabidopsis, these effects were drastically reduced and flowers, siliques and inflorescence were found to accumulate five-fold higher level of glycine betaine. Thus, it was suggested that increase in glycine betaine could lead to tolerance even at the reproductive phase which in general, is more sensitive to stress. In another study, transgenic rice has been produced by overexpressing the choline monooxygenase gene from spinach (Shirasawa et al., 2006). The transgenic plants accumulated glycine betaine at the level of 0.29–0.43 μM/g d.wt. and had enhanced tolerance to salt stress and temperature stress at the seedling stage.

4.3. Ectoine

Ectoine (1,4,5,6-tetrahydro-2-methyl-4-pyrimidine carboxylic acid) was identified as a compatible solute in Ectothiorhodopsira halochloris, an extremely halophilic phototrophic eubacterium (Galinski et al., 1985). The usefulness of ectoine as an enzyme protectant against heat, freezing, and drying has earlier been demonstrated (Lippert and Galinski, 1992). The biosynthetic pathway of ectoine comprises of three step enzymatic reaction involving ectA, ectB and ectC genes encoding L-2,4-diaminobutyric acid acetyltransferase, L-2,4-diaminobutyric acid transaminase and L-ectoine synthase respectively. To investigate the function of ectoine as a compatible solute in plant cells, the three genes were individually placed under the control of cauliflower mosaic virus 35S promoter and introduced together in cultured tobacco cells. The transgenic cells accumulated small quantity of ectoine and showed increased tolerance to hyperosmotic shock (900 mOsm). The transgenic cells also showed normal growth pattern under hyperosmotic conditions in which growth of the untransformed cells was delayed indicating that ectoine accumulation results in hyperosmotic stress tolerance (Nakayama et al., 2000). In a recent study, Rai et al. (2006) developed transgenic tobacco plants using genes cloned from Marinococcus halophilus in such a way that that all the three enzymes were targeted to chloroplast which in turn showed higher levels of ectoine. Such plants showed high level of salinity and temperature tolerance. Interestingly, these plants also showed an increase in proline, ABA, phenol as well as in the activities of enzymes such as PAL, catalase, polyphenol oxidase and only slight increase in the enzymes involved in ammonia assimilation like GS-GOGAT and a decrease in GDH. The mechanism by which ectoine increase leads to salinity tolerance was also studied by Moghaieb et al. (2004) who developed transgenic tobacco plants overexpressing ect ABC genes taken from Halomonas elongata. Their studies showed that ectoine helped in maintaining root functions so that there is no hindrance in the uptake of water and it also maintains adequate, infact, increased supply of nitrogen. The rate of photosynthesis is also maintained in ectoine producing plants under stress conditions.

4.4. Proline

Many plants and lower organisms accumulate free proline in response to osmotic stress, particularly drought and salinity. Although various studies have focused on the ability of proline as a compatible osmolyte involved in osmotolerance, its specific role throughout plant growth is still unclear. It has been speculated that overproduction of proline in plants may confer tolerance to such stresses. In plant systems, pyrroline-5-carboxylate synthetase (P5CS) catalyses the production of proline from glutamate. Overexpression of P5CS gene in tobacco, rice and potato showed an increase in proline content and improved tolerance to salinity (Kavi Kishore et al., 1995; Zhu et al., 1998). Antisense suppression of proline degradation improved tolerance to freezing and salinity stress in transgenic Arabidopsis plants (Nanjo et al., 1999). Not only this, removal of feedback inhibition of a gene which is involved in the accumulation of proline resulted in increased proline accumulation and protection of plants from osmotic stress (Hong et al., 2000). Thus, these reports strongly suggest the potential application of these osmolytes as suitable candidates for improving stress tolerance.

4.5. Polyamines

Polyamines (spermidine and spermine) are ubiquitous cellular polycations. They are known to play essential role in a variety of plant cellular processes such as the regulation of growth and development, membrane stability, synthesis and functioning of nucleic acids and proteins and protein-DNA interactions. Apart from this, they are also involved in plant responses to abiotic stresses. It has been documented that polyamines accumulate under various stress conditions like water and mineral deficiency, salinity, extreme temperatures, low pH etc. (Galston and Kaur-Sawhney 1990; Rajam 1997). The polycationic nature of polyamines results in their strong binding to active sites of nucleic acids, plasma membrane and phospholipids, thereby stabilizing them. Polyamines may also act as free-radical scavengers and stabilizers of RNAase, protease and other enzymes (Tiburcio et al., 1993; Rajam 1997). Also, excess synthesis of polyamines under stress conditions may be useful to maintain the ionic balance in the cell (Galston and Kaur-Sawhney, 1990). These properties make polyamines a potential candidate for engineering abiotic stress tolerance. A spermidine synthase cDNA that was cloned from Cucurbita ficifolia was overexpressed under the control of 35S promoter and the confirmed transgenic Arabidopsis plants were tested for increased enzyme activity. The T2 and T3 plants showed increased spermidine content, exhibited enhanced tolerance to drought and salinity and in addition, were also tolerant to chilling, freezing, hyperosmosis and paraquat toxicity (Kasukabe et al., 2004). A cDNA microarray analysis of transgenic plants having higher spermidine revealed up-regulation of a number of genes, including DREBs, indicating its role as an important signaling regulator in abiotic stress responses.

4.6. Mannitol

Mannitol-1-phosphate dehydrogenase (mtl1D) is an enzyme that catalyzes the biosynthesis of mannitol from fructose. The gene from E.coli has been used in model plants like tobacco and Arabidopsis to increase mannitol levels and the resulting transgenic plants were found to be tolerant to high salinity and oxidative stress (Tarczynski et al., 1992, 1993; Thomas et al., 1995; Shen et al., 1997). In a recent study, the mtl1D gene was overexpressed in poplar (*Populus tomentosa*) and the transgenic plants were found to accumulate more mannitol as quantified by GC/MS and capillary GC (Hu et al., 2005). In comparison to the untransformed plants, which could tolerate upto 25 mM NaCl in hydroponic cultures, the transgenic plants survived in 75 mM NaCl. However, the transgenic plants showed a decrease in height by about 50% in absence of salt thus indicating towards the unsuitability of this strategy towards genetic modification programs.

5. TRANSGENIC PLANTS FOR TRANSPORTERS/ PUMPS/ CHANNELS

5.1. NHX

The compartmentation of excess Na^+ ions away from the cytosol into the vacuoles is mediated through the vacuolar Na^+/H^+ antiporter. This antiport transports Na^+ into the vacuole using the electrochemical gradient of protons generated by vacuolar H^+ translocating enzymes. Overexpression of the Na^+/H^+ antiporter from Arabidopsis thaliana (AtNHX1) in Arabidopsis promoted sustained growth and development in soil watered with upto 200 mM sodium chloride (Apse et al., 1999). Engineering of the same gene in tomato plants enabled them to grow, flower and produce fruit in the presence of 200 mM sodium chloride without affecting the fruit quality (Zhang and Blumwald, 2001). Overexpression of plasma membrane Na^+/H^+ antiporter from Suaeda salsa in rice markedly enhanced tolerance of transgenic plants to salt stress and water deprivation. The transgenic rice accumulated less sodium in shoots but accumulated more K^+, Ca^{++} and Mg^{++} and showed higher photosynthetic activity and reduced ROS generation (Zhao et al., 2006a).

5.2. Vacuolar Proton Pyrophosphatase (AVP)

It was shown earlier that transgenic plants overexpressing Arabidopsis vacuolar H^+- pyrophosphatase (AVP1) showed higher tolerance to salinity and drought conditions (Gaxiola et al., 2001). In an interesting study, Gaxiola's group (Li et al., 2005) showed that in addition to regulating vacuolar pH, AVP1 also affects auxin transport and fluxes, and hence plant development. In fact, avp1 null mutants showed decreased root and shoot development. With this observation in mind, Park et al. (2005a) overexpressed AVP1 in tomato and found that the transgenic plants had higher root biomass and the plants could recover better from drought stress. These plants maintained higher leaf water potential and take up more water during water deficit period.

5.3. Aquaporins

Aquaporins or water channels are specialized transmembrane proteins found at the vacuolar and the plasma membrane that have a special role to play as far as water movement in the cell is concerned and hence the name. When opened, they facilitate the passive movement of water molecules down a water potential gradient. Aquaporins have been found to be important for the cytosolic osmoregulation. Plants contain a large number of aquaporin isoforms with distinct cell type- and tissue-specific expression patterns. Some of these are constitutively expressed, whereas the expression of others is regulated in response to environmental factors, such as drought and salinity (Johanson et al., 2000). The effectiveness of overexpression of Brassica juncea aquaporin in tobacco was tested for drought tolerance. The overexpressing lines showed tolerance to water stress at the whole plant level and their seeds germinated in soil containing 20% PEG.

5.4. CAX

Cation transport is important for maintaining ion homeostasis in plants. Ca^{++}levels in plants are controlled in part by H^+/Ca^{++} exchangers. A putative cation/proton antiporter was cloned from soybean and overexpressed in Arabidopsis. The transgenic plants accumulated less Na^+ and hence were more tolerant to Na^+ (Luo et al., 2005). Similarly another CAX gene similar to CBF was overexpressed in Arabidopsis and the transgenic plants did not show higher tolerance to salinity or drought but were more tolerant to cold stress (Catala et al., 2003) indicating that CAX also plays important role in processes related to cold adaptations.

6. OTHERS

Apart from the various genes discussed above which have been tested as candidate genes for genetic modification studies, there are other genes which are little less-understood for their involvement in stress responses. The following section present details about some of the recent reports where genetic modification using such genes has been attempted.

A glyceraldehyde-3 phosphate dehydrogenase (GPD) cDNA cloned from an oyster mushroom (*Plerotus sajor-caju*) was found to confer stress resistance in yeast. When the cDNA was introduced in potato and expressed under the constitutive 35S promoter, the transgenic plants showed greater tolerance to salinity (Jeong et al., 2001).

A gene (OSISAP1) was cloned from rice that was found to be upregulated in response to multiple stresses and to ABA. This gene was found to have homology to Zn-finger protein A20 of mammalian origin. When overexpressed in tobacco, it conferred tolerance towards cold, dehydration and salinity stress (Mukhopadhyay et al., 2004). The exact function of this protein is not yet known.

In eukaryotic cells, regulatory proteins such as 14-3-3 are reported to bind to a large number of target proteins phosphorylated at Ser/Thr site (Palmgren, 2001). In Arabidopsis, 14-3-3 family has been documented to have 12 members possessing different affinities for a given target within a cell. An Arabidopsis gene that encodes a 14-3-3 protein was overexpressed in cotton and the transgenic plants analyzed for stress tolerance. Transgenic lines that showed higher expression demonstrated 'stay–green' phenotype and exhibited water stress tolerance (Yan et al., 2004). It was found that higher photosynthetic activity of stay-green plants was due to stomatal conductance probably by regulating the activity of H^+/ATPase.

Ubiquinones (UQ) are electron carriers that show antioxidative property. A yeast gene coq2, encoding p-hydroxybenzoate:polyprenyltransferase, involved in UQ biosynthesis, was overexpressed in tobacco. Transgenic plants showing 3–6 fold increase in UQ showed salinity tolerance and oxidative stress tolerance as caused by methyl viologen (Ohara et al., 2004). This effect was seen irrespective of the fact if the protein was targeted to endoplasmic reticulum or to mitochondria and is attributable to better ROS scavenging ability of the transgenic plants.

A barley gene HVA1 was overexpressed in barley. The third generation transgenic plants exhibited better growth and increase in tolerance to salt stress at 200 mM NaCl. This tolerance was seen for various traits like days to heading, plant height, flag leaf area, root length, panicle length and number of tillers and kernels and kernel yield (Oraby et al., 2005).

The gene encoding small heat shock protein HSP17.6A from Arabidopsis was found to be induced by heat and osmotic stress. Overproduction of this protein could increase salt and drought tolerance in Arabidopsis (Sun et al., 2001). A Nicotiana HSP-70 (NtHSP70–1) was found to be a drought and ABA-inducible gene. The transgenic tobacco plants overexpressing NtHSP70 were found to be tolerant to water stress and with the progression of drought, the retention of optimum water was correlated with the level of the expressed protein (Cho and Hong, 2006).

LEA-type (late embryogenesis abundant) proteins accumulate in wide range of plant species in response to water deficit resulting from desiccation, cold and osmotic stress (Wang et al., 2003; Goyal et al., 2005). Hydrophilicity and heat stability are the notable features of these stress proteins. A model has been recently proposed for LEA proteins where their role as molecular chaperone has been indicated helping the plant to prevent the formation of damaging protein aggregates during water stress (Goyal et al., 2005). Chinese cabbage expressing *B. napus* LEA gene showed enhanced ability to grow under salt and drought stress conditions and also recorded improved recovery upon removal of stress conditions (Park et al., 2005b).

A soybean antiquitin homologue gene, designated as GmTP55, when overexpressed in Arabidopsis and tobacco conferred tolerance to salinity during germination and to water deficit during plant growth (Rodrigues et al., 2006). These transgenic plants also exhibited an enhanced tolerance to H_2O_2 suggesting that antiquitin may be involved in adaptive responses mediated by a physiologically relevant detoxification pathway in plants.

In plant system, RNA helicases play an important role as molecular motors that rearrange RNA secondary structure, potentially performing roles in any cellular process involving RNA metabolism. The role of helicases in response to abiotic stress is only beginning to emerge (Owttrim, 2006). One of the helicase – PDH45, when overexpressed in tobacco, imparted improved tolerance towards salinity stress (Sanan-Mishra et al., 2005). Similarly another helicase, PDH47 has been documented to be upregulated in plants under both salinity and low temperature (Vashisht et al., 2005). However, the exact mechanism of the functioning of helicases in improvement of stress tolerance is not completely understood (Owttrim, 2006)

7. TRANSGENICS WITH GENES FROM SALINE AND DROUGHT TOLERANT PLANTS

A serine –rich protein encoding gene was cloned from Porteresia coarctata that grows under high saline areas and is a relative of rice. The gene was overexpressed in finger millet under the control of rice actin-1 promoter. The transgenic plants were found to grow and set seeds even when grown under 250 mM NaCl (Mahalakshmi et al., 2006).

As mentioned above, AVP gene has been shown to play an important role in stress tolerance. At the Key laboratory of Plant Stress Research, Jinan, China, Guo et al. (2006) cloned and characterized an AVP gene (SsVP) from a halophyte, Suaeda salsa. This plant does not have salt glands or salt bladders and hence it sequesters sodium into vacuoles. The transgenic Arabidopsis overexpressing SsVP showed that both vacuolar ATPase and vacuolar pyrophosphatase activities were higher when plants were grown under 200 mM NaCl and drought stress and the plants showed higher levels of tolerance to both the stresses. In another study, SsNHX 1 gene, encoding vacuolar membrane Na^+/H^+ antiporter, was co-expressed with ArabidopsisAVP1 gene in rice and the resultant plants showed enhanced tolerance to NaCl and recorded higher K^+/Na^+ratio in their shoots (Zhao et al., 2006b). It has now been well demonstrated that one of the key factors for salinity tolerance in plants is the sodium transporter, which was detected as a QTL and then cloned from a salt tolerant rice cultivar named nonabokra (Ren et al., 2005).

A gene (PcINO) encoding inositol synthase was cloned from Porteresia and characterized in comparison to the gene (OsINO) that was cloned from salt sensitive rice variety. The PcINO showed enzyme activity in vitro even in the presence of NaCl and when overexpressed in tobacco and other evolutionary diverse organisms, conferred salt tolerance to the transgenic plants (Majee et al., 2004; Das-Chatterjee et al., 2006). It was later found that PcINO protein had a stretch of 37 amino acids, which was responsible for retaining its enzyme activity in the presence of NaCl (Ghosh-Dastidar et al., 2006).

8. TRANSGENICS TOLERANT TO BOTH ABIOTIC AND BIOTIC STRESSES: CROSS TALK

Plants growing under field conditions continuously experience a multitude of stresses which involve both biotic as well as abiotic agents. Thus, it is quite logical to assume that plants signaling machinery has evolved parallel components to respond to each of these signals. Rapidly accumulating data, resulting from large-scale transcriptome analyses with DNA microarray technology, strongly support the existence of such crosstalk between signaling networks. Biotic and abiotic stresses regulate the expression of different but overlapping suites of genes. It has now been well conceived that there are multiple stress perception and signaling pathways, some of which are specific and others may cross talk at various steps to regulate the expression of genes in response to varied stress signals (Fujita et al., 2006). It is now also being increasingly shown that many of the components involved in abiotic stresses may also be involved in biotic stresses (Chinnusamy et al., 2004).

An activated disease resistance gene (ADR1), that codes for coiled–coil –nucleotide –binding site –leucine rich repeat protein, when overexpressed was found to confer significant drought tolerance (Chini et al., 2004). These plants showed an increase in DREB2A but not DREB1A expression and showed increased sensitivity to thermal and salinity stresses. Microarray also revealed an enhancement in many other known drought tolerant genes.

While a lot of work has been done with transcription factors, the role of transcription co-activators associated with abiotic stresses is not well studied. Recently, a multiprotein bridging factor (MBF1c) was overexpressed in Arabidopsis. It enhanced the tolerance of transgenic plants to osmotic stress, heat stress and also to bacterial infection (Suzuki et al., 2005). It seems that MBF may be involved in ethylene signaling pathway and these can be used to enhance tolerance of plants to different abiotic and biotic stresses.

A ERF/AP2 family of transcription factors was cloned from pepper (CaPF1) and overexpressed in Pinus virginiana. The overexpression led to an increase in the activities of many antioxidant enzymes like APOX, GR, SOD and plants were protected from oxidative damage caused by abiotic factors as well as by pathogens (Tang et al., 2005). A case of reverse cross talk kind of situation was reported by Xiong and Yang, (2003). Transgenic rice plants were generated by overpressing rice MAP kinase (OsMAPK5). Suppressed lines were also generated by using dsRNAi. It was found that suppression of OsMAPK5 led to the induction of PR related genes and plants showed tolerance to fungal and bacterial infection and at the same time had reduction in drought, salt and cold tolerance. The overexpressed lines showed reverse trend and the plants were tolerant to all the three stresses.

Recently, it has been reported that transgenic plants overexpressing chitinase genes which are of fungal origin, show enhanced resistance to both biotic (fungus) as well as abiotic stresses (salinity and heavy metals) supporting that plant machineries taking part in these diverse responses have perhaps co-evolved (de Las Mercedes et al., 2006).

9. TRANSGENICS FOR ABIOTIC STRESSES: FIELD TRIALS

During the last decade, a large body of information has been generated on the molecular biology of stress tolerance. The mechanism of stress perception, signaling pathways and role of participatory components and transcription factors have been elucidated in plants like Arabidopsis. This information is now widely used for other plants, and attempts are being made to develop transgenic crop plants with the genes validated in Arabidopsis (Zhang et al., 2004). As has been discussed in this chapter, in many cases overexpression in crop plants has given reasonable level of tolerance

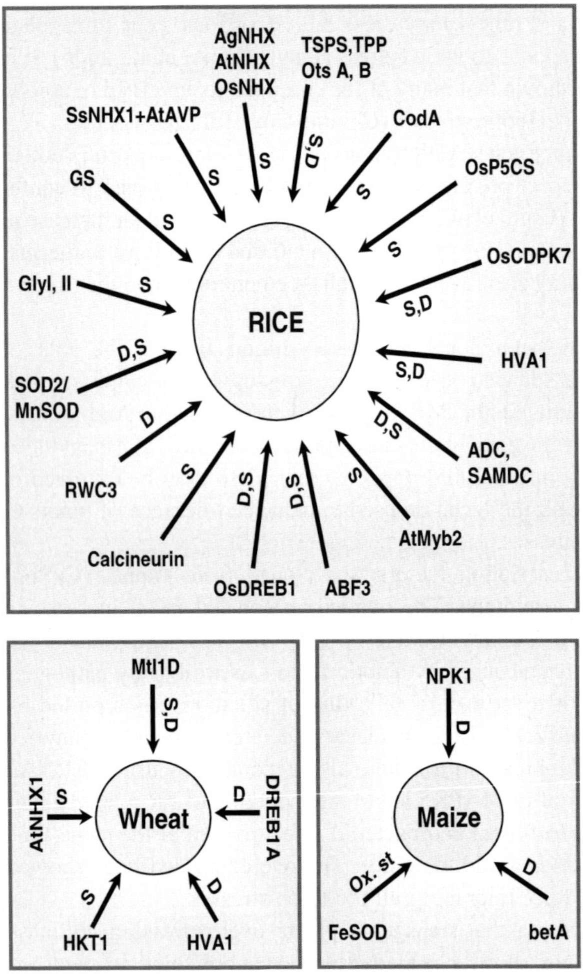

Figure 1. Cartoon summarizing the reports where genetic engineering of monocot plant species has been attempted employing genes belonging to diverse cellular functions. These transgenic plants showed improved tolerance to salinity (S) and/or desiccation (D) or oxidative (Ox. st) stress

towards various abiotic stresses. The contribution of this battery of genes (taking part in diverse cellular functions) towards imparting tolerance to both salinity and drought has been shown in several monocots (Figure 1) and dicot plant species (Figure 2). However, in majority of these reports, the data has been validated at the laboratory or at the best extended to green house level. Presently, field testing of these plants has been carried out only in a limited cases (see Dhlamini et al., 2005 (FAO) and no transgenic crop variety that tolerate abiotic stress has been released for wide scale cultivation with success yet.

Transgenic wheat (*Triticum aestivum*) expressing barley LEA was field evaluated for four seasons. T4 progeny from six independent lines were tested in nine field conditions over six cropping seasons. While some variations were seen between these lines, the grain yield of line 111/1 was significantly higher as also plant height and total biomass under dry land conditions in two of the four locations. Broadly, the experimental data suggests that HVA1 gene has potential to confer drought protection in spring wheat (Bahieldin et al., 2005). In another experiment, Xue et al. (2004) overexpressed AtNHX1 in wheat and found that the transgenic lines showed improved biomass and increased germination rates in severe saline condi-

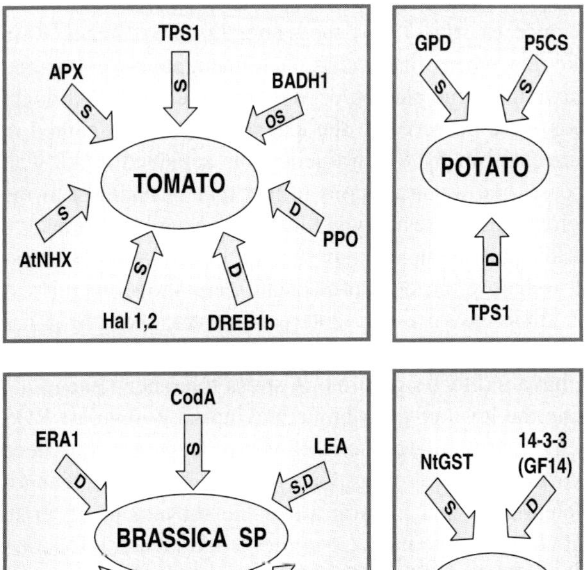

Figure 2. Cartoon summarizing the reports where genetic engineering of dicot plant species has been attempted employing genes belonging to diverse cellular functions. These transgenic plants showed improved tolerance to salinity (S) and/or desiccation (D) or osmotic (OS) stress

tions. When field trials were conducted in soils with electric conductivity of 10.6 and 13.7dSm^{-1}, the transgenics showed higher grain yields. It is not clear why these plants accumulated less Na^+ and higher K^+ since vacuolar NHX was overexpressed.

10. LESSONS LEARNT: FUTURE PERSPECTIVE

10.1. Different Genes: The Same End Result

Research in the last decade has moved extensively to use genetic engineering as a tool to develop stress tolerant crop plants. This followed soon after the molecular understanding of stress tolerance began to be revealed and many potential genes of consequence in abiotic stress tolerance were identified in different systems, especially, Arabidopsis, bacteria and yeast. In fact, a large number of genes from bacteria and yeast have been deployed and overexpressed using 35S promoter to generate transgenic plants that showed tolerance to salinity and drought conditions (Table 1). Similarly, genes from the same or different plant, when overexpressed also conferred stress tolerance as discussed earlier in this chapter. One of the striking observations accrued from this analysis is that despite using different genes (from various pathways or regulatory genes such as transcription factors), the resultant physiological trait (i.e. tolerance to stress) was the same. Different genes were targeted to different crops like rice, wheat, maize (Fig. 1), tomato, potato, cotton and Brassica (Fig. 2) etc. to achieve the same phenotype viz tolerance towards drought or salinity, (though variations were observed to the extent of tolerance gained in the transgenic plants). The mechanism by which tolerance is achieved is not well worked out in most of the cases. However, it does reflect that in addition to the intended effect, other biochemical processes may get influenced which also lead to stress tolerance. To support this statement, we may take an example where transgenic plants over producing trehalose, an osmoprotectant, were shown to possess improved tolerance towards stress (Garg et al., 2002). However, recently it has been found that an increase in trehalose can also induce the expression of many other genes, some of which have indirect/direct role in stress tolerance (Bae et al., 2005). Similarly, manipulating the level of glutathione may not only regulate ROS but can also act as a messenger to regulate other cellular processes. It has been shown recently that overexpression of pyrophosphate gene can lead to changes in the plasma membrane potentials which in turn affect the movement of auxin thus affecting developmental changes, like increase in root length, leading to stress tolerance (Li et al., 2005; Park et al., 2005a). Recently, it has been reported that overexpression of a AP-2 transcription factor from Cicer sp. (CAP2) in tobacco, besides altering the stress response, also showed changes in the expression of genes which are phytohormone regulated (Shukla et al, 2006). These observations clearly indicated possible crosstalk between pathways operating for growth and development along with the stress response. These results suggest that there are different metabolic pathways leading to stress tolerance and overexpression/manipulation of one of the gene can feed into other regulatory and biochemical pathways thus leading to the same 'end result'. It is therefore essential to find out the overall changes that a plant undergoes following overexpression of a gene.

10.2. Overexpression of Stress Responsive Genes May Not Always Impart Stress Tolerance

In most of these reports mentioned in the previous sections, it appears that overexpression of any of the stress inducible gene in transgenic plant may lead to enhanced stress tolerance. However, sometimes it becomes intriguing to find that the gene in question being altered is involved in stress tolerance or itself is an effect of stress? Thus, it is not surprising to find some examples in literature where overexpression of a given gene makes the system more vulnerable to stress. For example, an Arabidopsis glycine rich RNA binding protein (GR-RBP4) was found to be upregulated in response to cold and down regulated under salt and drought conditions. When overexpressed in Arabidopsis, the transgenic plants showed retarded germination under salt and dehydration stress and did not show any cold tolerance either (Kwak et al., 2005). In another recent study, it was found that overexpression of tobacco glutathione-S-transferase (GST) in cotton (*Gossypium hirsutum*) did not show improved tolerance to salinity, chilling or even to herbicides. Infact, these transgenic plants showed higher levels of oxidized glutathione when exposed to salt stress (Light et al., 2005), cautioning that the genetic background of the recipient plant plays an important role in adjusting to the levels of expression of foreign protein and/or the metabolite and in responding to stress conditions.

In another report, an Arabidopsis clone encoding for a nuclear localized calmodulin binding protein (AtCaMBP 25) was found to be upregulated in response to drought, salinity and cold. Transgenic plants that overexpressed AtCaMBP 25 showed increased sensitivity during seed germination under salinity and osmotic stresses. Infact, the antisense plants were found to be more tolerant to mannitol and NaCl stress (Perruc et al., 2004). Further detailed studies need to be carried out to get a possible mechanism leading to these results. The reason for availability of only few reports where overexpression of a gene has been shown to be not leading to tolerance could be due to the fact that many such reports may not have been published.

10.3. Is Threshold Level of Protein Essential for Imparting Tolerance?

Scientific literature is full of reports where a stress-responsive gene from a given plant when put under the control of 35S promoter could lead to improved stress tolerance. This shows that level of a specific protein in a cell is the most crucial factor determining its survival under given set of conditions. On the other hand, excess production of a protein may also lead to heavy energy drain and hence, be unsuitable for genetic modification studies. Keeping this in mind, serious efforts are being channeled to clone and characterize promoters which are inducible by stress signals. It is suggested that strength of the promoter used to derive the gene of interest may be critical in contributing towards the ultimate goal of achieving the desired level of tolerance in the transgenic plants. Differences in promoter strength and specificity could also be the cause for adaptation of plants under stress conditions. This is important especially in cases where the trait being addressed is directly quantitative in nature. Protection from dehydration stress via accumulation of osmolytes is one such case where desired level(s) of osmolytes are very critical for stress tolerance. The fine tuning has to be achieved so that enough desired products are available in the cell without compromising the energy drain or the yield ultimately.

10.4. Allelic Differences May Be Important

Wild species of crop plants represent a potential source of new alleles for improving yield, quality, and stress resistance in cultivated plants. Due to the favourable agricultural practices, the cultivated varieties tend to loose their resistance genes (and hence, their ability to fight with conditions they have not been exposed to) with each generation. Thus these uncultivated relatives of crop plants have become a favourite source in 'gene hunts'. This may also be important in overcoming criticism of the opponents of development of transgenic plants. Recently, it has been argued that transfer of genes from the same species pool could be similar to traditionally bred plants and such plants could be called cisgenic and may receive wider acceptance (Schouten et al., 2006). In a recent report, the PcINOI from local wild salinity resistant rice (Porteresia) has been found to possess a short stretch of 37 amino acids which seems to make the protein more tolerant towards salinity stress (Ghosh-Dastidar et al., 2006). This wild homologue was tested in a range of species and usefulness of the same for conferring stress tolerance was shown (Das-Chatterjee et al., 2006). Overexpression of a serine-rich-protein from Porteresia (PcSrp) in yeast and finger millet improved salinity tolerance (Mahalakshmi et al., 2006). With the similar objectives, comparative analysis between different rice genotypes has also been attempted employing salinity tolerant (CSR27 and Pokkali) and sensitive (PB1) cultivars of rice (Sahi et al., 2003). This study highlighted that genes such as SalT, glycine rich RNA binding proteins, ADP ribosylation factor, NADP dependent malic enzyme, Mub ubiquitin fusion protein, tumor suppressor genes, wound inducible genes, ethylene response element binding protein, alanine aminotransferase, copper chaperone, aspartate aminotransferase, ripening regulated protein, metallothionine and Zn finger transcription factor are important constituents of the rice salt stress response. In another study of almost similar nature, comparative analysis between salt-sensitive rice cultivar IR64 and naturally salt tolerant Pokkali revealed several ESTs specifically induced in higher amounts in the stress tolerant Pokkali rice (Pareek et al., unpublished).

Thellungiella halophila is closely-related to *A. thaliana*. In sharp contrast with Arabidopsis, Thellungiella tolerates extreme cold, drought, and salinity (Bressan et al., 2001; Inan et al., 2004; Taji et al., 2004; Amtmann et al., 2005). It has been noted that this naturally-occurring wild plant remains always "ready" to handle stress by keeping, in anticipation, the levels of stress responsive transcripts higher which are otherwise induced by stress signal in *A. thaliana* (Amtmann et al., 2005).

This again reflects that differences may be due to promoter functioning as discussed earlier. Besides this, differences in transcript stability may also be important. In a recent study, the transcriptome of Yukon ecotype of Thellungiella has been analyzed employing 6578 ESTs, which represented 3628 unigenes from cDNA libraries of cold-, drought-, and salinity stressed plants (Wong et al., 2005). In-depth analysis indicated that of the 140 common unigenes which are present in all the three libraries, 70% have no known functions demonstrating that Thellungiella can be a rich source of genetic information on environmental responses. Thellungiella orthologs of some stress-related Arabidopsis genes showed higher base levels of expression. Thus it is clear from above examples that allelic differences in stress related genes might prove to be useful in achieving our goal towards raising stress tolerant plants. With this view, work has already begun to find out useful differences in the structural aspects of stress induced genes as well as their regulatory machinery.

10.5. Gene Pyramiding: Is It Required?

It has been observed that in many cases, the overall stress tolerance imparted by overexpression of a single gene may not sustain itself under field conditions. Though the transgene would express faithfully, yet in many cases there was only a limited level of tolerance which is sometimes also accompanied by morphological abnormalities. In other cases, the intended effect on the changes in the metabolites was not sufficient to give durable tolerance. In contrast, a recent report has indicated the usefulness of the strategies based on co-expression of more than one gene for improving stress tolerance of plants (Zhao et al., 2006b). Simultaneous expression of the Suaeda salsa Na^+/H^+ antiporter (SsNHX1) and Arabidopsis vacuolar H^+PPase (AVP1) conferred greater tolerance to the transgenic plants than that of the single gene clearly establishing the need for gene pyramiding in this endeavor. In our own study, plants overexpressing both glyoxalaseI and glyoxalaseII performed better under stress conditions than either of the single gene transgenic plants (Singla-Pareek et al., 2003, 2006).

Keeping the above observations in mind, an urgent need for gene stacking is being felt where combination of various different modes and ways of stress tolerance should be sought after. In fact, Claire Halpin, (2005) has emphasized gene stacking as one of the major challenges of the 21st century to improve the growth and yield of plants under abiotic stresses. Moreover, under natural environmental conditions, plants have to face many stresses at the same time or at different developmental stages. The use of multiple stress mechanisms for one or more of the abiotic stresses through stepwise co-transformation or via classical breeding and backcrossing programs may help to achieve high levels of tolerance for commercial cultivation of crop plants. More studies are needed to identify novel and key genes in stress regulation through comparative genomics approaches, or by QTL mapping of stress tolerance. Using this information, one need to develop vectors with right genes whose expression can give tolerance under stress conditions under field environment without major morphological changes and yield losses. Once this is achieved, it will be a major 'step forward' in contributing to increase in food yields and meeting the demands of the growing population.

ACKNOWLEDGEMENTS

The work of the authors referred to in the chapter was supported by grants from the Department of Biotechnology, Govt. of India.

REFERENCES

Agarwal PK, Agarwal P, Reddy MK and Sopory SK. (2006) Role of DREB transcription factors in abiotic and biotic stress tolerance in plants. Plant Cell Rep. 25:1263–1274.

Amtmann A, Bohnert HJ and Bressan RA. (2005) Abiotic stress and plant genome evolution. Search for new models. Plant Physiol. 138:127–130.

Apse MP, Aharon GS, Snedden WA and Blumwald E. (1999) Salt tolerance conferred by overexpression of a vacuolar Na+/H+ antiport in Arabidopsis. Science 285:1256–1258.

Arabidopsis Genome Initiative. (2000) Analysis of the genome sequence of the flowering plant Arabidopsis thaliana. Nature 408:796–815.

Badawi GH, Kawano N, Yamauchi Y, Shimada E, Sasaki R, Kubo A and Tanaka K. (2004) Over-expression of ascorbate peroxidase in tobacco chloroplasts enhances the tolerance to salt stress and water deficit. Physiol. Plant. 121:231–238.

Bae H, Herman EM and Sicher Jr RC. (2005) Exogenous trehalose I induces chemical detoxification and stress response proteins and promotes nonstructural carbohydrate accumulation in Arabidopsis thaliana grown in liquid Culture. Plant Sci.168:1293–1301.

Bahieldin A, Mahfouz HT, Eissa HF, Saleh OM, Ramadan AM, Ahmed IA, Dyer WE, El-Itriby HA and Madkour MA. (2005) Field evaluation of transgenic wheat plants stably expressing the HVA1 gene for drought tolerance. Physiol. Plant. 123: 421–427.

Bajaj S and Mohanty A. (2005) Recent advances in rice biotechnology – towards genetically superior transgenic rice. Plant Biotechnol. J. 3: 275–307.

Bohnert HJ, Gong Q, Li P and Ma S. (2006) Unraveling abiotic stress tolerance mechanisms—getting genomics going. Curr. Opin. Plant Biol. 9:180–188.

Bressan RA, Zhang C, Zhang H, Hasegawa PM, Bohnert HJ and Zhu JK. (2001) Learning from the Arabidopsis experience. The next gene search paradigm. Plant Physiol. 127:1354–1360.

Catala R, Santos E, Alonso JM, Ecker JR, Martinez-Zapater JM and Salinas J. (2003) Mutations in the Ca^{2+}/H^{+} transporter CAX1 increase CBF/DREB1 expression and the cold-acclimation response in Arabidopsis. Plant Cell 15:2940–2951.

Chini A, Grant JJ, Seki M, Shinozaki K and Loake GJ. (2004) Drought tolerance established by enhanced expression of the CC-NBS-LRR gene, ADR1, requires salicylic acid, EDS1 and ABI1. Plant J. 38:810–822.

Chinnusamy V, Schumaker K and Zhu JK. (2004) Molecular genetic perspectives on cross-talk and specificity in abiotic stress signaling in plants. J. Exp. Bot. 55:225–236.

Cho EK and Hong CB. (2006) Over-expression of tobacco NtHSP70–1 contributes to drought-stress tolerance in plants. Plant Cell Rep. 25:349–358.

Cortina C and Culianez-Macia FAB. (2005) Tomato abiotic stress enhanced tolerance by trehalose biosíntesis. Plant Sci. 169:75–82.

Cuartero J, Bolarin MC, Asins MJ and Moreno V. (2006) Increasing salt tolerance in the tomato. J Exp. Bot. 57:1045–1058.

Das-Chatterjee A, Goswami L, Maitra S, Dastidar KG, Ray S and Majumder AL. (2006) Introgression of a novel salt-tolerant L-myo-inositol 1-phosphate synthase from Porteresia coarctata (Roxb.) Tateoka (PcINO1) confers salt tolerance to evolutionary diverse organisms. FEBS Lett. 580: 3980–3988.

De Las Mercedes Dana M, Pintor-Toro JA and Cubero B. (2006) Transgenic tobacco plants overexpressing chitinases of fungal origin show enhanced resistance to biotic and abiotic stress agents. Plant Physiol. Epublished ahead of print.

Dezar CA, Gago GM, Gonzalez DH and Chan RL. (2005) Hahb-4, a sunflower homeobox-leucine zipper gene, is a developmental regulator and confers drought tolerance to Arabidopsis thaliana plants. Trans. Res. 14:429–440.

Dhlamini Z, Spillane C, Moss JP, Ruane J, Urquia N and Sonnino A. (2005) FAO research and technology development service status of research and application of crop biotechnologies in developing countries. Preliminary assessment: Food and Agriculture Organization of the United Nations.

Ellul P, Rios G, Atares A, Roig LA, Serrano R and Moreno V. (2003) The expression of the Saccharomyces cerevisiae HAL1 gene increases salt tolerance in transgenic watermelon [Citrullus lanatus (Thunb.) Matsun. & Nakai.]. Theor. Appl. Genet. 107:462–469.

Foyer CH and Noctor G. (2005) Redox homeostasis and antioxidant signaling: a metabolic interface between stress perception and physiological responses. Plant Cell 17:1866–1875.

Fujita M, Fujita Y, Maruyama K, Seki M, Hiratsu K, Ohme-Takagi M, Tran LP, Yamaguchi-Shinozaki K and Shinozaki K. (2004) A dehydration-induced NAC protein, RD26, is involved in a novel ABA-dependent stress-signaling pathway. Plant J. 39: 863–873.

Fujita M, Fujita Y, Noutoshi Y, Takahashi F, Narusaka Y, Yamaguchi-Shinozaki K and Shinozaki K. (2006) Crosstalk between abiotic and biotic stress responses: a current view from the points of convergence in the stress signaling networks. Curr. Opin. Plant Biol. 9:436–442.

Fukushima E, Arata Y, Endo T, Sonnewald U and Sato F. (2001) Improved salt tolerance of transgenic tobacco expressing apoplastic yeast-derived invertase. Plant Cell Physiol. 42:245–249.

Galinski EA, Pfeiffer HP and Truper HG. (1985) 1,4,5,6-Tetrahydro-2-methyl-4-pyrimidinecarboxylic acid. A novel cyclic amino acid from halophilic phototrophic bacteria of the genus Ectothiorhodospira. Eur. J Biochem. 149:135–139.

Galston AW and Kaur-Sawhney R. (1990) Polyamines in plant physiology. Plant Physiol.94:406–410.

Gapper C and Dolan L. (2006) Control of plant development by reactive oxygen species. Plant Physiol. 141:341–345.

Garg AK, Kim JK, Owens TG, Ranwala AP, Choi YD, Kochian LV and Wu RJ. (2002) Trehalose accumulation in rice plants confers high tolerance levels to different abiotic stresses. Proc. Natl. Acad. Sci. USA 99:15898–15903.

Gaxiola RA, Li J, Undurraga S, Dang LM, Allen GJ, Alper SL and Fink GR. (2001) Drought- and salt-tolerant plants result from overexpression of the AVP1 H^+-pump. Proc. Natl. Acad. Sci. USA. 98:11444–11449.

Ghosh-Dastidar, Maitra S, Goswami L, Roy D, Das KP and Majumder AL. (2006) An insight into the molecular basis of salt tolerance of L-myo-inositol 1-P synthase (PcINO1) from Porteresia coarctata (Roxb.) Tateoka, a halophytic wild rice. Plant Physiol. 40:1279–1296.

Gisbert C, Rus AM, Bolarin MC, Lopez-Coronado JM, Arrillaga I, Montesinos C, Caro M, Serrano R and Moreno V. (2000) The yeast HAL1 gene improves salt tolerance of transgenic tomato. Plant Physiol. 123:393–402

Goddijn OJM and VanDun K. (1999) Trehalose metabolism in plants. Trends Pl. Sci. 4:315–319.

Goff SA, Ricke D, Lan TH, Presting G, Wang R, Dunn M, Glazebrook J, Sessions A, Oeller P, Varma H, Hadley D, Hutchison D, Martin C, Katagiri F, Lange BM, Moughamer T, Xia Y, Budworth P, Zhong J, Miguel T, Paszkowski U, Zhang S, Colbert M, Sun WL, Chen L, Cooper B, Park S, Wood TC, Mao L, Quail P, Wing R, Dean R, Yu Y, Zharkikh A, Shen R, Sahasrabudhe S, Thomas A, Cannings R, Gutin A, Pruss D, Reid J, Tavtigian S, Mitchell J, Eldredge G, Scholl T, Miller RM, Bhatnagar S, Adey N, Rubano T, Tusneem N, Robinson R, Feldhaus J, Macalma T, Oliphant A and Briggs S. (2002) A draft sequence of the rice genome (Oryza sativa L. sp. japonica). Science 296:92–100.

Goyal K, Walton LJ and Tunnacliffe A. (2005) LEA proteins prevent protein aggregation due to water stress. Biochem. J. 388:151–157.

Grennan AK. (2006) Abiotic stress in rice. An "omic" approach. Plant Physiol. 140:1139–1141.

Grover A, Agarwal PK, Kapoor A, Katiyar-Agarwal S and Agarwal M. (2003) Production of abiotic stress tolerant transgenic crops: present accomplishments and future needs. Curr. Sci. 84: 355–367.

Guo Y, Qiu QS, Quintero FJ, Pardo JM, Ohta M, Zhang C, Schumaker KS and Zhu JK. (2004) Transgenic evaluation of activated mutant alleles of SOS2 reveals a critical requirement for its kinase activity and C-terminal regulatory domain for salt tolerance in Arabidopsis thaliana. Plant Cell 16:435–449.

Guo S, Yin H, Zhang X, Zhao F, Li P, Chen S, Zhao Y and Zhang H. (2006) Molecular cloning and characterization of a vacuolar H^+ -pyrophosphatase gene, SsVP, from the halophyte Suaeda salsa and its overexpression increases salt and drought tolerance of Arabidopsis. Plant Mol. Biol.60:41–50.

Hayashi H, Alia, Mustardy L, Deshnium P, Ida M and Murata N. (1997)Transformation of Arabidopsis thaliana with the cod A gene for choline oxidase:accumulation of glycine betaine and enhanced tolerance to salt and cold stress. Plant J.12:133–142.

He XJ, Mu RL, Cao WH, Zhang ZG, Zhang JS and Chen SY. (2005) AtNAC2, a transcription factor downstream of ethylene and auxin signaling pathways, is involved in salt stress response and lateral root development. Plant J. 44:903–916.

Hong Z, Lakkineni K, Zhang Z and Verma DP. (2000) Removal of feedback inhibition of delta(1)-pyrroline-5-carboxylate synthetase results in increased proline accumulation and protection of plants from osmotic stress. Plant Physiol. 122:1129–1136.

Hu L, Lu H, Liu Q, Chen X and Jiang X. (2005) Overexpression of mtlD gene in transgenic Populus tomentosa improves salt tolerance through accumulation of mannitol. Tree Physiol. 25:1273–1281.

Inan G, Zhang Q, Li P, Wang Z, Cao Z, Zhang H, Zhang C, Quist TM, Goodwin SM, Zhu J, Shi H, Damsz B, Charbaji T, Gong Q, Ma S, Fredricksen M, Galbraith DW, Jenks MA, Rhodes D, Hasegawa PM, Bohnert HJ, Joly RJ, Bressan RA and Zhu JK. (2004) Salt stress. A halophyte and cryophyte Arabidopsis relative model system and its applicability to molecular genetic analyses of growth and development of extremophiles. Plant Physiol. 135:1718–1737.

International Rice Genome Sequencing Project (2005) The map-based sequence of the rice genome. Nature 436:793–800.

Jang IC, Oh SJ, Seo JS, Choi WB, Song SI, Kim CH, Kim YS, Seo HS, Choi YD, Nahm BH and Kim JK. (2003) Expression of a bifunctional fusion of the Escherichia coli genes for trehalose-6-phosphate synthase and trehalose-6-phosphate phosphatase in transgenic rice plants increases trehalose accumulation and abiotic stress tolerance without stunting growth. Plant Physiol. 131: 516–524.

Jeong MJ, Park SC and Byun MO. (2001) Improvement of salt tolerance in transgenic potato plants by glyceraldehyde-3 phosphate dehydrogenase gene transfer. Mol. Cell 12:185–189.

Jin Y, Weining S and Nevo E. (2005) A MAPK gene from Dead Sea fungus confers stress tolerance to lithium salt and freezing-thawing: Prospects for saline agriculture. Proc. Natl. Acad. Sci. USA 102:18992–18997.

Johansson I, Karlsson M, Johanson U, Larsson C and Kjellbom P. (2000) The role of aquaporins in cellular and whole plant water balance. Biochim. Biophys. Acta 1465:324–342.

Kasuga M, Liu Q, Miura S, Yamaguchi-Shinozaki K and Shinozaki K. (1999) Improving plant drought, salt, and freezing tolerance by gene transfer of a single stress-inducible transcription factor. Nat. Biotechnol. 17:287–291.

Kasuga M, Miura S, Shinozaki K and Yamaguchi-Shinozaki K. (2004) A combination of the Arabidopsis DREB1A gene and stress-inducible rd29A promoter improved drought and low-temperature stress tolerance in tobacco by gene transfer. Plant Cell Physiol. 45:346–350.

Kasukabe Y, He L, Nada K, Misawa S, Ihara I and Tachibana S. (2004) Overexpression of spermidine synthase enhances tolerance to multiple environmental stresses and up-regulates the expression of various stress-regulated genes in transgenic Arabidopsis thaliana. Plant Cell Physiol. 45:712–722.

Kavi Kishore PB, Hong Z, Miao GH, Hu CA and Verma DPS. (1995) Overexpression of pyrroline-5-carboxylate synthetase increases proline production and confers osmotolerance in transgenic plants. Plant Physiol. 108:1387–1394

Kim S, Kang JY, Cho DI, Park JH and Kim SY. (2004) ABF2, an ABRE-binding bZIP factor, is an essential component of glucose signaling and its overexpression affects multiple stress tolerance. Plant J. 40:75–87.

Ko JH, Yang SH and Han KH. (2006) Upregulation of an Arabidopsis RING-H2 gene, XERICO, confers drought tolerance through increased abscisic acid biosynthesis. Plant J. 47:343–355.

Koh Eun-Ji, Won-Yong Song, Youngsook Lee, Kyoung Heon Kim, Kideok Kim, Namhyun Chung, Kwang-Won Lee, Suk-Whan Hong and Hojoung Lee. (2006) Expression of yeast cadmium factor 1 (YCF1) confers salt tolerance to Arabidopsis thaliana. Plant Sci. 170: 534–541.

Kwak KJ, Kim YO and Kang H. (2005) Characterization of transgenic Arabidopsis plants overexpressing GR-RBP4 under high salinity, dehydration, or cold stress. J Exp. Bot. 56:3007–3016.

Kwon SY, Choi SM, Ahn YO, Lee HS, Lee HB, Park YM and Kwak SS. (2003) Enhanced stress-tolerance of transgenic tobacco plants expressing a human dehydroascorbate reductase gene. J Plant Physiol. 160:347–353.

Li J, Yang H, Peer WA, Richter G, Blakeslee J, Bandyopadhyay A, Titapiwantakun B, Undurraga S, Khodakovskaya M, Richards EL, Krizek B, Murphy AS, Gilroy S and Gaxiola R. (2005) Arabidopsis H^+-PPase AVP1 regulates auxin- mediated organ development. Science 310:121–125.

Light GG, Mahan JR, Roxas VP and Allen RD. (2005) Transgenic cotton (Gossypium hirsutum L.) seedlings expressing a tobacco glutathione S-transferase fail to provide improved stress tolerance. Planta 222:346–354.

Lilius G, Holmberg N and Bulow L. (1996) Enhanced NaCl stress tolerance in transgenic tobacco expressing bacterial choline dehydrogenase. BioTechnol. 14:177–180

Lippert K and Galinski EA. (1992) Enzyme stabilisation by ectoine-type compatible solutes: protection against heating, freezing and drying. Appl. Microbiol. Biotechnol. 37:61–65.

Liu Q, Kasuga M, Sakuma Y, Abe H, Miura S, Yamaguchi-Shinozaki K and Shinozaki K. (1998) Two transcription factors, DREB1 and DREB2, with an EREBP/AP2 DNA binding domain separate two cellular signal transduction pathways in drought- and low-temperature-responsive gene expression, respectively, in Arabidopsis. Plant Cell 10:1391–1406.

Luo GZ, Wang HW, Huang J, Tian AG, Wang YJ, Zhang JS and Chen SY. (2005) A putative plasma membrane cation/proton antiporter from soybean confers salt tolerance in Arabidopsis. Plant Mol. Biol. 59:809–820.

Luo M, Gu SH, Zhao SH, Zhang F and Wu NH. (2006) Rice GTPase OsRacB: potential accessory factor in plant salt-stress signaling. Acta Biochim. Biophys. Sin. (Shanghai). 38:393–402.

Ma X, Qian Q and Zhu D (2005). Expression of a calcineurin gene improves salt stress tolerance in transgenic rice. Plant Mol. Biol. 58:483–495.

Mahalakshmi S, Christopher GS, Reddy TP, Rao KV and Reddy VD (2006) Isolation of a cDNA clone (PcSrp) encoding serine-rich-protein from Porteresia coarctata T. and its expression in yeast and finger millet (Eleusine coracana L.) affording salt tolerance. Planta 224:347–359.

Majee M, Maitra S, Dastidar KG, Pattnaik S, Chatterjee A, Hait NC, Das KP and Majumder AL. (2004) A novel salt-tolerant L-myo-inositol-1-phosphate synthase from Porteresia coarctata (Roxb.) Tateoka, a halophytic wild rice: molecular cloning, bacterial overexpression, characterization, and functional introgression into tobacco-conferring salt tolerance phenotype. J Biol. Chem. 279:28539–28552.

Malik V and Wu R. (2005) Transcription factor AtMyb2 increased salt-stress tolerance in rice (Oryza sativa L.) Rice Genet. Newslett. 22:63.

Marin-Manzano MC, Rodriguez-Rosales MP, Belver A, Donaire JP and Venema K. (2004) Heterologously expressed protein phosphatase calcineurin downregulates plant plasma membrane H^+-ATPase activity at the post-translational level. FEBS Lett. 576:266–270.

Matityahu I, Kachan L, Bar Ilan I and Amir R. (2006) Transgenic tobacco plants overexpressing the Met25 gene of Saccharomyces cerevisiae exhibit enhanced levels of cysteine and glutathione and increased tolerance to oxidative stress. Amino Acids 30:185–194.

Mittler R. (2002) Oxidative stress, antioxidants and stress tolerance. Trends Plant Sci. 7:405–410.

Mittler R. (2006). Abiotic stress, the field environment and stress combination. Trends in Plant Sci. 11: 15–19.

Moghaieb REA, Tanaka N, Saneoka H and Kounosuke F. (2004) Expression of ectoine biosynthetic genes in tobacco plants (Nicotiana tabaccum) leads to the maintenance of osmotic potential under salt stress. 4th International Crop Science Congress.

Mohanty A, Kathuria H, Ferjani A, Sakamoto A, Mohanty P, Murata N and Tyagi AK. (2002) Transgenics of an elite indica rice variety Pusa Basmati 1 harbouring the codA gene are highly tolerant to salt stress. Theor. Appl. Genet. 106:51–57.

Mukhopadhyay A, Vij S and Tyagi AK. (2004) Overexpression of a zinc-finger protein gene from rice confers tolerance to cold, dehydration, and salt stress in transgenic tobacco. Proc. Natl. Acad. Sci. USA 101: 6309–6314.

Nakayama H, Yoshida K, Ono H, Murooka y and Shinmyo A. (2000) Ectoine, the compatible solute of Halomonas elongata, confers hyperosmotic tolerance in cultured tobacco cells. Plant Physiol. 122:1239–1247.

Nanjo T, Kobayashi M, Yoshiba Y, Kakubari Y, Yamaguchi-Shinozaki K and Shinozaki K. (1999) Antisense suppression of proline degradation improves tolerance to freezing and salinity in Arabidopsis thaliana. FEBS Lett.461:205–210.

Ogawa K. (2005) Glutathione-associated regulation of plant growth and stress responses. Antioxid. Redox Signal 7:973–981.

Oh SJ, Song SI, Kim YS, Jang HJ, Kim SY, Kim M, Kim YK, Nahm BH and Kim JK. (2005) Arabidopsis CBF3/DREB1A and ABF3 in transgenic rice increased tolerance to abiotic stress without stunting growth. Plant Physiol. 138:341–351.

Ohara K, Kokado Y, Yamamoto H, Sato F and Yazaki K. (2004) Engineering of ubiquinone biosynthesis using the yeast coq2 gene confers oxidative stress tolerance in transgenic tobacco. Plant J. 40:734–743.

Olsen AN, Ernst HA, Leggio LL and Skriver K. (2005) NAC transcription factors: structurally distinct, functionally diverse. Trends Plant Sci. 10:79–87.

Olsson P, Yilmaz JL, Sommarin M, Persson, S and Bulow L. (2004) Expression of bovine calmodulin in tobacco plants confers faster germination on saline media. Plant Sci. 166:1595–1604.

Oraby HF, Ransom CB, Kravchenko AN and Sticklen MB.(2005) Barley HVA1 gene confers salt tolerance in R3 transgenic oat. Crop Sci. 45:2218–2227.

Owttrim GW. (2006) RNA helicases and abiotic stress. Nuc. Acids Res. 34:3220–3230.

Palmgren MG. (2001) Plant plasma membrane H^+-ATPases: Powerhouses for nutrient uptake. Annu. Rev. Plant Physiol. Plant Mol. Biol. 52:817–845.

Pandey GK, Reddy VS, Reddy MK, Deswal R, Bhattacharya A and Sopory SK. (2002) Transgenic tobacco expressing Entamoeba histolytica calcium binding protein exhibits enhanced growth and tolerance to salt stress. Plant Sci. 162:41–47.

Pardo JM, Reddy MP, Yang S, Maggio A, Huh GH, Matsumoto T, Coca MA, Paino-D'Urzo M, Koiwa H, Yun DJ, Watad AA, Bressan RA and Hasegawa PM. (1998) Stress signaling through Ca^{2+}/calmodulin-dependent protein phosphatase calcineurin mediates salt adaptation in plants. Proc. Natl. Acad. Sci. USA. 95:9681–9686.

Park S, Li J, Pittman JK, Berkowitz GA, Yang H, Undurraga S, Morris J, Hirschi KD and Gaxiola RA. (2005a) Up-regulation of a H^+-pyrophosphatase (H^+-PPase) as a strategy to engineer drought-resistant crop plants. Proc. Natl. Acad. Sci. USA. 102:18830–18835.

Park BJ, Liu Z, Kanno A and Kameya T. (2005b) Transformation of radish (Raphanus sativus L.) via sonication and vacuum infiltration of germinated seeds with Agrobacterium harboring a group 3 LEA gene from B. napus. Plant Cell Rep. 24:494–500.

Perruc E, Charpenteau M, Ramirez BC, Jauneau A, Galaud JP, Ranjeva R and Ranty B. (2004) A novel calmodulin-binding protein functions as a negative regulator of osmotic stress tolerance in Arabidopsis thaliana seedlings. Plant J. 38: 410–420.

Pilon-Smits EAH, Terry N, Sears T, Kim H, Zayed A, Hwang S, Van Dun K, Voogd E, Verwoerd TC, Krutwagen RWHH and Goddijn OJM. (1998)Trehalose-producing transgenic tobacco plants show improved growth performance under drought stress. J Plant Physiol. 152:525–532.

Pitzschke A and Hirt H. (2006) Mitogen-activated protein kinases and reactive oxygen species signaling in plants. Plant Physiol. 141:351–356.

Prasad KVSK, Sharmila P, Kumar PA and Pardha Saradhi P. (2000) Transformation of Brassica juncea L. Czern with bacterial codA gene enhances its tolerance to salt stress. Mol. Breed. 6:489–499.

Rai M, Pal M, Sumesh KV, Jain V and Sankaranarayanan A. (2006) Engineering for biosynthesis of ectoine (2-methyl 4-carboxy tetrahydro pyrimidine) in tobacco chloroplasts leads to accumulation of ectoine and enhanced salinity tolerance. Plant Sci.170:291–306.

Rajam MV. (1997) Polyamines in Plant Ecophysiology pp. 343–374 ed MNV Prasad (New York: John Wiley).

Ren ZH, Gao JP, Li LG, Cai XL, Huang W, Chao DY, Zhu MZ, Wang ZY, Luan S and Lin HX. (2005) A rice quantitative trait locus for salt tolerance encodes a sodium transporter. Nat. Genet.37:1141–1146.

Rodrigues SM, Andrade MO, Gomes AP, Damatta FM, Baracat-Pereira MC and Fontes EP. (2006) Arabidopsis and tobacco plants ectopically expressing the soybean antiquitin-like ALDH7 gene display enhanced tolerance to drought, salinity, and oxidative stress. J Exp. Bot. 57:1909–1918.

Romero C, Belles JM, Vaya JL, Serrano R and Culianez-Macia FA. (1997) Expression of the yeast trehalose-6-phosphate synthase gene in transgenic tobacco plants: pleiotropic phenotypes include drought tolerance. Planta 201:293–297.

Roxas VP, Smith RK, Allen ER and Allen RD. (1997) Overexpression of glutathione S-transferase/glutathione peroxidase enhances the growth of transgenic tobacco seedlings during stress. Nat. Biotechnol. 15:988–991.

Sahi C, Agarwal M, Reddy MK, Sopory SK and Grover A. (2003) Isolation and expression analysis of salt stress-associated ESTs from contrasting rice cultivars using a PCR-based subtraction method. Theor. Appl. Genet. 106:620–628.

Saijo Y, Hata S, Kyozuka J, Shimamoto K and Izui K. (2000) Over-expression of a single Ca^{2+}-dependent protein kinase confers both cold and salt/drought tolerance on rice plants. Plant J. 23:319–327.

Sakamoto A, Alia and Murata N. (1998) Metabolic engineering of rice leading to biosynthesis of glycinebetaine and tolerance to salt and cold. Plant Mol. Biol. 38:1011–1019.

Sanan-Mishra N, Pham XH, Sopory SK and Tuteja N. (2005) Pea DNA helicase 45 overexpression in tobacco confers high salinity tolerance without affecting yield. Proc. Natl. Acad. Sci. USA. 102:509–514.

Schouten HJ, Krens FA and Jacobsen E. (2006) Cisgenic plants are similar to traditionally bred plants: international regulations for genetically modified organisms should be altered to exempt cisgenesis. EMBO Rep. 7:750–753.

Shen V, Jensen RG, and Bohnert HJ. (1997) Increased resistance to oxidative stress in transgenic plants by targeting mannitol biosynthesis to chloroplasts. Plant Physiol. 113:1177–1183.

Shi H, Lee BH, Wu SJ and Zhu JK. (2003) Overexpression of a plasma membrane Na^+/H^+ antiporter gene improves salt tolerance in Arabidopsis thaliana. Nat. Biotechnol. 21:81–85.

Shirasawa K, Takabe T, Takabe T and Kishitani S. (2006) Accumulation of glycinebetaine in rice plants that overexpress choline monooxygenase from spinach and evaluation of their tolerance to abiotic stress. Ann. Bot. (Lond). 98:565–571.

Shukla RK, Raha S, Tripathi V and Chattopadhyay D. (2006) Expression of CAP2, an AP2-family transcription factor from chickpea enhances growth and tolerance to dehydration and salt stress in transgenic tobacco. Plant Physiol. Epublished ahead of print.

Singla-Pareek SL, Reddy MK and Sopory SK. (2001) Transgenic approach towards developing abiotic stress tolerance in plants. Proc. Indian Natn. Sci. Acad. **67**: 265–284.

Singla-Pareek SL, Reddy MK and Sopory SK. (2003) Genetic engineering of the glyoxalase pathway in tobacco leads to enhanced salinity tolerance. Proc. Natl. Acad. Sci. USA 100:4672–14677.

Singla-Pareek SL, Yadav SK, Pareek A, Reddy MK and Sopory SK. (2006) Transgenic tobacco overexpressing glyoxalase pathway enzymes grow and set viable seeds in zinc-spiked soils. Plant Physiol. 140:613–623.

Sreenivasulu N, Altschmied L, Radchuk V, Gubatz S, Wobus U and Weschke W. (2004) Transcript profiles and deduced changes of metabolic pathways in maternal and filial tissues of developing barley grains. Plant J. 37:539–553.

Stockinger EJ, Gilmour SJ and Thomashow MF. (1997) Arabidopsis thaliana CBF1 encodes an AP2 domain-containing transcriptional activator that binds to the C-repeat/DRE, a cis-acting DNA regulatory element that stimulates transcription in response to low temperature and water deficit. Proc. Natl. Acad. Sci. USA. 94:1035–1040.

Sulpice R, Tsukaya H, Nonaka H, Mustardy L, Chen TH and Murata N. (2003) Enhanced formation of flowers in salt-stressed Arabidopsis after genetic engineering of the synthesis of glycine betaine. Plant J. 36:165–176.

Sun W, Bernard C, van de Cotte B, Van Montagu M and Verbruggen N. (2001) At-HSP17.6A, encoding a small heat-shock protein in Arabidopsis, can enhance osmotolerance upon overexpression. Plant J. 27: 407–415

Suzuki N, Rizhsky L, Liang H, Shuman J, Shulaev V and Mittler R. (2005) Enhanced tolerance to environmental stress in transgenic plants expressing the transcriptional coactivator multiprotein bridging factor 1c. Plant Physiol. 139:1313–1322.

Taji T, Seki M, Satou M, Sakurai T, Kobayashi M, Ishiyama K, Narusaka Y, Narusaka M, Zhu JK and Shinozaki K. (2004) Comparative genomics in salt tolerance between Arabidopsis and a Arabidopsis-related halophyte salt stress using Arabidopsis microarray. Plant Physiol. 135:1697–1709.

Tang W, Charles TM and Newton RJ. (2005) Overexpression of the pepper transcription factor CaPF1 in transgenic Virginia pine (Pinus Virginiana Mill.) confers multiple stress tolerance and enhances organ growth. Plant Mol. Biol. 59:603–617.

Tang W, Newton RJ, Lin J and Charles TM. (2006a) Expression of a transcription factor from Capsicum annuum in pine calli counteracts the inhibitory effects of salt stress on adventitious shoot formation. Mol. Gen. Genomics 276:242–253.

Tang L, Kwon SY, Kim SH, Kim JS, Choi JS, Cho KY, Sung CK, Kwak SS and Lee HS. (2006b) Enhanced tolerance of transgenic potato plants expressing both superoxide dismutase and ascorbate peroxidase in chloroplasts against oxidative stress and high temperature. Plant Cell Rep. [Epublished ahead of print]

Tarczynski MC, Jensen RG and Bohnert HJ. (1993) Stress protection of the transgenic tobacco by production of the osmolyte mannitol. Science 259:508–510.

Tarczynski MC, Jensen RG, and Bohnert HJ. (1992)Expression of a bacterial mtlD gene in transgenic tobacco leads to production of accumulation of mannitol. Proc. Natl. Acad. Sci. USA89:2600–2604.

Tausz M, Sircelj H and Grill D. (2004) The glutathione system as a stress marker in plant ecophysiology: is a stress-response concept valid? J Exp. Bot. 55:1955–1962.

Thomas JC, Sepahi M, Arendall B, and Bohnert HJ. (1995) Enhancement of seed germination in high salinity by engineering mannitol expression in Arabidopsis thaliana; Plant Cell Environ. 18:801–806.

Tiburcio AF, Campos JL, Figueras X, and Besford RT.(1993)Recent advances in the understanding of polyamine functions during plant development. Plant Growth Regul.12:331–340.

Umezawa T, Fujita M, Fujita Y, Yamaguchi-Shinozaki K and Shinozaki K. (2006) Engineering drought tolerance in plants: discovering and tailoring genes to unlock the future. Curr. Opin. Biotechnol. 17:113–122.

Valliyodan B and Nguyen HT. (2006) Understanding regulatory networks and engineering for enhanced drought tolerance in plants. Curr. Opin. Plant Biol. 9:189–195.

Van Breusegem F and Dat JF. (2006) Reactive oxygen species in plant cell death. Plant Physiol. 141:384–390.

Vashisht AA, Pradhan A, Tuteja R and Tuteja N. (2005) Cold- and salinity stress-induced bipolar pea DNA helicase 47 is involved in protein synthesis and stimulated by phosphorylation with protein kinase C. Plant J. 44:76–87.

Veena, Reddy VS and Sopory SK (1999). Glyoxalase I from Brassica juncea: molecular cloning, regulation and its over-expression confer tolerance in transgenic tobacco under stress. Plant J. 17: 385–395.

Vinocur B and Altman A. (2005) Recent advances in engineering plant tolerance to abiotic stress: achievements and limitations. Curr. Opin. Biotechnol. 16:123–132.

Waie B and Rajam MV. (2003) Effect of increased polyamine biosynthesis on stress responses in transgenic tobacco by introduction of human S-adenosylmethionine gene. Plant Sci. 164:727–734.

Wang FZ, Wang QB, Kwon SY, Kwak SS and Su WA. (2005) Enhanced drought tolerance of transgenic rice plants expressing a pea manganese superoxide dismutase. J Plant Physiol. 162:465–472.

Wang H, Huang Z, Chen Q, Zhang Z, Zhang H, Wu Y, Huang D and Huang R (2004) Ectopic overexpression of tomato JERF3 in tobacco activates downstream gene expression and enhances salt tolerance. Plant Mol. Biol. 55:183–192.

Wang W, Vinocur B and Altman A. (2003) Plant responses to drought, salinity and extreme temperatures: towards genetic engineering for stress tolerance. Planta 218:1–14.

Wong CE, Li Y, Whitty BR, Diaz-Camino C, Akhter SR, Brandle JE, Golding GB, Weretilnyk EA, Moffatt BA and Griffith M. (2005) Expressed sequence tags from the Yukon ecotype of Thellungiella reveal that gene expression in response to cold, drought and salinity shows little overlap. Plant Mol. Biol. 58:561–574.

Xiong L and Yang V. (2003) Disease resistance and abiotic stress tolerance in rice are inversely modulated by an abscisic Acid-inducible mitogen-activated protein kinase. Plant Cell15:45–59.

Xue, ZY, Zhi, DY, Xue, GP, Zhang, H, Zhao, YX, and Xia, GM (2004) Enhanced salt tolerance of transgenic wheat (Triticum aestivum L.) expressing a vacuolar Na^+/H^+ antiporter gene with improved grain yields in saline soils in the field and a reduced level of leaf Na^+. Plant Sci. 167:849–859.

Yadav SK, Singla-Pareek SL, Reddy MK and Sopory SK. (2005) Transgenic tobacco plants overexpressing glyoxalase enzymes resist an increase in methylglyoxal and maintain higher reduced glutathione levels under salinity stress. FEBS Lett. 579:6265–6271.

Yamaguchi T and Blumwald E. (2005) Developing salt-tolerant crop plants: challenges and opportunities. Trends Plant Sci. 10:615–620.

Yamaguchi-Shinozaki K and Shinozaki K. (2006) Transcriptional regulatory networks in cellular responses and tolerance to dehydration and cold stresses. Annu. Rev. Plant Biol. 57:781–803.

Yamamoto A, Bhuiyan MN, Waditee R, Tanaka Y, Esaka M, Oba K, Jagendorf AT and Takabe T. (2005) Suppressed expression of the apoplastic ascorbate oxidase gene increases salt tolerance in tobacco and Arabidopsis plants. J Exp. Bot. 56:1785–1796.

Yan J, He C, Wang J, Mao Z, Holaday SA, Allen RD and Zhang H. (2004) Overexpression of the Arabidopsis 14-3-3 protein GF14 lambda in cotton leads to a "stay-green" phenotype and improves stress tolerance under moderate drought conditions. Plant Cell Physiol. 45:1007–1014.

Yang SX, Zhao YX, Zhang Q, He YK, Zhang H and Luo D. (2001) HAL1 mediate salt adaptation in Arabidopsis thaliana. Cell Res. 11:142–148.

Yeo ET , Kwon HB, Han SE, Lee JT, Ryu JC and Byu MO. (2000) Genetic engineering of drought resistant potato plants by introduction of the trehalose-6-phosphate synthase (TPS1) gene from Saccharomyces cerevisiae. Mol. Cell 10: 263–268.

Yoo JH, Park CY, Kim JC, Heo WD, Cheong MS, Park HC, Kim MC, Moon BC, Choi MS, Kang YH, Lee JH, Kim HS, Lee SM, Yoon HW, Lim CO, Yun DJ, Lee SY, Chung WSand Cho MJ. (2005) Direct interaction of a divergent CaM isoform and the transcription factor, MYB2, enhances salt tolerance in Arabidopsis. J. Biol. Chem.280:3697–3706.

Yoshimura K, Miyao K, Gaber A, Takeda T, Kanaboshi H, Miyasaka H and Shigeoka S. (2004) Enhancement of stress tolerance in transgenic tobacco plants overexpressing Chlamydomonas glutathione peroxidase in chloroplasts or cytosol. Plant J. 37:21–33.

Yu J, Hu SN, Wang J, Wong GKS, Li SG, Liu B, Deng YJ, Dai L, Zhou Y, Zhang XQ, Cao ML, Liu J, Sun JD, Tang JB, Chen YJ, Huang XB, Lin W, Ye C, Tong W, Cong LJ, Geng JN, Han YJ, Li L, Li W, Hu GQ, Huang XG, Li WJ, Li J, Liu ZW, Li L, Liu JP, Qi QH, Liu JS, Li L, Li T, Wang XG, Lu H, Wu TT, Zhu M, Ni PX, Han H, Dong W, Ren XY, Feng XL, Cui P, Li XR, Wang H, Xu X, Zhai WX, Xu Z, Zhang JS, He SJ, Zhang JG, Xu JC, Zhang KL, Zheng XW, Dong JH, Zeng WY, Tao L, Ye J, Tan J, Ren XD, Chen XW, He J, Liu DF, Tian W, Tian CG, Xia HG, Bao QY, Li G, Gao H, Cao T, Wang J, Zhao WM, Li P, Chen W, Wang XD, Zhang Y, Hu JF, Wang J, Liu S, Yang J, Zhang GY, Xiong YQ, Li ZJ, Mao L, Zhou CS, Zhu Z, Chen RS, Hao BL, Zheng WM, Chen SY, Guo W, Li GJ, Liu SQ, Tao M, Wang J, Zhu LH, Yuan LP and Yang HM. (2002) A draft sequence of the rice genome (Oryza sativa L. sp. indica). Science 296: 79–92.

Zhang HX and Blumwald E. (2001) Transgenic salt-tolerant tomato plants accumulate salt in foliage but not in fruit; Nat. Biotechnol. 19:765–768.

Zhang JY, Broeckling CD, Blancaflor EB, Sledge MK, Sumner LW and Wang ZY. (2005) Overexpression of WXP1, a putative Medicago truncatula AP2 domain-containing transcription factor gene, increases cuticular wax accumulation and enhances drought tolerance in transgenic alfalfa (Medicago sativa). Plant J. 42:689–707.

Zhang JZ, Creelman RA and Zhu JK. (2004) From laboratory to field. Using information from Arabidopsis to engineer salt, cold, and drought tolerance in crops. Plant Physiol. 135:615–621.

Zhao F, Wang Z, Zhang Q, Zhao Y and Zhang H. (2006a) Analysis of the physiological mechanism of salt-tolerant transgenic rice carrying a vacuolar Na^+/H^+ antiporter gene from Suaeda salsa. J. Plant Res. 119:95–104.

Zhao F, Zhang X, Li P, Zhao Y and Zhang H. (2006b) Co-expression of the Suaeda salsa SsNHX1 and Arabidopsis AVP1 confer greater salt tolerance to transgenic rice than the single SsNHX1. Mol. Breed. 17:341–354.
Zhu B, Su J, Chang MC, Verma DPS, Fan YL and Wu R. (1998) Overexpression of a pyrroline-5-carboxylate synthetase gene and analysis of tolerance to water and salt stress in transgenic rice. Plant Sci. 139: 41–48.
Zhu JK. (2002) Salt and drought stress signal transduction in plants. Annu. Rev. Plant Biol. 53:247–273.
Zielinski RE. (1998) Calmodulin and calmodulin-binding proteins in plants. Annu. Rev. Plant Physiol. Plant Mol. Biol. 49:697–725.

CHAPTER 21

BREEDING FOR DROUGHT AND SALT TOLERANT RICE (*ORYZA SATIVA* L.): PROGRESS AND PERSPECTIVES

ZHI-KANG LI[1,2] AND JIAN-LONG XU[2]
[1] *International Rice Research Institute,DAPO Box 7777, Metro Manila, Philippines*
[2] *Institute of Crop Sciences/National Key Facility for Crop Gene Resources and Genetic Improvement, Chinese Academy of Agricultural Sciences*
E-mail: lizhk@caas.net.cn or z.li@cgiar.org

Abstract: Water shortage and salinity are the most important factors limiting rice production worldwide. No drought tolerant (DT) or salt tolerance (ST) rice varieties have been commercially released in the past, due largely to the lack of breeding efforts and partially to the complexity of genetics and physiology underlying DT/ST in rice. The real challenge facing plant breeders is how to efficiently develop high yield and DT or ST cultivars for varied stress scenarios of different rice ecosystems. Progress has been recently made in developing DT/ST rice cultivars using the conventional breeding approach at IRRI and hybrid rice cultivars tend to show high yield potential and good levels of water use efficiency or DT. Tremendous QTL mapping efforts in the past decade have identified numerous QTLs affecting DT/ST in rice, but the results have not let to any successful MAS. A new and promising strategy combining BC breeding with designed QTL pyramiding have been practiced at IRRI and in China, in which exploiting useful genetic diversity for DT/ST from the primary gene pool of rice by BC breeding and developing DT/ST introgression lines in elite genetic backgrounds, discovery, allelic mining and characterization of QTL networks for DT/ST, and directed trait improvement by designed QTL pyramiding are well designed and integrated. Many promising DT and ST rice lines have been developed using this strategy, even though the theoretical aspects underlying this strategy remain to be fully established

Keywords: QTL pyramiding, allelic mining, introgression lines, backcross breeding, drought and salinity tolerance, yield, rice

1. INTRODUCTION

Rice is the staple food for more than 3 billion people in Asia, where more than 90% of the world's rice is produced and consumed. As a semi-aquatic plant species originated from the tropic swamps, rice loves water. Rice production in Asia has

M.A. Jenks et al. (eds.), Advances in Molecular Breeding Toward Drought and Salt Tolerant Crops, 531–564.

been more than tripled in the past three decades, resulting primarily from the "Green Revolution" which dramatically increased the rice productivity in the high-input irrigated systems (Khush, 1999 and 2001). However, rice production requires the use of large amounts of water. Although current high yielding semidwarf rice cultivars have approximately the same water productivity with respect to transpiration efficiency as other C_3 cereals such as wheat, at about 2 kg grains m^{-3} water transpired (Bouman and Tuong, 2001), the total seasonal water input to rice fields is 2–3 times more than for other cereals because of additional water required for land preparation and higher evaporation rates from the water layer in rice fields (Tuong et al., 2005). Thus, water deficiency or drought has been the single largest factor limiting rice yields in approximately 46 million ha of rainfed rice fields of Asia (Pandey 2000). Most modern semi-dwarf cultivars are not adapted to the rainfed systems which are characterized with many abiotic stresses. As a result, low-yielding traditional varieties are still grown in about 50% of the rainfed area of Asia with an average yield of 1–2 t ha^{-1} because of their better adaptation to different stresses and favored grain quality (Mackill et al., 1996). Also, a significant portion of traditionally irrigated rice growing areas in Asia has become rainfed fields as water resources are rapidly diminishing due to rapid population growth, water pollution, industrialization and urbanization. High investment is required to achieve high yields in these areas, reducing rice farmer's already very low income from rice production.

Drought stress is a complex phenomenon and can occur at any time during a cropping season and fluctuates considerably across years and locations in the rainfed areas of Asia. Drought at the early (germination, seedling and tillering) stages causes delayed transplanting (in the rainfed lowlands) or delayed germination (in the uplands) and slowed growth, resulting in poor crop establishment and thus reduces number of panicles per unit area and panicle size. Drought at the reproductive stage (panicle initiation, flowering, and grain filling) causes varied degrees of spikelet sterility and poor grain filling. This latter type of terminal stress tends to cause more severe yield loss because rice is extremely sensitive to drought at the reproductive stage (Cruz and O'Toole, 1984). In the rainfed systems, rice crops may encounter either or both types of drought in a single season, but the occurrence frequency of each drought type tends to show specific patterns in different geographic ecosystems, providing the robust target environments for breeders.

Similarly, salinity is a major growing threat to rice production secondary only to drought. It is estimated that ~10% of the world's croplands are affected by salinity. In Asia, about 49 million ha of lands suitable for rice production remain uncultivated due to saline (Ponnamperuma and Bandyopadhya, 1980). Rice plants are sensitive to salt, particularly at the seedling stage. A low level of salinity at EC 5–6 dSm^{-1} can cause significant yield loss in susceptible rice lines (Pearson et al 1966, Akbar and Yabuno, 1974). In addition, saline soils are characterized by an array of properties such as drought, mineral deficiencies (Zn, P) and toxicities (Fe, Al), etc (Gregorio et al. 2002). Drought often goes hand in hand with salinity in many areas of Asia where irrigation is used to reduce soil salt of rice paddy fields. However, because of the shortage of irrigation water, salinity has become increasingly severe as salt is moving up to soil surfaces in these areas.

The effect of salinity on rice growth depends on many factors such as the plant development stage, salt concentration, duration of salt exposure, soil pH, temperature, humidity and solar radiation. Dissolved salts depress the external water potential and make water less readily available to plants, causing osmotic effect or 'physiological drought'. The ions of salinity also have specific toxic effects, which disturb metabolisms and mineral nutrition or nutrient acquisition (Greenway and Munns 1980). Accumulated evidence indicates that rice is relatively tolerant to salt during germination, becomes very sensitive at the early seedling stage (2 leaf stage), gains tolerance during vegetative growth, becomes sensitive during pollination and fertilization, and then becomes increasingly more tolerant during grain filling (Maas and Hoffman, 1977; Shannon, 1985; Rood, 2000). These results suggest tolerance of a variety may vary considerably at different developmental stages and salinity at the reproductive stage depresses grain yield much more than at the vegetative growth stage (Akbar and Ponnamperuma 1982; Pearson et al. 1966).

2. MECHANISMS OF DROUGHT TOLERANCE AND SALT TOLERANCE IN RICE

To most plant breeders and physiologists, the final and meaningful definition of drought tolerance (DT) or salt tolerance (ST) is the yield loss under stress and this definition will be used throughout the text of this chapter. Rice plants may achieve their adaptation to either drought or salinity by complex mechanisms in both physiology and phenology, which is important to identify specific traits related to DT or ST and develop appropriate screening techniques in breeding programs. For DT, these systems include better water uptake system such as deep and thicker roots (Lafitte et al. 2002), traits that reduce transpiration or nonproductive water loss from shoots such as curticular resistance to water vapor/leaf surface wax (Haque et al. 1992; O'Toole and Cruz, 1983), high water use efficiency/rapid stomatal closure and leaf rolling (Dingkhun et al. 1989), and rapid osmotic adjustment and dehydration tolerance (Lilley and Ludlow 1996). Drought escape by accelerated or delayed flowering under stress may also contribute significantly to rice adaptation to drought depending on specific situations (Lafitte et al. 2004; Xu et al. 2005; Lafitte et al. 2006).

The mechanisms of ST in rice may take place at three levels: the whole plant (Jeschke and Hartung, 2000, Munns et al., 1983), cellular (Munns et al., 1983), and molecular levels (Blumwald, 2000; Munns et al., 2002), and include (1) salt exclusion — plants do not take up excess salt by selective absorption; (2) salt reabsorption — tolerant varieties absorb excess salt but it is reabsorbed from the xylem and Na^+ is not translocated to the shoot; (3) root-shoot translocation — ST is associated with a high electrolyte content in the roots and a low content in the shoot; (4) salt translocation — tolerant plants have the ability to translocate a lesser proportion of Na^+ to the shoot; (5) salt compartmentation — excess salt is transported from younger to older leaves; (6) tissue tolerance — plants absorb salt but are properly compartmentalized in vacuoles within leaves in order to lower the harmful effects on plant growth; and (7) salt dilution — plants take up salt but dilute it by fast growth rate and high water content in the shoot (Yeo and Flowers 1984).

The above mentioned morph-physiological mechanisms underlying DT and ST are important to understand how rice plants adapt to stressful conditions of drought and salinity, but important questions remain regarding how to determine correct target traits and develop appropriate screening techniques in breeding programs to improve DT or ST because there is no convincing evidence that any single DT or ST mechanisms (traits) would be sufficient to confer rice plants' ability to adapt well to very stressful conditions.

3. GENETIC BASES OF DT AND ST IN RICE

3.1. Genetic Basis of DT in Rice

It is well known that significant variation exists among different rice genotypes for DT and its components (Babu et al. 1999, 2001; Price et al. 1997; Lafitte et al. 2006). The complex physiology and phenology involved in DT already imply a complicated genetic basis for this variation in DT. The complex and quantitative nature of DT explains, at least partially, the frustration of breeders in developing DT rice cultivars resulting from the substantial genotype x environment interaction (Fukai and Cooper, 1995; Pantuwan et al. 2002; Lafitte and Courtois, 2002; Kamoshita et al. 2002b). Over the past decade, there have been tremendous efforts to genetically dissect DT and its component traits in rice using QTL mapping approaches. Table 1 shows the reported results in mapping QTLs affecting DT and its components in rice from 31 independent studies on 12 different rice populations which could be summarized in the following four points. First, the number of loci affecting DT and each of its components are very large and widely distributed across the rice genome, but only a few QTLs are detectable in any specific population/environment. Second, most QTLs tend to have varied and inconsistent effects on DT and its components and QTLs with large and consistent effects on DT and related traits are few. Third, individual component traits each contributes little to DT and so for most QTLs affecting DT component traits. Fourth, epistasis, or interactions among QTLs affecting DT and its components, has not been addressed adequately in most studies. Thus, it remains unclear how to apply QTL information from mapping populations to improving DT in breeding populations unrelated to the reference mapping populations because of possible epistasis and QTL-by-environment interaction, uncertain relationships between secondary traits and grain yield under drought, and unknown allelic diversity at identified DT QTLs in parental lines of breeding populations (Li et al. 2000).

Table 1. Summarized results in mapping QTLs affecting DT and its component traits in rice

Trait	N[1]	Pop.[2]	Env.[3]	Type	QTL #	Reference
Tiller and root traits	4	1	2	RIL	18	Champoux et al. 1995
Tiller and root traits	4	1	1	RIL	29	Ray et al. 1996
Osmotic adjustment and DT	1	1	1	RIL	7	Lilly et al. 1996
Root traits	10	2	1	DH	39	Yadav et al. 1997
Root traits	4	2	1	DH	12	Zheng et al. 2000
Drought score	1	2	2	DH	2	Hemamalini et al. 2000
Shoot traits	3	2	2	DH	16	Hemamalini et al. 2000
Root traits	5	2	2	DH	23	Hemamalini et al. 2000
Shoot traits	4	2	3	DH	42	Courtois et al. 2000
Yield and root traits		2	2	DH	?	Venuprasad et al. 2002
Root traits	4	2	2	DH/NIL	9	Shen et al. 2001
Plant height and tillering	2	2	2	DH/NIL	3	Shen et al. 2001
Leaf rolling/stomatal conductance	2	3	1	F_2	8	Price et al. 1997
Root traits	8	3	1	F_2	24	Price and Tomos 1997
Root traits and tillering	4	3	1	RIL	18	Price et al. 2000
Root traits	8	3	2	RIL	24	Price et al. 2002
Dehydration avoidance traits		3	2	RIL	17	Price et al. 2002
Grain yield	1	3	2	RIL	3	Lafitte et al. 2004
Yield components	6	3	2	RIL	48	Lafitte et al. 2004
Heading date, plant height	2	3	2	RIL	15	Lafitte et al. 2004
Root thickness	1	3	2	RIL	2	Lafitte et al. 2004
Biomass	1	3	2	RIL	4	Lafitte et al. 2004
Harvest index	1	3	2	RIL	5	Lafitte et al. 2004
Root traits	5	4	2	RIL	28	Ali et al. 2000
Root traits and shoot biomass	7	4	1	RIL	22	Kamoshita et al. 2002a
Root traits	7	5	1	DH	35	Zhang et al. 2001
Osmotic adjustment	1	5	1	DH	5	Zhang et al. 2001
Cellular membrane stability	1	5	1	DH	9	Tripathy et al. 2000
Shoot biomass and root traits	7	5	4	DH	15	Kamoshita et al. 2002b
Root traits	7	5	1	DH	37	Nguyen 2004
Osmotic adjustment	1	5	1	DH	5	Nguyen 2004
Grain Yield	1	5	1	DH	5	Babu et al. 2003
Relative Yield	1	5	1	DH	2	Babu et al. 2003
Yield components	3	5	1	DH	12	Babu et al. 2003
Heading date and plant height	2	5	1	DH	14	Babu et al. 2003
Shoot traits	4	5	1	DH	8	Babu et al. 2003
Grain yield (DT)	1	5	5	DH	7	Lanceras et al. 2004
Biomass	1	5	5	DH	8	Lanceras et al. 2004
Harvest index	1	5	5	DH	6	Lanceras et al. 2004
Yield components	3	5	5	DH	40	Lanceras et al. 2004

(Continued)

Table 1. (Continued)

Trait	N[1]	Pop.[2]	Env.[3]	Type	QTL #	Reference
Heading date, plant height	2	5	5	DH	16	Lanceras et al. 2004
Relative yield	1	7	2	RIL	4	Yue et al. 2005
Relative spikelet fertility	1	7	2	RIL	5	Yue et al. 2005
Drought respond index	1	7	2	RIL	7	Yue et al. 2005
Leaf traits	3	7	2	RIL	16	Yue et al. 2005
Heading date	1	7	2	RIL	7	Yue et al. 2005
Yield components	6	7	2	RIL	27	Yue et al. 2006
Grain yield	1	7	2	RIL	5	Zou 2005
Yield components	4	7	2	RIL	27	Zou 2005
Relative yield	1	7	2	RIL	3	Yue et al. 2006
Relative yield components	6	7	2	RIL	15	Yue et al. 2006
Root traits	11	7	2	RIL	38	Yue et al. 2006
Leaf drying and rolling	2	7	2	RIL	10	Yue et al. 2006
Osmotic adjustment	1	8	1	BC_3F_3	14	Robin et al. 2003
Heading date and plant height	2	9	2	NIL	26	Xu et al. 2005
Grain Yield	1	9	2	NIL	10	Xu et al. 2005
Yield per plant	1	10	2	BC_2F_2	2	Moncada 2001
Yield components	4	10	2	BC_2F_2	13	Moncada 2001
Heading date and plant height	2	10	2	BC_2F_2	10	Moncada 2001
Leaf size/ABA accumulation	3	11	1	F_2	17	Quarrie et al. 1997
Root traits	7	12	2	RIL	40	Li et al. 2005

[1] N = the number of component traits studied;
[2] Populations: 1 = CO39 × Moroberekan, 2 = IR64 × Azucena, 3 = Azucena × Bala, 4 = IR58821 × IR52561, 5 = CT9993 × IR62266, , 6 = Kalinga III × Azucena, 7 = Zhenshan97 × IRAT109, 8 = IR62266 × IR60080-46A, 9 = Teqing × Lemont, 10 = Caiapo × *O. rufipogon* L., 11 = IR20 × 63–83, 12 = Yuefeng × IRAT109, 13 = IR1552 × Azecena;
[3] the number of environments in which the studies were conducted;
[4] DHL, RIL and NIL represent doubled haploid lines, recombinant inbred lines and near isogenic lines.

3.2. Genetic Basis of ST in Rice

ST of rice is also genetically complex even though there is tremendous genotypic variation for ST and its components in rice germplasm accessions. Classical genetic studies indicate that both genetic (additivity and dominance) and environmental effects are important in the inheritance of ST and related traits in rice (Moeljopawiro and Ikehashi 1981; Akbar et al., 1985; Gregorio and Senadhira, 1993; Lee, 1995; Mishra et al. 1996). Recent results from QTL mapping studies indicate that ST and its components in rice at the seedling stage are involved multiple QTLs (Gong et al., 1998; Gu et al., 2000; Takehisa et al., 2004; Prasad et al., 2000; Flowers et al., 2000; Koyama et al., 2001; Lang et al., 2003; Lin et al., 2004), but single genes/QTLs with large effects on ST were reported in several cases (Zhang et al., 1995; Guo et al., 1997;Fukuda et al., 1999; Bonilla et al. 2002; Lin et al. 2004). One major ST QTL, SKCL, from a japonica line, Nona Bokra, was cloned. This gene turns out to be a protein in the HKT family that exclusively mediates K^+ and Na^+ translocation between roots and shoots, thereby regulates K^+/Na^+ homeostasis in the shoots, resulting in improved ST (Ren et al. 2005).

3.3. Breeding for Improved DT and ST in Rice

3.3.1. Improving DT by the conventional breeding approach

Developing DT rice varieties has long been recognized as the most efficient way to overcome the problem of drought. However, progress in developing DT rice cultivars has been slow. For example, most rice cultivars grown in the rainfed areas of Asia today remain traditional landraces (Pandey, personal communication). There are two major reasons for this. First, rice cultivation has been historically accompanied with steadily improved irrigation and yield potential. Thus, past rice breeding efforts worldwide were largely devoted to increase yield potential under the high input conditions. In other words, breeding for improved DT has largely been neglected in most Asian breeding programs in the past. As a result, most modern high yielding semidwarf rice varieties were poorly adapted to the water-limited conditions of the rainfed systems where low yielding traditional landraces are still widely grown because of their better DT. This situation is changing as breeding for improved DT has recently become the research priority of the International Rice Research Institute (IRRI) (see IRRI Mid-Term Plan of 2007) and many national breeding programs of Asian countries to reduce poverty in the fragile rainfed systems. Second, the complex physiology, phenology and genetics as well as large environmental effects and genotype x environment interaction involved in DT make it difficult to combine high yield potential of the modern rice cultivars with a desirable level of DT through the conventional breeding strategy. For example, different levels of DT are required to achieve the yield stability for different target environments. For most shallow rainfed lowlands of Asia that are characterized with high (or potentially high) productivity but become increasingly in water-deficit or drought prone, a new variety should have a good level of water use efficiency (WUE)/DT at the reproductive stage combined with high yield potential and some other desirable properties such as grain quality and biotic stress resistance in order to be beneficial to farmers in these areas. On the other hand, a good level of tolerance to delayed transplanting plus a high level of DT at the reproductive stage are required for a variety to adapt well to the upper rainfed areas of South/Southeast Asia where drought occurs more frequently. For the upland ecosystem with frequent and severe drought, a new variety should have a high level of DT during the whole life cycle plus excellent resistance to rice blast.

The conventional breeding approach based on line crossing and mass selection remains the predominant method in all rice breeding programs worldwide. The two key elements for success to improve rice DT using the conventional breeding method is to generate sufficient genetic variation for DT in breeding populations and develop a reliable and feasible screening protocol to identify individuals with target traits from large segregating populations. Although yield under stress is used as the target trait for DT by most rice breeders today, this trait itself can result from a wide range of adaptive strategies in breeding populations segregating for flowering time, different types of DT including drought avoidance and drought escape, and the general adaptability to specific environments. Lafitte et al. (2006) tested 166 rice germplasm accessions from worldwide under mild terminal drought in the lowland conditions which reduced, on average, grain yield to 84% of the control value. The tested varieties showed a wide range of yield response to the stress — some lines produced up to 150% as much grain yield under stress as in the control, while others suffered a yield reduction over 90% (Table 2). They found that yield under full irrigation was positively correlated with yield under stress ($r = 0.55$, $P < 0.001$), even though cultivars with greater yield potential tended to be affected more by stress than low potential or poorly adapted cultivars.

In practical breeding, breeders tend to use upland rice landraces, the only ecotype that adapts well to the more extreme drought in the rainfed uplands of Asia, as donor parents for DT in their breeding programs. Line crossing between DT upland ecotypes and high yielding lowland varieties do create tremendous segregation for DT and related traits in breeding populations, as seen in most QTL mapping populations (Table 1), but it is also difficult to break undesirable linkage between DT and poor yield potential associated with most upland landraces, particularly when breeders are targeting at developing high yielding and WUE/DT varieties for the shallow rainfed lowlands.

Table 2. Summarized statistics of the performance of 166 parental lines under continuously flooded lowland conditions (L irr), lowland conditions but with stress imposed near heading (L stress), under upland (aerobic soil) conditions with frequent irrigation (U irr), or under upland conditions with restricted irrigation to impose stress (U stress). Grain yield in the upland experiment is the average measured across the two irrigation regimes (Lafitte et al. 2006)

Water level	Grain yield (g/m^2)		Plant height (cm)		Flowering date (d)	
	Mean ± SD	Range	Mean ± SD	Range	Mean ± SD	Range
Lowland irr	304 ±135	42–667	103 ± 22	40–157	82 ± 12	48–106
Lowland stress	236 ± 114	3–547	99 ± 20	45–145	83 ± 13	50–106
Upland irr			75 ± 15	38–135	86 ± 12	52–116
	50 ± 30*	0–141*				
Upland stress			66 ± 14	45–123	88 ± 12	54–112
% change L	−16 ± 42	−94–152	−3 ± 10	−28–37	1 ± 15	−18–18
% change U**	−83 ± 21	−100–58	−11 ± 10	−30–30	4 ±7	−15–27

*Grain yield reported for the upland experiments is the average of both irrigation levels.

**For yield, this is % change relative to the lowland irrigated lowland. For other traits, this is the % change from the upland irrigated treatment to the upland stress treatment.

At IRRI, efforts have been taken to develop ‘aerobic rice’ that can be grown as a dryland crop in unpuddled fields without standing water, targeting at the large areas of shallow and upper rainfed systems because growing aerobic rice can minimize water losses from seepage and percolation and avoid evaporation from the ponded surface layer (Bouman and Toung 2005). Segregating populations have been created from crosses between high yielding lowland varieties and DT upland varieties, which are typically screened under aerobic conditions during dry-seasons to select progeny that combine the input responsiveness and high harvest index of modern irrigated varieties with the deep roots and vigorous seedling growth of traditional upland types. Table 3 shows the mean grain yields of 40 varieties evaluated under severe, intermittent water stress in replicated yield trials in an upper field and fully irrigated conditions in the transplanted lowland field at IRRI during the 2003 dry-season (Atlin et al. 2004). These lines include 5 check varieties (IR20 and IR64 are two typical lowland drought sensitive varieties; PSBRc 80 is a high yielding line from the IRRI irrigated program with moderate lowland DT; Magat, an IRRI hybrid that released in the Philippines; and Apo, a DT variety released from IRRI upland breeding program), 6 new hybrid lines (names are ended with H) from the IRRI's hybrid rice breeding program, and 29 advanced breeding lines from IRRI's rainfed breeding program. Several interesting results representing the progress in breeding for DT were obtained. First, most advanced lines selected for DT indeed had significantly improved DT (less yield reduction under stress) and on average out-yielded the lowland checks (IR64 and IR20) by 46.2% under stress and 6.8% under non-stress conditions. Of these, IR74963-262-5-1-3-3 is the most promising inbred line which had significantly higher yields than the lowland checks under both stress and non-stress conditions. Second, the overall productivity (averaged over stress and non-stress conditions) of the 7 hybrids out-yielded the selected inbreds by 21.2% under stress and by 20.9% under non-stress conditions, even though their average DT (0.61) was about the same as the selected inbreds. Two hybrid lines, IR80227H and IR80228H are the best for yield under stress and overall productivity. This is interesting because all these hybrids were never selected for improved DT under stress, but for high yield potential under fully irrigated conditions, suggesting the presence of a significant level of heterosis for DT in rice. This result is consistent with the empirical observation in China that hybrid rice cultivars tend to have better DT than most inbreds. Third, phenotypic selection for improved DT did not necessarily incur a yield penalty as the mean yield under stress was positively correlated ($r = 0.49^{**}$) with that under the non-stress conditions (Atlin et al. 2004).

Based on a study on 1 $BC_1F_{2:4}$ population and 4 $F_{2:4}$ populations derived from upland/lowland line crosses, Venuprasad et al. (2006) obtained several important results regarding the efficiency of the conventional breeding approach for improving DT under aerobic and upland conditions (Tables 4 and 5): First, upland rice yield under the reproductive-stage drought stress is a moderately heritable trait, and therefore direct selection for yield under stress is effective. Second, selection under severe managed drought stress in the dry season resulted in yield gains under both

Table 3. Mean grain yield of 42 varieties screened under transplanted lowland conditions at IRRI, 2003 dry-season: severely stressed (stress) and fully irrigated treatments (control) (Singh et al. 2004)

Entry	Yield in stress		Yield in control		Yield reduction (%)		Total productivity	
	kg ha^{-1}	Rank	kg ha^{-1}	Rank	kg ha^{-1}	Rank	kg ha^{-1}	Rank
IR80227H	3967	1	7186	5	0.45	1	11153	1
IR80228H	3461	2	7519	3	0.54	6	10980	2
IR78386H	2437	18	7915	1	0.69	33	10352	3
IR77843H	2010	33	7373	4	0.73	39	9383	6
IR77266H	2494	16	6602	8	0.62	19	9096	11
IR80224H	2511	14	6469	10	0.61	15	8980	13
Magat (hybrid check 5)	2586	12	7659	2	0.66	29	10245	4
Mean	**2780.9**		**7246.1**		**0.61**		**10027**	
IR74963-262-5-1-3-3	2607	11	7053	6	0.63	23	9660	5
IR72176-140-1-2-2-3	2975	4	6372	13	0.53	5	9347	7
IR77298-14-1-2	3143	3	6020	23	0.48	3	9163	8
IR72894-35-2-2-2	2702	7	6434	12	0.58	12	9136	9
IR77298-5-6	2624	9	6494	9	0.60	13	9118	10
IR75298-59-3-1-3	2500	15	6461	11	0.61	16	8961	14
IR72903-121-2-1-2	2254	25	6679	7	0.66	30	8933	15
IR73943-120-5-3-2	2658	8	6165	20	0.57	9	8823	16
IR71604-4-1-4-7-10-2-1-3	2431	19	6353	14	0.62	20	8784	17
IR73008-138-2-2-2	2857	5	5839	29	0.51	4	8696	18

IR74371-46-1-1	2451	17	6205	19	0.60	14	8656	19
IR75298-98-2-3-2	2357	21	6141	21	0.62	21	8498	20
IR69715-72-1-3	2260	24	6222	18	0.64	26	8482	21
IR74293-95-1-1-2-2	2207	28	6260	15	0.65	28	8467	22
IR71726-18-2-1-2	2517	13	5859	27	0.57	10	8376	24
IR73718-3-1-3-3	2252	26	6088	22	0.63	24	8340	25
IR72862-27-3-2-3	2312	23	5902	26	0.61	17	8214	26
IR73013-95-1-3-2	2140	30	6007	24	0.64	27	8147	27
IR73005-23-1-3-3	2246	27	5809	30	0.61	18	8055	28
IR72875-94-3-3-2	2420	20	5529	35	0.56	8	7949	29
IR71700-247-1-1-2	2187	29	5741	31	0.62	22	7928	30
IR73014-59-2-2-2	1886	34	5987	25	0.69	34	7873	31
IR77298-12-7	2314	22	5397	36	0.57	11	7711	32
IR74286-55-2-3-2-3	2067	32	5609	32	0.63	25	7676	33
IR68098-B-78-2-2-B-1	2622	10	4875	40	0.46	2	7497	34
IR74052-54-3-2	1755	35	5554	34	0.68	32	7309	36
IR74714-141-3-3-2-3	1578	37	5379	37	0.71	35	6957	37
IR73009-3-1-1-3	1525	38	5364	38	0.72	38	6889	38
IR74053-144-2-3	859	40	5561	33	0.85	40	6420	40
Mean	**2300.2**		**5977.9**		**0.62**		**8278.1**	
IR20 (check 1)	1525	39	5336	39	0.71	36	6861	39
IR64 (check 2)	1622	36	5857	28	0.72	37	7479	35
PSBRc 80 (check 3)	2772	6	6247	17	0.56	7	9019	12
Apo (DT check 4)	2130	31	6260	16	0.66	31	8390	23
$LSD_{0.05}$	683							

artificially-imposed stress in the dry season and natural stress in the wet season. Third, it is possible to develop genotypes combining good yield potential with improved DT, but the correlation was not high enough that selection under non-stress conditions would result in significant gains in stress environments. Four, potential "spillovers" from selection for yield potential under non-stress conditions

Table 4. Mean yields (g m^{-2}) of random $F_{2:4}$ lines from each of 5 upland/lowland populations and yields expressed as a percentage of the yield of the parents in stress and non-stress environments: IRRI, 2003 dry season (Venuprasad et al. 2006)

Population	Population size	Mean of random lines	Population mean as percentage of:		
			Lowland parent	Upland parent	Relative yield reduction
Stress					
Apo/IR64	215	46	151	49	0.83
Apo/IR72	215	77	314	81	0.67
Vandana/IR64	215	86	200	66	0.24
Vandana/IR72	215	67	237	81	0.64
IR64*2/Azucena	400	39	61	79	0.84
Non-stress					
Apo/IR64	215	265	116	88	
Apo/IR72	215	237	92	93	
Vandana/IR64	215	112	85	449	
Vandana/IR72	215	186	90	323	
IR64*2/Azucena	400	244	91	137	

Table 5. Grain yield (g m^{-2}) of stress-selected (25 highest yielding), non-stress-selected (25 highest yielding), and random lines and checks in Apo/IR64 and Vandana/IR64 populations in selection response trials: IRRI, 2004 dry and wet season (Venuprasad et al. 2006)

	Evaluation environment					
	Non-stress (DS 2004)		Stress (DS 2004)		Natural stress (WS 2004)	
	Apo/IR64	Vandana/IR64	Apo/IR64	Vandana/IR64	Apo/IR64	Vandana/IR64
Lines						
Stress-selected	236	185	17	70*	101*	105
Non-stress-selected	255	217*	13	56	74	92
Random	234	186	18	56	75	98
Parents						
Apo	322	235	18	13	179	73
IR64	346	284	6	0	78	26
Vandana	93	50	83	103	192	206
Trial mean	**242**	**196**	**17**	**60**	**86**	**98**

* Significant difference of a selected set from random set, p = 0.05 level.

are likely to be limited to environments where stress is relatively mild. Five, using a highly tolerant donor appears to be critical to achieve gains in the most stressful environments, at least effective for improving drought tolerance of upland rice. Six, screening breeding lines for yield under both stress and non-stress conditions based on an index combining information from both environments is an effective strategy for developing rice cultivars combining improved DT with acceptable yield potential under favorable conditions.

3.3.2. Improving ST by the conventional breeding approach

Early studies have clearly shown the significant differences for ST among different rice genotypes, even though few rice accessions are known to have a high level of ST (Moeljopawiro and Ikehashi 1981; Akbar et al., 1985; Gregorio and Senadhira, 1993; Lee, 1995; Mishra et al. 1996; Fang et al. 2004). When ST measured as the yield loss under stress, there are two practical problems in accurately assessing rice ST. First, it is difficult to screen ST under the field conditions because environments significantly affect salinity levels under natural conditions. These include seasonal and climate changes, and fine scaled soil heterogeneity (Malcolm 1969; Richards 1983). The idea of using physiological criteria for screening ST has been embraced by many researchers (Epstein et al. 1980; Greenway 1973; Shannon 1985; Tal 1985; Yeo 1994; Rajanaidu and Zakri 1988), yet no ST varieties have been developed and released by using this approach. Thus, screening for ST has been considered to be more reliable and efficient in controlled than field conditions (Chaubey and Senadhira 1994) and a highly efficient technique for screening ST at the seedling stage has been developed at IRRI (Gregorio et al. 1997). However, a second complication arises from the fact that different rice varieties have varied levels of ST at different developmental stages. Zaidem et al. (2004) compared ST of 10 rice varieties at different growth stages under the controlled conditions at IRRI and found that ST at the seedling stage appeared to be independent from ST at the reproductive stage (Table 6). For example, two inbred varieties, IR64 and PSBRc 86, are moderately susceptible to salt at the seedling stage, but relatively more tolerant at the reproductive stage. They further found that the Na-K ratio of rice seedlings contributed only partially to the seedling ST and was independent from yield under stress (Table 7). Thus, screening of ST in breeding programs should be taken in two steps: (1) to screen seedling ST for large segregating populations (early generation screening) under the controlled conditions; and (2) to test ST of promising lines from the first round screening at the reproductive stage, preferably under the field conditions. At IRRI, a natural testing site in Iloilo of Philippines with seawater intrusion during high tides was identified, which offers a variable levels of salinity (between EC 12–30 dSm^{-1}) for screening ST during the whole growth period of rice (Ali et al. 2006).

Progress has been made in developing ST rice varieties using the conventional approach at IRRI and many ST breeding lines have been developed based primarily on ST screening at the seedling stage (Adorada et al. 2004). Evaluation of selected ST lines in a replicated experiment under the non-stress conditions indicates that

Table 6. Comparison of salinity tolerance between seedling, vegetative, and reproductive stages of rice based on the standard evaluation system (SES) and plant height percent reduction (Zaidem et al. 2004)

Variety	Plant height reduction (%)			SES		
	Seedling	Vegetative	Reproductive	Seedling	Vegetative	Reproductive
IR66946-3R-78-1-1	33.51	22.78	29.32	5.67	3.00	5.33
IR66946-3R-178-1-1	30.39	11.32	15.47	4.67	3.00	5.67
IR65192-4B-10-3	34.59	18.16	17.48	5.33	3.33	6.67
PSBRc 50	37.12	11.21	17.06	5.00	4.33	8.00
PSBRc 86	41.05	7.47	6.45	5.33	3.67	5.33
IR63295-AC209-7	25.90	12.83	13.26	3.67	2.33	6.33
IR63307-4B-24-2	11.90	17.81	1.30	5.00	3.00	6.33
IR29	56.50	13.14	31.97	7.67	8.00	8.00
IR64	35.94	11.20	0.00	6.33	5.67	5.33
PSBRc 28	38.78	8.98	6.29	5.33	6.33	6.67

Table 7. Comparison of different rice varieties and breeding lines for salinity tolerance based on dead leaves and shoot Na-K ratio, panicle length, and grain weight at the vegetative and reproductive stages (Zaidem et al. 2004)

Variety	Vegetative stage		Reproductive stage	
	Dead leaves	Na-K ratio	Panicle length (cm)	Grain weight (g)
IR66946-3R-78-1-1	74.00	0.29	22.00	13.56
IR66946-3R-178-1-1	73.00	0.29	22.80	14.25
IR65192-4B-10-3	86.00	0.32	19.75	7.25
PSBRc 50	63.50	0.24	19.80	8.91
PSBRc 86	75.67	0.24	20.33	10.46
IR63295-AC209-7	48.33	0.15	21.87	11.21
IR63307-4B-24-2	49.00	0.22	25.67	14.43
IR29	51.00	0.36	14.10	0.50
IR64	89.33	0.34	21.47	13.18
PSBRc 28	102.00	0.31	13.00	6.00

the selected ST lines show some interesting characteristics (Table 8). For example, the majority of the ST lines showed improved seedling vigor, which is expected from the fact that they were selected for ST at the seedling stage under stress. Also, almost all ST lines had high tiller numbers, resulting probably from the fact that high tiller numbers would allow the ST lines to survive better under stress by reducing their tillers. Otherwise, the ST lines showed considerable variation in plant height, growth duration, panicle exertion, grain type, etc, which allow identification of different ST genotypes to meet phenotypic requirements in different locations and seasons in the target environments.

Table 8. Frequency distribution of agronomic trait ratings on 780 salinity-tolerant lines based on the IRRI Standard Evaluation System (SES)

Agronomic trait[a]	Entries per SES rating[a] (no.)				
	1	3	5	7	9
Vigor (Vg)	278	349	113	37	3
Tillering ability (Tl)	146	277	357	0	0
Plant height (Ht)	582	–	169	–	29
Panicle exertion (Exs)	38	331	332	79	0
Panicle threshability (Thr)	41	73	274	341	51
Maturity (Mat)	181	–	299	–	300
Heading (HD)	Ranging from 34 to 123 days after seeding				
Lodging incidence (Lg)	No lodging incidence (dry season)				

[a] Vg: 1=extra vigorous; 3 = vigorous; 5 = normal; 7 = weak; 9 = very weak. Tl: 1 = more than 25 tillers/plant; 3 = 20–25 tillers; 5 = 10–19 tillers; 7 = 5–9 tillers; 9 = less than 5 tillers. Ht: 1 = semidwarf, less than 110 cm; 5 = intermediate, 110–130 cm; 9 = tall, more than 130 cm. Exs: 1 = well exerted; 3 = moderately exerted; 5 = just exerted; 7 = partly exerted; 9 = Enclosed. Thr: 1 = difficult; 3 = moderately difficult; 5 = intermediate; 7 = loose; 9 = easy. Mat: 1 = ≤115 early; 5 = 116–125 medium; 9 = ≥126 late (Adorada et al. 2004)

3.3.3. Marker-assisted selection(MAS) for improving DT/ST in rice

As mentioned above, the past efforts in identifying QTLs affecting DT/ST and their components were primarily targeting at improving DT/ST by MAS if QTLs affecting secondary traits of DT/ST can be accurately mapped and characterized (Lafitte and Courtois 2000). To date, no DT or ST rice varieties have been developed and released to farmers by MAS even though a few attempts have been made in applying MAS to improving DT or ST in rice. Shen et al. (2001) reported an effort at IRRI to introgress a large segment of rice chromosome 1 containing a putative QTL for deep and thick roots from an upland cultivar, Azucena, into IR64 using MAS and found that the majority of BC progeny carrying the desired introgressions failed to show expected deeper roots than IR64. In a six-year effort, Steele et al. (2006) were able to put 4 QTLs for deep roots from Azucena into an elite cultivar, Kalinga III by MAS, but only 1 of the 4 target QTL expressed the expected effect for increased root length. All these introgressed QTLs have yet to be verified to be associated with DT. Similarly, a major QTL on chromosome 1, *Saltol* which has a large effect on ST at the seedling stage (Bonilla et al. 2002), has been fine-mapped and is being used in MAS to improve ST of important rice cultivars (G. Gregorio, personal communication).

Today, most rice breeders are still reluctant to apply MAS to improving complex traits such as DT and ST in their breeding programs. This is not surprising because, in addition to a relatively high costs, most information is missing for breeders to choose appropriate target QTLs, which includes the magnitudes and consistency of identified QTL(s) in the target genetic backgrounds and environments, and the possible genetic drag associated with target QTL(s).

4. IMPROVING RICE DT AND ST BY BC BREEDING AND DESIGNED QTL PYRAMIDING

Recently, a new strategy- ‘trait improvement by designed QTL pyramiding’, has been successfully applied to combining high yield potential with significantly improved DT/ST in rice as part of the International Rice Molecular Breeding Network coordinated at IRRI (Yu et al. 2003; Lafitte et al. 2006; Li et al. 2005; Li 2006). Technically, this strategy includes 3 major steps: (I) developing introgression lines (ILs) for DT/ST by BC breeding; (II) identifying genes/QTLs and genetic networks for DT/ST by using ILs and DNA markers; and (III) developing DT or ST rice cultivars by designed QTL pyramiding (Figure 1), which are described separately as follows:

4.1. Developing Introgression Lines (ILs) for DT and ST by BC Breeding

In the first step, a large scale BC breeding program was taken to introgress useful genes/QTLs from the primary gene pool of rice into elite genetic backgrounds and develop large numbers of ILs with improved DT/ST. At IRRI, three elite rice lines, IR64 and Teqing (high yielding and widely adaptable indica varieties), and a new plant type (NPT, a high yielding tropical japonica line), were used as the recurrent parents (RPs) and crossed with 195 diverse donors, and backcrossed twice to the RPs to create large numbers of BC_2F_2 bulk populations (Figure 1, Ali et al. 2006). The parental lines of the BC breeding program are originated from 34 countries worldwide and represent a significant portion of the genetic diversity in the primary gene pool of rice according to a survey with 101 well distributed SSR markers (Yu et al. 2003).

For DT, 362 BC_2F_2 bulk populations were screened under 2 types (lowland and upland stress) of severe drought that killed the RPs (Lafitte et al. 2006), resulting in 4669 selected BC_2F_2 plants that showed better DT than the RPs (Table 9). Progeny testing indicated that most selected BC progeny indeed had improved DT as compared with the respective RPs (data not shown). Interestingly, transgressive DT plants were identified from 83.7% of the screened BC populations, including 99.3% of the IR64 populations, 81.0% of the Teqing populations and 65.5% of the NPT populations. The number of survival plants selected from each bulk ranged from 0 to 110 with an average selection intensity of 6.8% (10.6% for the IR64 populations, 3.5% for the Teqing populations, and 4.0% for the NPT populations).

Similarly, 175 BC_2F_2 populations were screened for seedling ST under EC 24–30 d Sm^{-1} in the growth chamber at IRRI, resulting in a total of 1292 surviving BC_2F_2 plants under the salinity stress that killed the RPs (Table 10, Ali et al. 2006). The average selection intensity was 3.95%, 3.69% and 3.40% for IR64, Teqing and NPT BC populations, respectively. Although ST BC progeny were identified in all BC populations, some donors produced more ST plants in all

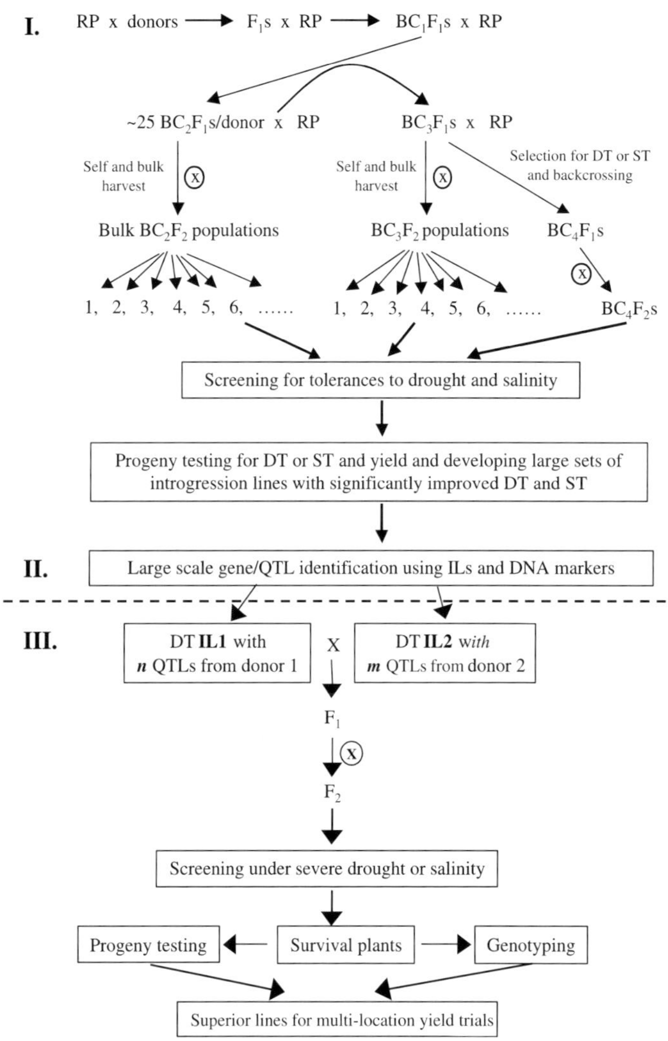

Figure 1. Procedures of backcross breeding procedure (**I**) for developing DT and ST introgression lines (ILs), identifying QTLs/networks (**II**) and developing DT and ST rice cultivars by designed QTL pyramiding (**III**), RP = recurrent parent

three RPs. These donors are OM1706, OM1723, FR13A, Nan29–2, Babaomi, and Khazar. Some donors (BG300, Cisedane, Pahk Maw Peun Meuang and TKM 9) produced more ST plants in the *indica* (IR64 and Teqing) genetic background than the *japonica* background (NPT). Progeny testing of the selected ST BC_2F_3 lines from 68 BC populations indicated that over 90% of the single plant selections from the BC_2F_2 populations indeed showed improved ST (Ali et al, 2006).

Table 9. Summary of the selection experiments of the BC_2F_2 progeny for drought tolerance in two dry seasons at IRRI

Year	Stress	RP	N	No. of survival (selected) plants/population			
				n	Range	Total	SI (%)
2000	LL	IR64	64	64	2–23	583	4.55
2000	UL	IR64	25	25	5–29	326	7.41
2001	LL	IR64	60	59	0–110	2192	18.26
2000	LL	Teqing	31	31	3–15	279	4.50
2000	UL	Teqing	22	22	2–22	182	4.70
2001	LL	Teqing	47	28	0–30	210	2.24
2000	LL	NPT	62	32	0–9	62	0.50
2001	LL	NPT	51	42	0–66	835	8.18
		Total	362	303		4669	6.8

RP, N, n and SI are the recurrent parents, the total number of BC_2F_2 bulk populations screened, the number of BC_2F_2 populations with survival plants, and the selection intensity; LL and UL are the lowland and upland stress conditions. In the lowland condition, the stress was applied at the reproductive stage, and in both seasons, the recurrent parents did not survive. In the upland condition of 2000, IR64 did not survive but Teqing did (those plants performed better than Teqing were selected) (Lafitte et al. 2006).

Table 10. Summary results of BC populations for screening salinity tolerance

Details	BC_2F_2screening				BC_2F_3 progeny testing			
	IR64	Teqing	NPT	Total	IR64	Teqing	NPT	Total
Total BC_2F_2 populations	62	58	55	175	24	34	10	68
No. of selected plants/population	4–12	4–13	1–14	1–14	0–43	0–49	0–11	
Total selected ST BC_2F_3 lines	490	428	374	1292	448	392	21	861
Selection intensity (%)	3.95	3.69	3.40	3.69				
Number of indica donors	47	47	42	136	20	27	7	54
Selected lines	369	345	289	1003	372	269	21	662
Selection intensity (%)	4.39	3.67	3.44	3.69				
Number of japonica donors	9	9	7	25	3	6	1	10
Selected lines	70	66	44	180	43	123	0	166
Selection intensity (%)	3.89	3.67	3.14	3.60				
Number of intermediate donors	4	1	3	8	1	1	2	4
Selected lines	35	5	19	59	33	0	0	33
Selection intensity (%)	4.38	2.50	3.16	3.69				

When tested under the natural field stress of EC 16–18 in Iloilo, Philippines, most selected ST BC progeny survived and set seeds under the field stress conditions. Of these, 22 BC_2F_5 ILs with a ST score of 2–3 at the seedling stage showed a good level of ST at the reproductive stage when the tolerant check, Bicol, all donors and RPs were severely damaged or dead (with SES scores ranging from 7 to 9) (Table 11).

The unique feature of the BC breeding program is that many donors of divergent origins were used and all BC populations were screened for DT or ST regardless the donor performances for the target traits. Progeny testing indicated that most selected BC progeny indeed had improved DT or ST, even though individual selected BC progeny did vary for the level of tolerance. Several important results were obtained. First, there are tremendous amounts of 'hidden' diversity in the primary gene pool of rice for DT and ST, reflected by the fact that BC progeny showing transgressive performance of DT or ST over the parental lines in most BC populations regardless of performances of their donors. In other words, DT or ST genes appear to be widely and randomly distributed in the primary gene pool of rice. Thus, the common practice of selecting donor parents based on phenotype practiced by most breeders is a poor way to exploit this hidden diversity. Second, it was common to identify BC progeny with extreme phenotypes, particularly for ST. Third, the selection efficiency, defined as the number of superior progeny identified per BC population, is highly dependent upon (1) the recipient genetic background (Table 12); (2) the recipient by donor combinations; and (3) the levels of stress applied. More severe stresses could significantly increase the accuracy of selection and reduce the number of total selected plants to a manageable size. Third, the first round selection (screening) should be done at BC_2 instead of BC_3 generation because much reduced selection efficiency was observed in the latter. Fourth, BC breeding combined with direct selection for yield under severe stress is a highly effective way to improve DT/ST in rice because most individuals in a BC population have the same genetic background, and are less affected by the genetic 'noise' from co-segregating non-target traits, such as flowering time and plant size. It is also easy to apply a uniform severe stress at the critical developmental stage(s) and to identify superior BC progeny in direct comparison with RPs.

One potential limitation of this approach is that the level of stress needed to expose genetic variation in single plant screens may be unrealistically severe. Further evaluation of the selected lines has to be conducted to establish gains resulting from the first round screening. In addition, applying an appropriate level of stress for DT/ST remains a major challenge under the field conditions. Nevertheless, the high probability of being able to identify large numbers of DT/ST progeny in our advanced BC populations demonstrated that despite the complex genetics and diverse physiological mechanisms underlying DT and ST, introgression of genes from a diverse source of donors into elite genetic backgrounds through BC breeding and efficient selection, is a powerful way to exploit the hidden diversity for genetic improvement of DT and ST in rice.

Table 11. Performance of 22 promising IR64 introgression lines with significantly improved ST under the saline (EC 16–18) field condition at Iloilo, Philippines (Ali et al. 2006)

ILs	Donor		Seed set (%)	Spikelets/panicle	Grain yield (g)	1000-grain weight (g)	Salinity damage score	
	Name	Origin					Seedling	Maturity
SAT2	Y134	China (I)	84.3	119.8	21.26	21.05	5	5
SAT4	Yue-Xiang-Zan	China (I)	89.0	48.9	9.91	22.78	4	7
SAT5	Zhong413	China (I)	96.0	48.1	8.57	19.66	4	5
SAT9	TKM9	India (I)	89.3	42.1	7.88	20.96	4	3–5
SAT17	TKM 9	India (I)	88.0	35.9	6.10	21.03	4	5
SAT36	STYH	Myanmar (I)	87.9	43.8	7.63	19.82	5	3
SAT39	Bg300	Sri Lanka (I)	86.4	47.1	7.52	18.48	5	3
SAT42	OM997	Vietnam (I)	93.0	35.1	6.31	19.91	5	3–5
SAT43	M401	USA (J)	86.4	40.4	6.79	19.46	5	3–5
SAT50	M401	USA (J)	87.9	41.2	7.97	22.02	4	5
SAT51	M401	USA (J)	81.0	39.7	5.63	17.70	5	3
SAT55	PMPM	Thailand (I)	82.9	34.5	5.53	19.34	5	1–3
SAT56	PMPM	Thailand (I)	81.8	42.9	6.37	18.15	4	1–3
SAT57	PMPM	Thailand (I)	82.0	34.4	5.51	19.54	4	1–3
SAT58	PMPM	Thailand (I)	87.1	41.7	6.97	19.20	3	1–3
SAT59	PMPM	Thailand (I)	83.9	38.6	6.78	20.93	3	1–3
SAT60	PMPM	Thailand (I)	86.7	39.1	7.41	21.86	3	3
SAT61	PMPM	Thailand (I)	88.6	35.9	6.97	21.92	5	3
SAT62	PMPM	Thailand (I)	88.3	37.7	7.10	21.32	4	3–5
SAT63	PMPM	Thailand (I)	85.1	48.9	7.72	18.56	5	3–5
SAT85	93072	China (I)	88.0	41.1	7.20	21.69	5	1–3
SAT87	93072	China (I)	81.1	79.7	12.06	18.67	5	1–3

[a] The recurrent parent, IR64, and all donors had a salinity damage score of 9 at both seedling and final stages of evaluation, and none of them survived the stress. Grain yield was the mean grain weight per plant harvested from 10 plants in the field plot. PMPM = Pahk Maw Peun Meuang, STYH = Shwe-Thwe-Yin-Hyv; I and J are indica and japonica.

Table 12. Genetic background effects on the selection of drought tolerant BC_2F_2 plants under the lowland water stress during the 2000–2001 dry-season (Lafitte et al. 2006)

Donor	Performance (%)[a]	Recurrent parent			Donor	Performance (%)	Recurrent parent		
		IR64	Teqing	NPT			IR64	Teqing	NPT
B4122	−47.6	37	2	1	Hei Mi Chan	46.6	52	2	
Shwewartun	−41.5	5	0	6	Tek Si Chut	−50.0	13	0	
Pokhreli	152.4	119	7	30	Sadajira 19	−100.0	55	0	
Khole marshi	−39.2	84	30	14	Dacca 6	−58.0	20	19	
UPR191-66	−40.6	68	0	24	Zale	−9.8	2	0	
ASD18	20.3	59	0	54	Giza 14	– –	29	4	
IRBB60	−36.0	110	0	66	M202	86.3	46		45
SML242	−81.6	6	0	5	Jumli Marshi	82.6	72		40
Rusty Late	−31.5	38	2	4	Rasi	– –	63		9
CHIPDA	−19.1	47	0	85	Moroberekan	71.8	13		11
Ziri	6.4	10	5	0	TGMS29	33.5	22		3
Vary Lava 16	58.1	24	0	5	Palung 2	−51.5	33		36
LA 110	−13.3	25	5	47	SLG-1	−100.0	26		0
Khumal 4	21.2	0	0	0	Dhan4	−51.5		1	0
Pusa	−33.8	15	4		ASD 16	−9.8		10	0
Guang122	13.7	52	0		Jalmagna	– –		0	28
Minghui63	−27.5	23	0		TKM 6	63.8		11	5
MR 77	0.5	31	0		UP 15	−41.7		22	24
Budda	−18.9	75	11		UZ-Ros 275	50.0		6	37
Doddi	−19.3	81	2		Chorofa	−48.7		1	20
Gajale	–	61	22						
					Mean		**41.6**	**4.9**	**21.4**

[a] Population size was 250 plants per population and donor performance was the percentage of yield reduction under the lowland stress condition.

4.2. Discovering and Mining QTL Alleles for DT and ST using ILs and DNA Markers

As mentioned above, three large sets of ILs with significantly improved DT and ST have been developed at IRRI, which are unique in two aspects. First, all sister lines within a single set of ILs are in the same elite genetic background but each has a few introgressed genomic segments associated with DT or ST from a known donor. These ILs are valuable genetic materials to characterize the effect of specific introgressions on DT/ST and related traits. Second, the three sets of ILs together contain a wide range of DT/ST types and QTLs from many donors of diverse origin, providing a unique set of genome-wide genetic stocks for large-scale QTL/allele discovery and functional genomic research of DT/ST in rice (Li et al. 2005). Third, further improvement of DT/ST can be achieved by designed trait/QTL pyramiding using populations derived from crosses between promising sister ILs carrying different sets of target QTLs, which will be described in the next section.

The second step is to characterize the genomewide introgression patterns in the ILs and identify DT or ST QTLs (donor segments that are responsive to selection for DT or ST) using DNA markers (Figure 1). The principle of using selected ILs and DNA markers to identify and map QTLs affecting DT or ST is straightforward and takes advantage of both linkage mapping and linkage disequilibrium (LD) mapping (Li et al. 2005). Figure 2 shows the introgression pattern and the X^2 profiles along chromosomes 1 and 2, based on SSR marker genotypes, of 38 DT ILs selected under severe lowland (at the reproductive stage) and upland drought from the IR64/Type3 BC_2F_2 population to demonstrate the methodology of QTL identification using ILs. A total of 36 DT QTLs were detected in which the Type3 (donor) allele and genotypic frequencies deviated significantly from the expectations (Table 13). These included 34 QTLs with excessive introgression in ILs selected under both lowland and upland conditions and 2 QTLs on chromosomes 2 and 4 (bins 2.9 and 4.5) specifically detected in the 15 upland selected ILs. Most DT QTLs appeared to be additive because the donor homozygote at these loci was apparently favored by selection, and only four loci (bins 3.8, 4.5, 6.7, and 6.9) appeared to be dominant or partially dominant with both the donor homozygote and heterozygote favored at these loci. Nineteen of the QTLs appeared to have large effects on DT with introgression frequencies >0.55 (Table 13). In addition, 8 QTLs (underlined) were detected in two association loops from LD analyses (Figure 3).

Gametic LD analyses revealed large numbers of non-random associations between or among the introgressed donor loci in the 38 selected DT ILs from the IR64/Type3 BC2F2 population and most of these non-random associations occurred between unlinked loci (Li et al. 2005). Fig. 3 shows two high-confidence genetic networks constructed based on the principle of hierarchy and complete genetic overlap between loci. Fig. 3A is the genetic network constructed based on 244 non-redundant significant Ds between 36 loci detected in the 23 lowland-selected ILs and Fig. 3B is the one built upon 270 non-redundant significant LDs between

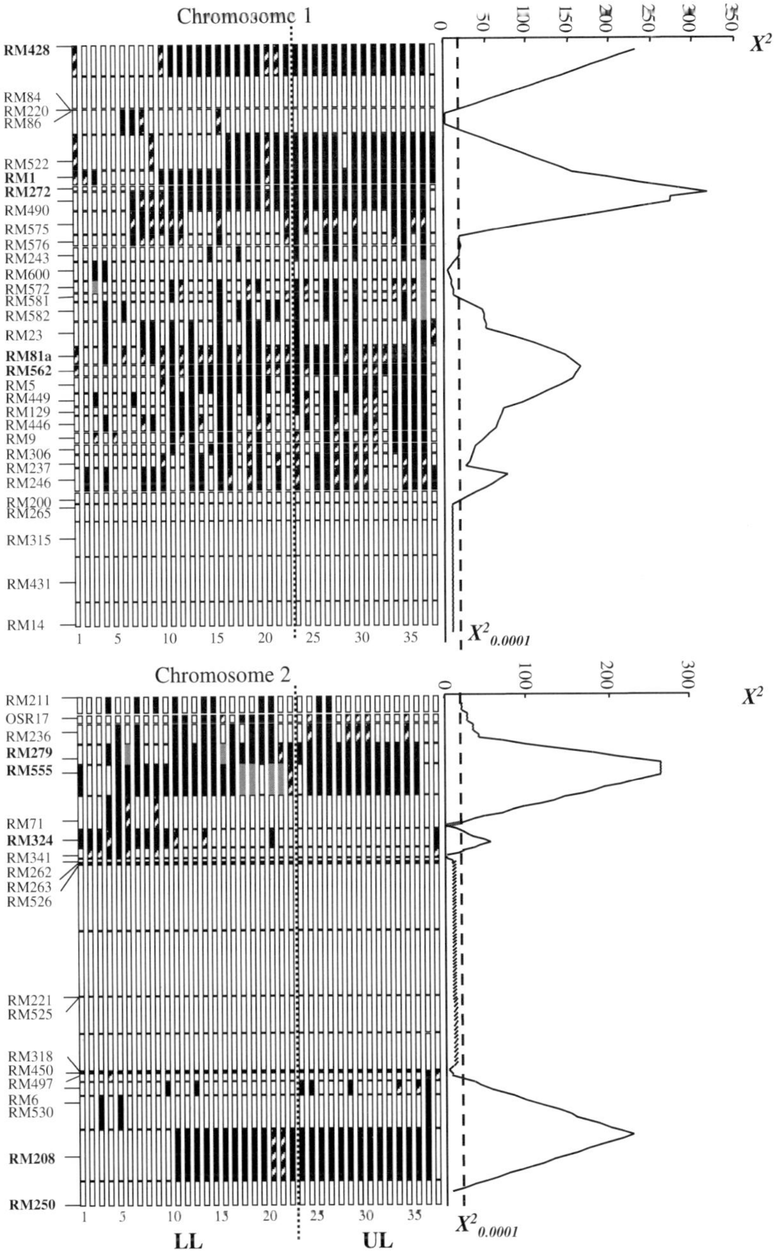

Figure 2. Introgression patterns and chromosomal locations of QTLs for drought tolerance detected by linkage and linkage disequilibrium of SSR markers in 38 introgression lines selected from the IR64 × Type3 BC^2F^2 population which survived very severe lowland drought (LL) and upland drought (UL). X^2 values are obtained from the deviations of the observed genotypic frequencies from the expectations (Table 2)

Table 13. Forty-six QTLs for drought tolerance detected by linkage or linkage disequilibrium in 38 introgression lines selected from the IR64/Type3 BC_2F_2 population under the lowland and upland drought (Li et al. 2005)

QTL[a]			Lowland (n = 23) [b]		Upland (n = 15)		Total (n = 38)		
Marker	Bins	AL	Frequency	X_2^2	Frequency	X^2	Frequency	X_2^2	GA[c]
RM428	1.2	AL1	0.565	70.2	0.933	194.1	0.711	231.1	A
RM490	1.4	AL1	0.652	117.4	1.000	225.0	0.789	318.5	A
RM562	1.6	AL1	0.523	49.6	0.833	118.9	0.649	152.5	A
RM5	1.7	AL1	0.500	68.0	0.667	93.7	0.566	156.2	A
RM246	1.8	–	0.455	45.1	0.533	32.2	0.486	69.1	A
OSR17	2.1	AL1	0.326	23.5	0.300	7.9	0.316	20.8	A
RM555	2.3	AL1	0.826	204.3	0.867	165.6	0.842	368.3	A
RM324	2.5	AL2	0.500	55.0	0.067		0.500	55.0	A
RM6	2.9	–			0.333	10.8	0.333	10.8	A
RM208	2.11	AL1	0.522	67.9	0.933	194.1	0.684	229.9	A
RM81B	3.1	–	0.391	43.8	0.400	30.0	0.395	73.9	A
RM231	3.2	AL1	0.630	100.3	0.933	194.1	0.750	272.9	A
RM411	3.8	–	0.348	23.4	0.500	27.1	0.408	43.6	CD
RM504	3.9	–	0.182	7.8	0.643	80.8	0.361	57.2	A
RM293	3.10	AL1	0.565	99.9	0.867	165.6	0.684	251.3	A
RM85	3.12	–	0.674	118.4	0.500	27.1	0.605	131.4	A
RM307	4.1	AL1	0.435	55.6	0.667	93.7	0.526	140.9	A
RM471	4.4	I	0.565	82.9	0.933	194.1	0.711	250.7	A
RM241	4.5	–			0.500	29.5	0.500	29.5	CD
RM349	4.8	AL1	0.674	118.4	0.933	194.1	0.776	295.6	A
RM122	5.1	AL1	0.413	26.3	0.767	96.8	0.553	102.6	A
RM13	5.2	AL3	0.043						A
RM509	5.4	AL1	0.870	229.7	0.967	194.6	0.908	422.6	A
RM161	5.5	AL3	0.043						A
RM276	6.3	AL1	0.696	136.8	0.933	194.1	0.789	318.5	A
RM527	6.5	AL1	0.261	9.5	0.567	56.8	0.382	50.7	A
RM528	6.7	–	0.391	22.0	0.500	31.2	0.434	46.7	PD
RM340	6.8	AL1	0.326	12.1	0.643	51.9	0.446	51.3	A
RM494	6.9	AL1	0.283	7.3	0.467	23.2	0.355	26.1	PD
RM432	7.3	AL1	0.391	43.8	0.867	165.6	0.579	174.1	A
RM346	7.4	AL2		0.067					A
RM172	7.7	AL1	0.630	117.1	0.933	194.1	0.750	294.9	A
RM408	8.1	AL2	0.130	0.067					A
RM38	8.2	–	0.262	6.1	0.700	93.5	0.433	58.4	A
RM126	8.3	AL1	0.761	159.1	1.000	225.0	0.855	369.0	A
RM331	8.4	–	0.283	16.1	0.433	18.4	0.342	27.4	A
RM223	8.5	AL1	0.478	44.3	0.333	10.8	0.421	52.4	A
RM264	8.8	AL2	0.478	68.9	0.067		0.478	68.9	A
RM321	9.3	AL2		0.067					A
RM242	9.6	AL3	0.043						A
RM271	10.4	AL1	0.370	23.3	0.567	44.1	0.447	62.6	A
RM171	10.5	AL3	0.043						A
RM228	10.6	AL2		0.067					A

RM202	11.3	AL2	0.130		0.067				A
RM206	11.6	AL1	0.500	42.0	0.700	93.5	0.579	116.6	A
									A
RM19	12.2	AL2	0.130		0.067				L

[a] Markers located at the peaks of the X^2 statistics (Figure 2), and the bins each represents a genomic region of approximately 20 cM, in which the number before dot indicating the chromosome and the number after the dot indicating the position of the bin, starting from the top of the chromosome. Grouping of the QTLs were based on the results of LD analyses and underlined markers are those involved in association loops or ALs (Figure. 3A and 3B).

[b] The frequency indicates the frequency of introgression at each marker and X^2 statistics were obtained based on the deviations of the observed genotypic frequencies in the selected ILs from the expectations in a BC_2F_2 population. X_2^2 values at significance levels of P = 0.05, 0.01, 0.001, and 0.0001 are 6.0, 9.2, 13.8 and 18.4.

[c] Gene action is inferred based on the observed genotypic frequencies of the selected ILs, in which additivity (A) is suggested by excess donor homozygote, complete or partial dominance (CD or PD) by excess of both donor homozygote and heterozygote, and overdominance (OD) by excess heterozygote. The underlined markers are those detected by LD analyses at a threshold of $P < 0.0000001$ (Figure. 3).

33 DT loci detected in the 15 upland-selected ILs from the same population. The genetic overlap between the two networks from the independent selection experiments of the same BC population in the lowland and upland drought, measured as the percentage of the same loci in both networks, was 85.1%. These results indicate that large numbers of QTLs are acting in a hierarchical manner in response to the strong directional selection for DT. The strong and positive associations between unlinked DT loci within each of the QTL association loops (ALs) indicate that QTLs are acting in groups and loci within a QTL group were possibly co-regulated in response to selection. This type of multilocus structure and similar genetic networks were detected in the 793 DT ILs selected from 67 BC2 populations (Gao et al. 2007). Further data analyses from progeny testing have detected large effects on multiple phenotypes associated with individual ALs, indicating that these ALs were indeed the targets of selection (data not shown). Thus, identification and characterization of the genetic networks associated with DT and ST should be an important task in future QTL mapping studies.

4.3. Developing DT or ST Rice Cultivars by Designed QTL Pyramiding

In the third step (Figure 1), promising ILs which have the same or better yield potential and unrelated DT/ST QTLs from different donors are identified based on results from steps I and II and used as parents for QTL pyramiding. Crosses are designed and made between promising sister ILs to produce segregating F_2 populations, which are then screened for DT or ST under severe stress to identify superior individuals that have desirable QTLs for DT or ST from 2 different donors and good yield potential. This third step can be repeated to pyramid multiple QTLs from 4 and 8 different donors in the 2nd and 3rd round QTL pyramiding to develop superior DT rice cultivars. At IRRI, 10 F_2 populations developed this way

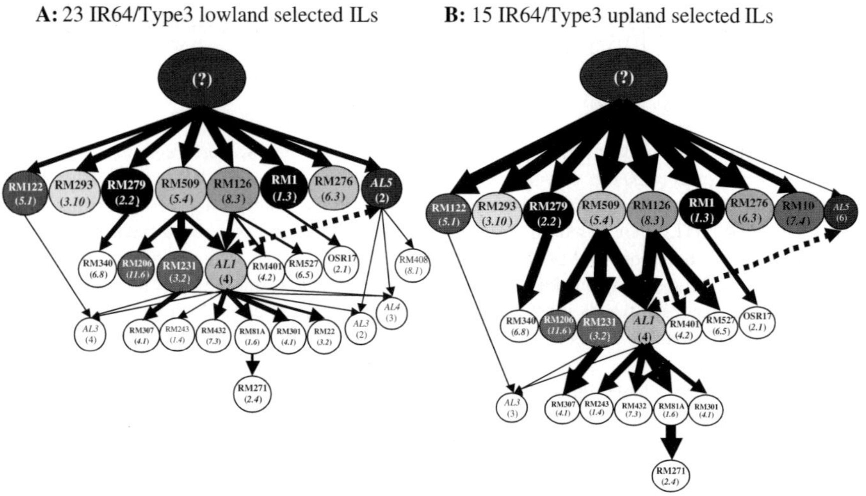

Figure 3. Genetic networks constructed based on the principle of hierarchy and the level of introgression of the DT QTLs and association loops (ALs) detected in 23 lowland selected DT ILs (A) and 15 upland selected ILs B) from the IR64/Type3 BC2F2 population. Here, an AL is defined as a group (2-6) of unlinked but perfectly and positively associated loci in the selected ILs from the BC population (Li et al. 2005). The hierarchy of each network was determined based on the level of introgression in which a hypothetical locus/AL with 100% introgression was put on the top (presumably as genes for signal transduction), followed by loci or ALs with successively decreasing introgression. The arrowed lines in the networks each has three meanings: (1) the level of introgression of the locus or AL it pointed; (2) the level of the absolute overlap (D' = 1.00) between the two loci/ALs it linked; and (3) the inclusive relationship of the locus/AL at the lower level in the upper one. The number under each marker represents the bin it locates (Table 13) and the one under an AL indicates the number of loci it contains (Gao et al. 2007)

were screened under very severe lowland drought, resulting in 560 DT F_3 progeny (Table 14). The selection efficiency of 22.8% in the designed QTL pyramiding F_2 populations was much higher than that (6.8%) in the first round screened BC_2 populations under less severe stress (see Table 9).

Figure 4 shows the genotypic frequencies of 25 DT QTLs segregating in the 25 survival plants from the first F_2 population in Table 14, in which the female IL has 16 DT QTLs from a Bangladesh upland variety, BR24 and the male IL has 9 DT QTLs from a commercial variety, Shwe-Thwe-Yin, from Myanmar. At all QTLs, the observed allelic and genotypic frequencies deviated significantly from the expectations with one allele was significantly more frequent than the other, including 6 loci (QTL # 4, 7, 9, 11, 13, and 14) at which one of the alleles was fixed (Figure 4). All QTLs, except QTL 20 on chromosome 10, showed excess homozygosity and the heterozygote was virtually eliminated at most loci. On average, each of the selected F_2 lines had 22.5 DT QTLs, ranging from12 to 25. LD analyses again detected many highly significant ($P < 0.001$) non-random associations between the segregating DT QTLs, which resulted in a genetic network

Table 14. Ten pyramiding F_2 populations from crosses between 14 promising DT IR64 introgression lines from which 560 surviving F_2 plants carrying DT QTLs from 2 different donors were selected under severe lowland drought during the 2002–2003 dry season

Cross	Female introgression line			Male introgression line			F_2 population	
	Code	Donor name[a]	Origin	Code	Donor	Origin	N	N
1	1	STY (I)	Myanmar	5	BR24 (I)	Bangladesh	237	25
2	1	STY (I)	Myanmar	6	BR24 (I)	Bangladesh	190	55
3	2	BR24 (I)	Bangladesh	10	Zihui100 (I)	China	299	30
4	3	BR24 (I)	Bangladesh	11	Binam (J)	Iran	318	90
5	3	BR24 (I)	Bangladesh	12	OM1723 (I)	Vietnam	305	105
6	4	BR24 (I)	Bangladesh	12	OM1723 (I)	Vietnam	248	55
7	4	BR24 (I)	Bangladesh	11	Binam (J)	Iran	154	30
8	7	Type3 (I)	India	13	Haoannong (J)	China	255	70
9	8	Type3 (I)	India	10	Zihui100 (I)	China	235	70
10	9	Zihui100 (I)	China	14	Haoannong (J)	China	219	30

[a] STY = Shwe-Thwe-Yin, and I and J are indica and japonica, respectively.

consisting of all 25 loci in a hierarchial way, indicating the presence of complex epistatic relationships among many of the DT QTLs (Figure 5).

Progeny testing in replicated experiments in 2004 indicated that on average, the 25 pyramiding F_4 lines were flowering 3 days earlier and yielded 2.84 times as much as IR64 under severe terminal stress. Under the non-stress conditions, these lines were 2 days earlier heading and 2.5 cm taller than IR64, and yielded the same as IR64 (Table 15). However, there was a considerable variation among the 25 lines under both stress and non-stress conditions, which led to identification of 4 promising lines that on average, out-yielded IR64 by 37.8% under the non-stress condition and by 238% under stress (Table 16). In fact, many promising lines have been developed from the 10 pyramiding populations in Table 14 which had significantly improved DT and yield potential as compared to the recurrent parent, IR64, the most widely grown variety in South and Southeast Asia (data not shown).

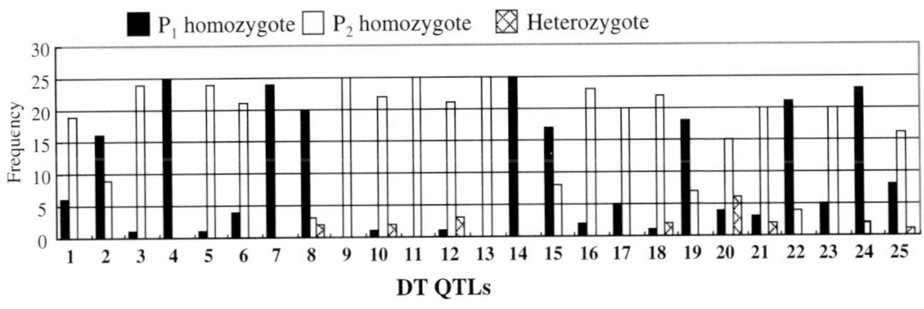

Figure 4. Genotypic frequencies of the 25 segregating DT QTLs in the 25 selected DT F_2 plants

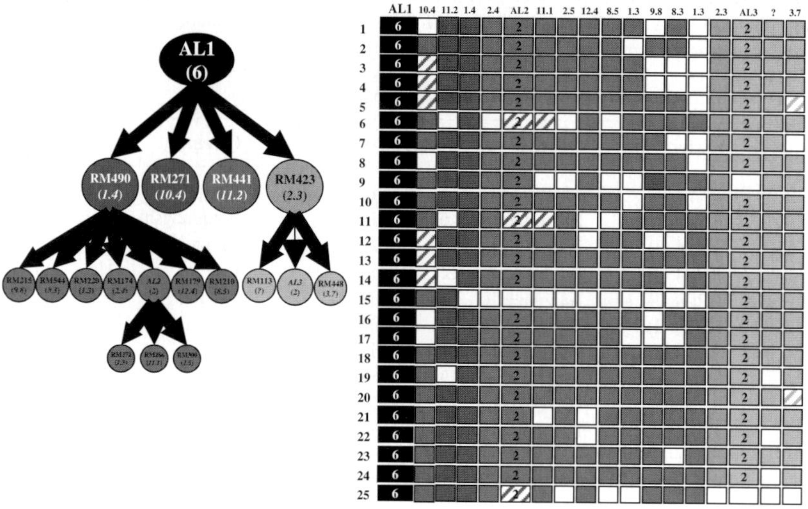

Figure 5. Graphical genotypes of 25 segregating DT loci detected in 25 progeny selected under severe drought from an F2 population derived from a cross between two ILs, DGI21 (IR64/STY) and DGI 60 (IR64/BR24) and the genetic network consisting of all 25 loci identified the population

Table 15. Performance of the 25 selected drought tolerant F_2 plants with pyramided QTLs in replicated experiments under the terminal stress and non-stress conditions during the 2004 dry season

Traits		QTL pyramiding F_4 lines				IR64	
		Mean	SD	Min	Max	Mean	SD
Days to heading	(S)	87.2	2.9	79.3	91.0	89.7	1.8
(days)	(N)	80.1	2.1	75.0	83.5	82.0	2.3
S − N		7.1**		4.3	7.5	7.7**	
Plant height (in cm)	(S)	60.7	3.7	50.6	69.1	58.4	3.3
	(N)	79.9	5.6	67.8	88.4	77.4	3.2
S − N		−19.1**		−17.2	−19.4	−19.0**	
Tillers/plant	(S)	22.6	1.5	20.0	26.0	23.3	2.5
	(N)	23.3	2.3	19.3	28.3	23.5	3.2
S − N		−0.7		0.7	−2.3	−0.2	
Panicles/plant	(S)	14.4	2.2	9.0	18.7	14.6	3.5
	(N)	22.3	2.0	18.6	26.3	21.9	3.0
S − N		−7.9**		−9.6	−7.7	−7.4**	
Fertility (%)	(S)	58.0	11.3	29.7	76.9	33.8	16.0
	(N)	87.5	5.3	73.5	96.6	89.0	7.6
S − N		−29.5		−43.8	−19.7	−55.3**	
Grain yield/plant (g)	(S)	16.5	9.8	3.4	35.9	7.2	4.8
	(N)	179.4	38.0	109.4	246.9	167.7	46.4
S − N		−162.9**		−106.1	−210.9	−160.5**	

*,** Indicate the significance levels of $p<0.05$ and $p<0.01$ for the difference between trail values obtained under stress and non-stress conditions, respectively.

Table 16. Performances of top 5 pyramided F_4 lines in replicated experiments under the terminal stress and non-stress conditions during the 2004 dry season

Lines #	HD (S)	PH (S)	TN (S)	PN (S)	FT% (S)	GY (S)	HD (N)	PH (N)	TN (N)	PN (N)	FT% (N)	GY (N)
6	87.0	58.8	21.0	14.7	60.8	21.4	80.0	86.9	19.3	18.6	94.6	246.9
8	87.7	61.5	25.0	17.0	58.7	18.1	80.0	77.6	24.2	23.8	91.3	220.2
17	84.7	61.3	23.2	18.1	71.4	25.4	77.3	75.3	22.6	22.1	85.5	215.9
23	85.3	65.4	24.1	18.7	72.0	32.3	79.7	86.2	23.2	22.1	91.6	227.9
Mean	**86.2**	**61.8**	**23.3**	**17.1**	**65.7**	**24.3**	**79.3**	**81.5**	**22.3**	**21.7**	**90.8**	**227.7**
SD	1.2	2.4	1.5	1.5	6.0	5.3	1.1	5.1	1.8	1.9	3.3	11.9
IR64	**89.7**	**58.4**	**23.3**	**14.6**	**33.8**	**7.2**	**82.0**	**77.4**	**23.5**	**21.9**	**89.0**	**167.7**
SD	1.8	3.3	2.5	3.5	16.0	4.8	2.3	3.2	3.2	3.0	7.6	46.4

5. SUMMARY

In summary, there is an urgent need to develop DT and ST rice varieties because of the increasing threats of water shortage and salinization of arable lands worldwide. Few DT or ST rice varieties have been commercially released in the past decades, which could largely attributed to the lack of breeding efforts specifically targeting at improving DT and ST, and partially to the complexity of genetics and physiology associated with DT and ST in rice. There have been very limited efforts in applying MAS to improving DT/ST in rice despite the numerous studies in genetically dissecting DT/ST in rice using the QTL mapping approach. Progress has been made recently in developing DT and ST rice cultivars at IRRI, which indicates that the conventional breeding approach is effective for breeding DT for the upper rainfed ecosystems of Asia because of the high level of heterosis for WUE/DT in rice. However, developing hybrid rice cultivars should be an effective strategy to combine high yield potential with a good level of WUE/DT for most shallow rainfed lowlands of Asia because of the high level of heterosis for WUE/DT in rice. The BC breeding and designed QTL pyramiding appears to be a new and promising breeding strategy, in which development of large numbers of DT/ST ILs in elite genetic backgrounds by large scale BC breeding, deep exploitation of useful genetic diversity for DT/ST from the primary gene pool of rice, effective selection, discovery, allelic mining and characterization of QTL networks for DT/ST, and directed trait improvement by designed QTL pyramiding based on accurate genetic information of QTL networks are well designed and integrated. Many promising DT and ST lines have been developed in the program, even though the theoretical aspects underlying the genetic networks underlying the target traits and QTL pyramiding by design remain to be fully established. However, care should be taken in selecting recurrent parents that already have most desirable traits in target environments in each BC breeding program.

REFERENCES

Adorada DL, RD Mendoza, and GB Gregorio. 2004. Agronomic characterization of saline-tolerant elite breeding lines with multiple tolerance for abiotic stresses, in PBGB 2003 Annual Report, p. 29, International Rice Research Institute, Los Banos, Philippines

Akbar M, and FN Ponnamperuma. 1982. Saline soils of South and Southeast Asia as potential rice lands. In Rice Research Strategies for the Future. IRRI, Los Banos, Philippines. pp. 265–281

Akbar M, and T Yabuno. 1974. Breeding for saline-resistant varieties of rice. II. Comparative performance of some rice varieties to salinity during early development stage. Jap. J. Breed. 25:176–181

Akbar M, GS Khush and D Hillerislambers. 1985. Genetics of Salt Tolerance. In: Rice Genetics, IRRI, Philippines, pp. 399–409

Ali AJ, JL Xu, AM Ismail, BY Fu, CHM Vijaykumar, YM Gao, J Domingo, R Maghirang, SB Yu, G Gregorio, S Yanaghihara, M Cohen, B Carmen, D Mackill, ZK Li. 2006. Hidden diversity for abiotic and biotic stress tolerances in the primary gene pool of rice revealed by a large backcross breeding program. Field Crops Research 97: 66–76

Ali ML, MS Pathan, J Zhang, G Bai, S Sarkarung, HT Nguyen. 2000. Mapping QTLs for root traits in a recombinant inbred population from two indicaecotypes in rice. Theor Appl Genet 101:756–766

Atlin G, P Virk, SS Virmani, and M Amante. 2004. Identification of drought-tolerant genotypes for shallow rainfed lowland production, 2003 Annual Report of Plant Breeding, Genetics and Biotechnology Division, pp. 18–19, the International Rice Research Institute, Los Banos, The Philippines

Babu RC, BD Nguyen, V Chamarerk, P Shanmugasundaram, P Chezhian, P Jeyaprakash, SK Ganesh, A Palchamy, S Sadasivam, S Sarkarung, LJ Wade and HT Nguyen. 2003. Genetic analysis of drought resistance in rice by molecular markers: association between secondary traits and field performance. Crop Sci. 43:1457–1469

Babu RC, MS Pathan, A Blum, and HT Nguyen. 1999. Comparison of Measurement Methods of Osmotic Adjustment in Rice Cultivars. Crop Sci. 39:150–158

Babu RC, HE Shashidhar, JM Lilley, ND Thanh, JD Ray, S Sadasivam, S Sarkarung, JC O'Toole and HT Nguyen. 2001. Variation in root penetration ability, osmotic adjustment and dehydration tolerance among accessions of rice adapted to rainfed lowland and upland ecosystems. Plant Breeding 120:233–238

Blumwald E. 2000. Sodium transport and salt tolerance in plant cells. Curr. Opin. Cell Bidi. 12:431–434

Bonilla P, J Dvorak, D Mackill, K Deal, and G Gregorio. 2002. RFLP and SSLP mapping of salinity tolerance genes in chromosome 1 of rice (*Oryza sativa* L.) using recombinant inbred lines. Philippine Journal of Agricultural Science 85:68–76

Bouman BAM, and RP Tuong. 2001. Field water management to save water and increase its productivity in irrigated rice. Agricultural Water Management 49(1):11–30

Champoux MC, G Wang, S Sarkarung, DJ Mackill, JC O'Toole, N Huang and SR McCouch. 1995.Locating genes associated with root morphology and drought avoidance in rice via linkage to molecular markers. Theor. Appl. Genet. 90:969–981

Chaubey CN, and D Senadhira. 1994. Conventional plant breeding for tolerance to problem soils. In: AR Yeo and TJ Flowers (eds) Soil fylineral stresses: Approaches to Crop Improvement. Springer-Verlag, Berlin, pp. 11–29

Courtois B, G McLaren, PK Sinha, K Prasad, R Yadav and L Shen.2000. Mapping QTLs associated with drought avoidance in upland rice. Molecular Breeding 6:55–66

Cruz RT, and JC O'Toole. 1984. Dry land rice response to an irrigation gradient at flowering stage. Agron J 76: 178–183

Dingkuhn M, RT Cruz, JC O'Toole, K Doerffling. 1989. Net photosynthesis, water use efficiency, leaf water potential, and leaf rolling as affected by water stress in tropical upland rice. Aust. J. Agric. Res. 40:1171–1181

Epstein E, JD Nbrlyn, DW Ruh, RW Kingsbury, DB Kelly, and GA Cunningham. 1980. Saline culture of crops; a genetic approach. Science 210: 399–404

Fang XW, LH Tang, and YP Wang. 2004 Identification of rice germplasm tolerant to salt stress. J. Plant Genetic Resources 5(3): 295–298

Flowers TJ, ML Koyama, SA Flowers, C Sudhakar, KP Singh, and AR Yeo. 2000. QTL: their place in engineering tolerance of rice to salinity. Journal of Experimental Botany/the Society or Experimental Biology 51(342): 99–106

Fukai S, M Cooper. 1995. Development of drought-resistant cultivars using physio- morphological traits in rice. Field Crops Res. 40: 67–86

Fukuda A, A Nakamura, Y Tanaka. 1999. Molecular cloning and expression of the Na^+/H^+ exchanger gene in *Oryza sativa*. Biochim Biophys Acta 1446: 149–155

Gao YM, BY Fu, R Lafitte, TQ Zheng, JL Xu, CHM Vijayakumar, YZ Jiang, MF Zhao, SB Yu, D Dwivedi, J Domingo, J Ali, R Maghirang, S Veruka, LH Zhu, J O'Toole, AH Paterson, HQ Zhai, GS Khush, JM Ribaut, and ZK Li 2007 Genome-wide response to selection and genetic basis of cryptic genetic variation of drought tolerance in rice (*Oryza sativa L.*). (in preparation)

Gong Ji-ming, He Ping, Qian Qian, Shen Li-shuang, Zhu Li-huang, Chen Sou-yi. 1998. QTL mapping of salt tolerance in rice. Chinese Science Bulletin (17):1847–1850

Greenway H. 1973. Salinity, plant growth and metabolism. Journal of Australian Institute of Agricultural Science 39: 24–34

Greenway H, and R Munns. 1980. Mechanism of salt tolerance in non-halophytes. Annual Review of Plant Physiology, 31: 149–190

Gregorio GB, D Senadhira, RD Mendoza. 1997. Screening rice for salinity tolerance. IRRI, Los Banos, Laguna, Philippines

Gregorio GB, D Senadhira, RD Mendoza, NL Manigbas, JP Roxas, CQ Guerta. 2002. Progress in breeding for salinity tolerance and associated abiotic stresses in rice. Field Crops Research 76: 91–101

Gregorio GB, and D Senadhira. 1993. Genetic analysis of salinity tolerance in rice. Theor. AppJ. Gen. 86:333–338

Gu Xing-you, Mei Man-tong, Yan Xiao-long, Zherng Shao-ling, Lu Yong-gen. 2000. Preliminary detection of quantitative trait loci for salt tolerance in rice. Chinese J Rice Sci 14 (2):65–70.

Guo Yan, Chen Shaolin, Zhang Gengyun, Chen Shouyi. 1997. Effect gene were obtained by cell engineering technique. Acta Genetica Sinica 24(2): 122–126

Lafitte HR, CHM Vijayakumar, YM Gao, Y Shi, JL Xu, BY Fu, SB Yu, AJ Ali, J Domingo, R Maghirang, R Torres, D Mackill and ZK Li. 2006. Improvement of rice drought tolerance through backcross breeding: evaluation of donors and results from drought nurseries. Field Crops Research 97:77–86

Haque MM, DJ Mackill and T Ingram. 1992. Inheritance of leaf epicuticular wax content in rice. Crop Sci 32: 865–868

Hemamalini GS, HE Shashidhar, S Hittalmani. 2000. Molecular marker assisted tagging of root morphological traits under two contrasting moisture regimes at peak vegetative stage in rice (*Oryza sativa* L.). Euphytica 112: 69–78.

Jeschke WD, and W Hartung, 2000. Root-shoot interactions in mineral nutrition. Plant Soil 226: 310–314

Kamoshita A, J Wade, L Ali, S Pathan, J Zhang, SS arkarung, T Nguyen. 2002a. Mapping QTLs for root morphology of a rice population adapted to rainfed lowland conditionsTheor Appl Genet. 104(5):880–893

Kamoshita A, J Zhang, J Siopongco, S Sarkarung, HT Nguyen, LJ Wade. 2002b. Effects of phenotyping environment on identification of quantitative trait loci for rice root morphology under anaerobic conditions. Crop Sci. 42(1):255–265

Khush GS. 1999. Green revolution: preparing for the 21st century. Genome 42: 646–655

Khush GS. 2001. Green revolution: the way forward. Nat Rev Genet. 2: 815–22

Koyama ML, A Levesley, RMD. Koebner, TJ Flowers, AR Yeo. 2001. Quantitative Trait Loci for Component Physiological Traits Determining Salt Tolerance in Rice. Plant Physiol 125:406–422

Lafitte HR, B Courtois, M Arraudeau. 2002. Genetic improvement of rice in aerobic systems: progress from yield to genes. Field Crops Res. 75:171–190

Lafitte HR, B Courtois. 2000. Genetic variation in performance under reproductive stage water deficit in a doubled-haploid rice population in upland fields. In: Ribaut JM, Poland D (eds) Molecular approaches for the genetic improvement of cereals for stable production in water-limited environments. A strategic planning workshop held on 21–25 June 1999. CIMMYT, El Batan, pp 97–102

Lafitte HR, AH Price, B Courtois. 2004. Yield response to water deficit in an upland rice mapping population: associations among traits and genetic markers. Theor Appl Genet. 109(6):1237–46
Lafitte HR, B Courtois. 2002. Interpreting cultivar – environment interactions for yield in upland rice: assigning value to drought-adaptive traits. Crop Sci. 42, 1409–1420
Lanceras JC, G Pantuwan, B Jongdee, T Toojinda. 2004. Quantitative trait loci associated with drought tolerance at reproductive stage in rice. Plant Physiol 135:384–399
Lang NT, S Masood, S Yanagihara, BC Buu. 2003. Mapping QTLs for salt tolerance in rice. In: Khush GS, DS Brar, B Hardy. Advances in Rice Genetics. Supplement to Rice Genetics IV. Proceedings of the Fourth International Rice Genetics Symposium, 22-27 October 2000. Los Banos, Philippines. International Rice Research Institute. pp. 294–298
Lee M. 1995. DNA markers and plant breeding programs. Advances in Agronomy 55:265–344
Li ZK, BY Fu, YM Gao, JL Xu, J Ali, HR Lafitte, YZ Jiang, JD Rey, CHM Vijayakumar, R Maghirang, TQ Zheng and LH Zhu. 2005. Genome-wide introgression lines and a forward genetics strategy for functional genomic research of complex phenotypes in rice. Plant Molecular Biology 59:33–52
Li Z, LS Shen, B Courtois, R Lafitte. 2000. Development of nearisogenic introgression line (NIIL) sets for QTLs associated with drought tolerance in rice. In: Ribaut JM, Poland D (eds) Molecular approaches for the genetic improvement of cereals for stable production in water-limited environments. A strategic planning workshop held on 21–25 June 1999. CIMMYT, El Batan, pp 103–107
Li ZK, BY Fu, YM Gao, JL Xu, J Ali, HR Lafitte, YZ Jiang, JD Rey, CHM Vijayakumar, R Maghirang, TQ Zheng and LH Zhu. 2005. Genome-wide introgression lines and a forward genetics strategy for functional genomic research of complex phenotypes in rice. Plant Molecular Biology 59:33–52
Lilley JM, and MM Ludlow. 1996. Expression of osmotic adjustment and dehydration tolerance in diverse rice lines. Field Crops Res. 48: 185–197
Lilley JM, MM Ludlow, SR McCouch, MC Champoux and JC O'Toole. 1996. Locating QTL for osmotic adjustment and dehydration tolerance in rice. J. Expt Bot 47: 1427–1436
Lin HX, MZ Zhu, M Yano, JP Gao, ZW Liang, WA Su, XH Hu, ZH Ren, DY Chao. 2004. QTLs for Na^+ and K^+ uptake of the shoots and roots controlling rice salt tolerance. Theor Appl Genet 108(2):253–260
Maas EV, and GJ Hoffman. 1977. Crop salt tolerance current assessment. ASCE J. Irrig. and Drainage Div. 103:115–134
Mackill D, W Coffman, D Garrity. 1996. Rainfed Lowland Rice Improvement. International Rice Research Institute, Manila, Philippines
Malcolm CV. 1969. Use of halophytes for forage production on saline wastelands. Journal of Australian Institute of Agricultural Sciences 35: 38–49
Mishra B, M Akbar, DV Seshu, and D Senadhira. 1996. Genetics of salinity tolerance and ionic uptake in rice. IRRN 21:38–39
Moeljopawiro S and H Ikehashi. 1981. Inheritance of salt tolerance in rice. Euphytica 30:291–300
Moncada MP, CP MartÃnez, J Tohme, E Guimaraes, M Chatel, J Borrero, H Gauch, and SR McCouch. 2001. Quantitative trait loci for yield and yield components in an *Oryza sativa* × *Oryza rufipogon* BC_2F_2 population evaluated in an upland environment. Theor Appl Genet 102:41–52
Munns R, H Greenway and GO Kirst. 1983. Halotolerant eukaryotes: In: Physiological Plant Ecology. III. Responses to the Chemical and Biological Environment. Eds. O.L. Lange, P.S. Nobel, C.B Osmond and H. Zeigler. Encycl. Plant Physiol., New Series, Vol. 12C. Springer, Berlin. pp. 59–135
Munns R, S Hussain, AR Rivelli, RA James, AG Condon, MP Lindsay, ES Lagudh, DP Schachtman and RA Hare. 2002. Avenues for increasing salt tolerance of crops, and the role of physiologically based selection traits. Plant and Soil 247: 93–105
Nguyen TTT, N Klueva, V Chamareck, A Aarti, G Magpantay, ACM Millena, MS Pathan, HT Nguyen. 2004. Saturation mapping of QTL regions and identification of putative candidate genes for drought tolerance in rice. Mol Gen Genomics 272: 35–46
O'Toole JC, and RT Cruz. 1983. Genotypic variation in epicuticular wax of rice. Crop Sci 23: 392–394
Pandey S, D Behura, R Villano, D Naik, 2000. Economic Cost of Drought and Farmers' Coping Mechanisms: A Study of Rainfed Rice in Eastern India. IRRI Discussion Paper Series, pp. 1–35

Pantuwan G, FS ukai, M Cooper, S Rajatasereekul, JC O'Toole, 2002. Yield response of rice (Oryza sativa L.) genotypes to different types of drought under rainfed lowlands. Part 3. Plant factors contributing to drought resistance. Field Crops Res. 73: 181–200

Pearson GA, AD Ayers, DL Eberhard. 1966. Relative salt tolerance of rice during germination and early seedling development. Soil Sci. 102: 151–156

Poonamperuma FN, AK Bandyopadhya, 1980 Soil salinity as constraints on food production in the humid tropics. In: Soil Related Constraints to Food Production in the Tropics, IRRI, Los Banos, Philippines, p. 203–216.

Prasad SR, PG Bagali, S Hittalmani, HE Shashidhar. 2000. Molecular mapping of quantitative trait loci associated with seedling tolerance to salt stress in rice (*Oryza sativa* L.). Curr Sci 78: 162–164

Price AH, AD Tomos.1997. Genetics dissection of root growth in rice (Oryza sativa L.) II: Mapping quantitative trait loci using molecular markers. Thero Appl Genet 95:143–152

Price AH, J Townend, MP Jones, A Audebert, B Courtois. 2002. Mapping QTLs associated with drought avoidance in upland rice grown in the Philippines and West Africa. Plant Mol Biol. 48(5–6):683–95

Price AH, AD Tomos and DS Virk. 1997. Genetic dissection of root growth in rice (*Oryza sativa* L.) I: a hydrophonic screen. Theor Appl Genet 95: 132–142

Price AH, KA Steele, BJ Moore, PB Barraclough, LJ Clark. 2000. A combined RFLP and AFLP linkage map of upland rice (*Oryza sativa* L.) used to identify QTLs for root-penetration ability. Theor Appl Genet 100:49–56

Quarrie SA, DA Laurie, J Zhu, C Lebreton, A Semikhodskii, A Steed, H Witsenboer, C Calestani. 1997. QTL analysis to study the association between leaf size and abscisic acid accumulation in droughted rice leaves and comparisons across cereals. Plant Mol Biol. 35(1–2):155–65

Rajanaidu N, and AH Zakri. 1988. Breeding for morpho-physiological traits in crop plants. In: Zakri A.H.(ed) Plant breeding and Genetic Engineering. SABRAO publishers. pp. 116–139

Ray JD, L Yu, SR McCouch, MC Champoux, G Wang, HT. Nguyen. 1996. Mapping quantitative trait loci associated with root penetration ability in rice (*Oryza sativa* L.). Theor. Appl. Genet. 92:627–636

Ren Zhong-Hai, Gao Ji-Ping, Li Le-Gong, Cai Xiu-Ling, Huang Wei, Chao Dai-Yin, Zhu Mei-Zhen, Wang Zong-Yang, Luan Sheng, Lin Hong-Xuan. 2005. A rice quantitative trait locus for salt tolerance encodes a sodium transporter. Nature Genetics 37(10):1141–1146

Richards RA. 1983. Should selection for yield in saline regions be made in saline or non-saline soils. Euphytica 32:431–438

Robin S, MS Pathan, B Courtois, R Lafitte, S Carandang, S Lanceras, M Amante, HT Nguyen, Z Li. 2003. Mapping osmotic adjustment in an advanced back-cross inbred population of rice. Theor Appl Genet. 107(7):1288–96

Rood MA. 2000. Monitoring levels in tail water recovery system can save yields. Rice Journal 103(1):12–13

Shannon MC. 1985. Principles and strategies in breeding for higher salt tolerance. Plant Soil 89: 227–241

Shen L, B Courtois, K McNally, S Robin, ZK Li. 2001. Evaluation of near-isogenic lines of rice introgressed with QTLs for root traits through marker-aided selection. Theor Appl Genet 103:70–83

Singh AK, D Singh, SP Rathi, JL Dwivedi, PK Sinha, NP Mandal, D Tao, and G Atlin. 2004. International Aerobic Rice Variety Trial, 2003 Annual Report of Plant Breeding, Genetics and Biotechnology Division, pp. 16–17, the International Rice Research Institute, Los Banos, the Philippines

Steele KA, AH Price, HE Shashidhar, JR Witcombe. 2006. Marker assisted selection to introgress into an Indian upland rice variety, Theor Appl Genet 112: 208–221

Takehisa H, T Shimodate, Y Fukuta, T Ueda, M Yano, T Yamaya, T Kameya and T Sato 2004. Identification of quantitative trait loci for plant growth of rice in paddy field flooded with salt water. Field Crops Research 89:85–95

Tal M. 1985. Genetics of salt tolerance in higher plants: theoretical and applied considerations. Plant and soils 89: 199–226

Tripathy JN, J Zhang, S Robin and HT Nguyen. 2000. QTLs for cell-membrane stability mapped in rice (Oryza sativa L.) under drought stress. Theor. Appl. Genet. 100: 1197–1202

Tuong TP, BAM Bouman, and M Mortimer. 2005. More rice, less water – integrated approaches for increasing water productivity in irrigated rice-based systems in Asia. Plant Production Science 8(3): 231–241

Venuprasad R, HE Shashidhar, S Hittalmani and GS Hemamalini.2002. Tagging quantitative trait loci associated with grain yield and root morphological traits in rice (*Oryza sativa* L.) under contrasting moisture regimes. Euphytica 128:293–300

Venuprasad R, HR Lafitte, and GN Atlin. 2006. Response to direct selection for grain yield under drought stress in rice. Crop Sci. (in press)

Xu JL, HR Lafitte, YM Gao, BY Fu, R Torres, and ZK Li. 2005. QTLs for drought avoidance and tolerance identified in a set of random introgression lines of rice. Theor Appl Genet 111:1642–1650

Yadav R, B Courtois, N Huang, and G McLaren. 1997. Mapping genes controlling root morphology and root distribution in a doubled-haploid population of rice. Theor Appl Genet (1997) 94: 619–632

Yeo AR. 1994. Physiological criteria in screening and breeding. In: A. R. Yeo and T. J. Flowers (eds) Soil mineral stresses: Approaches to crop improvement. Springer-Verlag, Berlin. pp. 37–57

Yeo AR, and TJ Flowers. 1984. Mechanisms of salinity resistance in rice and their roles as physiological criteria in plant breeding. In: Staples RC, GH Toenniessen eds. Salinity tolerance in plants strategies for crop improvemet. New York: John wiley-interscience. Pp.151–170

Yu SB, WJ Xu, CHM Vijayakumar, J Ali, BY Fu, JL Xu, R Marghirang, J Domingo, YZ Jiang, C Aquino, SS Virmani, ZK Li. 2003 Molecular diversity and multilocus organization of the parental lines used in the International Rice Molecular Breeding Program. Theor. Appl. Genet. 108:131–140

Yue Bing, Lizhong Xiong, Weiya Xue, Yongzhong Xing, Lijun Luo, Caiguo Xu. 2005 Genetic analysis for drought resistance in field with different types of soil. Theor Appl Genet 111: 1127–1136

Yue Bing, Weiya Xue, Lizhong Xiong, Xinqiao Yu, Lijun Luo, Kehui Cui, Deming Jin, Yongzhong Xing, and Qifa Zhang. 2006. Genetic basis of drought resistance at reproductive stage in rice: separation of drought tolerance from drought avoidance. Genetics 172: 1213–1228

Zaidem ML, RD Mendoza, EBTumimbang, IA Duka, and GB Gregorio. 2004. Genetic variability of salinity tolerance at different growth stages of rice, in PBGB 2003 Annual Report: pp. 19–20, International Rice Research Institute, Las Banos, Philippines

Zhang GY, G Yan, SL Chen, SY Chen. 1995. RFLP tagging of a salt tolerance gene. Plant Science 110 (2):227–234

Zhang J, HG Zheng, A Aarti, G Pantuwan, TT Nguyen, JN Tripathy, AK Sarial, S Robin, RC Babu, BD Nguyen, S Sarkarung, A Blum, HT Nguyen. 2001. Locating genomic regions associated with components of drought resistance in rice: comparative mapping within and across species. Theor Appl Genet 103:19–29

Zheng HG, RC Babu, P MS athan, L Ali, N Huang, B Courtois, HT Nguyen. 2000. Quantitative trait loci for root-penetration ability and root thickness in rice: comparison of genetic backgrounds. Genome 43(1):53–61

Zou GH, HW Mei, HY Liu, GL Liu, SP Hu, XQ Yu, MS Li, JH Wu, LJ Luo. 2005. Grain yield responses to moisture regimes in a rice population: association among traits and genetic markers. Theor Appl Genet. 112(1):106–13

CHAPTER 22

RECENT ADVANCES IN BREEDING WHEAT FOR DROUGHT AND SALT STRESSES

RANA MUNNS AND R.A. RICHARDS

CSIRO Plant Industry, GPO Box 1600, Canberra, ACT 2601, Australia

Abstract: Substantial advances have been made in breeding wheat for dry environments that will also improve performance in saline environments. These genetic gains have been made by conventional breeding. Further gains in productivity will come from the addition of traits that increase the efficiency of water use in dry soils, and control the uptake of salts from saline soils. Conventional breeding methods will continue to be important to provide farmers with higher yielding varieties in dry or saline soils which are resistant to current diseases and which have the grain quality demanded by competitive markets. Trait-based breeding approaches, which often utilize molecular markers to improve selection efficiency, are starting to deliver new and significant gains. To target the most important traits, it is important to know how they will influence yield. Is it through more water use, more efficient use of water or a higher harvest index? For example, the trait of 'early vigour' may be an advantage in some years but in others may lead to the exhaustion of soil water and a low yield. The challenge for breeders will be to efficiently integrate trait-based and molecular methods to increase yield in dry and saline environments

Keywords: Drought tolerance, salinity tolerance, trait-based breeding, marker-assisted selection, yield, wheat

1. INTRODUCTION

Wheat (*Triticum aestivum* L.) is the most widely grown crop in the world. It is considered a moderately salt-tolerant and drought-tolerant crop and, with barley, it is the preferred cereal in most arid and semi-arid agricultural regions. Wheat is hexaploid and allopolyploid (AABBDD), and is used primarily for baking bread. Wheat has extraordinary adaptation, growing successfully in cool moist regions from Finland to southern Chile to tropical regions such as parts of Asia and Africa. Much of the world's wheat is produced under irrigation; however, the area sown to rainfed wheat is very substantial and is expected to grow further as water for irrigation declines globally. The rainfed wheat crops are often in semi-arid environments which range from the centre of origin of wheat in the Fertile Crescent, where crops

M.A. Jenks et al. (eds.), Advances in Molecular Breeding Toward Drought and Salt Tolerant Crops, 565–585.

rely on current rainfall, to dry regions of north eastern Australia and South Africa, where little rain falls during the growing season and crops rely mostly on stored soil water.

The grain yield of wheat is a function of rainfall (or evapotranspiration) and several possible relationships between yield and rainfall (or evapotranspiration) are shown in Figure 1. The broken line represents average yields typically found on-farm in semi-arid environments (rainfed). The thin line represents the yield potential of current well-adapted cultivars. An average farm yield that falls below the potential can be due to management practices (eg, late sowing, insufficient fertilizer), hostile soils (eg, salinity, acidity) or biotic factors (soil-borne or foliar diseases, insect pests). The thick line represents a new yield potential that may be achievable through breeding for dry environments using conventional breeding methods. A new yield potential will come from increases in water productivity; ie, greater or more efficient use of the water resource, than from improvements in drought tolerance or drought resistance.

Figure 1 shows that the minimum rainfall required to produce harvested grain is about 100 mm. In other words, if there is no water stored in the soil, then a minimum of 100 mm of rain falling during the growing season is necessary to compensate for the evaporative loss. The slope of the line in Figure 1 is the transpiration efficiency, and is close to 20 kg $ha^{-1}mm^{-1}$. Thus, if 100 mm of water were stored in the soil at sowing and there was no further precipitation or evaporative loss, this could result in a wheat yield of 2 t ha^{-1}. Barley typically yields more than wheat when crops are reliant on rainfall during the growing season (Lopéz-Casteñeda and Richards

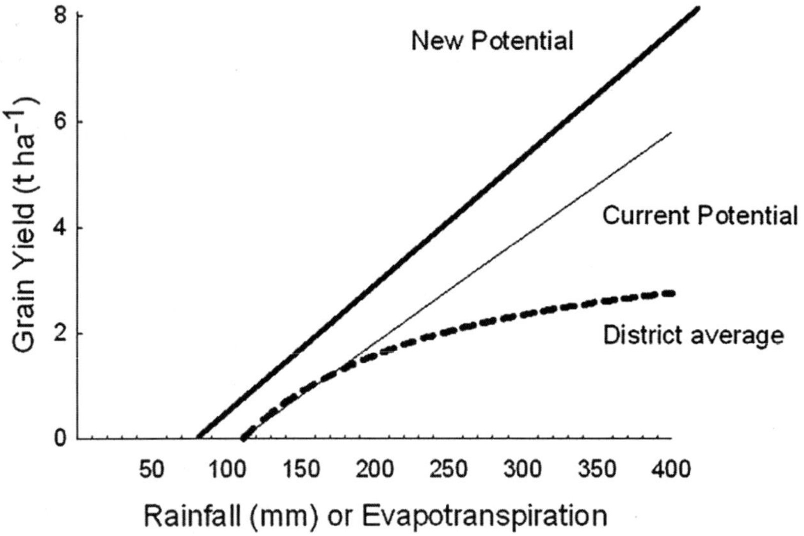

Figure 1. Relationship between grain yield and rainfall or evapotranspiration. Adapted from French and Schultz (1984)

1994a), whereas barley achieves a yield equal to or less than wheat when crops are reliant on stored soil moisture. This difference between wheat and barley is attributed to variation in their early leaf area growth which in turn regulates crop water use and not to any inherent variation in their drought resistance.

Breeding targets that should improve the grain yield of wheat (thick line in Figure 1) are also evident from this figure. Breeding to increase crop water use will increase yield. This may be achieved, for example, by a more effective root system. Reducing direct evaporation from the soil surface achieved by breeding for more vigorous plants that shade the soil surface will reduce the intercept in Figure 1 and also result in more crop water use and more yield. Breeding to increase the transpiration efficiency, i. e. the slope in Figure 1, will also result in a higher yield. These improvements in yield are due to increases in "water productivity", a term that has more meaning for breeding in dry environments compared with "drought tolerance" or "drought resistance" (see Passioura et al., this volume).

As wheat is grown in many arid or semi-arid regions of the world, it is likely to incur salinity, whether caused by irrigation, land clearing, or natural processes. This is because of large amounts of salt in the soil profile, deposited there by weathering of rocks or by deposition of oceanic salt in wind and rain. Where wheat is grown under irrigation it is likely to suffer from salinity because the irrigation water contains salt or because of salt already deposited in the soil. Where wheat is grown without irrigation, and the rainfall exceeds the crop's water use (400–500 mm), and recharges the groundwater, water tables will rise and move salt to the surface. If the rainfall does not exceed the crop's water use, salt will remain below the surface but possibly move up into the root zone by capillary action as the soil dries (Rengasamy, 2002).

Bread wheat is moderately salt tolerant, compared to other cereals. In a field where the salinity rises to 10 dS m^{-1} (about 100 mM NaCl), rice (*Oryza sativa* L.) will die before maturity, while bread wheat will produce a reduced yield. Even barley (*Hordeum vulgare* L.), the most tolerant cereal, dies after extended periods at salt concentrations higher than 250 mM NaCl (equivalent to 50% seawater). Durum wheat (*Triticum turgidum* L. ssp. *durum* [Desf.]; tetraploid, AABB), which is used for making pasta and couscous, is less salt tolerant than bread wheat (Maas and Hoffman, 1977; Munns et al., 2006).

Increases in drought tolerance would also increase the salt tolerance of both bread and durum wheat. For example, faster early growth characteristics are expected to be important in saline soils as well as in dry environments when the soil is not saline, for the reasons described above. A saline soil reduces plant growth for two main reasons: an osmotic effect due to the salt outside the root, and a toxic effect if the salt accumulates in leaves to excessive amounts (Munns, 2002). The osmotic effect is similar to a drought stress, and adaptive responses are in common. The salt-specific effect is greater in species that do not have tight control of salt uptake by roots, and salt transport within the plant, and may explain why durum wheat is more sensitive to salinity than bread wheat (Munns et al., 2006).

2. EMPIRICAL SELECTION AND BREEDING

2.1. Drought

So-called conventional breeding for grain yield in dry environments is a proven and successful way to improve productivity in dry environments (Trethowan et al. 2002). This involves sowing large numbers of bordered plots in numerous environments and measuring grain yield. Yield progress is often slower than where there is irrigation, as heritability for yield is lower and large genotype by environment interactions (g x e) make selection progress difficult. The low heritability and high g x e in dry environments is due to the large and unpredictable variation in the amount of rainfall, and the timing of this rainfall on a seasonal basis. However, despite these difficulties, conventional breeding for yield is expected to remain the most widespread method of wheat improvement in dry environments. It has been assisted by significant advances in experimental designs that can account for spatial variability which have improved the effectiveness of yield selection in conventional breeding programs (Gilmour et al., 1997). It has also been assisted by the flexibility in breeding methods that vary from breeder to breeder and from season to season allowing breeders to put greater emphasis on particular traits depending on the season. Even in dry environments, there is unlikely to be any selection pressure for drought-related traits. Breeders will usually select for flowering time, plant height, and their assessment of agronomic type, in early generations. It is then likely that selection for grain yield in field plots will be the only other cull made for performance in dry environments.

The success of conventional breeding in dry environments is evident from the large number of wheat varieties released. For example, in Australia where almost no wheat is grown under irrigation and rainfall limits yield every year, in the last 20 years an average of more than six new wheat cultivars were released each year (Whiting, 2004). Each new cultivar is expected to be an improvement on previous ones and a commercial success, although each variety may not yield more grain than the last in many environments, as improvements in disease resistance and grain quality characteristics are also important determinants of variety acceptance. Thus, many of the new cultivars released may not have a greater yield in dry environments or during a drought than previous cultivars, but instead may have additional protection against current diseases, or have an improvement in grain quality that will confer a market advantage.

It is significant that apart from flowering time most of the yield progress made in dry environments has been due to improvements in traits that initially may seem unrelated to yield performance during drought. A major factor has been yield potential under irrigation. Clearly factors that contribute to high yield under favourable conditions may also contribute to yield under less favourable conditions. This is evident from the germplasm selected at CIMMYT under irrigation also being successful under dry conditions in many countries (Laing and Fischer, 1977) and is also evident in other studies (Blum and Pnuel, 1990). Other factors that have been important to increase yields in dry environments have been associated with overcoming biotic and abiotic constraints in the soil. Thus tolerance to abiotic factors such as acidity or high boron content have been important, as have tolerance to biotic factors such as nematodes and other soil-borne pathogens. Genetic improvements in these other factors has resulted in greater crop water use and thereby increased yields.

2.1.1. *Use of wild relatives for breeding for drought tolerance*

Tolerance to drought is more complex than tolerance to soil salinity in that for salinity the critical factors are the control of salt uptake by roots and transport within the plant. There is no such target for drought, and drought is also more unpredictable in its timing and intensity. For this reason the use of wild relatives has been of less significance to breeding in dry than saline environments. However, wild relatives have often been explored for new sources of drought tolerance (Valkoun, 2001). Perennial grasses related to wheat that survive desiccating conditions could be useful sources for dehydration tolerance (Tabaei-Aghdaei et al., 2000). As yet however, they have not been of value in wheat improvement. There are multiple reasons for this, one being that breeders are reluctant to introduce totally different germplasm into their breeding gene pool as it can 'disturb' the favourable combinations of traits which are required by farmers and consumers. A second is that alien sources used in wheat breeding often reduce yields or carry unfavourable characteristics (The et al., 1988) which are difficult to eliminate. Thirdly, although alien species have been shown to contribute to survival in a drought they have not been shown to contribute to improved productivity.

An important source of variation for performance under drought may come from synthetic wheats (Mujeeb-Kazi et al., 1998). These are hexaploid wheats reformed from their ancestral genomes, such as the synthetic wheats formed from improved durum wheats (AB genome) hybridized with *Aegilops tauschii* (syn. *Triticum tauschii*) which contributes the D genome. It is thought that *Ae. tauschii* will be an important donor of drought-adaptive traits as it occurs naturally in dry environments. Selected synthetic wheats have shown improved performance under drought (Trethowan et al., 2005). However, in a recent study where selected lines were extensively evaluated in dry environments in Mexico and Australia, in general, the yield of the best local check in Mexico and Australia continued to outyield the best synthetic wheats (Dreccer et al., 2007). In addition, the magnitude of g x e was found to be large, and few elite synthetic wheats performed well across all sites.

2.2. Salinity

2.2.1. *Successful examples of conventional breeding*

Targeted breeding for salt tolerance has been successful in India, Pakistan and Egypt (reviewed by Munns et al., 2006). The most successful cultivars have been the Indian KRL1-4, released by the Central Soil Salinity Research Institute (CSSRI) at Karnal, the Pakistani LU26S and SARC-1, released by the Saline Agriculture

Research Cell (SARC) at Faisalabad, and the Egyptian Sakha 8, released by the Agricultural Research Centre at Giza.

In India, almost all salt tolerant wheat germplasm is derived from Kharchia 65. The Central Soil Salinity Research Institute (CSSRI) at Karnal released KRL1-4 for saline areas, a cross of Kharchia 65 with WL711 (Hollington, 2000). KRL1-4 has done well on the saline soils of northern India, but not in Pakistan, possibly because of the heavier soils and greater problems of waterlogging (Hollington, 2000). Another derivative of Kharchia 65 was developed in the UK by S.A. Quarrie and A. Mahmood: a doubled haploid line, KTDH 19, from a cross of Kharchia 65 with a line identified with exceptional sodium exclusion, TW161. This derivative performed well in Spain (Hollington et al., 1994) but in India and Pakistan, although highly tolerant in terms of total dry matter, its grain yield was very low due to its late maturity (Hollington, 2000).

Little is known about the physiology of the Indian landrace Kharchia 65, universally regarded as highly salt tolerant (Ashraf, 2002; Sharma et al., 1984; Wang et al., 2003), apart from an observation by Sharma et al., (1984) that it combined low Na^+ uptake rates with successful osmotic adjustment, and the finding of Richards and Lukacs (2002) that it has unusually high specific leaf area and early vigour. Yet it is still the mainstay of the Indian wheat breeding program.

The Pakistan selection LU26S showed improved yields on saline soils in Pakistan (Qureshi et al., 1990), but is now susceptible to rust (Hollington, 2000). LU26S was crossed with Kharchia, and two salt tolerant genotypes, S24 and S36 were selected (Ashraf and O'Leary, 1996). S24 had high salt tolerance, as high as Kharchia and SARC-1, possibly due to its low leaf Na^+ accumulation (Ashraf, 2002).

2.2.2. *Use of large collections for breeding for salt tolerance*

Large international collections have been screened in hydroponic or sand culture (summarised by Colmer et al., 2005), but no new cultivars have resulted from the best genotypes identified. The most extensive screen for salt tolerance in the field has been done by Jafari-Shabestari et al. (1995), who evaluated 400 Iranian wheats in one site in California over two seasons, irrigated with water at three salinity levels (1, 5 and 8 dS m^{-1}). They identified several accessions that were consistently high for grain yield in both low and high salinity treatments, but no cultivar was developed as a consequence (pers. com. C.O. Qualset). This is possibly because of the low correlation found between grain yield at high salinity with relative yield (yield in saline soil relative to non-saline soil), biomass, or harvest index. Moreover, Jafari-Shabestari et al. (1995) noted that some genotypes with high relative yield had low yield potential (yield in non-saline soils). They concluded that the calculation of relative yield is highly subject to experimental errors, especially with small plots, and questioned its use. A lack of correlation between relative yield and absolute yield was also noted by Richards et al. (1987), who concluded that the most efficient way to increase yields at high salinity was to select for the highest yielding lines at low salinity.

2.2.3. *Use of wild relatives for breeding for salt tolerance*

An extensive review of the use of wild relatives to improve the salt tolerance of wheat is given by Colmer et al. (2006), and is summarised below. In most cases, there was no knowledge of traits or genes, and the physiological reason for the improved salt tolerance in the progeny is not known.

Aegilops tauschii (DD) has been hybridized with durum wheat (AABB) to produce synthetic hexaploid wheat (Schachtman et al., 1992; Mujeeb-Kazi and Diaz de Leon, 2002). Variation in the salt tolerance of the D genome was shown to influence the salt tolerance of synthetic hexaploids (Schachtman et al., 1992), however a direct comparison was not made with modern bread wheat and the value of this approach has not yet been established.

Ae. cylindrica (CCDD) has been used to introduce salt tolerance into bread wheat. Backcrossed lines were produced from hybrids between *Ae. cylindrica* and the Pakistani cultivars LU26 and Pak81 (Farooq et al., 1995). These were tested in both saline and non-saline fields and shown to be both salt and drought-tolerant (Farooq et al., 1995; Farooq, 2004). Wheat lines WL1076 and WL41 out-yielded LU26, the salt-tolerant parent, and require less irrigation water and fertilizer than other genotypes. They are being used for rotation with cotton, and also inter-planted within a cotton crop, with higher yield than the current wheat variety Inqlab (Farooq, 2004; Farooq and Azam, 2005).

Tall wheatgrasses (E or J genomes) are very salt tolerant. At low to moderate salinity, they have a similar decline in biomass as does barley and the more tolerant bread wheat varieties (Figure 2), but continue to grow at high salinity even up to seawater concentrations and beyond. The diploid E genome species *Lophopyrum elongatum* (syn. *Agropyron elongatum* and *Elytrigia elongata*) was hybridised with bread wheat, and disomic addition and substitution lines were produced by J. Dvorák and colleagues as a source of novel genes for improving the salt tolerance of bread wheat (summarised by Colmer et al., 2006). Field studies showed that the amphiploid had a higher salt tolerance but a lower yield than Chinese Spring, and that chromosome 3E had a major effect on salt tolerance. *Thinopyrum ponticum* (decaploid, E genome) is the "tall wheatgrass" commonly used as a forage crop in saline land, and is very salt tolerant (Figure 2). Somatic hybridization techniques were used to transfer *Th. ponticum* chromosomes into bread wheat, and field experiments were conducted with F_4 and F_5 generation lines grown in a soil of moderately high salinity(Chen et al., 2004). The bread wheat parent died before maturity, whereas the two hybrids produced a good yield. Thus, salt tolerance of *Th. ponticum* appears to have been introgressed into bread wheat, with the *Th. ponticum* chromatin stably inherited (Chen et al., 2004). *Thinopyrum junceum* (hexaploid, mixed E and J genomes, $J_1J_1J_2J_2EE$) has been utilised by Wang et al. (2003) to produce recombinant lines of wheat containing segments of chromosome 5J. The data in Wang et al. (2003) show that yield of these lines was little affected by moderate salinity, and had a yield equal or better than that of Kharchia 65. However there were no data for yield in non-saline soil, so it remains possible that lines carried a significant yield penalty.

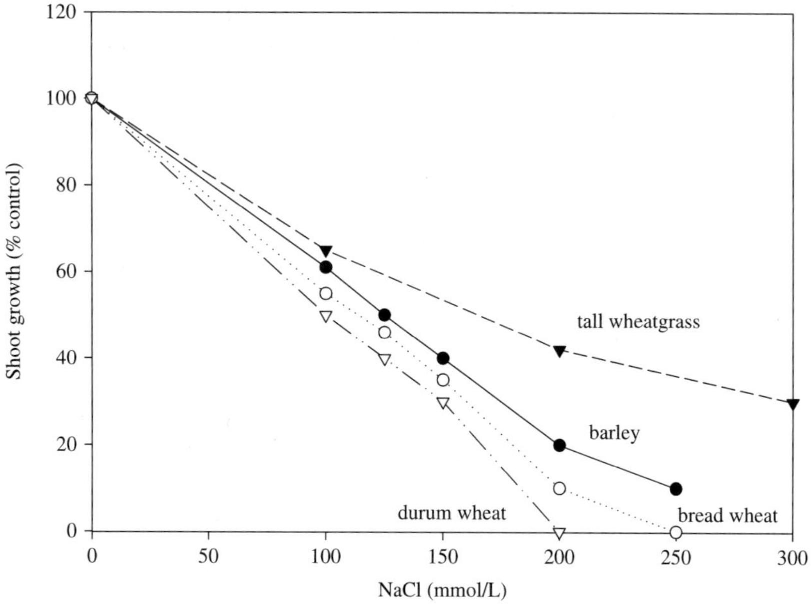

Figure 2. Effect of NaCl for 4–5 weeks on shoot biomass of wheat, barley and tall wheatgrass (*Thinopyrum ponticum*). Adapted from Colmer et al. (2005)

3. TRAIT-BASED BREEDING FOR YIELD UNDER DROUGHT AND MARKER-ASSISTED SELECTION

Targeted breeding for improvements in tolerance to biotic and abiotic factors has lifted yields in many environments, and has improved water use and thereby grain yield in water-limited environments. The best examples are breeding for resistance to nematodes, and tolerance of soil acidity and boron toxicity. Breeding for physiological traits associated with performance under drought has not been widely adopted by wheat breeders. Accordingly, there are only a small number of examples where trait based approaches have been successful in breeding for improved yield under drought.

Successful examples of yield improvement under drought are related more to improved water productivity, as discussed earlier and shown in Figure 1, than to improved drought tolerance or to traits associated with plant water relations. All traits fit the framework proposed by Passioura (1977) whereby improved yield under drought must be a function of (i) soil water extracted (ii) the efficiency of use of this water and (iii) harvest index. This is true for traits in all species that have been successful in improving yield under drought (Richards 2006). For wheat, the successful examples are breeding for (i) a reduced xylem vessel diameter in the seminal roots that slows water use when the soil is dry thereby saving more water for grain-filling to result in a higher harvest index (Richards and Passioura,

1989), (ii) osmotic adjustment associated with improved water relations about the time of flowering and hence improved fertility (Morgan, 2000), and (iii) a reduced carbon isotope discrimination (CID) to improve transpiration efficiency (Rebetzke et al. 2002). New wheat varieties combining improved performance under drought with disease resistance and stringent grain quality were released for CID (cultivars Drysdale and Rees) and osmotic adjustment (cultivar Mulgara) whereas for xylem vessel diameter a backcrossing program resulted in improved performance in dry conditions but a new rust strain appeared which was virulent on advanced breeding lines which made them unacceptable to farmers. Figure 3 shows the yield advantage of lines with narrow xylem vessels whereas Figure 4 shows the yield advantage of lines selected for improved transpiration efficiency.

A trait which has been universally regarded as important under drought is earlier flowering, as it leads to drought-avoidance and a higher harvest index and yield. It is a widespread practice in breeding programs to select for it in early generations. Earlier flowering was associated with yield progress in Australian breeding programs in dry areas for over 100 years until the 1970s. This is no longer evident and the same is likely to be true in other countries as well. However, genes controlling flowering time may still be important. Earlier sowing, made

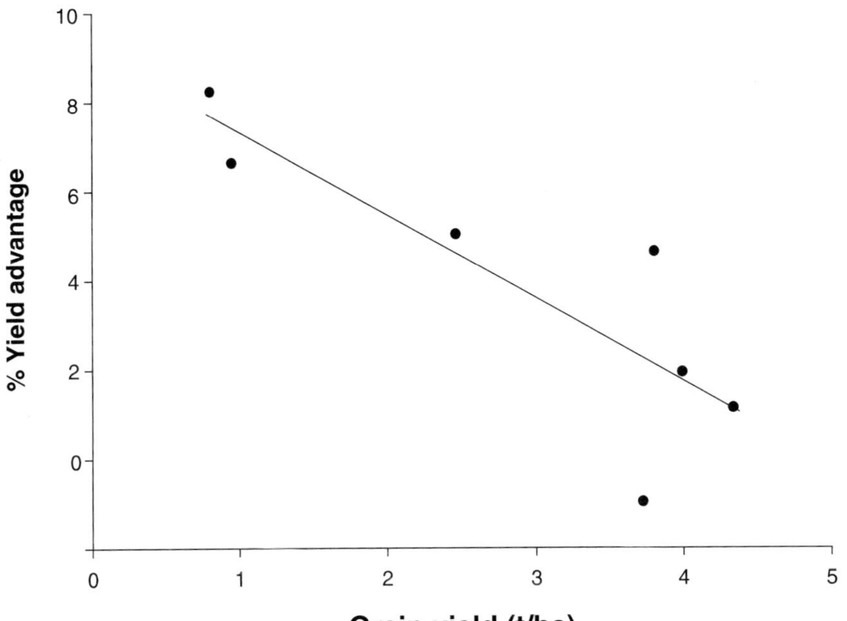

Figure 3. Yield advantage of lines selected for narrow xylem vessels. Values shown are the yield advantage of lines selected for narrow xylem vessels compared with the unselected controls and averaged over two genetic backgrounds (cultivars Kite and Cook). Data adapted from Richards and Passioura (1989)

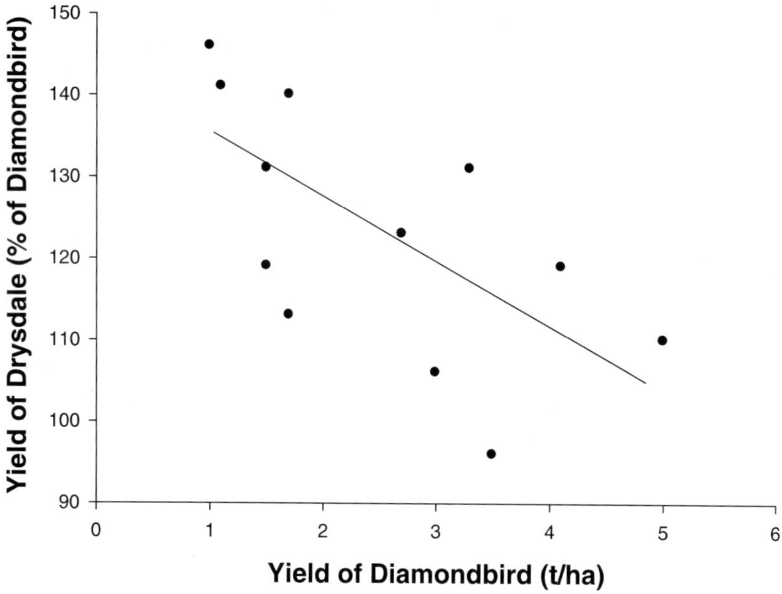

Figure 4. Yield advantage of Drysdale (high transpiration efficiency) wheat compared with Diamondbird (low transpiration efficiency) in 12 environments in southern New South Wales in 2003. Diamondbird is the current recommended wheat variety in the region where trials were sown. Data from Agritech Services, Young, NSW Australia, 2003

possible by flowering genes that are regulated by photoperiod or cold (vernalisation), can improve water use and water-use efficiency (Gomez-Macpherson and Richards, 1995; Richards, 2006). Numerous wheat varieties with these genes have been released in Australia from CSIRO and NSW Agriculture (Whiting, 2004). In addition to improving water-use efficiency and productivity these varieties can be grazed by animals providing farmers with an additional source of income achieved by better use of the limited rainfall. Similar successes have been achieved in the Great Plains of the USA (Winter and Musick, 1991).

Several additional traits are being selected in breeding programs to improve performance under drought. Tillering is being reduced in wheat by selecting for the *tin* gene, which inhibits tillering in wheat, and thereby reduces the production of wasteful tillers (Richards, 1988). Field studies contrasting near-isogenic lines with and without the *tin* gene over numerous years and locations show either a yield gain or neutral effect in most dryland field environments (Duggan et al., 2005). This trait results in a small reduction in leaf area development and an improved harvest index. In addition, a larger root system tends to be associated with this gene (Richards et al., 2006). Wheats with long coleoptiles and larger seedling leaf areas are being selected primarily in Mediterranean-like environments where current rainfall during early crop growth results in significant losses of soil water by direct evaporation (Lopéz-Castañeda and Richards 1994a). Such wheats could emulate barley, which has greater early vigour than wheat and a higher yield in dry environments(Cooper et al., 1987). Most wheats also possess the GA insensitive dwarfing genes *Rht1* (*Rht-B1b*) and *Rht2* (*Rht-D1b*). These genes have been associated with the 'Green Revolution', primarily in high yielding irrigated environments, but they reduce coleoptile length and slow growth (Allen, 1980; Richards, 1992). Both of these characteristics are undesirable in rainfed Mediterranean-like environments, especially if deep sowing is necessary to access moisture below the soil surface. A range of alternative dwarfing genes that retain the semidwarf stature but allow the development of long coleoptiles are now being used in breeding programs instead of *Rht1* and *Rht2* (Richards et al., 2002; Ellis et al. 2005). These genes not only improve crop establishment, particularly when soil water is receding or there is a heavy stubble load, (Rebetzke et al. 2005) but also improve early vigour (Ellis et al., 2004). These traits are also being incorporated into durum wheat (A. G. Condon and D. Mullan, unpublished).

In summary, trait-based breeding has delivered improved cultivars for dry environments, and there remains a range of additional traits that are likely to be important for further improvement. Some of these traits are complex and are controlled by many genes (eg, xylem vessel diameter, carbon isotope discrimination, seedling vigour), whereas others are simple and are under the control of single genes (tiller inhibition, dwarfing genes) although significant additional variation is often evident which cannot be accounted for by the major genes.

Knowledge of the target environment is very important in trait-based breeding as the benefit from a trait may be confined to specific environments. Good examples of this are enhanced early vigour and transpiration efficiency (TE). Greater early vigour will be very important to reduce the evaporation from the soil surface when crops are reliant on within-season rainfall so that more water is used by the crop for transpiration and growth. But when the crop is reliant on summer rainfall, greater early vigour may reduce yield as a vigorous crop may use the soil water too fast leaving little available for grain filling. Figure 5 shows a summary of yield outcomes from crop simulations that (a) increase early vigour by doubling the size of the first seedling leaf and (b) increase TE by 25% but at the same time imposing a growth penalty of 10% associated with greater TE (A. G. Condon and M. Stapper, unpublished; cited in Condon et al., 2002). The crop simulations were run in two representative low rainfall environments using long term weather data (~30 years). In the environment where the crop is reliant on growing season rainfall the simulations show yields may increase 11% if crops have greater early vigour, but there would be no increase associated with an improvement in TE. The converse was true when simulations were run in the environment reliant on stored soil water. When simulations were conducted by combining TE and vigour in the two contrasting environments, larger yield gains were found in both environments.

Marker-assisted selection has not contributed to improved selection efficiency in the examples given above. However, molecular markers have the potential to select for osmotic adjustment (Morgan and Tan, 1996), and have greatly improved

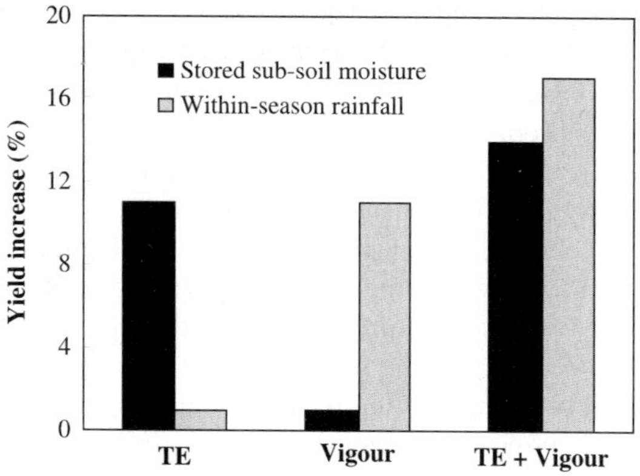

Figure 5. Wheat yield summary from simulations over about 30 years in two contrasting environments following breeding to increase (i) early seedling vigour (ii) transpiration efficiency (TE) and (iii) the combinations of (i) and (ii). Adapted from Condon et al., 2002

the selection efficiency for tiller inhibition (Spielmeyer and Richards, 2004), GA responsive dwarfing genes (Ellis et al., 2002, 2005), and early vigour (Spielmeyer et al., 2007). For other biotic or abiotic factors that limit water use there are several molecular markers that have been important in plant selection (see Passioura et al., this volume). Quantitative trait loci (QTL) have been identified for carbon isotope discrimination CID (G. J. Rebetzke et al., unpublished) but these are unlikely to replace the direct measurement of CID in breeding programs. Challenges associated with the use of QTLs in marker-assisted selection for yield in dry environments are discussed by Passioura et al. (this volume).

4. TRAIT-BASED BREEDING FOR SALINITY TOLERANCE, AND MARKER-ASSISTED SELECTION

Many traits for salt tolerance are in common with drought tolerance. Obvious examples are osmotic adjustment, and the production of osmoprotectants including compounds and enzymes that de-toxify reactive oxygen species. These would help to maintain turgor as the soil water potential falls, and prevent oxidative damage in leaves as stomates close and photosynthetic rate slows. Less obvious traits are those for improved efficiency of water use, such as early vigour, long coleoptiles and transpiration efficiency, as described above. Early vigour would maximize growth when conditions are favourable for growth early in the season or when water is more available and salt concentrations are lower. Transpiration efficiency would optimize the water use when soil moisture is less available.

Two traits that are specific to salinity relate to the prevention of Na^+ toxicity in wheat, namely (1) Na^+ exclusion by roots, and the associated high discrimination for K^+ over Na^+ in leaves, and (2) tolerant of high internal Na^+ concentrations in leaves. To date there have been no releases of salt tolerant wheat based on a particular trait. Successes have come from empirical selection, as described above. However several breeding programs are in progress in durum and bread wheat using trait-based selection.

4.1. Breeding for the Na^+ Exclusion Trait

Amongst wheat and its relatives, salt tolerance is associated with low rates of transport of Na^+ to shoots, and high selectivity for K^+ over Na^+ (Gorham et al. 1987). Na^+ exclusion in bread wheat is associated with the D genome. Durum wheats (AABB) have higher rates of Na^+ accumulation and poor K^+/Na^+ discrimination (Gorham et al., 1987) and are less salt tolerant than bread wheat. Bread wheats (AABBDD) have a low rate of Na^+ accumulation and enhanced K^+/Na^+ discrimination, a character located on the long arm of chromosome 4D (Gorham et al., 1987). This character is controlled by a single locus, *Kna1* (Dvorák et al., 1994). There have been two major efforts to improve the salt tolerance of durum wheat by enhancing its ability to exclude Na^+.

The first approach to improve the salt tolerance of durum wheat was to transfer the *Kna1* locus from the D genome of hexaploid wheat into tetraploid wheat (Dvorák et al., 1994). Recombination of the distal part of the long arm of chromosome 4D with chromosome 4B was obtained using the pairing mutant *ph1c* which inhibits the normal suppression of pairing between homoeologous chromosomes. This has created novel tetraploid germplasm with enhanced K^+/Na^+ discrimination. However the absolute biomass was lower in lines containing *Kna1* than in current durum cultivars, indicating that the recombined chromosomal fragment was bringing undesirable genes. Subsequent work reduced the size of the chromosomal fragment via another round of homoeologous recombination using the *ph1c* mutant, but has not yet produced an agronomically acceptable plant (Gorham et al., 1997).

The second approach was to select for natural variation in the A or B genome. From a screen of a number of tetraploids, a particular durum wheat, Line 149, was selected as having exceptionally low rates of Na^+ accumulation in leaves (Munns et al., 2000). This phenotype is controlled by two dominant genes of major effect (Munns et al., 2003), which have been named *Nax1* and *Nax2* (for $\underline{Na}^+$ e$\underline{x}$clusion). One gene retrieves Na^+ from the xylem in the roots, and enhances K^+ loading of the xylem, while the other retrieves Na^+ from the xylem in both the root and the shoots (James et al., 2006). Both genes appear to derive from a *T. monococcum* accession, and to be absent in modern tetraploid and hexaploid wheat (James et al., 2006). Selected progeny of the original cross of Line 149 with the Australian durum cultivar Tamaroi were backcrossed into Tamaroi and other Australian cultivars, and are being evaluated in the field in saline and non-saline soil. Initial trials indicate a significant yield improvement in saline soil (R. Munns, R. Hare and A. Rathjen, unpublished data).

The transfer of *Nax* genes from the durum wheat Line 149 into durum cultivars has been assisted by molecular markers. *Nax1* was mapped to the long arm of chromosome 2A (Lindsay et al. 2004) and one very tightly linked marker, *gwm312*, is being used routinely to select low Na^+ progeny in the durum breeding program. *Nax2* has recently been mapped (Byrt et al., 2007) and a tightly linked marker is being used for selection of lines containing *Nax2*. In the field, the *Nax* genes reduce the Na^+ concentration in the leaves to very low concentrations, and also reduce the Na^+ in the grain (R. Munns and R. Hare, unpublished data). The markers for *Nax1* and *Nax2* are being used to pyramid the *Nax* genes with traits associated with improved performance under drought (A.G. Condon and D. Mullan, unpublished data).

The *Nax* genes have also been transferred into hexaploid wheat cultivars by inter-specific crosses, and progeny selected using the markers described above. The character of Na^+ exclusion was expressed in hexaploid wheat, *Nax1* halving the Na^+ concentration in leaves of hexaploid wheat and *Nax2* also reducing it but not as markedly (R. Munns and R. James, unpublished data). The material is being evaluated in field trials.

4.2. Breeding for the Tissue Tolerance Trait

Most durum wheats have high rates of Na^+ uptake, but do not tolerate these high internal Na^+ concentrations. In order to introduce this trait of tissue tolerance into modern cultivars, a number of accessions of durum and durum-related tetraploid subspecies were screened. Some landraces were identified with high degree of salt tolerance, despite having very high leaf Na^+ levels (Munns and James, 2003). Barley was included as a benchmark, because of its established reputation for salinity tolerance coupled with high rates of salt accumulation, and previous observations that it was slow to develop leaf injury. Significant variation in percent dead leaf (weight of dead leaf as % of total leaf dry weight) was found between individual tetraploid lines, the percent dead leaf ranging from 2 to 29% (Munns and James, 2003). The barley cultivar had a low degree of leaf injury as expected, only 3%. Five tetraploid genotypes with an exceptional combination of high Na^+ accumulation and low leaf injury have been crossed with Australian durum cultivars. Recombinant inbred lines are being developed from these crosses for marker development and breeding.

5. GENE-BASED BREEDING

5.1. Drought

'Perfect' markers have been developed for the GA insensitive dwarfing genes (*Rht-B1b*, *Rht-D1b*) and these are being used to select against these genes so as to replace them with GA-responsive dwarfing genes which allow the selection for desirable agronomic characters such as long coleoptiles (Peng et al., 1999: Ellis et al., 2005). The latter are expected to be more robust in dry environments if sown deep and where early seedling vigour is important. This is the only example known in wheat where a perfect marker is being selected to enhance performance in dry environments. However, there are two other known examples where perfect markers are also being used to select for genes which indirectly improve plant growth and yield in dry environments. Both are associated with a more effective root system and a longer duration of root growth. One is for genes that contribute to resistance to cereal cyst nematode (Ogbonnaya et al., 2001), a soil-borne organism that stunts roots, the other is for a gene resulting in healthier and more effective roots in acid soils (Sasaki et al., 2004) – a common feature of dry environments.

Genes for vernalisation response have also been isolated in wheat (Yan et al., 2003). Vernalization genes are responsible for delaying the transition from vegetative to reproductive development until a period of extended low temperatures is experienced by the plant. These genes enable crops to be sown earlier, possibly allowing grazing by animals, making them dual-purpose as grain can also be harvested at maturity (Davidson et al., 1990). If they are not grazed, biomass productivity (and water-use efficiency) can be doubled (Gomez-Macpherson and Richards, 1995). Marker-based selection is useful but not essential because the time of flowering can be easily assessed.

The above examples have been drawn from an agronomic understanding and experience in dry environments and for this reason have contributed to improved on-farm productivity increases. A vast number of genes are expressed in response to dry conditions (eg, Ozturk et al., 2002). However, few of these are likely to be relevant to field grown crops as they are often laboratory phenomena or are related more to survival than productivity. Survival may be important for natural plant communities but is unlikely to contribute to greater grain yield nor water productivity.

Public information is available on two genes which have been introduced into wheat using transformation techniques and then tested in the field. Dehydration responsive element binding gene (DREB1A), a stress responsive transcription factor, has been expressed in hexaploid wheat at CIMMYT, Mexico, but further evaluation is not proceeding. No results are available from CIMMYT on the field trials. In a glasshouse study where plants were grown in small (5 cm) pots the transgenic plants survived drought longer than controls but no information is available on the size, productivity, water-use or water-use efficiency of lines (Pellegrineschi et al., 2004). Comprehensive information is available with the *HVA1* gene from barley aleurone layers (Bahieldin et al., 2005). This gene belongs to the LEA (late embryogenesis abundant) group of proteins which accumulate during seed desiccation. Transgenic wheats with *HVA1* constitutively expressed were tested in nine field experiments over six seasons in Egypt. Considerable variability was detected in performance across the field experiments between different transgenic events. However, the best transformed lines had higher yield than the non-transformed parent cultivar (Hi-Line) in most of the dryland trials conducted (Bahieldin et al., 2005).

5.2. Salinity

A comprehensive discussion of candidate genes for salt tolerance is given by Tester and Davenport (2003), particularly in relation to roots, and by Munns (2005) particularly in relation to leaves and an expected phenotype. The three individual genes that have attracted the most interest are the Arabidopsis *SOS1*, *NHX1* and *HKT1*.

Little transformation work has been done with wheat, because of the slowness to develop an efficient transformation technology, and the reduced chance of an introduced gene causing a distinctive phenotype because of the three genomes. Two breeding programs have evaluated transformed wheat in saline fields.

A Chinese commercial wheat has been transformed to constitutively over-express the vacuolar Na^+/H^+ antiporter gene *AtNHX1* (Xue et al. 2004). Grain yield of the best T_3 transgenic wheat line grown in moderately saline (EC_e 10.6 dS/m) field plots was 50% of that under non-saline conditions, whereas in the non-transformed control the yield was only 34% of that under non-saline conditions (Xue et al. 2004). At a higher salinity (EC_e 13.7 dS/m), yield of the best transgenic line was 18% of its non-saline control, while yield of the non-transformed control was still lower at 11%. These gains in salt tolerance appear modest and convincing.

Two Australian commercial wheats have been transformed with ornithine-d-aminotransferase (OAT) by Grain Biotech Australia (Scott McNeil, pers. com.), the aim being to increase the levels of the osmoprotectant, proline. OAT is the first enzyme in the conversion of ornithine to proline and not subject to feedback regulation. Glasshouse studies showed two lines that had 2–5 fold higher yields and 2-fold higher levels of free proline than the control varieties when the plants were salt stressed. The transgenic lines were assessed for yield on a salt gradient under field conditions. They had significantly higher yields at the high salinity levels than the commercial parents, although of course not as great as when grown in non-saline soil.

6. TRAITS IN COMMON FOR DROUGHT AND SALT TOLERANCE

There are significant synergies in breeding for drought that are likely to result in higher yields in salt-affected areas. Indeed maximizing both water use and water-use efficiency in saline soils will significantly enhance crop biomass production and yield. Specific breeding for traits to improve salt tolerance, such as excluding salt from the most photosynthetically active organs and increasing tissue tolerance, will further enhance water use and thereby productivity.

Early biomass growth is a good example of a trait that has a significant benefit in both dry and saline environments. Rawson et al. (1988) report two aspects of salt tolerance – absolute salt tolerance and physiological salt tolerance. The former is the amount of biomass under saline conditions and this is primarily dependent on the intrinsic growth rate of the genotype. In general if growth is fast under favourable conditions it is also fast in saline conditions. The latter is the specific benefit derived from a salt tolerance trait. Fast biomass growth is also very important in dry regions that are largely reliant on current rainfall. Lopéz-Castañeda and Richards (1994a) report that the faster early growth of barley compared with wheat largely accounts for the 20% yield advantage of barley over wheat in dry environments of south-eastern Australia. This is attributed to the more vigorous barley shading the soil surface thereby reducing water loss by direct evaporation from the soil surface and increasing crop water use and growth (Lopéz-Castaneda and Richards, 1994b). Barley was also more vigorous than wheat in saline conditions and accounted for much of its salt tolerance (Rawson et al., 1988). The exceptional early growth of Kharchia noted earlier may be partly responsible for its salt tolerance in the field.

A high transpiration efficiency which is acknowledged as important in dry conditions should also be important in saline soils. New selection methods using carbon isotope discrimination which has resulted in the release of new wheat cultivars (Richards et al., 2002) for dry conditions should also be important in saline environments. Indeed the identity developed by Passioura (1977) for dry environments, described earlier in this chapter, is equally valid for saline environments. Additional selection for genes responsible for salt exclusion or tolerance to tissue salinity will provide additional specific benefits to wheat in saline soils, although the exclusion of Na^+ needs to be balanced by the requirement to generate turgor with other solutes, preferably potassium.

7. CONCLUDING REMARKS

Important genetic gains have been made in wheat to improve its performance in dry and saline environments using conventional breeding. Conventional breeding methods will continue to be important as farmers demand not only higher yielding varieties in dry and/or saline environments but also robust varieties which are resistant to current diseases, and consumers demand a product suitable for animal feed or the very demanding food market where such properties as flour colour and stability, milling yield and dough properties are important. This array of characteristics will continue to be most effectively integrated by conventional breeding. The challenge for breeders will be to efficiently integrate trait-based and molecular methods to increase yield progress in dry and/or saline environments. Trait-based breeding approaches, which often utilize molecular markers for key traits, are starting to deliver new and significant gains. However, important traits are often complex and controlled by a number of genes, and yield is the ultimate selection criterion.

The challenge for physiologists, molecular biologists and those involved in pre-breeding will be to convince breeders that adopting yet another trait will be significant. This may be achieved by validation studies conducted in the target environment using near-isogenic lines or populations varying in trait expression. Breeders will also require assurance that the trait they are selecting for is highly heritable using molecular markers or phenotypic selection. It is also important to have a clear understanding of how a particular trait will influence yield. Is it through more water use, more efficient use of water or a higher harvest index? When breeders adopt new traits or use new parent lines to improve performance in dry and/or saline environments they will also want information on possible trade-offs or pleiotropic effects. For example, fast early vigour may be an advantage in some years but in others may lead to the exhaustion of soil water and a low yield.

Continued gains will not be easy to achieve in dry environments. This is partly due to the large seasonal variability from year to year which often result in large g x e effects and a low heritability for yield. This also makes trait validation difficult. However, trait based breeding may be advantageous and complement empirical breeding. Some advantages of trait based breeding are that it can introduce important new variation into breeding programs, and result in greater progress in yield improvement if traits have a higher heritability than yield. This enables out-of-session selection or early generation selection, and so accelerates breeding programs.

In summary, substantial advances have been made in breeding wheat for dry environments, that will also improve performance in saline environments. Further gains in productivity will come from the addition of traits that increase the efficiency of water use in dry soils, and control the uptake of salts from saline soils.

REFERENCES

Allan, R. E., 1980, Influence of semidwarfism and genetic background on stand establishment of wheat, Crop Sci., 20, 634–638.

Ashraf, M., 2002, Exploitation of genetic variation for improvement of salt tolerance in spring wheat, in 'Prospects for Saline Agriculture' (Eds R Ahmad, KA Malik) 113–121, Kluwer Academic Publishers, Dordrecht.

Ashraf, M., O'Leary, J. W., 1996, Responses of some newly developed salt-tolerant genotypes of spring wheat to salt stress: 1. Yield components and ion distribution, J. Agron. Crop Sci., 176:91–101.

Bahieldin, A., Mahfouz, H. T., Eissa, H. F., Saleh, O. M., Ramadan, A. M., Ahmed, I. A., Dyer, W. E., El-Itriby, H. A., and Madkour, M. A., 2005, Field evaluation of transgenic wheat plants stably expressing the *HA1* gene for drought tolerance, Physiol. Plant., 123:421–427.

Blum, A. and Pnuel, P., 1990, Physiological attributes associated with drought resistance of wheat cultivars in a Mediterranean environment, Aust. J. Agric. Res., 41: 799–810.

Byrt, C., Platten, J.D., Spielmeyer, W., James, R.A., Lagudah, E.S., Dennis, E.S., Tester, M. and Munns, R., 2007, HKT1; 5-like cation transporters linked to Na^+ exclusion loci in wheat, Nax2 and Kna1, Plant Physiol., 143: 1918–1928.

Chen, S. Y., Xia, G. M., Quan, T. Y., Xiang, F. N., Yin, J., Chen, H. M., 2004, Introgression of salt-tolerance from somatic hybrids between common wheat and *Thinopyrum ponticum*, Plant Sci., 167:773–779.

Colmer, T. D., Flowers, T. J., Munns, R., 2006, Use of wide crosses and wild relatives to improve salt tolerance of wheat, J. Exp. Bot., 57:1059–1078.

Colmer, T. D., Munns, R., Flowers, T. J., 2005, Improving salt tolerance of wheat and barley: future prospects, Aust. J. Exp. Ag., 45: 1425–1443.

Condon, A. G., Richards, R. A., Rebetzke, G. J. and Farquhar, G. D., 2002, Improving intrinsic water-use efficiency and crop yield, Crop Sci., 42:122–131.

Cooper, P. J. M., Gregory, P. J., Keatinge, J. D. H. and Brown, S. C., 1987, Effects of fertilizer, variety and location on barley production under rainfed conditions in northern Syria. 2. Soil water dynamics and crop water use, Field Crops Res., 16:67–84.

Davidson , J. L., Jones, D. B. and Christian, K. R., 1990, Winter feed production and grain yield in mixtures of spring and winter wheats, Aust. J. Agric. Res., 41:1–18.

Dreccer, M. F., Borgognone, G., Ogbonnaya, F. C., Trethowan, R. M. and Winter, B., 2007, CIMMYT-selected derived synthetic bread wheats for rainfed environments: yield evaluation in Mexico and Australia. Field Crops Res. 100: 218–228.

Duggan, B. L., Richards, R. A. and van Herwaarden, A. F., 2005, Agronomic evaluation of a tiller inhibition gene (*tin*) in wheat. II. Growth and partitioning of assimilate, Aust. J. Ag. Res., 56: 179–86.

Dvořák, J., Noaman, M. M., Goyal, S., and Gorham, J., 1994, Enhancement of the salt tolerance of *Triticum turgidum* L by the *Kna1* locus transferred from *Triticum aestivum* L chromosome 4D by homoeologous recombination, Theor. Appl. Genet., 87:872–877.

Ellis, M. H., Rebetzke G. J, , Chandler, P., Bonnett, D., Spielmeyer, W., Richards, R. A., 2004, The effect of different height reducing genes on the early growth of wheat, Funct. Plant Biol., 31: 583–589.

Ellis, M. H., Spielmeyer, W., Rebetzke G. J. and Richards, R. A., 2002, "Perfect" markers for the Rht-B1b and Rht-D1b dwarfing genes in wheat. Theor. Appl. Genet., 105: 1038–1042.

Ellis, M. H., Rebetzke, G. J., Spielmeyer, W. I., Richards, R. A., and Bonnett, D. G., 2005, Molecular Mapping Gibberellin-Sensitive Dwarfing Genes in Wheat (Triticum Aestivum L.), Theor. Appl. Genet., 111: 423–30.

Farooq, S. 2004, Salt tolerance in *Aegilops* species: A success story from research and production to large-scale utilization of salt tolerant wheat. In: Taha, F. K., Ismail, S. and Jaradat, A., eds, *Prospects of Saline Agriculture in the Arabian Peninsula*, Massachusetts, Amherst Scientific Publishers, 121–134.

Farooq, S., Asghar, M., Iqbal, N., Askari, E., Arif, M., and Shah, T. M., 1995, Production of salt-tolerant wheat germplasm through crossing cultivated wheat with *Aegilops cylindrica* .2. Field evaluation of salt-tolerant germplasm, Cer. Res. Commun., 23:275–282.

Farooq, S. and Azam, F., 2005, Salinity tolerance in *Triticeae*, Czech J. Genet. Plant Breed., 41:252–262.

French, R. J. & J. E. Schultz, 1984. Water use efficiency of wheat in a Mediterranean-type environment. I. The relation between yield water use and climate. Aust. J. Ag. Res. 35:743–764.

Gilmour, A. R., Cullis, B. R. and Verbyla, A. P., 1997, Accounting for natural and extraneous variation in the analysis of field experiments, J. Agric., Biol. and Environ. Stats., 3:269–293.

Gomez-Macpherson, H. and Richards R. A., 1995, Effect of sowing time on yield and agronomic characteristics of wheat in south-eastern Australia, Aust. J. Ag. Res., 46:1381–1399.

Gorham, J., Hardy, C., Wyn Jones, R. G., Joppa, L. R. and Law, C. N., 1987, Chromosomal location of a K/Na discrimination character in the D genome ofwheat, Theor. Appl. Genet., 74, 584–588.

Gorham, J., Bridges, J., Dubcovsky, J., Dvořák, J., Hollington, P. A., Luo, M. C., and Khan, J. A., 1997, Genetic analysis and physiology of a trait for enhanced K^+/Na^+ discrimination in wheat, New Phytol, 137:109–116.

Hollington, P. A., 2000, Technological breakthroughs in screening/breeding wheat varieties for salt tolerance, in: Proceedings of the National Conference Salinity management in agriculture (Eds. Gupta, S. K. Sharma, S. K., and Tyagi, N. K.), December 1998, Karnal, India: Central Soil Salinity Research Institute, 273–289.

Hollington, P. A., Royo, A., Miller, T. E., Quarrie, S. A., Mahmood, A., and Aragüés, R, 1994, The use of doubled haploid breeding techniques to develop wheat varieties for saline areas. Proceedings of the 3rd Congress of the European Society of Agronomy, 156–157.

Hollington, P. A., Aktar, J., Aragüés, R., Hussain, Z., Mahar, A. R., Quarrie, S. A., Qureshi, R. H., Royo, A., and Saqib, M., 2002, Recent advances in the development of salinity and waterlogging tolerant bread wheats, in 'Prospects for Saline Agriculture' (Eds Ahmad, R. and Malik, K. A.), 83–99, Kluwer Academic Publishers, Dordrecht.

Jafari-Shabestari, J., Corke, H., and Qualset, C. O., 1995, Field evaluation of tolerance to salinity stress in Iranian hexaploid wheat landrace accessions, Genetic Resources and Crop Evaluation, 42:147–156.

James, R. A., Davenport, R., and Munns, R., 2006, Physiological characterisation of two genes for Na^+ exclusion in wheat: *Nax1* and *Nax2*, Plant Physiol., 142:1537–1547.

Laing, D. R. and Fischer, R. A., 1977, Adaptation of semidwarf wheat cultivars to rainfed conditions, Euphytica, 26:129–39.

Lindsay, M. P., Lagudah, E. S., Hare, R. A., and Munns, R., 2004, A locus for sodium exclusion (*Nax1*), a trait for salt tolerance, mapped in durum wheat, Funct. Plant Biol., 31: 1105–1114.

Lòpez-Castañeda, C. and Richards, R. A., 1994a, Variation in temperate cereals in rainfed environments. I. Grain yield, biomass and agronomic characteristics, Field Crops Res., 36:51–62.

Lòpez-Castañeda, C. and Richards, R. A., 1994b, Variation in temperate cereals in rainfed environments. III Water use and water-use efficiency, Field Crops Res., 39:85–98.

Maas, E. V., Hoffman, G. J., 1977, Crop salt tolerance - current assessment, J. Irrig. Drain. Div. Amer. Soc. Civil Eng., 103:115–134.

Morgan, J. M., 2000, Increases in grain yield of wheat by breeding for an osmoregulation gene: relationship to water supply and evaporative demand. Aust. J. Ag. Res., 51:971–78.

Morgan, J. M. and Tan, M. K., 1996, Chromosomal location of a wheat osmoregulation gene using RFLP analysis, Aust. J. Plant Physiol., 23:803–806.

Mujeeb-Kazi, A., Diaz de Leon, J. L., 2002, Conventional and alien genetic diversity for salt tolerant wheats: focus on current status and new germplasm development. In: Ahmad, R. and Malik, K. A., eds, *Prospects for Saline Agriculture*, vol. 37, Dordrect, Kluwer Academic Publishers, 69–82.

Mujeeb-Kazi, A., Gilchrist, L. I., Fuentes-Davila, G. and Delgado, R., 1998, Production and utilization of D genome synthetic hexaploids in wheat improvement. Proceedings of the Third International Triticeae Symposium, ICARDA, 369–374. Enfield, New Hampshire, USA: Science Publishers.

Munns R., 2002, Comparative physiology of salt and water stress. Plant Cell Environ., 25: 239–250.

Munns, R. 2005, Genes and salt tolerance: bringing them together, New Phytol., 167: 645–663.

Munns, R., Hare, R. A., James, R. A., Rebetzke, G. J., 2000, Genetic variation for improving the salt tolerance of durum wheat, Aust. J. Ag. Res., 51: 69–74.

Munns, R. and James, R. A., 2003, Screening methods for salinity tolerance: a case study with tetraploid wheat, Plant Soil, 253: 201–218.

Munns, R., James, R. A., and Läuchli, A., 2006, Approaches to increasing the salt tolerance of wheat and other cereals, J. Exp. Bot. 57:1025–1043.

Munns, R., Rebetzke, G. J., Husain, S., James, R. A., and Hare, R. A., 2003, Genetic control of sodium exclusion in durum wheat, Aust. J. Ag. Res., 54:627–35.

Ogbonnaya, F. C., Subrahmanyam, N. C., Moullet, O., de Majnik, J., Eagles, H. A., Brown, J. S., Eastwood, R. F., Kollmorgen, J., Appels, R., and Lagudah, E. S., 2001, Diagnostic DNA markers for cereal cyst nematode resistance in bread wheat, Aust. J. Ag. Res., 52: 1367–74.

Ozturk, A. N., Talamé, V., Deyholos, M., Michalowski, C. B., Galbraith, D. W., Gozukirmizi, N., Tuberosa, R., and Bohnert, H. J., 2002, Monitoring large-scale changes in transcript abundance in drought- and salt-stressed barley, Plant Mol. Biol., 48:551–573.

Passioura, J. B., 1977, Grain yield, harvest index, and water use of wheat, J. Aust. Inst. Agric. Sci., 43:117–120.

Passioura, J. B., Spielmeyer, W. I. and Bonnett, D. G., 2006, Requirements for success in marker-assisted breeding for drought-prone environments (this volume).

Pellegrineschi, A., Reynolds, M., Pacheco, M., Brito, R. M., Almeraya, R., Yamaguchi-Shinozaki, K., and Hoisington, D., 2004, Stress-induced expression in wheat of the *Arabidopsis thaliana* DREB1A gene delays water stress symptoms under greenhouse conditions, Genome, 47:493–500.

Peng, J., Richards, D. E., Hartley, N. M., Murphy, G. P., Devos, K. M., Flintham, J. E., Beales, J., Fish, L. J., Worland, A. J., Pelica, F., Sudhakar, D., Christou, P,, Snape, J, W,, Gale, M, D,, Harberd, N, P, 1999, 'Green revolution' genes encode mutant gibberellin response modulators. Nature, 400:256–261.

Qureshi, R. H., Rashid, A., Ahmad, N., 1990, A procedure for quick screening of wheat cultivars for salt tolerance. In : El Bassam, N., Dambroth, M. and Loughman, B. C. (eds): Genetic aspects of plant mineral nutrition. Kluwer, 315–324.

Rawson, H. M., Richards, R. A., and Munns, R., 1988, An examination of selection criteria for salt tolerance in wheat, barley and triticale genotypes, Aust. J. Agric. Res., 39:759–772.

Rebetzke, G. J.; Bruce, S. E.; 2005, Longer coleoptiles improve emergence through crop residues to increase seedling number and biomass in wheat (*Triticum aestivum* L., Plant and Soil, 272:87–100.

Rebetzke, G. J., Condon, A. G., Richards, R. A., and Farquhar, G. D., 2002, Selection for reduced carbon-isotope discrimination increases aerial biomass and grain yield of rainfed bread wheat, Crop Sci., 42:739–745.

Rengasamy, P., 2002, Transient salinity and subsoil constraints to dryland farming in Australian sodic soils: an overview. Aust. J. Exp. Ag., 42:351–361.

Richards R. A., 1988, A tiller inhibitor gene in wheat and its effect on plant growth, Aust. J. Agric. Res., 39:749–757.

Richards R. A., 1992, The effect of dwarfing genes in spring wheat in dry environments. II. Growth, water use and water use efficiency, Aust. J. Agric. Res., 43:529–539..

Richards, R. A., 2006, Physiological traits used in the breeding of new cultivars for water-scarce environments., Agricult. Water Manag., 80:197–211.

Richards, R. A., Bennett, C. W., Qualset, C. O., Epstein, E., Norlyn, J. D., and Winslow, M. D., 1987, Variation in yield of grain and biomass in wheat, barley, and triticale in a salt-affected field, Field Crops Res., 15:277–287.

Richards, R. A., and Lukacs, Z., 2002, Seedling vigour in wheat - sources of variation for genetic and agronomic improvement, Aust. J. Agric. Res., 53:41–50.

Richards, R. A., and Passioura, J. B., 1989, A breeding program to reduce the diameter of the major xylem vessel in the seminal roots of wheat and its effect on grain yield in rain-fed environments, Aust. J. Agric. Res., 40:943–50.

Richards, R. A., Rebetzke, G. J., Condon, A. G., and van Herwaarden, A. F., 2002, Breeding opportunities for increasing the efficiency of water use and crop yield in temperate cereals, Crop Sci., 42:111–121.

Richards, R. A., Watt, M. and Rebetzke, G. J., 2006, Physiological traits and cereal germplasm for sustainable agricultural systems, Euphytica, in press.

Sasaki, T., Yamamoto, Y., Ezaki, B., Katsuhara, M., Ahn, S. J., Ryan, P., Delhaize, E., and Matsumoto, H., 2004, A wheat gene encoding an aluminum-activated malate transporter. Plant J., 37: 645–653.

Schachtman, D. P., Lagudah, E. S., and Munns, R., 1992, The expression of salt tolerance from *Triticum tauschii* in hexaploid wheat, Theor. Appl. Genet., 84:714–719.

Sharma, S. K., Joshi, Y. C. and Bal, A. R., 1984, Osmotic and ionic effects in salt sensitive and resistant wheat varieties, Indian J. Plant Physiol., 27:153–158.

Spielmeyer, W. and Richards, R. A., 2004, Comparative mapping of wheat chromosome 1AS which contains the tiller inhibition gene (*tin*) with rice chromosome 5S. Theor. Appl. Genet., 109:1303–10.

Spielmeyer, W., Hyles, J., Joaquim, P., Azanza, F., Bonnett, D., Ellis, M. E., Moore, C., and Richards, R. A., 2007, A QTL on chromosome 6A in bread wheat (*Triticum aestivum*) is associated with longer coleoptiles, greater seedling vigour and final plant height, (in press).

Tabaei-Aghdaei, S. R., Harrison, P. and Pearce, R. S., 2000, Expression of dehydration-stress-related genes in the crowns of wheatgrass species [*Lophopyrum elongatum* (Host) A. Love and *Agropyron desertorum* (Fisch. ex Link.) Schult.] having contrasting acclimation to salt, cold and drought, Plant Cell Environ., 23:561–571.

The, T. T., Latter, B. D. H., McIntosh, R. A., Ellison, F. W., Brennan, P. S., Fisher, J., Hollamby, G. J., Rathjen, A. J. and Wilson, R. E., 1988, Grain yields of near-isogenic lines with added genes for stem rust resistance, (Eds) Miller, T. E. and Koebner, R. M. D., Proceedings of the seventh international wheat genetics symposium, Cambridge, UK, 13–19 July 1988, 901–906.

Tester, M., and Davenport, R., 2003, Na^+ tolerance and Na^+ transport in higher plants, Ann. Botany, 91:503–527.

Trethowan, R. M., Reynolds, M. P, Sayre, K. D. and Ortiz-Monasterio, I., 2005, Adapting wheat cultivars to resource conserving farming practices and human nutritional needs, Ann. Appl. Biol., 146:405–413.

Trethowan, R. M., van Ginkel, M. and Rajaram, S., 2002, Progress in breeding for yield and adaptation in global drought affected environments, Crop Sci., 42:1441–1446.

Valkoun, J., 2001, Wheat pre-breeding using wild progenitors, Euphytica, 119:17–23.

Wang, R. R. C., Li, X. M., Hu, Z. M., Zhang, J. Y., Larson, S. R., Zhang, X. Y., Grieve, C. M. and Shannon, M. C., 2003, Development of salinity-tolerant wheat recombinant lines from a wheat disomic addition line carrying a *Thinopyrum junceum* chromosome, Int. J. Plant Sci., 164:25–33.

Whiting, D. (ed.), 2004, Wheat varieties in Australia 1968–2001, Don Whiting:Snowtown S. A.

Winter, S. R. and Musick, J. T., 1991, Grazed wheat grain yield relationships, Agron. J., 83:130–135.

Xue, Z. Y., Zhi, D. Y., Xue, G. P., Zhang, H., Zhao, Y. X., and Xia, G. M., 2004, Enhanced salt tolerance of transgenic wheat (*Triticum aestivum* L.) expressing a vacuolar Na^+/H^+ antiporter gene with improved grain yields in saline soils in the field and a reduced level of leaf Na^+, Plant Sci., 167:849–859.

Yan, L., Loukoianov, A., Tranquillie, G., Helguera, M., Fahima, T. and Dubcovsky, J., 2003, Positional cloning of wheat vernalization gene *VRN1*, Proc. Natl. Acad. Sci. USA, 100:6263–6268.

CHAPTER 23

RECENT ADVANCES IN BREEDING MAIZE FOR DROUGHT AND SALINITY STRESS TOLERANCE

MARIANNE BÄNZIGER[1] AND JOSE-LUIS ARAUS[2]

[1] *CIMMYT (International Maize and Wheat Improvement Centre). P.O. Box 1041, Village Market-00621, Nairobi, Kenya*

[2] *CIMMYT (International Maize and Wheat Improvement Centre). Apdo. Postal 6-641, 06600 Mexico D.F. Mexico*

Abstract: Maize production losses due to drought and salinity prominently affect economies and the livelihoods of millions of people, given the global and regional importance of maize and its pronounced susceptibility to these stress factors. Climate change and accelerating competition for irrigation water are expected to further increase the need for adaptive strategies. There is vast evidence for genetic approaches being able to significantly improve the drought and salinity tolerance of maize. Field-based breeding approaches have resulted in average breeding gains of around 100 kg ha^{-1} yr^{-1} under drought conditions, and there are first reports on transgenic drought and salinity tolerance mechanisms increasing maize grain yields under laboratory and field conditions. Drought and salinity tolerance are based on complex genetic systems and successful genetic enhancement programs need to consider gene-by-gene, gene-by environment and gene-by-developmental stage interactions. In the case of drought, field-based and transgenic approaches have resulted in the improvement of diverse and potentially additive tolerance mechanisms. Increasing yields and yield stability of maize in the face of climate change and scarcity of irrigation water will therefore likely be the most successful if complementary investments in field-based and transgenic breeding approaches are being made

Keywords: Drought and salinity tolerance, conventional and marker-assisted selection, transgenic approaches, yield, maize

1. INTRODUCTION

No exact figures on yield and economic losses in maize due to drought and salinity are available. They can be assumed extensive and significantly greater for drought than salinity. Heisey and Edmeades, (1999) estimated that 20–25% of the global maize area is affected by drought in any given year. Drought in major maize

M.A. Jenks et al. (eds.), Advances in Molecular Breeding Toward Drought and Salt Tolerant Crops, 587–601.

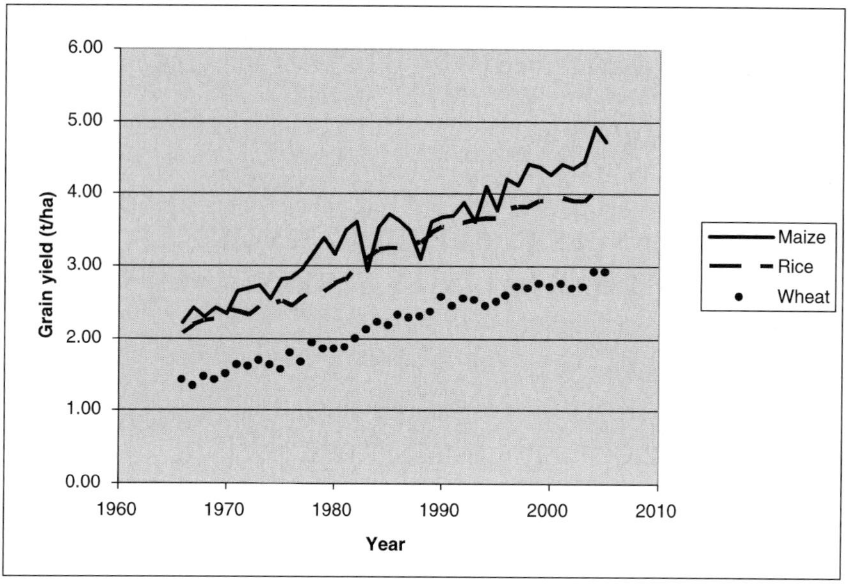

Figure 1. World maize, rice and wheat grain yields between 1966 and 2005 (FAOSTAT, 2006)

producing countries such as the United States or China routinely affects world maize yields, much more so than for rice or wheat (Figures 1 and 2). Most of the total world maize area of 150 million ha is grown under rainfed conditions, and maize is more susceptible to drought than all other cereals except rice.

Even though yield fluctuations in the main maize producing countries and more developed economies have the greatest influence on world maize production, impacts of drought on the economies and human well-being in developing economies is likely much greater. In eastern and southern Africa, where maize is the most important staple food for over 300 million people, a close correlation between rainfall and maize yields can be observed (Heisey and Edmeades, 1999), and total maize production can result in close to two-fold variation between two years (FAOSTAT, 2006; 12.5 million ton in 1992; 23.5 million ton in 1993). Drought in these countries can result in wide-spread maize crop failure, affecting the livelihood of millions of people. Between 2003 and 2005 alone, the World Food Program spent USD 1.5 billion to meet food deficiencies due to drought and crop failure in Africa (World Food Program, 2006).

As for the future Jones (2003) estimated that up to 10 million tons of maize may be lost in the developing world each year as climate change increases temperature, decreases water use efficiency and changes precipitation patterns which could eventually affect 140 million people. Increasing costs and scarcity of irrigation water, degradation of soil water holding capacity due to erosion or soil compaction, and shift of maize cropping into less favorable areas, due to population pressure or

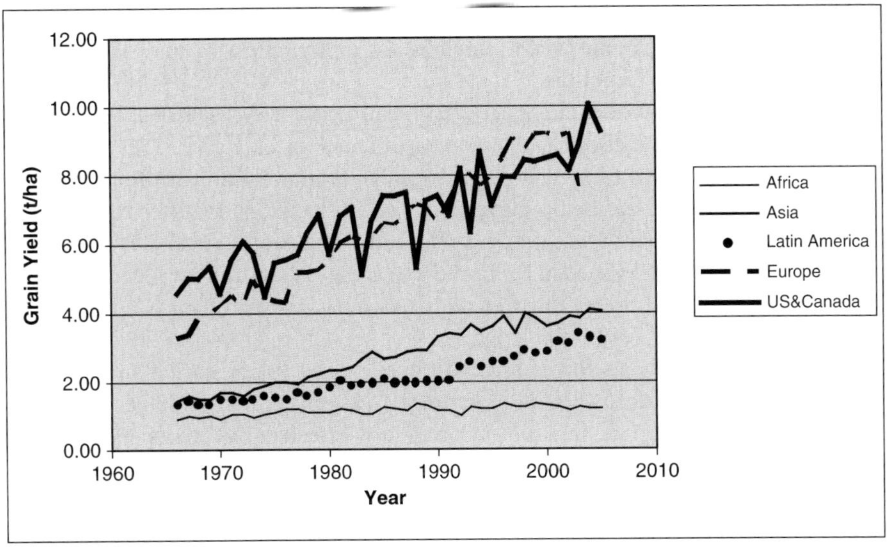

Figure 2. Maize grain yields by region between 1966 and 2005 (FAOSTAT, 2006). Arrows indicate reports of droughts. Maize production in the US and Canada in 1995 was likely affected by floods and drought

the use of more favorable areas for higher value crops or land uses, are other factors which will likely increase the frequency of drought in maize (Edmeades et al., 2006).

The greater susceptibility of maize to drought, compared to other cereals except rice, is often associated with its separation of male and female flowers and pronounced susceptibility to environmental stresses at flowering time (Grant et al., 1989; Prine 1971). The dominance of the apical tassel results in protandry which is enhanced under drought due to decreased allocation of assimilates to ears, ovules and silks resulting in reduced ear and silk growth rate, and increased kernel and ear abortion (Edmeades et al., 1993; Westgate and Boyer, 1986). Underlying processes were recently reviewed by Westgate and Boyer (2004). As a cross pollinating species which was propagated for thousands of years in highly variable landrace populations (see e.g. Reif et al., 2004), natural selection may have exerted little selection pressure for increased female survival or productivity under drought, given that cross pollination with lesser stressed individuals within a population could ensure survival of inferior alleles.

There are other characteristics which seem to disadvantage maize under drought. Maize has no noteworthy compensation mechanism through tillering, and its use of pre-flowering stem reserves for grain yield formation under drought seems less important than that of small grain cereals (Chapman and Edmeades, 1999). Lesser depth of water extraction, larger leaf area, a greater transpiration rate, slower grain grow rate and longer grain filling duration seem to disadvantage maize compared to sorghum in drought environments (Sinclair and Muchow, 2001). Many of these sorghum-like traits, however, are negative for grain yield formation in favorable years and hence undesirable when searching for maize cultivars adapted to a wide range of water supply situations.

The unpredictability of drought, geographically and across seasons, has emphasized the importance of drought tolerance as a breeding objective. Among various abiotic and biotic stress factors, drought stress is an important cause for genotype-by-environment interactions in maize across years, locations (Löffler et al., 2005; Setimela et al 2005) and most likely within individual fields (Bruce et al., 2002). Drought tolerance is thus needed for farmers to achieve high and stable maize yields and for seed companies to be able to widely market a maize cultivar (Campos et al., 2004).

Salinity is a significant factor affecting current and future agricultural productivity. FAO estimates that 45 million ha (or 19.5%) of the irrigated and 32 million ha (2.1%) of the rainfed agricultural areas are affected by secondary (human-induced) salinity with greatest areas in Argentina, Australia, Bangladesh, Brazil, Canada, China, Indonesia, India, Iran, Mexico, Pakistan, Spain, Uzbekistan (FAO, 2000). Several of these countries are major maize producers, even though irrigation may not be proportionally allocated to maize. Given the accelerated demand for maize as feed in particular in Asia and associated expansion of irrigated maize production (Falconer and Naylor, 1998), salinity tolerance could become an increasingly important breeding objective in maize.

Crops differ in their threshold level for salinity below which there is no reduction in yield. Above the threshold value, the yield decreases as a linear function of salinity until the plants die. With a threshold value of 1.7 dS/m and a slope of 12.0 % per dS/m, maize is moderately sensitive, more tolerant than rice but less tolerant than barley, sorghum, bread and durum wheat (Mass and Hoffman 1977; US Salinity Laboratory, 2006). Nitrogen fertilization and evaporative demand may influence the salinity threshold value and the sensitivity in maize (Beltrão and Asher, 1997; Katerji et al., 2000) to the extent that Katerji et al. (2000) classified maize as moderately tolerant to salinity when adjusting for evaporative demand.

Salinity is limiting crop production primarily through reductions in the expansion and photosynthetic capacity of the leaves, and accelerated senescence of older leaves. First symptoms of salinity in maize are droughty appearance and poor growth. As the stress becomes more severe, plants become stunted and develop short, thick stems and erect, gray appearing foliage (Jones, 2003). Although the number of ears developed in maize may not be affected, ear and kernel sizes are reduced with consequent reduction in grain yield. In general, however, grain yield in maize seems not as sensitive to salinity as maize forage yield (Beltrão and Asher, 1997), probably because there is little phloem transport of Na^+ and Cl^- to reproductive structures (Munns, 2002).

A wide variety of physiological, morphological and molecular traits have been suggested for use in improving the drought and salinity tolerance of crops, many of them potentially applicable to maize. Several recent reviews are available (e.g. Barker et al., 2004; Cushman and Bohnert, 2000; Flowers, 2004;

Hasegawa et al., 2000, Holmberg and Bulow, 1998; Ingram and Bartels, 1996; Munns, 2002) and additional information on potential mechanisms has been provided in other parts of this book. This chapter focuses mostly on maize-specific conventional and transgenic breeding approaches and their recent advances.

2. DROUGHT TOLERANCE IMPROVEMENT THROUGH CONVENTIONAL SELECTION

Earliest attempts to increase the drought tolerance in maize indicated low genetic variance and heritability of grain yield under drought, and large genotype-by-environment effects which fostered the notion that breeding progress for drought tolerance would be difficult to achieve (Johnson and Geadelmann, 1989), or that drought tolerance may even be negatively associated with yield potential (Quisenberry, 1982). Over the past decade, however, experiences from selection experiments confirmed significant grain yield increases under drought in both temperate and tropical maize.

2.1. Progress in Temperate Maize

Selection for yield and yield stability has been at the core of most temperate maize breeding programs (Duvick and Cassman, 1999; Troyer, 1996). In spite of earlier pessimism, it is now well established that rainfed breeding nurseries with high plant densities and large scale multi-location testing contributed to significant breeding gains in temperate maize under drought and other stress conditions (Bruce et al., 2002). Multi-environment trials, conducted at >100 to >1000 locations, exposed new hybrids frequently to drought conditions, and selection for yield stability applied consistent selection pressure on drought tolerance related traits.

Using trials exposed to different weather conditions in different years, Duvick (1997) estimated the rate of breeding progress under mild drought at 73 kg ha^{-1} $year^{-1}$ (0.85% $year^{-1}$) for hybrids released between 1930 and 1990, slightly less than under optimal conditions (89 kg ha^{-1} $year^{-1}$). In a later study, which used managed and more severe drought stressed conditions imposed at different stages of development, Campos et al. (2004) estimated rate of breeding progress at 146 kg $ha^{-1}year^{-1}$ under flowering drought and as compared to 189 kg $ha^{-1}year^{-1}$ under unstressed conditions, or about 2.0–2.5 % $year^{-1}$. Breeding progress was less if drought stress was imposed during the second half of grain filling (76 kg $ha^{-1}year^{-1}$).

Compared to hybrids from previous decades, recent and more drought tolerant hybrids showed increased interception of seasonal incident radiation through increased leaf longevity and more erect leaves, a greater uptake of nutrients and water through a more active root system, decreased apical dominance as tassel size decreased and flowering synchronization increased, and an increased grain sink size through decreased plant-to-plant variability and fewer barren plants (Campos et al., 2004; Duvick and Cassman, 1999; Tollenaar and Wu, 1999). Among the

18 hybrids evaluated by Campos et al. (2004), limited gains for tolerance to drought stress were observed in the latter part of grain filling, which was interpreted as a possible lack of genetic variation in functional stay-green. In one report, drought susceptibility of an older hybrid was associated with faster water extraction in the upper soil layers (Campos et al., 2004).

2.2. Progress in tropical maize

Tropical maize has a shorter breeding history, and a different and often broader genetic basis than temperate maize. In the 1970s, CIMMYT initiated a unique selection experiment for drought tolerance, using a lowland tropical maize population, Tuxpeño Sequía, which was subsequently improved through eight selection cycles of full-sib recurrent selection using three water levels, flowering drought stress, grain-filling drought stress, and well-watered conditions. Drought stress levels were managed by growing progenies during the dry season and using irrigation to manage timing and intensity of the drought stress. Selection was for yield and a range of secondary traits including improved flowering synchronization, increased leaf and stem extension rates, delayed leaf senescence, and reduced canopy temperature (Bolaños and Edmeades, 1993a).

Analysis of original and advanced cycles of selection of Tuxpeño Sequía established average breeding gains of 108 kg ha^{-1} yr^{-1} under drought, at yield levels ranging from 1 to 8 t ha^{-1} (Bolaños and Edmeades, 1993a). Reevaluation of similar selection procedures in several additional unrelated breeding populations produced selection gains of 80 to 144 kg ha^{-1} yr^{-1} (3.8% to 6.3% $year^{-1}$) under drought at yield levels of 1.0 to 4.5 t ha^{-1}, and slightly less under well-watered conditions (38 to 88 kg ha^{-1} yr^{-1}) where mean yields ranged from 5.8 to 10.4 t ha^{-1} (Edmeades et al., 1999).

Selection gains in tropical maize were associated with increased flowering synchronization (i.e. a reduced anthesis-silking interval), fewer barren plants, a smaller tassel size, a greater harvest index, delayed leaf senescence and, in one population, a reduced root length density in the upper soil profile, but no changes in water uptake or biomass (Bolaños et al; 1993; Bolaños and Edmeades, 1993a, b; Chapman and Edmeades 1999). Gene effects for grain yield were mostly additive (Betran et al., 2003) and polygenic (Ribaut et al., 2002).

The CIMMYT group examined a significant number of morphological and physiological traits for their value as selection criteria for drought tolerance, either by estimating their broad-sense heritabilities and genetic correlations with grain yield (Bolaños and Edmeades, 1996) or by assessing realized heritability and indirect response of grain yield to divergent selection for one particular secondary trait (Edmeades et al., 1997). Consistently, physiological or morphological traits indicative of improved water status (including leaf rolling, leaf senescence, leaf angle, leaf extension rate, canopy temperature, leaf chlorophyll concentration) or radiation interception (including leaf senescence, leaf angle, leaf chlorophyll concentration) showed relatively little impact on increasing grain yield under drought, whereas increased flowering synchronization and reduced kernel and ear abortion were closely related to grain yield improvements.

Application of what had become an established drought breeding methodology (Bänziger et al, 2000) in a product-oriented maize breeding program in southern Africa resulted more recently in significant larger selection gains under random stress conditions than those expressed by equivalent genotypes selected through multi-location testing (Bänziger et al., 2006). Differences between hybrids those selection included managed drought stress conditions and conventionally selected hybrids averaged 19% at average grain yields of 3 t ha^{-1}, and more in best hybrids. The authors concluded that inclusion of managed drought screening at early breeding stage when genetic variance is large, careful and uniform management of timing and intensity of drought stress during selection to keep heritability high and express genotype-by-drought stress interactions optimally (Bolaños and Edmeades, 1996), and use of secondary traits in addition to grain yield, i.e. classical factors of selection progress (Falconer, 1989), were the main reasons for progress.

Limited research has been directed towards improving tropical maize for seedling drought tolerance. After three selection cycles for improved biomass production and survival under seedling drought stress using field-based selection protocols, Bänziger et al. (1997) recorded only insignificant selection gains. Environmental variation was high and more success may be obtained by screening large numbers of diverse maize genotypes in more controlled systems. However, it should also be considered that natural selection in maize likely put a much higher selection pressure on survival related traits than it did on traits related to production.

3. SALINITY TOLERANCE IMPROVEMENT THROUGH CONVENTIONAL SELECTION

Salinity imposes primarily four types of stresses on plants: osmotic stress, specific ion toxicities (e.g. Na^+ and Cl^-), ionic imbalance (e.g. Na^+ versus K^+; Na^+ versus Ca^{2+}) and developmental disturbance (Grattan and Grieve, 1999; Munns, 2002). Despite intense research effort, it remains unclear whether osmotic or ionic effects dominate. Munns (1993) proposed a 'two-phase growth response to salinity' model, in which water deficits inhibit growth shortly after salinisation and then ionic effects occur later. Recently Sumer et al. (2004) found that both osmotic and ion effects were involved in the first phase of the reduction in maize growth under saline conditions.

Direct selection of superior salinity tolerant genotypes under field conditions has been hindered by the significant influence of environmental factors. Soil conditions may vary strongly not just from site to site but more importantly within a site (Richards et al., 1983). Salinity is often accompanied by changes in other soil physical and chemical properties such as sodicity, high pH, and boron (FAO, 2000), and interactions between these stresses with salinity can occur, stimulating genotype-by-environment interactions and making breeding progress more difficult.

Apart from more refined control of water application and use of saline waters for selection purposes, a significant amount of research has therefore focused on finding and understanding genetic variation in the salinity tolerance of maize seedlings. Even though salinity tolerance in maize and other cereal crops (to direct effects of salt) tends to increase with the age of the plant (Flowers, 2004; Yamaguchi and Blumwald 2005), salinity tolerance measured at maize seedling stage may persist through to mature plants (Ashraf and McNeilly, 1989; Maiti et al., 1996), indicating that solution culture at seedling stage could provide a screen for selecting for enhanced salinity tolerance in maize (Khan et al., 2003).

There is little evidence of selection for salinity tolerance having been systematically applied in applied maize breeding programs. Flowers and Yeo (1995) estimated that no more than thirty salinity tolerant crop cultivars have been developed. Among the Crop Science registrations of salt resistant cultivars listed by Flowers (1994), none included maize. Flowers (1994) concluded that "*although salinity might be of profound local importance, it had not yet had sufficient impact on regional agricultural production to warrant the effort necessary to produce new salt-tolerant cultivars*".

Genetic studies are available on the salinity tolerance of maize seedling, and they indicate that breeding progress for seedling salinity tolerance could well be made in maize. Progenies realizing greater seedling biomass and shoot length under salinity showed a narrow-sense heritability of 0.54 (Ashraf and McNeilly, 1990), and a broad-sense heritability of 0.4 in the study of Maiti et al. (1996). Additive and non-additive effects were found for root length under salinity stress, with broad and narrow sense heritability estimates approximating 0.6–0.8, and 0.4, respectively (Khan et al., 2003; Rao and McNeilly, 1999), and complex genetic systems appearing to control component traits for root growth (Khan et al., 2003) and salinity tolerance at large (Flowers, 2004).

Mechanisms of salinity tolerance between individual maize cultivars were found to be associated with different rates of salt accumulation and leaf senescence, indicating better Na+ exclusion by the more resistant maize cultivar (Fortmeier and Schubert, 1995). In other studies, varietal differences in shoot growth (Cramer et al., 1994), and shoot and root growth (Mladenova, 1990) were likely due to osmotic effects (Neumann, 1997). Maize cultivars can also differ in their response to supplemental Ca when salinized (Cramer, 2002). Across crops, characteristics of a salinity tolerant varieties include Na+ 'exclusion' from the plant or cytoplasm, K+/Na+ discrimination, decreased loading into and removal of Na + from the xylem, retention of ions in the leaf sheath, tissue tolerance, ion partitioning into different-aged leaves, and processes that promote fast growth despite the osmotic stress of the salt outside the roots including osmotic adjustment, transpiration efficiency, early vigor and early flowering (Colmer et al. 2005; Flowers, 2004; Munns, 2002). The individual genes that regulate these processes have been reviewed recently (Munns, 2005).

Computer simulation models have been used to define plant ideotypes better adapted to salinity, avoiding difficulties of field testing. Feng et al. (2003) combined the outputs of the ENVIROGRO model along with experimental data and concluded that a deeper root distribution would increase yield, particularly as longer irrigation intervals are imposed and therefore water storage capacity within the root zone becomes more important.

4. DROUGHT AND SALINITY TOLERANCE IMPROVEMENT THROUGH MARKER-ASSISTED SELECTION

First attempts to apply QTL analysis to get genetic insights into the drought tolerance response in maize were reported by Lebreton et al. (1995), but up to now few applications have emerged in practical maize breeding programs. Reasons include the complex genetic basis and influence of genetic background, developmental stage and environment on QTL effects (Tuberosa et al., 2002), limitations for precise phenotyping of components traits, time and cost considerations in fine mapping QTLs, and gene-by-gene effects (Campos et al., 2004).

Ribaut et al. (1996) identified six putative QTLs for anthesis-silking interval under drought on chromosomes 1, 2, 5, 6, 8 and 10, together accounting for 47% of the phenotypic variance. Agrama and Moussa (1996) found QTLs for drought on chromosomes 1, 3, 5, 6, and 8, explaining 50% of the phenotypic variance of grain yield under drought and expressing different types of gene action. Sanguineti et al., (1999) found 16 of 17 QTL regions influencing leaf ABA concentration also mapping to QTLs stomatal conductance, leaf temperature, relative leaf water content, anthesis-silking interval or grain yield.

Given strong QTL-by-environment effects and low explanation of phenotypic variation by individual QTLs, Ribaut et al. (1997) concluded that marker-assisted selection would have to take into account best QTLs for grain yield and secondary traits. Proof of concept was delivered in a marker-assisted backcrossing experiment which increased the drought tolerance of a recipient maize line based on the incorporation of five chromosome segments (Ribaut et al., 2002). Through co-localization of QTLs for morphological traits, related physiological parameters, and candidate genes, a consensus map was generated, including 11 genomic regions of key importance for drought tolerance in tropical maize (Ribaut et al., 2004a). Their application in selecting for drought tolerance in unrelated crosses, however, did not prove to be successful (Ribaut et al., 2004b). Campos et al. (2004) concluded that QTL information would have to be used selectively and based on the specific maize breeding situation to which they are to be applied as many QTLs identified for complex traits such as drought tolerance in maize are likely to be context-dependent.

5. TRANSGENIC APPROACHES

Recently, genomic approaches have increased our understanding of stress adaptation and stimulated drought and salinity stress related research in a wide range of areas including osmo-protectants, stress proteins, salt shock proteins, ion/proton transporters, water status, signaling components, control of transcription, growth regulators (Cushman and Bohnert, 2000). Even though most research is being conducted in model plants, there are now several accounts where transgenic approaches have led to increased drought or salinity tolerance in maize.

Overexpression of the C_4 phosphoenolpyruvate carboxylase activity in maize increased water use efficiency by 30% and dry weight by 20% under moderate drought conditions (Jeanneau et al., 2002). The overexpression of AtNHX1, a vacuolar Na+/H+ antiporter, resulted in enhanced salinity tolerance in transgenic maize (Yin et al. 2004). Quan et al. (2004) reported that transformation of maize with the betA gene from *Escherichia coli* encoding choline dehydrogenase resulted in higher glycine betaine accumulation, tolerance to drought stress at germination, young seedling stage and increased grain yields. The transformation provided greater protection of the integrity of the cell membrane and greater activity of enzymes under drought, and could also enhance salinity tolerance (Saneoka et al., 1995) Shou et al. (2004) expressed a tobacco mitogen-activated protein kinase kinase kinase constitutively in maize. As a result, transgenic maize plants maintained in pot experiments had significantly higher photosynthesis rates and produced 40–60% higher kernel weights than the non-transgenic controls. Again, the underlying protection of photosynthesis from dehydration could potentially be effective under both salinity and drought stress. More recently, information is emerging about constitutively expressed transcription factors showing a yield advantage in maize exposed to drought under multi-location field conditions (Warner et al., 2005; Heard et al., 2005).

In spite of these promising results, the development and deployment of transgenic drought tolerant maize cultivars will likely take some more time and research efforts. Drought tolerant orthologs may not produce the same results in a model plant and in an elite cultivar (Zhang et al., 2000), as experimental conditions between laboratory and field experiments often greatly vary, and as gene-by-crop effects may be expected between model plants and maize given its long history of improvement (Campos et al., 2004). The genetics underlying drought and salinity tolerance is extremely complex which may imply that manipulation of individual genes or even pathways may not result in desirable field results (Flowers 2004; Munns et al. 2006), as alteration of a single process may be compensated or damped out (Sinclair and Purcell, 2005), too small to result in significant phenotypic effects (Edmeades et al., 2004), or because tolerance mechanisms may differ between developmental stages (Flowers, 2004; Yamaguchi and Blumwald 2005), and gene and event-specific effects may be dependant on the genotype and the target environment of deployment. Given these complexities, Sinclair and Purcell (2005) concluded that genetic improvements would have to come from the integration and concurrent improvement in several traits.

6. CONCLUSIONS

Drought and salinity are important factors affecting maize production, economies and livelihoods. As for other crops, many traits, most prominently those associated with plant water status and regulation, have been proposed to influence maize drought and salinity tolerance. While the tolerance of individual maize genotypes may be ascribed to a wide range of traits, few have led so far to deliberate breeding progress in maize.

Traditional breeding approaches, nevertheless, were quite successful in increasing the drought tolerance in temperate and tropical maize at a rate averaging about 100 kg ha^{-1} yr^{-1}. Analysis of the underlying elements of breeding success has resulted in a converging and increasingly fine-tuned selection methodology for improving maize for drought tolerance based on the use of managed drought stress environments, selected secondary traits, environmental characterization, genotype-by-environment analysis, and crop simulation (Bänziger et al., 2006; Löffler et al., 2005). Realized heritablilities measured for the salinity tolerance of maize seedlings imply that progress could be achieved through conventional breeding, if implemented. In this regard, lack of application in applied maize breeding programs of traits and selection strategies much more than the lack of understanding of the underlying complex mechanisms has prevented breeding progress for drought and salinity tolerance in maize.

A significant proportion of the breeding progress in maize seems linked to reverting the pronounced susceptibility of reproductive structures to drought during the 2–3 weeks bracketing flowering time, and much less to changes in plant water status. Given that many transgenic approaches target other mechanisms and first successes are emerging, conventional and transgenic drought tolerance mechanism could turn out complementary, giving rise to significantly improved yield stability of maize under drought and salinity stress in future. At this stage, selection progress for drought tolerance from conventional breeding approaches still outperform those from transgenic approaches, raising the question why not more investment is directed towards increasing the drought and salinity tolerance of maize in applied maize breeding programs.

REFERENCES

Agrama, H.A.S. and M. E. Moussa. 1996 Mapping QTLs in breeding for drought tolerance in maize (Zea mays L.). Euph. 91: 89–97.

Ashraf, M., and T. McNeilly, 1989. Effect of salinity on some cultivars of maize. Maydica 34: 179–189.

Ashraf, M. and T. McNeilly, 1990. Improvement of salt tolerance in maize by selection and breeding. Plant Breed 104: 101–107.

Bänziger, M., G.O. Edmeades, and S. Quarrie. 1997. Drought stress at seedling stage - are there genetic solutions? pp 348–354 *In* G.O. Edmeades, M. Bänziger, H. R. Mickelson, and C.B. Peña-Valdivia (eds.) Developing Drought and Low N-Tolerant Maize. Proceedings of a Symposium, March 25–29, 1996, CIMMYT, El Batán, Mexico. Mexico D.F., CIMMYT.

Bänziger, M., G.O. Edmeades, D. Beck, and M. Bellon. 2000. Breeding for drought and nitrogen stress tolerance in maize. CIMMYT Special Publication. Mexico, D.F.: CIMMYT. 68 p.

Bänziger, M., P.S. Setimela, D. Hodson, and B. Vivek. 2006. Breeding for improved drought tolerance in maize adapted to southern Africa. Agricultural Water Management 80: 212–224.

Barker, T. , H. Campos, M. Cooper, D. Dolan, G.O. Edmeades, J. Habben, J. Schussler, D. Wright, C. Zinselmeier, 2005. Improving drought tolerance in maize. Plant Breed. Rev. 25: 173–253.

Beltrão, J. and J.B. Asher. 1997. The effect of salinity on corn yield using the CERES-maize model. Irrig. Drain. Syst. 11: 15–28.

Betran, F.J., D. Beck, M. Bänziger, and G. O. Edmeades. 2003. Genetic analysis of inbred and hybrid grain yield under Stress and nonstress environments in tropical maize. Crop Sci. 43:807–817.

Bolaños, J., and G.O. Edmeades. 1993a. Eight cycles of selection for drought tolerance in tropical maize. I. Responses in grain yield, biomass and radiation interception. Field Crops Res. 31:233–252.

Bolaños, J., and G.O. Edmeades. 1993b. Eight cycles of selection for drought tolerance in tropical maize. II. Responses in reproductive behavior. Field Crops Res. 31:253–268.

Bolaños, J., G.O. Edmeades and L. Martinez. 1993. Eight cycles of selection for drought tolerance in tropical maize. III. Responses in drought-adaptive physiological and morphological traits. Field Crops Res. 31:269–286.

Bolaños, J., and G.O. Edmeades. 1996. The importance of the anthesis-silking interval in breeding for drought tolerance in tropical maize. Field Crops Res. 48: 65–80.

Bruce, W.B., G.O. Edmeades and T.C. Barker. 2002. Molecular and physiological approaches to maize improvement for drought tolerance. J Exp Bot. 53: 13–25.

Campos, H., M. Cooper, J.E. Habben, G.O. Edmeades, J.R. Schussler. 2004. Improving drought tolerance in maize: a view from industry. Field Crops Res. 90: 19–34.

Chapman, S.C., and G.O. Edmeades. 1999. Selection improves drought tolerance in tropical maize populations: II. Direct and correlated responses among secondary traits. Crop Sci. 39:1315–1324.

Colmer T.D., R. Munns, and T.J. Flowers. 2005. Improving salt tolerance of wheat and barley: future prospects. Austr. J. Exp. Agric. 45: 1425–1443.

Cramer, G.R. 2002. Sodium-calcium interactions under salinity stress. p. 205–228. *In* A. Lauchli and U. Luttge (eds) Salinity: Environment - Plants – Molecules. Kluwer Academic Publishers, Dordrecht, The Netherlands.

Cramer, G. R. and D.C. Bowman. 1991. Kinetics of maize leaf elongation. I. Increased yield threshold limits short-term, steady-state elongation rates after exposure to salinity, J. Exp. Bot. 42: 1417–1426.

Cramer G.R., G.J. Alberico, and C. Schidt. 1994. Salt tolerance is not associated with the sodium accumulation of two maize hybrids. Austr. J. of Plant Physiol. 21: 675–692.

Cushman, J.C. and H.J. Bohnert. 2000. Genomic approaches to plant stress tolerance. Current Opinion Plant Biol 2000, 3:117–124.

Duvick, D.N. 1997. What is yield? p. 332–335. *In* G.O. Edmeades, M. Bänziger, H. R. Mickelson, and C.B. Peña-Valdivia (eds.) Developing Drought and Low N-Tolerant Maize. Proceedings of a Symposium, March 25–29, 1996, CIMMYT, El Batán, Mexico. Mexico D.F., CIMMYT.

Duvick, D.N. and K. G. Cassman. 1999. Post–green revolution trends in yield potential of temperate maize in the North-Central United States. Crop Sci. 39:1622–1630.

Edmeades, G.O., J. Bolaños, M. Hernandez, and S. Bello., 1993. Causes for silk delay in lowland tropical maize. Crop Sci. 33: 1029–1035.

Edmeades, G.O., J. Bolaños, and S.C. Chapman. 1997. Value of secondary traits in selecting for drought tolerance in tropical maize. *In* G.O. Edmeades, M. Bänziger, H.R. Mickelson, and C.B. Peña-Valdivia, (eds.). Developing Drought- and Low N-Tolerant Maize. Proceedings of a Symposium, March 25–29, 1996, CIMMYT, El Batán, Mexico. Mexico, D.F.: CIMMYT.

Edmeades, G.O., J. Bolaños, S.C. Chapman, H.R. Lafitte, and M. Bänziger. 1999.Selection improves drought tolerance in tropical maize populations: I. Gains in biomass, grain yield, and harvest index. Crop Sci. 39:1306–1315.

Edmeades, G.O., G.S. McMaster, J.W. White, and H. Campos. 2004. Genomics and the physiologist: bridging the gap between genes and crop response. Field Crops Res. 90: 5–18.

Edmeades, G.O., M. Bänziger, H. Campos, and J.R. Schussler. 2006. Improving tolerance to abiotic stresses in staple crops: a random or planned process? p. 293–309. In: K.R. Lamkey and M. Lee (Eds) Plant Breeding: The Arnel R. Hallauer International Symposium. Blackwell Publishing, Ames IA.

Falcon, W.P. and R.L. Naylor. 1998. The maize transition in Asia: Unlocking the controversy. Amer. J Agri. Econ. 80: 960–968.

Falconer, D.S. 1989. Introduction to Qunatitative Genetics. 3rd ed. John Wiley and Sons, New York NY.

FAO, 2000. Extent and causes of salt-affected soils in participating countries. FAO Rome, Italy. www.fao.org.

FAOSTAT. 2006. Food and Agriculture Organization of the United Nations. FAO Rome, Italy. http://faostat.fao.org.

Feng, G.L., A. Meiri, and J. Letey. 2003. Evaluation of a model for irrigation management under saline conditions: II. Salt distribution and rooting pattern effects. Soil Sci. Soc. Am. J. 67: 77–80.

Flowers, T.J. 2004. Improving crop salt tolerance. J. Exp. Bot. 55: 307–319.

Fortmeier, R., and S. Schubert. 1995. Salt tolerance of maize (Zea mays L.): the role of sodium exclusion. Plant, Cell and Environment 18: 1041–1047.

Grant, R. F., B.S. Jackson, J.R. Kiniry, and G.F. Arkin. 1989. Water deficit timing effects on yield components in maize. Agron. J. 81: 61–65.

Grattan, S.R., and C.M. Grieve. 1999. Mineral nutrient acquisition and response by plants grown in saline environments. pp. 203–229. *In* M. Pessarakli (ed.) Handbook of plant and crop stress. Marcel Dekker, New York.

Hasegawa, P.M., R. A. Bressan, J.K. Zhu, and H.J. Bohnert. 2000. Plant cellular and molecular responses to high salinity. Ann. Rev. Plant Physiol. Plant Mol. Biol. 51: 463–499.

Heard, J., T.R. Adams, G. Anstrom, R. Benson, D. Nelson, D. Warner, O. Ratcliffe, R. Creelman, and S. Dotson. 2005. Increasing yield stability in corn under drought conditions: new insights from transgenic studies. Abstract L 8.02 *In* InterDrought-II The 2nd International Conference on Integrated Approaches to Sustain and IMporve Plant Production under Drought Stress: Final Program and Abstract Book. Avenue media, Bologna, Italy.

Heisey, P.W., and G.O. Edmeades. 1999. Maize production in drought-stressed environments: Technical options and research resource allocation. Part 1 of CIMMYT 1997/1998 World Facts and Trends; Maize Production in Drought-Stressed Environments: Technical Options and Research Resource Allocation. Mexico, D.F.: CIMMYT.

Holmberg, N., and N. Vulgo. 1998. Improving stress tolerance in plants by gene transfer. Trends in Plant Science 3: 61–66.

Ingram, J., and D. Bartels. 1996. The molecular basis of dehydration tolerance in plants. Ann Rev Plant Physiol. Plant Mol. Biol. 47: 377–403.

Jeanneau, M., D. Gerentes, X. Foueillassar, M. Zivy, J. Vidal, A. Toppan and P. Perez. 2002. Improvement of drought tolerance in maize: towards the functional validation of the Zm-Asr1 gene and increase of water use efficiency by over-expressing C4–PEPC. Biochimie. 84:1127–35.

Johnson, S.S., and J. L. Gaedelmann. 1989. Influence of water stress on grain yield response to recurrent selection in maize. Crop Sci. 29:558–564.

Jones, J.B. 2003. Agronomic Handbook: Management of Crops, Soils and Their Fertility. CRC Press. Boca Raton, London, New York, Washington. p. 450.

Jones, P.G. and P.K. Thornton. 2003. The potential impact of climate change on maize production in Africa and Latin America in 2055. Global Environmental Change 13:51–59.

Katerji, N., J.W. van Hoorn, A. Hamdy, and M. Mastrorilli. 2000. Salt tolerance classification of crops according to soil salinity and to water stress day index. Agric. Water Manag. 43: 99–109

Khan, A.A., S.A. Rao, and T. McNeilly. 2003. Assessment of salinity tolerance based upon seedling root growth response functions in maize (Zea mays L.). Euphyt. 131: 81–89.

Lebreton C, Laziejancie V., Steed A, Pekic S, Quarrie, S.A. 1995. Identification of QTL for drought responses in maize and their testing casual relationships betweentraits. Journal of Experimental Botany 46:853–865.

Löffler, C.M., J. Wei, T. Fast, J. Gogerty, S. Langton, M. Bergman, B. Merrill, and M. Cooper. 2005. Classification of Maize Environments Using Crop Simulation and Geographic Information Systems. Crop Sci. 45:1708–1716.

Maas, E.V., and G.J. Hoffman. 1977. Crop salt tolerance, current assessment. J. Irrig. Drain. Div. ASCE 103, 115–134.

Maiti, R.K., L.E. Delgado Amaya, S. Ibarra Cardona, A.M. Ontiveros Dimas, M. De La Rosa-Ibarra, and H. De Leon Castillo. 1996. Genotypic variability in maize cultivars (Zea mays L.) for resistance to drought and salinity at the seedling stage. J. Plant Physiol. 148 741–744.

Mladenova, Y.I. 1990. Influence of salt stress on primary metabolism of Zea mays L. seedlings of model genotypes. Plant and Soil 123: 217–222.

Munns, R. 2002. Comparative physiology of salt and water stress. Plant Cell and Env. 25: 239–250.
Munns, R. 2005. Genes and salt tolerance: bringing them together. New Phytol. 167: 645–663.
Munns, R. 1993. Physiological processes limiting plant growth in saline soil: some dogmas and hypotheses. Plant Cell Env. 16: 15–24.
Munns, R., and H.M. Rawson. 1999. Effect of salinity on salt accumulation and reproductive development in the apical meristem of wheat and barley. Austr. J. Plant Physiol. 26: 459–464.
Munns, R., R. A. James, and A. Läuchli. 2006. Approaches to increasing the salt tolerance of wheat and other cereals. J. Exp. Bot. 57: 1025–1043.
Neumann, P. 1997. Salinity resistance and plant growth revisited. Plant Cell Env. 20: 1193–1198.
Prine, G.M., 1971. A critical period for ear development in maize. Crop Sci. 11: 782–786.
Quan, R., M. Shang, H. Zhang, Y, Zhao, and J. Zhang. 2004. Engineering of enhanced glycine betaine synthesis improves drought tolerance in maize. Plant Biotech. J. 2: 477
Quisenberry, J.E. 1982. Breeding for drought resistance and plant water use efficiency. p. 193–212. *In* M.N. Christiansen and C.F. Lewis (ed.) Breeding plants for less favorable environments. John Wiley and Sons, New York.
Rao, S.A., and T. McNeilly 1999. Genetic basis of variation for salt tolerance in maize (Zea mays L). Euphyt. 108: 145–150.
Reif, J. C., X. C. Xia, A. E. Melchinger, M. L. Warburton, D. A. Hoisington, D. Beck, M. Bohn and M. Frisch. 2004. Genetic Diversity Determined within and among CIMMYT Maize Populations of Tropical, Subtropical, and Temperate Germplasm by SSR Markers. Crop Sci. 44:326–334.
Richards RA. 1993. Should selection for yield in saline regions be made on saline or non-saline soils? Euphytica 32, 431–438.
Ribaut, J.M. D. A. Hoisington, J. A. Deutsch, C. Jiang and D. Gonzalez-de-Leon. 1996. Identification of quantitative trait loci under drought conditions in tropical maize. 1. Flowering parameters and the anthesis-silking interval. Theor. App Gen: 92, 905–914.
Ribaut, J.M. C. Jiang, D. Gonzalez-de-Leon, G.O. Edmeades, and D. A. Hoisington. 1997. Identification of quantitative trait loci under drought conditions in tropical maize. 2. Yield components and marker-assisted selection strategies. Theor. App Gen: 94, 887–896.
Ribaut J.M., M. Bänziger, F.J. Betrán, C. Jiang, G.O. Edmeades, K. Dreher, and D. Hoisington. 2002. Use of molecular markers in plant breeding: drought tolerance improvement in tropical maize. pp. 85–99. *In* M.S. Kang (ed) Quantitative Genetics, Genomics and Plant Breeding. Wallingford, UK: CAB International.
Ribaut J.-M., M. Bänziger, T. Setter, G. Edmeades, and D. Hoisington. 2004a. Genetic Dissection of Drought Tolerance in Maize: A Case Study. *In* H. Nguyen and A. Blum (eds.), Physiology and Biotechnology Integration for Plant Breeding. New York: Marcel Dekker, Inc. Pp. 571–611.
Ribaut, J.M., M.C. Sawkins, M Bänziger, M. Vargas, E. Huerta , C. Martinez, and M. Moreno. 2004b. Marker-assisted selection in tropical maize based on consensus map, perspectives, and limitations. p. 267–268. *In* D. Poland, M. Sawkins, J.-M. Ribaut, and D. Hoisington (eds.). 2004. Resilient Crops for Water Limited Environments: Proceedings of a Workshop Held at Cuernavaca, Mexico, 24–28 May 2004. Mexico D.F.: CIMMYT.
Saneoka H, C. Nagasaka, D.T. Hahn, W.J. Yang, G.S. Premachandra, R.J. Joly, and D. Rhodes. 1995. Salt tolerance of glycinebetaine-defcient and containing maize lines. Plant Physiol. 107: 631–638.
Setimela, P. Z. Chitalu, J. Jonazi, A. Mambo, D. Hodson and M. Bänziger. 2005. Environmental classification of maize-testing sites in the SADC region and its implication for collaborative maize breeding strategies in the subcontinent. Euphytica 145. 123–132.
Shou, H., P. Bordallo, and K. Wang. 2004. Expression of the Nicotiana protein kinase (NPK1) enhanced drought tolerance in transgenic maize. J. Exp. Bot. 55: 1013–1019.
Sinclair, T.R. and R. C. Muchow. 2001. System analysis of plant traits to increase grain yield on limited water supplies. Agron. J. 93:263–270.
Sinclairl, T.R., and L. C. Purcell. 2005. Is a physiological perspective relevant in a 'genocentric' age? J. Exp. Bot. 56: 2777–2782.
Sumer A, C. Zörb, F. Yan and S. Schubert. 2004. Evidence of sodium toxicity for the vegetative growth of maize (Zea mays L.) during the first phase of salt stress. J. Appl. Bot. 78; 135–139.

Tollenaar, M., and J. Wu. 1999. Yield improvement in temperate maize is attributable to greater stress tolerance. Crop Sci. 39:1597–1604.

Troyer A.F. 1996. Breeding widely adapted, popular maize hybrids. Euphytica 92:163–174.

US Salinity Laboratory, 2006. Salt Tolerance Databases: Fiber, Grain and Special Crops. http://www.ussl.ars.usda.gov/pls/caliche/SALTT42A.

Warner, D.C., J. Heard, R. Bensen, and D. Nelson. 2005. Development of transgenes for improvement of drought stress tolerance in maize. Abstracts 2005 International Annual Meetings. ASA-CSSA-SSSA, Madison, WI.

Westgate, M.E., and J.S. Boyer. 1986. Reproduction at low silk and pollen water potentials in maize Crop Sci. 26: 951–956.

Westgate, M.E., and J.S. Boyer. 2004. Grain yields with limited water. J. Exp. Bot. 55: 2385–2394.

World Food Program, 2006. WFP in Africa: Facts, figures and partners. WFP Liaison Office to the AU and ECA, Addis-Ababa, Ethiopia.

Yamaguchi, T. and E. Blumwald. 2005. Developing salt-tolerant crop plants: challenges and opportunities. TRENDS in Plant Sci. 10: 615–620.

Yin, X.Y., A. F. Yang, K.W. Zhang and J. R. Zhang. 2004. Production and analysis of transgenic maize with improved salt tolerance by the introduction of AtNHX1 gene. Acta Botanica Sinica 46: 854–861.

Zhang, J., H Nguyen and A Blum. 2000. Genetic analysis of osmotic adjustment in crop plants. J. Exp Bot. 50: 291–302.

CHAPTER 24

RECENT ADVANCES IN BREEDING BARLEY FOR DROUGHT AND SALINE STRESS TOLERANCE

CHENGDAO LI[1], GUOPING ZHANG[2] AND REG LANCE[1]

[1]*Department of Agriculture and Food, Government of Western Australia, 3 Baron-Hay Court, South Perth WA6151, Australia*

[2]*Department of Agronomy, Zhejiang University, Hangzhou, 310029, China*

Abstract: Barley is the most tolerance cereal crop for drought and salinity and is an ideal model crop for genetic study of drought and salinity tolerance because of its early maturity, diploid and self-pollination. Selection for drought tolerance in convention breeding programs has achieved significant progress to improve yield and yield stability under drought through direct selection or indirect selection for early vigour, coleoptile length or "stay green". A large number of Quantitative Trait Loci (QTL) were mapped for drought and salinity tolerance related traits, including physiological /biochemical traits such as osmotic adjustment capacity, proline content, stomatal conductance, water-soluble carbohydrates, relative water content, leaf turgor, ABA content, transpiration efficiency, water use efficiency and carbon isotope discrimination; and developmental/ morphological traits such as height, leaf emergence, leaf area index, tiller development, flowering time, maturity rate and root characteristics. QTLs for yield and yield components were also identified under drought. Extensive research has been devoted to the characterization of genes induced or up-regulated by drought or salinity. Numerous candidate genes were identified to associate with tolerance to drought or salinity and some of the candidate genes co-located with the QTLs for drought tolerance. Wild barley *(Hordeum spontaneum)* was demonstrated as a key genetic resource for drought and salinity tolerance. QTLs from the wild barley increased yield by 12–22% under drought. New germplasm and molecular tools make it possible to develop better barley variety faster for drought or salinity tolerance, but challenges still remain due to complexity of drought and salinity tolerance.

Barley *(Hordeum vulgare L.)* is the fourth largest cereal crop in the world with annual production over 140 million tonnes. It has been used as a staple food for humans, feed for animals, and a key ingredient in beer and whiskey production. Barley has a wider ecological range than any other cereals and is widespread in temperate, subtropical and artic areas, from sea level to heights of more than 4,500 m in the Andes and Himalayas (Bothmer et al., 1995). Barley can be grown on soils unsuitable for wheat, and at altitudes unsuitable for wheat or oats. Because of its salt and drought tolerance, barley thrives in nearly every corner of the earth including extremely dry areas near deserts. Barley is a short-season, early maturing, diploid and self-pollinating crop, thus it is also an ideal model plant for genetic study of drought and salinity tolerance. Several

M.A. Jenks et al. (eds.), Advances in Molecular Breeding Toward Drought and Salt Tolerant Crops, 603–626.

papers have summarized research on barley abiotic stress tolerance including drought and salinity tolerance (Cattivelli et al., 2002; Stanca et al., 2003). In this chapter, we will review recently progress on molecular breeding for saline and drought tolerance in barley

Keywords: Drought and salinity tolerance, marker-assisted selection, QTLs, candidate genes, yield, barley

1. DROUGHT TOLERANCE

The mechanisms of drought tolerance are classified into three categories (1) drought escape, (2) drought avoidance and (3) drought tolerance. Drought escape is defined as the ability of a plant to complete its life cycle before serious soil and plant water deficits develop. Drought avoidance is the ability of plants to maintain relatively high tissue water potential despite a shortage of soil-moisture, whereas drought tolerance is the ability to withstand water-deficit with low tissue water potential. Traits that have been investigated for drought tolerance include (1) physiological/biochemical traits, such as osmotic adjustment capacity, proline content, stomatal conductance, plant water status, water-soluble carbohydrates, epidermal conductance, canopy temperature, relative water content, leaf turgor, ABA content, transpiration efficiency, water use efficiency and carbon isotope discrimination; and (2) developmental/ morphological traits, such as leaf emergence, leaf area index, leaf waxiness, stomatal density, tiller development, flowering time, maturity rate, cell membrane stability, and root characteristics. However, yield and yield stability under drought are still considered as the most important parameters for drought tolerance.

2. IDENTIFICATION OF QTLS CONTROLLING DROUGHT TOLERANCE

Large numbers barley mapping populations have been developed to map genes and quantitative trait loci (QTLs) controlling agronomic and quality traits. The results have been reviewed recently by Fox et al. (2003). Several barley populations have been developed to map the QTLs for drought tolerance in both controlled environments and Mediterranean field trials. These included Tadmor x (ER/Apm) RIL population (Teulat et al., 1998), Derkado x B83–12 DH population (Foster et al., 2004), Apex x ISR101–23 (Pillen et al., 2003) and Barke x Hor11508 populations (Talame et al., 2004).

Tadmor is a two-rowed barley variety selcted from a Syrian landrace and characterized by high yield stability (Grando, 1989) and a high osmotic adjustment capacity (Teulat et al., 1997). The Tadmor x ER/Apm population has been used extensively to map QTLs for osmotic adjustment traits (Teulat et al., 1997a, 1997b), plant water status, water-soluble carbohydrates (Teulat et al., 2001a) and grain carbon isotope discrimination (Teulat et al., 2002). It was also grown in fields to map QTLs for grain yield and agronomic traits under Mediterranean countries

(Foster et al., 2004; Teulat et al., 2001b), in which drought is frequently a major production limitation factor. Two other populations of Apex x ISR101–23 (Pillen et al., 2003) and Barke x Hor11508 were developed to map QTLs for drought tolerance from wild barley *Hordeum spontaneum* (Foster et al., 2004; Pillen et al., 2003; Talame et al., 2004).

Relative water content (RWC) was demonstrated to be a relevant screening tool of drought-tolerance in cereals, as well as a good indicator of plant water-status. QTLs for RWC were mapped in Mediterranean field trials. A total 6 different QTLs were found, for three of the five environments studied, on chromosomes 1H, 2H, 4H, 5H, 6H and 7H. The QTLs on 1H, 6H and 7H were detected across different environments (Teulat et al., 2003). A region on the long arm of chromosome 6H contains the most-consistent QTL. This region was previously identified as controlling RWC, as well as leaf osmotic potential under water stress and osmotic adjustment, from an experiment conducted in growth-chamber conditions (Teulat et al., 1998).

Transpiration efficiency, ratio of dry matter produced to water transpired, is considered as an important drought-adaptive trait in cereals. However, direct measurements of transpiration efficiency are difficult, slow, expensive and need uniform weather conditions on large populations of plants. Carbon isotope discrimination (CID) provides an integrated measurement of transpiration efficiency of C3 crop species (Farquhar and Richards, 1984). In an early research, CID on whole shoots was largely controlled by chromosome 4H in wheat/barley addition line (Handley et al., 1994). Ten QTLs for CID were identified from maturing grain growing in Mediterranean field conditions (Teulat et al., 2002): one was specific to one environment, two presented interaction with the environment, six presented main effects across three or two environments and one presented both effects. Eight regions controlling CID were concomitant with QTLs previously identified in the same population, either for agronomic traits (Teulat et al., 2001b), or for traits related to plant water status and/or osmotic adjustment (OA) (Teulat et al., 2001a). Six regions controlling agronomic traits co-located with QTLs for CID, which include QTLs associated with thousand-grain weight and plant height on chromosome 2H (Teulat et al., 2001b), plant height and harvest index on chromosome 7H, heading dates on chromosomes 3H and 5H and plant height, thousand grain weight and the number of fertile tillers on chromosome 6H. Four QTLs controlling CID also co-located with chromosomal regions where QTLs for physiological traits related to plant water status and/or OA have been mapped previously (Teulat et al., 2001a), including a chromosome 7H region for relative water content and leaf osmotic potential (Teulat et al., 1998, 2001a), chromosome 2H region for OA, chromosome 4H region for water soluble carbohydrates and chromosome 7H for OA. These regions are of interest in terms of plant breeding as they control both important drought-adaptive traits for barley and yield components. Confirmation of the influence of these genomic regions by refining the map or observing similar effects in different populations, could help to elucidate the biological processes underlying complex traits such as yield or yield stability under drought.

It is interesting to notice that some similar QTLs have also been mapped in other cereal crops. The QTLs on chromosome 7H for RWC and some other traits (Teulat et al., 1998) was shown to be colinear with a region of rice chromosome 8, where a QTL for OA at 70% of RWC was found by Lilley et al. (1996). Morgan and Tan (1996) also identified a major gene controlling osmoregulation on the same homeologous arm in wheat (chromosome 7A). However this gene is probably at a more distal position, corresponding by synteny to rice chromosome 6. Other osmoregulation genes in barley on 6H and on 2H (Teulat et al., 1998) also correspond to a homoeologous rice chromosome reported to be associated with osmotic potential in rice (Lilley et al., 1996). Thus rice can be used as model plant for barley and wheat to understand gene functions related to drought tolerance.

Several populations have been used to map the QTLs controlling grain yield and its components in Mediterranean field trials (Teulat et al., 1997; Foster et al., 2004; Pillen et al., 2003; Talame et al., 2004). Numerous loci have been detected to control grain yield in different environments. QTL analyses across environments have revealed QTLs that are specific for each testing environment, which reflects the diversity of environments among the test sites. QTLs on chromosome 2H (centromeric region), 3H, 4H (long arm), 6H (long arm) and 7H (centromeric) showed consistency across environments. Most of these loci were located in the chromosomal regions with genes for key developmental traits, which indicated that drought escape is an important mechanism in the germplasm pool used for mapping drought tolerance. The main effects detected in these studies can be attributed to single major genes: vernalisation genes (possibly *sgh1*) located on chromosome 4H), semi-dwarfing genes, *sdw1* (3H) and *ari-e*. GP (5H) and the six/two ear type gene, *vrs1* (2H). The segregation of major genes can be misleading and their effects to confound in QTL studies as they have strong pleiotropic effects on many traits (Forster et al., 2000b; Araus et al., 2003).

Wild barley *Hordeum spontaneum* has been recognized as an important source for drought tolerance. A QTL identified on chromosome 4H from *Hordeum spontaneum* consistently increased grain yield across 6 test environments with an average yield increase of 7.7% (Pillen et al., 2003). Talame et al. (2004) identified two QTLs on chromosomes 2H and 5H with relative yield increase ranged from 12–22% under dry conditions. These QTLs could be used as target chromosome regions for integration of wild barley genes for yield improvement under drought. Lu et al. (1999) suggested that drought tolerance in wild barley is related to their differing genetic abilities of osmotic adjustment under drought conditions. Thus, further genetic mapping and marker-assisted transfer of the osmotic-adjustment genes harboured in the wild progenitor could improve resistance of cultivated barley grown in water-limited environments.

3. GENES INVOLVED DROUGHT TOLERANCE IN BARLEY

In barley, more than 370,000 ESTs have been released (HvGI Release 9.0, September 15, 2004) and organised into more than 23,000 tentative consensus (TC) sequences. Among them, 132 TCs have been annotated by gene onthology

(GO) as 'osmotic stress-', 'cold-', and 'water deprivation-' response gene products. Candidate gene approaches have been used to study drought tolerance. These genes and the proteins that they encode can be divided into three categories: (1) signaling and transcriptional control; (2) the protection of membranes and proteins and (3) water and ion transport (Wang et al., 2003). Potential candidate genes have included those encoding; (1) transcription factors, (2) compatible solutes (e.g. proline), (3) antioxidants and detoxifying enzymes, (4) ion transport, and (5) heat shock and late embryogenesis abundant proteins. Ozturk et al. (2002) used 1463 DNA elements to study gene expression response to drought and salinity in barley (*Hordeum vulgare* L. cv. Tokak). Drought and salinity stresses affect largely different sets of transcripts. Over 100 genes were up- or down regulated by drought. Those transcripts significantly up-regulated under drought stress are jasmonate-responsive, metallothionein-like, late-embryogenesis-abundant (LEA) and ABA-responsive proteins. The most drastically down-regulated category was observed for photosynthesis-related functions. Up-regulation under both drought and salt stress was restricted to ESTs for metallothionein-like and LEA proteins, while increases in ubiquitin-related transcripts characterized salt stress. A number of functionally unknown transcripts from cDNA libraries of drought-stressed plants showed up regulation by drought but down-regulation by salt stress. Among the most notable biochemical traits, accumulation of proline has received considerable attention, though contrasting conclusions have been reached concerning its role in the adaptive response to drought (Blum, 1988). The gene pyrroline-5-carboxylate dehydrogenase, which catalyzes the second step in the conversion of proline to glutamate, is characterized in a number of cereal species (Ayliffe et al., 2005). The gene was up regulated by drought and located on barley chromosome 1HL.

A late-embryogenesis-abundant (LEA) gene family was induced by osmotic condition, dehydration, salt and ABA treatment. The gene family was up regulated by drought. Several genes from this gene family were regulated or induced by drought. These genes include HVA1 (Hong et al., 1992), ABA2 and ABA3 (Gulli et al., 1995), Paf93 (Grossi et al., 1995), dehydrins (Close, 1996) and a B19 gene family (Hollung et al., 1994). Cattivelli et al. (2002) have summarized this gene family isolated from barley.

Accumulation of glycinebetaine is one mechanism for barley response to drought or salt stress. Ishitani et al. (1995) reported a gene up-regulated by drought stress and encoding an enzyme of known function - betaine aldehyde dehydrogenase (BADH). This enzyme is the last step in the betaine synthesis pathway. The mRNA level of BADH increased significantly when barley was subjected to drought or saline conditions. Sorbitol has a role in osmoregulation. The enzyme aldose reductase involved in the accumulation of sorbitol was regulated by ABA (Bartels et al., 1991).

Transcripts for the biosynthesis of jasmonate were highly up regulated by drought, which is well-known as a signal in pathogen defence and under drought conditions. These genes included jasmonate induced proteins (JIPs) and a methyl-jasmonate inducible lipoxygenase. Lipoxygenase and fatty acid α-oxidase may be in the signal transduction pathway that is regulated by jasmonates. Two arginine decarboxylases are induced by drought, which may be involved in the synthesis of polyamines that are observed in many stressed plants (Ozturk et al., 2002).

One hundred sequenced probes regulated by drought (Ozturk et al., 2002) and 12 candidate genes were surveyed for polymorphism in the Tadmor x Er/Apm mapping population (Diab et al., 2004), in which 68 QTLs have been detected for drought tolerance relative traits (Teulat et al., 2001a, 2002). In total 33 loci were mapped and ESTs or candidate genes at 12 loci were co-segregated with 19 QTLs for drought tolerance. An ESTBM816463 encoding for a blue copper-binding protein co-segregated with QTLs for RWC, WSC, OP and DWSC on chromosome 3H. This gene may be involved in the generation of activated oxygen species. The Acl3 locus encoding barley acyl carrier protein III is associated with a QTL for RWC and WSC on chromosome 7H. This gene encodes a co-factor protein of the fatty acid synthetase involved in the de novo synthesis of the fatty-acyl chain, especially in chloroplasts. This gene could have a role in the protection of membranes or in membrane fluidity during stress. The gene bSS1B coding for sucrose synthase co-segregated with a QTL on chromosome 7H for RWC. The enzyme sucrose synthase is a key enzyme in carbohydrate metabolism, catalyzing the reversible conversion of sucrose uridine-diphosphate into fructose and UDP-glucose (Kleines et al., 1999). Synthesis of sugars or compatible solutes has widely been observed as a mechanism that may help plants cope with water deficit (Whittaker et al., 2001).

A cluster of 4 Dhn genes is located under a QTL for RWC and for OA on chromosome 6H (Teulat et al., 1998, 2001a, 2003), and the Dhn9 gene maps to another drought QTL on the long arm of chromosome 5H. Besides these, the locus containing the Dhn1 and Dhn2 genes is located under the cold tolerance QTL Fr-H1. Dehydrins are one of the families of proteins that are synthesized in plants in response to dehydration, low temperature, osmotic stress, seed drying, and exposure to abscisic acid. Thirteen barley Dhn genes have been identified and sequenced in barley (Close et al., 1989; Rodriguez et al., 2005).

The other genes co-segregated with the QTLs for drought tolerance were oxalate oxidase, glutathione S-transferase 1, cathepsin B, isocitrate dehydrogenase, endopeptidase Clp, and hypothetical protein C18B2.4. The role of these genes in drought tolerance is not yet known. Tondelli et al. (2006) mapped regulatory genes as candidates for cold and drought stress tolerance in barley. The genes were located on chromosomes 2H, 5H and 7H. Transcription factors and other regulators of cold and drought-induced genes that were mapped are not randomly dispersed in the barley genome. The most represented chromosome is 5H: 10 candidate genes at 5 loci, out of 16 CGs at 11 loci in total. This is also intriguing since most abiotic stress QTLs of the Triticeae are located on this chromosome (Cattivelli et al., 2002). HvCBF8 together with HvABI5 were mapped to the vicinity of a QTL that controls osmotic potential. HvMYB4 (Wissenbach et al., 1993) was mapped on the long arm of chromosome 2H in a region where multiple QTLs for drought-related traits—i.e. RWC, OA, and water soluble carbohydrates—were discovered by Teulat et al. (2002) and Diab et al. (2004). HvABI5 maps on the long arm of chromosome 5H, together with two overlapping drought tolerance QTLs controlling osmotic potential (Teulat et al., 2001). HvABI5 is a bZIP TF up-regulated by ABA and is responsible for ABA-dependent induction of the barley effector genes HVA1 and HVA22 (Casaretto and Ho, 2003). It has also been shown that constitutive expression of HVA1, which encodes a group 3 LEA protein, can confer dehydration tolerance to transgenic rice plants (Chandra Babu et al., 2004). On chromosome 7H, an association exists between the two drought tolerance QTLs and candidate genes TC147474 (FRY1) and TC143232 (ICE1). Rubisco activase was also reported to relate to drought tolerance.

There has been substantial progress in identifying genes for tolerance to various abiotic stresses. An alanine aminotransferase isolated from barley roots could increase drought tolerance of transformed tobacco (Muench and Good, 1994). Xu et al. (1996) has used a transgenic approach to investigate the function of the HVA1 protein in stress protection of rice. HVA1 is a group 3 LEA protein that is expressed in barley aleurone and embryo during late seed development correlating with the seed desiccation stage (Hong et al., 1988). The transgenic rice plants exhibited a high constitutive expression of HVA1 protein in leaves and roots. The progeny of three transgenic plants was used for evaluation of the growth performance under water deficit and salt stress treatment. The appearance and development of the major damage symptoms such as wilting, dying of old leaves and necrosis of young leaves caused by the stress conditions were delayed in the transgenic plants. The better performance of the transgenic lines under stress conditions was correlated with higher level of HVA1 protein accumulated in the plants. The general involvement of MYB transcript factors in the induction of drought responsive genes was demonstrated by AtMyb2 over expression in transgenic plants, resulting in an improved osmotic stress tolerance (Abe et al., 2003). These researches not only provide targets for crop improvement, but also provide insights into the physiological and biochemical mechanisms underlying field performance. However, it has to be emphasized that the function of most of the above genes related to barley drought tolerance are not clear. More detailed research is required in the future to elucidate their role under drought conditions.

4. BREEDING BARLEY FOR DROUGHT TOLERANCE

Drought is the single most important factor limiting yield. Yet, compared to other cereals, barley is well adapted due to better water-use efficiency and mechanisms of drought escape, avoidance and tolerance. Three strategies have been considered in relation to the optimum environment for selection (Byrne et al., 1995). The first strategy is based on selection where growing conditions are optimum or near-optimum. The second strategy assumes that the optimum environment(s) for selection should be as representative as possible of the target population of environments (Blum, 1988). The third strategy, the alternate use of optimum and stressed conditions has been used to select genotypes that yield well in both conditions

(Calhoun et al., 1994). Ceccarelli et al. (1998) demonstrated that the most effective way to improve productivity of barley grown in drought conditions is to use locally adapted germplasm and select in the target environment(s).

Jana and Wilen (2005) summarized previous research on breeding for abiotic stress tolerance in barley. Although breeding for drought resistance based on direct selection for grain yield in the target environment (empirical or pragmatic breeding) appears intuitively to be the most obvious solution, this approach faces two major problems; a) the precision of the yield trials conducted under drought conditions, and b) the existence of several target environments, each characterized by its own specific type of drought and combination of stress (Ceccarelli and Grando, 2002). Breeding for drought resistance based on putative traits (traits associated with drought resistance, but easier to select for than grain yield) has been very popular, but the progress is still slow. Traits that have been investigated include; 1) physiological/biochemical traits, such as proline content, stomatal conductance, epidermal conductance, canopy temperature, relative water content, leaf turgor, abscisic acid content, transpiration efficiency, water use efficiency, carbon isotope discrimination, and re-translocation of carbohydrates; and 2) developmental/morphological traits, such as leaf emergence, leaf area index, leaf waxiness, stomatal density, tiller development, flowering time, maturity rate, cell membrane stability, and root characteristics. In the case of barley, the traits more consistently associated with higher grain yield under drought include growth habit, early growth vigour, earliness, plant height under drought, long peduncle, and short grain filling duration. However, most of traits were controlled by multiple genes and environments played an important role in the expression of specific traits. Identification of molecular markers for these traits provided tools for directly selection of drought tolerance. In several studies, it has shown that the developmental genes are key factors in the determination of yield potential under drought condition (Baum et al., 2003; Foster et al., 2004; Teulat et al., 2001b). These genes include photoperiod response, basic vegetative period, earliness and vernalization. These genes have been well characterized and tagged using molecular markers (Boyd et al., 2003). However, the useful genes/alleles related to drought tolerance may have already been lost during domestication and modern breeding (Foster et al., 2004). Wild barley *Hordeum spontaneum* and landraces will provide a useful gene pool for drought tolerance. Molecular markers, especially the candidate genes regulated by drought can be used to characterize the germplasm (Maestri et al., 2002). In addition, genes controlling plant height on chromosome 3H and 5H are obviously target loci for selection of drought tolerance using molecular markers. The other candidate loci included the major QTLs on chromosome 6H and 7H for OA, RWU and DIS. Marker assisted breeding may significantly improve the breeding efficiency for drought tolerance as more markers and drought tolerance germplasm become available.

Extensive research has been devoted to the characterization of genes induced or up-regulated by drought. The up-regulation of a drought-induced barley gene (HVA1) improved tolerance to drought and salinity in rice grown under controlled conditions (Xu et al., 1996). Encouraging as these results are, there is widespread skepticism that up-regulation of one or more genes encoding structural proteins may not lead to meaningful results in terms of field tolerance to drought (Bajaj et al., 2000; Bohnert and Bressan, 2001).

5. SALINITY TOLERANCE

Soil salinity is one of the principal abiotic factors affecting crop yields in arid and semi-arid irrigated areas (Szabolcs, 1989). Almost three quarters of the surface of the earth is covered by salt water and so it is not surprising that salts affect a significant proportion of the world's land surface. Salt-affected soils contain sufficient concentrations of soluble salts, which cause toxicity to common crop plants. In agriculture, salt stress severely affects the growth and economic yield of many important crops (Maas and Hoffman, 1977). Compared with other cereal crops, including wheat, rice, rye and oat, barley is highly tolerant to salinity, thus offering a means for efficient utilization of saline soil and improvement of productivity in these environments. However, barley still suffers from salt toxicity in many areas of the world. On the other hand, dramatic differences can be found among and within the barley species, providing the potential for developing cultivars with improved salt tolerance. It is predicted that the genetic improvement in salt tolerance will be an important aspect of barley breeding in the future.

6. TOXIC EFFECT OF SALT STRESS AND TOLERANCE IN BARLEY

Salinity has three potential effects on plants: 1) lowering of the water potential; 2) direct toxicity of any Na and Cl absorbed; and 3) interference with the uptake of essential nutrients. Hence, the plant water potential should be lowered in order to maintain water uptake in salt environments. Meanwhile, higher accumulation of salt ion will cause toxic effect on cells and so must be separated from the metabolic machinery of the cells. This is achieved by compartmentation: salt-sensitive metabolic processes take place in the cytoplasm, while the salt necessary for osmotic adjustment is stored in vacuoles (Flowers and Yeo, 1986). Within the cytoplasm, osmotic adjustment is influenced by compatible solutes, such as glycinebetaine, mannitol and proline. When a plant is exposed to salt stress, it responds initially to the changed water conditions brought about by the lowering of the external water potential by the salt. These initial effects of salinity are likely to be the same for cultivars differing in salt tolerance. When ions accumulate over time, differences in salt tolerance appear (Munns, 1993). Sensitive cultivars accumulate ions more quickly than tolerant cultivars. Ions enter plant cells through a membrane-across protein, and the process is driven by energy-consuming ion pumps, which use the energy stored in ATP to move protons by generating a difference of hydrogen ion concentration (pH) and electric potential (ΔE). It is assumed that Na^+ is 'mistaken' for potassium by a K^+ carrier or channels, but it is also possible that Na^+ enter cells through non-selective cation channels (Maser et al., 2002).

A possible survival strategy of plants under saline conditions is to sequester absorbed Na^+ in the vacuole, thus maintaining a higher K^+/Na^+ ratio in the cytoplasm (Greenway and Munns, 1980). The K^+/Na^+ antiport in vacuolar membranes transports Na^+ from the cytoplasm to vacuoles using a pH gradient generated by H^+-ATPase and H^+-PPase, which was considered to be related to salt tolerance of plants (Atsunori et al., 1998). It has been demonstrated that the proton pump and K^+/Na^+ antiport in vacuolar membranes were important in ion selective absorption and compartmentation of Na^+ in barley seedlings (Garbarino and Dupont, 1988).

It is known that a specific phospholipid environment is required for optimal ATPase activity, and changes in phospholipids and free sterols of the cell membranes may contribute to salt tolerance (Norberg and Liljenberg, 1991; Mansour et al., 1994). Yamaguchi and Kasamo (2001) found that exogenously added tonoplast phospholipids would stimulate the activity of purified tonoplast H^+-ATPase. Meanwhile, fatty acids are considered to be important in salt tolerance of plants and micro-organisms (Somerville, 1995; Malkit et al., 2002). By using genetic mutants, it was demonstrated that unsaturated fatty acids in membrane lipids could protect the photosynthetic machinery against salt stress-induced damage (Allakhverdiev et al., 1999).

In addition, polyamines (PAs) have been found in all living organisms studied and are required for normal development of both prokaryotes and eukaryotes (Tabor and Tabor, 1984). The polycationic nature of PAs at physiological pH is one of the main properties believed to mediate their biological activity. They are able to bind negatively charged molecules, such as DNA (Basu et al., 1990), membrane phospholipids and proteins (Tassoni et al., 1996), and pectic polysaccharide (D'Oraci and Bagni, 1987). In addition to free forms, PAs can also be covalently bound to some specific proteins catalyzed by a class of enzymes known as trans-glutaminases (TGase) to form bound PAs (Serafini-Fracassini et al., 1995). Both free and bound PAs associated to the tonoplast vesicles from barley seedlings were detected, and their contents were found to be closely related to salt tolerance of the plants (Zhao et al., 2000).

Zhao and Qin (2005) found that linoleic acid at 1 mM in culture solution possessed protective effects on root tonoplast function against salt stress in the barley seedlings; this was accompanied with a significant suppression of the degradation of phospholipids and PAs in tonoplast vesicles. Moreover, these salt-ameliorating effects of linoleic acid on tonoplast function were also indicated by the increase in H^+-ATPase and H^+-PPase activities. An application of LA under saline condition resulted in an augmentation of the activity of a vacuolar K^+/Na^+ antiport. These findings suggested that the addition of linoleic acid resulted in a protective effect on tonoplast function in the barley seedlings under salt stress, perhaps due partly to suppress the degradation of phospholipids and PAs in tonoplast vesicles, thus leading to a partial restoration in the activities of vacuolar H^+-ATPase, H^+-PPase and the K^+/Na^+ antiport.

Farquhar and Richards (1984) reported that carbon isotope discrimination (CID) is linearly related to the ratio (pi/pa) of the intercellular (pi) and atmospheric (pa) partial pressure of CO_2 in C_3 plants. The ratio (pi/pa) is determined by leaf stomatal conductance and photosynthetic capacity, therefore, by genetic and environmental factors. Low CID is generally associated with low stomatal conductance (Farquhar et al., 1989), while the latter could be reduced under salinity stress (Shen et al., 1994; Benes et al., 1996). However, it has been reported that grain CID is negatively correlated with grain yield in barley (Romagosa and Araus, 1991; Craufurd et al., 1991). The inconsistency among the investigators could be attributed to the different cultivars or genotypes. Pakniyat et al. (1997) reported some barley mutants grown in hydroponic culture had a higher CID and a lower Na^+ shoot content, and less inhibition of shoot growth by salinity than their respective parental lines.

Ultra structural alterations to root cells in response to salt stress have been recorded in several species, including barley. Kramer (1984) suggested that the appearance of various alternations under salt stress might have a function in the adaptation of plant to salinity. Huang et al. (1990) observed the structural changes occurring in meristematic cells of barley in response to moderate salinity stress. In the apical region of the root, salt caused an increase in vacuolation, which may provide a means for accumulation of excess ions. Salt treatment also caused many plastids in the cortical cells in this region to adopt varying amoeboid shapes, often appearing to enclose part of the cytoplasm, which was less dense than the surrounding cytoplasm. It was suggested that plastid morphology may allow or alternatively result from adaptive change in protein synthesis or cytoplasmic composition.

Cramer et al. (1989) reported that salt-stressed plants often show symptoms of Ca deficiency. The transport and tissue concentrations of Na were significantly affected by supplemental Ca. Calcium transport and tissue concentrations were markedly inhibited by salinity. There were significant Na-Ca interactions with ion transport, ion accumulation, and growth. Lynch et al. (1988) proposed that leaf growth in salt-stressed barley plants was reduced by sub-optimal Ca availability in the leaf meristem. One cause of reduced Ca availability is that Na replaces Ca in the leaf apoplast (Zid and Grignon, 1985). Salinity stress has been shown to stimulate a release of Ca from intracellular compartments (Lynch and Lauchli, 1988). Calcium transport to the shoot is reduced in NaCl-stressed plants (Lynch and Lauchli, 1985; Wolf et al., 1990), and indeed, the ability to transport Ca to the shoot during salt stress has been proposed as an index of salt tolerance (Lahaye and Epstein, 1971).

7. EVALUATION AND IDENTIFICATION OF SALT-TOLERANT GENOTYPES

The reliable, convenient, inexpensive and quick screening techniques of salt-tolerant germplasm are the paramount in successful breeding. Unfortunately, lack of proper screening technique is still a bottleneck of salt-tolerant breeding programs

(Zhu, 2000; Munns and James, 2003). In most cases, field screening for salinity tolerance remains the main tool, despite its limitation of time required and environment dependence. Many potential criteria or traits have been proposed for screening. Examples include ranking of plants according to growth rate or yield (Greenway, 1962), plant survival at high salinity (Sayed, 1985), germination rate (von Well and Fossey, 1998), leaf or root elongation rate (Cramer and Quarrie, 2002), leaf injury and reduction of CO_2 assimilation (James et al., 2002), loss of chlorophyll and damage to the photosynthetic apparatus (Krishnaraj et al., 1993), Na exclusion (Garcia et al., 1995), K^+/Na^+ discrimination (Asch et al., 2000) and Cl^- exclusion (Nobel and Rogers, 1992).

Since the classical selection of genetic material based on their yield performance under saline conditions has been largely unsuccessful, particularly due to the high variability of naturally saline soils (Richards, 1983), several authors (Noble and Rogers, 1992; Flowers and Yeo, 1995) suggested the use of physiological traits as alternatives to screening for yield. Aragüés and Royo (1998) assessed the relationships between grain yield, carbon isotope discrimination, canopy temperature, stomatal temperature, stomatal conductance, and grain ash content in a set of barley cultivars grown in the soils with different salinity levels, and found that none of the studied characters would be useful in screening for high yield under salinity environments, and that grain yield under salt stress was the only trait which proved reliable for identifying higher salt tolerance.

Many of these criteria are often unrelated to each other, resulting in different estimates of salt tolerance. As a complex trait, salt tolerance involves responses to cellular osmotic and ionic stresses and their consequent secondary stresses and whole-plant co-ordination. Hundreds of different genes may be involved, either directly or indirectly. Some of these genes are expressed at very early stages, while others become crucial only at later stages of plant ontogeny. All this complicates plant screening for salt tolerance, and crop ranking made at one stage may be rather different from similar assessment made at another stage of plant ontogeny. Obviously, knowledge of underlying physiological mechanisms is of paramount importance for efficient screening methods (Zhu, 2000). Some researchers (Shannon and Noble, 1990; Flowers and Yeo, 1995) have suggested that screening for salt tolerance be carried out using physiological markers, or that physiological traits should be used as selection criteria, either singly or in combination, rather than selection being simply upon yield or yield components.

Discrimination between the stable atmospheric carbon isotopes ^{13}C and ^{12}C provides an integrated measure of stomatal control of internal CO_2 concentration. In theory, higher internal CO_2 concentration (Ci) implies higher carbon isotope discrimination for the heavier ^{13}C isotope. Ci is dependent on two main parameters: stomatal conductance (gs) and CO_2 assimilation capacity. Limitation of the former or the latter will lower or increase Ci, respectively. In a number of C_3 species subjected to salinity stress, a decline of CID values was reported (Brugnoli and Lauteri, 1991; Ouerghi et al., 2000; Rasmuson and Anderson, 2002).

Chlorophyll fluorescence is a rapid, extremely sensitive and non-intrusive measurement which can be performed on intact, attached leaves as well as isolated chloroplasts or sub-chloroplast particles. It has become an important tool in the study of photosynthesis, in particular the functioning of PS II (Schreiber et al., 1995), and also widely used to determine the influence of abiotic stress on plant growth. The in-vivo effects of salinity on chlorophyll fluorescence have been described for several crop species (Smillie and Nott, 1982; Sayed, 2003) and fluorescence parameters have been used to screen for salinity tolerance in barley, wheat and corn (Belkhodja et al., 1994; Shabala et al., 1998). However, Jiang et al. (2006) compared gas exchange, chlorophyll fluorescence parameters, above ground dry matter and carbon isotope discrimination among 14 barley genetic lines grown under control and saline treatments. The results showed that 2-week exposure to saline conditions decreased above-ground dry mass, net photosynthesis (A), stomatal conductance (g_s), internal CO_2 concentration (Ci), efficiency of light harvesting of photosystem II (Fv/Fm), photochemical quenching (qP), and carbon isotope discriminationrelative to control plants, and the measurement of g_s can provide the best information to assess genetic differences in barley for absolute performance when subjected to salinity stress. Lines with the highest g_s values under control conditions also showed some of the highest absolute values for A and Fv/Fm under saline conditions. They used salinity susceptibility indexes (SSI) to estimate the relative tolerance of lines to salinity.

One of the key features of plant salt tolerance is the ability of plant cells to maintain optimal K^+/Na^+ ratio in the cytosol (Maathuis and Amtmann, 1999; Tester and Davenport, 2003). Under salinity, the K^+/Na^+ ratio in the cytosol falls dramatically. This occurs as a result of both excessive Na accumulation in the cytosol (Leigh, 2001; Zhu, 2000) and enhanced K leakage from the cell (Shabala, 2000; Shabala et al., 2003), the latter resulting from NaCl-induced membrane depolarization under saline conditions (Cakirlar and Bowling, 1981; Shabala et al., 2003). Therefore, K^+/Na^+ ratio in plant tissues has often been suggested as a potential screening tool for plant breeders (Shannon, 1997; Poustini and Siosemardeh, 2004). However, there appears to be some confusion between cytosolic K^+/Na^+ ratios and K^+/Na^+ ratios in salinized plant tissues. The latter ratio might not explain for the fact that a significant part of accumulated Na may be compartmentalized in the vacuole. Vacuolar compartmentation is another key feature of plant salt tolerance (Blumwald, 2000). Unfortunately, traditional tissue analysis for Na content cannot account for such compartmentation, thus diminishing the predictive value of the K^+/Na^+ ratio in plant tissues to screen plants for salt tolerance.

Recently, Chen et al. (2005) tested the possibility of ion selection vibrating technique (MIFE technique) to screen salt-tolerant barley. According to them, a cell's ability to retain K is at least as important for plant salt tolerance as its ability to exclude or compartmentalize toxic Na (Shabala, 2000; Shabala et al., 2003). Thus K uptake measurement may provide a quick and reliable screening test on seedlings that will save field space and time. Experimental results showed MIFE is a relatively quick and reliable method to screen plants for salt tolerance using non-invasive K flux.

8. GENETICS OF SALT TOLERANCE IN BARLEY

The differences in salinity tolerance have been reported among barley varieties (Epstein and Norlyn, 1977; Rathore et al., 1977; Day et al., 1985; Forster et al., 2000) and barley species (Mano and Takeda, 1998). Mano and Takeda (1998) evaluated salt tolerance of 340 accessions of *Hordeum*, consisting of 41 brittle-rachis forms of *Hordeum vulgare L.* subsp. *vulgare (H. agriocrithon)* accessions, 154 *H. vulgare L. subsp. spontaneum (H. spontaneum)* accessions, and 145 accessions of ten other species or subspecies of wild *Hordeum*. They found the levels of salt tolerance for seed germination in wild *Hordeum* species were generally lower than those in cultivated barley and the NaCl tolerance level of the different species were as follows: *H. agriocrithon* > *H. spontaneum* > other wild *Hordeum* species. In addition, when leaf injury index was used to assess tolerance at the seedling stage, the levels of salt tolerance in wild *Hordeum* species were generally higher than those found in cultivated barley. Most wild Hordeum species showed high NaCl tolerance at the seedling stage and were considered good sources of germplasm for salt tolerance breeding.

Ramagopal (1988) found obvious difference between salt-tolerant barley genotypes, CM72 (California Mariout 72) and sensitive Prato in protein synthesis during seed germination. Salinity stress induced both quantitative and qualitative changes in the expression of some proteins in vivo. Around 8% of the nearly 400 resolved proteins in a tissue were affected this way. Some of the proteins in this category were specific to each genotype. About 1% of the total showed qualitative changes; these proteins were expressed only during salinity stress. In roots, two proteinswere detected in CM72 and five in Prato. In shoots, four proteins were found only in Prato and these were similar to those induced in roots. The four new proteins in germinating embryos were apparently induced only in CM72. It is indicated that ontogeny plays an important role in the expression of tissue-specific proteins during salinity stress in the salt tolerant and sensitive barley genotypes.

Wei et al. (2001) found a gene encoding the barley vacuole ATPase subunit B (BSVAP), which was differentially expressed between near isogenic barley cultivars, Golden Promise and Maythorpe. The gene was inducible under long-term salinity stress in the salt sensitive cultivar Maythorpe, but less so in the relatively salt tolerant Golden Promise and was highly expressed under control conditions in Maythorpe. It was concluded that that the short-term down-regulation of BSVAP under high salinity was an important mechanism contributing to Golden Promise. Dizetz et al. (2001) suggested the ability to respond to salinity stress with changes in gene expression of the vacuolar ATPase might be a prerequisite and a characteristic of salt tolerance in plants.

Forster (2001) reviewed the research made at the Scottish Crop Research Institute (SCRI) on the effects of semi-dwarfing genes on salt tolerance. The work was initiated in 1993 with the fortuitous and unexpected result that the cultivar 'Golden Promise' showed considerable tolerance to salt. Golden Promise is a gamma-ray induced semi-dwarf mutant of the cultivar 'Maythorpe'. The parent and mutant showed significant differences in their responses to salt stress. The positive and pleiotropic effects of the mutant gene Gpert were found to be effective in a number of genetic backgrounds. The Gpert mutation was allelic to the ari-e mutants in barley. The ari-e mutants were salt tested and found to show the same positive responses to salt stress as Golden Promise, supporting the allelism tests, and consequently the Gpert symbol was changed to ari-e. GP. The semi-dwarf mutant sdw1 and the erectoides semidwarf mutant, ert-k32 were also tested for their effects on salt tolerance, but did not show any positive effects. Salt tolerance was therefore not a general phenomenon of semi-dwarf stature but specific to mutations at the ari-e locus in these lines.

A differential display using randomly amplified polymorphic DNA (RAPD) primers and isolated salt-inducible cDNA clones found a gene encoding a putative methionine synthase in barley leaves (Muramoto, 1999; Shi et al., 2001). The gene was named as HvMS (Hordeum vulgare methionine synthase), and it was found that the expression of this gene is induced in barley leaves within 1 h by salt stress. This is one of the early responsive genes in barley leaves. It was assumed that the salt-inducible HvMS could play an important role as a member of this cycle for salt tolerance in barley plants (Shi et al., 2001). Eckermann et al. (2000) reported that the expression of methionine synthase (MS) is induced under salt stress. A MS gene from potato was cloned and characterized (Zeh, 2002). RNA transcription from this MS gene was regulated by a day/night rhythm, but protein levels did not alter. Therefore, MS was not considered as one of the important components in salt-stress tolerance of plants. However, Narita et al. (2004) reported HvMS protein levels were increased under salt stress and suggested that MS may indeed be important in salt tolerance in higher plants. Furthermore, they found that this gene complemented a yeast mutant lacking the ability to synthesize methionine under both non-stress and high-salinity conditions.

Plants respond to high salinity conditions by adjusting their physiological and metabolic processes (Rhode and Hanson, 1993). They have many genes for maintaining ion homeostasis, and metabolism including synthesis of compatible solutes (osmoprotectants), stress proteins for cell rescue and defense, proteins for signal transduction, components of protein synthesis and others affecting morphology (Bohnert and Jensen, 1996; Kasuga et al., 1999; Kawasaki et al., 2001). Concerning the genetics of salt tolerance in barley, Koval and Koval (1996) showed that the tolerance was controlled by semi-dominant additive genes. The number of genes coding for salt tolerance in barley is small (Kueh and Bright, 1982) and chromosomes 2H, 4H, and 5H contribute to the inheritance of this trait (Forster et al., 1990; Mano and Takeda, 1997). Salt tolerance is linked with certain morphological and biochemical markers (Mano and Takeda, 1996; Krestnikov et al., 1986).

9. BREEDING FOR SALT-TOLERANT BARLEY

As physiology and the genetics of salt tolerance are so complicated, is it going to be possible to breed for salt-tolerant crops? To date, there have been only limited successes. However, a variety of approaches have been advocated, including conventional breeding, wide crossing, the use of physiological traits and, more recently, marker-assisted selection and the use of transgenic plants. None of these approaches could be said to offer a universal solution. Conventional breeding programs have rarely delivered enhanced salt tolerance (Flowers and Yeo, 1995), while wide crossing generally reduces yield to unacceptably low levels (Yeo and Flowers, 1989). There has been success using physiological criteria as the basis of selection in rice (Dedolph and Hettel, 1997) and such an approach has recently been advocated for wheat (Munns et al., 2002). A recent analysis has shown that whilst it is possible to produce a wide range of transgenic plants where some aspect of a trait relating to salt tolerance was altered, none has been tested in the field and few claims for success meet even minimal criteria required to demonstrate enhanced tolerance (Flowers, 2004). With the development of molecular marker research, marker-assisted selection provides a more powerful tool in barley salt-tolerant breeding.

Combining the DNA technology and advanced statistical methods (Kearsey, 1998), chromosomal regions that contain the genes that determine quantitative traits can be identified. By crossing parents that differ in one or more aspects of salt tolerance (their phenotype), and then analyzing the phenotype and the genotype of their offspring, it has been possible to locate QTL for salt tolerance. For a plant breeder, such QTLs are particularly attractive, as they can, in principle, be developed to produce markers to aid selection. Such markers can be used in the selection of lines following a crossing program and without the need to determine their phenotype or to take all the lines to seed (Asins, 2002).

Mano and Takeda (1997) identified QLTs controlling salt tolerance at germination and the seedling stage in barley by interval mapping analysis using marker information from two doubled haploid (DH) populations derived from the crosses, Steptoe x Morex and Harrington x TR306. The results revealed that the QTLs for salt tolerance at germination in the DH lines of Steptoe x Morex were located on chromosomes 4H, 6H, and 5H, and in the DH lines of Harrington/TR306 on chromosomes 1H and 5H. In both DH populations, the most effective QTLs were found at different loci on chromosome 5H. Genetic linkage between salt tolerance at germination and ABA response was found from QTL mapping. The QTLs for the most effective ABA response at germination were located very close to those for salt tolerance on chromosome 5H in both crosses. The QTLs for salt tolerance at the seedling stage were located on chromosomes 2H, 1H, 6H, and 5H in the DH lines of Steptoe x Morex, and on chromosome 5H in the DH lines of Harrington x TR 306. Their positions were different from those of QTLs controlling salt tolerance at germination, indicating that salt tolerance at germination and at the seedling stage were controlled by different loci.

However, it has been reported that such QTL were dependent on the conditions under which the plants were grown (Foolad et al., 1999; Monforte et al., 1997). In addition, it was suggested that QTL associated with salt tolerance vary with the developmental stage, at which the analysis is performed in species as widely divergent as tomato, rice and barley, citrus and Arabidopsis. A further limitation to the use of QTL in plant breeding is the fact that QTLs may be specific to particular crosses. It was argued that whilst markers will be of value in using elite lines from the mapping population in backcrossing, the result cautions against any expectation of a general applicability of markers for physiological traits.

In selection of barley with salt tolerance, Koval et al. (2000) investigated the contribution of the gametophyte in the inheritance of salt tolerance by crossing F_3 and BC1 hybrids of the tolerant cultivars Rannii 1 and Pirkka with the sensitive cultivar K-30356, and found that the progenies of heterozygous plants grown in saline conditions show elevated salinity tolerance. A comparison of the BC_1 hybrid progenies showed that the male and female gametophytes contributed to the inheritance of salt tolerance. Gametic selection is maximally efficient during the formation of the female gametophyte and the germination of pollen grains on the stigma.

REFERENCES

Abe H., Urao T., Ito T., Seki M., Shonozaki K., Yammaguchi-Shinozaki K. 2003. Arabidopsis AtMYC (bHLH) and AtMYB2 (MYB) function as transcription activators in abscisic acid signalling. Plant Cell 15:63–78

Ayliffe M.A., Mitchell H.J., Deuschle K., Pryor A.J. 2005. Comparative analysis in cereals of a key proline catabolism gene. Mol Gen Genomics 274: 494–505

Allakhverdiev S. I., Nishiyama Y., Suzuki I., Tasaka Y., Murata N. 1999. Genetic engineering of the non-saturation of fatty acids in membrane lipids alters the tolerance of *Synechocystis* to salt stress. Proc Natl Acad Sci USA 96: 5862–5867

Aragüés R.I.R., Royo A. 1998. Validity of various physiological traits as screening criteria for salt tolerance in barley. Field Crops Res 58: 97–107

Araus J.L., Slafer G.A., Reynolds M.P., Royo C. 2002. Plant Breeding and Drought in C_3 Cereals: What Should We Breed For? Annals of Botany 89: 925–940

Asch F., Dingkuhn M., Dorffling K., Miezan K. 2000. Leaf K/Na ratio predicts salinity induced yield loss in irrigated rice. Euphytica 113: 109–118

Asins M.J. 2002. Present and future of quantitative trait locus analysis in plant breeding. Plant Breed 121: 281–291

Atsunori F., Yoshiaki Y., Ishikawa T., Setsuo K., Yoshiyuki T. 1998. Na+/K+ antiporter in tonoplast vesicles from rice roots. Plant Cell Physiol 39: 196–201

Bajaj S., Targolli J., Liu L.F., Ho T.-H.D., Wu R., 1999. Transgenic approaches to increase dehydration-stress tolerance in plants. Mol Breed 5: 493–503

Bartels D., Engelhardt K., Roncarati R., Schneider K., Rotter M., Salamini F.1991. An ABA and GA modulated gene expressed in the barley embryo encodes an aldose reductase related protein. EMBO J 10:1037–43

Basu H.S., Schwietert H.C.A., Feuerstein B.G., Marton L.J. 1990. Effect of variation in the structure of spermine on the association with DNA and the induction of DNA conformational changes. Biochem J 269: 329–334

Baum M., Grando S., Backes G., Jahoor A., Sabbagh A., Ceccarelli S.2003. QTLs for agronomic traits in the Mediterranean environment identified in recombinant inbred lines of the cross _Arta_ _ H. *spontaneum* 41–1. Theor Appl Genet 107:1215–225

Belkhodja R., Morales F., Abadia A., Medrano H., Abadia J. 1999. Effects of salinity on chlorophyll fluorescence and photosynthesis of barley *(Hordeum vulgare L.)* grown under a triple-line-source sprinkler system in the field. Photosynthetica 36: 375–387

Benes S.E., Aragüés R., Austin R.B., Grattan S.R. 1996. Brief pr- and post-irrigation with freshwater reduces foliar salt uptake in maize and barley irrigated with saline water. Plant and Soil 180: 87–95

Blum A. 1988. Plant breeding for stress environment. CRC, Boca Raton pp. 1–223

Blumwald E. 2000. Sodium transport and salt tolerance in plants. Current Opinion in Cell Biology 12: 431–434

Bohnert H. J., Jensen R. G. 1996. Strategies for engineering water-stress tolerance in plants. Trends Biotechnol 14: 89–97

Bohnert H.J., Bressan R.A. 2001. Abiotic stresses, plant reactions, and approaches towards improving stress tolerance. In: Nössberger J., ed. Crop Science: Progress and prospects. Wallingford, UK: CABI International, pp. 81–100

BothmerR. von, Jacobsen N., Baden C., Jorgensen R.B., Linde-Laursen I. 1995. An ecogeographical study of the genus Hordeum. Systematic and ecogeographic studies on crop genepools, 7. IPGRI, Rome, 2nd ed., pp. 129

Boyd W.J.R., Li C.D., Grime C., Cakir M., Potipibool S., Kaveeta L., Men S., Jala Kamali M.R., Barr A.R., Moody D.B., Lance R.C.M., Logue S.J., Raman H., Read B.J. 2003 Conventional and molecular genetic analysis of factors contributing to variation in the timing of heading among spring barley (*Hordeum vulgare L.*) genotypes grown over a mild winter growing season. Australian Journal of Agricultural Research 54: 1277–1301

Brugnoli E., Lauteri M. 1991. Effects of salinity on stomata l conductance, photosynthetic capacity, and carbon isotope discrimination of salt-tolerant *(Gossypium hirsutum L.)* and salt-sensitive *(Phaseolus vulgaris L.)* C_3 non-halophytes. Plant Physiol 95: 628–635

Byrne P.F., Bolaños J., Edmeades G.O., Eaton D.L. 1995. Gains from selection under drought versus multilocation testing in related tropical maize populations. Crop Science 35: 63–69

Cakirlar H., Bowling D.J.F. 1981. The effect of salinity on the membrane potential of sunflower roots. J Exp Bot 32: 479–485

Calhoun D.S., Gebeyehu G., Miranda A., Rajaram S., van Ginkel M. 1994. Choosing evaluation environments to increase wheat grain yield under drought conditions. Crop Science 34: 673–678

Casaretto J., Ho T.-H. D. 2003. The transcription factors hvabi5 and hvvp1 are required for the abscisic acid induction of gene expression in barley aleurone cells. Plant Cell 15: 271–284

Cattivelli L., Baldi P., Crosatti C., Grossi M., Vale G., Stanca A.M.2002 Genetic bases of barley physiological response to stressful conditions. In: GA Slafer, Molina-Cano JL, Savin R, Aruas JL, Romagosa I (eds) Barley science Recent advances from molecular biology to agronomy of yield and quality. Food Products Press, New York pp. 387–411

Ceccarelli S., Grando S., Impiglia A. 1998. Choice of selection strategy in breeding barley for stress environments. Euphytica 103: 307–318

Ceccarelli S., Grando S. 2002. Plant breeding with farmers requires testing the assumptions of conventional plant breeding: Lessons from the ICARDA barley program. In: Cleveland, David A. and Daniela. Soleri, (eds.). Farmers, scientists and plant breeding: Integrating Knowledge and Practice. Wallingford, Oxon, UK: CAB Publishing International. pp.297–332

Chandra Babu R., Zhang J.A., Blum J.A., Ho T.-H.D., Wu R., Nguyen H.T. 2004. HVA1, a LEA gene from barley confers dehydration tolerance in transgenic rice *(Oryza sativa L.)* via cell membrane protection. Plant Sci 166:855–862

Chen Z., Newman I., Zhou M., Mendham N., Zhang G., Shabara S. 2005. Screening plants for salt tolerance by measuring K+ flux: a case study for barley. Plant Cell and Environment 28: 1230–1246

Close T.J. 1996. Dehydrins: emergence of a biochemical role of a family of plant dehydration proteins. Physiol Plant 97:795–803

Close T.J., Kortt A., Chandler P.M. 1989. A cDNA-based comparison of dehydration-induced proteins (dehydrins) in barley and corn. Plant Mol Biol 13:95–108

Cramer G., Epstain E., Lauchli A. 1989. Na-Ca interactions in barley seedlings: relationship to ion transport and growth. Plant Cell and Environment 12: 551–558

Cramer G.R., Quarrie S.A. 2002. Abscisic acid is correlated with the leaf growth inhibition of four genotypes of maize differing in their response to salinity. Funct Plant Biol 29: 111–115

Craufurd P.Q., Austin R.C., Acevedo E., Hall M.A. 1991. Carbon isotope discrimination and grain-yield in barley. Field Crops Res 27: 301–313

Diab A.A., Teulat-Merah B., This D., Ozturk N.Z., Benscher D., Sorrells M.E. 2004. Identification of drought-inducible genes and differentially expressed sequence tags in barley. Theor Appl Genet 109: 1417–1425

D'Oraci D., Bagni N. 1987. In vitro interactions between polyamines and pectic substances. Biochem. Biophys. Res. Commun 148: 1159–1163

Day A.D., Ludeke K.L., Ottman M.J. 1985. Registration of Arizona 8501 Barley germplasm for disturbed land reclamation. Crop Sci 26: 387

Dedolph C., Hettel G. (Eds.) 1997. Rice varieties boost yield and improve saline soils. Partners Making a Difference. IRRI, Manila, p. 37

Dietz K.J., Tavakoli N., Kluge C., Mimura T., Sharma S.S., Harris G.C., Chardonnens A.N., Golldack D. 2001. Significance of the V-type ATPase for the adaptation to stressful growth conditions and its regulation on the molecular and biochemical level. J Exp Bot 52: 1969–1980

Eckermann C., Eichel J., Schröder J. 2000. Plant methionine synthase: new insights into properties and expression. Biol Chem 381: 695–703

Epstein E., Norlyn J.D., 1977. Seawater-based crop production: a feasibility study. Science 197: 249–251

Farquhar G..D., Ehleringer J.R., Hubick K.T. 1989. Carbon isotope discrimination and photosynthesis. Ann. Rev. Plant Physiol 40: 503–537

Farquhar G..D., Richards R.A. 1984. Isotopic composition of plant carbon correlates efficiency of wheat genotypes. Aust J Plant Physiol 11: 539–552

Flowers T.J. 2004. Improving crop salt tolerance. J Exp Bot 55: 1–13

Flowers T. J., Yeo A. R. 1986. Ion relations of plant under drought and salinity. Aust Plant Physiol 13: 75–91

Flowers T.J., Yeo A.R. 1995. Breeding for salinity resistance in crop plants-Where next? Aust J Plant Physiol 22: 875–884

Flowers, T.J., Yeo, A.R. 1986. Ion relations of plants under drought and salinity. Aust. J Plant Physiol 13: 75–91

Foolad M.R. 1999. Comparison of salt tolerance during seed germination and vegetative growth in tomato by QTL mapping. Genome 42: 727–734

Forster B.P., Phillips M.S., Miller T.E., Baird E., Powell W. 1990. Chromosomal location of genes controlling tolerance in salt (NaCl) and vigour in *Hordeum vulgare* and H. chilense. Heredity 65: 99–107

Forster B.P., Ellis R.P., Thomas W.T.B., Newton A.C., Tuberosa R., This D., El-Enein R.A., Bahri M.H., Salem M.B. 2000. The development and application of molecular markers for abiotic stress tolerance in barley. Journal of Experimental Botany 51:19–27

Forster B.P., Ellis R.P., ThomasW.T.B., Newton A.C., Tuberosa R., This D., El-Enein R.A., Bahri M.H., Ben-Salem M. 2000. The development and application of molecular markers for abiotic stress tolerance in barley. J Exp Bot 51: 19–27

Forster B.P. 2001. Mutation genetics of salt tolerance in barley: An assessment of Golden Promise and other semi-dwarf mutants. Euphytica 120: 317–328

Forster B.P., Ellis R.P., Moir J., Talamè V., Sanguineti M.C., Tuberosa R., This D., Teulat-Merah B., Ahmed I., Mariy S., Bahri H., Muahabi M., Zoumarou-Wallis N., El-fellah M., and Salem M.B. 2004. Genotype and phenotype associations with drought tolerance in barley tested in North Africa. Ann Appl Biol144:157–168

Fox G., Panozzo J.F., Li C.D., Lance R.C.M., Inkerman A., Henry R.J. 2003. Molecular basis of barley quality. Australian Journal of Agricultural Research. 54: 1081–1101

Garbarino J., Dupont F.M. 1988. NaCl induces a Na+/H+ antiport in tonoplast vesicles from barley roots. Plant Physiol 86: 231–236

Garcia E.S., Gonzalez M.S., Azambuja P., Baralle F.E., Frainderaich D., Torres H.N., Flawia M.M. 1995. Induction of *Trypanosoma cruzi* metacyclogenesis in the hematophagous insect vector by hemoglobin and peptides carrying globin sequences. Exp Parasitol 81: 255–261

Grando S. 1989. Breeding for low rainfall areas. In: Cereal improvement program annual report 1089, ICARDA, Aleppo, pp. 26–35

Greenway H. 1962. Plant response to saline substrates I. Growth and ion uptake of several varieties of *Hordeum during* and after sodium chloride treatment. Aust. J. Biol. Sci 15: 16–38

Greenway H., Munns R. 1980. Mechanisms of salt tolerance in nonhalophytes. Ann Rev Plant Physiol 31: 149–190

Grossi M., Gulli M., Stanca A.M., Cattivelli L. 1995. Characterization of two barley genes that respond rapidly to dehydration stress. Plant Science 105: 71–80

Gulli M., Maestri E., Hartings H., Raho C., Perrotta C., Devos K.M., Marmiroli N. 1995. Isolation and characterizationof abscisic acid inducible genes in barley seedlings and their responsiveness to environmental stress. Plant Physiology 14: 89–96

Handley L.L., Nevo E., Raven J.A., Martines-Carrasco R., Scrimgeour C.M., Pakniyat H., Foster B.P. 1994. Chromosome 4 controls potential water us efficiency ibarley. J Exp Botany 45: 1661–1663

Hong B., Uknes S., Ho T.H.D. 1988. Cloning and characterization of a cDNA encoding a mRNA rapidly induced by ABA in barley aleurone layers. Plant Molecular Biology 11: 495–506

Hong B., Barg R., Ho T.D. 1992. Developmental and organ-specific expression of an ABA- and stress induced protein in barley. Plant Molecular Biology 18: 663–674

Hollung K., Espelund M., JakobsenK.J. 1994 Another Lea B19 gene (Group 1 Lea) from barley containing a single amino acid hydrophilic motif. Plant Molecular Biology 25: 559–564

Huang C.X., Van Steveninck R.F.M. 1990. Salinity induced structural changes in meristematic cells of barley roots. New Phytol 115: 17–22

Ishitani M., Nakamura T., Han S.Y., Takabe T. 1995. Expression of the betaine aldehyde dehydrogenase gene in barley in response to osmotic stress and abscisic acid. Plant Molecular Biology 27: 307–315

James R. A., Rivelli A. R., Munns R., von Caemmerer S. 2002. Factors affecting CO2 assimilation, leaf injury and growth in salt-stressed durum wheat. Functional Plant Biology. 29: 1393–1403

Jana S., Wilen R.W. 2005. Breeding for abiotic stress tolerance in barley. In: M. Ashraf, P.J.C. Harris (eds), Abiotic stresses:plant resistance through breeding and molecular approaches. Haworth Press, pp. 491–511

Jiang Q., Roche D., Monaco T.A., Durham M. 2006. Gas exchange, chlorophyll fluorescence parameters and carbon isotope discrimination of 14 barley genetic lines in response to salinity. Field Crops Research 96: 269–278

Kasuga M., Liu Q., Miura S., Yamaguchi-Shinozaki K., Shinozaki K. 1999. Improving plant drought, salt, and freezing tolerance by gene transfer of a single stress-inducible transcription factor. Nat. Biotechnol 17: 287–291

Kawasaki S., Borchert C., Deyholos M., Wang H., Brazille S., Kawai K., Galbraith D., Bohnert H.J. 2001. Gene expression profiles during the initial phase of salt stress in rice. Plant Cell 13: 889–905

Kearsey M.J. 1998. The principles of QTL analysis (a minimal mathematics approach). J Exp Bot 49: 1619–1623

Kleines M., Ralph-Cyrus E., Maria-Jesus R., Anne-Sophie B., Francesco S., Dorothea B., Max P. 1999. Isolation and expression analysis of two stress-responsive sucrose-synthase genes from the resurrection plant *Craterostigma plantagineum* (Hochst.). Planta 209:13–24

Koval V. S. 2000. Male and female gametophyte selection of barley for salt tolerance. Hereditas 132: 1–5

Koval V.S., Koval S.F. 1996. Genetic analysis of salt tolerance in barley: identification of the number of genes. Russian J Genet 32: 954–958

Kramer D. 1984. Cytological aspects of salt tolerance in higher plants. In: Salinity Tolerance in Plants (Ed. by Staples R.C. and Toenniessen G.H.). John Wiley and Sons, New York. pp. 3–5

Krestnikov I.S., Netsvetaev V.P., Biryukov S.V. 1986. Genotypic variability of root superoxide dismutase in spring barley. *Nauch.- Tekhn. Bull. VSGI (Odessa)*. 62: 35–40

Krishnaraj S., Mawson B.T., Yeung E.C., Jhorpe J.A. 1993. Utilizationof induction and quenching kinetics of chlorophyll fluorescencefor in vivo salinity screening studies in wheat (*Triticum aestivumvass.* Kharcia-65 and Fielder). Canadian Journal of Botany 71: 87–97

Kueh J.S.H., Bright S.W.J. 1982. Biochemical and genetic analysis of three proline accumulating barley mutants. Plant Sci Lett 27: 233–241

Lahaye P.A., Epstein E. 1971. Calcium and salt toleration by bean plants. Physiologia Plantarum 25: 223–218

Leigh R.A. 2001. Potassium homeostasis and membrane transport. J Plant Nutr Soil Sci 164: 193–198

Lilley J.M., Ludlow M.M., McCouch S.R., O'Toole J.C. 1996. Locating QTLs for osmotic adjustment and dehydration tolerance in rice. J Exp Bot 47:1427–1436

Lu Z., Tamar K., Neumann P.M., Nevo E. 1999. Physiological characterization of drought tolerance in wild barley (Hordeum spontaneum)from the Judean Desert. Barley Genetics Newsletter 29: 36–39

Lynch J., Lauchli A. 1985. Salt stress disturbs the calcium nutrition of barley (*Hordeum vulgare L.*). The New Phytologist 99: 345–354

Lynch J., Lauchli A. 1988. Salinity affects intracellular calcium in corn root protoplasts. Plant Physiology 87: 351–356

Lynch J., Thiel G., Lauchli A. 1988. Effects of salinity on the extensibility and Ca availability in the expanding region of growing barley leaves. Botanica Acta 101: 355–361

Maas E.V., Hoffman G. J. 1977. Crop salt tolerance–current assessment. Journal of the Irrigation and Drainage Division 103: 115–134

Maathuis F. J.M., Amtmann A. 1999. K^+ nutrition and Na^+ toxicity: the basis of cellular K^+/Na^+ ratios. Ann Bot 84: 123–133

Maestri E., MalcevschiA., Massari A., Marmiroli N. 2002. Genomic analysis of cultivated barley (*Hordeum vulgare*) using sequence-tagged molecular markers. Estimates of divergence based on RFLP and PCR markers derived from stress-responsive genes, and simple-sequence repeats (SSRs). Mol Genet Genomics 267: 186–201

Malkit A., Sadka A., Fisher M., Goldshlag P., Gokhman I., Zamir A. 2002. Salt induction of fatty acid elongase and membrane lipid modifications in the extreme halotolerant Alga *Dunaliella salina*. Plant Physiol 129: 1320–1329

Mano Y., Takeda K. 1996. Genetical studies on salt tolerance at germination in recombinant, inbred, iso-genic and doubted haploid lines of barley (*Hordeum vulgare L.*). Bull Res. Ins Okayama Univ 4: 79–88

Mano Y., Takeda K. 1997. Mapping quantitative trait loci for salt tolerance at germination and the seedling stage in barley *(Hordeum vulgare L.)*. Euphytica 94: 263–272

Mano Y., Takeda K. 1998. Genetic resources of salt tolerance in wild *Hordeum* species. Euphytica 103: 137–141

Mansour M.M.F., van Hasselt P.R., Kuiper P.J.C. 1994. Plasma membrane lipid alterations induced by NaCl in winter wheat roots. Physiol. Plant 92: 473–478

Maser P., Eckelman B., Vaidyanathan R., Horie T., Fairbairn D. J., Kubo M., Yamagami M., Yamaguchi K., Nishimura M., Uozumi N., Robertson W., Sussman M. R., Schroeder J. I. 2002. Altered shoot/root Na+ distribution and bifurcating salt sensitivity in Arabidopsis by genetic disruption of the Na+ transporter AtHKT1. FEBS Lett 531: 157–161

Monforte A.J., Asins M.J., Carbonell E.A. 1997. Salt tolerance in Lycopersicon species 6. Genotype-bysalinity interaction in quantitative trait loci detection: constitutive and response QTLs. Theor Appl Genet 95: 706–713

Morgan J.M., Tan M.K. 1996. Chromosomal location of a wheat osmoregulation gene using RFLP analysis. Aust J Plant Physiol 23:803–806

Muench D.G., Good A.G. 1994. Hypoxically inducible barley alanine aminotransferase: cDNA cloning and expression analysis. Plant Mol Biol 24: 417–427

Munns R. 1993. Physiological processes limiting plant growth in saline soils: some dogmas and hypotheses. Plant Cell and Environ 16: 15–24

Munns R. 2002. Comparative physiology of salt and water stress. Plant Cell Environ 25: 239–250

Munns R., James R.A. 2003. Screening methods for salinity tolerance: a case study with tetraploid wheat. Plant and Soil 253: 201–218

Muramoto Y., Watanabe A., Nakamura T., Takabe T. 1999. Enhanced expression of a nuclease gene in leaves of barley plants under salt stress. Gene 234: 315–321

Narita Y., Taguchi H., Nakamura T., Ueda A., Shi W., Takabe T. 2004. Characterization of the salt-inducible methionine synthase from barley leaves. Plant Science 167: 1009–1016

Noble C.L., Rogers M.E. 1992. Arguments for the use of physiological criteria for improving the salt tolerance in crops. Plant and Soil 146: 99–107

Norberg P., Liljenberg C. 1991. Lipids of plasma membranes prepared from oat root cells: effect of induced water deficit tolerance. Plant Physiol 96: 1136–1141

Ouerghi Z., Cornic G., Roudani M., Ayadi A., Brulfert J. 2000. Effect of NaCl on photosynthesis of two wheat species (*Triticum durum and T. aestivum*) differing in their sensitivity to salt stress. J Plant Physiol 156: 335–340

Ozturk Z.N., Talame V., Deyholos M., Michalowski C.B., Galbraith D.W., Gozukirmizi N., Tuberosa R., Bohnert H.J. 2002. Monitoring large-scale changes in transcript abundance in drought- and salt-stressed barley. Plant Molecular Biology 48: 551–573

Pakniyat H., Handley L.L., Thomas W.T.B., Connolly T., Macaulay M., Caligari O.D.S., Forster B.P. 1997. Comparison of shoot dry weight, Na^+ content and $d^{13}C$ values of Ali-E and other-dwarf mutants under salt-stress. Euphytica 94: 7–14

Pillen K., Zacharias A., Lèon J. 2003. Advanced backcross QTL analysis in barley (*Hordeum vulgare L.*). Theoretical and Applied Genetics 107:340–352

Poustini K., Siosemardeh A. 2004. Ion distribution in wheat cultivars in response to salinity stress. Field Crops Research 85: 125–133

Ramagopal S. 1988. Regulation of protein synthesis in root, shoot and embryonic tissues of germinating barley during salinity stress. Plant, Cell and Environment 11: 501–515

Rasmuson D.E., Anderson J.E. 2002. Salinity affects development, growth, and photosynthesis in cheatgrass. Journal of Range Management 55: 80–87

Rathore A.K., Sharma R.K., Lal P. 1977. Relative salt tolerance of different varieties of barley (*Hordeum vulgare L.*) at germination and seedling stage. Ann Arid Zone 16: 53–60

Rhode D., Hanson A.D. 1993. Quaternary ammonium and tertiary sulfonium compounds in higher plants. Ann Rev Plant Physiol Plant Mol Biol 44: 357–384

Richards R.A. 1983. Should selection for yield in saline regions be made on saline or non-saline soil. Euphytica 32: 431–438

Rodriguez E.M., Svensson J.T., Malatrasi M., Choi D.W., Close T.J. 2005. Barley Dhn13 encodes a KS-type dehydrin with constitutive and stress responsive expression. Theor Appl Genet 110: 852–858

Romagosa I., Araus J.L. 1991. Genotype-environment interaction for grain yield and 13C discrimination in barley. Barley Genetics VI 563–567

Sayed J. 1985. Diversity of salt tolerance in a germplasm collection of wheat (*Triticum aestivum*). Theor Appl Genet 69:651–657

Sayed O.H. 2003. Chlorophyll fluorescence as a tool in cereal crop research. Photosynthetica 41: 321–330

Serafini-Fracassini D., Del Duca S., Beninati S. 1995. Plant transglutaminases. Phytochemistry 40: 355–365

Shabala S. 2000. Ionic and Osmotic components of salt stress specifically modulate net ion fluxes from bean leaf mesophyll. Plant, Cell & Environment 23: 825–838

Shabala S., Shabala L., Van Volkenburgh E. 2003. Effect of calcium on root development and root ion fluxes in salinised barley seedlings. Functional Plant Biology 30: 507–514

Shabala S.I. 2002. Screening plants for environmental fitness: chlorophyll fluorescence as a ''Holy Grail'' for plant breeders. In: Hemantaranjan, A. (Ed.), Advances in Plant Physiology, vol. 5. Scientific Publishers, Jodhpur, India, pp. 287–340

Shannon M.C. 1997. Adaptation of plants to salinity. Adv Agron 60: 76–199

Shannon M.C., Noble C.L. 1990. Genetic approaches for developing economic salt-tolerant crops. In: K.K. Tanji (ed.) Agric. Salinity Assessment and Management. Manuals and Reports on Engineering Practice No. 71. Am Soc Civil Eng, New York. pp.161–185

Shen Z., Shen Q., Liang Y., Liu Y. 1994. Effect of nitrogen the growth and photosynthetic activity of salt stress barley. J Plant Nutri 17: 787–799

Shi W.M., Muramoto Y., Ueda A., Takabe T. 2001. Cloning of peroxisomal ascorbate peroxidase gene from barley and enhanced thermotolerance by overexpressing in *Arabidopsis thaliana*. Gene 273: 23–27

Smillie R.M., Nott R. 1982. Salt tolerance in crop plants monitored by chlorophyll fluorescence in vivo. Plant Physiol 70: 1049–1054

Somerville C. 1995. Direct tests of the role of membrane lipidcomposition in low-temperature-induced photoinhibition andchilling sensitivity in plants and cyanobacteria. Proc Natl Acad Sci USA 92: 6215–6218

Stanca A.M., Romagosa I., Takeda K., Lundborg T., Terzi V., Cattivelli L. 2003. Diversity in abiotic stress tolerance. In: RV Bothmer, Hintum TV, Knupffer H and Sato K (eds), Diversity in barley. ELSEVIER, pp.307–361

Szabolcs I. 1989. Salt-affected soil. CRC Press

Tabor C.W., Tabor, H. 1984. Polyamines. Annu. Rev Biochem 53: 749–790

Tassoni A., Antognoni F., Bagni N. 1996. Polyamine binding to plasma membrane vesicles from zucchini hypocotyls. Plant Physiol 110: 817–824

Talame V., Sanguineti M.C., Chiapparino E., Bahri H., Salem M.B., Forster B.P., Ellis R.P., Rhouma S., Zoumarou W., Waugh R., Tuberosa R. 2004. Identification of *Hordeum spontaneum* QTL alleles improving field performance of barley grown under rainfed conditions. AnnAppl Biol144:309–319

Teulat B., Rekika D., Nachit M.M., Monneveux P. 1997a. Comparative osmotic adjustments in barley and tetraploid wheats. Plant Breed 116:519–523

Teulat B., Monneveux P., Wery J., Borries C., Souyris I., Charrier A., This D. 1997b. Relationships between relative water-content and growth parameters under water stress in barley: a QTL study. New Phytol 137:99–107

Teulat B., This D., Khairallah M., Borries C., Ragot C., Sourdille P., Leroy P., Monneveux P., Charrier A. 1998. Several QTLs involved in osmotic-adjustment trait variation in barley (*Hordeum vulgare L.*). Theor Appl Genet 96:688–698

Teulat B., Borries C., This D., 2001a. New QTLs identified for plant water-status, water soluble carbohydrate and osmotic adjustment in a barley population grown in a growth-chamber under two water regimes. Theor Appl Genet 103:161–170

Teulat B., Merah O., Souyris I., This D. 2001b. QTLs for agronomic traits from a Mediterranean barley progeny grown in several environments. Theor Appl Genet 103:774–787

Teulat B., Merah O., Sirault X., Borries C., Waugh R., This D. 2002. QTLs for grain carbon-isotope discrimination in field-grown barley. Theor Appl Genet 106:118–126

Teulat B., Zoumarou-Wallis N., Rotter B., Ben Salem M., Bahri H., This D. 2003. QTL for relative water content in field-grown barley and their stability across Mediterranean environments. Theor Appl Genet 108:181–188

Tester M., Davenport R. 2003. Na^+ tolerance and Na^+ transport in higher plants. Ann Bot 91: 503–527

Tondelli A., Francia E., Barabaschi D., Aprile A., Skinner J.S., Stockinger E.J., Stanca A.M., PecchioniN. 2006. Mapping regulatory genes as candidates for cold and drought stress tolerance in barley. Theor Appl Genet 112: 445–454

von Well E., Fossey A. 1998. A comparative investigation of seed germination, metabolism and seedling growth between two polyploidy *Triticum* species. Euphytica 101:83–89

Wang W., Vinocur B., Altman A. 2003 Plant responses to drought, salinity and extreme temperatures: towards genetic engineering for stress tolerance. Planta 218: 1–14

Wei W.X., Bilsborrow P., Hooley P., Fincham D., Foster B. 2001. Variation between two near isogenic barley (*Hordeum vulgare*) cultivars in expression of the B subunit of the vacuolar ATPase in response to salinity. Hereditas 135: 227–231

Whittaker A., Bochicchio A., Vazzana C., Lindsey G., Farrant J. 2001. Changes in leaf hexokinase activity and metabolite levels in response to drying in the desiccation-tolerant species Sporobolus *stapfianus and Xerophyta viscosa*. J Exp Bot 52:961–969

Wierstra I., Kloppstech K. 2000. Differential effects of methyljasmonateon the expression of the early light-inducible proteins and other light-related genes in barley. Plant Physiol 124: 833–844

Wissenbach M., Uberlacker B., Vogt F., Becker D., Salamini F., Rohde W. 1993. Myb genes from Hordeum vulgare: tissuespecific expression of chimeric Myb promoter/Gus genes in transgenic tobacco. Plant J 4:411–422

Wolf O., Munns R., Tonnet M.L., Jeschke W.D. 1990. Concentrations and transport of solutes in xylem and phloem along the leaf axis of NaCl-treated *Hordeum vulgare*. Journal of Experimental Botany 41: 1133–1141

Xu D., Duan X., Wang B., Hong B., Ho T.-H.D., Wu R. 1996. Expression of a late embryogenesis abundant protein gene, HVAJ, from barley confers tolerance to water deficit and salt stress in transgenic rice. Plant Physiol 110:249–257

Yamaguchi M., Kasamo K. 2001. Modulation in the activity of purified tonoplast H+-ATPase by tonoplast glycolipids prepared from cultured rice (*Oryza sativa* L. var. Boro) cells. Plant Cell Physiol 42: 516–523

Yeo A.R., Flowers T.J. 1989. Selection for physiological characters – examples from breeding for salt tolerance. In: Jones, H.G., Flowers, T.J., Jones, M.B. (Eds.), Plants under Stress Biochemistry, Physiology and Ecology and their Application to Plant Improvement. Cambridge University Press, Cambridge, pp. 217–234

Zeh M., Leggewie G., Hoefgen R., Hesse H. 2002. Cloning and characterization of a cDNA encoding a cobalamin-independent methionine synthase from potato (*Solanum tuberosum L.*), Plant Mol Biol 48: 255–265

Zhao F.G., Qin P. 2005. Protective effects of exogenous fatty acids on root tonoplast function against salt stress in barley seedlings. Environmental and Experimental Botany 53: 215–223

Zhao F.G., Sun C., Liu Y.L., Liu Z.P. 2000. Effects of salinity stress on the levels of covalently and noncovalently bound polyamines in plasma membrane and tonoplast isolated from leaves and roots of barley seedlings. Acta Bot Sin 42: 920–926

Zhu J.K. 2000. Genetic analysis of plant salt tolerance using Arabidopsis. Plant Physiol 124: 941–948

Zid E., Grignon C. 1985. Sodium-calcium interactions in leaves of *Citrus aurantium* grown in the presence of NaCl. Physiologie Vegetale 23: 895–203

CHAPTER 25

RECENT ADVANCES IN BREEDING CITRUS FOR DROUGHT AND SALINE STRESS TOLERANCE

GOZAL BEN-HAYYIM[1] AND GLORIA A. MOORE[2]

[1]*Institute of Plant Sciences, Agricultural Research Organization, The Volcani Center, Bet Dagan, Israel*

[2]*Department of Horticultural Sciences, Plant Molecular and Cellular Biology Program, University of Florida, Gainesville, USA*

Abstract: Citrus is a major world horticultural commodity, and most of its world-wide production depends on irrigation, which is inevitably associated with the deterioration of water quality from run-off or ground water. Citrus, like most fruit trees, is relatively salt sensitive. The deleterious effects of salt stress lead to reduction in fruit yield and quality. In recent years, only a few relatively salt-tolerant rootstocks have been obtained through selection and conventional breeding, due to a rather limited existing genetic pool and the long period of time required for experiments. Attempts to regenerate salt-tolerant citrus plants via in vitro production of salt-tolerant callus or mutagenesis have been rather limited and as of yet not in use. Therefore, efforts should be invested to identify traits/genes that have a key role in tolerance to salt in order to speed up the process and to enlarge these genetic resources.

QTL analyses revealed that response to salt in citrus is a multigenic trait, as has been shown in other species, but some genes probably exist that have a major impact on salt tolerance and (or) mineral accumulation. Several robust EST databases now exist and are growing, the first microarray chips have been manufactured, and an initial genome sequencing effort is underway. These tools should allow citrus physiologists, biochemists, and geneticists to make much more rapid progress in understanding salt and water stress in the future and to design strategies to ameliorate their effects

1. INTRODUCTION

Citrus is a major world horticultural commodity, and most of its world-wide production depends on irrigation for economic production (Shalhevet and Levy, 1990). Irrigation is inevitably associated with the deterioration of water quality from run-off or ground water especially due to increases in soluble salts. Poor water quality unavoidably leads to increased soil salinity (Levy and Syversten, 2004).

M.A. Jenks et al. (eds.), Advances in Molecular Breeding Toward Drought and Salt Tolerant Crops, 627–642.

Citrus, like most fruit trees, is relatively salt sensitive (Bernstein, 1969). The deleterious effects of salt stress lead to reduction in fruit yield and quality. The common citrus rootstocks differ in their tolerance to salinity (Bernstein, 1969; Wutscher, 1979) and citrus trees can withstand relatively moderate salinity levels depending on the climate, scion cultivar, rootstock, and irrigation-fertilizer management. The response of different citrus species to different salt(s) further brings differential responses, when fruit quality is concerned. In young trees, salt damage is usually manifested as leaf burn and defoliation, which is associated with accumulation of toxic levels (Na^+ and/or Cl^-) in leaf cells. In many studies Cl^- exclusion from leaves served as a reliable criterion for salt tolerance leading to a decreasing order of salinity tolerance in rootstocks: Cleopatra mandarin > Sour orange > Sweet orange = Swingle citrumelo > Rough lemon > *Poncirus trifoliata* (Chapman, 1968; Newcomb, 1978). Differences in salt tolerance have also been shown to depend on the nature of the citrus scions (Cooper et al., 1952). Unraveling the mechanisms by which plants adapt to sustain salt stress, might provide an indication to plant breeders and biotechnologists as to how to proceed further in crop improvement.

Present day scion-rootstock combinations represent outcomes of human selection over the last 1500 years, with an especially intense selection pressure during the last century, enabling citriculture in environments far removed from source habitats of citrus species. While citrus is commonly grown in regions where the salinity of the irrigation water is relatively low, e.g. in Australia, typically <0.5 dSm^{-1}, it is also grown in regions where the salinity of the irrigation water is significantly higher, e.g. 1.4 dSm^{-1}. Continual improvement of rootstocks and/or scions will be necessary to sustain irrigated citrus in increasingly salinized environments (Storey and. Walker, 1999). A few relatively salt-tolerant rootstocks have been obtained through selection and conventional breeding, which is rather limited in the existing genetic pool and requires a long period of time for experiments. In the last twenty years, efforts have been invested in adopting modern biochemical and genetic tools to speed up the process and to enlarge the genetic resources.

This chapter summarizes the efforts of utilizing modern biotechnological tools to understand the physiological, biochemical and genetic basis for tolerance to salinity in citrus, and to apply this knowledge for production of enhanced salt-tolerant rootstocks.

2. PHYSIOLOGICAL AND BIOCHEMICAL PARAMETERS IDENTIFYING SALT/DROUGHT TOLERANCE

In many parts of the world, the citrus industry is limited by the availability of good quality water and salt/drought tolerance is one of the major traits that determine its ability to expand. In addition to conventional breeding programs, studies have emerged during the last twenty years adopting modern tools to analyze and enhance salt tolerance. These include tissue culture techniques and identification of major parameters, physiological, biochemical and molecular, that have a crucial impact on tolerance, to be later used in genetic manipulation.

2.1. In Vitro Systems and Salt Tolerance

The utilization of in vitro techniques and production of cultured cells has enabled researchers to gain better insight into the cellular mechanisms underlining salt response and adaptation to salt tolerance. Singh et al (2004) prepared callus from internodal stem segments of Rangpur lime (*C. limonia* Osbeck), rough lemon (*C. jambhiri* Lush.) and *Poncirus trifoliata* (L.) Raf., and confirmed that after 6 weeks of salt stress levels of Cl^- uptake and Na/K ratio in the calli corresponded to those found in the respective leaves, i.e. those of Rangpur lime were lower than those of rough lemon, which were lower than those of trifoliate orange. Thus, it was concluded that Cl^- uptake and Na/K ratio are reliable markers and that in vitro techniques can be a useful tool for screening a large citrus germplasm for salt tolerance. In a different study, using embryo callus prepared from Troyer citrange [*C. sinensis* (L.) Osbeck X *P. trifoliata*], sour orange (*C. aurantium*) and *P. trifoliata*, the authors claim that *P. trifoliata* is moderately salt-tolerant, Troyer citrange is the most salt-sensitive genotype and *C. aurantium* is moderately salt-sensitive (Rochdi et al., 2003). Their conclusion is based on measurements including Cl^- accumulation and loss of K^+ and it confirms the degree of sensitivity determined for these rootstocks by Loussert (1989).

In vitro techniques were used to select callus with increased salt tolerance via prolong sub-culturing on salt. Salt-adapted ovular cultured cells of Shamouti orange (*C. sinensis* L. Osbeck) exhibited reduction in Cl^- and Na^+ accumulation, but no change in the K content (Ben-Hayyim and Kochba, 1983). Salt-adapted ovular cultured cells of sour orange (*C. aurantium*), on the other hand, displayed higher levels of Cl^- and Na^+, but also showed a dramatic ability to maintain high K^+ under salt stress, which corresponded well with their ability to grow in the presence of salt (Ben-Hayyim et al., 1985). A similar feature was observed in *P. trifoliata* cv Pomeroy embryo-derived callus sub-cultured on salt, where higher levels of K^+ and Ca^{2+} were maintained in the salt-selected cell line compared with the original cell line (Beloualy and Bouharmont, 1992). In agreement, shoots and plants regenerated from salt-selected cell lines showed better growth on media containing salt than those derived from non-selected cells. Likewise, calli derived from these plants, whether regenerated in the absence or presence of salt, were markedly more salt-tolerant. The persistence of the acquired trait was also demonstrated in callus derived from plantlets regenerated from salt-tolerant Shamouti orange callus culture (Ben-Hayyim and Goffer, 1989). However, it should be noted that at least in the case of Shamouti orange, regenerated plantlets were abnormal and could not develop beyond the in vitro stage.

Salt-tolerant Troyer citrange plants were developed following chemical mutagenesis of unfertilized ovules (Garcia-Agustin and Primo-Millo, 1995). Plants regenerated from ovules cultured in vitro were subjected to selection by irrigation with saline nutrient solution, and plants obtained from vegetative propagation of a selected plant showed faster growth, less leaf damage and lower leaf concentrations of Cl^- and Na^+ than the original clone when subjected to salt stress.

Although there is a strong relationship between salt and drought/osmotic stress, no direct selection for osmotic stress has been reported in citrus. Salt-adapted cultured cells of Shamouti orange and sour orange were evaluated for their tolerance to osmotic stress produced by the addition of polyethylene glycol (Ben-Hayyim, 1987). While salt-tolerant Shamouti orange cells also exhibited higher tolerance to osmotic stress, salt-tolerant sour orange cells did not. This difference is probably related to two types of mechanisms for salt tolerance acquired by these cells, e.g. partial exclusion of salt by Shamouti orange cells and its accumulation by sour orange cells (Ben-Hayyim and Kochba, 1983; Ben-Hayyim et al., 1985).

2.2. Ion Content and Salt Tolerance

In the question which of the ions, Na^+ or Cl^- is the more toxic one, many reports point to the fact that Cl^- exclusion, rather than Na^+ exclusion, is a reliable and desirable marker for salt tolerance. Na^+ exclusion is characteristic for trifoliate orange and its hybrids, such as Troyer and Carrizo citrange, whereas Cl^- exclusion is characteristic for Cleopatra mandarin (*C. reticulata* Blanco) and Rangpur lime, both serving as a good rootstock in marginal lands. Many experiments show a good correlation between growth and Cl^- accumulation under salt stress. Testing ungrafted rootstocks for salt tolerance revealed that Cleopatra mandarin grew better than sour orange and accumulated less Cl^- (Zekri, 1991). Greenhouse experiments showed that saline irrigation of Valencia orange (*C. sinensis* L. Osbeck) budded on either Cleopatra mandarin or *Poncirus trifoliata* resulted in better growth of the former as well as less Cl^- accumulation (Banuls et al., 1997). In addition, it was shown that NaCl and KCl reduced growth whereas $NaNO_3$ was less harmful, suggesting that Cl^- is indeed the toxic ion. In a different greenhouse experiment, Sunburst mandarin was grafted on similar rootstocks, namely, Cleopatra mandarin and Carrizo citrange, which resulted in reduced Cl^- accumulation in the mandarin leaves grafted on Cleopatra mandarin compared with those grafted on Carrizo citrange (Garcia-Sanchez et al., 2002). These results suggest that increased salt tolerance induced by the rootstock Cleopatra mandarin is associated with better sequestering of Cl^- in the roots and its lower transport to the shoots. In a field experiment, where Star Ruby grapefruit (*C. paradisi* Macf.) trees were grafted either on Cleopatra mandarin or Carrizo citrange, essentially the same pattern was obtained (Garcia-Sanchez et al., 2002). Under saline conditions, the trees grafted on Cleopatra mandarin had higher fruit yield and less Cl^- accumulated in the leaves. On the other hand, those trees grafted on Carrizo citrange had impaired fruit quality due to a decrease in juice percentage and an increase in pulp and peel percentage. Screening for salt-tolerant rootstocks was also tested in hydroponic cultures, where five rootstocks were treated with various irrigation solutions containing salt (Garcia et al., 2002). In the ungrafted rootstocks, the highest survival was exhibited by Flying Dragon (*Poncirus trifoliata*) and Carrizo citrange and the performance of Cleopatra mandarin was rather poor. These results seem to contradict the above described experiments, and may reveal an important role for Na^+ exclusion. Another explanation for the ability of Flying Dragon and its hybrid to survive better under salt stress may lie in its ability to defoliate and thus remove the excess ions (Garcia et al., 2002). The performance of Flying Dragon and Carrizo citrange as relatively salt-tolerant rootstocks was further shown by the lower salt-induced leaf drop of Satsuma mandarin grafted on these rootstocks compared with Cleopatra mandarin. Similar results were reported by Vardi et al., (1988), where the mortality of Shamouti orange grafted on *P. trifoliata* was much lower than when it was grafted on sour orange at the end of 3 years of salinization, in spite of the fact that the Shamouti leaves accumulated high levels of Cl^-. Thus, although many studies utilize Cl^- accumulation under salt stress as a marker for salt tolerance, this parameter maybe reliable in grafted trees, but not necessarily in leaves of the ungrafted rootstocks. The importance of leaf parameters in determining salt tolerance was also exhibited in an experiment where pummelo (*C. grandis*) and *P trifoliata* were salinized (Tozlu et al., 2000b). Pummelo was actually more affected than Poncirus by salinity in terms of root and net (overall) growth, but it grew for a longer period of time during salinization. It also increased leaf mass as a percentage of total mass during the salinization period, although total mass went down. In contrast, *Poncirus* uniquely appeared to respond to salinity by increasing the production and turnover of fine roots (Tozlu et al., 2000a). So, probably in citrus types, as opposed to *Poncirus* which shed its leaves, leaves are more important for tolerance. Relatively recent data provide evidence for the emphasis on the importance of increasing salt tolerance in rootstocks. Leaves of the Cl^--sensitive Carrizo citrange (*C. sinensis* L. Osb. X *Poncirus trifoliata* L. Raf.) grafted on the Cl^--tolerant Cleopatra mandarin (*C. reshni* Hort. Ex Tan.) were more salt-tolerant in terms of shoot growth and lower accumulation of Cl^- than the non-grafted leaves (Moya et al., 2002). The reciprocal grafts exhibited essentially the same pattern, where Cleopatra mandarin grafted on Carrizo citrange was slightly less tolerant to salt than non-grafted Cleopatra.

It is really difficult to make clear cut conclusions following the results of the experiments described above. Different experiments included different genotypes and the conditions of salinization were varied. In many of then *P. trifoliata*, which shed its leaves, was included, making the comparisons of leaf parameters quite difficult. Nevertheless, it seems that there is a general agreement that Cl^- exclusion is a desired parameter for a good rootstock and that leaf Cl^- accumulation is a useful tool for screening salt tolerance.

2.3. Metabolites and Enzymatic Activities as Markers for Salt Tolerance

Several attempts have been made to correlate salt tolerance with biochemical functions which might eventually lead to improvement of tolerance via genetic engineering. Most of these studies have been performed in cultured cells. In salt-tolerant callus culture selected from lemon (*C. limon* L. cv Verna), the levels of proline, as well as glycine-betaine and choline were elevated compared with the non-selected callus (Piqueras et al., 1996). Salt tolerance of Shamouti orange cell culture, which was adapted to salt for a long period of time, was associated with salt-induced reduction in the unsaturation of fatty acids, exhibited by the ratio of linolenic to linoleic acids, while no such change was observed in the original cell culture (Gueta-Dahan et al., 1997). The authors suggested that the ability of salt-tolerant cells to reduce unsaturation reflects a protective mechanism, where by minimizing the content of linolenic acid, the fatty acid most prone to oxidation, these cells reduce the deleterious effect of salt-induced oxidative stress. In these salt-tolerant cells, a transient induction of a 9-lipoxygenase was observed under salt stress accompanied by a very fast reduction of the product hydroperoxides to the corresponding hydroxy derivatives (Ben-Hayyim et al., 2001). No such activity was found in the salt-sensitive cells, suggesting that the activity of 9-lipoxygenase in the salt-tolerant cells leads to a production of metabolites playing a key role in triggering defense mechanism against salt stress.

The activity of some of the anti-oxidant enzymes has been correlated with salt tolerance. In extracts of lemon cell cultures, an additional superoxide dismutase (SOD) isozyme was observed in the selected salt-tolerant culture as compared with the salt-sensitive one, following a separation on isoelectric focused gels (Piqueras et al., 1996). This isozyme was characterized as a Mn-SOD isozyme. It should be noted, however, that this comparison was made between the sensitive cells cultured in the absence of salt and the tolerant cells cultured in the presence of salt, and no data is provided cells cultured under similar conditions. In contrast, the isozyme pattern of SOD was rather similar in salt-sensitive and salt-tolerant Shamouti orange cultured cells, where no change was observed in the Mn-SOD activity, constitutively or under salt stress stress (Guata-Dahan et al., 1997). Salt-induced reduced activity of Mn-SOD and Fe-SOD was observed in leaves of rough lemon (*C. volkameriana*) (Gueta-Dahan et al., 1997) and in leaves of lemon trees (*C. limonium*) grafted on the salt-sensitive rootstocks *C. reticulata* and *C. macrophylla*, but not on sour orange (Almansa et al., 2002). On the other hand, salt increased Cu/Zn-SOD activity was detected in leaves of rough lemon, in leaves of lemon trees grafted on the salt-tolerant rootstock *C. aurantium* or the salt-sensitive rootstocks *C. reticulata* and *C.macrophylla* and in Shamouti cultured cells (Gueta-Dahan et al., 1997; Almansa et al., 2002). Salt adaptation of Shamouti culture cells resulted in slightly higher constitutive activity of this isozyme, namely increased activity in salt-tolerant cells in the absence of salt (Gueta-Dahan et al., 1997).

The constitutive activity of ascorbate peroxidase (APX) was found to exhibit the most striking difference between salt-sensitive and salt-tolerant Shamouti orange cultured cells (Gueta-Dahan et al., 1997). In the absence of salt, the activity of APX in the salt-tolerant cell line was about 10-fold higher than in the salt-sensitive one, and although salt reduced APX activity in both salt-sensitive and salt-tolerant cell lines, the salt-tolerant cell line could maintain higher activity. Thus, it was suggested that APX activity plays a crucial role in the defense against salt-induced oxidative stress. This idea was further supported by the observation that APX activity was much higher in leaves of Cleopatra mandarin than in those of Troyer citrange, a rootstock known to be rather sensitive to salt (Ben-Hayyim, personal communication).

Phospholipid hydroperoxide glutathione peroxidase (PHGPX) has a lipid hydroperoxide scavenging activity and was shown to be differentially induced by salt in salt-sensitive and salt-tolerant citrus cells (Avsian-Kretchmer et al., 1999). In the presence of salt, both transcript and protein levels were earlier induced in salt-sensitive cells, indicating that the rate of induction of this gene/protein could be a useful marker for the degree of stress imposed by salt.

In conclusion, attempts to regenerate salt-tolerant citrus plants via in vitro production of salt-tolerant callus or mutagenesis have been rather limited and as yet not in use. Therefore, in addition to conventional breeding based on the availability of limited salt-tolerant rootstocks, efforts should be invested to identify traits/genes that have a key role in tolerance to salt. Taken together the data available on ion accumulation under salt stress raise a question on its reliability to serve as a good marker for salt tolerance. This is a rather complex trait differing among tissues and it is not consistent in grafted and ungrafted leaves. Thus, the currently preferred effort should be directed towards identification of genes conferring stress tolerance. Once such genes are identified, their genetic manipulation could dramatically shorten the period required for achieving the goal of salt-tolerant rootstock, and production of fruits which are not genetically modified.

3. QTL MAPPING OF EFFECTS OF SALINIZATION IN CITRUS

As the development of polymorphic DNA markers of numerous types has progressed, the past 20 years has witnessed the production of genetic linkage maps of many species, many of them are now quite saturated. By crossing parents that differ in one or more aspects of a trait, such as salt or drought tolerance (their phenotype), and then analyzing the phenotypes and molecular genotypes of their progeny, it has been possible to locate quantitative trait loci (QTLs), molecular markers that appear to segregate with some aspect of the trait in question. Plant breeders in particular, found the concept of QTLs particularly attractive, as they could, in principle, be developed to produce markers to aid selection for the trait using marker-assisted selection (MAS). The idea of MAS was that the DNA genotype that was correlated with the positive expression of trait could be used to select promising progeny plants without the need to determine their phenotype (for example, without the need to evaluate performance of the plant under salt stress) and to discard unpromising plants, saving time and space (Flowers et al., 2000; Asins, 2002; Flowers and Flowers, 2005). Studies for the identification of QTLs have been done in a wide variety of species, although tomato, rice, and maize predominate, and perennial plants are little represented (Lexer et al., 2004; Ronnberg-Wastljung et al., 2005; Tschaplinski et al., 2006). Early analyses of QTLs in many species established that salt tolerance and drought tolerance were undoubtedly polygenic traits. However, an increasing number of analyses have demonstrated that the expression of QTLs is highly dependent not only on initial plant genotypes but also on plant stage of development, phenotypic characteristics examined, and genotype/environment interactions. Thus a MAS scheme that was successful in one cross will not necessarily be applicable to other situations (Campos et al., 2004). Positional cloning of promising genes at QTL is difficult because of lack of genetic information in many species and the sheer genetic distances between the marker loci and candidate genes. With the advent of genomic sequencing in model species and microarrays of defined ESTs becoming increasingly available in still more species, QTL mapping may be seen as an increasingly inefficient way to detect loci affecting traits of interest unless QTL maps can be linked to other genomic tools (Sawkins et al., 2004). None the less, QTL analyses have shown in many species, including citrus (see below), that salt tolerance and drought tolerance are heritable, although they are multigenic traits, where many genes are involved, but that some of the genes may be of large effect.

In the case of citrus, we conducted an intensive genetic study with a population that we expected would segregate for various parameters of salinity sensitivity/tolerance (Tozlu et al., 2000a,b). The major experiment consisted of evaluating the performance of an intergeneric BC_1 population [(*C. grandis)* x (*C. grandis* x *P. trifoliata* selected F_1 17–40), consisting of 54 individuals, relative to its parents for different growth and mineral accumulation related traits under both saline (40 mM NaCl) and non-saline environments (Tozlu et al., 1999a, b). A total of 36 traits related to growth (six traits) and tissue or whole plant dry mass production (30 traits) and 38 traits related to different tissue or whole-plant Na^+ and (or) Cl^- accumulation were evaluated. The difference between the responses of the original parents and the heterotic F_1 plant led to wide segregation in the BC_1 progeny in response to 16 weeks of salinization. Many traits showed transgressive segregation in both directions that may yield extreme values to breed salt hardy citrus genotypes. For example, 15% of the population performed better than the best performing parent, the F_1, for growth parameters, 21% of the population displayed less leaf damage than any parent for leaf symptom responses, and some progenies accumulated less Na^+ and/or Cl^- in different tissues and/or in whole plants than did any parent. Superior genotypes were identified for overall performance as well. Mapping analyses of these quantitative traits resulted in the identification of a number of potential quantitative trait loci (PQTL) with LOD scores greater than or equal to 3.0 located on a previously generated linkage map (Durham et al., 1992; Cai et al., 1994). The small progeny population size used made further analyses of these PQTLs necessary. Correlation analyses and locations of PQTLs indicated that many traits were controlled by fewer genes than the actual number of QTLs mapped for them. For example, 21 PQTLs mapped for Na^+ accumulation and Cl^-/Na^+ ratios were located in a cluster at the beginning of one linkage group (LG), while 10 PQTLs mapped for Cl^- accumulation and Cl^-/Na^+ ratios were located in a cluster at the beginning of another LG. However, the analyses revealed that, as has subsequently been shown in other species by microarray analyses, that response to salt in citrus is a multigenic trait, but some genes probably exist that have a major impact on salt tolerance and (or) mineral accumulation. Similar results were obtained with a genetically similar but larger population when freezing tolerance, another desiccation stress, was studied (Weber et al., 2003).

4. CANDIDATE GENES FOR RESPONSE TO WATER DEFICIT STRESS IN CITRUS – CLONING AND TRANSFORMATION

Several genomic studies in other plant species have indicated similarities in gene expression patterns following water deficit stresses of various kinds, including high or low temperatures, salinity and drought (Iba, 2002; Seki et al., 2002; Sung et al., 2003). Therefore, citrus genes that have been studied in any of these contexts are discussed here.

There have not as yet been any global gene expression studies in citrus on response to salinity or drought, but such studies will probably occur in the near future because resources to perform them are becoming available; a number of EST libraries have been established and sequenced, including libraries from environmentally stressed tissue (Forment et al., 2005), a convenient database of all publicly available DNA sequences has been set up and is maintained by Timothy Close and Steve Wanamaker, University of California, Riverside, and a microarray chip based on the sequences in that database is now commercially available. One study has been done where a subtracted cDNA library was constructed from cold-acclimated leaf tissue of the cold-hardy citrus relative *Poncirus trifoliata* (L.) Raf (Sahin-Cevik and Moore, 2006b). This library appeared to be successfully enriched for genes upregulated in response to cold. In some cases, for instance with two AP2 domain-containing genes, the genes were differentially induced in cold-hardy *Poncirus* and cold-sensitive pummelo in response to cold temperatures and were not induced in either type by drought. On the other hand, a novel RING-H2 finger gene isolated from the library was induced to a greater extent by drought than cold in both species ((Sahin-Cevik and Moore, 2006). Thus genes obtained from this library and characterized as cold-regulated may also respond to other kinds of abiotic stress. Similar studies have been done using a differential display technique (Lang et al., 2005; Zhang et al., 2005a; Zhang et al., 2005b)

In the past, the genes that were studied most extensively to characterize and ameliorate abiotic stresses were those encoding functional proteins of various kinds (Zhang et al., 2004; Umezawa et al., 2006). Some of these studies have been done in citrus.

One set of genes frequently implicated in plant response to dehydration stress are those that encode heat shock proteins (HSPs). HSPs are a diverse group of proteins, ranging in molecular weight from 15 to 115kD that are expressed in all organisms in response to elevated temperatures. In plants, HSPs function as molecular chaperones, assisting in protein folding, assembly, and transport, minimizing the aggregation of proteins, and targeting aggregated or degraded proteins for degradation. The expression of many HSPs is regulated by temperature stress, either high or low, although some are constitutively expressed and some are developmentally regulated. HSPs have been little studied in citrus. However, four

HSP cDNAs that shared high homology with plant HSP18–1, HSP18–2, HSP22, and HSP70 genes were cloned from grapefruit flavedo ((Rozenzvieg et al., 2004). Expression of these genes in fruit flavedo was briefly up-regulated by hot water or hot air treatments, but more stably up-regulated by a hot water treatment followed by prolonged chilling, leading to a hypothesis that they could be involved in the heat-induced chilling-tolerance response under study. In addition, two HSP70s were isolated from a cold acclimation subtracted cDNA library from *Poncirus trifoliata* and were shown to be highly induced in response to cold (Sahin-Cevik and Moore, in press).

The plant hormone abscisic acid (ABA) also plays a significant role in the adaptation of plants to various environmental stresses. In higher plants, ABA is derived from C_{40}-*cis*-epoxycarotenoids 9'-*cis*-neoxanthin and/or 9-*cis*-violaxanthin, which are cleaved by a 9-cis-epoxycarotenoid dioxygenase (NCED) to produce xanthoxin, the C_{15} precursor of ABA ((Nambara and Marion-Poll, 2005). In all plant species examined, NCEDs comprise a small gene family. Accumulation of ABA and increased expression of NCEDs have been found to be correlated with increasing water stress and transgenic plants overexpressing a NCED gene were more resistant to drought stress (Qin and Zeevaart, 2002). Rodrigo et al. (2006) cloned two full-length NCED cDNAs from peel of sweet orange fruits. Expression of one of them (*CsNCED1*) increased during natural and induced fruit maturation and in water-stressed leaves in a pattern consistant with accumulation of ABA, while expression of the other one was limited to fruit tissue. Thus, *CsNCED1* might be a candidate for engineering citrus to be more drought tolerant.

Biosynthesis of a second plant hormone, ethylene, is also associated with environmental stress, and a key enzyme in the biosynthetic pathway is 1-aminocyclopropane-1-carboxylate synthase (ACC synthase), also typically encoded by a multigene family. Two chilling regulated ACC synthase genes have been isolated from sweet orange fruit (Wong et al., 1999). Both genes were also induced by wounding. When the chilling inducible CS-ACS1 gene was transformed in an antisense orientation into Carrizo citrange and *Poncirus trifoliata*, the resulting transgenic plants producing ACS antisense RNA did not increase ACC content following chilling (Wong et al., 2001). Ethylene and ACC synthase are probably also important factors in leaf abscission in rehydrated citrus trees following water stress, with water stress promoting ACC synthesis in plant roots and rehydration leading to ACC transport to shoots, where it is oxidized to ethylene, inducing leaf abscission (Tudela and Primomillo, 1992).

Another gene isolated from citrus during the study of heat-induced chilling tolerance in grapefruit was cNHX1, a vacuolar membrane Na^+/H^+ antiporter gene involved in the compartmentalization of sodium ions into the vacuole (Porat et al., 2002a). Overexpression of a homologous gene in Arabidopsis increased salt tolerance in transgenic plants (Apse et al., 1999). In citrus, gene expression was transiently increased by a heat treatment, more markedly and stably increased in heat-treated and then chilled fruit, and also markedly induced by salt stress (Porat et al., 2002a).

The gene group that has been studied most in citrus in response to water stress is comprised of the dehydrins, which encode a subgroup of LEA (Late Embryo Abundant) proteins known as LEA-D11 or LEA type II. This is an immunologically distinct family of proteins, members of which have been shown to be induced during periods of water deficit imposed by extreme temperatures, drought, and salinity and during certain developmental events such as seed maturation (Close, 1997). The dehydrins have in common extreme hydrophilicity, solubility at high temperature, and a conserved lysine-rich 15 amino acid motif (the K-segment) present in one or more copies. The K-segment is predicted to form a class A amphipathic α-helix with the potential for both binding water and hydrophobic interaction. Other structural features of dehydrins include a tract of serine residues (the S-segment), a conserved amino acid sequence in the N-terminus (DEYGNP) and a φ-segment rich in polar amino acids and either glycine or proline and alanine. Although the functions of dehydrins are not completely understood, there is a great deal of evidence suggesting that they may act as structural stabilizers, protecting nuclear, cytoplasmic, and membrane macromolecules from dehydration-induced damage, thus maintaining cell structure and integrity.

Several dehydrin genes have now been isolated from citrus and its relatives and analyzed. Two dehydrin genes, *cor11* (a KS type) (Close, 1997) and *cor19* (a K_3S type), were first identified froma cDNA library constructed from cold acclimated leaf tissue of the cold-hardy citrus relative *Poncirus trifoliata* (Cai et al., 1995). Both genes were induced in response to cold temperatures, more so in cold-hardy *Poncirus* than in cold-sensitive *C. grandis*, but expression of *cor19* was repressed in response to drought and flooding. A *cor19* homologue (*CuCOR19*) was isolated from the flavedo of the *Citrus unshui* mandarin fruit and was shown to be induced in leaf tissue in response to cold, but not following treatment with salt or ABA (Hara et al., 1999; Hara et al., 2001). Overexpression of the gene enhanced cold tolerance in transgenic tobacco (Hara et al., 2003). *cor15*, a K_2S dehydrin type, from the flavedo of grapefruit has been characterized (Porat et al., 2002b). A highly homologous gene, designated *Crcor15*, was isolated from the flavedo of chilling-sensitive Fortune mandarin ((Sanchez-Ballesta et al., 2004). In contrast to grapefruit *cor15*, however, *Crcor15* was highly and constitutively expressed in fruit flavedo during fruit development and maturation and expression was depressed by a treatment that conferred chilling tolerance to the fruit. In addition, while expression was barely detectable in non-stressed leaf tissue, expression was rapidly and highly induced in response to both cold and water stress in leaves.

The genes described above constitute a unique dehydrin gene family for an angiosperm in that their K-segment(s) is similar in sequence to that of gymnosperms and their S-segment is located in an unusual position at the c-terminus of the protein. Two additional dehydrin genes, that have the angiosperm K-segment consensus sequence and the S-segment at the usual location in the n-terminus of the protein have also been isolated, one from Navel orange (*csDHN*) and one from Star Ruby grapefruit (*cpDHN*), (Porat et al., 2004). Expression of these genes in fruit peel is down-regulated by environmental stresses such as wounding, UV irradiation, water stress, and ethylene exposure, but expression was maintained in chilled fruit subsequent to a short exposure to heat.

Few of the genes described above as being affected by water deficit stress have yet been used in genetic transformation studies. However, a couple of such studies have been done. Carrizo citrange, perhaps the easiest type of citrus to genetically transform, was engineered to express a mutant Δ^1-pyrroline-5-carboxylate synthetase gene, which encodes the rate-limiting step of the proline biosynthetic pathway (Molinari et al., 2004). The transgenic plants displayed superior osmotic adjustment and significantly higher photosynthetic rates than control plants when water was withheld. Carrizo citrange has also been transformed with a yeast halotolerance gene, *HAL2*, in an effort to produce plants more tolerant to salinity (Cervera et al., 2000).

In contrast to exploring the individual effects of functional genes on various abiotic stresses, more recently there have been efforts to examine the actions of transcription factors (TFs) on suites of genes involved in these processes. Transcriptome analyses in other species have revealed that dozens of TFs are involved in response to, for instance, drought stress (Umezawa et al., 2006). Most of these fall into several large TF families, such as AP2/ERF, bZIP, NAC, MYB, MYC, or WKKY. Overexpression of TFs that upregulate stress-responsive genes has been used to engineer increased tolerance to environmental stresses such as salt, drought, and cold in a number of plant species (Zhang, 2003; Zhang et al., 2004; Umezawa et al., 2006), although this has not yet been accomplished in citrus. However, transcriptome analyses of *Citrus* and its relative *Poncirus* have revealed that similar TFs exist in the citrus genome and such studies are underway (Sahin-Cevik and Moore, 2006) (Sahin-Cevik and Moore, 2006; Champ et al., 2006a, c).

5. CONCLUDING REMARKS

Large size, long generation times, high heterozygosity, reproductive barriers, and commercial expectations for particular citrus types makes the study of citrus physiology, biochemistry, and genetics challenging, and the multigenic nature of salt tolerance and tolerance to other water deficit stresses adds greatly to the challenge. Nevertheless, citrus is an extremely important fruit crop grown in many tropical and semitropical parts of the world, so citrus scientists have made efforts in analyzing water deficit stresses and improving tolerance to these stresses. Many of these experiments in the past have been specifically on salinity, perhaps because this water deficit is relatively easy to quantitate and because salinity is a problem in so many parts of the world where citrus is grown. The recent sequencing of the genomes of model species such as *Arabidopsis* and rice, the increasing availability of large EST databases, and the rapid increase of information generated in high-throughput analyses such as microarrays will reveal increasing information on the genes involved in all types of water deficit that will be generally relevant in plants, so much knowledge gained should be applicable to citrus. However, there will probably be genes/mechanisms involved that are unique to perennial plants and to citrus specifically. Therefore, it is fortunate that citrus specific resources are being generated, although admittedly not with the speed and completeness as those in model species. Several robust EST databases now exist and are growing, the first microarry chips have been manufactured, and an initial genome sequencing effort is underway. These tools should allow citrus physiologists, biochemists, and geneticists to make much more rapid progress in understanding salt and water stress in the future and to design strategies to ameliorate their effects.

REFERENCES

Almansa, M.S., Hernandez, J.A., Jimenez, A., Botella, M.A., and Sevilla, F., 2002, Effect of salt stress on the superoxide dismutase activity in leaves of Citrus limonum in different rootstock-scion combinati. *Biol. Plant.* **45**:545 549.

Apse, M.P., Aharon, G.S., Snedden, W.A., and Blumwald, E., 1999, Salt tolerance conferred by overexpression of a vacuolar Na^+/H^+ antiport in Arabidopsis, *Science* **285**:1256 1258.

Arbona, V., Flors, V., Jacas, J., García-Agustí, P., and Gómez-Cadenas, A., 2003, Enzymatic and nonenzymatic antioxidant responses of Carrizo citrange, a salt-sensitive Citrus rootstock, to different levels of salinity, *Plant Cell Physiol.* **44**:388 394.

Asins, M.J., 2002, Present and future of quantitative trait locus analysis in plant breeding, *Plant Breeding* **121**:281 291.

Avsian-Kretchmer, O., Eshdat, Y., Gueta-Dahan, Y., and Ben-Hayyim, G., 1999, Regulation of stress-induced phospholipid hydroperoxide glutathione peroxidase expression in citrus, *Planta* **209**:469 477.

Banuls, J., Serna, M. D., Legaz, M. and Primo-Millo, E., 1997, Growth and gas exchange parameters of Citrus plants stressed with different salts, *J. Plant Physiol.* **150**:194 199.

Beloualy, N., and Bouharmont, J., 1992, NaCl-tolerant plants of Poncirus trifoliate regenerated from tolerant callus lines, *Theor. Appl. Genet.* **83**:509 514.

Ben-Hayyim, G., 1987, Relationship between salt tolerance and resistance to polyethylene glycol-induced water stress in cultured citrus cells, *Plant Physiol.* **85**:430 433.

Ben-Hayyim, G., and Goffer, Y., 1989, Plantlet regeneration of NaCl-selected salt-tolerant callus culture of Shamouti orange (*Citrus sinensis* L. Osbeck), *Plant Cell Rep.* **7**:680 683.

Ben-Hayyim, G., and Kochba, J., 1983, Aspects of salt tolerance in a NaCl selected stable cell line of *Citrus sinensis*, *Plant Physiol.* **72**:685 690.

Ben-Hayyim, G., Gueta-Dahan, Y., Avsian-Kretchmer, O., Weichert, H., and Feussner, I., 2001, Preferential induction of a 9-lipoxygenase by salt in salt-tolerant cells of *Citrus sinensis* L. Osbeck, *Planta* **212**:367 375.

Ben-Hayyim, G., Spiegel-Roy, P., and Neumann, H., 1985, Relation between ion accumulation of salt-sensitive and isolated salt-tolerant cell lines of *Citrus aurantium*, *Plant Physiol.* **78**:144 148.

Bernstein, L., 1969, Salinity factors and their limits for citrus culture, *Proc. First Int. Citrus Symposium* **3**:1779 1782.

Cai, Q., Guy, C.L., and Moore, G.A., 1994, Extension of the Linkage Map in Citrus Using Random Amplified Polymorphic DNA (Rapd) Markers and Rflp Mapping of Cold-Acclimation-Responsive Loci, *Theor. Appl. Genet.* **89**:606 614.

Cai, Q.Y., Moore, G.A., and Guy, C.L., 1995, An Unusual Group-2 Lea Gene Family in Citrus Responsive to Low-Temperature, *Plant Mol. Biol.* **29**:11 23.

Campos, H., Cooper, A., Habben, J.E., Edmeades, G.O., and Schussler, J.R., 2004, Improving drought tolerance in maize: a view from industry, *Field Crops Res.* **90**:19 34.

Cervera, M., Ortega, C., Navarro, A., Navarro, L., and Pena, L., 2000, Generation of transgenic citrus plants with the tolerance-to-salinity gene HAL2 from yeast, *J. Hort. Sci. Biotechnol.* **75**:26 30.

Champ, K.I., Febres, V.J., and Moore, G.A., 2006, The role of CBF transcriptional activators in two Citrus species (Poncirus and Citrus) with contrasting levels of freezing tolerance. *Physiol. Plant.*, 129:529–541.

Chapman, H. D., 1968, Salinity and alkali. In: *The Citrus Industry*, Univ. of Calif. Press, Berkeley, Vol **2**, pp. 243 266.

Close, T.J., 1997, Dehydrins: A commonality in the response of plants to dehydration and low temperature, *Physiol. Plant.* **100**:291 296.

Cooper, W. C., Gorton, B. S., and Olson, E. O, 1952, Ionic accumulation in citrus as influenced by rootstock and scion and concentration of salts and boron in the substrate, *Plant Physiol.* **27**:191 203.

Durham, R.E., Liou, P.C., Gmitter, F.G., and Moore, G.A., 1992, Linkage of restriction-fragment-length-polymorphisms and isozymes in citrus, *Theor. Appl. Genet.* **84**:39 48.

Flowers, T.J., and Flowers, S.A., 2005, Why does salinity pose such a difficult problem for plant breeders? *Agr. Water Manage.* **78**:15 24.

Flowers, T.J., Koyama, M.L., Flowers, S.A., Sudhakar, C., Singh, K.P., and Yeo, A.R., 2000, QTL: their place in engineering tolerance of rice to salinity, *J. Exp. Bot.* **51**:99 106.

Forment, J., Gadea, J., Huerta, L., Abizanda, L., Agusti, J., Alamar, S., Alos, E., Andres, F., Arribas, R., Beltran, J.P., Berbel, A., Blazquez, M.A., Brumos, J., Canas, L.A., Cercos, M., Colmenero-Flores, J.M., Conesa, A., Estables, B., Gandia, M., Garcia-Martinez, J.L., Gimeno, J., Gisbert, A., Gomez, G., Gonzalez-Candelas, L., Granell, A., Guerri, J., Lafuente, M.T., Madueno, F., Marcos, J.F., Marques, M.C., Martinez, F., Martinez-Godoy, M.A., Miralles, S., Moreno, P., Navarro, L., Pallas, V., Perez-Amador, M.A., Perez-Valle, J., Pons, C., Rodrigo, I., Rodriguez, P.L., Royo, C., Serrano, R., Soler, G., Tadeo, F., Talon, M., Terol, J., Trenor, M., Vaello, L., Vicente, O., Vidal, C., Zacarias, L., and Conejero, V., 2005, Development of a citrus genome-wide EST collection and cDNA microarray as resources for genomic studies, *Plant Mol. Biol.* **57**:375 391.

Garcia, M.R., Bernet, G.P., Puchades, J., Gomez, I., Carbonell, E.A., and Asins, M.J., 2002, Reliable and easy screening technique for salt tolerance of citrus rootstocks under controlled environments, *Aust. J. Agric. Res.* **53**:653 662.

Garcia-Agustin, P., and Primo-Millo, E., 1995, Selection of NaCl-tolerant *Citrus* plant, *Plant Cell Rep.* **14**:314 318.

Garcia-Sanchez, F., Jifon, J.L., Carvajal, M., and Syversteen, J.P., 2002, Gas exchange, chlorophyll and nutrient contents in relation to $Na^+$and Cl^- accumulation in 'Sunburst' mandarin grafted on different rootstocks, *Plant Sci* **162**: 705 712.

Gueta-Dahan, Y., Yaniv, Z., Zilinskas, B.A., and Ben-Hayyim, G., 1997, Salt and oxidative stress: similar and specific responses and their relation to salt tolerance in citrus, *Planta* **203**:460–469.

Hara, M., Terashima, S., and Kuboi, T., 2001, Characterization and cryoprotective activity of cold-responsive dehydrin from Citrus unshiu, *J. Plant Physiol.* **158**:1333 1339.

Hara, M., Terashima, S., Fukaya, T., and Kuboi, T., 2003, Enhancement of cold tolerance and inhibition of lipid peroxidation by citrus dehydrin in transgenic tobacco, *Planta* **217**:290 298.

Hara, M., Wakasugi, Y., Ikoma, Y., Yano, M., Ogawa, K., and Kuboi, T., 1999, cDNA sequence and expression of a cold-responsive gene in Citrus unshiu, *Biosci. Biotechnol. Biochem.* **63**:433 437.

Iba, K., 2002, Acclimative response to temperature stress in higher plants: Approaches of gene engineering for temperature tolerance, *Annu. Rev. Plant Biol.* **53**:225 245.

Lang, P., Zhang, C.K., Ebel, R.C., Dane, F., and Dozier, W.A., 2005, Identification of cold acclimated genes in leaves of Citrus unshiu by mRNA differential display, *Gene* **359**:111 118.

Levy, Y., and Syvertsen, J.P., 2004, Irrigation water quality and salinity effects in citrus trees, in: *Horticultural Reviews*. J. Janick, ed., Vol, 30, pp. 37 82.

Lexer, C., Heinze, B., Alia, R., and Rieseberg, L.H., 2004, Hybrid zones as a tool for identifying adaptive genetic variation in outbreeding forest trees: lessons from wild annual sunflowers (Helianthus spp.), *Forest Ecol. Manag.* **197**:49 64.

Loussert, R., 1989, Les argumes, *Techniques agricoles mediterraneennes*, Lavoisier, Paris.

Molinari, H.B.C., Marur, C.J., Bespalhok, J.C., Kobayashi, A.K., Pileggi, M., Leite, R.P., Pereira, L.F.P., and Vieira, L.G.E., 2004, Osmotic adjustment in transgenic citrus rootstock Carrizo citrange (Citrus sinensis Osb. x Poncirus trifoliata L. Raf.) overproducing praline, *Plant Sci.* **167**:1375 1381.

Moya, J.L., Tadeo, F.R., Gómez-Cadenas, A., Primo-Millo, E., Talón, M., 2002, Transmissible salt tolerance traits identified through reciprocal grafts between sensitive Carrizo and tolerant Cleopatra citrus genotypes, *J. Plant Physiol.* **159**:991 998.

Nambara, E., and Marion-Poll, A., 2005, Abscisic acid biosynthesis and catabolism, *Annu. Rev. Plant Biol.* **56**:165 185.

Newcomb, D. A., 1978, Selection of rootstocks for salinity and disease resistance, *Proc. Intern. Soc. Citriculture* **1**:117 120.

Piqueras, A., Hernandez, J.A., Olmos, E., Helfin, E., and Sevilla, F., 1996, Changes in antioxidant enzymes and organic solutes associated with adaptation of citrus to salt stress, *Plant Cell Tissue and Organ Culture*, **45**:53 60.

Porat, R., Pasentsis, K., Rozentzvieg, D., Gerasopoulos, D., Falara, V., Samach, A., Lurie, S., and Kanellis, A.K., 2004, Isolation of a dehydrin cDNA from orange and grapefruit citrus fruit that is specifically induced by the combination of heat followed by chilling temperatures, *Physiol. Plant.* **120**:256 264.

Porat, R., Pavoncello, D., Ben-Hayyim, G., and Lurie, S., 2002a, A heat treatment induced the expression of a Na^+/H^+ antiport gene (cNHX1) in citrus fruit, *Plant Sci.* **162**: 957 963.

Porat, R., Pavoncello, D., Lurie, S., and McCollum, T.G., 2002b, Identification of a grapefruit cDNA belonging to a unique class of citrus dehydrins and characterization of its expression patterns under temperature stress conditions, *Physiol. Plant.* 115:598 603.

Qin, X.Q., and Zeevaart, J.A.D., 2002, Overexpression of a 9-cis-epoxycarotenoid dioxygenase gene in Nicotiana plumbaginifolia increases abscisic acid and phaseic acid levels and enhances drought tolerance, *Plant Physiol.* **128**:544 551.

Rochdi, A., El Yacoubi, H., Rachidai, A., 2003, Responses to NaCl stress of *Citrus aurantium*, *Citrange troyer and Poncirus trifoliata* in callus cultures: assessment of characters for evaluating salt stress responses in citrus rootstocks, *Agronomie* **23**:643 649.

Rodrigo, M.J., Alquezar, B., and Zacarias, L., 2006, Cloning and characterization of two 9-cis-epoxycarotenoid dioxygenase genes, differentially regulated during fruit maturation and under stress conditions, from orange (*Citrus sinensis* L. Osbeck), *J. Exp. Bot.* **57**:633 643.

Ronnberg-Wastljung, A.C., Glynn, C., and Weih, M., 2005, QTL analyses of drought tolerance and growth for a Salix dasyclados x Salix viminalis hybrid in contrasting water regimes, *Theor. Appl. Genet.* **110**:537 549.

Rozenzvieg, D., Elmaci, C., Samach, A., Lurie, S., and Porat, R., 2004, Isolation of four heat shock protein cDNAs from grapefruit peel tissue and characterization of their expression in response to heat and chilling temperature stresses, *Physiol. Plant.* **121**:421 428.

Sahin-Cevik, M., and Moore, G.A., 2006a, Isolation and characterization of a novel RING-H2 finger gene induced in response to cold and drought in the interfertile Citrus relative *Poncirus trifoliate*, *Physiol. Plant.* **126**:153 161.

Sahin-Cevik, M., and Moore, G.A., 2006b, Identification and expression analysis of cold-regulated genes from the cold-hardy *Citrus* relative *Poncirus trifoliata* (L.) Raf., *Plant Mol. Biol.*, 129:529–541.

Şahim-çevik, M., and Moore, G.A. 2006c. Two AP2 domain containing genes isolated from the Cold-Hardy Citrus relative Poncirus trifoliata (L.) Raf. are induced in responce to cold. Funtional Plant Bilogy. 33:863–875.

Sanchez-Ballesta, M.T., Rodrigo, M.J., LaFuente, M.T., Granell, A., and Zacarias, L., 2004, Dehydrin from citrus, which confers in vitro dehydration and freezing protection activity, is constitutive and highly expressed in the flavedo of fruit but responsive to cold and water stress in leaves, *J. Agri. Food Chem.* **52**:1950 1957.

Sawkins, M.C., Farmer, A.D., Hoisington, D., Sullivan, J., Tolopko, A., Jiang, Z., and Ribaut, J.M., 2004, Comparative Map and Trait Viewer (CMTV): an integrated bioinformatic tool to construct consensus maps and compare QTL and functional genomics data across genomes and experiments, *Plant Mol. Biol.* **56**:465 480.

Seki, M., Narusaka, M., Ishida, J., Nanjo, T., Fujita, M., Oono, Y., Kamiya, A., Nakajima, M., Enju, A., Sakurai, T., Satou, M., Akiyama, K., Taji, T., Yamaguchi-Shinozaki, K., Carninci, P., Kawai, J., Hayashizaki, Y., and Shinozaki, K., 2002, Monitoring the expression profiles of 7000 Arabidopsis genes under drought, cold and high-salinity stresses using a full-length cDNA microarray, *Plant J.* **31**:279 292.

Shalhevet, J., and Levy, Y., 1990.Citrus trees, in:*Irrigation of agricultural crops*, A. R. Stewart, and D. R. Nielsen, eds., Am. Soc. Agron., Crop Sci. Soc. Am. Soil Sci. Soc. Am., Madison, WI, Vol. **30**, pp. 951 986.

Singh, A., Saini, M. L., Behl, R. K. 2004, In vitro screening of citrus rootstocks for salt tolerance, *Indian J. Genet. Plant Breed.* **64**:54 57.

Storey, R., and Walker, R.R., 1999, Citrus and salinity, *Scientia Horticulturae* **78**:39 81.

Sung, D.Y., Kaplan, F., Lee, K.J., and Guy, C.L., 2003, Acquired tolerance to temperature extremes, *Trends Plant Sci.* **8**:179 187.

Tozlu, I., Guy, C.L., and Moore, G.A., 1999a, QTL analysis of morphological traits in an intergeneric BC1 progeny of Citrus and Poncirus under saline and non-saline environments, *Genome* **42**:1020 1029.

Tozlu, I., Guy, C.L., and Moore, G.A., 1999b, QTL analysis of Na^+ and Cl^- accumulation related traits in an intergeneric BC1 progeny of Citrus and Poncirus under saline and nonsaline environments, *Genome* **42**:692 705.

Tozlu, I., Moore, G.A., and Guy, C.L., 2000, Regulation of growth and differential tissue dry mass accumulation by *Citrus gandis*, *Poncirus trifoliata*, and their F_1 under salinized and non-salinized environments, *Aust. J. Plant Physiol.* **27**:27 33.

Tozlu, I., Moore, G.A., and Guy, C.L., 2000a, Effects of increasing NaCl concentration on stem elongation, dry mass production, and macro- and micro-nutrient accumulation in *Poncirus trifoliata*, *Aust. J. Plant Physiol.* **27**:35 42.

Tozlu, I., Moore, G.A., and Guy, C.L., 2000b, Regulation of growth and differential tissue dry mass accumulation by *Citrus grandis*, *Poncirus trifoliata*, and their F-1 under salinized and non-salinized environments, *Aust. J. Plant Physiol.* **27**:27 33.

Tschaplinski, T.J., Tuskan, G.A., Sewell, M.M., Gebre, G.M., Donald, E.T.I., and Pendleyi, C., 2006, Phenotypic variation and quantitative trait locus identification for osmotic potential in an interspecific hybrid inbred F-2 poplar pedigree grown in contrasting environments, *Tree Physiol.* **26**:595 604.

Tudela, D., and Primomillo, E., 1992, 1-Aminocyclopropane-1-Carboxylic Acid Transported from Roots to Shoots Promotes Leaf Abscission in Cleopatra Mandarine (Citrus-Reshni Hort Ex Tan) Seedlings Rehydrated after Water-Stress, *Plant Physiol.* **100**:131 137.

Umezawa, T., Fujita, M., Fujita, Y., Yamaguchi-Shinozaki, K., and Shinozaki, K., 2006, Engineering drought tolerance in plants: discovering and tailoring genes to unlock the future, *Curr. Opin. Biotechnol.* **17**:113 122.

Vardi, A., Spiegel-Roy, P., Ben-Hayyim, G., Newmann, H., and Shalhevet, J., 1988, Response of Shamouti orange and Minneola tangelo on six rootstocks to salt stress, *Proceeding of the sixth international Society of Citriculture,* Balaban publishers, Rehovot, pp.75 82.

Weber, C.A., Moore, G.A., Deng, Z., and Gmitter, F.G., 2003, Mapping freeze tolerance quantitative trait loci in a *Citrus grandis* x *Poncirus trifoliata* F-1 pseudo-test cross using molecular markers, *J. Am. Soc. Hort. Sci.* **128**:508 514.

Wong, W.S., Li, G.G., Ning, W., Xu, Z.F., Hsiao, W.L.W., Zhang, L.Y., and Li, N., 2001, Repression of chilling-induced ACC accumulation in transgenic citrus by over-production of antisense 1-aminocyclopropane-1-carboxylate synthase RNA, *Plant Sci.* **161**:969 977.

Wong, W.S., Ning, W., Xu, P.L., Kung, S.D., Yang, S.F., and Li, N., 1999, Identification of two chilling-regulated 1-aminocyclopropane-1-carboxylate synthase genes from citrus (Citrus sinensis Osbeck) fruit, *Plant Mol. Biol.* **41**:587 600.

Wutscher, H. K., 1979, Citrus rootstocks, in: *Horticultural Reviews*, A.V.I. Publishers, Westport, CT, Vol. **1**, pp. 237 269.

Zekri, M., 1991, Effects of NaCl on growth and physiology of sour orange and Cleopatra mandarin seedlings, *Scientia Horticulturae*, 47:305 315.

Zhang, C.K., Lang, P., Dane, F., Ebel, R.C., Singh, N.K., Locy, R.D., and Dozier, W.A., 2005a, Cold acclimation induced genes of trifoliate orange (*Poncirus trifoliata*), *Plant Cell Rep.* **23**:764 769.

Zhang, C.K., Lang, P., Ebel, R.C., Dane, F., Singh, N.K., Locy, R.D., and Dozier, W.A., 2005b, Down-regulated gene expression of cold acclimated *Poncirus trifoliata*, *Can. J. Plant Sci.* **85**:417 424.

Zhang, J.Z., 2003, Overexpression analysis of plant transcription factors, *Curr. Opin. Plant Biol.* **6**:430 440.

Zhang, J.Z., Creelman, R.A., and Zhu, J.K., 2004, From laboratory to field. Using information from Arabidopsis to engineer salt, cold, and drought tolerance in crops, *Plant Physiol.* **135**:615 621.

CHAPTER 26

INTEGRATING FUNCTIONAL GENOMICS WITH SALINITY AND WATER DEFICIT STRESS RESPONSES IN WINE GRAPE - *VITIS VINIFERA*

JÉRÔME GRIMPLET, LAURENT G. DELUC, GRANT R. CRAMER AND JOHN C. CUSHMAN
MS 200, Department of Biochemistry & Molecular Biology, University of Nevada, Reno, NV 89557-0014 USA

Abstract: Wine grape (*Vitis vinifera*) is the world's most important fruit crop both in terms of crop production and economic value. For most crops, water deficit stress has negative implications for production and quality. For wine grape, however, vegetative growth is more sensitive to water-deficit stress than fruit growth. Thus, moderate water-deficit can positively influence the quality of wine produced from grapes harvested from vines grown under regulated (water) deficit irrigation (RDI) and partial root zone drying (PRD) conditions. However, the interaction between water deficit stress and berry composition is complicated by the ability to accurately measure water potential under field conditions and can be influenced by many parameters including timing of stress application within a season and across several seasons, grapevine variety and canopy, leaf to fruit ratio, and soil structure. The mechanistic basis for observed quality differences are poorly understood. However, recent studies using integrated transcriptome and metabolome data sets have revealed potential underlying changes in gene expression and determinants of fruit characteristics that explain the major effects that water deficit treatment can be expected to have on wine quality. Major responses include gene expression changes resulting in alterations in sugar content, anthocyanin accumulation, and decreased organic acid accumulation

Keywords: *Vitis vinifera*; water deficit; cold; salinity; abiotic stress

1. INTRODUCTION

Berries from wine and table grapes (*Vitis vinifera*) are the most widely cultivated and economically important fruit crop worldwide. *V. vinifera* has been domesticated for approximately 8,000 years (Aradhya et al., 2003). Cultivated germplasm apparently originated from at least two subspecies, *V. vinifera* ssp. *sativa*, which gave rise to many Western European cultivars, and ssp. *sylvestris*, which gave rise to current Eurasian cultivars in the near east and the Iberian peninsula (Arroyo-Garcia

M.A. Jenks et al. (eds.), Advances in Molecular Breeding Toward Drought and Salt Tolerant Crops, 643–668.

et al., 2006). Today, most of the world's wines are made from the fruit of these two subspecies, which are represented by more than 5,000 different cultivars (This et al., 2006). Like many other important crop species, grape vine is considered moderately sensitive to salinity stress, but highly tolerant to water deficit stress. In contrast to tomato, a well-studied climacteric fruit model (Giovannoni, 2001, 2004; Tanksley, 2004), relatively little is known about the molecular genetic mechanisms that govern grape berry development and ripening under conditions of environmental stress. In most row crops, with the exception of cotton, lack of water typically reduces vegetative growth and crop yield. In contrast, lack of water does not necessarily reduce fruit yield in trees and vines (Goldhamer and Fereres, 2005). Therefore, it is possible to conduct regulated water-deficit irrigation (RDI) or partial root zone drying (PRD) regimes that can save substantial amounts of water with little negative influence on fruit production. However, the physiological and developmental processes yielding fruits, such as berries, cannot be studied directly in popular plant model species that lack the ability to produce economically valuable fruits. In the last two years, the genomic resources in grape (*Vitis vinifera*) have increased dramatically, setting the stage for systematic functional genomic studies of shoot and berry development under various environmental stress conditions. In contrast to model fruit crop species within the Solanaceae and Rosaceae, *Vitis* is the only agriculturally important genus in the family Vitaceae. Improvements in wine grape production efficiency and fruit quality will be possible with a better understanding of the molecular genetic basis of berry development and environmental stress responses (Bisson et al., 2002; Vivier and Pretorius, 2002).

2. ECONOMIC AND HEALTH IMPORTANCE OF GRAPES

Grapes from the genus *Vitis* rank first among fruit crops in the world in terms of both production and economic importance. Of the approximately 60 species within the family *Vitaceae*, fruit from cultivars of *Vitis vinifera* L. are used mainly for the production of wine and distilled liquor production. However, many *Vitis* species are also used for table grapes, grape juice (concentrate), dried fruit (raisins), and processed for use as pharmaceutical and food supplements. Worldwide, grapes are the most widely cultivated fruit crop, encompassing about 8 million hectares of arable land (Vivier and Pretorius, 2002) with about 67,000 kilotonnes produced in 2004 (FAOSTATdata 2006). The United States is ranked 4th internationally in grape production or 10% of the world's production on 5% of land area or about 405,000 hectares with 92% of this area located in California. In the U.S. grape production is the 6th most economically important crop behind corn, soy, hay, wheat, and cotton, valued at $3 billion in 2000. Although the grape berry is used for multiple purposes, wine and distilled liquor produced from cultivars of *V. vinifera* have the highest economic value (Mullins et al., 1992). The annual economic impact of the U.S. grape and wine industry is approximately $50 billion, employing over 500,000 people.

In addition to the economic importance, consumption of grape products, such as red wine, provides numerous benefits to human health, including the reduction of risk of cardiovascular disease, stroke and cancer (German and Walzem, 2000; Middleton et al., 2000). The >200 polyphenolic compounds located primarily in the skin and seeds of the grape berry contribute to these health benefits (Dixon et al., 2005). A large-scale study recently indicated that moderate consumption of wine improves cognitive function in women (Stampfer et al., 2005), and daily moderate consumption of alcoholic beverages like wine is now included in the dietary guidelines of the U.S. government (Peregrin, 2005). The health promoting effects of wine consumption has an indirect economic impact by reducing health care costs.

3. ABIOTIC STRESS TOLERANCE OF GRAPE VINE

Vitis vinifera L. is considered as having medium salinity tolerance and high drought tolerance (McKersie and Leshem, 1994). Once established in deep soil with adequate water retention characteristics, grapevines will produce root systems several meters deep enabling the vines to survive all but the most severe drought conditions. However, such conditions will invariably result in reduced crop quality and yield. Drought conditions can be overcome, in part, by irrigation. However, depending upon soil and climate conditions, irrigation can lead to salinity build-up resulting in crop quality and yield reductions.

3.1. Salinity Tolerance of Grapevine

Vitis vinifera varieties are ranked moderately sensitive to salinity stress by most researchers (Hawker and Walker, 1978; Maas and Hoffman, 1977; Shani et al., 1993; Walker et al., 2002). However, one study ranked *Vitis vinifera* varieties as sensitive (Prior et al., 1992). Growth reduction occurs within 10 days of exposure to 90–100 mM NaCl (Walker et al., 1981; Fisarakis et al., 2001). Vines can survive at these concentrations of NaCl for at least 60 days. However, prolonged (3 years) irrigation with water containing 75 mM Cl^- resulted in the death of more than 90% of *V. vinifera* vines (Shani and Ben-Gal, 2005). Salt sensitivity is mainly a result of chloride ion accumulation. Following treatment with NaCl, Cl^- is the principle ion accumulated (Downton, 1977). Chloride accumulation in the leaves is a greater problem than Na^+, as Na^+ is generally retained in the roots (Tester and Davenport, 2003). Thus, grapevine appears to be more sensitive to Cl^- toxicity than Na^+ toxicity (Walker et al., 2004). Overall growth and crop yields are reduced by NaCl concentrations in soil greater than 25 mM (Downton, 1977). Exposure to Cl- concentrations as low as 20 mM can reduce fruit yield (Shani and Ben-Gal, 2005). Accumulation of low concentrations of leaf Cl^- (150 mM) leads to disruptions in metabolism including reductions in photochemical efficiency and increases in photorespiration (Walker et al., 1981). Under greenhouse conditions, exposure to Cl^- concentrations in excess of 200 mM lead to irreversible physiological damage.

3.2. Water Deficit Tolerance of Grapevine

Like most plants, *V. vinifera* fruit yields are correlated with the availability of water. Shoot growth is reduced at relatively moderate water deficits of about -0.3 MPa (Cramer et al., 2006) with stomatal closure occurring at about -1.2 MPa (Smart, 1974). Thus, monitoring changes in shoot growth is a very sensitive indicator of water deficit and can reveal water deficit stress even before changes in leaf water potential can be detected. Other measures of water deficit include changes in leaf angle and trunk swelling (Smart, 1974). Crop yields are reduced at leaf water potentials at or below -0.9 MPa (Grimes and Williams, 1990). Grapevines mainly utilize inorganic ions for osmotic adjustment to water-deficit stress, which is expected to have a lower energetic cost than using organic solutes (Patakas et al., 2002).

Increasing irrigation in dry, Mediterranean climates generally increases overall fruit yield at a rate of approximately 24% per 100 mm of irrigation supplied (McCarthy et al., 1983; Ginestar et al., 1998a). Grape berries are most susceptible to water deficit stress for a period of approximately four weeks after flowering (Alexander, 1965). Withholding irrigation of field-grown grape vines between budburst and flower (Smart et al., 1974) or between flowering and the beginning of the lag phase of berry development (Van Zyl, 1984) resulted in significant reduction in berry weight compared with non-stressed vines. Long-term studies conducted on field grown vines over the course of four years have revealed that water deficit applied during a one month period after flowering resulted in the greatest reduction in berry weight compared to well watered vines particularly in years with higher summary temperatures (McCarthy, 1997, 2000). Sensitivity to water deficit coincides with the early period of approximately 25 days after flowering when both cell division and cell expansion contribute to berry growth (Harris et al., 1968). Thereafter, berry growth is due to cell expansion alone, which can account for a more than 300-fold increase in the volume of mesocarp cells during berry enlargement (Coombe, 1976). This sensitivity is consistent with the observation that smaller berry size from early season (prevéraison) water deficit was due to a reduction in the number of cells per berry (Matthews et al., 1987). However, subsequent cell enlargement does not seem to be impaired when stressed vines are provided with adequate water supply during the late season up to harvest (McCarthy, 2000).

4. WATER EFFECTS ON GRAPE PRODUCTION

Soil, climate, and cultivar, collectively termed 'terroir' by the French, influence grape and wine quality. Of these components, soil and climate, have a greater impact on vine behavior and berry composition than that of cultivar (van Leeuwen et al., 2004). Although the precise contribution of each factor is not well understood, a good 'terroir' can be defined as conditions in which climactic extremes are limited from year-to-year with 1) adequate, but not excessive, soil fertility, especially with respect to nitrogen, 2) conditions that ameliorate the effects of heavy rain, especially after véraison, which can be found in deep, but well-drained soils in which deep roots can prevent excessive dehydration, and 3) ability to survive drought in very dry years, which can result from deep roots, good moisture-holding capacity of the soil or an appropriate water table (Sequin, 1986). Excessive rain and associated humidity, especially after véraison, can promote berry splitting and increased incidence of fungal diseases, which can have adverse effects on wine quality.

The effects of soil and climate are most likely mediated by vine water status, which accounted for a larger percent of total variance than any other single factor (van Leeuwen et al., 2004). The major effects of vine water deficit were studied over a five-year period in the context of three different soil types and cultivars. Prevéraison water deficits caused an early or accelerated shoot growth cessation and reduced berry size. Postvéraison water deficit accelerated berry ripening, increased anthocyanin and sugar content, and reduced total acidity and malate content (van Leeuwen et al., 2004). The best vintages occurred on soils where water deficit occurs early in the season, but are moderate.

Most quality wine growing regions of the world receive between 70 to 80 cm of annual precipitation with excessive rainfall or irrigation being associated with a reduction in wine quality (Jackson and Lombard, 1993). Until the 1970's many of the vineyards in the world's Mediterranean climates relied solely on winter rains typically stored in relatively deep root zones or a combination of winter and growing-season rainfall to supply the water needs of wine grapes. Early studies indicated that irrigation could increase yields over non-irrigated controls particularly in shallow soils, but cutting off irrigation early in the season reduced yield and berry size (Christensen, 1975; Neja et al., 1977). However, excessive and prolonged irrigation actually reduced grape quality as assessed by acidity and Brix measurements and yield due to late season vine growth (Neja et al., 1977). Concentrations of abscisic acid (ABA) increase in response to salinity (Downton and Loveys, 1978) or water deficit stress (Coombe and Monk, 1979). Excessive irrigation slows ripening, increases yield, in part, by berry enlargement, elevates pH and acid content of berries, and reduces anthocyanins from shading due to continuous and excessive shoot growth (Smart and Coombe, 1983). In contrast, although inadequate irrigation can enhance ripening, it reduces yield, berry weight, and malic acid content.

4.1. Effects of Water Abundance

In some European countries, such as Spain, irrigation of winegrapes was illegal due to real or perceived negative irrigation-related impacts on wine quality (Goldhamer and Fereres, 2005). High rainfall can lower ripening capacity more so than predicted by temperature indices (Jackson and Cherry, 1987). Excessive irrigation can have similar effects. Greater irrigation rates tend to delay fruit ripening and prevent berries from reaching full maturity and this can reduce wine quality. Indeed, several studies have shown that abundant water supply generally delays berry growth, development and ripening (Hofäcker et al., 1976; Alleweldt et al., 1984; Bravdo et al., 1985). This effect can be offset by exogenous application of ethylene (applied as Ethephon (trade name Ethrel)), which enhances the accumulation of soluble solids and berry size (Hardie et al., 1981). When applied in concert with moderate water deficit stress (about -0.3 MPa), ethephon accentuated the accumulation of soluble solids and anthocyanins. These results suggested that ethephon treatment can complement the use of moderate water deficit stress as a useful practical tool to enhance ripening and fruit characteristics such as color. Conversely, water-deficit stress has not always been found to enhance ripening and can have limited effects on the onset of véraison or duration of fruit ripening depending on the magnitude of the water deficit (Matthews and Anderson, 1988). Although supplemental irrigation can improve yield (Neja et al., 1977; Hepner and Bravdo, 1985; Morris et al., 1983), excess irrigation can lower berry sugar levels, but increase total acids and arginine levels and thereby reduce fruit quality for wine production. An over supply of water to cv. Carignane increased berry and pruning weight, must proline and arginine levels, pH and color compared to unirrigated control (Kliewer et al., 1983; Freeman, 1983). However, moderate irrigation, particularly in dry years, can increase sugar levels (Morris et al., 1982). Irrigation resulted in a significant increase in cv. Riesling fruit weight, a slight increase in berries per cluster, a significant increase in the number of berries per vine, and a lowering of Brix suggesting that irrigation delayed ripening slightly (McCarthy and Coombe, 1985). Greater irrigation rates tend to delay berry ripening regardless of crop load and is associated with reduced color content (Rankine et al., 1971), low anthocyanins content, and with low wine quality as measured by low color and high pH in cv. Cabernet Sauvignon (Bravdo et al., 1985).

Abundant water availability from irrigation often increases potassium and pH level in the must and wine (Hepner and Bravdo, 1985; Freeman and Kliewer, 1983), although one study showed a decrease in juice pH and soluble solids as a result of supplemental irrigation (Neja et al., 1977). However, in multi-year studies, significant differences were observed in only one out of two (Smart and Coombe, 1983) or three years (Freeman and Kliewer, 1983). Increased watering can also reduce color (Rankine et al., 1971) and anthocyanins content (Bravdo et al., 1985; Freeman, 1983; Matthews and Anderson, 1988; Morris and Cawthon, 1982).

4.2. Effects of Water Deficit

A large number of studies have reported that water status can influence vine growth and berry development depending on the amount of water application. In addition to the quantity of irrigation applied to a vineyard, the timing of the water deficit stress is critically important and may explain a great deal of the variation described within such studies. The effects of regulated water-deficit irrigation (RDI) or partial root zone drying (PRD) can also be influenced greatly by vineyard management practices such as the extent and type (manual *versus* mechanical) of pruning, the water-holding capacity of the soil, use of cover crops to remove soil water, planting density of vines, and management of the irrigation practices themselves (FAO, 2000), which, in turn, affect the number of clusters and the number and maturation of berries per cluster. Thus, the effects of water deficit irrigation are not always consistent depending on the complex interactions between crop load and soil composition and drainage. However, certain trends on the effects of water deficit irrigation on fruit and wine quality are beginning to emerge. With modern irrigation systems, it has become possible to manipulate soil water availability to precisely influence vegetative and fruit growth in desirable ways.

To study and compare the effects of water-deficit stress during berry development stages, water was withheld from four periods of berry development after flowering of cv. Shiraz using a modern irrigation system that supplied water on demand and where soil water content was monitored throughout the growing season (McCarthy 1997, 1999, 2000). In summary, these experiments concluded that berry growth was most sensitive to water stress during pericarp cell division, high levels of water-deficit stress are needed to reduce berry size compared with vegetative growth, reductions in berry size (and cropping level) resulted in earlier fruit maturity, smaller berries resulted in higher anthocyanin concentration, application of water-deficit stress during the early stages of fruit ripening (prevéraison) may enhance anthocyanin concentration, application of water-deficit stress during the late stages of fruit ripening (postvéraison) reduced solute accumulation in berries, and accumulation of flavor compounds occurred relatively late in the ripening process and was sensitive to water-deficit stress (Coombe and McCarthy, 2000). These authors recommend the application of water-deficit stress after fruit set to minimize berry size and to control vegetative growth and then to restore irrigation during berry ripening to encourage more rapid and complete ripening and the development of flavor compounds. Maintenance of higher soil water content postharvest may also be beneficial to post-harvest root growth and to ensure that vines do not enter dormancy under water stress, which may result in susceptibility to cold weather damage. This more refined water-deficit irrigation strategy has been termed 'strategic irrigation management' (SIM) and the use of this terminology is encouraged instead of RDI.

RDI and SIM practices have been applied primarily to red varieties, such as Shiraz, Cabernet Sauvignon, Merlot, and Grenache, for reducing berry size and skin-to-pulp ratios for optimal color extraction and early control of growth of the vine canopy. RDI/SIM practices are less important for white wine grapes as berry size and skin-to-pulp ratios are not as relevant in white varieties. However, RDI/SIM may still be useful for reducing excess vegetative growth, which can have beneficial effects on berry ripening. Another important advantage of using RDI/SIM is to reduce water consumption. In New Zealand, water consumption could be reduced 40% without causing differences in yield or Sauvignon Blanc fruit quality parameters (Greven et al., 2005).

Partial root zone drying (PRD) is a more recently developed irrigation technique that improves water use efficiency during wine grape production without significant crop reductions (FAO, 2000). PRD involves permitting one part of the root system to dry out while keeping another part well watered. Switching the wet and dry sectors of the root zone on a regular basis overcomes transient responses to partial root zone drying (Dry and Loveys, 1998; Dry et al., 2000). The effect of PRD is to stimulate stomatal closure via abscisic acid signaling to restrict water loss and thereby improve water use efficiency. A number of long-term, large-scale field studies have concluded that PRD can reduce water usage by half, achieve a balance between vegetative and fruit development, but without yield reductions sometime associated with RDI (Loveys et al., 1997, 1998, 1999; Dry et al., 1990). Under PRD with half the amount of irrigation applied to control vines, there was no apparent decrease in berry size, in contrast to a significant decrease in berry size in response to a substantial reduction in the amount of irrigation applied using RDI particularly when water deficit was applied between flowering and véraison (Smart and Coombe, 1983; Williams and Matthews, 1990).

4.2.1. Measuring water-deficit stress

Given the sensitivity of grape yield and quality traits based on compositional changes within the berry, careful scheduling of irrigation is needed to maximize the use of often limiting water resources. Although various measurements can be used to estimate when to apply irrigation, such as canopy size, climatic conditions, and soil moisture content, direct measurement of plant responses is likely to be more useful for scheduling irrigation. Several different methods of measuring plant responses improve the precision of measuring water-deficit stress (Ginestar et al., 1998a,b). Direct measure of stem water potential is a common method for measuring plant water status (McCutchan and Shackel, 1992). Alternatively, measuring sap flow, which is a direct measure of transpiration, using sap-flow sensors can provide an accurate measure of applied stress that can be used to monitor plant water status when using irrigation to manipulate canopy size, yield, and fruit composition (Ginestar et al., 1998a,b). Measurement of fruit stable carbon isotope composition ($\delta^{13}C$) has also been used as a convenient and reliable predictor of vine water status under natural conditions (Gaudillere et al., 2002; de Souza et al., 2005). During berry ripening, sucrose is translocated from leaves to fruit and is then rapidly converted to glucose and fructose and $\delta^{13}C$ values from the juice of mature berries and water-soluble leaf extracts are very similar (De Marco et al., 1977). Therefore, carbon isotope ratio in the sugars of mature berries should integrate leaf photosynthetic isotopic discrimination of carbon during berry ripening (Gaudillere et al., 2002). Surveys conducted over four growing seasons indicate that berry must sugar $\delta^{13}C$ at harvest correlates well with predawn leaf water potential and can be used to characterize the soil water holding capacity of a vineyard. More importantly, $\delta^{13}C$ values can be used in canopy management when inducing mild water stress in order to improve wine quality (Gaudillere et al., 2002). The $\delta^{13}C$ values from berry pulp showed the best correlation with intrinsic water use efficiency and cumulative integral of leaf water potential (de Souza et al., 2005).

Many studies have shown that supplemental irrigation during dry seasons can increase crop yields, but the size of the yield response can vary dramatically between different experiments and between seasons in the same vineyard (Kliewer et al., 1983; Matthews et al., 1987). In general, the occurrence of water deficit stress after véraison affects yield less than deficits applied before véraison (Hardie and Considine, 1976; Matthews and Anderson, 1989). Other studies found that late season or postvéraison water deficit had a greater impact on yield (Goodwin and Jerie, 1992). Similarly, some studies have concluded that irrigation can improve the quality of wine (Hardie and Considine, 1976; Freeman and Kliewer, 1983), whereas other studies have concluded that irrigation reduces quality due to delays in fruit ripening or attainment of desirable levels of sugar or berry weight (Sinton et al., 1978; Bravdo et al., 1985; Hepner and Bravdo, 1985). Such variability in the effects of water deficit is likely attributable to the complex interactions of vegetative growth, resultant canopy architecture, and crop load (Van Zyl, 1984). Part of the sensitivity to water stress is likely dependent upon a critical ratio between leaf area and fruit weight ratio that mediate, for example, exposure of clusters to sunlight (Kliewer and Lider, 1968; Lakso, 1990). Another likely reason for variability in reported yield responses to water deficit regimes for different studies is that the degree of water stress experienced by the grapevines under investigation is unlikely to be absolutely uniform over each treatment block (Ginestar et al., 1998a,b). Thus, reliable and accurate methodologies for predicting and measuring levels of water-deficit stress are critical to vineyard irrigation management strategies that lead to improvements in wine quality.

4.2.2. *Stress impacts on vines and berries*

In contrast to its effects on shoot growth, water deficit treatment has a far lesser effect on berry growth (McCarthy, 1997). However, berries are sensitive to water deficit stress during the post flowering period (Hardie and Considine, 1976; Matthews and Anderson, 1989; McCarthy, 2000) and less sensitive to water deficit after véraison with only minor decreases in apparent berry weight (McCarthy, 1997). In some early studies, irrigation had no effect on berry composition (Neja et al., 1977). In other studies, postvéraison water deficits caused the greatest reductions in yield and total soluble solids, while prevéraison water deficit stress had little effect on total soluble solids (Goodwin and MacRae, 1990; Goodwin and Jerie, 1992). Application of water stress in Shiraz grapevine resulted in increased berry anthocyanin and phenolic content, but no significant changes in juice total soluble solids and pH (Ginestar et al., 1998b). Application of moderate water deficit (–1.1 MPa) before véraison resulted in a significant reduction in malate concentrations, whereas application of water deficit stress after véraison (-1.3 MPa) increased proline concentration significantly (Matthews and Anderson, 1988). Withholding water pre- or postvéraison or over the entire time of Cabernet Franc berry development resulted in a 15% and 30% increase in anthocyanin content in skin extracts and phenolics in juice, respectively, compared with control vines maintained at higher water status (Matthews and Anderson, 1988). However, the moderate water deficit irrigation applied to this north coast region of California vineyard either before or after véraison did not have a significant effect on the onset of véraison, ripening rate, juice pH, titratable acidity, or berry sugar content (Matthews and Anderson, 1988). Color development changes were more sensitive to vine water status in the early rather than the late stages of fruit ripening. This study suggested the importance of response to water deficit may facilitate improved wine grape production for cultivars and environments where color development is a concern. Water deficits did not change the timing of the onset of véraison or the duration of the ripening period suggesting that vine water status directly effects berry metabolism during ripening.

In a three-year study the effect of irrigation on cv. Tempranillo, one of the most important red grape cultivars in Spain, showed that water deficit irrigation reduced titratable acidity (TA), organic acids (tartaric, malic, and citric), and berry total soluble solids (°Brix) or sugar content, but did not significantly effect the glucose to fructose ratio relative to irrigated controls (Esteban et al., 1999). However, phenolic and tannin content were found to be higher in irrigated vines in this variety (Esteban et al., 2001). Anthocyanin concentrations were found to be higher in irrigated vines on most sampling dates, but were sometimes higher in non-irrigated vines. Application of postvéraison water deficit irrigation in cv. Cabernet Sauvignon caused small increases in anthocyanins and decreases in flavonols (Kennedy et al., 2002).

The timing of water-deficit irrigation has important consequences for leaf area development and berry size, growth, and composition. Water-deficit stress causes a reduction in the total amount of leaf area developed and photosynthetic activity (Gomez-del-Campo et al., 2002). Non-stressed grapevines produced more dry matter after véraison, whereas water-deficit stressed vines accumulated a greater amount of total dry matter between fruit set and véraison (Gomez-del-Campo et al., 2002).

Biosynthesis of flavonols in cv. Shiraz berries increased as a result of applying either of pre- and postvéraison water-deficit stress (Ojeda et al., 2002). However, biosynthesis of flavan-3-ols (total tannins) was increased by the application of prevéraison water-deficit stress, whereas biosynthesis of proanthocyanins and anthocyanins increased only in response to postvéraison stress (Ojeda et al., 2002). Water-deficit stress reduced berry size regardless of the timing of stress application and this increased the skin-to-pulp weight ratio resulting in a consistent relative increase in berry skin phenolic concentrations (Ojeda et al., 2002).

Postvéraison water deficit can impact fruit yield and composition during the current and subsequent season (Petrie et al., 2004). Application of water deficit either pre- or postvéraison reduced berry and cluster weight and yield, as well as reduced sugar concentrations, whereas phenolic concentrations were increased without a significant change in anthocyanin content when assessed within a single season (Petri et al., 2004). However, despite restoration of irrigation in the following season, vines that were subjected to deficit irrigation in the previous year showed reduced yield, which was mainly the result of fewer clusters per vine – a direct consequence of fewer shoots per vine (lower budburst). Significantly, this lower crop load resulted in higher sugar and anthocyanin concentrations in fruit the following season (Petri et al., 2004). In contrast, minimal pruning of cv. Shiraz vines delayed fruit maturation as measured by sugar accumulation. Severe chronic water deficit stress can delay berry maturation. In other cases, the degree of water deficit was not severe enough to cause carry-over effects from one year to the next (Poni et al., 1994). The interaction between water deficit stress and crop level can be significant. In one study using cv. Concord, reductions in sugar accumulation due to water deficit stress was aggravated in vines that had a heavy canopy (Poni et al., 1994). In several studies, the magnitude of changes to berry composition due to water-deficit stress tended to be much less than changes to vegetative growth resulting from heavy vegetative crop loads (Jackson and Lombard, 1993; Poni et al., 1994). Low vine vigor, as measured by trunk cross sectional area, average shoot length, and leaf chlorophyll, correlated with significant increases in skin proanthocyanins and average mass of proanthocyanins, relative proportion of (-)-epigallocatechin extension units, and pigmented polymer content in berries from vineyard zones with a reduction in vine vigor (Cortell et al., 2005). These differences in proanthocyanidins have possible implications for wine quality as skin proanthocyanidins and pigmented polymers are considered to have an effect on proanthocyanidin perception (Cheynier et al., 1998). Because this study was performed with georeferenced data, reduced vine vigor could be associated with shallow soils and reduced soil water-holding capacity (Cortell et al., 2005).

In order to assess the effects of postvéraison water deficit, the relative fresh mass components or proportion of seed, skin, and flesh of six different berry sizes of cv. Cabernet Sauvignon were compared among vines exposed to control, low and high water status on mature fruit (Roby and Matthews, 2004). Berry growth was much less sensitive to water deficit than grapevine shoot growth. However, midday water potentials of around -1.5 MPa inhibited berry growth by 13–18% relative to well-watered controls (-1.0 MPa), whereas water potentials of -1.2 MPa had no effect on berry growth reduction. Inhibition of berry growth by water deficit stress was attributed almost exclusively to reduced growth of the mesocarp tissue (flesh or pulp), which increased the proportion of whole-berry fresh mass represented in seeds and skin (Roby and Matthews, 2004). Seed tannin content was influenced to a greater extent by berry size than vine water status (Roby et al., 2004). In contrast, water deficit resulted in reduced berry size and increased skin tannins and anthocyanins per berry and the concentrations of skin tannins and anthocyanins. These effects were, however, independent of berry size and attributed to the differential growth sensitivity of inner mesocarp and exocarp tissues rather than direct effects on phenolic biosynthesis (Roby et al., 2004).

The effect of water deficit irrigation on juice and winc composition is not merely caused by the decrease in berry size caused by water deficit. Anthocyanin and phenolic concentrations in fruit were greater from water deficit treated vines even when expressed on the basis of berry surface area (Matthews and Anderson, 1988). Differences in sensory attributes were also unlikely to be due to fruit maturity as sources compared had the same degree of fruit maturation (Matthews et al., 1990). Lack of irrigation had no effect on free terpenes in juice from cv. Riesling berries, but did increase the amount of bound terpenes suggesting that such grapes contained greater levels of potential volatile terpenes (McCarthy and Coombe, 1985).

Water deficit treatment typically increases the proportion of fruit mass of seed/skin to pulp compared to well watered vines (Ojeda et al., 2002; Roby et al., 2004), decreases berry size and number (Matthews and Anderson, 1989) and increases the amount of skin tannin and anthocyanins, but does not appear to affect the quantity or polymerization state of seed tannins (Geny et al., 2003; Roby et al., 2004). Water-deficit stress treatment reduced overall sugar content and titratable acidity (malic acid decreases, tartaric acid increases) (Salón et al., 2005), but increased total phenolic, anthocyanin (colored pigments), and proanthocyanin (tannin) content (Wildman et al., 1976; Hardie et al., 1981; Esteban et al., 1999; Ginestar et al., 1998b; Ojeda et al., 2002; Petrie et al., 2004; Keller 2005; Reynolds et al., 2005; Salón et al., 2005; Sivilotti et al., 2005; Koundouras et al., 2006), and the rates of accumulation of the compounds as well as increasing the degree of tannin polymerization (Ojeda et al., 2002; Geny et al., 2003; Cortell et al., 2005; Sivilotti et al., 2005; Koundouras et al., 2006), increasing proline content (Matthews and Anderson, 1988), and improving microbial disease resistance (Keller, 2005).

Water-deficit stress also has a close relationship with vine mineral nutrition. The application of water-deficit stress before fruit set may reduce cluster and berry number, especially if combined with nitrogen deficiency (Keller, 2005). Furthermore, the combination of regulated water-deficit irrigation and low-to-moderate rates of nitrogen application between flowering and véraison reduced canopy size, berry size, and yield, yet accelerates ripening, improved fruit color and microbial disease resistance (Keller, 2005). Severe water deficit stress, as did nitrogen deficiency, appeared to limit aroma potential for grapes. Optimal aroma potential can be attained when mild water-deficit is applied in combination with moderate nitrogen supply (des Gachons et al., 2005).

5. WATER DEFICIT STRESS EFFECTS ON WINE QUALITY

Relatively few studies examining the effect of regulated deficit irrigation has moved beyond must or wine parameter measurements to actually performing tasting trials to assess changes via organoleptic quality traits. However, water-deficit stress conditions have long been recognized as an important factor in affecting grape quality and can have a marked influence on the sensory attributes of the resulting wine (Esteban et al., 1999; Chapman et al., 2005; Koundouras et al., 2006). Regulated-deficit irrigation has been used advantageously to inhibit vine growth without fruit yield reductions and to make measurable improvements in grape quality (Matthews and Anderson, 1988; Matthews et al., 1990; Sipiora and Granda, 1998; Esteban et al., 1999, 2001). In an early study, tasting trials were used to examine the effect of intensive irrigation on cv. Cabernet Sauvignon vines (Bravdo et al., 1985). Intensive irrigation treatment, which delayed ripening regardless of low or high crop load, resulted in reduced wine quality as expressed as low tasting scores, along with low color and high pH. The timing of the application of water deficit stress was critically important in determining the extent of changes in wine composition. Application of water deficit early in the season before véraison resulted in greater water deficit and greater concentrations of anthocyanins and phenolics than in vines exposed to late season water deficit treatments (Matthews and Anderson, 1988; Matthews et al., 1990). This suggested that the developmental period near véraison may be more sensitive to water deficit for fruit ripening. In a later study, the effects on wine composition and color of stopping irrigation pre- or postvéraison were investigated in cv. Cabernet Sauvignon along with the effects of skin contact time on anthocyanin concentrations in resultant wines (Sipiora and Granda, 1998). Prevéraison irrigation cutoff resulted in a reduction in berry size suggesting that berry growth is sensitive to vine water deficit during one-to-two weeks before the onset of véraison consistent with earlier studies performed with potted grapevines (Creasy and Lombard, 1993). However, the smaller berry size resulting from the prevéraison or non-irrigated control water stress treatments did not result in either an increased total anthocyanin or total phenolic content of the finished wines and reduced berry yield by up to 22%. Berries from non-irrigated vines and prevéraison water deficit stress vines had significantly lower soluble sugars (Brix), higher titratable acidity, and lower potassium concentration than fruit from postvéraison stress treatments. Extended skin contact (30 d vs. 5 d) resulted in a greater extraction of total phenols regardless of berry size effects brought about by pre- or postvéraison water deficit stress treatments. In summary, prevéraison wine from only 5 d skin contact had the highest color density and the lowest concentration of total phenols, whereas wines from postvéraison water deficit treatment from 30 d skin contact had the highest concentrations of total phenolics and the lowest color density. These results suggest that enological practices, such as extended skin contact or adjustment of skin/juice ratio during fermentation could have a greater impact than the use of irrigation management for the manipulation of anthocyanin and total phenol content, anthocyanin equilibria, and wine color (Sipiora and Granda, 1998).

Sensory evaluation of wines produced from cv. Cabernet Franc vines were determined following early, late or full season water deficit irrigation and continually irrigated vines over the course of two seasons (Matthews et al., 1990). The concentrations of anthocyanins and total soluble phenolics were greater in wines from water-deficit treated vines than continually irrigated vines although other levels of residual sugar, titratable acidity, pH, and ethanol were similar to wines made from fully irrigated vines (Matthews et al., 1990). Sensory differences in wine appearance, flavor, taste, and aroma could be detected by non-professional judges between wines produced from continually irrigated versus early, late or full season water-deficit treated vines. A majority of professional wine tasters were able to detect visual color (hue, color density) differences in paired wine comparisons, but not flavor differences. Color differences were most likely the result of increased anthocyanin synthesis brought about by water-deficit stress applied either early or late in the season (Matthews and Anderson, 1988). Wine preference tends to be positively correlated with an increase in wine color intensity (Somers and Evans, 1977). However, comparisons among judges showed that differences in aroma were easier to detect than those in taste, suggesting that changes in volatile constituents tended to be greater than changes in soluble constituents (Matthews et al., 1990). In particular, late season water deficits led to significantly greater detection of "black currant" aroma than continually watered vines.

In addition to saving irrigation water in dry climates, mild water stress is generally recognized to improve the organoleptic properties of wine produced from berries harvested from water deficit treated vines (Matthews et al., 1990; Esteban et al., 1999; Chapman et al., 2005; Koundouras et al., 2006). Aromatic components perceived as fruity flavors and aromas are enhanced, whereas vegetal aromas and astringency are reduced (Chapman et al., 2005; des Gachons et al., 2005; Koundouras et al., 2006). In addition to increasing the concentration of anthocyanins and total phenolics in berry skins, application of early (prevéraison) water deficit appeared to increase the level of bound volatile compounds present in wines produced (Koundouras et al., 2006). Greater polymerization of polyphenolics also results in improved color stability and mouth feel properties (Sivilotti et al., 2005). Furthermore, wines produced from grapes harvested from water-deficit treated vines were also preferred in tasting trials (Koundouras et al., 2006). Specifically, analysis of variance (ANOVA) and principal component analysis (PCA) showed that wines made from vines with minimal irrigation treatment were significantly higher in red/blackberry, jam/cooked berry, and dried fruit/raisin aroma, and fruit than wines produced from vines that had been irrigated (Chapman et al., 2005). In contrast, wine from irrigated vines were rated significantly higher than minimally irrigated vines in vegetal, bell pepper, black pepper aroma and astringency (Chapman et al., 2005). Interestingly, wines produced from vines pruned to low bud numbers or "low yield" mimicked in organoleptic characteristics wines produced from fully-irrigated vines, whereas wines produced from vines pruned to high bud numbers or "high yield" had quality traits that resembled wines from minimally irrigated vines (Chapman et al., 2004). Other researchers have found little or no influence of berry cluster thinning on shoot growth, leaf area, pruning weight, berry number, berry weight, and fruit composition (soluble solids, titratable acidity, pH, and color) in a mature, deficit-irrigated vineyard (Keller et al., 2005). Although interpretation of such results can be extremely difficult due to the complex interaction between pruning and cluster-thinning treatments and perturbations in light and water availability, one possible explanation is that vines with low bud numbers may experience a lower degree of water stress due to less evaporative water loss from a reduced canopy. Direct measurements of vine water potential and/or soil water content could provide additional information in the context of such experiments.

6. FUNCTIONAL GENOMICS IN GRAPE VINE

Functional genomic resources for *Vitis vinifera* and related species have proliferated rapidly within the last several years mainly in the form of large, publicly available expressed sequence tag (EST) databases (da Silva, et al., 2005; Moser et al., 2005). The availability of such information has permitted large-scale mRNA expression profiling studies of gene expression profiles in flowers and during berry skin development using cDNA or oligonucleotide microarrays (Terrier et al., 2005; Waters, 2005, 2006). A high-density, oligonucleotide microarray containing approximately one-third of the expected gene content of the *Vitis vinifera* genome with some bias towards leaf and berry tissues, has recently been developed (Cramer et al., 2007). Experiments can now be conducted that yield a broader view of the gene expression changes that occur in response to a wide variety of environmental treatments. Here we provide some initial observations about the qualitative and quantitative changes in gene expression and metabolites brought about by water-deficit stress.

6.1. Water Deficit Stress Effects on the *Vitis* Transcriptome

Comprehensive transcript profiling using high-density microarrays has recently been used to investigate the effect of long-term, water-deficit and isoosmotic salinity stress effects on grapevine shoot tissues of greenhouse-grown vines (Cramer et al., 2007), and in berry tissues harvested from field-grown plants exposed to water-deficit stress irrigation. Tissue-specific expression patterns of berries from field-grown vines exposed to well watered and water-deficit irrigation have also been investigated.

6.1.1. Stress impacts on vegetative shoots

A microarray experiment was designed to differentiate water-deficit and salinity responses that occur over a 16-day time course in which greenhouse-grown vines in pots were allowed to dry out naturally over time. Vine stem water potentials were measured every other day and an isoosmotic saline solution was used to salinize vines to a level that mimicked the drop in stem water potential of the water-deficit stressed plants. Vine stem water potentials barely decreased in the initial phase of the experiment, but dropped significantly relative to irrigated controls after eight days. There were not any significant differences in stem water potential between water-deficit-treated and salt-stressed plants at any time during the course of the experiment ($p \leq 0.29$ and treatment x day interaction was 0.45). Even though stem water potentials were equal on day 16, water-deficit-treated plants began to wilt, but no wilting was observed in salinized plants. Relative elongation rates of shoots were inhibited by the stress treatments within a day after treatment indicating that shoot elongation was very sensitive to changes in soil water availability.

A microarray experiment was then performed from shoot samples taken every 4 days over a time course of 16 days. ANOVA identified 10,251 unigenes with significant F-statistics based on differences between water-deficit stress and control. Similarly, 8,687 unigenes showed significant differences between salt stress and control conditions across all time points. An additional and partially overlapping set of 8,632 unigenes showed significant differences in response to both stresses over time. For each of these two gene sets, genes with $\log_2$-transformed expression ratios versus control (day 0) of at least 1 or at most -1 during any time point were extracted for further inspection. This resulted in 2,497 unigenes associated with water-deficit

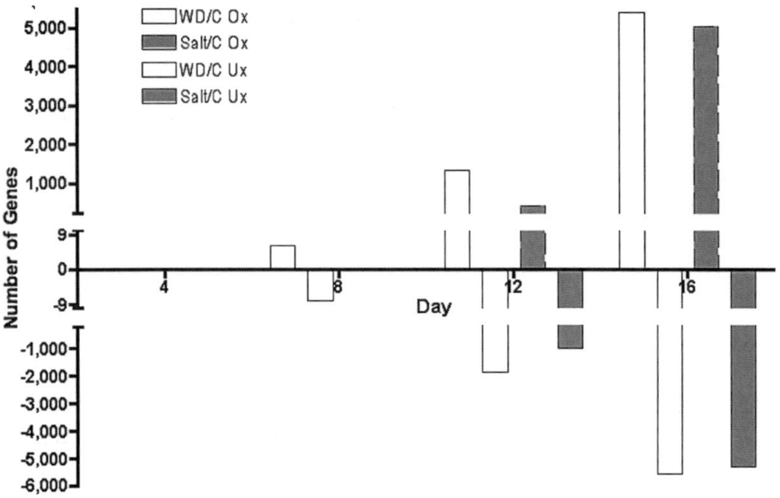

Figure 1. Significant changes in steady-state transcript abundance over time in response to salinity and water-deficit stress treatments in *Vitis vinifera* cv. Cabernet Sauvignon shoots relative to the control treatment. Ox = over-expression; Ux = under-expression. (From Cramer et al., 2007)

stress, and 2,260 with salt stress (Figure 1). A few genes had significantly different expression from controls on day 4, at a time when stress levels were very small. As the stress increased with time, so did the number of stress-responsive genes. After day 8, when stem water potentials began to decline significantly, gene expression changes increased dramatically. By day 16, there were massive changes in gene expression with more than 5,000 unigenes exhibiting significant changes in steady-state transcript abundance in response to both stresses (Figure 1). Water-deficit had a greater effect on growth and this was reflected in the observation that water-deficit affected more genes than salinity (Figure 1). A majority of unigenes (88%) were coordinately up-regulated or down-regulated by both water-deficit and salinity stress, however, 373 unigenes (12%) exhibited differential expression between the treatments with a greater number of these unigenes being affected more by water-deficit stress than by salinity stress (164 vs. 77 for up-regulated unigenes and 209 vs. 114 for down-regulated unigenes, respectively) (Cramer et al., 2007).

6.1.2. Stress impacts on berry development

To better understand the process of grape berry development, mRNA expression profiling has been conducted on two varieties of *V. vinifera* (cv. Cabernet Sauvignon (CS) and Chardonnay (CH)) using the Affymetrix GeneChip® *Vitis* genome array from stages 31–38 according to the developmental stages as defined (Coombe, 1995). Raw data were processed via Robust Multi-Array Average (RMA), normalized, and subjected to ANOVA (Didier et al., 2002). Across berry development, in CS, 7,804 unigenes (53%) were differentially expressed ($P<0.05$: Multiple-Test Correction). Among this gene set, 3,901 unigenes (27%) displayed a two-fold ratio or higher change in their expression. In CH, 8,163 unigenes were found significantly expressed (55.7%), whereas 5,126 unigenes (35%) had a ratio change in their expression equal or up to two-fold. Principal component analysis (PCA) was used to simplify this complex multidimensional data set to lower dimensions in order to visualize variances within the data set. PCA transforms the data to a new coordinate system such that the greatest variance by any projection of the data comes to lie on the first coordinate (called the first principal component), the second greatest variance on the second coordinate, and so on, resulting in dimensionality reduction in a dataset while retaining those characteristics of the dataset that contribute most to its variance, by keeping lower-order principal components and ignoring higher-order ones. The PCA of berry development of two cultivars exposed to well watered and water deficit stress conditions across berry development revealed that the most important aspects of the data were the cultivar, the water deficit stress treatment, and the developmental status of the berries at the 3rd and 4th axes (Figure 2). Indeed, when one compared both CS and CH in well-watered conditions, 6,162 unigenes (42%) exhibited significant differences in their expression that were attributable to the cultivar. With regard to vine water status, in CS, 2,668 unigenes (18%) were found to be differentially expressed among well watered and water-deficit treated plants along berry development whereas, in CH, 4,195 unigenes (29%) displayed differential expression. Of the differen-

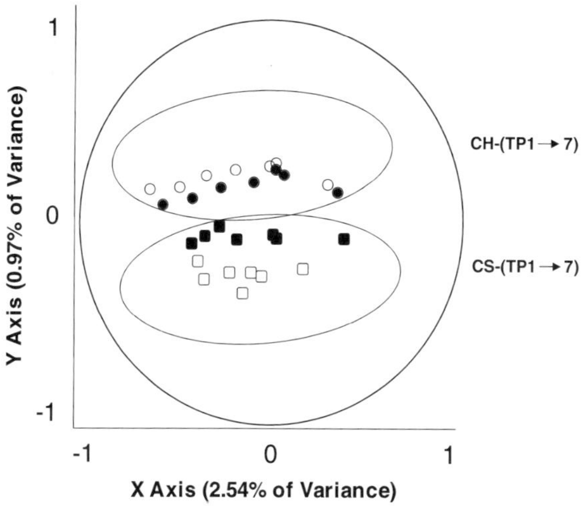

Figure 2. Principal component analysis (PCA) of mRNA expression profiles in cv. Chardonnay (circles) and cv. Cabernet-Sauvignon (squares) berries during fruit development in well watered (closed symbols) and water-deficit stress (open symbols) vines during time points (TP) 1–7, which correspond to developmental stages 31–38 (Coombe, 1992). Each symbol represents the average of three biological replicates for each cultivar/condition indicated. The X-axis (3rd axis of the model) explains 2.54% of the variance associated with developmental stage. The Y-axis (4th axis of the model) explains 0.97% of the variance and indicates the difference between cultivars and water status conditions. Analysis was performed using GeneANOVA software (Didier et al., 2002)

tially expressed genes across berry development, 61% and 62.5% of genes in CS and CH, respectively, shared homology (E-value $< 1 \times 10^{-10}$) with known gene products or known protein domains. Clustering analysis defined 21 gene expression profiles (e.g., increasing, decreasing, transiently increasing or decreasing, etc.,) of steady-state mRNA abundance throughout berry development that encompassed the majority of gene expression patterns observed.

To identify factors controlling berry quality, we focused on genes involved in secondary metabolism. Of differentially expressed genes, 45 unigenes in CS and 73 unigenes in CH had biosynthetic or regulatory functions in the phenylpropanoid pathway. In CS, approximately two-third of these genes exhibited a constant reduction in steady-state transcript abundance across berry development, whereas one-third displayed various patterns of increased transcript abundance. In CH, an approximately equal number of genes showed decreased or increased transcript abundance. Taken together, it appears that the berry undergoes many changes in gene expression during the course of its ripening and that this development can be greatly affected by water deficit stress. In addition, the significant number of unigenes that are differentially expressed between the two cultivars suggests that further investigations are needed to improve our knowledge with regard to intra-cultivar variability.

6.1.3. Stress impacts on berry tissues

Microarray analysis was also used to identify 2,947 genes differentially expressed among seed, pulp, and skin tissues within stage 38 berry tissues (Grimplet et al., 2007). Among berry tissues, 482 CS genes were differentially expressed significantly between well watered and water-deficit stress conditions (94 in the pulp, 96 in the skin and 17 in the skin). Principal component analysis (PCA) was used to visualize variances within the data set. The PCA of the gene expression of three berry tissues (skin, pulp, and seeds) of cv. Cabernet Sauvignon exposed to well-watered and water deficit stress conditions revealed that the most important differences in the data mainly reflecting berry tissue-specific expression patterns, were mainly due to differences between seed and skin tissues and the water deficit stress treatment status of the berry at the 2nd and 3rd axes of the model, respectively (Figure 3).

Of the 2,947 identified genes, 65% share homology (E-value $< 1 \times 10^{-10}$) with proteins of known function. Genes with functions in the phenylpropanoid and flavonoid biosynthetic pathways are specifically associated with the skin and the seed, with anthocyanin biosynthesis pathway genes being expressed specifically in the skin and tannin biosynthesis genes being preferentially expressed in the seed. Genes with cell wall expansion-related functions were specifically expressed in pericarp tissues. Interestingly, genes with ABA, auxin and jasmonate biosynthesis-related functions and with signal transduction functions also presented tissue-specific expression pattern. Genes with sugar and malate transport functions were differentially expressed between the pericarp and seed. Finally, genes involved

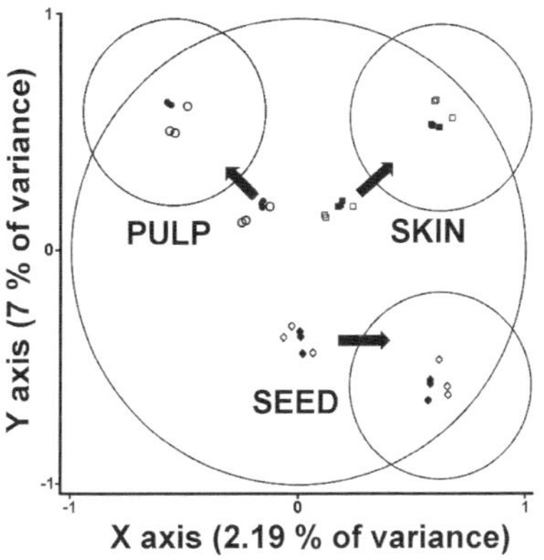

Figure 3. Principal component analysis (PCA) of mRNA expression profiles in different tissues of cv. Cabernet-Sauvignon (squares) berries. Berries were harvested at developmental stage 38 (Coombe, 1992) from well watered (closed symbols) and water-deficit stress (open symbols) and dissected into pulp (circles), skin (squares), and seed (diamonds) tissues prior to RNA extraction and microarray analysis. PCA analysis was performed on the entire data set (center) or for each tissue (outlying circles) to visualize the differences between well watered and water deficit stress berry tissues. The first axis of the model, explaining 88.32% of the variance indicating that most of the probe sets have similar expression values, is not presented. X-axis (3rd axis of the model) explains 2.19% of the variance associated with water status. The Y-axis (2nd axis of the model) explains 7.0% of the variance and indicates the difference between tissues. Analysis was performed using GeneANOVA software (Didier et al., 2002)

in the biosynthesis of aroma compounds were differentially expressed among tissues and in response to water deficit stress.

6.2. Water Deficit Stress Effects on the *Vitis* Metabolome

To complement ongoing mRNA expression profiling, procedures for the identification of metabolites by gas chromatography (GC)-MS in polar extracts of leaves and berries have been optimized. Comparisons of leaves subjected to control, water-deficit, and salinity stress conditions and in berries subjected control and water-deficit conditions are well underway. Of ~250 metabolites identified to date, 191 metabolites can be identified reproducibly. Of these, 143 were common to all three leaf treatments, 5 were unique to water-deficit, 7 were unique to salinity and 6 were unique to controls. Significant effects of water deficit on berry metabolites in both cv. Cabernet Sauvignon and Chardonnay have also been identified. Some organic acid concentrations were reduced by water deficit, particularly malate concentrations. More than half of the 19 measured amino acids were affected by water deficit, but the profiles differed depending on the cultivar analyzed. Water-deficit stress also affects polyphenol accumulation in wines produced from control and water-deficit stressed vines (Table 1) resulting in less acidic characteristics and more color, intensity, and tannin relative to control (well-watered) wines. Volatile analysis of

Table 1. Impacts of water-deficit stress on important wine quality components. Grapes were subjected to irrigation deficits in the field over the course of the 2004 summer season. Stem water potentials of well-watered vines ranged from −0.5 to −0.7 MPa and water-deficit-treated vines ranged from −1.0 to −1.25 MPa

Wine component	Well-watered cabernet sauvignon	Water-deficit stressed cabernet sauvignon	Well-watered chardonnay	Water-deficit stressed chardonnay
pH	3.87	3.99	3.42	3.37
Titratable Acidity	5.88	5.69	6.75	6.13
Hue (Color)	14.9	16.7	96.3	97.3
Chroma (Intensity of Hue)	21.0	36.6	6.7	10.8
Luminosity (Lightness of Hue)	81.6	59.1	97.7	96.5
Tannin (Catechin Equivalents)	35.7	234.8	21.5	25.0

the wines indicates that water-deficit increased the number of volatile compounds in the wine consistent with earlier suggestions from organoleptic sensory analyses (Matthews et al., 1990; Esteban et al., 1999; Chapman et al., 2005; Koundouras et al., 2006). There was a 2-fold increase in the number of components in cv. Chardonnay and a 1.3-fold increase in cv. Cabernet Sauvignon. Component identification and correlation of metabolite abundance with steady-state mRNA abundance changes in response to water deficit stress is in progress.

7. CONCLUSION

Wine grape is moderately sensitive to salinity stress and extremely tolerant to water-deficit stress. This latter trait has been exploited through the development and application of SIM practices that use both RDI and PRD to grow grapevine with relatively low water inputs without having a significant yield reduction in fruit production. Furthermore, such water-deficit stress treatments have the added benefit of actually improving wine quality, mainly through alterations in berry composition that increase the relative accumulation of proanthocyanins and anthocyanins and volatile flavor compounds. In order to more fully understand the complex changes that are occurring under water-deficit stress conditions, functional genomics tools, including large-scale oligonucleotide microarray analysis for high-throughput mRNA expression profiling and GC-MS analysis for high-throughput metabolite profiling, have now been developed and optimized. Initial experiments with shoot and berry tissues have revealed that there are thousands of gene expression changes ongoing in these tissues in response to salinity and water-deficit stress treatments. Future experiments will reveal a detailed picture of the complex, hierarchical regulatory networks that control tissue-specific and abiotic stress-responsive changes in mRNA, protein, and metabolite expression patterns. Ultimately, such information will provide a useful framework for understanding and manipulating the functions of individual regulatory proteins and enzymes that contribute to desirable organoleptic qualities in wine.

ACKNOWLEDGEMENTS

This work was supported by NSF Plant Genome Program DBI-0217653, the American Vineyard Foundation, and the Nevada Agricultural Experiment Station, publication number 03077033. The genomic and proteomic equipment used in this study was acquired with NIH/NCRR grants 5P20RR16464 and 1S10RR15919, and NSF Grant EPS-0132556.

REFERENCES

Alexander, D. M., 1965, The effect of high temperature regimes or short periods of water stress on development of small fruiting Sultana vines, *Aust. J. Agric. Res.* **16**:817–823.

Alleweldt, G., During, H., and Jung, K.H., 1984, Zum einfluss des klimas auf beerentwicklung, ertrag, und qualitt bei reben: Ergebnisse einer siebenjahringen factorenanalyse, *Vitis* **23**:127–142.

Aradhya, M.K., Dangl, G.S., Prins, B.H., Boursiquot, J.M., Walker, M.A., Meredith, C.P., and Simon, C.J., 2003, Genetic structure and differentiation in cultivated grape, *Vitis vinifera* L., *Genet Res.* **81**:179–192.

Arroyo-Garcia, R., Ruiz-Garcia, L., Bolling, L., Ocete, R, Lopez, M.A., Arnold, C., Ergul, A., Soylemezoglu, G., Uzun, H.I., Cabello, F., Ibanez, J., Aradhya, M.K., Atanassov, A., Atanassov, I., Balint, S., Cenis, J.L., Costantini, L., Goris-Lavets, S., Grando, M.S., Klein, B.Y., McGovern, P.E., Merdinoglu, D., Pejic, I., Pelsey, F., Primikirios, N., Risovannaya, V., Roubelakis-Angelakis, K.A., Snoussi, H., Sotiri, P., Tamhankar, S., Thiss, P. Troshin, L., Malpica, J.M., Lefort, F., and Martinez-Zapater, J.M., 2006, Multiple origins of cultivated grapevine (*Vitis vinifera* L. ssp. Sativa) based on chloroplast DNA polymorphisms, *Molec. Ecol.* **15**: 3707–3714.

Bisson, L.F., Waterhouse, A.L., Ebeler, S.E., Walker, M.A. and Lapsley, J.T., 2002, The present and future of the international wine industry, *Nature* **418**: 696–699.

Bravdo, B., Hepner, Y., Loinger, C., Cohen, S., and Tabacaman, H., 1985, Effect of irrigation and crop level on growth, yield and wine quality of Cabernet Sauvignon, *Am. J. Enol. Vitic.* **36**:132–139.

Chapman, D.M., Matthews, M.A., and Guinard, J-X., 2004, Sensory attributes of Cabernet Sauvignon wines made from vines with different crops yields, *Am. J. Enol. Vitic.* **55**:325–334.

Chapman, D.M., Roby, G., Ebeler, S.E., Guinard, J-X., and Matthews, M.A., 2005, Sensory attributes of Cabernet Sauvignon wines made from vines with different water status, *Aust. J. Grape Wine Res.* **11**:329–347.

Cheynier, V., Fulcrand, H., Brussaud, F., Asselin, C., Moutounet, M., 1988, Phenolic composition as related to red wine flavor. In. Chemistry of Wine Flavor; A.L. Waterhouse and S. E. Ebeler, eds. Oxford University Publishing: New York, pp. 125–141.

Christensen, P., 1975, Response of 'Thompson Seedless' grapevines to the timing of preharvest irrigation, *Am. J. Enol. Vitic.* **26**:188–194.

Coombe, B.G., 1976, The development of fleshy fruits, *Ann. Rev. Plant Physiol.* **27**:507–528.

Coombe, B.G., 1995, Adoption of a system for identifying grapevine growth states, *Aust. J. Grape Wine Res.* **1**:104–110.

Coombe, B.G., and McCarthy, M.G., 2000, Dynamics of grape berry growth and physiology of ripening, *Aust. J. Grape Wine Res.* **6**:131–135.

Coombe, B.G., and Monk, P.R., 1979, Proline and abscisic acid content of the juice of ripe Riesling grape berries: effect of irrigation during harvest, *Am. J. Enol. Vitic.* **30**:64–67.

Cortell, J.M., Halbleib, M., Gallagher, A.V., Righetti, T.L., and Kennedy, J.A., 2005, Influence of vine vigor on grape (*Vitis vinifera* L. Cv. Pinot Noir) and wine proanthocyanidins, *J. Agric. Food Chem.* **53**:5798–5808.

Cramer, G.R., Ergül, A., Grimplet, J., Tillett, R.L., Tattersall, E.A.R., Bohlman, M.C., Vincent, D., Sonderegger, J., Evans, J., Osborn, C., Quilici, D., Schlauch, K.A., Schooley, D.A., and Cushman, J.C., 2007, Transcript and metabolite profiling of grapevines exposed to gradually increasing, long-term water-deficit or isoosmotic salinity, *Funct Integr Genomics* **7**:111–134.

Creasy, G.L., and Lombard P.B., 1993, Vine water stress and peduncle girdling effects on pre- and postvéraison grape berry growth and deformability, *Am. J. Enol. Vitic.* **44**:193–197.

da Silva, F.G., Iandolino, A., Al-Kayal, F., Bohlman, M.C., Cushman, M.A., Lim, H., Ergul, A., Figueroa, R., Kabuloglu, E.K., Osborne, C., Rowe, J., Tattersall, E.A.R., Leslie, A., Xu, J., Baek, J.-M., Cramer, G.R., Cushman, J.C., and Cook, D.R., 2005, Characterizing the grape transcriptome: analysis of ESTs from multiple *Vitis* species and development of a compendium of gene expression during berry development, *Plant Physiol.* **139**:574–597.

de Souza C.R., Maroco J.P., dos Santos T.P., Rodrigues M.L., Lopes C.M., Pereira J.S., and Chaves M.M., 2005, Impact of deficit irrigation on water use efficiency and carbon isotope composition (δ13C) of field-grown grapevines under Mediterranean climate, *J. Exp. Bot.* **56**:2163–2172.

des Gachons, C.P., Van Leeuwen, C., Tominaga, T., Soyer, J-P., Gaudillère J-P., and Dubourdieu, D., 2005, Influence of water and nitrogen deficit on fruit ripening and aroma potential of *Vitis vinifera* L cv. Sauvignon blanc in field conditions, *J. Sci. Food Agric.* **85**:73–85.

Didier, G., Brezellec, P., Remy, E., and Henaut, A., (2002) GeneANOVA-gene expression analysis of variance, *Bioinformatics* **18**: 490–491.

De Marco, G., Grego, S., Tricoli, D., and Turi, B., 1977, Carbon isotope ratios ($^{13}C/^{12}C$) in fractions of field grown grapes, *Physiol. Plant.* **41**: 139–141.

Dixon, R.A., Xie, D.Y., Sharma, S.B., 2005, Proanthocyanidins–a final frontier in flavonoid research? *New Phytol* **165**:9–28.

Downton, W.J.S., 1977, Salinity effects on the ion composition of fruiting Cabernet Sauvignon vines, *Am. J. Enol. Vitic.* **28**:210–214.

Downton, W.J.S., and Loveys, B.R., 1978, Compositional changes during grape berry development in relation to abscisic acid and salinity, *Aust. J. Plant Physiol.* **5**:415–423.

Dry, P.R. and Loveys, B.R., 1998, Factors influencing grapevine vigour and the potential for control with partial rootzone drying, *Aust. J. Grape Wine Res.* **4**: 140–148.

Dry, P.R., Loveys, B.R., Stoll, M., Stewart, D. and McCarthy, M.G., 2000, Partial rootzone drying - an update, *Aust. Grapegrower Winemake*r **438a**: 35–39.

Esteban, M.A., Villanueva, M.J. and Lissarrague, J.R., 1999, Effect of irrigation on changes in berry composition of Tempranillo during maturations. Sugars, organic acids, and mineral elements, *Am. J. Enol. Vitic.*, **50**:418–434.

Esteban, M.A., Villanueva, M.J. and Lissarrague, J.R., 2001, Effect of irrigation on changes in the anthocyanin composition of the skin of cv Tempranillo (*Vitis vinifera* L) grape berries during ripening, *J. Sci. Food Agric.* **81**: 409–420.

FAO, 2000, Regulated deficit irrigation and partial rootzone drying as irrigation management techniques for grapevines, By McCarthy, M.G., Loveys, B.R., Dry, P.R., Stoll, M. 2000, In: Deficit Irrigation Practices. Water Rep. Vol 22, *FAO Corp. Doc. Deposit.* **http:// www.fao.org/ docrep/ 004/Y3655E/y3655e11.htm.**

FAOSTATdata, 2006, http://faostat.fao.org/.

Fisarakis, I., Chartzoulakis, K., and Stavrakas, D., 2001, Response of Sultana vines (V. vinifera L.) on six rootstocks to NaCl salinity exposure and recovery, *Agricultural Water Management* **51**:13–27.

Freeman, B.M., 1983, Effect of irrigation and pruning on Shiraz grapevines on subsequent red wine pigments, *Am. J. Enol. Vitic.* **34**: 23–26.

Freeman, B.M., and Kliewer, W.M., 1983, Effect of irrigation, crop level and potassium fertilization on Carignane vines. II. Grapes and wine quality, *Am. J. Enol. Vitic.* **34**: 197–207.

Gaudillere, J-P., Van Leeuwen, C., and Ollat, N., 2002, Carbon isotope composition of sugars in grapevine, an integrated indicator of vineyard water status. *J. Exp. Bot.* **53**:757–763.

Geny, L., Saucier, C., Bracco, S., Daviaud, F., and Glores, Y., 2003, Composition and cellular localization of tannins in grape seeds during maturation, *J. Agric. Food Chem.* **51**:8051–8054.

German, J.B., and Walzem R.L., 2000, The health benefits of wine. *Annu. Rev. Nutr.* **20**:561–593.

Giovannoni, J., 2001, Molecular biology of fruit maturation and ripening, *Annu Rev Plant Physiol Plant Mol Biol.* **52**:725–749.

Giovannoni, J.J., 2004, Genetic regulation of fruit development and ripening, *Plant Cell.* **16** Suppl:S170–180.

Ginestar, C., Eastham, J., Gray, S., and Iland, P., 1998a, Use of sap-flow sensors to schedule vineyard irrigation. II. Effects of post-veraison water deficits on water relations, vine growth, and yield of Shiraz grapevines, *Am. J. Enol. Vitic.* **49**:413–420.

Ginestar, C., Eastham, J., Gray, S., and Iland, P., 1998b, Use of sap-flow sensors to schedule vineyard irrigation. II. Effects of post-veraison water deficits on composition of Shiraz grapes, *Am. J. Enol. Vitic.* **49**:421–428.

Goldhamer, D., and Fereres, E., 2005. The promise of regulated deficit irrigation in California's orchards and vineyards. In: Crop water Use. *California Water Plan Update*, **4**, 207–210.

Gomez-del-Campo, M., Ruiz, C., and Lissarrague, J.R., 2002, Effect of water stress on leaf area development, photosynthesis, and productivity in Chardonnnay and Airén grapevines, *Am. J. Enol. Vitic.* **53**:138–143.

Goodwin, I., and MacRae, I., 1990, Regulated deficit irrigation of Cabernet Sauvignon grapevines, *Austral. NZ Wind Ind. J.* **5**:131–133.

Goodwin, I., and Jerie, P., 1992, Regulated deficit irrigations: from concept to practice, *Austral. NZ Wind Ind. J.* **5**:258–261.

Greven, M., Green, S., Neal, S., Clothier, B., Neal, M., Dryden, G., and Davidson, P., 2005, Regulated deficit irrigation (RDI) to save water and improve Sauvignon Blanc quality, *Water Sci. Tech.* **51**:9–17.

Grimes, D.W., and Williams, L.E., 1990, Irrigation effects on plant water relations and productivity of Thompson seedless grapevines, *Crop Sci.*, **30**:255–260.

Grimplet, J., Deluc L.G., Schlauch, K.A., Wheatley, M., Cramer, G.R., and Cushman, J.C., 2007, Tissue-specific mRNA Expression Profiling in Grape Berry Tissues, BMC Genomics **8**:187.

Hardie, W.J., and Considine J.A., 1976, Response of grapes to water-deficit stress in particular stages of development, *Am. J. Enol. Vitic.* **27**:55–61.

Hardie, W.J., Johnson, J.O., and Weaver, R.J., 1981, The influence of vine water regime on ethephon-enhanced ripening of Zinfandel, *Am. J. Enol. Vitic.* **32**:115–121.

Harris, J.M., Kriedemann, P.E., and Possignham, J.V., 1968, Anatomical aspects of grape berry development, *Vitis* **7**:106–119.

Hawker, J.S. and Walker, R.R., 1978. The effect of sodium chloride on the growth and fruiting of Cabernet Sauvignon vines, *Am. J. Enol. Vitic.* **29**:172–176.

Hepner, Y., and Bravdo, B., 1985, Effect of crop level and drip irrigation scheduling on the potassium status of Cabernet Sauvignon and Carignane vines and its influence on must and wine composition and quality, *Am. J. Enol. Vitic.* **36**:140–147.

Hofäcker, W., Alleweldt, G., and Khader, S., 1976, Zum einfluss unweltfacktoren beerenwachstum und mostqualität bei der rebe, *Vitis* **15**:96–112.

Jackson, D.I. and Lombard, P.B., 1993, Environmental and management practices affecting grape composition and wine quality- A review, *Am. J. Enol. Vitic.* **44**:409–430.

Jackson, D.I. and Cherry, N.J., 1987, Prediction of a district's grape-ripening capacity using a latitude-temperature index (LTI), *Am. J. Enol. Vitic.* **39**:19–28.

Keller M., 2005, Deficit Irrigation and Vine Mineral Nutrition, *Am. J. Enol. Vitic.* **56**:267–283.

Keller M., Mills, L.J., Wample, R.L., and Spayd, S.E., 2005, Cluster thinning effects on three deficit-irrigated *Vitis vinifera* cultivars, *Am. J. Enol. Vitic.* **56**:91–103.

Kennedy, J.A., Matthews, M.A. and Waterhouse, A.L., 2000, Changes in grape seed polyphenols during fruit ripening, *Phytochemistry* **55**:77–85.

Kennedy, J.A., Matthews, M.A. and Waterhouse, A.L., 2002, Effect of maturity and vine water status on grape skin and wine flavonoids, *Am. J. Enol. Vitic.* **53**:268–274.

Kliewer, W.M., Freeman, B.M., and Hossom C., 1983, Effect of irrigation, crop level and potassium fertilizer on Carignane vine. I. Degree of water stress and effect on growth and yield. *Am. J. Enol. Vitic.* **34**:186–196.

Kliewer, W.M. and Lider, L.A., 1968, Influence of cluster exposure to the sun on the composition of 'Thompson Seedless' fruit, *Am. J. Enol. Vitic.* **19**:175–184.

Koundouras S., Marinos V., Gkoulioti A., Kotseridis Y., and van Leeuwen C., 2006, Influence of vineyard location and vine water status on fruit maturation of nonirrigated cv. Agiorgitiko (*Vitis vinifera* L.). Effects on wine phenolic and aroma components, *J. Agric. Food Chem.* **54**:5077–5086.

Lakso, A.N., 1990, Interactions of physiology with multiple environmental stresses in horticultural crops, *HortScience* **25**: 1365–1369.

Loveys, B.R., Dry, P.R., and McCarthy, M.G., 1999, Using plant physiology to improve the water use efficiency of horticultural crops, *Acta Horticulturae* **537**: 187–199.

Loveys, B.R., Grant, W.J.R., Dry, P.R. and McCarthy, M.G., 1997, Progress in the development of partial root-zone drying, *Aust. Grapegrower Winema*ker **403**: 18–20.

Loveys, B.R., Stoll, M., Dry, P.R. and McCarthy, M.G., 1998, Partial rootzone drying stimulates stress responses in grapevine to improve water use efficiency while maintaining crop yield and quality, *Aust. Grapegrower Winemak*er **414a**: 108–113.

Maas, E.V. and Hoffman, G.J., 1977, Crop salt tolerance - current assessment, *J. Irrig. Drain. Div. ASCE.* **103**:115–134.

Matthews, M.A. and Anderson, M.M., 1988, Fruit ripening in *Vitis vinifera* L.: responses to seasonal water deficits, *Am. J. Enol. Vitic.* **39**: 313–320.

Matthews, M.A., and Anderson M.M., 1989, Reproductive development in grape (*Vitis vinifera* L.): Responses to seasonal water deficits, *Am. J. Enol. Vitic.* **40**:52–60.

Matthews, M.A., Anderson M.M., and Schultz, H.R., 1987, Phenological and growth responses to early and late season water deficits in Cabernet Franc, *Vitis* **26**:147–160.

Matthews, M.A., Ishii, R., Anderson, M.M. and O'Mahony, M., 1990, Dependence of wine sensory attributes on vine water status, *J. Sci. Food Agr.*, **51**: 321–335.

McCarthy, M.G., 1997, The effect of transient water deficit on berry development of cv. Shiraz (*Vitis vinifera* L.), *Aust. J. Grape Wine Res.* **3**:102–108.

McCarthy, M.G., 1999, Weight loss from ripening berries of cv. Shiraz grapes (*Vitis vinifera* L. cv. Shiraz), *Aust. J. Grape Wine Res.* **5**:10–16.

McCarthy, M.G., 2000, Developmental variation in sensitivity of *Vitis vinifera* L. (Shiraz) berries to soil water deficit, *Aust. J. Grape Wine Res.* **6**:136–140.

McCarthy, M.G., and Coombe, B.G., 1985, Water status and grapevine quality, *Acta Hort.* **171**:447–456.

McCarthy, M.G., Cirami, R.M., and McCloud, P., 1983, Vine and fruit responses to supplementary irrigation and canopy management, *S. Afric. J. Enol. Vitic.* **4**:67–76.

McCutchan, J., and Shackel, K.A., 1992, Stem water potential as a sensitive indicator of water stress in prune trees (*Prunus domestica* L., cv. French), *J. Amer. Soc. Hort. Sci.* **117**: 607–611.

McKersie, B.D., and Leshem, Y.Y., 1994, Stress and stress coping in cultivated plants. Dordrecht, The Netherlands, Kluwer Academic Publishers.

Medrano, H., Escalona, J.M., Bota, J., Gulías, J., and Flexas, J., 2002, Regulation of photosynthesis of C3 plants in response to progressive drought: stomatal conductance as a reference parameter, *Ann. Bot. (Lond).* **89**:895–905.

Middleton, E., Kandaswami, C., and Theoharides, T.C., 2000, The effects of plant flavonoids on mammalian cells: implications for inflammation, heart disease, and cancer, *Pharmacol . Rev.* **52**: 673–751.

Morris, J.R., and Cawthon, D.L., 1982, Effects of irrigation, fruitload, and potassium fertilization on yield, quality, and petiole analysis of Concord (*Vitis labrusca* L.) grapes, *Am. J. Enol. Vitic.* **33**:145–148.

Morris, J.R., Spayd, S.E., and Cawthon, D.L., 1983, Effects of irrigation, pruning severity and nitrogen levels on yield and juice quality of Concord grapes, *Am. J. Enol. Vitic.* **34**:229–233.

Moser, C., Segala, C., Fontana, P., Salakhudtinov, I., Gatto, P., Pindo, M., Zyprian, E., Toepfer, R., Grando, M.S., and Velasco, R., 2005, Comparative analysis of expressed sequence tags from different organs of *Vitis vinifera* L., *Funct. Integr. Genomics* **5**:208–217.

Mullins, M.G., Bouquet, A. and Williams, L.E., 1992, *Biology of the grapevine*. Cambridge University Press, Cambridge, New York, pp. 239.

Neja, R.A., Wildman, W.E., Ayers, R.S., and Kasimatis, A.N., 1977, Grapevine response to irrigation and trellis treatments in the Salinas valley, *Am. J. Enol. Vitic.* **28**:16–26.

Ojeda, H., Andary, C., Kraeva, E., Carbonneau, A., and Deloire, A., 2002, Influence of pre- and postveraison water deficit on synthesis and concentration of skin phenolic compounds during berry growth of *Vitis vinifera* cv. Shiraz, *Am. J. Enol. Vitic.* **53**:261–267.

Patakas, A., Nikolaou, N., Zioziou, E., Radoglou, K. and Noitsakis, B., 2002, The role of organic solute and ion accumulation in osmotic adjustment in drought-stressed grapevines, *Plant Science*, **163**:361–367.

Patakas, A. and Noitsakis, B., 1999, Osmotic adjustment and partitioning of turgor responses to drought in grapevines leaves. *American Journal of Enology and Viticulture*, 50:76–80.

Peregrin T., 2005, Wine-A Drink to Your Health? *J. Amer. Diet. Assoc.* **105**:1053–1054.

Petrie, P.R., Cooley, N.M., and Clingeleffer, P.R. (2004) The effect of post-veraison water deficit on yield components and maturation of irrigated Shiraz (*Vitis vinifera* L.) in the current and following season, *Aust. J. Grape Wine Res.* **10**:203–215.

Poni, S., Lakson, A.N., Turner, J.R., and Melious, R.E. 1994, Interactions of crop level and late season water stress on growth and physiology of field-grown concord grapevines. *Am. J. Enol. Vitic.* **45**: 252–258.

Prior, L.D. Grieve, A.M., Cullis, B.R., 1992, Sodium chloride and soil texture interactions in irrigated field grown Sultana grapevines. I. Yield and fruit quality, *Aust. J. Agric. Res.* **43**:1051–1066.

Rankine, B.C., Fornachon J.C.M., Boehm, E.N, and Cellier, K.M., 1971, the influence of grape variety, climate and soil on grape composition and quality of table wines, *Vitis* **10**: 33–50

Reynolds, A.G., Lowrey, W.D., and De Savigny, C., 2005, Influence of irrigation and fertigation on the composition, vine performance, and water relations of Concord and Niagara Grapevines, *Am. J. Enol. Vitic.* **56**:110–128.

Roby G., and Matthews M.A., 2004, Relative proportions of seed, skin and flesh, in ripe berries from Cabernet Sauvignon grapevines grown in a vineyard either well irrigated or under water deficit, *Aust. J. Grape Wine Res.* **10**:74–82.

Roby, G., Harbertson, J.F., Adams, D.A., and Matthews, M.A., 2004, Berry size and vine water deficits as factors in winegrape composition: anthocyanins and tannins, *Aust. J. Grape Wine Res.* **10**:100–107.

Salón, J.L., Chirivella, C., and Castel, J.R., 2005, Response of cv. Bobal to timing of deficit irrigation in Requena, Spain: Water relations, yield, and wine quality, *Am. J. Enol. Vitic.* **56**: 1–8.

Shani, U., Waisel, Y., Eshel, A., Xue, S. and Ziv, G., 1993, Responses to salinity of grapevine plants with split root systems, *New Phytol*, **124**: 695–701.

Shani, U., and Ben-Gal, A., 2005, Long-term responses of grapevine to salinity: osmotic effects and ion toxicity, *Amer. J. Enol. Viticult.* **56**:148–154.

Sequin, G. 1986, 'Terriors' and pedology of wine growing, *Experientia* **42**:861–873.

Sinton, T.H., Ough, C.S., Kissler, J.J., and Kasimatis, A.N., 1978, Grape juice indicators for prediction of potential wine quality. I. Relationship between crop level, juice, and wine composition and wine sensory ratings and scores, *Am. J. Enol. Vitic.* **29**:267–271.

Sipiora, M.J. and Granda, M.J.G., 1998, Effects of pre-veraison irrigation cutoff and skin contact time on the composition, color, and phenolic content of young Cabernet Sauvignon wines in Spain, *Amer. J. Enol. Viticult.* **49**:152–162.

Sivilotti, P., Bonetto, C., Paladin, M., and Peterlunger, E., 2005, Effect of soil moisture availability on Merlot: from leaf water potential to grape composition, *Am. J. Enol. Vitic.* **56**:9–18.

Schultz, H.R. and Matthews, M.A., 1988, Resistance to water transport in shoots of *Vitis vinifera* L.: relation to growth at low water potential, *Plant Physiol.* **88**:718–724.

Smart, R.E., 1974, Aspects of water relations of the grapevine (*Vitis vinifera*), *Amer. J. Enol. Vitic.* **25**:84–91.

Smart, R.E., Turkington, C.R., and Evans, J.C., 1974, Grapevine response to furrow and trickle irrigation, *Amer. J. Enol. Vitic.* **25**:62–66.

Smart, R.E. and Coombe, B.G., 1983, Water relations of grapevines. In: Water Deficits and Plant Growth. Vol. VII Additional woody crop plants. T.T. Kozlowski (Ed.), pp. 137–196. Academic Press, Inc. New York.

Somers, T.C., and Evans, E.E., 1977, Spectral evaluation of young red wines: anthocyanin equilibria, total phenolics, free molecular SO_2, and 'chemical age', *J. Sci. Food Agric.* **28**: 279–287.

Stampfer, M.J., Kang, J.H., Chen, J., Cherry, R., and Grodstein, F., 2005, Effects of moderate alcohol consumption on cognitive function in women, *New Engl. J. Med.* **352**:245–253.

Tanksley, S.D., 2004, The genetic, developmental and molecular bases of fruit size and shape variation in tomato, *Plant Cell* **16**:S181–189.

Terrier, N., Glissant, D., Grimplet, J., Barrieu, F., Abbal, P., Couture, C., Ageorges, A., Atanassova, R., Leon, C., Renaudin, J.-P., Dedaldechamp, F., Romieu, C., Delrot, S., and Hamdi, S., 2005, Isogene specific oligo arrays reveal multifaceted changes in gene expression during grape berry (*Vitis vinifera* L.) development, *Planta* **222**:832–847.

Tester, M., and Davenport R., 2003. Na^+ tolerance and Na^+ transport in higher plants, *Annals Bot.* **91**:503–527.

This, P., Lacombe, T., and Thomas, M.R., 2006, Historical origins and genetic diversity of wine grapes, *Trends Genet.* **22**:511–519.

van Leeuwen, C., Friant, P., Choné, X., Tregoat, O., Koundouras, S., and Dubourdieu, D., 2004, Influence of climate, soil, and cultivar on terroir, *Am. J. Enol. Vitic.* **55**:207–217.

Van Zyl, J.L., 1984, Response of Columbard grapevines to irrigation as regards quality and aspects of growth, *South African J. Enol. Vitic.* **5**:19–28.

Vivier, M.A., and Pretorius I.S., 2002, Genetically tailored grapevines for the wine industry, *Trends Biotechnol.* **20**:472–478.

Walker, R.R., Torokfalvy, E., Steele Scott, N. and Kriedemann, P.E., 1981, An analysis of photosynthetic response to salt treatment in *Vitis vinifera*, *Aust. J. Plant Physiol.* **8**:359–374.

Wildman WE, Neja RA, and Kasimatis AN., 1976, Improving grape yield and quality with depth-controlled irrigation, *Am. J. Enol. Vitic.* **27**:168–175.

Walker, R.E., Torokfalvy, N., Scott, N. and Kriedemann, P., 1981, An analysis of photosynthetic response to salt treatment in *Vitis vinifera*, *Aust. J. Grape Wine Res.* **8**:359–374.

Walker, R.R., Blackmore, D.H., Clingeleffer, P.R. and Correll, R.L., 2002, Rootstock effects of salt tolerance of irrigated field-grown grapevines (*Vitis vinifera* L. cv. Sultana) I. Yield and vigour inter-relationships, *Aust. J. Grape Wine Res.* **8**:3–14.

Walker, R.R., Blackmon, D.H., Clingeleffer, P.R. and Correll, R.L., 2004, Rootstock effects on salt tolerance of irrigated field-grown grapevines (*Vitis vinifera* L. cv. Sultana) II. Ion concentration in leaves and juice, *Aust. J. Grape Wine Res.* **10**:90–99.

Waters, D.L., Holton, T.A., Ablett, E.M., Lee, L.S., and Henry, R.J., 2005, cDNA microarray analysis of developing grape (*Vitis vinifera* cv. Shiraz) berry skin, *Funct Integr Genomics* **5**:40–58.

Waters, D.L, Holton, T.A., Ablett, E.M., Lee, L.S., and Henry, R.J., 2006, The ripening wine grape berry skin transcriptome, *Plant Science* **171**:132–138.

Williams, L.E. and Matthews, M.A. 1990, Grapevine. In: B.A. Stewart and D.R. Nielsen, eds. Irrigation of Agricultural Crops. Madison, Wisconsin, United States of America, America Society of Agronomy.

CHAPTER 27

CURRENT STATUS OF BREEDING TOMATOES FOR SALT AND DROUGHT TOLERANCE

MAJID R. FOOLAD

Department of Horticulture, The Pennsylvania State University, University Park, PA 16802, USA
E-mail: mrf5@psu.edu

Abstract: Salinity and drought are among the most challenging environmental constraints to crop productivity worldwide. The cultivated tomato, *Lycopersicon esculentum* Mill., is moderately sensitive to both of these stresses throughout its ontogeny, including during seed germination, seedling emergence, vegetative growth and reproduction. Limited variation exists within the cultivated tomato for abiotic stress tolerance, however, the related wild species of tomato is a rich source of genetic variation which can be used for crop improvement. During the past several decades this variation has been utilized for characterization of physiological and genetic bases of tolerance to different abiotic stresses, including salinity and drought. Abiotic stress tolerance is a complex phenomenon, controlled by more than one gene and influenced by uncontrollable environmental factors. Furthermore, tomato stress tolerance is a developmentally-regulated state-specific phenomenon, such that tolerance at one stage of plant development is independent of tolerance at other stages. This has been demonstrated by analysis of response and correlated response to selection as well as identification of quantitative trait loci (QTLs) conferring tolerance at different stages. Transgenic approaches also have been employed to gain a better understanding of the genetic and physiological bases of salt and, to a lesser degree, drought tolerance in tomato, and to develop transgenic plants with improved stress tolerance. However, despite considerable traditional genetics and physiological research as well as contemporary molecular marker and transgenic studies in tomato, there is yet no report of any commercial cultivar of tomato with salt or drought tolerance. To achieve this goal, cooperation among plant geneticists, physiologists, molecular biologists and breeders engaged in tomato stress tolerance is imperative. In this chapter, I review the recent progresses in genetics and breeding of salt and drought tolerance in tomato and discuss the prospects for developing commercial cultivars with stress tolerance

Keywords: breeding, drought stress, drought tolerance, gene mapping, genetic engineering, genetic transformation, quantitative trait loci (QTL), salt stress, salt tolerance, transgenic plants

Abbreviations: BC: backcross; DS: drought stress; DT: drought tolerance; DW: dry weight; FW: fresh weight; h^2: heritability; MAS: marker-assisted selection; PS: phenotypic selection; QTL: quantitative trait loci; RIL: recombinant inbred line; SG: seed germination; SS: salt stress; ST: salt tolerance; TI: tolerance index; VS: vegetative stage; WUE: water use efficiency

M.A. Jenks et al. (eds.), Advances in Molecular Breeding Toward Drought and Salt Tolerant Crops, 669–700.

1. INTRODUCTION

1.1. The Tomato

The cultivated tomato, *Lycopersicon esculentum* Mill., is the 2nd most important vegetable crop in the world in terms of consumption per capita and is the most popular garden vegetable. In addition to tomatoes that are eaten directly as raw vegetable or added as ingredient to other food items, a variety of processed products have gained popularity. Although a tropical plant, tomato is grown in almost every corner of the planet. It is grown in greenhouses where summers are too cool for pollination or fruit set to occur in outdoors. Worldwide, a total of 4,528,519 ha of tomato were harvested in 2005 with a total production of 124,748,292 Mt (FAOSTAT 2005). Major production countries in descending orders include China, U.S.A., Russia, Turkey, India and Italy. In the U.S., it is the 3rd most economically important vegetable crop (with a total farm value of $2.062 B) after potato ($2.564 B) and lettuce ($2.064 B) (http://www.usda.gov/nass/pubs/agr05/agstats2005.pdf). In the U.S., total harvested area in 2004 was 170,808 ha (505,60 ha fresh-market tomatoes valued $1.34 B and 120,248 ha processing tomatoes valued $0.72 B) (http://www.nass.usda.gov:8080/QuickStats/index2.jsp). California is by far the leading producer of processing tomatoes followed by Florida, which is also the leading state in producing fresh tomatoes (USDA 2005). Per capita consumption in the U.S. includes 31.7 kg of processing and 8.7 kg of fresh tomatoes (http://www.ars.usda.gov/). Although tomatoes do not rank high in nutritional value, they contribute significantly to the dietary intake of vitamins A and C and essential mineral and nutrients. In the U.S. diet, tomato ranks first among all fruits and vegetables as a source of vitamins and minerals (Rick 1980). Also, tomatoes are the richest source of lycopene, a phytochemical that protects cells from oxidants that have been linked to cancer (Giovannucci 1999).

Tomato belongs to the nightshade family *Solanaceae*, which is the most variable of all crop families in terms of agricultural utility, the third most economically-important after grasses and legumes, and the most valuable in terms of vegetable crops. The genus *Lycopersicon* is one of the smallest genera in *Solanaceae*, though the centerpiece for genetic and molecular research in the family. There are 9 known species within *Lycopersicon*, including the cultivated type *L. esculentum* and its wild form *L. esculentum* var. *cerasiforme* (Dun.) Gray, and the 8 wild species *L. pimpinellifolium* (Jusl.) Mill., *L. cheesmanii* Riley, *L. chmielewskii* Rick, Kes., Fob. & Holle, *L. chilense* Dun., *L. parviflorum* Rick, Kes., Fob. & Holle, *L. peruvianum* (L.) Mill., *L. hirsutum* Humb. and Bonpl. and *L. pennellii* (Corr.) D'Arcy (Rick 1976a; Rick 1979b). All species are native to western S. America, between Ecuador and Chile (Rick 1976b). However, their natural habitat is variable, from very dry to very wet, and from coastal to mountainous areas of more than 3300 m elevations (Warnock 1988). Among the 9 species, only *L. esculentum* has become a domesticated crop (Rick 1978), which includes the common fresh-market and processing tomatoes, land races, primitive cultivars, and the wild cherry, *L. esculentum* var. *cerasiforme*.

All tomato species are diploid (2N = 2X = 24) and have the same chromosome number and structure. Tomato is one of the most genetically characterized higher plant species and an excellent model system for basic and applied research. This is due to many reasons, including ease of culture, short life cycle, high self-fertility and homozygosity, great reproductive potential, ease of use for controlled pollination and hybridization, availability of a wide array of mutants and genetic stocks (http://tgrc.ucdavis.edu/; http://www.sgn.cornell.edu/), diploid with a rather small genome (0.86 pg, 950 kb) (Arumuganathan and Earle 1991), and amenability to asexual propagation and protoplast, cell and tissue cultures and whole plant regeneration thereof (McCormick et al. 1986). Members of *Lycopersicon* are easily transformed and transgenic tomatoes are routinely produced using *Agrobacterium tumefaciens* (McCormick et al. 1986). Recent availability of high MW insert genomic libraries of tomato has facilitated map-based gene cloning, and advances in EST databases and genome sequencing have added additional tools for further expansion of basic and applied research in tomato.

1.2. Sources of Genetic Variation and Response to Environmental Stresses

The cultivated tomato has a narrow germplasm base, largely because of several genetic bottlenecks that occurred during domestication and evolution of modern cultivars (Rick 1976b). Although higher levels of variability can be found in primitive cultivars in the native regions of tomato, it is estimated that only about 5% of the total genetic variation within *Lycopersicon* is within the cultivated species (Miller and Tanksley 1990; Rick and Fobes 1975). As a consequence, genes for many desirable agricultural characteristics, including environmental stress tolerance, are not found within *L. esculentum*. Fortunately, however, the related wild species of tomato are a rich source of desirable genes and characteristics for tomato crop improvement, all of which can be hybridized with the cultivated species, though with different degrees of difficulty (Rick 1976a, 1979a; Rick et al. 1987). The species with the greatest genetic variability are *L. chilense*, *L. hirsutum*, *L. peruvianum* and *L. pennellii* whereas the least variable species are *L. cheesmanii* and *L. pimpinellifolium* (Breto et al. 1993; Miller and Tanksley 1990). During the past several decades, tomato wild species have been extensively utilized for tomato crop improvement, in particular for improving disease resistance. Comparatively, however, only a superficial assessment of the extent of the genetic variation for environmental stress tolerance within *Lycopersicon* has been made. Nonetheless, some accessions with tolerance to abiotic stresses have been identified and used for characterization of physiological and genetic bases of stress tolerance as well as for improving crop stress tolerance. In this chapter, the existing variation in *Lycopersicon* in relation to salt and drought tolerance and the recent advancements in genetics and breeding of stress tolerance are reviewed and discussed.

1.3. Production Environments

Tomato is grown under wide varieties of climates ranging from tropics to within a few degrees of the Artic Circle. However, despite its global distribution, a major portion of the world tomato production is concentrated in a number of warm and dry regions, in particular areas around the Mediterranean Sea, southern and western parts of the U.S., and Mexico. These climates on the other hand are prone to drought and/or salinity stress during tomato production. For various reasons, nearly all tomato-breeding programs have largely focused their breeding activities on developing cultivars with high yield potential under favorable (i.e., nonstress) conditions. This is similar to the situation in many other crop species, where such breeding efforts have resulted in improved efficiency of crop production per unit area (Duvick 1986). In case of processing tomato, for example, the average-yield per unit area in the U.S.A. increased by seven fold between 1920s and 1990s (Warren 1998). However, with the rapid increase in human population and a greater demand for food, and with an increasing diminution in natural resources and arable lands, greater efforts must be devoted to increasing crop productivity in stressful agricultural environments as well as bringing marginal lands under cultivation. Although soil reclamation and deliberate irrigation management could alleviate stresses due to salinity or drought, development of cultivars with stress tolerance is considered a complementary approach to achieve higher yields in stressful environments. This approach has been suggested as an effective and economic solution to crop production in stress environments (Blum 1988). Toward this goal, within the past few decade considerable research has been undertaken and significant information has been obtained regarding the physiology, genetics and breeding of tomatoes for stress tolerance. In this chapter, the current information on tomato response to salt and drought stress and the available genetic resources for stress tolerance breeding are reviewed and the prospects for developing commercially acceptable, stress-tolerant tomato cultivars through conventional breeding and genomic approaches are discussed. In the following sections, each of the two stresses is dealt with separately.

2. GENETICS OF AND BREEDING FOR SALT TOLERANCE IN TOMATO

2.1. Background

Commercial cultivars of tomato are moderately sensitive to salinity at all stages of development, including seed germination, vegetative growth, and reproduction (Jones et al. 1988; Maas 1986). Genetic resources for salt tolerance (ST), however, have been identified within tomato related wild species. Attempts to find sources of genes for ST in tomato were first made by Lyon (Lyon 1941), who suggested that ST of the cultivated tomato might be improved by introgression of genes from *L. pimpinellifolium*, the most closely related wild species of tomato. Later investigations resulted in identification of other salt-tolerant accessions within this and other wild species, including *L. peruvianum* , *L. cheesmanii* , *L. hirsutum* and *L. pennellii* (Foolad and Lin 1997b; Jones 1986a; Phills et al. 1979; Rush and Epstein 1976; Sarg et al. 1993; Tal 1971; Tal and Shannon 1983). However, it is expected that more salt-tolerant accessions can be found within the wild species of tomato if more comprehensive screenings were conducted (Foolad 2004; Foolad and Lin 1997b).

In tomato (Asins et al. 1993a; Foolad 1999; Foolad and Lin 1997a; Jones and Qualset 1984) as well as many other plant species (Ashraf and McNeilly 1988; Johnson et al. 1992; Mano and Takeda 1997; Quesada et al. 2002) ST at each stage of plant development is often independent of tolerance at other stages. Also, in general ST of a plant is increased with its age in many species, including tomato (Bolarin et al. 1993), barley (*Hordeum* spp.), corn (*Zea mays* L.), rice (*Oryza sativa* L.) and wheat (*Triticum* spp.) (Maas 1986). Therefore, to facilitate a better understanding of the genetics of ST, in tomato often individual developmental stages have been studied for assessment of tolerance and the identification, characterization and utilization of useful genetic components. Below, recent findings on genetics of ST in tomato during different developmental stages are briefly reviewed and discussed.

2.2. Salt Tolerance During Seed Germination

Commercial cultivars of tomato are most vulnerable to salt stress (SS) during seed germination (SG) and early seedling growth stages (Cook 1979; Foolad and Jones 1991; Foolad and Lin 1997b; Jones 1986b; Maas 1986), when they exhibit sensitivity even to low concentrations (~75 mM) of salt (Cuartero and Fernandez-Munoz 1999; Foolad and Lin 1997b; Jones 1986a). Surface soils, however, may have salinities several fold that of the subsoil, presenting a serious problem during SG and seedling emergence. High salinity delays the onset, reduces the rate and final percentage of germination, and increases the dispersion of SG events in tomato. This sensitivity has important biological and applied significance. The costly operations of greenhouse seedling production and transplantation into the field are good reasons for tomato producers to consider growing direct-seeded crops. However, the dependence upon mechanization in modern cultivation systems and the use of costly hybrid seed, requires rapid, uniform and complete SG. Genetic resources for ST during SG have been identified within primitive cultivars and related wild species of tomato, including *L. pennellii*, *L. pimpinellifolium* , and *L. peruvianum* (Cuartero and Fernandez-Munoz 1999; Foolad and Lin 1997b; Jones 1986a). Salt-tolerant accessions have been utilized for investigation of the physiology and genetics of ST during SG in tomato.

2.2.1. *Physiology of seed germination under salt stress*

Salt tolerance during SG is a measure of the seed's ability to withstand the effects of salts in the medium. Excessive salt depresses the external water potential, making water less available to the seed. Slower SG under SS compared to nonstress conditions, however, could be due to osmotic and/or ionic effects of the saline germination medium. Physiological investigations to distinguish between the two types of effects have been scarce. However, accumulating evidence in different crop species suggests that low water potential of the external medium, rather than ion toxicity effects, is the major limiting factor to germination under SS (Bliss et al. 1986; Bradford 1995; Haigh and Barlow 1987; Kaufman 1969; Ungar 1978), although a few reports have indicated otherwise (Choudhuri 1968; Redmann 1974; Younis and Hatata 1971). In a recent investigation, germination responses of eight tomato genotypes were evaluated in iso-osmotic (water potential ≈ -700 kPa or ≈ 15 dSm^{-1}) medium of NaCl, $MgCl_2$, KCl, $CaCl_2$, sorbitol, sucrose, or mannitol (JR Hyman and MR Foolad, unpubl. data). Comparison of germination in SS treatments with those in osmotic-stress treatments indicated that all genotypes responded similarly to these two types of stresses. Also, comparison of germination among the SS treatments indicated that different types of salt generally affected germination of all genotypes similarly. The results supported the suggestion that the delay in germination of tomato seed under SS was mainly due to osmotic rather than ion-toxicity effects.

2.2.2. *Inheritance of salt tolerance during seed germination*

Most studies which examined the inheritance of ST during SG in tomato concluded that the heritability (h^2) for this trait was in the range of medium to high and the trait could be improved by directional phenotypic selection (PS). For example, generation means analysis of parental, filial and backcross (BC) populations of a cross between a salt-sensitive breeding line and a salt-tolerant *L. esculentum* plant introduction (PI174263) indicated that the ability of tomato seed to germinate rapidly under SS was genetically controlled with a narrow-sense h^2 of 0.75 ± 0.03 (Foolad and Jones 1991). This conclusion was confirmed in a subsequent study using F_2:F_3 and F_3:F_4 regression analysis of the progeny of the same cross (Foolad and Jones 1992). In a later study, the effectiveness of PS in improving tomato SG under SS was demonstrated using F_2, F_3 and F_4 progeny (Foolad 1996b). This study indicated that directional PS for rapid SG under SS significantly improved progeny ST, indicating a realized h^2 of 0.67–0.76. The overall conclusion from these studies was that ST during SG in tomato was controlled by a few major genes with largely additive effects. However, to verify this and to facilitate marker-assisted breeding of this trait a few genetic mapping studies were pursued, as discussed in below.

2.2.3. *QTL analysis of salt tolerance during seed germination*

Studies to identify quantitative trait loci (QTLs) for ST during SG in tomato have generally employed interspecific crosses, where presence of molecular marker polymorphisms allowed such studies. In one study, a trait-based marker analysis (a.k.a. selective genotyping) of an F_2 population of a cross between a salt-sensitive tomato breeding line and a salt-tolerant accession (LA716) of *L. pennellii* resulted in the identification of five QTLs on chromosomes 1, 3, 7, 8 and 12 (Foolad and Jones 1993). The validity of these QTLs was examined in a few subsequent studies, using populations derived from the same or different interspecific crosses, including

BC_1, BC_1S_1 and recombinant inbred line (RIL) populations of crosses between *L. esculentum* and *L. pimpinellifolium* (Foolad and Chen 1998; Foolad et al. 1998; Foolad et al. 1997; M.R. Foolad et al., unpubl. data). These studies validated the previously-identified QTLs and detected a few additional QTLs on chromosomes 2 and 9. The combined results supported the notion that ST during SG in tomato was a quantitative trait controlled by more than one gene. Notably, however, in all of these studies it was determined that this trait was controlled by a few QTLs with major effects in addition to several QTLs with smaller effects. A comparison of QTLs indicated that some QTLs were stable across populations/generations whereas others were population-specific. Further comparisons of QTLs across interspecific populations, including those derived from *L. esculentum* × *L. pennellii* (Foolad and Chen 1998; Foolad et al. 1997) and *L. esculentum* × *L. pimpinellifolium* crosses (Foolad et al. 1998; M.R. Foolad, unpubl. data), indicated that some QTLs were conserved across species whereas others were species-specific. Most studies also suggested absence of significant epistatic interactions among QTLs. The overall results from these studies indicate that, in comparison to ST at later stages of development (described in below), ST during SG in tomato is less affected by variation in genetic and environmental backgrounds and thus, it should be feasible to transfer this trait to commercial cultivars by PS or marker-assisted selection (MAS). However, because in most cases tolerance QTLs are found within the wild species of tomato and often more than one gene resource is utilized during the life of a breeding project, the use of MAS may be beneficial.

2.2.4. *Comparison of salt tolerance at different stress levels during seed germination*

A successful cultivar would be one which exhibits ST at a wide range of SS levels and whose performance would not decline in the absence of salts. This is because in many saline soils the concentration of salts varies across the soil horizon, ranging from low to moderate and high (Richards and Dennett 1980). Practically, however, in a breeding program it may not be feasible to conduct selections under different SS levels. It is, therefore, important to determine whether there is a critical salt concentration at which selections could be made to develop cultivars with ST at most other SS levels. Several studies have been conducted to examine relationships among germination responses under different SS levels in tomato. Evaluation of 56 tomato genotypes for ST during SG at 75 mM (low), 150 mM (intermediate) and 200 mM (high) salt indicated that generally genotypes that germinated rapidly at the low SS level also germinated rapidly at the moderate and high concentrations (Foolad and Lin 1997b). Linear correlation analysis indicated the presence of a strong phenotypic correlation ($r = 0.90$, $P < 0.01$) between germination response at 75 mM and 150 mM salts. The results suggested that the same genes might control the rate of tomato SG under different SS levels. This suggestion was subsequently confirmed by an analysis of response and correlated response to selection for ST, where selections were made separately under low (100 mM), medium (150 mM) or high (200 mM) salt concentration and progeny responses were examined at all three levels (Foolad 1996b). The results indicated that selection for rapid SG at any SS level led to progeny with enhanced germination rate at all three SS levels, suggesting that similar or identical genes with additive effects were responsible for rapid SG response at different SS levels. This suggestion was consistent with the finding of similar QTLs for ST during SG at different SS levels (Foolad and Jones 1993). The combined results suggest that to develop tomato cultivars with improved ST during SG, it is sufficient to conduct selections at a single SS level, preferably at a medium SS level (Foolad 1996b).

2.2.5. *Physiological genetics of salt tolerance during seed germination*

Although QTLs for ST during SG in tomato have been identified, their genetic nature or the physiological mechanisms that they modulate have not been determined. However, based on the current knowledge of the physiology of ST during SG, some speculations can be made as to their roles. The tomato seed is comprised of a seed coat that encloses the embryo and an endosperm that practically fills the lumen of the seed not occupied by the embryo (Esau 1953). For germination to occur, the hydraulic extension force of the embryo must exceed the opposing force of the seed coat and the living endosperm tissues (Bradford 1986; Groot and Karssen 1987; Hegarty 1978; Liptay and Schopfer 1983). Embryo genotype was suggested to play a major role in determining the time to germination of tomato seed under nonstress or stress conditions (Liptay and Schopfer 1983). According to this hypothesis, differences in salt sensitivity of tomato seeds during germination reside either in the osmotic potential or pressure potential of the germinating embryo. However, osmotic stress can also negatively affect seed imbibition, and thus retard (or prevent) weakening of the restrictive forces of the endosperm and seed coat, resulting in reduced rate (or inhibition) of germination (Dahal et al. 1990; Groot and Karssen 1987; Liptay and Schopfer 1983). Thus, the rate of SG may be influenced by the physical, chemical, and thus, genetic composition of the embryo, endosperm and/or the seed coat. The identified QTLs for ST during SG in tomato could therefore affect germination rate by affecting the vigor of the germinating embryo, the variation in the thickness of the endosperm, the physical and permeability properties of the endosperm cell walls, the time of onset or rate of activity of enzymes which modify the properties of the endosperm cell wall, the release of gibberellin by the embryo, the base water potential required for SG, the hydrotime constant (Bradford 1995), the rate of metabolic activities in the embryo or endosperm under osmotic stress, osmoregulation during germination, or any other physiological or metabolic processes which are essential for the initiation of germination. However, isolation, characterization and comparison of functional genes which facilitate rapid SG under SS would be necessary to determine the actual roles of the identified QTLs.

2.3. Salt Tolerance During Vegetative Stage

For tomato production under saline conditions, ST during vegetative stage (VS) is more important than ST during SG because most tomato crops are established by seedling transplantation. ST during VS may also be more important than ST during reproduction (flowering and fruit set) as tomato ST generally increases with plant age and plants are usually most tolerant at maturation (Bolarin et al. 1993). During flowering and fruiting stages, for example, tomato plants can withstand salt concentrations that can kill them at the seedling stage. Most commercial cultivars of tomato are moderately sensitive to SS during VS (Foolad and Lin 1997b; Maas 1986; Tal and Shannon 1983). At low concentrations of salt (EC = 3–5 dSm^{-1}), tomato growth is mainly restricted by nutritional imbalances, as nutrients become the limiting factor under such conditions (Cuartero and Fernandez-Munoz 1999). At moderate to high levels of salt (EC $\geq$ 6 dSm^{-1}), in addition to nutrient imbalances, osmotic effects and ion toxicity contribute to reductions in growth. Phenotypic variation for ST during VS has been identified within the cultivated (Cuartero et al. 1992; Foolad 1997; Sarg et al. 1993) and wild species of tomato, including *L. peruvianum* (Tal and Gavish 1973), *L. pennellii* (Cano et al. 1998; Dehan and Tal 1978; Perez-Alfocea et al. 1994; Saranga et al. 1991), *L. cheesmanii* (Asins et al. 1993a; Rush and Epstein 1976), and *L. pimpinellifolium* (Asins et al. 1993a; Bolarin et al. 1991; Cuartero et al. 1992; Foolad and Chen 1999). This variation has been utilized for investigation of the physiology and genetic basis of ST during VS in tomato.

2.3.1. *Physiology of salt tolerance during vegetative stage*

Most salt-tolerant genotypes within the cultivated tomato and closely-related wild species *L. pimpinellifolium* generally exhibit a glycophytic response to salinity, that is, exclusion of toxic ions (e.g. Na^+) at the root or shoot level and synthesis and accumulation of compatible organic compounds (e.g., sugars and amino acids) for osmoregulation (Bolarin et al. 1993; Caro et al. 1991; Cuartero et al. 1992; Foolad 1997; Perez-Alfocea et al. 1993b; Santa-Cruz et al. 1998). In contrast, salt-tolerant accessions within the tomato wild species *L. pennellii*, *L. cheesmanii* and *L. peruvianum* generally exhibit a halophytic response to salinity, in which osmotic adjustment is achieved by uptake of inorganic ions from the soil and compartmentalization in cell vacuoles (Bolarin et al. 1991; Perez-Alfocea et al. 1994; Sacher et al. 1983; Tal and Shannon 1983). However, differential accumulation of ions has not always been identified as a major factor in determining tomato ST or sensitivity. For example, analysis of BC populations of a cross between a salt-sensitive cultivar and a salt-tolerant *L. pennellii* accession (LA716) indicated that tissue ion content was not likely to provide an efficient selection criterion for ST, as no direct relationship was observed (Saranga et al. 1992). In another study, analysis of the relationship between ST and leaf ion compositions in the cultivated and three wild species of tomato prompted Saranga et al. (1993) to conclude that dry matter production under SS was positively correlated with K^+/Na^+ ratio in the stem and negatively correlated with Cl^- concentration in leaves and stems. The authors suggested that tissue ion content and ion selectivity were good selection criteria for ST breeding in tomato. Potassium selectivity over Na^+ was also reported as a good indicator of ST in a study of several genotypes of the cultivated and wild species of tomato (Cuartero et al. 1992). Further studies of wild species of tomato, including *L. peruvianum* (Tal 1971), *L. cheesmanii* (Rush and Epstein 1981b) and *L. pimpinellifolium*, *L. hirsutum* and *L. pennellii* (Bolarin et al. 1991), related elevated concentrations of Na^+ in the leaf to plant ST. Other studies suggested that the ability to regulate Na^+ concentration in the leaf tissue was more closely correlated with ST than Na^+ concentration per se (Sacher et al. 1983) and that the distribution of Na^+ in young and mature leaves were important part of such regulation (Shannon et al. 1987). In a more recent study, however, no relationship was observed between tissue ion content and plant ST in BC populations of a cross between a tomato breeding line and a salt-tolerant accession (LA722) of *L. pimpinellifolium* (Foolad and Chen 1999). The overall conclusion from the various studies is that tissue ion content per se may not be a universal indicator of ST across tomato genotypes.

In tomato genotypes with glycophytic response to salinity, as ion concentration increases beyond a threshold level the exclusion mechanism fails and further increases in ion concentration in the root zone would result in fading plant growth and gradual death (Foolad 1997; Perez-Alfocea et al. 1993a). Thus, such genotypes may only be useful for cultivation under low to moderate levels of salt. At higher SS levels, genotypes that exhibit a halophytic response may be more advantageous. Unfortunately, however, many salt-tolerant wild accessions of tomato that exhibit a halophytic response to salinity often grow extremely slowly under SS with limited fruit production (Foolad 1996a; Tal 1997). Whether these associations are due to pleiotropic effects of the same genes or undesirable linkage between different genes is unknown. Several studies in tomato and other plant species have suggested that genes contributing to plant vigor are different from those conferring ST, and when breeding for efficient production under saline conditions genes for both plant vigor and ST are important (Foolad 1996a; Forster et al. 1990). This may limit the utility of wild accessions with halophytic response to salinity for breeding tomatoes with enhanced ST. However, further studies are needed to verify this conclusion.

2.3.2. Inheritance of salt tolerance during vegetative stage

Genetics research on tomato ST during VS started about 3 decades ago, when Emanuel Epstein proposed exploitation of gene resources within the wild *Lycopersicon* species to increase ST of the cultivated tomato (Epstein et al. 1980; Rush and Epstein 1976). Subsequently, hybridizations were made between a salt-tolerant accession (LA1401) of *L. cheesmanii* and a salt-sensitive tomato cultivar and filial and BC progeny were produced (Rush and Epstein 1981a). The authors reported that selection in the segregating populations led to progeny with enhanced ST, suggesting that ST of LA1401 could be transferred to the cultivated tomato. Although no salt-tolerant cultivar was derived from these materials, this study led to other investigations of genetics and breeding of ST in tomato. (Saranga et al. 1992) developed BC populations of a cross between a salt-sensitive tomato line and a salt-tolerant accession (LA716) of *L. pennellii* and evaluated them for tolerance under saline field conditions. Estimates of h^2 for total dry matter and total fruit yield under saline conditions as well as total dry matter under salt relative to control conditions were moderate (0.3–0.45), suggesting that ST of the cultivated tomato could be improved by using LA716 as a gene resource. However, there has not been any report of a salt-tolerant cultivar derived from these materials. By evaluating F_2 progeny of a cross between a salt-sensitive tomato and a salt-tolerant accession of *L. pimpinellifolium* under SS, (Asins et al. 1993b) concluded that total fruit yield and fruit number were useful selection criteria for improving tomato ST; estimates of broad-sense h^2s for these traits were 0.53 and 0.73, respectively. In a greenhouse hydroponics study, using parental, filial and BC populations of an intraspecific cross between a salt-sensitive tomato breeding line and a salt-tolerant primitive cultivar (PI174263), it was determined that growth under SS relative to control, the most widely used index in physiological investigation of ST in tomato, was under additive genetic control and could be a possible selection criterion for improving tomato ST (Foolad 1996a). In none of the aforementioned studies, however, was any empirical selection made to verify the suggestion that ST of tomato could be improved by directional PS. Nonetheless, these and other studies (Bolarin et al. 1991; Foolad 1996a) have suggested that shoot growth under salinity relative to control (a.k.a. relative growth under SS) should be the best indicator of ST, which may be useful in ST breeding in tomato.

2.3.3. *Physiological genetics of salt tolerance during vegetative stage*

Direct selection for ST under field conditions is generally difficult because of confounding effects of numerous other environmental factors (Richards 1983; Yeo and Flowers 1990). A suggested approach to improve the efficiency of selection for ST has been the adoption of new selection criteria based on knowledge of physiological processes which limit crop production under saline conditions (Flowers and Yeo 1988, 1997; Tal 1985; Yeo and Flowers 1990). Physiological criteria that have been suggested as potential indicators of ST in tomato include tissue water potential, tissue ion content, K^+/Na^+ ratio, osmoregulation, succulence, and water use efficiency (WUE) (Asins et al. 1993b; Foolad 1996a, 1997; Guerrier 1996; Martin and Thorstenson 1988; Perez-Alfocea et al. 1993b; Romero-Aranda et al. 2001; Saranga et al. 1993). However, whether these physiological parameters are good indicators of ST in tomato, or if there are genetic variations in these responses, must be determined before assessing their utility as indirect selection criteria for improving tomato ST.

Genetic research to examine the value of physiological parameters for breeding for ST in tomato has been scarce. In one study, analysis of the parental, filial and BC generations of an intraspecific cross between a salt-sensitive tomato line and a salt-tolerant primitive cultivar (PI174263) indicated that growth under SS was positively correlated with leaf Ca^{2+} content and negatively correlated with leaf Na^+ content (Foolad 1997). Generation means analysis of these populations indicated that accumulations of both Na^+ and Ca^{2+} in the leaf under SS were genetically controlled with additivity being the major genetic component. Tissue ion concentration was therefore suggested as a useful selection criterion when breeding for improved ST of tomato using PI174263 as a genetic source (Foolad 1997). As discussed in section 2.3.1., a few other studies have speculated on the utility of physiological parameters as indirect selection criteria for breeding salt-tolerant tomatoes (Asins et al. 1993b; Cuartero et al. 2006; Foolad 1997; Saranga et al. 1993; Tal and Gavish 1973; Tal et al. 1979). However, despite these studies, there is yet no consensus on what might be the best physiological or morphological characteristic(s) that should be employed as indirect selection criteria when breeding tomatoes for ST. Most likely a combination of different characteristics should be considered if salt-tolerant genotypes with commercial values are expected. This, by itself, indicates the complexity of ST and the need for identifying better approaches for characterizing genetic bases of tolerance components to facilitate development of commercial cultivars with enhanced ST. Recent advances in molecular marker technology, QTL mapping, MAS, and genetic transformation have provided some promising approaches.

2.3.4. QTL analysis of salt tolerance during vegetative stage

A few studies have identified QTLs for ST during VS in tomato. In one study, a BC_1S_1 population of a cross between a tomato breeding line and a salt-tolerant accession of *L. pimpinellifolium* (LA722) was screened for ST (Foolad and Chen 1999). The two parents were distinctly different in ST: while 80% of LA722 survived after two weeks under a salt concentration of 700 mM NaCl + 70 mM $CaCl_2$ (equivalent to $\sim$64 dSm^{-1}), only 25% of the *L. esculentum* line remained alive. The BC_1S_1 population exhibited a continuous variation, with survival rate ranging from 9% to 94% across families. Interval mapping identified five QTLs for ST on tomato chromosomes 1, 3, 5 and 9. All QTLs had the positive alleles from *L. pimpinellifolium*. The results supported the previous suggestion (Foolad 1996a, 1997) that ST during VS in tomato was controlled by more than one gene. However, the involvement of only a few QTLs, which accounted for a large portion of the total phenotypic variation, suggested utility of MAS for transferring ST QTLs from LA722 to the cultivated tomato. Analyses of leaf Na^+, K^+, Mg^{2+}, Ca^{2+}, Cl^-, $NO3^-$, $SO4^{2-}$ and $PO4^{3-}$ contents indicated the absence of a correlation between ST and tissue ion content in this population; no QTL was identified for tissue ion content under SS. Using a different BC population of the same cross, a selective genotyping approach was used to verify the previously-identified QTLs and possibly identify new QTLs (Foolad et al. 2001). In this study, from a population of 792 BC_1 plants grown under SS, 37 most salt-tolerant individuals were selected and grown to maturity and produced BC_1S_1 seeds. The 37 selected BC_1S_1 families and 119 nonselected (random) BC_1S_1 families were evaluated for ST and their performances compared. A realized h^2 of 0.46 was obtained for ST during VS, consistent with a previous estimate of h^2 for this trait obtained from an intraspecific cross of tomato (Foolad 1996a). A trait-based marker analysis (selective genotyping) led to the detection of five QTLs for ST on chromosomes 1, 3, 5, 6 and 11 (Foolad et al. 2001). Except for one, all QTLs had positive alleles contributed from the salt-tolerant *L. pimpinellifolium* parent. Three of the five QTLs were at the same locations as those identified in the first study (Foolad and Chen 1999). The high level of consistency between results of the two studies indicated the genuine nature of the detected QTLs and their potential utility for ST breeding using MAS. In each of these two studies, a few individuals were identified with most or all of the QTLs and with a ST comparable to that of the salt-tolerant *L. pimpinellifolium* accession for future ST breeding.

In a more recent study, 145 F_9 recombinant inbred lines (RILs) of a *L. esculentum* × *L. pimpinellifolium* cross were evaluated in replicated trials for ST during VS. The RILs were genotyped for 129 RFLP and 62 resistance gene analog (RGA) markers, covering 1,505 cM of tomato genome with an average marker distance of 7.9 cM. Interval analysis identified 7 QTLs for ST during VS on tomato chromosomes 3, 4, 5, 7, 8, 9 and 12 (M.R. Foolad et al., unpubl. data). The QTLs detected on chromosomes 3, 5, and 9 were the same as those identified in the previous studies and exhibited larger effects than the newly identified QTLs on chromosomes 4, 7, 8 and 12. The overall results from these three studies indicated that the stable QTLs on chromosomes 3, 5 and 9 should be useful for introgression into the cultivated tomato via MAS to improve tomato ST during VS. However, further studies are needed to verify these QTLs in other genetic backgrounds or identify new QTLs for gene pyramiding and development of tomatoes with enhanced ST during VS.

2.4. Salt Tolerance During Reproduction

Much less research has been conducted on tomato ST during reproduction than earlier stages. In particular, little effort has been devoted to determine pollen viability or stigma receptivity, and/or the ability of the plant to produce flowers or set fruit under SS. This may be due in part to a higher level of ST generally observed during reproduction than earlier stages in tomato. For example, increasing salinity to 10 dSm^{-1} did not significantly affect fruit set in tomato, which was reduced only at 15 dSm^{-1} (Adams and Ho 1992). Also, it was reported that salinity did not affect tomato pollen viability, though the number of pollen grains per flower decreased with the duration of salinity (Grunberg et al. 1995). In a recent study, 13 tomato accessions from 3 different species were grown under saline (300 mM NaCl + 30 mM $CaCl_2$; equivalent to ~ 28 dSm^{-1}) and control conditions and their pollen production and *in-vitro* pollen germination were examined (S Prakash and MR Foolad, unpubl. data). For most accessions, there was no significant reduction in pollen production (per flower) in response to SS. Pollens from both salt-grown and control-grown plants were cultured at different SS levels, including 0, 0.2, 0.4 and 0.8% NaCl, and evaluated for percentage germination after 4 or 8 h of incubation. In all accessions, pollen germinability was decreased under salt compared to control treatment, and the reduction was greater at higher (0.8%) than lower (0.2%) salt concentrations. However, in most accessions, *in-vitro* pollen germinability of salt-grown plants was generally higher than that of the control-grown plants, suggesting that pollen ST was increased by growing plants under SS.

In the cultivated tomato, fruit yield generally starts decreasing when the EC of the saturated soil extract exceeds 2.5 dSm^{-1} (Maas 1990; Saranga et al. 1991), though there are reports of higher thresholds for yield reduction in tomato (Adams 1991). A 10% reduction in fruit yield is expected per additional dSm^{-1} beyond the threshold level (Saranga et al. 1991). The major cause of yield reduction in tomato under low to moderate levels of salinity (EC = 3–9 dSm^{-1}) is the reduction in the average fruit size, and not a reduction in fruit number (van Ieperen 1996). A 10% reduction in fruit size is caused following irrigation with 5–6 dSm^{-1} water, a 30% reduction with 8 dSm^{-1}, and about 50% reduction at 9 dSm^{-1} (Cuartero and Fernandez-Munoz 1999). Thus, small-fruited genotypes, including cherry tomatoes, may be more successful than large-fruited ones when grown under low to moderate salinity (Caro et al. 1991). However, at higher levels of salinity, or prolonged exposure to salinity, a reduction in the total number of fruits per plant is the major cause of yield reduction, thus affecting both large-fruited and small-fruited genotypes (Cuartero and Fernandez-Munoz 1999; van Ieperen 1996). It is notable that the potential of tomato wild species as sources of ST during reproduction has not been assessed critically, mainly because most of the wild accessions are self-incompatible and/or produce very small fruits and thus cannot be easily compared with the cultivated tomato. However, progenies derived from interspecific crosses have often been used for salt tolerance studies.

Limited research has been conducted to identify genes or QTLs for ST during reproduction in tomato. In one study, using 14 genetic markers and an F_2 population of a cross between a salt-sensitive cultivar and a salt-tolerant *L. pimpinellifolium* accession, a few QTLs were detected affecting fruit yield, fruit number and fruit size under SS. However, because of the extreme difference in fruit size between the parents of the F_2 population, it is likely that QTL effects were confounded by effects of genes controlling fruit size. Similar studies were conducted in F_2 populations of different crosses between *L. esculentum* and either *L. pimpinellifolium* or *L. cheesmanii*, and several other QTLs were reported for the same fruit-related traits (Monforte et al. 1996, 1997, 1999). However, large differences between parental lines of these populations, including differences in flowering habits, maturity time, fruit size, fruit number and total fruit yield, would have adversely affected the power of the experiments in detecting true QTLs affecting ST. Therefore, the identified QTLs should be validated using advanced generations before they are employed in MAS. In conclusion, more comprehensive studies are needed to carefully identify genetic factors (QTLs) which truly contribute to ST during reproduction in tomato and which could be used for marker-assisted breeding.

2.5. Relationship Among Salt Tolerance at Different Developmental Stages

Knowledge of genetic relationships among tolerance at different developmental stages is necessary to facilitate development of cultivars with enhanced ST throughout the plant ontogeny. Early studies had suggested absence of phenotypic relationships among different stages of plant development in regard to ST in various plant species (Abel and Mackenzie 1963; Greenway and Munns 1980; Johnson et al. 1992). In tomato, recently systematic approaches were taken to examine phenotypic as well as genetic relationships among tolerance to salinity in different developmental stages. In one study, an F_4 population of a cross between a salt-nsensitive tomato breeding line and a primitive cultivar (PI174263) with ST during both SG and vegetative stages was evaluated for tolerance during both stages. In the F_4 population, there were significant variation among families in terms of ST during both SG and VS, however, there was no significant correlation ($r_p = -0.10$, $P > 0.05$) between ST during the two stages (Foolad and Lin 1997a). To examine the genetic correlation between ST during SG and VS, selection was made for rapid SG under SS in an F_2 population of the same cross and the selected F_3 progeny were evaluated for ST separately during both SG and VS. The results indicated that while selection improved germination ST of the F_3 progeny significantly, it did not affect ST of the F_3 progeny during VS, suggesting that genetic and physiological mechanisms that contributed to ST during SG were different from those conferring ST during VS (Foolad and Lin 1997a). This relationship was further examined by comparison of QTLs affecting ST during each of the two stages (Foolad 1999). Using a BC_1S_1 population of a cross between a salt-sensitive tomato line and a *L. pimpinellifolium* accession (LA722) with ST during both SG and VS, it was determined that QTLs for ST during SG were different from QTLs for ST during VS. A similar QTL study was recently conducted using 145 F_9 RILs of the same cross, and the results supported the previous finding of absence of a genetic relationship between ST during SG and VS (MR Foolad et al., unpubl. data). The overall results indicated that ST during SG in tomato was independent of ST during VS, consistent with earlier reports that ST of young tomato plants did not correlate with that of mature plants (Shannon et al. 1987) and that ST ranking of tomato genotypes based on vegetative characteristics differed from the ranking based on fruit yield (Caro et al. 1991).

Absence of genetic relationships in ST among different developmental stages have also been reported in other plant species, including alfalfa, *Medicago sativa* L. (Johnson et al. 1992), barley (Mano and Takeda 1997), Arabidopsis (Quesada et al. 2002), wheat, *Triticum aestivum* L. (Ashraf and McNeilly 1988), triticale, *Triticale hexaploide* Lart. (Norlyn and Epstein 1984), and slender wheatgrass, *Elymus trachycalus* spp. *Trachycalus* (Link) Malte (Pearen et al. 1997). Findings from different studies suggest that when breeding for improved ST, each stage of plant development must be evaluated separately for assessment of tolerance and identification, characterization and utilization of useful genetic components. However, identification of QTLs for ST at different developmental stages may facilitate pyramiding of tolerance factors and development of cultivars with improved ST at all stages.

2.6. Transgenic Approaches to Develop Salt Tolerant Tomatoes

Many genes are involved in a plant's response to SS, which may lead to a wide variety of biochemical and physiological changes. These include expression of genes that facilitate compartmentalization of toxic ions in the vacuoles, activation of detoxifying enzymes, synthesis of late-embryogenesis-abundant (LEA) proteins, and accumulation of compatible solutes. Genetic engineering approaches to developing stress-tolerant plants are considered an attractive alternative to conventional breeding protocols. Recently, transgenic approaches have been employed to produce plants with enhanced tolerance to various abiotic stresses, including salinity, by overexpression of genes controlling different tolerance-related physiological mechanisms (Bajaj et al. 1999; Bartels and Sunkar 2005; Chinnusamy et al. 2005; Rontein et al. 2002; Seki et al. 2003; Serrano et al. 1999; Wang et al. 2003; Yamaguchi and Blumwald 2005; Zhang et al. 2004). For example, plants have been engineered with genes encoding enzymes that enhance the synthesis of compatible solutes such as mannitol (Thomas et al. 1995), glycine betaine (Lilius et al. 1996), proline (Zhu et al. 1997) and polyamines (Galston et al. 1997), which contribute to osmoregulation and improving plant stress tolerance (Rathinasabapathi 2000; Rontein et al. 2002). Compatible solutes may also contribute to stress tolerance through other functions such as protection of enzyme and membrane structure and scavenging of radical oxygen species (Bohnert and Shen 1999; Rathinasabapathi 2000; Shen et al. 1997; Wang et al. 2003). Transgenic plants also have been produced with overexpression of different vacuolar antiport proteins, which facilitate exclusion of toxic ions from the cell cytosol (Apse et al. 1999; Apse and Blumwald 2002; Serrano et al. 1999; Wang et al. 2003; Yamaguchi and Blumwald 2005; Zhang and Blumwald 2001; Zhang et al. 2001a). Furthermore, transgenic plants have been developed with increased expression of detoxification enzymes, which reduce oxidative stress (Tanaka et al. 1999). Although in almost all cases growth of transgenic plants were examined under controlled conditions and their performance under field were unknown, the transgenic approach has facilitated a better understanding of the mechanisms leading to stress tolerance.

Despite considerable efforts in the area of genetic transformation, limited attempts have been made to develop transgenic tomatoes with enhanced ST. A notable progress has been development of tomato plants overexpressing *AtNHX1*, a single-gene controlling vacuolar Na^+/H^+ antiport protein, introduced from *Arabidopsis thaliana* (Apse and Blumwald 2002; Yamaguchi and Blumwald 2005; Zhang and Blumwald 2001). The overexpression of this gene was previously shown to improve ST in *Arabidopsis* (Apse et al. 1999). Transgenic tomato plants overexpressing this gene were reported to have the ability to grow, set flower and produce fruit in the presence of 200 mM NaCl in greenhouse hydroponics whereas the control plants did not survive the saline conditions. The transgenic plants were reported to have acquired a halophytic response to SS, accumulating salts in the vacuoles. This is unlike the normal response of the cultivated tomato to SS, which is exclusion of salts from cells at the root shoot level, a glycophytic response. Accordingly, under high salinity conditions, transgenic tomato plants accumulated high concentrations of Na^+ and Cl^- in their leaves (Apse et al. 1999). The overproduction of the vacuolar Na^+/H^+ antiport protein enhanced the ability of the transgenic plants to sequester Na^+ in their vacuoles, averting its toxic effects in the cell cytosol. At the same time Na^+ was used to maintain an osmotic balance to drive water into the cell, and thus used salty water for cell expansion and growth. This was the first reported example of a single-gene transformation in any crop species that resulted in such a significant enhancement in plant ST. Subsequently, however, transfer and overexpression of the same gene in canola, *Brassica napus* (Zhang et al. 2001a), corn (Yin et al. 2004) and wheat (Xue et al. 2004) resulted in transgenic plants with enhanced ST under controlled saline conditions. However, the transgenic plants are yet to be evaluated for ST under field conditions and examined for their commercial value. To date, there is no report of such studies. Obviously, much more research is needed to gain a better understanding of the genetics, biochemical, and physiological basis of plant ST using the transformation technology. However, knowledge of various tolerance components and identification, cloning and characterization of responsible genes may allow development of plants harboring multiple transgenes and production of highly salt-tolerant transgenic plants. With the recent advances in molecular biology of stress tolerance in tomato, this expectation may not be unlikely.

3. GENETICS OF AND BREEDING FOR DROUGHT TOLERANCE IN TOMATO

3.1. Background

Drought, defined as the occurrence of a substantial water deficit in the soil or in the atmosphere, is an increasingly important constraint to crop productivity and yield stability worldwide (Ceccarelli and Grando 1996). It is by far the leading environmental stress in agriculture. The worldwide losses in yield due to drought probably exceed the losses from all other causes combined (Blum 1988; Kramer 1980; Schonfeld et al. 1988). In the U.S., up to 45% of the land surface is subject to continuous or frequent water stress (Boyer 1982; Tanji 1990) and a drought occurs somewhere in the country every year, costing billions of dollars in damage to crops and businesses (Ross and Lott 2000).

Most crop plants, including tomato, are sensitive to drought stress (DS) throughout the ontogeny of the plant, from SG to harvest (Hsiao 1973). Plant response to DS can be generally classified into three categories, drought escape, dehydration avoidance, and dehydration tolerance (Blum 1988; Kramer 1983). Drought escape includes situations where plants with short growth cycle and early maturity avoid experiencing drought. Breeding for drought escape should therefore be directed toward developing cultivars with early maturity so that by the time drought occurs the plant has already completed its life cycle. Dehydration avoidance is defined as the ability of the plant to retain a relatively higher level of "hydration" during the period of water stress (Blum 1988). In this situation, the plant protects its various growth related physiological, biochemical, and metabolic processes from the external water stress. A common measure of dehydration avoidance is the maintenance of a higher tissue water or turgor potential under conditions of water stress. Osmotic adjustment, as a means for retaining a higher turgor at a given tissue water potential, is an example of dehydration avoidance at the cell level. When the tissue is not protected by any of the avoidance mechanisms, cells lose turgor and dehydrate, resulting in various cellular physicochemical injuries (Hsiao and Bradford 1983). Complete loss of free water will result in desiccation or dehydration. In general, however, different genotypes exhibit different responses to cellular and whole plant stresses caused by dehydration, and there are varying levels of dehydration tolerance. It should also be noted that characteristics of the three categories of plant response to DS are not generally independent of each other, and some plants may exhibit a combination of characteristics (Blum 1988).

A complementary approach in agricultural methods currently followed is to minimize losses incurred by water stress and develop "drought tolerant" cultivars with the ability to escape, avoid, and/or tolerate effects of water stress. However, despite many decades of research on drought tolerance (DT), till date drought stress continues to be a major challenge to plant breeders. This is in part due to the complexity of the trait. Accumulating evidence suggests that plant response to DS is controlled by many genes and physiological mechanisms (Blum 1988; Subudhi et al. 2000; Zhang et al. 2001b; Zhu et al. 1997) and varies depending on the influence of other environmental factors (Ceccarelli and Grando 1996; Richards 1996). Selection and breeding for DT is also difficult because tolerance appears to be a developmentally-regulated, stage-specific phenomenon (Blum 1988; Ludlow and Muchow 1990; Mitchell et al. 1998; Richards 1996). Each stage may be considered as a separate trait and may require a different evaluation method. Furthermore, no reliable evaluation procedure is known that can effectively and efficiently be employed to identify drought-tolerant plants at different stages of development. These and other complexities have led to a limited success in developing drought-tolerant crop plants, including tomato.

In tomato, most commercial cultivars are sensitive to DS throughout the ontogeny of the plant, yet genotypic variation for DT exists within the cultivated (Wudiri and Henderson 1985) and related wild species such as *L. cheesmanii*, *L. chilense*, *L. pennellii*, *L. pimpinellifolium*, and *L. esculentum* var. *cerasiforme* (Martin et al. 1989; Pillay and Beyl 1990; Richards and Phills 1979; Rick 1973, 1978; Rick 1979b; Rick 1982; Yu 1972). The latter species, being native of coastal deserts of western South America, witness rainless long periods except for the occasional El Niño episodes of heavy rains. These species grow at habitats where condensation of dew and fog drip at night are the main source of moisture (Rick 1973). They are also remarkably capable of overcoming brief wilting. Only few formal studies have been conducted to screen for DT in tomato. In one study, (Rana and Kalloo 1990) evaluated 150 lines of cultivated and wild species of tomato under water-deficit conditions and identified a few *L. esculentum* genotypes and a few accessions of *L. pimpinellifolium* and *L. chilense* with DT attributes. In a recent study, Husain and Foolad (unpubl. data) screened over 120 tomato genotypes and identified a few wild accessions exhibiting considerable DT (described in below). However, very limited effort (Kahn et al. 1993; Martin et al. 1999; Pillay and Beyl 1990) has been devoted to characterization of the physiology or genetics of DT in tomato to warrant breeding activities toward development of drought-tolerant tomatoes. This is unlike extensive research that has been conducted on DT in many other crop species, including rice (Nguyen et al. 1997; Zhang et al. 2001b), corn (Ribaut et al. 1997), sorghum, *Sorghum bicolor* L. Moench (Subudhi et al. 2000) and lettuce, *Lactuca sativa* L. (Johnson et al. 2000). Also, comparatively less research has been done on tomato DT than tomato tolerance to other abiotic stresses such as salinity and extreme temperatures. Here, the available information on germplasm resources and genetics of DT in tomato is reviewed and the prospect for developing drought-tolerant tomatoes is discussed.

3.2. Drought Tolerance During Seed Germination

The ability of the seed to germinate rapidly and uniformly under DS is a desirable trait for direct seeding tomato crops. Successful establishment of direct-seeded crops, however, depends on successful SG and seedling emergence. Most commercial cultivars of tomato are sensitive to DS during SG, however, sources of tolerance have been identified within the related wild species of tomato, including *L. pennellii* and *L. pimpinellifolium* (M.R. Foolad et al, unpubl. data), and some studies have been undertaken to discern the genetic basis of DT during SG in tomato.

3.2.1. Inheritance of drought tolerance during seed germination

The genetic basis of DT during SG in tomato has recently been studied using interspecific crosses between *L. pimpinellifolium* and *L. esculentum* (Foolad et al. 2003a; Foolad et al. 2003b; Subbiah 2001). In one study, for example, a BC_1 population (N = 1000) from a cross between a drought-tolerant *L. pimpinellifolium* accession (LA722) and a drought-sensitive tomato breeding line was evaluated for SG under DS (14% PEG, $\psi_w \approx -680$ kPa), and the most rapidly germinating seeds (first 3% germinated) were selected. The 30 selected BC_1 individuals were grown to maturity and self-pollinated to produce BC_1S_1 progeny seeds. Select BC_1S_1 progeny families were evaluated for germination under DS and their average performance was compared with that of a nonselected BC_1S_1 population of the same cross. Results indicated that selection for rapid SG under DS was effective and significantly improved progeny SG rate under DS; a realized h^2 of 0.41 was obtained for DT during SG in this population. The results indicated that DT during SG in tomato was genetically controlled and could be improved by PS.

3.2.2. Mapping of QTLs for drought tolerance during seed germination

A few recent studies have identified QTLs for DT during SG in tomato. In one study, a trait-based marker analysis, using BC_1 individuals of a cross between a drought-sensitive tomato breeding line and a drought-tolerant *L. pimpinellifolium* accession (LA722), detected four QTLs on chromosomes 1, 4, 8, 9, and 12 for DT (Foolad et al. 2003b). The results indicated that DT during SG in tomato was a quantitative trait, controlled by more than one gene. A few BC_1S_1 families were identified with most or all of the QTLs and with a DT comparable to that of LA722. These families should be useful for developing germination drought-tolerant tomato lines using MAS. In another study, 145 F_9 RILs of the same cross were evaluated for germination rate under DS and, by using composite interval mapping analysis, several QTLs for DT during SG were identified on tomato chromosomes 1, 2, 3, 4, 8, 9, and 12 (MR Foolad et al., unpubl. data). The results of this study were consistent with those of the previous one and suggested the presence of stable QTLs for DT during SG in populations derived from the *L. esculentum* × *L. pimpinellifolium* cross. These QTLs should be useful for improving tomato DT during SG using MAS.

3.3. Drought Tolerance During Vegetative Growth and Reproduction

Potential sources of DT during vegetative growth and later stages in tomato have been identified among accessions of the wild species *L. chilense* and *L. pennellii* (Rick 1973, 1978; Rick 1979b; Rick 1982). Different tolerance indices (TIs) have been suggested or employed to characterize physiological and genetic bases of DT in tomato, including dry weight (DW) of shoot and root, root length, root morphology, leaf rolling, flower and fruit set, fruit weight, fruit yield, WUE, recovery after re-watering, stomatal resistance, plant survival, leaf water potential, leaf osmotic potential, osmoregulation, transpiration rate, photosynthetic rate, enzymatic activities (e.g. superoxide dismutase and Rubisco), and pollen viability and germination (Blum 1988; Cohen et al. 1991; Kalloo 1991; Lutfor-Rahman 1998; Martin and Thorstenson 1988; Pillay and Beyl 1990; Rana and Kalloo 1989; Richards and Phills 1979). In a germplasm screening study, for example, tomato cultivar Saladette was considered drought tolerant as determined by a smaller reduction in fruit set compared to other cultivars, which in turn was attributed to its ability to roll up leaves under a high evaporative demand and maintain a high leaf water potential (Wudiri and Henderson 1985). The physiological basis of DT in *L. chilense* was attributed to its deep vigorous root system (Rick 1978), similar to those reported for cultivar Red Rock (Stoner 1972) and a few accessions of *L. pimpinellifolium* (Rana and Kalloo 1989). In contrast to these findings, the "drought-tolerant" *L. pennellii* accession LA716 has a limited and shallow root system and the basis for its DT is largely due to the ability to conserve moisture in succulent leaves during periods of limited rainfall. Also, LA716 has been characterized as having a greater WUE under DS than *L. esculentum*, as measured by g DW produced per Kg of water consumed (Martin and Thorstenson 1988). A high WUE in this accession was attributed to smaller leaf conductance due to fewer and smaller stomata, longer trichomes, lower chlorophyll content and Rubisco activity per unit leaf area, and larger mesophyll cell surface exposed to intercellular air space (Martin et al. 1999). However, though WUE may be a good indicator of DT in tomato, its measurement under field condition is not without inherent difficulties. Thus, attempts have been made to determine the relationship between WUE and stable carbon isotope discrimination (Δ), a measure of proportion of ^{13}C relative to ^{12}C in plant organic matter, which is easier to measure when dealing with large number of plants. (Martin et al. 1999) suggested that WUE in progeny of crosses between *L. esculentum* and *L. pennellii* LA716 could be increased by selecting for low Δ, however, this could lead to the selection of smaller plants, an agriculturally undesirable characteristic. The authors suggested that the small plant size could be corrected by conventional breeding following selection for DT, but no such effort has been reported.

Most recently a systematic study was conducted to identify sources of DT during vegetative stage in tomato (S Husain and MR Foolad, upbubl.). In this study, over 120 accessions from the cultivated tomato and wild species *L. pimpinellifolium, L. chilense, L. peruvianum and L. pinnellii* were screened in two treatments of control (no stress) and drought (stress) under greenhouse conditions. The growth parameters measured were shoot length, fresh and dry weight as well as root length and DW. TIs were also calculated as the ratio of growth under DS to growth under control conditions. The greenhouse experiments were repeated 3 times and similar parameters were measured. Based on absolute shoot DW under drought stress, *L. esculentum* genotypes exhibited the least DT. *L. pennellii* accessions were found to be the most drought tolerant, exhibiting greater shoot DW under stress and greater TIs, followed by *L. pimpinellifolium* accessions. As to the root DW under stress, *L. pimpinellifolium* accessions had the most root biomass accumulated followed by *L. peruvianum* accessions. Similar trend was observed as to the root length. An interesting observation was that *L. chilense* accessions showed the best performance as to TIs for the root length and root DW, followed *L. pennellii, L. peruvianum, L. pimpinellifolium* and *L. esculentum.* Overall this study identified some new accessions within the wild species of tomato with DT, which deemed to be better than those previously reported. These accessions should be useful for physiological and genetic studies, including mapping of tolerance-related genes/QTLs and their use in marker-assisted breeding.

3.3.1. Inheritance and QTL mapping for drought tolerance during vegetative growth and reproduction

Very limited research has been conducted to characterize genetic controls of DT or develop tomatoes with improved tolerance. In one study, three QTLs associated with low Δ were identified using F_3 and BC_1S_1 progeny of a cross between a *L. esculentum* breeding line and *L. pennellii* accession LA716 (Martin et al. 1989). However, it was not determined whether selection for these QTLs would increase WUE in tomato. Other related studies on genetics of tomato DT during vegetative growth include identification of several genes or mRNAs whose expressions were reportedly elevated in response to DS. For example, four drought-induced genes, *le4, le16, le25* and *le20,* were identified and characterized in tomato (Cohen et al. 1991; Kahn et al. 1993; Plant et al. 1991). It was determined that the increase in expression of these genes occurred after a longer period of water deficit in

L. pennellii than in the cultivated tomato, although these genes did not appear to be responsible for DT in *L. pennellii* (Kahn et al. 1993). Overall, in tomato too few studies have been undertaken to characterize genetic controls of DT post germination stage and/or to warrant any type of breeding activities. Obviously DT has not been a pressing issue for tomato breeders or its complexity has deterred them of breeding attempts.

3.4. Transgenic Approaches to Tomato Drought Tolerance

Very limited transgenic research has been done on tomato DT. This is unlike considerable research conducted in other plant species to identify, characterize and transfer genes toward development of drought-tolerant transgenic plants (Bajaj et al. 1999; Bartels and Sunkar 2005; Cherian et al. 2006; Grover et al. 1999; Kasuga et al. 2004; Oh et al. 2005; Serrano et al. 1999; Shou et al. 2004; Zhang et al. 2004). However, a few basic studies have been undertaken to investigate effects of selected foreign genes, in particular from Arabidopsis, on tomato response to water deficit. In one study, for example, transfer of an Arabidposis DNA cassette containing C repeat / dehydration-responsive element binding factor 1 (*CBF1*), under the control of CAMV35S promoter, resulted in transgenic tomatoes with water-deficit resistance greater than normal plants (Hsieh et al. 2002). In this study, however, it appeared that the introduced DNA had negative pleiotropic effects on plant growth under normal conditions such that fresh weight and fruit and seed numbers in transgenic plants were less than the isogenic wild-type plants. Further studies demonstrated that such negative effects were reversible by application of exogenous GA, which did not have any effect on plants' DT. In another study, transgenic tomato plants expressing CBF1 driven by an ABA-responsive complex (ABRC1) from the barley *HAV22* gene exhibited tolerance to DS, low temperature and SS; these plants maintained normal growth and yield under nonstress conditions (Lee et al. 2003). The results of this study suggested the potential benefit of using ABRC1-CBF1 transgenic tomato plants for production under stressful conditions. In a more recent study, (Na 2005) investigated the possibility of developing drought-tolerant tomatoes by developing transgenic plants containing either a tomato type I inositol 5 polyphosphatase (5PTse) or an ABRE binding factor ABF4 derived from Arabidopsis. While transgenic tomatoes containing the former gene exhibited some resistance to water deficit, they were retarded in growth. However, transgenic tomatoes expressing Arabidopsis ABF4/AREB2 exhibited more DT than non-transgenic plants, which was demonstrated to be due to lower water loss per unit leaf area. In another recent study, it was determined that transgenic tomato plants harboring the yeast trehalose-6-phosphate synthase (TPS1) gene under the control of CAMV35S promoter were more drought tolerant than the wild-type plants, though the transgenic plants exhibited some undesirable pleiotropic changes in plant morphology (Cortina and Culianez-Macia 2005). Overall, the results of these studies clearly demonstrate the potential utility of transgenic approaches to develop drought-tolerant tomatoes, though none of these investigations has led to development of any agriculturally-acceptable stress-resistant cultivar. While there is a good prospect for developing transgenic tomato cultivars with improved DT, it seems fine-ntuning of this approach necessitates a lot more basic and applied research efforts.

4. CURRENT STATUS AND FUTURE PROSPECTS FOR DEVELOPING TOMATOES WITH SALT AND/OR DROUGHT TOLERANCE

Most commercial cultivars of tomato are sensitive to salt and drought stresses during all stages of plant development, thus restricting tomato production in environments with such stresses. Occurrence of several genetic bottlenecks during tomato domestication and evolution, led the cultivated tomato to be depauparate in genetic diversity, including genes for abiotic stress tolerance. Fortunately, however, the related wild species of tomato are a rich source of desirable genes for tomato crop improvement. Although thus far only a superficial assessment of the extent of genetic variation for abiotic stress tolerance within *Lycopersicon* species has been made, some accessions with tolerance to salt or drought stress have been identified. Such resources have been utilized in physiological and genetic studies of salt and drought tolerance in tomato. However, more research is needed before commercial cultivars of tomatoes with the ability to grow and produce economic yield under saline or drought conditions will be available.

Absence of any tomato cultivar with proven field tolerance to salinity can be attributed to several factors including complexity of the trait, multifaceted interactions of ST with other agronomically important traits, insufficient understanding of the basic physiological and genetic mechanisms of ST, lack of efficient selection criteria, and, most importantly, limited efforts that has been devoted to identification, characterization and utilization of genetic resources for ST breeding. However, with the advent of new tools of plant molecular biology, including molecular marker technology and genetic transformation, the focus has largely been shifted to discerning genetic and physiological bases of ST in tomato, and some notable progress has been made. Recently, some tolerance components have been defined and their genetic controls characterized, and several controlling QTLs or genes with major effects have been identified and/or cloned. The new technology of gene transfer has provided opportunities to engineer tomatoes with enhanced ST using genes from unrelated species. Although transgenic plants have only been subjected to artificial laboratory tests of ST, the prospect for engineering tomato plants with field tolerance is improving. Furthermore, with our improved understanding of the significance of ST breeding in tomato, it is not unexpected to witness tomato cultivars with improved field ST in a near future. Notably, several research programs around the world, which are equipped with traditional and/or modern technologies of crop improvement, are currently working on development of tomatoes with enhanced ST.

Comparatively, however, much less progress has been made in genetics and breeding of tomatoes for DT. From the preceding discussion in this chapter, it is evident that currently there is limited physiological and/or genetic information on tomato DT to warrant development of cultivars with improved tolerance. Primarily, very limited knowledge is available as to genetic resources in *Lycopersicon* with DT attributes. Ironically, most tomato studies on DT have employed a single accession (LA716) of *L. pennellii* as a source of tolerance. However, due to various undesirable characteristics of this accession, in particular its extremely slow growth rate under DS, its usefulness as a genetic source for DT breeding in tomato is questionable. Although this accession can survive long periods of dryness, it lacks many other characteristics needed for use as a gene resource for DT breeding. Thus, initially larger germplasm screening experiments must be carried out including different wild species of tomato to identify useful sources of DT. In particular, collections from torrid areas should be examined for DT at different developmental stages.

Selection criteria for screening or breeding tomatoes for DT are also less clear than those available for ST breeding. More comprehensive studies are needed to identify and validate useful selection criteria, including morphological, agronomical, physiological, biochemical and molecular characteristics. In general, considering the normal climatic conditions for growing tomatoes, where short periods of drought may occur intermittently throughout the growing season, it seems that the ability of the tomato plant to survive transient periods of water stress and to recover rapidly upon re-availability of water is far more important than the ability to survive long-term water stress. Rather limited investigation has been done in this area in tomato, which deserve more attention. From a practical point of view, the most reliable criteria for breeding for DT are agronomic characteristics such as yield, and absolute and relative plant growth under stress and nonstress environments. Such criteria, however, may not be efficient or feasible to apply because in most initial germplasm evaluation or breeding projects often a large number of individuals, families or populations are screened, many of which may have wild genetic backgrounds. Alternative criteria based on physiological characteristics such as photosynthetic rates, stomatal resistance and leaf water potential might be more efficient. These characteristics are easier to measure, compared to yield, and generally show good correlations with agronomic characteristics. However, such characteristics must be identified and verified for specific sources of tolerance. Other selection criteria include biochemical characteristics such as enzyme activities and protein contents. These characteristics, however, often show weak correlations with agronomic traits and are expensive to measure. Additional options include identification and utilization of molecular markers associated with tolerance-related physiological, morphological or agronomic characteristics. Limited research has been conducted in this area in tomato. Transgenic approaches, which have been employed in several other plant species to increase DT, may also be useful for developing tomatoes with improved DT. This approach may require identification, examination and utilization of DT-related genes or proteins across species. In general, however, if tomato cultivars for commercial production under DS conditions are desired, it may be necessary to create and employ innovative combinations of germplasms, trait characteristics, tolerance criteria, and technologies at different stages of the breeding process.

In summary, to facilitate development of tomatoes with improved salt or drought tolerance, the following recommendations are made:

1. Conduct large screening experiments to identify highly desirable sources of genetic tolerance, in particular in relation to DS.
2. Identify and characterize major components of tolerance at different developmental stages. Often it is not only one physiological mechanism or genetic factor that contributes to plant stress tolerance throughout its ontogeny. Also, different physiological or genetic mechanisms of tolerance may be involved in different genetic backgrounds. Identification and characterization of individual components of genetic tolerance may simplify the breeding process and allow pyramiding of tolerance components across developmental stages and genetic backgrounds.
3. Extend the search for identification and utilization of potential tolerance components, including genes and proteins, beyond the limits of species within *Lycopersicon*, and possibly include other genera, including model plants and microbial organisms.
4. Establish interdisciplinary collaborations among plant physiologists, geneticists, breeders and molecular biologists interested in stress tolerance. Successful development of commercial cultivars with proven tolerance under field conditions is beyond the capabilities of one individual scientist or laboratory.

REFERENCES

Abel GH, Mackenzie AJ (1963) Salt tolerance of soybean varieties (*Glycine max* L. Merrill) during germination and later growth. Crop Sci 3:159–161

Adams P (1991) Effects of increasing the salinity of the nutrient solution with major nutrients or sodium chloride on the yield, quality and composition of tomatoes grown in rockwool. J Hort Sci 66:201–207

Adams P, Ho LC (1992) The susceptibility of modern tomato cultivars to blossom-end rot in relation to salinity. J Hort Sci 67:827–839

Apse MP, Aharon GS, Snedden WA, Blumwald E (1999) Salt tolerance conferred by overexpression of a vaculolar Na/H anitort in *Arabidopsis*. Science 285:1256–1258

Apse MP, Blumwald E (2002) Engineering salt tolerance in plants. Curr Opin Biotech 13:146–150

Arumuganathan K, Earle ED (1991) Nuclear DNA content of some important plant species. Plant Mol Biol Rep 9:208–218

Ashraf M, McNeilly T (1988) Variability in salt tolerance of nine spring wheat cultivars. J Agron Crop Sci 160:14–21

Asins MJ, Breto MP, Cambra M, Carbonell EA (1993a) Salt tolerance in *Lycopersicon* species. I. Character definition and changes in gene expression. Theor Appl Genet 86:737–743

Asins MJ, Breto MP, Carbonell EA (1993b) Salt tolerance in *Lycopersicon* species. II. Genetic effects and a search for associated traits. Theor Appl Genet 86:769–774

Bajaj S, Targolli J, Liu LF, Ho THD, Wu R (1999) Transgenic approaches to increase dehydration-stress tolerance in plants. Mol Breed 5:493–503

Bartels D, Sunkar R (2005) Drought and salt tolerance in plants. Crit Rev Plant Sci 24:23–58

Bliss FA, Platt-Aloia KA, Thomson WW (1986) Osmotic sensitivity in relation to salt sensitivity in germinating barley seeds. Plant Cell Environ 9:721–725

Blum A (1988) Plant Breeding for Stress Environment. CRC Press, Boca Raton

Bohnert HJ, Shen B (1999) Transformation and compatible solutes. Scientia Hort 78:237–260

Bolarin MC, Fernandez FG, Cruz V, Cuartero J (1991) Salinity tolerance in four wild tomato species using vegetative yield-salinity response curves. J Am Soc Hort Sci 116:286–290

Bolarin MC, Perez-Alfocea F, Cano EA, Estan MT, Caro M (1993) Growth, fruit yield, and ion concentration in tomato genotypes after pre- and post-emergence salt treatments. J Am Soc Hort Sci 118:655–660

Boyer JS (1982) Plant Productivity and environment. Science 218:443–448

Bradford KJ (1986) Manipulation of seed water relations via osmotic priming to improve germination under stress conditions. HortScience 21:1105–1112

Bradford KJ (1995) Water relations in seed germination. In: Kigel J, Galili G (eds) Seed Development and Germination. Marcel Dekker, Inc., New York, pp 351–396

Breto MP, Asins MJ, Carbonell EA (1993) Genetic variablility in *Lycopersicon* species and their genetic relationships. Theor Appl Genet 86:113–120

Cano EA, Perez-Alfocea F, Moreno V, Caro M, Bolarin MC (1998) Evaluation of salt tolerance in cultivated and wild tomato species through *in vitro* shoot apex culture. Plant Cell, Tissue and Organ Cult 53:19–26

Caro M, Cruz V, Cuartero J, Estan MT, Bolarin MC (1991) Salinity tolerance of normal-fruited and cherry tomato cultivars. Plant and Soil 136:249–255

Ceccarelli S, Grando S (1996) Drought as a challenge for the plant breeder. Plant Growth Regul 20:149–155

Cherian S, Reddy MP, Ferreira RB (2006) Transgenic plants with improved dehydration-stress tolerance: progress and future prospects. Biol Plant 50:481–495

Chinnusamy V, Jagendorf A, Zhu J-K (2005) Understanding and improving salt tolerance in plants. Crop Sci 45:437–448

Choudhuri GN (1968) Effect of soil salinity on germination and survival of some steppe plants in Washington. Ecology 49:465–471

Cohen A, Plant AL, Moses MS, Bray EA (1991) Organ-specific and environmentally regulated expression of two absicisic acid-induced genes of tomato. Plant Physiol 97:1367–1374

Cook RE (1979) Patterns of juvenile morbidity and recruitment in plants. In: Solbrig OT, Jain S, Johnson GB, Raven PH (eds) Topics in plant population biology. Columbia University Press, Los Angeles, pp 207–301

Cortina C, Culianez-Macia FA (2005) Tomato abiotic stress enhanced tolerance by trehalose biosynthesis. Plant Sci 169:75–82

Cuartero J, Bolarin MC, Asins MJ, Moreno V (2006) Increasing salt tolerance in the tomato. J Exp Bot 57:1045–1058

Cuartero J, Fernandez-Munoz R (1999) Tomato and salinity. Scientia Hort 78:83–125

Cuartero J, Yeo AR, Flowers TJ (1992) Selection of donors for salt-tolerance in tomato using physiological traits. New Phytol 121:63–69

Dahal P, Bradford KJ, Jones RA (1990) Effects of priming and endosperm integrity on seed germination rates of tomato genotypes: I. Germination at suboptimal temperature. J Expt Bot 41:1431–1440

Dehan K, Tal M (1978) Salt tolerance in the wild relatives of the cultivated tomato: Responses of *Solanum pennellii* to high salinity. Irrig Sci 1:71–76

Duvick DN (1986) Plant breeding: past achievement and expectations for the future. Econ Bot 40:289–294

Epstein E, Norlyn JD, Rush DW, Kingsbury RW, Kelly DB, Gunningham GA, Wrona AF (1980) Saline culture of crops: A genetic approach. Science 210:399–404

Esau K (1953) Plant Anatomy. John Wiley, New York

FAOSTAT (2005) FAO Statistical Databases. Food and agriculture organization of the United Nations, Statistics Division

Flowers TJ, Yeo AR (1988) Salinity and Rice: a physiological approach to breeding for resistance. The International congress of Plant Physiology, New Delhi, India, pp 953–959

Flowers TJ, Yeo AR (1997) Breeding for salt resistance in plants. In: Jaiwal PK, Singh RP, Gulati A (eds) Strategies for improving salt tolerance in higher plants. Science Publishers, Inc., U.S.A., pp 247–264

Foolad MR (1996a) Genetic analysis of salt tolerance during vegetative growth in tomato, *Lycopersicon esculentum* Mill. Plant Breed 115:245–250

Foolad MR (1996b) Response to selection for salt tolerance during germination in tomato seed derived from P.I. 174263. J Am Soc Hort Sci 121:1006–1011

Foolad MR (1997) Genetic basis of physiological traits related to salt tolerance in tomato, *Lycopersicon esculentum* Mill. Plant Breed 116:53–58

Foolad MR (1999) Comparison of salt tolerance during seed germination and vegetative growth in tomato by QTL mapping. Genome 42:727–734

Foolad MR (2004) Recent advances in genetics of salt tolerance in tomato. Plant Cell, Tiss Org Cult 76:101–119

Foolad MR, Chen FQ (1998) RAPD markers associated with salt tolerance in an Interspecific cross of tomato (*Lycopersicon esculentum* ∞ *L. pennellii*). Plant Cell Rep 17:306–312

Foolad MR, Chen FQ (1999) RFLP mapping of QTLs conferring salt tolerance during vegetative stage in tomato. Theor Appl Genet 99:235–243

Foolad MR, Chen FQ, Lin GY (1998) RFLP mapping of QTLs conferring salt tolerance during germination in an interspecific cross of tomato. Theor Appl Genet 97:1133–1144

Foolad MR, Jones RA (1991) Genetic analysis of salt tolerance during germination in *Lycopersicon*. Theor Appl Genet 81:321–326

Foolad MR, Jones RA (1992) Parent-offspring regression estimates of heritability for salt tolerance during germination in tomato. Crop Sci 32:439–442

Foolad MR, Jones RA (1993) Mapping salt-tolerance genes in tomato (*Lycopersicon esculentum*) using trait-based marker analysis. Theor Appl Genet 87:184–192

Foolad MR, Lin GY (1997a) Absence of a relationship between salt tolerance during germination and vegetative growth in tomato. Plant Breed 116:363–367

Foolad MR, Lin GY (1997b) Genetic potential for salt tolerance during germination in Lycopersicon species. HortScience 32:296–300

Foolad MR, Stoltz T, Dervinis C, Rodriguez RL, Jones RA (1997) Mapping QTLs conferring salt tolerance during germination in tomato by selective genotyping. Mol Breed 3:269–277

Foolad MR, Subbiah P, Kramer C, Hargrave G, Lin GY (2003a) Genetic relationships among cold, salt and drought tolerance during seed germination in an interspecific cross of tomato. Euphytica 130:199–206

Foolad MR, Zhang L, Subbiah P (2003b) Genetics of drought tolerance during seed germination in tomato: Inheritance and QTL mapping. Genome 46:536–545

Foolad MR, Zhang LP, Lin GY (2001) Identification and validation of QTLs for salt tolerance during vegetative growth in tomato by selective genotyping. Genome 44:444-454

Forster BP, Phillips MS, Miller TE, Baird E, Powell W (1990) Chromosome location of genes controlling tolerance to salt (NaCl) and vigor in *Hordeum vulgare* and *H. chilense* . Heredity 65:99–107

Galston AW, Kaur-Sawhney R, Altabella T, Tiburcio AF (1997) Plant polyamines in reproductive activity and response to abiotic stress. Bot Acta 110:197–207

Giovannucci E (1999) Tomatoes, tomato-based products, lycopene, and cancer; Review of the epidemiologic literature. J Natl Cancer Inst 91:317 331

Greenway H, Munns R (1980) Mechanism of salt tolerance in non-halophytes. Ann Rev Plant Physiol 31:149–190

Groot SPC, Karssen CM (1987) Gibberellins regulate seed germination in tomato by endosperm weakening: A study with gibberellin-deficient mutant. Planta 171:525–531

Grover A, Sahi C, Sanan N, Grover A (1999) Taming abiotic stresses in plants through genetic engineering: current strategies and perspective. Plant Sci 143:101–111

Grunberg K, Fernandez-Muñoz R, Cuartero J (1995) Growth, flowering, and quality and quantity of pollen of tomato plants grwon under salline conditions. Acta Hort 412:484–489

Guerrier G (1996) Fluxes of Na+, K+ and Cl-, and osmotic adjustment in *Lycopersicon pimpinellifolium* and *L. esculentum* during short- and long-term exposures to NaCl. Physio Plant 97:583–591
Haigh AH, Barlow EWR (1987) Water relations of tomato seed germination. Austral J Plant Physiol 14:485–492
Hegarty TW (1978) The physiology of seed hydration and dehydration, and the relation between water stress and the control of germination: A review. Plant, Cell Environ 1:101–119
Hsiao TC (1973) Plant responses to water stress. Annu Rev Plant Physiol 24:519–570.
Hsiao TC, Bradford KJ (1983) Physiological consequences of cellualar water deficits. In: Taylor HM, Jordan WR, Sinclair TR (eds) Limitations to efficient water used in crop production. American Society of Agronomy, Madison, Wis., pp 227–
Hsieh T-H, Lee J-T, Charng Y-Y, Chan M-T (2002) Tomato plants ectopically expressing Arabidopsis CBF1 show enhanced resistance to water deficit stress. Plant Physiol 130:618–626
Johnson DW, Smith SE, Dobrenz AK (1992) Genetic and phenotypic relationships in response to NaCl at different developmental stages in alfalfa. Theor Appl Genet 83:833–838
Johnson WC, Jackson LE, Ochoa O, Wijik Rv, Peleman J, Clair DAS, Michelmore RW (2000) Lettuce, a shallow-rooted crop, and *Lactuca serriola*, its wild progenitor, differ at QTL determining root architecture and deep soil water exploitation. Theor Appl Genet 101:1066–1073
Jones RA (1986a) High salt-tolerance potential in *Lycopersicon* species during germination. Euphytica 35:576–582
Jones RA (1986b) The development of salt-tolerant tomatoes: breeding strategies. Acta Hort 190:101–114
Jones RA, Hashim M, El-Beltagy AS (1988) Developmental responsiveness of salt-tolerant and salt-sensitive genotypes of *Lycopersicon*. In: Whitehead E, Hutchison F, Timmeman B, Varazy R (eds) Arid Lands: Today and Tomorrow. Westview Press, Boulder, pp 765–772
Jones RA, Qualset CO (1984) Breeding crops for environmental stress tolerance. In: Collins GB, Petolino JF (eds) Application of Genetic Engineering to Crop Improvement. Nijihoff/Junk, The Hague, pp 305–340
Kahn TL, Fender SE, Bray EA, O'Connell MA (1993) Characterization of Expression of Drought- and Abscisic Acid-Regulated Tomato Genes in the Drought-Resistant Species *Lycopersicon Pennellii*. Plant Physiol 103:597–605
Kalloo G (1991) Breeding for environmental resistance in tomato. In: Kalloo G (ed) Genetic Improvement of Tomato. Springer-Verlag, Berlin Heidelberg, Germany, pp 153–165
Kasuga M, Miura S, Shinozaki K, Yamaguchi-Shinozaki K (2004) A combination of the arabidopsis *DREB1A* gene and stress-inducible *rd29A* promoter improved drought- and low-temperature stress tolerance in tobacco by gene transfer. Plant Cell Physiol 45:346–350
Kaufman MR (1969) Effects of water potential on germination of lettuce, sunflower, and citrus seeds. Can J Bot 47:1761–1764
Kramer PJ (1980) Water Relations of Plants. Academic Press, New York
Kramer PJ (1983) Water Relations of Plants. Academic Press, New York
Lee J-T, Prasad V, Yang P-T, Wu J-F, Ho T-HD, Charng Y-Y, Chan M-T (2003) Expression of Arabidopsis CBF1 regulated by an ABA/stress inducible promoter in transgenic tomato confers stress tolerance without affecting yield. Plant, Cell Environ 26:1181–1190
Lilius G, Holmberg N, Bulow L (1996) Enhanced NaCl stress tolerance in trangenic tobacco expressing bacterial choline dehydrogenase. Bio/technology 14:177–180
Liptay A, Schopfer P (1983) Effect of water stress, seed coat restraint, and abscisic acid upon different germination capabilities of two tomato lines at low temperature. Plant Physiol 73:935–938
Ludlow MM, Muchow RC (1990) A critical evaluation of traits for improveing crop yields in water-limited environments. Adv Agron 43:107–153
Lutfor-Rahman SM (1998) Eco-physiological study on tomato drought tolerance. Division of Environmental Science and Technology. Kyoto University, Kyoto, Japan, p 80
Lyon CB (1941) Responses of two species of tomatoes and the F_1 generation to sodium sulphate in the nutrient medium. Bot Gaz 103:107–122
Maas EV (1986) Salt tolerance of plants. Appl Agric Res 1:12–26

Maas EV (1990) Crop salt tolerance. In: Tanji KK (ed) Agricultural salinity assessment and management. ASCE Mannuals and Reoprts on Engineering No. 71, New York, pp 262–304

Mano Y, Takeda K (1997) Mapping quantitative trait loci for salt tolerance at germination and the seedling stage in barley (*Hordeum vulgare* L.). Euphytica 94:263–272

Martin B, Nienhuis J, King G (1989) Restriction fragment length polymorphisms associated with water use efficiency in tomato. Science 243:1725–1728

Martin B, Tauer CG, Lin RK (1999) Carbon isotope discrimination as a tool to improve water-use efficiency in tomato. Crop Sci 39:1775–1783

Martin B, Thorstenson YR (1988) Stabel carbon isotope composition (delta ^{13}C), water use efficiency and biomass productivity of *Lycopersicon esculentum*, *Lycopersicon pennellii*, and the F_1 hybrid. Plant Physiol 88:213–217

McCormick S, Niedermeyer J, Fry J, Barnason A, Worsch R, Fraley R (1986) Leaf disk transformation of cultivated tomato (*L. esculentum*) using *Agrobacterium tumifaciens*. Plant Cell Rep 5:81–84

Miller JC, Tanksley SD (1990) RFLP analysis of phylogenetic relationships and genetic variation in the genus *Lycopersicon*. Theor Appl Genet 80:437–448

Mitchell JH, Siamhan D, Wamala MH, Risimeri JB, Chinyamakobvu E, Henderson SA, Fukai S (1998) The use of seedling leaf death score for evaluation of drought resistance of rice. Field Crops Res 55:129–139

Monforte AJ, Asins MJ, Carbonell EA (1996) Salt tolerance in *Lycopersicon* species. IV. Efficiency of marker-assisted selection for salt tolerance improvement. Theor Appl Genet 93:765–772

Monforte AJ, Asins MJ, Carbonell EA (1997) Salt tolerance in *Lycopersicon* species. V. Does genetic variability at quantitative trait loci affect their analysis? Theor Appl Genet 95:284–293

Monforte AJ, Asins MJ, Carbonell EA (1999) Salt tolerance in *Lycopersicon* spp. VII. Pleiotropic action of genes controlling earliness on fruit yield. Theor Appl Genet 98:593–601

Na JK (2005) Genetic approaches to improve drought tolerance
of tomato and tobacco. Horticulture and Crop Science. The Ohio State University, Wooster, p 119

Nguyen HT, Babu RC, Blum A (1997) Breeding for drought resistance in rice: Physiology and molecular genetics considerations. Crop Sci 37:1426–1434

Norlyn JD, Epstein E (1984) Variablility in salt tolrance of four Triticale line at germination and emergence. Crop science 24:1090–1092

Oh S-J, Song SI, Kim YS, Jang H-J, Kim S-Y, Kim M, Kim Y-K, Nahm BH, Kim J-K (2005) Arabidopsis CBF3/DREB1A and ABF3 in transgenic rice increased tolerance to abiotic stress without stunting growth. Plant Physiol 138:341–351

Pearen JR, Pahl MD, Wolynetz MS, Hermesh R (1997) Association of salt tolerance at seedling emergence with adult plant performance in slender wheatgrass. Can J Plant Sci 77:81–89

Perez-Alfocea F, Estan MT, Caro M, Bolarin MC (1993a) Response of tomato cultivars to salinity. Plant and Soil 150:203–211

Perez-Alfocea F, Estan MT, Caro M, Guerrier g (1993b) Osmotic adjustment in *Lycopersicon esculentum* and *L. pennellii* under NaCl and polyethylene 6000 iso-osmotic stresses. Physiologia Plantarum 87:493–498

Perez-Alfocea F, Guerrier G, Estan MT, Bolarin MC (1994) Comparative salt responses at cell and whole-plant levels of cultivated and wild tomato and their hybrid. J Hort Sci 69:639–644

Phills BR, Peck NH, McDonald GE, Robinson RW (1979) Differential responses of *Lycopersicon* and *Solanum* species to salinity. J Am Soc Hortic Sci 104:349–352

Pillay I, Beyl C (1990) Early responses of drought-resistant and -susceptible tomato plants subjected to water stress. J Plant Growth Regul 9:213–219

Plant AL, Cohen A, Moses MS, Bray EA (1991) Nucleotide sequence and spatial expression pattern of drought- and ABA-induced gene for tomato. Plant physiol 97:900–906

Quesada V, Garcia-Martinez S, Piqueras P, Ponce MR, Micol JL (2002) Genetic architecture of NaCl tolerance in Arabidopsis. Plant Physiol 130:951–963

Rana MK, Kalloo G (1989) Morphological attributes associated with the adaptation under water deficit conditions in tomato (*L. esculentum* Mill.).12th Eucarpia Congress 1989, Vortrage Pflanzenzucht, pp 23–27

Rana MK, Kalloo G (1990) Evaluation of tomato genotypes under drought conditions (Abstr.).23rd International Horticultre Congress, Firenze, Italy

Rathinasabapathi B (2000) Metabolic engineering for stres tolerance: Installing osmoprotectant synthesis pathways. Anna Bot 86:709–716

Redmann RE (1974) Osmotic and specific ion effects on the germination of alfalfa. Can J Bot 52:803–808

Ribaut JM, Jiang C, Gonzalez-de-Leon D, Edmeades GO, Hoisington DA (1997) Identification of quantitative trait loci under drought conditions in tropical maize. 2. Yield components and marker-assisted selection strategies. Theor Appl Genet 94:887–896

Richards MA, Phills BR (1979) Evaluation of Lycopersicon species for drought tolerance (Abstr.). HortScience 14:121

Richards RA (1983) Should selection for yield in saline regions be made on saline or non-saline soils? Euphytica 32:431–438

Richards RA (1996) Defining selection criteria to improve yield under drought. Plant Growth Regul 20:157–166

Richards RA, Dennett CW (1980) Variation in salt concentration in a wheat field. Soil and Water 44:8–9

Rick CM (1973) Potential genetic resources in tomato species: clues from observation in native habitats. In: Srb AM (ed) Genes, Enzymes, and Populations. Plenum Press, New York, USA, pp 255–269

Rick CM (1976a) Natural variability in wild species of *Lycopersicon* and its bearing on tomato breeding. Genet Agraria 30:249–259

Rick CM (1976b) Tomato, *Lycopersicon esculentum* (Solanaceae). In: Simmonds NW (ed) Evolution of Crop Plants. Longman, London, UK, pp 268–273

Rick CM (1978) The Tomato. Sci Amer 23:76–87

Rick CM (1979a) Biosystematic studies in *Lycopersicon* and closely related species of *Solanum*. In: Hawkes JC, Lester RN, Skelding AD (eds) The Biology and Taxnomy of the Solanaceae. Academic Press, New York, USA, pp 667–678

Rick CM (1979b) Potential improvement of tomatoes by controlled introgression of genes from wild speies. Proc Conf Broadening Genetic Base of Crops. Pudoc, Wageningen, pp 167–173

Rick CM (1980) Tomato. Hybridization of Crop Plants. Am. Soc. Agron./Crop Sci. Soc. Am., Madison, WI, USA, pp 669–680

Rick CM (1982) The potential of exotic germplasm for tomato improvement. In: Vasil IK, Scowcroft WR, Frey KJ (eds) Plant Improvement and Somatic Cell Genetics. Academic Press, New York, pp 1–28

Rick CM, DeVerna JW, Chetelat RT, Stevens MA (1987) Potential contributions of wild crosses to improvement of processing tomatoes. Acta Hort 200:45–55

Rick CM, Fobes JF (1975) Allozyme variation in the cultivated tomato and closely related species. Bul Torrey Bot Club 102:376–384

Romero-Aranda R, Soria T, Cuartero J (2001) Tomato plant-water uptake and plant-water relationships under saline growth conditions. Plant Sci 160:265–272

Rontein D, Basset G, Hanson AD (2002) Metabolic engineering of osmoprotectant accumulation in plants. Metabolic Engin 4:49–56

Ross R, Lott N (2000) A climatology of recent extreme weather and climate events. U.S. Department of Commerce, Technical Report 2000–02, NOAA/NESDIS, National Climatic Data Center, Asheville, NC

Rush DW, Epstein E (1976) Genotypic responses to salinity: differences between salt-sensitive and salt-tolerant genotypes of the tomato. Plant Physiol 57:162–166

Rush DW, Epstein E (1981a) Breeding and selection for salt tolerance by the incorporation of wild germplasm into a domestic tomato. J Am Soc Hort Sci 106:699–704

Rush DW, Epstein E (1981b) Comparative studies on the sodium, potassium, and chloride relations of a wild halophytic and domestic salt-sensitive tomato species. Plant Physiol 68:1308–1313

Sacher RF, Staples RC, Robinson RW (1983) Ion regulation and response of tomato to sodium chloride: A homeostatic system. J Amer Soc Hort Sci 108:566–569

Santa-Cruz A, Perez-Alfocea F, Caro M, Acosta M (1998) Polyamines as short-term salt tolerance traits in tomato. Plant Sci 138:9–16

Saranga Y, Cahaner A, Zamir D, Marani A, Rudich J (1992) Breeding tomatoes for salt tolerance: Inheritance of salt tolerance and related traits in interspecific populations. Theor Appl Genet 84:390–396

Saranga Y, Zamir D, Marani A, Rudich J (1991) Breeding tomatoes for salt tolerance: Field evaluation of *Lycopersicon* germplasm for yield and dry matter production. J Am Soc Hort Sci 116:1067–1071

Saranga Y, Zamir D, Marani A, Rudich J (1993) Breeding tomatoes for salt tolerance: variation in ion concentration associated with response to salinity. J Am Soc Hort Sci 118:405–408

Sarg SMH, Wyn-Jones RG, Omar FA (1993) Salt tolerance in the Edkawy tomato. In: Lieh H, Al-Masoom A (eds) Towards the rational use of high salinity tolerant plants. Kluwer Academin Publishers, The Netherlands, pp 177–184

Schonfeld MA, Johnson RC, Carver BD, Mornhinweg DW (1988) Water relations in wheat as drought resistance indicators. Crop Sci 28:526–531

Seki M, Kamei A, Yamaguchi-Shinozakiz K, Shinozaki K (2003) Molecular responses to drought, salinity and frost: Common and different paths for plant protection. Curr Opin Biotech 14:194–199

Serrano R, Culiañz-Macia FA, Moreno V (1999) Genetic engineering of salt and drought tolerance with yeast regulatory genes. Scientia Hort 78:261–269

Shannon MC, Gronwald JW, Tal M (1987) Effects of salinity on growth and accumulation of organic and inorganic ions in cultivated and wild tomato species. J Am Soc Hort Sci 112:416–423

Shen B, Jensen RG, Bohnert JJ (1997) Mannitol protects against oxidation by hydroxul radicals. Plant Physiol 115:527–532

Shou H, Bordallo P, Wang K (2004) Expression of the *Nicotiana* protein kinase (NPK1) enhanced drought tolerance in transgenic maize. Exp Bot 55:1013–1019

Stoner AK (1972) Merit, Red Rock and Potomac-tomato varieties adapted to mechanical harvesting. USDA Prod. Res. Rep.

Subbiah P (2001) Genetic Investigation of Abiotic Stress Tolerance in Lycopersicon Species.Genetics. The Pennsylvania State University, University Park, p 104

Subudhi PK, Rosenow DT, Nguyen HT (2000) Quantitative trait loci for the stay green trait in sorghum (*Sorghum bicolor* L. Moench): consistency across genetic backgrounds and environments. Theor Appl Genet 101:733–741

Tal M (1971) Salt tolerance in the wild relatives of the cultivated tomato: Responses of *Lycopersicon esculentum*, *L. peruvianum*, and *L. esculentum minor* to sodium chloride solution. Aust J Agric Res 22:631–638

Tal M (1985) Genetics of salt tolerance in higher plants: Theoretical and practical considerations. Plant and Soil 89:199–226

Tal M (1997) Wild germplasm for salt tolerance in plants. In: Jaiwal PK, Singh RP, Gulati A (eds) Strategies for improving salt tolerance in higher plants. Science Publishers, Inc., U.S.A., pp 291–320

Tal M, Gavish U (1973) Salt tolerance in the wild relatives of the cultivated tomato: Water balance and abscisic acid in *Lycopersicon esculentum* and *L. peruvianum* under low and high salinity. Aust J Agric Res 24:353–361

Tal M, Katz A, Heikin H, Dehan K (1979) Salt tolerance in the wild relatives of the cultivated tomato: proline accumulation in *Lycopersicon esculentum* Mill., *L. peruvianum* Mill., and *Solanum pennellii* Cor. treated with NaCl and polyethylene glycol. New Phytol 82:349–355

Tal M, Shannon MC (1983) Salt tolerance in the wild relatives of the cultivated tomato: Responses of *Lycopersicon esculentum*, *L. cheesmanii*, *L. peruvianum*, *Solanum pennellii* and F_1 hybrids to high salinity. Aust J Plant Physiol 10:109–117

Tanaka Y, Hibino T, Hayashi Y, Tanaka A, Kishitani S, Takabe T, Yokota S, Takabe T (1999) Salt tolerance of trangenci rice overexpressing yeast mitochondrial Mn-SOD in chloroplasts. Plant Science 148:131–138

Tanji KK (1990) Nature and extent of agricultural salinity. In: Tangi KK (ed) Agricultural Salinity Assessment and Management. Am. Soc. Civil Engineers, New York, pp 1–13

Thomas JC, Sepahi M, Arndall B, Bohnert HJ (1995) Enhancement of seed germination in high salinity by engineering mannitol expression in *Arabidopsis thaliana*. Plant Cell Envr 18:801–806

Ungar IA (1978) Halophyte seed germination. The Bot Rev 44:233–264

USDA (2005) Agricultural statistics 2005. United State Department of Agtriculture, National Agricultural Staticstics Service

van Ieperen W (1996) Effects of different day and night salinity levels on vegetative growth, yield and quality of tomato. J Hort Sci 71:99–111

Wang W-X, Vinocur B, Altman A (2003) Plant responses to drought, salinity and extreme temperatures: towards genetic engineering for stress tolerance. Planta 218:1–14

Warnock SJ (1988) A review of taxonomy and phylogeny of the genus *Lycopersicon*. HortScience 23:669–673

Warren GF (1998) Spectacular increases in crop yields in the twentieth century. Weed Technol 12:752–760

Wudiri BB, Henderson DW (1985) Effects of water stress on flowering and fruit set in processing tomatoes. Sci Hortic 27:189–198

Xue Z-Y, Zhi D-Y, Xue G-P, Zhang H, Zhao Y-X, Xia G-M (2004) Enhanced salt tolerance of transgenic wheat (*Triticum aestivum* L.) expressing a vacuolar Na+/H+ antiporter gene with improved grain yields in saline soils in the field and a reduced level of leaf Na+. Plant Sci 167:849–859

Yamaguchi T, Blumwald E (2005) Developing salt-tolerant crop plants: Challenges and opportunities. Trends Plant Sci 10:615–620

Yeo AR, Flowers TJ (1990) Screening of rice (*Oryza sativa* L.) genotypes for physiological characters contributing to salinity resistance, and their relationship to overall performance. Theor Appl Genet 79:377–384

Yin XY, Yang A-F, Zhang K-W, Zhang J-R (2004) Production and analysis oftransgenic maize with improved salt tolerance by the introduction of *AtNHZ1* gene. Acta Bot Sin:854–861

Younis AF, Hatata MA (1971) Studies on the effects of certain salts on germination, on growth of root, and on metabolism. I. Effects of chlorides and sulphates of sodium, potassium, and magnesium on germination of wheat grains. Plant and Soil 13:183–200

Yu TT (1972) The genetics and physiology of water usage in *Solanum pennellii* Corr. and its hybrids with *Lycopersicon esculentum* Mill. University of Cal., Davis

Zhang HX, Blumwald E (2001) Transgenic salt-tolerant tomato plants accumulate salt in foliage but not in fruit. Nature biotechnology 19:765–768

Zhang HX, Hodson JN, Williams JP, Blumwald E (2001a) Engineering salt-tolerant *Brassica* plants: Characterization of yield and seed oil quality in transgenic plants with increased vacuolar sodium accumulation. Proc Natl Acad Sci USA 98:12832–12836

Zhang J, Zheng HG, Aarti A, Pantuwan G, Nguyen TT, Tripathy JN, Sarial AK, Robin S, Babu RC, Nguyen BD, Sarkarung S, Blum A, Nguyen HT (2001b) Locating genomic regions associated with components of drought resistance in rice: comparative mapping within and a cross species. Theor Appl Genet 103:19–29

Zhang JZ, Creelman RA, Zhu J-K (2004) From laboratory to field. Using information from Arabidopsis to engineer salt, cold, and drought tolerance in crops. Plant Physiol 135:615–621

Zhu JK, Hasegawa PM, Bressan RA (1997) Molecular aspects of osmotic stress in plants. Crit Rev Plant Sci 16:253–277

CHAPTER 28

RECENT ADVANCES IN MOLECULAR BREEDING OF CASSAVA FOR IMPROVED DROUGHT STRESS TOLERANCE

TIM L. SETTER[1] AND MARTIN A. FREGENE[2]

[1]*Department of Crop and Soil Sci, Cornell University, Ithaca, NY USA*
[2]*Centro Internacional de Agricultura Tropical (CIAT), Cali, Colombia*
E-mail: TLS1@cornell.edu and M.Fregene@cgiar.org

Abstract: Cassava is an important tropical starchy root crop that is used extensively in drought prone tropical regions. It responds to water deficit with a dehydration avoidance and growth arrest syndrome. Carbohydrate is supplied from stems via remobilization. It is very limited in its use of osmotic adjustment, compatible solute synthesis, dehydrin accumulation and other tolerance mechanisms for low water potential. Given the difficulties of conventional breeding of cassava due to its long breeding cycle, heterozygousity, and difficulties in producing seed, an important recent development is the use of molecular markers and marker assisted selection (MAS). MAS is also contributing to the introgression of traits from wild relatives

Keywords: leaf retention, marker assisted selection, storage root, drought, water deficit, stomatal conductance, leaf growth, stem carbohydrate remobilization

Cassava (*Manihot esculenta*, Crantz) is an important tropical crop that ranks sixth among crops as a source of calories in the human diet worldwide. Cassava produces starchy storage roots which are processed for direct human consumption, and it is also used to make refined starch products such as tapioca. It is particularly important as a staple crop to subsistence farmers in the tropics between 30° N and 30° S latitude, many of whom utilize low-fertility and stress-prone soils. Although cassava is grown in a wide range of climates from drought-prone to well-watered, it is commonly cultivated in areas receiving less than 800 mm rainfall per year with a dry season of 4 to 6 months, where tolerance to water deficit is an important attribute. Such farmers value cassava as a food security crop that can be depended upon to provide sustenance in years when other crops fail.

Paradoxically, although cassava is well regarded for its performance in stress-prone environments, it is one of the most highly productive crops available for

M.A. Jenks et al. (eds.), Advances in Molecular Breeding Toward Drought and Salt Tolerant Crops, 701–711.

favorable environments as well. For example, in an evaluation of 1400 accessions, Kawano et al. (1978) found that with adequate moisture in a non-irrigated environment the best lines produced greater than 20 Mg storage root biomass ha^{-1} yr^{-1}, while several genotypes showed consistently superior performance over a wide range of environments. This potential for a broad range of adaptation and reliable performance are attributes that are prized, but the underlying bases of this combination of traits are only partially understood, and the tools for selecting and genetically manipulating them are only beginning to be developed.

1. CHARACTERISTICS OF CASSAVA STRESS RESPONSE

A trait that contributes to cassava's productivity in favorable environments is its high leaf photosynthetic rate. Studies have indicated that under high-input conditions rates exceed 40 μmol (CO_2) m^{-2} s^{-1}, competing with other high-productivity plants, including those of both C3 and C4 types (El Sharkawy et al. 1990; El Sharkawy et al. 1992; El Sharkawy et al. 1993; El-Sharkawy 2004; Periera et al. 1986). Photosynthesis under high input conditions increases with increasing sunlight irradiance with a tendency to saturate only at high photon flux densities above 1500 μmol m^{-2} s^{-1} (Angelov et al. 1993; Periera et al. 1986; El Sharkawy and Cock 1990; El Sharkawy et al. 1993). Studies indicate that even when grown in elevated atmospheric [CO_2] of 680 μmol/mol, photosynthesis was not down regulated, suggesting that cassava creates sufficient sink capacity to maintain a favorable source:sink ratio (Fernandez et al. 2002). Correspondingly, studies have indicated that storage root yield and total plant biomass among a wide range of cultivars is significantly correlated to single leaf photosynthetic rate when light interception is not limiting (El Sharkawy and Cock 1990) or with seasonal average canopy photosynthesis of upper canopy leaves when canopy leaf area is taken into account (El Sharkawy et al. 1993).

Photosynthetic rate is one of the most affected processes in water deficit environments, due largely to sensitivity of cassava's stomata to lowered water status (Palta 1984). This was demonstrated in a study of plants subjected to 45 days of water limitation, where photosynthetic gas exchange of young intact leaves that were exposed to intermediate irradiance was decreased 66%, whereas CO_2-saturated photosynthetic rates, measured with the oxygen electrode, were comparable in well-watered and water-limited plants (Calatayud et al. 2000). Also, photochemical quenching of Chlorophyll-a fluorescence and the quantum efficiency of PSII photochemistry in young leaves were comparable in both control and stressed plants. In addition to stomatal closure, leaves droop at early phases of water limitation such that incident light flux densities are lessened and photoinhibition avoided (Calatayud et al. 2000). During water deficit, stomatal opening tends to be limited to the early part of the day when temperature and vapor pressure differences (VPD) are lower, thereby permitting photosynthesis to occur with higher water use efficiency (Itani et al. 1999).

The sensitivity of cassava stomata to slight decreases in leaf water potential (ψ_w) is such during initial periods of water deficit, leaves drastically limit water loss and maintain leaf ψ_w at values near those of well-watered controls for relatively extended periods of drought (Ike et al 1982; Itani et al. 1999; Palta et al 1984). Stomatal closure also occurs in a highly sensitive response to leaf-air VPD (Cock et al. 1985; El Sharkawy and Cock 1984; El Sharkawy and Cock 1990; Oguntunde 2005). In a comparison with four other tropical species including rice (*Oryza sativa* L.) and eucalyptus (*Eucalyptus deglupta* Blume), cassava stomata were the most responsive to VPD over the range from 1.0 to 4.0 kPa (El Sharkawy et al. 1984). This highly homeohydric behavior (tendency to limit water loss to the extent that ψ_w is kept at about the same values as well watered plants; Wood 2005) places cassava in the category of plants described as isohydric, along with maize, cowpea, and poplar, among others (Jones 1998; Tardieu and Simonneau 1998).

Although the high sensitivity of cassava stomata to water status is potentially advantageous, there is also some evidence that it is excessively conservative of water use with respect to optimal crop performance. In field trials of irrigated and rain-fed plots, Cock et al. (1985) found that when relative humidity in the crop canopy was increased by artificial misting, crop biomass production increased by 27%, and storage root yield by 91%. The effect was neither associated with changes in soil water nor with changes in leaf area indices, but rather with enhanced photosynthetic rate per unit leaf area of the misted plants due to stomatal opening. It is possible that the underlying cause of stomatal sensitivity to VPD and the misting benefit is low hydraulic conductance in the pathway from roots to leaves, such as has been reported for certain genotypes of grape (Schultz et al. 2003).

The sensitivity of cassava stomata to incipient water deficit is associated with large increases in [ABA] (Alves and Setter 2000). Also, leaves droop such that radiant heating and water loss is lessened (Calatayud et al. 2000), and as stress progresses, a substantial fraction of leaves abscise, thereby decreasing transpirational surface area (Ike and Thurtell 1981b; Lenis et al. 2006; Ramanujam 1990). Nevertheless, studies have indicated that genotypes with greater leaf retention have better yield performance in stress environments. In an evaluation of 1350 clones at a field site with water stress, clones with the leaf retention trait produced more total fresh biomass and yielded 33% more root dry matter than plants without the trait (Lenis et al. 2006). Furthermore, the trait is highly heritable (h^2=55%), and genetic correlation with yield is high (0.46) under irrigated conditions as well (Lenis et al. 2006).

Leaf growth in cassava is also highly sensitive to water stress (Alves and Setter 2000; 2004a; Conner and Cock 1981; Yao et al. 1988). Moreover, despite the severity of leaf abscission (described above), the loss of leaf area during water stress is dominated by restricted leaf area development and not by leaf loss (Conner and Cock 1981). Studies of leaf growth have indicated that both the rate of leaf expansion in existing leaves and the rate of new leaf appearance are drastically affected (Alves and Setter 2004a; Yao et al. 1988). Cell expansion, cell proliferation, and new leaf appearance are essentially halted during short-term stresses lasting a few days – a response that conserves photosynthate resources. However, while well watered leaves advance beyond their developmental window for growth, leaves in water stressed plants remain arrested at a young developmental stage, and when rainfall returns, are able to rapidly resume growth from where they left off (Alves and Setter 2004a; El-Sharkawy and Cadavid 2002). Such quiescence conserves resources during stress, but permits affected tissues to respond rapidly upon renewed rainfall. This ability may be ideal for environments with numerous water deficit episodes interspersed with brief rainfalls. This response along with stomatal closure and leaf fall, helps diminish transpirational surface area during water scarcity.

In response to more extended and severe water deficits that entail substantial leaf senescence and abscission, leaf buds grow slowly to develop numerous young shoots in the vicinity of abscised leaf scars (Duque and Setter, unpublished), but recovery of leaf growth is much slower, thereby limiting storage root bulking and yield (Baker et al. 1989). Nevertheless, at late stages, storage root growth is sustained relatively better than leaf regrowth, as the proportion of assimilates imported by storage roots is higher at this stage, perhaps due to utilization of stored carbohydrates in stems and other vegetative organs by roots (Yao et al. 1988).

Growth of the fine roots (those other than the starch-storing roots) is of particular importance in relation to drought. In general studies have indicated that cassava's fine root system is sparse, although some roots can reach two meters in depth (El Sharkawy 2004). Water deficit tends to decrease the number and length of adventitious and lateral roots and the total fine-root dry weight (Pardales and Esquibel 1996). This could be a favorable alteration in carbon partitioning away from excess root branching if it is coupled to partitioning of deep root growth and enhanced access to deep soil water.

Given the drastic decrease in stomatal conductance and attendant decrease in photosynthetic CO_2 assimilation during stress, the carbon economy of the plants represents an additional challenge. While cessation of leaf and stem growth limit demands for assimilate, sources of assimilate to meet respiratory and other tissue maintenance requirements must also be provided to withstand long periods of stress (Gent 1994). Recent studies have indicated that cassava stores large quantities of starch in stems and leaf petioles, which is remobilized during water deficit (Duque and Setter 2005). This showed that the amount of starch stored in stems is considerable, representing about 35% of the total non-structural carbohydrate in a plant at the initial period of storage root growth, and 6% of total plant dry mass at that stage. Additional reserves are in the petioles, whereas the leaf lamina has minimal reserves. While stomatal closure essentially stops photosynthesis during water deficit, petiole and stem carbohydrate reserves are gradually translocated and utilized throughout the plant to sustain tissue metabolism and viability. In contrast, small storage roots retain carbohydrates during stress. This ability to accumulate substantial reserves of carbohydrate, and to make them available during stress, may be one of the key adaptations by which this crop performs well in water, salinity, and numerous other stress environments.

Despite its favorable yield performance in drought environments, studies have indicated that cassava is very limited in its use of several stress tolerance traits that confer ability to withstand low water potentials, such as osmotic adjustment, and accumulation of proline and dehydrin-like proteins. In the case of osmotic adjustment (OA) studies have reported negligible OA in leaves of plants exposed to several days of water deficit (Itani et al. 1999), or minor OAs of about 0.15 MPa (Alves and Setter 2004b; Ike and Thurtell 1981a), and OA was not correlated with genotypic differences in stress resistance (Ike and Thurtell 1981c). Although proline accumulates in response to stress, studies have shown that it had a negligible contribution to OA (Alves and Setter 2004b), and was negatively correlated with cultivar stress tolerance (Sundaresan and Sudhakaran 1995).

Relatively few studies have investigated cassava's response to salinity. Hawker and Smith (1982) found that cassava was moderately salt tolerant in evaluations with hydroponic culture containing 0 to 75 mM NaCl. Tuber weight was reduced by 50% with 30 to 50 mM NaCl and there was some burning of apical leaves at 50 and 75 mM NaCl. Tolerance is expected to be influenced by cassava's highly sensitive stomatal response to lowered water potential which would diminish NaCl uptake via the transpiration stream. In plantlet culture for 8 weeks where evaporation was minimized, cassava tolerated up to 100 mM NaCl, and demonstrated an ability to selectively accumulate K^+ and proline (Potluri and Prasad 2001).

In summary, cassava responds to water deficit with a stress avoidance syndrome involving highly sensitive stomatal closure, leaf drooping, leaf loss and halt of leaf growth. These responses keep water potential from decreasing substantially such that tissues are not exposed to injurious low water potential stress. Although this also limits photosynthetic carbon assimilation, growth and carbon consumption is kept to a minimum, and metabolic needs are supplied via remobilization from reserves in petioles and stems. When rainfall occurs, cassava rapidly resumes growth of leaves and storage roots, such that full advantage is made of available moisture. Studies have indicated that cassava is very limited in its use of osmotic adjustment, compatible solute synthesis, dehydrin accumulation and other tolerance mechanisms for low water potential. However, it is not known whether cassava's performance could be improved if such mechanisms were introduced into plants subjected to environments with extreme, extended droughts that inevitably deplete soil water to the point of severely low water potential. To explore these possibilities, investigators have begun to employ modern genetic approaches toward identifying diverse germplasm and creating plants with new stress tolerance traits.

2. GENETIC IMPROVEMENT

Despite its importance to world agriculture, cassava has received relatively little breeding attention until recently (Ceballios et al. 2004; Kawano 2003). In part, this is due to its long breeding cycle (18–24 months), the difficulties in production of recombinant seed, and its highly heterozygous nature which (a) masks allelic differences in segregating populations, (b) permits a sizable genetic load of deleterious or undesirable alleles to persist in populations subjected to selection, and (c) creates difficulties in transferring desirable traits from one genotype to another (Ceballios et al. 2004). Nevertheless, breeding has made sizable gains when intense, sustained programs have been applied (Kawano 2003). Such efforts have also provided valuable lessons to guide future work, such as the finding that tests of response to selection are valuable for evaluating the contribution of particular traits (Kawano 2003), and that it would be valuable to introduce inbreeding into the process (Perez et al. 2005). It has also provided the impetus to develop rapid methods by which inbred lines can be created, such as through doubled haploids (Woodward and Puonti-Kaelas 2001).

Given the difficulties of conventional breeding in cassava, there is considerable interest in augmenting breeding with molecular marker approaches (Fregene et al. 2001) to accelerate the rate of genetic gain. The relative efficiency of molecular marker-assisted selection (MAS) compared to phenotypic selection is very high if a trait in question has low narrow sense heritability (h^2) under field-based evaluation methods and the ratio of variance explained by the marker compared to total additive genetic variance (V_M/V_A) is high. The ratio V_M/V_A is high if the marker is associated with a major gene or markers for quantitative trait loci (QTLs) that control a large proportion of additive genetic variance. There are many instances when traits of importance in cassava breeding have low h^2. Some examples are:

1) Evaluation based upon a single plant, particularly for quantitative traits and resistance to several pests and diseases.
2) Disease resistance traits where the pathogen pressure is absent or low, such as cassava mosaic disease resistance in the Neo-tropics or cassava green mite during the wet season.
3) Highly variable experimental fields and/or poor management
4) Traits that are often affected by stage of plant growth, e. g. dry matter content.

Markers may enhance narrow sense heritability in the examples mentioned above but they also help in the reduction of breeding populations via the elimination of undesirable genotypes at the seedling stage. For example, the number of genotypes at the seedling stage can be reduced by 50% if a trait is controlled by a single gene, or by 87.5% if controlled by three genes. This is one of the most crucial selection stages, since it contains the highest level of genetic diversity for the breeder to find the trait combinations of interest. Often, up to 90% of genotypes are discarded in the seedling stage. Furthermore molecular markers could provide an efficient way to transfer genes conferring traits that are difficult to phenotype in large scale trials, such as those for deep rooting or osmotic adjustment.

A MAS program normally involves three basic steps. The first is genome (linkage) mapping where markers are placed on a molecular genetic framework map on the basis of their segregation in a mapping population. In the second step, genome linkage mapping is followed by QTL mapping. In this step the genome location of markers that co-segregate with the traits of interest are located on the linkage map. The third stage involves the selection of molecular markers at such QTL during the evaluation and selection.

Progress has been made in developing genetic linkage maps using RFLP, RAPD, and microsatellite markers (Fregene et al. 1997), and using them in identifying QTLs for desired traits (Jorge et al. 2000; Jorge et al. 2001; Okogbenin and Fregene 2002; Okogbenin and Fregene 2003). Markers developed have also being used to study genetic diversity for traits of agronomic interest (Fregene et al; 2000; Fregene et al. 2003). Recently initiated work is addressing water deficit and other associated stresses (Fregene 2006, unpublished). For example, water deficit in cassava creates serious co-occurring stress caused by mealybug (*Phenacoccus herreni*) (Calatayud et al. 2002) and mites (*Mononychellus tanajoa*) (A. Bellotti, personal communication). The research efforts seek to tag QTLs controlling traits related to drought tolerance in cassava and to develop molecular markers to improve the efficiency of breeding cassava for drought tolerance which in itself is a trait with low heritability in cassava given the difficulty of reproducing water stress environments. Eight contrasting genotypes were crossed to develop drought tolerant mapping populations. The populations obtained provide an important resource to create a consensus map of genes across the different cassava populations that may represent universal genetic "hot spots" in those genomic regions that confer drought tolerance in diverse settings to varying degrees. Quantitative trait locus (QTL) mapping of these populations require marker systems that are highly polymorphic and amenable to high-throughput genotyping.

On MAS of water stress associated traits, sources of resistance to mites and whiteflies have been identified from wild *Manihot* relatives of cassava, and marker assisted approaches are being use to introduce multiple traits into desirable germplasm (Fregene and Mba 2004). But the use of wild relatives in regular breeding programs is complicated by the long reproductive breeding cycle of cassava, high genetic load that is released on backcrossing, and linkage drag associated with the use of wild relatives in crop improvement. A project was initiated recently to accelerate the introgression of useful genes from wild relatives into cassava via a modified Advance Back Cross QTL (ABC-QTL) (Tanksley and Nelson 1996) breeding scheme. The traits include resistance to green mites, whiteflies, and hornworm, delayed post harvest physiological deterioration (PPD), and high protein and dry matter content (DMC) (Fregene et al. 2006).

Another approach to overcome the difficulties of conventional breeding is to use transformation technology. Work has succeeded in establishing transformation protocols for cassava that use either particle bombardment or Agrobacterium systems (Li et al. 1996; Raemakers et al. 1996; Schopke et al. 1997; Taylor et al. 2004). For example, a transformation approach has been used to address the problem of excess leaf abscission in response to stress by introducing an expression construct consisting of the cytokinin synthesis gene, ipt, under the control of the senescence-associated promoter SAG12 (Zhang and Gruissem 2004). Preliminary trials have shown that transgenes have improved retention of green leaf color in water deficit and field trials are underway to evaluate the impact of the transgene on performance under water deficit (Zhang and Gruissem 2005).

SUMMARY

Marker assisted selection (MAS) can contribute to the efficient reduction of large breeding populations at the seedling stage based upon a 'minimum selection criteria'. This is particularly important given the length of the growing cycle of cassava and the expenses involved in the evaluation process. The selection of progenies based on genetic values derived from molecular marker data substantially increases the rate of genetic gain, especially if the number of cycles of evaluation or generations can be reduced. Another application of MAS in cassava breeding is reducing the length of time required for the introgression of traits from wild relatives. Wild relatives are an important source of genes for pest and disease resistance in cassava but the need to reduce or eliminate undesirable donor genome content, linkage drag, can lengthen the process making it unrealistic for most breeders.

REFERENCES

Alves, A. A. C. and Setter, T. L., 2000, Response of cassava to water deficit: Leaf area growth and abscisic acid, *Crop Science* **40**:131–137.

Alves, A. A. C. and Setter, T. L., 2004a, Abscisic acid accumulation and osmotic adjustment in cassava under water deficit, *Environmental and Experimental Botany* **51**:259–271.

Alves, A. A. C. and Setter, T. L., 2004b, Response of cassava leaf area expansion to water deficit: Cell proliferation, cell expansion and delayed development, *Annals of Botany* **94**:605–613.

Angelov, M. N., Sun, J., Byrd, G. T., Brown, R. H. and Black, C. C., 1993, Novel characteristics of cassava, *Manihot esculenta* Crantz, a reputed C-3-C-4 intermediate photosynthesis species, *Photosynthesis Research* **38**:61–72.

Baker, G. R., Fukai, S. and Wilson, G. L., 1989, The response of cassava to water deficits at various stages of growth in the subtropics, *Australian Journal of Agricultural Research* **40**:517–528.

Calatayud, P. A., Llovera, E., Bois, J. F. and Lamaze, T., 2000, Photosynthesis in drought-adapted cassava, *Photosynthetica* **38**:97–104.

Calatayud, P. A., Polania, M. A., Seligmann, C. D. and Bellotti, A. C., 2002, Influence of water-stressed cassava on *Phenacoccus herreni* and three associated parasitoids, *Entomologia Experimentalis et Applicata* **102**:163–175.

Ceballos, H., Iglesias, C. A., Perez, J. C. and Dixon, A. G. O., 2004, Cassava breeding: opportunities and challenges, *Plant Molecular Biology* **56**:503–516.

Cock, J. H., Porto, M. C. M. and El Sharkawy, M. A., 1985, Water use efficiency of cassava *Manihot esculenta* 3. Influence of air humidity and water stress on gas exchange of field grown cassava, *Crop Science* **25**:265–272.

Connor, D. J. and Cock, J. H., 1981, Response of cassava to water shortage 2. Canopy dynamics, *Field Crops Research* **4**:285–296.

Duque, L. O. and Setter, T. L., 2005, In *Interdrought II, The 2nd International Conference on Integrated Approaches to Sustain and Improve Plant Production Under Drought Stress. Rome, Italy, September 24 to 28, 2005*, Avenue media, Bolgna, Italy, pp. L 5.09.

El Sharkawy, M. A. and Cock, J. H., 1984, Water use efficiency of cassava *Manihot esculenta* 1. Effects of air humidity and water stress on stomatal conductance and gas exchange, *Crop Science* **24**:497–502.

El Sharkawy, M. A. and Cock, J. H., 1990, Photosynthesis of cassava *Manihot esculenta*, *Experimental Agriculture* **26**:325–340.

El Sharkawy, M. A., Cock, J. H. and Held, K. A. A., 1984, Water use efficiency of cassava *Manihot esculenta* 2. Differing sensitivity of stomata to air humidity in cassava and other warm climate species, *Crop Science* **24**:503–507.

El Sharkawy, M. A., Cock, J. H., Lynam, J. K., Del Pilar Hernandez, A. and Fernando Cadavid, L. L., 1990, Relationships between biomass root yield and single-leaf photosynthesis in field-grown cassava, *Field Crops Research* **25**:183–202.

El Sharkawy, M. A., Del Pilar Hernandez, A. and Hershey, C., 1992, Yield stability of cassava during prolonged mid-season water stress, *Experimental Agriculture* **28**:165–174.

El-Sharkawy, M. A., 2004, Cassava biology and physiology, *Plant Molecular Biology* **56**:481–501.

El-Sharkawy, M. A. and Cadavid, L. F., 2002, Response of cassava to prolonged water stress imposed at different stages of growth, *Experimental Agriculture* **38**:333–350.

El-Sharkawy, M. A., De-Tafur, S. M. and Cadavid, L. F., 1993, Photosynthesis of cassava and its relation to crop productivity, *Photosynthetica* **28**:431–438.

Fernandez, M. D., Tezara, W., Rengifo, E. and Herrera, A., 2002, Lack of downregulation of photosynthesis in a tropical root crop, cassava, grown under an elevated CO_2 concentration, *Functional Plant Biology* **29**:805–814.

Fregene, M., Angel, F., Gómez, R., Rodríguez, F., Chavarriaga, P., Roca, W., Tohme, J. and Bonierbale, M., 1997, A molecular genetic map of cassava (*Manihot esculenta* Crantz), *Theoretical and Applied Genetics* **95**:431–441.

Fregene, M., Bernal, A., Duque, M., Dixon, A. and Tohme, J., 2000, AFLP analysis of African cassava (*Manihot esculenta* Crantz) germplasm resistant to the cassava mosaic disease (CMD), *Theoretical and Applied Genetics* **100**:678–685.

Fregene, M. and Mba, C., 2004, In *Cassava Breeding*, (Ed, Hershey, C.) FAO, Via Caravelle, Rome, Italy.

Fregene, M., Morante, H., Sanchez, T., Marin, J., Ospina, C., Barrera, E., Gutierrez, J., Guerrero, J., Bellotti, A., Santos, L., Alzate, A., Moreno, S. and Ceballos, H., 2006, Molecular markers for introgression of useful traits from wild Manihot relatives of cassava, marker-assisted selection (MAS) of disease and root quality traits., *Root Crop Journal*:(in press).

Fregene, M., Okogbenin, E., Mba, C., Angel, F., Suarez, M.-C., Janneth, G., Chavarriaga, P., Roca, W., Bonierbale, M. and Tohme, J., 2001, Genome mapping in cassava improvement: Challenges, achievements and opportunities, *Euphytica* **120**:159–165.

Fregene, M. A., Suarez, M., Mkumbira, J., Kulembeka, H., Ndedya, E., Kulaya, A., Mitchel, S., Gullberg, U., Rosling, H., Dixon, A. G. O., Dean, R. and Kresovich, S., 2003, Simple sequence repeat marker diversity in cassava landraces: Genetic diversity and differentiation in an asexually propagated crop, *Theoretical and Applied Genetics* **107**:1083–1093.

Gent, M. P. N., 1994, Photosynthate reserves during grain filling in winter wheat, *Agronomy Journal* **86**:159–167.

Hawker, J. S. and Smith, G. M., 1982, Salt tolerance and regulation of enzymes of starch synthesis in cassava *Manihot esculenta* cultivar Maus-7, *Australian Journal of Plant Physiology* **9**:509–518.

Ike, I. F., 1982, Effect of water deficits on transpiration photosynthesis and leaf conductance in cassava *Manihot esculenta* cultivar Llanera, *Physiologia Plantarum* **55**:411–414.

Ike, I. F. and Thurtell, G. W., 1981a, Osmotic adjustment in indoor grown cassava Manihot esculenta in response to water stress, *Physiologia Plantarum* **52**:257–262.

Ike, I. F. and Thurtell, G. W., 1981b, Response of indoor grown cassava *Manihot esculenta* cultivar Llanera to water deficits and recovery of leaf water potential and stomatal activity after water stress, *Journal of Experimental Botany* **32**:1029–1034.

Ike, I. F. and Thurtell, G. W., 1981c, Water Relations of Cassava Manihot-Esculenta Water Content Water Osmotic and Turgor Potential Relationships, *Canadian Journal of Botany* **59**:956–964.

Itani, J., Oda, T. and Numao, T., 1999, Studies on mechanisms of dehydration postponement in cassava leaves under short-term soil water deficits, *Plant Production Science* **2**:184–189.

Jones, H., 1998, Stomatal control of photosynthesis and transpiration, *Journal of Experimental Botany* **49**:387–398.

Jorge, V., Fregene, M., Velez, C.-M., Duque, M. C., Tohme, J. and Verdier, V., 2001, QTL analysis of field resistance to *Xanthomonas axonopodis* pv. manihotis in cassava, *Theoretical and Applied Genetics* **102**:564–571.

Jorge, V., Fregene, M. A., Duque, M. C., Bonierbale, M. W., Tohme, J. and Verdier, V., 2000, Genetic mapping of resistance to bacterial blight disease in cassava (*Manihot esculenta* Crantz), *Theoretical and Applied Genetics* **101**:865–872.

Kawano, K., 2003, Thirty years of cassava breeding for productivity: Biological and social factors for success, *Crop Science* **43**:1325–1335.

Kawano, K., Daza, P., Amaya, A., Rios, M. and Goncalves, W. M. F., 1978, Evaluation of cassava germ-plasm for productivity, *Crop Science* **18**:377–380.

Lenis, J. I., Calle, F., Jaramillo, G., Perez, J. C., Ceballos, H. and Cock, J. H., 2006, Leaf retention and cassava productivity, *Field Crops Research* **95**:126–134.

Li, H. Q., Sautter, C., Potrykus, I. and Puonti-Kaerlas, J., 1996, Genetic transformation of cassava (*Manihot esculenta* Crantz), *Nature Biotechnology* **14**:736–740.

Munyikwa, T. R. I., Kreuze, J., Fregene, M., Suurs, L., Jacobsen, E. and Visser, R. G. F., 2001, Isolation and characterisation of cDNAs encoding the large and small subunits of ADP-glucose pyrophosphorylase from cassava (*Manihot esculenta* Crantz), *Euphytica* **120**:71–83.

Oguntunde, P. G., 2005, Whole-plant water use and canopy conductance of cassava under limited available soil water and varying evaporative demand, *Plant and Soil* **278**:371–383.

Okogbenin, E. and Fregene, M., 2002, Genetic analysis and QTL mapping of early root bulking in an F1 population of non-inbred parents in cassava (*Manihot esculenta* Crantz), *Theoretical and Applied Genetics* **106**:58–66.

Okogbenin, E. and Fregene, M., 2003, Genetic mapping of QTLs affecting productivity and plant architecture in a full-sib cross from non-inbred parents in cassava (*Manihot esculenta* Crantz), *Theoretical and Applied Genetics* **107**:1452–1462.

Palta, J. A., 1984, Influence of water deficits on gas-exchange and the leaf area development of cassava *Manihot esculenta* cultivars, *Journal of Experimental Botany* **35**:1441–1449.

Pardales, J. R., Jr. and Esquibel, C. B., 1996, Effect of drought during the establishment period on the root system development of cassava, *Japanese Journal of Crop Science* **65**:93–97.

Pereira, J. F., Splittstoesser, W. E. and Ogren, W. L., 1986, Photosynthesis in detached leaves of cassava, *Photosynthetica* **20**:286–292.

Perez, J. C., Ceballos, H., Jaramillo, G., Morante, N., Calle, F., Arias, B. and Bellotti, A. C., 2005, Epistasis in cassava adapted to midaltitude valley environments, *Crop Science* **45**:1491–1496.

Potluri, S.-D.-P. and Prasad, P.-V.-D., 2001, In-vitro studies on the effects of varying levels of sea-salt on two cassava cultivars, *Tropical Agriculture* **78**:62–65.

Raemakers, C. J. J. M., Sofiari, E., Taylor, N., Henshaw, G., Jacobsen, E. and Visser, R. G. F., 1996, Production of transgenic cassava (*Manihot esculenta* Crantz) plants by particle bombardment using luciferase activity as selection marker, *Molecular Breeding* **2**:339–349.

Ramanujam, T., 1990, Effect of moisture stress on photosynthesis and productivity of cassava, *Photosynthetica* **24**:217–224.

Schopke, C., Taylor, N. J., Carcamo, R., Beachy, R. N. and Fauquet, C., 1997, Optimization of parameters for particle bombardment of embryogenic suspension cultures of cassava (*Manihot esculenta* Crantz) using computer image analysis, *Plant Cell Reports* **16**:526–530.

Schultz, H. R., 2003, Differences in hydraulic architecture account for near-isohydric and anisohydric behaviour of two field-grown *Vitis vinifera* L. cultivars during drought, *Plant Cell and Environment* **26**:1393–1405.

Sundaresan, S. and Sudhakaran, P. R., 1995, Water stress-induced alterations in the proline metabolism of drought-susceptible and -tolerant cassava (*Manihot esculenta*) cultivars, *Physiologia Plantarum* **94**:635–642.

Tanksley, S. D. and Nelson, J. C., 1996, Advanced backcross QTL analysis: A method for the simultaneous discovery and transfer of valuable QTLs from unadapted germplasm into elite breeding lines, *Theoretical and Applied Genetics* **92**:191–203.

Tardieu, F. and Simonneau, T., 1998, Variability among species of stomatal control under fluctuating soil water status and evaporative demand: modelling isohydric and anisohydric behaviours, *Journal of Experimental Botany* **49**:419–432.

Taylor, N., Chavarriaga, P., Raemakers, K., Siritunga, D. and Zhang, P., 2004, Development and application of transgenic technologies in cassava, *Plant Molecular Biology* **56**:671–688.

Wood, A.J., 2005, Eco-physiological adaptations to limited water environments, In Plant Abiotic Stress, M.A. Jenks and P.M. Hasegawa (eds.), Blackwell Publishing, Oxford, UK, pp. 1–13.

Woodward, B. and Puonti-Kaerlas, J., 2001, Somatic embryogenesis from floral tissue of cassava (*Manihot esculenta* Crantz), *Euphytica* **120**:1–6.

Yao, N. R., Goue, B., Zeller, B. and Monteny, B., 1988, Effect of drought on leaf development and dry matter production of the cassava *Manihot esculenta* Crantz plant, *Tropical Agriculture* **65**:84–88.

Zhang, P. and Gruissem, W., 2004, In *Sixth International Scientific Meeting of the Cassava Biotechnology Network.*, CIAT/CBN, Cali, Colombia, March 8–14, 2004, pp. 99.

Zhang, P. and Gruissem, W., 2005, In *The Global Food & Product Chain. Dynamics, Innovations, Conficts, Strategies*, Deutscher Tropentag, October 11–13, 2005, Univ. Hohenheim, Stuttgart, Germany.

CHAPTER 29

RECENT ADVANCES IN GENETIC ENGINEERING OF POTATO CROPS FOR DROUGHT AND SALINE STRESS TOLERANCE

MYUNG-OK BYUN[1], HAWK-BIN KWON[2], AND SOO-CHUL PARK[1]

[1]*Department of Molecular Physiology and Biochemistry, National Institute of Agricultural Biotechnology, RDA, Suwon, 441–707, Korea*

[2]*Division of Applied Biological Sciences, Sunmoon University, Asan, 336–708, Korea*

Abstract: Defense systems are triggered when plants encounter environmental stresses such as high salinity or drought. Many studies have shown that these defense systems depend on protective mechanisms created by altering the expression levels of stress genes. The agricultural species *Solanum tuberosum* is an autotetraploid with a highly complicated, quantitative inheritance pattern. Thus, breeding new potato cultivars that are tolerant of saline and drought stress by conventional methods is tedious, difficult and time-consuming, and generally requires between 10 and 15 years. Genetic engineering techniques represent a faster and more reliable way to improve potato cultivars. As a first step towards developing drought- and saline-tolerant potato plants by molecular breeding methods, numerous potato stress genes, including those that code for functional and regulatory proteins, have been isolated and characterized by homologue gene screening, differential screening, microarray analysis and proteome analysis. There have been many attempts around the world to create drought- and saline-tolerant potato plants by introducing abiotic stress genes for functional proteins, such as proline synthesis protein, osmotin-like protein, GPD, trehalose synthesis protein, and regulatory proteins such as StEREBP, CBF and StRD22

Keywords: potato; drought stress; salt stress; stress related genes; transgenic potato

1. INTRODUCTION

Potato (*Solanum tuberosum*) is the world's 4th major crop after rice, wheat and corn in terms of yield, and 8th in terms of area under cultivation (FAO statistics). The potato tuber is a high-energy staple food in many countries around the world and since it provides high productivity per unit area, it can be cultivated intensively. Thus, potato represents one of the best candidates for alleviating food shortages. The most commonly cultivated potato is *S. tuberosum*, an autotetraploid ($2n = 4x = 48$)

M.A. Jenks et al. (eds.), Advances in Molecular Breeding Toward Drought and Salt Tolerant Crops, 713–737.

species in which the commercial cultivars are often sterile. Potatoes originate from Peru and Bolivia and various related wild-type species continue to grow throughout the Andes, with the range of some varieties extending into the USA (Hooker, 1981). This broad geographical distribution has resulted in a diversity of potato ecotypes, many of which are adapted to their specific environments. Thus, one of the main aims in potato breeding programs is to introduce useful traits from wild-type varieties into agricultural species.

Cultivation of potato began 3,500 to 4,500 years ago; it was introduced to Europe, Asia and North America in the late 16th and early 17th centuries (Kim, 2005). At present, the major potato-producing countries are China, Russia, the USA, India and the Ukraine, accounting for 22, 12, 7, 7 and 6 % of world production, respectively. Thus, potato is not only an important crop worldwide but also serves as a model plant for other members of the Solanaceae family such as tomato, tobacco and pepper.

Potatoes grow optimally under relatively cool conditions and the formation and enlargement of the tuber depends upon a sufficient difference in temperature between day and night, to enable metabolites produced during daytime to accumulate in the tuber during the night (Hooker, 1981). In general, potato is relatively vulnerable to salt stress (1.7 dSm^{-1}, EC) and is classified as a moderately saline-sensitive crop (Katerji *et al.*, 2003); salt sensitivity represents a major limitation to cultivation area. In comparison to other crops, its resistance to salt stress is greater than pepper or corn, but weaker than tomato, rice, soybean or barley (Chinnusamy *et al.*, 2005). Another factor that limits cultivation area is susceptibility to drought. Water availability determines yield in potato plants, as it is required for tuber formation (Harries, 1978, Deblonde *et al.*, 1999). Thus, improvements in the resistance of potato to abiotic stresses such as salinity and drought could increase both the cultivation area and yield, and such cultivars would be valuable for areas in which there is a need to increase food production.

Breeding new potato cultivars by conventional methods is tedious, difficult and time-consuming, generally requiring between 10 and 15 years. The agricultural species *S. tuberosum* is autotetraploid and has a highly complicated quantitative inheritance pattern. However, as many quantitative traits are difficult to distinguish from environmentally-induced variation, the field trials required to breed specific traits into new cultivars often take many years to complete. Genetic engineering techniques represent a faster and more reliable way to improve potato cultivars. Here, we provide a brief description of potato physiology and molecular biology with respect to drought and saline stress. We then review some of the defense mechanisms against these stresses and discuss current developments in molecular breeding of drought- and saline-tolerant potato plants.

2. PHYSIOLOGY AND MOLECULAR BIOLOGY OF DROUGHT AND SALT STRESS IN POTATO PLANTS

Stresses such as cold, drought, high salinity and freezing damage have been shown to induce dehydration or water stress-resistance mechanisms (Shinozaki and Yamaguchi-Shinozaki, 1997; Thomashow, 1998). Under water stress, plant cells lose water and reduce turgor pressure (Kopka *et al.*, 1997; Holmberg and Bulow, 1998), and levels of the plant hormone abscisic acid (ABA) increase; it has been established that ABA levels play an important role in the stress-tolerance of plants (Plant and Bray, 1999; Choi *et al.*, 2000). Both early growth and tuber formation require large amounts of water and once a potato plant experiences water deficit, it does not usually recover (Harris, 1978; Deblonde and Ledent, 2001). Plants undergoing drought conditions during the tuber formation stage are susceptible to scab (*Streptomyces scabies*) and soil cracking can leave tubers vulnerable to insect pests such as the potato tuber moth (*Phthorimaea opercullella*; Hide and Lapwood, 1978).

Drought stress affects crop yields directly due to decreased respiration and photosynthesis, and indirectly through evaporation from the soil and transpiration from the leaves, resulting in elevation of soil and plant temperatures, respectively. Increased temperatures are harmful to tuber formation, and drought and heat stress acting in tandem, during the late growth stage can cause problems such as brown spots inside the tubers (Hide and Lapwood, 1978). Deblonde and Ledent (2001) studied the effects of drought conditions on morphology and tuber yield in six different drought-sensitive and -tolerant cultivars and found that in the former, drought reduced the numbers of green leaves, stem length, stem height, tuber number and average tuber dry weight.

Salt stress also decreases growth and crop yield and causes serious problems for the cultivation of tetraploid potatoes with damage thresholds ranging from 1.5 to 3.0 dSm^{-1}(EC) NaCl (Mass and Hoffman, 1977), particularly in some of the most potentially productive regions of the world such as the Mediterranean, California and South East Asia. Saline stress decreases potato growth by altering metabolic processing, resulting in decreased stomatal conductance and respiration, decreased water potential, ion imbalances and toxicity of specific ions (Fidalgo *et al.*, 2004). Salinity affects tuberization and stolon growth *in vitro* and decreased tuber yields have been attributed to changes in the partitioning of assimilate, with tuber development being more affected by salinity than canopy growth; at high salt concentrations (80 mM NaCl) tuber development was inhibited completely (Silva *et al.*, 2001; Zhang *et al.*, 2005). In addition, salinity is known to reduce canopy expansion and advance senescence.

Higher salinity also brings about oxidative stress, which in turn damages the photosynthetic apparatus and cell membranes (Benavides *et al.*, 2000; Fidalgo *et al.*, 2004; Rahnama and Ebrahimzadeh, 2005). This is because levels of reactive oxygen species (ROS) increase when plants experience relatively strong salt stress or pathogen attack, resulting in membrane peroxidation, protein denaturation and DNA damage (Fidalgo *et al.*, 2004). Potato plants treated with 100–200 mM NaCl for 30 d exhibited several negative effects, including decreased relative water content, stomatal cell conductance and respiration rate, as well as decreased ascorbate and protein levels and decreased superoxide dismutase (SOD) and catalase activities. In contrast, proline levels increased, and ascorbate peroxidase activity remained unchanged in comparison to untreated control plants. When chloroplasts were examined under electron microscopy, it was observed that the thylakoids bulged and grana stacking had decreased (Fidalgo *et al.*, 2004).

3. DEFENSE MECHANISMS AGAINST SALINE AND DROUGHT STRESS IN POTATO PLANTS

Under conditions of water stress, plant cells lose water and decrease turgor pressure (Holmberg and Bulow, 1998). If plants encounter environmental stresses such as high salinity or drought (both which alter the cell water balance), defense systems are triggered (Figure 1). For example, ABA levels increase, and many studies have shown that such protective mechanisms are regulated via alteration of the expression levels of stress genes. The protein products of these genes can be separated into two categories according to function. The first category comprises the functional proteins involved in direct protection activities; the second category comprises regulatory proteins involved in stress signal transduction pathways and control of

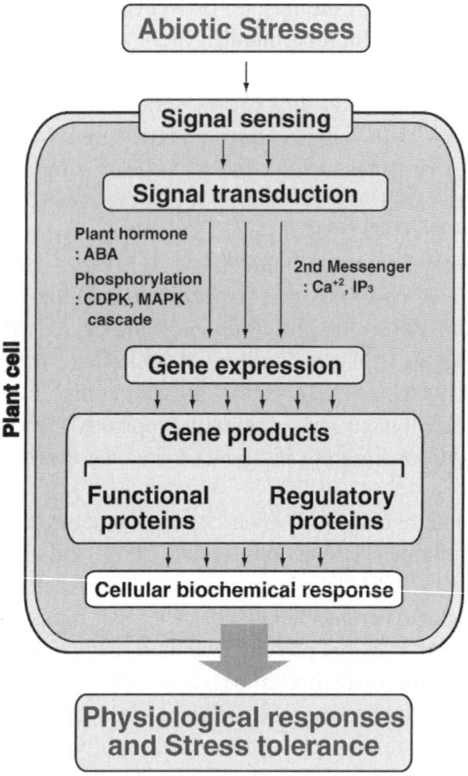

Figure 1. Abiotic signal perception, transduction and induction of stress genes. (Modified from Shinozaki and Yamaguchi-Shinozaki, 1977)

the expression of stress tolerant genes. Genes regulated by abiotic stresses have been identified using DNA microarrays (Schenk *et al.*, 2000; Bohnert, 2001; Seki *et al.*, 2001, 2002; Kreps *et al.*, 2002). Many of these stress genes are regulated at the transcriptional level and their mRNA levels alter in response to abiotic stress (Shinozaki and Yamaguchi-Shinozaki, 1997; Jaglo-Ottosen *et al.*, 1998; Cheong *et al.*, 2002; Shinozaki *et al.*, 2003). These genes encode metabolic proteins that are important for cellular protection as well as those that are involved in stress response signal transduction (Shinozaki and Yamaguchi-Shinozaki, 1997; Kasuga *et al.*, 1999; Seki *et al.*, 2002; Xiong *et al.*, 2002).

The first category of stress gene is represented by those that encode proteins involved in plant abiotic stress-tolerance, such as water channel proteins, enzymes for biosynthesis of osmoprotectant metabolites, chaperones, late embryogenesis abundant (LEA) proteins, proteinases and enzymes involved in detoxification (Shinozaki and Yamaguchi-Shinozaki, 1997).

The compatible osmolytes (also called compatible solutes or osmoprotectants) include proline, glycerol, betaines and sugars such as mannitol and trehalose, which accumulate under water stress (Knipp and Honermeier, 2006); the accumulation of these osmolytes in plants is regarded as a general adaptation to such stress (Hmida-Sayari *et al.*, 2005). In addition, calcium plays an important role in tolerance to salt stress (Shaterian *et al.*, 2005) and a study in which grafting was used to investigate the control of calreticulin (a calcium storage protein) expression, determined that this protein was involved in ABA-induced salt tolerance (Shaterian *et al.*, 2005). Recently, the biogenic amine catecholamine was implicated as a stress agent in potato plants (Swiedrych *et al.*, 2004). Proteins that protect macromolecules and membranes (LEA proteins, antifreeze protein, chaperone and osmotin) play important roles in plant abiotic stress tolerance, as do thiol protease and ubiquitin.

Strong salt stress or pathogen attack will also result in an increase in ROS levels (Wu *et al.*, 1995; Rahnama and Ebrahimzadeh, 2005) and plants use antioxidant defense systems to avoid damage from ROS accumulation. These systems involve enzymes such as ascorbate peroxidase, glutathione-S-transferase (GST), SOD and catalase, and salt stress has been shown to increase SOD activity (Bendavides *et al.*, 2000; Fidalgo *et al.*, 2004). However, catalase activity was either reduced or unaffected by salt stress. Thus, it appears that under salt stress, salt-tolerant cultivars may effect better protection against ROS by increasing the activity of antioxidant enzymes such as SOD (Rahnama and Ebrahimzadeh, 2005).

The second category of stress genes comprises the regulatory genes involved in plant tolerance to abiotic stresses, which include transcription factors, protein kinases, phospholipase C and 14-3-3 protein (Shinozaki and Yamaguchi-Shinozaki, 1997). Recently, it was reported that expression of 14-3-3 protein (20R isoform) from potato was induced by metal ions and NaCl, and its promoter activity was found to be responsive to ABA (Aksamit *et al.*, 2005). We isolated the binding protein for the ABA responsive element from potato, and using Northern blot analysis, demonstrated its induction in response to salt, cold and drought stress (Byun *et al.*, unpublished data).

The genes encoding the potato ethylene responsive element binding protein (EREBP) and CIP353 were isolated and characterized from a cDNA library prepared from potato tubers stored at 4°C for 7 months. These genes appeared to contain an AP2/ERF domain and their expression increased in response to cold treatment, especially in potato tubers (Mine *et al.*, 2003). Thus, the authors suggested that CIP353 was an AP2/ERF domain regulatory factor that regulates genes expressed in tubers stored at low temperature

4. ISOLATION AND CHARACTERISTICS OF ABIOTIC STRESS-RELATED GENES IN POTATO PLANTS

The results of many studies have indicated that the mechanisms which protect the plant from various abiotic stresses are regulated by alterations in the expression levels of a suite of genes, collectively referred to as stress genes. Thus, the first step required for the genetic study and engineering of abiotic stress-resistant potato plants was isolation and characterization of their abiotic stress-related genes. These genes were isolated by a variety of methods including screening libraries with homologues, cDNA microarray and proteomic analyses, screening of functional cDNA and EST analysis (Kim *et al.*, 2003; Rensink *et al.*, 2005a, b).

4.1. Isolation of Abiotic Stress-Related Genes Using Homologue Screening and Differential Screening

In general, abiotic stress-specific libraries have been used to isolate abiotic stress-related genes from potato plants (Kim *et al.*, 2003; Rensink *et al.*, 2005a). These libraries were constructed from leaves, tubers, flowers and stems that were treated with various abiotic stresses. mRNA from stress-treated organs was then isolated and a cDNA library generated. The abiotic stresses that have been used include drought, salt, high temperature, cold, and stress response effectors such as pathogens, salicylic acid, jasmonic acid (JA) and hormones such as ABA. To avoid wounding roots, plants are usually grown hydroponically in liquid culture medium (Rensink *et al.*, 2005b). For examination of abiotic stresses in guard cells, cDNA libraries were constructed from epidermal cells (Kopka *et al.*, 1998).

In potato the transcriptional response (transcriptome) to abiotic stress was analyzed using 20,756 expressed sequence tags (ESTs) isolated from a cDNA library comprising mRNA pooled from leaves and roots that were treated with high temperature, cold, salt and drought (Rensink *et al.*, 2005a). When compared to 78,825 ESTs from potato cDNA libraries derived from untreated root, leaf, stolon, tuber, germinating eye, and callus tissues, 1,476 ESTs were found to be unique to abiotic-stressed potato leaf and root tissue. These ESTs were compared to the *Arabidopsis* transcriptome of genes which responded to abiotic stresses and similar genes were identified. These included genes encoding sensors, transcription factors and defense proteins: RD19; RD21; RD 22; RD28; LEA14; DREB2A and 2B;

PLC1; CDL, ETR1; LTI65; CBF1,2,3,4; ABF1,2,3,4; ERD6,7,13,14; ICE; NHX; and myb2 (Rensink *et al.*, 2005a).

Several transcription factors that were regulated by cold, salt, drought and *Phytophthora infestans* treatment were identified from the 12,000 ESTs isolated from cold stress-treated potato plants. Expression of genes encoding potato EREBP increased in response to cold stress, salt and ABA treatment, and ABF levels increased in response to drought and ABA. Expression of the gene encoding myb increased with salt stress and the expression of genes encoding LEA, SRP, COR and StRD22 increased following salt and drought treatment (Kim *et al.*, 2003). In addition, several reports have considered the phenomenon of cross-talk between genes induced by cold, drought and salt stresses, all of which result in water stress. Potato *StDS2* expression was elevated in leaves, flowers and tubers that were subjected to drought stress, whereas it did not respond to cold, high temperature, salt, hypoxia or ABA treatment (Doczi *et al.*, 2005). *StDS2* expression also increased following treatment with polyethylene glycol (PEG) and mannitol, which affect osmotic pressure.

The genes for the phophatidylinositide-specific phospholipase C (PI-PLC) isoforms StPLC1, StPLC2 and StPLC3, were isolated from the guard cells of *S. tuberosum*, then expressed in various tissues including leaves, flowers, tubers and roots. Although *StPLC2* and *StPLC3* mRNAs accumulated following wilting, which occurred as a result of air drying the root system for 6 h, *StPLC1* transcript levels were reduced (Kopka *et al.*, 1998). Using reverse northern blot analysis, we isolated the gene encoding a cold-inducible potato lipid transfer protein (*St*LTP; Byun *et al.*, unpublished data) and its expression was induced by treatment with drought and salt stress, ABA, JA, cold and wounding (Figure 2). In addition, Table 1 indicates several salt- and drought-related genes that have been isolated from potato.

Genes that were differentially expressed in potato were cloned using differential display reverse transcription PCR (DDRT-PCR) and cDNA-amplified fragment length polymorphism (AFLP; Bachem *et al.*, 1996; Leone *et al.*, 1999). In order to identify water stress-induced genes, potato cells that had been treated with PEG or ABA were subjected to DDRT-PCR. This approach led to the identification of

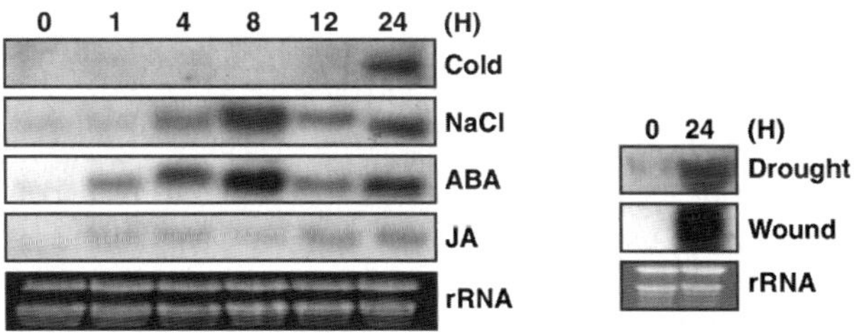

Figure 2. Northern blot analysis showing expression of potato *StLTP* mRNA under various abiotic stresses. Cold (4°C); 250 mM NaCl; 100 μM ABA; 100 μM JA; drought (air-dried on filter paper); wounding (cut with a razor blade)

Table 1. Salt- and drought-related genes isolated from potato

Gene name	Source	Isolation methods	Description	Reference
EREBP (Ethylene responsive element binding protein)	*S. tuberosum*	cDNA library screening	Drought, salt, and pathogen tolerance	Lee et al., 2004
ABF (Abscisic acid binding factor)	*S. tuberosum*	cDNA library screening	Drought tolerance	Unpublished data
Rd22 (Response to dehydration)	*S. tuberosum*	cDNA library screening	Drought tolerance	Unpublished data
Thioredoxin like protein CDS32	*S. tuberosum*	2D analysis of chloroplast	Tolerance to water deficit	Rey et al., 1998
CDS34 (Fibrillin: plastid associated protein)	*S. tuberosum*	2D analysis of chloroplast	Tolerance to water deficit	Pruvot et al., 1996
StGCPRP (Guard cell proline rich protein)	*S. tuberosum* Guard cell	Differential screening	Drought tolerance	Menke et al., 2000
StPLC2, 3 (Phospholipase C)	*S. tuberosum*	Epidermal fragment cDNA library screening	Drought tolerance	Kopka et al., 1998
Osmotin like protein	*S. commersonii*	Genomic library screening	Salt tolerance	Zhu et al., 1995
DS2 (Dehydration specific)	*S. chacoense/ tuberosum*	Genomic library screening	Drought tolerance	Silhavy et al., 1995/ Doczi et al., 2005

transcripts that accumulated during low water potentials, and included transcripts for proteins such as α-1-elongation factor and myosin, as well as a variety of proteins of unknown function.

Salt stress-induced transcripts were identified from potato leaves using AFLP and of the 5,000 bands identified, 154 and 120 were up- and down-regulated, respectively (Hmida-Sayari *et al.*, 2005a). These transcripts encoded proteins involved in cell wall structure and turnover such as β-galactosidase, stress response proteins such as glyceraldehyde dehydrogenase and wound-induced protein, and proteins involved in the pathogen response.

4.2. Isolation of Abiotic Stress-Related Genes Using Microarray Analysis

Temporal and spatial monitoring of the plant transcriptome is required for understanding of the abiotic stress response. Therefore, expression profiling using microarray analysis, is the preferred method for large-scale identification of stress-induced changes. Rensink *et al.*, (2005b) screened a 12,000 clone potato cDNA microarray prepared from plants that had been treated with low temperature (4°C), heat (35°C), or salt (100 mM NaCl). Following 3, 9, and 27 h of stress treatment, the expression profiles of the roots and aerial parts of the plant were identified and 3,314 genes were shown to respond significantly to at least one of the stress conditions. As with the abiotic stress response genes of *Arabidopsis* and rice, these genes encoded transcription factors, signal transduction and chaperone proteins, prompting Rensink *et al.* (2005b) to suggest that a similar abiotic stress response signal transduction pathway to that found in *Arabidopsis*, was present in potato. By categorizing these genes, stress-specific and shared-response genes were identified. The salt-inducible genes included those encoding protein phosphatase 2C, class II chitinase, LTP, branched chain amino acid aminotransferase, ABA and environmental stress inducible protein, homeoprotein Athb-7, and LEA-like protein.

Using microarray analysis, Kim *et al.* (2003) analyzed 100 potato genes involved in abiotic stress responses and demonstrated the presence of crosstalk in signal transduction of abiotic stresses such as drought (air drying of leaves), cold (4°C), salt (250 mM NaCl) and pathogen attack (*P. infestans*). These genes included StRD22, COR, EREBP, Myb and LEA (Figure 3).

In order to examine the salt tolerance of transgenic plants overexpressing StEREBP, Lee et al., (2007) compared root growth in transgenic and control plants (transformed with empty pBI121 vector) grown on MS plates with or without Nacl and observed tolerance to Nacl stress in the former (Figure 4). Using microarray analysis, the StEREBP signal transduction pathway was investigated by isolating genes that were up-regulated in transgenic lines over-expressing StEREBP. mRNA was isolated from these transgenic lines and used to screen for StEREBP-regulated genes using a 12,000 clone TIGR microarray chip of potato cDNA. Among the genes exhibiting a >2-fold change in expression compared to non-transgenic

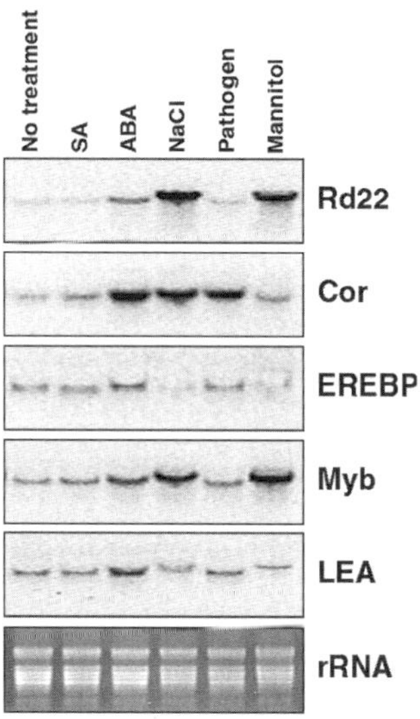

Figure 3. Northern blot analysis showing expression of some stress-related potato genes under abiotic and biotic stress, and phytohormones. C, no treatment; 100 M salicylic acid (SA); 100 M ABA; 250 mM NaCl; Pathogen (*P. infestans*); and 200 mM Mannitol

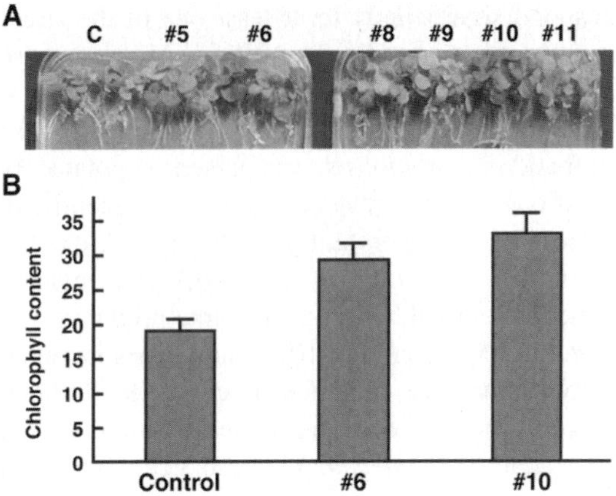

Figure 4. Transgenic potato plants expressing StEREBP exhibit tolerance to Nacl stress. To progeny were transferred and grown on MS medium supplemented with (top) or without (bottom) 75mM Nacl. Root growth was assessed after 3 weeks. C; control plant transformed with empty pBI121 vector. 2, 3 and 10; StEREBP over expressing potato lives

control plants, were 401 genes that appeared to be homologous to those found in *Arabidopsis*. We examined the upstream regions of these genes (−1,000 bp) for the presence of a GCC (AGCCGCC) or DRE core element (CCGAC), which represents the binding site for EREBP. Five genes appeared to contain the GCC element, 79 genes contained the DRE core element and 5 genes had both elements. (Figure 5; Byun *et al.*, unpublished data). Several StEREBP-regulated genes are shown in Table 2.

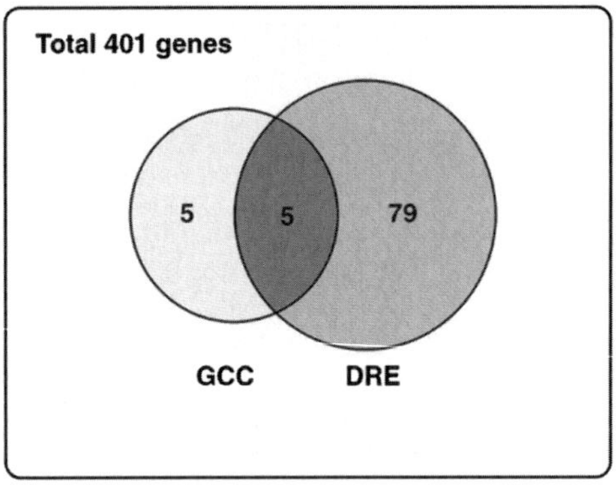

Figure 5. Genes downstream of StEREBP were up-regulated in StEREBP transgenic potato plants. A 12,000 clone TIGR potato cDNA microarray chip was screened using transcripts from potato StEREBP-expressing transgenic plants. Among 400 genes showing >2-fold expression compared to non-transgenic control plants, 5 genes appeared to contain the GCC element (AGCCGCC), 79 genes contained the DRE core element (CCGAC) and 5 genes contained both elements

Table 2. Examples of StEREBP up-regulated genes in StEREBP transgenic potato plants.

AGI No.	Description	Category	GCC	DRE	Potato (TC)	EREBP regulation
AT4G22690	Cytochrome P450 family protein, flavonoid 3',5'-hydroxylase Hf1	Metabolism	1	1	TC46322	8.93
AT1G22170	Phosphoglycerate mutase	Metabolism		1	TC55957	6.50
AT4G15510	Photosystem II reaction center PsbP family protein	Metabolism		1	TC53944	5.58
AT1G27980	Pyridoxal-dependent decarboxylase family protein	Metabolism		1	TC49063	4.89
AT1G29280	WRKY 65 transcription factor	Cell rescue/defense		1	TC54604	5.11
AT1G19940	Endo-β-1,4-D-glucanse	Cell rescue/defense		1	TC54158	4.64
AT5G52060	Apoptosis regulator Bcl-2 protein	Cell rescue/defense	1		TC49742	4.21
AT5G16710	Glutathione dehydrogenase	Cell rescue/defense		1	TC41316	4.04
AT1G18150	AtMPK8	Signal transduction		1	TC49283	2.62
AT4G27300	S-locus protein kinase	Signal transduction		1	TC55477	2.73
AT2G07180	Protein kinase	Signal transduction	1		TC53776	2.90
AT4G17560	Ribosomal protein L19 family protein	Prtein biosynthesis		2	TC45161	7.06
AT4G02930	Elongation factor Tu	Protein biosynthesis		1	TC45719	3.73
AT1G64520	26S proteasome regulatory subunit	Protein catabolism		1	TC42610	2.90
AT3G08690	Ubiqitin-conjugating enzyme 11	Protein catabolism		1	TC55438	2.41
AT1G50640	AtERF3	Regulation of transcription		1	TC46552	2.43
AT5G52510	Transcription factor SCL8	Regulation of transcription		1	TC46670	2.84
AT1G53190	Zinc finger (C3HC4-type RING finger)	Protein binding		1	TC54792	2.75

4.3. Isolation of Abiotic Stress-Related Genes Using Proteome Analysis

Only a few studies have considered the effects of abiotic stress on protein expression patterns in potato and these have used two dimensional electrophoresis to observe drought-induced changes in chloroplast proteins (Gillet *et al.*, 1998; Rey *et al.*, 1998; Langenkämper *et al.*, 2001). Chloroplast proteins induced by water deficit were CDSP34 (chloroplast drought- induced stress protein, thylakoid protein,

Table 3. Drought- and salt-resistant potato plants generated by the introduction of abiotic stress response genes

Induced genes	Source of the genes	Potato Cultivar	Function	Reference
ScTPS	*S. cerevisiae*	Superior	Drought resistance	Yeo et al., 2000
StRD22	*S. tuberosum*	Superior	Drought resistance	Unpublished data
PsGPD	*P. sajor-caju*	Dejima	Salt resistance	Jeong et al., 2001
Osmotin like protein	*A. thaliana*	Bintje	Salt resistance	Evers et al., 1999
AtCBF1	*A. thaliana*	Superior	Salt resistance	Unpublished data
AtCBF3	*A. thaliana*	Desiree	Salt resistance	Celebi-Toprak et al., 2005
SST/FFT (Fructan biosynthetic genes)	*C. cardunculus*	Desiree	Proline increase	Knipp and Honermeier, 2005
Repression of FDH (Formate dehydrogenase)	*S. tuberosum*	Desiree	Proline increase	Ambard-Bretteville et al., 2003

34 kDa) and CDSP32 (thioredoxin-like protein, 32 kDa). Interestingly, although the protein CDSP34 accumulated in the thylakoids following drought stress, its levels reduced after the plants were watered (Gillet *et al.*, 1998). In contrast, CDSP32 did not accumulate under conditions of mild water deficit (relative water contents of leaves, RWC 85%) whereas, high level of CDSP32 mRNA were observed in response to severe water deficit (RWC 75 %). Thus, it has been suggested that CDSP32 is involved in the preservation of the thiol-disulfide redox potential of chloroplast proteins during water deficit (Rey *et al.*, 1998).

5. GENERATION OF STRESS-TOLERANT POTATO PLANTS

5.1. Conventional Breeding Methods and Tissue Culture

Currently, the potato species cultivated globally is the heterozygotic, tetraploid *S. tuberosum* ($2n = 4x = 48$). Thus, difficulties are experienced when using conventional breeding methods because crossing and inbreeding depression result in a significant decrease in the efficiency of selection of a superior line. In addition, the tetraploid genetics and quantitative nature of the breeding targets increase the difficulty of selection of a pure breeding line. Potato plants range from diploid to hexaploid and although most cultivated potato species are tetraploid, over 74% of naturally-occurring species are diploid. Thus, the disadvantages of breeding tetraploid potato plants can be circumvented by using wild-type diploid species.

S. tuberosum lacks genetic diversity and introduction of useful traits is a way to increase the genetic diversity of cultivated potatoes. In order to breed a tetraploid potato, a diploid that is generated from a tetraploid plant is crossed with a closely-related wild-type diploid potato, and the resultant hybrid is re-crossed with a tetraploid plant to increase the genetic diversity. Jefferies (1996) evaluated the seed germination and survival of *S. tuberosum* L. seedlings that had been selected for tolerance to salinity and demonstrated genetic variation between the cultivars. In *in vitro* experiments, a stable salt-tolerant potato cell line that could grow in media containing 60–450 mM NaCl, was generated (Ochatt *et al.*, 1999) and 20 salt-resistant potato varieties were screened using culture medium containing different concentrations of NaCl (0.05 to 0.5 M; Kim *et al.*, 1995). However, as potato improvement using traditional breeding methods is slow and unpredictable, the faster and more reliable techniques that are available through genetic engineering are now being used to improve the resistance of potato to abiotic stresses.

5.2. Breeding Using Molecular Engineering Techniques

Transformation of potato plants using the *Agrobacterium* Ti plasmid was established relatively early (An *et al.*, 1986) and more recently, techniques such as the bacterial artificial chromosome (BAC), have been developed; these can introduce very large DNA fragments into plants (Ercolano *et al.*, 2004). Using these molecular engineering techniques, cultivars have been developed and are in use, which are resistant to insects, viruses and late blight disease (*P. infestans*). However, breeding of potato plants that are resistant to abiotic stresses is still at an early stage.

Selection of useful traits through visible phenotypes requires vast effort and is time consuming because it requires large areas for cultivation and testing of progeny. In the field of abiotic stress studies, potato plants exhibit very complicated genetic patterns due to their polyploid nature and thus, the use of molecular markers can increase the efficiency of selection of useful genetic traits (Watanabe, 2002). Marker-assisted selection is rapid, accurate, convenient and inexpensive (Watanabe, 2002) and recent developments in molecular markers have made it possible to develop crop plants with polygenic traits (Celebi-Toprak *et al.*, 2005a). The first RFLP molecular map of potato was constructed using the tomato genetic map (Bonierbale *et al.*, 1988). Prior to this, it had not been possible to construct a genetic map of potato by conventional methods due to its tetrasomic inheritance and heterozygosity that had resulted from inbreeding depression following repeated selfing and heteropolyploidy (Gebhardt C., and Valkonen J.P., 2001). Thus, the linkage map of diploid potato plants was constructed using molecular markers (Bonierbale *et al.*, 1988 Valkonen J.P.; Tanksley *et al.*,1992) and currently, markers that are related to pathogen tolerance are under active development. The techniques using these molecular markers include RFLP, RAPD, AFLP, CAPS and SCAR (Watanabe, 2002; Celebi-Toprak *et al.*, 2005a). In addition, PCR markers have become increasingly useful in potato breeding due to their simplicity of use (Celebi-Toprak, 2005a; Colton *et al.*, 2006).

In order to improve stress tolerance, genetic engineering techniques have been used to introduce genes involved in different abiotic stress responses into a variety of plants including potato, *Arabidopsis*, tobacco and rice, (Kishor *et al.*,

1995; Pilon-Smits *et al.*, 1995; Xu *et al.*, 1996; Goddijn *et al.*, 1997; Holmberg and Bulow, 1998; Kasuga *et al.*, 1999; Huang *et al.*, 2000; Yeo *et al.*, 2000; Maqbool *et al.*, 2002). Here, we briefly review some improvements that have been achieved through the introduction of stress-related genes using genetic engineering techniques.

5.3. Genetic Engineering of Salt-Tolerant Potato Plants

Potato is more sensitive to salt stress than rice, corn or barley (Chinnusamy *et al.*, 2005) and in general, two main genetic engineering approaches are used in the improvement of salt tolerance. One approach reduces physiological and biochemical metabolic damage by decreasing absorption of Na^+ or Cl^- ions or by compartmentalizing ions absorbed into the cell vacuole. The other approach increases the availability of compatible solutes such as sugars, sugar-alcohols and amines, which improve membrane stability. Thus, genes have been introduced into potato plants which encode proteins that are related to salt stress such as proline synthase, osmotin-like protein, glyceraldehyde-3-phosphate dehydrogenase, CBF1, CBF3 and EREBP (Zhu *et al.*, 1995; Jeong *et al.*, 2001; Hmida-Sayari *et al.*, 2005b).

5.3.1. Introduction of genes encoding functional proteins

Proline is recognized as a compatible osmolyte and its levels increase in response to water stress brought about by salt or drought. In plants, proline is synthesized from either glutamate or ornithine and in *Arabidopsis*, Δ^1-pyrroline-5-carboxlyate synthetase (AtP5CS) plays a key role in osmotic stress-induced proline biosynthesis. When *AtP5CS* was introduced into potato and over-expressed under the control of the 35S promoter, proline accumulated and plants demonstrated improved salt tolerance although tuber yield and weight were lower than for non-transgenic controls (Hmida-Sayari *et al.*, 2005b).

Osmotin-like protein is a pathogenesis-related (PR) protein, which is expressed in response to the presence of a pathogen or osmotic stress, although the response to the former stress is the greater of the two. The gene encoding osmotin-like protein from *Arabidopsis* was introduced into potato plants and over-expressed from the 35S promoter. Following growth in 100 mM NaCl, no difference was observed in the number of nodes, root formation or aerial parts of the plant between transgenic and wild-type control plants. However, there was an increase in the length and number of roots, as well as in the dry weight and biomass of the aerial parts of transgenic plants, compared to the wild-type controls (Evers *et al.*, 1999). In addition, transformants expressing osmotin-like protein exhibited elevated proline levels in response to salt stress. Thus, recombinant osmotin-like protein conferred salt-tolerance to potato transformants.

The gene encoding glyceraldehyde-3-phosphate dehydrogenase (GPD) from the oyster mushroom (*Pleurotus sajor-caju*) was also shown to increase salt tolerance and biomass when it was introduced and over-expressed in transgenic potato plants.

In addition, the authors suggested that *P. sajor-caju GPD* might function in other stress responses (Jeong *et al.*, 2001).

5.3.2. *Introduction of genes encoding regulatory proteins*

The *Arabidopsis* cold stress-related gene *CBF3 (DREB1A)* has been over-expressed in *Arabidopsis* under the control of the 35S promoter. Although the resultant transgenic plants exhibited increased tolerance to various abiotic stresses such as salt, cold and drought, growth retardation was also observed (Kasuga *et al.*, 1999). In contrast, when *Arabidopsis CBF1* was introduced into potato plants and over-expressed under the control of 35S promoter, no growth retardation was observed. In addition, when the transgenic potato plants were grown in 50 mM NaCl for 4 weeks, they exhibited normal growth of both the aerial and ground parts, demonstrating increased salt tolerance (Figure 6).

Potato *CBF3* was introduced into potato plants and expressed under the control of the rd29A promoter. Following 2 M NaCl treatment, *CBF3* mRNA accumulated and reached a maximum level within 5 min, then decreased thereafter. Notably,

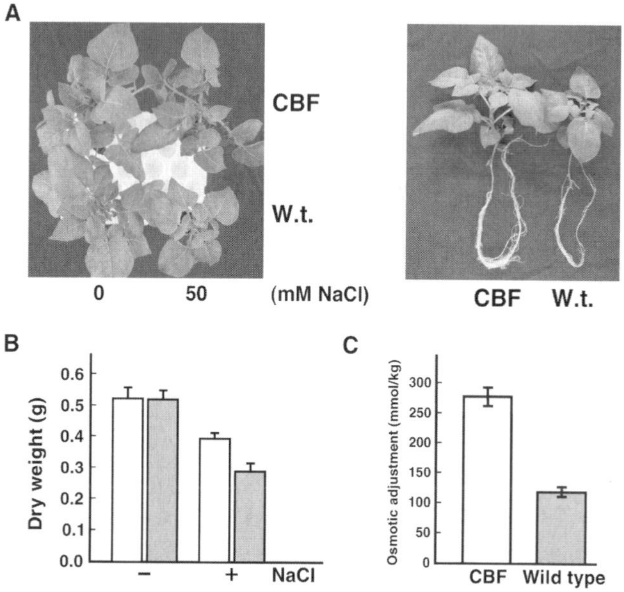

Figure 6. Increased salt tolerance of CBF transgenic potato plants. (A) When plants were treated with 50 mM NaCl for 3 weeks, the biomass of both the aerial and ground parts of the CBF transgenic potato plants was less damaged by NaCl treatment, whereas wild-type (W.t.) control plants showed clear damage. (B) Dry weights of control and CBF transgenic plants were compared after salt treatment. The transgenic plants demonstrated a higher dry weight than control plants following NaCl treatment as compared with the plants grown in a nutrient solution without the stress. unshaded, CBF; shaded, wild type (C) Osmotic adjustment of the CBF transgenic plant and control plants was examined by measuring solute concentration in cell sap. The CBF transgenic plant exhibited greater osmotic adjustment ability in the presence of the salt stress

transgenic plants which were treated with 2 M NaCl for 24 h, then returned to normal conditions and grown for a further 6 d, still survived (Celebi-Toprak *et al*., 2005b). When *CBF3* was over-expressed in *Arabidopsis* under control of the 35S promoter, the resulting transgenic plants also showed retarded growth. However, when *CBF3* was expressed under the control of the rd29A promoter from *Arabidopsis*, the resulting plants exhibited normal growth and increased tolerance to salt, drought and cold stresses (Celebi-Toprak *et al*., 2005b). Thus, the improved salt-tolerance would appear to have resulted from the lower level of *CBF3* expressed when under the control of the rd29A promoter. Notably, two transgenic lines exhibited tolerance to 2 M NaCl (Celebi-Toplak *et al*., 2005b). Thus, when *Arabidopsis CBF* genes were introduced into potato, the transgenic plants exhibited improved tolerance to abiotic stress and especially salt tolerance. Figure 6 shows the increased salt tolerance of transgenic plants treated with 50 mM NaCl for 3 weeks.

Arabidopsis CBF3 was over-expressed under the control of the stress-inducible rd29A promoter, conferring salt tolerance to the resulting transgenic potato plants (Celebi-Toprak *et al*., 2005b). In addition, transgenic potato expressing the *DREB* transcription factors and associated genetic components, exhibited multiple abiotic stress tolerances (e.g. drought, salinity and freezing; Kasuga *et al*., 1999). Transgenic potato plants over-expressing oxalate oxidase also showed increased salt tolerance (Turhan, 2005).

5.4. Genetic Engineering of Drought-Tolerant Potato Plants

5.4.1. Introduction of genes encoding functional proteins

Compatible osmolytes such as proline, glycerol, betaines, mannitol and trehalose are accumulated under water stress (Hmida-Sayari *et al*., 2005b) and their biosynthetic genes have been introduced into potato plants in order to improve drought tolerance (Goddijn *et al*., 1997; Yeo *et al*., 2000; Hmida-Sayari *et al*., 2005b). Plants that grow in arid regions such as deserts are known to have high levels of trehalose (Adams *et al*., 1990) and salt stress increases the levels of this osmolyte in yeast (Wiemken,1990). Thus, trehalose is proposed to function as a compatible solute for increasing drought tolerance.

In order to generate drought-tolerant potato plants using trehalose accumulation, the yeast (*Saccharomyces cerevisiae*) gene encoding trehalose phosphate synthase (*tps1*) was introduced into potato plants and over-expressed under the control of the 35S promoter. However, although the resultant plant exhibited increased drought resistance (Figure 7), it showed growth retardation and was phenotypically aberrant, producing small leaves and numerous branches. Interestingly, Satoh-Nagasawa *et al*., (2006) found that inflorescence branching in maize was regulated by three genes including trehalose-6-phosphate phosphatase. In addition, there was no accumulation of trehalose in the transgenic potato plants expressing *tps1*, although mRNA and protein expression was confirmed. Thus, these findings might indicate the presence of an endogenous trehalose degradation enzyme, trehalase (Goddijn *et al*., 1995; Yeo *et al*., 2000).

W C 2-3 5-3 5-5 8-1 10-1

Figure 7. Drought stress treatment of TPS1 transgenic potato plants. Transgenic potato plants were grown at room temperature, in a soil mixture for 2 weeks. They were not watered for 15 d after the 2 weeks growth. W: wild-type. C: control potato plant transformed with empty pBI121 containing the Gus gene. 2–3, 5–3, 5–5, 8–1 and 10–1: TPS1 transgenic potato plants

Proline levels increased in genetically modified potato crops that generated fructan and suppressed expression of formate dehydrogenase, under water deficit. In potato, fructan synthesis influences carbohydrate partitioning in microtubers and raises the total nonstructural carbohydrate content (starch, glucose, sucrose, fructose and fructan levels) in leaves. In addition, fructan performs a stabilizing function for membranes and proteins, by reducing free radical activity and stabilizing proline (Fricke and Pahlich, 1990). Transgenic potato plants that generated a high level of the soluble carbohydrate fructan were developed in order to investigate whether or not water stress could induce proline synthesis in transgenic potato plants (Knipp and Honermeier, 2006). Interestingly, transgenic potato lines generating fructans did not accumulate proline under water stress, suggesting that modification of carbohydrate metabolism and high soluble carbohydrate contents, might affect water stress-induced proline accumulation.

Transgenic potato plants that produced low levels of formate dehydrogenase (FDH; a mitochondrial enzyme that oxidizes formate into CO_2), were generated using antisense *FDH* mRNA and exhibited a reduced ability to use formate as respiratory substrate (Ambard-Bretteville *et al.*, 2003). Suppression of *FDH* expression resulted in rapid accumulation of proline in the leaves under drought stress conditions.

5.4.2. *Introduction of genes encoding regulatory proteins*

The ABA-dependent signal transduction pathway is responsible for binding of the transcription factors myb or myc to the *StRD22* promoter and activating its expression in response to dehydration. *StRD22* mRNA accumulated within 1 h, and peaked after 4 h following treatment with ABA or salt. In addition, expression of StRD22 was induced by cold and drought, and exogenous application of ABA resulted in the accumulation of *StRD22* mRNA (Figure 8).

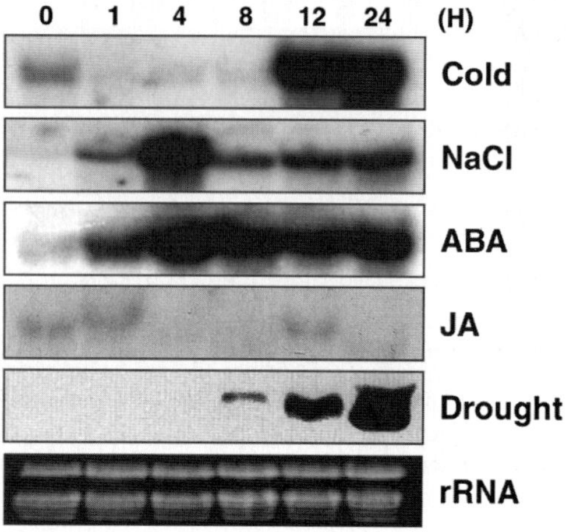

Figure 8. Temporal expression pattern of potato StRD22 following various abiotic stress treatments. C: Un-treated control potato (*S. tuberosum* cv Superior). Cold, 4°C; NaCl, 250 mM NaCl; ABA, 100 μM ABA; JA, 100 mM JA; Drought, air drying. Time after treatment: 0 (untreated control potato), 1, 4, 8, 12 and 24 h

In order to analyze the role of potato StRD22 in the stress response, we generated transgenic *StRD22* tobacco plants in which expression of this gene was controlled by the *A. thaliana* abiotic stress-specific promoter rd29A. We subjected these plants to drought and observed wilting in both transgenic plants over-expressing StRD22 and controls containing vector alone. However, following watering, the former group recovered more quickly than the latter (Figure 9). Thus, potato StRD22 was shown to improve recovery following drought stress rather than conferring drought tolerance. In our lab, we have generated and are characterizing transgenic potato plants harboring *StRD22*.

Figure 9. Recovery of StRD22 over-expressing tobacco plant following dehydration. StRD22 over-expressing transgenic tobacco plants that had been cultured in Jiffy pots for 3 weeks, were subjected to drought stress until the wilting point. Plants were watered and the recovery compared between a control plant and transformants containing empty vector and transformants over-expressing StRD22. C: untransformed control. V: transformant containing empty vector. #1 and #2: rd29A::StRD22 transgenic tobacco lines

5.5. Considerations for Genetic Engineering of Abiotic Stress-Resistant Potato Plants

Due to the quantitative nature and multiple loci of genes involved in plant stress tolerance, it is possible that crop growth and yield may not simply be improved through over-expression of a single gene. Thus, the use of genes encoding stress-related transcription factors or signal transduction pathway components may represent a means to overcome problems arising from the polygenic nature of stress tolerance and regulation of a network of genes (Jaglo-Ottosen *et al.*, 1998; Kasuga *et al.*, 1999; Celebi-Toprak *et al.*, 2005b). For example, transgenic potato plants, which over-expressed MAP kinase under the control of a pathogen-specific promoter, exhibited late blight (*P. infestans*) resistance and alteration in the expression pattern of genes downstream of MAPK (e.g. *hsr203J*, *StrbohC* and *StrbohD*; Yamamizo *et al.*, 2006). Thus, it was proposed that MAP kinase conferred pathogen resistance by up-regulating its downstream genes.

Another consideration in the genetic engineering of abiotic stress-resistant potato plants is to avoid causing detrimental effects to the constitutive plant metabolism. Thus, it is necessary to effect both appropriate and efficient expression of an introduced gene. Such requirements may be met by using promoters that are tissue-specific, developmental stage-specific or stress-inducible (Kasuga *et al.*, 1999). Abiotic and biotic stress-responsive potato promoters that exhibit tissue- or developmental stage-specific expression are being studied and the promoters of vetispiradiene synthase (PVS) and one of the sesquiterpene cyclases, have demonstrated pathogen responsiveness (Yamamizo *et al.*, 2006).

Other promoters, such as the cold-inducible *C17* promoter, exhibit responses to cold, drought, ABA and salt stresses (Kirch *et al.*, 1997). The expression of *ci21A* is induced in potato tubers by cold (Schneider *et al.*, 1997), whereas the promoter of the desiccation-specific gene *StDS2* is induced in response to dehydration (Doczi *et al.*, 2002) and the promoter of the *14-3-3* gene exhibits a strong response to metal ions and NaCl (Aksamit *et al.*, 2005). Moreover, the expression of genes encoding class I patatin is normally tuber-specific, but can be induced in leaves by high concentrations of sucrose. One report has indicated that expression of genes involved in starch synthesis could be controlled using the tuber-specific patatin promoter (Kok-Jacon *et al.*, 2005). Furthermore, regulation of MAP kinase expression by the pathogen-specific potato PVS promoter enabled transformed plants to exhibit tolerance to late blight (*P. infestans*; Yamamizo *et al.*, 2006).

One of the most common problems encountered in the genetic engineering of transgenic plants is that the introduced gene products are unable to function properly due to difficulties in achieving an active formation. Reasons for incorrectly formed gene products include post-translational modification, cofactor acquisition and inhibitory cellular environments.

In some cases introduced gene products may have to compete with endogenous gene products for required precursors, and precursor availability can present a serious problem to active product synthesis. In addition, introduced gene products can sometimes exhibit negative feedback control. For example, although potato plants usually accumulate proline upon water deficit, fructan-generating transgenic potato lines demonstrated lower proline levels than wild-type potato plants. The reason proposed for this finding was that a modification of carbohydrate metabolism, especially at high concentrations of soluble carbohydrate, might affect water stress-inducible proline accumulation (Knipp and Honermeier, 2006).

A further consideration is the unpredictable nature of metabolic changes that can be caused by the introduced gene. Such changes could result in desired gene products being degraded and/or toxic compounds produced. For example, transgenic potato plants that were transformed with the *FDH* antisense construct showed no detectable FDH activity. However, upon drought-stress these under-expressing FDH potato plants exhibited a much more rapid accumulation of proline in the leaves than the non-transgenic control plants (Ambard-Brettevile *et al.*, 2003).

6. CONCLUSIONS AND FUTURE PERSPECTIVES

Due to its tetraploid genetics and the quantitative nature of its traits, improvement of the cultivated potato species by traditional breeding methods was slow and unpredictable. Thus, genetic engineering provides a faster and more reliable means for crop improvement and these techniques are especially applicable to development of resistance to biotic and abiotic stresses such as pathogens, salt, cold and drought.

Several potato varieties have been developed using molecular tools and these include insect-resistant (Colorado beetle, *Leptinotarsa decemlineata*; Perlak *et al.*, 1993) and late blight (*P. infestans*)-resistant plants, the latter being developed using an *RB* gene cloned from the wild potato species *Solanum bulbocastanum* (Song *et al.*, 2003). In addition, virus-resistant potato plants have been developed using molecular engineering techniques and these are commercially available e.g. potato leaf roll virus (PLRV)-resistant potato (NewLeaf Plus™, Monsanto) and potato virus Y (PVY)-resistant potato (NewLeaf Y™, Monsanto).

Genetically modified potato plants have been developed and some that have altered starch profiles for better tuber quality, are undergoing field tests (Mullins *et al.*, 2006). In addition, a hepatitis B surface antigen (HBsAg)-expressing potato has been developed as a potato-based edible vaccine. Following ingestion, 57% of individuals tested developed a hepatitis B immunogenic response. This experiment demonstrated that a non-replicating vaccine, which was delivered by an oral route in the absence of adjuvant, could provide a meaningful immunogenic response (Thanavala *et al.*, 2005).

However, transgenic potato plants that are resistant to abiotic stresses, and in particular to salt and drought, have yet to be developed. Given the developments in the molecular genetics of this species, abiotic stress-resistant potato cultivars could be achieved within the near future. Abiotic stress-related genes, especially those involved in regulatory pathways, such as transcription factors are being isolated and characterized, as are stress-inducible promoters. Many labs around the world, including our lab, have developed some salt-, cold- and drought-resistant transgenic potato plants by introducing stress genes and thus, we foresee that abiotic stress-resistant transgenic potato plants may be commercially available within the near future.

ACKNOWLEDGEMENTS

This material is based upon work supported by the Crop Functional Genomics Project of the 21C Frontier Program governed by the Ministry of Science and Technology and Rural Development Administration.

REFERENCES

Adams, R. P., Kendall, E., and Kartha, K.K., 1990, Comparison of free sugars in growing and desiccated plants of *Selaginella lepidophylla*, *Biochem. Syst. Ecol.* **18**:107–110.

Aksamit, A., Korobczak, A., Skala, J., Lukaszewicz, M., and Szopa, J., 2005, The 14-3-3 gene expression specificity in response to stress is promoter-dependent, *Plant Cell Physiol.* **46**(10):1635–1645.

Ambard-Bretteville, F., Sorin, C., Rebeille, F., Hourton-Cabassa, F., and des Francs-Small, C., 2003, Repression of formate dehydrogenase in *Solanum tuberosum* increases steady-state levels of formate and accelerates the accumulation of proline in response to osmotic stress, *Plant Mol. Biol.* **52**:1153–1168.

An, G., Watson, B. D., and Chiang, C. C., 1986, Transformation of tobacco, tomato, potato and *Arabidopsis thaliana* using a binary Ti vector system, *Plant Physiol.* **81**:301–305.

Bachem, C. W. B., van der Hoeven, R. S., de Bruijn, S. M., Vreugdenhil, D., Zabeau, M., and Visser, R. G. F., 1996, Visualization of differential gene expression using a novel method of RNA fingerprinting based on AFLP: Analysis of gene expression during potato tuber development, *Plant J.* **9**:745–753.

Benavides, M. P., Marconi, P. L., Gallego, S. M., Comba, M. E., and Tomaro, M. L., 2000, Relationship between antioxidant defense systems and salt tolerance in *Solanum tuberosum*, *Aust. J. Plant Physiol.* **27**:273–278.

Bohnert, H. J., 2000, What makes desiccation tolerable? *Genome Biol.* **1**(2):1010.1–1010.4.

Bonierbale, M., Plaisted, R. L., and Tanksley, S. D., 1988, RFLP maps based on a common set of clones reveal modes of chromosomal evolution in potato and tomato, *Genetics* **120**:1095–1103.

Celebi-Toprak, F., Watanabe, J. A., and Watanabe, K. N., 2005a, Molecular markers in identification of genotypic variation, in genetic improvement of Solanaceous crops, Vol 1:Potato, edited by Razdan M, and Mattoo AK, Science publishers, Inc, USA.

Celebi-Toprak, F., Behnam, B., Serrano, G., Kasuga, M., Yamaguchi-Shinozaki, K., Naka, H., Watanabe, A. J., Yamanaka, S., and Watanabe, K. N., 2005b, Tolerance to salt stress of the transgenic tetrasomic tetraploid potato, *Solanum tuberosum* cv. Desiree appears to be induced by the DREB1A gene and rd29A promoter of *Arabidopsis thaliana*, *Breeding Sci.* **55**:311–319.

Cheong, Y. H., Chang, H. S., Gupta, R., Wang, X., Zhu, T., and Luan, S., 2002, Transcriptional profiling reveals novel interactions between wounding, pathogen, abiotic stress, and hormonal responses in Arabidopsis, *Plant Physiol.* **129**:661–677.

Chinnusamy, V., Jagendorf, A., and Zhu, J. K., 2005, Understanding and improving salt tolerance in plants, *Crop Sci.* **45**:437–448.

Choi, H. I., Hong, J. H., Ha, J. O., Kang, J. Y., and Kim, S. Y., 2000, ABFs, a Family of ABA-responsive Element Binding Factors, *J. Biol. Chem.* **275**:1723–1730.

Colton, L. M., Groza, H. I., Wielgus, S. M., and Jiang, J., 2006, Marker-assisted selection for the broad-spectrum potato late blight resistance conferred by gene RB derived from a wild potato species, Crop Sci. 46:589–594.

Deblonde, P. M. K., Haverkort, A. J., and Ledent, J. F., 1999, Responses of early and late potato cultivars to moderate drought conditions: agronomic parameters and carbon isotope discrimination, *Europ. J. Agronomy* **11**:91–105.

Deblonde, P. M. K., and Ledent, J. F., 2001, Effects of moderate drought conditions on green leaf number, stem height, leaf length and tuber yield of potato cultivars, *Europ. J. Agronomy* **14**:31–41.

Dóczi, R., Csanaki, C., and Bánfalvi, Z., 2002, Expression and promoter activity of the desiccation-specific *Solanum tuberosum* gene, *StDS2*, *Plant Cell Environ.* **25**:1197–1203.

Doczi, R., Kondrak, M., Kovacs, G., Beczner, F., Banfalvi, Z., 2005, Conservation of the drought-inducible DS2 genes and divergences from their ASR paralogues in Solanaceous species, *Plant Physiol Biochem.* **43**(3):269–276.

Ercolano, M. R., Ballvora, A., Paal, J., Steinbiss, H. H., Salamini, F., and Gebhardt, C., 2004, Functional complementation analysis in potato via biolistic transformation with BAC large DNA fragments, *Mol. Breed.* **13**:15–22.

Evers, D., Overney, S., Greppin, H., and Hausman, J. F., 1999, Salt tolerance of *Solanum tuberosum* L. overexpressing an heterologous osmotin-like protein, *Biologia plantarum* **42**(1):105–112.

Fidalgo, F., Santos, A., Santos, I., and Salema, R., 2004, Effects of long-term salt stress on antioxidant defence systems, leaf water relations and chloroplast ultrastructure of potato plants. *Annals of Applied Biology* **145**: 185–192.

Fricke, W., and Pahlich, E., 1990, The effect of water stress on the vacuole-extravacuole compartmentation of proline in potato cell suspension cultures, Physiologia Plantarum 78:374–378.

Gebhardt, C., and Valkonen J.P. 2001, Organization of genes controlling disease resistance in the potato genome, Annu. Rev. Phytopathol. 39:79–102.

Gibson, R. W., 1978, Pest aspects of potato production, in The potato crop, edited by Harries PM, Chapman and Hall, London

Gillet, B., Beyly, A., Peltier, G., and Rey, P., 1998, Molecular characterization of CDSP 34, a chloroplastic protein induced by water deficit in *Solanum tuberosum* L. plants, and regulation of CDSP 34 expression by ABA and high illumination, *Plant J.* **16**(2):257–262.

Goddijn, O. J., Verwoerd, T. C., Voogd, E., Krutwagen, R. W., de Graaf, P. T., van Dun, K., Poels, J., Ponstein, A. S., Damm, B., and Pen, J., 1997, Inhibition of trehalase activity enhances trehalose accumulation in transgenic plants, *Plant Physiol.* **113**(1):181–190.

Harris, P. M, 1978, Water, in The potato crop, edited by Harris PM, Chapman and Hall, London

Hide, G. A., and Lapwood, D. H., 1978, Disease aspects of potato production, in The potato crop, edited by Harris PM, Chapman and Hall, London

Hmida-Sayari, A., Costa, A., Leone, A., Jaoua, S., and Gargouri-Bouzid, R., 2005a, Identification of salt stress induced transcripts in potato leaves-AFLP, *Mol. Biotechnol.* **30**:31–39.

Hmida-Sayari, A., Gargouri-Bouzid, R., Bidani, A., Jaoua, L., Savoure, A., and Jaoua, S., 2005b, Overexpression of Δ1-pyrroline-5-carboxylate synthetase increases proline production and confers salt tolerance in transgenic potato plants, *Plant Sci.* **169**:746–752.

Holmberg, N., and Bulow, L., 1998, Improving stress tolerance in plants by gene transfer, *Trends Plant Sci.* **3**:61–66.

Hooker, W. J., 1981, Compendium of potato diseases, The American phytopathological society press, St.Paul, USA.

Huang, J., Hirji, R., Adams, L., Rozwadowski, K. L., Hammerlindl, J. K., Keller, W. A., and Selvaraj, G., 2000, Genetic engineering of glycinebetaine production toward enhancing stress tolerance in plants: metabolic limitations, *Plant Physiol.* **122**:747–756.

Jaglo-Ottosen, K. R., Gilmour, S. J., Zarka, D. G., Schabenberger, O., and Tomashow, M. F., 1998, Arabidopsis CBF1 overexpression induces COR genes and enhances freezing tolerance, *Science* **280**:104–106.

Jefferies, R. A., 1996, Evaluation of seedling selection for salinity tolerance in potato (*Solanum tuberosum* L), *Euphytica* **88**:207–213.

Jeong, M. J., Park, S. C., and Byun, M. O., 2001, Improvement of salt tolerance in transgenic potato plants by glyceraldehyde-3-phosphate dehydrogenase gene transfer, *Mol. Cells.* **12**(2):185–189.

Kasuga, M., Liu, Q., Miura, S., Yamaguchi-Shinozaki, K., and Shinozaki, K., 1999, Improving plant drought, salt, and freezing tolerance by gene transfer of a single stress-inducible transcription factor, *Nat. Biotechnol.* **17**(3):287–291.

Katerji, N., van Hoorn, J. W., Hamdy, A., and Mastrorilli, M., 2003, Salt tolerance classification of crops according to soil salinity and to water stress day index, *Agricultural water management* **43**:99–109.

Kim, D. Y., Lee, J. E., Yi, K. W., Han, S. E., Kwon, H. B., Go, S. J., and Byun, M. O., 2003, Expression pattern of potato (*Solanum tuberosum*) genes under cold stress by using cDNA microarray, *Kor. J. Genetics* **25**(4):345–352.

Kim, H. Y., 2005, Potato cytogenetics, GEO Book, Seoul, Korea.

Kim, H. S., Jeon, J. H., Jeung, Y. H, and Joung, H., 1995, In vitro selection of salt resistant *Solanum tuberosum* L. varieties, *J. Kor. Soc. Hort. Sci.* **36**(2):172–178.

Kirch, H. H., van Berkel, J., Glaczinski, H., Salamini, F., and Gebhardt, C., 1997, Structural organization, expression and promoter activity of a cold-stress-inducible gene of potato (*Solanum tuberosum* L.), *Plant Mol. Biol.* **33**(5):897–909.

Kishor, P., Hong, Z., Miao, G. H., Hu, C., and Verma, D., 1995, Overexpression of [delta]-pyrroline-5-carboxylate synthetase increases proline production and confers osmotolerance in transgenic plants, *Plant Physiol.* **108**:1387–1394.

Knipp, G. and Honermeier, B., 2006, Effect of water stress on proline accumulation of genetically modified potatoes (*Solanum tuberosum* L.) generating fructans, *J. Plant Physiol.* **163**(4):392–397.

Kok-Jacon, G. A., Vincken, J. P., Suurs, L. C., Wang, D., Liu, S., and Visser, R. G., 2005, Production of dextran in transgenic potato plants, *Transgenic Res.* **14**(4):385–395.

Kopka, J., Pical, C., Gray, J. E., and Muller-Rober, B., 1998, Molecular and enzymatic characterization of three phosphoinositide-specific phospholipase C isoforms from potato, *Plant Physiol.* **116**:239–250.

Kopka, J., Provart, N. J., and Muller-Rober, B., 1997, Potato guard cells respond to drying soil by a complex change in the expression of genes related to carbon metabolism and turgor regulation, *Plant J.* **11**(4):871–882.

Kreps, J. A., Wu, Y., Chang, H. S., Zhu, T., Wang, X., and Harper, J. F., 2002, Transcriptome changes for Arabidopsis in response to salt, osmotic, and cold stress, *Plant physiol.* 130(4):2129–2141.

Langenkämper, G., Manac'h, N., Broin, M., Cuiné, S., Becuwe, N., Kuntz, M., and Rey, P., 2001, Accumulation of plastid lipid-associated proteins (fibrillin/CDSP34) upon oxidative stress, ageing and biotic stress in Solanaceae and in response to drought in other species, *J Exp. Bot.* 52:1545–1554.

Lee, H. E., Shin D. J., Park, S. R., Han, S. -E., Jeong, M. -J, Kwon, T. R., S. K., Park, S. C, Yi, B. Y., Kwon, H. B., and Byun, M. O (2007) Ethylene responsive element binding protein 1 (StEREBP1) from *Solanum tuberosum* increases tolerances to abiotic stress in transgenic potato plants. *Biochem. Biophys. Res. Commun.* 353(2007) 863–869.

Leone, A., Costa, A., Consiglio, F., Massarelli, I., Dragonetti, E., De Palma, M., and Grillo, S., 1999, Tolerance to abiotic stresses in potato plants: a molecular approach, *Potato research* **42**:333–351.

Luo, Z. W., Zhang, Z., Leach, L., Zhang, R., Bradshaw, J., and Kearsey, M., 2006, Constructing genetic linkage maps under a tetrasomic model, *Genetics* **172**:2635–2645.

Maqbool, B., Zhong, H., El-Maghraby, Y., Ahmad, A., Chai, B., Wang, W., Sabzikar, R., and Sticklen, B., 2002, Competence of oat (Avena sativa L.) shoot apical meristems for integrative transformation, inherited expression, and osmotic tolerance of transgenic lines containing hva1, *Theor Appl. Genet.* **105**:201–208.

Mass, E. V., and Hoffman, G. K. J., 1977, Crop salt tolerance-current assessment, *J. Irrig. Drain Div.***103**:115–134.

Menke, U., Renault, N., and Mueller-Roeber, B., 2000, StGCPRP a potato gene strongly expressed in stomatal guard cells, defines a novel type of repetitive proline-rich proteins, *Plant Physiol.* **122**: 677–686.

Mine, T., Hiyoshi, T., Kasaoka, K., and Ohyama, A., 2003, CIP353 encodes an AP2/ERF-domain protein in potato (*Solanum tuberosum* L.) and responds slowly to cold stress, *Plant cell physiol.* **44**(1):10–15.

Mullins, E., Milbourne, D., Petti, C., Doyle-Prestwich, B. M., and Meade, C., 2006, Potato in the age of biotechnology, *Trends Plant Sci.* **11**(5):254–260.

Ochatt, S. J., Marconi, P. L., Radice, S., Arnozis, P. A., and Caso, O. H., 1999, In vitro recurrent selection of potato: production and characterization of salt tolerant cell lines and plants, *Plant. Cell, Tissue and Organ Culture* **55**: 1–8.

Park, S., Kang, T. S., Kim, C. K., Han, J. S., Kim, S., Smith, R. H., Pike, L. M., and Hirschi, K. D., 2005, Genetic manipulation for enhancing calcium content in potato tuber. *J Agric Food Chem.* **53**(14):5598–5603.

Perlak,F. J., Stone, T. B., Muskopf, Y. M., Petersen, L. J., Parker, G. B., McPherson, S. A., Wyman, J., Love, S.,Reed, G., Biever, D., and Fischhoff, D. A., 1993, Genetically improved potatoes: protection from damage by Colorado potato beetles, *Plant Mol. Biol.* **22**:313–321.

Pilon-Smits, E., Ebskamp, M., Paul, M., Jeuken, M., Weisbeek, P., and Smeekens, S., 1995, Improved performance of transgenic fructan-accumulating tobacco under drought stress, *Plant Physiol.* **107**:125–130.

Plant, A. L., and Bray, E. A., 1999, Regulation of gene expression by abscisic acid during environmental stress, in plant responses to environmental stress, edited by Lerner HR, Printed by Marcel Dekker, New York.

Pruvot, G., Cuiné, S., Peltier, G., and Rey, P., 1996, Characterization of a novel drought-induced 34-kDa protein located in the thylakoids of *Solanum tuberosum* L. plants, *Planta* **198**:471–479.

Rahnama H., and Ebrahimzadeh, H., 2005, The effect of NaCl on antioxidant enzyme activities in potato seedling, *Biologia Plantarum* **49**:93–97.

Rensink, W. A., Hart, A., Liu, J., Ouyang, S., Zismann, V., and Buell, C. R., 2005a, Analyzing the potato abiotic stress transcriptome using expressed sequence tags, *Genome.* **48**(4):598–605.

Rensink, W. A., Iobst, S., Hart, A., Stegalkina, S., Liu, J., and Buell, C. R., 2005b, Gene expression profiling of potato responses to cold, heat, and salt stress, *Funct. Integr. Genomics* **5**(4):201–207.

Rey, P., Pruvot, G., Becuwe, N., Eymery, F., Rumeau, D., and Peltier, G. A., 1998, Novel thioredoxin-like protein located in the chloroplast is induced by water deficit in *Solanum tuberosum* L. plants, *Plant J.* **13**(1):97–107.

Satoh-Nagasawa, N., Nagasawa, N., Malcomber, S., Sakai, H., and Jackson, D., 2006, A trehalose metabolic enzyme controls inflorescence architecture in maize, *Nature* **441**: 227–230.

Schenk, P. M., Kazan, K., Wilson, I., Anderson, J. P., Richmond, T., Somerville, S. C., and Manners, J. M., 2000, Coordinated plant defense responses in Arabidopsis revealed by microarray analysis, *Proc. Natl. Acad. Sci. USA.* **97**(21):11655–11660.

Schneider, A., Salamini, F., and Gebhardt, C., 1997, Expression patterns and promoter activity of the cold-regulated gene ci21A of potato, *Plant Physiol.* **113**(2):335–345.

Seki, M., Narusaka, M., Abe, H., Kasuga, M., Yamaguchi-Shinozaki, K., Carninci, P., Hayashizaki, Y., and Shinozaki, K., 2002, Monitoring the expression pattern of 1300 Arabidopsis genes under drought and cold stresses by using a full-length cDNA microarray, *Plant Cell* **13**(1):61–72.

Seki, M., Narusaka, M., Ishida, J., Nanjo, T., Fujita, M., Oono, Y., Kamiya, A., Nakajima, M., Enju, A., Sakurai, T., Satou, M., Aiyama, K., Taji, T., Yamaguchi-Shinozaki, K., Carninci, P., Kawai, J., Hayashizaki, Y., and Shinozaki, K., 2002, Monitoring the expression profiles of 7000 Arabidopsis genes under drought, cold and high-salinity stresses using a full-length cDNA microarray, *Plant J.* **31**(3):279–292.

Shaterian, J., Georges, F., Hussain, A., Waterer, D., Jong H. D., and Tanino, K. K., 2005, Root to shoot communication and abscisic acid in calreticulin(CR) gene expression and salt-stress tolerance in grafted diploid potato clones, *Env. Exp. Bot.***53**:323–332.

Shinozaki, K., Yamaguchi-Shinozaki, K., and Seki, M., 2003, Regulatory network of gene expression in the drought and cold stress response, *Curr. Opin. Plant Biol.* **6**:410–417.

Shinozaki, K., and Yamaguchi-Shinozaki, K., 1997, Gene expression and signal transduction in water-stress response, *Plant Physiol.* **115**:327–334.

Silhavy, D., Hutvágner, G., Barta, E., and Bánfalvi, Z., 1995, Isolation and characterization of water stress inducible cDNA clone from *Solanum chacoense*, *Plant Mol. Biol.* **27**:587–595.

Silva, J. A. B., Otoni, W. C., Martinez, C. A., Diasm L. M., and Silvam M. A. P., 2001, Microtuberization of Andean potato species (Solanum spp.) as affected by salinity, *Scientia Horticulturae* **89**:91–101.

Song, J., Bradeen, J. M., Naess, S. K., Raasch, J. A., Wielgus, S. M., Haberlach, G. T., Liu, J., Kuang, H., Austin-Phillips, S., Buell, C. R., Helgeson, J. P., and Jiang, J., 2003, Gene *RB* cloned from *Solanum bulbocastanum* confers broad spectrum resistance to potato late blight, *Proc.Natl. Acad. Sci. USA*. **100**:9128–9133.

Swiedrych, A., Lorenc-Kukula, K., Skirycz, A., and Szopa, J., 2004, The catecholamine biosynthesis route in potato is affected by stress, *Plant Physiol. Biochem.* **42**:593–600.

Tanksley, S. D., Ganal, M. W., Prince, J. P., de Vicente, M. C., Bonierbale, M. W., Broun, P., Fulton, T. M., Giovannoni, J. J., Grandillo, S., Martin, G. B., Messeguer, R., Miller, J. C., Miller, L., Paterson, A. H., Pineda, O., Roder, M. S., Wing, R. A., Wu, W., and Young, N. D., 1992, High density molecular linkage maps of the tomato and potato genomes, *Genetics* **132**(4):1141–1160.

Thanavala, Y., Mahoney, M., Pal, S., Scott, A., Richter, L., Natarajan, N., Goodwin, P., Arntzen, C. J., and Mason, H. S., 2005, Immunogenicity in humans of an edible vaccine for hepatitis B, *Proc. Natl. Acad. Sci. U S A*. **102**:3378–3382.

Thomashow, M. F., 1998, Role of cold-responsive genes in plant freezing tolerance, *Plant Physiol.* **118**(1):1–8.

Turhan, H., 2005, Salinity response of transgenic potato genotypes expressing the oxalate oxidase gene, *Turk. J. Agric.* **20**:187–195.

Valverde, R., Chen, T. H. H., and Li, P. H., 1997, Frost hardiness and cold acclimation in Solanum species. in Plant cold hardiness: molecular biology, Biochemistry and physiology, edited by Li PH and Chen THH, Plenum Press, NewYork.

Van Breusegem, F., Slooten, L., Stassart, J. M., Moens, T., Botterman, J., Van Montagu, M., and Inze, D., 1999, Overproduction of *Arabidopsis thaliana* FeSOD confers oxidative stress tolerance to transgenic maize, *Plant Cell Physiol.* **40**(5):515–523.

Watanabe, K. N., 2002, Challenges in biotechnology for abiotic stress tolerance on roots and tubers, JIRCAS Working Report 75–83.

Wiemken, A., 1990, Trehalose in yeast, stress protectants rather than reserve carbohydrate, *Antonie van Leeuwenhoek* **58**:209–217.

Wu, G., Shortt, B. J., Lawrence, E. B., Levine, E. B., Fitzsimmons, K. C., and Shah, D. M., 1995, Disease resistance conferred by expression of a gene encoding H_2O_2-generating glucose oxidase in transgenic potato plants, *Plant Cell* **7**(9):1357–1368.

Xiong, L., Schumaker, K. S., and Zhu, J. K., 2002, Cell signaling during cold, drought, and salt stress, *Plant Cell* **14** Suppl:S165–183.

Xu, D., Duan, X., Wang, B., Hong, B., Ho, T., and Wu, R., 1996, Expression of a late embryogenesis abundant protein gene, HVA1, from barley confers tolerance to water deficit and salt stress in transgenic rice, *Plant Physiol.* **110**(1):249–257.

Yamamizo, C., Kuchimura, K., Kobayashi, A., Katou, S., Kawakita, K., Jones, J. D., Doke, N., and Yoshioka, H., 2006, Rewiring mitogen-activated protein kinase cascade by positive feedback confers potato blight resistance, *Plant Physiol.* **140**(2):681–692.

Yeo, E. T., Kwon, H. B., Han, S. E., Lee, J. T., Ryu, J. C., and Byun, M. O., 2000, Genetic engineering of drought resistant potato plants by introduction of the trehalose-6-phosphate synthase (TPS1) gene from *Saccharomyces cerevisiae*. *Mol. Cells.* **10**(3):263–268.

Zhang, Z., Mao, B., Li, H., Zhou, W., Takeuchi, Y., and Yoneyama, K., 2005, Effect of salinity on physiological characteristics, yield and quality of microtubers *in vitro* in potato, *Acta Physiologiae Plantarum*, **27**:481–489.

Zhu, B., Chen, T. H. H., and Li, P. H., 1995, Activation of two osmotin-like protein genes by abiotic stimuli and fungal pathogen in transgenic potato plants, *Plant Physiol.* **108**:929–937.

CHAPTER 30

RECENT ADVANCES IN BREEDING FOR DROUGHT AND SALT STRESS TOLERANCE IN SOYBEAN

MD S. PATHAN[1], JEONG-DONG LEE[2], J. GROVER SHANNON[2], AND HENRY T. NGUYEN[1]

[1]*National Center for Soybean Biotechnology and Division of Plant Sciences, University of Missouri-Columbia, Missouri 65211, USA*
[2]*Delta Center, University of Missouri-Columbia, Portageville, Missouri 63873, USA*

Abstract: Drought and salinity are two important abiotic factors limiting soybean production worldwide and drought alone accounts for about 40% crop loss. Irrigation and soil reclamation are not economically viable options for soybean production under drought and salinity. Hence, genetic improvement for drought and salt tolerance are cost effective. Conventional breeding has made a significant contribution to soybean improvement in the last 50 years. Through conventional breeding, it is easy to manipulate simply inherited qualitative traits which are less sensitive to environmental variation, but quantitative traits like yield or tolerance to abiotic stress are significantly influenced by environment. Most agronomically important traits are quantitatively inherited and are difficult to improve through conventional breeding. Molecular marker technologies can dissect quantitative traits into individual components, known as quantitative trait loci enabling marker assisted selection of desired traits in much shorter time avoiding labor intensive, conventional, phenotypic selection. A molecular breeding approach can supplement the conventional breeding system. Well developed molecular genetic maps, functional genomic resources, and other molecular tools are available for soybean. Effective use of these resources will allow a greater understanding of basic mechanisms of tolerance to abiotic stress. Integration of these genomic tools coupled with well-designed breeding strategies will help to develop soybean varieties with higher tolerance to drought and salt

Keywords: Drought and salinity tolerance, conventional breeding, marker assisted selection, genomics, candidate genes, genetic engineering, yield, soybean

1. INTRODUCTION

Differences in soybean [*Glycine max* (L.) Merr.] yield from year to year on the same farm are often due to abiotic stress. Globally these factors are more important to the farmer than diseases or insects (Carter et al., 1999). Abiotic stress in soybean is

M.A. Jenks et al. (eds.), Advances in Molecular Breeding Toward Drought and Salt Tolerant Crops, 739–773.

primarily due to too little (drought) or too much water (waterlogging/submergence), salinity, nutrient deficiency or toxicity and low or high temperatures. Abiotic stress are responsible for more than 50% yield losses worldwide (Boyer, 1982; Bray et al., 2000). Drought and salinity are major abiotic stress that adversely affect soybean production and quality. Drought is the single most economically important abiotic stress affecting crop productivity, thus drought tolerance is often a major goal of most soybean breeding programs (Tuberosa and Salvi, 2004). About 20% of irrigated agricultural land is adversely affected by salinity (Flowers and Yeo, 1995). Salt damage to soybean occurs as a result of storm surges and seawater intrusion from the ocean, over-fertilization and from irrigation water with high salt content.

When plants are subjected to abiotic stress, they activate different physiological, cellular, metabolic and defense mechanisms to survive and sustain growth until maturity. Therefore, understanding genetic mechanisms for stress tolerance is crucial in the development of tolerant varieties. Plant traits associated with stress tolerance are often controlled by several genes or quantitative trait loci (QTL) and are difficult to improve through conventional breeding.

1.1. Origin, Genome, Importance, and Production of Soybean

China is the primary center of origin of soybean and the crop was domesticated during 1500–1100 B. C. It was introduced into East Asian countries during first century A. D. to the age of discovery (15th–16th century). Soybean was introduced in Europe during the 16th and 17th centuries and was brought into North America in 1765 (Hymowitz, 1970, 1990, 2004). Soybean is a member of the genus *Glycine* willd., which is a member of the legume family *Leguminosae,* subfamily *Papilionoideae* and tribe *Phaseoleae*. *Phaseoleae*, which includes common bean, lima bean, mungbean and cowpea, is the most important tribe of the *Leguminosae* with members that have great importance for food and feed. Soybeans are divided into two subgenera, *Glycine* (perennials) and *Soja* (Moench) F. J. Herm. (annuals). The subgenus *Glycine* contains 22 species including *G. tabacina* and *G. tomentella*. The subgenus *Soja*, includes *Glycine max,* the cultivated soybean, and *Glycine soja,* the wild annual soybean. Wild soybean grows in China, Japan, Korea, Russia, and Taiwan in fields, hedgerows, along roadsides and riverbanks. *G. soja* plants are annual, procumbent with slender twining growth and generally have purple flowers and tawny pubescence (Hymowitz, 2004). Soybean plants are diploid with 20 pairs of chromosomes on which a genetic map for each of the 20 linkage groups has been constructed (Song et al., 2004). The soybean genome consists of ~1.1 Mbp, which is relatively larger than those of Arabidopsis (7.5 times; Mbp/C) or rice (2.5 times; Mbp/C), but smaller than corn (2.4 times; Mbp/C) or wheat (14 times; Mbp/C) (Arumuganathan and Earle, 1991). The soybean genome evolved from two rounds of polyploidization or duplication (Shoemaker et al., 1996; Blanc and Wolfe, 2004; Schlueter et al., 2004), and 35% of the soybean genome is diploidized (Shultz et al., 2006).

Soybean is the world's primary source of protein and oil and is often called the miracle crop because of its numerous uses. Soybean seeds contain an average 40% protein, 35% carbohydrate, 20% oil, and 5% ash (Liu, 1997). Soybean is now an essential and dominant source of protein and oil with over 200 uses in feed, food and industrial applications. Recent studies indicate that consumption of soybean reduces cancer, blood serum cholesterol, osteoporosis and heart disease (Birt et al., 2004). Also soybeans are a good source of minerals, vitamins, folic acid, and isoflavones which are credited with slow development of these diseases (Wilson, 2004). Thus, the demand for many edible soybean products has increased dramatically. Also, the desire for more meat in diets among the world's population has increased, consequently the demand for soybean protein for livestock and poultry feed has also increased. In addition to feed and food, soybean has numerous industrial applications such as building materials, plastics, printing inks, paints, hydraulic fluids, cosmetics, pharmaceuticals and soy-diesel fuel that burns cleaner and pollutes less than petroleum derived fuels.

The increased importance of soybeans as a world crop has led to a huge expansion in world soybean production (http://www.soystats.com). In the last twenty years, world soybean production has increased steadily from 70 million tons in 1984 to 217 million tons in 2004 (Figure 1). About 80% of the world soybean was produced in North and South America. The United States, Brazil and Argentina were the major producers and exporters of soybean. Among these countries, the United States was the leading soybean producer at 86 million tons or about 40% of the total world production. At the same time Brazil and Argentina produced about 50 and 38 millions tons (24% and 18 % of the total), respectively (www.soystats.com). Although soybean is native to China, China produced 17 millions tons (8.5% of the

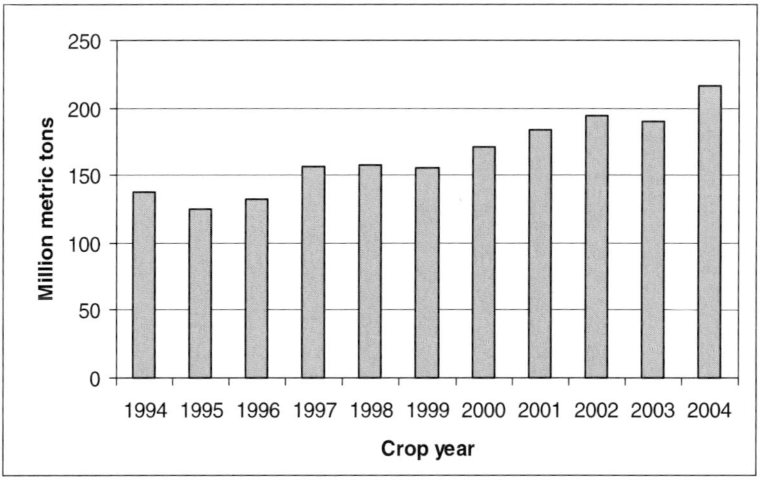

Figure 1. World soybean production in million metric tons from 1984 to 2004 (USDA & Soy Stats-www.fas.usda.gov, www.soystats.com)

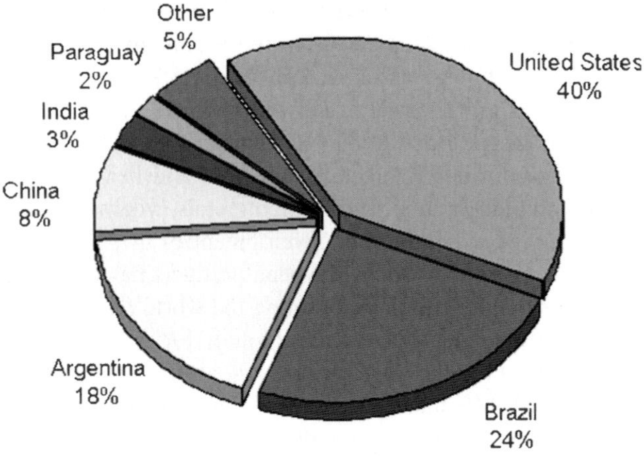

Figure 2. World soybean production, 2004 (Soy Stats-www.soystats.com)

total) and India produced about 7 millions tons (3% of the total). The remaining 5% was produced in countries of Asia and Europe (Figure 2). In 2005 total oilseed production was 380 million tons of which 57% was from soybean making it the world's number one oil seed crop followed by rapeseed and cotton seed at 12% each (Figure 3).

Genetic variability is a key resource for varietal improvement. The genetic base of North American soybean cultivars is narrow (Singh and Hymowitz, 1999; Cui et al., 2000). Improved soybean cultivars are depend on the addition of

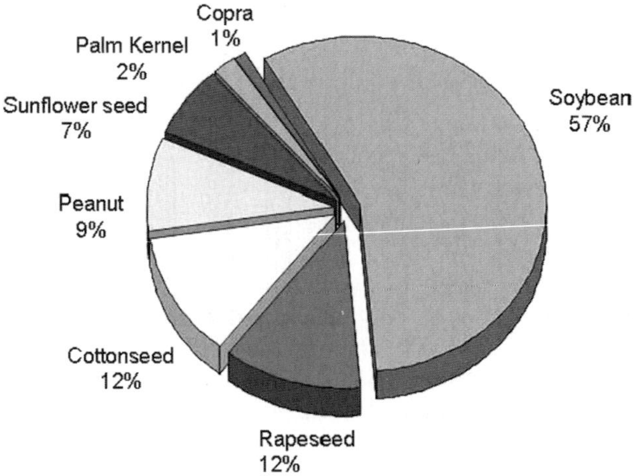

Figure 3. World oilseed production, 2004 (Soy Stats-www.soystats.com)

genes from exotic germplasm for higher yield and resistance to biotic and abiotic stress. Exploitation and utilization of diverse germplasm is essential to widen the genetic base for the development of new soybean cultivars. Beneficial alleles have been identified in wild germplasm in other crops and utilized for cultivar improvement, but so far a little progress has been made in soybean. More than 170,000 *G. max* accessions are maintained by 160 institutions in 70 countries. In addition, there are also 10,000 accessions of *G. soja* and 3,500 accessions of *glycine* species (www.ipgri.org). China has the largest collection of soybean germplasm in the world with about 26,000 accessions of *G. max* followed by United States with about 17,000 accessions of *G. max* in the USDA soybean germplasm collection (Chang et al., 1999; Carter et al., 2004).

2. SOYBEAN MOLECULAR GENETIC AND PHYSICAL MAPS

During the last two decades, different molecular markers have been developed and used for genetic map construction, QTL analysis, and marker-assisted selection in soybean. A number of soybean molecular linkage maps have been constructed using important mapping populations like recombinant inbred lines (RILs) or F_2 derived lines. The first soybean genetic linkage map was constructed with 150 RFLPs markers using 59 F_2 lines derived from the inter-specific cross between *Glycine max* (A81-356022) and *Glycine soja* (PI468916) (Keim et al., 1990). Shoemaker and Olson (1993) added more than 300 markers on the initial map. A new integrated genetic linkage map of soybean was constructed by integrating five mapping populations comprised of 20 consensus linkage groups that spanned 2,523 cM of Kosambi map distance (Song et al., 2004). Population sizes of each of the integrated five maps range from 57 to 240 lines. This map consists of 20 linkage groups with 1849 markers, including 1015 SSRs, 709 RFLPs, 73 RAPDs, 6 AFLPs, 24 classical traits, 10 isozymes, and 12 other markers. Recently, Cregan et al. (2006) placed about 1,183 SNP markers on the pre-existing RFLP/SSR-based soybean genome map. Availability of integrated SSR/RFLP/SNP soybean genetic linkage maps facilitates the precise dissection of specific genetic loci of interest. As of October 2006, a total of 480 soybean genes and 1174 QTLs have been reported in the USAD-ARS soybase database (http://soybase.ncgr.org). A high density genetic map is a prerequisite for soybean genome studies and genetic analysis of genes related to any important soybean agronomic trait. The National Center for Soybean Biotechnology at the University of Missouri-Columbia has initiated the construction of a high density genetic linkage map using more than 750 F_2 lines developed from the cross between Forrest x Williams 82.

Physical maps are a powerful resource for large-scale genome sequencing marker development, positional cloning and EST mapping (Adams et al., 2000; Wu et al., 2004). Whole-genome physical maps have been constructed for different species including *Arabidopsis thaliana* (Mozo et al., 1999; Chang et al., 2001) and rice (Tao et al., 2001; Chen et al., 2002). Wu et al. (2004) reported a genome-wide, BAC and plant-transformation competent binary large-insert plasmid clone (BIBAC)-based physical map of the soybean genome. The map was developed from five BAC and BIBAC libraries representing 9.6 haploid genomes, with 2,905 BAC/BIBAC contigs, in an estimated span of about 1,400 Mb. To accelerate the soybean genomics research, the soybean research community at the University of Missouri-Columbia has constructed a BAC-based physical map of cultivar Williams 82, since this genotype has been commonly used for soybean genomic research and whole genome sequencing. Anchoring genetically mapped molecular markers to the physical map will help to understand soybean genome structure and function (Wu et al., 2006). The soybean genome database, (SoyGD: http://soybeangenome.siu.edu), is an important resource for the soybean physical map, BAC fingerprint database and genetic map (Shultz et al., 2006).

3. ASSOCIATION MAPPING/ LINKAGE DISEQUILIBRIUM MAPPING

Association mapping or linkage disequilibrium (LD) mapping studies the association of a molecular marker with a phenotypic trait of interest in unrelated individuals of a population rather than a mapping population of known pedigree. Association mapping does not require any crossing and is suitable for fine scale mapping with a greater possibility for recombination to take place than traditional pedigree studies (Nordborg and Tavare, 2002). This approach is mainly used for the study of marker-trait association followed by MAS, and the study of genetic diversity in a natural population and development of germplasm for crop improvement. Association mapping has made significant progress in human genetics and recently applied in plant genetics. Gupta et al. (2005) in their review summarized a list of LD studies in plants. Aranzana et al. (2005) first studied genome wide association mapping for flowering time and pathogen resistance in 95 accessions of *Arabidopsis thaliana*. In soybean, LD has been used for the studies of genetic diversity and SNP frequency detection (Cregan et al., 2002; Zhu et al., 2003; Hyten et al., 2004). Recently, Hyten and his group (Hyten et al., 2006) used 96 *Glycine max* landrace to assess genome-wide LD in soybean using 345 SNPs. They also reiterated the need of genome-wide LD map to determine the optimum marker coverage to detect most QTL present in a genome wide association analysis. Linkage analysis is suitable for general QTL mapping and alternately, LD mapping gives a more precise location of the QTL that controls the trait of interest (Glazier et al., 2002; Gupta et al., 2005). A SNP based genome map, genome-wide sequence information and integrated linkage and LD map of certain QTL/genes of interest will revolutionize soybean improvement through MAS.

4. FUNCTIONAL GENOMIC TOOLS AND RESOURCES

Functional genomics has become an important discipline to identify genes, gene structure and function, and to elucidate the biochemical pathways operating in a cell or tissue to define that specific cell or tissue type. The availability of complete genome sequence information of model plants (Arabidopsis, rice and Medicago), has enabled scientists to focus on understanding the relationship between genotype and phenotype through an integrated functional genomics approach. The US Department of Energy Joint Genome Institute (DOE-JGI) in collaboration with soybean research community has launched a program to sequence the soybean genome. Important functional genomic tools and resources currently used are expressed sequence tags (ESTs), full length cDNA sequences (FL-cDNA), gene expression profiling through microarray and serial analysis of gene expression (SAGE), proteomics, metabolomics, and bioinformatics.

Expressed sequence tags (ESTs) are random sequences of gene transcripts, and are a novel genomic tool for gene identification. EST sequence information is largely used for making a catalogue of expressed genes through microarray and SNP detection. More than 390,000 soybean ESTs are available in the GenBank and are derived from different tissue and organ systems including developing seeds, seed coats, leaves, pods, roots, and numerous stages of plants regenerated via tissue culture (http://www.ncbi.nlm.nih.gov/entrez), and out of these, about 21,000 are unigenes. This resource includes about 15,000 ESTs derived from drought stressed soybean roots generated by the scientists of National Center for Soybean Biotechnology (NCSB) at the University of Missouri-Columbia. Tian et al. (2004) analyzed 314,000 ESTs (284,000 ESTs from GenBank and 30,000 ESTs from their lab) and detected 61 genes regulated by salicylic acid, 326 disease resistance genes and 1,322 transcription factors. Salicylic acid genes respond to abiotic stress like salt and osmotic stress (Borsani et al., 2001). Recently, through the cDNA-amplified fragment length polymorphism (cDNA-AFLP), several salt-induced genes and one novel gene, *GmTDF-5* involved in water potential changes under salt stress (Aoki et al., 2005) have been reported in soybean

DNA microarray analysis has provided a unique opportunity for transcript profiling at the whole genome level to study gene expression patterns and discovery of gene function under certain conditions. In soybean, currently three different array platforms are available for gene expression profiling studies. Among them, GeneChip®, the soybean genome array designed and marketed by Affymetrix (www.affymetrix.com), contains about 60,000 transcripts. This short oligo array (25-mer) contains about 36,000 transcripts from *Glycine max*, 16,000 transcripts from the important pathogen *Phytophthora sojae* (a water mold that commonly attacks soybean crops) and 7,500 transcripts from the world's most devastating pest, soybean cyst nematode *Heterodera glycines*. The spotted cDNA microarray contains 36,000 elements constructed from soybean cDNAs derived from a variety of soybean EST libraries representing a wide source of tissues and organs, developmental stages and stress or pathogen infected plants (Vodkin et al., 2004). Recently, Vodkin et al. (2006) developed a set of 70-mer long oligo arrays representing 38,000 unigenes. Only a few reports are available on soybean gene expression analysis. Maguire et al. (2002) used soybean cDNA microarray containing about 4,100 unigene ESTs derived from axenic roots to evaluate tissue specific differentiation. Transcript profiling was also done for somatic embryogenesis in soybean (Thibaud-Nissen et al., 2003), soybean cyst nematode (Khan et al., 2004), *Pseudomonas syringae* (Zou et al., 2005), and nitrogen fixation symbiosis (Brechenmacher et al., 2006). To date there are no reports on soybean abiotic stress and related gene expression profiling. Recently, expression profiling of drought stressed soybean leaf, root and seed protein work is in progress at the author's laboratory using both Affymetrix and cDNA arrays.

Gene expression profiling measures mRNA expression at the genome level, but it is not always proportionate to the amount of protein derived from the expressed gene. Proteomics deals with the analysis of protein content in a biological unit at a specific developmental stage and under various biotic and abiotic conditions. Most of the proteomic analysis is based on two-dimensional electrophoresis (2-DE) and matrix-assisted laser desorption/ionization time of flight mass spectrometry (MALDI-TOF MS). Plants produce different proteins and protective compounds to resist biotic and abiotic stress and proteomic analysis helps to identify stress-related proteins (Horvath-Szanics et al., 2006; Rampitsch and Srinivasan, 2006). A proteomic approach has been successfully applied in different crops to identify stress related proteins. Stress related proteins have been identified in drought stressed rice leaves (Salekdeh et al., 2002), in rice roots exposed to salt (Yan et al., 2005) and in drought stressed wheat (Horvath-Szanics et al., 2006). Hajduch et al. (2005) did a systemic study to determine the expression pattern and to identify proteins during seed filling in soybean and they have also developed a user-intuitive database (http://oilseedproteomics.missouri.edu). Kim et al. (2006) used a proteomic approach to study flooding stress in soybean. Proteomic analysis of drought stressed soybean root, leaf and seed is in progress in author's lab at the University of Missouri-Columbia.

Metabolomics measures low molecular weight endogenous metabolites present in a sample and profiles of these compounds in response to specific conditions, like biotic and abiotic stress. Since metabolites play a significant role in regulating cellular processes, transcriptomic and proteomic data are not sufficient to fully explain the complex biological systems. Lin et al. (2006) noted that a metabolomics approach provides the most functional measure of cellular status and helps to describe a genotype of an organism. This approach has been used for metabolic engineering to increase isoflavone biosynthesis in soybean seed (Yu et al., 2003), and profiling for biotic and abiotic elicitors on metabolism in *Medicago truncatulla* (Broeckling et al., 2005). Research work is in progress at the University of Missouri on metabolic engineering for soybean seed sterol biosynthesis (Neelakandan et al., 2006) and metabolic profiling of soybean seed under drought stress.

A large amount of data has been generated by high throughput functional genomics techniques. A computational tool known as bioinformatics being used to analyze, integrate, deposit, and make these data available and user friendly accessible to these resources.

5. APPLICATION OF MOLECULAR BREEDING IN SOYBEAN

The goal of soybean breeding programs is to develop superior soybean cultivars with improved yield, seed composition, and resistance to biotic and abiotic stress. Plant breeding requires incorporation of important traits into adapted varieties through conventional or molecular breeding. Most of the soybean improvement so far has been achieved through conventional breeding approach. Development of a soybean cultivar through conventional breeding requires considerable time and labor and also takes a large amount of space in the greenhouse and field for evaluation. A limited number of soybean lines can be evaluated in the field per season and generally it takes about 8 to 10 years to develop a cultivar. However, molecular applications are being used as a tool to improve accuracy and efficiency, and to reduce the time from the cross between two selected parents to variety/germplasm release.

Through conventional breeding, it is easy to manipulate simply inherited or qualitative traits which often are less sensitive to environmental variation. However, it is difficult to manipulate quantitative traits like yield, or tolerance to abiotic stress, since these traits generally have low heritability and sensitivity to environmental variations. Most of the agronomically important traits are quantitatively inherited and it is difficult for breeders to improve these traits using conventional breeding methods. With the development of molecular marker technologies, it is possible to dissect quantitative traits into individual components, known as quantitative trait loci (QTL) (Tanksley, 1993; Quarrie, 1996; Beavis, 1998; Tuberosa et al., 2002). Thus, the use of molecular markers to select desired traits can be done in much shorter time by avoiding labor intensive and expensive conventional phenotypic selection in the greenhouse or field. Earlier there was concern about accuracy of QTLs in tagging genes. But recently, Price (2006) reported that even though few plant QTLs have been cloned or accurately mapped to within 2 cM or less, QTL mapped to within 20 cM of genes in initial mapping studies are accurate enough to be useful in identifying genes for specific traits. However, fine mapping to detect QTL to within 2 cM of genes for various traits is highly desirable. Availability of high-density soybean genetic maps, progress in QTL mapping, and application of marker–assisted selection (MAS) has increased significantly. Orf et al. (2004) summarized the potential uses of MAS in soybean breeding, such as selection of parents with highest potential for breeding programs, to monitor gain or loss of genetic regions during backcrossing, recovery of recurrent parents, selection in segregating populations, and mining new beneficial alleles from the exotic/wild germplasm. MAS coupled with the conventional breeding techniques have increased efficiency and reliability in soybean breeding programs. MAS is being used both in the public and private sector, but on a limited scale such as the improvement of a few traits like resistance to soybean cyst nematode. Pioneer Hi-Bred International (Johnston, IA) has successfully used MAS for development of disease resistant soybean varieties. Monsanto has also integrated molecular marker based breeding tools in their breeding programs and have successfully improved their selection efficiency in cultivar development (Kruger, 2006).

6. BREEDING FOR DROUGHT TOLERANCE

Drought is the most economically important abiotic stress affecting soybean. A basic understanding of physiological, biochemical and gene regulatory networks is important to develop plants with drought tolerance. In soybean, drought reduces yield about 40% (Muchow and Sinclair, 1986; Specht et al., 1999). Crop loss depends on plant growth stages and duration of drought. Drought stress during flowering and early pod development stages significantly increases the rate of flower and pod abortion, ultimately decreasing grain yield (Boyer, 1983; Westgate and Paterson, 1993; Liu et al., 2004). Irrigation is not a viable option for most of the soybean growing regions in the USA (Boyer, 1983). In environments where water is limited, genetic improvement of a crop for drought tolerance is an economically feasible option (Blum, 2002). Even though large resources are committed to soybean breeding, progress has been slow for drought tolerance. Carter et al. (1999) mentioned several reasons for slow progress in breeding for drought tolerance in soybean. These reasons include: 1) Breeding in high yielding environments results in more progress and more return than breeding in low yielding environments e.g. drought prone. Additionally, low yield environments provide data that are suspect due to soil heterogeneity. Data from low yielding environments are often ignored because small yield differences among lines fail to adequately separate high yielding lines from low yielding lines. Identifying lines with the highest yield potential is of utmost importance in soybean breeding and is normally done by conducting tests where moisture is optimum and high yields can be achieved. 2) Most of the varieties released in the early days of soybean breeding were selected for disease resistance, shatter resistance and other factors, but not for resistance to abiotic stress such as drought. Thus, little emphasis was placed on utilization of germplasm to achieve a broader genetic base in breeding programs for drought tolerance in the past. 3) The study of drought tolerance is high risk and difficult in that drought is unpredictable as to when and where it will occur. Little progress for drought tolerance can be made without the ability to impose stress year after year. Thus, a field with poor moisture holding capacity, good soil uniformity, and a reasonable probability of drought each year is important in selecting genotypes for drought tolerance. Unfortunately, such a specialized environment is rare at universities and federal field experiment stations.

Plants use different mechanisms to cope with drought stress, namely drought escape and drought resistance (Levitt, 1980). Drought escape allows the plant to complete its life cycle before the onset of drought during the period of maximum water supply via short life cycle. The Early Soybean Planting System (ESPS), now widely used in the southern USA, is an example of drought escape. In this system, short season cultivars are planted in March or early April in zones where later maturing cultivars have traditionally been grown. These early maturing cultivars begin blooming in late April to early May; start setting seed in late-May to early June and reach full seed setting by mid-July to early August. In the southern US, rainfall is often plentiful from April to early July allowing the soybean crop to reach the critical reproductive stage with ample water prior to July and August where conditions often favor drought stress (Heatherly and Elmore, 2004).

Drought resistance is generally divided into drought avoidance and drought tolerance. Drought avoidance helps plants maintain relatively high leaf water potential during water stress by extracting more water from the soil through a well-developed root system and/or by leaf rolling, reducing water loss through stomatal closure and thick leaf cuticles (O'Toole and Bland, 1987; Ludlow and Muchow, 1990). Drought tolerance allows plants to maintain turgor and cell volume at low leaf water potential to continue metabolic activity longer under water stress through osmotic adjustment (OA), antioxidant capacity, and cell membrane stability (CMS). However, so far, no data support the positive contribution of OA towards yield under drought stress (Blum, 2005).

In soybean, a widely accepted equation for grain yield (Y) under water-limited environments is a function of three independent components, i.e. amount of water transpired (T), water-use efficiency (WUE), and harvest index (HI); that is, Y= T x WUE x HI (Passioura, 1977, 1994, 1996; Ludlow and Muchow, 1990; Turner et al., 2001). Ludlow and Muchow (1990), and Purcell and Specht (2004) suggested that the following 8 traits are related to increase or possibly maintain T during drought. These traits are phenology, photoperiod sensitivity, developmental plasticity, leaf area maintenance, heat tolerance, osmotic adjustment, early vigor, and rooting depth and density. Two traits, transpiration efficiency and leaf reflectance are related to WUE.

Blum (2005) reviewed the association among yield potential (YP), drought resistance (DR), and water-use efficiency (WUE). He explained that higher WUE is expressed in yield improvement only when there is limited and known soil moisture reserve to balance crop water demand. But, lower WUE is expressed under dry land conditions where crop production mostly depends on unpredicted rainfall. High axial root resistance, mobilization of pre-anthesis dry matter and nitrogen fixation during drought is considered to be related to increased harvest index. Liu et al. (2005) detected significant correlations between drought tolerance and different root traits like dry root weight, total root length, and root volume in soybean. They proposed using these traits as root indicators for drought tolerance. Benjamin and Nielsen (2006) compared root distribution of soybean, field pea and chickpea under a water deficit situation. In soybean, under both irrigated and water deficit conditions, there was no effect on root distribution and about 97% of the total roots were distributed in the surface of 0.23 m. These results support earlier findings of Mitchell and Russell (1970). But, Merrill et al. (2002) and Hoogenboom et al. (1987) reported that soybean roots grow deeper under water stress. Robertson et al. (1980) found greater root mass for irrigated than non-irrigated soybean, but Mayaki et al. (1976) found no change in root mass. Sponchiado et al. (1980) suggested that soybean root response under stress would be cultivar and climate dependent as was found in dry bean. It is well documented that plants with deep roots have the ability to extract water from a deeper soil depth during water stress. Thus, soybean genotypes with deeper root systems may help to avoid drought stress. Ohashi et al. (2006) reported that soybean stem diameter decreased after sunrise and increased after sunset under drought. This suggests using stem diameter as an effective parameter to characterize water status under drought stress. More research is needed to confirm this hypothesis.

Approaches to breed for drought tolerance in soybeans have involved several mechanisms and include slow canopy wilting, prolific rooting, sustained nitrogen fixation, and selecting for higher yield under drought conditions. Over 2000 Plant Introductions (PIs) from the USDA-ARS national soybean germplasm collection have been evaluated over the past 20 years in North Carolina to search for drought tolerance utilizing special fields where drought occurs each year. PI's and breeding lines have been identified or developed which wilt more slowly than existing varieties. Two PIs, PI416937 and PI471938 are slow wilting and exhibit drought tolerance. These lines were among the best drought resistant sources identified to date and have been used in most of the breeding programs for drought tolerance in the North Carolina program. BR-4 and Ocepar 4 are Brazilian cultivars that are drought tolerant (Newmaier et al., 1995). Pedigrees of the Brazilian cultivars trace to US breeding lines without tolerance to drought. Thus, genetic control of drought response is not clear since drought tolerant lines may be derived from drought sensitive parents. Other sources of soybean germplasm for drought tolerance have been identified from research in China (Xu et al., 1999). Although identity of lines was not given, 463 strains from the 7023 evaluated were listed as having a high level of drought resistance. These Chinese strains could be a very valuable source of germplasm in breeding soybeans for improved tolerance to drought.

PI416937 was the first slow wilting line identified and is the most studied among drought tolerant lines in the US to identify mechanisms for tolerance to drought. Several possible explanations have been given as to why this line wilts more slowly under drought than other soybean strains. Hudak and Patterson (1996) reported that PI416937 had more highly-branched roots in the upper soil profile than drought sensitive lines. Pantalone et al. (1996a, 1996b) demonstrated that the prolific rooting trait with heritability similar to yield can be effectively manipulated in the field. Busscher et al. (2000) compared soybean genotypes for root penetration in soil hard pans and suggested that PI416937 possesses the genetic capability to continue root growth in compacted soils. Although, the PI roots grow deeper, water use efficiency was not better than drought sensitive lines (Purcell et al., 1997). However, PI416937 has other positive traits like tolerance to high soil aluminum (Carter and Rufty, 1993) and salt (Abd-Alla et al., 1998) which may increase adaptability to drought conditions.

The drought tolerant lines PI416937 and PI471938 are being extensively utilized to develop slow wilting varieties that perform well relative to other varieties regardless of water regime. N98-9683, a group VII line from the USDA program at North Carolina State University with 25% of its pedigree from PI416937 yielded 10% more than Benning in two years of tests across 20 environments in the southern US (Paris, 2003). G00-3209, a University of Georgia soybean line also with 25% of its pedigree derived from PI416937 yielded 12% more than Benning across two years and 15 environments (Paris and Shelton, 2006). The high yielding varieties with slow wilting capacity are being developed with improved tolerance to drought, yet broad adaptation to all environments. The genetically diverse slow

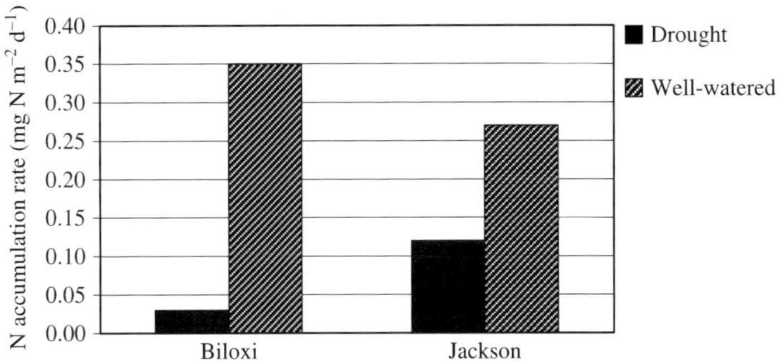

Figure 4. Nitrogen fixation of Biloxi (drought sensitive) and Jackson (drought tolerant) under drought and well-watered conditions

wilting lines listed above are showing value as parents by combining new yield genes, drought tolerance, and adding much needed diversity to a very narrow genetic base.

Data support that crop growth and yield under water deficient conditions are limited by extreme sensitivity of nitrogen fixation to drought. Nitrogen fixation in the drought tolerant line Jackson was less affected by water deficits as compared to the cultivar Biloxi (Figure 4). Serraj et al. (1997) and King et al. (2001) have shown that Jackson has larger nodules and accumulates ureides less in leaf petioles versus a drought sensitive genotype Biloxi. Jackson avoids the high ureide concentration in the leaves that triggers reduced nitrogen fixation during the initial stages of drought. Several soybean genotypes in addition to Jackson in maturity groups V-VIII have been identified with nitrogen fixation tolerance to water deficits. They are PI227547 (VII), PI374163 (VIII), PI423886 (VI), PI429328 (VIII), PI507039 (VII,), PI227557 (V), PI507414 (VI), and PI578315B (VIII) (Sinclair et al., 2000).

6.1. QTL Mapping and Molecular Breeding for Drought Stress

Identification of genes/QTLs related to drought resistance traits is the first critical step in molecular breeding. Tuberosa and Salvi (2004) mentioned that it is important to select QTLs with limited interaction with water regime or other environmental variables, and to consider the effect of the beneficial QTL in the genetic background to be improved. Other factors, like size of the segregating population, multiple studies across environments, selection of traits based on both morphological and agronomic traits and confounding effects of morpho-physiological traits and plant water status under drought stress should also be considered during QTL studies for drought. The accurate characterization and validation of a QTL is often done through the development of near isogenic lines (NILs) via MAS. Once a QTL is consistently detected in different field trials and confirmed in different populations, then this QTL may be selected for introgression into cultivars for drought improvement.

Although a large number of QTLs (1174) have been mapped in soybean for agronomic, physiological, seed composition traits, biotic and abiotic factors (www.soybase.ncgr.org), only a handful QTLs have been reported for drought and salt (Table 1). Almost half of the reported QTLs have explained less than 10% of the phenotypic variation. In most cases single and small population sizes have been used for QTL detection which may lead to faulty estimation and precision in QTL detection as suggested by Beavis (1998).

WUE is an important trait related to drought tolerance in soybean. Mian et al. (1996, 1998) mapped QTL for WUE in two mapping populations, Young x PI416937 and S-100 x Tokyo. They detected seven QTLs for WUE. Among them, two QTLs linked to RFLP markers, cr392-1 of linkage group (LG)-J and A489H of LG-L explained 13 and 14% phenotypic variation, respectively. These two loci, however, have not been confirmed yet in any other genetic background or across other environments. Interestingly, the marker locus A489H (WUE2-1) also found associated with soybean leaf length, leaf weight, leaf width, leaf shape, leaf area, yield, plant height and oil content (www.soybase.ncgr.org). Another QTL linked to RFLP marker A063E for WUE was common in both the populations, but the phenotypic effect was less than

Table 1. Reported Quantitative trait loci (QTL) related to soybean drought and salinity tolerance

Mapping populations	Trait reported	Marker linked to QTL, linkage group (LG) and contribution (R^2)	Reference
Drought related traits			
Hutcheson× PI471938, 140 F_4	yield (3), wilting (3)	Satt226, LG-D2; Sat_375, LG-F1; Sat_074, LG-F2;	Monteros et al. (2006)
Jackson× KS4895, 81 RILs	leaf wilting (1)	Sat_044, LG-K, R^2 = 17	Bhatnagar et al. (2005)
Minsoy× Noir 1, 256 RILs	yield (1)	Satt205-satt489, LG-C2, R^2=7	Specht et al. (2001)
S-100× Tokyo, 116 F_2	water use efficiency (2) (WUE)	A489H, LG-L, R^2 = 14 A063-1, unlinked, R^2 =8	Mian et al. (1998)
Yong× PI416937, 120 F_4	water use efficiency (5) (WUE)	B031-1, LG-G, R^2 = 8.5 A089-1, LG-H, R^2 = 8.7 cr497-1, LG-J, R^2 = 13.2 K375-1, LG-J, R^2 = 7.5 A063-1, LG-C1, R^2 =5	Mian et al. (1996)
Salt tolerance			
S-100× Tokyo, 100 F_2	salt tolerance (1)	Sat_091, LG-N, R^2 = ~41–79 (field and greenhouse screening)	Lee et al. (2004)

10%. Specht et al. (2001) used 236 RILs developed from a cross between Minsoy x Noir 1 to determine the genetic basis of beta and carbon isotope discrimination (CID). They reported a QTL for CID on LG C2 with phenotypic contribution of <10% and with no effect on beta. Recently, Bhatnagar et al. (2005) identified a major QTL for a slow wilting trait related to increased drought tolerance in soybean. The major QTL linked to SSR marker Sat_044 on linkage group K explained 17% phenotypic variation. Recently, Wood et al. (2006) detected a number of QTLs related to water stress tolerance in soybean. They reported three QTLs for root architecture of basal root, Satt509 (LG A2), Sat_083 (LG B2), and Satt316 (LG 316); one QTL for root dry weight, Satt554-CAA19 (LG F); and one QTL Satt214 in (LG G) for root and shoot dry weight ratio. They also detected the presence of Trigonelline, a low molecular weight compatible solute in soybean, shown to stabilize enzyme activity during water and salt stress. Also Monteros et al. (2006) identified three QTLs associated with seed yield and slow wilting in a mapping population of 140 F_4 lines from Hutcheson x PI471938 (drought tolerant). One QTL from the P1 mapped to LG D2 near the SSR marker Satt226 and two QTL to LG F_1 near Sat_375 and Sat_074. The PI471938 QTL on LG D2 and LG F1 were associated with yield and linked with slow wilting. Eight F_5 derived homozygous lines were among the highest yielding in the field across the environments studied with Satt226 on LG D2 having lowest values in carbon isotope discrimination.

7. BREEDING FOR SALT TOLERANCE

Salt stress or salinity is also an important abiotic factor limiting crop production. Soil with electric conductivity greater than 40mM NaCl (about 4ds/m) is considered saline (Stoddard et al., 2006). About 7% of the earths land (Akoi et al., 2005) and 20% of irrigated land (Flowers and Yeo, 1995) is affected by salt stress. Salinity problems in soybeans are most prevalent in coastal areas where tides from hurricanes inundate farm land; in some of the irrigated fields pumping water high in salt content and in over-fertilized fields or fields naturally high in salt. Every year, more land is brought under irrigation to increase crop production. Expansion of irrigated land and high salt content in irrigation water, coupled with poor drainage has increased salt stress. Blumwald and Grover (2006) predicted about 50% of the arable land will be affected by salt stress by 2050. Irrigation management, improved drainage, and the development and use of salt tolerant cultivars are a feasible solution to increase crop production in saline soils.

Salt tolerance is the ability of plants, to grow and complete their life cycle with good yield potential under saline conditions. Halophytes are the most salt tolerant plants and can grow in high concentrations of sodium chloride, but the majority of the crop species are glycophytes and unable to tolerate higher salt stress. There are two mechanisms of salt tolerance; ionic effects that minimize intercellular toxicity due to presence of higher concentration of salt and osmotic effect that minimizes entry of salt from root to leaf (Greenway and Munns, 1980; Zhu, 2001; Munns,

2002). Compared to glycophytes like soybean, halophytes are salt tolerant because they can exclude and compartmentalize salt in the cell vacuoles. Salt tolerance among genotypes can be measured by comparing biomass production between salt treated and control plants in annual crops and by monitoring survival ratings (Munns, 2002). The total saline area where most of the world's soybeans are grown including sodic areas of North America, Central America, and South America is 15.8, 2.0, and 129.0 million hectares, respectively. Saline areas are affected by amount of rainfall, rock weathering, wind-transported materials from soil or lake surfaces, quality of irrigation water, seawater intrusion onto land, climatic features, and human activity (Rengasamy, 2006). All potential crop production areas have a chance to be affected by salt stress in the future. Thus, salt tolerant cultivars are desirable for various crops including soybean.

7.1. Soybean Response to Salt Stress

Like many other species, soybean growth in hypersaline environments results in plants that suffer hyperosmotic effects. The consequences of hyperosmotic stress include membrane disorganization, metabolic toxicity, disruption of photosynthesis, and in extreme cases plant death (Malhotra and Blake, 2004). Salinity stress in soybean results in increased plant mortality, leaf necrosis, and accumulation of chloride in stems and leaves, and reduced green leaf color. It results in decreased plant biomass, plant height, leaflet size, seed yield, seed quality, and field emergence (Able and MacKenize, 1964; Parker et al., 1983; Yang and Blacnchar, 1993; Pannneerselvam et al., 1998; Wang and Shannon, 1999; An et al., 2002; Essa, 2002). Salinity also decreases root growth, root osmotic adjustment, root pressure, sodium ion exclusion and water extraction (An et al., 2001, 2002).

Soybean nodulation is also adversely affected by salt stress. Studies have shown salinity significantly decreases nodule number and dry-weight (Bernstein and Ogata, 1966; Singleton and Bohlool, 1984). Availability of oxygen to nodules is reduced and fermentative pathways are stimulated (Serraj et al., 1994). The ability of tissues to supply water to root cells under salt stress is reduced (Joly, 1989). In other studies nuclear deformation of the meristematic root cells occurred and was followed by degradation of nuclei in the apical region of the root tip (Liu et al., 2000). Salinity stress induced a significant increase in soybean leaf sodium and chloride and reduced the accumulation of potassium, calcium, and magnesium (Abel, 1969; An et al., 2001; Essa, 2002). Soybean yield is dramatically decreased under salt stress. Soybean yield was 80% at 4.0 dS m^{-1}, and 44% at 6.7 dS m^{-1} versus 100% at 0.8 dS m^{-1}. The effect of soil salinity on the relative biological nitrogen contribution of the soil was 77% at 4.0 dS m^{-1}, and 28% at 6.7 dS m^{-1} verse 100% at 0.8 dS m^{-1} (Katerji et al., 2003).

The mechanism of salt tolerance is one of the most important subjects in plant science, because the mechanism is complicated, and is thought to consist of two principle components (Greenway and Munns, 1980; Zhu, 2001). One component is an 'osmotic effect' which limits water absorption due to salinity in the rhizosphere.

The other component is an 'ionic effect' which is able to overcome intercellular toxicity from excess ions. Soybean salt tolerance is thought to be primarily from the ionic component in which tolerant plants limit the accumulation of excess ions to reduce injury (Läuchli and Wieneke, 1979; Umezawa et al., 2000; Umezawa et al., 2002). Salt tolerant soybean genotypes 1) prevent salt ions from moving from the roots to other plant parts, 2) do not accumulate as much salt in leaves and stems, and 3) have better osmatic adjustment in plant cells.

Abel (1969) reported that the chloride concentration of tolerant soybeans in soybean leaves was 10 times less than salt susceptible soybeans. Transport of chloride ions in salt tolerant cultivars from the root to stems and leaves of was exceedingly low (Able and MacKenize, 1964). Plants of the tolerant cultivar Lee were taller, maintained lower Na^+ and Cl^- concentrations, a higher K^+ concentration and a higher K^+/Na^+ ratio at higher salinity levels than salt sensitive cultivars Colquitt and Clark 63 (Essa, 2002). Dare another tolerant cultivar showed a higher relative shoot and root growth, water extraction ability, root pressure, better root osmotic adjustment and less sodium accumulation in plant tissue than salt sensitive cultivars. The salt tolerance of Dare was associated with high water uptake, and $Na^+$and Cl^-exclusion from being transported from roots to upper portions of plants (An et al., 2002). Tolerance in soybean was also shown to be related to maintenance of stable water content in shoots and the accumulation of soluble saccharides, soluble proteins, the amino acid, proline, and K^+ and Ca^+ for osmotic adjustment (Abd El-samad and Shaddad, 1997).

Differences in salt tolerance among *Glycine* species were reported (Pantalone et al., 1997; Luo et al., 2005; Kao et al., 2006). Greater variation in sodium chloride tolerance was shown among the perennial *Glycine* accessions than among the *G. max* cultivars. The sodium chloride tolerance thresholds ranged from 3.0 to 17.5 g/L NaCl for the perennial accessions, but only ranged from 5.2 to 8.0 g/L NaCl in the *G. max* cultivars based on the Weibull model for measuring leaf chlorosis (Pantalone et al., 1997).

Luo et al. (2005) reported that the salt tolerance mechanism of cultivated soybean (*G. max*) was different from wild soybean (*G. soja*). In a recent study comparing differential sensitivity of *G. max* and *G. soja* to chloride and sodium ions, *G. max* genotypes were more heavily damaged in the Cl^-solution than Na^+ solution. Salt tolerance in *G. max* was mainly due to prevention of Cl^- ion transport from the roots to the upper portion of the plant preventing toxic accumulation in stems and leaves. In contrast leaves of salt tolerant *G. soja* strains or wild soybeans were not as susceptible as *G. max* to Cl^-toxicity as that of Na^+. Salt tolerance in *G. soja* was primarily from exclusion of sodium ions from the roots preventing accumulation at toxic concentrations in stems and leaves. The descendants of a cross between tolerant wild and cultivated soybean strains were more tolerant to salt stress of NaCl and Cl^- salts than those of cross between tolerant *G. max* cultivars (Luo et al., 2005). This indicates that interspecific crosses between *G. max* and *G. soja* offer the possibility of improving salt tolerance in soybean cultivars.

7.2. Genetic Sources for Salt Tolerance in Soybean

Numerous soybean genetic resources are reserved in 70 countries. The genetic base of Northern American soybean cultivars is narrow because they trace to few parental ancestors. Analysis of US public cultivars showed that 80 ancestors accounted for 99% of the parentage in cultivars. Only 26 ancestors accounted for nearly 90% of the total ancestry, with the remaining ancestors each contribution <1% of the total ancestry to US cultivars. Comparing soybean cultivars within succeeding decades of releases from 1960 through the 1990s, coefficient of parentage analysis indicate less diversity among southern US lines than northern US lines for most decades in North America. This is not surprising because nearly 40% of southern parentage of North America is derived from two ancestors, CNS and S-100 (salt tolerant). The cultivar Lee (salt tolerant) and three sibs derived from S-100 x CNS are the primary conduit for 37% of the parentage in southern soybean cultivars (Carter et al., 2004). Ancestral cultivars, Lee, Manokin, Centennial, and many others derived from ancestors tracing to S-100 X CNS have been used in several salt studies (Abel, 1969; Wang et al., 1983; Yang and Blanchar, 1993; Ragab et al., 1994a, 1994b; Pantalone et al., 1997; Wang and Shannon, 1999; Wood, 1999; Essa 2002; Umezawa et al., 2000, 2002; Lee et al., 2004; Parker et al., 2005). S-100 was a selection from Illini. Illini and A. K. (Harrow) were selected from A. K. A. K was collected from northeast China in 1912 (Bernard et al., 1988). So, it is possible to assume that the salt tolerance genes in cultivars grown in U. S. could be from the same source or A. K. soybean.

This was demonstrated by the pedigree tracking using flanking SSR markers of salt tolerance QTL which derived from S-100 (Lee et al., 2004).

Genetic diversity is one of the most important elements to make genetic progress in crop breeding. If gene pools derived from various germplasm sources have target genes, breeders have a higher chance to develop improved cultivars. In four independent studies, 32 of 65 (Parker et al., 1986), 10 of 15 (Parker et al., 1983), 19 of 60 (Yang and Blanchar, 1993) and 10 of 257 (Shao et al., 1995), U. S. cultivars and breeding lines were identified as tolerant to high chloride. The USDA Germplasm Resources Information Network shows that 151 soybean genotypes have salt tolerance (USDA). About 10,000 Chinese accessions have been screened for tolerance to salt; 176 lines were indicated as having excellent tolerance when evaluated during germination in soil with high salt concentration. Eight accessions were identified as having very high tolerance and 456 lines showed good tolerance when evaluated at the seedling stage (Xu et al., 1999). Several salt tolerant soybean genotypes were collected from different gene pools. These include Flambeau from Russia, Bilomi #3 from Philippines, Fiskeby III and Fiskeby-840-7-3 collected from Sweden (Carter et al., 2006), and Dare (An et al., 2002)., Only a few accessions from wild soybean species have been reported as salt tolerant. Pantalone et al. (1997) reported five tolerant perennial *Glycine* accessions, *G. argyrea* 1626, *G. clandestina* 1388 and 1389, and *G. microphylla* 1143 and 1195, were significantly lower in leaf chlorosis score than that of the *G. max* cultivars at the 10 g/L NaCl treatment. Several wild soybean (*G. soja)* plant introductions PI378701A, PI468916, PI483463, PI483468A, and PI549048 were determined to have salt tolerance compared to cultivar Hutcheson, a sensitive variety (J. D. Lee and J. G. Shannon, University of Missouri-Delta Center, unpublished data). Because there is genetic diversity in both cultivated and wild species, it is likely that different genes condition tolerance which could be used to improve salt stress in soybean cultivars.

7.3. Inheritance of Salt Tolerance

To characterize the inheritance of salt tolerance, genotypes were screened for chloride inclusion (sensitive) and exclusion (tolerant). Parents, F_1, and F_2 progenies of 2 combinations of includer x includer, one combination of excluder x excluder, and 4 combinations of excluder x includer were planted in a low-salt soil. Salt solution was added to screen F_2 populations from parents differing in chloride accumulation, populations segregated in ratios of 3 non-necrotic plants very low in chloride to 1 necrotic plant very high in chloride with a 1:1 segregation ratio from a test cross. It was concluded that the factor to exclude or include chlorides in soybean leaves and stems is controlled by a single dominant gene. Gene symbols *Ncl* and *ncl* were proposed as the dominant chloride excluder allele for tolerance and the recessive chloride includer for sensitivity, respectively (Able, 1969)

Lee et al. (2004) studied salt response in the green house in a nutrient solution and in the field with a salt content having an electrical conductivity $\geq$4.0 dS m^{-1}. Parents and 106 F_2-derived recombinant inbred lines derived from S-100 (salt tolerant) x Tokyo (salt sensitive) were used. Lines were scored from 0 to 5 with 0 representing plant death and 5 no apparent salt injury. Visual ratings were 3.1, 5.0 and 4.1 for S-100 and 0.1, 1.0, and 0.6 for Tokyo for screenings in the field, greenhouse, and the combined trials, respectively. The progeny mean for visual salt ratings were 1.7, 2.4 and 2.0 for ratings in the field, greenhouse, and combined trials, respectively. They estimated heritabilities by entry-means which were 0.85, 0.48, and 0.57 for the field, greenhouse, and combined trials, respectively. They concluded that more than one gene controls salt tolerance in soybean. Future genetic studies are needed to determine the inheritance of salt tolerance among unrelated genotypes including wild soybean species.

7.4. Screening for Salt Tolerance Genotypes

Appropriate screening techniques are important for the successful development of soybean cultivars with salt tolerance. Three types screening techniques have been reported for selection of salt tolerant soybeans, 1) soil with high salt content, 2) hydroponically with high salt added to a nutrient solution or 3) DNA markers.

Screening genotypes in fields with high salt content has been used in some breeding programs. However the evaluation of genotypes in fields with high salt content can be difficult because of variability of salt levels across field locations and the potential for interactions with other environmental factors, including soil fertility, temperature, light intensity and water loss due to transpiration. Parker et al.

(1983) planted 15 soybean genotypes in two fields with a history of leaf scorch symptoms from Cl^- toxicity from KCl fertilizer. Soil types were a Leefield sand (arenic Plinthaquic Paleudults) a Alapaha sand (arenin Plinthic Paleaquults). They found 10 of 15 soybean cultivars were tolerant to high chloride by rating for leaf scorch and leaf Cl^- concentration. Yang and Blanchar (1993) used Mexico silt loam soil with and without added Cl^-. A total of 673 Kg Cl^- ha^{-1} was added as a CaCl2 solution for the high Cl^- plot. Nineteen of 60 soybeans were salt tolerant based on leaf scorch ratings and leaf Cl^- concentration. Shao et al. (1995) used a field of salinized soil located in Shandong province, China. The soil was a sandy loam with moderate saline content. They screened soybean lines at germination and the V2 or V3 seedling stage using 15-17 ds m^{-1} saline water made from a mixture of fresh water and underground salt water. These studies show that different soil types can be used for screening for phenotyping soybean genotypes for salt tolerance.

Hydroponics modified by Johnson et al. (1957) and Hoagland and Arnon (1953) with added NaCl to the nutrient solution is widely used to screen soybeans for tolerance. Many genotypes can be screened in limited space like the greenhouse and the salt rate can be easily controlled. Several studies for evaluating salt tolerance have used the following procedure: 1) Germinating soybean in sand and placing 5 seedlings per replication in nutrient solution 14–21 days after emergence; 2) adding NaCl to the solution for 14–31 days, and 3) scoring genotypes for leaf scorch by assessing the relative proportion of visual symptoms of scorch induced toxicity (Figure 5) on a scale of 0 = healthy (no apparent symptoms of scorch), 1 = slight scorch (25% of the leaf area showed scorch symptoms), 2 = moderate scorch (50% of the leaf area showed scorch symptoms), 3-severe (75% of the leaf area showed scorch symptoms), 4 = dead (plants were brown and withered) (Ragab et al., 1994a, 1994b; Pantalone et al., 1997; An et al., 2001; Lee et al., 2004). The results of these procedures depend on genotype, salt concentration, and other environmental factors such as temperature and light. Salt concentration is the most critical factor for phenotyping genotypes for tolerance.

Threshold values to detect salt tolerance in soybean have varied. Chinnusamy et al. 2005 determined in their study that the threshold salinity to detect tolerance was 3.2 dS m^{-1}. The cultivar Lee (tolerant) produced more twice the relative shoot fresh weight and was significantly lower in chlorosis score than other genotypes at a 6.0 dSm^{-1} salt level. Soybean genotypes were able to differentiate for tolerance or sensitivity at 7.5 dSm^{-1}, but all genotypes were sensitive at 10.9 dSm^{-1} salt content and had similar leaf chloride levels (Ragab et al., 1994a, 1994b). The most significant phenotypic differences for salt tolerance between Dare (tolerant) and 'Tachiyutaka' (sensitive) were obtained at 40 mM NaCl (An et al., 2002). Lee et al. (2004) used 100 mM NaCl to screen 106 recombinant inbred lines. The sodium chloride tolerance thresholds were estimated from a Weibull model of leaf chlorosis and ranged from 3.0 to 17.5 g L^{-1} NaCl for the perennial accessions but only ranged form 5.2 to 8.0 g L^{-1} for the cultivars (Pantalone et al., 1997). This salt tolerance threshold showed different values among *Glycine* species thus, the degree of salt tolerance is different among species and genotypes. Both field and hydroponic screening are affected by environmental

Figure 5. Soybean leaf symptoms caused by salinity (A), tolerant accessions (center of tray) and susceptible accessions (sides of tray) (B), leaf scorch scores (C); 1 healthy (no apparent symptoms), 2 slight chlorosis (not shown), necrosis (25% of leaf area show symptoms), 3 moderate chlorosis, necrosis (50% of leaf area show symptoms), 4 severe chlorosis, necrosis (75% of leaf area show symptoms), and 5 dead (plants wilted and dead)

conditions such as temperature, plant growth stage, and salt concentrations (Li et al., 2000; Xu et al., 1999; Pantalone et al. 1997; Able and MacKenize, 1964). Therefore, researchers need to set critical levels of salt concentration to screen soybean genotype for tolerance for their conditions.

7.4.1. *DNA markers for salt tolerance*

Use of DNA markers such as RAPD, AFLP, RFLP, SSR, and SNP markers for selection is a powerful tool in modern breeding programs. Marker assisted selection allows screening numerous genotypes in less time with less effort, greatly improved efficiency of selection for specific traits. Several studies were conducted to determine markers that distinguish soybean genotypes for salt tolerance.

Eleven RAPD markers for salt tolerance were obtained from 148 polymorphic RAPD bands from a wild soybean population with high salt tolerance. Six of markers, OPF05 (213), OPF19 (4361), OPF19 (1727), OPF19 (14000), OPF19 (700), and OPH02 (1350), were significantly associated with salt tolerance. These markers were present in each of the salt-tolerant individuals and absent in all the salt-sensitive lines in the study. Lines with intermediate tolerance had only some markers present (Zhang et al., 1999). Guo et al. (2000) studied different crosses of soybean to screen and identify PCR markers associated with salt-tolerant genes.

Three populations of salt-tolerant and susceptible cultivars were used and two dominant PCR markers were identified via analysis of the segregation of F_2 plants. The markers were closely linked with salt-tolerant/susceptible alleles, and are now being utilized to select salt-tolerant, high yielding lines.

A major salt tolerant QTL was found from 106 recombinant inbred lines derived from soybean S-100 (salt tolerant) x Tokyo (sensitive) (Lee et al., 2004). This QTL was discovered near the Sat_091 on linkage group N, accounting for 41, 60, and 70% of the total genetic variation for salt tolerance trials from the field, greenhouse, and from the combined screening trials, respectively. Pedigree tracking for 27 U. S. soybean cultivars descending from the ancestors S-100 and 'Tokyo' was used to examine the association between the salt tolerance QTL and flanking SSR marker alleles. The presence of alleles from S-100 at the Sat_091 (159bp) and Satt237 (240bp) marker loci was always associated in tolerant descendants. Alleles from Tokyo for these same markers was generally associated in sensitive descendents. A strong relationship was evident between the Sat_091 marker and tolerance so it could be readily be used for marker-assisted breeding.

Recent research has focused on developing MAS techniques. Breeding programs for tolerance of salinity should emphasize developing MAS techniques that can increase selection efficiency. The development of high-density DNA maps that incorporate SNP, SSR, RFLP, and AFLP, and advances in marker-assisted selection techniques will facilitate pyramiding genes to improve soybean salt tolerance. Numerous subgenus *Soja* (*G. max* and *G. soja*) accessions are in worldwide collections. Many of these accessions have not been evaluated for salt tolerance. It is likely that undiscovered salt stress genes are available from the soybean germplasm. Genetic and DNA mapping studies soybean will be required in the future to find and confirm new markers for high levels of salt tolerance.

7.5. Candidate Genes for Salt Tolerance

All living things have a response to detrimental situations such as abiotic stress. This response stimulates gene action in defense of abiotic stress. Several studies have been conducted to determine gene action under salt stress in soybean. Zhong et al. (1997) used DAF (DNA Amplification Finger printing) for two salt-tolerant (Morgan and Wenfeng No. 7) and two salt-sensitive (Hark and Jackson) soybean cultivars and found three polymorphic markers (8.6f/350bp, 8-27/240bp and 8-15/215bp) which only appeared in the salt-tolerant cultivars. The amplified DNA fragment at marker 8-27/240bp was cloned. The cloned DNA sequence had significant homology to the *Oryza sativa* MADS-box protein (MADS3) mRNA (length-1316bp) that encodes regulatory proteins and plays an important role in flower morphogenesis (Kang et al., 1995). They suggested that the queried sequence is not a gene but part of a regulatory factor which may play an important role in regulating the transcription and expressing of salt-tolerant genes (Qin et al., 2000).

Umezawa et al. (2002) obtained 140 expressed cDNA-AFLP fragments induced by salt stress (100 mM NaCl or 12% PEG for 24 h) in soybean. After sequencing,

140 individual clones were determined and were designated as *Glycine max Stress Responsive* genes (GSR). Approximately 80% of the GSR genes matched to reported soybean EST sequences. However, 14% of the GSR genes did not show homology to any nucleotide or amino acid sequences in the GenBank data base. They found several GSR genes, GSR-8, 98, 110, and 112, induced by the NaCl treatment, and these genes showed tissue-specific expression. They suggested that salt tolerance of soybean is achieved from the response to both ionic effects and osmotic effects. The gene expression was abundant in soybean under salt stress. Transcripts which could be determined from ionic (NaCl-specific) and osmotic effects (common from NaCl with PEG) were 44 and 40 %, respectively. The gene expression dependent ionic effects was more abundant in roots indicting a greater response to ionic stress in roots than shoots. On the other hand, GSR gene expression from osmotic effects was more in shoots than roots.

A cation/proton antiporter beneficial for regulation of ion homeostasis in soybean for salt tolerance was reported. A putative *GmCAX1* was expressed in all tissue of the plants, but at a lower level in roots under PEG, ABA, Ca^{2+}, Na^+ and Li^+ stress. Transgenic *Arabidopsis* plants over expressing *GmCAX1* accumulated less Na^+, $K^+$and Li^+, and were more tolerant to elevated Li^+ and Na^+ levels during germination. Thus, *GmCAX1* may function as an antiporter for Na^+, K^+ and Li^+ (Luo et al., 2005).

Three DREB (dehydration-responsive element binding) homologue genes have function specifically in response to abiotic stress in soybean, *GmDREBa*, *GmDREBb*, and *GmDREBc*. They were isolated from soybean under salt, dehydration, and abscisic acid (ABA) treatments. The transcriptions of *GmDRBa* and *GmDRBb* in leaves of soybean seedlings were induced by salt, drought, and cold stress. The expression of *GmDRBc* was induced in roots by salt, drought, and abscisic acid treatments (Li et al., 2005).

A novel gene, *GmPAP3* that plays a role in the adaptation of soybean to NaCl stress, was identified from salt-stressed soybeans. NaCl stress causes a general induction of *GmPAP3* expression in both roots and leaves of various cultivated and wild (*Glycine soja*) soybeans (Liao et al., 2003). Protein sequence alignment studies and phylogenetic analysis suggested that *GmPAP3* belongs to the group of plants purple acid phosphatases (PAPs)-like proteins.

Among 106 salt-inducible soybean genes designated *GmTDFs*, a soybean gene *GmTDF-5* was characterized as a novel cytosolic leucine-zipper-like protein functioning in mature organs of soybean shoots against water-potential changes. The *GmPTF-5* was induced in the stem and lower-expanded leaf. The amount of mRNA increased 5.1-fold and 2.0-fold up to 72 h by a 100 mM NaCl treatment, respectively (Aoki et al., 2005). The full length of cDNA coding for a novel vacuolar Na^+/H^+ antiporter, *GmNHX1*, was cloned from soybean. Northern blot analysis demonstrated that the expression of the *GmNHX1* was tissue-specific. Expression was increased by ABA treatment, NaCl, KCl, LiCl and dehydration stress. *GmNHX1* expression was lower in leaves, but higher in roots and hypocotyls of salt tolerant than salt sensitive cultivars. The *GmNHX1* was over expressed under the control of a tandem cauliflower mosaic virus (CaMV) 35S promoter in the model leguminous plant *Lotus corniculatus* L. and conferred salt-tolerance of the transgenic plants. Measurements of Na^+ and K^+ contents in both roots and shoots demonstrated that the plantlets of lines overexpressing *GmNHX* 1 had lower Na^+ and K^+ content, and higher K^+/Na^+ that the control lines, which indicates that salt-tolerance conferred by *GmNHX1* is closely related with decreased accumulation of Na^+ in the transgenic plants (Sun et al., 2006).

8. GENETIC ENGINEERING FOR TOLERANCE TO DROUGHT AND SALT IN CULTIVAR IMPROVEMENT

When plants are exposed to abiotic stress like drought and salinity, several fold changes occur in gene expression. Recent advancement in the areas of functional genomics and development of analytical tools help in understanding the physical, biochemical and molecular aspects of gene regulatory networks for abiotic stress tolerance. Integrated molecular and genomic approaches have facilitated selection of functional and regulatory candidate genes related to stress tolerance in plants (Umezawa et al., 2006; Valliyodan and Nguyen, 2006; Yamaguchi-Shinozaki and Shinozaki, 2006). In different crops, transcriptional profiling through microarray analysis have identified a number of drought and salt stress related genes and functional analysis of some of these genes were performed (Oono et al., 2003; Rabbani et al., 2003; Takahashi et al., 2004; Buchanan et al., 2005; Hazen et al., 2005; Walia et al., 2006). Yamaguchi-Shinozaki and Shinozaki (2006) indicated that about 50% of the drought-inducible genes are also induced by salt stress, showing a significant cross talk between the two stress systems. In last the ten years, several successful attempts have been made to genetically engineer drought and salt tolerance in plants (summarized by Zhang et al., 2004; Bartels and Sunkar, 2005; Umezawa et al., 2006). So far, a large number of genes have been identified and a portion of them being used to develop drought and salt tolerant transgenic plants. They are classified as functional genes and regulatory genes. Functional genes are involved in the synthesis of osmotically active compounds, transporters, chaperons and reactive oxygen species quenchers thus, protecting cells from stress effects. To develop drought tolerant transgenic plants, genes are involved in synthesis of proline, polyamines, glycine beatine, trehalose, and late embryogenesis abundant (LEA) proteins have been used (Bartels and Sunkar, 2005; Umezawa et al., 2006). On the other hand, regulatory genes involved in signal transduction and gene regulation include transcription and signaling factors. The most important and most frequently used transcription factors are dehydration-responsive element binding (DREB) protein/C-repeat; basic region leucine zipper (bZIP) proteins, Myb-like proteins and stress responsive NAM, ATAF, and CUC family transcription factor NAC1 (SNAC1). Abiotic stress responsive important signaling factors are calcium dependent protein kinases (CDPK), mitogen-activated protein kinase (MAPK), and farnesyl transferase (ERA1) (Bartels and Sunkar, 2005; Umezawa et al., 2006).

Important events in genetic engineering for drought and salt tolerance in plants include Trehalose and (LEA) proteins. Trehalose a nonreducing disaccharide that functions as a compatible solute to protect biological structure under stress, has been used to increase drought tolerance in tobacco, rice and tomato. Garg et al. (2002) have successfully engineered rice plants with a trehalose-6-phosphate synthase/phosphatase (TPSP) gene. Under drought stress, transgenic rice plants showed increased amounts of trehalose and also high levels of tolerance to salt, drought, and low temperature stress as compared to controlled plants without any negative effect on plant growth and grain yield. Jang et al. (2003) confirmed this using trehalose-6-phosphate phosphatase (TPP) and trehalose-6-phosphate synthetase (TPS) in rice plants. These results suggest potential to use a transgenic approach to other crops including soybeans to increase tolerance to abiotic stress. Late embryogenesis abundant (LEA) proteins are a group of stress protective proteins expressed during embryo maturation and synthesized by *lea* gene. HVA1 gene derived from barley, encodes for synthesis of LEA protein has been used to develop transgenic rice and wheat. During drought stress, transgenic rice with HVA1 protein has performed better by protecting cell membranes from injury (Babu et al., 2004). Wang et al. (2005) demonstrated in field tests that the protein farnesyl transferase enhanced drought tolerance in Arabidopsis and canola maintaining seed yield and oil composition. In a field trial, Hu et al. (2006) demonstrated that the over expression of SNAC1 gene significantly increased drought tolerance in transgenic rice plants as compared to non-GMO plants. These transgenic rice plants showed drought and salt tolerance at the vegetative growth stage. Most of the above genes are also induced by salt stress. Recently, Meng et al. (2006) cloned and characterized NAC-like genes in soybean which may provide information to determine the role of NAC-like genes in soybean seed development and other physiological processes. Other important salt stress inducible genes are AtNDPK2 (vacuolar Na^+/H^+ antiporter) SOS1 (plasma membrane Na^+/H^+ antiporter) and HAL1. Most of the transgenic research work involving abiotic stress has been done in crops other than soybean. De Ronde and his group (2000, 2004a, 2004b) have generated antisense soybean plants with a pyrroline-5-carboxylate reductase (P5CR) gene. They demonstrated that transgenic soybean lines accumulated more proline and were more drought tolerant than non-transgenic lines. Research is in progress at the University of Missouri to generate transgenic soybean lines with genes for drought tolerance.

9. CONCLUSIONS AND FUTURE PERSPECTIVES

Soybean is a very important world oil and food crop. However, very little progress has been made to develop soybean cultivars with enhanced drought and salt tolerance due to the complexity in understanding genetic and physiological mechanisms of these traits. More concentrated efforts are needed to screen germplasm, determine new genes, and to combine genes for higher levels of tolerance to these stresses in soybean. Identification and use of traits related to drought and salt tolerance and development of suitable screening techniques are the prime criteria for cultivar development. Significant efforts have been made to understand molecular and physiological aspects of drought and salt tolerance. However, a better understanding of the networks regulating root growth, water use efficiency and nitrogen fixation under drought; and ion transport, ion effects, osmotic effects, and the mode of action for ion exclusion under salt stress will facilitate the ability to develop tolerant cultivars. Moreover, a basic understanding of these mechanisms under water deficits and salt stress conditions will open new avenues for genetic engineering for drought and salinity tolerance in soybean.

A significant number of QTLs/genes have been identified for different traits in soybean, but only a few are for drought and salt tolerant related traits. More studies are needed to elucidate and determine novel genes and their mode of action for high tolerance. Although there are numerous cultivated and wild soybean accessions in the soybean germplasm collections of the world, little of this germplasm has been screened for drought or salt tolerance. Combining genes from both wild and cultivated species show promise to obtain genotypes with higher levels of tolerance (Luo et al., 2005). Mapping for new QTL/gene and determination of gene action under drought and salt stress will provide key resources to improve tolerance to drought and salt stress in soybean.

Marker assisted selection (MAS) will be important for pyramiding genes at two or more loci to elevate drought and salt tolerance in soybean. Limitations of molecular markers have been surpassed with the discovery and use of gene-based abundant SNP markers. SNP and other markers have helped to develop a high density soybean genetic map for the identification and characterization of QTLs/genes conditioning drought and salt tolerance related traits facilitating MAS programs. The United States Department of Energy and Joint Genome Institute (DOE-JGI) have lunched a program to sequence the entire soybean genome. With the availability of soybean genome sequence information, integrated soybean genetic and physical map and traits specific SNP markers will make a significant and successful contribution in molecular breeding for abiotic stress including drought and salinity.

Genetic engineering technology is an attractive approach to improve soybean for drought and salt stress tolerance. Genes have been identified and some are being used successfully to develop genetically engineered drought and salt tolerant rice and canola. Introduction of drought and salt tolerance genes from soybean and other crops into elite soybeans that have high yield, enhanced resistance to pathogens and improved seed quality is highly desirable. Molecular techniques like QTL mapping, gene cloning, gene transformation and DNA microarray and gene expression analysis related to specific QTL regions are rapidly advancing and will play a vital role in the development of stress tolerant soybeans. Effective use of available genetic resources, understanding of tolerance mechanisms, construction of a fine map of the genome, development of marker assistant selection techniques and well-designed breeding strategies will advance the development of soybean varieties with significantly greater tolerance to drought and salt.

REFERENCES

Adb-Alla, M. H., Voung, T. D., and Harper, J. E., 1998, Genotypic differences in dinitrogen fixation response to NaCl stress in intact and grafted soybean, *Crop Sci.* **38**:72–77.

Abd El-Samad, H. M., and Shaddad, M. A. K., 1997, Salt tolerance of soybean cultivars, *Biologya Plantarum,* **39**:263–269.

Abel, G. H., 1969, Inheritance of the capacity for chloride inclusion and chloride exclusion by soybeans, *Crop Sci.* **9**:697–698.

Able, G. H., and MacKenzie, A. J., 1964, Salt tolerance of soybean varieties (*Glycine max* L. Merrill) during germination and later growth, *Corp Sci.* **4**:157–161.

Adams, M. D., Celniker, S. E., Holt, R. A., Evans, C. A., Gocayne, J. D., Amanatides, P. G., Scherer, S. E., Li, P. W., Hoskins, R. A, and others, 2000, The genome sequence of *Drosophila melanogaster*, *Science*, **287**:2185–2195.

An, P., Inanaga, S., Cohen, Y., Kafkafi, U., and Sugimoto, Y., 2002, Salt tolerance in two soybean cultirvars, *J. Plant Nutrn.* **25**:407–423.

An, P., Inanaga, S., Kafkafi, U., Lux, A., and Sugimoto, Y., 2001, Different effect of humidity on growth and salt tolerance of two soybean cultivars, *Biologa Plantarum*, **44**:405–410.

Aoki, A., Kanegami, A., Mihara, M., Kojima, T., Shiraiwa, M., Takahara, H., 2005, Moleclar characterization of a novel soybean gene encoding a leucine-zipper-like protein induced to salt stress, *Gene*, **356**:135–145.

Aranzana, M. J., Kim, S., Zhao, K., Bakker, E., and other 13 co-authors, 2005, Genome-wide association mapping in Arabidopsis identifies previously known flowering time and pathogen resistance genes, *PloS Genet.* **1**:531–539 (e60).

Arumuganathan, K., and Earle, E. D., 1991, Nuclear DNA content of some important plant species, *Plant Mol. Biol. Rep.* **9**:208–219.

Babu, R. C., Zhang, J., Blum, A., Ho, T. D-H., Wu, R., and Nguyen, H. T., 2004, HVA1, a LEA gene from barley confers dehydration tolerance in transgenic rice (*Oryza sativa* L.) via cell membrane protection, *Plant Sci.* **166**:855–862.

Bartels, D., and Sunkar, R., 2005, Drought and salt tolerance in plants, *Crit. Rev. Plant Sci.* **24**:23–58.

Beavis, W. D., 1998, QTL analysis: power, precision, and accuracy, in: *Molecular dissection of complex traits*, A. H. Paterson, ed., CRC Press, Boca Raton, Florida, pp. 145–162.

Benjamin, J. G., Nielsen, D., C., 2006, Water deficit effects on root distribution of soybean field pea and chickpea, *Field Crop Res.* **97**:248–253.

Bernard, R. L., Juvik, G. A., Hartwig, E. E., and Edwards, C. J. Jr., 1988, *Origins and Pedigrees of Public Soybean Varieties in the Unites States and Canada*, U. S. Department of Agriculture, Technical Bulletin No. 1746.

Bernstein, L., and Ogata, G., 1966, Effects of salinity on nodulation, nitrogen fixation, and growth of soybeans and alfalfa, *Agron. J.* **58**:201–203.

Bhatnagar, S., King, C. A., Purcell, L., Ray, J. D., 2005, Identification and mapping of quantitative trait loci associatiated with crop responses to water-deficit stress in soybean [*Glycine max* (L.) Merr.], The ASA-CSSA-SSSA International annual meeting poster abstract, November 6–10, 2005.

Birt, D. F., Hendrick, S., Alekel, D. L., 2004, Soybean and the prevention of chronic human disease, in: *Soybeans: Improvement, Production, and Uses*, Agronomy Monographs 3rd ed. No. 16, H. R. Boerma, and J. E. Specht, eds., ASA-CSSA-SSSA, Madison, WI, USA, pp.1047–1117.

Blanc, G., and Wolfe, K. H., 2004, Wide spread paleopolyploidy in model plant species inferred from age distributions of duplicate genes, *Plant Cell*, **16**:1667–1678.

Blum, A., 2002, Drought tolerance - is it a complex trait? Field screening for drought tolerance in crop plants with emphasis on rice, in: *Field screening for drought tolerance in crop plants with emphasis on rice*, International Workshop on Field Screening for Drought Tolerance in Rice, N. P. Saxena and J. C. O'Toole, eds., ICRISAT, Patancheru, India, pp. 17–22.

Blum, A., 2005, Drought resistance, water-use efficiency, and yield potential- are they compatible, dissonant, or mutually exclusive? *Aust. J. Agri. Res.* **56**:1159–1168.

Blumwald E., Grover, A., 2006, Salt tolerance, in: *Plant Biotechnology: Current and future uses of genetically modified crops*, Nigel G. Halford, eds., John Wiley and Sons Ltd, UK, pp. 206–224.

Borsani, O., Valpuesta, V., and Botella, V., 2001, Evaluation for a role of salicylic acid in the oxidative damage generated by NaCl and osmotic stress in *Arabidopsis* seedlings, *Plant Physiol.* **126**:1024–1030.

Boyer, J. S., 1982, Plant productivity and environment, *Science*, **218**:443–448.

Boyer, J. S., 1983, Environmental stress and crop yields, in: *Crop reaction to water and temperature stresses in humid, temperate climates*, C. D. Raper and P. J. Kramer, eds., Westview Press, Boulder, CO. pp. 3–7.

Bray, E. A., Bailey-Serres, J., Weretilnyk, E., 2000, Responses to abiotic stresses, in: *Biochemistry and molecular biology of Plants*, Gruissem, W., Buchannan, B., Jones, R., eds., American Society of Plant Physiologists, pp. 1158–1249.

Broeckling, C. D., Huhman, D. V., Farag, M. A., Smith, J. T., May, G. D., Mendes, P., Dixon, R. A., and Sumner, L. W., 2005, Metabolic profiling of *Medicago truncatula* cell cultures reveals the effects of biotic and abiotic elicitors on metabolism, *J. Expt. Bot.* **56**:323–336.

Brechenmacher, L., Kim, M.Y., Govindarajulu, M., Benitez, M., Li, M., Lee, M.P., Libault, M., Lee, S.H., Taylor, C., Clough, S.J., Stacy., G, 2006, Transcript profiling of soybean during nitrogen fixation symbiosis, The 11th Biennial conference on the molecular and cellular biology of the soybean, abstract, August 5–8, 2006, Lincoln, Nebraska.

Buchanan, C. D., Lim, S., Salzman, R. A., Kagiampakis, I., Morishige, D. T., Weers, B. D., Klein, R. R., Pratt, L. H., Cordonnier-Pratt, M-M., Klein, P. E., and Mullet, J. E., 2005, *Sorghum bicolor's* transcriptome response to dehydration, high salinity and ABA, *Plant Mol. Biol.* **58**:699–720.

Busscher, W. J., Lipiec, J., Bauer, P. J., and Carter, T. E., 2000, Improved root penetration of soil hard layers by a selected genotypes, *Comm. Soil Sci. Plant Anal.* **31**:19–20.

Carter Jr, T. E., Boerma, H. R., Lee, G. J., Zhou, X., Villagarcia, M. R., Cardinal, A., Shannon, J. G., 2006, On-farm QTL mapping of salt tolerance in the genetic base of North American soybean, The 11th Biennial conference on the molecular and cellular biology of the soybean, abstract, August 5–8, 2006, Lincoln, Nebraska.

Carter Jr, T. E., Nelson, R. L., Sneller, C. H., Cui, Z., 2004, Genetic diversity in soybean, in: *Soybeans: Improvement, Production, and Uses*, Agronomy Monographs 3rd ed. No. 16, H. R. Boerma, and J. E. Specht, eds., ASA-CSSA-SSSA, Madison, WI, USA, pp. 303–416.

Carter, T. E. Jr., and Ruffy, T. W., 1993, Soybean plant introduction exhibiting drought and aluminum tolerance, in: *Adaptation of food crops to temperature and water stress,* G. C. Kuo, ed., Proceedings of an international symposium, Asian Vegetable Research and Development Center, Taipei, Taiwan, 13–18 August, 1992, pp. 335–346.

Carter, T. E. Jr., D Souza, P. I., and Purcell, L. C., 1999, Recent advances in breeding for drought and aluminum resistance in soybean, in: Proceedings at the World Soybean Research Conference VI Chicago, IL. Superior Printing, Champagne, IL, pp. 106–125.

Chang, Y. l., Tao, Q., Scheuring, C., Meksem, K., and Zhang, H-B., 2001, An integrated map of *Arabidopsis thaliana* for functional analysis of its genome sequence, *Genetics*, **159**:1231–1242.

Chang, R., Qiu, L., Sun, J., Chen, Y., Li, X., and Xu, Z., 1999, Collection and conservation of soybean germplasm in China, in: *Proceedings at the World Soybean Research Conference VI*, H. E. Kauffman, ed., Chicago, Illinois. Superior Printing, Champagne, Illinois. pp. 172–176.

Chen, M., Presting, G., Barbazuk, W. G., et al., 2002, An integrated physical and genetic map of the rice genome, *Plant Cell,* **14**:537–545.

Chinnusamy, V., Jagendorf, A., and Zhu, J. K., 2005, Understanding and improving salt tolerance in plants, *Crop Sci.* **45**:437–448.

Cregan, P., Randall, N., and Youlin, Z., 2002, Sequence variation, haplotype diversity and linkage diseqilibrium in cultivated and wild soybean, in: *First International Conference on Legume genomics and Genetics: translation to Crop Improvement,* June 2–6, Minneapolis-St. Paul, MN.

Cregan, P. B., et al., 2006, A SNP based soybean genome map and applications in soybean breeding and genetics, The 11th Biennial conference on the molecular and cellular biology of the soybean, abstract, August 5–8, 2006, Lincoln, Nebraska.

Cui, Z., Carter Jr., T. E., and Burton, J. W., 2000, Genetic base of 651 Chinese soybean cultivars released during 1923 to 1955, *Crop Sci.* **40**:1470–1481.

De Ronde, J. A., Cress, W. A., Kruger, G. H. J., Strasser, R. J., and Staden, J. V., 2004a, Photosynthetic response of transgenic soybean plants, containing an *Arabidopsis P5CR* gene, during heat and drought stress, *J. Plant Physiol.* **161**:1211–1224.

De Ronde, J. A., Laurie, R. N., Caetano, T., Greyling, M. M., and Kerepesi, I., 2004b, Comparative study between transgenic and non-transgenic soybeans lines proved transgenic lines to be more drought tolerant, *Euphytica*, **138**:123–132.

De Ronde, J. A., Spreeth, M. H., and Cress, W. A., 2000, Effect of antisense L-Δ^1-pyrroline-5-carboxylate reductase transgenic soybean plants subjected to osmotic and drought stress, *Plant Growth Reg.* **32**:13–26.

Essa, T. A., 2002, Effect of salinity stress on growth and nutrient composition of three soybean (*Glycine max* L. Merrill) cultivars, *J. Agron. Crop Sci.* **188**:86–93.

Flowers, T. J., Yeo, A. R., 1995, Breeding for salinity resistance in crop plants. Where next ? *Aust. J. Plant Physiol.* **22**:875–884.

Garg, A. K., Kim, J-K., Owens, T. G., Ranwala, A. P., Choi, Y. D., Kochian, L. V., and Wu, R. J., 2002, Trehalose accumulation in rice plants confers high tolerance levels to different abiotic stresses, *Proc. Natl. Acad. Sci.* **99**:15898–15903.

Glazier, A. M., Nadeau, J. H., and Aitman, T. J., 2002, Finding genes that underlie complex traits, *Science*, **298**:2345–2349.

Greenway, H., and Munns, R., 1980, Mechanisms of salt tolerance in nonhalophytes, *Ann. Rev. Plant Physiol.* **31**:149–190.

Guo, B., Lijusn, Q., Guihua, S., Ruzhen, C., Lihong, L., Zhanyou, Z., Xianghua, L., Jinaying, S., Guo, B., and Qiu, L. J., 2000, Tagging salt tolerant gene using PCR markers in soybean, *Scientia Agri. Sinica.* **33**:10–16.

Gupta, K. P., Rustgi, S., and Kulwal, L., 2005, Linkage disequilibrium and association studies in higher plants: present status and future prospects, *Plant Mol. Biol.* **57**:461–485.

Hajduch, M., Ganapathy, A., Stein, J. W., Thelen, J. J., 2005, A systemic proteomic study of seed filling in soybean- establishment of high-resolution two dimensional reference maps, expression profiles, and an interactive proteome databse, *Plant Physiol.* **137**:1397–1419.

Hazen, S. P., Pathan, M. S., Sanchez, A., Baxter, I., Dunn, M., Estes, B., Chang, H-S., Zhu, T., Kreps, J., and Nguyen, H. T., 2005, Expression profiling of rice segregating for drought tolerance QTLs using a rice genome array. *Funct. Integ. Genom.* **5**:104–116.

Heatherly, L. G., and Elmore, R. W., 2004, Managing inputs for peak production, in: *Soybeans: Improvement, Production, and Uses*, Agronomy Monographs 3rd ed. No. 16, H. R. Boerma, and J. E. Specht, eds., ASA-CSSA-SSSA, Madison, WI, USA, pp. 451–536.

Hoagland, D. R., and Arnon, D. I., 1938, The water-culture method for growing plants without soil, *Calif. Agric. Expt. Circ.* **347**:1–39.

Hoognenboom, G., Nuck, M. G., Peterson, C. M., 1987, Root growth rate of soybean as affected by drought stress, *Agron. J.* **79**:607–614.

Horvath-Szanics, E., Szabo, Z., Janaky, T., Pauk, J., Hajos, Gy., 2006, Proteomics as an emergent tool for identification of stress-induced proteins in control and genetically modified wheat lines, *Chromatographia*, **63**: S143-S147.

Hu, H., Dai, M., Yao, J., Xiao, B., Li, X., Zhang, Q., and Xiong, L., 2006, Overexpressing a NAM, ATAF, and CUC (NAC) transcription factor enhances drought resistance and salt tolerance in rice, *Proc. Natl. Acad. Sci.* **103**:12987–12992.

Hudak, C. M., and Patterson, R. P., 1996, Root distribution and soil moisture depletion pattern of a drought-resistant soybean plant introduction, *Agron. J.* **88**:478–485.

Hymowitz, T., 1970, On the domestication of the soybean, *Eco. Bot.* **24**:408–421.

Hymowitz, T., 1990, Soybean: The success story, in: *Advances in new crops*, J. Janick and J. E. Simon, eds., Timber Press, Portland, OR, USA, pp. 159–163.

Hymowitz, T., 2004, Speciation and cytogenetics, in: *Soybeans: Improvement, Production, and Uses*, Agronomy monographs 3rd ed., No. 16, H. R. Boerma, and J. E. Specht, eds., ASA-CSSA-SSSA, Madison, WI, USA, pp. 97–136.

Hyten, D. L., Song, Q., and Cregan, P. B., 2004, Linkage disequilibrium in four soybean populations, in: Plant and Animal Genome Conference, 10–14 January 2004, San Diego, CA, p. 534.

Hyten, D., Choi, I-Y., Yoon, M-S., Song, Q., Specht, J., Nelson, R. l., Chase, K., Young, N, Lark, K. G., Shoemaker, R., and Cregan, P., 2006, An initial assessment of genome-wide linkage disequilibrium in soybean, abstract, ASA-CSSA-SSSA 2006 International meetings, 12–16 November, Indianapolis, IN. (http://a-c-s.confex.com/crops/2006am/techprogram/P2109.HTM)

Jang, I-C., OH, S-J., Seo, J-S., Choi, W-B., Song, S. I., Kim, C. H., Kim, Y. S., Seo, H-S., Choi, Y-D., Nahm, B. H., and Kim, J-K., 2003, Expression of a bifunctional fusion of the Escherichia coli genes for trehalose-6-phosphate synthase and trehalose-6-phosphate phosphatase in transgenic rice plants increases trehalose accumulation and abiotic stress tolerance without stunting growth, *Plant Physiol.* **131**:516–524.

Johnson, C. M., Stout, P. R., Broyer, T. C., and Carlton, A. B., 1957, Comparative chlorine requirements of different plant species, *Plant Soil*, **8**:337–353.

Joly, R. J., 1989, Effect of sodium chloride on the hydraulic conductivity of soybean root systems, *Plant Physiol.* **91**:1262–1265.

Kang, H. G., Noh, Y. S., Chung, Y. Y., Costa, M. A., An, K., and An, G., 1995, Phenotypic alterations of petal and sepal by ectopic expression of a rice MADS box gene in tobacco, *Plant Mol. Biol.* **29**:1–10.

Kao, W. Y., Tsai, T. T., Tsai, H. C., and Shih, C. N., 2006, Response of three *Glycine* species to salt stress, *Environ. Expt. Bot.* **56**:120–125.

Katerji, N. Hoorn, J. W., Hamdy, A., and Mastrorilli, M., 2003, Salinity effect on crop development and yield, analysis of salt tolerance according to several classification methods, *Agril. Water Mangt.* **62**:37–66.

Keim, P., Diers, B. W., Olson, T. C., and Shoemaker, R. C., 1990, RFLP mapping in soybean: Association between marker loci and variation in quantitative traits, *Genetics*, **126**:735–742.

Khan, R., Alkharouf, N., MacDonald, M., Chouikha, I., Meyer, S., Grefenstette, J., Knap, H., and Mathews, B., 2004, Microarray analysis of gene expression on soybean roots susceptible to the soybean cyst nematode two days post invasion, *J. Nemat.* **36**:241–248.

Kim, T. S., Cho, S. W., Lee, J. E., Kim, y. H., Cho, G., Choi, J. S., Chung, K. Y., Song, B. H., Jong, S. K., Kim, H. S., Woo, S. H., 2006, Proteomic approach to analyzing flooding stress in soybean [*Glycine max* (L.) Merr.], Plant and Animal Genomes XIV Conference abstract, January 14–18, 2006, San Diego, CA. p.930.

King, C. A., and Purcell, L. C., 2001, Soybean nodule size and relationship to nitrogen fixation response to water deficit, *Crop Sci.* **41**:1099–1107.

Kruger, W. M., 2006, Application of molecular marker to improve grain yield potential in soybean, The 11th Biennial conference on the molecular and cellular biology of the soybean, abstract, August 5–8, 2006, Lincoln, Nebraska.

Läuchli, A., and Wieneke, J., 1979, Studies on growth and distribution of Na+, K+, and Cl- in soybean varieties differing in salt tolerance, *Zeitschrift Fur Pflanzenernahrung und Bodenkunde*, **142**:3–13.

Lee, G. J., Boerma, H. R., Villagarcia, M. R., Zhu, X., Carter, T. E. Jr., Li, Z., and Gibbs, M. O., 2004, A major QTL conditioning salt tolerance in S-100 soybean and descendent cultivars, *Theor. Appl. Genet.* **109**:1610–1619.

Levitt, J., 1980, Response of plants to environmental stresses. In. *Water, radiation, salt and other stresses* vol. II, Academic Press, New York, pp. 93–128.

Li, D., 2006, Soybean QTL for yield and yield components associated with *Glycine soja* alleles, ASA-CSSA-SSSA 2006 International meetings, (http://a-c-s.confex.com/crops/2006am/techprogram/P21109.HTM)

Li, X. P., Tian, A. G., Luo, G. Z., Gong, Z. Z., Zhang, J. S., and Chen, S. Y., 2005, Soybean DRE-binding transcription factors that are responsive to abiotic stresses, *Theor. Appl. Genet.* **110**:1355–1362.

Li, Y. B., Hu, Z. A., and Wang, H. X., 2000, Further study on genotypic variation of salt tolerance to wild soybean (*Glycine soja* Sieb. & Zucc.), *Soybean Genetic Newsletter* 27 [Online journal]. URL http://www.soygenetics.org/articles/sgn2000–016htm.

Liao, H., Wong, F. L., Phang, T. H., Cheung, M. Y., Francisca Li, W. Y., Shao, G., Yan, X., and Lam, H. M., 2003, *GmPAP3*, a novel purple acid phosphatase-like gene in soybean induced by NaCl stress but not phosphorus deficiency, *Gene*, **318**:103–111.

Lin, C. Y., Viant, M. R., and Tjeerdema, R. S., 2006, Metabolomics: methodologies and applications in the environmental sciences, *J. Pestic. Sci.* **31**:245–251.

Liu, K., 1997, *Soybeans: Chemistry, Technology and Utilization*, Aspen Publishers, Gaithersburg, Maryland, USA, p. 532.

Liu, T., Staden, J., and Cress, W. A., 2000, Salinity induced nuclear and DNA degradation in meristematic cells of soybean (*Glycine max* L.) roots, *Plant Growth Regul.* **30**:49–54.

Liu, F. L., Anderson, M. N., and Jensen, C. R., 2004, Root signal controls pod growth in drought-stressed soybean during the critical abortion-sensitive phase of pod development, *Field Crops Res.* **85**:159–166.

Liu, Y., Gai, J. Y., Lu, H. N., Wang, Y. J., Chen, S. Y., 2005, Identification of drought tolerant germplasm and inheritance and QTL mapping of related root traits in soybean [*Glycine max* (L.) Merr.], *Yi Chuan Xue Bao*, **32**(8):855–863. (article in Chinese and abstract in english, PMID:16231741).

Ludlow, M. M., and Muchow, R. C., 1990, A critical evaluation of traits for improving crop yields in water-limited environments, *Adv. Agro.* **43**:107–153.

Luo, G. Z., Wang, H. W., Huang, J., Tian, A. G., Wang, Y. J., Zhang, J. S., and Chen, S. Y., 2005, A putative plasma membrane cation/proton antiporter from soybean confers salt tolerance in *Arabidopsis*, *Plant Mol. Biol.* **59**:809–820.

Luo, Q., Yu, B., Liu, Y., 2005, Differential sensitivity to chloride and sodium ions in seedlings of *Glycine max* and *G. soja* under NaCl stress, *J. Plant Physiol.* **162**:1003–1012.

Maguire, T. L., Grimmond, S., Forrest, A., Iturbe-Ormaetxe, I., Meksem, K., and Gresshoff, P., 2002, Tissue specific gene expression in soybean (*Glycine max*) detected by cDNA microarray analysis, *J. Plant physiol.* **159**:1361–1374.

Malhotra, R. S., and Blake, T., 2004, Breeding for salinity tolerance, in: *Abiotic Stresses, Plant Resistance Through Breeding and Molecular Approaches*, M. Ashraf, and P. J. C. Harris, eds., Food Products Press, Binghamton, NY, pp.125–143.

Mayaki, W. C., Stone, L. R., Teare, I. D., 1976, Irrigated and non-irrigated soybean, corn, and grain sorghum root systems. *Agron. J.* **68**:532–534.

Meng, Q., Zhang, C., Gai, J., and Yu, D., 2006, Molecular cloning, sequence characterization and tissue-specific expression of six NAC-like genes in soybean (*Glycine max* L. Merr.), *J. Plant Physiol.* On line published (doi:10.1016/j.jplph.2006.05.019).

Merrill, S. D., Tanaka, D. L., Hanson, J. D., 2002, Root length growth of eight crop species in Haplustoll soils, *Soil Sci. Soc. Am. J.* **66**:913–923.

Mian, M. A. R., Ashley, D. A., Boerma, H. R., 1996, An additional QTL for water use efficiency in soybean. *Crop Sci.* **38**:390–393.

Mian, M. A. R., Mailey, M. A., Ashley, D. A., Wells, R., Carter, T. E., Parrot, W. A., Boerma, H. R., 1998, Molecular markers associated with water use efficiency and leaf ash in soybean, *Crop Sci.* **36**:1252–1257.

Mitchell, R. L., and Russell, W. J., 1970, Root development and rooting patterns of soybean (*Glycine max* L. Merrill) evaluated under field conditions, *Agron. J.* **63**:313–316.

Monteros, M. J., Lee, G., Missaoui, A. M., Carter Jr., T. E., Boerma, H. R., 2006, Identification and confirmation of QTL conditioning drought tolerance in Nepalese soybean PI471938, The 11th Biennial conference on the molecular and cellular biology of the soybean, abstract, August 5–8, 2006, Lincoln, Nebraska.

Mozo, T., Dewar, K., Dunn, P., Ecker, J. R., et al., 1999, A complete BAC-based physical map of the *Arabidopsis thaliana* genome, *Nat. Genet.* **22**:271–275.

Muchow, R. C., and Sinclair, T. R., 1986, Water and nitrogen limitation in soybean grain production. II. Field and model analysis, *Field Crops*, **15**:143–156.

Munns, R., 2002, Comparative physiology of salt and water stress, *Plant Cell Environ.* **25**:239–250.

Neelakandan, A., Valliyodan, B., Nes, D., and Nguyen, H., 2006, Metabolic engineering of sterol biosynthesis pathway in soybean (abstract), in: Second Annual Soybean Biotechnology Symposium,

March 22, 2006, University of Missouri-Columbia, MO.

Neumaier, N., Farias, J. R. B., and Nepomuceno, A. L., 1995, Indice de tolerancia a seca de quarto cultivares de soja. Sociedade Brasileria de Agronmeteorologia. UFP. Congresso Brasileiro de Agrometeorologia. 80–82.

Nordborg, M., and Tavare, S., 2002, Linkage disequilibrium: what history has to tell us, *Trends Genet.* **18**:83–90.

Ohashi, Y., Nakayama, N., Saneoka, H., Fujita, K., 2006, Effects of drought on photosynthetic gas exchange, chlorophyll fluorescence and stem diameter of soybean plants, *Biologia Plantarum*, **50**:138–141.

Oono, Y. Seki, M., Nanjo, T., Narusaka, M., Fujita, M., et al., 2003, Monitoring expression profiles of Arabidopsis gene expression during rehydration process after dehydration using ca. 7000 full-length cDNA microarray, *Plant J.* **34**:868–887.

Orf, J. H., Diers, B. W., Boerma, H. R., 2004, Genetic improvement: conventional and molecular-based strategies, in: *Soybeans: Improvement, Production, and Uses*, Agronomy Monographs 3rd ed. No. 16, H. R. Boerma, and J. E. Specht, eds., ASA-CSSA-SSSA, Madison, WI, USA, pp. 417–450.

O'Toole, J. C., and Bland, W. L., 1987, Genotypic variation in crop plant root systems, *Adv. Agron.* **41**:91–145.

Panneerselvam, R., Muthukumarasamy, M., and Rajan, S. N., 1998, Amelioration of NaCl stress by triadimefon in soybean seedlings, *Biologia Plantarum*, **41**:133–137.

Pantalone, V. R. Rebezke, G. J., Burton, J. W., and Carter, T. E. Jr., 1996a, Phenotypic evaluation of root traits in soybean and applicability to plant breeding, *Crop Sci.* **36**:456–459.

Pantalone, V. R., Butron, J. W., and Carter, T. E. Jr., 1996b, Soybean fibrous root heritability and genotypic correlations with agronomic and seed quality traits, *Crop Sci.* **36**:125–136.

Pantalone, V. R., Kenworthy, W. J., Slaughter, L. H., and James, B. R., 1997, Chloride tolerance in soybean and perennial *Glycine* accessions, *Euphytica*, **97**:235–239.

Paris, R. L., 2003, Uniform soybean tests southern states 2002, pp. 127–142.

Paris, R. L., and Shelton, G. W., 2006, Uniform soybean tests southern states 2005, pp. 147–162.

Parker, M. B., Gaines, T. P., and Gascho, G. J., 1986, Sensitivity of soybean cultivars to soil chloride, Research Bulletins 347, The Georgia Agricultural Experiment Stations, University of Georgia, pp. 1–14.

Parker, M. B., Gascho, G. J., and Gaines, T. P., 1983, Chloride toxicity of soybeans grown on Atlantic cost flatwoods soils, *Agron. J.* **75**:439–443.

Price, A. H., 2006, Believe it or not, QTLs are accurate! *Trends Plant Sci.* **11**:213–216.

Passioura, J. B., 1977, Grain yield, harvest index and water use of wheat, *J. Aust. Inst. Agric. Sci.* **43**:117–120.

Passioura, J. B., 1994, The yield of crops in relation to drought, in: *Physiology and determination of crop yield*, K. J. Boote et al. eds., American Society of Agronomy, Madison, WI, pp.343–359.

Passioura, J. B., 1996, Drought and drought tolerance, in: *Drought tolerance in higher plants: Genetical, physiological, and molecular biological analysis*, E. Belhassen, ed., Kluwer Academic Publications, The Netherlands, pp.1–5.

Purcell, L. C., DeSiva, M., King, C. A., and Kim, W. H., 1997, Biomass accumulation and allocation in soybean associated with genotypic differences in tolerance of nitrogen fixation to water deficits, *Plant Soil*, **196**:101–113.

Purcell, l. C., and Specht, J. C., 2004, physiological traits for ameliorating drought stress, in: *Soybeans: Improvement, Production, and Uses*, Agronomy Monographs 3rd ed. No. 16, H. R. Boerma, and J. E. Specht, eds., ASA-CSSA-SSSA, Madison, WI, USA, pp. 569–620.

Qin, Z., Rui, Y., Zhiang, H., Min, Z., and Xin, D., 2000, Cloning and characterization of a molecular marker associated with salt tolerance from soybean cultivars, *Soybean Genetics Newsletter* 27 [Online journal]. URL http://www.soygenetics.org/articles/sgn2000–008.htm.

Quarrie, S. A., 1996, New molecular tools to improve the efficiency of breeding for increased drought resistance, *Plant Growth Regulator*, **20**:167–178.

Rabbani, M. A., Maruyama, K., Abe, H., Khan, M. A., Katsura, K., et al., 2003, Monitoring expression profiles of rice genes under cold, drought, and high-salinity stresses and abscisic acid application using cDNA microarray and RNA gel-blot analyses, *Plant Physiol.* **133**:1755–1767.

Ragab, A. S., Pantalone, V. R., Kenworthy, W. J., and James, B. R., 1994a, Salt tolerance of soybean in solution culture experiments, I. Evaluation of screening technique, *Soybean Genet. Newsl.* **21**: 274–276.

Ragab, A. S., Pantalone, V. R., Kenworthy, W. J., and James, B. R., 1994b, Salt tolerance of soybean in solution culture experiments, II. Reaction of 19 genotypes, *Soybean Genet. Newsl.* **21**:277–279.

Rampitsch, C., Srinivasan, M., 2006, The application of proteomics to plant biology: a reviw, *Can. J. Bot.* **84**: 883–892.

Rengasamy, P., 2006, World salinization with emphasis on Austrialia, *J. Expt. Bot.* **57**:1017–1023.

Robertson, W. K., Hammond, L. C., Johnson, J. T., and Boote, K. J., 1980, Effects of plant-water stress on root distribution of corn, soybeans and peanuts in sandy soil, *Agron. J.* **72**:548–550.

Salekdeh, G. H., Siopongco, J., Wade, L. J., Ghareyazie, B., Bennett, J., 2002, Proteomic analysis of rice leaves during drought stress and recovery, *Proteomics*, **2**:1131–1145.

Schlueter, J. A., Dixon, P., Granger, C., Grant, D., Clark, L., Doylee, J. J., and Shoemaker, R. C., 2004, Mining EST databases to resolve evolutionary events in major crop species, *Genome*, **47**:868–876.

Serraj, R., Bona, S., Purcell, L. C., and Sinclair, T. R., 1997, Nitrogen accumulation and nodule activity of field-grown 'Jackson' soybean in response to water deficits, *Field Crops Res.* **52**:111–118.

Serraj, R., Roy, G., and Drevon, J. J., 1994, Salt stress induces a decrease in the oxygen uptake of soybean nodules and in their permeability to oxygen diffusion, *Physiologia Plantarum*, **91**:161–168.

Shao, G., Chang, R., and Chen, Y., 1995, Screening for salt tolerance to soybean cultivars of the United States, *Soybean Genet. Newsl.* **22**:32–42.

Shoemaker, R. C., lzin, K., Labate, J., Specht, J., Brummer, E. C., Olson, T., Young, N., Concibido, V., Wilcox, J., Tamulonis, J. P., Kochert, G., Boerma, H. R., 1996, Genome duplication in soybean (*Glycine* subgenus *soja*), *Genetics,* **144**:329–338.

Shoemaker, R. C., and Olson, T. C., 1993, Molecular linkage map of soybean [*Glycine max* (L.) Merr.], in: *Genetic Maps: Locus Maps of Complex Genomes*, S. J. O'Brien, ed., Cold Spring Harbor Press, Cold Spring Harbor, New York, pp. 6131–6138.

Shultz, J. L., and others, 2006, The soybean genome database (SoyGD): a browser for display a duplicated, polyploidy, regions and sequence tagged sites on integrated physical and genetic maps of *Glycine max*, *Nucleic Acid Res.* **34**: D758-D765.

Sinclair, T. R., Purcell, L. C., Vadez, V., Serraj, R., King, C. A., and Nelson, R., 2000, Identification of soybean genotype with N_2 fixation tolerance to water deficits, *Crop Sci.* **40**:1803–1809.

Singh, R. J., Hymowitz, T., 1999, Soybean genetic resources and crop improvement, *Genome*, **42**: 605–616.

Singleton, P. W., and Bohlool, B. B., 1984, Effect of salinity on nodule formation by soybean, *Plant Physiol.* **74**:72–76.

Song, Q. J., Marek, L. F., Shoemaker, R. C., Lark, K. G., Concibido, V. C., Delannay, X., Specht, J. E., Cregan, P. B., 2004, A new integrated genetic linkage map of the soybean, *Theor. Appl. Genet.* **109**:122–128.

Specht, J. E., Hume, D. J., and Kumudini, S. V., 1999, Soybean yield potential- A genetic and physiological perspective, *Crop Sci.* **39**:1560–1570.

Specht, J. E., Germann, M., Markwell, J. P., Lark, K. G., Orf, J. H., Macrander, M., Chase, K., Chung, J., Graef, G. L., 2001, Soybean response to water: a QTL analysis of drought tolerance, *Crop Sci.* **41**:493–509.

Sponchaido, B. N., White, J. W., Castillo, J. A., and Jones, P. G., 1980, Root growth of four common bean cultivars in relation to drought tolerance in environments with contrasting soil types, *Exp. Agric.* **25**:249–257.

Stoddard, F. L., Balko, C., Erskine, W., Khan, H. R., Link, W., and Sarker, A., 2006, Screening techniques and sources of resistance to abiotic stresses in cool-season food legumes, *Euphytica*, **147**:167–186.

Sun, Y., Wang, D., Bai, Y., Wang, N., and Wang, Y., 2006, Studies on the over expression of the soybean *GmNHX1* in *Louts corniculatus*: The reduced Na^+level is the basis of the increased salt tolerance, *Chinese Sci. Bullt.* **51**:1306–1315.

Takahashi, S., Seki, M., Ishida, J., Satou, M., Sakura, T., Narusaka, M., Kamiya, A., Nakajima, M., Enju, A., Akiyama, K., Yamaguchi-Shinozaki, K., and Shinozaki, K., 2004, Monitoring the expression profiles of genes induced by hyperosmotic, high salinity, and oxidative stress and abscisic acid treatment in Arabidopsis cell culture using a full-length cDNA microarray, *Plant Mol. Biol.* **56**:29–55.

Tanksley, S. D., 1993, Mapping polygenes, *Ann. Rev. Genet.* **27**:205–233.

Tao, Q., Chang, Y. L., Wang, J., Chen, H., Islam-Faridi, M. N., Scheuring, C., Wang, B., Stelly, D. M., and zhang, H-B., 2001, BAC-based physical map of the rice genome constructed by restriction fingerprint analysis, *Genetics*, **158**:1711–1724.

Thibaud-Nissen, F., Shealy R. T., khanna, A., and Vodkin, L. O., 2003, Clustering of microarray data reveals that transcript patterns associated with somatic embryogenesis in soybean, *Plant Physiol.* **132**:118–136.

Tian, A-G., Wang j., Cui, P., Han, Y-J., Xu, H., Cong, L-J., Huang, X-G., Wang, X-L., Jiao, Y-Z., Wang, B-J., Wang, Y-J., Zhang, J-S., and Chen, S-Y., 2004, Characterization of soybean genomic features by analysis of its expressed sequence tags, *Theor. Appl. Genet.* **108**:903–913.

Tuberosa, R., Gill, B. S., and Quarrie, S. A., 2002, Cereal genomics: ushering in a brave new world, *Plant Mol. Biol.* **48**:445–449.

Tuberosa, R., Salvi, S., 2004, Markers, genomics and post-genomics approaches – will they assist in selecting for drought tolerance? in: *New directions for a diverse planet*, proceedings of the 4th International Crop Science Congress, 26 September – 1 October, 2004, Brisbane, Australia.

Turner, N. C., Wright, G. C., and Siddique, K. H. M., 2001, Adaptation of grain legumes (pulses) to water limited environments, *Adv. Agron.* **71**:193–231.

Umezawa, T., Mizuno, K., and Fujimura, T., 2002, Discrimination of genes expressed in response to the ionic or osmotic effect of salt stress in soybean with cDNA-AFLP, *Plant Cell Environ.* **25**: 1617–1625.

Umezawa, T., Shimizu, K., Kato, M., and Ueda, T., 2000, Enhancement of salt tolerance in soybean with NaCl pretreatment, *Physiologya Plantarum*, **110**:59–63.

Umezawa, T., Fujita, M., Fujita, Y., Yamaguchi-Shinozaki, K., and Shinozaki, K., 2006, Engineering drought tolerance in plants: discovering and tailoring genes to unlock the future, *Curr. Opin. Biotech.* **17**:113–122.

USDA, ARS, National Genetic Resources Program, *Germplasm Resources Information Network - (GRIN)*, [Online Database] National Germplasm Resources Laboratory, Beltsville, Maryland. Available: http://www.ars-grin.gov/cgi-bin/npgs/html/findtaxon.pl (24 May 2006).

Valliyodan, B., and Nguyen, H. T., 2006, Understanding regulatory networks and engineering for enhanced drought tolerance implants, *Cur. Opn. Plant Biol.* **9**:1–7.

Vodkin, L. O., Khanna, A., Shealy R., Clough S. J., Gonzalez, D. O., Philip, R., Zabala, G., thibaud-Nissen, F., sidarous M., stromvik, M., et al., 2004, Microarrays for global expression constructed with a low redundancy set of 27,500 sequenced cDNAs representing an array of developmental stages and physiological conditions of thesoybean plant, *BMC Genom.* **5**:73.

Vodkin, L. O., Zabala, G., Hunt, M., Boone, A. M., Gonzalez, D. O., and Tuteja, J., 2006, The 11th Biennial conference on the molecular and cellular biology of the soybean, Speaker abstract, August 5–8, 2006, Lincoln, Nebraska.

Walia, H., Wilson, C., Wahid, A., Condamine, P., Cui, X., and Close, T. J., 2006, Expression analysis of barley (*Hordeum vulgare* L.) during salinity stress, *Funct. Integr. Genom.* **6**:143–156.

Wang, D., and Shannon, M. C., 1999, Emergence and seedling growth of soybean cultivars and maturity groups under salinity, *Plant Soil.* **214**:117–124.

Wang, D., Grieve, C. M., and Suarez, D. L., 2005, Composition of irrigation water salinity affects growth characteristics and uptake of selenium and salt ions by soybean, *J. Plant Nutrn.* **28**:1073–1088.

Wang, Y., Ying, J., Kuzma, M., Chalifoux, M., Sample, A., Mcarthur, C., Uchacz, T., Sarvas, C., Wan, J., Dennis, D. T., 2005, Molecular tailoring of farnesylation for drought tolerance and yield protection, *Plant J.* **43**:413–424.

Westgate, M. E., and Peterson, C. M., 1993, Flower and pod development in water-deficient soybeans (*Glycine max* L. Merr.), *J. Expt. Bot.* **44**:109–117.

Wilson, R. F., 2004, Seed Composition, in: *Soybeans*: *Improvement, Production, and Uses*, Agronomy Monographs 3rd ed. No. 16, H. R. Boerma, and J. E. Specht, eds., ASA-CSSA-SSSA, Madison, WI, USA, pp. 621–677.

Wood, A. J., 1999, Comparison of salt-induced osmotic adjustment and trigonelline accumulation in two soybean cultivars, *Biologia Plantarum*, **42**:389–394.

Wood, A. J., Kassem, A. M., Lightfoot, D. A., 2006, Genetic components of water stress tolerance in soybean, the 11th Biennial Conference on the molecular cellular biology of the soybean, poster abstract, August 5–8, 2006, Lincoln, Nebraska.

Wu, C., Sun, S., Nimmakayala, P., Santos, F. A., Meksem, K., Springman, R., Ding, K., Lightfoot, D. A., Zhang, H-B., 2004, A BAC- and BIBAC-based physical map of the soybean genome, *Genome Res.* **14**:319–326.

Wu, X., Zhong, G., Luo, M-C., Dvorak, J., Cregan, P., Stacey, G., and Nguyen, H., 2006, Anchoring the soybean physical map to the genetic map, The 11th Biennial conference on the molecular and cellular biology of the soybean, poster abstract, August 5–8, 2006, Lincoln, Nebraska.

Xu, Z., Chang, R., Que, L., Sun, J., and Li, X., 1999, Evaluation of soybean germplasm in China, in: *Proceedings of World Soybean Research Conference VI*, Chicago, IL, Superior Printing, Champagne, IL, pp. 156–165.

Yamaguchi-Shinozaki, K., and Shinozaki, K., 2006, Transcriptional regulatory networks in cellular responses and tolerance to dehydration and cold stress, *Annu. Rev. Plant Biol.* **57**:781–803.

Yan, S., Tang, Z., Su, W., Sun, W., 2005, Proteomic analysis of salt stress-responsive proteins in rice root, *Proteomics*, **5**:235–244.

Yang, J., and Blanchar, R. W., 1993, Differentiation chloride susceptibility in soybean cultivars, *Agron. J.* **85**:880–885.

Yu, O., Shu, J., Hession, A. O., Maxwell, C. A., McGonigle, B., and Odell, J. T., 2003, Metabolic engineering to increase isoflavone biosynthesis in soybean seed, *Phytochem.* **63**:753–763.

Zhang, Q., Wang, H., and Hu, Z., 1999, RAPD markers associated with salt tolerance in wild soybean populations, *Soybean Genetic Newsletter.* 26 [Online journal]. URL http://www.soygenetics.org/articles/sgn1999–101.html.

Zhang, J. Z., Creelman, R. A., and Zhu, J-K., 2004, From laboratory to field. Using information from Arabidopsis to engineer salt, cold, and drought tolerance in crops, *Plant Physiol.* **135**:615–621.

Zhong, M., Hu, Z. A., and Gresshoff, P. M., 1997, Search for molecular markers of salt tolerance of soybean by DNA amplification fingerprinting, *Soybean Genet. Newsl.* **24**:81–82.

Zhu, J. K., 2001, Plant salt tolerance, *Trends Plant Sci.* **6**:66–71.

Zhu, Y. L., Song, Q. J., Hyten, D. L., Van Tassell, C. P., Matukumalli, L. K., Grimm, D. R., Hyatt, S. M., Fickus, E. W., Young, N. D., and Cregan, P. B., 2003, Single-nucleotide polymorphisms in soybean, *Genetics*, **163**:1123–1134.

Zou, J., Rodriguez-zas, S., Aldea, M., Li, M., Zhu, J., Gonzalez, D. O., Vodkin, L. O., DeLucia, E., and Clough, S. J., 2005, Expression profiling soybean response to *Pseudomonas syringae* reveals new defense-related genes and rapid HR-specific down regulation of photosynthesis, *Mol. Plant microbe Interact.* **18**:1161–1174.

CHAPTER 31

RECENT ADVANCES AND FUTURE PROSPECTIVE IN MOLECULAR BREEDING OF COTTON FOR DROUGHT AND SALINITY STRESS TOLERANCE

EDWARD L. LUBBERS[1], PENG W. CHEE[1],
YEHOSHUA SARANGA[2], AND ANDREW H. PATERSON[3]

[1]*Cotton Molecular Breeding Lab, NESPAL, University of Georgia, 2356 Rainwater Road, P.O. Box 748, Tifton, GA 31793*

[2]*Department of Field Crops, Vegetables and Genetics, Faculty of Agricultural, Food and Environmental Quality Sciences, The Hebrew University of Jerusalem, P.O. Box 12, Rehovot76100, Israel*

[3]*Department of Crop and Soil Science, Dept. Botany, and Dept. Genetics, University of Georgia, Athens, GA30602, USA*

Abstract: Fiber from cotton (*Gossypium hirsutum* and *G. barbadense*) is a major product in the world economy. It is a botanically unique plant as it is a perennial allotetraploid derived from diploid *Gossypium* species, one of which does not produce lint, which is grown as an annual row crop. Cotton is an especially appropriate system for research into the molecular basis of plant response to water deficit and salinity, as it originates from wild perennial plants adapted to semi-arid, sub-tropical environments which experienced periodic drought and temperature extremes that are associated with soils with high salt content. The current primary molecular breeding approaches include transgenic modification and quantitative trait mapping with marker-assisted selection. The preliminary work in QTL mapping for drought response and the relationships of the QTLs with the drought-associated measurements is developing a foundation for understanding and using the molecular basis of drought tolerance. QTL mapping for salt tolerance is not moving apace. Using and/or regulating transgene effects on the plant responses to drought and salinity has shown success and will continue to increase our understanding of the complexity of plant's physiological pathways. Improvements in all areas of molecular breeding are almost certain, but the most effective improvements will come from exploiting our improved understanding of the genetic architecture

Keywords: QTL, MAS, abiotic stress, water-use efficiency, stable carbon isotope ratio/ discrimination

M.A. Jenks et al. (eds.), Advances in Molecular Breeding Toward Drought and Salt Tolerant Crops, 775–796.

1. INTRODUCTION

The domesticated members of the *Gossypium* (cotton genus) are among the most important field crops in the world primarily due to the intrinsic value of their lint (fiber). Cotton lint as soft, breathable textile products has added greatly to the comfort, style, and culture of human civilization. The cotton plant also is an important source of vegetable oil used extensively in foodstuffs for baking and frying and in spreads such as margarine and mayonnaise. The seed bagasse is used as raw materials in livestock feed, fertilizer, paper, and biofuel. Despite the importance of cotton's secondary products, 90% of cotton's value resides in the lint fiber.

While *Gossypium* species are endemic to the tropics and subtropics, about 33 million hectares of cotton are planted annually in almost 70 countries (ICAC 2004) ranging from latitudes 47°N in Ukraine (UNCTAD 2006) to about 34°S in Australia (AOGTR 2002). The top six cotton producing countries in the 2000 to 2005 seasons include, by the average order of importance, China, the United States, India, Pakistan, Uzbekistan, and Brazil, and collectively account for almost three quarters of the world's cotton production (NCC 2006). Annual cotton fiber production has now reached about 25 million metric tons with a farm-gate value of about US$20 billion, contributing about 40% to the world fiber market (ICAC 2006) and making cotton the single most important natural fiber in the global economy. The value of the processing of cotton adds US$10 billion to the farm-gate for an aggregate value estimated to be US$30 billion. More than 350 million people are engaged in jobs related to the production and processing of cotton.

Cotton is unique among crop plants in that four separate species in the genus *Gossypium* (Malvaceae) have been independently domesticated and cultivated for commercial lint fiber production. *Gossypium* appears to have arisen 10 to 20 million years ago and now includes eight genome types labeled A to G and K. Most cotton fiber production around the world is concentrated in a pair of species, *G. hirsutum* and *G. barbadense*, with small roles for another pair, *G. arboreum* and *G. herbaceum*. Both *G. hirsutum* and *G. barbadense* are tetraploids derived from a hypothetical common ancestor that formed in the New World from the combination of an invasive diploid A-genome species and a native diploid D-genome species. These diploids appear to have diverged from a common ancestor about 4–11 million years ago then rejoined in a common nucleus as an allotetraploid ($2n = 4x = 52$) about 1–2 million years ago (Wendel 1989). Both *G. arboreum* and *G. herbaceum* are A-genome diploid species that supplied all of the cotton to the Old World (Zaitzev, 1928; as cited by Fryxell, 1979) until being supplanted by the New World cotton.

G. hirsutum 'Upland cotton' is the primary source of cotton fiber, accounting for about 97% of the world production (NCC 2006). *G. hirsutum* race 'yucatanense' appears to be the wild progenitor of Upland cotton (Brubaker and Wendel 1994) and is native to the Yucatan peninsula in Mexico, with race 'punctatum' considered the first version of a domesticated form that is found in a ring surrounding the habitat of the race 'yucatanense'. Additional forms that appear to be versions of further domesticated cotton (Brubaker et al. 1999) are found on the Mexican highlands and in southern Mexico and Guatemala (races 'latifolium', 'palmeri', 'richmondi', and 'morilli'). The earliest cotton cultivars in the US cottonbelt were a collection of diverse cultivars that were native to the islands of the Caribbean Sea and Mesoamerica (Ware 1951), but these were essentially replaced by 'latifolium' cultivars from southern Mexico/Guatemala via the Mexican highlands (Ware 1951; Wendel et al. 1992; Brubaker and Wendel 1994). Additional cultivars directly from southern Mexico/Guatemala were introduced as a response to the devastation caused by the boll weevil (Ware 1951). This 'melting pot', which has included some introgressions from *G. barbadense*, has been the basis of the main groupings of US Upland cultivars, Acala, Delta, Plains, and Eastern, and can be considered the genetic foundation of most of the *G. hirsutum* cotton cultivars throughout the world (Niles and Feaster 1984).

G. barbadense, the other domesticated allotetraploid, yields an extra-long staple or extra-fine quality cotton fiber that makes about 3% of the total world cotton market (ICAC 2005). Even though the fiber quality is better than Upland cotton, neither is the yield as high nor the normal production environment as widespread, thus, breeding efforts in this species have not been as extensive as in Upland cotton. Using molecular markers, Percy and Wendel (1990) and Westengen et al. (2005) report northwest South America as the center of diversity for *G. barbadense* as well as the subsequent dispersal across the Andes to northeast South America and then following the islands up the Caribbean into Central America. The material in the Caribbean developed into Sea Island cotton that was subsequently combined in Egypt with another *G. barbadense*, reportedly Jumel's tree cotton (Balls 1912), that was the basis for Egyptian cotton which was then brought back to the new world as Pima cotton. Interestingly, since it appears that there was one original interspecies cross (Wendel 1989), there has to be some movement of these protoallotetraploids since the centers of diversity for *G. hirsutum* and *G. barbadense* don't overlap.

2. NEED FOR DROUGHT AND SALT TOLERANCE IN COTTON

Almost 2/3 (64%) of Earth's land is desert or drylands, and of that, 57% of the world's potentially productive area is located in drylands (FAO 2000). Even though this includes areas that are to cold for cotton and removes irrigated desert production areas, it still illustrates the overall impact of drought on agriculture. In virtually all agricultural regions, crop yields are periodically reduced by drought (Kramer 1980; Boyer 1982) and global climatic trends may accentuate this problem (Le Houerou 1996). Areas that are prone to drought are interrelated with salinity throughout the world since suitable amounts of precipitation will wash the salt into the seas or deep into the soil profile. Efficient irrigation technologies help to reduce the gap between potential and actual yield; however, diminishing water supplies in many regions drive programs to improve inherent crop genetics for productivity under arid conditions (Blum 1988) that are an affordable and sustainable resolution to this need. Even though cotton is usually grown during the summer in dryland regions where water availability is often limited, the high value of cotton justifies irrigation and makes cotton a major consumer of agricultural water. Regardless of whether it is irrigated or not, cotton is often exposed to drought stress, which adversely affects both yield and fiber quality.

In order to improve any trait such as drought or salinity tolerance by selective breeding, heritable variability that affects the plant's response to these conditions is required. Cotton is an especially appropriate system for research into the molecular basis of plant response to water deficit, as it originates from wild perennial plants adapted to semi-arid, sub-tropical environments which experienced periodic drought and temperature extremes (Kohel 1974), and adaptations to heat and drought stress are known to exist (Peterschmidt and Quisenberry 1981, Quisenberry et al. 1982). However, modern cotton cultivars are the result of intensive selection to facilitate mechanical harvesting and processing as well as to produce large amounts of specific types of fibers, often under irrigated conditions. Since selection has unintentionally narrowed the genetic variability for drought tolerance (Rosenow et al. 1983), increasing demands on limited water supplies makes this an urgent priority for improvement. Considerable variation does persist within and between *G. hirsutum* and *G. barbadense* in physiological traits such as water-use efficiency (Yakir et al. 1990; Saranga et al. 1998) and photosynthetic rate (Pettigrew and Meredith 1994). Additional characterization of other accessions of these species and exploiting the other wild tetraploid cottons (*G. tomentosum, G. darwinii*, and *G. mustelinum*) will likely yield additional valuable alleles.

In cotton, cultivar development has usually involved breeding for yield and fiber quality while the plants are stressed by water scarcity or salinity, and selecting based on relative performance between stressed and non-stressed conditions. Much of the success of this approach has involved improving harvest index, the ratio of yield weight to total plant weight, which should work for cultivars growing under abiotic stress or not. Advanced molecular techniques such as quantitative trait loci (QTLs) and marker-assisted selection (MAS) can provide valuable tools. Since cotton yield and fiber quality are quantitative traits, using transgenic approaches may be limited in this end product-focused approach, however, see the section Transgenics to Study and Improve Drought and Salinity Tolerance.

Selection for the end product of greater yield and enhanced quality in stress environments is commonly inefficient because the heritability is reduced by large phenotypic variation from season to season and from year to year. Larger genotype by environment (G x E) interactions also confuse the effectiveness of selection (Richards 2006). Much of the interaction can be traced to variability of the timing and amount of precipitation throughout the growing season. It is easy to see that the effect of a terminal water deficit could easily require different physiological adaptations than water scarcity during the beginning of the season. Stiller et al. (2004) suggests that given G x E crossover interactions for yield between well-watered and drought-stress, breeding programs must select in the stressed environments or they will not likely to be effective. This approach will ultimately require developing cultivars that are regional when positive G x E interactions begin to be utilized.

Addressing additional challenges needed for commercial cotton production (for example, biotic restraints such as disease and nematodes, nutrient deficiencies, and inferior soil properties) would make the breeding even more complex.

3. PHYSIOLOGICAL RESPONSES OF COTTON RESPONSE TO DROUGHT AND SALINITY

Developing a physiological understanding of the plants' responses to abiotic stresses and utilizing it in a stepwise breeding program may help to address shortcomings of the classical breeding approach. Any heritable improvement in yield must be the result of an underlying physiological change. Appropriate targets for physiological breeding will develop from an understanding of the factors regulating growth, development, and yield of the crop (Richards et al. 2002; Chaves et al. 2003). This should include a detailed understanding of the relevant production system and regional climatic conditions. The research can range from the underlying basic sciences of chemistry and physics, to understanding the plant as a part of an ecological system which includes interactions with other individuals of the population and with the biotic and abiotic environment.

Understanding of the physiology can guide breeding for specific components of the crop's responses that increase yield and enhance quality, or breeding to avoid the stress by changing things like reproductive timing or increasing the volume of soil that roots exploit. Developing the requisite physiological understanding is the key, and, of course, is not simple. It increases the scope of breeding programs by the need to include plant physiology as a preliminary and vital part of the program, first selecting and characterizing a physiological trait that affects the production of the end product. Also necessary is determination that genetic variability exists for the trait upon which selection might be based. Secondly, techniques to quickly and inexpensively measure the trait in single plants are required since plant breeding is a statistical endeavor.

There are few early examples in which physiological understanding has identified traits that limit yield under drought which have subsequently been used in breeding programs to enhance the yield (Richards 2006). Some of the previous work reviewed by Richards (2006) has focused on survival of the plants which does not necessarily correspond with yield. Also, traits shown by individual plants may not necessarily be important in field populations. And as mentioned before, traits that ease the effect of one type of drought may not work on other types of drought, i.e. terminal drought versus early season drought. On the whole, this can be a very complex system.

Of the examples that Richards (2006) mentioned as successful physiological breeding, some are being studied in cotton such as enhancing the efficiency of water uptake by increasing the volume of soil explored by the roots. Basal et al. (2003) evaluated rooting traits at the seedling stage for 68 converted cotton race stocks and compared them with a modern elite line TAM94L-25 and Lankart 142, an obsolete cultivar. Considerable genetic variation occurred for root length, lateral root number, root fresh weight, lateral root dry weight, and total root dry weight, but none of the converted race stocks were superior to TAM94L-25. Two BC_2F_2 populations were derived by crossing TAM94L-25 to two of the converted race stock donor parents that were considered robust and nonrobust rooting, respectively. Variation in each population for the rooting traits was substantial; the ratio of the standard deviation to the range was more than 4.5 for all of the traits. These results suggest that the converted race stocks have useful genetic variability for improving root growth.

Later flowering and maturity has been strongly, positively associated with lint yield (Stiller et al. 2004) and also can be consistent with development of more roots. Later flowering also permits the plant to accumulate more biomass that may be remobilized during later growth and reproductive stages. Dumka et al. (2004) looked at delaying initiation of cotton fruiting to enhance root growth and attempt to avoid episodic drought. Although fruiting was delayed, no improvement for drought tolerance was noted as determined by boll number or yield. Singh et al. (2006) showed that increased P uptake from a drying soil leads to an increased supply of osmotically-active inorganic solutes for the cells in the growing leaves. This appears to lead to the accumulation of both free- and bound water, ultimately leading to increased leaf expansion rates. Root traits that emphasize P uptake could be a powerful mechanism to help tolerate drought.

Water loss (transpiration) is an unavoidable consequence of photosynthesis (Cowan 1986), whereby the energy of solar radiation is used for carbon fixation. Both transpiration and photosynthesis are interdependent to changes in stomatal aperture allowing gas exchange, evaporation of water to the surrounding air and diffusion of CO_2 into the leaf. Water could be conserved by closing the stomates, which decreases the rate of CO_2 fixation by the photosynthetic apparatus. Conversely, photosynthesis could be enhanced by opening the stomates so that CO_2 would not be limiting which would also increase water loss. Physiological water-use efficiency (WUE), the ratio between the rate of carbon fixation by photosynthesis and transpiration rate, can also estimated as a ratio of the photosynthesis rate and stomatal conductance since water loss through the stomates is the primary water movement pathway out of the plant (Baker 1984).

Crop WUE can be defined in a practical sense as the ratio of yield (or total biomass) produced to the water used (water applied plus the change of the water stored in the soil). It is a key factor that can be improved to either produce greater yield for a given amount of water or for having a stable yield with less water. Much of the significant improvement of crop WUE has been mainly from modern irrigation and management techniques (Stanhill 1991). Further improvement may come from exploiting the physiology and genetics of the cotton plant. Genetic variability for components of WUE is available in cotton for stomatal resistance (Voloudakis et al 2002) and in the rate of stomatal closure in adjusting the transpiration rate during changes in water availability (Basal et al. 2005).

Reliable evaluations of either crop or physiological WUE are complicated, difficult to obtain, and not inexpensive. A large drawback for crop WUE is that it realistically cannot be done on single plants. Physiological WUE is a standard measurement of the moment that can be performed on single plants, but it fails to incorporate seasonal effects of WUE. Measures of WUE that represent a single plant over an entire season are essential to make breeding selections.

Since neither the physiological WUE nor crop WUE is directly suitable for selecting plants from segregating populations, utilizing the carbon isotope ratio ($\delta^{13}C$) or stable carbon isotope discrimination (Δ) as an indirect measure (Farquhar et al. 1989) has been shown to be of potential value in several crops (Ehleringer et al. 1993). Lu et al. (1996) found that Δ in *G. barbadense* was positively correlated with yield and stomatal conductance. Since there was no difference in photosynthetic rate in this experiment (Lu and Zeiger 1994), the physiological WUE would of formulaic necessity be negatively correlated with Δ. Saranga et al. (1998) determined that $\delta^{13}C$ was positively correlated with crop WUE and yield using *G. barbadense*, *G. hirsutum*, and an interspecific hybrid. Leidi et al. (1999) found Δ negatively associated to physiological WUE in *G. hirsutum* cultivars. Since Δ and $\delta^{13}C$ are essentially different sides of the same coin (Farquhar et al. 1989), these test all showed the same general relationship between WUE and yield. Ulloa et al. (2000) confirmed a hypothesis that higher stomatal conductance allows greater cotton yields for genotypes that experience above optimal temperatures under irrigated environments. The advantages of this physiological trait would be small or nonexistent in moderate temperate zones where enhanced evaporative cooling would not be expected to enhance lint yield.

Since $\delta^{13}C$ and Δ are assessed in plant tissues that developed over a relatively long time, this measurement reflects WUE integrated over time that can be performed on individual plants. Further studies relating $\delta^{13}C$ and Δ with production and water use remain important. $\delta^{13}C$ showed more sensitivity by finding significant differences between cultivars than did any of the other physiological measures [net photosynthesis and leaf conductance to water vapor (WUE components) and the ratios of A/g and intercellular CO_2 concentration (C_i) /ambient CO_2 concentration (C_a) (alternate WUE formulas)] by more than a factor of two (Stiller et al. 2005). However, even though the estimates of broad-sense heritability for A, C_i/C_a, and $\delta^{13}C$ were good, none were much better than the broad-sense heritability for lint yield and none were better than the broad-sense heritability of the fiber quality measures of length, strength, micronaire, uniformity, or elongation. As the interaction of the processes of carbon fixation and transpiration within the plant becomes clearly understood, the heritability issue may be resolved.

Additional traits may be of value to further understand the relationship between WUE and carbon assimilation and transpiration. Nonstomatal effects can affect internal leaf CO_2 concentration and thereby affect $\delta^{13}C$ (Ennahli and Earl 2005). There are also artifacts that affect measurements of internal leaf CO_2 concentration during stress conditions. Chloroplastic CO_2 concentration, calculated from gas exchange measurements and chlorophyll fluorometry, eliminates the possible artifacts. It also allows differentiation between the effects that stomates and mesophyll have on reducing carbon fixation by reducing CO_2 concentration in the leaf and the effects of stress damage or inhibition of the chloroplast (Ennahli and Earl 2005). WUE that is found by using chloroplastic CO_2 concentration can also separate the levels of water-stress, as compared to the check, significantly better than gas exchange measurements alone. Severe stress was shown to decrease net leaf photosynthetic carbon assimilation from both lasting chloroplast-level injury and decreased chloroplastic CO_2 concentration; the decreased chloroplastic CO_2 concentration was transient, recovering after rewatering.

WUE can be considered part of resource management under moderate drought stress much like the consideration of mineral nutrition. Beyond increasing the efficiency of production against the amount of water used, another level of stress protection is applicable in drought as well as salt stress. Ahmad et al. (2002) and Ashraf (2002) review salt tolerance specifically in cotton, but there are a number of more general reviews discussing approaches to understanding possible tolerance mechanisms against abiotic stresses as drought and salt, including osmotic and structural adjustment, damage prevention and repair, metabolic modification, regulated ion uptake and compartmentalization, and Late-Embryogenesis-Abundant (LEA) proteins (McKersie and Leshem 1994; Bohnert et al. 1995; Ingram and Bartels 1996; Smirnoff 1998; Toppi and Pawlik-Skowrońska (eds.) 2003; Ashraf 2004; Ashraf and Harris 2004; Hirt and Shinozaki (eds), 2004; Bartels and Sunkar 2005; Jenks and Hasegawa (eds.) 2005).

The accumulation of compatible solutes (i.e. which do not interfere with normal biochemical reactions (Bohnert et al. 1995)) associated with putative osmotic adjustments has been shown to be initiated by heat stress, salt, and drought in cotton (Kuznetsov et al. 1999; Meloni et al. 2001; Showler 2002). These three studies in cotton illustrate the first step of physiological breeding for a trait which affects drought and salt tolerance in cotton: developing an understanding of the physiology. Kuznetsov et al. (1999) in two *G. hirsutum* cultivars found that ion concentration accounted for up to 90% the osmotic pressure in non-stressed plants with the free amino acids and amides (compatible solutes) contributing only 3%. In heat shocked plants under water stress, the contribution of the free amino acids and amides increased 5- to 7-fold over the control. The change in arginine, proline, and asparagine concentrations (240-, 160-, and 150-fold increases, respectively over the control) were specifically highly correlated (0.98 to 0.99) with changes in osmotic pressure. Meloni et al. (2001) in two *G. hirsutum* cultivars did not find any significant increase in proline in NaCl-stressed treatments compared to the non-stressed control. Different profiles of the various free amino acids were found for different stress-related treatments (Showler 2002). In drought stress, all free amino acids were significantly different from the control, except for glutamic acid, with proline increasing almost 50-fold. Arginine had the greatest concentration in the greatest stress treatment which follows the response by arginine in the study by Kuznetsov et al. (1999).

The production of reactive oxygen species in chloroplasts and mitochondria is another change detected in plants under drought or salt stress. The activities of antioxidative enzymes superoxide dismutase, peroxidase, and glutathione reductase increased in the cotton cultivar "Pora" as the salt level increased but were not affected by the salt increase in "Guazuncho" (Meloni et al. 2003). Guazuncho had higher membrane damage, lower photosynthetic rate, lower stomatal conductance, and lower chlorophyll content under salt stress but these measures did not further increase with increasing salt levels. Lower photosynthetic rate and lower stomatal conductance was also found for Pora but there was no difference in chlorophyll content. The membrane damage for Pora also increased (except for a small decrease from unstressed control to the lowest salt stressed treatment) but it was lower for all salt levels. Unlike Guazuncho, Pora's responses were not a single step but followed the change in salt. The activity of the antioxidative enzymes and the measurement of membrane damage in Pora fit the hypothesis that increases in antioxidative enzyme activity decreased the amount of membrane damage in cotton.

4. GENE REGULATION AND STRESS-RESPONSIVE PROTECTION MECHANISMS

Regulation of gene expression may be even more important than the mechanisms such as antioxidative enzymes or osmotic adjustments. Besides up-regulation that occurs in response to abiotic stress; promoters, second messengers and signaling molecules, posttranscriptional control, and down-regulation of genes can further influence stress responses (Ingram and Bartels 1996). As with WUE, the effect of such variables as timing and severity of drought along with production practices must be understood before efficient breeding for stress tolerance using osmoticants can take place. Adding an understanding of the physiological and biochemical activities associated with the responses of the plant to salinity and drought reveals the complexity of selecting for tolerance to salt and water stress. Stress-responsive control of the induction and/or the amount of osmoticants reduces/eliminates the energy requirement to produce an osmoticant in favorable conditions.

In cotton, genes of osmotin, a PR5 (pathogenesis-related) protein, are clustered as two functional genes and two pseudogenes (Wilkinson et al. 2005). Osmotin genes have a wide variety of potential promoter elements that occur in their 5′-flanking regions and are possible promoter elements for inducible gene expression during osmotic stress. A related tobacco gene is up-regulated by drought, salinity, and other environmental indicators as well as ethylene and abscisic acid (Kitajima and Sato 1999). Cotton plants can be induced to express the osmotin proteins upon treatment with ethephon and hydrogen peroxide (Wilkinson et al. 2005). The osmotin system in cotton might be an excellent model to study stress signaling.

Trehalose is a non-reducing disaccharide that serves as a carbohydrate reserve and stress protectant, stabilizing and protecting proteins and membranes from denaturation (Crowe et al. 1992). In cotton, production of trehalose is induced by a number of stresses including drought and salt (Kosmas et al. 2006) but it does not accumulate in high amounts. Trehalose-6-phosphate synthase RNA was found in all tissues tested in both water-stressed and well-watered plants with increased levels of expression in stressed leaves and roots as compared to the well-watered controls. Kosmas et al. (2006) speculates that the trehalose-6-phosphate synthase gene is involved in osmotic stress signal transduction. In *Gossypium*, *trehalose-6-phosphate synthase* is found as a single copy in each of the genomes. Further work is planned to identify and analyze the promoter, study subfunctionalization effects (Adams and Wendel 2004), and explore more drought tolerance mechanisms at the transcriptional level.

Some heat-shock proteins are also found in plants subject to water stress (Kuznetsov et al. 1999). They likely function as molecular chaperones that assist in protein folding and prevent protein denaturation. A heat shock protein was found (Lu et al. 1995) that would bind with calmodulin and was named Heat shock protein calmodulin binding (HSPCB). Calmodulin senses nanomolecular changes in Ca^{++} and acts as a molecular switch to regulate other proteins and enzymes. These target proteins and enzymes, called calmodulin-binding proteins, are thought to be the response elements through which the Ca^{++}/calmodulin second messenger system affects signal transduction. HSPCB is likely involved in the regulation of Ca^{++} mediated processes. A couple of possibilities could be a chaperone to stabilize calmodulin to maintain activity or participating in the stress response system as a component enzyme. The Ca^{++}/calmodulin complex may be involved in key regulatory roles in plant metabolism such as fluctuation in cytosolic Ca^{++} controls stomatal closure as a response of guard cells to ABA. Voloudakis et al. (2006) found that the *HSPCB* is mainly expressed in leaves of drought-tolerant cotton varieties under high water-stressed conditions. Gene-specific DNA probes were used to detect reverse transcription-PCR products that came from total plant RNA isolated at the end of the stress period. This suggests that this gene could be used as a selection marker for cotton drought tolerance in plants grown in a water-stressed condition and was found in all tissues tested. The *HSPCB* has also been isolated and characterized (Soitiros et al. 2006) as a preliminary to study the stress tolerance mechanisms at the transcriptional level in cotton through promoter identification and analysis.

5. TRANSGENICS TO STUDY AND IMPROVE DROUGHT AND SALINITY TOLERANCE

The use of transgenes is an excellent tool to confirm the effects of the genes of suspected traits that affect the response of plants to salt and drought and gain further information to better understand particular mechanisms controlling stress (Bajaj et al. 1999; Umezawa et al. 2006). Overexpression with a strong constitutive promoter of a cDNA that encodes for dehydration responsive element (DRE)-binding proteins triggered the expression of many genes for stress tolerance under normal growing conditions (Kasuga et al. 1999). This improved the tolerance to drought and salt in these transgenic plants, but it also resulted in severe growth retardation under normal growing conditions. Expression using a stress inducible promoter gave rise to even greater tolerance to stress conditions while minimizing the effects on plant growth. A cDNA encoding *G. hirsutum* DRE-binding protein 1 (GhDBP1) was found to act as a transcriptional repressor for DRE-mediated gene expression (Huang and Liu 2006). A next step in order to make this useful in breeding would be to determine if there is genetic variation in the activity of this system or other promoters that can be used to slightly adjust the responses.

A *G. hirsutum* cDNA clone, *GhNHX1*, which showed high sequence identity with plant vacuolar-type Na^+/H^+ antiporters, was isolated via differential hybridization in response to salt stress in cotton seedlings (Wu et al. 2004). Analysis by northern blot showed that mRNA accumulation of *GhNHX1* in cotton seedlings was strongly induced by salt stress and abscisic acid. *GhNHX1* activity in a mutant for yeast tonoplast Na^+/H^+ antiporter showed function complementation thereby proving that the antiporter is in the vacuolar membrane. Tobacco plants that overexpressed *GhNHX1* had higher salt tolerance than the wild-type plants. The mRNA level of *GhNHX1* was 3 to 7 times higher in a salt-tolerant cotton cultivar than in the two salt-sensitive cotton cultivars under salt treatment. Almost concurrently, a transgenic cotton plant was developed to overexpress *AtNHX1*, an *Arabidopsis* vacuolar Na^+/H^+ antiporter (He et al. 2005). *GhNHX1* and *AtNHX1* share 77.6% sequence identity. Cotton plants with *AtNHX1* had more biomass and produced more fibers when grown in the presence of high NaCl. Overexpression of both of the tonoplast Na^+/H^+ antiporters increases sodium transfer into vacuoles, which leads to higher vacuolar salt concentration and therefore higher salt tolerance. The sequestering of Na^+ in the vacuoles gives two advantages: (1) reduced toxic levels of Na^+ in cytosol; and (2) increased osmotic potential of the vacuole and therefore a more negative water potential that aids water uptake by the cells and water retention under high salt conditions. Also, the *AtNHX1*-expressing cotton plants yielded more and higher quality lint in the field than the controls.

Arabidopsis gene GF14λ, that encodes a 14-3-3 protein, was introduced into cotton and showed a "Stay-Green" phenotype and improved stress tolerance under moderate drought conditions (Yan et al. 2004). The 14-3-3 proteins are a group of regulatory proteins that can bind to over 100 proteins. In plants, 14-3-3 proteins can regulate primary metabolism, ion transport, cellular trafficking, enzyme activities, and gene expression. The *Arabidopsis* gene GF14λ interacts with several proteins that include ascorbate peroxidase 3 and ankyrin repeat-containing protein 2 (Zhang et al 1997; Yan et al. 2002). Ascorbate peroxidase 3 plays an important role in protecting plants under oxidative stress and water-deficit conditions (Wang et al. 1999; Yan et al. 2003) and ankyrin repeat-containing protein 2 is involved in both disease resistance and antioxidation metabolism (Yan et al. 2002). Overexpression of GF14λ in cotton conferred a "stay-green" phenotype under well-watered conditions. These cotton plants also displayed increased water-stress tolerance and maintained higher photosynthetic rates under conditions of low-water availability.

6. INTEGRATING QTL MAPPING INTO COTTON IMPROVEMENT

Crop breeding requires selecting high yielding, high quality cultivars. The cotton industry must have high lint yield per hectare with quality that is required by the cotton mills to make fabric in order to compete with synthetic fiber. There are many traits that are selected in a breeding program; some are controlled by single genes whereas others are quantitative traits that are controlled by many genes. These quantitative trait loci (QTLs) are associated with both yield and quality of cotton as well as numerous other desirable traits. There are several good reviews and books at a basic level describing markers and QTL mapping (for e.g. Paterson et al. 1991; Paterson 1995; Paterson 1996; Bernardo 2002; Collard et al. 2005). Using genetic mapping to dissect inheritance can help understand complex traits in the same population by distinguishing common heredity from casual associations and thereby help develop a logical progression of discrete factors that affect/control the trait (Paterson et al. 1988).

There are traits that are associated with stress tolerance that are not biochemical or osmotic in character, but might be described as traits that physically affect stress tolerance such as leaf boundary layer or leaf morphology. These traits can be considered components of yield since they do affect photosynthesis by influencing transpiration and carbon fixation. Stiller et al. (2004) determined that cotton with the okra leaf morphology was effective for increased yield in dryland conditions, however, additional research is required to make a more complete story to ensure that the okra leaf trait was not simply part of a suite of traits that were selected together. There is no doubt that the leaf morphology will be valuable in a number of production scenarios. Jiang et al. (2000) mapped and characterized 40 QTLs determining cotton leaf morphology in 180 F_2 plants from an interspecific cross between a *G. hirsutum* genotype carrying four morphological mutants, and a wild-type *G. barbadense*. The region characterized by the largest cluster of QTLs affecting leaf-lobe length and width was found at the lower end of chromosome 15, corresponding approximately to the location of the "Okra-leaf" mutation on the classical map (Endrizzi et al. 1984). The prominent effects of this putative cluster were modified by QTLs on several other chromosomes affecting leaf size and shape. Manipulating leaf architecture can be a useful tool in breeding for drought tolerance with effects in traits such as early maturity (Andries et al. 1969; Heitholt 1993), reduced leaf area index and higher canopy CO_2-uptake per unit leaf area (Kerby et al. 1980), higher light-saturated, single-leaf photosynthesis per unit leaf area (Pettigrew et al. 1993), a shorter sympodial plastochron (Kerby and Buxton 1978), and increased numbers of flowers per season (Wells and Meredith 1986). Meredith (1984) compared super-okra, okra- and sub-okra leaf types vs. normal-leaf types and found sub-okra types that yielded greater than normal-leaf types. The discovery that the *G. hirsutum* and *G. barbadense* genotypes each contribute some alleles that increase and others that decrease leaf length, width, sublobe angles, and other attributes suggests that there exists considerable opportunities to breed cotton that go beyond the present range of leaf phenotypes.

The locus "hair-Chr.6", that affects trichome density on the leaves, is another trait with potential in breeding for drought tolerance and was mapped on chromosome 6 (Jiang et al. 2000). Trichome density affects leaf boundary layer thus affecting transpiration rate (Schuepp 1993). Wright et al. (1999) reported on a QTL in a different cotton population that affected trichome density which mapped to the same region of chromosome 6 that exhibited many of the expected features of the classical t*1* locus.

Shen et al. (2006), while analyzing DNA markers for QTLs in a fine mapping project for resistance to nematodes, also found a major QTL for root weight on chromosome 7 which accounted for almost 30% of the phenotypic variability of root weight, with the Pima S-6 parent conferring the increased root weight. Work is presently in progress to identify QTLs for root characters across the entire genome and not just the area that was under scrutiny for the root-knot nematode resistance (Chee, pers. comm.).

In a series of studies, Saranga and Paterson et al. (2001, 2003, 2004) developed an F_2 and thence an $F_{2:3}$ population from a *G. hirsutum*/*G. barbadense* interspecific hybrid to detect QTLs that were associated with water stress. Phenotypic measurements in water-stressed and well-watered environments for fiber quality [fiber span length, length uniformity, fineness (micronaire value), strength, elongation, and color components (reflectance and yellowness)], plant productivity [dry matter, seed cotton yield, harvest index, boll weight, and boll number] and physiological traits [osmotic potential, carbon isotope ratio ($\delta^{13}C$), canopy temperature, and chlorophyll *a* and *b* content] were made to dissect out their genotype by environment interactions for these traits using QTL analyses.

Overall, the water-limited conditions were responsible for yield reduction of ~ 50% relative to well-watered conditions. Of the 161 QTLs detected for the above traits, 33 QTLs (20%) influenced the traits only in water-limited treatment and therefore these QTLs can be used as markers to improve the traits with which the QTLs are associated. A subset of 63% showed no significant difference in their effects between well-watered and water-limited conditions which indicates that adaptation to both arid and favorable conditions can be combined in the same genotype. This finding along with Tuinstra et al. (1997) indicates that genetic potential for productivity under arid conditions may be improved with little or no penalty under irrigated conditions.

At face value, these results seem contradictory to the long-held notion that selection for stress tolerance will generally result in reduced productivity under favorable environments and a decrease in average overall production (Finley and Wilkinson 1963; Rosielle and Hamblin 1981; Acevedo and Fereres 1993). These findings might be reconciled with this classical expectation in that simultaneous improvement of productivity (and/or quality) for both arid and irrigated conditions will reduce the expected rate of genetic gain, because of the need to manipulate larger numbers of genes and conduct more extensive field testing (Falconer and Mackay 1996). These difficulties may be alleviated by the efficiencies gained through marker-assisted selection (MAS). Another realistic possibility would be a scenario during the introgression of the desired alleles in which a number of the QTLs would likely be replaced by less valuable alleles from the donor parent which, in turn, would cause a temporary decrease in performance until the more desirable alleles were restored via additional breeding cycles. Again, this could be rectified by efficiencies gained through identification and use of diagnostic DNA markers.

Fiber length, length uniformity, elongation, strength, fineness, and color (yellowness) were influenced by 6, 7, 9, 21, 25, and 11 QTLs (respectively) that could be detected in one or more treatments. The genetic control of cotton fiber quality was markedly affected both by general differences between growing seasons and by specific differences in water management regimes. Seventeen QTLs were detected only in the water-limited treatment while only two were specific to the well-watered treatment, thus suggesting that improvement of fiber quality under water stress may be even more complicated than improvement of this already complex trait under well-watered conditions. In crops such as cotton with widespread use of both irrigated and rainfed production systems, the need to manipulate larger numbers of genes to confer adequate quality under both sets of conditions will reduce the expected rate of genetic gain. Once again, these difficulties may be relieved by using MAS.

Testing the extent to which different complex traits share common genetic control provides a means to distinguish associations between the physiological variables and the measures of crop productivity that are truly diagnostic of genetic potential for improved adaptation to abiotic stress from those that are incidental phenotypic correlations. Of the 33 QTLs detected for the five physiological variables and the 46 QTLs for the five measures of crop productivity, only reduced plant osmotic potential was clearly implicated in improved cotton productivity under arid conditions. QTL likelihood intervals for high seed cotton yield and low osmotic potential corresponded in three genomic regions, two of which mapped to homoeologous locations on the two subgenomes of tetraploid cotton. Other studies of osmotic adjustment have been largely based on phenotypic associations with yield under drought stress (Ludlow et al. 1990; Morgan 1995; Tangpremsri et al. 1995; El Hafid et al. 1998; and Kumar & Singh 1998). These results add a new dimension to previously reported relationships between these traits; this shows that there appears to exist not only a phenotypic correlation but also a partly common genetic basis of osmotic adjustment and productivity. The importance of osmotic adjustment as an effective mechanism of crop drought resistance is receiving growing attention (Zhang et al. 1999), but cautionary reviews are also in evidence (Serraj and Sinclair 2002).

An obvious application of QTL mapping is using marker-assisted selection (MAS) as a tool in a breeding program (Lande and Thompson 1990). Reviews of MAS and QTLs show the potential and practical value of these tools to develop improved cultivars and to determine the genetic basis of phenotypic expression (Stuber 1999; Asins 2002; Bernardo 2002; Slafer 2003; Charcosset and Moreau 2004). MAS is selecting DNA markers as a surrogate for the genotype and not the usual selection of a desired phenotype. MAS can either be a direct selection if the gene itself is marked or indirect selection for a marker that is close to the desired gene. An example of direct selection is currently used with a transgenic cotton line since the sequence of the transgene is known. By developing a QTL map of traits such as drought or salt tolerance, the markers are not statistically likely to be within the sequence of the desired gene, therefore the closest marker is used and it is indirect selection.

Although MAS has received a lot of attention from the breeders in the last decade or so, applying it is still rare (Lacape et al. 2003). In cotton, MAS has mostly been used in backcrossing transgenes from transformed cultivars to the elite cultivars and then, at times, into cultivars that are further advanced. Introgressing fiber quality traits is the next most obvious program since, from the work cited earlier in the chapter, QTL mapping has provided considerable information for markers that are associated with fiber quality.

Lubbers et al (2006) is developing a series of near-isogenic introgression lines (NIILs) to contain small fragments of the genome of the *G. barbadense* donor parent within the genetic background of a specific *G. hirsutum* cultivar. This series will be useful to determine the phenotypic effect of the individual QTLs more precisely. Fine mapping (Paterson et al. 1990) will be more efficient without the clutter of the original cross and it will easier to correctly target a chosen region for fine mapping. It is expected that using these NIILs will provide a powerful tool to identify numerous QTLs that can be used to increase the tolerance of cotton to drought and salt among many other characters desired by the cotton industry.

7. SYNTHESIS

The preliminary work in QTL mapping for drought response and the relationships of the QTLs with the drought-associated measurements have developed a foundation for understanding and using the molecular basis of drought tolerance. This work is moving the timetable forward to release tolerant cotton cultivars, but further effort is needed. Testing of further traits is needed to correlate unlinked QTL alleles to their physiological basis. For example, from Saranga and Paterson et al. (2001, 2003, 2004), some QTLs showed no association with any of the measured physiological parameters but were associated with higher harvest index in the arid environment. This could indicate drought responses that either favor photoassimilation to the lint and seeds or hinder dry matter accumulation or some of both. There were a relatively large number of QTLs in this map set that were associated with $\delta^{13}C$ that may help identify the important physiological relationships between traits affecting stomatal conductance and photosynthetic capacity. The availability of cotton bacterial artificial chromosome libraries (Tomkins et al. 2001) and established transformation methods for cotton (Da et al. 2006), together with the possibility of using comparative approaches (Paterson et al. 1996) to exploit complete sequence data from botanical models such as *Arabidopsis*, may help to address the complexities of cloning QTLs. Clues as to the physiological roles of the underlying genes may help in designing appropriate probes for parallel high-throughput gene expression studies (Schena et al. 1995; De Risi et al. 1997; Hieter and Boguski 1997; Ruan et al. 1998) and/or mutation searches (Underhill et al. 1997) to identify high-probability candidate genes.

QTL mapping for salt tolerance is needed but it is not moving apace. Replicated experiments within the same genetic background as well as understanding the differences between different genetic backgrounds are needed to use QTL mapping in breeding confidently (Bernardo 2002) and will also aid our understanding of the genetics of salt and drought tolerance. With the numerous possible mechanisms and regulatory controls that have been suggested as associated with stress tolerance in the previously cited reviews and articles, a large population of recombinant inbred lines (RILs) may be required to have the resource to make confident progress in determining the genetic relationships of all these traits (Asins 2002). One drawback is that RILs will not detect the dominance component of a QTL. An intriguing mapping tool, multiple-interval mapping (reviewed by Zeng et al. 1999), can identify epistasis which other analytical tools are not designed to do. Given QTL epistasis, MAS can then be used to select parents based on the predicted genotypic values of the offspring. Recent research that demonstrated extensive epistasis in the genetics of salt tolerance for yield, yield components, and fiber quality (Bhatti 2006) indicates that a tool such as multiple-interval mapping will likely be required to effectively breed for salt and drought tolerance.

A review by Wilson et al. (2003) gives a quick tour of recently developed technologies that will be used to facilitate molecular breeding cotton for drought and salt tolerance. Transcriptomics, high throughput differential gene expression technologies, will efficiently improve research that studies the changes in mRNAs that are responses to changing environments such as drought and salt. It is necessary to not only find when and where genes are expressed but also what other genes are being regulated by the same signals. Gene knockout technology will assist in understanding the function of the numerous genes that are part of the stress tolerance pathways. SNPs (single nucleotide polymorphisms) with their genome wide distribution and the availability of high throughput systems are very good candidates for mapping and associating phenotypes with genotypes which are key parts of the research described in this chapter. With these technologies the quantity of data is going to be even greater than it is presently. Bioinformatics will be imperative to keep the information available and to help assimilate it.

Improvements in all areas of molecular breeding are almost certain, particularly in genotyping and mapping, but the most effective improvements will come from exploiting our improved understanding of the genetic architecture, perhaps from utilizing the polyploid nature of this crop. Virtually all genes in tetraploid cotton are represented by one or more copies in each sub-genome, in similar (although not identical) chromosomal orders in the two subgenomes (Reinisch et al. 1994, Rong et al. 2004) and their diploid ancestors (Brubaker et al. 1999, Desai et al. 2006). Favorable alleles in stress tolerance can come from either *G. hirsutum* or *G. barbadense*, thus recombination of favorable alleles from each of these species may form novel genotypes that are better-adapted to arid conditions than either of the parental species. The relationship of osmotic potential and seed cotton yield in the work of Saranga et al. (2004) showed a *G. hirsutum* allele that was favorable on chromosome 6 and a *G. barbadense* allele that was favorable on the homoeologous region on chromosome 25. Of course, it would require more work to determine if these are truly homoeologous genes, but it leads to an intriguing possibility that, in principle, we might be able to assemble complementary alleles across genomes in polyploids.

REFERENCES

Acevedo E, Fereres E (1993) Resistance to abiotic stresses. Chapman and Hall, London

Adams KL, Wendel JF (2004) Exploring the genomic mysteries of polyploidy in cotton. Biol J Linn Soc 82:573–581

Ahmad S, Khan N-u-I, Iqbal MZ, Hussain A, Hassan M (2002) Salt tolerance of cotton (*Gossypium hirsutum* L.). Asian J Plant Sci 1:715–719

Andries JA, Jones JE, Sloane LW, Marshall JG (1969) Effects of okra leaf shape on boll rot, yield, and other important characters of Upland cotton, *Gossypium hirsutum* L. Crop Sci 9:705–710

AOGTR (2002) The biology and ecology of cotton (*Gossypium hirsutum*) in Australia. Australian Office of the Gene Technology Regulator, www.agbios.com/docroot/decdocs/06-059–003.pdf Cited 8 Oct 2006

Ashraf M (2002) Salt tolerance of cotton: Some new advances. Crit Rev Plant Sci 21:1–30

Ashraf M (2004) Some important physiological selection criteria for salt tolerance in plants. Flora 199:361–376

Ashraf M, Harris PJC (2004) Potential biochemical indicators of salinity tolerance in plants. Plant Sci 166:3–16

Asins MJ (2002) Present and future of quantitative trait locus analysis in plant breeding. Plant Breeding 121:281–291

Bajaj S, Targolli J, Liu LF, Ho THD, Wu R (1999) Transgenic approaches to increase dehydration-stress tolerance in plants. Mol Breed 5:493–503

Baker DA (1984) Water relations. In: Wilkins MB (ed) Advanced plant physiology. Pitman, London; Marshfield, MA

Balls WL (1912) The cotton plant in Egypt. Macmillan and Co. Ltd, London

Bartels D, Sunkar R (2005) Drought and salt tolerance in plants. CritRev Plant Sci 24:23–58

Basal H, Bebeli P, Smith CW, Thaxton P (2003) Root growth parameters of converted race stocks of upland cotton and two BC_2F_2 populations. Crop Sci 43:1983–1988

Basal H, Smith CW, Thaxton PS, Hemphill JK (2005) Seedling drought tolerance in upland cotton. Crop Sci 45:766–771

Bernardo R (2002) Breeding for quantitative traits in plants. Stemma Press, Woodbury, MN

Bhatti MA (2006) Genetics of salt tolerance in *Gossypium hirsutum* L. Dept of Plant Breeding and Genetics. University of Agriculture, Faisalabad, Pakistan

Blum A (1988) Plant breeding for stress environment. CRC, Boca Raton, FL

Bohnert HJ, Nelson DE, Jensen RG (1995) Adaptations to environmental stresses. Plant Cell 7: 1099–1111

Boyer JS (1982) Plant productivity and environment. Science 218:443–448

Brubaker CL, Paterson AH, Wendel JF (1999) Comparative genetic mapping of allotetraploid cotton and its diploid progenitors. Genome 42:184–203

Brubaker CL, Wendel JF (1994) Reevaluating the origin of domesticated cotton (*Gossypium hirsutum*; Malvaceae) using nuclear Restriction-Fragment-Length-Polymorphisms (RFLPs). Am J Botany 81:1309–1326

Charcosset A, Moreau L (2004) Use of molecular markers for the development of new cultivars and the evaluation of genetic diversity. Euphytica 137:81–94

Chaves MM, Maroco JP, Pereira JS (2003) Understanding plant responses to drought – From genes to the whole plant. Funct Plant Biol 30:239–264

Collard BCY, Jahufer MZZ, Brouwer JB, Pang ECK (2005) An introduction to markers, quantitative trait loci (QTL) mapping and marker-assisted selection for crop improvement: The basic concepts. Euphytica 142:169–196

Cowan IR (1986) Economics of carbon fixation in higher plants. In: Givnish TJ (ed) On the economy of plant form and function. Cambridge University Press, Cambridge, pp 133–170

Crowe JH, Hoekstra FA, Crowe LM (1992) Anhydrobiosis. Annu Rev Physiol 54:579–599

Da K, McCurdy J, Chee PW (2006) Development of plant regeneration and transformation protocols for elite Georgia cotton lines. Beltwide Cotton Conferences. National Cotton Council of America, San Antonio, Texas

DeRisi JL, Iyer VR, Brown PO (1997) Exploring the metabolic and genetic control of gene expression on a genomic scale. Science 278:680–686

Desai A, Chee PW, Rong J, May OL, Paterson AH (2006) Chromosome structural changes in diploid and tetraploid A genomes of *Gossypium*. Genome 49:336–345

Dumka D, Bednarz CW, Maw BW (2003) Delayed initiation of fruiting as a mechanism of improved drought avoidance in cotton. Crop Sci 44:528–534

Ehleringer JR, Hall AE, Farquhar GD (eds) (1993) Stable isotopes and plant carbon-water relations. Academic Press, San Diego, CA

El Hafid R, Smith DH, Karrou M, Samir K (1998) Physiological attributes associated with early-season drought resistance in spring durum wheat cultivars. Can J Plant Sci 78:227–237

Endrizzi JE, Turcotte EC, Kohel RJ (1984) Qualitative genetics, cytology, and cytogenetics. In: Kohel RJ, Lewis CF (eds) Cotton. ASA/CSSA/SSSA, Madison, WI, pp 81–129

Ennahli S, Earl HJ (2005) Physiological limitations to photosynthetic carbon assimilation in cotton under water stress. Crop Sci 45:2374–2382

Falconer DS, Mackay TFC (1996) Introduction to quantitative genetics, 4th edn. Longman, Essex, England

FAO (2000) Land resource potential and constraints at regional and country levels. Food and Agricultural Organization of the United Nations, Land and Water Development Division, Rome ftp://ftp.fao.org/agl/agll/docs/wsr.pdf Cited 8 Oct 2006

Farquhar GD, Ehleringer JR, Hubick KT (1989) Carbon isotope discrimination and photosynthesis. Annual Review of Plant Physiology and Plant Molecular Biology 40:503–537

Finlay KW, Wilkinson GN (1963) The analysis of adaptation in a plant-breeding programme. Aust J Agric Res 14:742–754

Fryxell PA (1979) The natural history of the cotton tribe. Texas A&M University Press, College Station, TX

He CX, Yan JQ, Shen GX, Fu LH, Holaday AS, Auld D, Blumwald E, Zhang H (2005) Expression of an *Arabidopsis* vacuolar sodium/proton antiporter gene in cotton improves photosynthetic performance under salt conditions and increases fiber yield in the field. Plant Cell Physiol 46:1848–1854

Heitholt JJ (1993) Cotton boll retention and its relationship to lint yield. Crop Sci 33:486–490

Hieter P, Boguski M (1997) Functional genomics: It's all how you read it. Science 278:601–602

Hirt H, Shinozaki K (eds) (2004) Plant responses to abiotic stress. Springer, Berlin

Huang B, Liu JY (2006) A cotton dehydration responsive element binding protein functions as a transcriptional repressor of DRE-mediated gene expression. Biochem Biophys Res Commun 343:1023–1031

ICAC (2004) Cotton: Review of the World Situation. Vol 58 (2), November-December 2004 International Cotton Advisory Committee www.icac.org/cotton_info/publications/samples/reviews/erev_november_04.pdf Cited 8 Oct 2006

ICAC (2006). Townsend, Terry, pers. comm.

Ingram J, Bartels D (1996) The molecular basis of dehydration tolerance in plants. Ann Rev Plant Phys Plant Mol Bio 47:377–403

Jenks MA, Hasegawa PM (eds) (2005) Plant abiotic stress. Blackwell Publishing Ltd., Oxford; Ames, Iowa

Jiang C, DelMonte TA, Paterson AH, Wright RJ, Woo SS (2000) QTL analysis of leaf morphology in tetraploid *Gossypium* (cotton). Theor Appl Genet 100:409